A TEXTBOO

PRODUC

ENGINEERING

Dr. P.C. SHARMA

B.Sc. Engg. (Mech.) (Hons.)
M.Sc. Engg. (Mech.) (Distinction)
Ph.D., (IIT Delhi)

LMISME, MISTE

Ex. Principal, SUSCET,
Tangori (Mohali), Punjab
Formerly from
Punjab Engineering College
Chandigarh

S Chand And Company Limited
(ISO 9001 Certified Company)

S Chand And Company Limited

(ISO 9001 Certified Company)

Head Office: Block B-1, House No. D-1, Ground Floor, Mohan Co-operative Industrial Estate, New Delhi – 110 044 | Phone: 011-66672000

Registered Office: A-27, 2nd Floor, Mohan Co-operative Industrial Estate, New Delhi – 110 044 Phone: 011-49731800

www.**schandpublishing.com**; e-mail: **info@schandpublishing.com**

Branches

Chennai	:	Ph: 23632120; chennai@schandpublishing.com
Guwahati	:	Ph: 2738811, 2735640; guwahati@schandpublishing.com
Hyderabad	:	Ph: 40186018; hyderabad@schandpublishing.com
Jalandhar	:	Ph: 4645630; jalandhar@schandpublishing.com
Kolkata	:	Ph: 23357458, 23353914; kolkata@schandpublishing.com
Lucknow	:	Ph: 4003633; lucknow@schandpublishing.com
Mumbai	:	Ph: 25000297; mumbai@schandpublishing.com
Patna	:	Ph: 2260011; patna@schandpublishing.com

First Edition 1982
Subsequent Editions and Reprints 1986, 88, 89, 90, 92 (Twice), 94, 95, 96, 97 (Twice), 99, 2000, 2002, 2003, 2005, 2006 (Twice), 2008, 2009, 2010, 2013 (Twice), 2014, 2015, 2016, 2017, 2018 (Twice), 2019
Reprint 2021

ISBN : 978-81-219-0111-6 **Product Code :** H3PRE67PRDE10ENAK0XO

PRINTED IN INDIA

By Vikas Publishing House Private Limited, Plot 20/4, Site-IV, Industrial Area Sahibabad, Ghaziabad – 201 010 and Published by S Chand And Company Limited, A-27, 2nd Floor, Mohan Co-operative Industrial Estate, New Delhi – 110 044.

PREFACE TO THE ELEVENTH EDITION

The author feels very happy to present the revised edition of the book to the readers. This standard treatise on "Production Engineering" was first publishing in 1982. During the last about 26 years, it has kept a close rapport with the readers. The publication of the new edition has given an opportunity to me to incorporate the latest developments in the field in the text. Additional material has been included in chapters : 1, 2, 4, 5, 9, 10, 11, 14, 15, 20, 21 and 24.

While every attempt has been made to ensure that no errors (printing or otherwise) enter the text, the possibility of these creeping into the text is always these. I shall be grateful to the readers to bring these errors to my notice so that these may be rectified in the subsequent editions.

AUTHOR

PREFACE TO THE TENTH EDITION

The author is grateful to the readers for the tremendous response to the Ninth edition of the book.

The author has done his best to remove all the errors in the text. The author shall feel grateful if the readers point out the errors in the text, which might have been overlooked.

In the present revised edition of the book, about 200 problems from various competitive examinations (GATE, IES, IAS) have been included. The author does hope that with this, the utility of the book will be further enhanced.

AUTHOR

CONTENTS

1

JIGS AND FIXTURES

1.1. GENERAL

Jigs and fixtures are special purpose tools which are used to facilitate production (machining, assembling and inspection operations) when workpieces are to be produced on a mass scale. The mass production of workpieces is based on the concept of interchangeability according to which every part will be produced within an established tolerance. Jigs and fixtures provide a means of manufacturing interchangeable parts since they establish a relation, with predetermined tolerances, between the work and the cutting tool. They eliminate the necessity of a special set up for each individual part. Once a jig or fixture is properly set up, any number of duplicate parts may be readily produced without additional set up. Hence jigs and fixtures are used :

1. To reduce the cost of production, as their use eliminates the laying out of work and setting up of tools.
2. To increase the production.
3. To assure high accuracy of the parts.
4. To provide for interchangeability.
5. To enable heavy and complex-shaped parts to be machined by being held rigidly to a machine.
6. Reduced quality control expenses.
7. Increased versatility of machine tool.
8. Less skilled labour.
9. Saving labour.
10. Their use partially automates the machine tool.
11. Their use improves the safety at work, thereby lowering the rate of accidents.

A jig may be defined as a device which holds and positions the work, locates or guides the cutting tool relative to the workpiece and usually is not fixed to the machine table. It is usually lighter in construction.

A fixture is a work holding device which only holds and positions the work, but does not in itself guide, locate or position the cutting tool. The setting of the tool is done by machine adjustment and a setting block or by using slip gauges. A fixture is bolted or clamped to the machine table. It is usually heavy in construction.

Jigs are used on drilling, reaming, tapping and counterboring operations, while fixtures are used in connection with turning, milling, grinding, shaping, planning and boring operations.

Jigs and fixtures, because of their functions and advantages are also called "Production Devices". To facilitate interchangeability, we need "Inspection Devices, *i.e.*, the different types of Gauges (Refer to Chapter 9 and 10)".

To fulfil their basic functions, both jigs and fixtures should possess the following components or elements:

1. A sufficiently rigid body (plate, box or frame structure) into which the workpieces are loaded.

2. Locating elements.
3. Clamping elements.
4. Tool guiding elements (for jigs) or tool setting elements (for fixtures).
5. Elements for positioning or fastening the jig or fixture on the machine on which it is used.

Locating pins are stops or pins which are inserted in the body of jig or fixture, against which the workpiece is pushed to establish the desired relationship between the workpiece and the jig or fixture. To assure interchangeability, the locating elements are made from hardened steel. The purpose of clamping elements is to exert a force to press a workpiece against the locating elements and hold it there in opposition to the action of the cutting forces. In the case of a jig, a hardened bushing is fastened on one or more sides of the jig, to guide the tool to its proper location in the work. However, in the case of a fixture, a target or set block is used to set the location of the tool with respect to the workpiece within the fixture. Most jigs use standard parts such as drill bushings, screws, jig bodies and many other parts. Fixutres are made from grey cast iron or steel by welding or bolting. Fixtures are usually massive bodies because they have to withstand large dynamic forces. Because the fixtures are in between the machine and the workpiece, their rigidity and the rigidity of their fastening to the machine table are most important. Jigs are positioned or supported on the machine table with the help of feet which slide or rest on the machine table. If the drill size is quite large, either stops are provided or the jig is clamped to the machine table to withstand the high drilling torque. Fixtures are clamped or bolted to the machine table. A simple jig and a fixture are shown in Fig. 1.1.

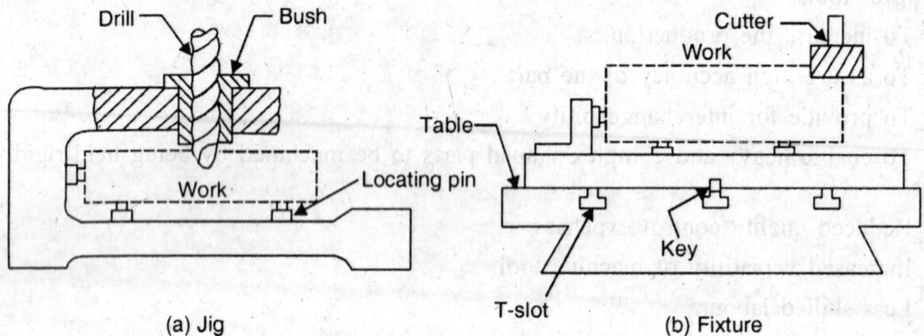

Fig. 1.1. A Simple Jig and a Fixture.

According to the degree of mechanization and automation, jigs and fixtures are classified as : (*a*) hand operated (*b*) power (*c*) semi-automatic (*d*) automatic.

1.2. LOCATING AND CLAMPING

The question of properly locating, supporting, and clamping the work is important since the overall accuracy is dependent primarily on the accuracy with which the workpiece is consistently located within the jig or fixture. There must be no movement of the work during machining. Locating refers to the establishment of a proper relationship between the workpiece and the jig or fixture. The function of clamping is to exert a force to press the workpiece against the locating surfaces and hold it there against the action of cutting forces.

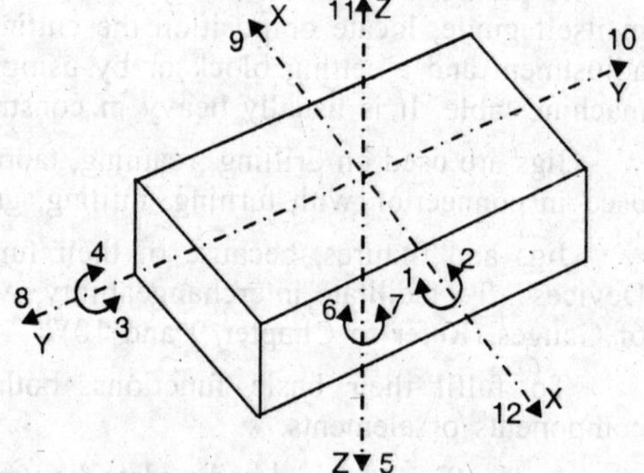

Fig. 1.2. Workpiece in Space.

1.2.1. Principle of Location. In order to study the complete location of a workpiece within a jig or fixture, let us consider a workpiece in space (Fig. 1.2). The workpiece is assumed to have true and flat faces. In a state of freedom, it may move in either of the two opposed directions along three mutually perpendicular axes, *XX*, *YY* and

and *ZZ*. These six movements are called "movements of translation". Also, the workpiece can rotate in either of two opposed directions around each axis, clockwise and anticlockwise. These six movements are called "rotational movements". The sum of these two types of movements gives the twelve degrees of freedom of a workpiece in space. To confine the workpiece accurately and positively in another fixed body (jig or fixture), the movement of the workpiece in any of the twelve degrees of freedom must be restricted. For this, let us refer to Fig. 1.3.

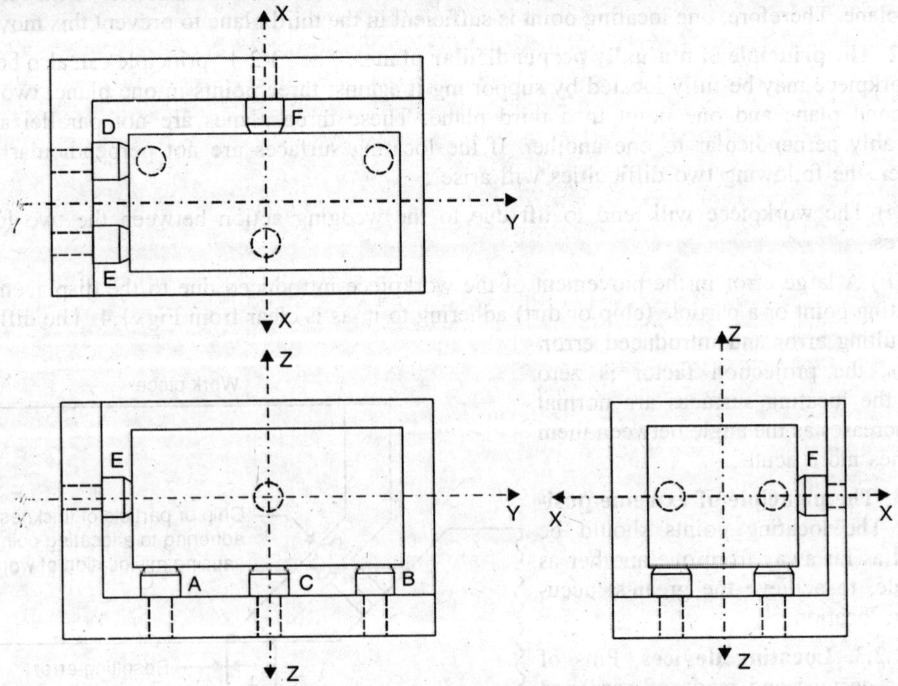

Fig. 1.3. Workpiece Located in a Fixed Body.

(*a*) The workpiece is resting on three pins *A*, *B* and *C* which are inserted in the base of the fixed body.

The workpiece cannot rotate about the axes *XX* and *YY* and also it cannot move downward. In this way, the five degrees of freedom 1, 2, 3, 4 and 5 have been arrested.

(*b*) Two more pins *D* and *E* are inserted in the fixed body, in a plane perpendicular to the plane containing the pins *A*, *B* and *C*. Now the workpiece cannot rotate about the *Z*-axis and also it cannot move towards the left. Hence, the addition of pins *D* and *E* restrict three more degrees of freedom, namely 6, 7 and 8.

(*c*) Another pin *F* in the second vertical face of the fixed body, arrests degree of freedom 9.

Thus, six locating pins, three in the base of the fixed body, two in a vertical plane and one in another vertical plane, the three planes being perpendicular to one another, restrict nine degrees of freedom. Three degrees of freedom, namely, 10, 11 and 12 are still free. To restrict these, three more pins, one for each of these degrees of freedom are needed. But this will completely enclose the workpiece making its loading and unloading into the jig or fixture impossible. Due to this, these remaining three degrees of freedom may be arrested by means of a clamping device. The above method of locating a workpiece in a jig or a fixture is called the "3-2-1" principle or "six point location" principle.

1.2.2. Principles of pin location. 1. **The principle of minimum locating points.** According to this principle, only the minimum locating points should be used to secure location of the workpiece in any one plane. Considering the "3-2-1" principle, there pins are used in the base of the fixed body.

This is due to the reason that this is the minimum number of locating points through which a plane can be drawn on which the workpiece will seat. The workpiece may rock and get strained if more than three locating points are provided..Now considering the second plane, it is clear that if one locating point is provided, the workpiece will swivel about this point, but not if there are two such location points. These two locating points establish a line parallel to the first plane. With the workpiece located against a plane and a line, it has only one direction of movement in third plane. Therefore, one locating point is sufficient in the third plane to prevent this movement.

2. **The principle of mutually perpendicular planes.** The "3-2-1" principle can also be put as : a workpiece may be fully located by supporting it against three points in one plane, two points in second plane and one point in a third plane. These three planes are not parallel and are preferably perpendicular to one another. If the locating surfaces are not perpendicular to one another, the following two difficulties will arise :

(*i*) The workpiece will tend to lift due to the wedging action between the two locating surfaces.

(*ii*) A large error in the movement of the workpiece introduced due to the displacement of a locating point or a particle (chip or dirt) adhering to it, as is clear from Fig. 1.4. The difference of resulting error and introduced error, that is, the projection factor is zero when the locating surfaces are normal and increases as the angle between them becomes more acute.

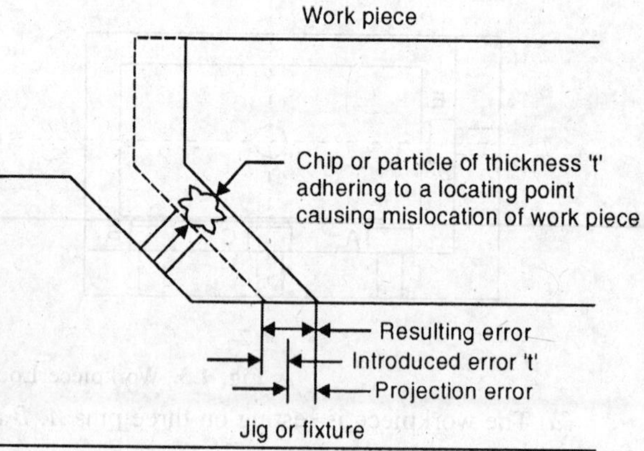

Fig. 1.4. Principle of Mutually Perpendicular Planes.

3. **The principle of extreme positions.** The locating points should be placed as far away from one another as possible, to achieve the greatest accuracy in location.

1.2.3. Locating devices. Pins of various designs and made of hardened steel are the most common locating devices used to locate a workpiece in a jig or fixture. The shank of the pin is press fitted or driven into the body of the jig or fixture. The locating diameter of the pin is made larger than the shank to prevent it from being forced into the jig or fixture body due to the weight of the workpiece or the cutting forces. Depending upon the mutual relation between the workpiece and pin, the pins may be classified as :

1. Locating pins 2. Support pins 3. Jack pins

1. **Locating pins.** When reamed or finelly finished holes are available in the workpiece, these can be used for locating purposes in the manner shown in Fig. 1.5. Depending upon their form, the locating pins are classified as :

(*i*) *Conical locating pins.* These pins are used to locate a workpiece which is cylindrical and with or without a hole as shown in Fig. 1.5 (*a*) and (*b*). Any variation in the hole size will be easily accommodated due to the conical shape of the pin.

(*ii*) *Cylindrical locating pins, 1.5* (*c*), (*d*) *and* (*e*) *:* In these pins, the locating diameter of the pin is made a push fit with the hole in the workpiece, with which it has to engage. The top portion of these pins is given a sufficient lead either by chamfering [Fig. 1.5 (*c*) and 1.5 (*d*)] or by means of radius [Fig. 1.5 (*e*)] to facilitate the loading of the workpiece.

2. **Support pins, Fig. 1.6.** With these pins (also known as rest pins, buttons or pads), workpieces with flat surfaces can be supported at convenient points. In the fixed type of support

pins, the locating surface is either ground flat [Fig. 1.6 (*a*)] or is curved [Fig. 1.6 (*b*)]. Support pins with flat head are usually employed to provide location and support to machined surfaces, because more contact area is available during location. It would ensure accurate and stable location and would not indent the work. The spherical head or rounded-head rest buttons are conventionally used for supporting rough surfaces (unmachined and cast surfaces), because they provide a point support which may be stable under these circumstances. Adjustable type support pins [Fig. 1.6 (*c*) and Fig. 1.6 (*d*)] are used for workpieces whose dimensions can vary, *e.g.*, sand castings, forging or unmachined faces.

If the component is to be located in the jig/fixture body, without the aid of these support pins, then the surface of the jig/fixture body where the component will be supported, will have to be machined. This will involve unnecessary machining time. The use of support pins saves machining time as only seats for the pins can be machined instead of the entire body of a large fixture. For small workpieces, however, no support pins are necessary. The fixture body itself can be machined suitably to provide the locating surfaces. An ample recess should be provided in corners so that burr on the workpiece corners or dirt and swarf do not obstruct proper location through positive contact of the workpiece with the locating surface. Support pins in large fixtures automatically provide similar recesses.

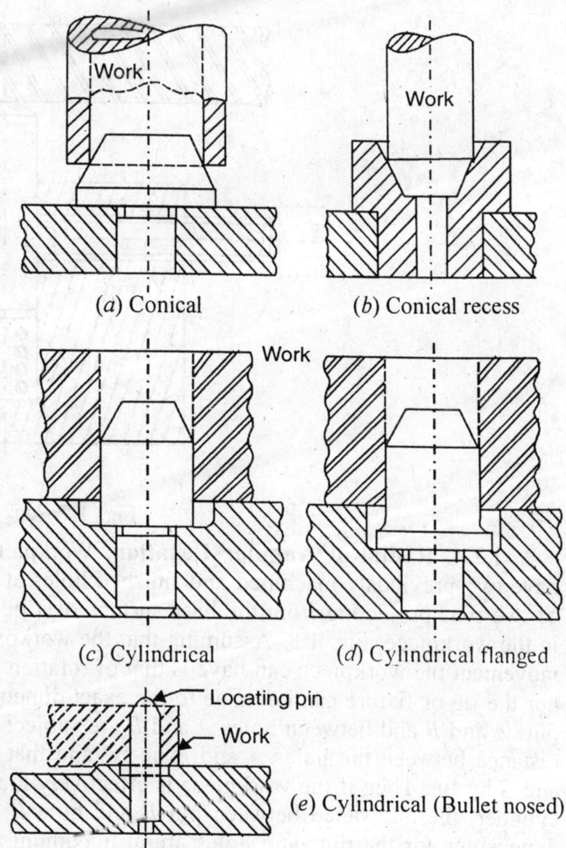

(*a*) Conical (*b*) Conical recess

(*c*) Cylindrical (*d*) Cylindrical flanged

(*e*) Cylindrical (Bullet nosed)

Fig. 1.5. Locating Pins.

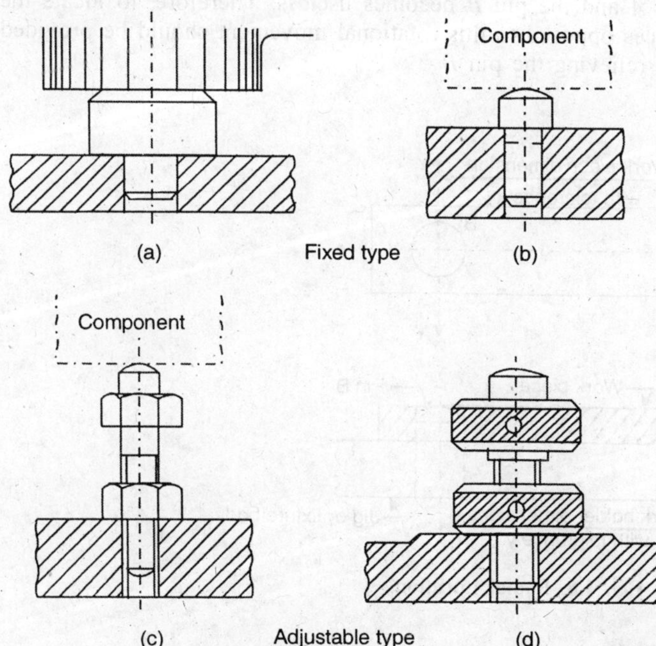

Fig. 1.6. Support Pins.

3. **Jack pins.** Jack pins or spring pins are also used to locate the workpieces whose dimensions are subject to variation, Fig. 1.7. The pin is allowed to come up under spring pressure or conversely is pressed down by the workpiece. When the location of the workpiece is secured, the pin is locked in this position by means of the locking screw.

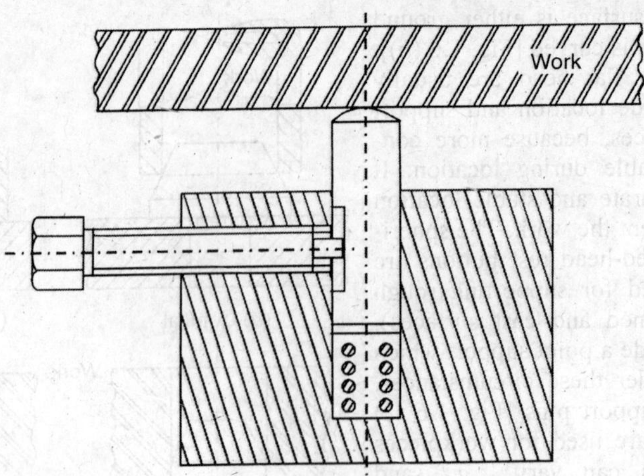

Fig. 1.7. Jack Pin.

1.2.4. Radial or Angular Location. Workpieces such as connecting rod or lever, which have two previously machined and finished holes at the two ends, may be located with the help of two pins projecting from the base surface of a jig or fixture, which will fit into the two holes in the workpiece, Fig. 1.8. Assuming that the workpiece is effectively located on pin *A*, the only movement the workpiece can have is that of rotation about the pin *A*. Now, neither the workpiece nor the jig or fixture can be made to the exact dimensions. It means the centre distances between pins *A* and *B* and between holes *A* and *B* are subject to variation. Let the tolerance for the centre distance between the holes *A* and *B* be '*x*' and that for the centre distance between the pins, *A* and *B* by '*y*'. Then if the workpiece is effectively located on pin *A* and if the pin *B* is a complete cylinder, the allowance between pin *B* and hole *B* will be *x* plus *y*. When the centre distance dimensions for the pins and holes are at maximum and minimum conditions, a large allowance will result between the hole and pin at *B* in the *Y* direction. Due to this, the workpiece will have undesirable rotation about the pin *A* and the pin *B* becomes useless. Therefore, to locate the component completely, location faces opposed to this rotational movement should be provided at the hole *B*. This is achieved by relieving the pin *B*

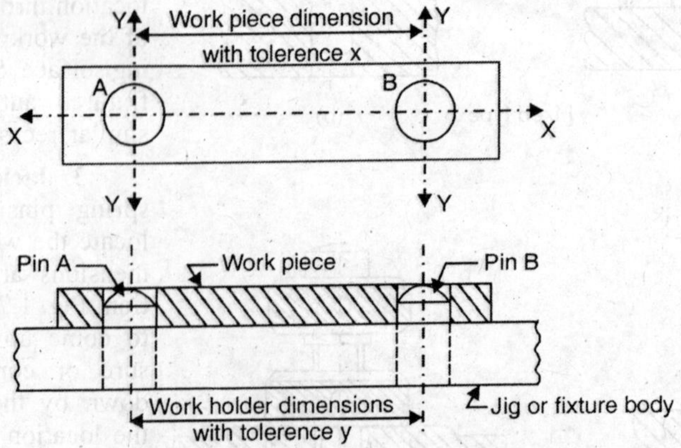

Fig. 1.8. Radial Location.

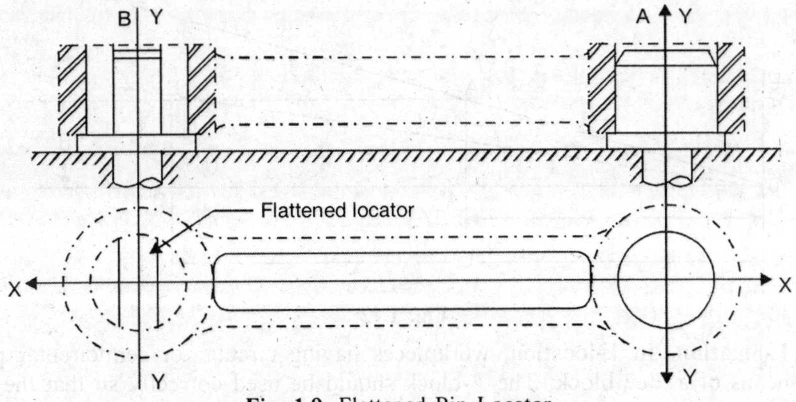

Fig. 1.9. Flattened Pin Locator.

on two sides perpendicular to the X-axis. This will allow for variations in the X-direction but will provide cylindrical locating surfaces in the Y-direction. This will result in a flattened or diamond pin locator as shown in Fig. 1.9 and Fig. 1.10 respectively.

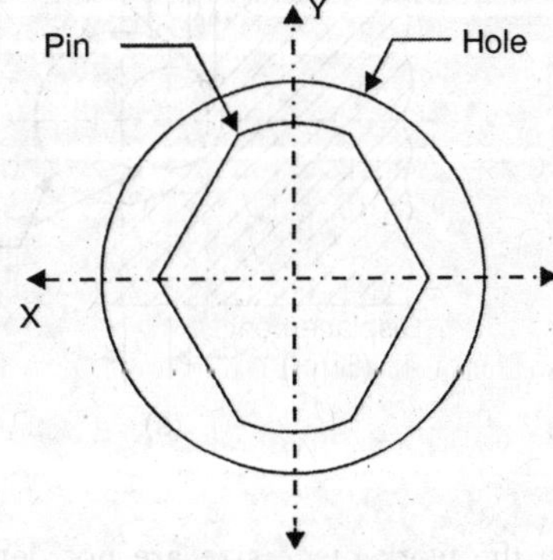

Fig. 1.10. Diamond Pin Locator.

The important and accurate hole of the two holes should be used for principal cylindrical location with a full cylindrical pin. The diamond pin is used to constrain the pivoting of the workpiece around the principal location. The principal locator should be longer than the diamond pin so that the workpiece can be located and pivoted around it before engaging with the diamond pin. This simplifies and speeds up locating of the workpiece.

A workpiece with only one hole can also be fully located as shown in Fig. 1.11. The principal location is secured from pin A. The radial movement in both the directions of Y-axis is restricted by providing two pins B confining the periphery of the workpiece. The basic principle for radial locations so as to minimize deviations from true locations is to position the radial locators as far from the axis of rotation as possible. This is clear in Fig. 1.12.

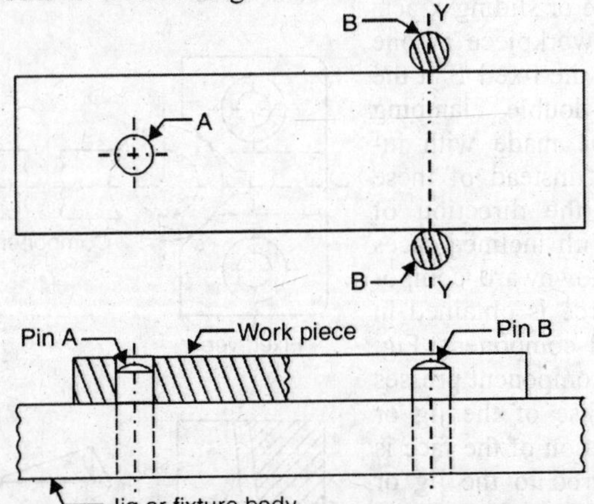

Fig. 1.11. Location of Workpiece with only one Hole.

A displacement 'd' at a distance 'a' from the axis O results in an angular error of AOA'. The same displacement 'd' at a greater distance 'b' gives an angular error of BOB' *which is smaller*.

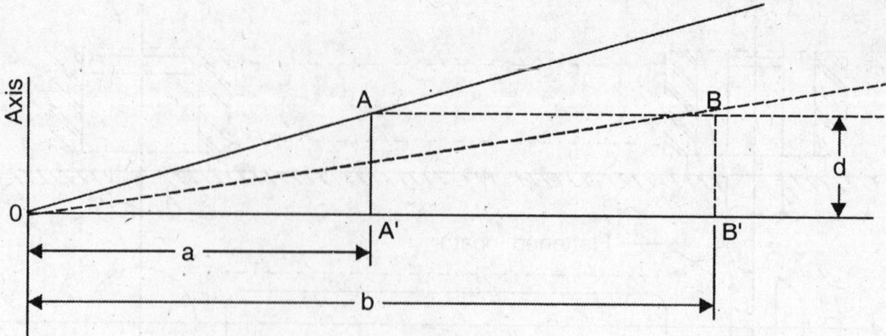

Fig. 1.12

1.2.5. *V*-location. In *V*-location, workpieces having circular or semicircular profile are located by means of a Vee block. The *V*-block should be used correctly so that the variations

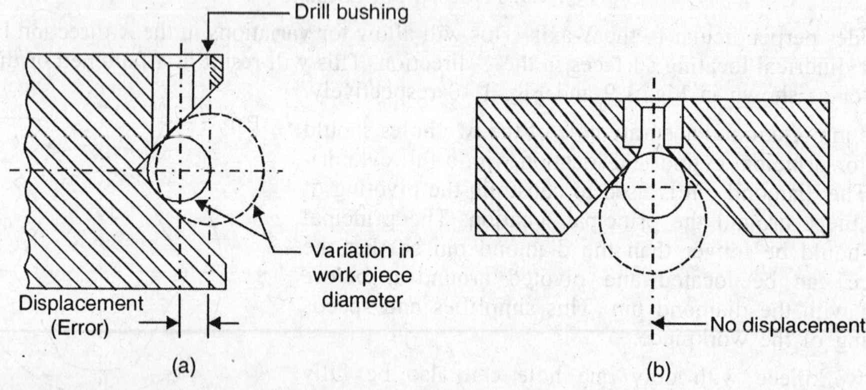

Fig. 1.13. *V*-location.

in the work piece size are not detrimental to location (Fig. 1.13). Vees can be used both for locating and clamping a workpiece. For this two Vees are employed, one fixed and the other sliding one. The fixed *V* acts to locate and the movable or sliding *V* acts to clamp and hold the workpiece at one end and forces it against the fixed *V* at the other end. To secure double clamping effect, the Vees may be made with inclined locating surfaces, instead of these being perpendicular to the direction of location of clamping. With inclined faces of the Vees a vertical downward component of the clamping force is obtained in addition to its horizontal component, Fig. 1.14. The vertical force component presses the workpiece on the base of the jig or fixture. The usual inclination of the face is 3°. The fixed *V* is secured to the jig or fixture body by means of caphead screws or dowel pins. The sliding *V* block may be actuated by means of a hand operated screw (Fig. 1.15) or a cam.

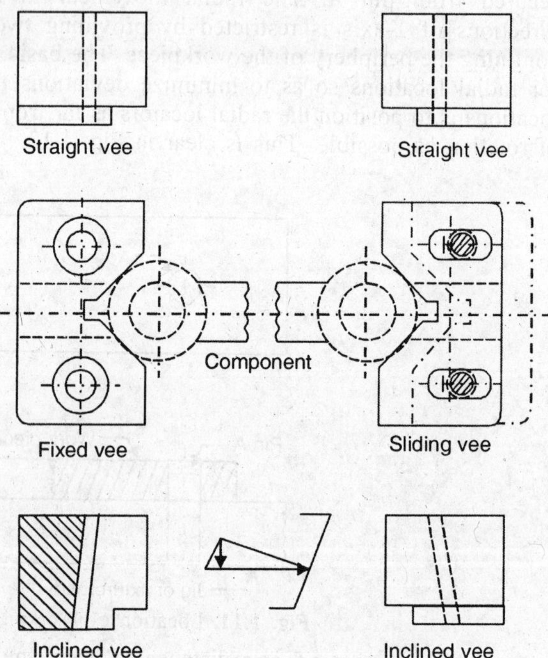

Fig. 1.14. Fixed and Sliding Vees

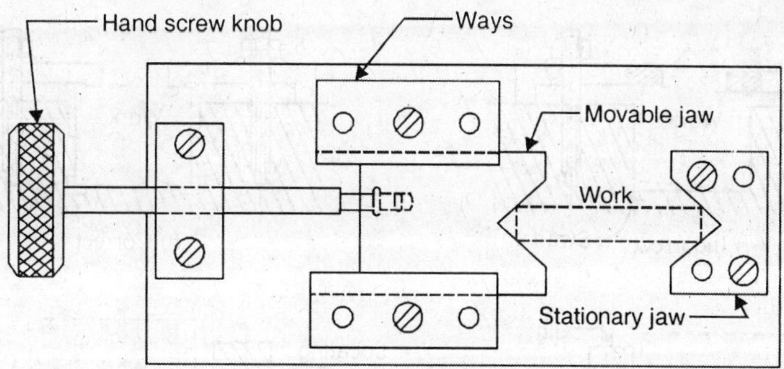

Fig. 1.15. *V*-block.

1.2.6. Bush location. Shaft type workpieces can be easily located in hardened steel bushes. Small and medium sized bushes are usually press fitted into the jig or fixture body whilst the large bushes are push fitted in the body and located by means of screws. The bushes can be plain or flanged type. A flange strengthens the bush and also prevents it from being driven into the jig body if it is left unlocked. In all the bushes, the entrance of the bush is chamfered, coned or bell mouthed to facilitate loading of the workpiece. A typical bush location is shown in Fig. 1.16.

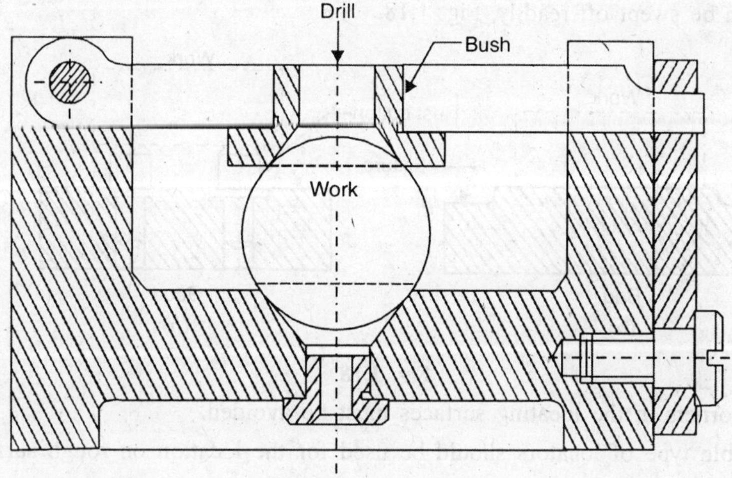

Fig. 1.16. Bush Location.

1.2.7. Design principles for location purposes. In addition to the principles discussed under Art. 1.2.2, the following principles should be followed while locating surfaces :

1. At least one datum or reference surface should be established at the first opportunity, from which subsequent machining will be measured.

2. For ease of cleaning, locating surfaces should be as small as possible consistent with adequate wearing qualities. Also, the location must be done from the machined surface.

3. The locating surfaces should not hold swarf and thereby misalign the workpiece. For this, proper relief should be provided where burr or swarf will get collected, as explained in Fig. 1.17.

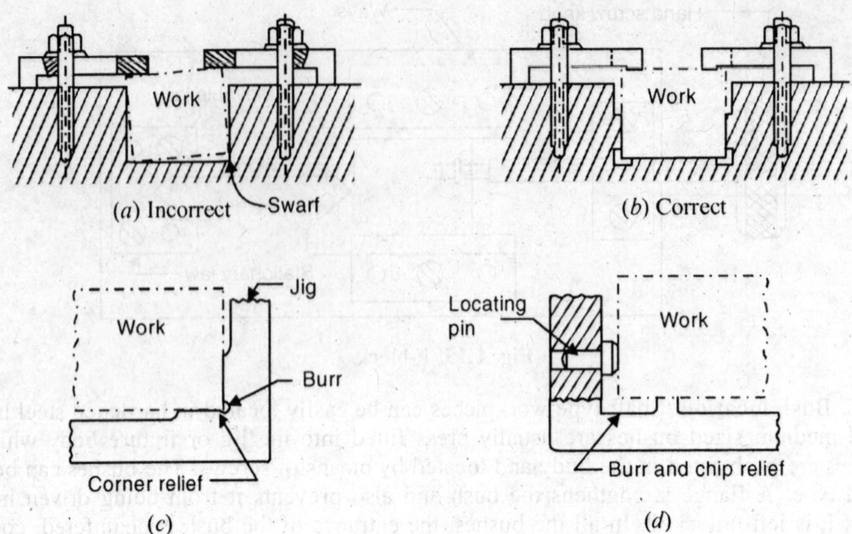

Fig. 1.17. Provision of Relief.

4. Locating surfaces should be raised above surrounding surfaces of the jig or fixture, so that chips fall or can be swept off readily, Fig. 1.18.

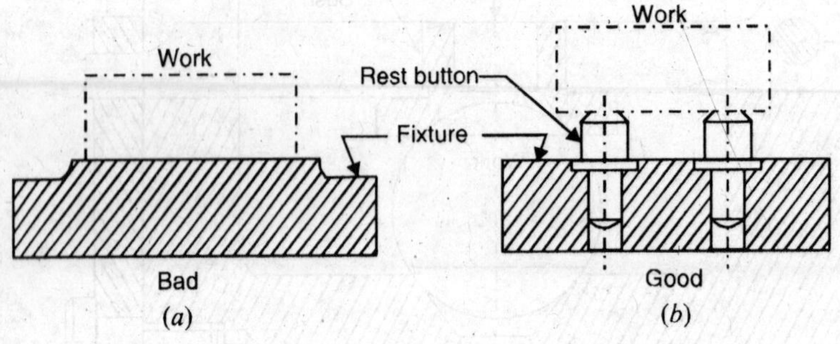

Fig. 1.18

5. Sharp corners in the locating surfaces must be avoided.
6. Adjustable type of locators should be used for the location on rough surfaces.
7. Locating pins should be easily accessible and visible to the operator.
8. To avoid distortion of the work, it should be supported as shown in Fig. 1.19.

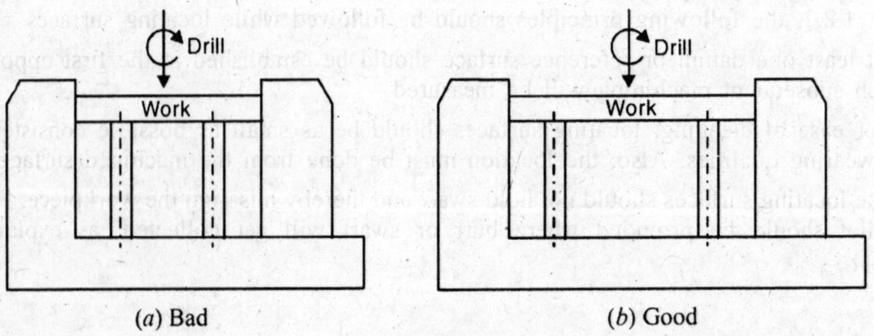

Fig. 1.19.

1.2.8. Clamping. If the workpiece cannot be restrained by the locating elements, it becomes necessary to clamp the workpiece in jig or fixture body. As already noted, the purpose of clamping is to exert a pressure to press a workpiece against the locating surfaces and hold it there in opposition to the cutting forces i.e to secure a reliable (positive) contact of the work with locating elements and prevent the work in the fixture from displacement and vibration in machining. The most common example of a clamp is the bench vise, where the movable jaw of the vise exerts force on the workpiece thereby holding it in the correct position of location in the fixed jaw of the vise.

Principles for clamping purposes. Since the proper and adequate clamping of a workpiece is very important, the following design and operational factors should be taken care of :

1. The clamping pressures applied against the workpiece must counteract the tool forces.

2. The clamping pressures should not be directed towards the cutting operation. Whenever possible, it should be directed parallel to it, Fig. 1.20.

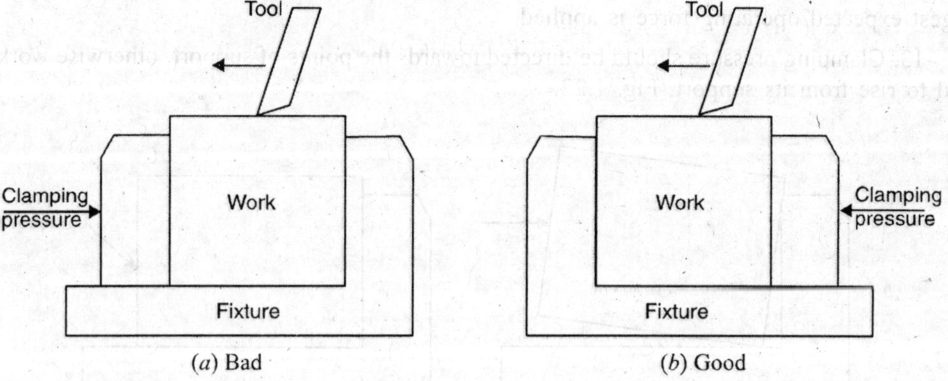

(*a*) Bad (*b*) Good

Fig. 1.20

3. The clamping pressure must only hold the workpiece and should never be great enough so as to damage, deform or change any dimensions of the workpiece.

4. The clamping and cutting forces should be directed towards the locating pins, otherwise the workpiece may get bent or forced away from the locating pins during machining.

5. Clamping should be simple, quick and foolproof. Complicated clamps lose their effectiveness as they wear.

6. The movement of a clamp should be strictly limited and if possible it should be positively guided.

7. Whenever possible, the lifting of the clamp by hand should be avoided if it can be done by means of a spring fitted to it.

8. Clamps should never be relied upon for holding the workpiece against the cutting force. The cutting force should be arranged against a fixed stop or a substantial part of the fixture body.

9. The clamps should always be arranged directly above the points supporting the work, otherwise the distortion of the work can occur, as illustrated in Fig. 1.21.

10. Fibre pads should be riveted to the clamp faces, otherwise soft and fragile workpiece can get damaged.

11. A clamp should be designed to deliver the required clamping force when operated by the smallest force expected.

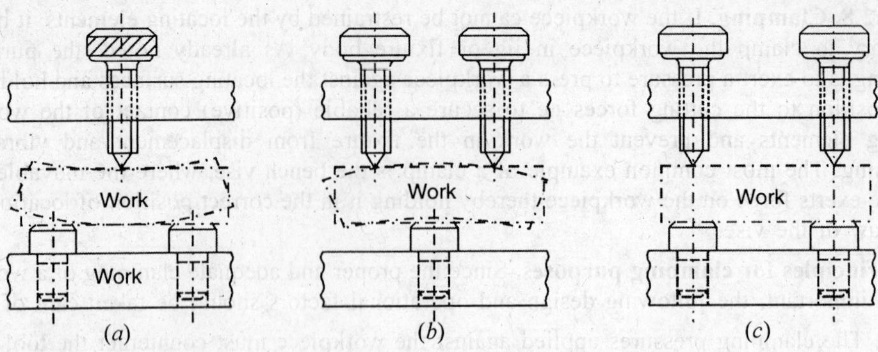

Fig. 1.21. Position of Clamp.

12. A clamp should be strong enough to withstand the reaction imposed upon it when the largest expected operating force is applied.

13. Clamping pressure should be directed towards the points of support, otherwise work will tend to rise from its support, Fig. 1.22.

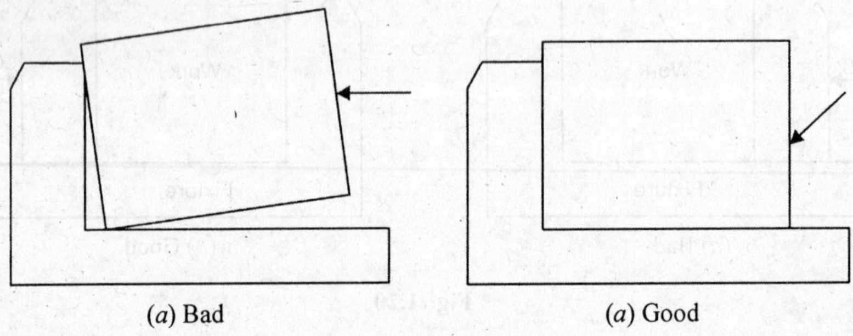

Fig. 1.22

1.2.9. Clamping Devices. The commonly used clamping devices are discussed below.

1. **Clamping screws.** Clamping screws are used for light clamping and typical examples are shown in Figs. 1.17 and 1.21.

2. **Hook bolt clamp.** This is very simple clamping device and is only suitable for light work and where the usual type of clamp is inconvenient. A typical hook bolt clamp is shown in Fig. 1.23.

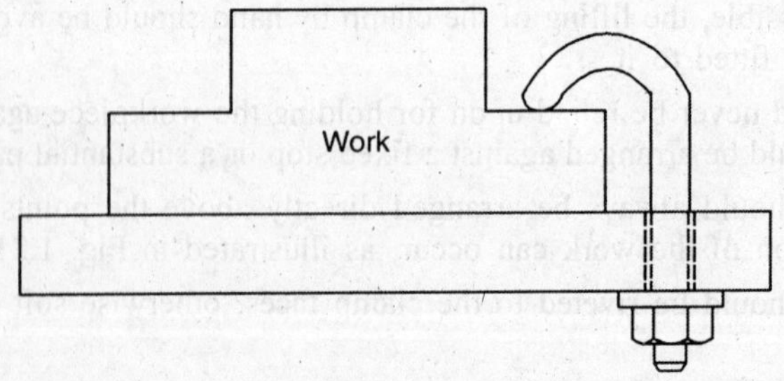

Fig. 1.23. Hook Bolt Clamp.

3. **Lever type clamps.** The various designs in the lever type clamp used with jigs and fixtures are discussed below.

(*i*) *Bridge clamp.* It is very simple and reliable clamping device. The clamping force is applied by the spring loaded nut Fig. 1.24 (*a*).

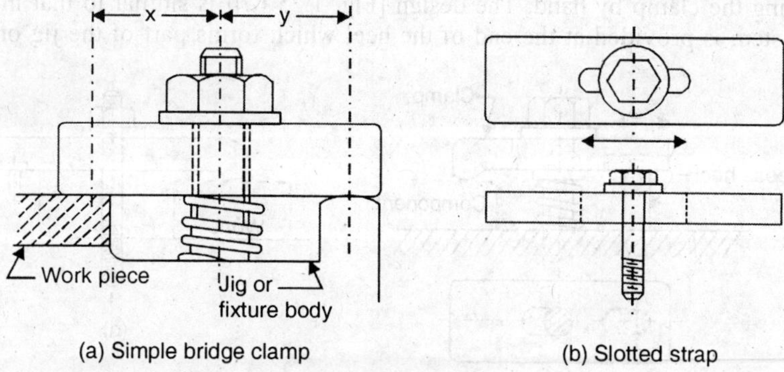

(a) Simple bridge clamp

(b) Slotted strap

Fig. 1.24. (*a, b*)

The relative positions of the nut, the point of contact of the clamp with the work and with outer support should be carefully considered, since the compressive force of the nut is shared between the workpiece and the clamp support inversely as the ratio of their distances from the nut. The distance '*x*' is less than or equal to but never greater than the distance '*y*'. The spring is fitted with the clamp for its automatic lifting when the nut is loosened to remove the workpiece from the jig or fixture. To avoid the complete removal of the nut every time a workpiece is changed the clamp may be slotted to draw it back as shown in Fig. 1.24 (*b*). A two way clamping can be obtained by the bridge clamp as shown in Fig. 1.24 (*c*).

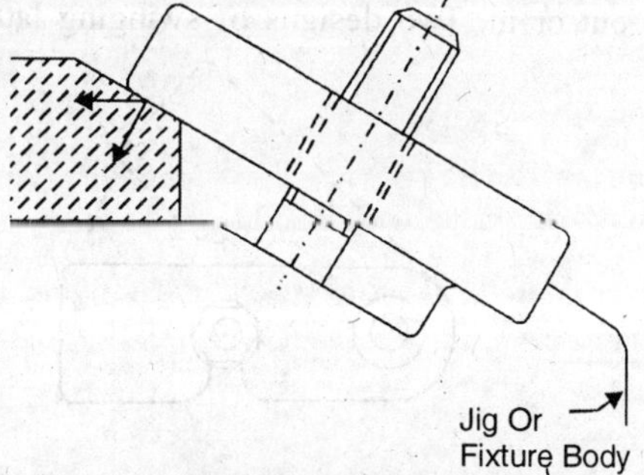

(c) Two way clamp

Fig. 1.24. Bridge Clamps.

(*ii*) *Heel clamps.* The various types of heel clamps are shown in Fig. 1.25. These consist of a robust plate or strap, centre stud and a heel. The strap should be strengthened at the point where the hole for the stud is cut out, by increasing the thickness around the hole. The design [Fig. 1.25 (*a*)] differs from the simple bridge clamp [Fig. 1.24 (*a*)] in that a heel is provided at the outer end of the clamp to guide its sliding motion for loading and unloading the workpiece. In design [Fig. 1.25 (*b*)], the heel is solid and one piece with the clamp. The workpiece is loaded into the jig or fixture or removed from these, by rotating the clamp. In design [Fig. 1.25 (*c*)], the clamp is guided by the loose heel which

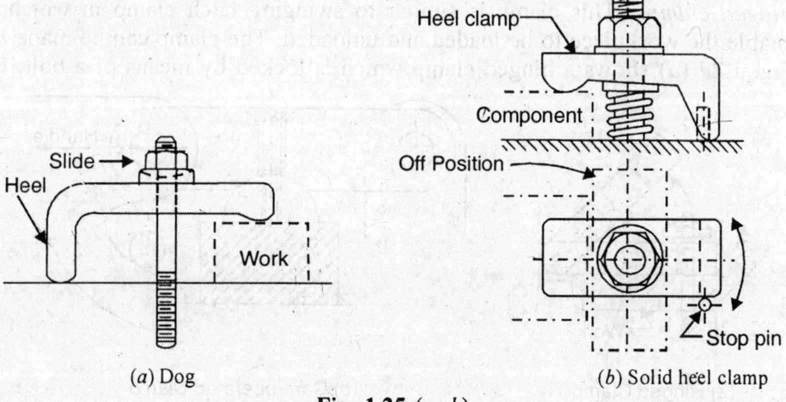

(*a*) Dog

(*b*) Solid heel clamp

Fig. 1.25 (*a, b*)

is driven into the jig or fixture body. A short stem is turned on the end of the heel which fits loosely into a keyway in the clamp strap. The loading and unloading of the workpiece is obtained by reciprocating the clamp by hand. The design [Fig. 1.25 (*d*)] is similar to that in Fig. 1.25 (*c*) but, here the stem is provided at the end of the heel which forms part of the jig or fixture body casting.

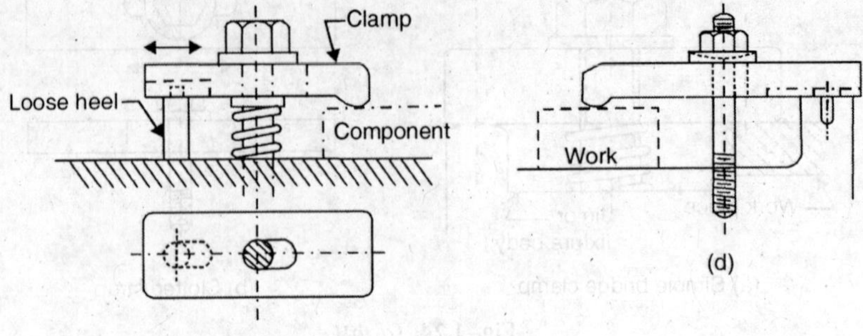

Fig. 1.25. (*c, d*) Heel Clamps.

(*iii*) *Swinging strap* (*latch*) *clamp.* This is a special type of clamp which provides a means of entry for loading and unloading the workpiece. For this, the strap (latch or lid) can be swung out or in. Two designs of swinging latch clamps are shown in Fig. 1.26.

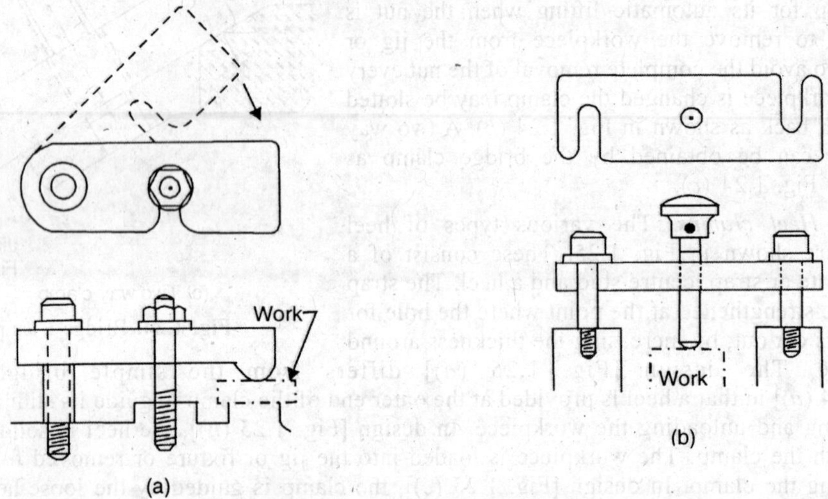

Fig. 1.26. Swinging Latch Clamp.

(*iv*) *Hinged clamps.* This clamp is similar to swinging latch clamp in which the latch is hinged to enable the workpiece to be loaded and unloaded. The clamp can be made integral with the latch. Fig. 1.27 (*a*) shows a hinged clamp which is locked by means of a bolt. Fig. 1.27 (*b*)

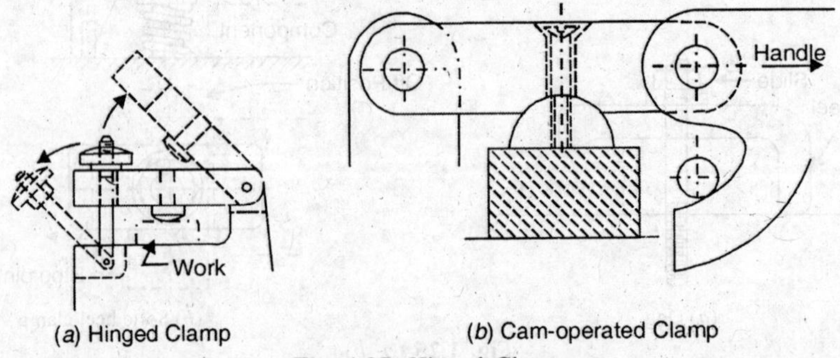

Fig. 1.27. Hinged Clamps.

shows a hinged clamp provided with a hook cam. This clamp is much quicker than the bolt type and is suitable for workpieces which maintain dimensional accuracy. The hooked end of the operating lever acts as a cam and engages a pin. Fig. 1.28 shows some other designs of the lids or straps which may be used for swinging latch or hinged clamps.

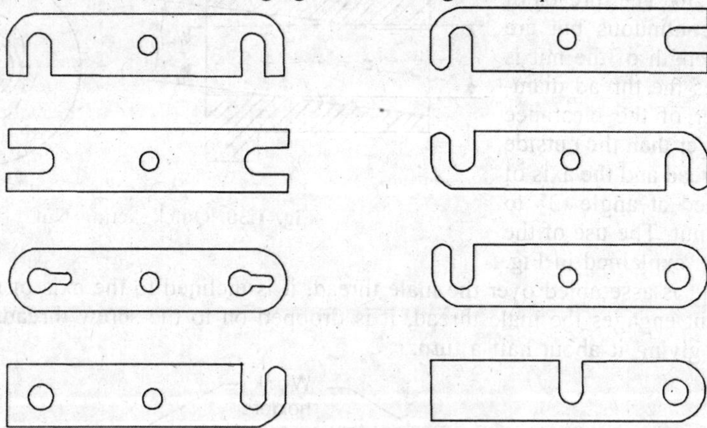

Fig. 1.28. Straps.

4. **Quick acting clamps.** There are many mechanical clamping devices (pneumatic and hydraulic devices will be discussed later), which can be termed as quick acting. These devices are costlier than the other types but ultimately prove economical since these help in reducing the total operating time. Some of the quick acting clamping devices are discussed below :

(*i*) *C-clamps.* The two types of C-clamps, free and captive are shown in Fig. 1.29. To unload the workpiece, the locking nut is unscrewed by giving it about one turn and this releases the *C*-clamp. When the clamp is removed or swung away, the workpiece can freely pass over the nut.

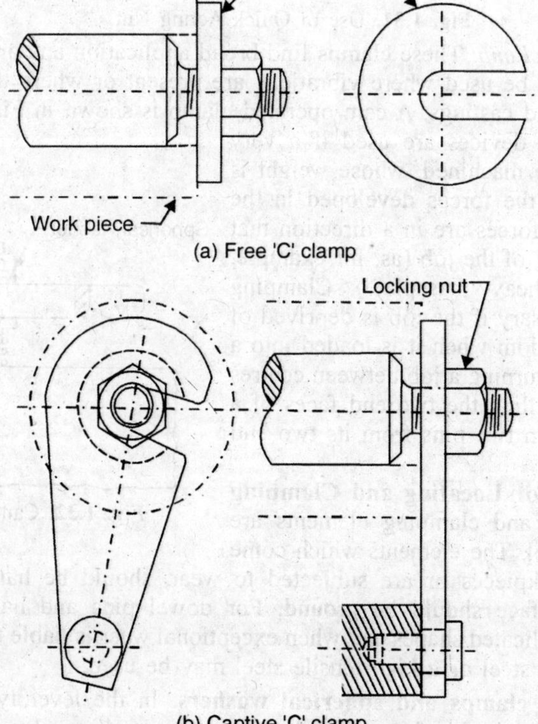

Fig 1.29. *C* - clamps.

The reverse procedure is adopted for loading the workpiece. The free *C*-clamp may be fastened to jig or fixture body to prevent it from being lost.

(*ii*) *Quick acting nut.* This nut is shown in Fig. 1.30. The threads of the nut are not continuous but are interrupted. The length of the nut is about 2 to 3 times the thread diameter. The diameter of the clearance '*D*' is slightly bigger than the outside diameter of the thread and the axis of the hole is inclined at angle (3° to 7°) to the axis of nut. The use of the quick acting nut is explained in Fig. 1.31. When the nut is assembled over the male thread, it is inclined to the axis of the clearance hole. When the nut engages the male thread, it is dropped on to the screw threads and is then tightly locked by giving it about half a turn.

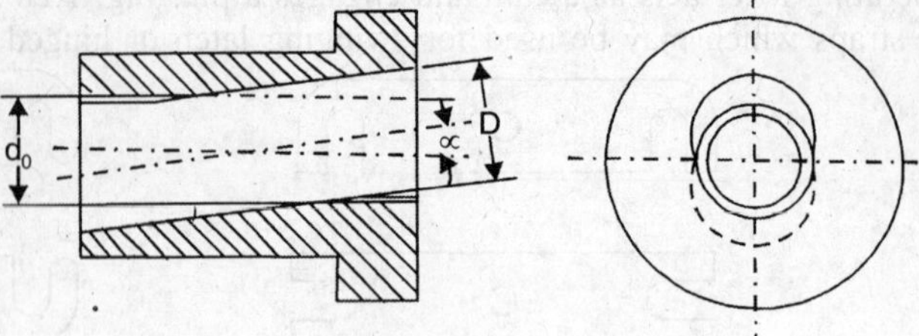

Fig. 1.30. Quick Acting Nut.

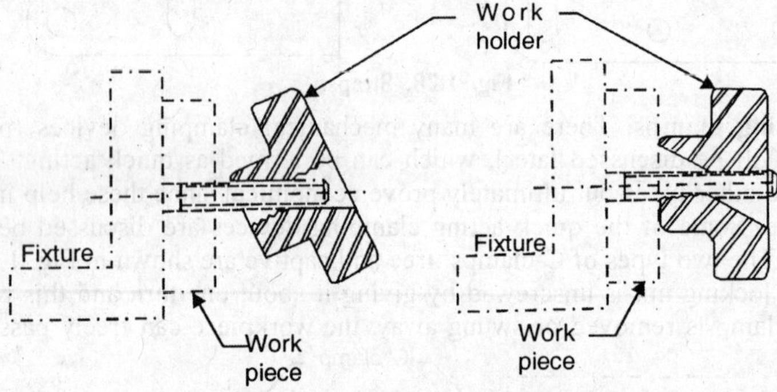

Fig. 1.31. Use of Quick Acting Nut.

(*iii*) *Cam-operated clamp.* These clamps find broad application and are fast and positive in action. These should not be used where vibrations are present or where the dimensions of the workpiece vary, *e.g.*, sand castings. A cam operated clamp is shown in Fig. 1.32.

Note. No clamping devices are used if a very heavy stable job is to be machined, whose weight is very great compared to the forces developed in the cutting process, if these forces are in a direction that cannot disturb the setting of the job (as, for example, in drilling holes in a heavy baseplate). Clamping devices are also unnecessary if the job is deprived of all of its degrees of freedom when it is loaded into a fixture (as. for example, turning a job between centres on a centre lathe and milling the two end faces of a connecting rod located on two pins from its two end holes, Fig. 1.9).

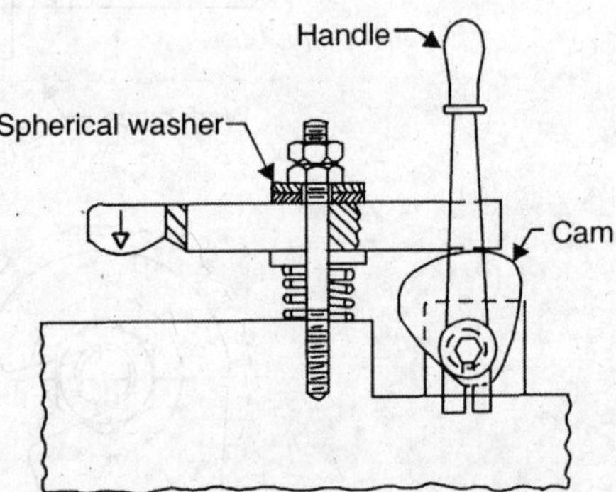

Fig. 1.32. Cam-operated Clamp

1.2.10. Materials for Locating and Clamping Elements. The locating and clamping elements are generally made from steel. The elements which come in contact with the workpieces or are subjected to wear, should be hardened and wherever necessary, the working face should be ground. For dowel pins and handles, silver steel is generally used. For complicated shapes and when exceptional wear is liable to occur, good quality case-hardened steel, tool steel or a high tensile steel may be used.

1.2.11. Lever type clamps and spherical washers. In the lever type clamps discussed above, it is seen that the clamping face of the lever is curved. This makes the clamp operatable even if there is variation in the workpiece. At the other end of the strap (pressure pad), the top of the bridge or the heel should also be in the shape of raised and rounded toes to permit some tilting of clamp. This design provides more effective clamping than a design having flat strap ends. This design will also take care of small variations in workpiece height.

Poor clamping conditions can result if there is a considerable variation in the workpiece or there is difference in workpiece and fulcrum block height. The misalignment between clamp surface and clamping nut due to tilting of clamp can be taken care of by interposing a pair (male-female) of spherical washers between the nut and the strap (instead of a plain washer), Fig. 1.33. The spherical bearing surfaces of the washers will allow the inclination of the strap caused by the difference in heights of the filcrum block and workpiece. The male washer (upper one) remains square with the nut while the female washer (lower one) tilts with the clamp; since the spherical bearing surfaces allow the pair of spherical washers to tilt with respect to each other. The angle of inclination of the strap that can be tolerated is limited by the clearance between the stud and the inside diameter of the washers. Spherical washers are thus commonly used for equalising clamping forces.

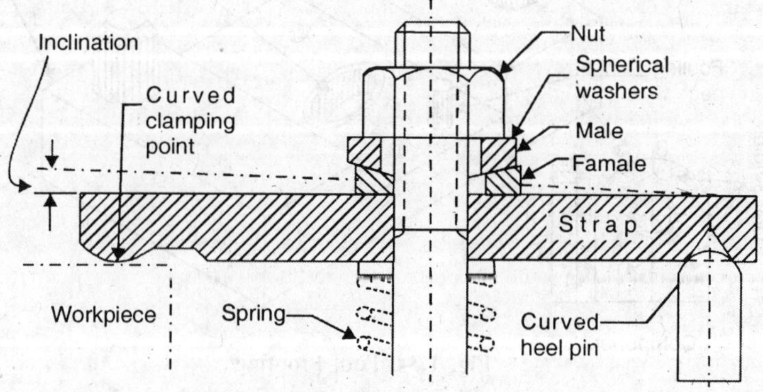

Fig. 1.33. Use of Spherical Washers.

1.3. DESIGN PRINCIPLES COMMON TO JIGS AND FIXTURES

1. Since the total machining time for a workpiece includes work-handling time, the methods of location and clamping should be such that the idle time is minimum. The various design principles regarding location and clamping have already been discussed.

2. The design of jig and fixture should allow easy and quick loading and unloading of the workpiece. This will also help in reducing the idle time to minimise.

3. The jig and fixture should be as open as possible to minimize chip or burr accumulation and to enable the operator to remove the chips easily with a brush or an air jet.

4. *Fool proofing.* It can be defined as the incorporation of design features in the jig or fixture, that will make it impossible to load the work into the jig or fixture in an improper position but will not interfere with proper loading and locating the workpiece. There are many foolproofing devices such as fouling pegs, blocks or pins which clear correctly positioned parts but prevent incorrectly loaded parts from entering the jig or fixture body.

Figure 1.34 explains this principle. Three holes are to be drilled in the component shown. The operator locates the component on the bottom plate of the drilling jig with the projection *C* on the component fitting the locating hole in the jig. Now if the component is being located incorrectly so that the end *A* of the component is oriented towards the left of the locating hole, the fouling peg in the body of the jig will obstruct the component.

For correct locations, the end *A* of the component is to be towards the right and the curved end *B* of the component is to be towards the left of the locating hole.

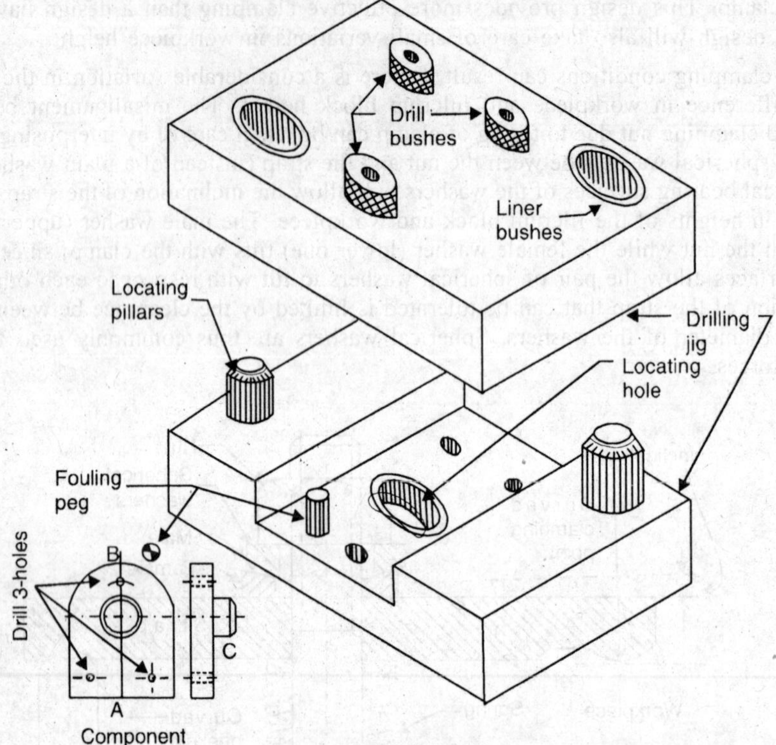

Fig. 1.34. Fool Proofing.

5. *Clearance.* Clearance is provided in the jig or fixture body for two main reasons :

(*i*) to allow for any variation in component sizes, especially castings and forgings.

(*ii*) to allow for hand movements so that the workpiece can easily be placed in the jig or fixture and removed after machining.

6. *Rigidity.* Jigs and fixtures should be sufficiently stiff to secure the preset accuracy of machining.

7. *Trunnions.* To simplify the handling of heavy jigs or fixtures, the following means can be adopted :

(*i*) Eye-bolts, rings or lifting lugs can be provided for the lifting of the jig or fixture.

(*ii*) If the workpiece is also heavy, then the jig design should allow for side loading and unloading by sliding the workpiece on the machine table.

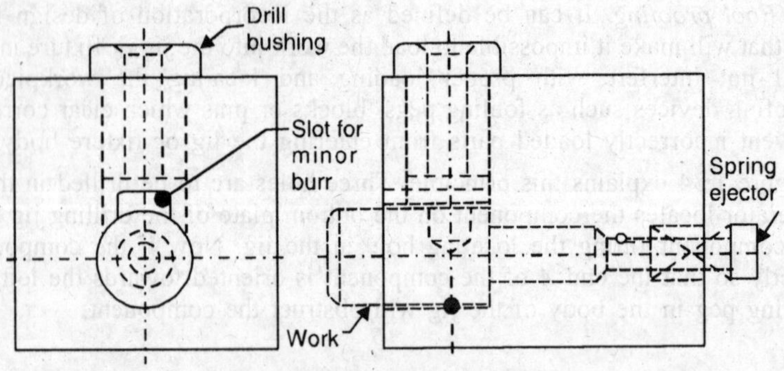

Fig. 1.35. Burr Grooves.

8. *Burr grooves.* A burr raised on the work at the start of a cut is termed a 'minor burr' and that at the end of a cut a 'major burr'. Jigs should be designed so that the removal of the workpiece is not obstructed by these burrs. For this, suitable clearance grooves or slots should be provided as shown in Fig. 1.35.

9. *Ejectors.* The use of ejection devices to force the workpiece out from the jig or fixture is important in two situations :

(*i*) the workpiece is heavy.

(*ii*) machining pressure forces the workpiece to the sides or base of the jig or fixture and the pressure and oil or coolant film will cause the work to stick and be difficult to remove.

On small jigs or fixtures, a pin located under the work will remove the part readily [Fig. 1.36 (*a*), (*b*)]. Hinged ejectors [Fig. 1.36 (*c*)] are also very useful and can be easily mounted.

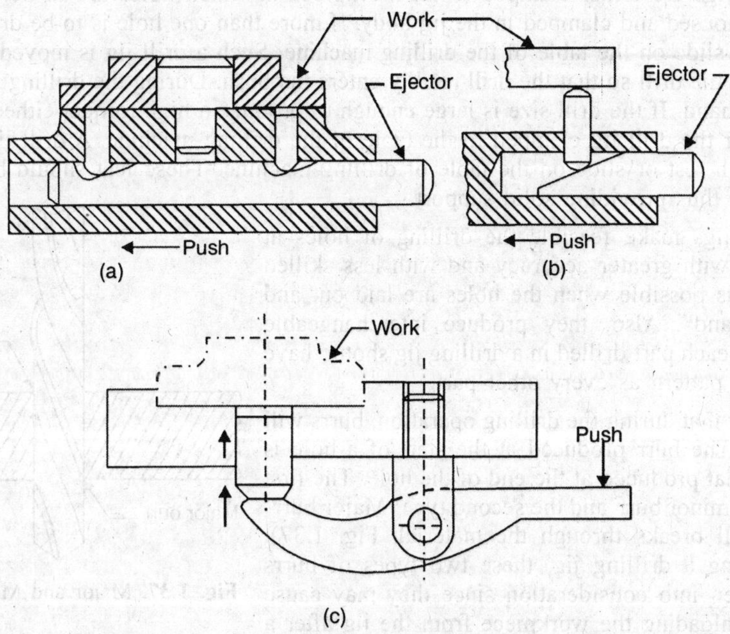

Fig. 1.36. Ejecting Devices.

10. **Inserts.** To avoid any damage to fragile and soft workpieces and also to the finished surfaces of a workpiece while clamping, inserts of some soft material such as copper, lead, fibre, leather, hard rubber, plastic or felt should be fitted to the faces of the clamps.

11. **Design for safety.** Jigs/fixtures must be safe and convenient in use. Following are some of the factors for the safety of the worker working with a jig/fixture :

(*i*) Sharp corners on the body of the jig/fixture should be avoided.

(*ii*) Sighting surfaces should be clear.

(*iii*) Bolts and nuts should be inside the body of the jig/fixture and not protrude on the surface.

12. **Sighting Surfaces.** Machining on a workpiece must be clearly visible to the worker. He should not be required to bend his neck for seeing the work surface.

13. **Simplicity in Design.** Design of the jig/fixtures should be a simple one. A complicated design requires a large maintenance. They should be cheap in manufacture and should lend themselves readily to maintenance and replacement of worn-out parts.

14. *Economical.* Jig/fixture should be simple in construction, give high accuracy, be sufficiently rigid and light in weight. To satisfy all these conditions, an economical balance has to be made.

15. They should be easy to *set in the* machine tool, which is so important in quatity production where jigs/fixtures are replaced at intervals.

1.4. DRILLING JIGS

Drilling jigs are used to machine holes in mechanical products. To obtain positional accuracy of the holes, hardened drill bushes or jig bushes are used to locate and guide drills, reamers etc., in relation to the workpiece. These guide bushes are not essential but these prove to be economical and technically desirable as will be discussed ahead. The portion of the jig into which the hardened bushes are fitted is called bush plate.

Drilling jigs are either clamped to the workpiece in which holes are to be drilled or the workpiece is housed and clamped in the jig body. If more than one hole is to be drilled, the drill jig is made to slide on the table of the drilling machine. Such a drill jig is moved by hand into position under the drill so that the drill readily enters the bush. During the drilling operation, the jig is held by hand. If the drill size is large enough to produce a high torque, either stops should be provided or the drill jig clamped to the table of the drilling machine. A drill jig is provided with feet which rest or slide on the table of drilling machine. These feet should be outside the cutting forces, thus providing solid support.

Drilling jigs make feasible the drilling of holes at higher speed, with greater accuracy and with less skilled workers than is possible when the holes are laid out and drilled "by hand". Also, they produce interchangeable parts, because each part drilled in a drilling jig should have the same hole pattern as every other part.

It is clear that during the drilling operation, burrs will be produced. The burr produced at the start of a hole is smaller than that produced at the end of the hole. The first type is called 'minor burr'and the second type 'Major burr' (when the drill breaks through the material, Fig. 1.37). When designing a drilling jig, these two types of burrs should be taken into consideration since they may cause difficulty in unloading the workpiece from the jig after a hole has been drilled.

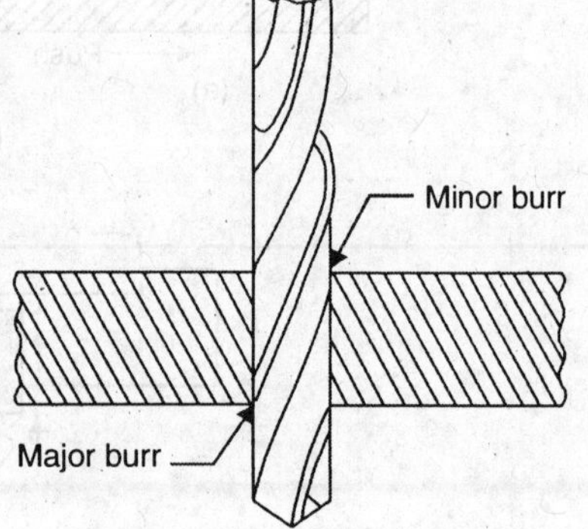

Fig. 1.37. Major and Minor Burrs.

1.4.1. Design Principles for Drilling Jigs. 1. A drilling jig should be of light construction consistent with adequate rigidity to facilitate its handling because it has to be handled frequently during the operation.

2. A drilling jig which is not normally clamped to the machine table should be provided with four feet so that it will rock if it is not resting square on the machine table and so warn the operator.

3. The stability of a drilling jig should be as good as possible since it is not usual to clamp it to the machine table and to ensure this, the feet or base of the jig should extend well outside the holes to be drilled.

4. Drill bushings should be fitted in fixed portion of the jig and not in clamps except for a few special cases (for example, leaf type jig).

1.4.2. Drill Bushes. Sometimes the stiffness of the cutting tool may be insufficient to perform certain machining operations. To eliminate the elastic spring back in machining and to locate the tool relative to the work, use is made of guiding parts, such as, jig bushings and templates. These must be precise, wear resistant and changeable.

Jig bushings are used in drilling and boring jigs. Their use permits giving up the marking out, reduces drill run-off and hole expansion (ovalization). The diametric accuracy of holes in jig drilling is 50 per cent higher on the average compared to that of holes drilled conventionally. Drill bushings can be classified as : Press fit, Renewable and Liner bushings.

(*i*) *Press Fit bushings.* These bushings are used when little importance is put on accuracy or finish, and the tool used is a twist drill. These bushings are installed directly in the jig body and are used mainly for short production runs not requiring bush replacement. These are also employed where the centre distances of holes are too close to permit the fitting of liners and renewable bushes. There are two designs of press fit bushings : (*a*) Plain or headless (*b*) Flanged or headed. A flanged or headed bush has a flange or head, Fig. 1.38. (*b*). It is employed when the jig plate into which it is installed is thin, the flanged or headed portion increasing the length of the bush which provides longer guiding portion to the bush than would otherwise be available. The flange or head also acts as a stop for the tool.

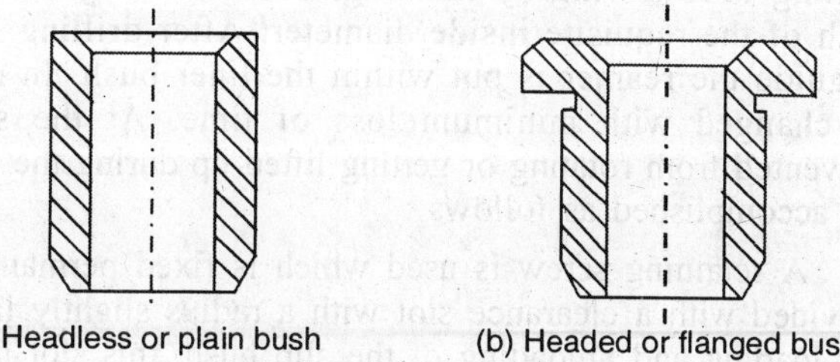

Fig. 1.38. Press Fit Bushings.

(*ii*) *Renewable bushes.* Figure 1.39 (*a*). When the guide bushes require periodic replacement (due to the wear of the inside diameter of the bush, in the case of continuous or large batch production), the replacement is simplified by using a renewable bush. These are of the flanged type and are sliding fit into the liner bush, which is installed (press fitted in the jig plate). The liner bush provides hardened wear resistant mating surface to the renewable bush.

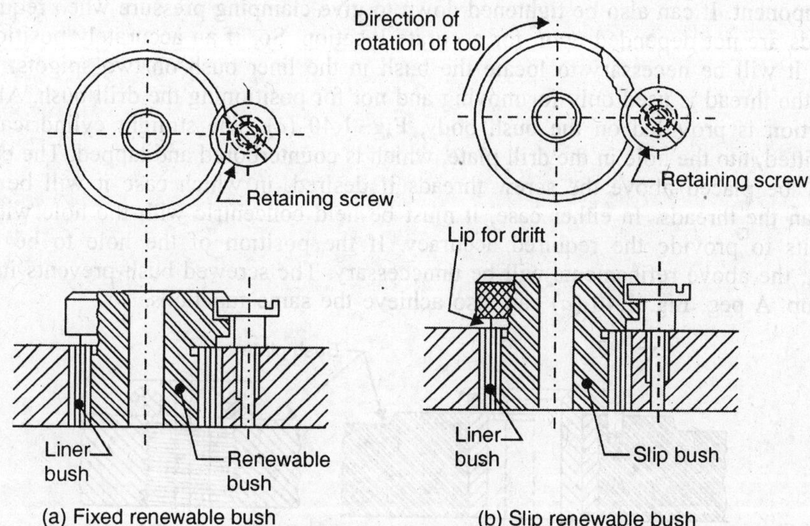

Fig. 1.39. Renewable Bushes.

The renewable bushes must be prevented from rotating or lifting with the drill. One common method is to use a retaining screw as shown in the Fig. 1.39 (*a*). The flange of the bush is provided with a flat. When the renewable bush is put into the liner bush and the retaining screw is tightened, the collar of the screw will press against the flat on the bush flange and prevent rotation of the bush and also its lifting up (when the drill is clearing the swarf or is being withdrawn at the end of the cut). When the renewable bush wears out, the retaining screw is removed and the worn bush is taken out. A new bush is then easily substituted. A normal press fit bush can be taken out only by driving it out of the jig plate. This usually damages the bore of the jig plate which has to be rebored for an oversize bush to be fitted.

(*iii*) *Slip Bushes.* Fig. 1.39 (*b*). Slip bushes are used when more than one bushings are to be interchanged in a given size of the liner, that is, where two or more operations in a job require different inside diameters of the guide bush in the same jig, for example, when drilling is following by reaming, counterboring etc. The hole is first drilled using a guide bush of the requisite inside diameter. After drilling, this bush is removed and another bush to guide the reamer is put within the liner bush. In mass production, these bushings should be changed with minimum loss of time. At the same time, the slip bushes should be prevented from rotating or getting lifted up during the machining process. Both these objectives are accomplished as follows :

A retaining screw is used which is fixed permanently in the bush plate. The slip bush is provided with a clearance slot with a radius slightly larger than that of the head of the screw. For loading and unloading of the slip bush, this slot is aligned with the collar of the retaining screw. The bush can be moved freely axially in this position. For loading, the bush can, therefore, be dropped over the screw. The bush is also provided with a step, which when the bush is rotated clockwise, will turn and lock under the flange of the screw. When the tool is withdrawn, the screw head prevents the bush from rising. For unloading, the bush is rotated anticlockwise to align the bush clearance slot with screw collar. Then the bush can be lifted up axially out of the liner. The slip bushings are also flange type and are sliding fit into the liner bush and for their easy loading/unloading, their heads are knurled.

(*iv*) *Screw bush.* The screwing of the bush into the jig body not only holds the bush in place, but it also makes the bush adjustable. This drill bush may also be used for locating purposes and is then invariably screwed into position; it can, therefore, be adjusted for length to suit the component. It can also be tightened down to give clamping pressure when required. The screw threads are not depended upon for accurate location. So, if an accurately positioned hole is required, it will be necessary to locate the bush in the liner bush on two spigots. This will ensure that the thread is used only for moving and not for positioning the drill bush. Alternately, guiding portion is provided on the bush body, Fig. 1.40 (*c*). The straight cylindrical guiding portion, is fitted into the hole in the drill plate, which is counterbored and tapped. The cylindrical portion may be placed above the screw threads if desired, in which case it will be of larger diameter than the threads. In either case, it must be held concentric with the hole within close enough limits to provide the required accuracy. If the position of the hole to be drilled is unimportant, the above refinements will be unnccessary. The screwed bush prevents its rotating and lifting up. A peg, Fig. 1.40 (*a*) can also achieve the same functions.

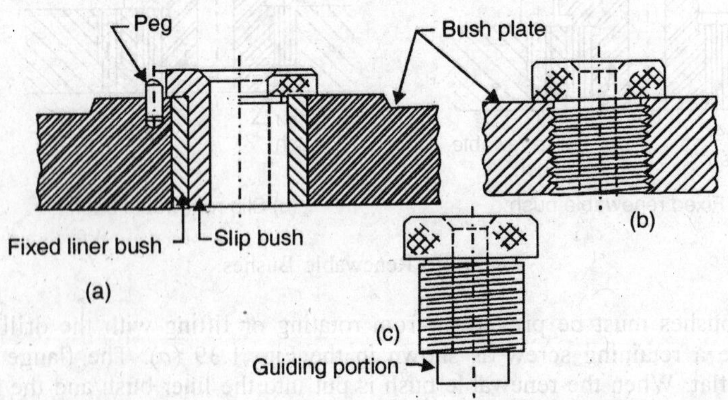

Fig. 1.40

(*v*) *Liner Bushings.* These bushings, also known as 'master bushings' are permanently fixed into the jig body. These act as guides for renewable type bushings. These bushings can be with or without heads, Fig. 1.41. A liner bush is always used in conjunction with a renewable bush, for example a slip bush. Slip bushes of different inside diameters but of the same outside diameters (same as the inner diameter of the liner bush) are used.

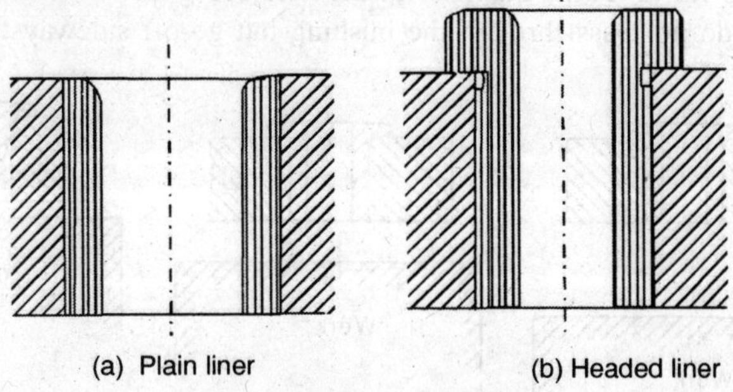

Fig. 1.41. Liner Bushings.

(*vi*) *Special drill bushings.* Some examples of special drill bushings are given in Fig. 1.42 :

(*a*) is used to drill a hole through an inclined surface.

(*b*) A long busing is used if the hole being drilled is in a recess. Since the bush is long, therefore, the drill friction is considerable. To reduce it, the bush can be counterbored (*i*). The other method is shown in (*ii*). The larger bush can be made of C.I. or a cheaper material, only the actual drill bush being of the normal material.

(*c*) If two holes are to be drilled close together, two bushings with flats can be used, or

(*d*) a bushing with two holes is used.

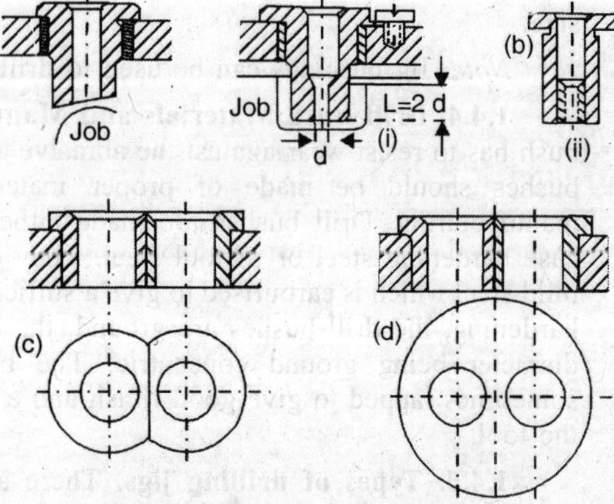

Fig. 1.42. Special Drill Bushings.

1.4.3. Design Principles for Drill Bushings. The following points should be kept in mind in the case of drill bushings :

1. To facilitate easy entry of drills, the entrances to drill bushes should be extremely smooth and well chamfered or rounded. Also, a suitable lead or chamfer should be on the outside of the bush at the lower end to facilitate its installation in the jig body (Fig. 1.38).

2. There should not be any sharp corners on the body of the bush.

3. Loose or screwed in solid bushes should not be used where accuracy is important.

4. The effective length of the drill bushing should be sufficient to guide and support the drill. A too short a bush will not be able to keep the drill in line as the drill can bend with the short bushing acting as a fulcrum. Due to this, out of line and oversize holes may be produced. On the other hand, if the bush is longer than necessary, the drill will wear out earlier. For drills having average helix angles, the length of the drill bushing should be from 1.75 to 2.50 times the drill diameter.

5. Adequate provisions must be made for the chips that are produced and for their easy removal. Insufficient clearance between the end of a drill bushing and the workpiece, Fig. 1.43 (*c*), will prevent the chips from escaping and these will come up through the drill bushing. This will result in the wear of the bushing due to the abrasive action of the chips. On the other hand, too much clearance, Fig. 1.43 (*a*), may not provide adequate drill guidance and can result in broken drills, due to bending between the bushing and the workpiece. Fig. 1.43 (*b*), shows the

correct design. The clearance between the lower end of the bushing and the workpiece should be from 1/3 to 1 times (lower values for C.I. and higher values for steel) and for drilling deep holes in steel, as much as 1.5 times the diameter of the hole being machined. This design reduces wear because the chips do not pass through the bushing but go-off sideways.

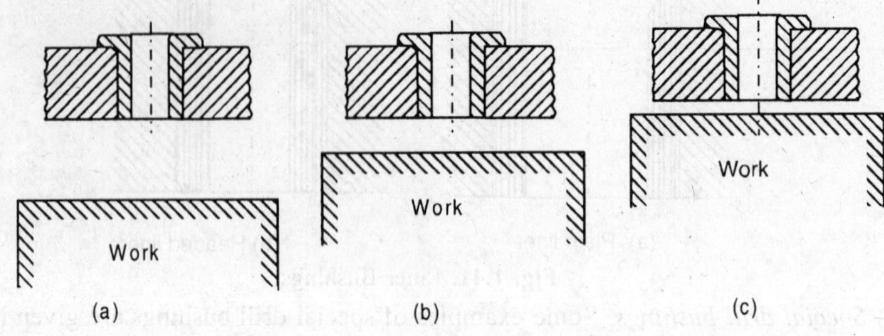

Fig. 1.43

6. The hole of the drill bushing should be from 0.00025 to 0.0025 cm larger than the drill size.

Note. Jig bushings can be used to drill from 10,000 to 15,000 holes.

1.4.4. Drill Bush Materials and Manufacture. The surface of the guide hole of the drill bush has to resist wear against the abrasive action of the chips and also of the drill. So, the drill bushes should be made of proper materials and good care should be taken for their manufacturing. Drill bushes are made either from good quality case hardening steel or of tool steel. They can also be made of mild steel which is carburised to give a sufficient case depth. After hardening, the drill bushes are ground, the bore and the outside diameter being ground concentric. The bore of the bush is sometimes lapped to give good finish and a fine running fit with the tool.

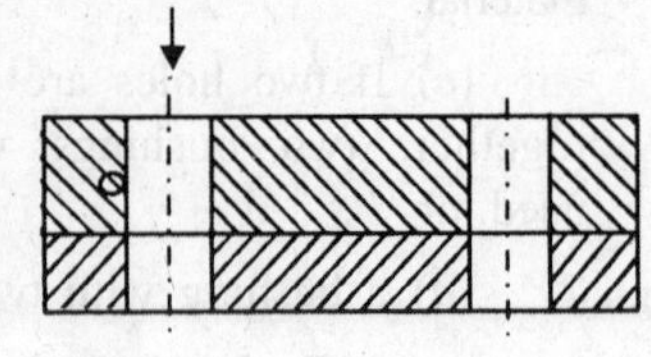

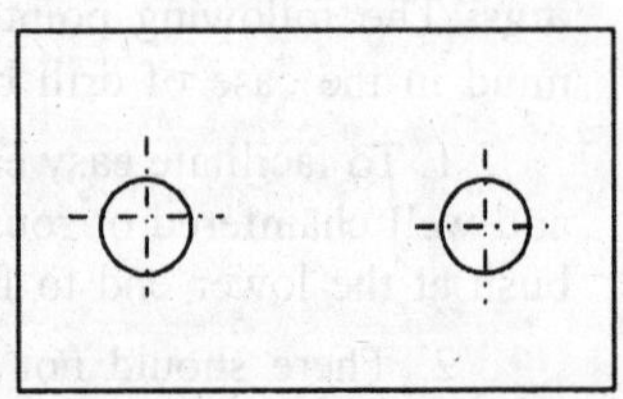

Fig. 1.44. Template Jig.

1.4.5. Types of drilling jigs. There are no hard and fast criteria for classifying the drilling jigs. However, the drilling jigs may be classified as follows :

1. *Template jig.* This is the simplest type of drilling jig. It is simply a plate made to the shape and size of the workpiece with the required number of holes made in it accurately. A simple template type of jig is shown in Fig. 1.44. It is placed on the workpiece and the holes in the workpiece will be made by the drill which will be guided through the holes in the template. The plate should be hardened to avoid its frequent replacement. This type of jig is suitable if only a few parts are to be made.

2. *Plate type jig.* This is an improvement over the template type of jig. In place of simple holes, drill bushes are provided in the plate to guide the drill. The workpiece can be clamped to the plate and the holes drilled, Fig. 1.45.

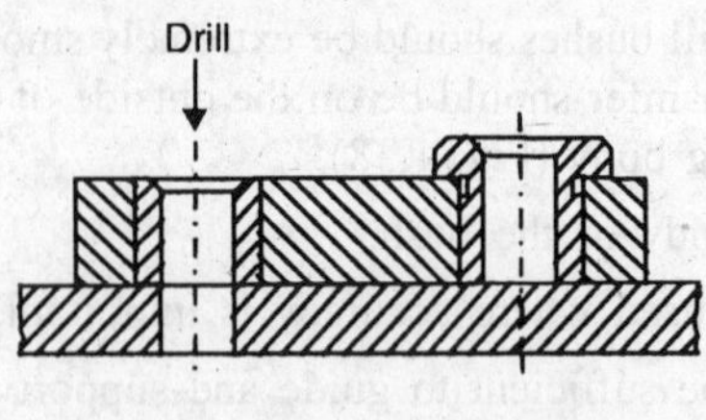

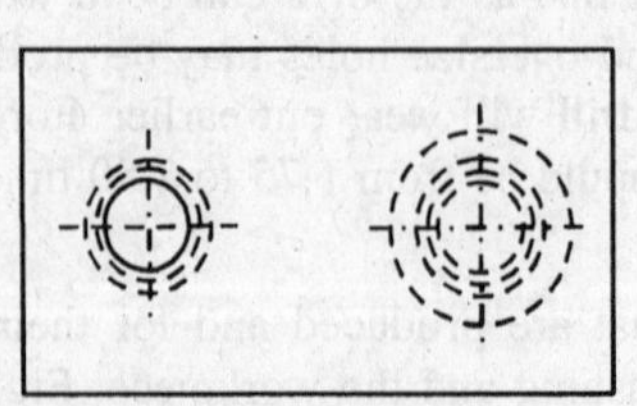

Fig. 1.45. Plate Type Jig

3. *Open type jig*. In this jig, Fig. 1.46, the top of the jig is open. The workpiece is placed on top.

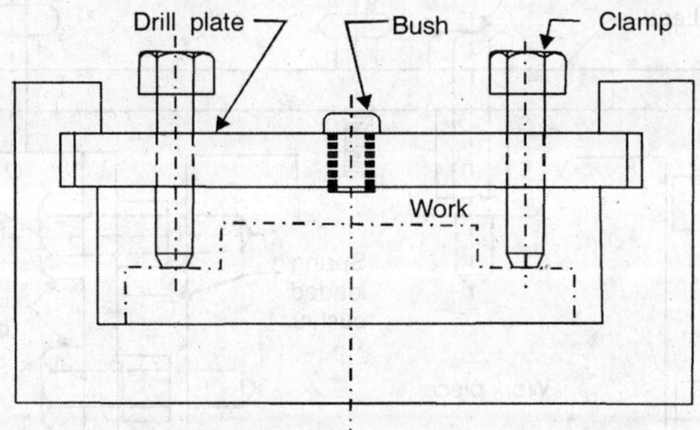

Fig. 1.46. Open Type Jig.

4. *Swinging leaf jig*. It is also a sort of open type jig in which the top plate is arranged to swing about a fulcrum point so that it completely clears the jig for easy loading and unloading of the workpiece, Fig. 1.47. The drill bushes are fitted into the plate which is also known as leaf, latch or lid.

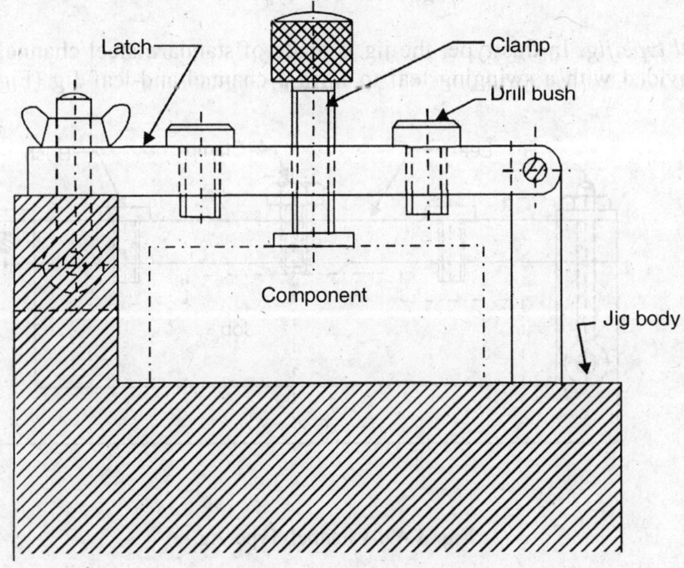

Fig. 1.47. Swinging Leaf Jig

5. *Box type jig*. When holes are to be drilled in more than one plane of a workpiece, the jig has to be provided with equivalent number of bush plates. For positioning the jig on the machine table, feet have to be provided opposite each drilling bush plate. One side of the jig will be provided with a swinging leaf for loading and unloading the workpiece. Such a jig would take the form of a box. (Fig. 1.48). The body of such a jig should be as light as possible since it will have to be lifted again and again. Fig. 1.48 is for a leaf-type box jig. When, one or more sides of the box jig are kept open for loading/unloading, it is known as Tumble type and Trunnion type box jig.

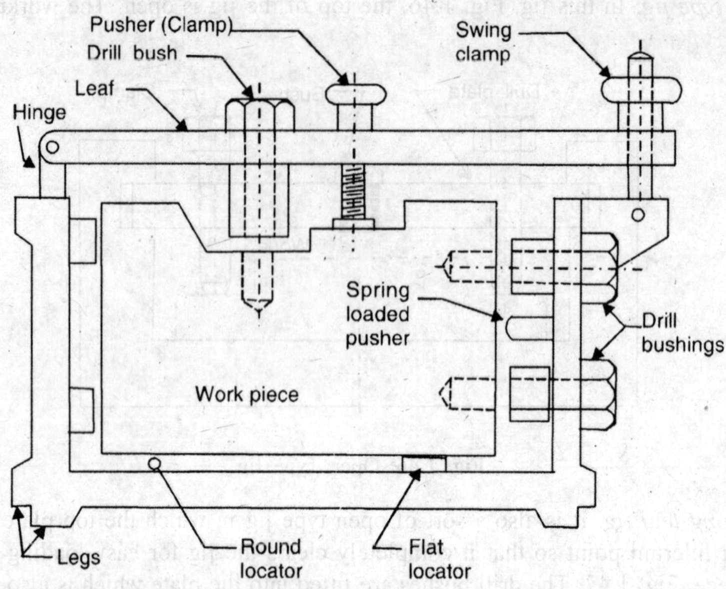

Fig. 1.48. Box Type Jig.

6. *Channel type jig*. In this type, the jig is made of standard steel channel section. This jig can also be provided with a swinging leaf to form a channel-and-leaf Jig (Fig. 1.49).

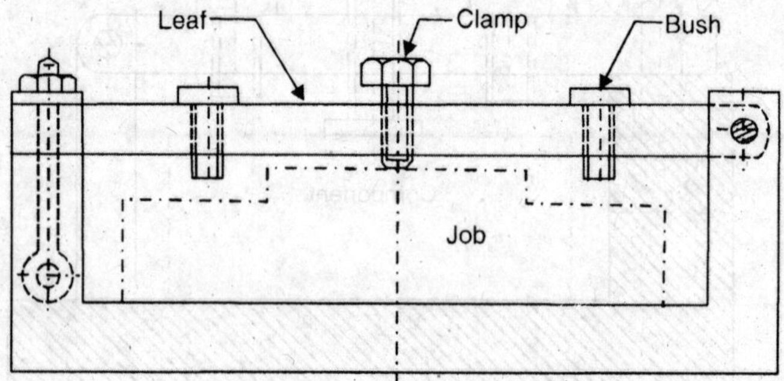

Fig. 1.49. Channel Type Jig.

7. *Sandwich jig:* A sandwich jig is a modification of a plate type of jig. The plate jig has a back-up plate. The job is held between the two plates. The jig is very useful for thin and ductile jobs which might get bent or warped on another type of jig.

8. *Angular jig:* This type of jig is used when a hole is to be drilled at an angle to the drilling bush axis. Fig. 1.50. This type of jig is used to drill holes in collars and hubs of pulleys and gears etc. Fig. 1.50 refers to a drilling jig for drilling oil holes in an I.C. engine connecting rod.

9. *Angle plate jig*: This type of jig is used to drill holes in parts at right angles to their mounting locations. Fig. 1.56 can be an example of angle plate jig.

10. *Pot jig:* This type of jig is used for drilling holes in circular components, which have both internal and external diameters. The body of the jig is in the form of a pot. The workpiece is located in the pot of the jig and is properly clamped with the help of a post type locating pin, a clamping plate and a clamping device, Fig. 1.51.

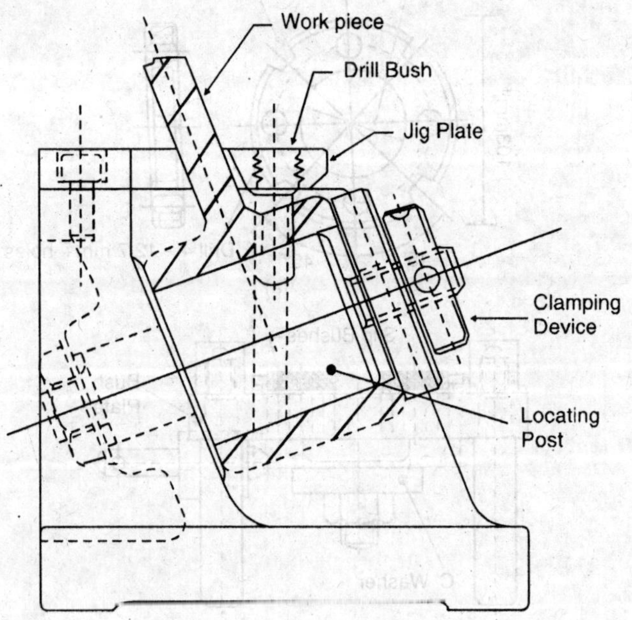

Fig. 1.50. Angular Jig

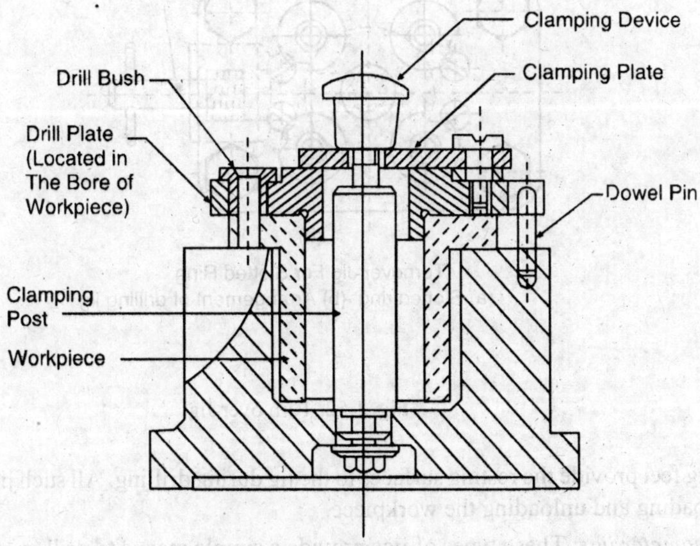

Fig. 1.51. Pot Jig

11. *Turn-over jig:* These jigs are used to drill holes in corresponding having no suitable resting surfaces. They are the modification of plate type jigs type jigs with jig feet, Fig. 1.52.

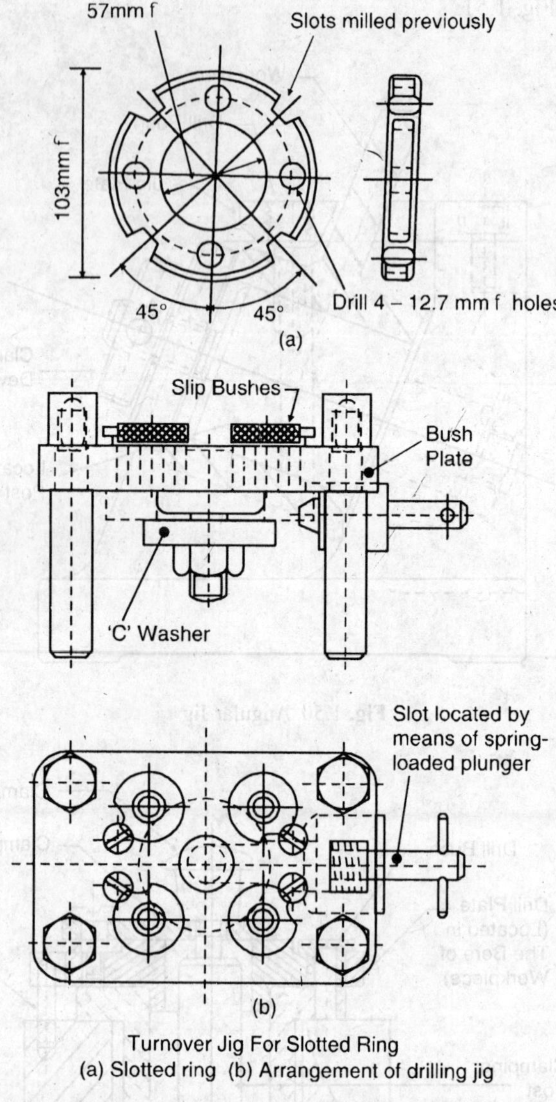

Fig. 1.52. Turn over Jig

The jig feet provide the resting surfaces to the jig during drilling. All such jigs should be "turned-over" for loading and unloading the workpiece.

12. *Diameter jigs*: These types of jigs provide a simple means to drill or ream radial holes on a diameter of cylindrical or spherical jobs. The job can be very conveniently located on a V-block and clamped by a clamping plate and a clamping bolt, Fig. 1.53.

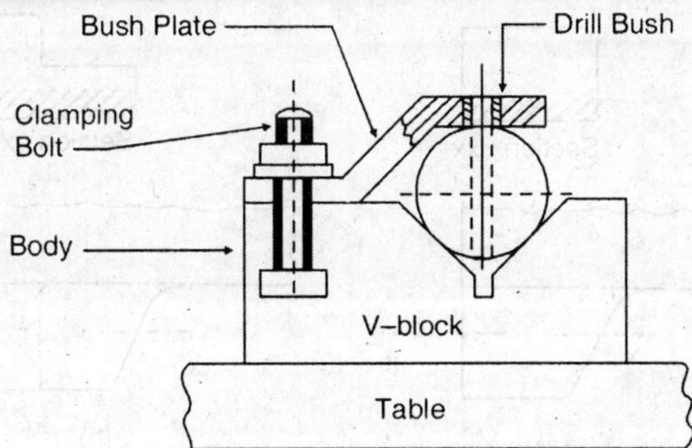

Fig. 1.53. Diameter Jigs

13. *Multi-station jig*: This type of jig has a circular indexable plate mounted on a circular base plate which is clamped to the machine table with the help of T-bolts and nuts. Such jigs are used on multi-spindle machines. Supposing the indexable circular plate has four stations. On station 1, drilling operation in being done, on station 2, reaming and on station 3, counterboring operation is being done. Station 4 is used for loading and unloading of the jobs. After every set of operations, the circular plate is indexed by 90°.

14. *Standard jigs*: There are many components that are similar in design, but different in dimensions (cylindrical pins of same diameter, but of different lengths of cylindrical pins of different diameters but of same length/different lengths, etc.) It is sometimes possible to drill several of these different components in one jig. The jig incorporates an adjustable end locator to accommodate a variety of lengths. When a jig is especially designed and fabricated for several similar parts, it is called a standard jig. The diameter jig, Fig. 1.53, provided with an adjustable end locator can be an example of such a jig.

15. *Universal jig*: This jig is first manufactured as a basic unit (just like unit heads for machine tools) to which a number of other elements and parts can be fitted to make it suitable for a specific job. Since the same basic unit can be adopted for different work-pieces and operations, simply by making additions and alterations, hence, the name "Universal Jig".

16. *Trunnion jigs*: The manual manipulation of heavy duty box type jigs is quite inconvenient and fatiguing. So, such jigs are mounted on trunnions to bring the different faces of the workpiece to the correct locations for drilling the holes.

17. *Ring jig*: Such a jig is suitable for drilling holes in round jobs, such as, flanges of pipes.

18. *Solid jig*: If holes are to be drilled in workpiece of simple shape and relatively of smaller size, then the jig body can be fabricated from a standard section of rolled steel. Such a jig can be named as a "solid Jig".

1.4.6. Jig Feet. Jig feet are needed to support the jig on the table of the machine. Jig feet should be provided opposite each working face of the jig. In normal design, the number of jig feet should be four. This is due to the reason that if any swarf or foreign matter gets under one of the jig feet, the jig with four feet will rock about this point. This will at once indicate that the jig is not resting properly on the machine table. This defect will go undetected if three feet are provided since any plane will pass through three points. So, when a jig with three feet is placed on the machine table, it will rest in almost any normal position even though swarf or some foreign matter be present under one or more of the feet. Such a jig will not be in perfect alignment with the machine spindle and the workpiece will be drilled or machined incorrectly. Hence a jig should have four feet.

Jig feet may be either cast integral with the body of the jig or fixture or built up. Cast feet are shown in Fig. 1.54. The built up feet (Fig. 1.55) can be installed into the jig body in many

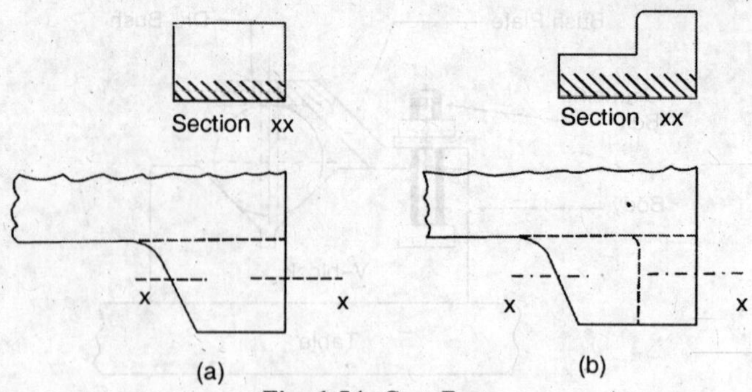

Fig. 1.54. Cast Feet.

ways. In Fig. 1.55 (*a*) the foot is press fitted into the jig body. A drift hole should be provided for forcing the foot out. In Fig. 1.55 (*b*) the foot is screwed into the jig body. Figs 1.55 (*c*) and 1.55 (*d*) show turnover feet. The short foot is used to support the jig when it is turned over.

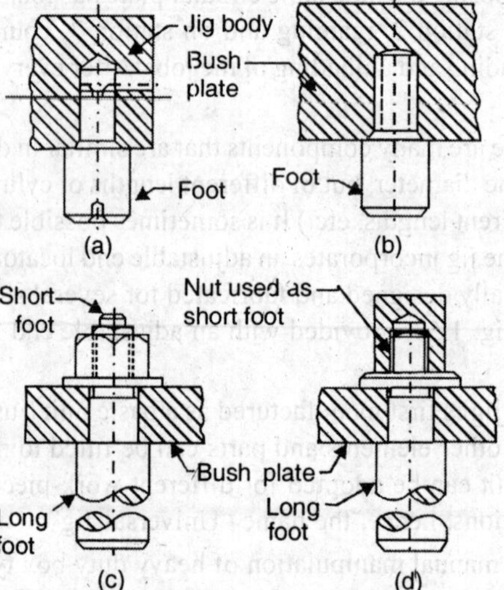

Fig. 1.55. Built up Feet.

In both cases, jig feet are held by a shoulder on the jig feet against the jig body by a nut. The supporting faces of the steel feet should be hardened and finished ground.

A drilling jig for machining a hole in a casting is shown in Fig. 1.56.

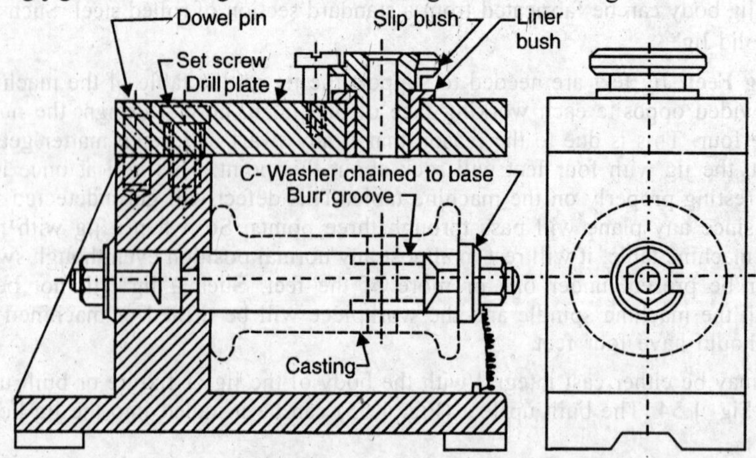

Fig. 1.56. A Drilling Jig.

1.5. MILLING FIXTURES

A milling fixture is a work holding device which is firmly clamped to the table of the milling machine. It holds the workpiece in correct position as the table movement carries it past the cutter or cutters. A milling fixture is usually a good solid casting, since cast iron has the important characteristic of absorbing and damping out the vibrations resulting from the cutting action of the milling cutters. The essential features of a milling fixture are : A heavy base, location and clamping elements and setting blocks.

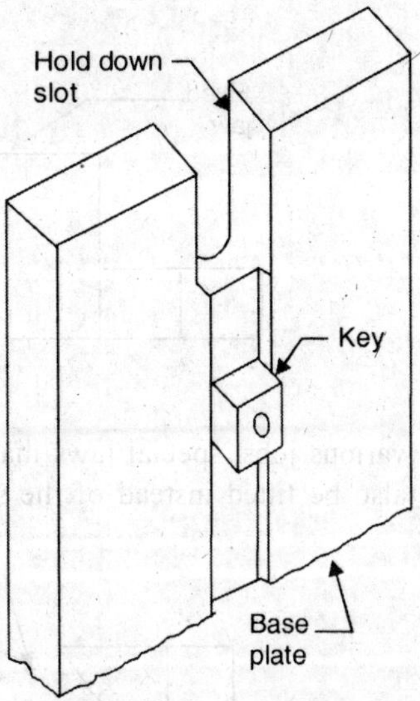

Fig. 1.57. Milling Fixture Base.

(*i*) *Base* : A heavy base is the most important element of a milling fixture. It is a plate with a flat and smooth underface. The complete fixture is built up from this plate. Keys are provided on the underface of the plate which are used for easy and accurate aligning of the fixture on the milling machine table by inserting them into one of the *T* slots in the table. These keys are usually set in keyways on the undersurface of the plate and are held in place by a socket head cap screw for each key Fig. (1.57). The fixture is fastened to the machine table with the help of two *T* bolts engaging in the *T*-slots of the table and protruding through the hold down slots or ears in the mill fixture base.

(*ii*) *Location and clamping elements*. The same design principles of location and clamping apply for milling fixtures as have been discussed earlier.

(*iii*) *Setting blocks*. After the fixture has been securely clamped to the machine table, the workpiece which is correctly located in the fixture, has to be set in correct relationship to the cutters. This is achieved by the use of a setting block and feeler gauges. The setting block is fixed to the fixture. Feeler gauges are placed between the cutter and reference planes on the setting block so that correct depth of cut and correct lateral setting is obtained (Fig. 1.58). The block is made of steel, hardened and with the reference planes (feeler surfaces) ground. Sometimes, if convenient, the feeler surfaces may be machined in a part of the fixture. These should be clearly marked so that the operator is certain as to which are the feeler surfaces. In its correct setting, the cutter should clear the feeler surfaces by at least 0.08 cm to avoid any damage to the block when the machine table is moved back to unload the fixture. The thickness of the feeler gauge to be used should be stamped on the fixture base near the setting block. It is common to keep a gap of 0.4 to 0.5 mm (0.8 mm noted above) between the cutter and the setting planes of the setting block. This gap facilitates the cutter setting with the help of feeler gauges, Fig 1.58.

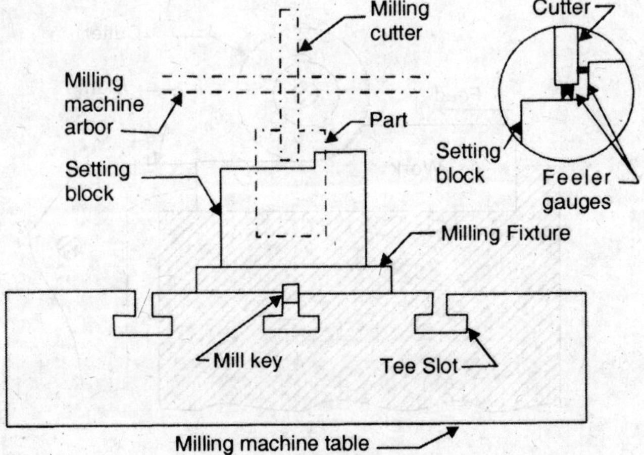

Fig. 1.58. Setting Block.

1.5.1. Milling Machine Vice. For many jobs, the standard machine vice can be used to clamp the workpiece under cutter, Fig. 1.59. The movable jaw can be operated either by a cam or a screw. To suit the contours of the various jobs, special jaws may be attached to the plain jaws of the standard vice, and vees may also be fitted instead of the standard plain jaws, Fig. 1.60.

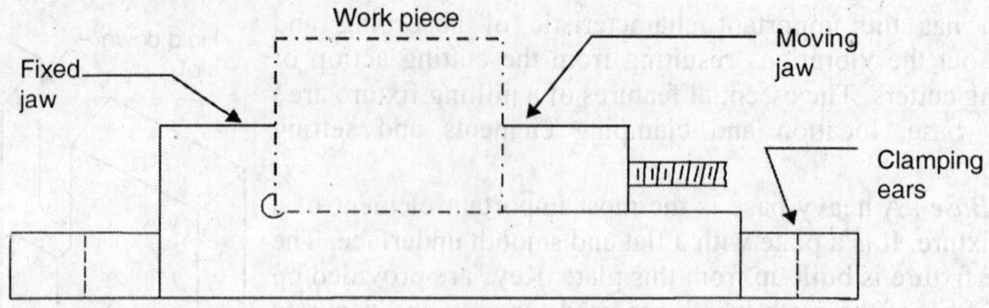

Fig. 1.59. Milling Vise.

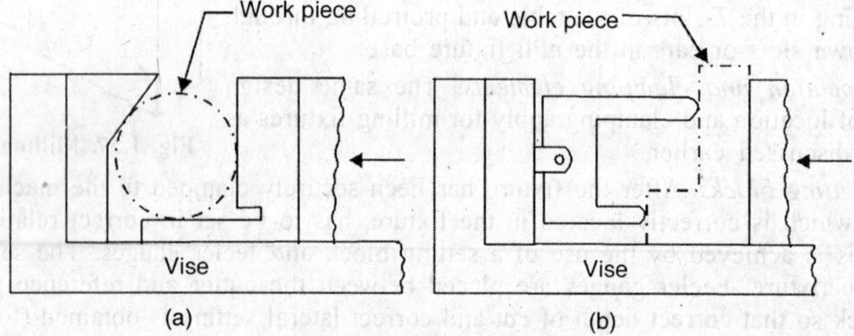

Fig. 1.60. Special Jaws for the Vice.

For short runs, the special jaws may be made of aluminium alloy or soft mild steel. For long runs, the wearing surfaces of the special jaws should be hardened and should be of renewable type. Cast iron jaws are also frequently used. Wherever possible and convenient, setting blocks should also be provided.

1.5.2. Some Design Principles for Milling Fixtures. 1. Pressure of cut should always be against the solid part of the fixture, Fig. 1.61.

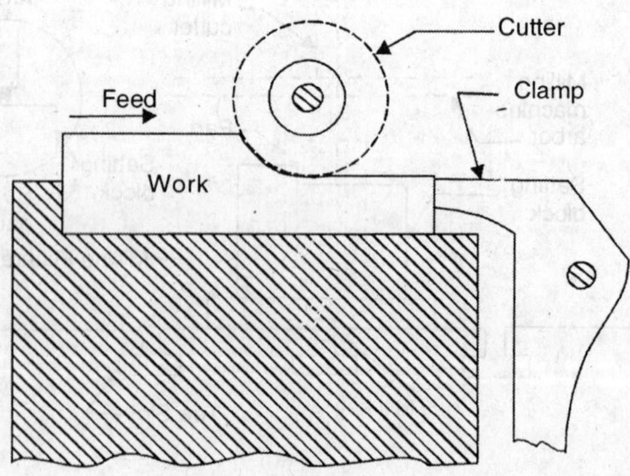

Fig. 1.61.

2. Clamps should always operate from the front of the fixture, Fig. 1.62.

3. The workpiece should be supported as near the tool thrust as possible, Fig. 1.63.

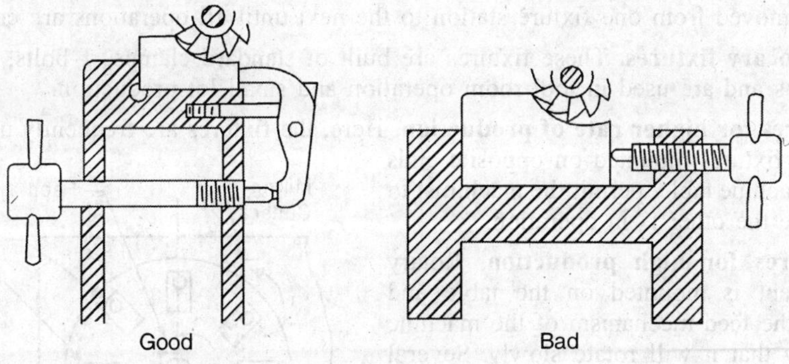

Fig. 1.62

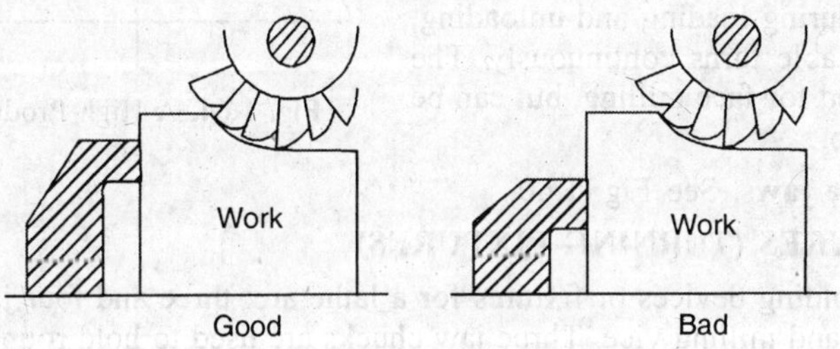

Fig. 1.63

1.5.3. Classification of Milling Fixtures. There is no standard method of classification of milling fixtures. This is because, each fixture is different and must be designed to meet certain requirements. However, these may be classified in the manner discussed below:

1. **According to the type of the milling operation performed on the work.** According to this criteria, the milling fixtures can be : Face milling fixtures, slab milling fixtures, slotting fixtures, straddle milling fixtures, gang milling fixtures, string or line milling fixtures, profile milling fixtures and so on.

2. **According to the way the workpiece is clamped.** Under this category, we have : Hand clamping fixture, power clamping fixtures, toggle fixtures or automatic fixtures.

3. **According to the way the workpiece is located**. Such as centre fixture, V-block fixtures and pin or stud fixtures.

4. **According to the method of presenting the workpeice to the cutter.** Here we have:

(*a*) *Cradle Fixtures*. Workpiece is rocked or rotated within a given angle during milling.

(*b*) *Rotary Fixtures*. Workpiece is rotated under the cutter.

(*c*) *Drum Fixtures*. Workpiece is mounted on the periphery of a rotating drum.

(*d*) *Indexing Fixtures*. Workpiece is indexed into the next position during the machining cycle of the mill.

(*e*) *Rise and Fall Fixtures*. These fixtures allow raising and lowering of the workpiece in conjunction with the mill feed.

5. **Universal fixture.** These fixtures are designed to hold a family of parts similar in shape but different in size.

6. **Progressive fixtures.** Here the workpiece can be located in different positions and progressively moved from one fixture station to the next until all operations are complete.

7. **Temporary fixtures.** These fixtures are built of standard clamps, T-bolts, heel blocks, jacks and stops and are used in tool room operation and small lot production.

8. **Fixtures for higher rate of production.** Here, the fixtures are frequently used in pairs. Two identical fixtures mounted on opposite ends of a milling machine table, permit the machinist to load one while the other is in operation.

9. **Fixtures for high production.** Rotary table attachment is mounted on the table and connected to the feed mechanism of the machine in such a way that it will rotate slowly. Several fixtures are mounted on the rotary table, Fig. 1.64, and a part is clamped in each. As a part is finished, the operator merely removes it and inserts an unfinished one. With this set-up, no cutting time is lost during loading and unloading, because, the rotary table turns continuously. The method is widely used for face milling, but can be used for slotting also.

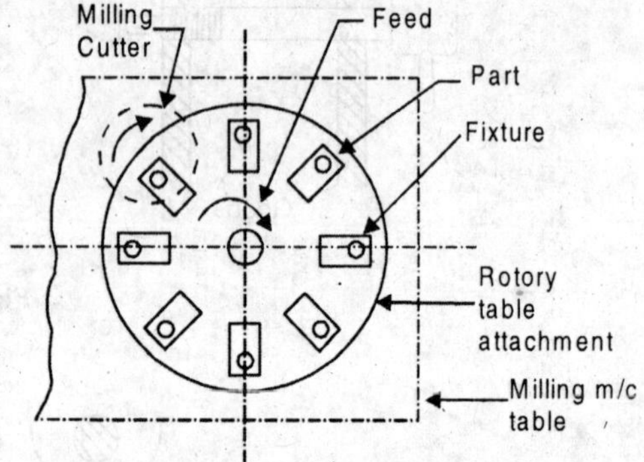

Fig. 1.64. A High Production Fixture.

10. **Special vise jaws.** See Fig. 1.60.

1.6. LATHE FIXTURES (TURNING FIXTURES)

The standard workholding devices or fixtures for a lathe are: three and four jaw chucks, collets, face plate, mandrels and milling vice. Three jaw chucks are used to hold round or hexagonal bar stock or other symmetrical work. Four jaw chucks are used for rough castings and square or octagonal work. When irregular shaped castings and forgings are to be machined, special chuck jaws may be used to suit the component. Collets are used primarily for bar stock. Some types of jobs can be mounted and supported on a face plate. Standard equipment such as angle brackets and flat plates may be screwed or bolted to the face plate and the work clamped to these. For example, jobs which are hard to hold such as pipe elbow may be bolted to an angle plate which in turn is bolted to the face plate. Hollow jobs such as gear blanks and pulleys can be supported on a mandrel between centres and rotated by a dog. A mandrel is a hardened bar held between centres. It has flats on each end for clamping a lathe dog. The mandrel is tapered about 0.5 mm per metre so that the job can be forced on it with a press fit and then unlocated after machining. A milling vice can be mounted on the lathe bed in place of the compound rest, to hold the work and the cutter is inserted in the lathe spindle.

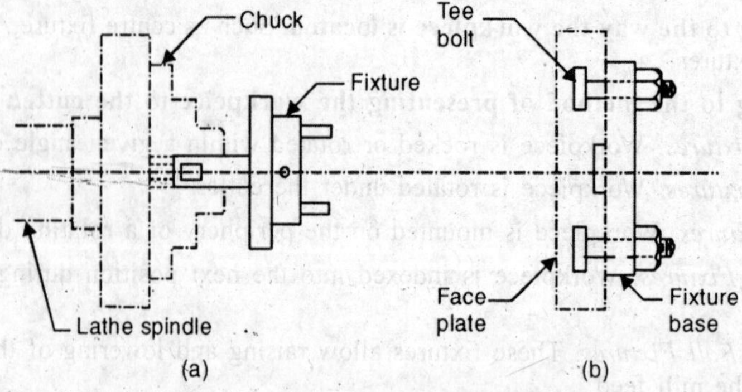

Fig. 1.65. Fixing of a Lathe Fixture.

If the job can be held easily and quickly in the above mentioned standard devices, then there is no need for a special workholding device. However, many jobs, particularly castings and forgings, because of their shape, cannot be conveniently held by any of the standard devices. It then becomes necessary to build a special work holding device for the job. Such a device is called a lathe fixture. A lathe fixture consists of a base, location and clamping devices as usual and an arrangement for locating and fastening the fixture securely and accurately in the lathe. A lathe fixture can be fixed to the lathe either by holding in the chuck jaws or fixing to a face plate, Fig. 1.65. The following design points should be considered for a lathe fixture:

1. To avoid vibrations while revolving, the fixtures should be accurately balanced.

2. There should be no projections of the fixture which may cause injury to the operator.

3. The fixture should be rigid and overhang should be kept minimum possible so that there is no bending action.

4. Clamps used to fix the fixture to the lathe should be designed properly so that they don't get loosened by centrifugal force.

5. The fixture should be as light weight as possible since it is rotating.

6. The fixture must be small enough so that it can be mounted and revolved without hitting the bed of the lathe for which it is designed. The designer should check the "swing" of the lathe.

7. The part must be mounted and held in such a way that there will be no obstructions in the way of the cutting tool or tools.

1.7. GRINDING FIXTURES

The workholding devices for grinding operations will depend upon the type of grinding operation and the machine used. We shall take each case one by one, as below :

1. **Fixtures for external grinding.** A mandrel is the most common fixture used for grinding external surface of a workpiece. As written earlier, a mandrel is hardened and is held between centres of a machine. The mandrel is used for internal chucking of round workpiece with bores. The workpiece is located and held on the mandrel with the help of the bore so that the external surface may be machined truly concentric to the bore. The various types of mandrel are:

(*i*) *Taper mandrel.* In this type of mandrel, the outer chucking surface is given a slender taper of about 0.5 mm per metre, Fig. 1.66. The diameter of the bore of the workpiece must be smaller than the largest diameter of the mandrel. The workpiece is forced endwise on to the mandrel. The gripping force between the workpiece and the mandrel is created due to the taper on the mandrel.

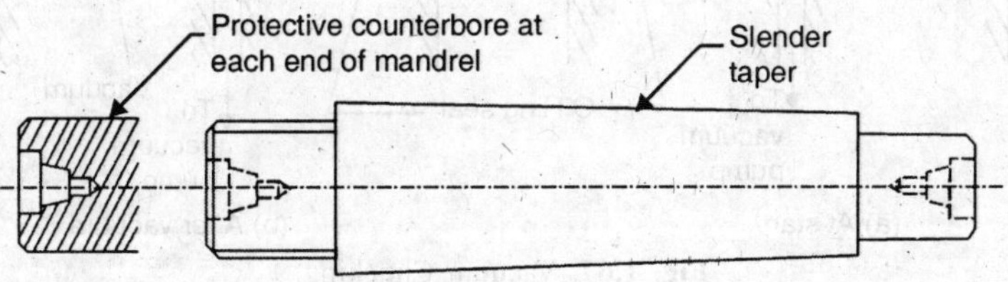

Fig. 1.66. Taper Mandrel.

(*ii*) *Straight mandrel.* It differs from the taper mandrel in that it has straight or untapered chucking surface. To press fit the workpiece on to the mandrel, the outer diameter of the mandrel is made bigger than the bore of the workpiece. The difference known as interference depends upon the wall thickness, diameter and the material of the workpiece. It should not be able to cause plastic deformation of the mandrel and/or of the workpiece.

(*iii*) *Combined taper and straight mandrel.* In this type of mandrel, a portion of the outer diameter of the mandrel is straight and the rest is tapered. The straight portion helps to prealign the bore of the workpiece and the tapered portion provides the driving arca on to which the workpiece is press fitted.

2. **Fixtures for internal grinding.** For grinding internal surfaces of simple circular workpieces, the chuck may be used as a standard workholding device. If required, special jaws can be provided for the chuck. However, for many components, special fixtures may have to be made which are designed on the same lines, as the lathe fixtures.

3. **Fixtures for surface grinding.** The workpiece can be held for machining on a surface grinder in the following ways:

(*i*) It may be clamped directly to the machine table or to an angle plate and so on.

(*ii*) It may be held in a vice.

(*iii*) The workpiece may be held by means of a magnetic chuck or a vacuum chuck. Here, the workpiece is held without any mechanical clamping.

(*iv*) The workpiece may be held in a special fixture.

(*a*) **Magnetic chuck.** A magnetic chuck holds a workpiece by the magnetic force which may be generated either by permanent magnets or by electromagnets powered by direct current. The gripping power of the chuck depends on the strength of the magnets and amount of magnetic flux which passes through the workpiece. The magnetic chuck is clamped to the machine table. Magnetic chucks are available in various shapes such as rectangular, circular and also as a *V*-block. A magnetic chuck has the advantage of being fast acting and causing minimum distortion of the workpiece. However, it can be used for holding only ferrous workpieces. Also, some residual magnetism is imparted to the workpiece which must be removed by demagnetising it.

(*b*) **Vacuum chucking.** This is a very convenient method of holding workpieces of non-magnetic materials. The principle of vacuum chucking is explained in Fig. 1.67. The workpiece is placed on the '*O*' ring seal inserted in the workholder. This forms a close chamber between the locating surface of the workpiece and the mating surface of the workholder. At start, the air pressure on the inside and outside of the chamber will be equal. Then a vacuum is created in the chamber. The outside pressure then holds the workpiece against the locating surface of the workholder.

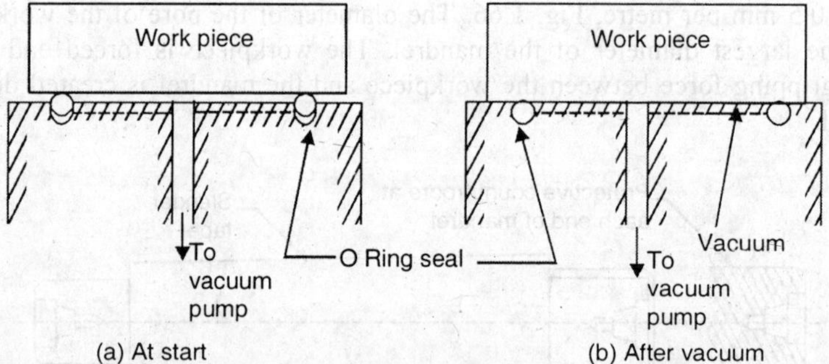

Fig. 1.67. Vacuum Chucking.

1.8. BROACHING FIXTURES

Broaching fixtures are must for workholding because of the high forces involved and because of the manner in which the operation is performed. Broaching fixtures are required to perform one or more of the following functions:

1. Hold the job rigidly.
2. Locate the job in correct position relative to the tool of the machine table.
3. Guide the broaching tool in relation to the job.

4. Move the job into and out of the cutting position.
5. Index the job between the cuts.

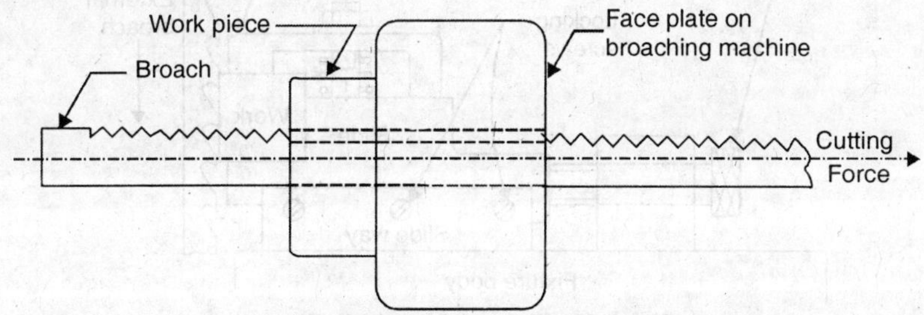

Fig. 1.68. Broaching Fixture.

Fixtures are needed for internal broaching and external broaching. The fixtures used for internal broaching are the simplest and for many operations may consist of a face plate or support plate on the broaching machine (Fig. 1.68). The fixtures for external broaching are made quite rigid so that the workpiece does not move during the broaching action.

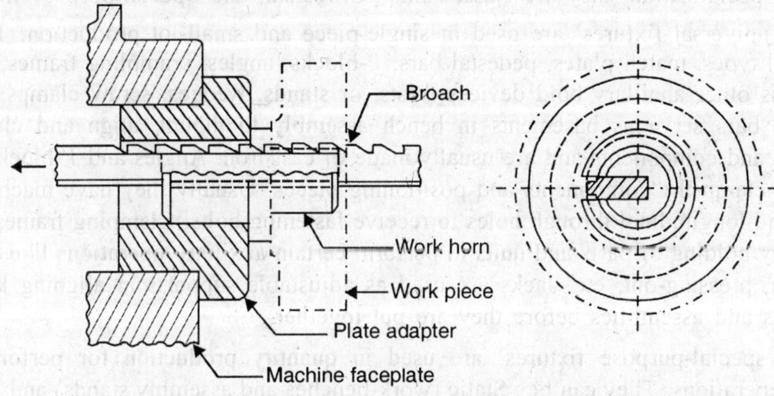

Fig. 1.69. A Simple Keyway Broaching Fixture.

A simple broaching fixture used to machine a keyway is shown in Fig. 1.69. The fixture is simply a plug or horn fixture which establishes the correct position of the workpiece in relation to the broach and the machine faceplate. The broaching force will hold the work in position of the fixture. Special work horn may be designed to broach keyways at an angle, Fig. 1.70.

In the case of external broaching, the broaching force will not hold the work in position on the fixture, but will push it away, Fig. 1.71. Therefore, surface broaching fixtures are more elaborate and require proper clamping arrangements. External broaching will usually require a special fixture for each job.

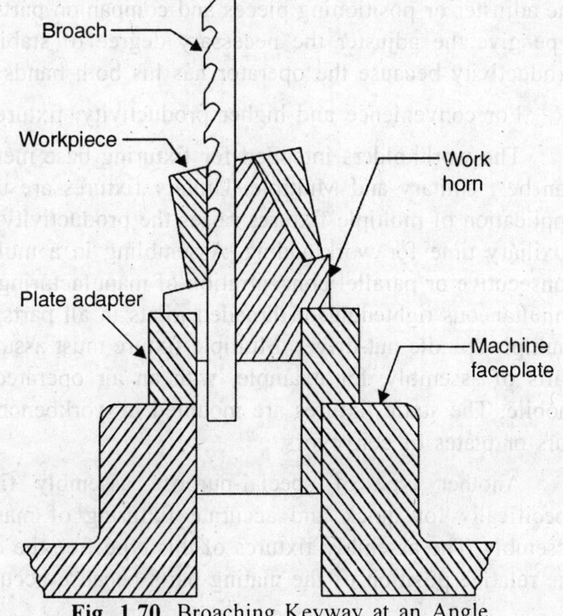

Fig. 1.70. Broaching Keyway at an Angle.

1.9. ASSEMBLY FIXTURES

Assembly fixtures are used to hold the various components in their correct position while they are assembled. This is particularly the case when the various parts are to

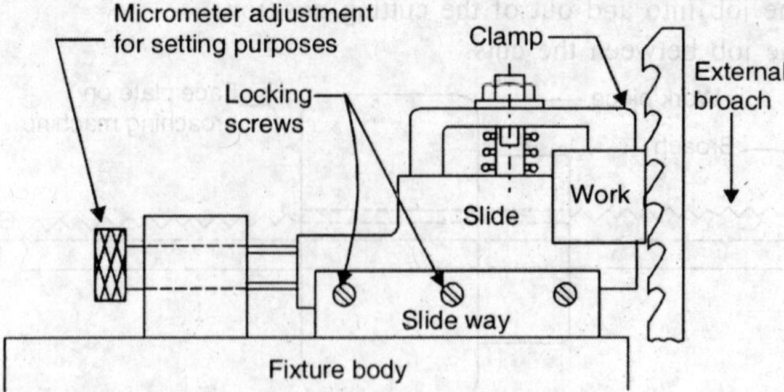

Fig. 1.71. External Broaching Fixture.

be put together for welding. They are indispensable in the component and final assembly of units. They are simple and effective means used to mechanize the manual assembly operations. Their use assures higher productivity and convenient performance of the assembling operations, securing a quick and accurate fixturing of the mating parts during assembly. According to the degree of specialization, they are classed into : Universal, and Special-purpose fixtures.

The 'universal fixtures' are used in single-piece and small-lot production. These include vices of all types, match plates, pedestal bars, *V*-blocks, angles, crampling frames, lifting jacks, and various other ancillary hold devices (rests, or stands, wedges, screw clamps, etc.). Match plates and bars serve as basements in bench assembly to mount, align and clamp machine assemblies and components and are usually made of cast iron. Angles and *V*-blocks are used to locate and clamp the components and positioning pieces. Usually they have machined locating surfaces and longitudinal through holes to receive fastening bolts. Clamping frames are used for a temporary holding of parts and units to perform certain auxiliary operations like straightening, pressing-in, pressing-out, etc. Jacks are used as adjustable supports in aligning large-size and heavy parts and assemblies before they are put together.

The 'special-purpose fixtures' are used in quantity production for performing special assembly operations. They can be : Static (work-benches and assembly stands) and mobile which are fixed to plates of chain conveyors, turntables, cars, dollies, etc.

The static fixtures are set up-clamping or fit-up fixtures used to fix in position and clamp the adjuster or positioning pieces and companion parts of the product assembly. Fixtures of this type give the adjuster the necessary degree of stability during the assembly operation, raise productivity because the operator has his both hands free to do the assembling.

For convenience and higher productivity, fixtures are often made turnable.

The workholders intended for fixturing base members and component parts of assemblies can be : Unitary and Multiple. Unitary fixtures are used to clamp one product assembly. The application of multiple fixtures raises the productivity of assembling due to the reduction of the auxiliary time for workloading. Assembling in a multiple fixture operates on the principle of consecutive or parallel concentration of manufacturing steps. An example of the later can be the simultaneous tightening of threaded joints in all parts, located in the fixture, with the help of a multiple-spindle nut-driver. Multiple fixture must assure a uniform and quick grasping of all the parts of assembly, for example, with an air operated clamp. Such fixtures can be static and mobile. The static fixtures are mounted in workbenches or assembly stands and the mobile, in cars or plates of conveyors.

Another type of special-purpose assembly fixtures includes work holders designed specifically for quick and accurate fixturing of mating parts and components of a product assembly. The assembly fixtures of this type free the operator from the need to check and align the relative position of the mating parts because accurate positioning is achieved automatically by bringing these parts into positive contact with the fixture supporting and guiding elements. Such fixtures are frequently used in welding, brazing and soldering, riverting glueing, expanding, fitting parts with interference into threaded and other assembly joints. These fixtures assure a considerable rise in productivity and are indispensable in automatic assembly. The special-purpose assembly fixture consists of a body (base member) on which mounted are the locating elements and clamping devices. The fixture need only be of light construction with adequate rigidity to ensure relative positional relationships of the various components. They may be built up from light castings, steel sections or completely from steel.

Some simple types of welding fixtures are shown in Fig. 1.72. The parts to be welded are placed in the fixture in their correct relationship and the whole arrangement tightened. The parts are then welded whilst in place. The locating and clamping principles described earlier, also apply to welding fixtures. However, a few design principles applicable only to welding fixtures are given below :

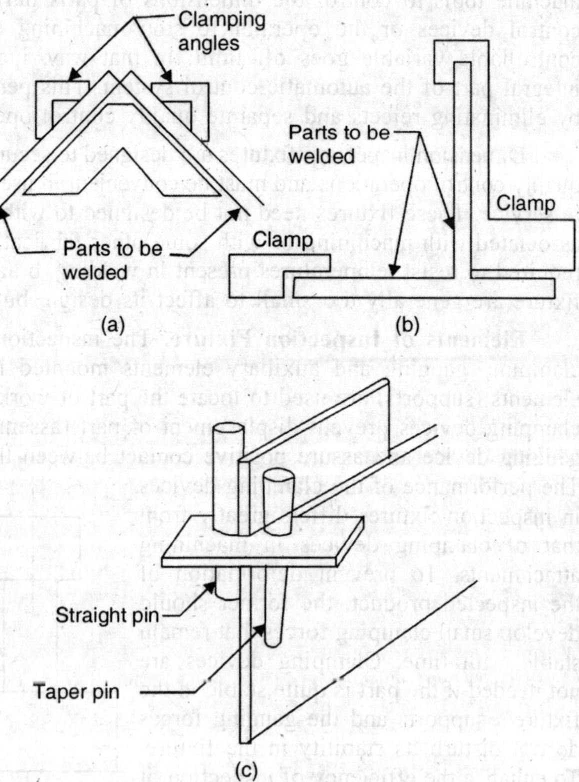

Fig. 1.72. Simple Welding Fixtures.

1. Due to intense heat produced during welding process, expansion will take place. Due to this, the welded assembly may get locked in the fixture, making its removal very difficult. To avoid this, the locators should be designed accordingly.

2. Clamping pressure should be light and clamps should be arranged in such a way that the workpiece does not get distorted.

3. Clamps should be kept clear of welding zone or be shielded.

4. Location and clamping should not make the welding zone inaccessible.

5. The welding fixture should be rigid and stable.

1.10. INSPECTION FIXTURES

An inspection (qualifying, gauging) operation is any examination of a workpiece that determines whether or not it meets the standards of quality. Inspection fixtures are used to check the quality of the workpieces, parts and components of machines. Dimension inspection or guaging fixtures raise the efficiency of the work of human inspectors, improve their working conditions, quality of workpieces, parts and components of machines. Fixtures for checking parts are usually employed at intermediate stages of machining (step-by-step inspection) and at the final stage of machining (acceptance inspection) to verify the accuracy of dimensions, relative position of surfaces and adequacy of surface geometry.

High precision of modern machines is conducive to the use of inspection fixtures of high sensitivity meters and gauges. Small and medium-size parts are checked in stationary fixtures and large-size parts and products, in portable fixtures. Gauging fixtures can be single-and multi-limit control devices. The last are used to measure several parameters in one set up.

Inspection fixtures are classed into in process and off-line gauging attachments. The latter are used for post-operation inspection. The in-process inspection attachments are mounted on machine tools to control the dimensions of parts during machining by signalling the machine control devices or the operator to stop machining or change the cutting conditions as the controllable variable goes off limit. In that way, independent inspection fixtures become an integral part of the automatic control system. This permits cutting down the cost of production by eliminating rejects and separate quality control operations.

Dimension inspection fixtures are designed to secure the pre-set accuracy and efficiency of the quality control operations and must be convenient in use, simple in construction, cheap and reliable in service. These fixtures need not be designed to withstand forces, such as shock and vibration, associated with machining or with some other fabricating and assembly processes. They are not required to resist temperatures present in welding, brazing etc. Clamping forces in an inspection fixture are generally too small to affect its design, but they should not distort the workpiece.

Elements of Inspection Fixture. The inspection or gauging fixture consists of locating, clamping, gauging and auxiliary elements mounted in the body of the fixture. The locating elements (supports) are used to locate the part or workpiece from its reference for gauging. The clamping devices prevent displacement of part (assembly) set in it for checking relative to the gauging device and assure positive contact between the part's locations and fixture's supports. The performance of the clamping devices in inspection fixtures differs greatly from that of clamping devices in machining attachments. To prevent deformation of the inspected product, the former should develop small clamping forces that remain stable with time. Clamping devices are not needed if the part is quite stable in the fixture's supports and the gauging forces do not disturb its stability in the fixture. To enhance the efficiency of inspection, it is advisable that the clamping device be quick-grasp and convenient in use.

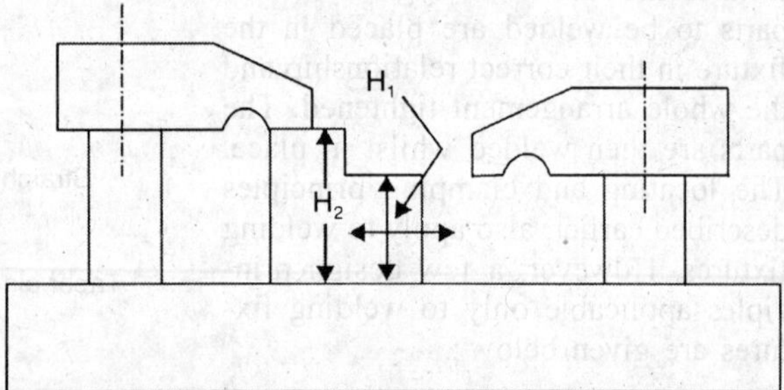

Fig. 1.73. An Inspection Fixture with Rigidly Fixed Limit Elements.

The clamping devices in inspection fixtures are usually hand-operated (lever and spring-actuated, screw and eccentric-type clamps) and power devices, such as pneumatic clamps whose power is also used to actuate the fixture's ancillary mechanisms (lifters, rotators, pushers).

Application of multiple gauging fixtures enhances the quality of control, frees human labour and reduces the cost of instrumentation. In order to make multiple gauging applicable, it is necessary that the controllable dimensions be measured against a common reference location.

The measuring instruments in inspection fixtures can be control-limit (non dial) and reading (dial) gauges. The devices which employ the principle of normal gauges constitute a special group of size-control instruments.

The simplest fixture arrangement is one which has rigidly fixed limit elements to check the height of the shoulders (dimensions H_1 and H_2) of a stepped part being moved along the baseplate of the fixture by hand, Fig. 1.73, the lower surface being the reference surface. The limit elements can also be extensible.

1.11. BORING FIXTURES

In boring operation, an already existing hole in the workpiece is enlarged. The boring tool is usually a single point tool and the size of the tool will depend upon the adjustment of the tool within the boring bar. The boring operation may be done by either of the two arrangements– boring bar is stationary and the workpiece moves into the bar or the workpiece is stationary and the boring bar moves into the workpiece. Boring fixtures can be divided into two general classes: one, where the fixture guides the boring bar as in drill jigs and it is more appropriately called a Boring jig, and the other in which fixture holds the workpiece in proper relation to the boring bar as in milling fixture. The various machine tools employed for boring operation can be : lathes, drill presses, milling machines, jig boring machines and vertical or horizontal boring mills. According to its length, a boring bar may be classified as : a stub bar, single piloted bar and double piloted bar. A stub boring bar is short in length and is not supported along its length. It is used for boring short holes and also blinds holes. A single piloted bar is longer than a stub bar and is supported at its leading end so that it does not spring under cutter thrust, Fig. 1.74. A double piloted bar is supported both at leading and trailing ends, Fig. 1.75. Sometimes, slip bushings fitted into the finished holes in the workpiece may act as support and guide for the boring bar, Fig. 1.76. In order to eliminate any trouble from misalignment of the jig and the spindle of the boring machine, the boring bar is connected to the spindle by a short intermediate shaft fitted with a universal joint at each end. The general principles of jig and fixture design are also applicable for boring jigs and boring fixtures.

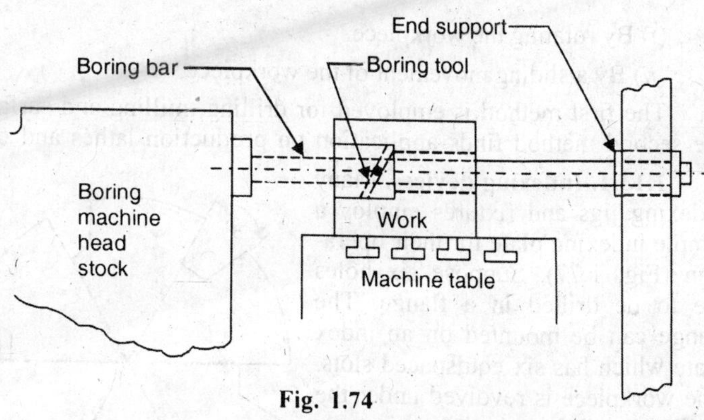

Fig. 1.74

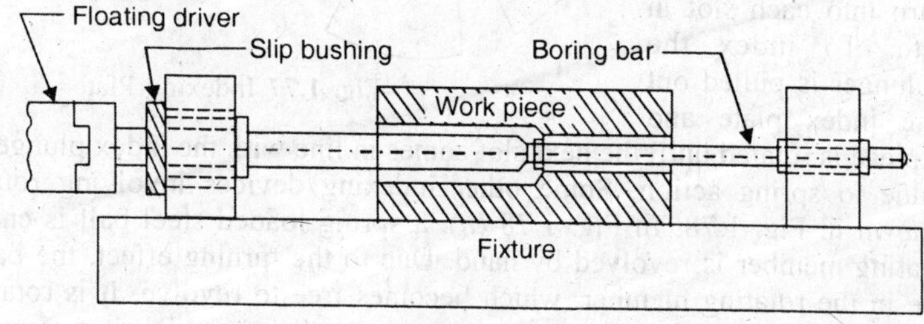

Fig. 1.75

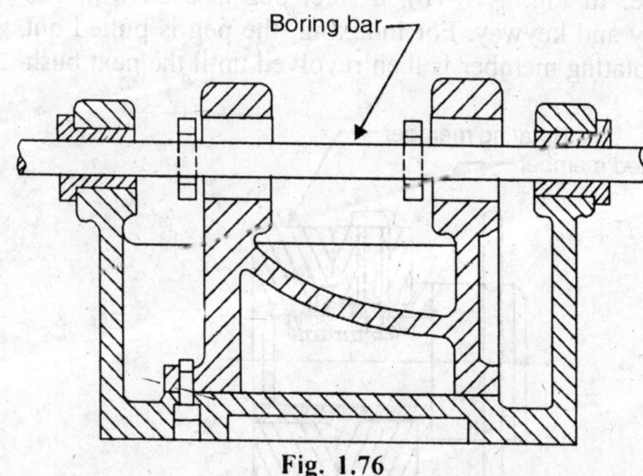

Fig. 1.76

1.12. PLANING AND SHAPING FIXTURES

We know that Planers and Shapers are not very efficient machine tools. Due to this, they are usually not recommended for mass production. However, jigs and fixtures are normally used on machine tools engaged in mass production. As such fixtures do not find much utility on Planers and Shapers. Most of the time, the general purpose work-holding devices are used on these machine tools, such as, T-bolts and clamps, stop pins and Toe dogs and Vises (See the book by the author: A T.B.O. Production Technology). However if a job can not be held in one of these devices, then a fixture will have to be designed and fabricated.

1.13. INDEXING JIGS AND FIXTURES

Indexing jigs and fixtures are used when holes or slots are to be machined to some specific relationship, in a workpiece. These are of particular interest for drilling, milling, and grinding operations. There are two ways in which a workpiece may be moved under the cutting tool for machining each hole or slot or for other operation:

(*i*) By rotating the workpiece.

(*ii*) By a sliding movement of the workpiece.

The first method is employed for drilling, milling and surface grinding operations, whereas the second method finds application on production lathes and on external grinding machines.

1.13.1. Indexing devices. Many indexing jigs and fixtures employ a simple indexing plate for their operation (Fig. 1.77). Suppose six holes are to be drilled in a flange. The flange can be mounted on an index plate which has six equispaced slots. The workpiece is revolved under the drill and each hole is drilled in turn. For this an index plunger is used which fits by turn into each slot in the index plate. To index the workpiece, the plunger is pulled out of the slot. The index plate and thereby the workpiece is rotated until the next slot comes in line with the index plunger into which it gets pushed due to spring action. Some other indexing devices involving rotation of the workpiece are shown in Fig. 1.78. In Fig. 1.78 (*a*), a spring loaded steel ball is employed. For indexing, the rotating member is revolved by hand. Due to the turning effect, the ball is pushed out of the groove in the rotating member, which becomes free to revolve. It is rotated until the next groove in the rotating member is encountered by the ball into which it gets pushed due to spring action. This device is not very accurate. In Fig. 1.78 (*b*), a steel peg is used which is retained in the fixed member by means of a key and keyway. For indexing, the peg is pulled out clear of the bush in the rotating member. The rotating member is then revolved until the next bush is encountered by the peg, into which it passes.

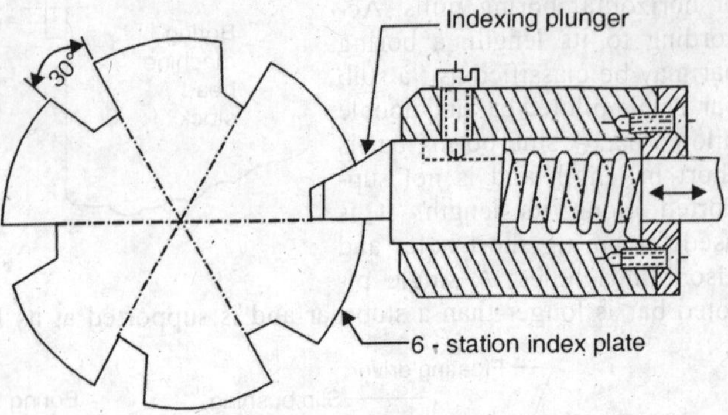

Fig. 1.77 Indexing Plate

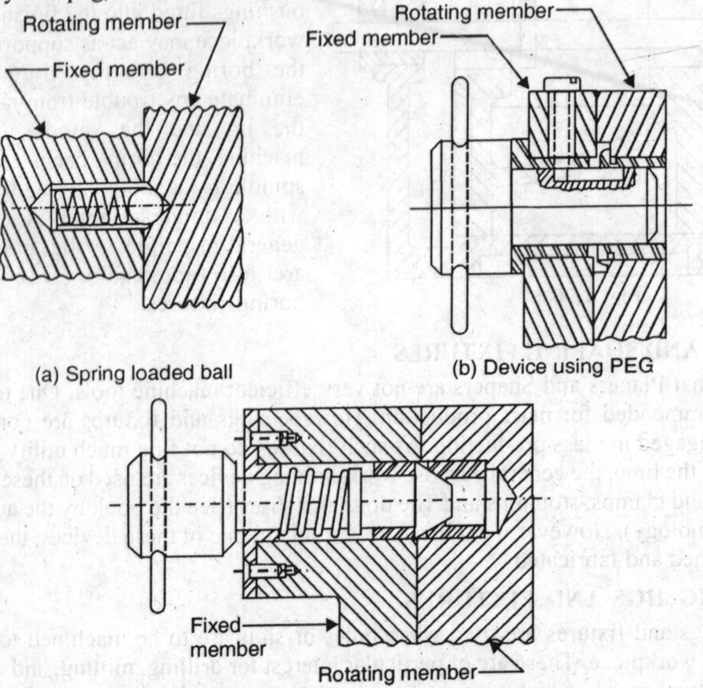

Fig. 1.78. Indexing Devices.

The same principle is employed in the device shown in Fig. 1.78 (*c*). Here the peg is spring loaded. The arrangement using a sliding member is shown in Fig. 1.79. The sliding member is provided with slot at suitable spacings and the fixed member carries a spring loaded lever, which fits the slots. For indexing, it is lifted off a slot and the sliding member is moved until the next slot comes under the lever.

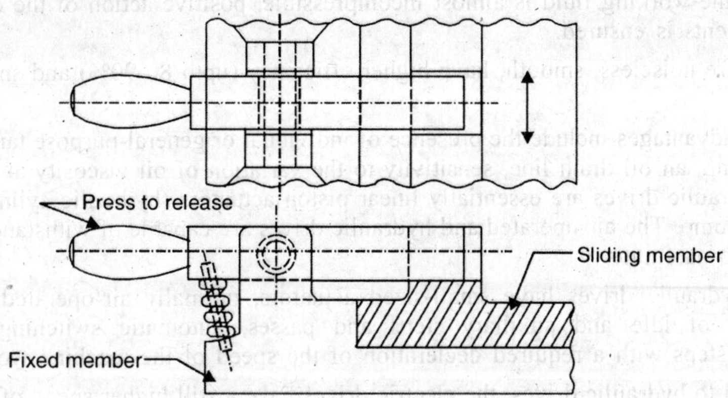

Fig. 1.79. Sliding Indexing Device.

1.14. AUTOMATED JIGS AND FIXTURES

In machining and assembling operations, the auxiliary time forms quite a large percentage of the standard time per piece. The auxiliary operations have to be done hundreds of times per day, and if done manually may cause considerable fatigue to the operator, thereby reducing his efficiency. Also, the time spent in this activity can seriously affect the production. These problems can be overcome by making the performance of jigs/fixtures automatic provided it is economically justified by the rate of production. This will assure higher efficiency and liberate the human labour considerably. Automation can be partial or full.

In 'partial automation', we automate one or several steps or passes : loading and unloading of jobs by various means : clamping and unclamping of workpieces; removal and push-out of workpieces from the workzone; rotation, indexing and locking of the rotating parts of multiple and mulit-place fixtures; in process inspection of workpieces. In 'fully automated jigs/fixtures' and machining cycles, the function of the operator is limited only to the loading of workpieces in hoppers (feeders) and control over the performance of jigs/fixtures and the machine tool.

Automated jigs/fixtures should exclude the possibility for incorrect location of workpieces. For this, they employ interlocking and safeguarding devices and clearance limits (in transfer lines) to the effect that the machine tool stops machining with the work in incorrect position (or absent) in the fixture.

The driving mechanisms in jigs/fixtures can be : (1) Mechanical (2) Pneumatic or Air operated (3) Hydraulic (4) Pneumohydraulic (5) Electrical, or a combination of these.

Air operated drives are more widely used due to the ready availability of the compressed air in most of the production shops. Their advantages are : relative simplicity and low cost, quick action and reliability in service, absence of return pipeline, insensibility to the changes of the ambient. The used air can be used again for swarf disposal and to transfer further small-sized machine parts or product assemblies. However, these drives are impracticable where large forces are required for clamping, because with pressures between 40 and 60 MPa, they have large overall dimensions and small efficiency, produce much noise by discharging the used air, and fail to assure smooth running of the fixture's working parts. Pneumatic drives are essentially air cylinders (linear piston actuators).

Hydraulic drives are used particularly with relatively large components and on hydraulic machine tools. They have the following advantages :

1. They are relatively small and compact due to high working pressures involved (400 to 600 MPa and higher).

2. As the working fluid is oil, there is no need of external lubrication. Also, there is no trouble of water condensation which may be there in a pneumatic system.

3. Since the working fluid is almost incompressible, positive action of the operating parts of the components is ensured.

4. They are noiseless, smooth, have higher efficiency (upto 80-90%) and small delay time (0.01-0.02s).

Their disadvantages include the presence of individual or general-purpose tank of sufficient capacity, a pump, an oil drain line, sensitivity to the variation of oil viscosity at heating, and a high cost. Hydraulic drives are essentially linear piston actuators (hydraulic cylinders) operated by a separate pump. The air-operated and hydraulic drives are capable of withstanding overloads and are easy to automate.

Pneumohydraulic drives have small-sized actuators, normally air-operated. They ensure quick motions of idle and auxiliary steps and passes, automatic switching over to the manufacturing steps with a required decleration of the speed of the machine working part.

Compared to hydraulic drives, the electric drives have a still higher speed of response, low power consumption, larger overall dimensions and mass, higher sensitivity to overload and overheating, and lower reliability in service. However, they have higher efficiency as compared to pneumatic and hydraulic drives.

Drives are controlled by cams and stops, master controllers, servo-valves and limit switches that operate in response to the action by the machine moving parts (tables in millers and spindles in boring and drilling machine tools).

1.14.1. Automatic Clamping Devices. The following advantages accrue from the use of automatic clamping devices :

1. Clamping is quick, thereby, considerably reducing the chucking time.

2. The clamping force is uniform and is independent of any personal variation caused by the operator.

3. More clamping force is obtainable than in manual clamping.

4. Two or more clamps can be operated simultaneously from a single control print.

5. As the working pressures are high, the automatic clamping devices (pneumatic and hydraulic) are small and compact.

In the simplest case, automatic clamping is achieved by applying clamps actuated by the feed mechanisms or cutting forces of the machine tool itself. Feed-actuated clamping devices are used largely in drilling machines. In a multi-spindle drilling machine, the jig plate is located in between the spindle head and the table and is supported on the pillars on which the spindle head is supported. The jig plate is spring loaded. As the spindle head is lowered, the jig plate moves towards the table. Subsequent lowering of the spindle head compresses the springs. This causes the clamping force to grow steadily to reach its maximum by the end of drilling. The drawback is the additional loading of the machine feed mechanism.

Electromagnetic, magnetic and vacuum clamping devices have already been discussed under 'Grinding Fixtures'.

Motor driven clamping devices, usually find applications in machine tools of the turret-lathe group, in transfer machines and transfer lines. They have screw clamps driven by a wrench which is a withdrawable device (manually or automatically) with a torque limiting sprial jaw clutch. Torque from the motor shaft is transferred through a reducer and a clutch to a screw which drives a nut linked to the rod of the clamping device. Upon reaching the desired clamping force, the clutch disengages. The torque is controlled by adjusting the strength of the torque-control spring. Unclamping is affected by reversing the motor, Fig. 1.80.

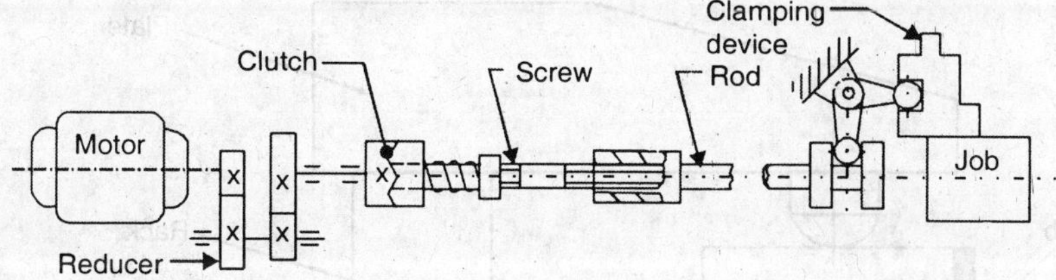

Fig. 1.80. Motor-driven Clamping Device

For pneumatically controlled clamping devices, compressed air is needed. For this, the installation includes a compressor, air receiver and a pipe line around the shop. Coupling points at suitable intervals are provided where a pipe may be plugged in. The working pressure is usually about 0.3 to 0.6 MPa. The operating equipment consists of a piston and cylinder arrangement with the piston rod connected to the levers or to some other mechanisms which comprise the method of clamping. To clamp and unclamp a workpiece, the piston must operate in both directions. For this, air inlets are provided at each end of the cylinder and a valve is provided to regulate the supply of air. The various arrangements of air operated clamping devices are shown in Fig. 1.81. The arrangements [Fig. 1.81 (*a*), (*b*)] are suitable where slight variations may be encountered since the toggel and cam clamp will remain securely locked under pressure.

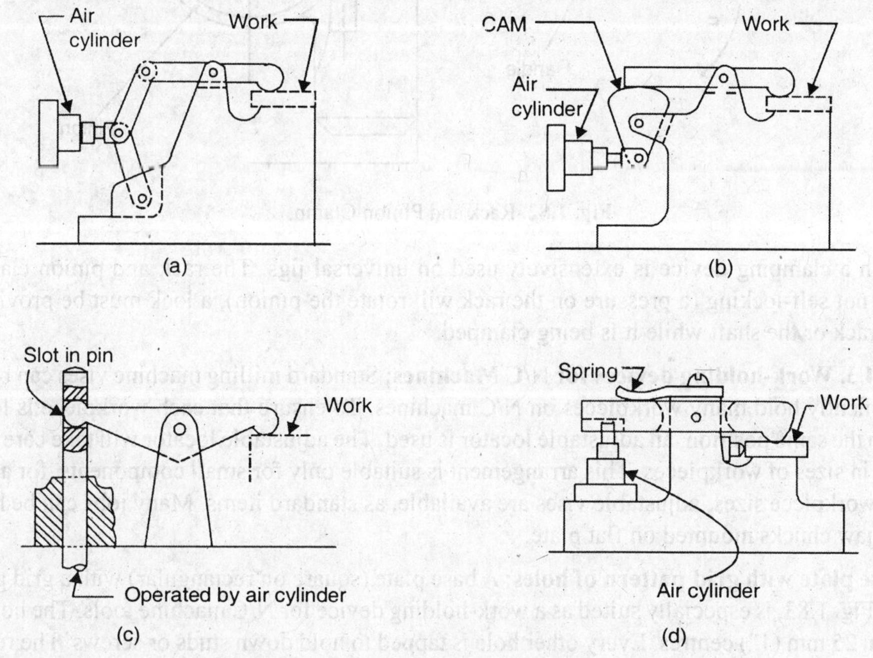

Fig. 1.81. Air Operated Clamping Devices.

Hydraulically operated clamps operate similarly to air operated clamps.

1.14.2. Rack and Pinion operated Clamping Device. Rack and pinion clamps, Fig. 1.82, consists of a rack, pinion mounted on a shaft and a handle. By turning the handle counterclockwise, the rack gets lowered resulting in the clamping of the job directly or through an intermediate piece (e.g., a plate). The clamping force depends on force F applied to the handle. To retain the developed clamping force after removing the applied force, the clamp has a lock to prevent the reversal of the rack pinion under the action of elastic forces due to arise in the links of the clamping system. For unloading the job at the end of the machining operation, the handle is turned clock-wise.

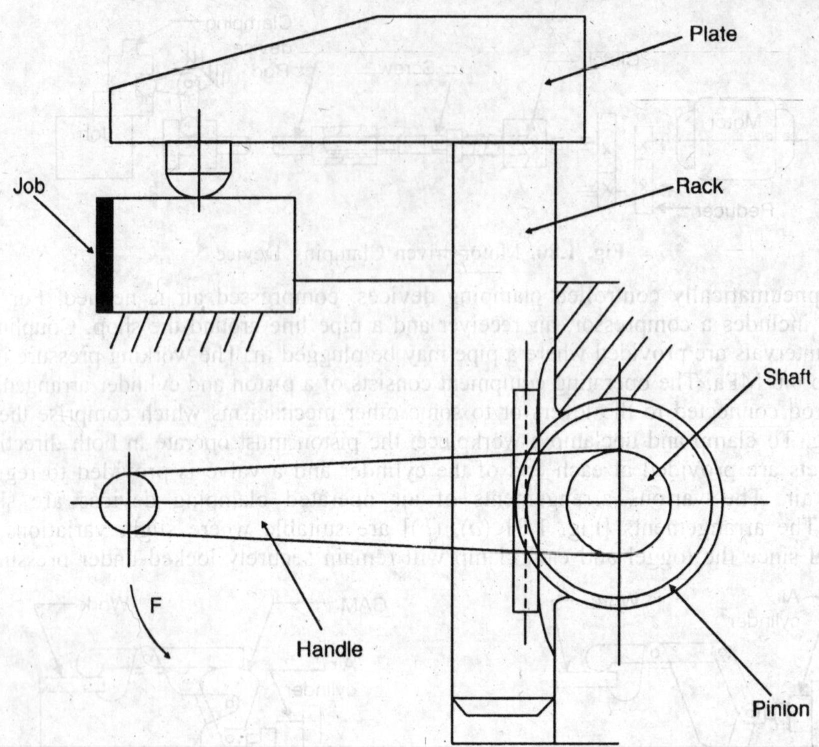

Fig. 1.82. Rack and Pinion Clamp.

Such a clamping device is extensively used on universal jigs. The rack and pinion clamping device is not self-locking (a pressure on the rack will rotate the pinion), a lock must be provided to hold the rack or/the shaft while it is being clamped.

1.14.3. Work-holding devices for N/C Machines: Standard milling machine vises can be used to conveniently hold many workpieces on N/C machines. To ensure that each workpiece is located exactly in the same position, an adjustable locator is used. The adjustable locator will take care of any variation in sizes of workpieces. This arrangement is suitable only for small components. for a larger range of workpiece sizes, adjustable vises are available, as standard items. Many jobs can be held in 3- and 4-jaw chucks mounted on flat plate.

Base plate with grid pattern of holes: A base plate (square on rectangular) with a grid pattern of holes, Fig. 1.83, is especially suited as a work-holding device for N/C machine tools. The holes are usually on 25 mm (1") centres. Every other hole is tapped to hold down studs or screws. The remaining holes are jig-bored to 0.005 mm accuracy, and these holes hold dowel pins. The dowel pins act as locators and the studs and screws are used for clamping the work-piece on the plate. The grid plate, in turn, is bolted directly on the machine table. The grid plate system can be used for an infinite varieties of workpiece.

The bottom-left most hole is taken as the zero reference point, Fig. 1.83. With this, the grid pattern and hence the work-piece will be placed in the first-quadrant (upper right-hand quadrant). All the dimensions measured from the reference point will be positive in magnitude. The N/C machine tool is set-up by bring the machine spindle directly over the reference hole with the help of a dial indicator mounted on the spindle. The workpiece is then located and held on the grid plate in reference to the set up point.

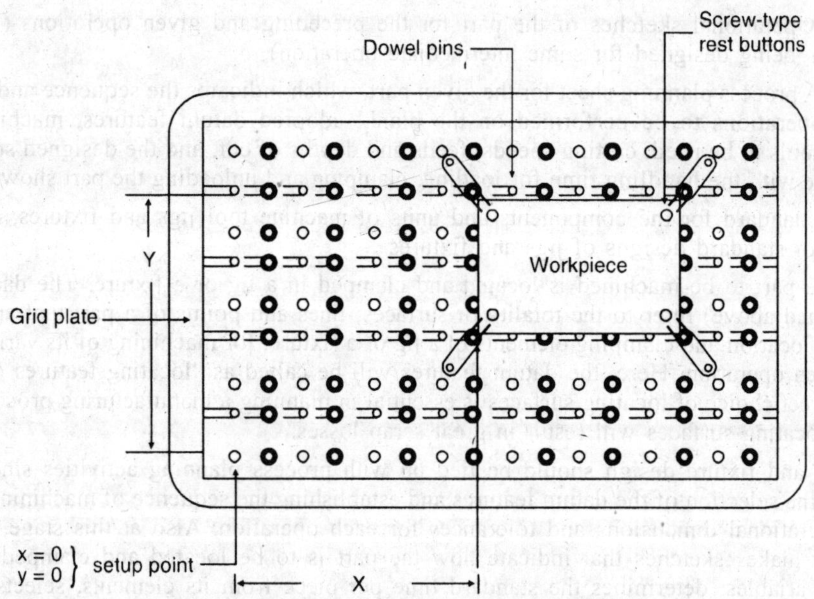

Fig. 1.83. A base plate with grid pattern of holes

One major drawback of grid plate is that the chips get accumulated in the holes. However, this can be avoided by keeping all the unused holes plugged.

Modular fixtures: These fixtures are commonly used on N/C machining centres. The fixtures comprise of modular components based on the grid principle (A module is an inter-changeable "plug in" item that may be combined with other inter-changeable items to form a complete unit, *i.e.*, a modular fixture is built up on the building block principle). The workpiece locators are located approximately on the grid pattern and then the workpiece is precision-machined under tape control. Dowel pins are not used. Fig. 1.84, shows a typical modular fixture.

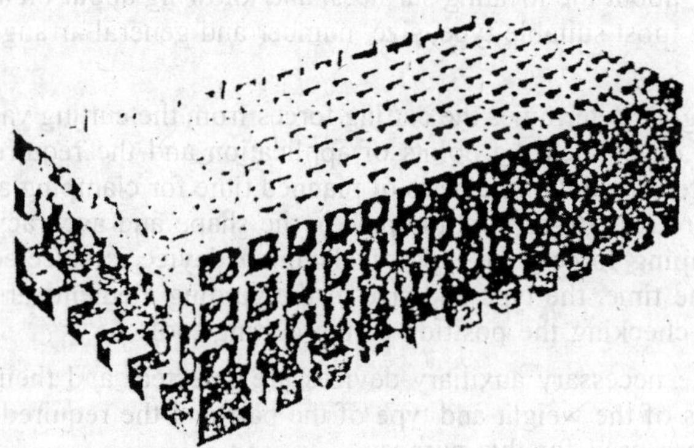

Fig. 1.84. Modular Fixture

1.15 Fundamentals of Jig and Fixture Design:

For designing a jig or a fixture the following data should be available to the designer:

1. Working drawing of the finished part, the specifications for its inspection and the working drawing of the blank. The term blank, here, refers to the starting work piece which will be machined to get the finished part. The blank may be in the form of: a casting, a stamping, a forging, a weldment, a blank cut from a rolled stock or a blank produced by Powder Metallurgy.

2. Operational sketches of the part for the preceding and given operations (if the jig or fixture is being designed for some intermediate operation).

3. A process planning sheet for the given part, which indicates the sequence and description of the operations to be performed on the blank, adopted datum features, machine tool and cutting tools to be used, cutting speeds, feeds and depths of cut, and the designed standard time per piece with the handling time for loading, clamping and unloading the part shown separately.

4. Standard for the components and units of machine tool jigs and fixtures, as well as a manual of standard designs of jigs and fixtures.

The part to be machined is located and clamped in a jig or a fixture. The datum features (mentioned above) refer to the totality of surfaces, lines and points of a part that are in contact with the location and clamping elements of a jig or a fixture, for machining of its various surfaces in a given operation. Here, the datum features will be called as "locating features or surfaces". The correct choice of locating surfaces is essential in planning a manufacturing process, because, wrong locating surfaces will result in great scrap losses.

Jig and fixture design should be tied up with process planning activities since the latter involve the selection of the datum features and establishing the sequence of machining operations with operational dimensions and tolerances for each operation. Also at this stage, the process engineer makes sketches that indicate how the part is to be located and clamped, selects the cutting variables, determines the standard time per piece from its elements, selects the cutting tools and the machine tools to be used.

5. In addition to the above information, it is necessary to know the basic dimensions of the machine tool with respect to the installation of the jig or fixture, that is, dimensions of the table, size and arrangement of the *T*-slots, minimum distance from the table to the spindle, size of spindle taper socket, etc., as well as the general condition of the machine tool.

The design of the jig or fixture is selected in accordance with the available production facilities and the required volume of production. Also, it is to be decided at this stage, whether the parts subjected to most wear in the jig or fixture are to be of renewable design.

With the above information available, the design of a jig or a fixture may follow the following steps:

(*i*) After deciding about the locating surfaces, and knowing about their accuracy and finish, the designer selects the most suitable type, size, number and general arrangement of the locating elements.

(*ii*) Next, the designer determines the cutting forces from the cutting variables (cutting speed and feed and depth of cut). Then the points of application and the required magnitudes of the clamping forces are determined. On the basis of planned time for clamping and releasing the part, the type of jig or fixture (Single-or multiple-piece), the shape and accuracy of the part, and the magnitude of the clamping forces, the type of clamping device is selected and its dimensions determined. At the same time, the type and size of the guiding elements are determined, as well as of the elements for checking the position of the cutting tool.

(*iii*) After that, the necessary auxiliary devices are selected, and their size and design are established on the basis of the weight and type of the part and the required machining accuracy. The proper standards are used for this purpose.

(*iv*) Next, to draw the general arrangement of the jig or the fixture a number of views of the part are drawn. The number of views will depend upon the complexity of the part. The elements of the jig or fixture are next consecutively drawn, where required, around the part. First, the locating elements are drawn then the clamping devices, tool guiding elements and the auxiliary devices. Finally the body is designed around the part as the element that links the other elements into an integral unit. The assembly drawing usually carries the overall dimensions and the dimensions that should be held to in assembling the jig or fixture and stipulates the assembly specifications. In making the assembly drawing and the part drawings, tolerances are established on the dimensions.

1.15.1. Standard Jig and Fixture Parts. We all know that the basic purpose of standardization of any manufactured items is to ensure interchangeability (which is the basis of mass production) and facilitate the manufacture of parts. It also facilitates maintenance and repair. Also, any item, mass produced, is much cheaper than the same item made as a single unit. A number of standardized component parts for the construction of jigs and fixtures are available commercially. Thus, when designing and manufacturing jigs and fixtures, maximum use should be made of standard, readily purchasable (or available in store) component parts. This will reduce the cost of design and manufacturing considerably. The designer should avoid, as far as possible, specifications of odd shaped or sized items that must be specially made.

The standardized component parts for jigs and fixtures include : drill bushings, washers, handles, various types of screws, studs and nuts, various types of knobs, jig straps, rest buttons or jig feet locating pins, fixture or mill keys, *T*-bolt and *T*-nut, Heel rest, brackets, toggles, came and wedges, etc. Many of these have already been sketched and discussed.

1.16. JIG AND FIXTURE CONSTRUCTION

The body of a jig or a fixture can be constructed by any of the following methods:

(1) Casting (2) Fabrication (3) Welding.

In the last two methods, the body is made up from separate parts. The advantages of the cast construction are :

(*i*) Jigs and fixtures with complicated shapes can be easily cast.

(*ii*) Cast iron has the property of absorbing and damping out the vibrations.

(*iii*) Any number of castings with the same characteristics can be made from one pattern.

(*iv*) If a cast jig or fixture drops down, it will probably break. It is not likely to bend and get out of alignment so as to result in the production of defective pieces.

Fabrication has the following advantages:

(*i*) Standard parts can be used to build up the body.

(*ii*) The jig or fixture can be built up quickly by using standard parts.

(*iii*) After use, the jig or fixture can be disassembled.

Welding has the following advantages:

(*i*) The jig or fixture can be constructed speedily.

(*ii*) The welded construction is cheaper.

(*iii*) Less machining is needed than for fabricated parts.

Comparison. Casting is the most popular method. It is costly both in time and money but it has the property of absorbing and damping out vibrations, a very useful characteristic for milling fixtures. Castings are usually heavy and where lightness is needed, welded or fabricated jig or fixture has the advantages. If a fabricated or welded jig or fixture gets dropped, it will get distorted. The defect will remain unnoticed until the device is used again when it produces defective piece. A cast construction will break rather than bend. For ease and quickness of manufacture, the order is : (*a*) Welded (*b*) Fabricated (*c*) Cast.

The welded construction must be "stress relieved" by heat treatment, to relieve all internal stresses induced during welding. Otherwise, the stresses will gradually distort the jig or fixture, as they relieve themselves. Only after stress relief, can finish machining be done on the construction, if accuracy is to be held. And, since machining must be done after assembly, it is sometimes difficult to machine an internal surface of a welded construction.

In a fabricated construction, all the component parts can be completely machined before assembly.

This is often an important enough advantage to offset the lesser cost of the welded construction. Often a combination of the two methods is successfully used. In the fabrication method, the construction is held together by screws and dowels. The screws serve only to hold the parts together with the dowels assuring accurate alignment. Two dowels are sufficient in any case, whereas the screws can be one or more as needed. A fabricated construction is easy to repair, because disassembly is comparatively easy. To facilitate disassembly and assembly, each dowel should be a press fit in one part and slip fit in the other.

Casting construction is sometimes used for the main body (especially for a jig) and other pieces are welded or assembled to it with screws and dowels.

1.17. MATERIALS FOR JIGS AND FIXTURES

S. No.	Description of the item	Recommended Material	Heat Treatment
	Jig bush	40Cr 1	Oil hardened and tempered 40–45 HRC
		T90	–do– 50–58 HRC
		C14	Case hardened and tempered to 55–56 HRC
		15 Cr 65	–do–
		T21 5Cr 12	Oil hardened and tempered 58–63 HRC
2.	Washer	St 42	
		C 45	
		55 Cr 70	Oil hardened and tempered to 40–45 HRC
3.	Palmgrip	C 20	
		C 45	
		40Cr 1	Oil hardened and tempered 40 to 50 HRC
		Cast Iron grade 20	
		Aluminium alloy type C	
4.	Taper handle	C45	
5.	Ball handle	C 45	
		Aluminium casting	
		Plastics	
6.	Tenon	C 45	Oil hardened and tempered 36–45 HRC
7.	Set collars	C 45	–do–
8.	Handle wheels	Gray cast Iron Grade 20	
		Aluminium Alloy Casting, A-24 M	
		Phenol Plastics	
9.	Strap clamp	C 45	
		40 Cr 1	Oil hardened and tempered 40 – 50 HRC
		C14	Case hardened and temepered 52–62 HRC
10.	'C' washers	15 NiCr 1 MO 12	–do–
11.	Jig Button	T 90	Oil hardened and tempered 54–58 HRC
12.	Locking dig (Catle)	C 14	Case hardened and tempered 56–60 HRC
13.	Jig Feet	T 90	Oil hardened and tempered 54–58 HRC
14.	Welded V-block	St 42	Stress relieving after welding
		C 45	–do–

15.	Base Plate	St 42	–do–
		C 45	–do–
		Cast Iron grade 20	
16.	Locating Pin	C 45	Oil hardened and tempered 55–61 HRC
		T 90	Case hardened and tempered 56–60 HRC
		15 Cr 65	–do–
17.	Locating plugs	C14	–do–
		T 215 Cr 12	Oil hardened and tempered 58–62 HRC
18.	Swing and Table clamps	40 Cr 1	Oil hardened and tempered 40–45 HRC
19.	Clamp plate	40 Cr 1	–do–
20.	Screw/wing, Knurled, Grub or socket head cap	C–45	
		St 42	
		40 Cr 1	Oil hardened and tempered 40– 45 HRC
		Brass	
		Bronze	
		Aluminium Alloy	
21.	Thrust pad	St 42	
		40 Cr 1	Oil hardened and tempered 55–60 HRC
22.	V Block	Cast iron grade 20	
		C 45	Surface flame hardened 40–45 HRC
		40 Cr 1	–do
		C 14	Case hardened and tempered 56–60 HRC
23.	T Nut	St 42	
		C 45	
		40 Cr 1	Oil hardened and tempered 42–45 HRC
24.	Plate, base, Housing	St 42	Stress relieving after welding if any
		C 45	–do–
		Cast iron Grade 20	
25.	Jaws for vice	T 90	Oil hardened and tempered 52–58 HRC
26.	Claming plates and levers	C 45	
		40 Cr 1	
27.	Key	C 45	
		St 42	
28.	Hinge pin drilled	T 90	
29.	Foot (positioned) Plate	T 90	Oil hardened and tempered 54–58 HRC
30.	Foot Round	T 90	–do–

Note: Cast Iron, because of its damping capacity, is a very important material for the manufacture of jigs and fixtures. For heavy bodies, it is always economical to use Cast iron. Again complicated and detailed machining can be avoided. Where lightness is the main consideration, the materials can be alloys of aluminium and magnesium. In the case of screw operated feeding and clamping devices, phosphor bronze replacable nuts are used, screws being costly items. Nuts will wear out early and can be replaced. They reduce the wear of the screws and also provide corrosion resistance.

1.17.1. Indian Standards for the Elements of Jigs and Fixtures:

S. No.	Element	IS.No
1.	Locating pins	5093–1969
2.	Locating plugs (end)	5095–1969
3.	Locating studs	5096–1969
4.	Diamond locating studs	5094–1969
5.	Diamond locating stud	5097–1969
6.	Jig Bushes Part I	666–1972
7.	Jig Bushes Part II	666–1972
8.	Jig buttons	4294–1967
9.	Jig feet	4299–1967
10.	Turning mandrels	7262–1974
11.	Grinding mandrels	7301–1974
12.	Tenons	2990–1965
13.	'V' blocks	2496–1964
14.	Welded 'V' blocks Dia. range from 300 to 1200 mm	4492–1968
15.	Dead Centres (carbide tipped)	2534–1963
16.	Dead centres for lathe 60°	2289–1963
17.	Live Centres	3793–1966
18.	Circular Base Plate	5991–1969
19.	Collets (spring)	6238–1971
20.	Feed Fingers	6105–1971
21.	Set collars	2995–1965
22.	Locking Dogs	4295–1967
	CLAMPING FEATURES	
23.	Cam operated clamps	6090–1971
24.	Double ended clamps	6080–1971
25.	Goose Neck clamps	4279–1975
26.	Parallel sided clamps	5998–1971
27.	Swing clamps	5250–1969
28.	Slotted clamps	6082–1971
29.	Strap clamps	4292–1967
30.	Swing latches	5999–1971
31.	Table clamps	5251–1969
32.	U–Type Clamps	4293–1975
33.	Wide strap clamps	5252–1969
34.	'T' bolts	2014–1962
35.	'T' Nuts	2015–1962
36.	'T' Slots	2013–1974
37.	Knurled Nuts	3460–1968
38.	Knurled Thumb nuts	6335–1971

HANDLES

39.	Ball handles	2904–1964
40.	Control lever with ball grips	2975–1964
41.	Cranked handles	2793–1964
42.	Handcranks	2908–1964
43.	Handwheels	3048–1965
44.	Palm grips	2804–1964
45.	Star grips	2909–1964
46.	Taper handles for machine tools	2890–1964

WASHERS

47.	'C' washers	4291–1967
48.	Swing 'C' washers	4298–1967
49.	Plain washers	4298–1967
50.	Spherical washers & conical seats	4297–1967

SCREWS

51.	Depth of holes for studs	4499–1968
52.	Clearance for metric bolts	1821–1967
53.	Thread centre holes	2540–1963
54.	Grub screws slotted	2388–1971
55.	Knurled thumb screws	3726–1966
56.	Wing screws	3729–1966
57.	Clamping screws	6388–1971
58.	Screws for thrust pads	6336–1971
59.	Thrust pads	6337–1971
60.	Pilots for Counter bore & C-sinks	5705–1970
61.	Counter sinks & counter bores Part I & Part II	3406–1975
62.	Hexagonal socket head cap screws	2269–1967
63.	Radial & chamfers for general Engg. purposes	3457–1966
64.	Dimensions for relief grooves	3428–1966
65.	Allowable deviations for dimension without specified tolerances	2102–1969
66.	Quick action drilling jigs	6440–1972
67.	Code of practice for general engineering drawings	696–1972
68.	Tolerances of form and position for engineering drawings	1800–1976
69.	Recommendations for limits and fits for engineering	919–1963
70.	Centre drills	664–1963
71.	Engineers parallels	4241–1967

1.18. TOLERANCE AND ERROR ANALYSIS

1.18.1. Theory of addition and subtraction of tolerances: If A and B are two linear measurements to be added/subtracted with tolerances a_2, a_1 and b_2, b_1 respectively to obtain a dimension C (where a_2, b_2 are higher limits and a_1, b_1 are lower limits of tolerances, and

$$\delta_A = \text{tolerance of } A = a_1 + a_2$$

$$\delta_B = \text{tolerance of } B = b_1 + b_2$$

Then tolerance of C, $\delta_C = \delta_A + \delta_B$

1.18.2. Calculation of tolerance on Centre distance between holes in Drilling Jigs:

If on a component, two holes are to be drilled and reamed at a distance L_0 within the limits $\pm \delta_0$ using a jig plate with slip bushes,

$L_1 \pm \delta_1$ = distance between the lines bushes in jig plate

c_1, c_2 = Clearance (around) between the liner bush and the slip bush for the first and second holes

e_1, e_2 = eccentricity of inside diameter of slip bush with respect to its outside diameter for the first and second slip bushes

$d - \delta d$ = diameter of drill

$d_B + \delta_B$ = Inside diameter of slip bush

Then maximum value of centre distance between holes obtained on component,

$$L_0 + \delta_0 \geq L_1 + \delta_1 + c_1 + c_2 + e_1 + e_2 + (d_B + \delta_B) - (d - \delta_d)$$

Minimum value of centre distance between holes obtained on component,

$$L_0 - \delta_0 \leq L_1 - \delta_1 - c_1 - c_2 + e_1 + e_2 - (d_B + \delta_B) + (d - \delta_d).$$

1.18.3. ERROR ANALYSIS

One of the causes of workpiece dimensional and geometrical errors is the error of locating the workpiece in a jig or fixture. A workpiece has the following surfaces:

(*i*) Surfaces to be machined.

(*ii*) Surfaces which orient the workpiece with respect to a cutting tool set to size. These surfaces (and also lines and points) of the workpiece are called "locating surfaces" or "locations".

(*iii*) Surfaces which contact work holding clamping elements.

(*iv*) Surfaces from which the size is to be held and checked in machining.

(*v*) Surfaces from from which the size is to be hold and checked in machining.

(*vi*) Free surfaces.

Locating features can be divided into: basic, functional, datum and production types.

The basic locations are selected when designing a product. These determine the normal position of its components and units relative to each other. On the drawings, these features are shown as geometric elements, such as, the axes of shafts and holes, the planes of symmetry, the bisector of angles etc.

The functional location features are the ones used to determine the position of a part in the product. The functional locations are provided by a actual material surfaces.

The datum locations are those to which the dimensions to be held are referred in machining and assembly, and from which the relative position of workpiece elements or product components is established for checking.

The production locating suraces are those used to determine the position of a workpiece being machined or assembled. These are the material surfaces that contact the fixture locating elements.

Work-piece set up errors: The set up error may arise as the work-piece to be machined is loaded in a jig/fixture. This error, ε, which is a component of the total machining error, comprises of locating error ε_{loc}, clamping error ε_{cl} and fixture error ε_{fix}.

(*a*) *Locating error,* ε_{loc} = Difference of maximum distance and minimum distance between the workpiece datum surface and the cutting tool set to size.

This error occurs when the locating surface is different from datum surface. In that case, this error will be designated by the symbol of a given dimension.

(*i*) A workpiece on which a step is to be milled is located as shown in Fig. 1.85. The workpiece is clamped by force F.

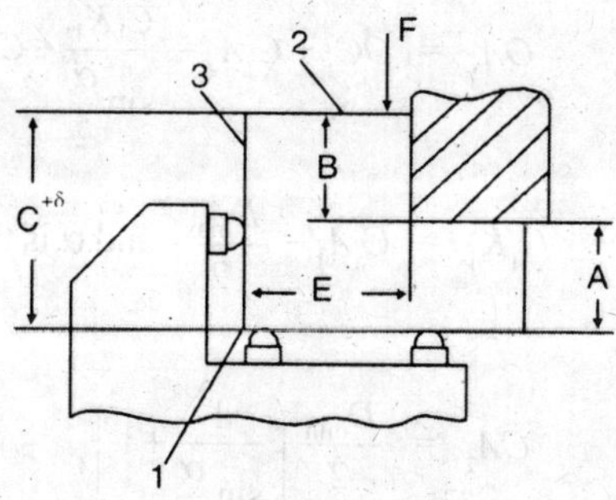

Fig. 1.85. Locating Errors

In this figure, the locating error with respect to dimension *A* is zero.

$$\varepsilon_{loc}\ A = 0$$

This is because, the surface 1 of the workpiece is both its production locating surface and datum surface.

$$\varepsilon_{loc\ B} = \delta \text{ (tolerance on dimension } C)$$

Here, surface 1 is the locating surface, and surface 2, the datum surface., *i.e.*, two different surfaces.

(*ii*) *Location on a V-block*: Fig. 1.86. (*a*), shows the location of a cylindrical job on a V-block. The maximum diameter and minimum diameter of jobs in a batch to be machined, are represented by the two circles with centres C_1 and C_2 resectively. When the dimension h_1 is to be held, the locating error is given as

= Difference between the limits of the distance from the datum (points A_1 and A_2) to the milling cutter set to size (point A_3).

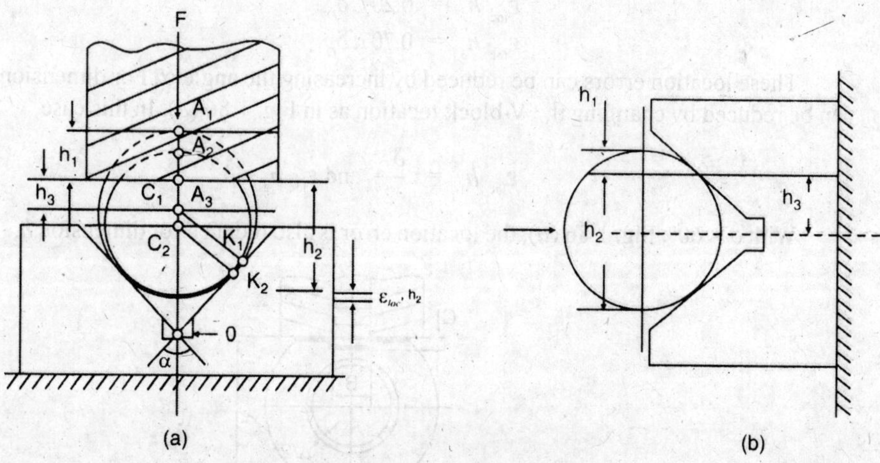

Fig. 1.86. Locating Errors

That is,

$$\varepsilon_{loc}\, h_1 = OA_1 - OA_2$$

Now,

$$OA_1 = OC_1 + C_1A_1 = \frac{C_1K_1}{\sin\frac{\alpha}{2}} + C_1A_1 = \frac{D_{max}}{2}\left(\frac{1}{\sin\frac{\alpha}{2}}+1\right)$$

Since,

$$C_1K_1 = C_1A_1 = \frac{D_{max}}{2} \text{ and } \alpha \text{ is the V-block angle.}$$

Similarly,

$$OA_2 = \frac{D_{min}}{2}\left(\frac{1}{\sin\frac{\alpha}{2}}+1\right)$$

$$\therefore \quad \varepsilon_{loc}\, h_1 = \frac{\delta_D}{2}\left(\frac{1}{\sin\frac{\alpha}{2}}+1\right),$$

where

$$\delta_D = \text{tolerance on workpiece diameter} = D_{max} - D_{min}$$

Similarly, for dimensions h_2 and h_3,

$$\varepsilon_{loc}\, h_2 = \frac{\delta_D}{2}\left(\frac{1}{\sin\frac{\alpha}{2}}-1\right)$$

and

$$\varepsilon_{loc}\, h_3 = \frac{\delta_D}{2}\,\frac{1}{\sin\frac{\alpha}{2}}$$

The commonly used value of $\alpha = 90°$. The other values are: 60°, 75° and 120°. With $\alpha = 90°$,

$$\varepsilon_{loc}\, h_1 = 1.207.\delta_D$$
$$\varepsilon_{loc}\, h_2 = 0.207.\delta_D$$
$$\varepsilon_{loc}\, h_3 = 0.707.\delta_D$$

These location errors can be reduced by increasing the angle α. For dimensions h_1 and h_3, error can be reduced by changing the V-block location as in Fig. 1.86 (*b*). In this case,

$$\varepsilon_{loc}\, h_1 = \frac{\delta_D}{2} \text{ and } \varepsilon_{loc}\, h_3 = 0$$

With $\alpha < 60°$, Fig. 1.86 (*a*), the location error is also reduced for dimension h_2.

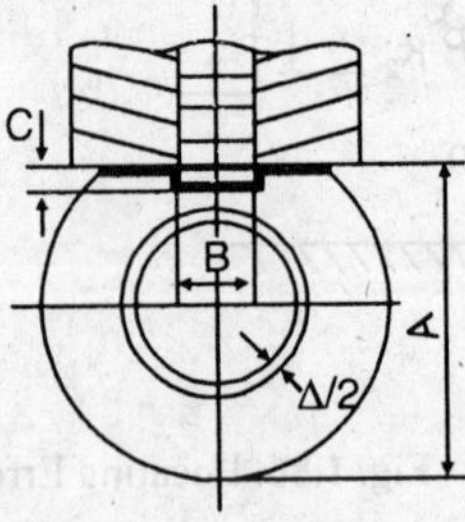

Fig. 1.87

A workpiece mounted by its locating hole on a cylindrical aligning pin of a jig/fixture and clamped on the end face (see Fig. 1.56) is shown in Fig. 1.87. With mounting without any clearance (on an expandable–mandrel type pin, for instance), the locating error for dimension A is equal to half the tolerance on the workpiece diameter. If the workpiece is mounted with a clearance (as a plain solid pin), the locating error will increase by the maximum amount of clearance Δ :

$$\varepsilon_{loc} A = \frac{\delta_D}{2} + \Delta$$

Where workpiece diameters or surface relationships are determined by the size or setting of the cutting tool, the locating error is neglected. Thus, in Fig. 1.87, $\varepsilon_{loc} B = 0$, and $\varepsilon_{loc} C = 0$, for dimensions *B* and *C*. The relative position of the workpiece surfaces machined in a single step is also not affected by locating errors.

The locating error is influenced by the error of form of the locating surface. For instance, the ovality of a cylindrical workpiece mounted on a V-block changes the position of the workpiece axis depending upon the angular position of the workpiece. That causes the error of dimension *E*, Fig. 1.88 *a*, which changes within the limits of $2c$, when c is the distance from the workpiece axis to the ellipse focus.

Macro-irregularities of the locating surface may cause the error of dimension *B*, Fig.1.88 *b*. In most cases, this error is small (with a substantial distance between the locating points, it may take a fraction of height H of macro-irregularities) and can be neglected. The error can be regarded as part of the locating error, since the actual locating points do not lie on the datum surface (plane I – I).

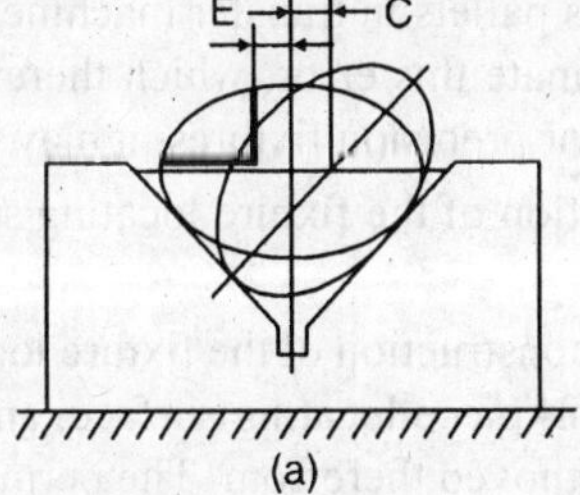

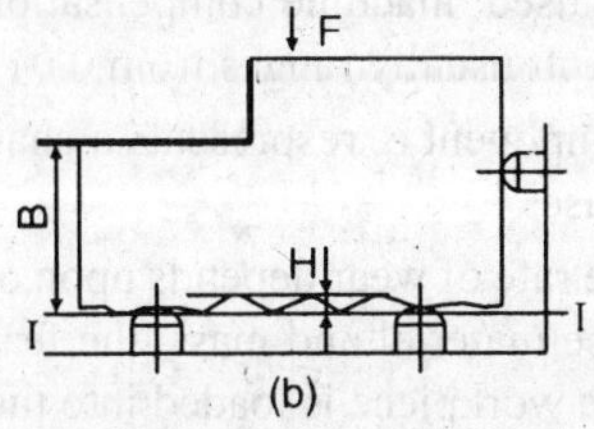

Fig. 1.88

Locating errors affect workpiece dimensional accuracy and surface relationships, but do not impair surface geometrical accuracy. For each locating method, the locating error can be found by geometrical calculations. The ways of eliminating or reducing locating errors are:- to combine locating and datum surfaces, to increases the accuracy of locating surfaces, to select the rational position of fixture locating elements and their proper size: to eliminate or reduce clearances when mounting work pieces on external or internal elements, *etc.*

Angular variation due to V-location of a cylindrical component:

If $\Delta\beta$ = error resulting over an angle β on pith circle diameter D, then

$$\tan\frac{\Delta\beta}{2} = \frac{\delta_D \cdot \sin\beta/2}{\left(D\sin\frac{\alpha}{2} - \delta_D \sin\beta/2\right)}$$

where δ_D = Tolerance on the diameter of the component

α = Included angle of V-block.

The clamping error, ε_{cl} can be defined as a difference between the maximum and minimum distances from the workpiece datum surface to the cutting tool set to size due to displacement of workpiece under the action of the clamping force.

If in machining a batch of workpieces, this displacement is constant, though large, it can be compensated for by the machine tool adjustment, whereby the clamping error is eliminated.

The clamping error is introduced in dimension A, Fig. 1.85, *i.e.*, $\varepsilon_{cl} A \neq 0$, whereas it does not affect dimension E ($\varepsilon_{cl} E = 0$), because the datum surface 3 has no horizontal displacement during the workpiece clamping. Datum surface 1 is displaced due to deformation of elements through which the clamping force is transmitted: the workpiece itself, the fixture body and locating elements etc. Here, the greatest displacements take place at the interface of the workpiece with the locating elements (points); in the other joints, the displacements are small with a rationally designed fixture.

The clamping error is often commensurable with the locating error. It can be reduced by using clamping devices (e.g;) pneumatic, hydraulic etc. which provide a constant clamping force, by improving the uni-formity of the workpiece material and surface layer structure and by rationally directing the clamping force..

The clamping force should press the workpiece locating surface against the fixture locating elements (see Fig. 1.22). A workpiece may shift or tilt where clamping force fails to hold it properly against the locating elements. As a result, the machined surface of the workpiece will be non-parallel to its bottom locating surface.

The fixture error, ε_{fix}, comprises machining and assembly errors ε_1 of the fixture locating elements, errors ε_2 due to their wear and error ε_3 arising due to the setting and location of jig/fixture on the machine tool. The component ε_1 is a constant systematic error.

With a single fixture being used, it can be compensated for by a corresponding machine tool adjustment. Where several identical fixtures (such as pallets in transfer machines) or multi-part fixtures are used, machine compensation can not eliminate this error, which therefore, fully, contributes to ε_{fix}. It usually ranges from 0.01 to 0.05 mm. For precision fixtures, it may be within 0.01 mm.

Component ε_2 respresents a change of the position of the fixture locating surfaces due to their wear in use.

The rate of wear depends upon on the size and construction of the fixture locating elements, the workpiece material and mass, the features of the workpiece locating surface, and the conditions in which the workpiece is loaded into the fixture and removed therefrom. The permitted extent of wear of the locating elements is specified depending upon the required accuracy of workpiece location and is checked during periodic inspection of fixtures. If the wear has reached the permissible limit, the locating elements are replaced. In machining medium size workpieces of the tolerance grade 2 or 3, the wear allowance must not exceed 0.0015 mm. For longer wear life, locating elements of fixtures are made of hardened steel. In some cases, they are chromium plated or carbide-tipped, which reduces wear by a factor of 3 and 10, respectively.

Component ε_3 is a result of misalignment of the fixture body on the machine table, face plate or spindle of a machine tool. In mass production, where a given fixture is set up and used for a long time, this error can be reduced to a minimum by careful fixture alignment during the setting; after that the error remains constant as long as the fixture is used. In addition, this error ε_3 can be compensated for by a machine adjustment. In batch production, the fixtures are periodically changed, so that the component ε_3 becomes a random value variable in certain limits and incapable of being corrected for. This also holds true for automatic transfer machines using pallet type work holders. Guiding elements (e.g., inserted tongues for worktable T-slots, locking devices, centring pilots etc.) introduced in fixtures and fitted to the mating machine tool surfaces with small clearances make it possible to reduce the value of ε_3 to 0.01 mm.

ε_1, ε_2 and ε_3 can be regarded as random variables following the normal frequency distribution.

Thus, $$\varepsilon_{fix} = \sqrt{\varepsilon_1^2 + \varepsilon_2^2 + \varepsilon_3^2}$$

The overall workpiece set up error is found as.

$$\varepsilon = \sqrt{\varepsilon_{loc}^2 + \varepsilon_{Cl}^2 + \varepsilon_{fix}^2}$$

The magnitudes of ε_{loc}, ε_{cl} and ε_{fix} are comensurable. For high accuracy machining applications, ε_{cl} and ε_{fix} must be reduced.

An analysis of these errors enables the selection of a fixture arrangement and formulation of accuracy requirements for fixture production, which is especially important designing pricision fixtures.

(*iii*) Location by two cylindrical pin locators or by one cylindrical locator and one diamond shaped locator. This has been discussed under Art. 1.2.4. (Radial or Angular location). Let (Refer Fig. 1.8)

L = Nominal value of the distance between the centres of the holes of the workpiece and also of the locating pins

x = Tolerance on the distance between hole centres of the workpiece

y = Tolerance on the distance between centres of the pins in the body of jig/fixture

$\delta_{A\ min}$ = Minimum gap between left pin and hole, A

$\delta_{B\ min}$ = Minimum gap between right pin and hole, B

It has been noted in Art. 1.2.4, that if the workpiece is effectively located on pin A (left one) and if the pin B (right one) is complete cylinder, the allowance between pin B and hole B will be atleast x plus y. In a general case, when the workpiece is not effectively located on one of the pins,

$$\delta_A \min + \delta_B \min = \geq (x + y) \qquad ...(1)$$

Thus, the condition for 100% mating of two holes with two round pins is:

Sum of the diametral clearances between holes and pins ≥ sums of tolerance on centre distance between pins and between holes.

The drawbacks of using both the pins as cylindrical have been discussed in Art. 1.2.4. The remedy is the use of one of the pins as a diamond pin. Let it be pin B. Let us calculate the width of its cylindrical portion. Fig. 1.89.

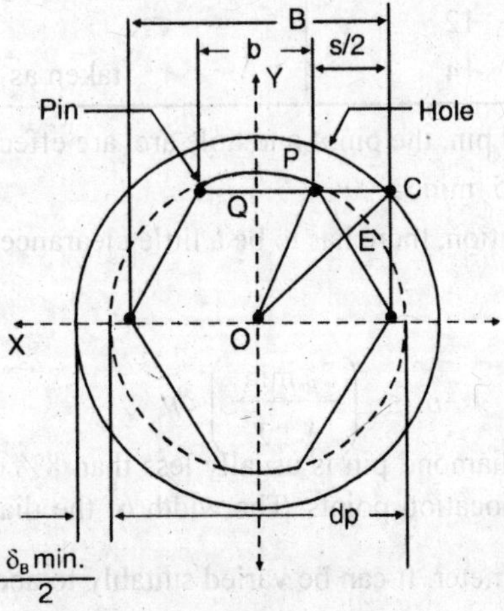

Fig. 1.89. Diamond pin locator.

Triangles OQP and PEC are similar.

$$\therefore \quad \frac{CE}{PQ} = \frac{PC}{OP}$$

$$\therefore \quad \frac{\delta_B \frac{\text{min.}}{2}}{\frac{b}{2}} = \frac{\frac{s}{2}}{\frac{dp}{2}}$$

where b = width of cylindrical portion of the diamond pin

dp = Diameter of pin

$\therefore \quad s = \delta_B \text{ min.}\left(\frac{dp}{b}\right)$, (It is clear that clearance $\frac{s}{2}$ is greater than the radial clearance $\frac{\delta_B \text{ min.}}{2}$).

Substituting s for δ_B min. in equ. (1), we have

$$\delta_A \text{ min.} + \delta_B \text{ min.}\left(\frac{dp}{b}\right) \geq (x+y)$$

$$b \leq \left(\frac{\delta_B \text{ min.}}{x+y-\delta_A \text{ min.}}\right) \cdot dp \quad \text{...(2)}$$

To avoid rapid wear of the pin, 'b' should be as large as possible. The included angle of the pin at the $x - x$-axis is usually 120°.

Recommended values of b for various pin diameters are given below:

***dp* (mm)**	***b* (mm)**	***B*, width of pin at hor. centre lim. mm**
4 to 6	2	$dp - 1$
6 to 10	3	$- 2$
10 to 18	5	$- 4$
18 to 30	8	$- 6$
30 to 50	12	$- 10$
50 and above	14	taken as per construction

When the pin B is diamond pin, the pin A and hole are are effectively located, that is,

$$\delta_A \text{ min} \cong 0$$

However, for accurate location, there has to be a little clearance (only a few microns) between hole A and pin A.

$\therefore$ From equ. (2),

$$b \leq \left(\frac{\delta_B \text{ min.}}{x+y}\right) \cdot dp \quad \text{...(3)}$$

The locating surface of a diamond pin is usually less than 8% of a full cylindrical pins. This provides more clearance at the location points. The width of the diamond locating pins is usually kept $\frac{1}{8}$th of workpiece hole diameter. It can be varied suitably to accommodate variation in centre distance of the workpiece and the desired fit between the hole and the diamond pin, which is usually Precision Running Fit, H7/F6.

1.19. ANALYSIS OF CLAMPING FORCES

The commonly used mechanical methods of applying clamping forces are:

1. Screw (V-threads)
2. Screw and Strap (lever type clamps)
3. Wedge
4. Eccentric Cam
5. Toggle linkage
6. Combined screw and wedge.

There are clamping devices in which clamping forces are applied through Pneumatic and Hydraulic means (Toggle linkage etc.).

1. Screw (V-threads): Fig. 1.90 (*a*), (Also See Fig. 1.21)

The tightening or loosening of a screw (clamping/unclamping the workpiece) is equivalent to the movement of a weight on a inclined plane. The actuating force P acts horizontally (tangent to the threads) to result in the holding or clamping force *F*. These two forces are connected as.

$$P = F.\tan(\alpha + \phi)$$

where α = Helix angle of the threads = $p/\pi\, dm$

p = pitch of the threads (lead in case of multistart threads)

dm = mean or pitch diameter of the threads

ϕ = angle of friction = $\tan^{-1}\mu$

μ = co-efficient of friction between nut and screw.

For unclamping, $P = F.\tan(\phi - \alpha)$

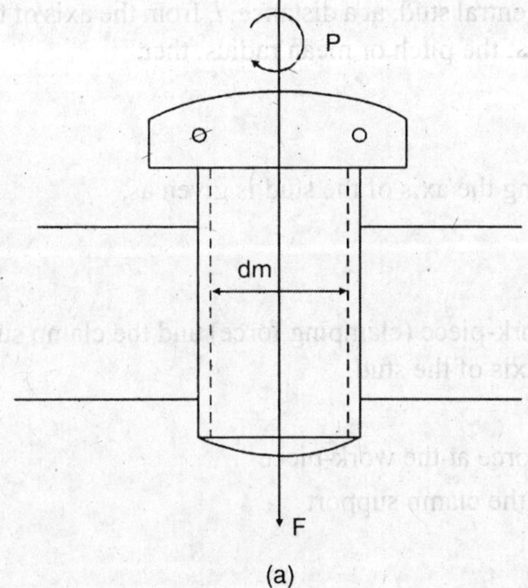

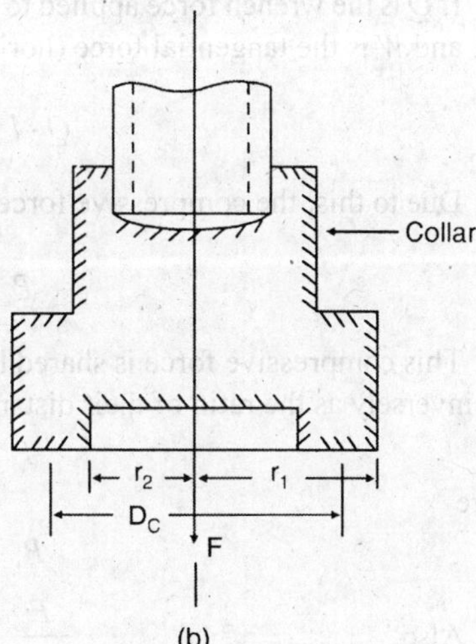

Fig. 1.90. Screw Clamp

If the end of the screw touches the work-piece directly, it can damage the surface of the work-piece, the work-piece can get displaced or the screw can get bent. To prevent these possibilities, a swivel pad or collar is provided, Fig. 1.90 (*b*), at the bottom end of the screw.

Total torrional moment needed to apply the clamping force

= Screw torsional moment + collar friction torisonal moment

$$= P.\frac{dm}{2} + \mu_C.F.\frac{D_C}{2}$$

where μ_C = co-efficient of friction between collar and workpiece

D_C = friction diameter of the collar

$= \frac{4}{3} \cdot \frac{r_1^3 - r_2^3}{r_1^2 - r_2^2}$, considering uniform pressure and when power lost in friction to be found out

$= (r_1 + r_2)$ considering uniform rate of wear and when power transmitted is to be determined.

∴ Total torsional moment, needed to clamp the workpiece

$$M_t = F \cdot \frac{dm}{2} \cdot \tan(\alpha + \phi) + \mu_C \cdot F \cdot \frac{D_C}{2}$$

The actuating force at the nut/screw head is usually applied with the help of a spanner or a wrench. If this force, T, is applied at a distance, *l*, from the axis of the screw, then

$$T.l = F \cdot \frac{dm}{2} \cdot \tan(\alpha + \phi) + \mu_C \cdot F \cdot \frac{D_C}{2}$$

∴ Clamping force,

$$F = \frac{T \cdot l}{\left[\frac{dm}{2} \cdot \tan(\alpha + \phi) + \mu_C \cdot \frac{D_C}{2}\right]}$$

The main drawback of screw clamps is that lot of time is wasted in clamping the job.

2. Screw and strap device: This type of clamping device (Fig. 1.24 to 1.26) is based on the lever or beam system. Refer Fig. 1.24 (*a*).

If Q is the wrench force applied to the nut of the central stud, at a distance, l, from the axis of the stud, and W is the tangential force (horizontal) acting at the pitch or mean radius, then

$$Q \cdot l = W \cdot \frac{dm}{2}$$

Due to this, the compressive force of the nut along the axis of the stud is given as,

$$P = \frac{W}{\tan(\alpha + \phi)}$$

This compressive force is shared between the work-piece (clamping force) and the clamp support inversely as the ratio of their distances from the axis of the stud.

$$P = F + P_1.$$

where F = clamping force at the work-piece

and P_1 = reaction at the clamp support

Also,

$$\frac{F}{P_1} = \frac{y}{x}$$

$$\therefore \quad \frac{F + P_1}{F} = \frac{x + y}{y}$$

∴ Clamping force,

$$F = P \cdot \frac{x}{x + y}$$

∴ The clamping force is dependent upon the distance y. Greater the distance y, greater will be the clamping force.

Also, $$\frac{F}{P_1} > 1$$

$$\therefore \quad \frac{y}{x} \geq 1$$

3. Wedge Clamps: Wedge clamps use the basic principle of inclined plane to hold the work in a manner similar to a cam. A wedge is a movable inclined plane which provides and should maintain the desired clamping force. It forces the workpiece against a fixed stop. The movement of the wedge should not require a large force for actuation. These requirements are controlled by the wedge angle, *i.e.*, the angle of inclination.

There are two general forms of wedge clamps:

– Using flat wedge (self releasing), Fig. 1.91

– Using conical wedge or mandrel used for holding work through a hole, Fig. 1.66.

Fig. 1.91 shows the forces acting on the plain wedge at the instant unclamping begins. F_2 is the clamping force when the wedge was inserted for clamping purposes. Taper angle of the wedge is α. If μ is the co-efficient of friction, between wedge and workpiece and fixture body, then

$$F_3 = \mu F_1$$

and $$F_4 = \mu F_2$$

Now, $$\Sigma F_V = 0$$

$$\therefore \quad F_2 = F_1 \cos\alpha + F_3 \sin\alpha$$

$$= F_1 (\cos\alpha + \mu \sin\alpha)$$

Also, $$\Sigma F_H = 0$$

$$\therefore \quad P = F_3 \cos\alpha + F_4 - F_1 \sin\alpha$$

$$= F_1[2\,\mu \cos\alpha + (\mu^2 - 1)\sin\alpha]$$

for small values of μ, μ^2 is much smaller and may be neglected, therefore,

$$P = F_1 (2\,\mu \cos\alpha - \sin\alpha)$$

P is + ve for small values of α and is negative for larger values of α. It means that as α increases, there is less and less tendency for the wedge to stay in place.

For $P = 0$, the wedge will not stay in place on its own accord.,

$$\tan\alpha = 2\mu$$

∴ For the wedge to stay tight α should be $< 2\tan^{-1}\mu$.

This is the condition for self-locking of a wedge clamp. If α is very small, the wedge is inclined to stick and the pull, P to disengage it becomes very large.

The taper angle of a plain wedge may range from 6° to 18°, depending upon μ (which depends upon wedge surface variables, such as the presence of oil and coolants, surface finish and hardness). Small values of α give the wedge more holding force but increases both the distance required to lock and the pulling force to disengage it. With the presence of oil or coolant, ∝ will still be be smaller unless an additional holding device is used. When α approaches the higher values (16 to 18°), the wedge tends to slip and to keep it in place, an auxiliary holding device can be used, Fig. 1.91 (b). A practical working angle is 7°. Even though a wedge is a quick-acting clamping device, it suffers from the drawbacks as discussed above. Also, they have the tendency to loosen under vibration. The working of a plain wedge can be improved by adding levers and links.

The force Q required to exert a clamping force F on the workpiece, is given as

$$Q = F[\tan(\alpha + \phi_1) + \tan\phi_2]$$

where ϕ_1 = Angle of friction between wedge and fixture body

ϕ_2 = Angle of friction between wedge and workpiece

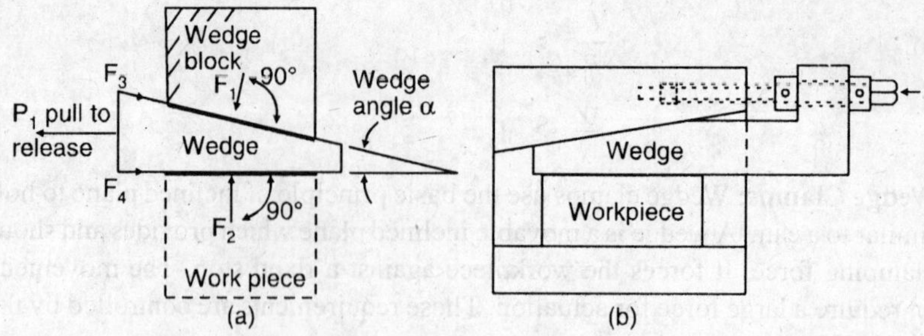

Fig. 1.91. Wedge Clamp

4. Eccentric Cam: The eccentric cam shown in Fig. 1.92 is a direct-acting cam with an outline in the form of a circle of diameter D, with centre at C. The lever is pivoted by the pin P, and the eccentricity is e.

When the lever (handle) is rotated clock–wise, the centre C will travel in a circle until the circumference contacts the workpiece at B. The clamping force will act through the pin, but there will be counteracting force resulting from the clamping force F. If this force is not too large and there is sufficient friction, the clamp will hold.

Note: If T_g = tolerance on the workpiece dimension, and

$L_{min.}$ = Minimum distance between the cam and the clamping face of the component

Then, $$e = \frac{3Tg + L_{min.}}{4}$$

F is the clamping force. At the contact point of cam and workpiece, the reaction F acs upwards

If dp = diameter of pin, then taking moments about the pin centre.

$$F_1.D/2 + F_z.dp/2 - F.e. = 0$$

$$\mu_1.F.\frac{D}{2} + \mu_2.F.\frac{dp}{2} - F.e = 0$$

where μ_1 = co-efficient of friction between cam surface and the work-piece

μ_2 = co-efficient of friction between pin and cam

e = eccentricity

For self-locking of the cam,

$$F \cdot e \leq \mu_1 \cdot F \cdot \frac{D}{2} + \mu_2 \cdot F \cdot \frac{dp}{2} \qquad \text{(Fig. 1.92(a))}$$

$$\therefore \quad e \leq \mu_1 \cdot \frac{D}{2} + \mu_2 \cdot \frac{dp}{2}$$

The pin surface and the internal bore of the cam are usually highly polished, so that μ_2 is very small and its effect can be neglected,

$$\therefore \quad e \leq \mu_1 \cdot \frac{D}{2}$$

or $$\frac{D}{e} \geq \frac{2}{\mu_1}$$

μ_1 normally varies from 0.1 to 0.2.

$\therefore \frac{D}{e}$ should range between 10 and 20.

The ratio $\frac{D}{e}$ is known as the characteristic of the eccentric cam, and it should lie between 14 and 16.

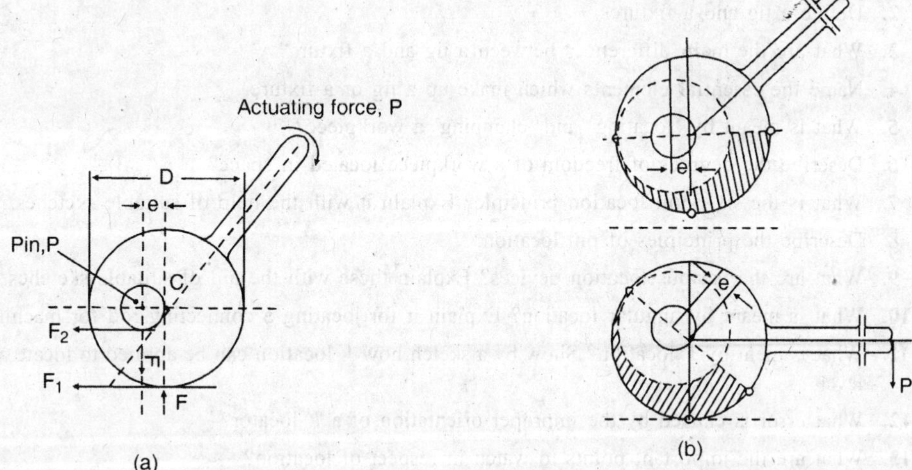

Fig. 1.92. Eccentric Cam

It is clear from the working of a cam-operated clamp that it is a kind of wedge (curvilinear wedge). It can be used as a sector, disk or cylinder whose working surface can be either a circle or a logarithmic or else Archemedian spiral. The circular cams are most common. In Fig. 1.27 (*b*), the inner (female) surface in contact with a pin is used as a cam. Such a clamp is called "Hook Cam Clamp". Fig. 1.92(*b*) shows the working of an eccentric cam.

5. Combined screw and wedge clamp: Fig. 1.93. Here, the clamping force is the combined action of a screw and wedge.

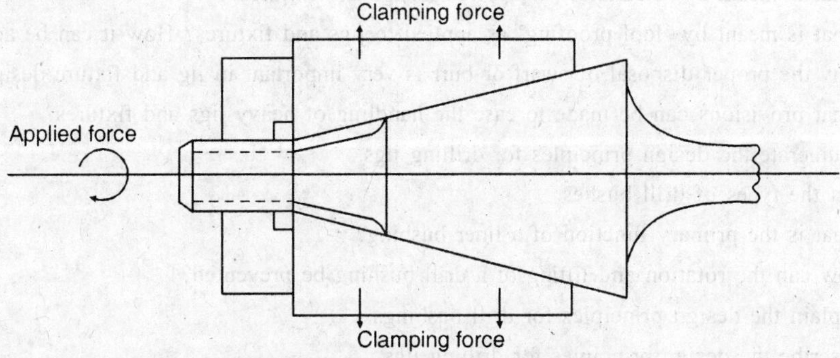

Fig. 1.93. Combined Screw and Wedge

6. Toggle linkage, Fig. 1.94: A toggle linkage clamp is a mechanical leverage device that multiplies a force exerted by an operator or through the effort of oil or air pressure (Fig. 1.81(*a*)).

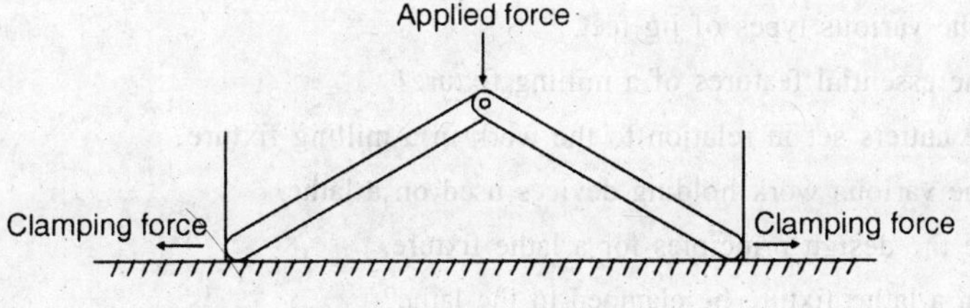

Fig. 1.94. Toggle Linkage

PROBLEMS

1. What are the function of jigs and fixtures?
2. Define a jig and a fixtures.
3. What are the main differences between a jig and a fixture?
4. Name the essential elements which make up a jig or a fixture.
5. What is meant by 'locating' and 'clamping' a workpiece?
6. Describe the degrees of freedom of a workpiece located in space?
7. What is the six-point location principle? Explain it with the help of suitable sketches.
8. Describe the principles of pin location.
9. What are the various location devices? Explain these with the aid of suitable sketches.
10. What is meant by angular location? Explain it for locating a connecting rod for machining.
11. What is ment by *V* location? Show by a sketch how *V* location can be applied to locate a small lever.
12. What error is caused by the improper orientation of a *V* locator?
13. What are the important points to watch in respect of location?
14. What are the important points to watch in respect of clamping? How should clamps be disposed of with respect to location points?
15. Name the more common types of clamps.
16. Sketch a two way clamp.
17. Sketch the various quick acting clamps.
18. What are the materials used for locating and clamping elements?
19. Explain with the aid of suitable sketches, the principles of jig and fixture design.
20. What is meant by 'clearance' as applied to jigs and fixtures?
21. What is meant by 'fool proofing' as applied to jigs and fixtures? How it can be achieved?
22. Whv the proper disposal of swarf or burr is very important in jig and fixture design?
23. What provisions can be made to ease the handling of heavy jigs and fixtures?
24. Enumerate the design principles for drilling jigs.
25. List the types of drill bushes.
26. What is the primary function of a liner bushing?
27. How can the rotation and lifting of a drill bushing be prevented?
28. Explain the design principles for drill bushings.
29. Describe the design principles for drilling jigs.
30. Write about the drill bush materials and their manufacture.
31. What are the main types of jigs? Discuss these with the help of suitable sketches.
32. Why should a jig have four feet and not three? Explain the reason.
33. Sketch the various types of jig feet.
34. Name the essential features of a milling fixture?
35. How are cutters set in relation to the work in a milling fixture?
36. Name the various work holding devices used on a lathe.
37. Describe the design principles for a lathe fixture.
38. How can a lathe fixture be clamped to the lathe?
39. Describe the various grinding fixtures.

40. Write short notes on "Broaching fixtures" "Assembly fixtures".

41. For what purposes are indexing jigs and fixtures used? List the operations suitable for use with indexing fixtures.

42. Explain with the aid of suitable sketches, the various indexing devices.

43. Explain the advantages to be obtained from the use of pneumatic and hydraulic clamping devices.

44. Compare compressed air and hydraulic power as a means for operating clamping devices.

45. What are the main types of jig and fixture construction? State the advantages of each and discuss the relative merits of each.

46. Design and draw drilling jigs for drilling the holes in the components shown in Fig. 1.95 to 1.99.

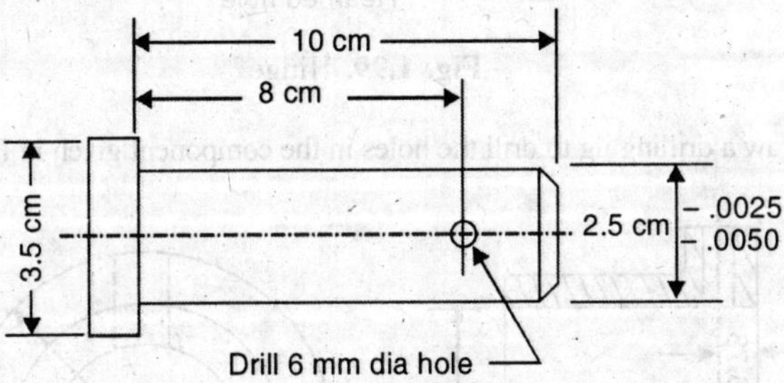

Fig. 1.95

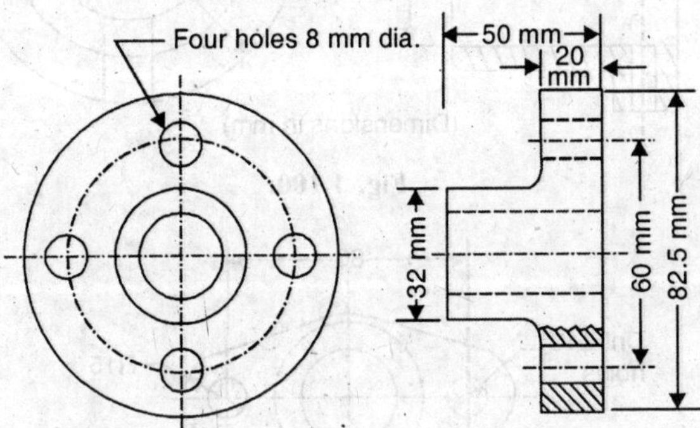

Fig. 1.96

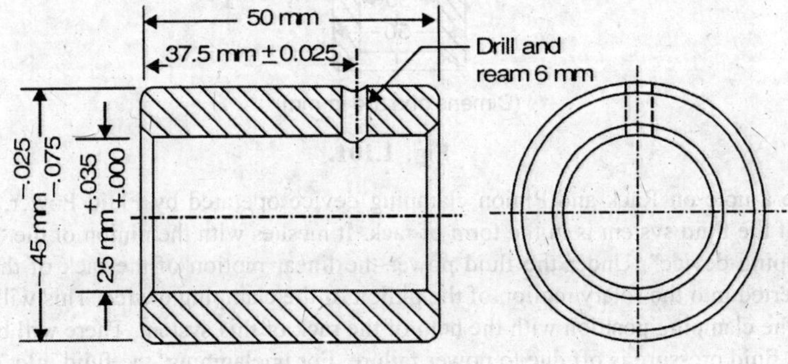

Fig. 1.97. Bush

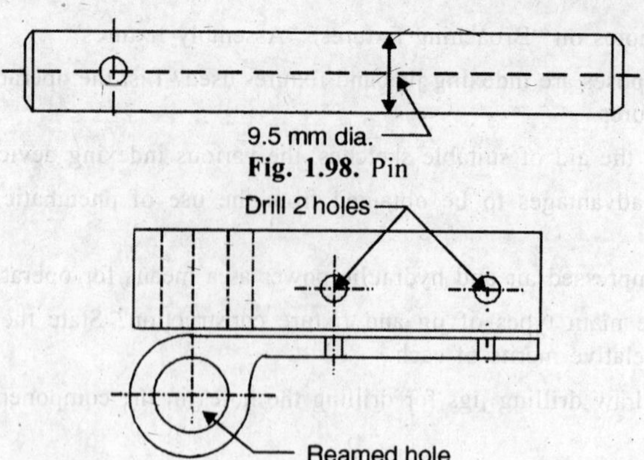

Fig. 1.98. Pin

Fig. 1.99. Hinge

47. Design and draw a drilling jig to drill the holes in the component given in Fig. 1.100 and in Fig. 1.101.

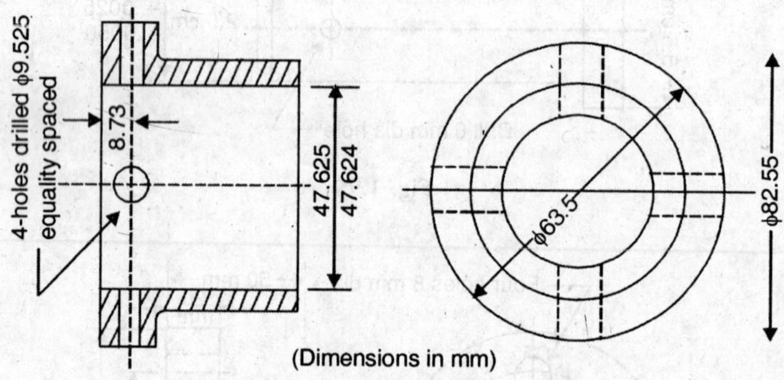

Fig. 1.100.

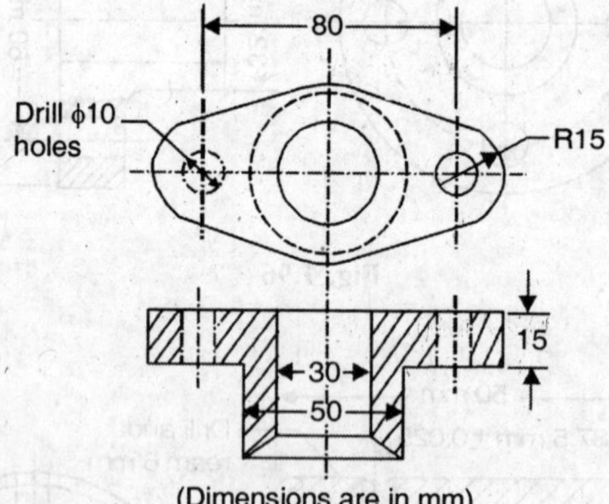

Fig. 1.101.

48. Write a note on Rack and Pinion clamping device operated by Fluid Power, Here, the piston rod of the fluid system is in the form of rack. It meshes with the pinion of the "Rack and Pinion Clamping device". Under the fluid power the linear motion of the rack of the fluid system is converted into the rotary motion of the pinion of the clamping device. This will move the clamp into the clamping position with the help of the rack of this system. There will be no unclamping if the fluid pressure is off due to power failure. For unclamping, the fluid inlet port is connected to the discharge line.

2

PRESS TOOL DESIGN

2.1. GENERAL

Press working may be defined as a chipless manufacturing process by which various components are made from sheet metal. This process is also termed as cold stamping. The machine used for press working is called a press. The main features of a press are : A frame which supports a ram or a slide and a bed, a source of mechanism for operating the ram in line with and normal to the bed. The ram is equipped with suitable punch/punches and a die block is attached to the bed. A stamping is produced by the downward stroke of the ram when the punch moves towards and into the die block. The punch and die block assembly is generally termed as a "die set" or simply as the "die". Press working operations are usually done at room temperature.

2.2. PRESS OPERATIONS

The sheet metal operations done on a press may be grouped into two categories, cutting operations and forming operations. In cutting operations, the workpiece is stressed beyond its ultimate strength. The stresses caused in the metal by the applied forces will be shearing stresses. In forming operations, the stresses are below the ultimate strength of the metal. In this operation, there is no cutting of the metal but only the contour of the workpiece is changed to get the desired product. The cutting operations include : blanking, punching, notching, perforating, trimming, shaving, slitting and lancing etc. The forming operations include : bending, drawing, redrawing and squeezing. The stresses induced in the metal during bending and drawing operations are tensile and compressive and during the squeezing operation these are compressive.

Below we give the definitions of the various press operations :

(*a*) *Blanking.* Blanking is the operation of cutting a flat shape from sheet metal. The article punched out is called the 'blank' and is the required product of the operation. The hole and metal left behind is discarded as waste. It is usually the first step of series of operations, Fig. 2.1 (*a*).

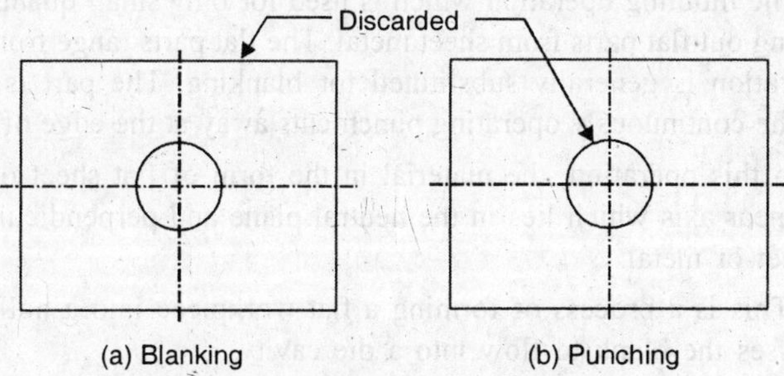

Fig. 2.1. Blanking and Punching.

Plain blanking is explained in Fig. 2.2.

(*b*) *Punching* (*Piercing*). It is a cutting operation by which various shaped holes are made in sheet metal. Punching is similar to blanking except that in punching, the hole is the desired product, the material punched out to form the hole being waste Fig. 2.1 (*b*).

(*c*) *Notching.* This is cutting operation by which metal pieces are cut from the edge of a sheet, strip or blank.

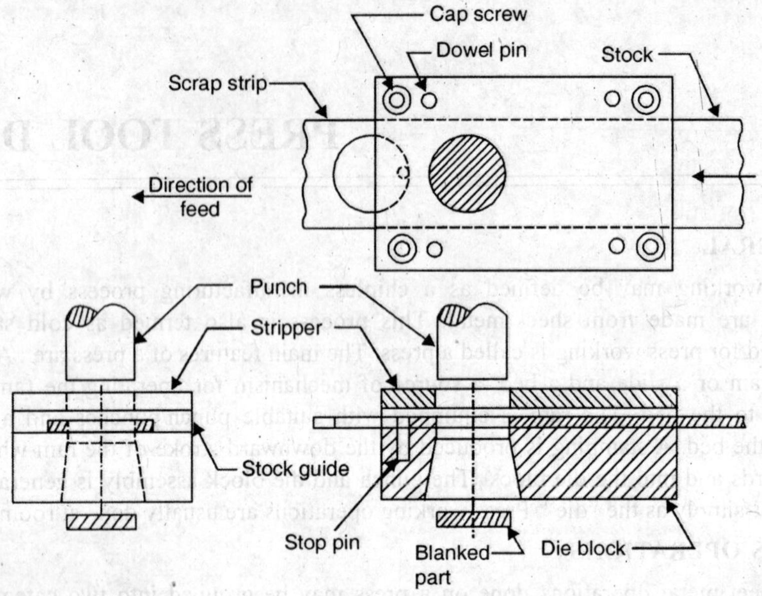

Fig. 2.2. Plain Blanking.

(*d*) *Perforating.* This is a process by which multiple holes which are very small and close together are cut in flat work material.

(*e*) *Trimming.* This operation consists of cutting unwanted excess material from the periphery of a previously formed component.

(*f*) *Shaving.* The edges of a blanked part are generally rough, uneven and unsquare. Accurate dimensions of the part are obtained by removing a thin strip of metal along the edges. The operation is termed as shaving.

(*g*) *Slitting.* It refers to the operation of making incomplete holes in a workpiece.

(*h*) *Lancing.* This is a cutting in which a hole is partially cut and then one side is bent down to form a sort of tab or louver. Since no metal is actually removed, there will be no scrap.

(*i*) *Nibbling.* The nibbling operation which is used for only small quantities of components, is designed for cutting out flat parts from sheet metal. The flat parts range from simple to complex contours. This operation is generally substituted for blanking. The part is usually moved and guided by hand as the continuously operating punch cuts away at the edge of the desired contour.

(*j*) *Bending.* In this operation, the material in the form of flat sheet or strip, is uniformly strained around a linear axis which lies in the neutral plane and perpendicular to the lengthwise direction of the sheet or metal.

(*k*) *Drawing.* This is a process of forming a flat workpiece into a hollow shape by means of punch which causes the blank to flow into a die cavity.

(*l*) *Squeezing.* Under this operation, the metal is caused to flow to all portions of a die cavity under the action of compressive forces.

Some of these operations are explained in Fig. 2.3.

As written above, the workpiece obtained after one or more press operations is called a metal stamping. A metal stamping can be defined as one which is usually made of, but not limited to, sheet or strip metal that has been cut, coined, pierced, bent, formed or drawn in one or more operations between matching dies under pressure. The size of a metal stamping, the thickness, the type of material and shape vary widely. Meta stampings range in size from tiny instrument part to freight car ends.

The materials amenable to cold stamping are low-carbon and plastic alloyed steels, non-ferrous metals including copper and its alloys (brass, bronze), aluminium, zinc, titanium and other materials. The starting material for press working is sheet metal in the form of bands or strips from tenth of a mm to about 6 or 8 mm thick. During the stamping process, the wall thickness of the parts remains almost constant and differs only slightly from the thickness of the initial sheet metal. The material must be free from scale before sheet metal working. Scale is removed by pickling and other methods.

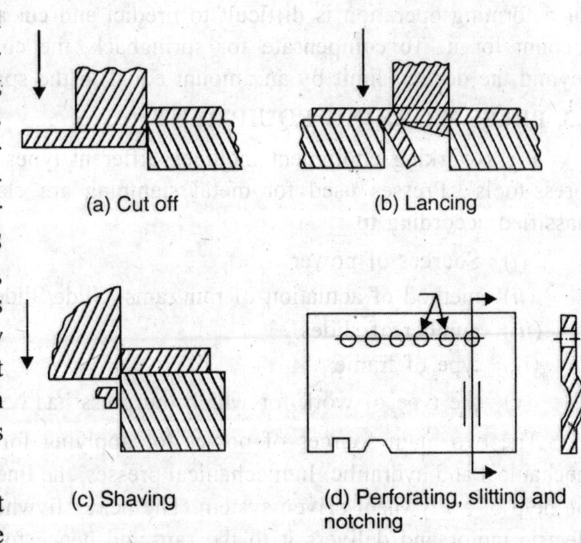

Fig. 2.3. Some Press Operations.

The use of metal stampings is not confined to any single industry. On the contrary, a large percentage of manufactured products of all types employ at least one or more metal stamping.

Advantages of metal stamping. Sheet stamping is an advanced technique of mechanical working. It has got the following advantages :

1. Small weight of fabricated parts.
2. High productivity of labour (upto 30,000 to 40,000 parts per 8 hour shift).
3. High efficiency of technique as regards the fabrication of items of diversified shapes, both simple and complex, such as washers, bushings, retainers (cages) of ball bearing, tanks and car bodies etc.
4. The parts made by cold sheet metal stamping are distinguished for their size accuracy (narrow tolerances and high surface finish). In many cases, they require no subsequent machining and are delivered to the assembly shop.
5. High-volume, low-cost production.
6. Predictable strength characteristics.
7. Uniformity of parts.
8. Low cost material
9. Less labour-consuming.

All the above advantages have made sheet stamping very attractive to a host of industries, particularly to automotive, aircraft, electrical engineering and others. The variety of products is immense, from aircraft skins and automobile bodies to appliance shells. Other products include : construction girders, truck frames, furniture legs, supertankers, bath tubes, beer cans, wheel rims, fan blades, watch gears and so on (list is very long).

Elastic Recovery or Springback

We know that in metal working proceses, the total deformation imparted to a workpiece will be the sum of elastic deformation and plastic deformation. We also know that elastic deformation is recoverable whereas plastic deformation is permanent. So, at the end of a metal working operation, when the pressure on the metal is released, there is an elastic recovery by the material and the total deformation will get reduced a little. This phenomenon is called as "Springback".

This phenomenon is more importance in cold working operations, especially in forming operations such as bending etc. Springback depends upon the yield point strength of a metal. The higher the yield point strength of a metal, the greater the springback. The amount of springback for a forming operation is difficult to predict and cut and try methods are most satisfactory to account for it. To compensate for springback, the cold deformation must always be carried beyond the desired limit by an amount equal to the springback.

2.3. PRESS-WORKING EQUIPMENT

Press-working equipment includes different types of presses and tools, known as dies or 'press-tools'. Presses used for metal stamping are classified in various ways. They may be classified according to :

(*i*) Sources of power

(*ii*) method of actuation of ram/rams (Slide/Slides)

(*iii*) number of slides

(*iv*) type of frame

(*v*) the type of work for which the press has been designed.

The two main sources of power for applying force to the ram or slide of a press are : mechanical and hydraulic. In mechanical presses, the linear movement of the ram is obtained with the help of a Flywheel driven system. The heavy flywheel absorbs energy continuously from an electric motor and delivers it to the ram and hence to the workpiece intermittently. Hydraulic presses are a large piston and cylinder arrangement coupled to a hydraulic pump. The piston and the press ram are one unit. The ram is actuated by oil pressure on the piston in the cylinder. Water and water-oil emulsion can also act as working mediums.

Mechanical presses have the following advantages over hydraulic presses :

(1) run faster (2) lower maintenance cost (3) lower capital cost

Advantages of hydraulic presses are :

1. Most versatile and easier to operate
2. Tonnage capacity adjustable from zero to maximum
3. Constant pressure can be maintained throughout the stroke
4. Force and speed can be adjusted throughout the stroke, if need be
5. More powerful than mechanical presses
6. Safe as it will stop to a pressure setting
7. Stroke can be varied to any length within the limits of hydraulic cylinder travel.
8. The press can exert its full tonnage at any position of the ram stroke.

The main disadvantage of the hydraulic press is that it is slower than a mechanical press.

The flywheel of the mechanical press drives the main shaft. The rotary motion of the main shaft is converted into the linear motion of the ram or slide. The most common drive is crankshaft which is used for longer strokes. The eccentric drive can be used for smaller strokes. The other drives include : Cam, toggles, rack and pinion, screws and knuckles which can be used to actuate the ram. Cam drive is similar to eccentric drive except that at the bottom of the stroke, dwell is obtained. Rack and pinion drive is used where a very long stroke is required. In this drive, the movement of the slide is uniform and much slower as compared to crank drive.

Most press working makes use of mechanically driven and less frequently, hydraulic presses.

According to the number of slides, there are three types of presses : single-, double, - or triple action press. A single action press has only one ram which acts against a fixed bed. A double action press has two slides (one inner and the other outer) moving in the same direction against the fixed bed of the press. Such presses are used in deep drawing operations. A triple action press has two slides moving in the same direction against the fixed bed as in double action press and the third or lower slide moves upward through the fixed bed. Such presses are used for large work such as motor car body panels.

According to the type of frame, there are two types of presses : open frame presses and closed frame presses. Due to their construction, the open frame presses are less rigid and strong and so are useful mainly for operations on smaller work. These presses are available upto 200-tonne capacity with strokes of 90 to 120 per minute. Closed frame presses are stronger, more rigid and balanced than open frame presses. These presses are available upto at least 3000 tonnes capacity.

The "Throat Depth" of a press is the distance from the slide centre line to the frame. It determines the maximum distance between the shank axis and the rearmost point of a die that can be mounted on the press.

Small presses often have inclinable press frames which facilitate removal of the stamped part by gravity.

Depending upon the type of work, the presses can be designed for doing the following types of work : Punching, Blanking, Drawing, Bending, Forming, Coining, Embossing etc.

According to the method of strip feeding the presses can be designed either for manual or automatic feed.

According to the method of removing finished parts, there are three types of presses :

1. Presses with dies with a hole through which the part drops.

2. The ones in which the component is ejected upwards and then removed by a positive knockout.

3. Those in which the finished parts are removed by compressed air or manually.

2.3.1. Rating of a press. A press is rated in Tonnes of force that can be applied to the slide without undue strain and without affecting the structural strength of the press.

The tonnage of a mechanical press is determined by the size of the bearing for the crankshaft or the eccentric. It is given approximately by the relation :

$$\text{Tonnage capacity} = \text{Shear strength of the crankshaft material} \times \text{the area of the crankshaft bearings}$$

The tonnage capacity of a mechanical press is always given when the slide is near the bottom of its stroke, because it will be maximum at that point, or with the crank turned through an angle of not more than 30° from the bottom zero position.

The tonnage capacity of a hydraulic press is given as :

$$\text{Rated capacity} = \text{Piston area} \times \text{oil pressure in the cylinder.}$$

As noted earlier, the capacity of a hydraulic press can be varied by changing oil pressure.

In double-action crank presses, the tonnage of the inner slide determines the maximum drawing pressure, while the maximum blank holding pressure depends upon the tonnage of the outer side. To keep the strains and the deflections of the press structure small, it is a usual practice to choose a press rated 50 to 100% higher than the force required for an operation.

In the case of single,- double,- and triple- action hydraulic presses, the rams may all be driven from a central hydraulic accumulator fed by pump, or have individual drives from one or more pumps.

The construction and working of the various types of presses mentioned above have been discussed in detail in the book "A Textbook of Production Technology" by the author.

2.3.2. Requirements of press tool design. The press tool design should suit the type of production it will be used for, that is, small batch, large batch, or mass production. The press tools should meet the following requirements :

1. The dimensional accuracy and surface finish of stampings should conform to the drawings and specification.

2. The working parts of the press tool (die and punch) must be adequately strong, durable in operation and easily replaceable when worn out.

3. The die should ensure the required hourly out put, easy maintenance, safe operation and reliable fastenings in the press.

4. The die should be designed in such a way that as far as possible standard components are used for its manufacture. As few special parts as possible should be used in its design.

5. The scrap in the stamping operation must be kept at a minimum by suitably designing the strip layout. The percent utilization of the material is given as :

$$\text{Percentage material utilization} = \frac{\text{Total area of blank cut}}{\text{Area of uncut strip}} \times 100$$

Normally, the value of this factor is 70 to 80%

2.3.3. Press tool components. The press tool components may be divided into the following types, see Fig. 2.4.

(*a*) Working components, which participate in the shaping of parts : dies, punches and their sections.

(*b*) Structural components, which serve for joining the working components to one another and to the press : Upper shoe (Punch holder), lower shoe (die holder) and shanks.

(*c*) Guiding components, which ensure accurate alignment of the upper shoe with the die shoe in operation : guide posts and bushings. Guide posts also facilitate tryout of the press tool in a press.

(*d*) Feeding components, which feed the stock strip or blanks to the stamping station.

(*e*) Locating and locking components, which provide for an accurate positioning of the stock or blank in the die and fix it in place while the operation is performed.

(*f*) Stripping components, which strip and remove the blanks and scrap from the working components after the operation is over : strippers, push off pins, knockouts.

(*g*) Fastening components, which join and hold together all parts and units of the press tool : punch plates, die-blocks and cases, and all fasteners.

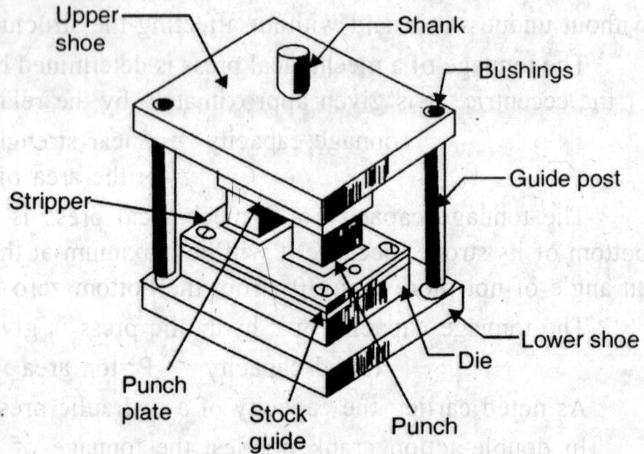

Fig. 2.4. Press Tool Components.

Fig. 2.4 is a schematic representation of a press tool comprising all principal components present in any typical press tool. The shank secures the press tool to the press ram. It fits in the clamping hole in the lower end of the press ram.

2.3.4. Building up of the press tool. Inspite of the great variety of press tools, most of them have certain components similar in purpose and design. To facilitate design, manufacture, maintenance and repair of press tools, similar units and components have been standardised.

The most important standardised component of a press tool is "Die-Set", Fig. 2.5, which consists of an upper shoe (punch holder), lower shoe (die-holder) and guide posts. Punch (or punches) is mounted on the upper shoe and the die components are mounted on the lower shoe (die shoe). The available space for mounting punch and die components is called the "die-area". The die area on the lower shoe should be at least 6.35 mm larger all around the die block. Ears are provided on the flange of the die shoe to fasten it to the bolster plate or the bed of the press, with the help of *T*-bolts.

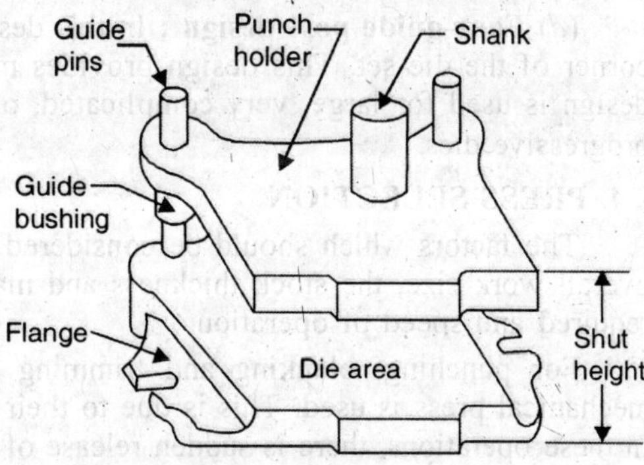

Fig. 2.5. A Commercial Die Set

In smaller die sets the upper shoes are made with a shank. On longer die sets (for cutting press tools) the standard shouldered shanks are press fitted into a specially bored hole of the upper shoe, Fig. 2.6 (*a*). In drawing and bending press tools, threaded shanks are used Fig. 2.6 (*b*). There are other arrangements also for fitting the shanks into the upper shoes of larger die sets. If the tool shank does not fit in the press slide hole, adapter shanks or split adapter bushings may be employed.

Arrangement of guide posts. The various arrangements of guide posts are given below :

(*a*) **Back-post design.** In this design, the die-set has two guide posts positioned in the rear, Fig. 2.7 (also see Fig. 2.5)

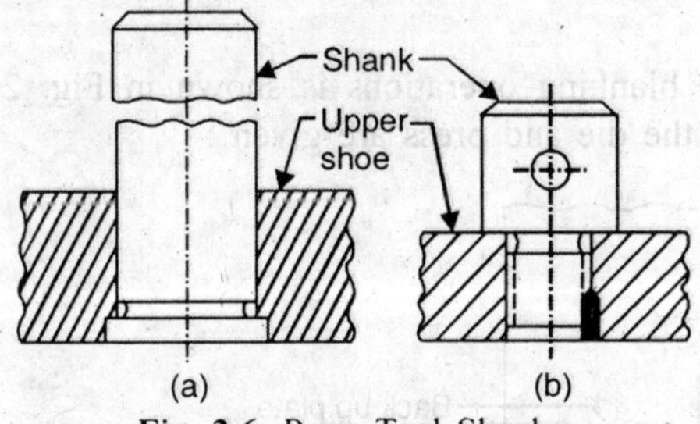

Fig. 2.6. Press Tool Shanks.

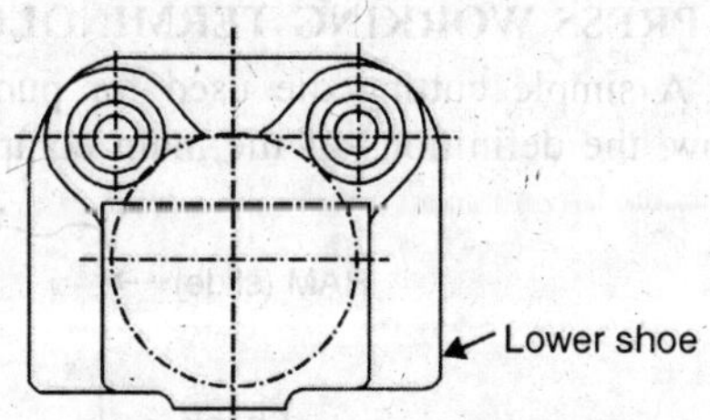

Fig. 2.7. Back Post Design.

This is the most frequently used design. This design gives clear view of the moving parts and there is no obstruction from guide posts. This helps the operator when feeding from left to right. This design also leaves a clear space for hand feeding blanks for second operations.

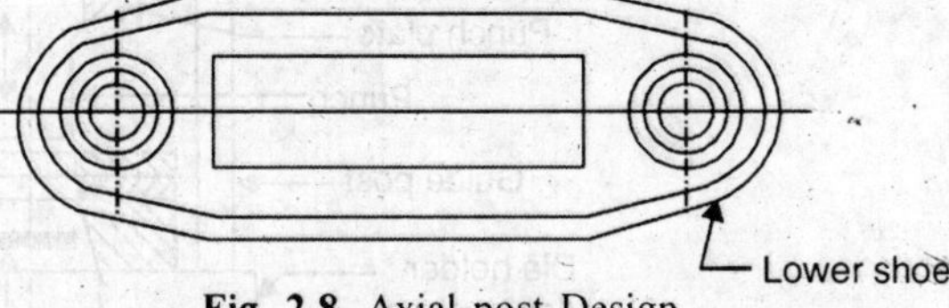

Fig. 2.8. Axial-post Design.

(*b*) **Centre-post or Axial-post design.** In this design, the two guide posts are positioned, along the transverse centre line of the die-set, Fig. 2.8. This design is suitable when the load is too heavy and the strip feeding is from the front. This design prevents end feeding.

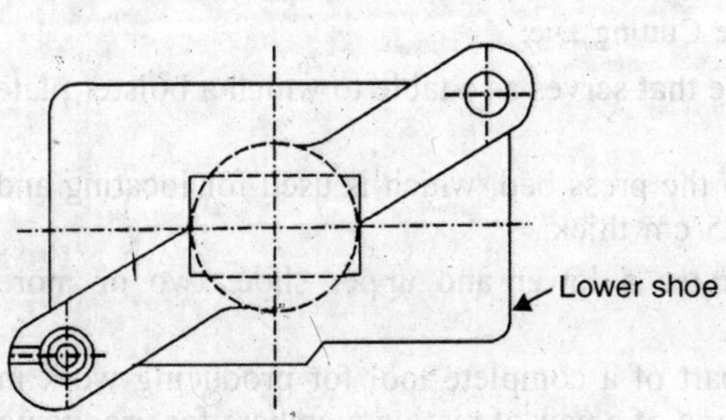

Fig. 2.9. Diagonal Post Design

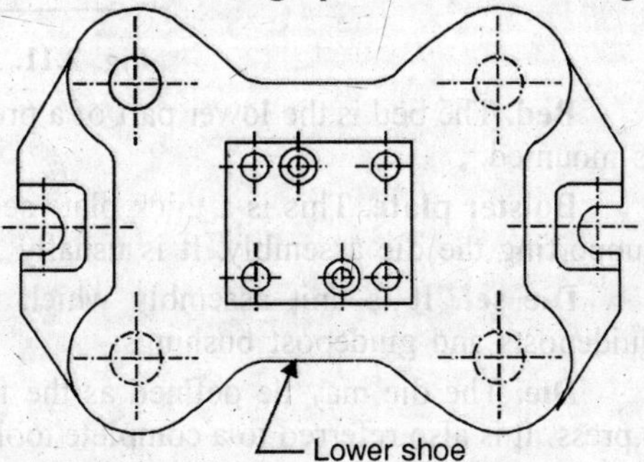

Fig. 2.10. Four-guide Post Design

(*c*) **Diagonal-post design.** This design, Fig. 2.9 is used when the load is very heavy and end feeding is required. The guide posts are along a diagonal line. Generally, the left guide post is in front, but if feed is to be from left to right, the right guide post is in front.

(*d*) **Four-guide post design :** In this design, Fig. 2.10, a guide post is positioned at each corner of the die-set. This design provides maximum rigidity and accuracy of alignment. This design is used for large, very complicated, or highly accurate stampings and for multi-station progressive dies.

2.4. PRESS SELECTION

The factors which should be considered while selecting a press for a given job, are : the overall work size, the stock thickness and material, kind of operation to be performed, power required and speed of operation.

For punching, blanking and trimming operations, usually the crank or eccentric type mechanical press is used. This is due to their small working strokes and high production rates. In these operations, there is sudden release of load at the end of the cutting stroke. This sudden release of load is not advisable in hydraulic presses. So, hydraulic presses are not preferred for these operations. If however these are inevitable, then some damping devices are incorporated in the press design. For coining and other squeezing operations, which require very large forces, knuckle joint mechanical press is ideally suited. Hydraulic presses, which are slower and more powerful, can also be used for these operations. Hydraulic presses are also better adapted to pressing, forming and drawing operations, which are basically slower processes.

2.5. PRESS WORKING TERMINOLOGY

A simple cutting die used for punching and blanking operations is shown in Fig. 2.11. Below, the definitions of the main components of the die and press are given :

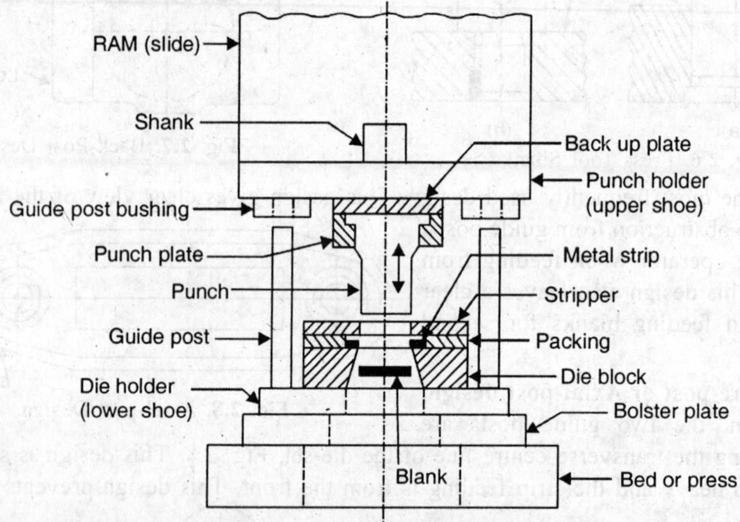

Fig. 2.11. A Simple Cutting Die.

Bed. The bed is the lower part of a press frame that serves as a table to which a bolster plate is mounted.

Bolster plate. This is a thick plate secured to the press bed, which is used for locating and supporting the die assembly. It is usually 5 to 12.5 cm thick.

Die set. It is unit assembly which incorporates a lower and upper shoe, two or more guideposts and guidepost bushings.

Die. The die may be defined as the female part of a complete tool for producing work in a press. It is also referred to a complete tool consisting of a pair of mating members for producing work in a press.

Die block. It is a block or a plate which contains a die cavity.

Lower Shoe. The lower shoe of a die set is generally mounted on the bolster plate of a press. The die block is mounted on the lower shoe. Also, the guide posts are mounted in it.

Punch. This is the male component of the die assembly, which is directly or indirectly moved by and fastened to the press ram or slide.

Upper Shoe. This is the upper part of the die set which contains guidepost bushings.

Punch plate. The punch plate or punch retainer fits closely over the body of the punch and holds it in proper relative position.

Back up plate. Back up plate or pressure plate is placed so that the intensity of pressure does not become excessive on punch holder. The plate distributes the pressure over a wide area and the intensity of pressure on the punch holder is reduced to avoid crushing.

Stripper. It is a plate which is used to strip the metal strip from a cutting or non-cutting punch or die. It may also guide the sheet.

Knockout. It is a mechanism, usually connected to and operated by the press ram, for freeing a workpiece from a die.

Pitman. It is a connecting rod which is used to transmit motion from the main drive shaft to the press slide.

Shut Height. It is the distance from top of the bed to the bottom of the slide, with its stroke down and adjustment up.

Stroke. The stroke of a press is the distance of ram movement from its up position to its down position. It is equal to twice the crankshaft throw or the eccentricity of the eccentric drive. It is constant for the crankshaft and eccentric drives but is variable on the hydraulic press.

2.5.1 Working of a cutting die (Fig. 2.11). As is clear in the figure, the punch holder (upper shoe) is fastened directly to the ram of the punch press, and the die shoe (lower shoe) is fastened to the bolster plate of the press. Guide posts may be used to better align the punch holder with the die shoe. These main components (punch holder, die shoe and guide posts) constitute what is known as the die set. A die set can be had with two guide posts located at the rear of the die set (known as back posts), diagonally, one at the back and the other in the front of the die set, or with four guide posts, one in each corner of the die set. The lower ends of the guide posts are press fitted into the die shoe. At the upper end, the guide posts have a slip fit with the guide bushings which are press fitted into the punch holder. With this, the guide posts have a free movement in the bushings. The punch is fastened to the punch holder and the die block is fastened to the die-shoe. The punch is aligned with the opening in the die-block. Since, both the punch and die block act as cutting tools, they are hardened.

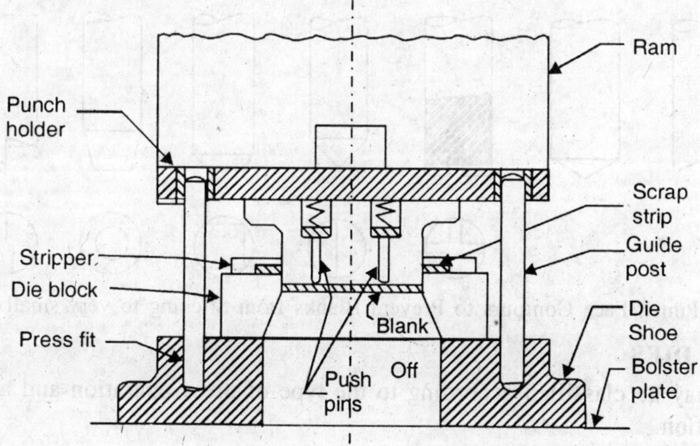

Fig. 2.12. A Cutting Die.

The cutting action takes place during the downward movement of the punch into the die block. After the cutting action, the elastic recovery in the strip material takes place. Due to this, the size of the blank (cut portion from the strip) increases and that of the hole in the strip decreases. So, at the end of the cutting action, when the punch starts to move upwards the scrap strip clings to the punch and the blank gets clogged in the die-opening. To remove the scrap strip from the punch surface, a stripper is used. In Fig. 2.11., a simplest type of stripper is used. It strips off the scrap strip from the punch surface when the scrap strip strikes the bottom surface of the stripper during the upward movement of the punch. To avoid clinging of the blank in the die opening, the walls of the die-opening are tapered.

In addition to the two above mentioned troubles, the blank may also adhere to the face of the punch. This usually happens with thin blanks which have been treated with a lubricant. To help the blank free itself from the punch face, push-off pins are provided which are fitted into the punch body, Fig. 2.12.

The various basic designs of push-off pins are shown in Fig. 2.13. The most popular design is a straight, round plunger with a sliding fit in the punch body, Fig. 2.13 (*a*). It is held in place by a light compression spring and a set screw. If at the end of the punch body, a pilot is to be provided, then the push-off pin is located off-centre, Fig. 2.13 (*b*). When the space available for a compressing spring is insufficient, a flat spring may be used, Fig. 2.13 (*c*).

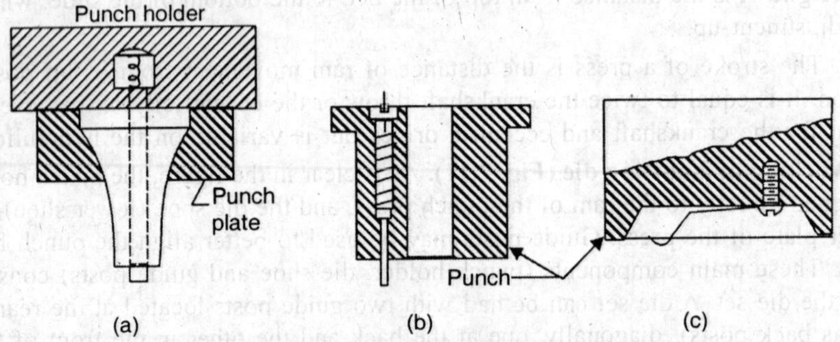

Fig. 2.13. Push-off Pins.

In the case of very small punches, no space is left to provide the push-off pin. In such cases, the sticking of the blank to the punch face may be avoided by changing the shape of the punch bottom, Fig. 2.14. With this, the oil seal or vacuum created by the oil film is broken and tendency of the blank to stick to the punch face is greatly decreased. The drawback of this method is that blank gets distorted. Due to this, this method is suitable for punching operation only.

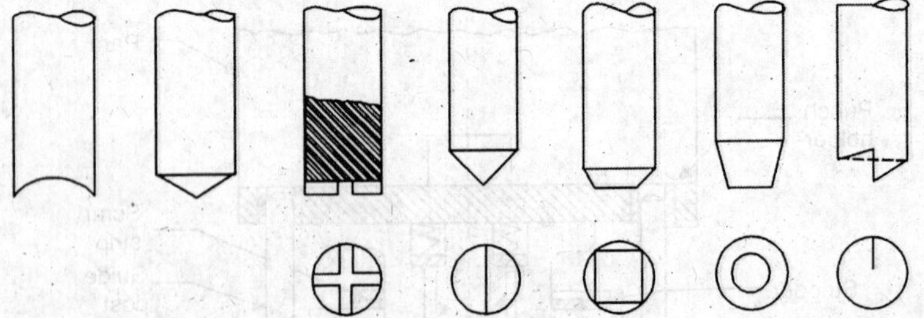

Fig. 2.14. Punch Face Contours to Prevent Blanks from Sticking to very small Punches.

2.6. TYPES OF DIES

This dies may be classified according to the type of press operation and according to the method of operation.

2.6.1 Type of press operation. According to this criterion, the dies may be classified as : cutting dies and forming dies.

Cutting dies. These dies are used to cut the metal. They utilize the cutting or shearing action. The common cutting dies are : blanking dies, piercing dies, perforating dies, notching, trimming, shaving and nibbling dies etc.

Forming dies. These dies change the appearance of the blank without removing any stock. These dies include bending dies, drawing dies, squeezing dies etc.

2.6.2. Method of operation. According to this criterion, the dies may be classified as : Single-operation dies or simple dies, compound dies, combination dies, progressive dies, transfer dies, and multiple dies.

Simple dies. Simple dies or single action dies perform single operation for each stroke of the press slide. The operation may be any of the operations listed under cutting or forming dies.

Compound dies. In these dies, two or more operations may be performed at one station. Such dies are considered as cutting tools since, only cutting operations are carried out. Fig. 2.15 shows a simple compound die in which a washer is made by one stroke of the press. The washer is produced by simultaneous blanking and piercing operations. Compound dies are more accurate and economical in mass production as compared to single operation dies.

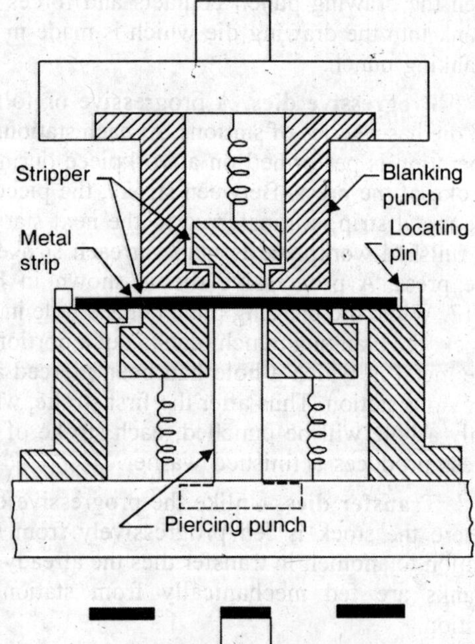

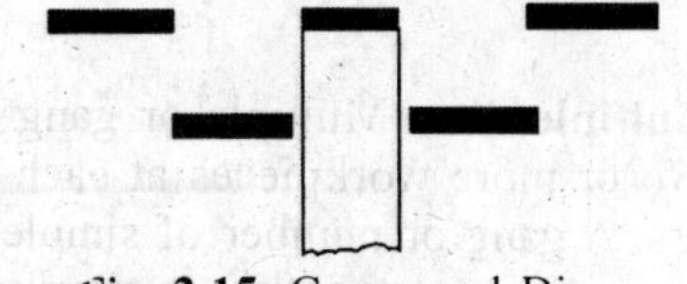

Fig. 2.15. Compound Die.

Combination dies. In this die also, more than one operations may be performed at one station. It differs from compound die in that in this die, a cutting operation is combined with a bending or drawing operation. Fig. 2.16 explains the working of a combination blank and draw die. The die ring which is mounted on the die-shoe, is counterbored at the bottom to allow the flange of a pad to travel up and down. This pad is held flush with the face of the die by a spring. A drawing punch of required shape is fastened to the die shoe. The blanking punch is secured to the punch holder. A spring stripper strips the skeleton from the blanking punch. A knockout extending through the centre opening

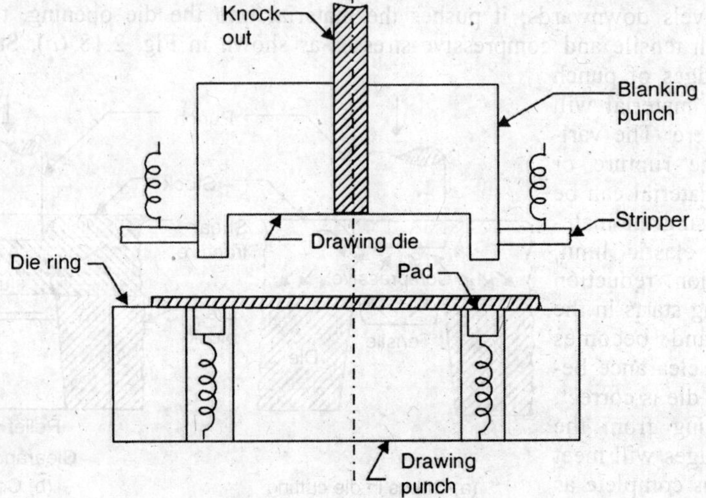

Fig. 2.16. Combination Die.

and through the punch stem ejects the part on the upstroke as it comes in contact with the knockout bar on the press. In operation, the blank holding ring descends as the part is blanked, then the drawing punch contacts and forces the blank into the drawing die which is made in the blanking punch.

Progressive dies. A progressive or follow on die has a series of stations. At each station, an operation is performed on a workpiece during a stroke of the press. Between stroke, the piece in the metal strip is transferred to the next station. A finished workpiece is made at each stroke of the press. A progressive die is shown in Fig. 2.17. While the piercing punch cuts a hole in the stock, the blanking punch blanks out a portion of the metal in which a hole had been pierced at a previous station. Thus after the first stroke, when only a hole will be punched, each stroke of the press produces a finished washer.

Transfer dies. Unlike the progressive dies where the stock is fed progressively from one station to another, in transfer dies the already cut blanks are fed mechanically from station to station.

Multiple dies. Multiple or gang dies produce two or more workpieces at each stroke of the press. A gang or number of simple dies and punches are ganged together to produce two or more parts at each stroke of the press.

Strip
Punch holder
Blanking punch
Piercing punch
Guide pin (pilot)
Stripper plate
Strip
Die
Bolster
Washer

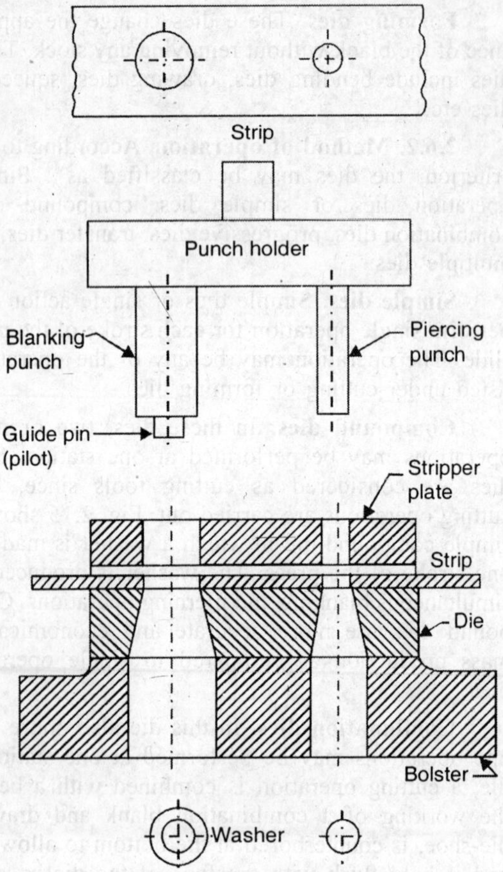

Fig. 2.17. Progressive Die.

2.7. PRINCIPLE OF METAL CUTTING

The cutting of sheet metal in press work is a shearing process. The cutting action is explained with the help of Fig. 2.18. The punch is of the same shape as of the die opening except that it is smaller on each side by an amount known as 'clearance'. As the punch touches the material and travels downwards, it pushes the material into the die opening, the material is subjected to both tensile and compressive stresses as shown in Fig. 2.18 (*a*). Stresses will be highest at the edges of punch and die and the material will start cracking there. The various steps in the rupture or fracture of the material can be written as : stressing the material beyond its elastic limit, plastic deformation, reduction in area, fracturing starts in the reduced area and becomes complete. If the clearance between punch and die is correct, the cracks starting from the punch and die edges will meet and the rupture is complete as shown in Fig. 2.18 (*b*). If the

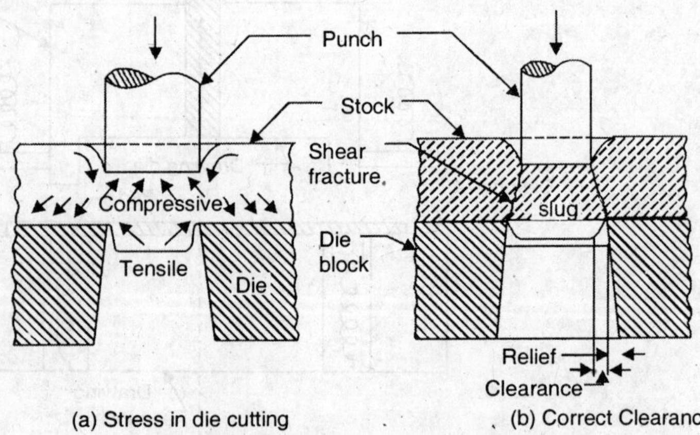

Fig. 2.18. Principle of Metal Cutting.

clearance is too large or too small, the cracks do not meet and a ragged edge results die to the material being dragged and torn through the die.

This is explained in Fig. 2.19.

2.8. CLEARANCE

As is clear in the previous article, the die opening must be sufficiently larger than the punch to permit a clean fracture of the metal. This difference in dimensions between the mating members of a die set is called 'clearance'. This clearance is applied in the following manner :

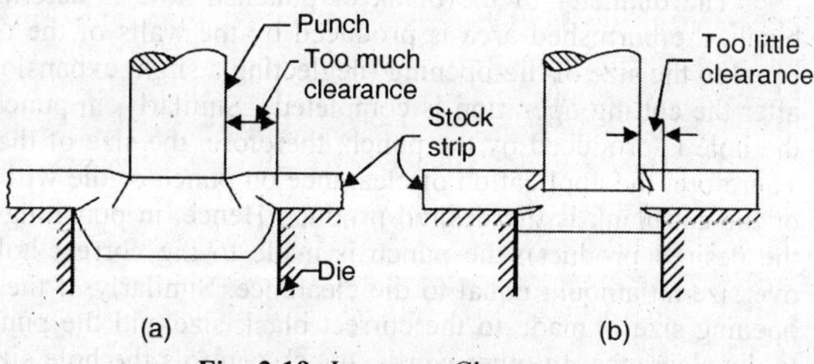

Fig. 2.19. Improper Clearance.

(*i*) When the hole has to be held to size, *i.e.*, the hole in the sheet metal is to be accurate (Punching operation), and slug is to be discarded, the punch is made to the size of hole and the die opening size is obtained by adding clearance to the punch size, Fig. 2.20 (*a*).

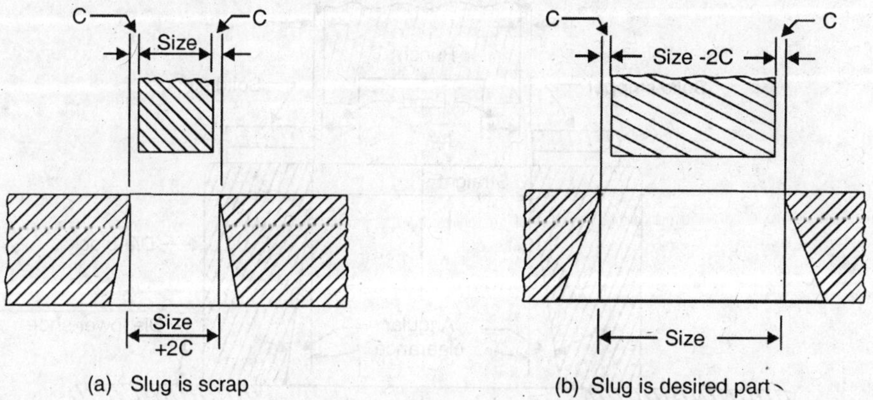

Fig. 2.20. Application of Clearance.

(*ii*) In blanking operation, where the slug or blank is the desired part and has to be held to size, the die opening size equals the blank size and the punch size is obtained by subtracting the clearance from the die-opening size, Fig. 2.20 (*b*).

In Fig. 2.20, C is the amount of clearance per side of the die opening. The clearance is a function of the kind, thickness and temper of the work material, harder materials requiring larger clearance than soft materials, the exception being Aluminium. The usual clearance per side of the die, for various metals, are given below in terms of the stock thickness, t :

For brass and soft steel, $c = 5\%$ of t

For medium steel, $c = 6\%$ of t

For hard steel, $c = 7\%$ of t

For aluminium, $c = 10\%$ of t

The total clearance between punch and die size will be twice these figures. These clearances are for blanking and piercing operations.

The clearance may also be determined with the help of the following relation :

$$c = 0.0032\ t \cdot \sqrt{\tau_s}\text{, mm}$$

where τ_s is the shear strength of the material in N/mm^2.

The reason behind the application of clearance in the manner as given above, is explained below :

The diameter of the blank or punched hole is determined by the burnished area. On the blank, the burnished area is produced by the walls of the die. Therefore, the blank size will be equal to the size of die-opening (neglecting a slight expansion of the blank due to elastic recovery after the cutting operation is completed). Similarly, in punching operation, the burnished area in the hole is produced by the punch, therefore, the size of the hole will be the same as the punch. Therefore, the application of clearance on punch or die will depend on whether the punched hole or the cut blank is the desired product. Hence, in punching operation (where hole in the strip is the desired product), the punch is made to the correct hole size and the die opening is made oversize an amount equal to die clearance. Similarly, if the blank is the desired product, the die opening size is made to the correct blank size and the punch is made smaller an amount equal to die clearance. In other words, punch controls the hole size and die opening controls the blank size.

A section through blanking die is given in Fig. 2.21, showing clearance, land, straight and angular clearance.

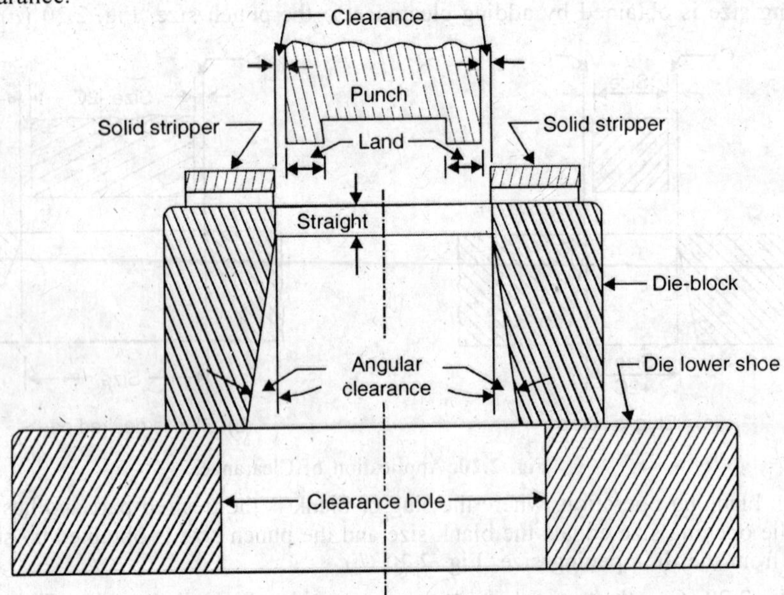

Fig. 2.21. Section through Blanking Die.

Land. It is the flat (usually horizontal surface contiguous to the cutting of a die which is ground and reground to keep the cutting edges of the punch sharp.

Straight. It is the surface of a cutting die between its cutting edge and the beginning of the angular clearance. This straight portion gives strength to the cutting surface of the die and also provides for sharpening of the die. This straight portion is usually kept at about 3 mm for all materials less than 2 mm thick. For thicker materials, it is taken to be equal to the metal thickness.

Angular clearance. Angular clearance or relief is provided to enable the slug to clear the die. It is provided below the straight portion of the die surface. It is usually $\frac{1^\circ}{4}$ to $1\frac{1^\circ}{2}$ per side but occasionally as high as 2°, depending mainly on stock thickness and frequency of sharpening. For round die opening (d ≤ 3 to 5 mm), instead of die angular clearance, die-clearance is maintained by slightly oversize drilling.

2.8.1 Punch and die clearance after considering the elastic recovery of the material. It has already been explained that after the cutting operation has been completed, elastic recovery of the strip material takes place. In blanking operation, after the release of blanking pressure, the blank expands slightly. The blanked part is thus actually larger than the die opening that has produced it. Similarly, in punching operation, after the strip is stripped off the punch, the material recovers and the hole contracts. Thus, the hole is actually smaller than the size of the punch which produced it. This difference in size due to elastic recovery will depend upon : blank size, stock thickness and stock material. It may be taken as between 0.0125 mm and 0.075 mm. If the stock thickness is upto 0.25 mm, this difference may be taken as zero. For stock thickness between 0.25 mm and 0.75 mm, it may be taken as equal to 0.025 mm and for stock thickness more than 0.75 mm, it may be taken as 0.050 mm. Thus to produce correct hole and blank sizes, the punch size should be increased and the die opening size should be decreased by an amount as explained above. Or, the swelling/shrinkage expected due to elastic deformation may be taken as = (0.003 to 0.01) × t; for softer to harder stocks.

2.9. CUTTING FORCES

In cutting operation, as the punch in its downward movement enters the material, it need not penetrate the thickness of the stock in order to affect complete rupture of the part. The distance which the punch enters into the work material to cause rupture to take place is called 'penetration' and is usually given as the percentage of the stock thickness.

The per cent penetration depends on the material being cut and also on the stock thickness. When a hard and strong material is being cut, very little penetration of the punch is necessary to cause fracture. With softer materials, the penetration will be greater. For example, for soft aluminium, it is 60% of 't'; for 0.15% carbon steel annealed, it is 38% of 't'; and only 24% of 't' for 0.5 per cent carbon steel annealed. The percentage penetration also depends upon the stock thickness, being smaller for thicker sheets and greater for thinner sheets, as shown in Table 2.1, i.e., it is inversely proportional to stock thickness.

Table 2.1 Penetration

Stock Thickness t.mm	25	20	15	12.5	10	8	6	5	3	2.5	1.6	Below 1.6
Penetration % of t	25	31	34	37	44	47	50	56	62	67	70	80

The maximum force F_{max} in newtons needed to cut a material is equal to the area to be sheared times the shearing strength, τ_s in N/mm² for the material. For a circular blank of diameter D mm and of thickness t, mm, the cutting force will be given as,

$$F_{max} = \pi D t\ \tau_s = P.t.\tau_s \ldots \qquad \ldots(2.1)$$

where 'P' is the perimeter of the section to be blanked.

For rectangular blanks with length L and width b, it is

$$F_{max} = 2\,(L + b)\, t.\tau_s \ldots \qquad \ldots(2.2)$$

The shear strengths of the various metals are given in Table 2.2.

Table 2.2. Shear Strengths of Various Metals

Metal	τ_s *N/mm²*
Carbon steels :	
0.10%C	245 to 311
0.20%C	308 to 385
0.30%C	364 to 469
High strength low-alloy steels	315 to 446
Silicon steels	420 to 490

Stainless steels	399 to 903
Aluminium alloys	49 to 322
Copper and bronze	154 to 490
Lead alloys	12.8 to 41
Magnesium alloys	119 to 203
Nickel alloys	245 to 812
Tin alloys	20.5 to 77.7
Titanium alloys	420 to 490
Zinc alloys	98.0 to 266

To allow for energy lost in machine friction and in pushing slugs through the die, etc.,

Press capacity will be taken as $= F_{max} \times C$,

where the factor C is

$= 1.1$ to 1.5 for normal to narrow profiles

$= 1.25$ to 1.75 for $d/t < 2$

Energy in press work, $E = F_{max} \times C \times$ Punch travel

$= F_{max} \times C \times K \times t$,

where K = Percentage penetration required to cause rupture.

Power in Press work $= \dfrac{E \times n_a}{60 \times \eta}$,

where n_a = actual number of strokes per minute

η = efficiency of the press (0.60 to 0.85), depending upon its condition

Shear force with non-parallel edges = $(0.67 \text{ to } 0.75) \times F_{max}$.

Horizontal force on Punch and die,

$F_{HD} = F_{HP} = (0.15 \text{ to } 0.25) \times F_{max}$, for normal clearance

$= (0.4 \text{ to } 0.6) \times F_{max}$, for narrow clearance, higher values for complicated contours.

$\cong \dfrac{1}{3} \times F$

Push through force for a blank = 10% of F_{max}.

2.10 METHODS OF REDUCING CUTTING FORCES

For calculating the cutting forces in the last article, it has been assumed that the bottom of the punch and the top of the die block lie in parallel planes and that the blank is severed from the sheet metal by shearing it simultaneously along the whole perimeter. This process is characterized by very high punch forces exerted over a very short time, resulting in shock or impulse conditions. It is usual, however, to reduce cutting forces and to smooth out the shock impact of heavy loads. This is achieved by arranging for a gradual cut instead of sudden cut of the stock. For this, two methods are generally used :

1. **Shear.** The working faces of the punch or die are ground off so that these don't remain parallel to the horizontal plane but are inclined to it. This angle of inclination is called 'shear'. This has the effect of reducing the area in shear at any one time and the maximum force is much less. It may be reduced by as much as 50%. This is made clear in Fig. 2.22, which shows the relation of cutting forces to amount of shear.

In Fig. 2.22 (*a*), the shear is zero, *i.e.*, the cutting edges are parallel. The material is cut at once on the entire perimeter resulting in maximum load. The force diagram shows a steep rise at maximum load and then sudden load release sometimes severe on both press and dies, as the cut is completed. In Fig. 2.22 (*b*), the face of the punch is ground off so that shear = $t/3$. The cutting action will start at the leading edge of the punch and then it will gradually spread to the rest of the punch. With this, only a part of the punch would be cutting at any one instant. While the maximum force will decrease, the energy needed to complete the cut is unchanged. So, the punch travel in this case will be more than in case (*a*). In case 2.22 (*c*), shear = $t/1$. When the leading edge has travelled through the stock a distance 't', the trailing edge will start making contact with the material. Maximum force would be at this position of the punch and since the cut is complete, the maximum load would be about half of that when shear is zero. The punch travel will be still greater.

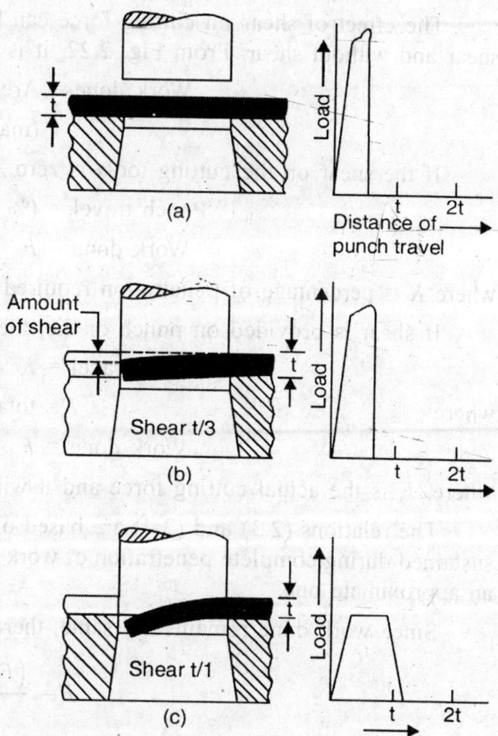

Fig. 2.22. Effect of Shear on Cutting Force.

The provision of shear distorts the material being cut. When shear is on the face of the punch, the blank cannot be flat and when shear is on the die, the piercing cannot be flat. So, for blanking operation, shear is provided on the die face and for punching or piercing operation shear is provided on the punch face. Fig. 2.23 shows the various methods of applying shear on the punch and die face. Wherever possible, double shear should be used so that the two shear faces neutralise the side thrusts which each sets up.

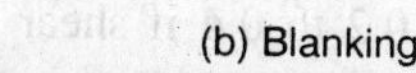

Fig. 2.23. Application of Shear.

It is clear from the above analysis that it is advisable to provide shear on punches and dies. The amount of shear to be applied is a matter of comprise. If the shear is quite big, say $2t$ or $3t$, then the cutting edges of the tools will become too acute and liable to break away easily. However, the shear must be at least equal to the percentage penetration.

The effect of shear on cutting force can be determined by comparing the work done with shear and without shear. From Fig. 2.22, it is clear that,

Work done = Area under curve

= (maximum punch force) × (punch travel)

If the shear on the cutting tools is zero, then

Punch travel = (% penetration) × (material thickness)

$$\text{Work done} = F_{max} \times K \times t \qquad ...(2.3)$$

where K = percentage of penetration required to cause rupture

If shear is provided on punch or die, then

Punch travel = $K \times t$ + Amount of shear in cm = $K \times t + I$

where I = total inclination or shear of punch or die, in cm

$$\therefore \quad \text{Work done} = F \times (K \times t + I) \qquad ...(2.4)$$

where, F, is the actual cutting force and it will be less than, F_{max}.

The relations (2.3) and (2.4) are based on the assumption that punch force F_{max} and F, are sustained during complete penetration of work material, which is not true. So, the analysis is only an approximate one.

Since work done remains the same, therefore, comparing equations (2.3 and 2.4), we get,

$$I = \frac{(F_{max} - F) \times K \times t}{F} \qquad ...(2.5)$$

and

$$F = \frac{F_{max} \times K \times t}{K \times t + I} \qquad ...(2.6)$$

The above is true for single or double shear.

It is clear that the provision of shear increases the punch travel to complete severance of the material. This results in a smoother cutting action and cleaner edges with a higher degree of squareness. However, the initial die cost increases and there is also greater chance for dimensional error and variation, since the dimensions of the cut tend to increase in proportion to the angle of shear. The shear angle chosen should provide a change in punch length of about $\frac{1}{2}$ to 2 times stock thickness. In piercing, the direction of shear angles must be such that the cut proceeds from the outer extremities of the contour towards the centre. This avoids stretching the material when it is cut free.

The application of shear reduces the cutting force according to the penetration before the breakthrough. The ratio of penetration to the shear gives the reduction factor. Thus, a simple relation can be

$$F = \frac{Kt}{I} \times F_{max} \text{ with } I > Kt \qquad ...(2.7)$$

or

$$F = P.t.\tau_s \times k$$

The factor k is

= 0.2 to 0.6 if shear $I = t$

= 0.2 to 0.4 if shear $I = 2t$

The work done or the shearing energy may also be approximately determined with the help of the following empirical relation :

E = Area under the force displacement curve, Fig. 2.22

$= C_1.F.t$

where the constant C_1 is,

$$C_1 = 0.5 \text{ for soft materials}$$
$$= 0.35 \text{ for hard materials}$$

As already noted, when the cutting edges (of punch and die) are parallel, the length to be cut is the entire length of the contour cut, that is, the perimeter of the blank. This can lead to very high cutting forces. These can be reduced by placing the two cutting edges at an angle to each other (called shear or rake) as discussed above. Here, only the instantaneous sheared length needs to be considered, which will be given as in the case of sheet cutting machine,

$$= \frac{t}{\sin \alpha}$$

where α = angle of shear.

2. *Staggering of punches.* As an effect similar to shear can be obtained by staggering two or more punches that all operate in one stroke of press. For staggering, the punches are arranged so that one does not enter the material until the one before it has penetrated through. In this manner, the cutting load may be reduced approximately 50 per cent, Fig. 2.24.

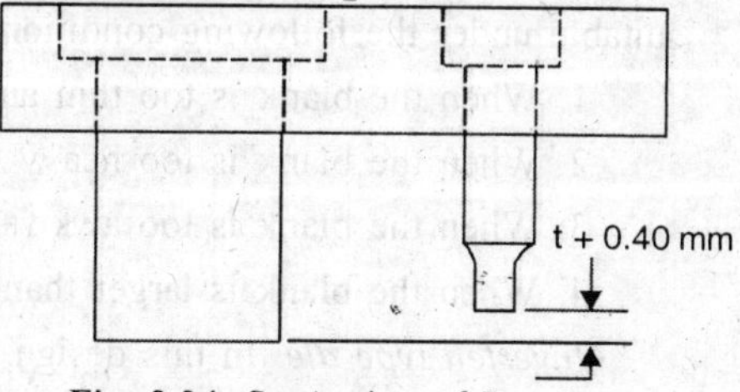

Fig. 2.24. Staggering of Punches.

2.11. MINIMUM DIAMETER OF PIERCING

To punch a hole without the failure of the punch, the compressive strength of the punch should be at least equal to the force necessary to fracture the material. The hole which will be punched under this condition will be the smallest. If τ_s is the shear strength of the work material σ_c is the compressive strength of the punch material, d is the smallest hole diameter to be pierced and t is the thickness, then,

$$\text{Piercing pressure} = \tau_s . \pi d.t$$

$$\text{Strength of punch} = \sigma_c . \frac{\pi}{4} d^2$$

$$\therefore \qquad \tau_s \pi d.t = \frac{\pi}{4} d^2 . \sigma_c$$

$$\therefore \qquad d = \frac{4\tau_s}{\sigma_c} \cdot t$$

If it is assumed that, $\sigma_c = 2\tau_s$ then,

$$d = 2t$$

i.e., the minimum diameter of hole that can be punched is twice the stock thickness. In actual practice, smaller holes than this are pierced successfully, showing to what extent the punch may be loaded without failure,

Minimum size of punched holes depending upon their shape (round, square, rectangular, oval) are given below :

$$= 0.7 \text{ to } 1.2\ t \text{ for soft steel}$$
$$= 0.9 \text{ to } 1.5\ t \text{ for steel}$$
$$= 1.75 \text{ to } 2\ t \text{ for Ti alloys}$$
$$= 0.6 \text{ to } 0.9\ t \text{ for brass and copper}$$
$$= 0.5 \text{ to } 0.8\ t \text{ for Zinc}$$

= 0.4 to 0.7 t for bakelite and textolite

= 0.3 to 0.6 t for cardboard and paper

Punching force for holes which are smaller than the stock thickness, may be estimated as follows :

$$F = \frac{\pi\, d t\, \sigma_t}{\sqrt[3]{d/t}}, \text{N}$$

where d and t are in mm and σ_t is the tensile strength in N/mm² or MPa.

2.12. BLANKING DIE DESIGN

2.12.1 Types of blanking dies. There are two general types of blanking dies : the drop through and the inverted.

Drop-through die. In this die, the die block assembly is mounted on the bolster plate or the press bed and the punch assembly on the press slide, Fig. 2.11. The blank drops of its own weight through the die opening and the clearance hole provided in the bolster plate and press bed. This design is economical to build and maintain and is fast in working. However, this design is not suitable under the following condition :

1. When the blank is too thin and fragile to be dropped very far.
2. When the blank is too heavy to be dropped for any appreciable distance.
3. When the blank is too awkward to be removed from below the press.
4. When the blank is larger than the press bed opening.

Inverted type die. In this design, the punch becomes the lower stationary part and the die is mounted on the ram. This type is somewhat more complicated, more costly and slower in operation. The scrap disposal is much easier but removal of blank from the die opening is troublesome and a device is needed to knock the blank out of the die opening. This type is widely used where the blank is large or heavy.

Fig. 2.25 shows an inverted type Blanking die. During the downward stroke of the ram, the blank is cut from the strip. The blank and the shedder (spring loaded knockout) are forced into the die opening, which compress a spring in the die opening. At the same time, a compression spring attached to the stripper is compressed. On the upstroke of the ram, the stripper strips the scrap off the punch and the shedder forces the blank out of the opening. The blank falls down or is blown out the rear of the press. Usually, the press is inclined to aid the discharge operation.

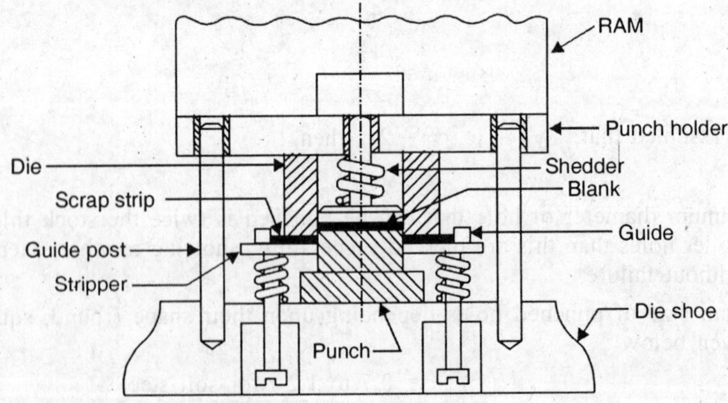

Fig. 2.25. Inverted Blanking Die.

2.12.2. Strip layout. In the design of a blanking die set, the first step is to prepare blanking layout, that is, to layout the position of the workpieces in the strip and their orientation with respect to one another. This is called 'Strip layout'. The factors which will influance the stock layout are :

1. Economy of material.
2. Direction of material grain or Fibre.
3. Strip or Coiled stock.
4. Direction of burr.
5. Press used
6. Production required.
7. Die cost.

1. **Economy of material.** In Fig. 2.26, the different ways of arranging to blank the given workpiece are shown. The arrangement in Fig. 2.26 (*a*) can be worked at single row, single pass with a single punch. For arrangement of Fig. 2.26 (*b*), the strip would either have to be fed twice, once for each row, or double blanking will have have to be employed. The per cent material utilization may be increased somewhat by the arrangement of Fig. 2.26 (*b*), that is, by having two (or even three) rows of blanks. Fig. 2.26 (*c*) shows a single row, double pass strip. This is called "stock nesting". Here, the strip will have to be passed through the dies once, turned over, and passed through dies a second time. Nesting considerably reduces the scrap. However, the strip layout with maximum material saving may not be the best strip layout, as the die construction may become more complex which would offset the savings due to material economy unless a large number of parts are to be produced.

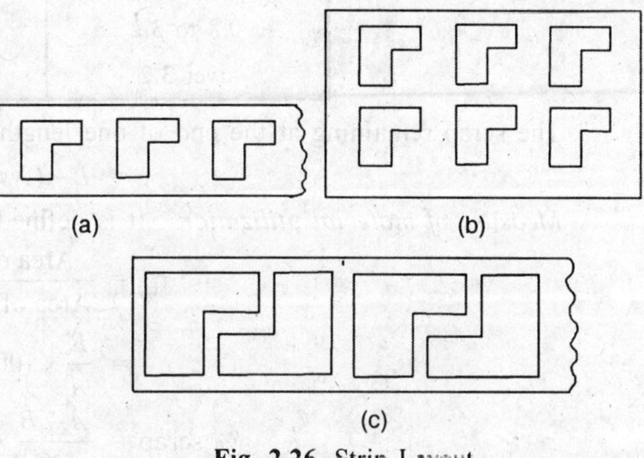

Fig. 2.26. Strip Layout.

Another important consideration in strip layout is the distance between the nearest points of blanks and between blanks and the edges of the strip. To prevent the scrap from twisting and wedging between the punch and the die, this distance must increase with material thickness. A general rule of thumb is to keep this distance, called as web, at least 1.5 times the material thickness. However, other factors such as strip thickness, hardness of material, type of operation, shape of blank etc. may allow the web to be thinner. The various terms connected with strip layout are shown in Fig. 2.27.

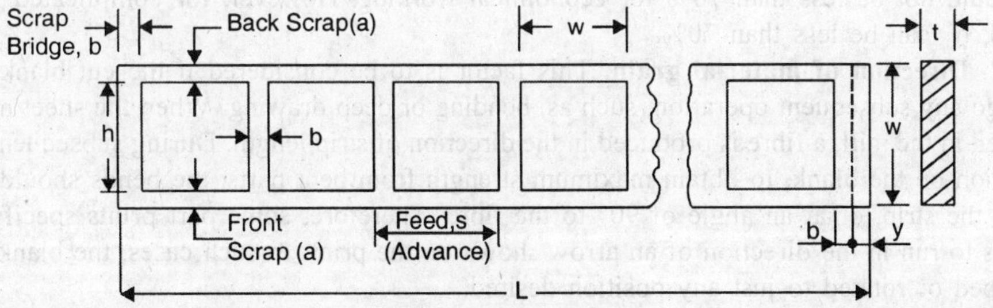

Fig. 2.27. Strip Layout.

The distance between the blank and edge of strip, known as back scrap (front scrap) may be determined by the equation,

$$a = t + 0.015\, h \qquad ...(2.8)$$

The distance between successive blanks and also the scrap bridge, *b*, is given in Table 2.3.

In general, softer metals require larger spacing and thinner metals require larger spacing.

The feed or advance or the length of one piece of stock needed to produce one blank is,

$$s = w + b \qquad ...(2.9)$$

The number of blanks which can be produced from one length of stock can be found out as,

$$N = \frac{L - b}{s} \qquad ...(2.10)$$

Table 2.3

Material Thickness	*b*
mm	mm
0.8	0.8
0.8 to 3.2	*t*
over 3.2	3.2

The scrap remaining at the end of one length of strip may be calculated from,

$$y = L - (Ns + b) \qquad ...(2.11)$$

Measure of material utilization : It is defined as,

$$\eta_m = \frac{\text{Area of blank to be cut}}{\text{Area of material available}}$$

$$= \frac{B}{A} \times 100$$

$$\therefore \quad \%\ \text{scrap} = \frac{A - B}{A} \times 100$$

Now area of material available per blank = Feed or advance × Stock width

Feed is also called "Progression".

$$\therefore \text{ For Fig. 2.27, } \%\ \text{utilization, } \eta_m = \frac{w \times h}{(w + b)(h + 2a)}$$

For a round blank of diameter d,

$$\eta_m = \frac{\frac{\pi}{4} \cdot d^2}{(d + b)(d + 2a)} \qquad ...(2.12)$$

η_m should not be less than 70% for economical working. However, for complicated shapes of blanks, η_m can be less than 70%.

2. **Direction of material grain.** This factor is to be considered if the cut blanks have to undergo any subsequent operation, such as, bending or deep drawing. When the sheet metal strip is rolled in the mill, a fibre is produced in the direction of strip length. During subsequent bending operation on the blank, to obtain maximum strength from bent parts, the bends should be made across the strip, or at an angle of 90° to the fibre. Therefore, some part prints specify that the fibre is to run in the direction of an arrow shown on the print. In such cases, the blanks can not be tipped or rotated to just any position desired.

3. **Strip or Coiled stock.** Another important consideration in the strip layout is whether the stock used will be in the form of a strip or coil. Whereas, the stock strip may be passed through the die more than once, the coiled stock is usually passed through the die only once.

When coiled stock is used, decoiling and recoiling of the stock is expensive (a decoiling reel, a stock straightner and feed rolls are necessary). Thus coiled stock is used when :

(*a*) Production is high.

(*b*) Thinner metal sheets are employed.

(*c*) The stock needs to be passed through the die only once.

Strip stock is used when :

(*i*) Production is low.

(*ii*) Thicker sheet metals are used.

(*iii*) The stock needs to be passed through the die more than once.

Here, the stock feed is manual.

4. **Direction of burr.** When sheet metal is cut in a die, a burr is produced on the die side of the scrap strip and on the punch side of the blank. If the burr has to be on the hidden side, then the expensive operation of removing the burr need not be done. For this, a note is often placed on the part drawing which reads "burr down". To control the position of the burr may limit the stock-layout arrangement. Scrap may not be reduced to a minimum.

5. **Press used.** During production planning, a press has been assigned to the operation and the die. Therefore, the stock layout has to be such that it allows the die to be designed within the press capacity. Shear may have to be provided on punch or die, to limit the maximum cutting force within the press capacity. Another factor is the bed area of the press. The relation of the press bed area to the blank area is a definite factor controlling the stock layout. The third factor is to have the cutting forces of the die evenly balanced around the centre line of the press ram.

6. **Production required.** The following guide lines may be followed when production is the main consideration :

(*a*) *Low production-thin material* :

(*i*) Strip stock and a single-pass layout.

(*ii*) Cutting of one or more blanks at a time.

(*b*) *Low production-thick material* :

(*i*) Strip stock and a single or double-pass layout.

(*ii*) Cutting one blank at a time.

(*c*) *High production-thin material* :

(*i*) Coiled stock and a single-pass layout.

(*ii*) Cutting of one or more blanks at a time.

(*d*) *High production-thick material* :

(*i*) Strip stock, and a single- or double-pass layout.

(*ii*) Cutting of more than one blank at a time.

7. **Die-cost.** Die-cost will be higher for :

(*i*) Higher productions.

(*ii*) Cutting more than one at a time, particularly when cutting extremely complicated blank shapes, or when cutting extremely accurate blank sizes.

However, for simple round or square-edged blanks, multiple cutting at one time is often practical. Also, double-pass dies are less expensive than cutting two at a time.

So, the designer has to decide while making the stock layout, as to which is preferred : more operator time per blank or more machine time per blank.

2.12.3. Die-block. The die block is the female half of the two mated tools which carry the cutting edges. It is subjected to extreme pressures and wear conditions. Hence the die block is made of a superior quality of tool steel. A simple layout of the

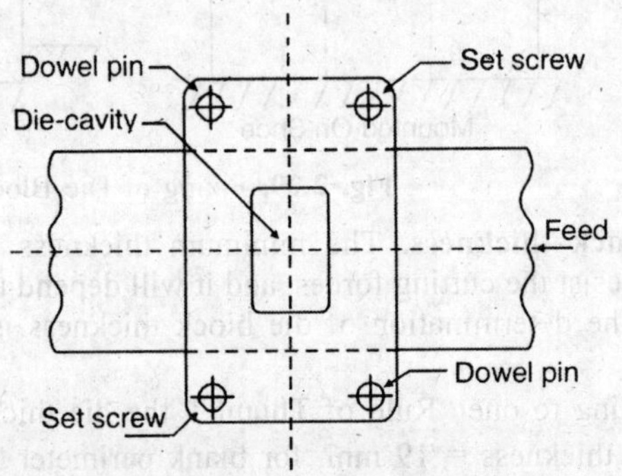

Fig. 2.28. Die Block Layout.

die block is shown in Fig. 2.28. The die block may be of solid or sectional construction, depending on the size and contour of the die opening. If the die opening is small and its contour is simple, a solid die block is the choice. Sectional dies are made up of accurately ground matching components which may be assembled together easily. Sectional dies have the following advantages :

1. Dies with long and complicated contours can be broken up into sections of simple geometrical forms which can easily and economically machined.

2. Building the die block in sections eliminate heat, treating, distortion and cracking as the separate pieces handled are more uniform in cross-section.

3. Grinding of the sections is more easily done.

4. Maintenance of the die is simple and less costly, because if one section happens to crack or chip, only that particular section will have to be replaced.

Sectional dies have the following disadvantages :

1. This design is not suitable when the stock thickness is quite large and there is great side pressure in blanking.

2. If the bed and press slide are not absolutely parallel, proper alignment between the punch and the sectional die will be extremely difficult to maintain.

For the high production of accurately shaved die products of complicated shapes, sectional dies have definitely an edge over solid dies. However, the availability of modern tooling equipment by which punch and die can be made in one cut, for example by contour sawing method, has extended the range of solid die construction. Generally, high carbon, high chromium steels are used for die sections which are hardened and ground. The die block is either mounted on or held in a die shoe, as shown in Fig. 2.29.

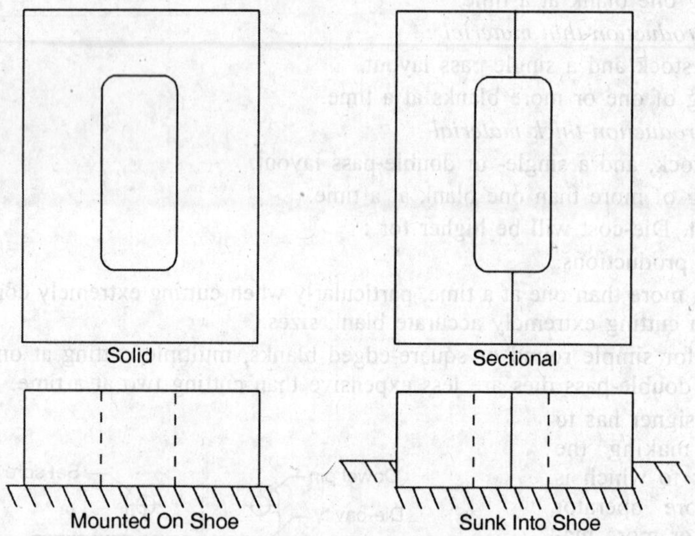

Fig. 2.29. Fixing of Die Block to Die Shoe.

Die block thickness. The minimum thickness of die block depends upon the strength required to resist the cutting forces, and it will depend upon the type and thickness of the material being cut. The determination of die block thickness is usually based on experience and thumb rules.

According to one "Rule of Thumb", the die thickness may be obtained as follows :

(*i*) Die thickness = 19 mm, for blank perimeter ≥ 75 mm

(*ii*) Die thickness = 25 mm, for blank perimeter = 75 mm to 250 mm.

(*iii*) Die thickness = 31 mm, for blank perimeter > 250 mm.

This rule of thumb is for die blocks made of tool steel.

The die thickness may also be obtained from table 2.4, in which it is given as a function of stock thickness and quantity of production. Allowances for polishing must be added to the values.

When the parts are made of lamination stock or other extremely hard or tough materials, the thickness given in table 2.4 can be increased by 3 mm. However, for bakelite, brass or similar soft materials, it can be reduced by 3 mm. Care must be taken not to make the die block unnecessary thick as it adds to the cost of manufacture and results in an excessive shut height.

Table 2.4. Die-Block Thickness.

Stock Thickness mm	*Production Quantity (million)*						
	5	25	50	100	500	*Unlimitted*	
0.375	9.375	12.5	15.625	19	25	28	*Die Block Thickness, mm*
0.775	12.5	15.625	19	25	28	28	
2.325	15.625	19	25	28	31	34	
3.125	19	25	28	31	31	34	
4.675	25	28	28	31	34	37.5	
6.25	31	31	31	34	37.5	37.5	

or, die-block thickness can be taken as

$$T = \sqrt[3]{F} \text{ cm; } F \text{ is in tonnes}$$

$$T_{min.} = 7.5 \text{ mm and } 10.5 \text{ mm}$$

for die-surface area above 3200 mm².

Die opening. A section of the die block is shown in Fig. 2.21. The side walls of the die block opening should be provided with sufficient relief, Fig. 2.18a or taper so that the blank drops clear through. This taper can either start from the top surface itself or after a straight land from the surface of the die, Fig. 2.2. Where the filing and grinding of the die is done by machine, the fully tapered cavity is quicker to produce and is therefore cheaper. But if the die has to be finished on bench, the straight land is easier to file and sandstone.

Soft metals such as copper, brass and aluminium tend to swell more rapidly after being cut. This may be due to a slight spring back or return of the material along the lines in which it which it has been stressed in compression. So, for such metals, the die cavity should be fully tapered. But for steel, the taper or relief should start after the straight land from the die surface. The chief advantage of the straight land is that the original dimensions of the die are retained after repeated regrinding. With fully tapered die cavity design, the die opening size increases after each regrinding. This increase, however, is very negligible to cause any appreciable effect on blank dimensions.

As already mentioned, the straight land should be at least equal to stock thickness and not less than 3 mm. The relief angle or taper may be taken as : $\frac{1^\circ}{4}$ to 1° for small dies, 1° to 2° for average dies and 2° to 3° on large dies. For softer materials, these values can be slightly increased. Also, the closer the blank tolerances, the less the relief.

The nominal size and outline of the die cavity is identical with the blank dimensions. At least a distance equal to stock thickness should be provided between blanks and the edge of the strip. The die wall thickness from the edge of the die opening to the die block border, that is, the distance *A* [Fig. 2.29 (*a*)], should be at least equal to as given below :

A = 1.5 to 2 times the die thickness (T) for small dies

= 2 to 3 times the die thickness for larger dies.

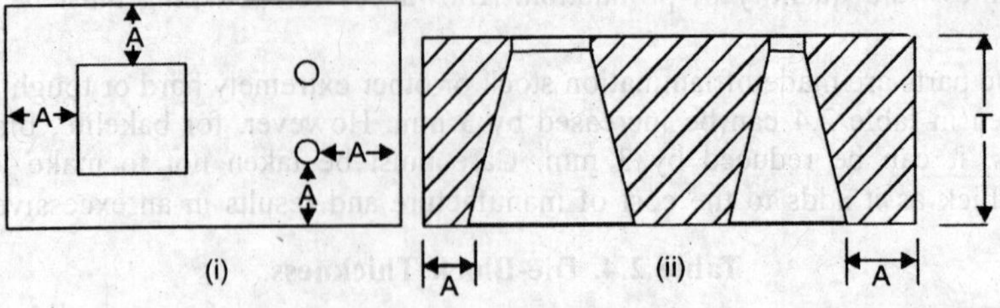

Fig. 2.29 (*a*)

After that, the die thickness T must be checked. According to an empirical rule, the cross-sectional area $A \times T$, [Fig. 2.29*a* (*ii*)] must bear a certain minimum relationship to the maximum cutting force ($P.t.\tau_s$), for a die put on a flat base. This is given in Table 2.4 (*a*). If the die thickness, as calculated above, does not give sufficient area for the critical distance A, the die thickness must be increased accordingly.

According to "Rule of Thumb" there should be a minimum of 32 mm margin around the opening of the die block (distance A). In general, on very thin materials, 12.7 mm die block thickness will be sufficient but, except for temporary tools, finished thickness (including a grinding allowance of 2.5 to 5 mm) is seldom less than 22 mm, which allows for blind screw holes. Sharp corners in the contour may lead to cracking in hat treatment, and so require greater wall thickness at such points.

Table 2.4 (*a*)

Maximum Cutting force, kN	*Area between die opening border (cm²)*
200	3.25
500	6.50
750	9.75
1000	13.00

Fastening of die block. The die block is secured either to the die shoe or bolster plate. The size of screws and bolts employed is usually not calculated. They are usually chosen by personal judgement and practical experience. According to one empirical rule.

Screw diameter = 0.5 t, for $T \leq 19$ mm.

= 0.4 t, for $T > 19$ mm.

Along with screws, dowel pins are also used for alignment purposes. They are usually located near diagonally opposite corners of the die block, for maximum locating effect. The diameter of the dowel pins is taken to be equal to the outside diameter of the fastening screws. Two, and only two, dowel pins should be provided in the die block for permanent positioning.

Dowel Pins. Any misalignment of die details may cause severe damage. Therefore, any details that must be located accurately are held in such a position by dowels. Two dowels are necessary for one detail to achieve complete location. Most dowels used in die construction we hardened and ground steel pins. They are press fitted in the detail and the mounting plate or shoe. Keeping the dowel diameters same as that of the screws facilitates the drilling and reaming of dowel holes.

The position of the dowel pins is so selected as to make through holes for them, Fig. 2.30, otherwise the removal of the pins during disassembly will be too difficult. If through holes are not available taper pins are driven, the holes for them being machined by a taper reamer.

Whenever possible, threading hardened components should be avoided. For example, the threads should be cut in the die shoe (which is made of C-steel and is not hardened) and die block should be clearance drilled and counter bored to accept the cap screws. The effective thread depth of the screws should be 1.5 to 2 times the screw diameter.

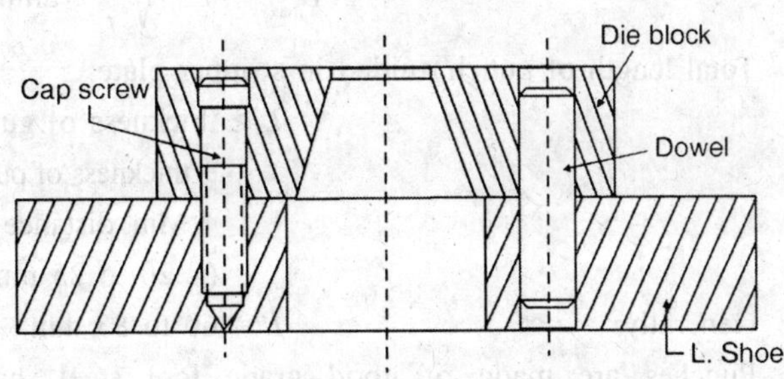

Fig. 2.30. Dowel Pin

1. On die blocks up to 18 cm square, use two 10 mm screws and two 10 mm dowel pins.

2. On sections upto 25 cm square, use three screws and two dowel pins.

3. For blanking heavy stock, use screw and dowels of 12.7 mm diameters. Counterbore the cap screws about 3 mm deeper than usual, to compensate for die sharpening.

Screws and dowels are preferably located about 1.5 to 2 times their diameter from the outer edges or the blanking contour. Min. permissible centre distance between the tapped hole and dowel pin $\geq 2 \times$ screw diameter.

The diameter of fastening screws can also be found by calculations. The basis for the computation is the stripping force, which depends upon many factors, but may be taken as 10 per cent of cutting force. With this force, the screws which fasten the stripper to the die plate or secure the punch holder to the punch plate can be easily calculated. Practically, the same values are then taken for screws which secure the die block to the press table.

$\therefore$ Stripping force, $\quad Fs = 0.1 \times P \times t \times \tau_s = \pi/4\ d^2c \times n \times \sigma_t$

where dc is the core diameter of screw, σ_t is the safe tensile stress of screw material and 'n' is the number of screws usually an even number; 2 for small dies, 4 for medium dies and 6, 8 or more for large dies.

Note. A minimum of 16 mm should be provided between punch and die blocks and guide pins and bushings to allow clearance for the grinding wheel when resharpening. When this is not possible, removable guide posts may be used for easy access to dies for maintenance purposes. The die holder or the lower shoe should be at least 6.5 mm larger all around than the die block.

2.12.4. Punch. The punch must be a perfect mate to the die block opening the sizeof the working surface of punch is obtained by subtracting the total clearance from the desired size of the blank. As already mentioned, shear is provided on the die surface for blanking operation. The punch is usually provided with a wide flange or shoulder to facilitate mounting and prevent its deflection under load. The minimum length of punch should be such that it extends far enough into the die block opening to ensure complete shearing of the blank. The punch length must also provide for the anticipated number of regrinds. The maximum length of the punch can be calculated from the formula, from the consideration of an axially loaded street,

$$L = \frac{\pi d}{8}\left(\frac{E}{\tau s}\cdot\frac{d}{t}\right)^{\frac{1}{2}} \qquad \text{...(2.13)}$$

where E is the modulus of elasticity, d is punch diameter.

or, the maximum permissible free length of circular punches to be safe against buckling can also be determined as,

$$L_{\text{max.}} = 7.5\sqrt{\frac{d^3}{t}}\text{; units are mm.}$$

Total length of punch guided in stripper plate,

$L \geq$ thickness of guides + thickness of stripper plate + thickness of punch holder plate + sharpening allowance + Min. distance between the stripper and punch holder ($\cong$ 20 to 25 mm)

Generally, L = 60 to 85 mm

Punches are made of good grade tool steel, hardened and ground. The hardness recommended is Rockwell C 60 to 62.

Punches with unguided length of more than 100 mm are avoided. A sharpening allowance of 6 to 12.5 mm is provided in punch length.

The strength of the punch is checked for the weakest cross-sectional area, by the formula,

$$\sigma_{\text{comp.}} = \frac{F_{\text{max.}}}{A} \leq |\, \sigma \,| \text{ permissible}$$

where A = cross-sectional area of punch, at its weakest section.

2.12.5. Back up plate. For small punches, back up plates or pressure plates are often provided between the punch plate and punch holder (Fig. 2.31). The 'punch plate' or 'punch retainer' fits closely over the body of the 'punch'and holds it in proper relative position. It is attached to the punch holder. The back up plate is provided to take the cutting force of the punch head, provide a base and insurance against punch deflection when punch does not have a flange or shoulder and prevent the hardened punch from being pushed into the softer punch holder, thus becoming loose.

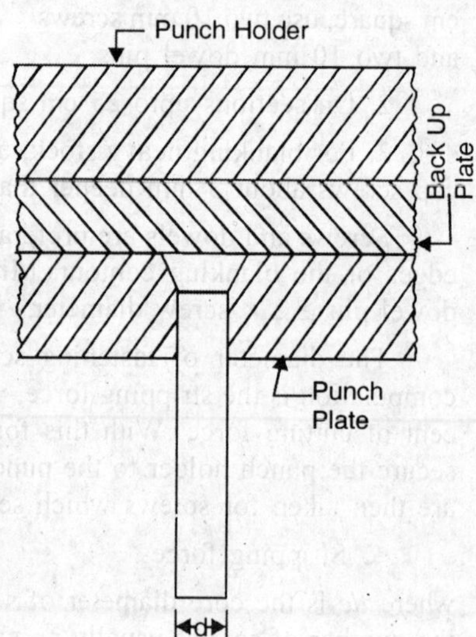

Fig. 2.31. Back up Plate.

The main criteria whether a back up plate should be provided or not is the unit of compressive stress on the punch, given as,

$$p = F/A$$

where A is the cross-sectional area of the punch.

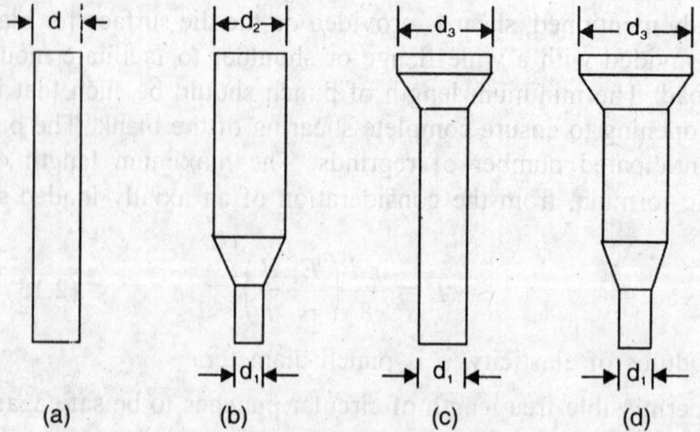

Fig. 2.32. Punch Construction.

Depending upon the punch construction, Fig. 2.32, the above equation will be.

For straight punches [Fig. 2.32 (*a*)], $p = \dfrac{F}{\frac{\pi}{4}d^2} = \dfrac{4F}{\pi d^2} = \dfrac{4t\tau_s}{d}$

For punches with a shoulder [Fig. 2.32 (*b*)],

$$P = 4t\tau_s \cdot d_1 / d_2^2$$

For punches with turned heads [Fig. 2.32 (*c*) and (*d*)],

$$P = 4t\tau_s \cdot d_1 / d_3^2$$

A back up plate is provided if p exceeds 245 N/mm². As a rule, a back up plate is to be used whenever the punch diameter is less than four times the stock thickness. A back plate is usually made from plain carbon steel, hardened and ground parallel. The thickness of the back up plate depends upon the stock thickness. For stock thickness upto 2 mm, the thickness of back up plate should be about 3 mm and for thicker stock, it should be about 6 mm.

2.12.6. Methods of holding punches. The mounting or securing of a blanking punch in the punch holder does not present any problems. Being relatively bigger, they are made with flanges that are doweled into position and directly fastened to the punch holder by screws without the use of punch plate and sometimes without even a back up plate (Fig. 2.33). When used, the thickness of punch plate should be 1.5 times the punch diameter.

For press fitted punches, a fit H_8/p_7 or H_7/p_6 is to be maintained.

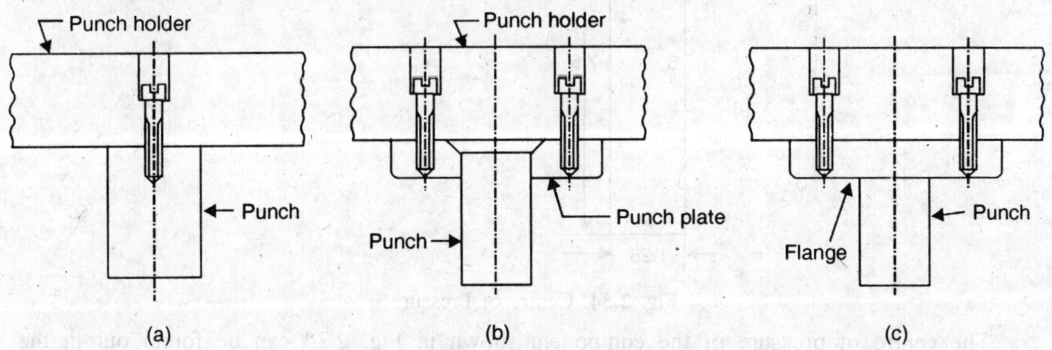

Fig. 2.33. Methods of Holding Punches.

2.12.7. Centre of pressure. When the shape of blank to be cut is irregular, the summation of shear forces about the centre line of press ram may not be symmetrical. Due to this, bending moments will be introduced in the press ram, producing misalignment and undesirable deflections. To avoid this the 'centre of pressure' of the shearing action of the die must be found and while laying out the punch position on the punch holder, it should be ensured that the centre line of press ram passes exactly through the centre of pressure of the blank. This 'centre of pressure' is the centroid of the line perimeter of the blank. It should be noted that it is not the centroid of the area of the blank.

The centre of pressure can be found out by the following procedure :

1. An outline of the piece part is drawn, Fig. 2.34.

2. The X and Y axes are placed on it in a convenient position.

3. The outline of the piece part is divided into convenient line elements. These are numbered as 1, 2, 3 and so on.

4. The length l_1, l_2, l_3 etc. of these line elements are determined.

5. The centroids of these line elements are determined.

6. The distance of the centroids from the X and Y axes is determined. Let x_1, x_2, x_3 etc. and y_1, y_2, y_3, etc., be the distance of centroids of line elements l_1, l_2, l_3 etc. from the Y and X axes respectively.

7. The distance of the centre of pressure from each axis is determined by the method of centroids, *i.e.*,

$$\bar{X} = \frac{l_1x_1 + l_2x_2 + l_3x_3 \ldots}{l_1 + l_2 + l_3 + \ldots} \qquad \ldots(2.14)$$

$$\bar{Y} = \frac{l_1y_1 + l_2y_2 + l_3y_3 \ldots}{l_1 + l_2 + l_3}$$

$\bar{X} = x$ distance to centre of pressure

$\bar{Y} = y$ distance to centre of pressure

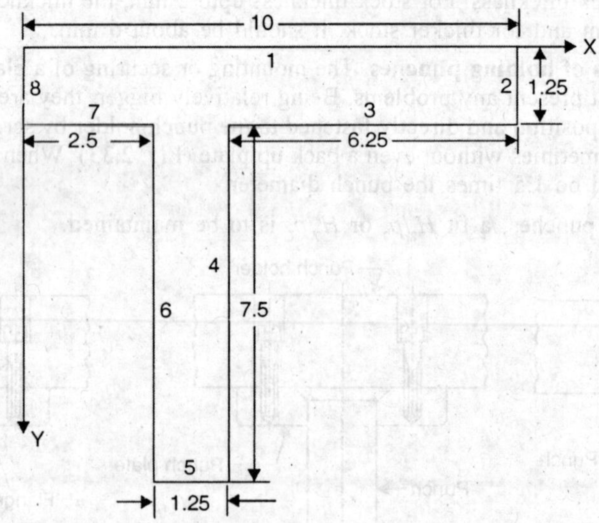

Fig. 2.34. Centre of Pressure.

The centre of pressure of the component shown in Fig. 2.34 can be found out in the following manner :

Element	*l*	*x*	*y*	*lx*	*ly*
1	10	5	0	50	0
2	1.25	10	0.625	12.5	0.78125
3	6.25	6.875	1.25	42.96875	7.8125
4	7.5	3.75	5	28.125	37.5
5	1.25	3.125	8.75	3.90625	10.9375
6	7.5	2.5	5	18.75	37.5
7	2.5	1.25	1.25	3.125	3.125
8	1.25	0	0.625	0	0.78125

$\Sigma\, l = 37.5$

$\Sigma\, lx = 159.375$

$\Sigma\, ly = 98.4375$

$$\therefore \quad \overline{X} = \frac{\Sigma lx}{\Sigma l} = \frac{159.375}{37.5} = 4.25 \text{ cm}$$

$$\overline{Y} = \frac{\Sigma ly}{\Sigma l} = \frac{98.4375}{37.5} = 2.625 \text{ cm}$$

Note. Life of die is usually 2 to 3 times the life of a punch.

2.12.8. Strippers. After a blank has been cut by the punch on its downwards stroke the scrap strip has the tendency to expand. On the return stroke of the punch, the scrap strip has the tendency to adhere to the punch and be lifted by it. This action interferes with the feeding of the stock through the die and some device must be used to strip the scrap material from the punch as it clears up the die block. Such a device is called 'stripper' or 'stripper plate'. Strippers are of two types :Fixed or stationary and spring loaded or movable, (Fig. 2.35). As is clear, it is a plate parallel with and above the die surface. An opening is cut through the stripper plate for free passage of the punch. By rule of thumb, this opening can be about 1.6 mm larger than the blank size on all sides.

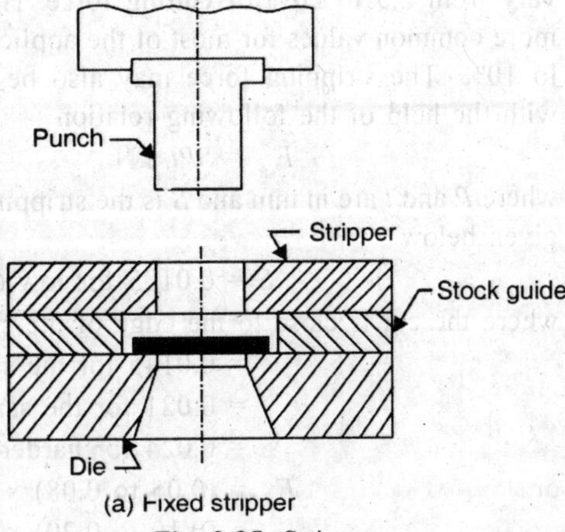

(a) Fixed stripper

Fig. 2.35. Strippers.

(*a*) **Fixed Stripper [Fig. 2.35 (*a*)].** This stripper is attached at a fixed height over the die block. The height should be sufficient to permit the sheet metal to be fed freely between the upper die surface and the under surface of the stripper plate. The stripper plate is usually of the same width and length as the die block. In simple dies, it is fastened with the same screws and dowels which are used for die block. In complex dies, the stripper fastening will be independent of diefastening. The thickness of the stripper plate should be sufficient to withstand the force needed to strip the scrap strip from the punch. The usual value is 9.5 mm to 16 mm. The following empirical formula may also be used for determining stripper plate thickness,

$$t_{str} = \frac{1}{8}(w/3 + 16t) \qquad \text{...(2.15)}$$

where w and t are the width and thickness of stock strip.

or $\quad t_{str} = 0.5 \times T$, if guided

$\quad = 0.75 \times T$, if un-guided

The thickness of the stripper plate may also be determined by the size of the socket-head cap screws used to hold it in place. The stripper plate must be thick enough to allow for the screw-head counter bores, which in most cases provides adequate stripper strength. In addition to screws, dowels are used to ensure accurate alignment of the die block. Stripper may be made of cold rolled M.S. The fixed stripper is also knows as 'channel stripper'.

For the stripping action, on the upward movement of the punch, the scrap strip will strike the underside of the stripper plate and get stripped off from the punch. The underside of the stripper plate which comes in contact with the strip should be machined and preferably ground. The height of the stock strip channel should be at least equal to 1.5 times the stock thickness. If the scrap trip is to be lifted over a fixed pin stop, this height should be increased. The width of the channel should be equal to the width of the stock plus adequate clearance. The disadvantages of fixed stripper are that it hides the work from the operator and it would interfere with removal of the scrap strip in large blanking operations.

(*b*) **Spring Loaded Stripper.** [Fig. 2.35 (*b*)]. This type is used on large blanking operations and also on very thin and highly ductile materials where it is desirable to utilize the pad pressure to hold the surrounding stock during the blanking operation. In this design the stripper plate is mounted over compression springs and suspended by bolts from the punch holder, with the lower surface of the stripper below the cutting end of the punch. As the punch travels downward for the blanking operation, the stripper plate contacts the stock strip first and holds it until the punch clears the strip on its return stroke. As the punch rises, spring pressure holds the strip, stripping it from the punch surface. The stripping force may vary from 2.5 to 20% of cutting force. However, the more common values for most of the applications are 5 to 10%. The stripping force may also be determined with the help of the following relation :

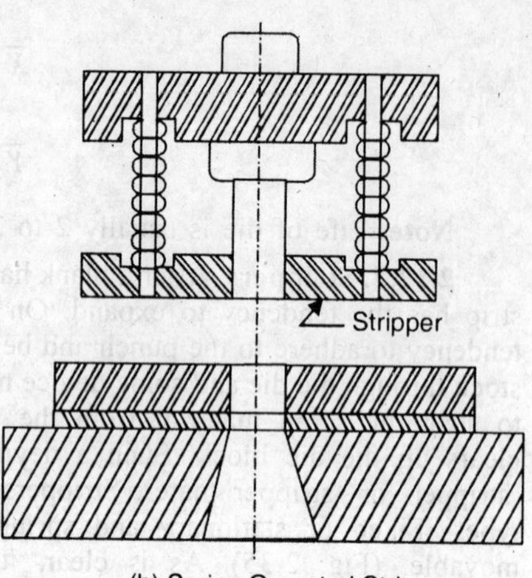

(b) Spring Operated Stripper

Fig. 2.35. Strippers.

$$F_{str} = SPt, \text{ kN}$$

where P and t are in mm and S is the stripping constant, given below :

$S = 0.0105$ for low C-steels with sheet metal under 1.6 mm thick where the cut is close to the edge of the strip or preceding cut.

$= 0.0145$ for the same materials but for other cuts.

$= 0.021$ for the same materials with $t > 1.6$ mm.

$= 0.024$ for harder materials.

or, $F_{str} = (0.05 \text{ to } 0.08) \times F$, for normal clearance

$= (0.10 \text{ to } 0.20) \times F$, for narrow clearance and complicated contours.

Comparison of fixed and spring loaded strippers :

1. Solid stripper is simple in construction and is cheap.

2. Fixed or solid stripper does not apply any hold-down pressure like a spring stripper. Due to this thinner sheets may buckle and jam in the stripper. So, it is mainly used for thick sheets.

3. Clearances in the fixed stripper channel allow the sheet to shift slightly. Additional edge stock is necessary to insure complete blanks being cut.

4. Fixed stripper is used mainly for strip sheet metal where feeding of the sheet metal is accomplished by hand. Coiled sheet metal uses a feed roll and spring stripper in most cases.

5. For very large stripping force, fixed stripper is preferred because it may not be possible to provide enough stripping force with springs in the space available.

6. The travel of spring strippers is limited by the compression allowance of the springs.

7. In spring strippers, the full stripping force is not available throughout the stripping action. It is build up as the springs are compressed.

8. Additional die space is required for placing the springs around the punch.

Design of Springs: The maximum force the spring can exert,

$$F_{smax} \geq 1.5 \times \frac{F_{str}}{i}, \; i = \text{no. of springs}$$

Now, movement of stripper,

$$y_{str} = t + 2, \text{ mm}$$

spring deflection, $y = (3 \text{ to } 4)\, y_{str}$ at F_{smax}.

Considering sharpening allowance,

$$y_{max} = (3 \text{ to } 4)(t + 2) + s; \text{ here } s \text{ is the sharpening allowance.}$$

Helical Springs: Wire diameter,

$$d = K \cdot \sqrt{\frac{8 F_{smax.}}{\tau_s} \cdot \frac{D}{d}}$$

K = Wahl factor

$$\frac{D}{d} = \text{spring index} \left(\frac{\text{mean diameter of coil}}{\text{wire diameter}} \right)$$

= 4 to 6 for shearing dies

= 12 to 15 for drawing dies

τ_s = Permissible shear stress, 600 to 700 N/mm²

$\therefore \quad d = ?$

Deflection of one free coil, $y_1 = \dfrac{8 \cdot F_{smax.} \cdot D^3}{Gd^4}$

$\therefore$ No. of free coils, $n = y_{max.} / y_1$

Compressed length of spring where max. force acts,

$$l_{min.} = 1.1\, nd + d$$

Free length of spring, $l_0 = l_{min.} + y_{max.}$

Pre-compressed length of spring,

$$l_1 = l_{min.} + y_{str} + s$$

Note: Belloville springs are more tough, *i.e.*, less deflection for higher loads, so, are good for shearing dies for stripping.

2.12.9. Stock stop. The strip of sheet metal is fed and guided through a slot in the stock guide, Fig. 2.35 (*a*), or through a slot in the stripper plate. After each blanking, the strip has to be advanced a correct distance. The device used to achieve this is called 'stock stop'. The simplest arrangement may be a dowel pin or a small block, against which an edge of the previously blanked hole is pushed after each stroke of the press. On its upward stroke, the punch carries the stock strip as far as the underside of the stripper plate. Due to this, the stock strip gets released from the stop. With constant pressure exerted pushing the stock strip to the left, the stock will move as it is lifted clear, then drop with the next hole over the stop as the scrap strip is stripped from the punch, Fig. 2.36.

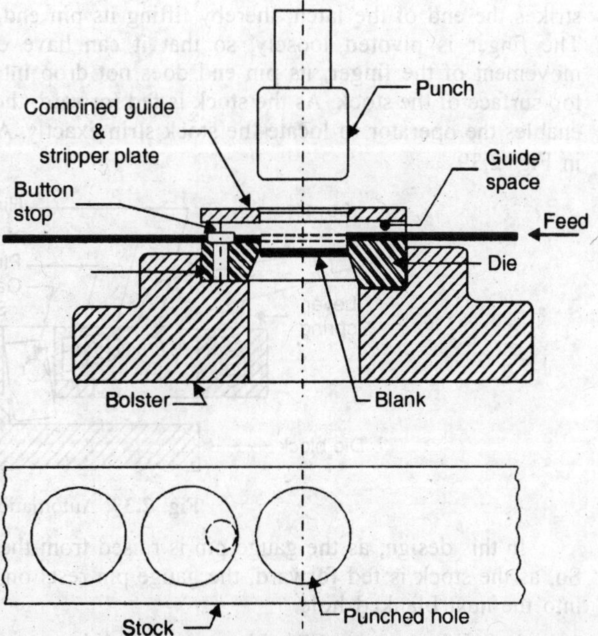

Fig. 2.36. Stock Stop.

This type of stock stop is suitable for only low and medium production dies and not for high production dies, since the operator has to force the stock over the shoulder to secure a desired feed length. Also, this needs considerable skill on the part of the operator. This stop is also, not suitable where frail die sections would be damaged by a misfeed. Other types of stock stops are discussed below :

1. **Latch stop or Trip stop or Pawl stop. Fig. 2.37.** In this type of stop, a latch or pawl is pivoted on a pin fitted into a block on stripper and is held down by a tension spring. The latch is lifted by the scrap bridge on the ratchet principle and drops into the blanked area, as the stock is fed forward manually into the die. The operator then pulls the stock back until the vertical surface of latch bears exactly against the scrap bridge. The design is also suitable for low production only.

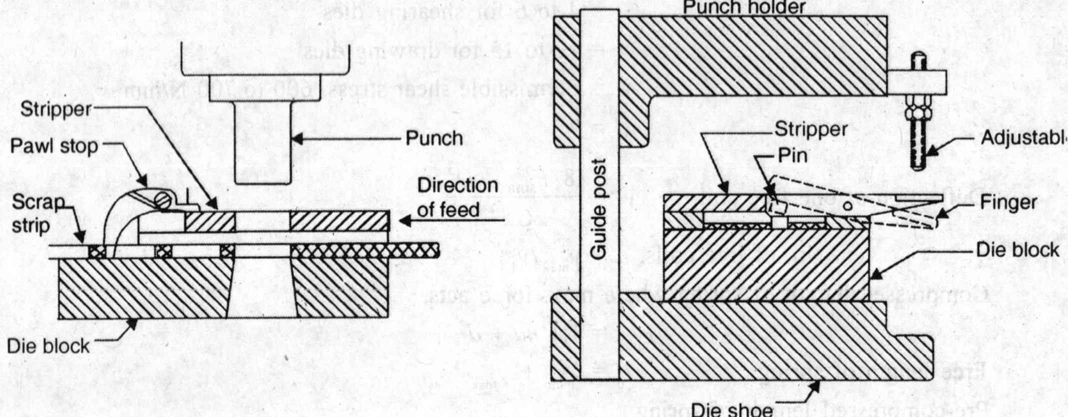

Fig. 2.37. Latch Stop. **Fig. 2.38.** Automatic Stop.

2. **Automatic stop.** In this design also, a hinged latch or lever (finger) is used. The working of an automatic stop is shown in Fig. 2.38. An adjustable strip screw is fastened to the punch holder. During the downward stroke of the slide, as the punch cuts the blank, this strip screw strikes the end of the latch, thereby lifting its pin end. On the return stroke, the pin end drops. The finger is pivoted loosely, so that it can have endwise movement. Due to this endwise movement of the finger, its pin end does not drop into its former position, but drops on to the top surface of the stock. As the stock is fed forward, the pin drops into the next blank space. This enables the operator to locate the stock strip exactly. Another design of automatic stop is shown in Fig. 2.39.

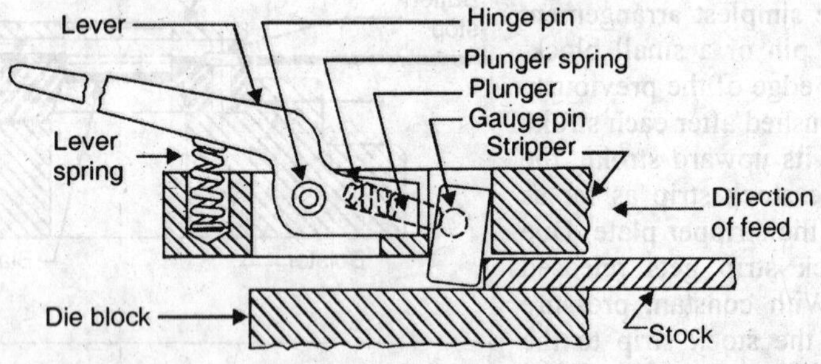

Fig. 2.39. Automatic Stop.

In this design, as the gauge pin is raised from the blank space due to lever action, it rocks. So, as the stock is fed forward, the gauge pin rests on the top surface of the stock until it drops into the next blanked hole.

3. **Solid stops or Shoulder stops.** Solid stops are extensively used on progressive dies, when the last operation is a cut off or trimming one, to position the end of the stock or the workpiece. Two designs are used in Fig. 2.40. In Fig. (*a*), the stop is fastened to the die, whereas in Fig. (*b*) it is fastened to the stripper.

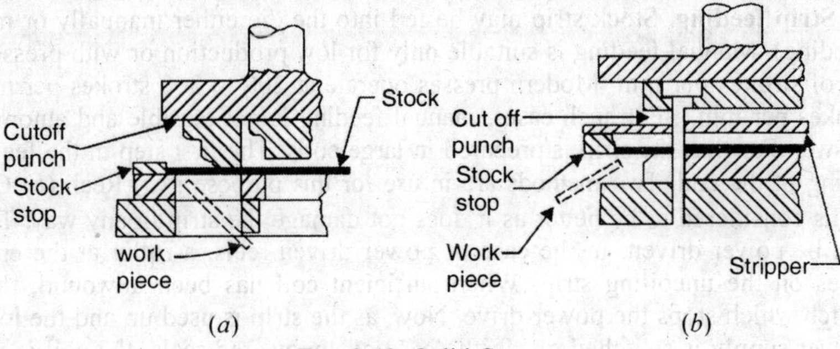

Fig. 2.40. Solid Stops.

4. **Starting stops.** A starting stop or a primary stop is used to position stock as it is initially fed into the die. It is mounted on the stripper plate, Fig. 2.41. It consists of a latch, which is pushed inward by the operator by one end until it contacts the stripper plate. The stock strip which is being fed by the operator with his other hand is made to locate against the latch. The latch is held in position until the first die operation is completed and then the latch is released.

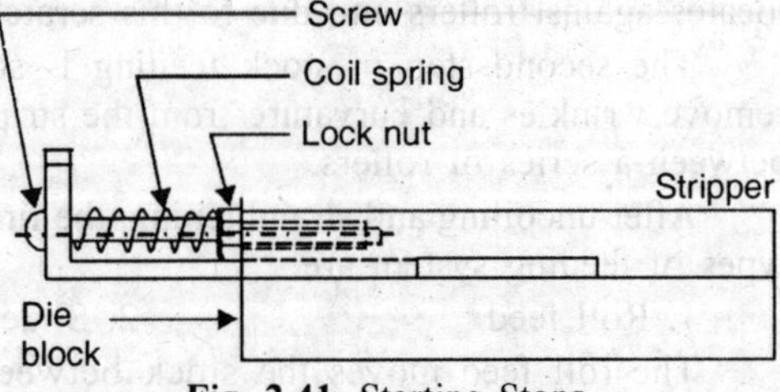

Fig. 2.41. Starting Stops.

Note. Stock stops are needed when the stock is fed manually. In the case of automatic feeding, stops are usually not needed as the strip can be advanced quite accurately.

2.12.10 Stock Guide. Stock guide is the space provided in the die-block, through which the stock strip is guided as it is fed into the die. The design of the stock guide will depend upon the type of stripper. For fixed strippers (channel strippers), the channel for guiding the stock is shown in Figs. [2.35 (*a*) and 2.36] and its design is given under "fixed stripper". For spring operated strippers, the various designs for guiding the stock are shown in Fig. 2.42 (*a, b, c*). In the figure, guide rails are mounted on the die block for guiding the stock. In Fig. 2.42 (*a*), the stripper acts as a pressure pad and presses directly against the stock strip. The stripper does not contact the guide rails. In Fig. 2.42 (*c*), hooks are provided on the guide rails in order to improve the guiding effect. When there is space limitations, button stock-strip guides may be used, Fig. 2.42 (*d*). At least three button guides should be used on each side of the stock strip. The guiding is not as efficient as with guide rails.

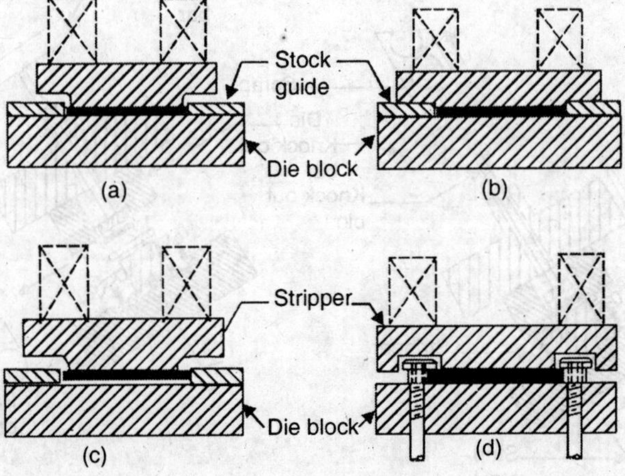

Fig. 2.42. Stock Guides with Spring Operated Strippers.

2.12.11. Strip feeding. Stock strip may be fed into the die either manually or mechanically (automatic feeding). Manual feeding is suitable only for low production or with presses operating at low values of strokes per min. Modern presses operate at 200 to 300 strokes per min. (may be up to 500 strokes per min.). For such cases, manual feeding is not feasible and automatic feeding is the only answer. For this, the strip is prepared in large coils. The first step in the feeding of strip is the unwinding of the coil. Two methods are in use for this purpose : (*a*) Reel, (*b*) Coil cradle.

The reel is considered to be better as it does not damage the strip in any way. The reel may be or may not be power driven. In the case of power driven reels, a roller at the end of a long loop arm, rides on the uncoiling strip. When sufficient coil has been unwound, the loop arm actuates a switch which stops the power drive. Now, as the strip is used up and the loop arm gets raised, the power supply is switched on. In the case of unpowered reels, the coil is unwound by an external power source, which may be feeding mechanism or straightening rolls. When enough coil has been unwound the reel is stopped from uncoiling by a manual or automatic brake.

In the case of cradle, the strip is supported on the outside diameter of the coil. The coil locates against rollers and due to this scratches may from on the coil.

The second step in stock feeding is straightening of the uncoiled strip. This is done to remove wrinkles and curvature from the strip. For straightening the coil, it is passed through in between a series of rollers.

After uncoiling and straightening, the final step is to feed the strip into the die. The two main types of feeding system are :

1. Roll feed 2. Slide or hitch feed

The roll feed moves the stock between a pair of rollers, which is driven through an overrunning clutch or ratchet mechanism timed from the press main shaft or ram. Roll feeds may be : single or double. In the case of single roll feed, rollers are provided only on one side of the press and they push or pull the strip through the die. In the case of double roll feed, feeding mechanism is mounted on each side of press bed with a drive connection between them. On feed pushes the stock and the other pulls stock through the die. This will keep strip tight and prevent its buckling.

In the case of slide or latch feed, grippers are used which grasp the strip mechanically and feed it into the die by a reciprocating mechanism which may be driven from the press crankshaft, a cam mounted on the punch holder, hydraulically or pneumatically.

2.12.12. Knockouts. As already stated, the function of a knockout is to shed or eject a workpiece from within the die cavity as the workpiece may get jammed in the die cavity due to friction. A knockout may be actuated by springs (Fig. 2.25) or by a positive acting knockout pin

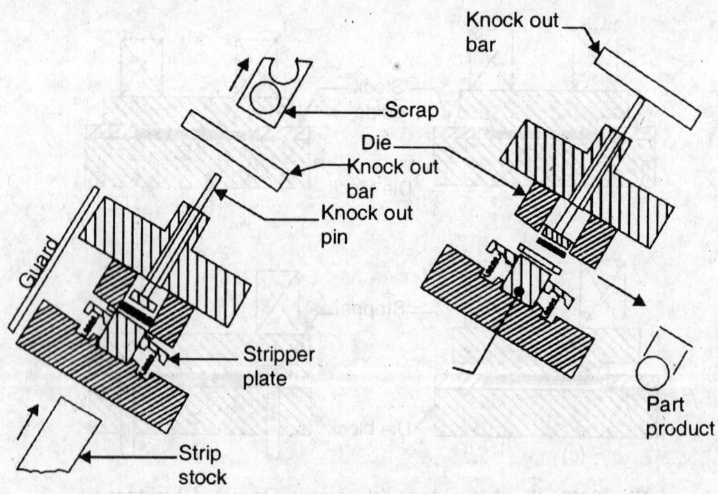

Fig. 2.43. Knockout.

and bar arrangement. The knockout pin usually leads through the shank. It may be a single pin or a double pin fastened to a pad or collar above the shank. The working of a positive knockout pin arrangement is shown in Fig. 2.43, for an inverted *OBI* press. On the return stroke of the press, the knockout pin strikes the knockout bar. This makes the knockout pin to actuate the knockout plate which forces the blank out of the die-opening. The function of the knockout plate is to support and guide fragile punches.

2.12.13. Die-set. The die shoe, the punch holder together with two or more guide posts constitute a die set, Fig. 2.44. The die shoe and the punch holder are made of *C. I.*, *C. S.* and rolled steel. For smaller dies, *C.I.* is used, whereas for larger and special die sets, *C. S.* and rolled tool steel are used. Bushings are assembled to the upper shoe (punch holder) by press fitting and guide posts are press fitted into the lower shoe (die shoe). The bushings and posts are sized to provide a slip fit. For average range of die sets, the diameter of guide posts varies from 2.5 cm to 7.5 cm. Larger pins may be used if extreme alignment is required. When die is fully closed, the upper ends of the guide posts should not project beyond the top surface of the upper shoe.

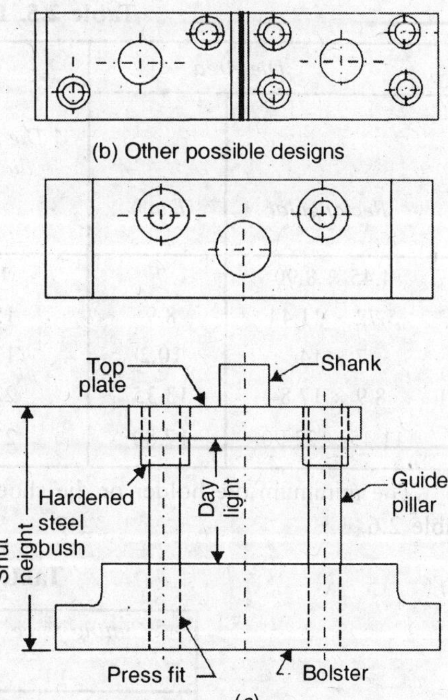

Fig. 2.44. Die Set.

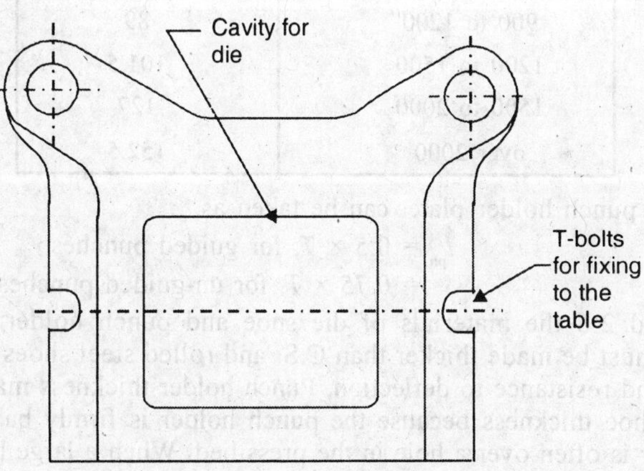

Fig. 2.45. Bottom Bolster.

A typical bottom bolster (die shoe) plan is shown in Fig. 2.45.

The dimensions of round diagonal-post die sets are given in Table 2.5. All dimensions are given in cm.

Table 2.5. Dimensions of Die Sets

Die Area		*Die holder diameter*	*Thickness*		*Min guide post diameter*
Rectangular	*Diam*		*Die holder*	*Punch holder*	
4.45 × 8.90	7	12.7	4.45	3.18	1.27
5.72 × 11.44	8.9	15.25	4.45	3.18	1.60
7 × 14	10.2	17.80	5.00	3.80	1.90
8.9 × 17.8	13.35	22.90	5.00	3.80	2.55
11.45 × 22.90	17.80	27.95	5.00	3.80	2.85

The minimum die holder or die shoe thickness in terms of cutting force is given below in Table 2.6.

Table 2.6. Die Shoe Thickness

Cutting force kN	*Die shoe thickness mm*
Upto 100	38
100 to 300	50
300 to 600	63.5
600 to 900	76
900 to 1200	89
1200 to 1500	101.5
1500 to 2000	127
over 2000	152.5

Thickness of the punch holder plate, can be taken as,

$$t_{ph} = 0.5 \times T, \text{ for guided punches}$$

$$= 0.75 \times T, \text{ for un-guided punches.}$$

In tables 2.5 and 2.6 the materials of die shoe and punch holder are not mentioned. However, C.I. shoes must be made thicker than C.S. and rolled steel shoes to provide the same mechanical strength and resistance to deflection. Punch holder thickness may be 12.5 to 25 mm thinner than the die shoe thickness because the punch holder is firmly backed up by the ram, whereas the die holder is often over a hole in the press bed. When a large hole is present in the bed of the press and no bolster plate is used, increase the die shoe thickness by approximately 50 per cent. The thickness of the shoe will be limited by the shut height of the press as discussed later.

Note : If a die set is not used, the working die details would have to be mounted on the press ram and bed. Normally, a die is used for a period of one or two years, and then replaced by a new die for the new part. On the other hand, the press must be used for several years.

If the die details are mounted directly on the ram and bed surfaces, the later will get damaged by the numerous screw and dowel pin holes used for attaching die details. The mounting of the new die details will become difficult. It will be very expensive to rework the press bed and the ram, these being the parts of the press frame. Due to this, the die details are mounted in a die set, which is relatively cheap. The same press can be used for different die sets.

Shut height of press. As already defined, the shut height of a press is the height of the opening between the ram and the bed, with the ram down and the adjustment up. Ram adjustment is provided by a special screw between the ram and connecting rod. One end of the screw is usually a ball that rides in a socket in the ram. The end of the screw is threaded and engages the connecting rod. By turning the screw clockwise, the overall length of the connecting rod and ram assembly is shortened. The total amount of adjustment is called the press "adjustment". Adjustment is used to change the closed opening of a press. Adjustment may decrease this opening from the shut height (maximum closed opening) down, but it does not increase upon the shut height.

Shut height of Die. Die shut height is found as follows: It is the distance from the top of the punch holder to the bottom of the die holder. Thus,

Die Shut Height = Punch Shoe thickness + Die Shoe thickness + Die Height + Punch Height Bypass of Die and punch (3.2. to 6.4 mm).

or, The die-shut height or the tool shut height, H, is given as,

$$H_{min} + 10 < H < H_{max} \cdot 5$$

where $H_{max} = H_u - \frac{1}{2}(s_{max} + s_{min})$

$H_{min} = H_u - l - \frac{1}{2}(s_{max} + s_{min})$

H_u = Distance between ram and press table in the up-stroke position with maximum length of stroke and the ram adjustment fully down

l Ram adjustment

s_{max} = max. stroke of the press

s_{min} = min. stroke of the press

If s = actual stroke length required for the tool, then $s_{min} < s < s_{max}$

The shut height of the die must be equal to or less than the shut height of the press, because as discussed above, due to the long life of presses and the short life of dies, the die usually must be fitted to the press. The die is first designed and the press capacity determined. Then a suitable available press is selected to meet the requirements. There can be three cases :

1. If the calculated die shut height is less than the press shut height by an amount less than the press adjustment, then the die may be designed as calculated.

2. If the calculated die shut height is less than the press shut height by an amount greater than the press adjustment, then a bolster plate is used to reduce the press closed height to approximately the die shut height. The press adjustment is then used for the remaining take up of height. Both the bottom bolster and top bolster (if need be) can be used for this purpose,

Thickness of top bolster = 1.25 (steel) to 1.5 (for C.I.) × T

3. If the calculated shut height is greater than the press shut height, then adjustments must be made in the die component heights or thicknesses. However, minimum design requirements should not be violated. If necessary, a new press must be found to suit the die rather than altering the original design.

2.12.14. Bolster Plate. When many dies are to run in the same press at different times, the wear occurring on the press bed is high. The bolster plate is incorporated to take this wear, plate (2.5 to 7.5 cm) is made from boiler plate or t_{bp} = 1.75 (for steel) to 2.00 (for C.I.) × T. It is attached to the press bed and the die shoe is then attached to it. It is machined so that its surfaces are flat and parallel. Bolster plates are relatively cheap and easy to replace. The other functions of a bolster plate are :

1. To provide attachment holes for the dies rather than drilling these holes in the press bed.

2. To support the die shoe when it is located over a large hole in the press bed.

3. To take up space in the press when the press shut height is too great for the die shut height.

4. To provide chutes for ejecting parts or scrap out the sides of the press.

Normally, bolster plate is taken to be the bottom bolster plate

2.13. PIERCING DIE DESIGN

The cutting action of piercing is essentially the same as in blanking. The piercing die is quite similar to the drop-through blanking die. But where a blanking die has one pair of cutting members, the piercing die may have many pairs of cutting tools which are usually much smaller in size. As already mentioned the size of the punch is the size of the hole and the clearance is provided on the die cavity. Also to reduce the cutting force, shear is applied to the punch rather than the die block. When the punch is withdrawn after cutting the hole, the hole has the tendency to shrink due to return flow of metal which was earlier pushed out by the penetrating punch. To compensate for this, the punch dimension is generally increased. For holes of 25 mm diameter, this allowance ranges from 0.025 mm on thin materials to 0.050 mm on thick stocks. For closely spaced holes, these should be positioned so as to leave a distance of at least equal to twice the stock thickness across nearest approach of a hole, to the sheet edge or to another hole.

2.13.1 Mounting piercing punches. The mounting of piercing punches presents some difficulties as their sizes can be very small and also holes may have to be pierced very close to one another.

1. **Peened-head punches.** Small punches (less than 19 mm diameter) are generally made from 19 mm or smaller drill rod and are left headless until assembly. A countersunk reamed hole is made in the punch plate. The punch is pressed lightly into this hole and then riveted over. To facilitate assembly, the shank or portion of the punch placed in the punch plate is always made circular and larger than the piercing portion, [Fig. 2.46 (*a*)]. For perforating operations, where a large number of small, round and closely spaced holes are to be pierced, the punch may be straight for its entire length. They are further supported and guided by making them a sliding fit in the stripper plate, Fig. 2.46 (*b*). To reduce the total cutting force, the adjoining punches are staggered as shown. For best results, the stripper plate should be hardened at the guide holes and/or fitted with bushing, Fig. 2.46 (*c*).

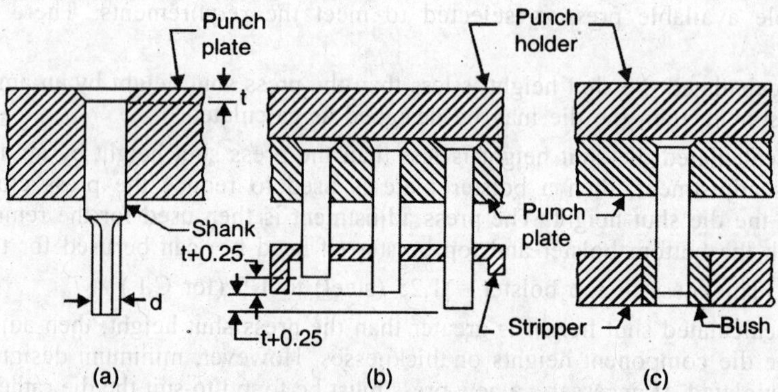

Fig. 2.46. Peened-Head Punches.

The peened head constructions have the following drawbacks :

(*a*) The punches are easily deformed, since the peening is done by hand and, therefore, is not completely uniform.

(*b*) These are unsuitable for comparatively high stripping pressures.

(*c*) Because of inadequate heat treatment, the head often breaks during use.

2. **Quilled Punches.** For punching very small holes in thick stock, the punch having a uniform point and shank diameter, is enclosed in quill, which is made of hardened tool steel,

Fig. 2.47. The quill is provided with a square-shouldered head and after fitting the punches, their heads are riveted over.

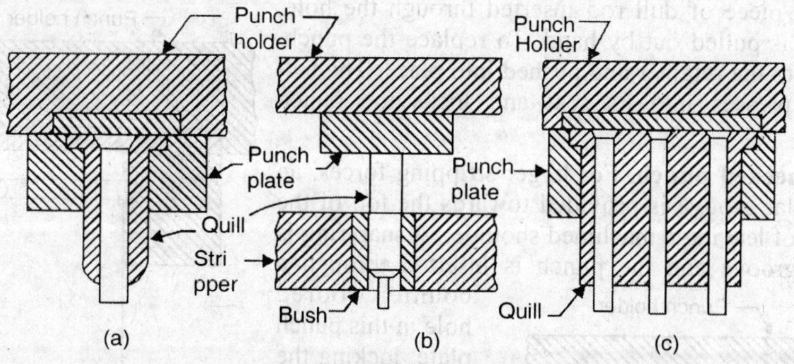

Fig. 2.47. Quilled Punches.

3. **Square headed punches.** Square headed punches, Fig. 2.48 (*a*), are usually used in high quality tools. They are machined from round bar stock of larger diameter than the punch body, leaving an integral head of sufficient size. The head prevents the punch from pulling out of the punch plate. The head thickness is usually 6 mm. The chief disadvantage of this design is an increase in labour and material costs.

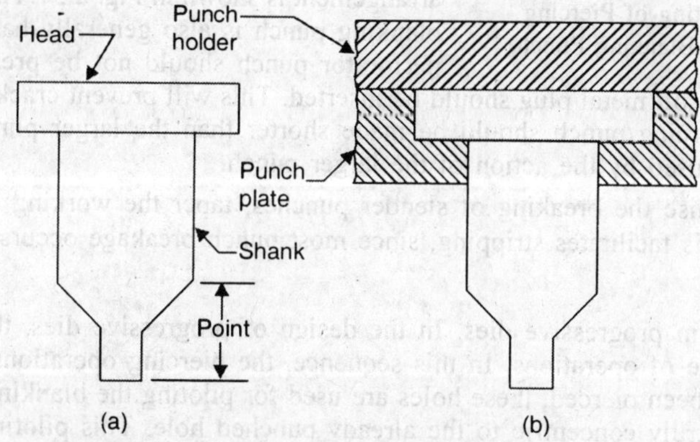

Fig. 2.48. Square Headed Punches.

(4) **Headless punches.** In the above mentioned designs of the punches, the punch plate has to be dismounted in order to change the punch. There is no such problem in the case of headless or straight-shanked punches. They can be frequently changed by loosening the set screw, Fig. 2.49 (*a*) and (*b*), or by the ball-spring mechanism. Fig. 2.49 (*c*).

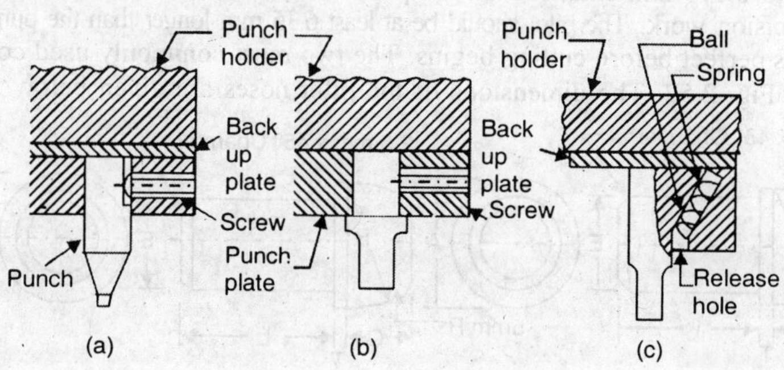

Fig. 2.49. Headless Punches.

The punch is firmly secured in the punch plate by the spring-loaded ball pressing into a recess in the shank of the punch. To remove the punch, the ball is pushed up against the spring by means of a piece of dull rod inserted through the hole, and the punch is pulled out by hand. To replace the punch, it is inserted into the retainer and pushed into place. The ball automatically presses into its seat and locks the punch securely.

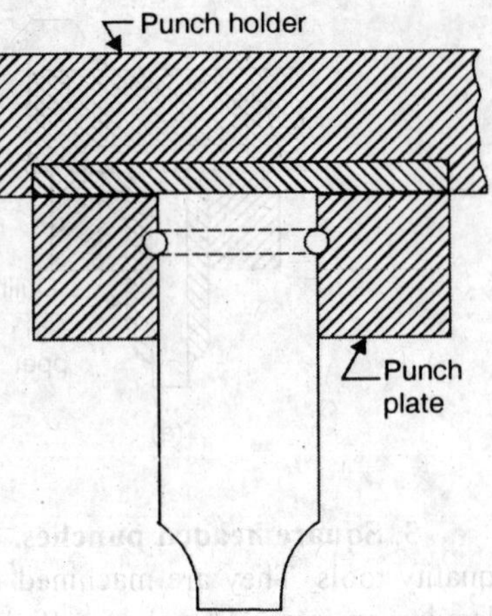

Fig. 2.50. Ring Necked Punch.

5. **Ring-necked punch.** For larger stripping forces, an adequate circular groove is provided towards the top of the punch instead of leaving a machined shoulder. A snap ring is fitted in this groove and the punch is inserted through a counter bored hole in this punch plate, locking the punch in position as shown in Fig. 2.50.

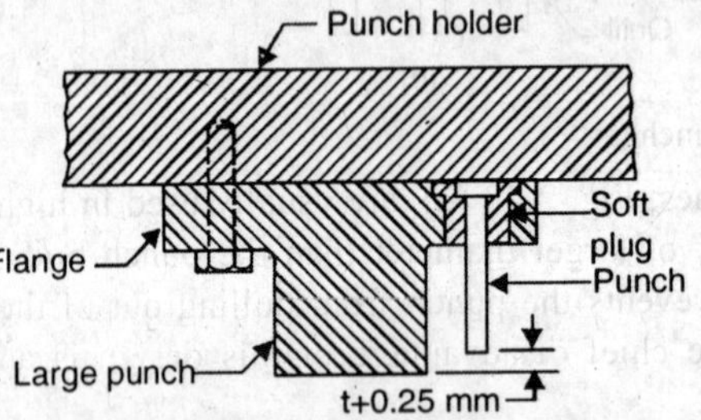

Fig. 2.51. Mounting of Piercing Punch Close to a Blanking Punch.

If a small perforator punch is to be mounted close to a blanking punch, the arrangement is shown in Fig. 2.51. As the flange of the blanking punch is also generally hardened; the small perforator punch should not be pressed directly into the flange. Instead a soft metal plug should be inserted. This will prevent cracking of the flange. Also, the slender piercing punch should be made shorter than the larger punch, as shown, to prevent deflection caused by the action of the larger punch.

Note. To minimise the breaking of slender punches, taper the working end of the punch 0.25 to 0.50 mm. This facilitates stripping, since most punch breakage occurs during stripping.

2.14. PILOTS

Pilots are used in progressive dies. In the design of progressive dies, the first step is to establish the sequence of operations. In this sequence, the piercing operations are placed first. After the holes have been pierced, these holes are used for piloting the blanking punches so that the blank formed is truly concentric to the already punched hole. This piloting is achieved by means of pilots secured under the blanking punch, which engage the already pierced holes. To be effective, the pilot must be strong enough to align the stock without bending. Pilots are made of good grade of tool steel heat treated to maximum toughness and to a hardness of 56 to 60 Rockwell *C*.

Pilot dimensions. The pilot should fit the already pierced hole with a close sliding fit. Its size is from 0.05 mm to 0.076 mm smaller than the punch size for normal work and from 0.0013 to 0.025 mm for precision work. The pilot should be at least 6.35 mm longer than the punches in order to ensure that registry is perfect before cutting begins. The two most commonly used contours of pilot nose are given in Fig. 2.52. The dimensions of the pilot noses are given below :

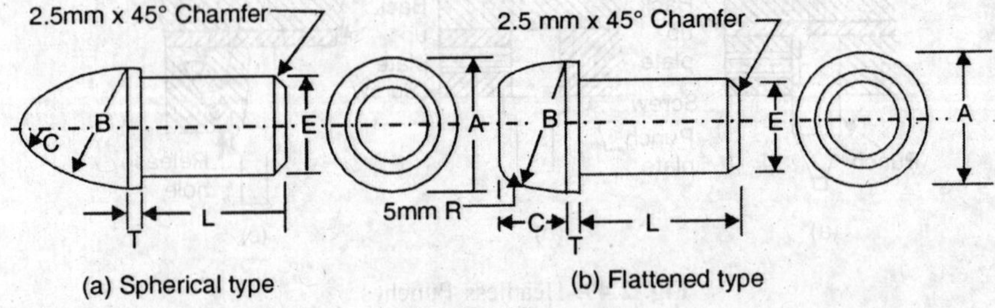

Fig. 2.52. Pilot noses

Spherical type	*Flattened type*
$B = A$	$B = A$
$C = 0.25\ A$	$C = 0.625\ A$
$L = A$ to $2\ A$	$L = 0.75\ A$
$E = 0.625\ A$	$E = 0.5\ A$

T = 1.6 mm or stock thickness whichever is greater. The nose of the pilot is hardened and then polished.

Types of Pilots. There are two types of pilots :

1. Direct pilots, and 2. Indirect pilots.

1. **Direct pilots.** Pilots which are mounted on the face of a punch are called direct pilots. The pilot holder is generally a block of steel which can be fastened to the punch holder. The simplest arrangement is a press fit pilot, Fig. 2.53 (*a*). This is often used for low speed dies. It is not recommended for high speed dies as it may drop out of the punch holder. Threaded shank design, Fig. 2.53 (*b*), is recommended for high speed dies. In Fig. 2.53 (*c*), the pilot is held by a socket screw. If the diameter of pilot is less than 6 mm, it may be headed and secured by a socket set screw 2.53 (*d*). The pilot can also be held by ball lock mechanism, 2.53 (*e*).

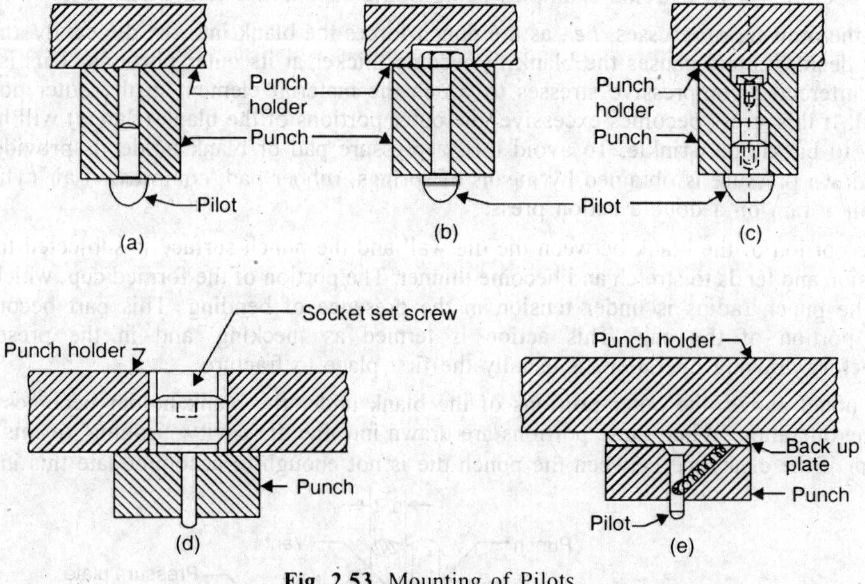

Fig. 2.53. Mounting of Pilots.

2. **Indirect pilots.** It is not always necessary to fix a pilot to the bottom of a blanking punch. The pilots may be independent of the blanking punch and directly retained in the punch holder. They are designed to enter previously punched holes in the strip some distance away from the blanking punch. In this way, they help to position the stock strip accurately and bring it into proper register for subsequent blanking and piercing operations. During manual feeding of stock strip, there is usually a slight overfeeding beyond the register position. The pilot then helps to bring the strip back into registry position in a direction away from the stock stop. This prevents buckling of the stock strip against the stock stop. Such pilots are well guided through the hardened bushes in the stripper plate.

2.15. DRAWING DIES

Drawing operation is the process of forming a flat piece of material (blank) into a hollow shape by means of a punch which causes the blank to flow into the die-cavity, Fig. 2.54.

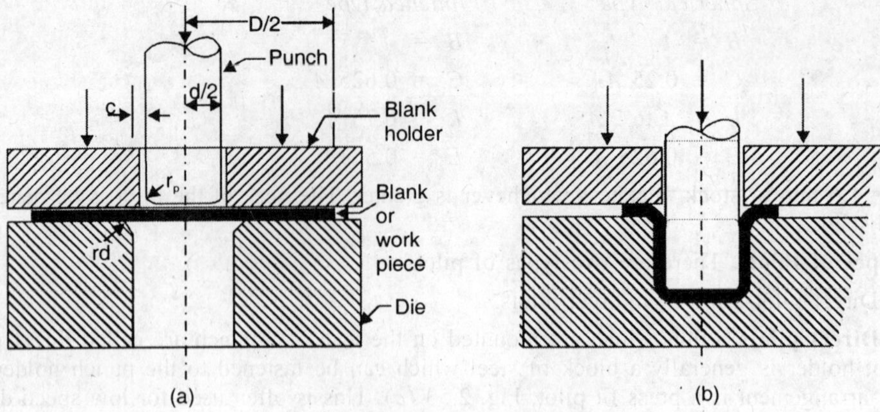

Fig. 2.54. Drawing Operation.

The depth of draw may be shallow, moderate or deep. If the depth of the formed cup is upto half its diameter, the process is called "Shallow drawing". If the depth of the formed cup exceeds the diameter, it is termed as "Deep Drawing". Parts of various geometries and sizes are made by drawing operation, two extreme examples being bottle caps and automobile panels.

As the drawing progresses, *i.e.*, as the punch forces the blank into the die cavity, the blank diameter decreases and causes the blank to become thicker at its outer portions. This is due to the circumferential compressive stresses to which the material element in the outer portion is subjected. If this stress becomes excessive, the outer portions of the blank (flange) will have the tendency to buckle or wrinkle. To avoid this, a pressure pad or blank holder is provided. The holding down pressure is obtained by means of springs, rubber pad, compressed air cylinder or the auxiliary ram on a double action press.

The portion of the blank between the die wall and the punch surface is subjected to nearly pure tension and tends to stretch and become thinner. The portion of the formed cup, which wraps around the punch radius is under tension in the presence of bending. This part becomes the thinnest portion of the cup. This action is termed as 'necking' and in the presence of unsatisfactory/drawing operation, is usually the first place to fracture.

As noted above, the outer portions of the blank under the blank holder becomes thicker during the operation. When these portions are drawn into the die cavity, 'ironing' of this section will occur if the clearance between the punch die is not enough to accommodate this increased

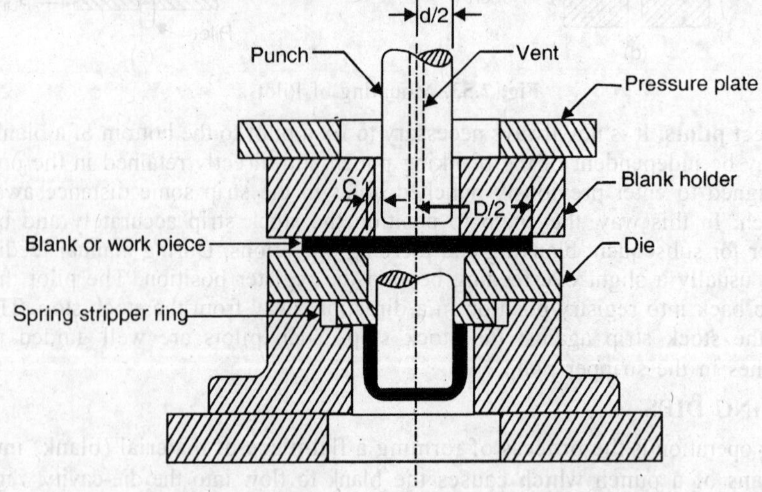

Fig. 2.55. A Simple Push through Drawing Die.

thickness of the workpiece. This ironing is useful if uniform thickness of the product is desired after the drawing operation.

Deep Drawability. Deep drawability or drawing ratio, β, of the metal is defined as the ratio of the maximum blank diameter to the diameter of the cup drawn from the blank (usually taken to be equal to the punch diameter), *i.e.*, *D*/*d*. For a given material there is a limiting drawing ratio (LDR) after which the punch will pierce a hole in the blank instead of drawing the blank. This ratio depends upon many factors, such as type of material, amount of friction present, etc. The usual range of the maximum drawing ratio is 1.6 to 2.3.

$$\text{Reduction ratio, } R = \frac{D-d}{D} = 1 - \frac{d}{D} = 1 - \frac{1}{\beta}$$

or

$$\beta = \frac{1}{1-R}$$

A simple push through drawing die is shown in Fig. 2.55. The drawing punch must be properly vented with drilled passages. Venting serves double purpose. It eliminates suction which would hold the cup on the punch and damage the cup when it is stripped from the punch. Secondly, venting provides passages for lubricants.

2.15.1. Design consideration. Radius of Draw Die. The edge radius of the first draw die, over which the blank is drawn, is very important. If the radius is too small, the resistance to the flow of metal increases resulting in cutting or tearing of the metal. On the other hand, too large a die radius may induce the formation of wrinkles in the metal as it is drawn into the die cavity from beneath the blank holder. This radius, r_d in Fig. 2.54, usually ranges from 4 to 10 times the blank thickness.

or

$$r_d = 0.035\,[50 + (D - d)] \times \sqrt{t}$$

Punch Radius. Like the radius of draw die, the punch radius is also critical. The edges of the punch must be rounded to avoid cutting or tearing the metal. An excessive punch radius increases tendency of material to buckle. Where more than one draw is needed to give the final shape to the part, the minimum punch radius, r_p in Fig. 2.54, should be 4 times the stock thickness. On subsequent draws, the punch radius can be reduced proportionately. That is,

$$r_p = (3 \text{ to } 6) \times t$$

Draw Clearance. The side clearance between the punch and the die should be more than the stock thickness to take into account the thickening of metal over the die radius when the flat blank is drawn into the die cavity, otherwise the blank may get jammed in the die cavity. The side clearance between the punch and the die is usually taken as 1.25 times the stock thickness, *i.e.*, Punch diameter = Die opening diameter − 2.5 *t*.

The average draw clearance is given in Table 2.7.

Table 2.7. Draw Clearance

Blank Thickness, mm	*First draw*	*Redraws*
Up to 0.38	1.07 *t* to 1.09 *t*	1.08 *t* to 1.1 *t*
0.40 to 1.27	1.08 *t* to 1.1 *t*	1.09 *t* to 1.12 *t*
1.30 to 3.18	1.1 *t* to 1.12 *t*	1.12 *t* to 1.14 *t*
3.20 and up	1.12 *t* to 1.14 *t*	1.15 *t* to 1.2 *t*

For sizing draws, the clearance is less.

Drawing Speed. The punch travels downward to force the blank into the die cavity, by the action of press ram. The ram speed is critical since to initiate plastic deformation, the internal inertia of the work material must be overcome. Sufficient time must be given for this to happen otherwise the metal can rupture instead of being drawn. The ideal drawing speeds are given in Table 2.8. Proper die design, proper lubrication etc. help to increase the permissible speed of ram travel.

Table 2.8. Ideal Drawing Speed

Work material	*Speed, mpm*
Aluminium	45–52.5
Brass	52.5–60
Copper	37.4–45
Steel	5.5–15
Stainless Steel	9–12
Zinc	37.5–45

Calculating Blank Sizes. The first step in the drawing process is to calculate the approximate size of the blank required for a particular drawn part. There are several methods to do so, but below we give the equations based on 'Area-methods' to determine the blank size for drawing plain cylindrical shells. In this method it is assumed that the surface area of the blank is equal to the surface area of the finished shell.

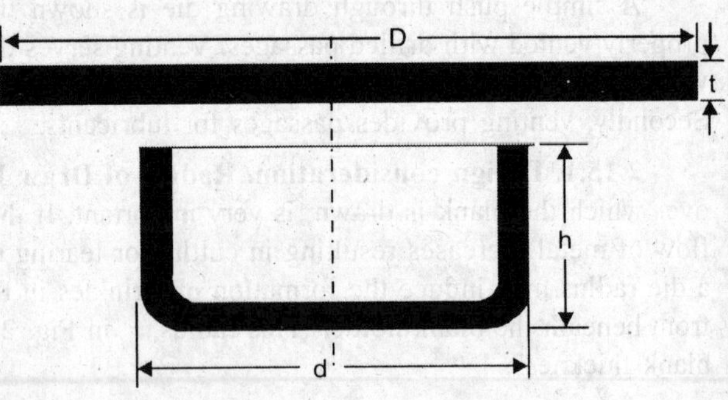

Fig. 2.56

(*i*) Where thin gauge stock is used and the shell has sharp inside corners, Fig. 2.56, the formula may be used.

$$D = \sqrt{d^2 + 4dh} \quad \text{...(2.16)}$$

where, D is the diameter of flat blank, 'd' is the diameter of finished shell and 'h' is the height of finished shell. This formula is valid for d/r 20 and more, where, r is the radius of bottom corner.

(*ii*) For a round corner cup, when the ratio d/r is between 15 and 20 the formula becomes

$$D = \sqrt{d^2 + 4dh} - 0.5\,r \quad \text{...(2.17)}$$

(*iii*) $D = \sqrt{d^2 + 4dh} - r$, when d/r is between 10 and 15 ...(2.18)

(*iv*) When d/r is < 10

$$D = \sqrt{(d - 2r)^2 + 4d\,(h - r) + 2\pi r\,(d - 0.7\,r)} \quad \text{...(2.19)}$$

(*v*) If thick-gauge stock is used, use net or mean diameter (outside diameter less average wal thickness) for 'd' in the above formulas.

(*vi*) On flanged cylindrical shells with relatively sharp corners, use the following formula :

$$D = \sqrt{d_2^2 + 4d_1 h} \quad \text{...(2.20)}$$

where 'd_2' is the outer diameter of flange on finished shell and 'd_1' is the diameter of cylindrica section on finished shell.

(*vii*) When deep drawing is done alongwith "ironing", so that the thickness of the wall o the shell is less than that of the bottom of the shell, the blank size can be approximatel determined as :

$$D = \sqrt{d^2 + 4dh.\frac{t}{T}} \quad \text{...(2.21)}$$

where t = wall thickness, and T = bottom thickness

Number of Draws. The stresses which can be imposed on the workmaterial during the drawing process limits the amounts of reduction in the blank diameter. If the reduction required is greater more than one draw will be needed to reduce the diameter and increase the height of the workpiece to the desired dimensions. There are several methods to determine the correct number of draws and the best reduction rate per draw.

(*i*) If the reduction is defined as $(D - d)/D$, it is clear that greater the difference between the blank and shell diameters, greater will be the area that must be made to flow, and therefore the greater the forces needed to make it flow. As per practice, for the first draw, the area of blank should not be more than 3.5 to 4 times the cross-sectional area of the punch, *i.e.*, a reduction of about 45 to 50 per cent. For the subsequent draws the practical reduction in diameter are : 30, 25, 16 and 13%. The metal work gets hardened due to repeated drawings. To prevent its failure, it must be annealed. After a total reduction of 60%, the metal must be annealed.

(*ii*) The number of draws can be determined from the ratio of inside shell height and mean shell diameter as in Table 2.9.

Table 2.9. Number of Draws

Height to Diameter (Ratio)	*Number of Draws*
Up to 0.7	1
0.7 to 1.5	2
1.5 to 3	3
3.4 to 7	4

(*iii*) Average permissible reductions for the various materials are given in table 2.10. These figures are for flangeless cylindrical shells. When the shell has a flange around the open end, these figures can be increased by 5 to 10%. Also, these figures correspond to stock thickness in a range average 3.2 mm.

Drawing Pressure. The force needed to draw a shell is equal to the product of the cross-sectional area, and the yield strength in tension of the work material. For a cylindrical shell, the drawing force can be determined by the empirical formula :

$$F = \pi dt\sigma yt\left(\frac{D}{d} - C\right), N \qquad ...(2.22)$$

d and t are in mm and σyt in N/mm^2

The term D/d takes into account the relation between the blank and shell diameters and C is a constant (0.6 to 0.7) and it accounts for friction and bending effects.

Table 2.10. Average Permissible Reduction

Material	*Reduction in Diameter*		
		Succeeding Draws	
	First draw	*Single action die*	*Double action die*
Brass	50	25	30
Deep Drawing Steel	45	25	30
Soft Steel	45	25	30
Aluminium (soft temper)	45	20	25
Stainless Steel (Cr-Ni)	45	15	20
Stainless Steel (Cr)	40	20	25
Duralumin	40	10	15
High Nickel Alloys	35	20	25
Zinc	35	10	15
Magnesium	30	10	15
		% of flat diameter	% of shell diameter to be reduced

Blank Holding Pressure. The amount of pressure to be applied by the blank holder is critical. It should be just sufficient to prevent wrinkling of the material and should permit the material to slip in the direction of the drawing force. The blank holding pressure is usually less than 40% of the drawing force. For general use, it is taken as 33% of the drawing force. The tendency to wrinkle decreases with increase in the stock thickness so the thick blanks may be drawn without a blank holder. Also if the ratio of punch diameter and stock thickness is quite small, it may be possible to draw ductle material without the help of a blank holder. In such cases, the tendency to wrinkle can be reduced by increasing the edge radius around the die opening. Usually the blank is lubricated to help it slide under the pressure pad and over the edge of the die.

For deep drawing without blank holder,

$$D - d < 18t$$

When 'ironing' of the drawn workpiece is also intended, then

$$\text{Ironing force} = \pi d\,(t - t_1).\sigma_c$$

where t_1 is the die clearance and σ_c is the compressive strength of metal.

or
$$F_i = 1.2\,\pi d_m \cdot t_i \cdot \sigma_m \cdot \log t/t_i$$

where d_m = mean diameter of shell after ironing

t_i = shell thickness after ironing

and σ_m = mean stress before and after ironing

Then Press Capacity = Drawing force + Blank holding force + Ironing force

Force is also needed to overcome friction in the die walls, which may be approximately taken to be 20% of drawing load.

2.15.2. Redraw dies. Redraw dies or secondary dies reduce the diameter of a drawn shell and differ in design from the simple draw die. Shells with thick side walls can be pushed through the die but the thin walled shells will tend to wrinkle in the simple draw die. These shells will need a holding force to prevent the material from buckling.

Push through redraw die. In the single-action push through or drop through die, a nest plate is used to position the drawn shell. The solid nest must be perfectly concentric with the inside diameter of the die cavity and it must be deep enough to hold the shell without wobbling.

Double action redraw die. In this die, a pilot punch or draw sleeve whose outside diameter closely fits the inside diameter of the drawn shell, is employed to hold down the already drawn shell, Fig. 2.57. The pilot punch is mounted on the outer or blank holding side of the press and leads the draw punch into the die cavity. The draw punch is actuated by the inner slide of the press and descends through the pilot punch into the die-cavity. The pilot punch also helps in stripping the shell from the draw punch on its upward stroke.

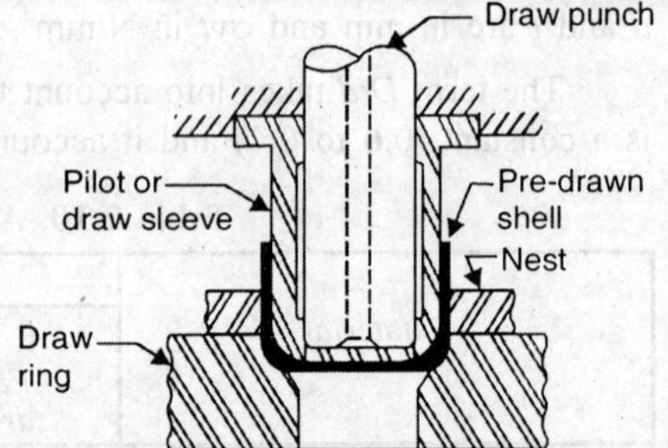

Fig. 2.57. Double Action Redraw Die.

2.15.3. Inverted draw die. Inverted draw die is used when : (1) flanged shells are to be drawn (2) shape and size of shell is such as to be impractical to drop through the die, and (3) more positive stripping pressures are required. In this die, the punch is mounted on the lower shoe and the draw die is mounted on the press slide, and moves with it. Within the die cavity, a knockout plate is used to strip the cup from the die-cavity. The working of an inverted draw die is explained in Fig. 2.58. The blank is positioned on the upper surface of the lower die plate. As the slide travels down, the face of the upper die comes in contact with the blank and depresses the pressure ring against the resistance offered by the air or rubber cushion, or springs, below. The descending die draws the blank into the cavity over the draw punch. On the upstroke, the pressure ascends as far as possible and strips the drawn shell from the punch. At or near the top of ram stroke, the knockout is actuated which strips the shell from the die surface.

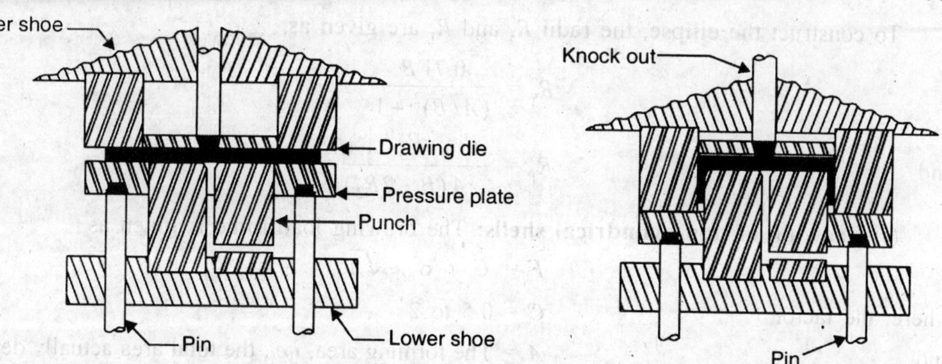

Fig. 2.58. Inverted Draw Die.

2.15.4. Blank development for non-cylindrical shells. As already noted, blank size for any shape of the shell is calculated on the basis of the constancy of volume of the material, and hence on the basis of the surface area of the material (assuming, the thickness of the material does not change during the drawing operation).

(*a*) *Square Shell*: The dimensions of the square shell (box) are $W \times W \times h$ (length × width × depth).

Approximate diameter of the circular blank for $h/W > 0.6 \sim 0.7$, will be

$$D = 1.13\sqrt{W^2 - 0.84\,r_p^{\,2} + 4\,(W - 0.43\,r_p)\,(h - 0.43\,r_f)}$$

where r_p = corner radius in plan

and r_f = corner radius in front view

Normally, $r_p = r_f = r$

so, $$D = 1.13\sqrt{W^2 - 0.84\,r^2 + 4\,(W - 0.43\,r)\,(h - 0.43\,r)}$$

(*b*) *Rectangular Shell*: Dimensions : $L \times W \times h$ (length, width, depth) and corner radius = r.

The blank will be approximately in the form of an ellipse, Fig. 2.59.

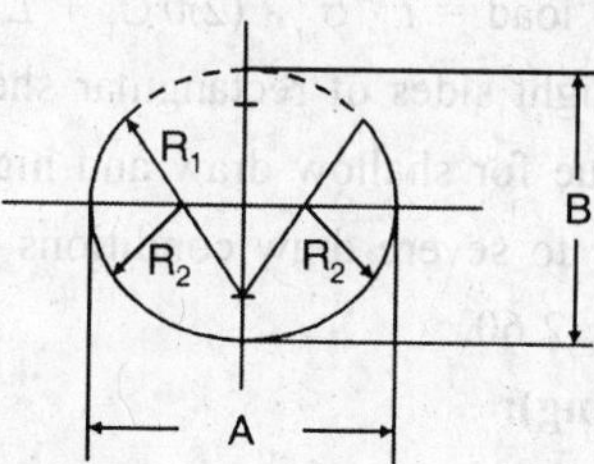

Fig. 2.59 Approximate form of blank for a rectangular shell

Minor axis of the blank, $B = \sqrt{1.27\,F - 0.5\,C^2}$

Major axis of the blank, $A = \sqrt{B^2 + C^2}$

where $C = \sqrt{(L - W)\,(L + W - 0.76\,r)}$

and F = Total area of the shell (box)

$= LW - 0.86\,r + 2\,(L + W - 0.86\,r)\,(h - 0.43\,r)$

To construct the ellipse, the radii R_2 and R_1 are given as,

$$R_2 = \frac{0.71\,B}{(A/B)^2 + 1}$$

and
$$R_1 = \frac{A^2 + B^2 - 4AR_2}{4(B - 2R_2)}$$

(*i*) **Drawing of non-cylindrical shells:** The drawing load/force is given as

$$F = C \cdot t \cdot \sigma_{ul} \cdot \sqrt{A}$$

where, the factor $C = 0.5$ to 2

and A = The forming area, *i.e.*, the total area actually deformed

Take lower value of C if the area is not loaded too much, i.e., if the elongation is not more than half the permitted elongation. Take higher values of C if the load is nearer to the tearing value.

(*ii*) **Drawing of rectangular shells:**

(*a*) **Relationship of corner radius, *r*, to height of draw, *h*:**

***r*, mm**	***h*, mm**
2.4 – 4.75	25
4.75 – 9.5	38
9.5 – 12.5	50
12.5 – 20	75

(*b*) **No. of draws:**

***h*/*r*, nominal**	**Allowable *h*/*r* Range**		**No. of draws**
	min.	**max**	
6	–	7	1
12	7	13	2
17	13	18	3
22	18	24	4

(*c*) **Drawing load:** The drawing load for rectangular shells is given as:

$$\text{Drawing load} = t \cdot \sigma_{ul} \cdot (2\pi r C_1 + L \cdot C_2), \text{ newtons,}$$

where L = Total length of straight sides of rectangular shell

C_1 = 0.5 to 2, lower value for shallow draw and higher value for $h/r > 6$

C_2 = 0.2 to 0.3, for easy to severe draw conditions

(*iii*) **Hole Flanging:** See Fig. 2.60.

90° hole flanging (for tapping):

$$b = a + \frac{5}{4} \times t, \text{ for } t < 1.2 \text{ mm}$$
$$= a + t, \text{ for } t > 1.2 \text{ mm}$$
$$h = t, \text{ for } t < 0.9 \text{ mm}$$
$$= \frac{4}{5}\, t \text{ for } t = 0.9 \text{ to } 1.25 \text{ mm}$$
$$= 0.6\, t \text{ for } t > 1.25 \text{ mm}$$
$$r = 0.25\, t \text{ for } t < 1.2 \text{ mm}$$
$$= \frac{t}{3} \text{ for } t > 1.2 \text{ mm}$$

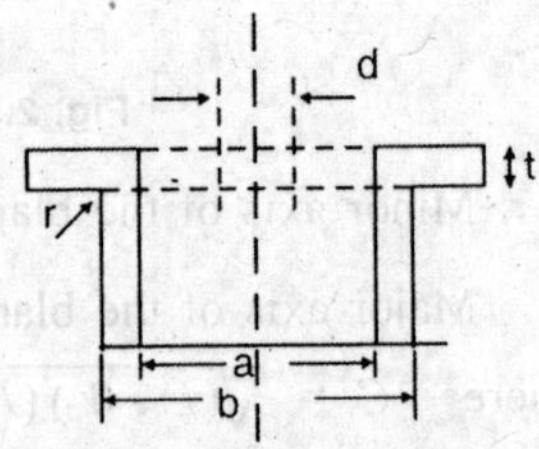

Fig. 2.60. Flanging

Pre-pierced hole diameter, $d = \sqrt{\dfrac{tb^2 + 4ta^2 + 4ha^2 - 4hb^2}{9t}}$

(*iv*) **Direct hole piercing and flanging (with stepped punch):**

$$\text{Force} = (2 \text{ to } 2.5) \times \pi d \times t \times \tau_s$$

(*v*) **Hole flanging after pre-piercing the hole:**

$$\text{Force} = (1.5 \text{ to } 2) \times \pi d \times t \times \tau_s$$

(*vi*) **Embossing/Beading:**

$$\text{Force} = t \cdot L \cdot \sigma_{ut}$$

where L = Height of embossing or bead

Bottoming force = $\sigma_y \cdot A$

where A = Plan area of bottoming zone

(*vii*) **Coining:**

$$\text{Force} = A \cdot p_C$$

where A = Total area of deformed surface

p_C = Surface pressure

It is given as,

= 1200 – 1500, N/mm², for Gold coins

= 1500 – 1800, N/mm², for Silver coins

= 1600 2000, N/mm?, for Ni coins

= 1200 – 3200, N/mm², for Steel coins

Depending upon C-content and hardness

= 1000 – 1700, N/mm², for Al and Alloys coins

(*viii*) **Flattening:**

$$\text{Force} = A \times p$$

where A = Surface area of the flattening portion

p = Surface pressure or specific pressure

It is given as:

Material	**p, N/mm²**
Al – soft	20 – 150
Al – hard	300 – 400
CRS – soft	100 – 150
CRS – Medium	250 – 400
CRS – Hard	500 – 600

2.16. BENDING DIES

Bending is the metal working process by which a straight length is transformed into a curved length. It is a very common forming process for changing sheet and plate into channels, drums, tanks etc. During the bending operation, the outer surface of the material is in tension and the inside surface is in compression. The strain in the bent material increases with decreasing radius of curvature. The stretching of the bending causes the neutral axis of the section to move towards the inner surface. In most cases, the distance of the neutral axis from the inside of the bend is 0.3 *t* to 0.5 *t*, where '*t*' is the thickness of the part. Bending terminology is illustrated in Fig. 2.61

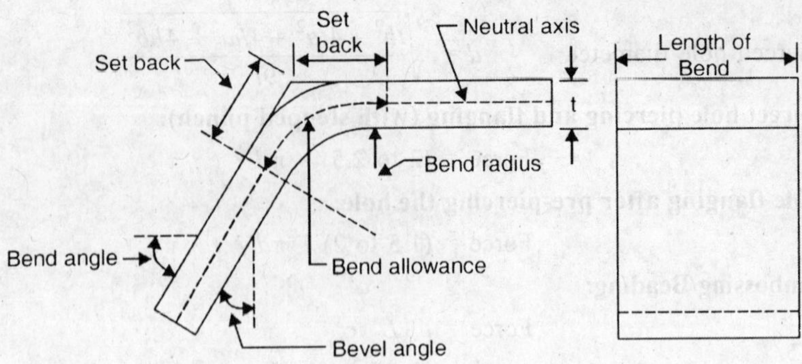

Fig. 2.61. Bending Terminology.

2.16.1. Bending methods. The two methods commonly used are : *V*-bending and Edge bending.

***V*-bending.** In *V*-bending, a wedge-shaped punch forces the metal sheet or strip into a wedge shaped die cavity, Fig. 2.62. The bend angle may be acute, 90°, or obtuse. As the punch descends,

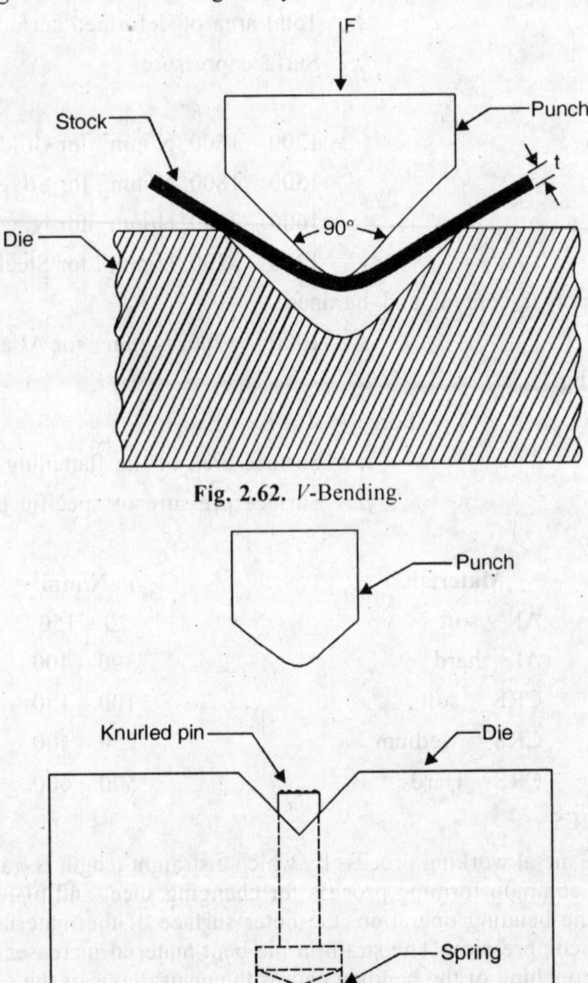

Fig. 2.62. *V*-Bending.

Fig. 2.63

the contract forces at the die corner produce a sufficiently large bending moment at the punch corner to cause the necessary deformation. To maintain the deformation to be plane-strain, the side creep of the part during its bending is prevented or reduced by incorporating a spring loaded knurled pin in the die. The friction between the pin and the part will help to achieve this, Fig. 2.63. Plane strain conditions will also be established in the centre of the sheet if its width is more than 10 times its thickness.

Edge bending. In edge bending, or cantilever bending, Fig. 2.64, a flat punch forces the stock against the vertical face of the die. The bend axis is parallel to the edge of the die and the stock is subjected to cantilever loading. To prevent the movement of the stock during bending, it is held down by a pressure pad before the punch contacts it. The die is called 'Wiping die'.

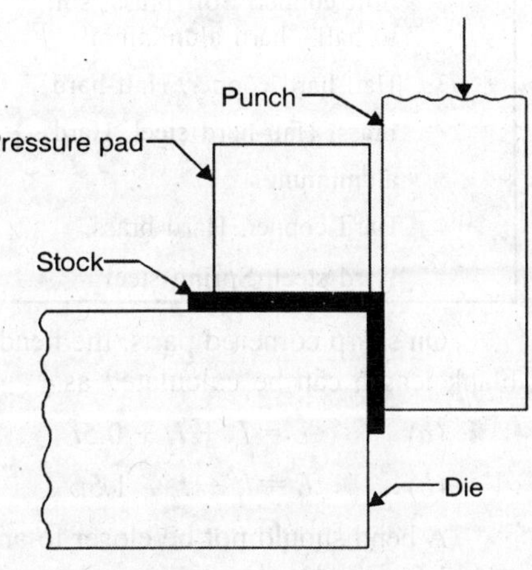

Fig. 2.64. Edge Bending.

2.16.2. Design Principles. Bend radius. It is defined as the radius of curvature on the inside or concave surface of the bend. To prevent the cracking of the material on the outer tensile surface, the bend radius cannot be made smaller than a certain value. Very ductile materials can have zero bend radius, that is, they can be folded upon themselves. However, to prevent any damage to the punch and die, the bend radius should not be less than 0.8 mm. In general, soft metals can be bent 180° with a radius equal to or less than the stock thickness. The bend radius must be larger and angle less for metals of high temper. For such materials, the bend radius is kept equal to or greater than five times the stock thickness. For magnesium the bend radius may be up to 20 times the stock thickness.

When bending to an angle of 90°, the minimum bend radius

= 2 to 5 t for C-steel

= 1 t for stainless steel

= 2 to 3.5 t for Ti alloys

= 0.3 to 0.5 t for brass

= 0.35 t for Al

Bend Allowance. To calculate the blank length for bending, the length of material in the curved section or bend area has to be calculated. This length in the bend area which will be more than the corresponding length of blank before bending, is called 'Bend Allowance'. The bend allowance added to the lengths of the straight legs of the part will give the length of blank. The bend allowance varies with the distance of the neutral axis from the inside surface of the bend. Bend allowance can be calculated from the following formula :

$$B = \frac{\alpha}{360} 2\pi (r + k)$$

where B = Bend allowance along neutral axis, cm.

α = Bend angle in degrees.

r = Inside radius of bend, cm.

k = Distance of neutral axis from the inside surface of the bend cm.

The values of 'k' are given in table 2.11.

$\therefore$ Blank length for bending = Length of the two arms upto the bend + B

Table 2.11. Value for 'k'.

Stock Material	*where $r < 2t$*	*where $r \geq 2t$*
1. Mild Steel		
(*i*) Edge Bending	$0.2t$	$0.33t$
(*ii*) *V*-bending, *U*-bending	$0.33t$	$0.50t$
2. Soft copper, Soft brass, soft to half—hard aluminium	$0.33t$	$0.50t$
3. Half-hard copper, Half-hard brass, Half-hard steel, Hard aluminium	$0.4t$	$0.5t$
4. Hard copper, Hard brass, Hard steel, Spring steel	$0.5t$	$0.5t$

On sharp cornered parts, the bend allowance may be kept as $0.5t$. Referring to Fig. 2.65, the blank length can be calculated as :

(*a*) $L = l_1 + l_2 + 0.5t$

(*b*) $L = l_1 + l_2 - 1.5t$

A bend should not be closer to an edge than $1.5t$ plus bend radius.

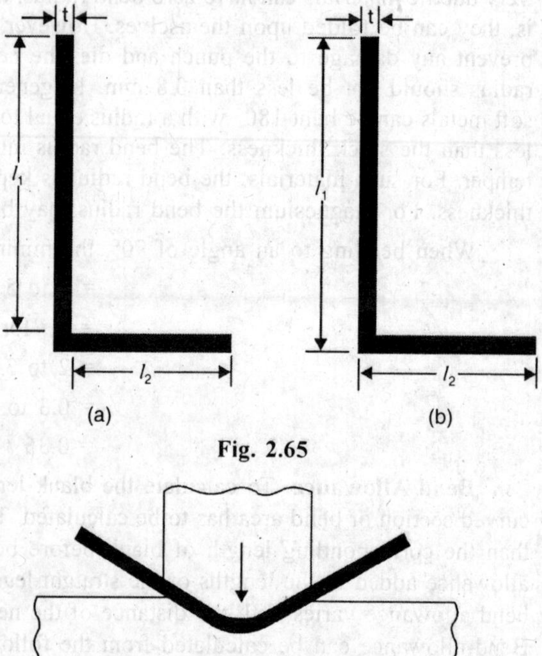

Spanking. During bending, the area of the sheet under the punch has a tendency to flow and form a bulge on the outer surface, Fig. 2.66 (*a*). This is prevented by having tool surfaces of sufficient area to restrain metal flow. Thus the nose radius of the punch is gradually blending into the punch faces, Fig. 2.66 (*b*). The lower die should be provided with mating surfaces, so that when the punch and die are completely closed on the blank, any bulging developed earlier will be completely pressed or "spanked" out.

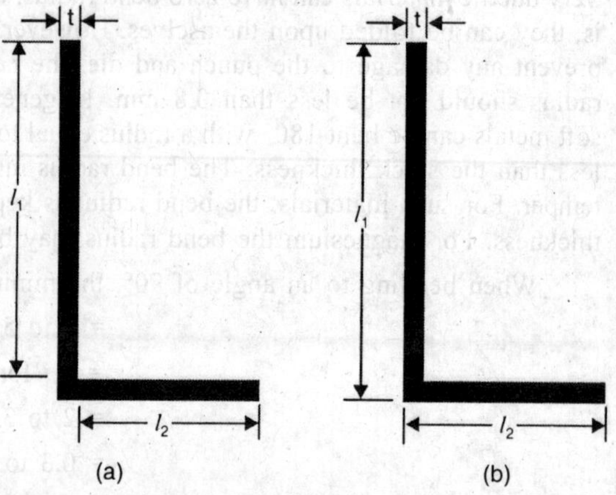

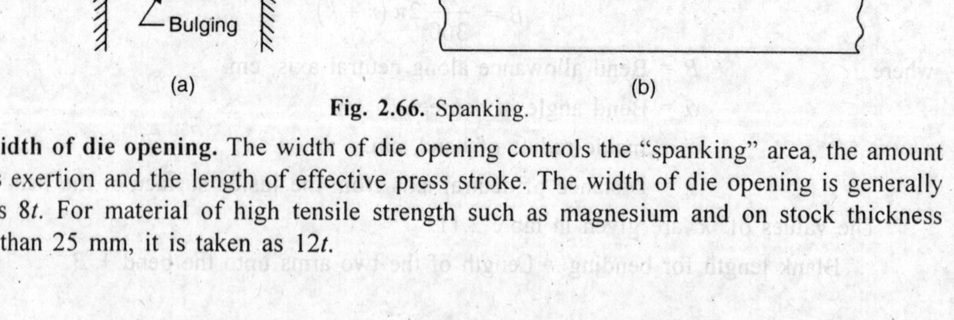

Fig. 2.65

Small nosed punch

Bulging

(a) (b)

Fig. 2.66. Spanking.

Width of die opening. The width of die opening controls the "spanking" area, the amount of press exertion and the length of effective press stroke. The width of die opening is generally taken as $8t$. For material of high tensile strength such as magnesium and on stock thickness greater than 25 mm, it is taken as $12t$.

Spring back. At the end of the bending operation, when the pressure on the metal is released, there is an elastic recovery by the material. This causes a decrease in the bend angle and this phenomenon is termed as spring back. For low carbon steel, it can be 1° to 2° and for medium carbon steel, it can be 3° to 4°, for phosphor bronze and spring steel the spring back can be from 10 to 15°. To compensate for spring back, the wedge-shaped punches and the mating dies are made with included angles somewhat less than required in the formed component. Due to this, the component will be bent to a greater angle than desired, but it will spring back to the desired angle. For other types of bending, the part is overbent by an angle equal to spring-back angle by having the face of the punch undercut or relieved. The values of spring back given above are for 90° bends and are usually greater for greater angles.

Bending pressure. Bending pressure depends upon the thickness of the stock, the length of the bend, the width of die opening and the type of bend. For '*V*' bending, the formula is,

$$F = \frac{K . l . \sigma_{ut} . t^2}{w}, N \qquad ...(2.23)$$

where l = length of bend (Fig. 2.61), mm, i.e., width of the stock.

σ_{ut} = ultimate tensile strength of material, MPa or N/mm²

w = width of die-opening, mm

t = blank thickness, mm

K = die-opening factor varies from 1.20 for a die opening of $\geq 16t$, to 1.33 for a die opening of $< 16t$.

or K is given as,

$$K = 1 + \frac{4t}{w}$$

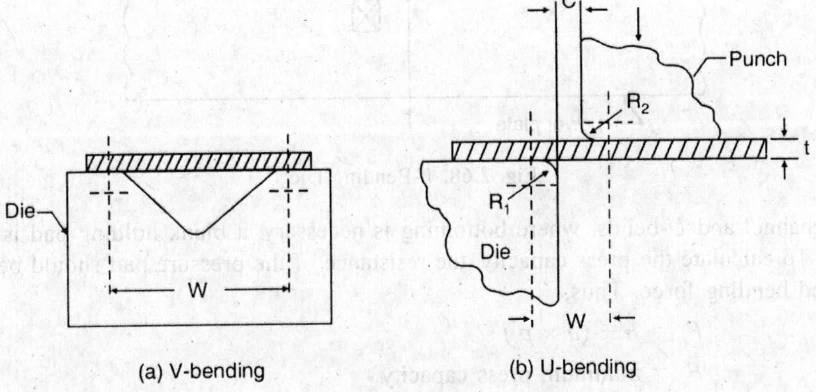

Fig. 2.67. Determining *w*.

For *U* or channel bending, the pressure will be twice as for *V*-bending and for edge bending it will be about one-half of that for *V*-bending, or, the values of *K* can be taken as

Condition	V Bending	U Bending	Edge Bending
$w < 16t$	1.33	2.67	0.67
$w \geq 16t$	1.20	2.40	0.60

Or we can use equation (2.23) with the values of K as given below :

$$K = 0.67 \text{ for } U\text{-bending or channel bending}$$
$$= 0.333 \text{ for edge bending or single wiping bending}$$

Here w will be the width between contact points on die and punch (span) and is determined as shown in Fig. 2.67 (*b*), and is given as :

$$w = R_1 + c + R_2, \text{ where}$$

R_1 = Punch edge radius

R_2 = Die edge radius

c = Clearance = t

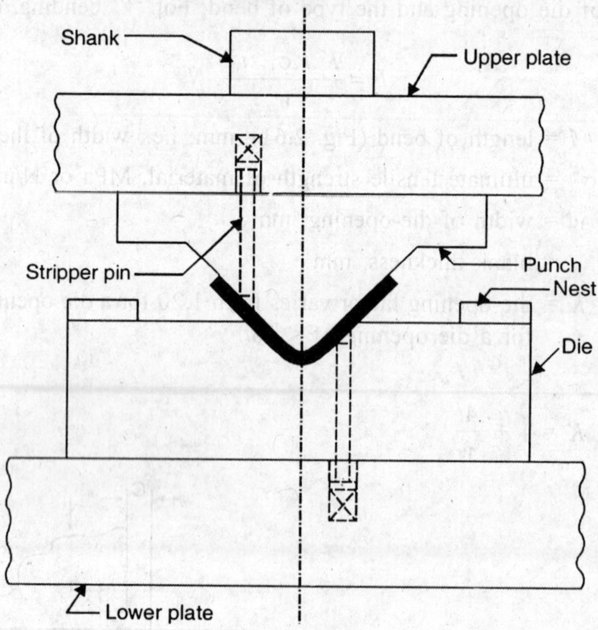

Fig. 2.68. *V*-Bending Die.

For channel and *U*-bends, where bottoming is necessary, a blank holding pad is generally employed. To calculate the press capacity, the resistance of the pressure pad should be added to the required bending force. Thus,

$$F_C = F + (a \times p_b) \qquad ...(2.24)$$

where F_C = minimum press capacity

a = Area of pad face

p_b = Average blank holding pressure, ranging from 1.05 N/mm² for the less ductile and thick materials to 2.8 N/mm² for more ductile and thin materials.

Or, the pad force may be takens as,

$$\text{Pad force} = 0.67\ \sigma_{ut}.l.t$$

A *V*-bending die and a *U*-bending die are shown in Fig. 2.68 and Fig. 2.69 respectively. Stripper pins are provided to shed the part from the punch.

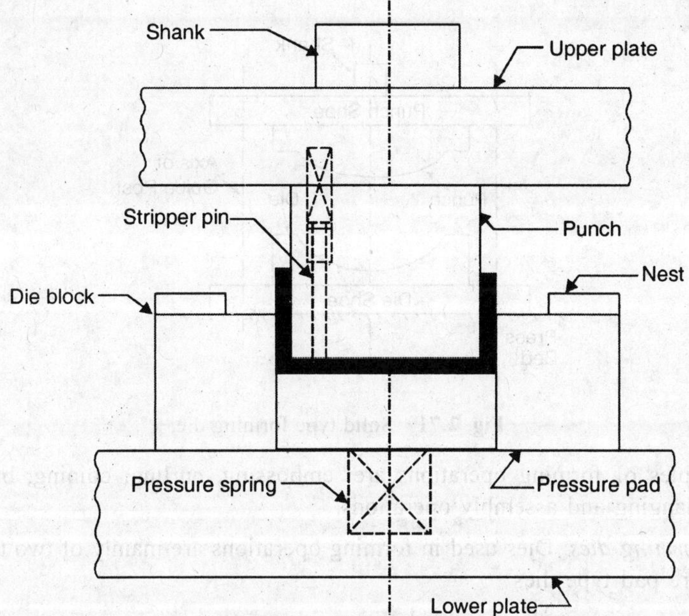

Fig. 2.69. *U*-Bending Die.

Bottoming force. To overcome springback, the bend area is often placed under high compressive stress to set the metal. This is called "Bottoming". It is achieved by having a projection or bead on the punch, Fig. 2.70. Bottoming force is calculated as,

$$\text{Bottoming force} = \sigma_c . l . b$$

where b = plan project width of bend or beads,

σ_c = compressive or setting pressure.

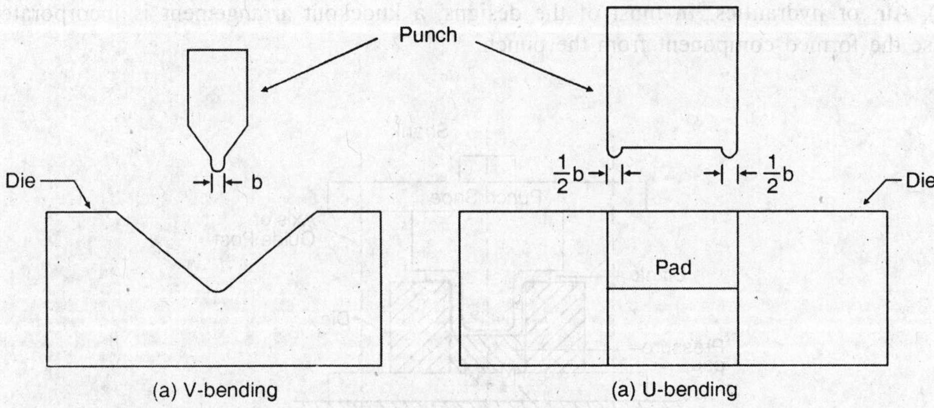

Fig. 2.70. Punch Beads.

2.16.3. Forming dies. Forming dies are much like bending dies, with the difference that, whereas, in bending dies, the blank is formed or bent along a straight axis, in forming dies, it is formed or bent along a curved axis. There is not enough metal flow in forming operation. So, no excessive thinning or tearing of the metal takes place.

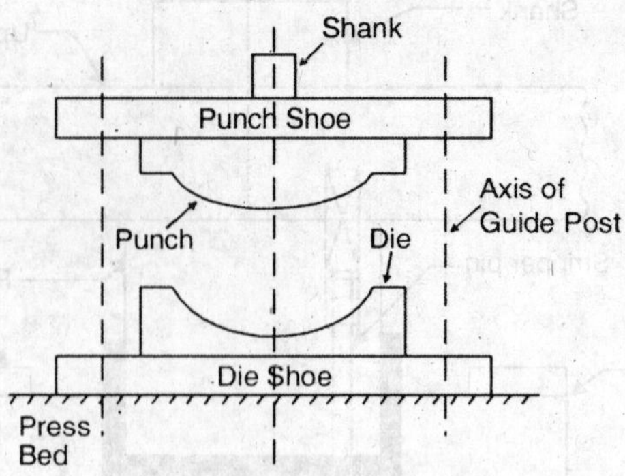

Fig. 2.71 Solid type forming die

The examples of forming operations are: embossing, curling, coining, bulging, beading, twisting, hole flanging and assembly operations.

Types of forming dies: Dies used in forming operations are mainly of two types: Solid type dies and pressure pad type dies.

Forming dies are normally provided with guide posts.

Solid type forming dies: These dies are usually of simple construction and design. A solid type forming die consists of male punch and female die shaped to the contour of the workpiece, Fig. 2.71. The shape of the punch and die is reproduced on the blank. Allowance for clearance equal to the thickness of the workpiece is provided between the punch and the die. These dies are used for forming blanks of soft grade metals, mostly strip stock.

Pressure pad type dies: A pressure pad may be necessary when the workpiece contains sharp radii or intricate details or when the workpiece may shift during the forming operations. Pressure pads are applied above (see Figs. 2.54 and 2.55) or below the workpiece (Fig. 2.72) depending upon the design of the die and type of bend. Pressure for the pad may be applied by springs (Fig. 2.72), Air, or hydraulics. In most of the designs, a knockout arrangement is incorporated to release the formed component from the punch.

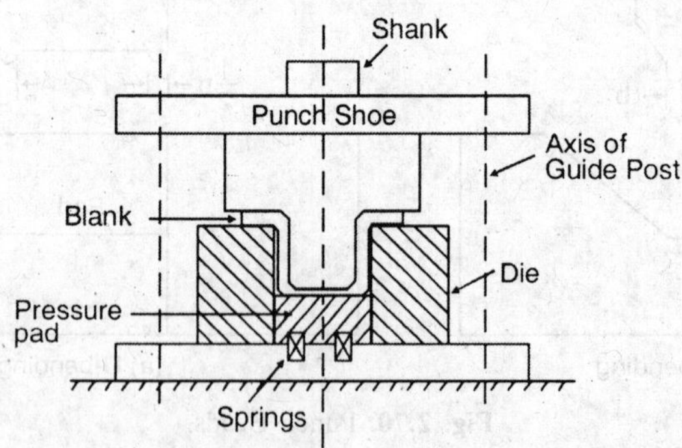

Fig. 2.72 Pressure pad type forming die

2.17. DESIGN PROCEDURE FOR PROGRESSIVE DIES

A progressive die, also known as cut-and-carry die, is a multi station die that performs several operations in succession in a single stroke of the press ram. The various operations may be all cutting operations or a combination of cutting and forming operations. The first step in the design of the die is the development of the blank. For this, the direction of metal grain is kept in mind. Normally, the grain of coiled strip is parallel to its length. To eliminate any tendency to fracture, the forming operating are usually performed perpendicular to the grain. Thus, while making a scale layout of the developed blank, the bend line is kept perpendicular to the strip. The next step is to plot the probable design of the scrap strip. In a progressive die, the individual operations (which are normally relatively simple) are combined in several stations. Due to this, it is usually difficult to evolve the most practical and economical strip design for optimum operations of the die. However, when establishing the sequence of operations for progressive dies, the following guide lines should be considered :

1. While evolving the strip design, the positioning and proper location of the stock in each station is of prime importance. For this, pilots are incorporated in the die. Advantage should be taken of any required holes in the workpiece for piloting. This method is known as "Direct piloting". Thus piering operations must be placed first in the sequence. However, sometimes (when direct piloting is not ideal or satisfactory) special piloting holes can be placed in the scrap part of the strip. This is called "Indirect piloting". Advantages of indirect piloting are : greater freedom of size or location and not affected by workpiece change. However, the material width and lead may increase. Care must be taken to avoid having pierced holes tool close to a bend.

2. Distribute pierced areas over several stations if they are close together or are close to the edge of die opening.

3. Check, if the blanked areas can be divided into simple shapes, so that commercially available punches of simple contours may be used. A blanked area will be partially cut at one station and the remaining area will be cut in later stations.

4. **Use idle stations.** They help the designer to distribute the total load uniformly over the complete length of the die. They also strengthen die blocks, stripper plates and punch retainer etc. and facilitate strip movement. Also, they permit later changes in workpiece design by providing for additional operations.

5. Bending and forming operations must be at the last stations.

6. Cutting and forming areas should be located to provide uniform loading of the press slide.

7. While designing strip layout, try for minimum scrap. Use a multiple layout if possible.

8. Design the strip so that scrap and workpiece can be ejected without interference.

The next step is to design the die. The design of the various die elements (die block, punches, strippers etc.) is done on the same lines as has been discussed for blanking die and piercing die. A progressive die is usually a heavy construction.

Note. Some hints on Press work :

1. The counterbores in the die-block, the tapped holes in the die shoe and the reamed dowel holes in the die shoe must all be made from 6.35 mm to 9.5 mm deeper than needed to allow for grinding of the die block.

2. If the punch diameter is twice the material thickness, it need not be guided.

3. If the punch produces a hole 40% deeper than its own diameter, the shank diameter must be at least twice the punch size.

4. Punches should be a tap fit in punch plates. For purposes of stability, the flange of the punch should never be smaller in diameter than the length of the punch.

5. Punches should be secured with pins, flats, keys etc. against rotating in the punch holders.

2.18. Table 2.12. Dimensions of Back-post Die Sets (cm)

Die Area		Thickness				Min. guide-post diam.
Rectangular	Diam (if circular)	Die Holder		Punch Holder		
		From	To	From	To	
7.5 × 7.5	7.5	2.5	3.2	2.5		1.9
10 × 10	10	3.5	4.45	3.2		2.5
10 × 15	–	3.8	7.0	3.2	5.7	2.5
12.5 × 10	–	3.5	4.45	3.2	–	2.5
12.5 × 12.5	12.5	3.8	5.0	3.2	4.45	2.5
12.5 × 20	–	3.8	7.5	3.2	5.7	2.5
15 × 7.5	–	3.8	5.0	3.2	4.45	2.5
15 × 10	12.5	3.8	7.0	3.2	5.7	2.5
15 × 15	16.25	3.8	6.25	3.2	5.7	2.5
15 × 22.5	–	3.8	8.25	3.2	5.7	2.5
17.5 × 12.5	14.4	3.8	7.6	3.2	5.7	2.5
17.5 × 17.5	18.75	3.8	6.25	3.2	5.7	2.5
17.5 × 25	–	4.1	8.2	3.5	5.7	3.2
20 × 10	–	3.8	6.25	3.2	5.7	2.5
20 × 15	17.5	3.8	7.5	3.2	5.7	2.5
20 × 20	21.25	3.8	6.25	3.2	5.7	2.5
22.5 × 30	–	4.45	8.90	3.8	5.7	3.8
25 × 12.5	–	3.8	6.25	3.2	4.45	3.2
25 × 17.5	–	3.5	7.0	3.5	5.7	3.2
25 × 25	25	4.1	8.2	3.5	5.7	3.2
25 × 35	–	4.75	9.5	4.1	8.2	3.8
27.5 × 22.5	25	4.45	8.9	3.8	5.7	3.2
30 × 10	–	4.45	5.7	3.8	5.0	3.2
30 × 15	–	3.8	6.25	3.8	5.0	3.2
30 × 30	31.25	4.45	8.9	4.45	5.7	3.8
30 × 40	–	5.0	9.5	4.45	8.2	3.8
35 × 20	–	4.45	8.25	4.1	8.2	3.8
35 × 25	28.75	4.45	8.25	4.1	8.2	3.8
35 × 35	35	4.45	8.25	4.1	5.7	3.8
37.5 × 12.5	–	3.8	6.25	3.8	5.0	3.8
45 × 20	–	3.8	6.25	3.8	5.0	3.8
45 × 25	–	3.8	6.25	3.8	5.0	3.8
45 × 35	37.5	5.0	7.5	4.45	5.7	3.8
45 × 40	42.5	5.0	7.5	4.45	5.7	3.8
50 × 12.5	–	4.45	6.25	3.8	5.0	3.8
55 × 15	–	4.45	6.25	3.8	5.7	3.8
55 × 30	–	5.0	7.5	3.8	5.0	3.8
62.5 × 17.5	–	4.45	7.5	3.8	5.7	3.8
62.5 × 35	–	4.45	7.5	3.8	5.7	3.8

The shut-height is established by the length of the guide-posts, which must be at least 12.5 mm shorter than the shut-height in order to allow for the reduced short-height due to resharpening.

2.19. MATERIALS AND MANUFACTURE OF SHEET METAL WORKING DIES

2.19.1. Materials. Tool materials are chosen mostly on the basis of the expected life of the production run. The efficiency, strength and dimensional stability of the working parts of the tool depend on the material for each component and its heat treatment. The tool steel used for punches and dies should possess the following properties :

High strength, High hardness, Adequate wear resistance, High toughness, Good hardenability, show little sensitivity to over heating, Good machineability and Good grindability.

Blanking tools are subjected to severe wear and are made from the various cold-working die steels. Bending and Drawing dies are made of similar materials, although C.I. and even hard zinc alloys or plastics are suitable for short production suns or softer workpiece materials. Below, we discuss the various materials used for making tools for press working :

1. *Blanking* :

(*a*) *For Al, Mg and Cu alloys*

(*i*) Zn alloy

(*ii*) Water hardening steel : 0.6-1.4C, Hardness = 62-66HRC

(*iii*) Oil hardening steel : 0.9 C, 1 Mn, 0.5Cr, Hardness = 57-62 HRC

(*iv*) Air hardening steel : 1C, 5Cr, 1Mo, Hardness = 57-62 HRC

(*v*) Cold working die steel : 1.5C, 12Cr, 1Mo, Hardness = 58-64 HRC

(*b*) *For Steels and Ni alloys*

(*i*) Mo H.S.S. : 0.85C, 4Cr, 5Mo, 6.25 W, 2V, Hardness = 60-66

(*ii*) WC

2. **Deep Drawing :**

(*a*) For Al, Mg, and Cu alloys

(*i*) Water hardening steel, Hardness = 60-62 HRC

(*ii*) Oil hardening steel, Hardness = 57-62 HRC

(*iii*) Air hardening steel, Hardness = 57.62 HRC

(*iv*) Cold working die steel, Hardness = 58-64 HRC

(*b*) *For Steels and Ni alloys*

(*i*) Water hardening steel, Hardness = 60-62 HRC

(*ii*) Mo H.S.S., Hardness = 60-65 HRC

(*iii*) WC

3. **Press Forming :**

(*a*) For Al, Mg and Cu alloys

(*i*) Epoxy/metal powder (*ii*) Zn alloy

(*iii*) Mild steel (*iv*) C.I.

(*v*) Oil hardening steel (*vi*) Air-hardening steel (*vii*) Cold working die steel

(*b*) For Steels and Ni alloys : Epoxy/metal Powder

Note. Die materials mentioned first are for lighter duties, shorter runs. Other basic die materials are :

1. *Al bronzes.* These have often been used for draw rings and other forming tools, especially those intended for working stainless steel. They possess : fair machineability, high dimensional

stability, high antigalling properties, high compression strength, good retention of high polish and no need for heat treating. Used for medium sized draw dies and irregular contours.

2. *Semi-steel.* It is a term often applied to C.I. alloyed with about 10% Si, 1% Ni, 2% Cr, 3% Mn. It is most commonly used for strippers, blank holders and secondary structural members of the die.

Flexible, Short Run and Alternative Materials

1. *Cerromatrix.* It is a special low melting point alloy of Bismith, Lead, tin and antimony with hard setting and expandable properties, designed specifically for die applications. It is widely used for the repair of dies and for the seating and holding of punches and die sections, especially long, slender members in intricate dies where precision machining would normally be time consuming and costly. The material is also used as the form block or die block member is short run forming.

2. *Kirksite.* A cast Zn base alloy with Al, Cu, and Mg is extensively used in the aircraft industry and to some extent in the automotive industry for large forming and drawing and stretch forming. It is most applicable on experimental and short production run. Kirksite dies are quickly made, being cast closely to shape and size required, are used without heat treating and are easily finished.

3. *Plastics.* Cast phenolics, cast or laminated epoxy and glass inforced polyester. They are proving of exceptional benefit to the aircraft, automotive, appliances and other industries for certain types of work. They are used principally for large forming and drawing dies, where their fast fabricating time, low cost and light weight are definite advantages. They are easily repaired and modified. Plastics are not necessarily limited to short run or experimental dies.

4. *Rubber.* Mostly used for forming dies.

2.19.2. Sintered carbide dies. A longer life of tools for different press working operations can be achieved by using inserts made from tungsten and cobalt carbides.

Carbide tools have found widest application in large lot and mass production, where conventional steel dies and punches do not prove stable enough, which calls for frequent duplication of dies. The life of carbide blanking dies rises by 8 or more times. This effect grows manifold when operating on high strength sheets of stainless, electrical and similar steels; carbide dies in such cases are 50 times as durable as conventional dies.

Carbide dies differ somewhat in design from steel dies, which is due to the physical and mechanical properties of carbides. The design and manufacture of such dies must ensure :

1. High stiffness of the die structure.
2. High wear resistance of guide posts, bushings, stock guides, stops and other die components.
3. Secure fastening of carbide inserts and their accurate fitting to the bearing surfaces.
4. Minimum penetration of the punch into the dies, ensured by application of limiting rests.
5. Elemination of the effects of in-accurate travel of press slide by using float type shanks and symmetrical arrangement of guide posts with respect to the contour of the cut.
6. Increased punch-to-die clearance.

2.19.3. Die Manufacture. The dies for the sheet metal working are manufactured by the same processes as used for the manufacture of forging dies, that is, mechanical machining methods (copying milling, NC milling), EDM, ECM and casting. Dies can be manufactured by these methods individually or these processes can be combined to manufacture a die at lower cost.

(*a*) **Manufacture of blanking and piercing dies.** Among the mechanical machining processes, in addition to the more common methods of milling, turning and grinding, processes such as sawing, shaping and planing are also employed. With EDM besides die sinking, cutting may also be performed using a NC wire electrode. The size of the work piece, the desired contour, the die material and the required accuracy all influence the choice of the machining process. Cutting edges are usually machined conventionally to required tolerances in the case of simple contours, such as those with cylindrical openings. More complicated contours are usually machined by conventional EDM or WEDM. This results in significant cost and time savings. For example, with dies made of WC, savings of 30-50% may be achieved. EDM die sinking and cutting, using a wire electrode, has lead to improvements in quality, (precision dies) because of elimination of distortion and of improved strength. This is because hardened tools can be machined by EDM. CAD/CAM techniques minimize the production costs involved in the design, manufacture, maintenance and modification of dies, especially for the design of progressive dies and dies of complex configuration. Additional heat treatment of the die surface, such as nitriding, surface coating of tools with a layer of WC, or ion implantation, leads to significant reduction in the wear of cutting edge.

(*b*) **Manufacture of dies for sheet-metal working (deep drawing etc.).** The single or double convex or concave working surfaces of sheet m.w, dies are largely machined by copy milling or EDM. In the case of medium and large-sized dies, the die block is generally cast with a 10-25 mm machining allowance. The choice of one or two process combinations is based primarily on the complexity of master models and electrodes and on the amount of hand finishing required in the form of spot grinding, lapping, polishing, and so on (about 60-80% of total machining time).

In conventional manufacture, using copy milling, individual dies are first rough and finish milled following a master pattern. After initial grinding to smooth down the milling edges, the surface is spot ground using a blueing model, smoothed and then polished. With complex die shapes, the finished male die may be used as a blueing aid in finish grinding the lower die. The desired gap width required for the sheet metal is checked with the aid of lead or Al wire.

EDM is best suited for finishing premachined (copy milled) dies to a high standard of accuracy and surface finish. The machining allowance may vary between 0.2 and 0.3 mm. The overall machining time may be reduced by 20-50%, and, depending on the die geometry, the time required for manual finishing by 40-90%.

The die blocks made by casting are first contour milled and then the remaining working surfaces are rough machined using copy milling. About 0.2-3 mm is left on the die surface and machined by EDM to tolerances of 0.1-0.3 mm and a surface finish of 1-3 mm Ra. The final manual finishing consists of grinding the projections left opposite the electrolyte supply channels as well as the edges formed between neighbouring elements in the case of multiple electrodes. This is best done by using a blueing model. In addition, minor shape corrections may be made where the electrodes are worn out and the drawing radii may be smoothed out and polished.

Dies of large working surface areas and gradually changing curvatures, such as dies for automobile body parts (doors, roof pannels etc.) are mainly machined by copy or NC milling and require little subsequent hand finishing. With large body panels, perfect stream lining is essential. This requires special precision machining. Heavily ribbed and ridged deep drawing dies may be machined advantageously using EDM.

Note. In general, working surfaces > 1 m^2 are considered large dies, while small dies have working areas $10^3 - 10^4$ mm^2.

2.19.4 Machining of punches. Normally, the following steps are involved :

1. Cutting of a blank
2. Rough machining
3. Rough grinding
4. Hardening
5. Finish grinding the profile
6. Finishing the profile
7. Sharpening the cutting face.

Straight punches for blanking round contours or piercing round holes are easy to manufacture. They are :

1. Turned on Lathe.
2. Hardened.
3. Ground on circular or universal grinders.
4. Then their working surface is finished by polishing and the cutting face is sharpened (ground).

The shaped punches are far more difficult to make. The following steps are involved in their manufacture :

1. The contour of a punch is first machined on metal cutting machines keeping an allowance for further machining.
2. Then an impression of the contour is made by the hardened and finished die.
3. Next, the punch is filed off according to the contour of the impression with due account for the required die-punch clearance.
4. Lastly, the punch is hardened and its working part is finished.

2.19.5. Increasing press tool service life. The life of a press tool may mean any of the following : The number of stampings made,

1. between sharpening and repair
2. between two sharpenings
3. between two repairs or
4. the total number of stampings made before the tool is completely worn out.

The life of press tool can be increased by :

1. Case hardening (nitriding).
2. Chrome plating. It is used for increasing the wear resistance of new parts and reconditioning worn parts made from carbon steels. The thickness of chromium layer can be from 3 μm upwards. It may be 50-60 μm in case of reconditioning. Chrome plated ports are subjected to heat treatment in an oil bath (2*h* at 170-180°C).
3. Cemented carbide inserts increase the life of press tools dozens of times (as discussed above under Art. 2.19.2.)
4. Hard facing of tools aims at increasing the wear resistance of new, mostly large size, press tools and at reconditioning worn tools. Hard facing is done by means of oxy-acetyline flame. It is done with special electrodes or hard alloys. Electrodes 6 to 7 mm in diameter and 400 to 500 mm long are used. The material of the electrodes (cast alloy) is satellite for blanking and cut off dies and chromite (based on chromium and iron) for bending and draw dies.

The hard faced press tool is placed in warm sand to allow it to cool slowly. The hard faced parts are annealed at a temperature of 800-900°C for 2-4 hours and subsequently tempered in ofl at a temperature of 450-600°C for one hour.

2.19.6. I.S. Code. I.S. Code for Tool and Die steels for cold work is : IS : 3749-1966. It specifies requirements for plain carbon and alloy tool and die steels in the form of bars, blanks, rings and other shapes for cold work, capable of being hardened and tempered. Chemical composition and hardness are among the requirements. See Appendix-I (C)

The I.S. code for High Speed tool steels is : I.S : 7291-1974.

2.20. SOLVED EXAMPLES

Example. 2.1. *Find the total pressure, dimensions of tools to produce a washer 5 cm. outside diameter with a 2.4 cm diameter hole, from material 4 mm thick, having a shear strength of 360 N/mm².*

Solution. The production of washer consists of two operations : Blanking the outside diameter and piercing the inner hole.

Blanking Pressure, $F_1 = \pi D t \tau_s$

$= \pi \times 50 \times 4 \times 360$

$= 226$ kN

Piercing Pressure, $F_2 = \pi d t \tau_s$

$= \pi \times 24 \times 4 \times 360$

$= 108.768$ kN

Total Pressure required, $F = F_1 + F_2$

$= 226.6 + 108.768$

$= 335.368$ kN

$= 335.4$ kN

Clearance is taken as 10% of stock thickness. Therefore,

Piercing punch diameter = 2.4 cm

Piercing die diameter = 2.4 + 0.04 = 2.44 cm

Blanking die diameter = 5 cm

Blanking punch diameter = 5 – 0.04 = 4.96 cm

The above calculations give the press capacity for tools without shear. When suitable shear is applied, the loads calculated can be reduced by 20%.

Press capacity = 0.8 × 335.4 = **268.32 kN**

The blanking punch will be flat and shear applied to the blanking die and for piercing, the punch will have shear and the die will be flat.

Example. 2.2. *A cup without flanges and of height 10 cm. and diameter 5 cm is to be made from sheet metal 2.5 mm thick. Find the suitable number of draws.*

Solution. First of all, blank diameter should be obtained. It is given as :

$$D = \sqrt{d^2 + 4dh}$$

Here, $d = 5$ cm and $h = 10$ cm

$\therefore$ $D = 15$ cm

The height diameter ratio of cup is 2. Therefore, from table 2.9, the tentative number of draws is 3. Let the reduction be : 45%, 25% and 20% respectively.

1st reduction = 45% = 15 × 0.45 = 6.75

$\therefore$ Diameter d_1 at first draw = 15 – 6.75 = 8.25 cm

2nd reduction = 25% = 8.25 × 0.25 = 2.06 cm

Diameter d_2 at second draw = 8.25 – 2.06 = 6.19 cm

3rd reduction = 20% = 6.19 × 0.2 = 1.238 cm.

Diameter d at final draw = 6.19 – 1.238 = 4.952 cm

To get exactly the value of 5 cm, the third reduction can be slightly lower than 20% (19.25 per cent exactly).

Example 2.3. *A 37.5 cm long, 19 mm wide and 2.5 mm thick strip is to be bent in a V-shaped die. Calculate the bending force necessary if the steel has 630 N/mm² tensile strength.*

Solution. Let the width of die opening be 16*t*, *i.e.*,

$w = 16 \times 2.5 = 40$ mm

Now Bending force is given as,

$$F = \frac{K \cdot l\sigma_{ut} t^2}{w}$$

Now $K_t = 1.20$, $l = 37.5$ cm, $t = 2.5$ mm

and $\sigma_{ut} = 630$ N/mm²

$$F = \frac{1.2 \times 37.5 \times 10 \times 630 \times (2.5)^2}{40}$$

$$= \mathbf{44.3\ kN}$$

Example 2.4. *Estimate the blanking force to cut a blank 25 mm wide and 30 mm long from a 1.5 mm thick metal strip, if the ultimate shear stress of the material is 450 N/mm². Also determine the work done if the percentage penetration is 25 percent of material thickness.*

Solution. (*a*) Perimeter of blank = 2 (25 + 30)

= 110 mm

∴ Blanking force = Area under shear × ultimate shear stress

= Perimeter × t × τ_s

= 110 × 1.5 × 450

= **74.25 kN**

(*b*) Punch travel = Percentage penetration

$$= \frac{1}{4} \times t = \frac{1.5}{4} = 0.375 \text{ mm}$$

Work done = Force × punch travel

$$= 74.25 \times 10^3 \times 0.375 \times \frac{1}{10^3}$$

= **27.84 Nm**

Example 2.5. *A washer with a 12.7 mm internal hole and an outside diameter of 25.4 mm is to be made from 1.5 mm thick strip of 0.2 per cent carbon steel. Considering the elastic recovery of the material, find : (a) the clearance (b) blanking die-opening size (c) the blanking punch size (d) the piercing punch size (e) the piercing die-opening size.*

Solution. (*a*) The clearance (one side) for soft steel

= 5% of t

∴ $c = 0.05 \times 1.5$

= **0.075 mm.**

(*b*) It is clear from Fig. 2.20 (*b*), the blanking die opening size is equal to the blank size. But to allow for the expansion of the blank, the die opening should be made smaller. Thus,

Blanking die opening = 25.40 – 0.05

= **25.35 mm**

(*c*) Blanking punch size = Blanking die opening – 2*C*

= 25.35 – 0.15

= **25.20 mm.**

(*d*) The punch size determines the size of the hole produced. To allow for the contraction of the hole due to elastic recovery, the punch size should be made larger by an amount 0.05 mm. (See article 2.8.1). So, the piercing punch size will be,

Piercing punch size = 12.70 + 0.05

= **12.75 mm**

(*e*) Piercing die size = Piercing punch size + 2*C*, [Fig. 2.22 (*a*)]

= 12.75 + 0.15

= **12.90 mm.**

Example 2.6. *The ultimate shearing strength of the material of the washer in example 2.5 is 280 N/mm². (a) Find the total cutting force if both punches act at the same time and no shear is applied to either punch or the die. (b) What will be the cutting force if the punches are staggered, so that only one punch acts at a time. (c) Taking 60% penetration and shear on punch of 1 mm, what will be the cutting force if both punches act together.*

Solution. (*a*) $F = \pi (D + d).t\tau_s$

$D = 25.4$ mm, $d = 12.7$mm, $t = 1.5$ mm

$\tau_s = 280$ N/mm²

$\therefore$ $F = \pi (25.4 + 12.7) \times 1.5 \times 280$

$\therefore$ = **50.27 kN**

(*b*) When the punches are staggered, the punch taking the largest cut will require the greatest force.

$\therefore$ $F = \pi D t \tau_s$

$= \pi \times 25.4 \times 1.5 \times 280$

= 33.5 kN

(*c*) $F = \dfrac{t \times K \times F_{max}}{(K \times t + I)}$, (Eqn. 2.6)

K = percentage of penetration = 0.60

I = shear on punch = 1 mm

F_{max} = 50.27 kN

$\therefore$ $F = \dfrac{1.5 \times 0.6 \times 50.27}{(0.9 + 1)}$

= **23.8 kN**

Example 2.7. *A hole of 60 mm diameter is to be produced in steel plate 2.5 mm thick. The ultimate shear strength of the plate material is 450 N/mm². If the punching force is to be reduced to half of the force using a punch without shear, estimate the amount of shear on the punch. Take percentage penetration as 40%.*

Solution. The punching force with non-sheared punch,

$F_{max} = \pi D t \tau_s$

$= \pi \times 60 \times 2.5 \times 450$

= 212 kN

Work done = F_{max} × penetration (punch travel)

$= F_{max} \times 0.4 \times t$

$= 212 \times 10^3 \times 0.4 \times 2.5 \times \dfrac{1}{10^3}$

= 212 Nm

Now work done remains the same with a sheared and a non-sheared punch.

For a sheared punch,

$$\text{Punch travel} = \text{penetration} + \text{shear}$$

$$= K.t + I$$

If F is the blanking force, then comparing the work done,

$$F_{max} \times K \times t = F\,(K.t + I)$$

$$\therefore \quad I = \frac{K \times t\left(F_{max} - F\right)}{F}$$

Now $$F = \frac{1}{2}\,F_{max} = 106 \text{ kN}$$

$$\therefore \quad I = \frac{0.4 \times 2.5 \times 106}{106}$$

$$= 1 \text{ mm}$$

∴ Amount of shear on punch = **1 mm.**

Example 2.8. *In Fig. 2.27, the strip thickness is 3.2 mm and the dimension 'h' of the blank is 10 cm and w is 2.5 cm. Find*

(*i*) *the value for back scrap.*

(*ii*) *the value for scrap bridge.*

(*iii*) *the width of strip.*

(*iv*) *the length of one piece of stock needed to produce one part.*

(*v*) *the number of parts which can be produced from a 2.4 m long strip.*

(*vi*) *the scrap remaining at the end of the strip.*

Solution. (*i*) Back scrap and front scrap are given as,

$$a = t + 0.015\,h$$

Here $$t = 3.2 \text{ mm and } h = 10 \text{ cm}$$

$$\therefore \quad a = 3.2 + 0.015 \times 100 = \mathbf{4.7 \text{ mm}}$$

(*ii*) The scrap bridge and the distance between successive blanks given as, see Table 2.1.

$$b = t = \mathbf{3.2 \text{ mm}}$$

(*iii*) $$W = h + 2a$$

$$= 10 + 0.94 = 10.94 \text{ cm, say } \mathbf{11.00 \text{ cm}}$$

(*iv*) The length of one piece of stock needed to produce one blank is,

$$S = w + b$$

$$= 2.5 + 0.32 = \mathbf{2.82 \text{ cm}}$$

(*v*) $$N = \frac{L - b}{s}$$

$$= \frac{240 - 0.32}{2.82} = \mathbf{84 \text{ blanks}}$$

(*vi*) Scrap remaining at the end,

$$y = L - (Ns + b)$$

$$\therefore \quad y = 240 - (84 \times 2.82 + 0.32)$$

$$= 240 - 237.2 = \mathbf{2.80 \text{ mm}}$$

Example 2.9. *The symmetrical-cup workpiece shown in Fig. 2.73, is to be made from cold rolled steel 0.8 mm thick. Make the necessary calculations for designing the drawing die for this part.*

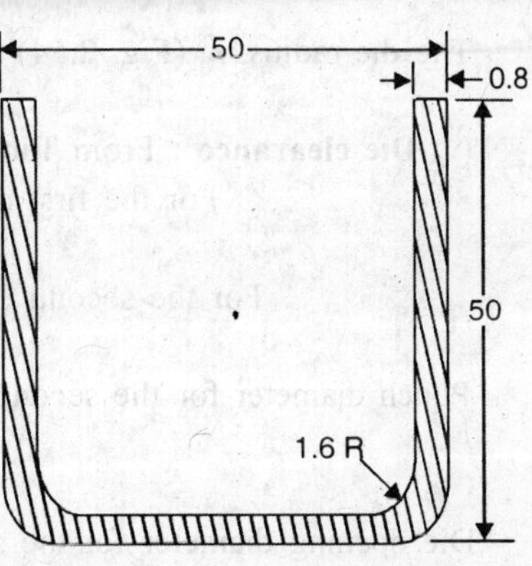

Fig. 2.73. Drawn Shell

Solution. The shell diameter d = 50 mm, radius of bottom corner, r = 1.6 mm

(*i*) **Size of blank** : The ratio

$$\frac{d}{r} = \frac{50}{1.6} = 31.25$$

So, the formula for determining the shell blank size will be, Eqn. 2.16.

$$D = \sqrt{d^2 + 4dh}$$

$$d = 50 \text{ mm}, h = 50 \text{ mm}$$

$$\therefore \quad D = 111.80 \text{ mm}$$

This is the theoretical blank size. However, to get a smooth edge, a small trimming is necessary. For this, it is necessary to add extra metal. Rule of Thumb is to add about 3.2 mm to the blank diameter for each 2.5 cm of cup diameter. Since cup diameter is 50 mm, 6.4 mm should be added for trimming.

$$\therefore \quad D = 111.80 + 6.40 = 118.2 \text{ mm} \approx 118.5 \text{ mm}$$

(*ii*) **Percentage reduction.** Percentage reduction is given as,

$$= 100\left(1 - \frac{d}{D}\right)$$

$$= 100\left(1 - \frac{50}{118.5}\right) = 57.8\%$$

As per practice, a reduction of about 45 to 50 per cent is permissible for first draw. Therefore, it is clear that the above cup cannot be drawn in one draw.

(*iii*) **Number of draw :**

$$\text{Height to diameter ratio} = \frac{50}{50} = 1.0$$

Therefore, from Table 2.9,

$$\text{number of draws} = 2$$

Let the first reduction be 45%

$$\therefore \quad \text{Diameter } d_1 \text{ at first draw} = 118.5 - 0.45 \times 118.5 = 65.175 \text{ mm}$$

Diameter at the end of second draw has to be 50 mm,

Therefore, reduction for second draw

$$= 100 \times \left(1 - \frac{50}{65.175}\right)$$

$$= 23.28\%$$

which is less than 30%, the permissible reduction for the second draw.

(*iv*) **Radius on punch and die :** For the first draw, the punch radius at the bottom should be at least = 4 × t

$$= 4 \times 0.8 = 3.2 \text{ mm, say 4 mm}$$

For the second draw, the punch radius, is determined by the corner radius of the finished shell, that is, it should be equal to 1.6 mm.

The die radius, r_d (Fig. 2.54) should be 4 to 10 times the stock thickness.

$\therefore$ r_d = 3.2 mm to 8 mm, say **6 mm**

(*v*) **Die clearance :** From Table 2.7, die-clearance is given as :

For the first draw = 1.08 t to 1.1 t, per side
= 0.864 to 0.88 mm, say 0.87 mm

For the second draw = 1.09 t to 1.12 t
= 0.872 to 0.896 mm, say 0.88 mm

Punch diameter for the second draw
= 50 − 2t
= **48.4 mm**

Die opening diameter for the second draw
= 48.4 + 2 × 0.88
= **50.16 mm**

Punch diameter for the first draw = 65.175 − 2t
= **63.575 mm**

Die opening diameter for the first draw
= 63.575 + 2 × 0.87
= **65.315 mm**

(*vi*) **Drawing pressure :** $F = \pi dt.\sigma_{yt}\left(\frac{D}{d} - c\right)$

d = 50 mm
t = 0.8 mm
c = constant (0.6 to 0.7)
σ_{yt} = 427 N/mm²
$F = \pi \times 50 \times 0.8 \times 427\,(2.37 - 0.65)$
= **92.30 kN**

(*vii*) **Blank holding pressure :**

Blank holding pressure may be taken as one-third of drawing pressure, *i.e.*, 30.8 kN.

Blank holding pressure can be found by the relation, for cylindrical draws,

$$F_{bh} = \frac{\pi}{4}(D^2 - d^2).p_{bh}, \text{ where}$$

p_{bh} is the specific blank holding pressure and can be taken as (N/mm²),

For M.S. (t < 0.5 mm) = 2.5 − 3.0; M.S. (t > 0.5 mm) = 2.0 − 2.5

Brass = 1.5 − 2.0; Cu = 1.0 − 2.0; Al = 0.8 − 1.2

(*viii*) **Press capacity :** A 150 kN capacity press will be alright for the total possible maximum pressure of about 120 kN.

Example 2.10. *Determine the developed length of the part shown in Fig. 2.74.*

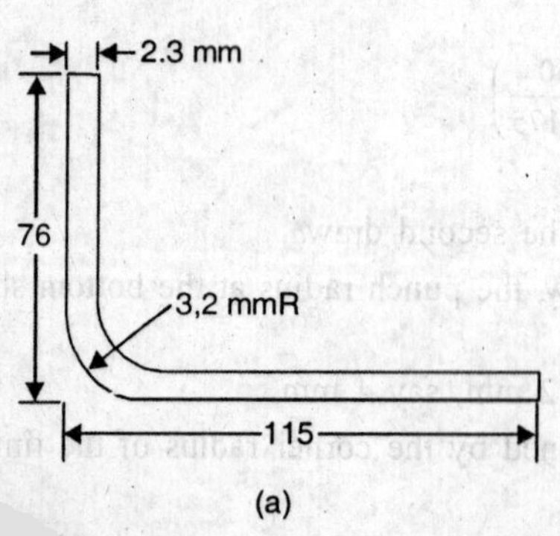

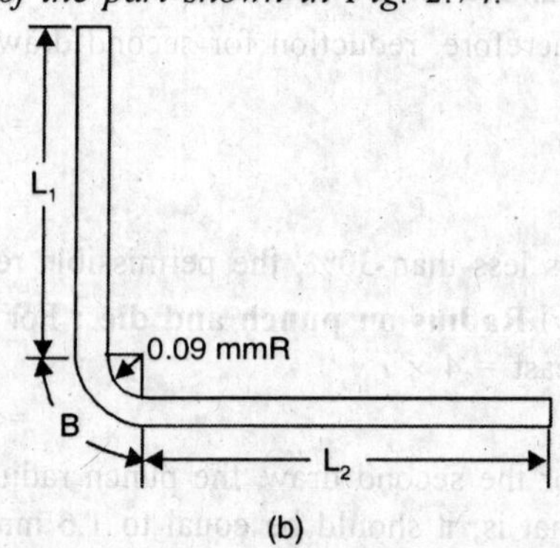

Fig. 2.74

Solution. While calculating the developed length for bending, external dimensions should be converted to internal dimensions, Fig. 2.74 (*b*).

$$\text{Developed length} = L_1 + L_2 + B$$

The inside radius of bend, $r = 3.2 - 2.3 = 0.90$ mm

$$\text{Length } L_1 = 76 - (2.3 + 0.90) = 72.8 \text{ mm}$$

$$\text{Length } L_2 = 115 - (2.3 + 0.90) = 111.8 \text{ mm}$$

Bend allowance, $B = \dfrac{\alpha}{360} \cdot 2\pi (r + K)$

Here $\alpha = 90°$

Let $K = t/3$

$$\therefore \quad B = \frac{90}{360} \cdot 2\pi \left(0.90 + \frac{2.3}{3}\right)$$

$$= 2.61 \text{ mm}$$

$$\therefore \quad \text{Developed length} = 72.8 + 111.8 + 2.61$$

$$= \mathbf{187.21\ mm}$$

Example 2.11. *Calculate the bending force for channel bending for the following data :*

Thickness of blank = *3.2 mm*

Bending length = *900 mm*

Die radius = *Punch radius* = *9.5* mm

Ultimate tensile strength of the material = *400 N/mm²*

Solution. Bending force is given as :

$$F = \frac{K \cdot l \cdot \sigma_{ut} \cdot t^2}{w}, N \qquad \text{(Eqn. 2.23)}$$

Now $K = 0.67$ for channel bending

$$l = 900 \text{ mm}$$

$$\sigma_{ut} = 400 \text{ N/mm}^2$$

$$t = 3.2 \text{ mm}$$

$$w = R_1 + R_2 + C \qquad \text{[Fig. 2.65 (b)]}$$

$$= 9.5 + 9.5 + 3.2 = 22.2 \text{ mm}$$

$$\therefore \quad F = \frac{0.67 \times 900 \times 400 \times 3.2 \times 3.2}{22.2}$$

$$= \mathbf{111.25\ kN}$$

Example 2.12. *Calculate the bending force for a 45° bend in aluminium blank. The following data is given :*

Blank thickness = *1.6 mm*

Bend length = *1200 mm*

Die-opening = *8 × metal thickness*

Ultimate tensile strength = *455 N/mm²*

Solution. Bending force is given as :

$$F = \frac{K \cdot l \cdot \sigma_{ut} \cdot t^2}{w}, N$$

$$K = 1.33$$

$$l = 1200 \text{ mm}$$

$$\sigma_{ut} = 455 \text{ N/mm}^2$$
$$t = 1.6 \text{ mm}$$
$$w = 8 \times t = 12.8 \text{ mm}$$
$$F = \frac{1.33 \times 1200 \times 455 \times 1.6 \times 1.6}{12.8}, \text{N}$$
$$= \mathbf{145.24 \text{ kN}}$$

Example 2.13. *Determine the capacity of the double bending die for the following data :*

Sheet metal thickness (cold rolled steel) = 1 mm

Sheet metal width at bend = 50 mm

Die radius = 3 mm

Punch radius = 1.5 mm

Die clearance = 1.25 mm

Tensile strength = 315 MPa

Setting pressure = 560 MPa

Beads on punch = 2

Projected width of each bead = 3 mm

Solution. (*i*) **Bending force :**
$$F = \frac{\left(0.67 l \cdot \sigma_{ut} \cdot t^2\right)}{w}, \quad w = R_1 + C + R_2$$
$$= \frac{0.67 \times 50 \times 315 \times 1 \times 1}{3 + 1.25 + 1.5} = 1835.2, \text{N}$$

(*ii*)
$$\text{Pad force} = 0.67 \cdot \sigma_{ut} \cdot l \cdot t$$
$$= 0.67 \times 315 \times 50 \times 1$$
$$= 10552.5, \text{N}$$

(*iii*)
$$\text{Bottoming force} = \sigma_c \cdot l \cdot b$$
$$= 560 \times 50 \times 2 \times 3$$
$$= 168{,}000 \text{ N.}$$

It is clear from above, that the actual bending force is minor as compared to other two forces. Now, bending force occurs above bottom of press stroke. Pad force occurs throughout the bending operation and Bottoming force occurs at bottom of press stroke. Mechanical presses are rated at the bottom of the stroke. So, in view of its magnitude, when bottoming is used to overcome springback, the bottoming force is used alone to determine the press capacity.

∴ Force required for the operation $= 10552.5 + 168{,}000$
$$= 178552.5, \text{N}$$
$$= \mathbf{18.2 \text{ tonnes}}$$

∴ use a 20 tonne press.

If bottoming had not been used, then, force required would be,

$$\text{Force required} = 1835.2 + 10552.5$$
$$= 12387.8, \text{N}$$
$$= \mathbf{1.263 \text{ Tonnes.}}$$

Example 2.14. *A M.S. sheet 5 mm thick and 2 m wide is cut in the width direction. Determine the cutting force for cutting :*

(*a*) *With parallel cutting edges.*

(*b*) *In a guillotine in which cutting edges are given a shear of 6°. Take ultimate stress in shear as 382.5 MPa.*

Solution. (*a*) Area to be cut = width × sheet thickness

$\therefore$ Cutting force = 2 × 1000 × 5 × 382.5, N

= 3.825 MN.

(*b*) In this case, the length to be cut is the instantaneous length to be cut, which is

$$= \frac{t}{\sin \alpha} = \frac{5}{\sin 6°}$$

= 47.83 mm

$\therefore$ F = 47.83 × 5 × 382.5, N

= 91.48 kN

It should be noted that there is a large drop in the cutting force.

Example 2.15. *A cup 105 mm inside diameter and 90 mm deep is to be drawn from steel sheet 1 mm thick. Determine the blank diameter and a suitable punch diameter for the first draw. Give the probable dimensions of the cup obtained from the first draw and estimate the press capacity.*

Solution : Blank diameter,

$$D = \sqrt{d^2 + 4dh} = [105^2 + 4 \times 105 \times 90]^{\frac{1}{2}} = 221 \text{ mm}$$

$\therefore$ Thickness ratio, $\frac{t}{D} \times 100 = \frac{1}{221} \times 100 = 0.453$

From the table below :

Thickness ratio $\frac{t}{D} \times 100$:	0.15	0.20	0.30	0.40	0.50	> 0.50
Max draw ratio, D/d :	1.43	1.54	1.67	1.82	1.91	2.00

(for sheet steel where pressure pad is used)

For thickness ratio of 0.453, the safe drawing ratio is 1.82

$\therefore$ Diameter for the first draw $= \frac{221}{1.82} = 121\text{mm}$.

The cup diameter is 105 mm, hence it can not be drawn by a single draw. If '*d*' is taken as 121 mm for the first draw, the reduction for the subsequent draw will be rather small. So, it will be better to have a somewhat lower reduction for the first draw.

Let the diameter of the first draw be d = 130 mm

$\therefore \frac{D}{d}$ for first draw is $= \frac{221}{130} = 1.7$

$\frac{D}{d}$ for second draw $= \frac{130}{105} = 1.24$, which is quite O.K.

$\therefore$ Depth of first draw : $130^2 + (4 \times 130) \times h = 221^2$

From here, h **= 61.43 mm**

Press capacity $= \pi dt\, \sigma_{yt} \left(\frac{D}{d} - C \right) = \pi \times 130 \times 1 \times 415 \left(\frac{221}{130} - 0.65 \right)$ = **177.45 kN**

Example 2.16. *A symmetrical-cup of 80 mm diameter and 250 mm height is to be fabricated on a deep drawing die. How many drawing operations will be necessary if no intervening annealing is done. Also find the drawing force.*

Solution : Blank diameter,

$$D = \sqrt{d^2 + 4dh}$$

$$= \sqrt{64 + 4 \times 8 \times 25} = 29.4 \text{ cm}$$

Let $$\frac{D - D_1}{D} = 0.5 \text{ for the first draw}$$

∴ Diameter after the first draw,

$$D_1 = D - 0.5\,D = 0.5\,D = 14.7 \text{ cm}$$

Let the reduction for the second draw be 40%.

∴ Diameter after the second draw

$$= 14.7 - 0.4 \times 14.7 = 8.82 \text{ cm}$$

Now, final diameter = 8 cm.

∴ Third draw is also necessary.

∴ Percentage reduction for the 3rd draw

$$= 8.82 - R \times 8.82 = 8$$

$$\therefore \quad 1 - R = \frac{8}{8.82} = 0.9$$

$$\therefore \quad R = 10\%$$

Now, $$\pi D_1 l_1 = \frac{\pi}{4} D^2 - \text{Area of base}$$

(neglecting corners and with no flanges)

$$= \frac{\pi}{4} \times (29.4)^2 - \frac{\pi}{4} \times (14.7)^2$$

$$= 509 \text{ cm}^2$$

∴ Height of cup after the first draw

$$= \frac{509}{\pi \times 14.7} = 11.02 \text{ cm}$$

Similarly, $$l_2 = 22.28 \text{ cm}$$

and $$l_3 = 25 \text{ cm}$$

Now, Drawing force, $$F = \pi \cdot d \cdot t \cdot \sigma_{yt} \left(\frac{D}{d} - C \right)$$

Let $t = 3$ mm, $\sigma_{yt} = 250$ N/mm² and $C = 0.66$.

$$\therefore \quad F = \pi \times 80 \times 3 \times 250 \left(\frac{294}{80} - 0.66 \right)$$

$$= \mathbf{56.556 \text{ kN}}$$

Example 2.17. *Determine the developed length of the put shown in Fig. 2.75(a).*

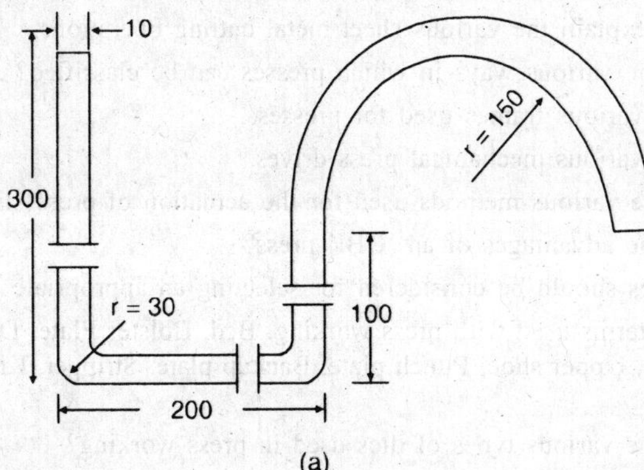

Solution : While calculating the developed length for bending, the external dimensions are converted into internal dimensions.

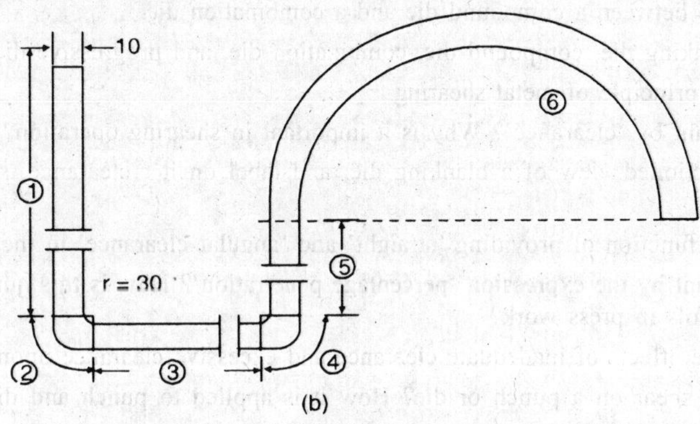

Fig. 2.75

Inside radius of the bends = 30 − 10 = 20 mm

$\therefore$ Length (1) = 300 − (10 + 20) = 270 mm

$$\text{Length (2)} = \frac{\alpha \cdot 2\pi}{360}(r + K)$$

Here $\alpha = 90°$, now $\frac{r}{t} = 2$. Let $K = 0.33\ t$.

$$\therefore \quad \text{Length (2)} = \frac{90}{360} \times 2\pi\,(20 + 0.33 \times 10) = 36.6 \text{ mm}$$

Length (3) = 200 − 2 (10 + 20) = 140 mm

Length (4) = Length (2) = 36.6 mm

Length (5) = 100 − (10 + 20) = 70 mm

For length (6), inner radius = 150 − 10 = 140 mm, $\alpha = 180°$

$$\therefore \quad \text{Length (6)} = \frac{180}{360} \times 2\pi\,(140 + 0.33 \times 10) = 450.25 \text{ mm}$$

$\therefore$ Total developed length = (1) + (2) + (3) + (4) + (5) + (6)

= 270 + 36.6 + 140 + 36.6 + 70 + 450.25

= **1003.45 mm**

PROBLEMS

1. Name and explain the various sheet metal cutting operations.
2. What are the various ways in which presses can be classified?
3. Sketch the various frames used for presses.
4. Sketch the various mechanical press drives.
5. Describe the various methods used for the actuation of press slides.
6. What are the advantages of an 'OBI' press?
7. What factors should be considered for selecting an appropriate press for a given job?
8. Define the terms used with press working. Bed, Bolster Plate, Die, Die block, Die set, Punch, Lower shoe, Upper shoe, Punch plate, Backup plate, Stripper, Knock out, Pitman, Shut height, Guide posts.
9. What are the various types of dies used in press working?
10. Differentiate between a cutting die and a forming die.
11. Differentiate between a Blanking die and a Piercing die.
12. Differentiate between a compound die and a combination die.
13. Sketch: Blanking die, compound die, combination die, and progressive die.
14. Explain the principle of metal shearing.
15. What is meant by 'clearance'? Why is it important in shearing operation?
16. Sketch a sectioned view of a blanking die, and label on it : clearance, straight and angular clearance.
17. What is the function of providing 'straight' and 'angular clearance' in the die block?
18. What is meant by the expression 'percentage penetration'? How is this quantity employed for designing tools in press work?
19. What are the effects of inadequate clearance and excessive clearance upon die-cut metals?
20. What is the shear on a punch or die? How it is applied to punch and die for blanking and piercing operations?
21. Sketch the various methods of applying shear to the punch and die.
22. What means should be employed to reduce the loading on the press during cutting operation?
23. Differentiate between a "Drop-through' and 'Inverted" blanking die.
24. Sketch the various methods of holding punches.
25. What are the various types of strippers ? Explain their function with the help of suitable sketches.
26. What is 'stock stop' and 'pilot'?
27. Sketch the various methods of mounting pilots in the punch.
28. Define "Deep Drawability". What is the usual range?
29. What is the purpose of pressure pad in drawing a cup?
30. How the size of a blank is calculated for drawing a cup?
31. Why more than one draw is needed to draw a cup?
32. What is the usual reduction for the first and the succeeding draws?
33. Sketch a 'push-through' and an 'inverted' draw die.
34. What are the various methods of bending? Sketch these.
35. Explain the 'Bending Terminology' with the help of a suitable sketch.
36. When is ironing done?
37. Define 'spring back' and explain how allowances may be made to compensate for its harmful effects.
38. Sketch : *V*-bending die and *U*-bending die.

39. How do hydraulic drives compare with mechanical drives for presses?
40. Sketch a die to manufacture cycle-chain links.
41. Find the pressure required to cut a circular blank 30 cm in diameter from 3 mm thick, if it has a shear strength of 376 N/mm^2.
42. A blank has a perimeter of 31.75 cm. The metal is 1 mm thick cold worked 0.15%. Carbon steel with a shear strength of 420 N/mm^2 and per cent penetration of 25%. Two holes of 1.25 cm diameter each are to be pierced during the same stroke when the piece is blanked. What are the forces required for blanking and for piercing? What is the maximum force the press must exert at any one time without shear? What energy is required per stroke?
43. A piece of stock 2.35 mm thick is bent to an angle of 120° with an inside radius of 6.25 mm. What is the original length of stock that goes into the bend?
44. A cup 5 cm in diameter and 7.5 cm deep is to be drawn from 1.5 mm thick drawing steel with a tensile strength of 315 N/mm^2. The corner radius is negligible. Determine
(*a*) Blank diameter.
(*b*) Least number of drawing operations.
(*c*) Force and energy for the first draw with 40% reduction.
45. Calculate the maximum punch force necessary to blank a steel washer 44 mm outside diameter 22.25 mm inside diameter and 1.60 mm thick, if τ_s = 400 N/mm^2. Estimate the work done if % penetration is 25%.
46. In the above problem, calculate the amount of shear which must be grounded upon the tool, if the maximum punch force is to be reduced to 60 kN.
47. Find the centre of pressure of the following blanked shapes :

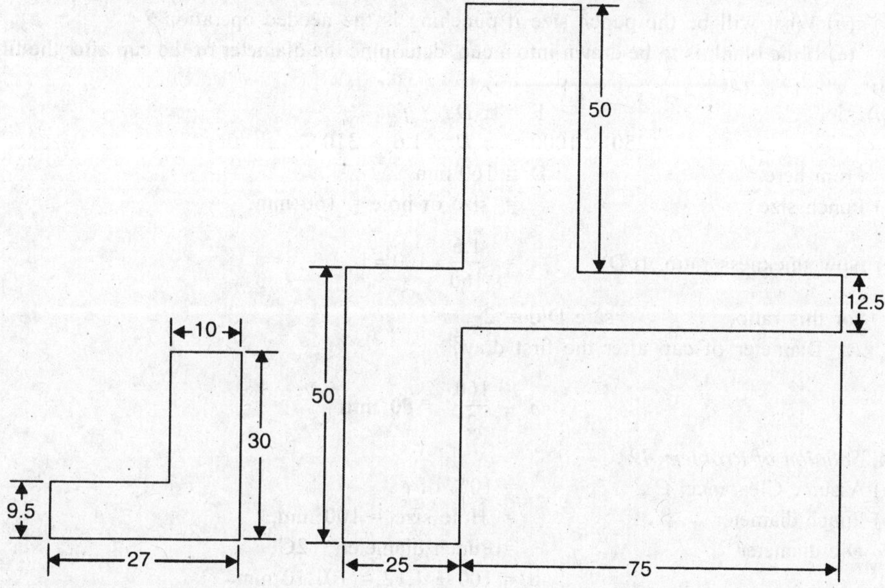

Fig. 2.76

48. Differentiate between blanking and piercing. Which tool size decides the blank size and which tool size that of the pierced hole?
49. A hole 100 mm diameter is to be punched in steel plate 5.6 mm thick. The material is cold-rolled 0.4% *C* steel for which ultimate shear strength is 550 MPa. With normal clearance on the tools, cutting is completed at 40% penetration of the punch. Give suitable diameter for punch and die and a suitable shear angle for the punch in order to bring the work within the capacity of a 30 tonne press (100 mm, 101.1 mm, 4.2°).
50. Design the press tool set to perform the piercing and blanking operations to manufacture bell crank lever shown in Fig. 2.77, from 2 mm thick mild steel.

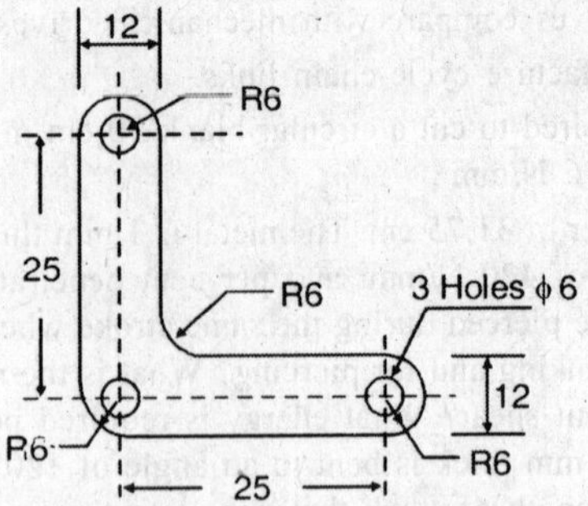

Fig. 2.77. A Bell Crank lever.

51. A steel cup of height 30 mm and internal diameter 40 mm with a flange width of 10 mm is to be deep drawn from a sheet 1 mm thick. Determine the diameter of the blank and the drawing force. What is the draw ratio? Can the cup be drawn in a single operation? **(GATE 1996)**

(Ans.: 94.25 mm, 32.54 kN, 2.35, No, in 2 draws)

52. In deep drawing of sheets, the value of limiting draw ratio depends upon :

(A) Percentage elongation of sheet metal (B) Yield strength of sheet metal (C) Type of press used (D) thickness of sheet. **(Ans. : D) (GATE 1994)**

53. (*a*) Determine the diameter of the hole that can be punched in a steel sheet of thickness 1.6 mm, for which the ultimate shear stress is 310 N/mm². Press Capacity is 250 kN.

(*b*) What will be the punch size if punching is the needed operation ?

(*c*) If the blank is to be drawn into a cup, determine the diameter of the cup after the first draw.

Sol.

(*a*) $F = \pi D t \times f_{su}$

∴ $250 \times 1000 = \pi D \times 1.6 \times 310$

From here, $D \cong 160$ mm

(*b*) Punch size = size of hole = 160 mm.

(*c*) Now thickness ratio, (t/D) $= \frac{1.6}{160} \times 100 = 1$

for this ratio, safe D/d = 2

∴ Diameter of cup after the first draw,

$$d \simeq \frac{160}{2} = \mathbf{80\ mm}$$

54. *Solution of Problem 49 :*

(*a*) Assume Clearance, C = 10% of t

(*b*) Punch diameter = Hole size = 100 mm.

Die diameter = Punch diameter + 2C

= 100 + 1.12 ≅ 101.10 mm

(*c*) Using equation (2.7),

$$F = \frac{kt}{I} \times F_{max}$$

$$F_{max} = \pi D t f_{su} = \pi \times 100 \times 5.6 \times 550 = 968 \; kN$$

$$k = 0.4, \quad F = 300 \; kN$$

∴ $I = \frac{968}{300} \times 0.4 \times 5.6 = 7.23$ mm ∴ $\tan\theta = \frac{7.23}{100}$

From here, θ = 4.2°.

3

FORGING DIE DESIGN

3.1. GENERAL

Forging process may be defined as a metal working process by which metals or alloys are plastically deformed to the desired shapes by a compressive force applied with the help of a pair of dies. One die is stationary and the other has a linear motion. Forging process can be carried out both in the cold and hot state of the metal. But, unless otherwise mentioned, forging process is considered to be "hot forging" process.

Forging improves the quality of steel, which becomes stronger after forging. Due to his, the parts which are subjected to heavy duty are generally made of forgings. Time of production is very often reduced. Much less steel is consumed in forging production. Hence, the cost of any given part is reduced. In forge shops, steel is received as ingots or as rolled sections. Ingots are used for manufacturing heavy forgings, while rolled billets are used for lighter forgings. Forgings which have to undergo subsequent machining are called 'blanks'. Those which do not need any further machining are called "finished" forgings.

Forgings may be produced in either open or closed dies. In open die forging which is also known as "Flat die forging", the hot metal is struck or pressed between two flat surfaces or simple contoured dies. The compressive force is progressively applied locally on different parts of the metal stock. The flow of the metal, that is, the changing of its dimensions and shape is controlled with the aid of various blacksmith's tools.

In closed die forging process, cavities or impressions are cut in the die block, the compressive force is applied to the entire surface and the metal is forced to take its final shape and dimensions as it flows into and fills the die cavities. The flow of the metal is limited by the surfaces of the recesses or cavities in the dies. When the pair of dies approach each other for completing the forging, the excess metal squirts out of the cavity as a thin ribbon of metal called "flash" or "fin". Because of flash, the term "closed-dies forging" is a bit of a misnomer. Closed-die forging means no flash. So, a better description of the process with recesses or cavities in the die blocks would be " Impression-die forging".

Open-dies are less costly than impression dies and so are used where number of components to be forged is too small to justify the cost of impression dies, or where the sizes are too large and too irregular to be contained in usual impression dies. Open-die forging can be used for simple shapes only such as : Bars, slabs or billets with rectangular, circular, hexagonal or octagonal cross-sections, welders rings, and many other components of simple shapes. On the other hand, for more complex and accurate parts and with increased production rates, impression dies are preferred.

In open die forging, the weight range of forging goes upto few tonnes, whereas in impression die forging, the weight range is limited upto few hundred newtons due to limitation of die size. In open die forging, the forgings are usually made on hydraulic presses designed for forging ingots, whereas impression die forgings are made on hammers or presses (mechanical/hydraulic). Open die forgings are required for heavy equipment and machinery such as for steel plants, power generation, shipping and defence, whereas impression die forgings are generally used in automobile sector. In open die forging, the simplicity of tooling is gained at the expense of the complexity of process control, whereas in impression die forging, the process is simplified to a sequence of simple compression strokes at the expense of complex die shape.

Limitations of open-die forging :

1. Less control in determining grain flow, mechanical properties and dimensions.
2. Restricted to short run production.
3. Poor material utilization.
4. Restricted to simple shapes.
5. Difficulty of maintaining moderately close tolerances.
6. Absolute need for skilled labour.
7. Final cost of production may be higher than other forging methods, because machining is often required.

Closed-die forgings have the following characteristics :

1. Saving of time as compared to open-die forging.
2. Makes good utilization of workpiece materials.
3. Excellent reproductivity with good dimensional accuracy.
4. Forgings are made with smaller machining allowances, thus reducing considerably the machining time and the consumption of metal required for the forging.
5. Forgings of complicated shapes can be made.
6. The equipment for closed-die forging does not require highly skilled workers.
7. The grain flow of the metal can be controlled ensuring high mechanical properties.
8. Method is suited for rapid production rate.
9. Cost of tooling is high, therefore, suitable for large production runs.

Classification of dies. The dies used in closed-die forging or impression-die forging may be classified into two groups :

(*i*) **Single Impression Die.** This die contains only one cavity or impression which is the finishing impression. The preliminary forging operations are done by hand or on forge hammers, forging rolls etc., and only final finishing operation is done in the die-cavity.

(*ii*) **Multi-Impression Die.** This die contains finishing operation and one or more auxiliary impressions for preliminary forging operations (discussed ahead). The final shape of part is progressively developed over a series of steps from one die impression to the next. Generally, multi-impression dies are very expensive to make and are employed only when the quantity to be made is sufficiently large, and for forging of intricate designs.

Advantage of multi-impression die

1. Complete sequence of forging operations can be carried out on one equipment only, avoiding the use of auxiliary forging equipment.
2. Use of multi-impression dies is particularly suited for production of small and medium sized forgings in large quantities as this method gives 2 to 3 times the production compared with the method of production using a single die. This is because the time of preparation of the 'use' on auxiliary equipment is reduced or eliminated.
3. All the preliminary operations can be performed on these dies with good ease. The 'use' can be prepared to fairly accurate dimensions. Besides this, more accurate forgings are prepared.
4. Wastage of forging metal is reduced.
5. 'Use' may not be reheated for the finishing impression.
6. Initial die cost becomes insignificant in case of high output.
7. Finishing impression lasts long, because, much of the load is taken by blocking impression.

Fig. 3.1 illustrates a multiple impression die and the forging sequence for forging a connection rod. The various steps are given below :

1. The heated forging stock is elongated by reducing its cross-section in impression 1, the operation is called as "fullering".

2. Next, the metal is redistributed, increasing the cross-section at certain places and reducing at others as required to fill the cavities of the die. The operation done in impression 2 is called as "edging" operation.

3. Then the stock is bent in bending impression 4.

4. Next, the general shape is given by forging in the "blocking" or semifinish impression 5.

5. The final shape is given to the forging in the finish impression 3.

Blocking operation reduces the wear of the finish impression. The flash gutter 6 is provided only around the finish impression.

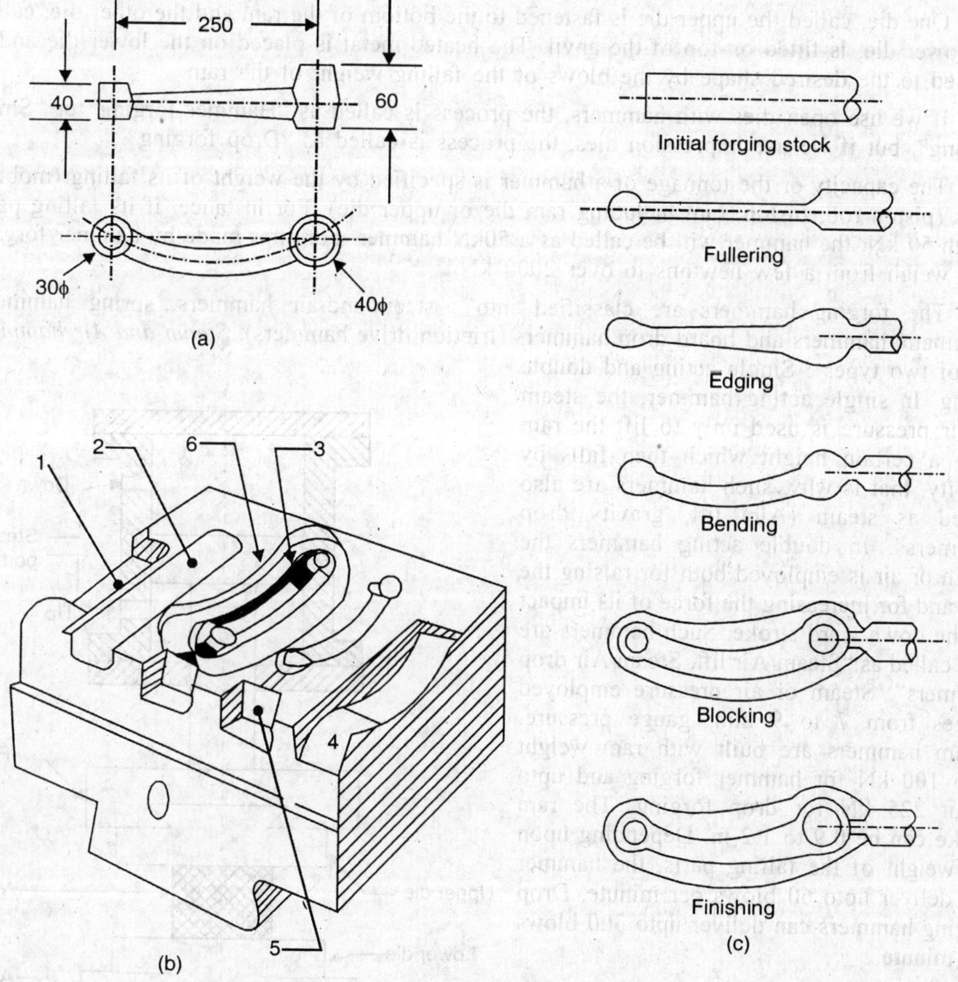

Fig. 3.1. A Multiple Impression Forging Die.

3.2. FORGING EQUIPMENT

Hot-forging equipment may be divided into three main categories :

1. **Hammers.** Hammers impart stress on the material by impact, and they operate in a vertical position. They are impact machines.

2. **Forging presses.** Here, the compressive force is applied continuously and the material is gradually pressed or squeezed into shape. They operate in a vertical position like the hammers.

3. **Forging machines or upsetters.** This unit also delivers its stress to the material by pressure or squeeze like a forging press, but it operates in a horizontal position.

3.2.1. Hammers

A hammer consists of four main parts :

1. Falling weight, *i.e.*, a ram or tup.
2. Frame or guide for the falling weight.
3. Base or anvil.
4. Mechanism to lift the ram.

One die, called the upper die is fastened to the bottom of the ram and the other die, called the lower die, is fitted on top of the anvil. The heated metal is placed on the lower die and is formed to the desired shape by the blows of the falling weight of the ram.

If we use open dies with hammers, the process is called as "Hammer forging" or "Smith forging". but if we use impression dies, the process is called as "Drop forging".

The capacity or the tonnage of a hammer is specified by the weight of its falling (mobile) parts (piston rod, piston, ram including ram die or upper die). For instance, if its falling parts weigh 50 kN, the hammer will be called as a 50kN hammer. Forgings made by hammer forging may weigh from a few newtons to over 2000 kN.

The forging hammers are classified into : steam-and-air hammers, spring hammers, pneumatic hammers and board drop hammers (friction drive hammers). *Steam and Air hammers* are of two types : Single acting and double acting. In single acting hammer, the steam or air pressure is used only to lift the ram upto a certain height which then falls by gravity, that is why, such hammers are also called as steam (Air) lift, gravity drop hammers". In, double acting hammers the steam or air is employed both for raising the ram and for increasing the force of its impact on the down ward stroke. Such hammers are also called as "Steam/Air lift, Steam/Air drop hammers". Steam or air pressure employed ranges from 7 to 9 atm, gauge pressure. Steam hammers are built with ram weight upto 100 kN for hammer forging and upto about 225 kN for drop forging. The ram stroke can be 0.9 to 1.2 m. Depending upon the weight of the falling parts, the hammer can deliver upto 60 blows per minute. Drop forging hammers can deliver upto 300 blows per minute.

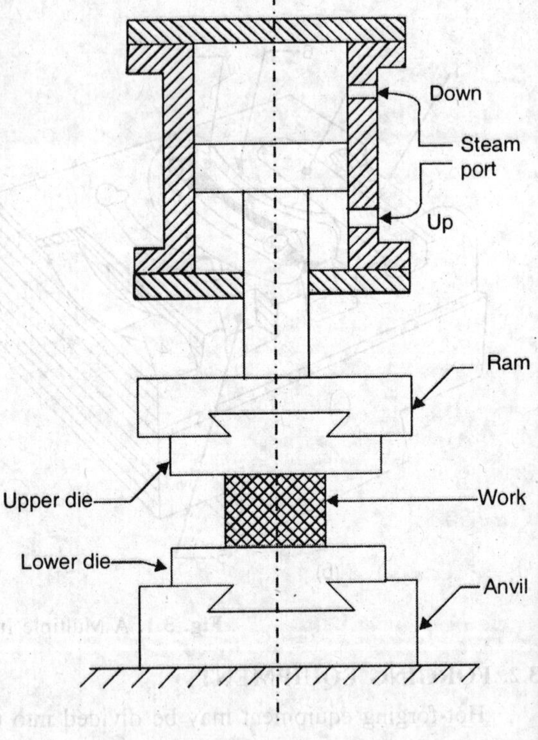

Fig. 3.2. Double acting Steam Hammer.

By the design of columns, steam-air hammers are divided into single column or Single frame hammer and Double column or Double frame hammer. Single frame hammers are usually built with mobile parts weighing 2.5 to 10 kN. Double frame hammers are built with falling parts weights from 10 to 50 and even 100 kN, but more frequently from 30 to 50 kN. A double frame hammer is more powerful than a single frame hammer, but it does not permit the operator more freedom of movement around the dies. In steam hammers, the anvil block normally weighs at least 15 to 20 times the total falling weight of the hammer.

Hammers are impact forging machines. The metal is plastically deformed by the kinetic energy, E, of the mobile parts, that is

$$E = \frac{1}{2} mv^2$$

where m = mass of mobile parts

v = velocity of impact.

It is clear that the power of a hammer can be increased by increasing the mass and the velocity of its mobile parts. When the mobile parts are raised to a height H, then released to fall by gravity, the velocity at impact is

$$v = \sqrt{2gH}$$

As there is a structural limit to height H, it is advantageous to accelerate the parts as much as feasible by the pressure of steam or compressed air (Double acting hammers). This will result in a great increase in the energy of impact E or reduction in the height of hammer. In steam-air hammers, the velocity of dies attain 7-8 m/s. Fig. 3.2 shows the working principle of a double acting steam hammer.

Difference in the construction of steam or air hammers for hammer forging and drop forging. Drop-forging hammers are very similar in design and arrangement to the steam or air hammers used for hammer forging. They differ from the latter chiefly in that the frames of steam forging hammers are secured to their anvil blocks, whereas the frames of drop-forging hammers are secured to special foundations independent of their anvil blocks. This is done to achieve rigidity of construction and precise alignment of the top and bottom dies during their impact, thereby ensuring the precision of the required shape and dimensions of the forging. Drop forging hammers have the added provisions of longer and more accurate guides for the vertical movement of the ram, so that the top die meets the bottom die in perfect alignment. Also, they have heavier anvils.

Another great advantage of drop-forging steam or air hammers is that their blows can be regulated over a much wider range than in hammer forging hammers from the lightest to the heaviest as desired. Their control and operation are much simpler. Drop forging hammers do not require a hammer operator- the die forger himself operates the hammer by manipulating the treadle with his foot.

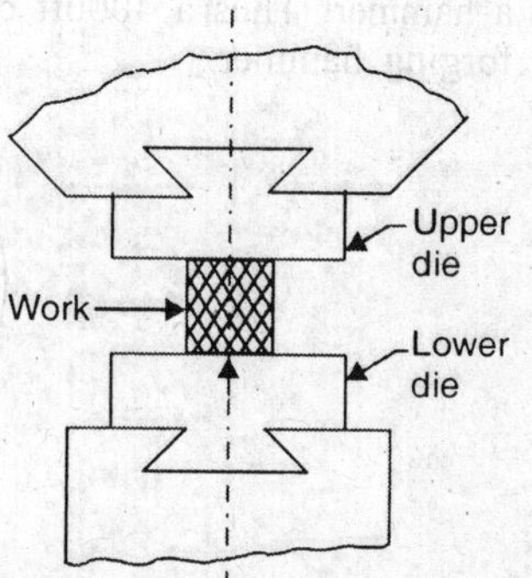

Fig. 3.3. Counter-blow Hammer.

Drop-forging hammers are most commonly employed for the mass or large-lot production of parts of steel or non-ferrous metals of a weight upto 3500 N.

The chief drawback of steam-air drop hammer is a need for large foundations, and despite the precaution, appreciable vibration of production building. This shortcoming has been eliminated in "Counter-blow hammers" in which two halves of the die move towards each other, Fig. 3.3. As this hammer produces a large impact, large forgings weighing upto about 45 kN can be made with it. The capacity of counter blow hammers is upto the range of 125 metric ton capacity (1250 kN).

Pneumatic hammer. Pneumatic power forging hammers are usually used to smith forge small parts. The capacity of pneumatic hammers usually varies from 0.5 to 10 kN. Such hammers operate at 80 to 200 blows per minute.

Board drop-hammer. A board-drop hammer or friction-drive hammer is a drop-forging hammer. It is a true drop-hammer, since, after the ram is lifted upto a certain height, it is allowed to fall freely under the force of gravity only. It comprises a ram carrying a wooden board which is compressed between two driving rollers. When actuated, the rollers raise the board with the ram and as they are drawn apart, the mobile parts fall to effect a blow upon a forging, Fig. 3.4. The board drop hammer sizes range from about 900 N to 45 kN falling weight. The height of fall ranges from about 0.75 to 1.5 m. The disadvantages of board hammer are : the boards are liable to frequent breakage, and the intensity of blow cannot be controlled during the stroke.

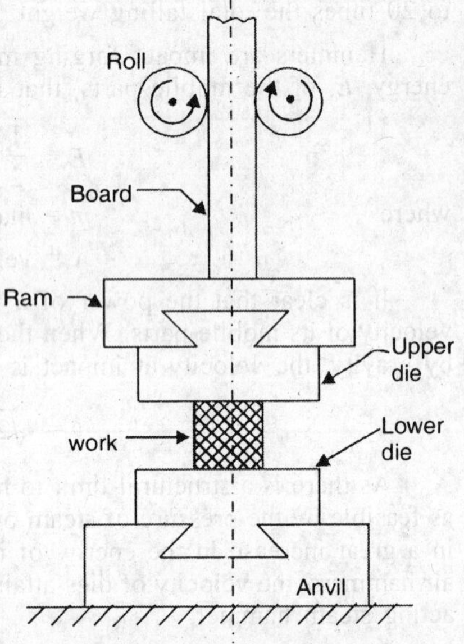

Fig. 3.4. Board Drop Hammer.

3.2.2. Presses

The various forging presses employed for the forging process are : Power-driven presses (mechanical presses) and the hydraulic presses. The common mechanical forging presses are : Crank-type forging presses and friction screw presses.

Crank-type forging presses. These presses employ the same operating principles as the cold stamping crank-type presses described in Chapter 2, Fig. 3.5. Unlike the latter, they have a frame of a more rigid design and a more powerful drive. These presses operate at about 25 to 100 strokes per minute. Their capacities range from 900 kN to 110 MN. Speed of upsetting in a crank-type press is 0.5 to 0.8 m/s. It may be of interest that every 10 MN (1000 tf) capacity of a forging press corresponds approximately to 1 ton of mobile parts of a hammer. Thus a 4000tf crank type forging press can be substituted for a 4tf capacity drop forging hammer.

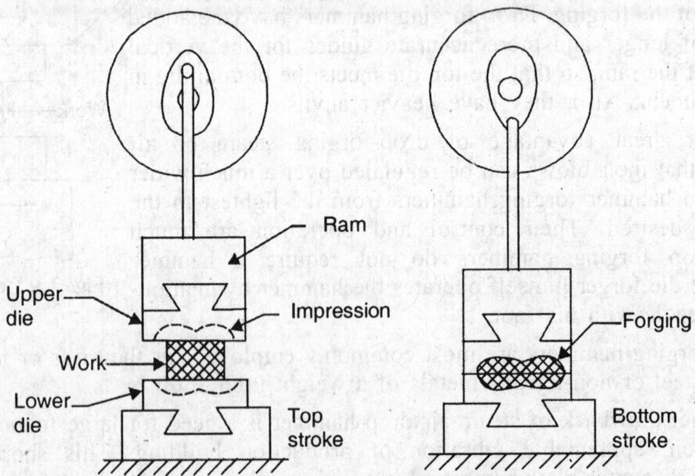

Fig. 3.5. Crank-type Forging Press.

Friction-screw press. Friction screw presses are used for the most part to produce small forgings. These presses are available in the capacity ranges from 630 kN to 2.5 MN and occasionally upto 6 MN. They can be of double, triple-, or no disc design. The principle of working of a double disc press is explained in Fig. 3.6.

Two massive discs (upstroke and downstroke discs) of C.I. are carried on a shaft which is driven from an electric motor via a *V*-belt drive or gearing. Arranged between the discs and set on the end of a working screw is a flywheel coated with leather or another friction material. The screw rotates in a nut fixed at the top of the press frame. The bottom end of the screw carries the slide. The distance between the discs is somewhat greater than the diameter of the flywheel. Because of this, when one of the discs is in contact with the flywheel, there will be a slight clearance between the second disc and the flywheel.

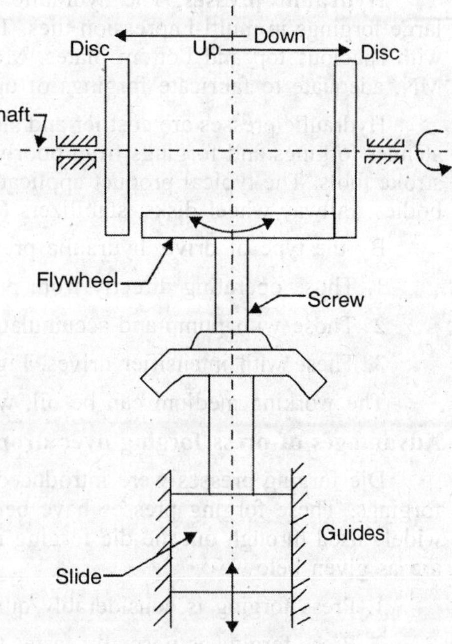

Fig. 3.6. Friction Screw Press.

By moving one or other of the discs to touch the flywheel, the screw can be made to rotate clockwise or anti-clockwise, thus raising or lowering the slide. As the flywheel is rotated by engaging the downstroke disc, it moves down and the slide motion is continuously accelerated. At the bottom of the stroke, the entire energy stored in the flywheel is consumed by the work. Due to this, the increase in pressure of the slide of a friction press is not so smooth as in a crank-type press. It is accompanied by a blow. This feature permits friction presses to be successfully employed for bending and straightening operations and also for upsetting bolt heads.

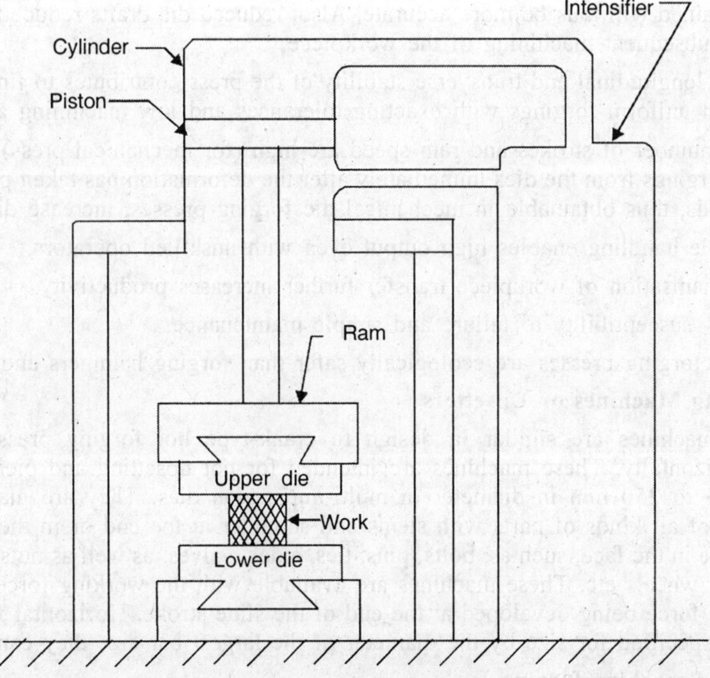

Fig. 3.7. Hydraulic Forging Press.

Hydraulic presses. The hydraulic presses are preferably used for stamping complicated large forgings in multi-impression dies. That is why they are made with four columns, provided with spacious top and bottom plates. Modern hydraulic presses develop a force of upto 1000 MN, adequate to fabricate forgings of upto 5 tons.

Hydraulic presses are costlier and slower than crank-type presses. They are used to produce slender forgings and forgings from poorly plastic alloys and to effect operations requiring large-stroke tools. The typical product applications are : large levers, flanges, toothed wheels, hollow bodies, railway wheel disks, stabilizers (fins) and other parts of aircraft and rocket bodies.

By the type of drive, hydraulic presses fall into one of the following three types :

1. Those operating directly from pumps.

2. Those with pump-and-accumulator drives.

3. These with intensifier drives, Fig. 3.7.

The working medium can be oil, water or oil-water emulsion.

Advantages of press forging over drop forging

Die forging presses were introduced in the early thirties, for large scale production of die forgings. These forging presses have been replacing the drop forging hammers and today are widely used through out the die forging industry. The special advantages of die forging presses are as given below :

1. Press forging is considerably quieter operation than drop forging.

2. Press forging is normally faster than drop forging since only one squeeze is needed at each die impresion.

3. Alignment of the two die halves can be more easily maintained than with hammering.

4. Structural quality of the product is superior to drop forging.

5. With ejectors in the top and bottom dies, it is possible to handle reduced die drafts. Forgings obtained will thus be more accurate. Also, reduced die drafts reduce the weight of the charge and subsequent machining of the workpiece.

6. High longitudinal and transverse stability of the press contributes to fine accuracies and this results in uniform forgings with exacting tolerances and low machining allowances.

7. The number of strokes and ram speed are high (for mechanical press) and the ejectors release the forgings from the dies immediately after the deformation has taken place. The shorter contact periods, thus obtainable in mechanical die forging presses, increase die life.

8. Simple handling enables high output even with unskilled operators.

9. Mechanisation of workpiece transfer further increases productivity.

10. Low susceptibility to failure and simple maintenance.

11. Die forging presses are ecologically safer than forging hammers and screw presses.

3.2.3. Forging Machines or Upsetters

These machines are similar in design to crank-type hot forging presses but they are mounted horizontally. These machines are intended for hot upsetting and piercing of forgings from bars 13 to 250 mm in diameter in multi-impression dies. They are mainly suitable for manufacture of all kinds of parts with stems and shoulder at the end or in the middle, with or without a hole in the face, such as, bolts, pins, ties, axles, valves, as well as nuts, bushings, blank caps for gear wheels etc. These machines are available with the working force of 1 to 30 MN and over, the force being developed at the end of the slide stroke. Horizontal forging machines are rated or specified for size by the diameter of the largest bar size they can handle.

Advantage of machine forging

1. The quality of the forging is better than obtained by drop forging or press forging, due to better orientation of grain.

2. Forging in forging machine is accompained by little or no flash, whereas in drop forging, the flash may amount from 10 to 75% or even upto 200% of the finished forging.

3. Forging machines have a higher productivity and their maintenance is much cheaper as compared to drop forging hammers.

4. There is saving in material and also machining expenses as no or little draft is needed on forging made by upsetters.

5. The upsetting process can be automated.

Limitations

1. It is not convenient to forge heavier jobs due to the material handling difficulties.

2. The maximum diameter of the stock which can be upset is limited (Max about 25 cm).

3. Intricate, non symmetric and heavy jobs are difficult to be forged on a forging machine.

4. The tooling cost may be high.

3.2.4. Comparison of Forging Equipment. All the three types of forging equipment, discussed above have their advantages and limitations with respect to the shapes to be produced and the materials to be used. Open die hammers are used for the production of simple shaped components, and for low outputs where the cost of complex impression or closed dies will not be justified. Drop-forging hammers are used for production of close-tolerance parts where the quantity of production is comparatively small (upto 2000 parts per die run). For close limit parts and for large quantity production (5000 to 6000 forgings per die run) presses and upsetters are employed. Presses are used for large jobs of symmetrical shape, whereas hammers and forging machines are used for smaller parts. When there is decrease in section, hammer is employed and when there is an increase in section forging machine or upsetter is used.

Whenever large changes in section size are required, such as in manufacture of turbine and compressor rotor discs, hammers or hydraulic presses are normally used.

The hammers and hydraulic presses employed in forging subject the metal to different deformation processes. The hammer operates by the application of successive blows, each deforming the material closer to the final form, while the press applies a continuous pressure to complete the deformation is one squeeze.

Mechanical presses are stroke-restricted machines since the length of the press stroke and the available load at various positions of the strokes represent their capability. Hydraulic presses are load-restricted machines because their capacity to do a forging process mainly depends upon their maximum rating. The ram stroke in mechanical presses (the most commonly used is eccentric press) is shorter than in a forging hammer or hydraulic press, therefore, these presses are best suited for low profile forgings.

Hammers are energy-restricted machines. During a working stroke, the deformation proceeds until the total kinetic energy of the hammer is dissipated by plastic deformation of the material and by elastic deformation of the ram and the anvil, when the die faces contact each other. Thus, the practice of rating the capacity of a hammer by the weight of its mobile parts is not useful. The weight of its mobile parts can be regarded only as a specification. The more appropriate method of rating the capacities of hammers is in terms of energy, e.g., kN.m or MN.m.

The initial cost of a hydraulic press is higher than that of a mechanical press of equivalent capacity. However, in a hydraulic press, pressure can be changed as desired at any point in the stroke by adjusting the pressure control valve. This will help in controlling the rates of deformation according to the metals being forged. But in hydraulic presses, the contact time between the workpiece and the dies is more as compared to a mechanical press. Due to this the die life in hydraulic presses is sometimes shortened because of heat transfer from hot work piece to the dies.

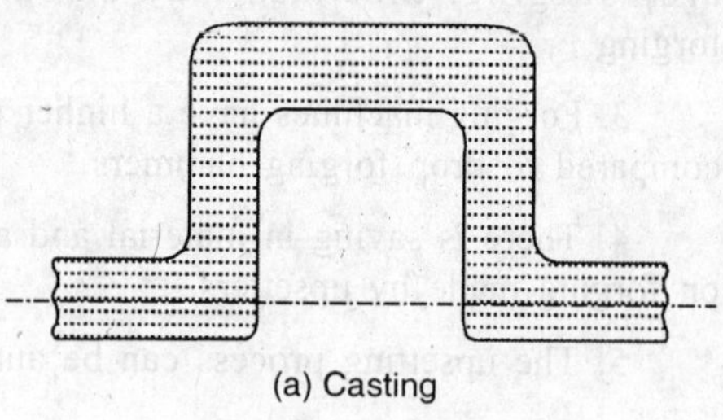

(a) Casting

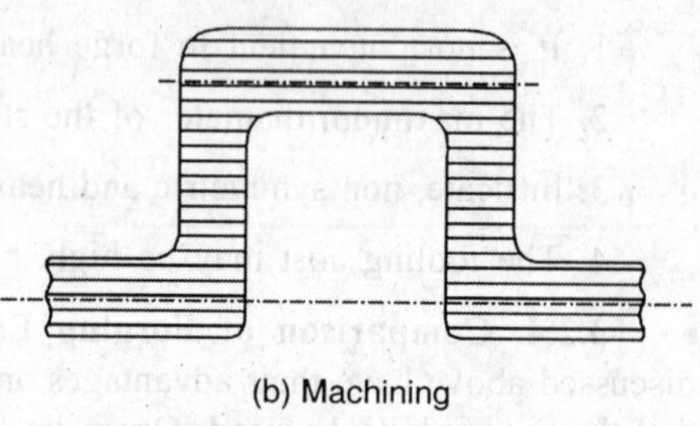

(b) Machining

Another basic difference between various types of forging equipment lies in their forging velocities or rates of deformation. Forging hammers, for instance, deform metals at rates of deformation of the order of 100 times faster than hydraulic presses, that is, their deformation velocities range from 2.5 to 10 m/s as compared to 0.025 to 0.35 m/s for hydraulic presses. In mechanical presses, the velocity ranges from 0.15 to 1.5 m/s. Screw presses have characteristics between mechanical presses and hammers.

3.3. DESIGN OF A FORGING

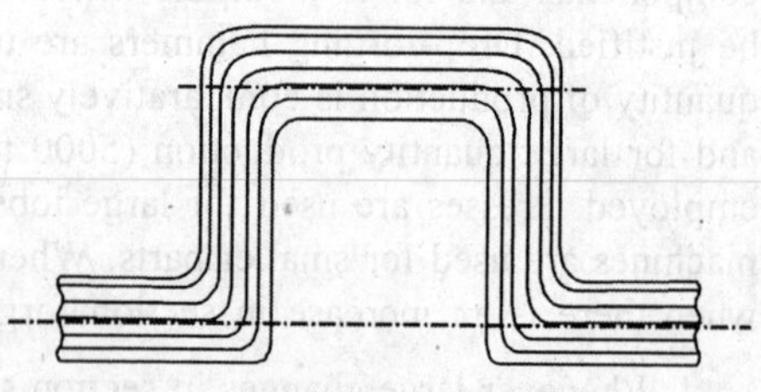

(c) Forging

Fig. 3.8. Comparison of Grain Flow.

The reason of high strength-to-weight ratio of forged parts is that these have fibrous structure and the grain structure or the flow lines of the metal are not interrupted but are made to follow the contour of the forged part. The main objective of good forging design is to control the lines of metal grain flow so that a part with greatest strength and resistance to fracture is produced. This is clear if we compare the three methods of manufacturing crankshaft : Casting, Machining and Forging, Fig. 3.8. The crankshaft produced by casting [Fig. 3.8 (*a*)] has no grain flow and so has the poorest mechanical properties. In Fig. 3.8 (*b*), the crankshaft has been made by machining from a bar stock and the fibre of the metal gets interrupted and for this

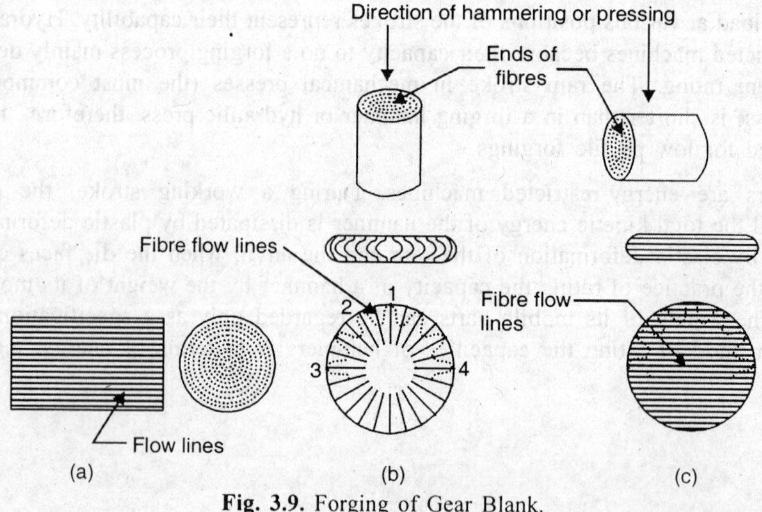

Fig. 3.9. Forging of Gear Blank.

reason the mechanical properties of this shaft will be poorer than those of the crankshaft made by forging [Fig. 3.8 (*c*)], where the fibre of the metal has not been interrupted and continues along the entire length of the shaft.

It is clear from Fig. 3.8 (*c*), that the effect of flow lines is to produce marked directional properties in the material and here (in the case of crankshaft), these qualities add toughness to be webs and where they join the round parts of the shaft. Controlling the directions of fibrous structure within forgings will develop the maximum mechanical properties for applications where shock and fatigue are encountered. For static loads, the direction of these fibre lines is not so important. Below, we will explain how flow lines are used to advantage in forging typical components.

(*i*) **Gear Blank.** Flat forgings of gear blanks are usually made by upsetting the bar stock. The pattern of the flow lines in the bar stock (which has been produced through forging or rolling) is shown in Fig. 3.9. (*a*). When this bar stock is upset on ends (that is, the bar stock is placed in dies such that the flow lines are vertical), the grain pattern produced will be radial and the gear blank will have flow lines concentrated where they will give greatest strength to the teeth, Fig. 3.9 (*b*), and all the teeth will be equally strong. This is desired because each tooth can be considered as a cantilever beam and the fibre flow lines will be parallel with the expected tensile and compressive stresses along the face of the gear teeth. However, if the bar stock is placed in dies for upsetting in such a manner that its flow lines are horizontal, then the grain pattern produced will be as shown in Fig. 3.9 (*c*). The teeth cut on such a blank will not be of the same strength. Teeth 1 and 2 in Fig. 3.9 (*b*) will withstand less stress than teeth 3 and 4, because they have fibres running perpendicular to the expected stresses.

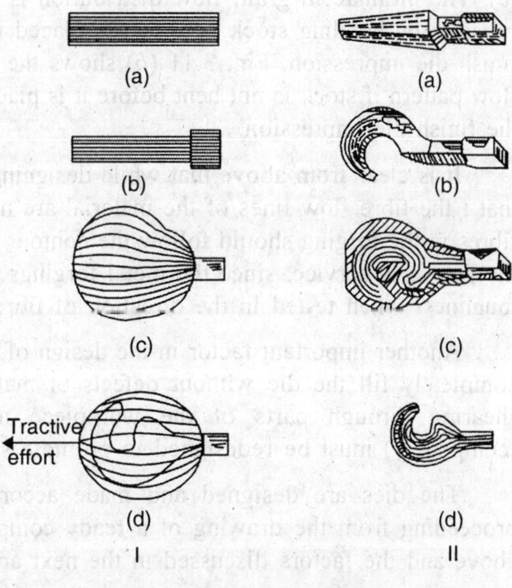

Fig. 3.10. Forging of Coupling Hook.

(*ii*) **Coupling Hooks.** Coupling hooks are subjected to severe service conditions and the safety of the railway traffic depends upon the strength of the hook. Therefore, the coupling hooks should be forged in such a manner that : (*a*) due consideration is given to the direction of the fibre of the material (*b*) the fibre of the material is not severed during the forging process. Also, it must be ensured that the highest possible tensile strength is developed in the hook. Below, we compare the two methods of forging a coupling hook.

Method A. In this method, Fig. 3.10-I, the sequence of operations is as follows :

(*i*) Cutting the bar stock to proper length Fig. 3.10 (*a*)

(*ii*) Drawing out the head, 3.10 (*b*)

(*iii*) Flattening the head, Fig. 3.10 (*c*)

(*iv*) Forging in dies, Fig. 3.10 (*d*)

(*v*) Cutting out jaws and trimming the flash.

It is clear from Fig. 3.10 (*d*), that in order to give bent shape to the hook, the metal is cut and the flow lines are severed at the nose of the hook which is the most dangerous section.

Since the tractive force is directed along and not across the flow lines, there is danger that failure will occur at the section.

Method *B*. In this method, Fig. 3.30-II, after drawing out the head of the hook, the material is bent and then forged in dies. In this way the flow lines will follow the contour of the hook which will result in maximum tensile strength of the hook. The mechanical properties of the hook produced by this method will be much higher than of the hook produced by the first method, because the direction of the attractive effort now coincides with that of the fibre.

(*iii*) **Bell crank forging.** The forging shown is Fig. 3.11 will look the same when completed, but the grain pattern may be detrimental to the strength of the part. The bell crank can be produced more economically without bending, Fig. 3.11 (*b*). However, if the part is to be highly stressed in the areas of transverse grain structure, it would be better to perform the bending operation, Fig. 3.11 (*c*). The ultimate in grain flow distribution is made possible by bending stock before it is placed in the finish die impression. Fig. 3.11 (*b*) shows the grain flow pattern if stock is not bent before it is placed in the finish die impression.

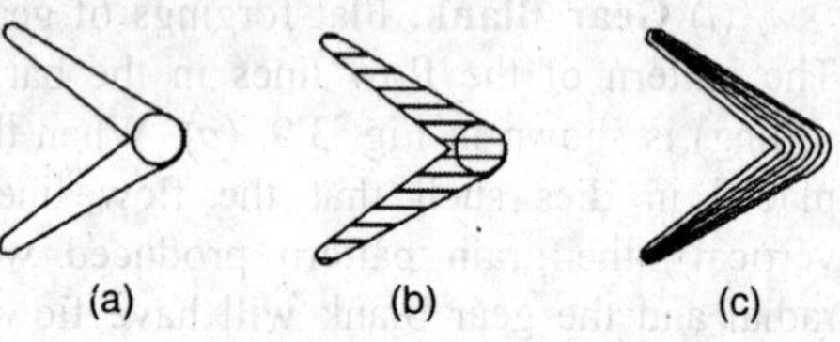

Fig. 3.11. Bell Crank Forging.

It is clear from above that while designing a forging and forging tools, it must be ensured that : the fibre flow lines of the material are not severed during the process of forging and the fibres in the forging should follow the contour or lie in the direction of maximum stresses when the part is in service, since the metal forgings with marked fibre have much more ductility and toughness when tested in the direction of fibre than when tested across the fibre.

Another important factor in the design of a forging is the material flow. The material must completely fill the die without defects of material flow, such as, pinching, folding down, or shearing through parts of the workpiece material. Therefore, the shape of the forging (component) must be redesigned to promote smooth material flow.

The dies are designed and made according to the forging design which is made up proceeding from the drawing of a ready component, taking into account the factors discussed above and the factors discussed in the next article.

3.3.1. Forging design factors

Before designing the tools to produce a given forging, the shape of the final forging has to be determined. There are certain underlying principles for achieving a practical and economical forging design. The tool engineer must have a clear understanding of these principles or factors, which are discussed below :

1. **Draft.** It is the angle of taper put on all sides of the forging to facilitate its quick removal from the die cavity after forging. In case of drop forging and press forging, the usual values of draft angle are :

3° to 7° for external surfaces

and 5° to 10° internal surfaces

If automatic ejection devices are employed to free the forging from the die cavity, the draft angle can be reduced to 1°. Draft is appreciably lowered in case of forging machine since the stock is firmly held by gripping dies. It is sometimes as small as 1/2°. The application of a draft on a forging is shown in Fig. 3.12.

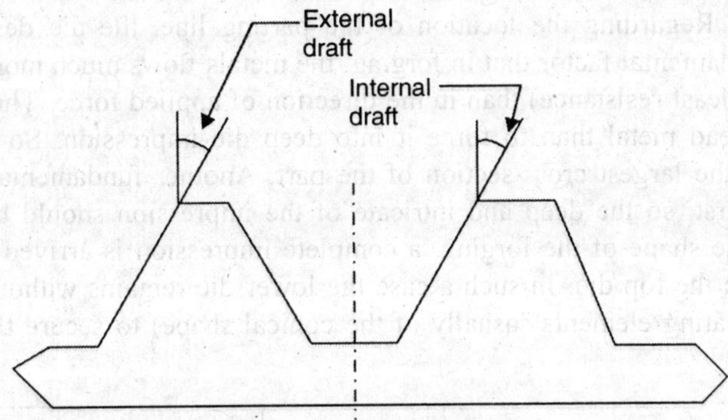

Fig. 3.12. Application of Draft.

2. **Fillet and corner radii.** A fillet means the rounding of the apex of an internal angle and corner radius means the rounding of the apex of the external angle, Fig. 3.13. Sharp edges on the body of the forging and hence the die cavity increases the tendency towards forging defects and accelerate die wear. Also sharp edges will hinder the complete filling of the die cavities. Therefore, generous fillet and corner radii are the most desirable features of a closed-dic forging bccausc it assists thc flow of hot metal and eliminates the possibility of forging laps or shuts. Also, the premature die-failures due to stress cracks and abrasions are prevented. Hence, larger the fillet and corner radii, longer will be the die life and better will be the forging quality. The usual values of fillet and corner radii are given in table 3.1.

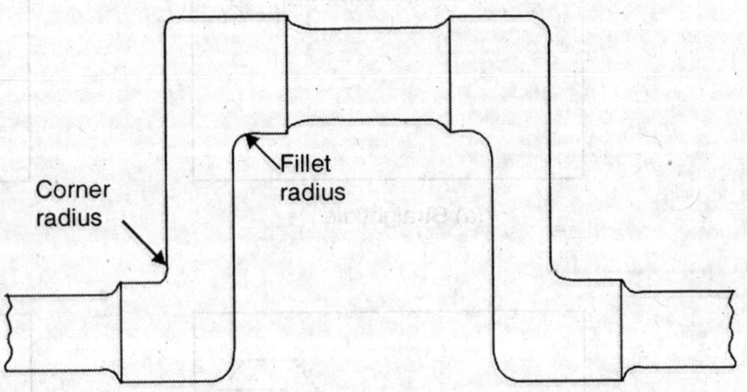

Fig. 3.13. Fillet and Corner Radii.

Table 3.1. Fillet and Corner Radii.

Net mass of Forging, kg	*Fillet radius mm*	*Corner radius mm*
Upto 2.25	4	2
2.25 to 4.5	6	3
4.5 to 13.5	8	4
13.5 to 22.5	10	5

Fillet radius can also be found as follows :

$$R_f = \frac{1}{3}\left(\frac{a+b}{2}\right),$$

where a and b are thicknesses of the two parts that are joined at the fillet.

3. **Parting line.** The parting line is the line along the forging where the two halves of a pair of forging dies meet. It divides the die impression into two parts from which one is made in the top die and the other in the lower die. The shape and location of the parting line is very important as these have considerable influence on the flow of metal, die cost and draft

requirements, etc. Regarding the location of the parting line, the die designer should always remember the fundamental factor that in forging, the metals flows much more easily in the lateral direction (path of least resistance) than in the direction of applied force. Thus in forging process, it is easier to spread metal than to force it into deep die impression. So in most forgings the parting line is at the largest cross-section of the part. Another fundamental factor is that metal fills the top die first, so the deep and intricate of the impression should be cut in the top die. When owing to the shape of the forging, a complete impression is arrived at in one part of the die, this should be the top die. In such a case the lower die remains without the impression and will have only locating elements (usually of the conical shape) to secure the proper location of the forging.

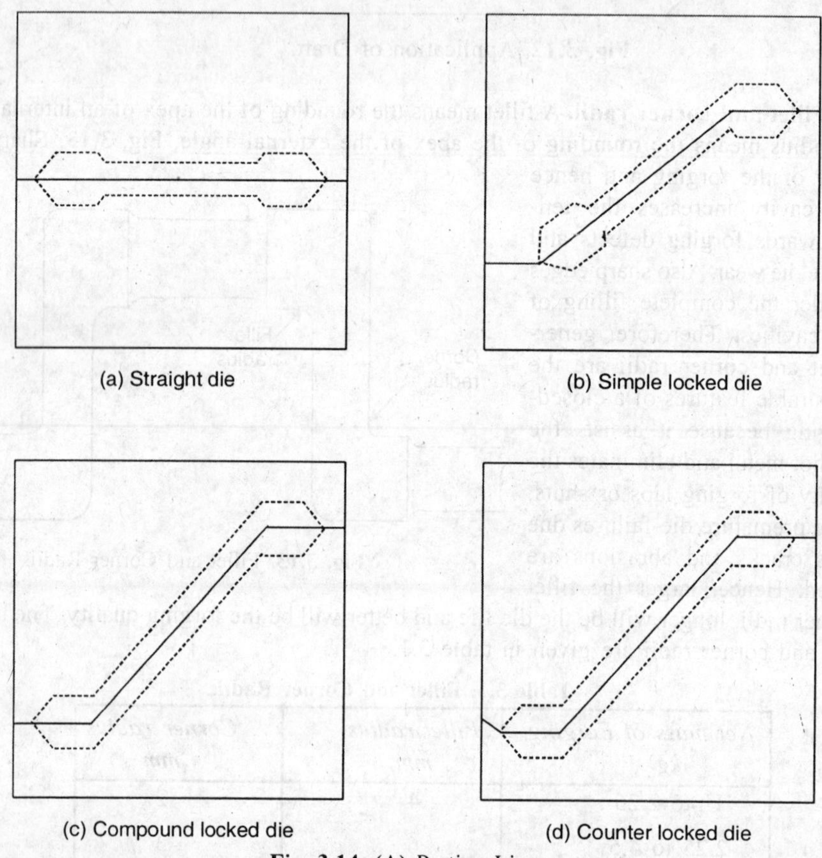

Fig. 3.14. (A) Parting Lines.

Regarding its shape the parting line may be perfectly flat (straight) or multidirectional. The straight die, Fig. 3.14 A (*a*) is preferable since no side thrust is present and it is cheaper to be produced. However, the forging produced may not be economical. Many forgings require a parting line that is not flat but is multidirectional or irregular, Fig. 3.14 A (*b, c, d*). Such dies will mesh or lock in vertical direction when closed and are called locked dies. The irregular parting lines tend to cause side thrust and add to the cost of the die. The side thrust during forging may cause mismatch of the dies and/or breakage of the forging equipment as the ram guides may not be fully adequate to ensure complete alignment of the dies, in the presence of the side thrust. The various types of locked dies are shown in Fig. 3.14A. Side thrust is present in simple locked dies and it may be present in compound locked die. There are several ways of eliminating or controlling side thrust. One way is to provide counterlocks or hedges by

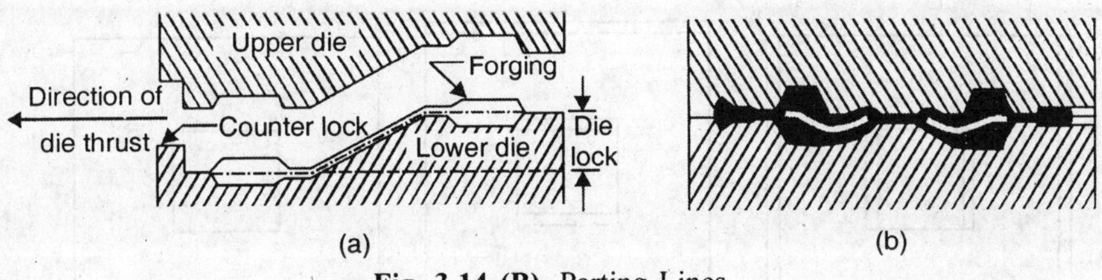

Fig. 3.14 (B). Parting Lines.

machining mating projections and recesses into the parting surfaces, Fig. 3.14A (*d*) and Fig. 3.14B (*a*). The height of the counterlock is usually equal to or slightly greater than the depth of the locking portion of the die. The thickness of the counterlock should be atleast 1.5 times the height so that it will have adequate strength to resist the side thrust. It is very difficult to maintain proper lubrication at the sliding parts of the counterlock due to high die temperature involved. Due to this the surfaces of the counterlock wear rapidly and need frequent reworking. So, counterlocks should be avoided as far as possible by inclining, rotating of individual forgings or by changing the design of forging. If the part is small in size and its production is large, forging two parts at a time, side thrust forces can be balanced in a multiple part die, Fig. 3.14B (*b*). From the above discussion, it is clear that a flat horizontal parting surface should be selected as far as possible.

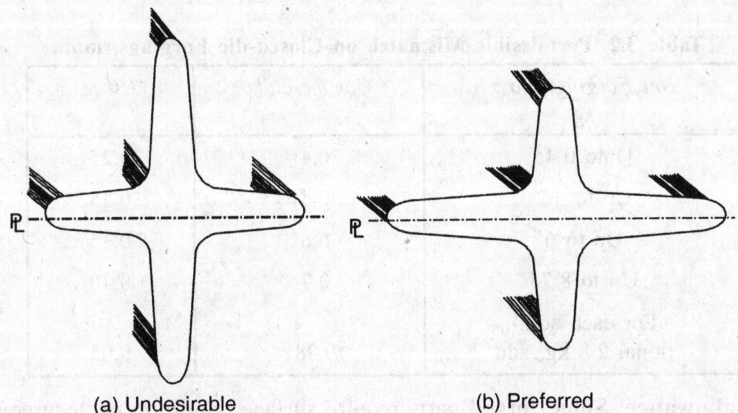

Fig. 3.14. (C) Parting Line.

Another factor in the selection of parting line is that it should avoid deeper die impressions to minimize die wear, Fig. 3.14(C). Deeper die impressions would require high forging pressure for complete filling and might lead to die breakage. A proper parting line may eleminate the chances of grain flow reversal and improve the mechanical properties. A parting line which may not require any additional draft provided for easy removal of forging from die, will result in large savings in machining cost and raw material.

4. **Shrinkage and die wear.** Before forging, the work material is heated to the forging temperature. During the forging operation, it is in the process of cooling and consequently shrinking. To take into account the expansion of the material at high forging temperatures, the die cavities are made correspondingly larger by using a shrink scale. For steel, this allowance is about 16 mm per metre.

Due to continuous use, the dimensions of the die do not remain the same. Die wear is the difference in dimensions which occurs due to abrasion of the die impression. This is accounted for in the forging design as follows :

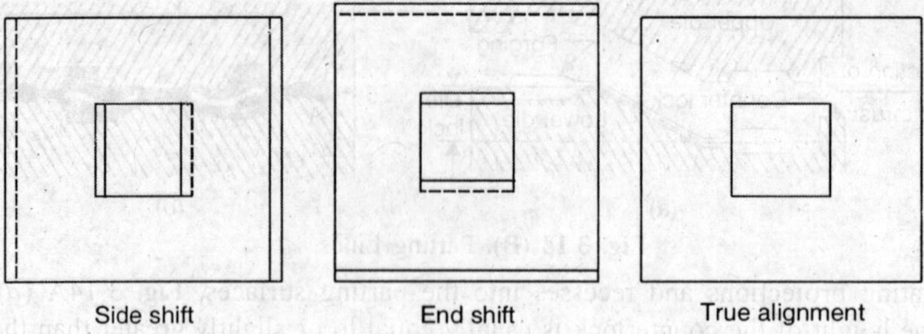

Fig. 3.15. Mismatch.

For forgins weighing up to about 45 N, the die wear allowance is taken as 0.4 to 0.8 mm for external as well as for internal surfaces. For forgings weighing from 90 to 225 N the allowance may increase from 1.6 mm to 2.4 mm.

5. **Mismatch.** In closed die forging it is very difficult to achieve perfect alignment of the two die halves and either the upper or the lower die may shift during forging. This shift may occur sideways or endways as shown in Fig. 3.15. The forging produced by shifted die will be mismatch. The mismatch should be kept minimum. The amount of permissible mismatch is given in Table 3.2.

Table 3.2. Permissible Mismatch on Closed-die Forgings, mm

Net Forging, mass kg	*Commercial*	*Close*
Upto 0.45	0.4	0.25
Up to 3.2	0.5	0.35
Up to 6	0.6	0.35
Up to 8.75	0.7	0.40
For each additional 2.8 kg, add	0.08	0.05

6. **Finish allowance.** Some forged parts require surface condition or accuracy which may not be possible to obtain during forging. For this, the parts will have to be subsequently machined. For this purpose extra material is provided on the forging which may vary from 0.8 mm to 3.2 mm depending upon the material and relative size of the forging.

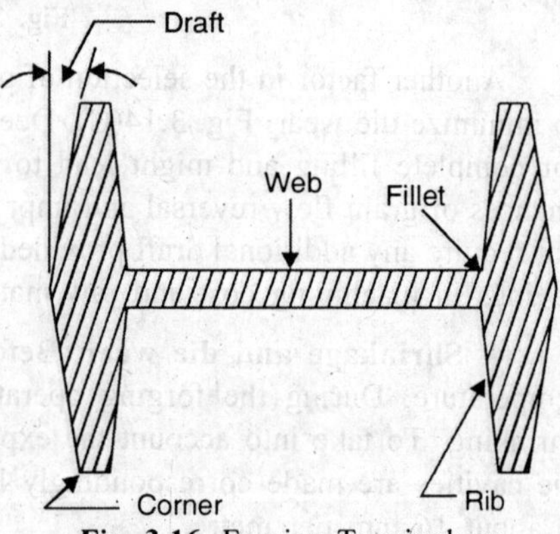

Fig. 3.16. Forging Terminology.

7. **Dimensional tolerances.** These are the variations permitted from the given or normal dimensions and these may include : thickness tolerance and length tolerance. Commercial tolerance for thickness are as given below :

On forgings upto 0.45 kg – 0.25 mm to + 0.80 mm

On forgings from 2.25 to 9 kg – 0.60 mm to + 1.60 mm

On forgings from 9 kg to 22.5 kg – 0.80 mm to + 2.40 mm

(8) **Webs and Ribs,** Fig. 3.16. A web is usually the thinnest portion of a forging. It will cool first and when it goes below the forging temperature, the forging pressure required increases rapidly. So, webs less than about 4.74 mm thick are not usually practical. The ribs should be proportionately low and wide. Their height should not be more than 8 times the width. The minimum recommended rib thickness is equal to that for webs, Fig. 3.16.

3.4. DIE-DESIGN FOR DROP FORGING AND PRESS FORGING

As already noted the dies are made in sets of halves. One-half of the die is attached to the ram and the other to the stationary anvil. The die halves may be having one or more than one impressions. In single impression dies, die impression is the finishing impression, the preliminary forging operations are done on other machines such as forging rolls, upsetter and benders etc. Multi-impression dies may have two (blocking and finishing) or more than two impressions. In these dies, the final shape of the forging is progressively developed over a series of steps from one die impression to the next. Each impression gradually distributes the flow of metal. and changes the shape of the work-piece as it is transferred from one impression to the next between stroke.

The art of forging die design aims at determining the minimum number of steps that lead from the starting material (usuallly a round or rectangular bar) to the finished shape.

3.4.1. Preliminary Forging Operations

For a multi-impression die, the preliminary forging operations or the preform operations that are usually required in shaping the part are generally classified as :

1. Fullering or swaging
2. Edging or Rolling
3. Bending
4. Drawing down or drawing out or cogging
5. Flattening
6. Blocking

The other operations on such a die are :

7. Finishing operation
8. Cut off

1. **Fullering.** Fullering is usually the first operation performed on the heated bar and its primary function is to reduce the cross-sectional area of the stock. The fullering impression looks like small mountain on each of the die block (Fig. 3.17). The stock is turned 90° between each

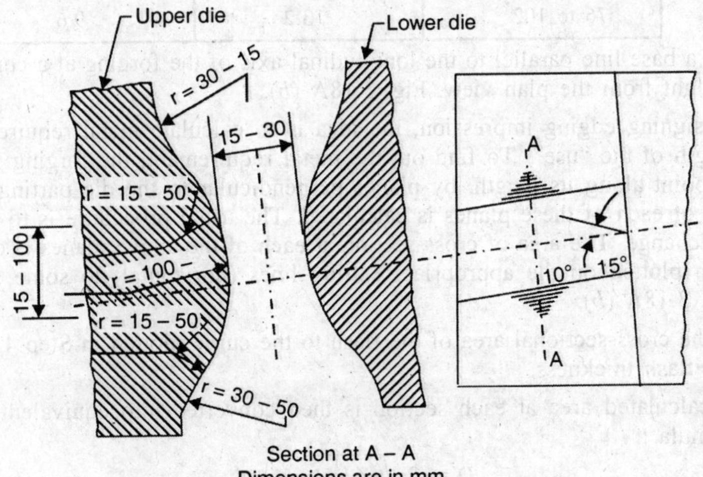

Fig. 3.17. Fullering Impression.

blow and is slowly progressed to obtain the required fullered length. As a rough rule, the gap between the two impressions should be kept equal to the thickness of forging less 1.6 mm to 5 mm. The length of the fullering impression depends upon the length of the fullered stock. Generally, it is taken as 1/4th the length of fullered stock or 1/2 times the length of the original stock required for fullering whichever is less. Fig. 3.17 gives fuller dimensions only for general guidance and are for bars of cross-section varying from 20 × 20 mm upto 60 × 60 mm.

2. **Edging or Rolling.** Edging or Rolling impression also called "pre-form impression" is probably the most important auxiliary impression. Its function is to distribute the metal longitudinally by moving metal from the portions where it is in excess to the portion which is deficient in metal and to iron out, possible imperfections in the fullered 'use'. Design of the edging impressions involves some calculations.

The following procedure can be used to design the edging impression. Refer Fig. 3.18 A :

1. The plan view of the forging is traced out on a tracing paper. If the forging is complicated, the side view may also be traced out.

2. Next, layout an estimated outline of the flash around the forging, Fig. 3.18 A (*a*). Flash or excess metal extruded from the finishing impression during forging acts as a cushion for impact blows and as a pressure relief valve for the almost incompressible work material. The flash leaves the die cavity through a narrow passage provided all round the die cavity (Fig. 3.19). Due to this it restricts the outward flow of the material and thereby helps in the complete filling of the die cavity including the ribs and the bosses. There are many empirical rules for calculating the dimension of flash (Refer Fig. 3.19) : Flash thickness or clearance = 0.5 mm (minimum) upto 9.525 mm for forging weighing upto 90 kg

or a reasonable guide for flash thickness = 3% of the maximum forging thickness

Minimum flash width per side = 6.35 mm

or the dimensions of flash can be taken from table 3.3 given below :

Table 3.3. Dimensions of Flash

Stock Size, mm	*Flash Dimensions, mm*	
	Thickness	*Width*
Upto 38	0.8	4.8
38 to 50	1.2	5.6
50 to 63.5	1.6	6.4
63.5 to 76	2.4	8.0
76 to 102	3.2	9.6

3. Draw a base line parallel to the longitudinal axis of the forging at a convenient distance towards the right from the plan view, Fig. 3.18A (*b*).

4. In designing edging impression, the idea is to calculate metal required at every point along the length of the "use". To find out the metal requirement, the forging is assumed to be cut at every point along its length, by planes perpendicular to the die parting surface and the metal content of each of these planes is calculated. The usual procedure is to take a section at every section change. The area of cross-section at each of the above planes is calculated. These areas are then plotted on the appropriate cutting lines (horizontal) to some suitable scale as shown in Fig. 3.18A (*b*).

5. Add the cross-sectional area of the flash to the curve plotted in Step 4. Area of flash = flash width × flash thickness.

6. The calculated area at each section is then converted into equivalent circular section using the formula :

$$D = 2\sqrt{(A/\pi)}$$

where A is the area of each section as plotted in step 5.

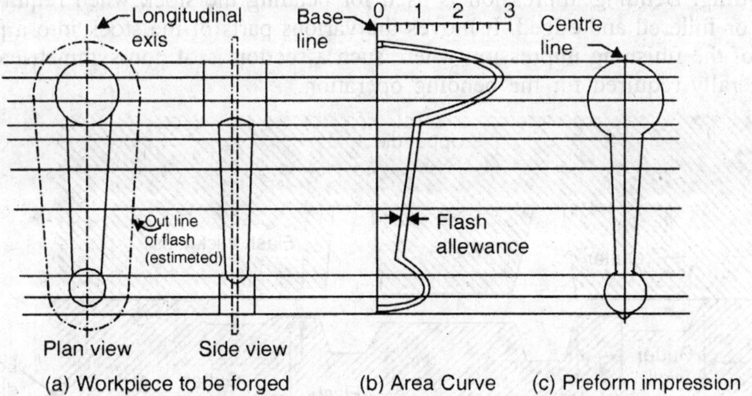

Fig. 3.18A. Designing Edging Impression.

7. Next, to trace the edging impression, the centre line of the impression is drawn in the extreme right corner.

8. Then corresponding to each section a point at a distance equal to the radius of the equivalent circular section, from the centre line of the edging impression is plotted on the tracing paper.

When all such points have been plotted, these are joined by a smooth curve. While joining the points, one should keep in mind the requirements of smooth metal flow, Fig. 3.18A (*c*).

From the edging impression, the dimensions of the blank needed to produce a forging can be calculated. Minimum diameter of blank will be equal to the maximum diameter of the preform and the length of the blank will be equal to the total volume of the preform impression divided by the area of the largest cross-section of the edging impression. The detailed procedure is given ahead under Art. 3.6.

The next impression is to design the front and back portion of the edging impression. In the front, a necking impression is provided. In the rear, the impression can be completely closed, but it is not desirable, as the operator may not be able to gauge the length exactly and some extra length may be there which may introduce a defect in the forging if space is not provided for its escape. This is usually obtained by providing 6.5 mm opening between the two dies in their closed positions. While tracing the edging impression, the centre line of the edging impression is shifted to 3.25 mm towards the points to allow for the opening between the dies in operation. If this is not provided, the metal obtained in the edging will be too much in excess and all that metal will be wasted in the form of flash.

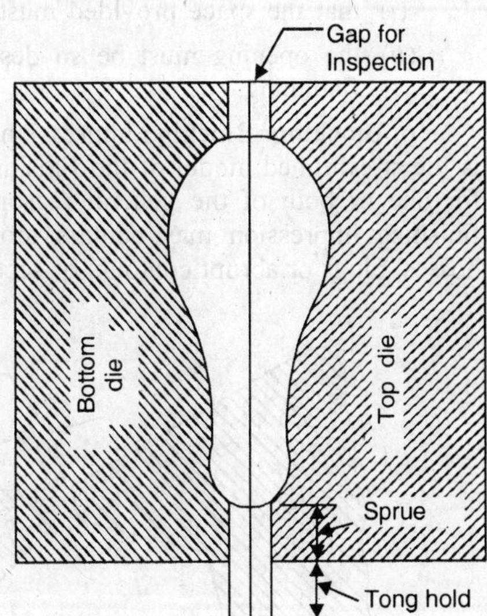

Fig. 3.18B. Edging Impression.

The edging impression is very important one and all the calculations must be done carefully because if the metal is less than what is required, the die cavity will not be filled up and the forging will have to be rejected. If on the other hand, the metal is more than the required, the metal will be wasted which will be very serious when large number of forgings are to be produced. The edging impression of the straight parting line job is not very difficult to design but when complicated jobs with parting lines in more than one plane are to be forged, the design of the edging impression is quite complicated. Fig. 3.18 B shows an edging impression for connecting rod.

3. **Bending.** Bending impression is used for bending the stock when required after it has been edged or fullered and edged. It moves the various parts of the stock into a proper relation with shape of the finishing impression where such a section is of non-symmetrical section. One blow is generally required for the bending operation.

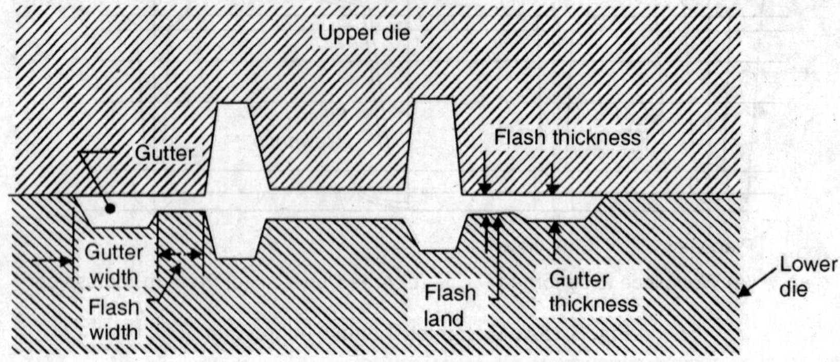

Fig. 3.19. Flash Land and Flash Gutter.

4. **Drawing out.** Drawing out is an operation similar to rullering with a difference that fullering reduces the stock between the two ends of the stock at a central place, whereas the drawing operation reduces the stock size only at one end.

5. **Flattening.** Sometimes there is a need for flattening the stock before passing it on to the final impression. This is done in flattening impression, usually very simple and situated in one of the front corners of the die block. In designing the flattening impression one should keep in mind two things :

(*i*) that the space provided must be large enough to accommodate the flattened stock.

(*ii*) the opening must be so designed as to give the required height of the stock after flattening.

6. **Blocking.** Blocking impression or the blocker also called as "semi-finishing impression" is the streamlined model of the finishing impression and is required on some types of forgings for one or both of the general reasons. The first and the more important reason is that the finishing impression may contain too many obstructions in the form of depressions, holes, bosses, plugs or abrupt contours or section changes, to permit a normal flow of metal to all parts

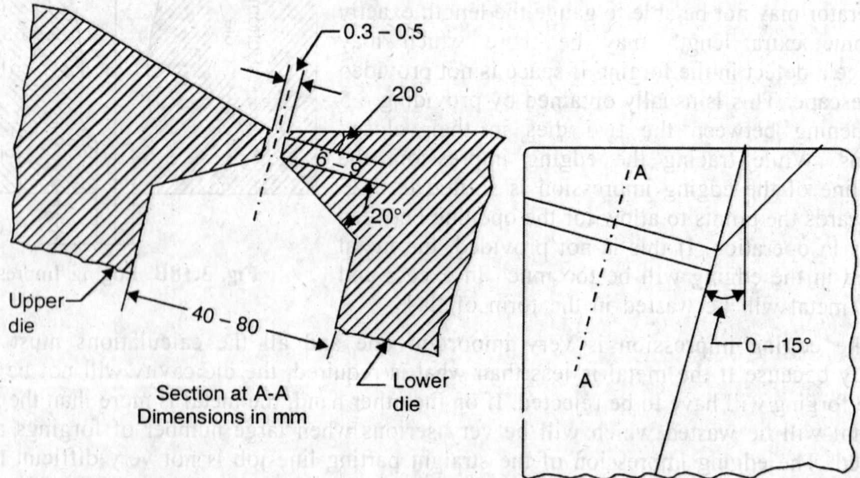

Fig. 3.20. Cut Off Impression.

of the impression without further preparations from the preceding operations. The blocking impression has the general shape of the finishing impressions. The blocking impression has the general shape of the finishing impression but with all the corners, holes and abrupt section changes thoroughly rounded so that plastic metal may be moved into suitable position for more exact shaping the finishing impression. Blocking impression aids in preventions of forging defects such as cold shuts. Second reason for the use of blocking impression is to reduce the wear of the finishing impression. Excessive wear of the finishing impression reduces the useful die life. In case of certain forgings which are symmetrical in shape, the blocking impression is only a preparatory impression. In blocking impression, the length and width are smaller, height or thickness is more, but the centre distance is the same as in the final impression. The difference may be 1 to 2 mm on each side, but higher clearance of 3 to 5 mm can be recommended where partial displacement of metal occurs.

7. **Finishing Impressions.** This impression represents the exact shape of the finished forging. The shape and size of the finishing impression is checked in the process of manufacture of the die by lead cast. The finishing impression is located in the middle of the die block, but it is not necessarily in its central axis. However, it is vital to locate the final impression in such a manner that there will be no horizontal forces which give a side thrust and may cause die shift. It is thus advisable to have the load centre of the forging directly below the axis of the hammer tup or ram.

Flash gutter. The flow of plastic metal under the blows of the drop hammer or the pressure of the forging press, proceeds first to fill up the finishing impression and then a small quantity of the extra metal moves into the shallow cavity provided around the finishing impression of the die. These small cavities which are directly outside the die impression are known as the flash gutter. The flash gutter is separated from the die impression by a narrow passage which is the flash land or flash pan. The volume of the flash land and flash gutter should be about 20 to 25% of the volume of forging, Fig. 3.19.

The amount of excess metal from the finishing impression may be too large to permit the complete closing of the dies. Gutter is provided to ensure the complete closing of the die. It acts as a storage for the excess material after it has passed through the flash land. Too large a gutter reduces the striking area for the die surfaces. Alternatively, too small a gutter will result in extrusion of flash between the striking surfaces of the dies, resulting in an over-size forging. The dimensions of gutter may be taken from table 3.4.

The flash cools rapidly and presents increased resistance to deformation. This builds up pressure inside the bulk of the workpiece (within the die-cavity) that helps in the complete filling of the die. Closed-die forging (Impression die forging without flash) does not depend upon the formation of flash to achieve complete filling of the die. Due to this, the workpiece volume has to be calculated very carefully so that it is completely contained in the dies without generating extreme pressures in the dies. One approach is to use roll-formed shapes or extruded preform shapes. These preform shapes can be sliced up and forged into individual components.

Table 3.4. Gutter Dimensions.

Stock Size, mm	*Gutter Dimensions, mm*	
	Thickness	*Width*
Upto 38	3.2	25.4
38 to 50	4.8	25.4 to 31.75
50 to 63.5	4.8	31.75 to 38
66 to 75	4.8	31.75 to 38
76 to 100	6.4	38 to 44.5

8. **Cut off.** When the forgings are made from the bar stock, they must be cut off after the forging operation is completed. This is done either by a special side cutter of the trimming press or by the cut off impression milled usually in the left back corner of the die block, Fig. 3.20.

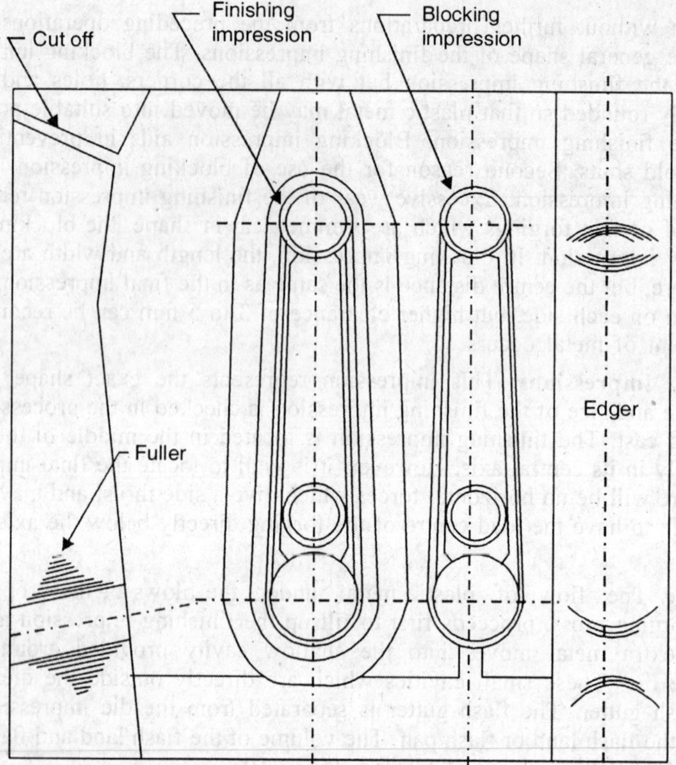

Fig. 3.21. Plan of a Multi-impression Die.

Fig. 3.21 shows the plan of a multi-impression die for the manufacture of I.C. Engine connecting rod.

3.4.2. Shape of the forging and preform design

The preliminary or preform operations (preform design) needed will depend upon the shape of the forging. In the case of forgings of simple shapes, preform design before finishing impression may not be necessary or economical. However, for forgings having wide variations in sections or of irregular shape, preforming operations before the finishing operation will be necessary for improving die life. While designing preform impressions, it must be ensured that minimum deformation is required to achieve the final shape in the finishing impression. The forgings can be classified as given below :

Class	*Shape*	*Dimensions*	*Typical examples*
1.	Compact	Length, width and height same	Hubs, small bevel gears etc.
2.	Flat or Pancake shape	Two dimensions same, third dimension very small.	Gear blank
3.	Irregular shape	Length greater than other dimensions	lever, Connecting rod, etc.

For class 1 and class 2 forgings, only blocker impression will be required, but for forgings of class 3, fullering, edging/rolling or bending impressions may be needed in addition to the blocker impression, depending upon the nature of the forging.

3.4.3. Die-block dimensions

The dimensions of the die-block depend upon the length of the finish forging impression, depth of the impression and the number of impressions in the die-block.

For a single-impression die, the length of the die-block may be taken as,

$L = l + 3h$ (minimum), Fig. 3.22, and breadth,

$B = c \times b$

where l = Total length of forging impression

h = maximum depth of the impression

b = maximum width of the impression

c = constant

= 3 for b upto 5 cm

= 2.5 for b upto 25 cm

= 2 for b above 25 cm

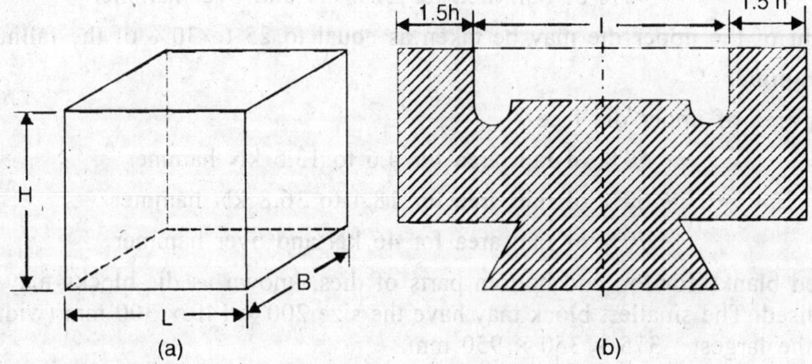

Fig. 3.22. Die-Block Dimensions.

The height of the die-block determines the maximum impression depth, since adequate die material must be there between the bottom of impression and bottom face of the die block to provide strength in the die.

From the strength point of view of the dies and the die wear, the ratio of h and b is given below :-

Table 3.5

Material	h/b	
	$l = b$	$l \geq 2b$
Al, Mg	1.0	2.0
Steel, Titanium	1.0	1.5

For multi-impression dies, a gap of at least 25 mm should be left between two adjacent impressions.

As per German handbooks, the inter impression distance, a_1, the distance of impression from the edge of the die block, a, and the height of the die block H, in terms of the maximum depth of the impression, h are given below (refer Fig. 3.22 A):

Table 3.6. Dimensions in mm

h	a	a_1	H
6	12	10	100
10	32	25	125
40	56	40	200
100	110	80	315

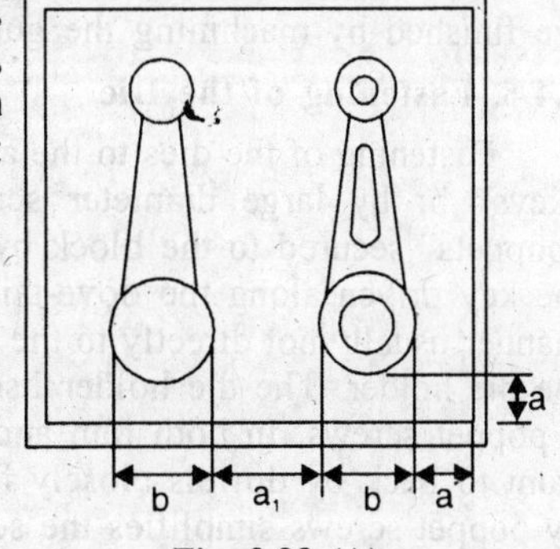

Fig. 3.22 (A)

So, once the length, width and depth of forging impression are known (from the design of the forging), the dimensions of the die-block can be determined as explained above.

The face area of the die-block may also be taken as :

500 mm^2 per 1000N of tup weight.

According to one manufacturer; for rams of standard forge alloy steel, the face area of the die block may be taken as given below :

Gravity drop hammer

Minimum area of upper die face

= 30% of ram area for 2.3 to 11.5 kN hammer

= 35% of ram area for 13.8 to 23 kN hammer

= 40% of ram area for 27.6 kN and over hammer

The weight of the upper die may be taken as equal to 25 to 30% of the falling weight.

Power drop hammer

Minimum area of upper die face

= 50% of ram area for 4.6 to 13.5 kN hammer

= 60% of ram area for 18.4 to 36.8 kN hammer

= 70% of ram area for 46 kN and over hammer

The forged blanks for top and bottom parts of dies, known as die-blocks may have their sizes standarddised. The smallest block may have the size 200 × 140 × 300 mm (width × height × length) and the largest : 375 × 350 × 950 mm.

3.4.4. Position of Impression in Dies

This question arises in multi-impression dies and not in single impression dies. The final impression in a multi-impression die should be as far as possible located at the centre of the die block. If it is not in the centre, then there is a tendency for the die to shift and it gives the shift of mis-match. The blocking and finishing impressions are cut in the block by highly skilled men who use the milling machine, specially designed for sinking dies. Cutters of various types are used in accordance with the shape of each section of the impression. But much of the accuracy of the die depends upon hand work performed after it is sunk. These days, E.D.M. (Electrical Discharge Machining and E.C.M. (Electro-Chemical Machining) are being increasingly employed to machine intricate cavities in the die-blocks. When the forging die impressions are completed, the die blocks are clamped together in the position in which they will meet in the forging operation and a lead antimony alloy is poured into the finishing impression. The resulting lead cast is used to check the accuracy of the forging dimensions and is sent to the customer for his approval. Since steel shrinks in cooling from its forging temperature and the lead alloy does not, it is necessary to allow for the shrinkage in checking the lead cast. The correction amounts to about 16 mm per metre. After the lead cast has been approved, the dies are finished by machining the gutter and flash land etc.

3.4.5. Fastening of the Die

Fastening of the dies to the anvil block is made by dovetail and long tapered wedges called "keys" or by large diameter screws passing through heavy forged steel blocks known as "poppets" secured to the block by steel cotter. The upper die is secured directly to the ram by the key driven along the dove-tailed shank of the die. The bottom die is fastened in a similar manner usually not directly to the base of the hammer or press but to the large steel block called the die holder. The die holder itself is fixed to the anvil either by dovetail and key or by 4 to 6 poppet screws. In both ram and die holder, the dies are secured against the movement from front to back by dowels closely fitted into recesses provided in ram and die holder. Fastening by poppet screws simplifies the setting and correct matching of the die but poppet screws have a tendency to loosen and it is necessary to tighten them up several times during the day. Fastening of the lower die by poppet screws directly to the base of the hammer without the die holder is used very seldom and only for big hammers. The dovetails of the dies together with long taper keys prevent side movement of the dies and the possibility of longitudinal movement is eliminated by cross keys.

3.4.6. Die Maintenance

The following main rules should be kept in mind for the proper maintenance of forging dies :

1. Before forging begins, warm the dies to a temperature of not less than 150-200°C by placing a heated slab of metal between the top and bottom dies.

2. Before starting to work, always remove scale from hot stock.

3. Do not forge metal which has cooled below the permissible minimum forging temperature.

4. Scale, which is knocked off of forging, should be removed from the die surfaces by blowing off with jets of compressed air.

5. To prevent forgings from sticking in die impressions, the die surfaces should be swabbed or sprayed lubricating oil before each new forging operation.

6. The forging dies should not get overheated (temperature above 400°C). If they do get overheated, cool them with compressed air.

3.5. DIE DESIGN FOR MACHINE FORGING

In completing a forging in the forging machine, one or more steps comprise the sequence so that metal may not fold upon itself producing a cold shut or cracks in the forgings. These steps are commonly termed as 'passes', 'blows', or 'shots'. To use good forging design, it is necessary to understand the practical laws that govern the action of metal in the forging machine and cover the majority of the forging machine die design problems.

For practical purposes, the laws have been translated into series of rules for use in forging machine practice so as to be certain that the forgings do not develop cold shuts and other injurious defects. The rules apply to all diameter stock to be upset.

Rule No. 1. The limit of length of unsupported stock that can be gathered or upset in one blow without injurious buckling is not more than 3 times the diameter of the bar, Fig. 3.23. If an attempt be made to upset a length of stock longer than 3 times the bar diameter, instead of upsetting uniformly, the stock will buckle at a point near the middle and result in more of the upset forming on one side of the centre of the stock. In practice, it is better than the length of the unsupported stock is within 2.5 times the bar diameter.

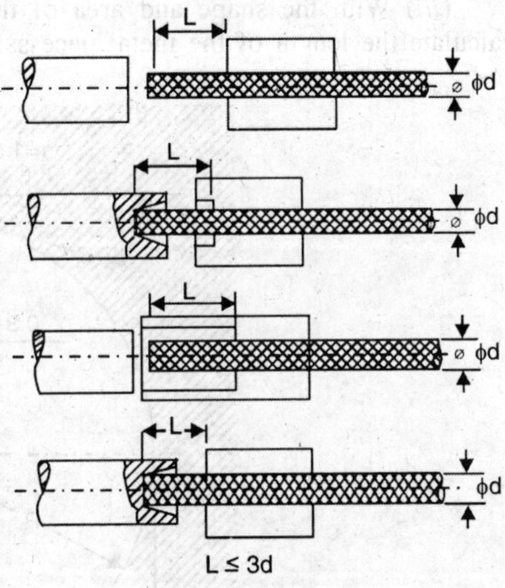

Fig. 3.23. Rule 1.

Rule No. 2. Length of stock more than 3 times bar diameter that is within the limits of the stroke of machine can be successfully upset in one blow, provided the diameter of the upset made is not more than 1.5 times bar diameter. If this is kept more than 1.5d, the buckling will be excessive and the stock will fold in. In practice, it is advisable not to exceed 1.3 times bar diameter Fig. 3.25.

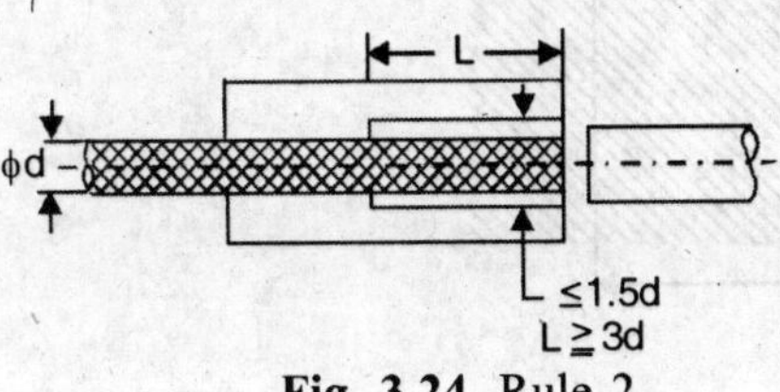

Fig. 3.24. Rule 2.

Rule No. 3. In an upset requiring more than 3d in length when the diameter of the upset is 1.5d, the amount of unsupported stock beyond the face of the die must not exceed one diameter of bar, Fig. 3.25. However, if the diameter of the hole in the die is reduced below 1.5d, then the length of unsupported stock beyond the face of the die can be correspondingly increased.

Rule No. 4. Avoid using head diameter greater than four times the stock diameter. A maximum of 63.5 mm diameter can be upset at one time.

Selection of correct size machine necessary to forge a part should be governed by the following factors :

1. Volume of stock required in the finished forging,
2. Size of stock used.
3. Maximum dimension of the finished forging.
4. Number of blows necessary to complete the forging.

Fig. 3.25. Rule No. 3.

The following sequences are formulated for the purpose of eliminating guess work and creating a more positive method of study for the forging problem. The steps are as given below :

(*i*) Calculate the volume of metal in the part to be forged.

(*ii*) Determine the proper cross-section of metal and shape of the metal to be used to make the forging.

(*iii*) With the shape and area of the cross-section as well as the volume of the upset, calculate the length of the metal necessary.

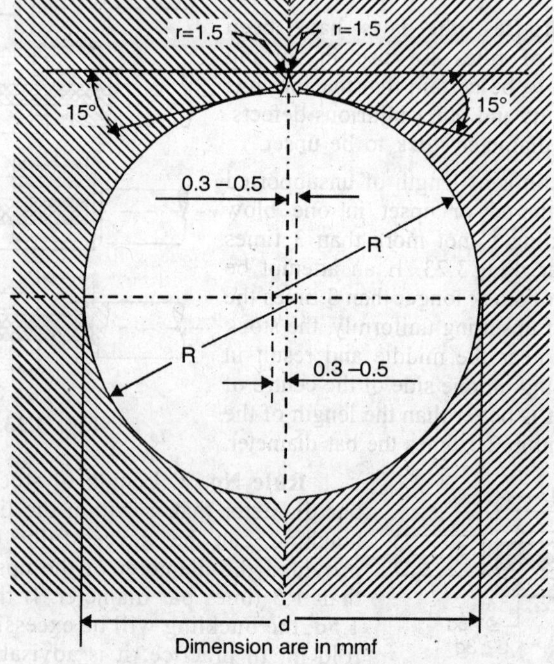

Fig. 3.26. Gripping Impression.

(*iv*) Calculate the number of blows necessary to complete the forging, using the general rules that eliminate folding or buckling.

(*v*) Make a die layout to determine the size of the die blocks and heading tools necessary to accommodate the required number of blows and cavity dimensions.

(*vi*) Determine the size of forging machine to be used, bearing in mind the size of the bar to be used, size of the forging to be made, size of the die blocks necessary, length of header slide, length of stroke, length of gather, length of die opening etc.

(*vii*) Use hot dimensions on all cavities.

(*viii*) Provide for clearance between heading tools and their mating dies when these tools enter the dies.

(*ix*) Provision should be made for proper grip of the stock. The length of this grip should not be less than 3d. Also, the cavity diameter in this area should measure approximately 0.30 to 0.50 mm smaller than the diameter of the bar to be forged. The cross-section of a well designed gripping impression is shown in Fig. 3.26.

(*x*) With the dies and tools designed and the impressions machined in the die blocks, it is advisable to place the tools in the tool holders and make a preliminary set up of the dies and tools in a face plate for final checking before placing them in the forging machine.

(*xi*) Setting of the dies and tools in the forging machine requires a check of parallelism for the die seats and also the travel of the header slide.

(*xii*) A small stream of coolant should be directed on the dies and tools to dissipate the heat and keep the dies free from scale, that may gather in the cavities. The best results may be obtained from a solution of soluble oil and water.

In addition to advantages mentioned under Art. 3.2.3, upsetters have the following plus points :

1. A high degree of accuracy in dimensional tolerances.
2. Die life is increased by minimising the contact time between the dies and the hot metal.
3. Reduction in the man power required as compared to drop forging.
4. Die setting time is less than drop forging for similar jobs.
5. Die manufacturing cost is less as insert technology is very suitable for upsetter dies.
6. Preparation of raw material (preform design) is not needed as the bar stock is directly used in dies.

3.5.1. Die-Block Dimensions

Die-block dimensions for coventary forging machines are given in Table 3.7.

Table 3.7. Die-Block Dimensions, cms

Nomenclature	*Size of Machine*							
	2.5	3.8	5.0	6.25	7.50	10.00	12.50	15
Opening of L.H.die	6.00	7.0	8.25	10.15	11.4	14.0	15.25	17.8
Length of die	22.5	25.0	30.5	35.0	40.0	45.0	55.9	69.9
Height of die	27.3	30.0	36.2	43.2	48.9	54.6	63.5	74.95
Thickness of die	9.5	10.8	13.0	14.6	17.15	19.0	21.6	26.65
Total Stroke	14.6	17.8	20.0	24.0	27.0	31.0	35.55	41.25
Stroke after Closing of die	8.9	11.0	13.0	15.0	17.15	19.0	21.6	26.65
Strokes per min	90	80	60	50	42	35	30	25
Power of motor, kW	7.35	11	14.7	18.35	25.7	33	44	55

The ratings of Upset forging machines as per Metals Handbook are given in Table 3.8.

Table 3.8. Upset Forging Machines

Rated Size, cms	*Nominal Rated Capacity kN*	*Average Strokes per min.*
2.5	...	90
3.1	1250	75
3.8	3000	65
5.0	4000	60
6.25	5000	55
7.50	6000	45
10.00	8000	35
12.50	10000	30
15.00	12000	27
17.50	15000	25
20.00	18000	23
22.50	22000	–

3.6. DETERMINATION OF STOCK SIZE

The factors in estimating the stock size include the size and shape of the forging, the method of heating and the method of forging. The consideration which draws the attention of the designer is the net weight of the forging, that is, the weight of the forging finished to the drawing dimensions. This can be easily found out by calculating the volume of forging and multiplying it by the density of the metal. The next step is to find the gross weight or the amount of metal required to fabricate the forging. The gross weight is the sum of the net weight and the losses due to flash, scale, tong hold, sprue and shear waste. The method of finding these losses is explained below :

1. **Flash loss.** Flash loss is a function of linear centimetres of flash, its width and thickness. The machine forged components generally will not have any flash and in such cases, flash loss should not be taken into account. This holds good for some of the press forged components where very little or no flash is formed. But in case of drop forgings, flash loss is unavoidable and is determined by flash thickness and area. In practice, a flash loss ranging from 15 to 20% of the net weight is taken into account.

2. **Scale loss.** Scale loss is due to the oxidation of the material and is a function of the surface area exposed, the temperature of the heated piece and the length of time exposed. In practice, the scale loss is considered as percentage of the net weight of the forging. For mild steel it is taken as :

For forgings with net weight less than 45N, scale loss is 7.5 per cent. For net weight from 45 to 90N, scale loss is 6 per cent and for forgings with net weight above 90 N, the scale loss is 5 per cent.

3. **Tong hold loss.** In die forging, a projection is often provided at one end of the forging to facilitate handling. In practice, about 50 to 60 mm projection is found ample. As a portion of this is also used up in the sprue, for calculation purposes, only 50% of this is taken into account.

4. **Sprue loss.** The connection between the tong hold and the forging is called the sprue and this should be heavy enough to permit the lifting of the forging out of the impression without any bending. Generally, 7.5% of the net weight is taken into account.

5. **Shear Waste.** The waste that occurs while the bar is cut, is termed as shear waste and this is equal to :

Size of round or square bar	*waste, %*
Up to 5 cm	3
5 cm to 7.5 cm	4
7.5 to 10 cm	5
above 10 cm	6

The procedure of determining the stock size is, now, given below :

(*a*) **Closed-die Forging.** The volume of metal required for closed-die forging is calculated from the following formula :

$$V_{stock} = V_{forg} + V_{flash} + V_{sc} + V_{tonghold \cdot sprue} + V_{shear}$$

where V_{stock} = volume of bar stock in cm^3

V_{forg} = volume of forging in cm^3

V_{flash} = volume of flash in cm^3

V_{sc} = volume of scale in cm^3

$V_{tonghold + sprue}$ = volume of tonghold and sprue in cm^3

V_{shear} = volume of shear waste in cm^3

The volume of the forging, V forg, is calculated from the drawing and includes all allowances. The volume of flash, V flash, is calculated from the formula :

$$V_{flash} = pwt \text{ cm}^3$$

where p = the perimeter of the forging along which the flash is located, cm.

w = average width of flash, cm

t = average thickness of flash, cm

The dimensions of flash are taken from standard tables and depend upon the forging method. However, it has been explained above how to take the flash loss. The shear waste occurs due to two factors. One is due to the factor that when forgings are made from rolled sections, the bar stock cannot be cut up in exact number of lengths for forging. The remainder is shear waste. However, this waste can be utilised for making other forgings. Secondly, the shear waste occurs equivalent to the thickness of saw while cutting individual lengths for each forging.

After determining all the losses, the volume of stock can be calculated. Then the length and cross sectional area of the bar stock can be calculated. When calculating the cross-sectional dimensions of the bar stock for drop-forging without upsetting, it is to be ensured that the cross-sectional area of the stock is 10 to 15 per cent greater than that of the finished forging.

(*b*) **Open-die Forging.** The initial material in open die forging may be ingots, blooms or rolled billets of various cross sections and lengths. The billet weight for a forging may be determined from the following formula :

$$W_{in} = W_f + W_s + W_c + W_t \text{ newtons}$$

where W_{in} = weight of the initial material

W_f = weight of the forging

W_s = weight of the scale loss

W_c = weight of the cropped ends

W_t = weight of the trimming scrap

The weight of the forging, W_n can be calculated from the part drawing as explained earlier. Scale loss is taken as 2 to 3 per cent of the ingot or billet weight for each heating and from 1.5 to 2 per cent for each subsequent heating. The dimensions and the weight of the croppings will depend upon the shape and cross-sectional area of the stock. The volume of the cropping may be calculated from the following formulas :

For press Forging

(*i*) For cylindrical section of diameter *d*,

$$V_c = 0.21\, d^3$$

(*ii*) For rectangular section of width *b* and height *h*

$$V_c = 0.28\, b^2h$$

For Hammer Forging

(*i*) For cylindrical section of diameter *d*,

$$V_c = 0.23\, d^3$$

(*ii*) For rectangular section of width *b* and height *h*

$$V_c = 0.30\, b^2h$$

In forging ingots, the weight of the cropped ends may be taken as 25 to 30 per cent of the ingot weight of the croppings may also be taken from Table 3.9, as a percentage of the weight of the forging, for various types of forgings.

The weight of the trimming scrap depends upon the complexity of the forging and the method of forging employed. For forging of simple form, it is taken as 5 to 8 per cent. For certain intricate forgings, it may reach 30 per cent of the billet weight.

Table 3.9 Loss due to Croppings

Type of forging	*Loss, percent*
Forgings of simple shape (shafts, discs)	5 to 10
Stepped and flanged shafts	10 to 20
Connecting rods	15 to 20
Levers	14 to 25
Small Crankshafts with flanges	25 to 30
Locomotive Connecting rods	20 to 30

Some loss of metal may also occur due to remainder. When making forgings, from rolled sections, remainders occur as shear was when the bar stock cannot be cut up in an exact number of lengths for forgings.

After calculating the total initial weight of the material, the next step is to establish the shape and dimensions of the billet, this will depend on the degree of working which is defined as the ratio of the cross-sectional area of the billet to that of the finished forging. The ratio should be 3 to 5 for steel ingots and from 1.1 to 1.5 for rolled billet.

3.7. SELECTION OF FORGING EQUIPMENT

Selection of forging equipment for a particular job depends upon many factors such as : Type of forging, type of metal to be forged, quantity of forging and cost of the equipment. These factors are discussed below :

1. **Type of Forging.** The shape and section of the part to be forged will determine the preference of the forging equipment for economy and quality of forging, as shown in Table 3.10.

Table 3.10

S.No.	*Type of forging*	*Preferred Forging Equipment*
1.	Larger parts of symmetrical shape	Presses
2.	Smaller parts	Hammers, Upsetters
3.	Circular and ring forging	Crank Press
4.	Asymmetrical forging	Hammers
5.	Larger circular forging	Hammers
6.	Thin plate type forging	Hydraulic Press
7.	Small simple forgings such as rivets, nuts, bolts, bearing balls	Upsetter

2. **Type of Metal.** The choice between hammer and press will also depend upon the metal to be forged, as explained below :

(*i*) Metals and alloys of good forgeability such as carbon and alloy steels can be forged under both hammer and presses. But for jobs which are intricate in shape, are of very small thickness and are heavy, hammers are preferred over presses.

(*ii*) Metals and alloys of poor forgeability, such as brasses, bronzes, magnesium, beryllium etc. are sensitive to deformation rate and if forged under hammers are liable to cracking. Such materials are preferred, therefore, to be forged under presses.

(*iii*) Metals having high thermal conductivity such as copper, aluminium etc. may cool during the forging operations if forged under presses. Therefore for such metals, hammers are preferred.

(*iv*) For metals and alloys with high forging temperature, such as molybdenum and its alloys, tungsten, zirconium, titanium etc., hammers are preferred over presses.

3. **Quantity of Forging.** The number of forging to be produced per die run is another important factor. Where the quantity of production is comparatively small (upto 2000 parts per die run) hammers are preferred. For larger production runs, presses are preferred.

4. **Cost of Equipment.** If the above technical factors give equal chance to hammer and press, the final selection will depend upon economic considerations. The total cost of production will include the capital cost of forging unit, foundation cost, building cost, die cost and labour cost. The capital cost including the installation cost of a press is higher than that of a hammer. This higher cost of press is justified only when the press capacity is utilized to a high degree above 80% and the annual production of parts is quite high (30,000 to 40,000 forgings). Presses and counter blow hammers require smaller and cheaper foundation, whereas fixed anvil block hammers (gravity drop and double acting) need costly heavy mass concrete foundations. The foundation cost and building cost also depend upon the vibrations produced during forging operation, being directly proportional to these. The forging equipment arranged in order of increasing ground vibrations are : Crank press, screw press, vertical or horizontal counter blow hammer and fixed anvil block hammer.

3.8. SELECTION OF SIZES OF FORGING EQUIPMENT

(*i*) **Hammers.** Practically for every forge shop, the hammer is often the determining factor between profit and loss. Hammer size and capacity and the relationship of this to the forging die design, the forging shape and its influence on the life of dies are today of increasing importance in drop forging practice. There are no known rules that govern the exact hammer size or capacity for the given job, as the factors influencing it are too many. Some of the factors that govern the slection of hammer size include : forging shape, its size, the number of forgings to be produced and the material of the forging. Consideration is also given to the die face area, total cubic centimetre displacement of the metal affected in a given time etc. Therefore, for the selection of hammer size, we must rely upon our fund of experience at our disposal. Table 3.11 gives data for determining the capacity of a hammer required for the production of forgings depending upon their weight and cross section.

The capacity of the die forging hammer (Mass of falling parts) may be calculated by using the following empirical relation :

$$C = 10\left(1 - 0.005\,D\right)\left(1.1 + \frac{2}{D}\right)^2 \left(0.75 + 0.001\,D^2\right) D\sigma_t \qquad ...(3.1)$$

where C is the capacity of the hammer, *kg*

D is the diameter of the round forging at the parting line

σ_t is the tensile strength of material at forging temperature.

Another empirical formula is :

$$C = (3.5 \text{ to } 5) \times A, \text{ kg}$$

where A = forging projection area, cm² ...(3.2)

(*ii*) **Presses.** For a given capacity of a press, the output increases and forging cost decreases with increase in section diameter to a maximum of optimum diameter. After that the output decreases and forging cost increases. The optimum diameter is given as

$$D = \sqrt{2.5\,C}, \text{ cm},$$

where C = Pressure developed by press or press capacity, tonnes

Table 3.11. Hammer Size (Hammer Forging)

Maximum cross-section of stock, mm (diameter or side of square)	*Weight of Forging, N*			*Hammer Size, kN*
	Shaped Forgings		*Smooth shafts*	
	Average Weight	*Maximum Weight*	*Maximum Weight*	
50	5	20	100	1
60	15	40	150	1.5
70	20	60	250	2
85	30	100	450	3
100	60	180	600	4
115	80	250	1000	5
135	120	400	1400	7.5
160	200	700	2500	10
225	600	1800	5000	20
275	1000	3200	7500	30
350	2000	7000	15000	50

The force or pressure necessary to forge a given part is selected according to the sam requirements which were discussed for selecting the capacity of a forging hammer. The pressur depends upon the cross-sectional area of the forging. The required pressure may be calculate by the relation,

$$C = A \times \sigma_t \qquad ...(3.3$$

where C = pressure developed by press, *kgf*

A = area of contact of the top die and forging cm² and

σ_t = tensile strength of the metal at the forging temperature, kgf/cm².

(*iii*) **Forging Machine.** Capacity of the horizontal forging machine depends upon th maximum diameter of bar stock it can handle. However, the capacity of the forging machine may be obtained by the following formula :

$$C = 5\,(1 - 0.001\,D)\,(D + 10)^2\sigma_t \qquad ...(3.4)$$

where C = capacity of forging machine, *kgf*

D = maximum diameter of upset forging cm and

σ_t = tensile strength of metal at forging temperature, kgf/cm².

There is another empirical formula;

$$C = \sigma_t \frac{\pi}{4} D^2 k \qquad ...(3.5)$$

where k = 1 to 6 = a constant found from practice.

There is another method of determining the capacities of the various forging equipments. The capacity fo the hydraulic press is first determined with the help of empirical relation. The capacity of the other forging equipment is then obtained in terms of capacity of the hydraulic press. The capacity of hydraulic press may be obtained by the following emprical relation

$$C = A \times \text{constant, } kgf \qquad ...(3.6)$$

where A is the projected plan area of forging including flash land, in mm², at die parting plane and constant ranges from 50 to 65 depending upon the material composition and complexity of the forging.

Another formula is

$$C = 8\,(1 - 0.001\,D)\,(1.1 + 20/D)^2 \times A \times \sigma_t\, kgf \text{ for circular forgings} \qquad ...(3.7)$$

For non-circular forgings,

$$C = C_1\left(1 + 0.1\sqrt{\frac{2L}{A}}\right), kgf \qquad ...(3.8)$$

where L is the maximum length of forging including flash land at die parting plane in mm and C_1 is calculated from (3.7) in which D is taken as $1.13\sqrt{A}$.

1. Capacity of Mechanical Press = capacity of Hydraulic Press (0.8 to 1.2).
2. Capacity of Forging Machine, *kgf* = capacity in tonnes by formula 3.6 or 3.7.
3. Capacity of Power Drop Hammer, *kgf* = capacity in tonnes by formula 3.6 or 3.7.

3.9. DIE INSERTS

In closed-die forging, the parts of the die which are subjected to excessive wear caused by flow of metal are usually designed with exchangeable inserts. that is, instead of one piece solid die block, the impressions are sunk in inserts which are then secured in die holders. The use of the inserts in closed-die forging has the following advantages :

1. Instead of replacing the complete die block, only the insert will be replaced when it is worn out.
2. They prolong the life of die block into which they fit.
3. The machine hours used for resinking are lesser.
4. Saving in die steel.
5. the die changing time is less as only inserts are to be taken out and replaced.
6. The same die holder can be used for many parts.
7. Handling of inserts is easier as compared to solid die blocks.

Against the above mentioned benefits, the use of inserts has the following limitations :

1. The use of inserts is suitable only for round and symmetrical parts with a single parting plane and which can be forged in a single die impression, for example, the gear blanks. In upset forging, the inserts are commonly used for heading tool faces.

2. The die life per sinking is lesser.

3. After resinking, the packings below the inserts have to be readjusted properly otherwise parts obtained will be outside the tolerance zone.

4. In drop forging, the inserts tend to loosen due to impact force and the keys have to be tightened frequently.

The parts requiring preform operations and having assymetric parting line are preferably forged with solid dies.

3.10. SOLVED EXAMPLE

Design suitable tooling for upset forging of the component shown in Fig. 3.27. The material is mild steel.

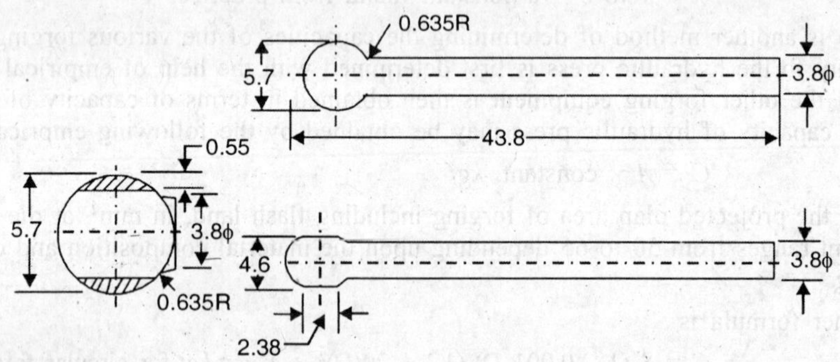

Fig. 3.27

Solution. The above figure shows the size and shape of the forging after giving 1.6 mm machining allowance on machined surfaces. All the dimensions shown above are at room temperatures. For the design of the die, all the dimensions will be taken on M.S. contraction scale.

Volume to be upset

$$\text{Volume of sphere} = \frac{4}{3}\pi(2.85)^3 = 96.97 \text{ cm}^3$$

Volume of two frustums of sphere $2 \times \frac{\pi}{3} h^2 (3R - h)$

Here, $h = 0.55$ cm, $R = 2.85$ cm

$$\therefore \quad \text{Volume of two frustums} = 2 \times \frac{\pi}{3} \times (0.55)^2 (3 \times 2.85 - 0.55)$$

$$= 5.07 \text{ cm}^3$$

$$\text{Fillet volume} = \frac{r^2}{5} \times 2\pi R'$$

Here $r = 0.635$ cm, and $R' = 1.9$ cm

$$\therefore \quad \text{Fillet volume} = \frac{(0.635)^2}{5} \times 2\pi \times 1.9 = 0.96 \text{ cm}^3$$

$$\therefore \quad \text{Total volume} = 96.97 + 0.96 - 5.07 = 92.86 \text{ cm}^3$$

Taking 3.8 cm dia. bar, the length of bar needed for upsetting only $= 92.86 / \frac{\pi}{4}(3.8)^2 = 8.188$cm.

$$\therefore \quad \frac{l}{d} = \frac{8.188}{3.8} = 2.15$$

Since l/d is less than 2.5, the forging can be upset in one blow. Since the length of the bar stock to be upset and kept unsupported is 2.15 d, *i.e.*, less than 2.5 d, so the first rule is satisfied.

Total length required for the component

$$= (43.8 - 5.7) + 8.188 + 0.16$$
$$= 46.45 \text{ cm}$$

Size of the machine. Since the bar size is 3.8 cm, so, 3.8 cm size machine of nominal rated capacity equal to 3000 kN (from Table 3.8) will be suitable.

Size of the die blocks. For 3.8 cm size of the machine, the die block sizes can be taken from Table 3.7.

Length of bar to be gripped = $3d$ = 11.4 cm.

Parting line of the job is taken along the diameter.

Half the impression is in the die block and the other half in the punch.

Justification for machine forging process. The following comparison justifies the above saving :

1. **With machining :**

(*a*) There will be too much machining because on a length of 38.1 cm, out of 43.8 cm; the dia. is to be turned from 5.7 cm to 3.8 cm.

(*c*) A form tool of 2.85 cm radius will be needed and such a big tool is not at all feasible.

(*d*) For component of this type, nearly 40% material will go as chips only and 60% will form the part of the component.

(*e*) Lot of milling will be done at the two faces.

(*f*) Cost of machining and material will be too heavy.

2. **Casting :**

(*a*) Smooth finish and close tolerances will not be obtained.

(*b*) Lot of machining will be required on 3.8 cm ϕ and on sphanical portion.

(*c*) The component will be weaker in strength.

In forging, only a limited amount of machining will be done and amount of scrap material will be only a few %. The shape of the component is such that it is easily adaptable to machine forging hence it can be fabricated very economically.

3.11. TOOLS FOR FLASH TRIMMING AND HOLE PIERCING

After a forging is produced in dies, the flash around the edges of the forging must always be removed, This operation of flash trimming or removal is done on special trimming dies of crank or eccentric type.

The set of flash trimming tools consists of a trimmer blade and a trimmer punch, (Fig. 3.28). A trimmer blade is designed and machined to the contour of the forging at the die parting line. The blade is made of tool-steel and has a cavity both in contour and dimensions to the finish impression of the forging die. For small and medium forgings, trimmer blades are usually made from one piece of steel, but for large

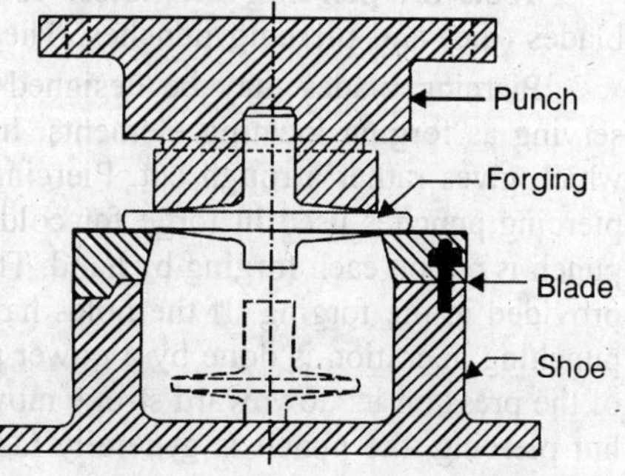

Fig. 3.28. A trimming Tool

forgings it is preferable to make trimmer blades built up from small segments mounted on the common base and accurately fitted together.

Fixing of the trimmer blades used for small medium sized forgings is usually made in special holders known as shoes or bolsters which are fastened by 4 bolts to the table of trimming press. Fastening part of the trimmer punches should be designed according to the nature of the fastening elements provided on the slide of the press. In small trimming presses, it is usually a round hole with a setting screw which takes the shank of the trimmer punch. Bigger presses have a rectangular slot with several setting screws. The large presses are usually made with a dove-tail slot and the fastening of the punch to the press slide is made exactly on the same principle as fixing the die block on the hammers.

For trimming the flash, the forging is placed over the cavity the trimmer blade, the flash serving as the support for the forging. When the press is started, punch travels downwards and cuts off the flash. The forging falls down into the slot and the flash remains on the top of the trimmer blade. Trimming can be done with or without previous heating. Forgings of aluminium alloys, copper brass, bronze and of mild steel (*C*-content 0.20 and 0.25 per cent) are usually trimmed in unheated state (cold work). However, the forgings of harder steels as well as those of larger dimensions, should always be heated before trimming (hot work).

The clearance between the trimmer blade cavity (die opening) and the trimmer punch is 0.8 to 1.5 mm. The angles of cutting edges of trimmer blade, that is taper, of the cavity, are different when provided for trimming the flash in hot or cold condition. It is :

10° for hot work, and

5° for cold work

Minimum thickness trimmer blade = 50 mm

Minimum width or length of trimmer blade

= Cavity diameter + 100 mm

Cavity diameter will be the outer dimension of the forging.

Capacity of trimming press can be found by the empirical relation :

$$C = \frac{1}{3} P \times t \times \frac{K}{1000}, \text{tonnes}$$

where P = Perimeter of the forging, mm

t = effective thickness of flash, mm

= 1.5 × flash thickness for normal trimming

= 2.5 × flash thickness for piercing

Factor K is 80 kgf/mm² for steel.

When tools are designed for hot trimming, they must be made with standard shrinkage allowance of 1/100th of the nominal length.

Tools for piercing the holes. Tools for piercing the holes in forgings consist of piercing blades (die) and piercing punches. They are similar to piercing die discussed in chapter 2.

Piercing blades may be designed with usual cutting edges but quite often they are only serving as forging locating elements. In this case, piercing operation is performed by a punch, which gives rather a rough cut. Piercing punches must always possess sharp cutting edges. The piercing punches used in forge for cold work are very often simple loose punches. This type of punch is put on each forging by hand. The punch locates itself on small cylindrical pegs, specially provided in the forging. If the holes have an irregular shape, two pegs are provided. The actual punching operation is done by a power press on which table a piercing blade is placed. The slide of the press on its downward stroke movement presses the punch and pushes it right through. For hot piercing, the punch is generally fastened to the slide of the press and the piercing blade is usually provided with a suitable stripper to strip forging from the punch.

Combined trimming and piercing tools

Combined trimming and piercing tools, which perform both operations simultaneously, are designed with the trimming blade and punch located on a suitable base with tapered cavity and central peg. When lifting up the punch and the trimming blade, it is very easy to remove from the tool the finished, trimmed and punched forging.

For the piercing operation :

Diameter of the Punch = Diameter of hole to be pierced and

Diameter of the Piercing blade (Die-cavity) = Diameter of hole + 0.8 to 1.5 mm

Minimum thickness of Piercing blade = 40 mm

Taper of Piercing blade recess = 5°

Thickness of Piercing punch = 25 mm

Total length of Piercing punch, min. = Depth of forging + 20 mm

Straightening. After the trimming operation, the forging may get slightly distorted or deformed, and must be straightened. This the trimming operation. Straightening can be done cold or hot as the case may be. Straightening under the hammer is done with one light blow in the finish forging impression of the die or in a special straightening impression.

3.12. MATERIALS AND MANUFACTURE OF FORGING DIES

3.12.1. Materials for Die-Blocks

The hot forging dies operates under very severe service conditions since the forging process is characterised by high interface pressures coupled with high temperatures. Therefore, the tool and die materials are selected and manufactured with the greatest care. The materials used for making dies must be heat resistant, possess adequate strength of low wear rate and lend themselves well for machining by cutting tools. A compromise between hardness and ductility must be struck since the dies are exposed to thermal shock.

Die blocks used for the production of forging dies are manufactured from high grade special tool steels. These steels are made by the open hearth process either basic or acid or by electric furnace. The ingots are worked under large forging presses, varying from 20 to 40 MN capacity, to achieve the utmost grain refinement and resistance to shock. After this, the die-blocks are normalised, and usually heat treated in the final stage by quenching and tempered to the required hardness. This double heat treatment insures a maximum resistance to wear.

Composition and heat treatment of die block steels depend on the type of die and its application. Typical steels used for the production of forging dies and tools are listed in Table 3.12. Applications of steels listed in Table 3.12 are briefly described in Table 3.13.

Heat treatment of die-blocks is carried out either before or after complete machining of the die. The most common practice is to purchase the blocks as hardened and tempered forgings and to machine the impressions into the forged blanks as recieved. Hardness of the die blocks vary from 269 to 477 BHN, the most common range being 341 to 375 BHN. At hardness higher than 302 BHN, these steels are difficult to machine and machining is done only at a sacrifice in cutting speed and tool life.

3.13. DIE MANUFACTURE

Below, we discuss the various methods (conventional and latest) for cutting impressions in the die-blocks.

3.13.1. Mechanical Machining Processes. These are the most commonly used methods for manufacturing dies. The processes include: turning, machining on planer, milling and grinding. Turning is used for rough and finish machining of rotary working surface, whereas for rectangular and square die-blocks, these operations are done on a planer. Grinding is mainly used for finishing the surfaces. The impressions are cut in the die block by highly skilled men who use the milling machine, specially designed for sinking dies. Cutters of various types are used in accordance with

Table 3.12. Typical Steels Used for Die Blocks, Die Inserts and Trimming Tools

Steel	*Type of steel*	*% C*	*% Mn*	*% Si*	*% Cr*	*% Ni*	*% Mo*	*%*	*Annealing °C*	*Hardening °C*	*Quenching medium*
DIE BLOCKS AND DIE INSERTS											
A	Carbon steel	0.55-0.65	0.50-0.80	0.80 max.	–	–	–	–	770-800	810-830	Water oil
B	Nickel-chromium, steel	0.50-60	0.50-0.80	0.30 max.	0.50-0.80	1.25-1.70	0.25-0.30	--	760-790	820-840	Oil
C	Nickel-chromium, molybdenum steel	0.50-0.60	0.50-0.80	0.30 max.	0.50-0.80	1.25-1.70	0.25-0.30	–	760-790	820-840	Oil
D	Chromium-molybd enum-vanadium-, steel	0.50-0.60	0.70-0.90	0.30 max.	0.85-1.05	–	0.45-0.50	*V* = 0.06-0.012	760-790	830-850	Oil
E	Chromium-molybd enium, steel	0.18-0.23	0.40-0.60	0.30 max.	1.40-1.60	–	0.45-0.55	–	770-800	850-870	Oil
F	Nickel-chromium-molybdenum-vanad-ium steel	0.50-0.60	0.50-0.80	0.30 max.	0.75-1.00	2.00-2.25	0.70-0.90	*V* = 0.08-0.012	760-790	820-840	Oil
G	Special steel for, nitriding	0.45-0.55	0.40-0.65	0.30 max.	1.40-1.80	2.25 max.	0.10-0.25	*Al* = 0.90-1.30	790-820	880-900	Oil
H	Tungsten-chromium-m-vanadium steel	0.35-0.45	0.20-0.40	0.30 max.	3.00-3.50	–	–	*W* = 17.00-18.00	860-890	slow-850	Air

I	Tungsten-chromium-nickel-molybdenum-vanadium steel	0.28-0.34	0.20-0.50	0.30 max.	3.00-4.00	3.00-4.00	0.30-0.60	$W = 7.25$-8.25 $V = 0.40$-0.60	860-890	slow-850	
				$V = 0.40$-0.6							
TRIMMING, PUNCHING, AND STRAIGHTENING											
J	Group of carbon, steel	0.50-1.50	0.20-0.40	0.25 max.	–	–	–	–	10-20 Above Ac	30-50 Above Ac	Water or Oil
K	Manganese steel,	0.85-0.90	1.80-2.00	0.35 max.	–	–	–	–	750-780	770-800	Oil
L	Chromium-vanadium, steel	0.75-0.85	0.20-0.40	0.30 max.	0.70-0.90	–	–	$V = 0.20$-0.30	750-780	770-800	Oil
M	Tungsten-chromium-vanadium steel	0.40-0.50	0.30-0.40	0.60-0.80	1.25-1.40	–	–	$W = 2.20$-2.40	770-800	850-870	Oil

Table 3.13. Application of Steels

A Carbon steel	Dies for drop hammers and forging machines when production is small (few hundreds of forgings only).
B Nickel-chromium steel	Dies for drop hammers when production is not exceeding 2,000-3,000 forgings.
C Nickel-chromium-molybdenum steel D Chromium-molybdenum vanadium steel	Most extensively used steels for drop hammers, and forging machine dies when production is large.
E Chromium-molybdenum steel	Dies for forging machines when production is large.
F Nickel-chromium-molybdenum-vanadium steel	Dies for mass production hammers and presses. Die inserts.
G Special steel for nitriding	Die inserts for hammers and presses, surface hardened by nitriding after complete machining and heat treatment
H Tungsten-chromium-vanadium steel	Small dies and die inserts for mass production on presses..
I Tungsten-chromium nickel-molybdenum and vanadium steel	Hot working dies and tools where resistance to wear and abrasion combined with toughness is necessary.
J Group of carbon steels	Cheap steels commonly used for trimming, blanking piercing, and straightening, press tools for hot and cold working operations.
K Manganese steel	Tools of intricate shape for trimming, piercing, straightening, coining etc. This steel is of non-shrinking type.
L Chromium-vanadium steel	Typical press tools for large production. The alloying element improves property and makes steel very suitable for cold working tools.
M Tungsten-chromium-vanadium steel	Tools for flash trimming and piercing and for hot and cold working operation. Specially suitable for mass production.

the shape of each section of the impression. But much of the accuracy of the die depends upon hand work performed after it is sunk. The impressions are machined either by manual sinking after layout and/or by copy milling using templates or patterns.

In copy milling, the following variations in the process are available :

1. Manual copy milling, where the feed of the spindle is controlled by hand.

2. Semi automatic copy milling, where the feed in the longitudinal and lateral axes is automated but th evertical movement is controlled manually.

3. Automatic copy milling, where the movement of the cutting tool is controlled automatically by the movement of a sensor over the surface of a $3D$ pattern. This method can produce intricate surfaces economically on both small and large dies.

Die sinking step :

1. Holes are drilled in the sides of the die blocks opposite each other. Handling bars may be inserted in these holes. This facilitates ease in lifting and handling heavy die blocks.

2. Next, the shanks (dovetails, 3.22 *b*) for the attachment of the dies in the forging hammer or press, are machined on a Planer.

3. Next, the die blocks are turned over the bed of the planer and the striking surfaces are machined to obtain clean and sound metal for the impressions which are to be sunk.

4. The die blocks are then squared which facilitate exact alignment of the die blocks when placed in the forming hammer or press.

5. *Laying out the outline of the impressions.* Metal templates are made from the blue print or model of the part to be forged. These templates are then used for laying out the outline of the impressions which are to be sunk in the die blocks. In the absence of a template, the layout is made from the forging or die drawing. The faces of the die blocks are given a colour back ground by coating them with copper sulphate or a similar purpose solution, to secure a convenient surface for the marking of the outline of the impression.

6. *Machine work.* The first impression to be sunk is the finishing impression. If the impressions are of simple shape, they are sunk on vertical milling machines, those of intricate shapes are machined on die sinkers.

7. *Bench Work.* After sinking the impression, the hand work on the bench includes operations such as scraping, filing, grinding and polishing the cavities. The finishing impression must be true for every dimension. They must be lapped and polished free of all tool marks and sharp corners, so that the impressions will allow the metal to move with least resistance in filling the cavities of the dies.

8. *Preparation of Lead Cast.* After the finishing impression has been completed, the die blocks are clamped together in exact alignment. A lead antimony alloy is pured in the finishing impression through a sprue which is machined into each die block from its outer edge and extends to the cavity of the finishing impression. The resulting lead cast or "proof" is now carefully checked for dimensional accuracy by the die maker, as well as by the engineer. The lead cast may also be checked by the producer and the user of the forgings, before final approval for release for production is obtained. Since steel shrinks on cooling from its forging temperature and the lead alloy does not, it is necessary to allow for the shrinkage in checking the lead cast. The correction amounts to about 16 mm per metre.

9. After the lead cast has been approved, the sinking of the other impressions is begun. The next impression to be sunk is the blocking impression. The rolling or edging impression can next be sunk into the die block and so on.

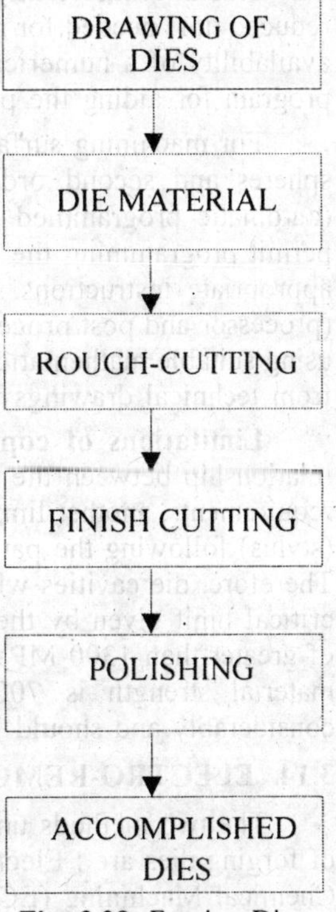

Fig. 3.29. Forging Die Manufacturing.

10. The "flash" may begin to form at the blocking impression, but most of it will develop at the finishing impression where the full impact of the hammer blows or full pressure of the press is utilized to the utmost. Thus, the blocking impressions are seldom flashed or guttered, whereas the finishing impression is usually flashed in both sides, and guttered only in the top die. By flashing both dies, we get a neater and more symmetrical appearance of the final forging. By guttering the top die only, the forging will sit flush in the trimming die which must be used to trim the excess flash. Machining of the flash gutter will complete the necessary machining work required for a given set of forging dies.

Flash must be thin to aid die filling and produce close tolerances. It also acts as "safety valve" for excess metal. As a general rule, flash thickness.

$$t = 0.015\sqrt{A}, \text{ mm},$$

where A = projected area of the forging, mm^2

A thin flash running out between parallel die surfaces would lead to very large length/ thickness ratios and thus to high die pressures (see equ. 13.17 and 13.18). Therefore, the length of flash is reduced by cutting a "flash gutter". This allows free flow of the flash and limits the minimum flash thickness to only a small width. The width of flash "flash land" is given as,

Flash land = 3 to $5t$, wide

Gutter depth and width should be sufficient to accommodate the extra material.

11. Lastly, the die is given a final dressing to fillet all sharp corners at the flash line to permit ease of metal flow.

12. Many dies are now surface treated for improved were resistance, by techniques similar to those described for metal cutting tools.

The various steps in the manufacture of a forging die are shown in Fig. 3.29.

3.13.2. Numerically Controlled Milling. This technology is gaining importance due to the reduced time needed for making dies. However, the application of this technology depends on the availability of a numerical description of the die geometry and on the development of suitable program for aiding the preparation of punched tapes.

For machining surfaces of relatively simple analytical discription such as cylinders, cones, spheres and second order surfaces, problem-oriented programming languages such as APT (automatic programmed tools) and EXAPT (extension of APT) have been developed. These permit programming the cutter paths required to machine the entire surface by means of a few appropriate instructions. The instructions are processed with the help of processing programs (processor and post processor) using a suitable computer. Complex surfaces may be described by using suitable mathematical surface models generated directly by CAD or indirectly by digitizing from technical drawings or from measurements made on 3D models.

Limitations of copy milling and NC milling. In addition to the limitations set by the relationship between the shape to be machined and the cutting tool as well as by the tolerance requirements, another limit is set by the fact that the shaft diameter and the tip radius of the sensor (stylus) following the pattern cannot be infinitely small because of the probing forces involved. Therefore, die cavities with sharp small corner radii can not be machined easily by milling. The critical limit given by the loading of the cutting tool is reached when materials with a strength of greater than 1300 MPa are machined. The machining of dies in the annealed state, where the material strength is 700 to 800 MPa, raises the performance of these milling methods considerably and should be preferred.

3.14. ELECTRO-REMOVAL PROCESSES

The two methods under this category, which are being used increasingly in the production of forging dies are : Electro Spark Erosion or Electro-Discharge machining (EDM) and Electro-Chemical Machining (ECM). These two methods have already been discussed in detail in the book "Production Technology". As already discussed, the metal removal rate is independent of the mechanical strength of the material being machined. Thus, these processes are particularly well suited to machining the hardened, tempered, high-temperature-resistant materials used in die making.

3.14.1. E.D.M. The method is used for manufacturing both dies with working surfaces of just a few mm2, common in precision engineering, and dies of several square metres of working surface, such as those required for sheet metal working in the automobile industry. As in conventional machining, the machining of 3D shapes is carried out in several stages of roughing and finishing. Important advantages of EDM over conventional die sinking processes are :

1. Since the metal removal rate of electrode wear is not dependent on the toughness and hardness of the die material, dies can be machined after heat treatment, thus avoiding distortion or heat cracks at critical areas.

2. The conventional machining of complex internal shapes is replaced by the easier machining of external shapes made from materials such as copper and graphite.

3. Hand finishing time, as compared to copy milling can be reduced by about 70%.

For the reasons mentioned above, die sinking applications that combine poor machineability, complex shape and close tolerances favour the use of EDM. On the other hand, dies that can be made readily produced by mechanical machining methods can seldom be made at lower cost by ordinary EDM technique, because EDM electrodes are short lived and the accurate electrodes (the finishing electrodes must be made to closer tolerances then those required of the die itself) are expensive. A particular adaptation of EDM "no wear" EDM-often is competitive for sinking large contoured cavities in hardened tool steel forging dies. No wear EDM use graphite electrodes at reverse (+) polarity, high current density and low pulse frequency. Maximum metal removal rate is 10 to 20% lower than for ordinary EDM roughing with graphite electrodes. When no wear technique is used, some of the molten or vaporised work metal solidifies on the electrode in a thin, adherent film that may have to be removed if it becomes thick enough to alter any of the processing conditions significantly. Because it does not produce low micro-metre surfaces, no wear EDM is not used for finishing. Instead, because the absence of machining wear on electrodes eleminates or greatly reduces the need for re-dressing or replacement of electrodes, no wear EDM is often used in roughing operations that require the removal of large amounts of metal from die cavity.

A special application of EDM is its combination with numerical path control. A continuously moving wire is used as the electrode and produces a cut equivalent to that produced by a band saw (WEDM discussed in the "Production Technology" book). In this way dies with any type of opening can be machined. The process is thus suitable for machining jigs, blanking tools and orifics. The necessary programs can be produced easily with the help of a small computer, using the lines and circles which define the contour to be machined.

3.14.2. ECM. Since late 1960s the process has been used for manufacturing dies with working surfaces upto approximately 250 cm². The main factors influencing the accuracy in EC die sinking is the machining gap. It becomes smaller when the electrode feed rate, supply voltage and conductivity are reduced and with this the accuracy increases. However, the danger of short circuits is also increased thereby. Tolerances achieved are : ± 0.05 mm for shallow cavity dies and ± 0.07 mm for deeper-cavity dies under appropriate conditions. Electrochemically machined forging dies may have a higher die life than conventionally machined dies due to the non mechanical removal process and the resulting improved fatigue resistance. The advantages of ECM over EDM (already discussed in "Production Technology" book) are :

1. Unlike EDM where separate roughing and finishing operations are needed to produce a good surface finish, in ECM, just a single machining operation is needed.

2. The workmaterial structure is not affected by thermal or mechanical stresses.

3. High metal removal rates may be achieved.

4. No wear of the electrode.

5. The time needed to machine a die impression by ECM may be reduced by 70 to 80% as compared with EDM.

However, ECM has got the following drawbacks :

1. Machining of working surfaces larger than 250 cm² is uneconomical.

2. Complex electrodes are required and more time is needed to produce and test the electrode.

3. Higher investment is needed for ECM equipment.

4. Greater disparity exists between electrode and cavity.

3.15. CAST DIES

In some plants, cast dies have proved satisfactory and economical for use in mechanical presses (cast dies are not used in forging hammers). They may be used where large tolerances are acceptable, where a large number of similar dies are required, or where the dies are to be finished by EDM or conventional machining methods. Their principal advantage is the saving in die making costs that can be affected by minimizing the amount of machining. When the shell moulding process is used, hot working tool steel can be cast to make dies that require only a polishing operation to finish the cavities. The saving can be considerable, especially on dies with complex cavities.

3.16. RESINKING OF DIES

Solid dies must be resunk after they have worn out of tolerance. For a block of given thickness, the number of resinkings depends mainly on the depth of the impression. In general, the thickness of the block remaining beneath or above the impression should be at least 3 times the depth of the impression. In practice, cast dies are not resunk when they become worn, but are melted and recast.

3.17. SIZE OF DIE BLOCKS

Once a forging has been designed for production and its size determined (length, breadth and thickness or height), the size of the die block can be determined. The method for this has already been discussed in Art. 3.4.3. Below, we shall further elaborate it.

For a single impression (finishing impression) die, the length and the breadth of the die block will be estimated by a simple rule of thumb : The distance between the outer periphery of the impression and the die edge should not be less than 1.5 times the maximum depth of impression. The height or thickness of die block below the deepest impression should be at least 4 times the maximum depth of the impression (as discussed above under resinking of dies.).

For a multiple impression die, the additional data needed is the minimum striking surface, that is, the distance between impressions and between outer edges of end impressions and die edges. As shown in Fig. 3.21, the edges or roller is placed on the extreme right side of the die block (it can be left side also). The roller width is taken as follows :

Roller width = Required Steel Size + 19 mm, for sizes upto 50 mm round
= Required Steel Size + 25 mm, for sizes from 50 to 75 mm round
= Required Steel Size + 32 mm, for sizes from 75 to 100 mm round
= Required Steel Size + 38 mm, for sizes from 100 to 125 mm round

The determination of required steel size has already been explained in Art. 3.6.

The next considerations are the distances between or around impressions. The area covered by these distances serve two purposes. One is the striking surface and the other being a provision for the excess metal. If the striking area between the impressions is too little, rapid upsetting of the die face will result, causing an undersize condition on the forging thickness. The striking surface between cavities is considerably reduced by the flashing and guttering of the dies.

Distance between finishing and blocking impression or other impression
= flash width + gutter width + 12.5 mm

The distance between impressions and also between edges of the impressions and the edges of the die block are kept equal to the value calculated above.

Blocker width = maximum width of forging
Finishing width = maximum width of forging
Length of fuller = (1 to 1.5) × d + (5 to 10) mm
Width of fuller = (1.25 to 1.50) × d + 20 mm, for inclined impression

where d = side of the blank.

Table 3.14, below gives the characteristics of Hammers and Presses.

As a rule of thumb, a hammer equipped with a 1 tonne-ram can do the work of a 1000 tonne-press, because it delivers the total energy required in several blows.

Table 3.14. Characteristics of Hammers and Presses

Equipment type	*Energy, kN m*	*Ram mass, kg*	*Force, + kN*	*Speed m/s*	*Strokes/ min*	*Stroke m*	*Bed area, m × m*	*Mechanical efficiency*
Hammers	0.5-40	30-5,000		4-5	350-35	0.1-1.6	0.1 × 0.1 to 0.4 × 0.6	0.2-0.5
Mechanical								
Steam and air	20-600	75-17,000 (25,000)		3-8	300-20	0.5-1.2	0.3 × 0.4 to (1.2 × 1.8)	0.05-0.3
Counterblow	5-200 (1250)				3-5	60-7	–	0.3 × 0.4 to (1.85 × 5)
Herf	15-750			8-20	<2			0.2-0.6
Presses								
Hydraulic, forging			100-8,000 (800 000)	< 0.5	30-5	0.3-1 (3)	0.5 × 0.5 to (3.5 8)	0.1-0.6
Hydraulic, sheet m.w			10-40 000	< 0.5	130-20	0.1-1	0.2 × 0.2 to 2 × 6	0.5-0.7
Hydraulic, extrusion			1000-50 000 (200 000)	< 0.5	<2	0.8-5	0.06 to 0.6, diam container	0.5-0.7
Mechanical, forging			10-80 000	< 05	130-10	0.1-1	0.2 × 0.2 to 2 × 3	0.2-0.7
Horizontal upsetter			500-30 000 (1-9 in diam)	< 0.1	90-15	0.05-0.4	0.2 × 0.2 to 0.8 × 1	0.2-0.7
Machanical, sheet m.w			10-20,000	< 1	180-10	0.1-0.8	0.2 0.2 to 2 × 6	0.3-0.7
Screw			100-80 000	< 1	35.6	0.2-0.8	0.2 × 0.3 to 0.8 × 1	0.2-0.7

Note : Numbers in parentheses indicate the largest sizes, available.

3.18. I.S. CODE

The I.S. code for Tool and Die steels for hot work is : IS : 3748-1966. It prescribes requirements for tool and die steels in the form of bars, blanks, rings and other shapes for hot working applications. Hardness and chemical composition are among the requirements. See Appendix- I (D)

PROBLEMS

1. Define forging process.
2. Name the various equipments used.
3. Compare : Smith forging, Drop forging, Press forging and Machine forging.
4. Compare open and closed die forging.
5. Compare cold forging, warm forging and hot forging.
6. Describe the advantages and limitations of forging process.
7. Describe the common types of forging hammers.
8. How the size of a hammer is specified?
9. Name the main parts of a hammer.
10. Compare single impression dies and multi-impression dies.
11. With the help of suitable sketch, explain the working of 'Steam drop hammer' and 'Board drop hammer'
12. What is counter blow steam hammer?
13. Enumerate the advantages of press forging over drop forging.
14. How the size of a press is specified?
15. What is upset forging and how it is done?
16. What are the advantages and limitations of upset forging?
17. How the size of a forging machine is specified?
18. Discuss the various factors of a good forging design.
19. Why the position of parting line is so important?
20. What is preform ? What is its main advantage?
21. Describe the various preliminary operations done on a multi-impression die.
22. What are the materials used for forging die blocks?
23. How the impressions are sunk in a die block?
24. How the dies are secured to the ram and anvil of a drop hammer or forging press?
25. What are the points which should be kept in mind for the proper maintenance of the dies? What type of work is upset forging particularly suited to?
27. Explain the basic rules for die design for upset forging.
28. How we arrive at the stock size required for forging?
29. How are the sizes of various forging equipment selected?
30. What are the advantages and limitations of using die inserts?
31. In what way the Blocking and the Finishing impressions differ?
32. In what may the drop hammers are different from forging hammers?
33. For what type and quantity of jobs, the following forging equipments are preferred : Forging hammers, drop hammers, presses, and forging machines.
34. Differentiate between Fullering and Drawing operations.
35. List the various factors upon which he selection of forging equipment for a particular job depend. Discuss these factors.
36. A drive shaft 25 mm diameter with one end upset to dimensions : diameter 75 mm and length 10 mm is to be forged on the upset forging machine. Design suitable tooling for upset forging

25. What are the points which should be kept in mind for the proper maintenance of the dies? What type of work is upset forging particularly suited to?

27. Explain the basic rules for die design for upset forging.

28. How we arrive at the stock size required for forging?

29. How are the sizes of various forging equipment selected?

30. What are the advantages and limitations of using die inserts?

31. In what way the Blocking and the Finishing impressions differ?

32. In what may the drop hammers are different from forging hammers?

33. For what type and quantity of jobs, the following forging equipments are preferred : Forging hammers, drop hammers, presses, and forging machines.

34. Differentiate between Fullering and Drawing operations.

35. List the various factors upon which he selection of forging equipment for a particular job depend. Discuss these factors.

36. A drive shaft 25 mm diameter with one end upset to dimensions : diameter 75 mm and length 10 mm is to be forged on the upset forging machine. Design suitable tooling for upset forging the drive shaft.

Sol. Volume of the head $= \frac{\pi}{4} \times 75^2 \times 10 = 44178.6 \text{ mm}^3$

Now, Area of Cross-section of the bar stock $= \frac{\pi}{4} \times 25^2 = 490.9 \text{ mm}^2$

$\therefore$ Length of the bar stock needed to be upset to make the head is = 44,178.6 / 490.9 = 90 mm

Now as per rule 1 of upset forging, the maximum length of the bar stock that can be upset to make the head without its getting buckled is 3d (Art 3.5), i.e. 75 mm Thus rule No. 1 is not valid here, Hence, an intermediate upset forging operation is needed to make the finished product. As per rule No. 2, the diameter of the intermediate blank will be = 1.5d = 37.5 mm.

$$\therefore \quad \text{Head length} = 44178.6 \Big/ \frac{\pi}{4} \times 37.5^2 = 40 \text{ mm}$$

For the final finishing operation, the ratio of length to diameter (40/37.5) is less than 3. Thus, the intermediate blank can be put straight in to the final finishing die to get the finished product.

37. Can all the known materials be forged? Justify your answer.

38. Compare a forged component with a cast component.

39. Write the steps of forging a hexagonal nut from a cylindrical bar stock.

40. A MS bar stock has the dimensions :- Length : 200 mm; Diameter : 25 mm

Write the steps to forge the following parts from this bar stock :-

(*a*) A round headed bolt of dimensions : Head diameter : 20 mm; Head length : 9 mm; Shank diameter : 10 mm; Shank length : 25 mm.

(*b*) Square headed nut : nut head length : 8 mm; outer diameter : 20 mm; Inner diameter : 10 mm.

4

COST ESTIMATING

4.1. DEFINITION

Cost estimating may be defined as the process of forecasting the expenses that must be incurred to manufacture a product. These expenses take into consideration all expenditures involved in design and manufacturing, with all the related service facilities such as pattern making, tool making, as well as a portion of the general administrative and selling costs.

Cost estimates are the joint product of the engineer and the cost accountant, and involves two factors : physical data and costing data. The engineer as part of his job of planning manufacturing determines the physical data. The cost accountant compiles and applies the costing data.

Purpose of Cost Estimating. Cost is the background of almost every decision the tool engineer makes in organizing manufacturing operations and in selecting materials, methods, tooling and facilities. An understanding of cost determination is essential to ensure that these decisions are based on sound and dependable estimates of cost.

Estimates of cost must be reasonably accurate if a venture is to be successful (realistic cost estimate). If a job is overpriced, it is lost to a competitor. If it is underestimated, it results in financial loss.

Detailed cost estimates are prepared to :

1. Determine the selling price of a product for a quotation or contract, so as to ensure a reasonable profit to the company.

2. Check the quotations supplied by the vendors.

3. Decide whether a part or assembly is economical to be manufactured in the plant or is to be purchased from outside.

4. Determine the most economical process or material to manufacture a product.

5. Initiate means of cost reduction in existing production facilities by using new materials which result in savings due to lower scrap loss and revised methods of tooling and processing.

6. To determine standards of production performance that may be used to control costs.

4.2. COST ACCOUNTING OR COSTING

It is the determination of an actual cost of a component after adding different expenses incurred in various departments. Or, it may be defined as a system which systematically records all the expenditures to determine the cost of manufactured products. The work of cost accounting begins with the pre-planning stage of the product and ends only after the whole lot of the product has been fully manufactured. Costing progresses with the progress of the product through the plant.

Difference between cost estimation and cost accounting. As discussed above, cost estimation is determining the anticipated or probable cost of a job much before the manufacturing of the job is undertaken, whereas cost accounting will be complete only after the job has been completed. Estimation as compared to costing is a higher technical job because an estimator must be familiar with factory methods and operation time etc., whereas costing only consists of compiling data from various sources by clerical staff. Cost estimation gives predicted or standard cost, whereas cost accounting gives actual or postmortem costs.

Purpose of Cost Accounting

1. To compare the actual cost with the estimated cost to know whether the estimate had been realistic or not.

2. Wastages and undesirable expenses are pointed out requiring corrective measures.

3. The costing data helps in changing the selling price because of change in material cost or labour cost etc.

4. It helps to locate the reasons for the increase or decrease of loss or profits of a company.

5. It helps in determining the discount on catalogue or market price of the product.

6. The actual cost helps the company to decide whether to continue with the manufacture of a product or to buy it from outside.

7. It helps the enterprise to prepare its budget.

8. The costing data helps to formulate policies and plans for the pricing of a new job.

9. It helps in regulating, from time to time, the production of a job so that it may be profitable to the company.

4.2.1. Classification of Costs. Costs can be classified in the following manner :

1. **Non-recurring costs.** These costs are also called "capital costs" and are one-time costs. These costs consist of two parts : fixed capital costs and non-depreciated capital costs. Fixed capital costs include depreciable items such as plant building, manufacturing equipment and tools. Nondepreciated capital cost includes land.

2. **Recurring cost.** These costs are a direct function of manufacturing process. These costs are also called 'operating cost' or 'manufacturing cost'.

Another classification of costs is :

(*i*) *Fixed costs.* Fixed costs associated with a productive unit are those costs which are independent of the rate of production of components. These costs will be there whether the facilities are being utilized or not .

(*ii*) *Variable costs.* These costs vary with the rate of production. If there is no production, variable costs will be nil.

A cost can be termed as 'direct cost' or 'indirect cost', as explained below :

(*a*) *Direct cost.* It is that cost which can be directly assigned to a product.

(*b*) *Indirect cost.* Indirect costs can not be directly assigned to a product but must be spread over an entire factory.

Working Capital. Working capital includes funds over and above the fixed capital and land investment, to get a facility started and to provide for the future obligations as and when they occur. It consists of :

Raw material on hand

Semi finished products in the process of manufacture

Finished products in the inventory

Accounts receivable

Cash in hand needed for day-to-day operation.

The working capital remains tied up during the useful life of the plant, but it is considered to be fully recoverable at the end of the life of the facility.

Turn over ratio. This concept provides a rough estimate of the investment cost of a new product. It is defined as :

$$\text{Turn Over Ratio} = \frac{\text{Annual Sales}}{\text{Total investment}}$$

The turn over ratio for steel industry is about 0.6. In chemical industry, it is nearly 1.0 for many products.

4.3. ELEMENTS OF COST

The constituents of cost of a product or the "cost elements" are : Material cost, Labour cost and Expenses. We shall discuss each element in turn.

1. **Material cost.** Material is divided into two basic categories : (*a*) material for fabricated parts (*b*) standard purchased parts. The total cost of these two will give the material cost. Again there are two kinds of materials which comprise the factory cost of a product. These are : Direct material and Indirect material.

(*i*) *Direct material.* The direct material is the raw material which is processed in the plant and finally forms the finished product. Any standard part which also becomes a part of the finished product will also come under the category of direct material.

(*ii*) *Indirect material.* Indirect materials are those which help in the processing of direct materials into the finished product. These materials don't form a part of the finished product. Indirect materials include : Shop supplies such as cotton waste, lubricating oil, cutting fluids, coal, oil, gas, shielding gases used in Arc welding, Emery paper used for polishing, quenching oils for heat treatment etc. Indirect materials form the part of oncost or overheads.

2. **Labour cost.** Labour which enter into the manufacture of a product is of two categories : Direct Labour and Indirect Labour.

(*i*) *Direct labour.* The operator or operators which actually process the raw materials either on machines or manually, form the direct labour.

(*ii*) *Indirect labour.* All the staff excepting administrative and sales office staff, which help in running the plant come under the category of indirect labour. Indirect labour includes : Foremen, supervisors, maintenance staff, stores personnel, time office staff, drawing office staff, etc. Indirect labour forms a part of overheads.

3. **Expenses.** Total cost of the product minus the costs of direct material and direct labour constitutes the 'Expenses'. Expenses may also be either direct or indirect.

(*i*) **Direct expenses.** These expenses like the direct material and direct labour are directly chargeable to the finished product. These are also known as "chargeable expenses". These include :

(*a*) Cost of patterns, jigs, fixtures, dies, drawings or designs specifically prepared for a particular product which cannot be used for other purposes.

(*b*) Cost of any experimental work done specially for a particular product.

(*c*) Cost of inward carriage or freight incurred on supply of special material needed for the particular product.

(*d*) Hire of special or single purpose tools or equipment for a particular product.

(*ii*) **Indirect expenses.** These are also called "oncosts" "overheads" or "burden". These include : cost of indirect material, cost of indirect labour and other expenses that cannot be conveniently charged directly to a particular job. Indirect expenses may be divided into :

(*a*) Factory expenses or overheads.

(*b*) Office and Administrative expenses or overheads.

(*c*) Selling and Distribution expenses or overheads.

Factory expenses. These expenses include : indirect materials, indirect labour, expenses, insurance, maintenance and depreciation of machine, power etc.

Office and Administrative expenses. These expenses consist of all expenses incurred in the direction, control and administration of an undertaking. These expenses include : rent and rates of office premises, salaries of office staff, printing and stationery, postage, salaries of high officers, depreciation of office equipment and insurance on office equipment.

Selling and distribution Expenses. These expenses include : salaries of sales staff, publicity and advertisement, catalogues, leaflets and price lists, packing and forwarding charges, godown rent, commission to salesmen etc.

The overheads may be grouped into two main categories :

1. **Fixed overheads or constant overheads.** These are such items of indirect expenses which remain constant or fixed irrespective of volume of production. These items include : salaries of higher officers (administrative and management executives), capital taxes, insurance charges, depreciation of building, plant machinery etc., rent of buildings.

2. **Variable or floating overheads.** These are such items of overheads which vary with the volume of production. Such items are : internal transport charges, power, fuel, stores expense, factory lighting and heating and sales office expenses and repairs of machine tools.

Since fixed overheads remain constant irrespective of volume of production, production should be increased to reduce the cost of the part. There should be some minimum production to meet the fixed expenses and start earning profit.

4.3.1. Cost structure. The elements of cost can be combined to give following types of cost :

1. Prime cost. Prime cost or direct cost is given as :

Prime cost = Direct material + Direct labour + Direct expenses (if any)

2. **Factory cost.** This cost is given as :

Factory cost = Prime cost + Factory expenses.

Factory cost is also called as 'Works cost'.

3. **Manufacturing cost.** Manufacturing cost or cost of production is given as :

Manufacturing cost = Factory cost + Administrative expenses.

4. **Total cost.** Total cost is given as :

Total cost = Manufacturing cost + Selling and Distribution expenses.

5. **Selling price.** Selling price is given as :

Selling price = Total cost + Profit

The above mentioned cost structure is explained with the help of a block diagram in Fig. 4.1.

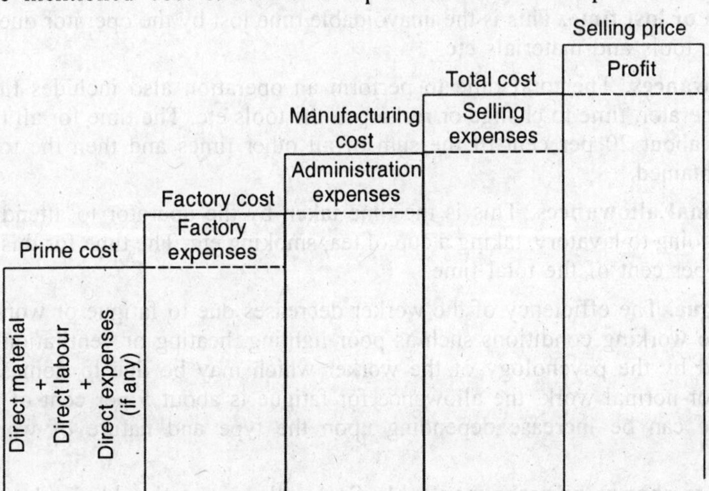

Fig. 4.1. Cost Structure.

4.4. ESTIMATION OF COST ELEMENTS

Direct material cost, direct !abour cost and direct expenses can be found out most accurately by the estimating procedure. The indirect expenses items which are so numerous are determined by cost accounting section only, which furnishes the figure department wise to the estimator. The various cost elements are estimated in the manner give below.

1. **Direct material cost.** The cost of standard purchased parts can be obtained from the purchasing section. The raw material chargeable to a product is that in the rough state and includes all scrap removed. Material can be in the form of sheet metal, bar stock, forgings or castings, plastic etc. The weight of the material can be determined from the drawing of the part. An irregular part is divided into simple sections to calculate its volume. Volume is multiplied by density of the material to find its weight. The weight of a part multiplied by the unit cost of the material gives the material cost per piece. If the unit cost covers only the purchase price of the material, the material cost is multiplied by one or more additional factors to account for bulk losses, purchasing and handling costs.

2. **Direct labour cost.** For estimating the direct labour cost of a product, the job is divided into operations needed to machine it, and then estimating the operation time for each operation. Total time multiplied by a labour rate gives the direct labour cost. The total time required to perform an operation may be divided into the following parts :

(*i*) **Set-up time.** This is the time needed to prepare for the operation and may include : time of study the blueprint or to do paper work, time to get tools from the crib and the time to install the tools also on the machine.

(*ii*) **Man or handling time.** This is the time the operator spends loading and unloading the work, manipulating the tools and the machine and making measurements during each cycle of operation.

(*iii*) **Machining time.** The elements comprising the machining time are those which are performed by the machine. This is the time during each cycle of operation that the machine is working or the tools are cutting.

(*iv*) **Tear down time.** This is the time required to remove the tools from the machine and to clean the tools and the machine after the last component of the batch has been machined.

Tear down time is usually small. It will seldom run over 10 minutes on the average machine in the shop. It may require only a few minutes to tear down a set up on a drilling press and 10 to 15 minutes on the average miller or turret lathe. In exceptional cases, it may go upto as high as 30 minutes on very large boring mills and large milling machines.

(*v*) **Down or lost time.** This is the unavoidable time lost by the operator due to breakdowns, waiting for the tools and materials etc.

(*vi*) **Allowances.** The total time to perform an operation also includes time for personal needs of the operator, time to change or resharpen the tools etc. The time for all these allowances is taken to be about 20 per cent of the sum of all other times and then the total time for the operation is obtained.

(*a*) **Personal allowances.** This is the time taken by the operator to attend to his personal needs such as going to lavatory, taking a cup of tea, smoking etc. The time for this is usually taken to be about 5 per cent of the total time.

(*b*) **Fatigue.** The efficiency of the worker decreases due to fatigue or working at a stretch and also due to working conditions such as poor lighting, heating or ventilation. The efficiency is also affected by the psychology of the worker which may be due to domestic worries, job security etc. For normal work, the allowance for fatigue is about 5 per cent of the other times. This allowance can be increase depending upon the type and nature of work and working conditions.

(*c*) **Time to change or resharpen tools.** Some allowances should also be provided for the time taken by the operator to get the tools changed or to resharpen the tools. This time varies from machine to machine.

(*d*) **Inspection or checking allowance.** To maintain the uniform quality of the parts, the dimensions of the parts should be checked or inspected at regular intervals depending upon the closeness of tolerances. The checking times for the various instruments are given below, to check one dimension :

With rule	0.10 minute
Vernier calliper	0.50 minute
Inside calliper	0.10 minute
Outside calliper	0.05 minute
Inside micrometer	0.30 minute
Outside micrometer	0.15 minute
Depth micrometer	0.20 minute
Dial micrometer	0.30 minute
Thread micrometer	0.025 minute
Plug gauge	0.20 minute
Snap guage	0.10 minute

Set up time and tear down time are performed usually once for each lot or batch of parts. Set up time per piece is obtained by dividing the set up time for the machine by the number of pieces produced in lot. Set up time, handling time and tear down time are estimated from previous performances on similar operations. All work on a particular type of maching tool consists of a limited number of elements. These elements can be standardized, measured and recorded. This is done under Time and Motion study. Standard data is available for set up time, tear down time and handling time. Machining time is obtained with the help of formulas for each machining operation which takes into account speeds of cutting, feeds and depth of cut and tool travel. The actual amount of down or lost time that will occur in a particular operation can scarcely be predicted. Some operations will run smoothly, others may be beset by troubles. The sum of machining time and the handling time is called 'run time' or 'unit of operation time'.

The total time to manufacture a product (from which the direct labour cost will be estimated) may be divided into the following major groups :

1. Set up time
2. Machining time
3. Non-machining time
4. Down time

The tear down time discussed earlier may be included in the set up time itself. The non-machining times will be man or handling time, personal needs, fatigue, cutter or tool sharpening and inspection. The man or handling time, as already discussed, includes : loading and clamping the part, unloading the part, advancing or retracting the cutting tool, tightening a chuck, a trial cut, trial gauging, deburring the machine, cleaning the fixture etc.

For a simple drilling operation of a drill press, the various elements can be listed as given below :

Handling elements	*Machining elements*	*Handling elements*
Pick up part	Drill	Clear drill from job
Place in jig	the	Move jig into clear
Fasten in jig	hole through	Release part from jig
Position under drill	the	Remove part from jig
Advance work to drill	job	Aside with part

Set up time and economic lot size. As discussed above, the set up time is taken once for each lot or batch of parts. The number of set ups, needed for a product will depend upon 'economic lot size'. The economic lot size (which will be discussed in detail in the next chapter) gives the number of parts to be produced for each set up that will result in lowest cost per part. It is usually determined by comparing the cost of set up, tooling, handling and carrying charge (interest, taxes, insurance and obsolescence etc.) with the cost of storing the finished product. The economic lot size may be found from the following approximate relation :

$$N = 5\sqrt{\frac{AS}{KC}} \qquad ...(4.1)$$

where A = Average monthly requirement of pieces as per schedule

S = set up cost per lot

C = cost per piece

K = carrying charge factor, %, 5 to 30%.

The carrying charge factor includes interest, taxes, insurance, storage cost, obsolescence etc.

The set up time, s, for each lot is then

$$s = \frac{NS}{A} \qquad ...(4.2)$$

Tool change and sharpening time. The frequency of sharpening of a tool will depend upon the amount of machining being done by the tool and the tool life. The formula for unit time to change tools in

$$t = \frac{T'cTk}{T} \qquad ...(4.3)$$

where T_C' = cutting time tool is in use during operation cycle.

Tk = Total time to change tool

T = Tool life.

Inspection or checking time. Inspection, checking measuring or guaging time depends upon : the type of operation performed; type of the measuring or checking device, the precision of the work surface etc. The time allowance for checking a specific dimension, may be obtained from the following equation :

$$tc = \frac{Tc}{N+1} \qquad ...(4.4)$$

where T_C = Total time required to check this dimension for all parts.

N = Number of parts

or $tc = F \times Tc$ ...(4.5)

where F is the frequency of checking a dimension.

If the value of Tc is computed for other dimensions, their sum will give the total time required to check or gauge all the dimensions for all the parts in a lot. Thus, from equ. (4.4).

$$Tc = tc\,(N + 1) \qquad ...(4.6)$$

Performance factor. The performance factor is considered to take into account the "downtime". The downtime factors are based on many series of case histories studies over a long period of time. These factors relate the actual time needed to do a job to the estimated time to do the same job. Since, downtime relates to the non-production time, these factors will always be more than 1 (between 1 to 2). This ratio of actual time to estimated time to do a job is known as "Performance factor", R, which may be written as,

$$R = \frac{Ta}{Te} \qquad ...(4.7)$$

where Ta = actual time needed to do a job

Te = estimated time to do a job.

3. **Indirect expenses.** Indirect expenses or overheads are those charges which vary in proportion to the production rate, but which are not easily attributable directly to a given operation or part. These expenses are apportioned among the operational units (machines, plants etc.) according to some weighting factor. These various methods for estimating the overheads are given below :

(*i*) **Percentage of direct labour cost.** Here all overhead costs are apportioned according to direct labour cost by means of a labour burden rate, which is the ratio of the annual total indirect expenses to the annual direct labour cost. It is then assumed that this burden rate also represents the ratio of hourly overhead costs to hourly labour costs. That is

$$\text{Overhead rate} = \frac{\text{Total overheads in Rs for a period}}{\text{Total Direct labour in Rs for a period}}$$

The overhead cost = overhead rate × direct labour cost per unit.

This method is simple and provides accurate results where labour is the main production element and wage rates are fairly uniform. On the other hand, it may not be an accurate indicator where machines of greatly different capacity and sizes are operated. Also, if two products uniform, then this method will give less overhead cost where labour is cheap and high overhead cost where labour is costly and therefore, this method increases the cost of a component which has already higher labour cost. Also, in many cases it gives very approximate results because some items of overhead such as depreciation and taxes have very little relationship to labour costs. However, if applied to such overhead items as arise more or less from the amount of direct labour, like supervision and personnel department costs, it can provide a reliable distribution.

(*ii*) **Percentage of direct labour hours.** Here overhead costs are estimated by multiplying the operation time by an overhead rate. This rate is obtained by dividing the total indirect costs applicable to a production unit for a period of time (say a month) by the total number of hours of direct labour in the same period.

This method is akin to the direct labour method and has most of its advantages and disadvantages. However, this method will give the same overhead cost for two components which take the same time for production. The overhead rate will be given as :

$$\text{Overhead rate} = \frac{\text{Total overheads for a period}}{\text{Total hours of direct labour for a period}}$$

(*iii*) **Machine hour method.** Here the overhead rate is obtained by dividing the total annual overhead costs chargeable to a machine by the number of hours machine is used during the year. That is,

$$\text{Overhead rate} = \frac{\text{Charges applied to a machine over a period}}{\text{Number of hours of operation of the machine during the same period}}$$

This method is suitable where work is done mostly by machines.

(*iv*) **Direct material method.** This method is based on the theory that the overhead expense is incurred in proportion to the value of the direct materials consumed. This method is simple but does not allow for the usual situation wherein some of the material is fabricated without the use of much equipment, whereas other material in the same plant requires extensive machinery, requiring considerably more labour, power, maintenance and floor space. However, for the allocation of material expenses such as purchasing, storage and handling, the method is useful. The method is also useful when major part of the cost is of material like foundries and mines. Here,

$$\text{Overhead rate} = \frac{\text{Total overhead expenses, Rs}}{\text{Total direct material cost, Rs}}$$

(*v*) **Unit of production method.** In this method, the overhead expenses are divided equally among all the parts produced. Here, overhead rate is Rs. for each unit will be,

$$\text{Overhead rate} = \frac{\text{Overhead expenses, Rs. over a period}}{\text{Quantity of production over a period}}$$

This method can be recommended for one product mass produced or a few products of close similarly.

(*vi*) **Space rate method.** The amount of space occupied by a machine has an evident relationship to certain overhead expenses. Typical of these are : building expense, heat, light, ventilation, and service equipment such as cranes and conveyors.

Department, Space rate, Rs per square meter

$$= \frac{\text{Total overheads assigned a department}}{\text{Total area of the production department in square metre}}$$

$\therefore$ Space charge to the individual machine for the defined period of time

$$= \text{Space rate} \times \text{total area with which the machine should be charged.}$$

(*vii*) **Combination of labour hour and machine hour method.**

(*viii*) **Percentage of Prime Cost.** Here,

$$\text{Overhead rate} = \frac{\text{Total overheads over a period}}{\text{Prime cost over a period}}$$

Then, overhead cost per unit = overhead rate × prime cost per unit.

This method will give the same overhead cost for two products with equal prime cost, even though their labour and material costs will be different. This method will be useful where only one type of product is being manufactured and when direct labour and direct material costs are nearly equal.

4. **Direct expenses.** The direct expenses are estimated in the following manner :

The engineering (preparation of drawings, blue prints, drafting etc.) and design cost of a product is calculated as a flat hourly rate for each estimated hour of design and engineering time. On the same lines, the cost of any experimental work done specifically for a particular product, can be estimated. The cost of patterns and special tools such as jigs, fixtures, dies and gauges can be estimated as outlined below :

Tool cost. Generally tool cost estimating is concerned with tool and other special equipment to be used for production. Much product cost estimating depends upon tooling cost estimating. Therefore, tool costs are often treated separately during a product cost study.

Tooling costs are estimated to :

1. Determine how much must be invested in tools and equipment to manufacture a product.

2. Determine the cost of alternate methods of tooling to help in selecting the more economical method.

3. Find the cost of a proposed machine or tool that promises to produce more economically method.

4. Determine the reasonable cost of a special machine or tool to gauge whether tool room performance is efficient or vendor's prices are reasonable.

The cost of cutting tools, both special and standard, jigs, fixtures, dies and guages and other special equipment is often separated from the cost of machine tools. This is done to determine what portion of the tooling cost is to be charged directly to the project or proposal and what portion is to be capitalized.

Tooling costs are of two categories : of perishable tools and of durable tools. Perishable tools consist of cutting tools that are used for a particular job such as drill bits, small cutters etc. Perishable tools that have a general application are considered as part of overhead. The cost of perishable tools is obtained in the same manner as materials, through the purchasing group. Where these costs cannot be obtained, however, without a drawing, they must be estimated, as are durable tools. Durable tools are usually estimated by the estimator. Durable tools or capital tools are : jigs, fixtures and ancillary support tooling that requires tool design and construction.

Estimating tool cost. Tooling cost is estimated in the same manner as the cost of manufacturing of a product is estimated (See Art. 4.3 and 4.4). There are four major factors in the cost of any tool. These are : material, labour and overhead or burden as in manufacturing costs. The fourth factor is the cost to engineer and design the tool, that is, the cost of designing and drafting the tool.

(*i*) **Engineering and design cost.** The engineering and design cost represents a large portion of the total cost of the tool, often as high as 20 to 30 per cent. This cost can be directly charged to the individual tool and as a consequence are always considered first in establishing an estimate, the engineering and design cost is applied to the tool-cost estimate as a flat hourly rate for each estimated hour of design and engineering time.

(*ii*) **Tool-materials cost.** The cost of the material for a proposed tool may be calculated and therefore becomes more of an actual cost thån an estimated cost. It is the most accurate item in the tool-cost estimate and is determined as discussed before. Standard parts such as knobs, hand wheels, bushings, bolts, screws, springs, and similar items that complete the tool bill of material can be accurately priced from catalogues, price lists or invoices.

(*iii*) **Tool-labour cost.** The labour involved in the machining, assembling, fitting, and tryout of tools is always difficult to estimate accurately, even for the most experienced estimator. The machining time can be calculated as explained in Art. 4.10. The other times cannot be calculated and must be estimated. It is difficult to foresee all the problems that may develop in fitting assembly and try out operations, even under most favourable conditions. Therefore, the estimate must include a liberal factor of safety for lost time, which is impossible to anticipate and will always be present. The sum of all the times will be the labour time for the tool. When multiplied by the toolmakers hourly rate, the estimated labour cost is determined.

(*iv*) **Burden or overhead.** The toolroom may be considered as a department, and therefore may have an established burden rate, just as production department has. In general, the man-hour or direct labour-cost method of burden distribution is applied as discussed under point (3) above.

Comparison of various methods of tooling set-ups :

Two tooling set-ups are compared in cost not upto cost of tooling equipment, but upto the tooling set-up stage. So the main aim of a good tooling set-up is to reduce the setting-up time of the total time to produce a component.

The various methods of reducing the setting-up time are :

(*a*) By using rapid-change facilities, for example, of the kind used in changing drills and reamers, interchangeable holders with preset tools, when the interchangeability method of maintaining the required accuracy is applied.

One example of such preset tooling is the interchangeable turret. It carries holders and tools set-up to machine one or several workpieces having the same purpose or, in some instance, even different purposes. Interchangeable turrets allow a high-production machine tool to be quickly changed over from one workpiece to another.

(*b*) The setting-up time per unit of the product can be reduced by increasing the size of the batch with a single set-up. The batch size can be increased in two ways :

(*i*) By increasing the number of identical items in the batch.

(*ii*) By increasing the number of workpieces by machining different parts closely resembling each other in their service function, construction, size, material, manufacturing specifications etc.

These are commonly called a family of workpiece or parts.

Consequently, the machining of workpieces of a family will require only slight adjustment of the tooling set-up in changing over from one workpiece to another.

For example, if screws of the same diameter but of different lengths are to be machined on a turret lathe, only the stop for limiting the travel of the turret needs to be readjusted, in changing over from a screw of one length to one of different length. Thus, the machining of such a group or family of workpieces has the effect of increasing the number of workpieces machined in a single set-up with a slight readjustment of the machine-fixture-tool-workpiece (MFTW) system.

(*iii*) Sometimes, in changing over from the machining of one workpiece to another, the readjustment of the MFTW system consists in changing the inserts of the fixture or chuck without changing themselves. Several interchangeable sets of jaws for a vise can be used. Workpieces of various shapes can be clamped and machined by changing the jaws of the same vise.

(*iv*) The same problem can be solved by the use of various kinds of universal fixtures that enable a family of work pieces to be handled since they can be quickly changed over from one workpiece to another.

(*v*) The same purpose may also be served by unified or combined tool set-ups for quickly changing over from one workpiece to another, for example, by changing a number of cutting tools without changing the holders.

(*c*) Another way of reducing tooling set-up time is with the use of permanent tool-set ups. Time can be saved on any type of machine, by issuing successive batches of similar work, which need the same types or size of tools and cutters. This fact can be utilised to save a considerable amount of time per week, on all types of turret lathes, by having certain machines permanently set-up with standard tools, to suit the types of work likely to be issued. Two types of permanent tool set-ups for turret lathes are :

Universal chucking equipment

Universal bar equipment

Drastic set-up changes are less likely on chucking machines than on bar machines. The tools should be mounted on the machine in the most logical order of use and the tools should be changed to other positions only when a large quantity of parts justifies the change.

A permanent set-up for a chucking lathe is shown in Fig. 7.11 (Chapter 7) and such a set up could be used without change, except for the sizes of the drills etc. for a wide range of jobs. Another permanent set-up for use with a bar machine is shown in Fig. 7.14.

4.5. METHODS OF ESTIMATING

The cost of a new product can be estimated by the following methods :

1. **Conference method.** Under this method, representatives of the various sections of the plant such as purchasing, process engineering, tool design and methods and time study sit together in a conference and estimate the costs of material, labour and tooling. A coordinator from either accounting or estimating section collects all the data and arrives at the manufacturing cost after applying some burden factors. This method does not involve detailed paper work, standard data or mathematical calculations and its accuracy will depend upon the availability of specifications, samples and drawings.

2. **Delphi method.** It is a special version of conference method, which involves cycles of questioning and feedbacks in which the opinions of individual participants are kept anonymous.

3. **Past experience.** In this method, the cost of the new product is approximately estimated through the use of past average actual costs of somewhat similar products. The cost is then adjusted to suit variations in the product, material and labour costs. In this method there is no

co-ordination between the various sections of the plant and the estimating clerk lacks the manufacturing and technical background to arrive at an accurate cost estimate.

4. **Detailed analysis method.** This is the most accurate and reliable method of cost estimating. However, this is also the most time consuming method, and the total cost of a new product is estimated in the manner explained under articles 4.3 and 4.4.

The complete analysis involves :

(*i*) calculating all raw material usage including scrap allowance and salvage material.

(*ii*) processing each individual component of the product, *i.e.*, writing the operation sheet.

(*iii*) calculating the production time and hence the direct labour cost for each operation.

(*iv*) the equipment required.

(*v*) tools, gauges and special fixtures or dies.

(*vi*) inspection and testing equipment.

(*vii*) packaging and shipping requirements.

4.6. DATA REQUIREMENTS FOR COST ESTIMATING

The following detailed data are required by the estimator to arrive at an accurate estimate of a new product :

1. General design specification, *i.e.*, a brief description of the product, its function, performance and purpose.
2. Quantity and rate of production.
3. Assembly or layout drawings.
4. List of sub-assemblies of the product.
5. Detailed drawings and a bill of material for the product.
6. Material release data.
7. Operation analysis.
8. Standard time data.
9. Machine tool and equipment required.
10. Tools, gauges and special fixtures, jigs or dies required.
11. Manufacturing routings.
12. Test and inspection equipment and procedures.
13. Packaging and transportation requirements.
14. Area and building requirements.

4.7. STEPS IN MAKING A COST ESTIMATE

The cost of a new product may be estimated by following the basic steps given below :

1. Make a complete and thorough analysis of the cost request to understand it fully.
2. Make an analysis of the part or product and make separate lists of standard parts and the parts to be fabricated within the plant.
3. Make a manufacturing process plan for the parts to be fabricated.
4. Determine the material costs for the standard and the fabricated parts.
5. Estimate the total production time for each operation listed in step 3.
6. Apply the labour and burden rates to each operation.
7. Add the material costs (step 4) and the labour and burden costs (step 6). This will give the total manufacturing cost.
8. Apply the profit factors to arrive at the selling price.

4.8. CHIEF FACTORS IN COST ESTIMATING

Each cost estimate may not be exactly the same as the actual manufacturing cost. The most significant causes for the cost deviations can be : Fluctuations in material and labour costs, incomplete design information at the time of estimate, unexpected delays resulting in premiums paid for overtimes and materials and the unexpected machining or assembly problems. However, the average of cost estimates over a period of time should be reasonably close to the actual manufacturing costs. For this, the following factors should be considered for arriving at an accurate and complete cost estimate :

1. Each estimate should contain complete costs of direct material, direct labour, factory overheads, spoilage, engineering, administration and selling.

2. If the cost of a new product is estimated on the basis of previous estimates of comparable parts, detailed estimating should be used. It is necessary to make substitutes in the past estimates for individual operations, individual parts or individual sub-assemblies.

3. The period of time between the cost estimating and the actual production of the part affects the determination of unit prices. During the intervening period, the basic material and labour rates may rise or fall. Therefore, an estimate of what the cost will be at the time of actual production is what is really needed. Thus the estimator should have the ability to project thinking and reasoning into the future.

4. The volume of the pieces to be produced also affects the costing rates since the time and therefore the cost of performing an operation decreases as the number of units produced is increased.

5. The addition of new types of equipment and special buildings require the development of new overhead rates etc.

4.9. NUMERICAL EXAMPLES

Example. 4.1. *From the following data, calculate the total cost and selling price for a job :*

Direct material = *Rs 5500*

Manufacturing wages = *Rs 3000*

Factory overheads to manufacturing wages = *100%*

Non-manufacturing overheads to factory cost = *15%*

Profit on total cost = *12%*

Solution. Direct material = Rs 5500

Manufacturing wages (Direct labour)

= Rs 3000

Factory overheads = 100% of Rs 3000

= Rs 3000

∴ Factory cost (Refer Fig. 4.1) = Direct material + Direct labour + Factory overheads

= Rs 5500 + 3000 + 3000

= Rs 11,500

Non-manufacturing overheads, *i.e.*, administrative and selling overheads

= 15% of Rs 11,500

= Rs 1725

∴ Total cost = Factory cost + Rs 1725

= Rs 11,500 + Rs 1725

= Rs 13,225

Profit = 12% of total cost

= Rs 1587

∴ Selling price = Total cost + Profit

= **Rs 14,812**

Example. 4.2. *From the records of a company, the following data are available :*

(*i*) **Raw materials**

Opening stock = Rs 20,000

Closing stock = Rs. 30,000

Total purchases during the year

= Rs 1,70,000

(*ii*) **Finished goods**

Opening stock = Rs 10,000

Closing stock = Rs 15,000

Sales = Rs 4,89,500

(*iii*) **Direct wages** = Rs 1,20,000

(*iv*) Factory expenses = Rs 1,20,000

(*v*) Non-manufacturing expenses

= Rs. 50,000

Find out what price should be quoted for a product involving an expenditure of Rs. 20,000 in material and Rs. 30,000 in wages

Solution. First of all, we shall determine the rates of factory expenses, non-manufacturing expenses and profit, from the given data above.

Direct material cost = opening stock + total purchases during the year – closing stock

= Rs 20,000 + 1,70,000 – 30,000

= Rs 1,60,000

Direct wages = Rs 1,20,000

Factory expenses = Rs 1,20,000

∴ Factory cost = Rs 1,60,000 + 1,20,000 + 1,20,000

= Rs 4,00,000

Non-manufacturing expenses = Rs 50,000

∴ Total cost = Rs 4,00,000 + Rs. 50,000

= Rs 4,50,000

Cost of finished goods sold = opening stock + cost of goods manufactured – closing stock

= Rs 10,000 + Rs 4,50,000 – Rs 15,000

= Rs 4,45,000

Total sales = Rs 4,89,500

∴ Profit = 4,89,500 – 4,45,000

= Rs 44,500

(*i*) Factory expenses (% of direct wages)

$$= \frac{1,20,000}{1,20,000} \times 100 = 100\%$$

(*ii*) Non-manufacturing expenses to factory cost

$$= \frac{50,000}{4,00,000} \times 100 = 12\frac{1}{2}\%$$

(*iii*) Profit to cost of sales $= \dfrac{44,500}{4,45,000} \times 100 = 10\%$

Now the cost of the product can be quoted as follows :

Direct material = Rs 20,000

Direct wages = Rs 30,000

Factory expenses (100% of wages)= Rs 30,000

∴ Factory cost = 20,000 + 30,000 + 30,000 = Rs 80,000

Non-manufacturing expenses ($12\frac{1}{2}$% of factory cost)

$$= \frac{80,000 \times 12.5}{100}$$

= Rs 10,000

∴ Total cost = Rs 80,000 + Rs 10,000

= Rs 90,000

Profit (10% of total cost) = Rs 9,000

∴ Selling price = Rs 90,000 + Rs 9,000

= **Rs 99,000**

Example 4.3. *Find the factory cost of a part made from solid brass bar 38 mm diameter and length of bar used being 25 mm. The machining time taken to finish the part is ninety minutes and the labour rate is Rs 2.00 per hour. Factory overheads are 50 per cent of direct labour cost. The density of the material is 8.6 gms. per cub. cm and its cost is Rs. 1.625 per newton.*

Solution. Weight of raw material $= \frac{\pi}{4} \times (3.8)^2 \times 2.5 \times 8.6 \times \frac{9.81}{1000}$, newtons

= 2.4 N

Material cost = 2.4 × 1.625 = Rs 3.90

Labour cost $= 2 \times \frac{90}{60} =$ Rs 3.00

Factory overheads = 50% of 3.00 = Rs 1.50

∴ Factory cost = 3.90 + 3.00 + 1.50

= **Rs. 8.40**

Example 4.4. *An electric immersion rod is being sold in the market for Rs 65.00. Find its production or manufacturing cost assuming 20% profit of the selling price and selling expenses to be 40% of production cost.*

If the cost of material used for rod is Rs 15.00 and the overheads of the department in which it is being made is 40% of labour cost, find the time taken for its manufacture if the labour rate is Rs 2.00 per hour.

Solution. Selling price = Rs 65.00

Profit = 20% of Rs 65.00 = Rs. 13.00

∴ Total cost = 65 – 13 = Rs. 52.00

If '*P*' is the production or manufacturing cost, then

Selling expenses = 0.4 P

But selling price = Production cost + selling expenses + profit

∴ 65 = P + 0.4 P + 13

$$P = \frac{52}{1.4} = \text{Rs. } 37.14$$

Now cost of raw material = Rs 15.00

If '*W*' is the direct labour cost, then

Overheads = 0.4 W

Now P = direct material + direct labour + overheads

∴ $37.14 = 15 + W + 0.4\,W$

$$\therefore \quad W = \frac{22.14}{1.4} = \text{Rs } 15.81$$

$$\therefore \quad \text{Time taken} = \frac{15.81}{2} = \textbf{7.905 Hours.}$$

Example. 4.5. *The market price of a machine is Rs 6000 and the discount allowed to distributors is 20% of the market price. For a certain period, it was found that the selling cost was half the factory cost. The material cost, labour cost and overhead charges of the factory are in the ratio 1 : 4 : 2. If the material cost is Rs. 400, what profit is made by the factory on each machine ?*

Solution. Market price of machine = Rs 6000

Discount = 20% of Rs 6000 = Rs. 1200

Selling price of the factory = 6000 – 1200 = Rs 4800

Now selling price = Factory cost + Administrative cost and Selling expenses + Profit

$= F + 0.5\,F + P = 1.5\,F + P$

Now F = Material cost + Labour cost + factory overheads

= 400 + 1600 + 800

= Rs 2800

∴ $4800 = 1.5 \times 2800 + P$

∴ P (Profit) = **Rs 600**

Example 4.6. *The monthly requirement of a company is 1500 components. The cost of each part is Rs 5 and the cost of each set up is Rs 30 per lot. If the carrying charge factor is 20% determine :*

(*a*) *Economic lot size*

(*b*) *Set up time for each lot*

Solution. From equation (4.1)

(*a*)
$$N = 5\sqrt{\frac{AS}{KC}}$$

$A = 1500$, $S = \text{Rs } 30$, $K = 0.20$, $C = \text{Rs } 6$

$$N = 5\sqrt{\frac{1500 \times 30}{0.20 \times 5}} = 1061 \text{ pieces}$$

(*b*) Time for each set up, equation (4.2)

$$s = \frac{NS}{A} = \frac{1061 \times 30}{1500} = \mathbf{21.22\ hours}$$

Example 4.7. *What is the unit time to change a milling cutter if it takes 10 minutes for the operator to change the cutter, the cutter works 2 minutes during each cycle and tool life is 4 hours ?*

Solution. From equ. (4.3)

$$\therefore \quad t = \frac{T'c.Tk}{T}$$

Here, $T_c = 2$ min, $Tk = 10$ min, $T = 240$ min

$$\therefore \quad t = \frac{2 \times 10}{240} = \mathbf{0.83\ min}$$

Example 4.8. *A tool will cut for 6 hours before it needed resharpening. It takes 20 minutes to change the tool. If the tool can be sharpened 12 times before it is discarded, determine the unit tool change time per cycle.*

Solution. Here $Tk = 6$ hours $= 360$ min

$T_c = 20$ min, $T = 12 \times 6 = 72$ hr $= 4320$ min

$$t = \frac{T'_c T_K}{T}$$

$$= \frac{20 \times 360}{4320} = \mathbf{1.67\ min}$$

Example 4.9. *A hole being bored on a lathe is to be checked by a Go, No-Go, plug gauge. The operator takes 10 seconds to check the hole. The guage is used twice for each piece and the number of pieces to be made are 200. Determine the measuring time allowance.*

Solution. From equation (4.5)

$$tc = Tc \times F$$

Hence $T_C = 10$ sec. $F = 2$

$\therefore$ tc (to check one piece) $= 10 \times 2 = 20$ secs.

$\therefore$ Total time needed to check this dimension for 200 pieces $= t_C \cdot (N + 1)$

$$= 20 \times 201 = \mathbf{67\ min}$$

Example 4.10. *40 forgings are to be machined in four set ups. Calculate the cost of production with the help of the following given data :*

Machining time	=	*12 min per forging*
Non-machining time	=	*21 min per forging*
Set up time	=	*45 min per set up*
Tool sharpening	=	*5 min per forging*
Fatigue	=	*20%*
Personal needs	=	*5%*
Tool change time	=	*10 min*
Tool life	=	*8 hours*
Checking time	=	*15 sec with 5 checks per forging*

Performance factor = *1.4*

Direct labour cost = *Rs 5 per hour.*

Sol. Machining + non-machining time = 33 min.

Fatigue = 20% of 33 min = 6.6 min

Personal needs = 5% = 1.65 min

$$\text{Tool sharpening time} = t = \frac{T'_c . T_k}{T}$$

$$= \frac{12 \times 10}{480} = 0.25 \text{ min. per forging}$$

$$\text{Measuring and checking time} = \frac{15 \times 5}{60} = 1.25 \text{ min per forging}$$

Sum of the above times = 12 + 21 + 1.65 + 6.6 + 0.25 + 1.25

= 42.75 min

∴ Time for 40 forgings = 42.75×40

= 1710 min

Set up time = 45 × 4 = 180 min.

∴ Total estimated time, *Te* = 1890 min

∴ Total actual time, *Ta* = *Te* × *R*

= 1890 × 1.4

= 2646 min.

∴ Direct labour cost = **Rs 220.5**

4.10. CALCULATION OF MACHINING TIMES

The basic relationship for determining the machining time for any machining operation is that the cutting time in minutes is equal to the distance the tool is fed, in mm, divided by the feed in mm per minute, *i.e.*,

$$T_m = \frac{L}{F}, \text{ per cut or per pass} \qquad ...(4.8)$$

where T_m = cutting time in minutes

L = Total tool travel in mm.

F = Feed of tool in mm per minute.

The distance a tool is fed to make a cut (L) is the sum of the distance the tool travels while cutting the material plus its approach distance plus its over travel. The "approach" is the distance a tool is fed from the time it touches the workpiece until it is cutting to the full depth. Approach distance for a drill is the length of its point which is about one-fourth the diameter of the standard drill. The approach of most of the single point tools is negligible. "Overtravel" is the distance the tool is fed while it is not cutting. It is the distance over which the tool idles before it enters and after it leaves the cut. This distance is calculated for face milling and slotting, but in other cases like drilling or turning it is taken as 0.8 to 6.0 mm.

Fig. 4.2 shows the cutting operation on a lathe. Advance of the tool is from position 1 to 3. At position 3, the feed is engaged. The tool travels to position 4, when it contacts the job. The distance 4-5 is the tool approach, *i.e.*, the tool travel before it starts cutting the required depth of material. So,

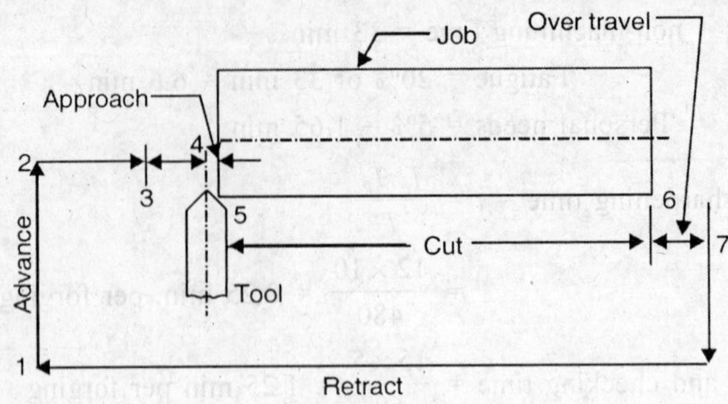

Fig. 4.2

Total cutting distance = (4 – 5) + (5 – 6)

The distance (6 – 7) is tool overtravel, so that the tool completely clears the job. The distance (3 – 4) is also the tool overtravel and this is provided so that the tool contacts the job smoothly and not with a jerk. So,

Total tool travel = Length of job + approach + Two overtravels

Now $F = f \times N$

and N = r.p.m. of work or cutter

$$= \frac{1000\,V}{\pi D} \quad \text{...(4.9)}$$

where V = cutting speed in metres per minute of work or cutter

and D = Diameter in mm of work or cutter.

∴ Relation (4.8) becomes,

$$T_m = \frac{L}{fN} \quad \text{...(4.10)}$$

Principal Elements of Metal Machining :

The principal elements of metal machining are :

(*a*) Cutting speed (*b*) Feed (*c*) Depth of cut

Cutting speed. The cutting speed can be defined as the relative surface speed between the tool and the job. It is a relative term, since either the tool or the job or both may be moving during cutting. It is expressed in metres per minute (mpm).

Feed. It may be defined as the relatively small movement per cycle of the cutting tool, relative to the work piece in a direction which is usually perpendicular to the cutting speed direction. It is expressed in millimetres per revolution (mm/rev) or millimetres per stroke (mm/str). It is more complex element as compared to cutting speed, since it is expressed differently for various operations. For example, in turning and drilling, the feed is the axial advance of the tool along or through the job during each revolution of the tool or job; for the shaper and planer, it is lateral offset between the tool and work for each stroke and for multitooth milling cutters, feed is the advance of the work or cutter between the cutting action of two successive teeth (expressed basically as mm/per tooth).

Depth of cut. The depth of cut is the thickness of the layer of metal removed in one cut, or pass, measured in a direction perpendicular to the machined surface. The depth of cut is always perpendicular to the direction of feed motion.

Selection of cutting speed. The cutting speed to be used will depend upon the following factors :

(*i*) **Work material.** Hard and strong materials require a lower cutting speed; whereas soft and ductile materials are cut at higher cutting speeds.

(*ii*) **Cutting tool material.** Special cutting tool materials, for example, cemented carbides, ceramics, Stellite and H.S.S. will cut at much higher cutting speeds than alloy or carbon steel tools.

(*iii*) **The depth of cut and feed.** A light finishing cut with a fine feed may be run at a higher speed than a heavy roughing cut.

(*iv*) **Desired cutting tool life.** The tool life is a direct function of cutting temperature which increases with increase in cutting speed. Thus as the cutting speed is increased, cutting tool life is decreased.

(*v*) **Rigidity and conditions of the machine and tool and the rigidity of the work.** An old, loose machine working with a poorly supported tool on a thin bar, will not cut at such a high speed, as a good machine with rigid tool operating on a well supported bar of reasonable dimensions.

Selection of feed. Feeds, to be used, will depend upon the following factors :

(*i*) **Smoothness of the finish required.** A coarse feed will give wider and deeper machining marks and an inferior finish to a fine feed. A blunt nosed tool will give a better finish than a sharp tool for the same feed.

(*ii*) **Power available, condition of the machine and its drive.** The product of the speed, feed and depth of cut gives the amount of metal being removed and hence the power necessary. A coarse feed on a poor or badly driven machine will be harmful both for the machine and the tool. This will also result in slipping of the drive or belt.

(*iii*) **Type of cut.** As a general rule, give coarest feed possible for a roughing cut because finishing is unimportant. For a finishing cut, the feed should be fine enough to give the class of finish required.

(*iv*) **Tool Life.** The cutting temperature increases with increase of feed, resulting in decreased tool life.

Selection of depth of cut. The depth of cut to be used will depend upon the following factors :

(*i*) **Type of cut.** Use large depths of cut for roughing operations than for finishing operations.

(*ii*) **Tool life.** The cutting temperature increases with increase of depth of cut, resulting in decreased tool life.

(*iii*) **Power required.** As discussed above, the cutting speed multiplied by area of cut (feed *x* depth of cut) gives the metal removal rate, which gives the power requirements. For a given area of cut, a large ratio of depth of cut to feed usually gives the most efficient performance as well as a better surface finish.

Procedure for assigning cutting variables

A definite order should be followed in selecting and assigning the cutting variables, that is, speed, feed and depth of cut. First, the depth of cut is established and then the rate of feed is established. After determining or assigning the values of depth of cut and rate of feed, the cutting velocity is established.

(*a*) *Depth of cut.* The depth of cut is determined primarily from the magnitude of the machining allowance. The machining allowance is the layer of metal (stock) that is to be removed from the surface of a blank in machining to obtain a finished part. Allowances are specified on one side, that is in the case of round parts, it equals one half of the difference in diameters of the blank and the finished part. The nearer is the blank to the finished part in shape and size, that is, the smaller the machining allowance, the less the amount of metal that will be converted into chips, the shorter the time required for machining, the higher the productivity in manufacturing the given part and the cheaper the machine of which it is a component. It is advantageous to remove the whole machining allowance in a single pass or cut as is done in rough machining. In that case, the depth of cut will equal the machining allowance. In removing large allowances or when the depth of cut is limited by : available power of the machine tool drive, lack of rigidity of the workpiece, insufficiently reliable clamping of the workpiece in the machine, and other factors, it becomes necessary to remove the total machining allowance in several passes, thereby reducing the depth of cut.

The depth of cut may be taken equal to 4 to 5 mm in rough turning 0.5 to 2 mm in semi finish turning and from 0.1 to 0.4 mm for finish passes.

The depth of cut is always associated with the type of machining being done. For example, a grinding allowance is always removed in several passes.

(*b*) *Feed.* In order to reduce the machining time, that is, to increase the productivity, it is advantageous to apply the maximum possible rate of feed, taking into account all the factors which may influence this value. In rough machining, when no special requirements are made to the surface finish of the workpiece, but the acting forces in the cutting process are of considerable magnitude, the maximum rate of feed may be limited by the available strength and rigidity of the cutting tool, rigidity of the blank and the capacity of the machine tool. Processing factors, namely machining accuracy and surface finish are of prime importance only for finish operations and especially when high classes of finish are required. The actual numerical values must be taken from those available on the machine tool. These are usually given in the service manuals and handbooks. For example, in rough turning with an ordinary carbide tipped lead angle lathe tool, with a depth of cut equal to 5 mm, the recommended rate of feed is maximum equal to 1.2 mm/rev. When higher quality of the machined surface is required, as in semi-finish and finish machining, the maximum rate of machining is restricted by the specified surface finish, since, the heavier the feed, the greater the roughness of the surface produced. For example, to obtain a surface finish of class between roughness grade number 7 to 8, with a turning tool on which tip radius = 1 mm, approach angle = 45° and end cutting edge angle ≤ 5°, the recommended rate of feed is maximum = 0.25 mm/rev.

(*c*) *Cutting velocity.* Having established the depth of cut and the rate of feed, it is possible to determine the cutting speed, V, from formulas given in metal cutting theory. Both these variables greatly affect the cutting speed permitted by the tool. The heavier the feed and depth of cut, the greater the forces acting on the tool, the higher the cutting temperature, the more intensive the tool wear will be and the lower the cutting speed permitted by the tool for the same tool life. The relationship between cutting speed, depth of cut and rate of feed, for a given tool life, can be expressed by the formula :

$$V = \frac{C}{d^m \cdot f^x}, \text{ m/ min} \qquad \text{...(4.11)}$$

where C = constant factor depending upon the work material, tool material, tool geometry, tool life, cutting fluid etc.

d = depth of cut, mm

f = rate of feed, mm/rev

m and x = exponents which differ for different work materials, tool materials and machining variables. In turning and boring work pieces of carbon steels with high-speed steel tools and with tool life of 60 min.

$$m = 0.25 \text{ and } x = 0.33 \text{ for } f \leq 0.25 \text{ mm/rev}$$

$$= 0.25 \text{ and } x = 0.66 \text{ for } f > 0.25 \text{ mm/rev}$$

The values for C can be taken from handbooks. For example, for turning with a single point tool tipped with cemented carbide and a tool life of 90 min, C = 141.5 for working material of carbon steel with a tensile strength of 750 N/mm^2.

It follows from the values for x and m that the exponent for feed is higher than for depth of cut, that is, an increase in feed leads to a greater reduction in cutting speed than an increase in the depth of cut. This is due to the more intensive tool wear with an increase in 'f' than with an increase in 'd'.

In equation (4.11), the values of the variables C, m and x will depend upon the tool life. A more general equation incorporating the tool life also is written as :

$$V = \frac{C}{T^{n} \cdot d^{m} \cdot f^{x}}, \text{ m/ min} \qquad \text{...(4.12)}$$

where T = tool life, min

For turning structural carbon and alloy steels with tensile strength equal to 750 N/mm^2 with a carbide-tipped tool, the various exponents and constant C have the values (from hand books) :

$$C = 273,\ n = 0.2,\ m = 0.15$$

$$x = 0.2 \text{ for } f \leq 0.3 \text{ mm/rev}$$

$$= 0.35 \text{ for } f \leq 0.75 \text{ mm/rev}$$

$$= 0.45 \text{ for } f > 0.75 \text{ mm/rev}$$

The common values of cutting speeds and feeds are given in Table 4.1 and also, for the individual operations their values have been recommended in the following pages.

Below, we drive relations to calculate the machining time for the various operations :

1. **Turning.** The turning operation is shown in Fig. 4.3.

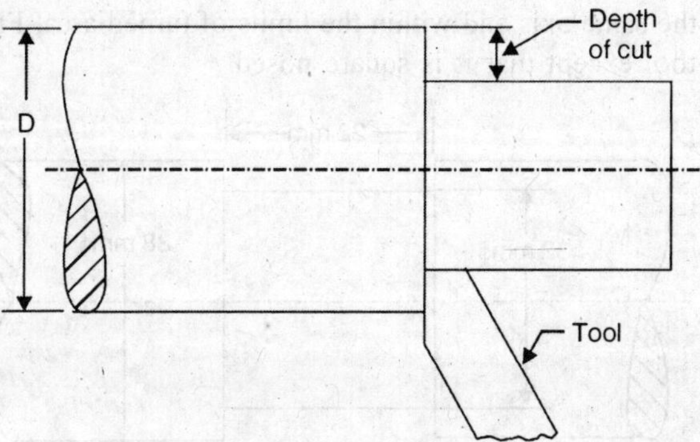

Fig. 4.3. Turning Operation.

Note the diameter '*D*' which will be used in equation (4.9) to obtain the r.p.m. of the cutter of work. To obtain the machining time for turning operation, equation (4.10) is used.

Example 4.11. *What is the machining time to turn the dimensions given in Fig. 4.4. The material is brass, the cutting speed with H.S.S tool being 100 m/min, and the feed is 7.5 mm/rev.*

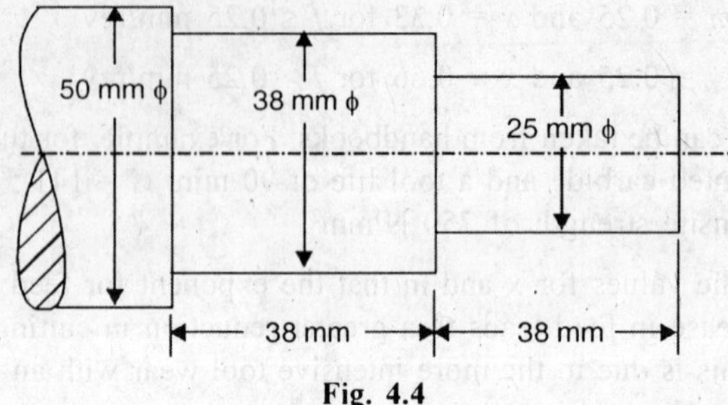

Fig. 4.4

Solution. First find the time to turn 38 mm diameter by 76 mm length of cut. Using the formula,

$$T_m = \frac{L}{fN}$$

Now $$N = \frac{1000\,V}{\pi D}$$

$$= \frac{1000 \times 100}{\pi \times 50} = 636.62\ r.p.m$$

$$\therefore \qquad T_m = \frac{76}{0.75 \times 636.62} = 0.16 \text{ min}$$

Next, to turn 25 mm diameter by 38 mm length,

$$N = \frac{1000 \times 100}{\pi \times 38} = 837.65\ r.p.m.$$

$$\therefore \qquad T_m = 380.75 \times 837.65 = 0.06 \text{ min}$$

$$\therefore \qquad \text{Total time} = 0.16 + 0.06 = \mathbf{0.22\ min.}$$

The turning is done in one pass only as the maximum depth of cut of 6 mm on a side in general shop practice for brass.

2. **External relief.** An external relief operation is the removal of material from a previously turned surface along the same axis and within the limits of turned area, Fig. 4.5. The cutting tool is similar to turning tool except that it is square nosed.

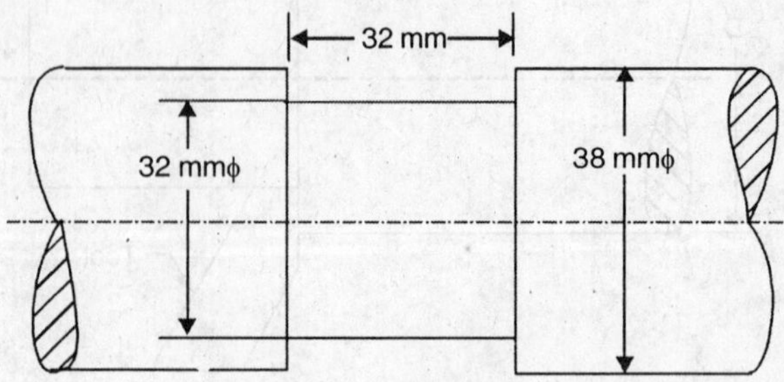

Fig. 4.5. External Relief.

Table 4.1

Cutting Speeds and Feed Rates

Work material	*Cutting Speed V in mpm*								*Feed rate f in mm/rev.*	
	H.S.S. tool				*Carbide tool*		*Stellite tool*			
	Turn		*Ream and* Thread	*Drill*	*Turn*		*Turn*			
	Rough	*Finish*			*Rough*	*Finish*	*Rough*	*Finish*	*Rough*	*Finish*
Mild Steel	40	60	7.5 to 15	30	90	180	50	75	0.625 to 2.0	0.125 to 0.75
Cast Steel	15	24	3.5	12	45	100	24	33	0.5 to 1.25	0.125 to 0.5
Stainless Steel	15	18	3	12	27	45	22	25	0.5 to 1.0	0.075 to 0.175
Grey C.I.	18	27	3.5	13	60	100	33	45	0.4 to 2.5	0.2 to 1.0
Aluminium	90	150	15	72	240	360	120	180	0.1 to 0.5	0.075 to 0.25
Brass	75	100	18	60	180	270	90	150	0.375 to 2.0	0.2 to 1.25
Phosphor Bronze	18	36	4.5	13	120	180	30	50	0.375 to 0.75	0.125 to 0.5

The time for doing external relief is calculated on the same lines as for plain turning.

Example 4.12. *Find the time to turn the external relief shown in Fig. 4.5. The material is mild steel and the feed is 0.379 mm/rev.*

Solution. From table 4.1, with H.S.S. cutting tool the cutting speed, $V = 60$ m/min.

$$N = \frac{1000 \times 60}{\pi \times 38} = 502.6 \text{ rev/min}$$

$$\therefore \quad T_m = \frac{L}{fN} = \frac{32}{0.375 \times 502.6}$$

$$= \mathbf{0.17\ min.}$$

3. **Point.** Pointing is the process of removing stock at the end of a bar to facilitate the approach of a cutter turner. It helps the cutter-turner rolls to find their work place. This operation which is usually hand feed operation may be done either from cross slide or turret of a lathe, Fig. 4.6. The time calculations are done on the same lines as for plain turning.

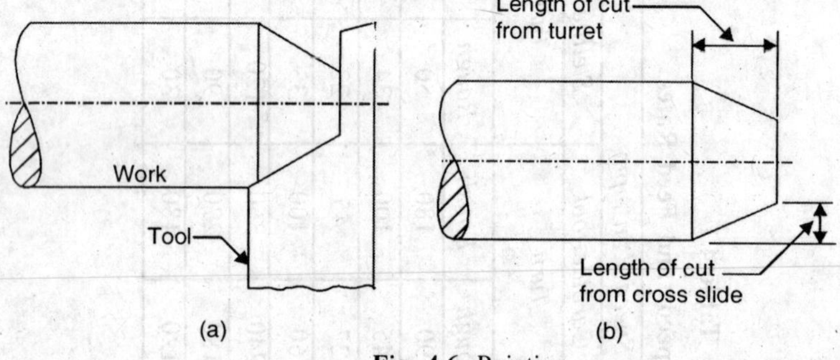

Fig. 4.6. Pointing.

4. **Chamfer.** Chamfering is the operation of removing material from the edges of external or internal diameters to facilitate the entering of mating parts, to form a seat or to remove sharp edges, Fig. 4.7. This is usually a hand feed operation and the machining time is relatively short. The time may be calculated by formula (4.10) as for plain turning.

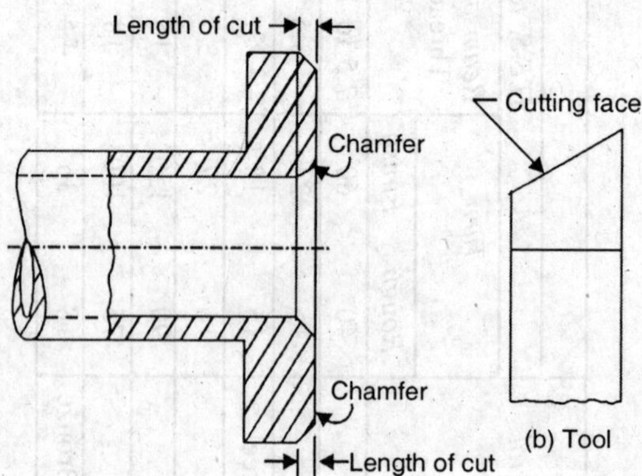

Fig. 4.7. Chamfering.

5. **Knurling.** Knurling is the operation of upsetting material so as to produce diamond-shaped or straight-lined patterns on the surface of the material. This is done to provide gripping

when the job is grasped by hand. The handles of gauges, hand screws, slip bushings etc., are often knurled. The knurling is obtained with the help of tools called 'knurls' and the operation may be performed either from cross slide or from turret of a lathe. In turret knurling, the tool is fed 'in' and 'out', the out feed being two to three times the infeed. The cutting time for knurling operation is estimated on the same lines as for plain turning, Fig. 4.8, *i.e.*,

$$\text{Time} = \frac{\text{Length of cut}}{(\text{feed})\,(\text{r.p.m.})}$$

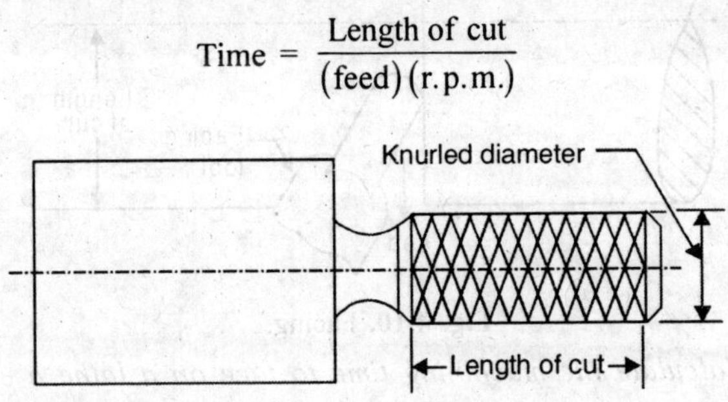

Fig. 4.8. Knurling.

6. **Forming.** Forming is the operation of producing surfaces that would be difficult to produce by the conventional methods of turning, boring or facing. The tool used is called a 'form tool' and carries the contour of the part to be formed, Fig. 4.9. The cutting time is estimated in same way as for plain turning, *i.e.*,

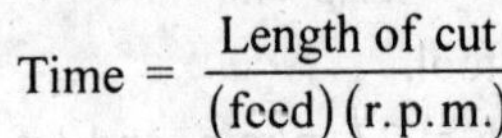

$$\text{Time} = \frac{\text{Length of cut}}{(\text{feed})\,(\text{r.p.m.})}$$

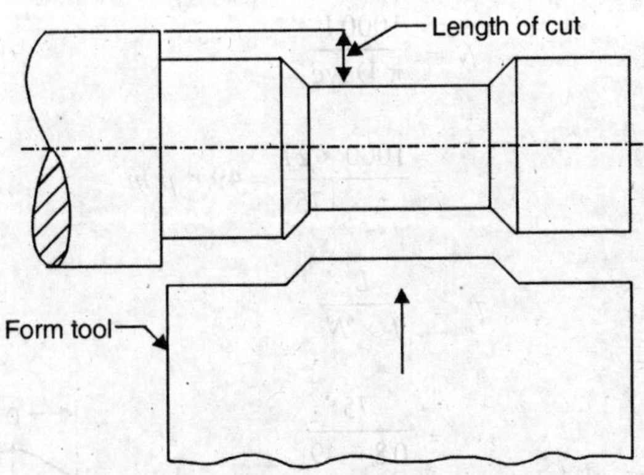

Fig. 4.9. Forming.

7. **Facing.** Facing is the operation of machining along some plane on a bar, casting or forging, Fig. 4.10. The purpose of this operation is to obtain a surface of good finish. The machining time is estimated with the help of formula (4.10) where r.p.m. is determined by using in the mean work diameter, *i.e.*,

$$N = \frac{1000\,V}{\pi\ \text{Dave}}$$

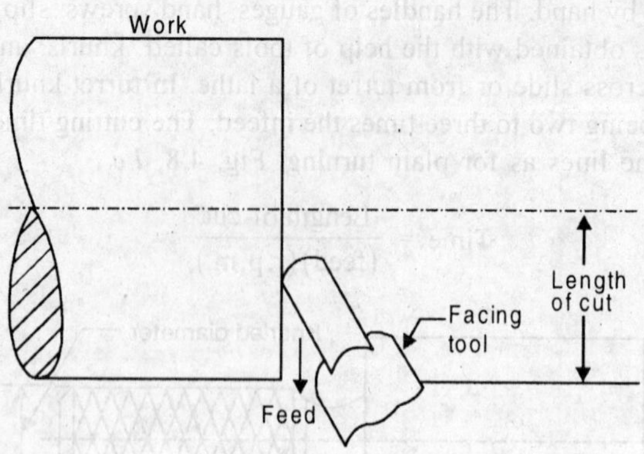

Fig. 4.10. Facing.

Example 4.13. *Calculate the machining time to face on a lathe a cast-iron flange shown in Fig. 4.11.*

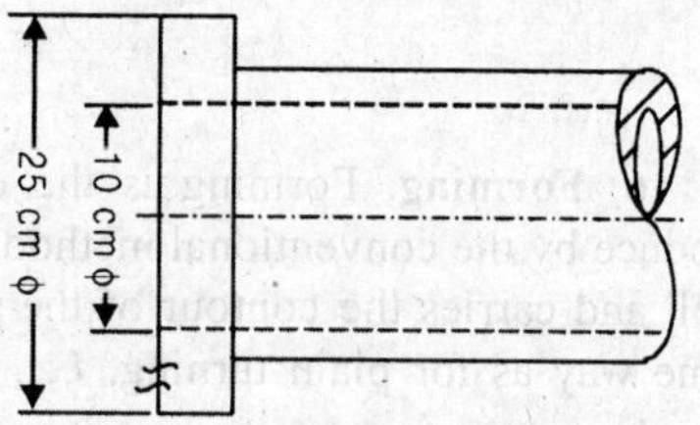

Solution. As is clear from Fig. 4.11.

$$\text{Length of cut} = 7.5 \text{ cm.}$$

$$\text{Average diameter, Dave} = \frac{1}{2}(25+10)$$

$$= 17.5 \text{ cm.}$$

Now from Table 4.1, for H.S.S. tool,

$$V = 27 \text{ m/min, and let } f = 0.8 \text{ mm/rev.}$$

Now
$$N = \frac{1000\,V}{\pi \text{ Dave}}$$

$$= \frac{1000 \times 27}{\pi \times 175} = 49\ r.p.m.$$

$\therefore$
$$T_m = \frac{L}{f \times N}$$

$$= \frac{75}{0.8 \times 49}$$

$$= \mathbf{1.91\ min.}$$

Fig. 4.11.

8. **Drilling.** Drilling is the operation of cutting holes in material, Fig. 4.12 in which the length of drill travel is

$$L = A + t + A$$

where $A = 0.29\,D$ for point angle of 120°

Drilling operation can be done on drill presses, turret lathes and engine lathe. On engine lathe, the drill is held in the tail stock spindle and the feed is given manually by rotating the hand wheel of the tail stock.

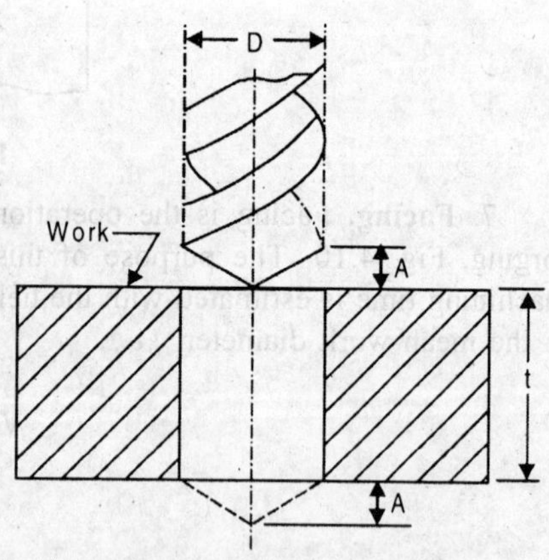

Fig. 4.12. Drilling Operation.

Here, the drilling time cannot be calculated with the help of a formula. Drill cutting speeds, in general, are lower than turning cutting speeds. The harder the material to be drilled, the lower should be the cutting speed. The feed varies with the diameter of the drill, increasing with the diameter. It is between about 0.05 mm/rev. for small drills of 1.5 mm diameter to about 0.25 mm/rev. for drills of 25 mm diameter. Table 4.2 and Table 4.3 give respectively cutting speeds and feed rates for drilling.

Table 4.2 — **Cutting Speed, m/min**; **Table 4.3** — **Feed rates, mm/rev**

Work material	*V*	*Drill size, mm*	*f*
Stainless Steel	9-12	0-3.2	0.025-0.05
C-Steels (0.4-0.5% C)	21-24	3.2-6.35	0.05-0.10
C-Steels (0.2-0.3% C)	24-33	6.35-12.7	0.10-0.175
Soft grey C.I	30-45	12.7-25.4	0.175-0.375
Brass and Bronze	60-90	> 25.4	0.375-0.625
Magnesium Alloys	75-120		

Example 4.14. *How long will it take a 12.7 mm drill to drill a hole 50 mm deep in brass.*

Solution. From Table 4.2 and Table 4.3.

$$V = 75 \text{ m/min. (say) and } f = 0.175 \text{ mm/rev.}$$

Length of drill travel, $L = 50 + 2 \times 0.29 \times 12.7$

$$= 57.366 \text{ mm.}$$

$$\therefore \quad T_m = \frac{L}{f \times N}$$

Now $$N = \frac{1000V}{\pi D} = \frac{1000 \times 75}{\pi \times 12.7} = 1880 \, r.p.m.$$

$$\therefore \quad T_m = \frac{57.366}{0.175 \times 1880} = \mathbf{0.174 \text{ min.}}$$

9. **Boring.** Boring is the operation of enlarging or finishing an internal hole which has been previously cored or drilled. The tool is usually a single point tool and it is mounted in a tool post, a square turret or a boring bar and is operated from a spindle, a cross slide, a hexagonal turret, and so on. The cutting speed relation and the relation to find machining time is similar to that used for simple turning.

10. **Undercutting (Internal Relief).** In undercutting operation, a previously bored hole is made larger along the same axis and within the longitudinal limits of the main bore, Fig. 4.13. Cutting time is obtained in the same manner as boring time.

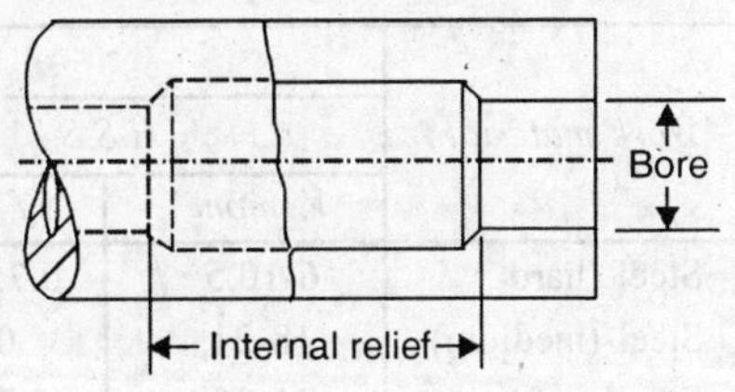

Fig. 4.13. Under cutting.

11. **Shaping, Planning and Slotting.** In all these operations, relative motion between the tool and the work piece is reciprocating. The shaper operation is shown in Fig. 4.14. Forward stroke is the cutting stroke and the return stroke is idle stroke, so it should be completed in minimum possible time, Usually,

$$\frac{\text{Return stroke}}{\text{Forward stroke}} = 2/3$$

i.e., time for forward stroke (cutting stroke)

$$= \frac{3}{5} \times \text{Total time (of both the strokes)}$$

Now cutting speed will be given as,

$$V = \frac{N \times L(1+K)}{1000} \text{ m/ min}$$

where N = Number of strokes per minute

L = Length of stroke, mm

(including tool clearance at each end).

and K = Ratio of return time to cutting time

Now, Time taken by cutting stroke

$$= \frac{L}{1000 \times V}$$

$\therefore$ Return time = $K \times$ cutting stroke time

$$= \frac{KL}{1000 \times V}$$

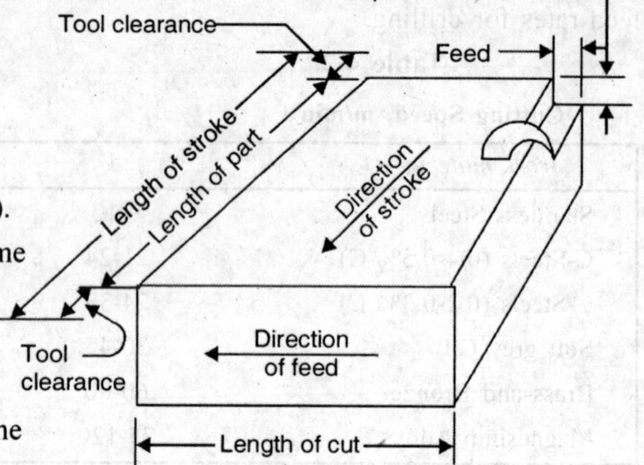

Fig. 4.14. Shaper operation.

$\therefore$ Time taken to complete one cycle (double strokes)

$$= \frac{L}{1000 \times V} + \frac{KL}{1000 \times V}$$

Now total number of double strokes required to complete one cut on full breadth, W, of the job = W/f

where f = feed/stroke

$$\therefore \text{Total time for completing one cut} = \frac{L(1+K)}{1000 \times V} \times \frac{W}{f}$$

Cutting speeds and feeds are given in Table 4.4.

Table 4.4

Cutting Speeds and Feeds for Shaping, Planning and Slotting

Work material	*Type of Tool*					
	H.S.S.		*Cast alloys*		*Carbides*	
	V, mpm	*f, mm*	*V, mpm*	*f, mm*	*V, mpm*	*f, mm*
Steel (hard	6-10.5	0.75-1.25	-	-	30-54	0.9
Steel (medium)	18-21	0.75	-	-	54-75	1.25
Steel (soft)	21-30	0.75-3.0	-	-	54-90	1.25
Cast steel	7.5-18	1.25	18-24	1.0	30-54	1.00
C.I. (hard)	9-15	1.50	15-24	1.25	30-60	1.25
C.I. (soft)	15-24	3.0	27-36	1.25	33-67.5	1.25
Malleable iron	15-27	2.25	14-36	1.25	45-75	1.0
Brass	45-75	1.25-1.50	-	-	-	-
Bronze	9-18	2.0	-	-	45-90	1.25
Aluminium	60-09	0.75-1.25	-	-	-	-

Example 4.15. *Find the time required on the shaper to complete one cut on a plate 600 × 900 mm, if the cutting speed is 6 m/min. The return time to cutting time ratio is 1 : 4 and the feed is 2 mm/stroke. The clearance at each end is 75 mm.*

Solution. Now, $V = \dfrac{N \times L(1+K)}{1000}$

Now $L = 900 + 2 \times 75 = 1050 \text{ mm}$

$$K = \frac{1}{4} = 0.25$$

$\therefore$ N, number of strokes/min. $= \dfrac{6 \times 1000}{1050 \times 1.25} = 5\,(\text{approx.})$

$\therefore$ $\text{Time} = \dfrac{W}{f \times N} = \dfrac{600}{2 \times 5} = \mathbf{60\ min.}$

12. **Broaching.** The cutting time for broaching operation may be calculated as,

$$T_m = \frac{\text{Length of stroke}}{\text{cutting speed}} + \frac{\text{Length of stroke}}{\text{return speed}}$$

The return stroke being idle is faster than cutting stroke and so, the return speed is usually twice the cutting speed. Cutting speeds for various material with H.S.S. broaches are given in Table 4.5.

Table 4.5 : Cutting Speeds for Broaching.

Material	*V, m/min.*
Steel (hard),	3
Steel (medium),	6
Steel (soft),	9
C.I. (hard),	6
C.I. (soft),	12
Bronze,	12
Brass,	12
Beryllium, copper,	3

For carbide broaches, the cutting speed ranges from 9 to 27 m/min. Feed is usually 0.075 mm or less between teeth.

Example 4.16. *How long will it take to broach a four-spline brass fitting if the cutting speed is 11 m/min, the return speed is 24 m/min., the length of the stroke is 70 cm.*

Solution. Using the formula,

$$T_m = \frac{L}{\text{cutting-speed}} + \frac{L}{\text{return speed}}$$

$$= \frac{70}{100 \times 11} + \frac{70}{100 \times 24} = 0.0636 + 0.0291$$

$$= \mathbf{0.0927\ min}$$

13. **Sawing.** For power hack saw,

$$T_m = \frac{\text{Length of cut (Thickness to be cut)}}{\text{(strokes per min.) (feed per stroke)}}$$

The recommended strokes per minute are as follows :

1. Cold-rolled brass, copper etc. 140
2. Annealed tool steels, medium steels, etc 100
3. Hard steels, special alloys, etc. 70

The positive feeds for power hack saw are given in Table 4.6.

Table 4.6. Positive Feeds for Power Hack Saw

Diameter of Stock, cm	*Recommended feed per stoke, mm*		
	Annealed tool steel	*Cold-rolled brass, copper*	*Standard pipe and tubing*
2.5	0.35	Max.	Max.
5.0	0.30	0.50	Max.
7.5	0.20	0.40	0.50
10.0	0.150	0.25	0.40
12.5	0.125	0.175	0.30
15.0	0.075	0.15	0.20

The formula for time to saw on a Band Saw is,

$$T_m = \frac{\text{Length of cut in cm.}}{\text{(cutting speed, mpm) 100 (teeth per mm) (Feed per tooth, mm)}}$$

Band Saw feeds and speeds are given in Table 4.7 and Table 4.8.

Table 4.7. Band-Saw Feeds (Carbon Steel)

	Material Thickness, cm				
	0.16 to 0.32	0.64	1.27	2.54	5 to 7.5
Feed per tooth, mm	0.0125	0.0075	0.0050	0.0025	0.00125
Saw pitch/cm	7-10	6-7	4-6	3-6	2-3

Table 4.8. Band-Saw Cutting Speeds (Carbon Steel), mpm

Material	*Material Thickness, mm,*				
	1.6	3.2	6.4	12.7	25.4
Aluminium	450	450	450	450	450
Bakelite	450	450	450	450	450
Beryllium Copper	120	105	90	82.5	75
Brass (soft)	450	360	225	165	105
Bronze	150	120	90	67.5	45
C.I. (soft)	60	52.5	45	37.5	30

Copper	330	270	210	180	150
Magnesium	450	450	450	450	450
Malleable iron	60	57	52.5	48	45
Monel	60	48	37.5	33	30
Phenolic	450	450	450	375	300
Plastic	450	450	450	375	300
Rubber (hard)	450	450	450	450	450
Steel (mild)	75	67.5	60	57	52.5
Steel (medium)	52.5	45	37.5	33	30
Steel (hard)	30	27	22.5	18	15

For band saws made of H.S.S., the cutting speed can be 50 to 140 per cent higher than those for carbon steel bands. The feed rate for H.S.S., band saws are given in table 4.9.

14. **Milling.** For milling also, the formula to calculate machining time will be,

$$T_m = \frac{\text{Length of cut}}{(\text{feed per rev.}) \times (\text{r.p.m.})}$$

Table 4.9. Feed rate : cm/min (H.S.S. Band Saw)

Work Thickness mm	*Cast Iron*	*Mild Steel cold-rolled*	*Tool Steel*	*High C-Steel High chrome steel*
6.25	80	52.5	30	18
12.5	40	27	15	12
25	17.65	11.43	7.93	6.65
37.5	11.43	7.16	4.11	4.87
75	5.70	3.35	1.67	2.08
150	2.54	1.76	0.76	1.00

Since the milling cutter is a multi-point cutter, the feed will be given as,

Feed per rev. = Feed per tooth × number of cutter teeth

Feed per min. (table feed) = Feed per rev. × r.p.m. of cutter

or $$F = f \times n \times N$$

where n is the number of cutter teeth.

The r.p.m. of cutter is obtained as,

$$N = \frac{1000 \times V}{\pi D}, \; D \text{ is cutter diameter in mm.}$$

The feed per tooth is usually 0.2 mm/tooth. However, the average cutting speeds and feed per tooth are given in tables 4.10 and 4.11 respectively.

Table 4.10. Cutting Speed (H.S.S. Cutter)

Material being cut	*Brass*	*C.I.*	*Bronze*	*Mild Steel*	*Hard C Steel*	*Hard alloy Steel*	*Alumi-nium*
V, mpm	45-60	21-30	24-45	21-30	15-18	9-18	150-300

Table 4.11. Feed Per Tooth (H.S.S. Cutter)

Type of cutter	*Slab Mill (Helix angle) up to 30°*	*Slab Mill (Helix angle) 30° to 60°*	*Face Mill*	*End Mill*	*Slot Mill*	*Form relieved cutter*
Feed per tooth, mm	0.10-0.25	0.075-0.20	0.125-0.50	0.025-0.20	0.075-0.125	0.075-0.20

Table 4.12. gives the cutting speed for carbide tipped cutters for a feed rate of 0.2 mm per tooth.

Table 4.12. Cutting Speed, Carbide Cutter

Work Material	*V, m/min*							
	Brazed Cutters				*Indexable inserts*			
	I.S.O. Carbide grade				*I.S.O. Carbide grade*			
	P.10	*P.30*	*P.40*	*K.20*	*P.10*	*P.30*	*P.40*	*K.20*
Aluminium	150	130	100	-	200	170	130	-
C-Steel, 0.7% C	120	90	75	-	150	90	75	-
Steel Casting	60	45	50	-	80	75	50	-
Stainless Steel	100	100	100	-	125	125	115	-
Grey C.I.	150	130	110	-	150	130	110	-
Aluminium Alloy	-	-	-	600	-	-	-	600

The total cutter travel will be more than the length of the job to be cut. The extra cutter travel will be the cutter overtravel and cutter approach and will depend upon the type of milling operation, as shown below :

(*i*) *Face Milling.* When a milling cutter has traversed the length of face, some portion of the face has yet to be milled as shown by shaded area in Fig. 4.15. In order to complete the milling, and additional distance must be travelled by the milling table. This is given as

$$\text{Added table travel} = \frac{1}{2}\left(D - \sqrt{D^2 - W^2}\right)$$

If the face milling cutter is of the same diameter as the width of the work, then,

$$\text{Added table travel} = D/2.$$

Marks are left on the job by the heel of the cutter. For the sake of finish, it is good to feed the cutter until it clears the job. For such cases, the additional table travel will equal the diameter of the cutter.

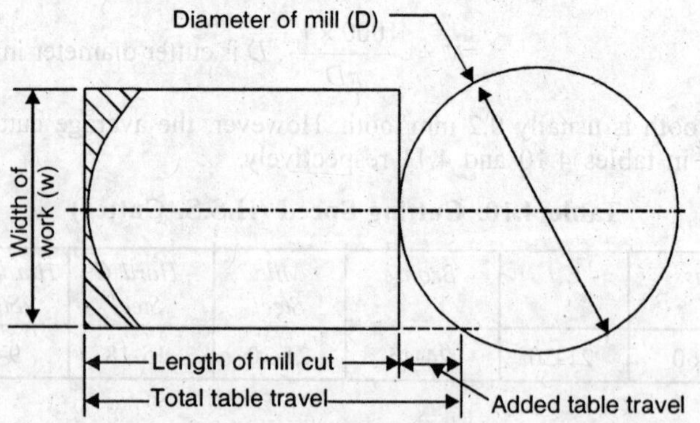

Fig. 4.15. Face Milling.

(*ii*) **Slab milling or Slot milling.** Referring to Fig. 4.16, the added table travel will be given as,

$$\text{Added table travel} = \sqrt{Dd - d^2}$$

where '*d*' is depth of cut.

The above formula is valid when the depth of cut is less than the radius of the cutter. If the depth of cut equals or exceeds the cutter radius the added table travel will be equal to radius of the cutter.

The above relations are valid for rough milling where cutter overtravel may not be required, and the cutter would stop at the dotted position shown in Fig. 4.16. For a finishing pass, the cutter is permitted to travel beyond the end of the workpiece so that the trailing edges give the same wiping action to the entire surface. So,

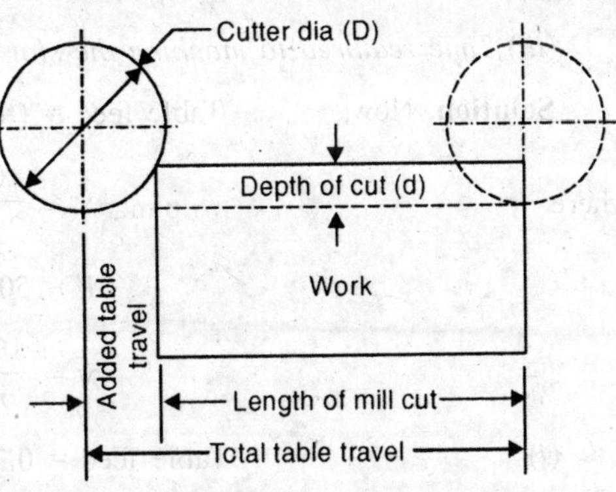

Fig. 4.16. Slab Milling.

$$\text{Over travel} = \text{Approach}$$

$$\therefore \quad \text{Added table travel} = \text{Approach} + \text{Overtravel}$$

$$= 2\sqrt{Dd - d^2}, \text{ when } d < \frac{D}{2}$$

$$= D, \text{ when } d \geq \frac{D}{2}$$

(*iii*) **Cutting a flat across round bar stock.** Refer Fig. 4.17, the length of cut will be the length of chord which is given,

$$L = 2\sqrt{dD_1 - d^2}$$

where D_1 is diameter of bar and '*d*' is depth of flat or chord.

The added table travel will be given as (Fig. 4.17b),

$$\text{Added Table Travel} = \sqrt{D_2 d + D_1 d - d^2} - \sqrt{D_1 d - d^2}$$

where D_2 is the cutter diameter

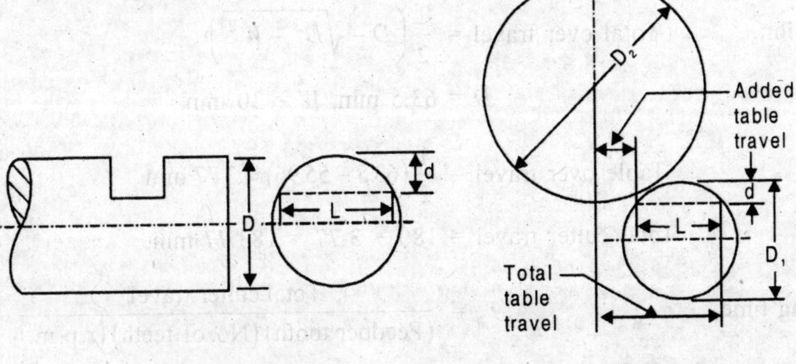

Fig. 4.17. Flat Across Round Bar.

Example 4.17. *A slot 25 mm deep is to be cut through a work piece 200 mm long with the help of H.S.S. side and face cutter whose diameter is 150 mm and that has 10 teeth. The cutting speed is 50 m/min. and feed is 0.25 mm per tooth. Determine :*

(*i*) *Table feed in mm/min*

(*ii*) *Total cutter travel*

(*iii*) *Time required to machine the slot.*

Solution. Now, Table feed = (feed per tooth) (number of teeth) (r.p.m.)

where r.p.m., $N = \dfrac{1000\,V}{\pi\,D}$

V = 50 m/min and D = 150 mm

$\therefore \quad N = \dfrac{1000 \times 50}{\pi \times 150} = 106$

(*i*) ∴ Table feed = 0.25 × 10 × 106 = 265 mm/min

(*ii*) Total cutter travel = Length of workpiece + total overtravel

Now, Table over travel = $\sqrt{Dd - d^2}$

where D = 150 mm and d = 25 mm

∴ Total over travel = $\sqrt{150 \times 25 - 625} = 55.9$ mm

(*ii*) Total cutter travel = 200 + 55.9 = 255.9 mm

(*iii*) $\text{Time} = \dfrac{\text{Total travel}}{\text{Feed/ min}}$

$= \dfrac{255.9}{265} = \mathbf{0.965\ min.}$

Example 4.18. *A 63.5 mm diameter plain milling cutter having 6 teeth is used to face mill a block of aluminium 18 cm long and 3 cm wide. The spindle speed is 1500 r.p.m. and the feed is 0.125 mm per tooth per rev. Find the cutting time.*

Solution. Total over travel = $\dfrac{1}{2}\left(D - \sqrt{D^2 - W^2}\right)$

Now D = 63.5 mm, W = 30 mm.

∴ Table over travel = $\dfrac{1}{2}(63.5 - 55.96) = 3.77$ mm

∴ Total cutter travel = 180 + 3.77 = 183.77 mm.

Cutting time, $T_m = \dfrac{\text{Total cutter travel}}{(\text{Feedper tooth})\,(\text{No. of teeth})\,(\text{r.p.m.})}$

$= \dfrac{183.77}{0.125 \times 6 \times 1500} = \mathbf{0.162\ min.}$

Example 4.19. *A flat is to be cut across a round brass bar 5 cm in diameter. The depth of cut is 19 mm. The diameter of cutter is 10 cm. Cutting speed is 50 m/min with a feed of 0.2 mm per tooth. The cutter has 8 teeth. Find the milling time.*

Solution. Refer Fig. 4.17

$$\text{Length of chord} = \text{Length of cut, } L = 2\sqrt{dD_1 - d^2}$$

where $d = 19 \text{ mm}, D_1 = 50 \text{ mm}$

$$\therefore \quad L = 2 \times \sqrt{19 \times 50 - 19^2} = 48.54 \text{ mm}$$

$$\text{Overrun} = \sqrt{D_2 d + D_1 d - d^2} - \sqrt{D_1 d - d^2}$$

$$D_2 = 100 \text{ mm}$$

$$\therefore \quad \text{Overrun} = \sqrt{100 \times 19 + 50 \times 19 - 361} - 24.27$$

$$= 25.62 \text{ mm}$$

$$\therefore \quad \text{Table travel} = 48.54 + 25.62 = 74.16 \text{ mm}$$

$$\therefore \quad T_m = \frac{\text{Table travel}}{(\text{Feed per tooth})(\text{No. of teeth})(\text{r.p.m.})}$$

Now,

$$\text{r.p.m.} = \frac{1000\,V}{\pi D}$$

$$= \frac{1000 \times 50}{\pi \times 100} = 159$$

$$\therefore \quad \text{Time} = \frac{74.16}{0.2 \times 8 \times 159} = \mathbf{0.29 \text{ min.}}$$

15. **Grinding.** Machine time for external cylindrical grinding is

$$T_m = \frac{\text{Length of part}}{(\text{Feed per rev.})(\text{r.p.m.})}, \text{ per pass}$$

The transverse feed is taken as,

For roughing cuts, $\text{feed} = \left(\frac{1}{2} \text{ to } \frac{3}{4}\right) \times \text{width of wheel}$

For finishing cuts, $\text{feed} = \left(\frac{1}{4} \text{ to } \frac{1}{2}\right) \times \text{width of wheel}$

The total grinding time equals the time per pass of cut multiplied by the number of cuts. The recommended depths of cut are 0.025 to 0.10 mm for roughing cuts. Use 0.025 mm for hard steels and upto 0.10 mm for soft materials. The recommended depths of cut are 0.00625 mm to

0.05 mm for finishing cuts. Use 0.00625 mm for hard steels and upto 0.05 mm for soft materials. Use cutting speeds as : 9 mpm on hard steel, 15 mpm on medium and mild steels, and 22.5 to 30 mpm on soft materials.

Example 4.20. *Rough grind medium steel shaft under the following conditions :*

Diameter	=	*38 mm*
Length of part	=	*200 mm*
Total stock	=	*0.25 mm*
Grinding wheel	=	*50 mm face*
Depth of cut	=	*0.025 mm*
Cutting speed	=	*15 m/min*

How much time will be required to grind the shaft ?

Solution.

$$\text{Feed per rev.} = \left(\frac{1}{2} \text{ to } \frac{3}{4}\right) \times \text{width of wheel}$$
$$= w/2 \text{ say.}$$

where '*w*' is width of grinding wheel,

$$\therefore \quad \text{Feed per rev.} = \frac{50}{2} = 25 \text{ mm}$$

$$\text{No. of cuts} = \frac{\text{Total stock}}{\text{Depth of cut per pass}} = \frac{0.25}{0.025}$$
$$= 10$$

$$\text{r.p.m.} = \frac{1000\,V}{\pi\,D}$$

$$= \frac{1000 \times 15}{\pi \times 38} = 125$$

$$T_m = \frac{200 \times 10}{25 \times 125} = \mathbf{0.64\ min.}$$

16. **Threading and tapping.** Threads can be cut on engine lathe with the help of a single point tool or on turret lathes with the help of a die head which operates from the nexagonal turret. The formula for time to cut threads is

$$T_m = \frac{\text{Length of cut}}{(\text{feed per rev.})\,(\text{r.p.m.})}$$

Feed per revolution = Pitch of threads for single-start threads

= Lead of thread for multi start threads

Tapping is the operation of cutting internal threads with the help of a special tool called a "tap". The cutting speed for threading and tapping are given in table 4.13. The number of cuts needed to complete the operation is also given in table 4.13.

Table 4.13. Cutting Speeds and No. of Cuts (Single-Point Tool)

Material	*V, m/min*	*No. of cuts*
Aluminium	9	4
Brass (commercial)	9	3
Brass (naval)	9	4
Bronze (ordinary)	9	5

Bronze (hard)	6	7
C.I. (Soft)	6	-
C.I. (Medium)	6	-
C.I. (Hard)	3	-
Copper	6	5
Drill Rod	3	8
Magnesium	9	4
Monel (bar)	3	8
Steel (Medium)	3	7
Steel (Hard)	3	8
Steel (Stainless)	3	8

Example 4.21. *An operator is cutting 44 mm - 2 threads per cm. 8.9 cm long in mild steel on a lathe with a single pointed tool. Find the time to cut the threads.*

Solution. From table 4.13,

$$V = 6 \text{ m/min and number of cuts} = 5$$

$$\text{r.p.m.} = \frac{1000 \times V}{\pi D}$$

$$= \frac{1000 \times 6}{\pi \times 44} = 43$$

$$\therefore \quad T_m = \frac{\text{Length of cut}}{(\text{feed per rev.})(\text{r.p.m.})} \times \text{No. of cuts}$$

$$\text{Now feed per rev.} = \text{Pitch} = \frac{1}{2} = 0.5 \text{ cm}$$

$$\therefore \quad T_m = \frac{89}{5 \times 43} \times 5 = \mathbf{2.07 \text{ min.}}$$

4.11. ESTIMATION OF TOTAL UNIT TIME

As already mentioned, the total time to produce one unit will consist of set up time, machining time, handling time, tear down time and allowances. Standard data are available for set up time, handling time and tear down time. Tear down time is usually very small and is considered along with set up time. After the machining time for individual operations has been calculated and the total obtained, including the set up time and handling time, a percentage of about 10% to 20% is added to allow for contingencies. That will give the total cycle time. The set up time which occurs once for a lot is given in table 4.14, for a new machine tools.

Table 4.14. Average Set Up Time

Machine Tool	*Time, hr*		*Machine Tool*	*Time, hr*	
	Min.	*Max.*		*Min.*	*Max.*
Engine Lathe (9″ × 48″)	0.30	1.50	Turret Lathe (3″ × 21″)	1.00	2.50
			Shaper	0.30	0.80
Engine Lathe (12″ × 102″)	0.50	2.00	Planer	0.50	1.50
Engine Lathe (24″ × 144″)	0.75	2.00	Surface Grinder	0.30	0.50
Turret Lathe (1″ × 10″)	0.30	1.50	Radial Drill	0.30	1.00
Turret Lathe (1.5″ × 21″)	1.00	1.00	Mills	0.75	1.00

The handling times for various operations on engine lathe and turret lathe are given in tables 4.15 and 4.16 respectively.

Table 4.15. Handling Times on the Engine Lathe

Operation	*Time, hr*
Use drill, reamer, etc., in tailstock-tail stock in fixed position-place and remove from tailstock spindle	0.006
Set tool to face-turn-bore, etc., - includes feed on and off	0.005
Clamp and release taper attachment on machine bed	0.008
Index square turret when used	0.003
Change speed-includes stop and start machine	0.002
Change feed	0.001
Pull out drill to relieve chip	0.002

Table 4.16. Handling Time on Turret Lathe

Operation	*Time, min*
Change speed	0.05
Change feed	0.05
Index Tool post	0.06
Engage feed	0.02
Feed to bar stop	0.06
Chuck in 3 Jaw chuck	0.75

Example 4.22. *Find the total time to produce one piece (Fig. 4.18) on a turret lathe. The material of the workpiece is mild steel and it is to be produced in lots of 100 pieces.*

Solution. To find out the total unit time, let us first of all write the various operations to be done on the job :

1. Feed to bar stop.
2. Turn 25 mm dia. to 15 mm dia. in two passes (max. depth of cut not to exceed 3 mm) a length of 50 mm.
3. Turn 15 mm dia. to 10 mm dia. in one pass, a length of 30 mm.
4. Form the end.
5. Cut screw thread.
6. Knurl.
7. Chamber head.
8. Part off.

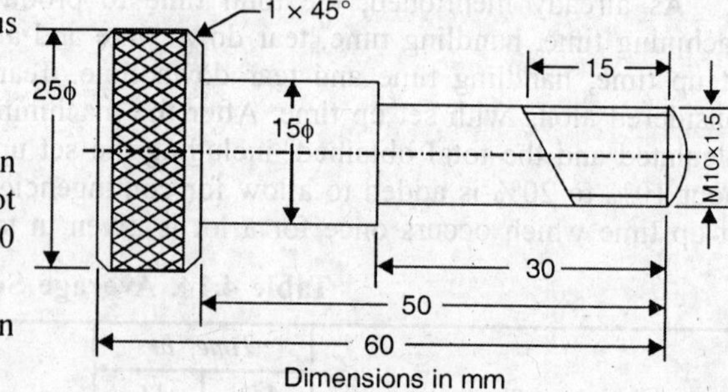

Fig. 4.18. Knurled Screw

The cutting speed for turning = 40 m/min.; screw cutting and knurling = 8 m/min. Feed : For turning = 0.4 mm/rev. For forming and part off = 0.2 mm/rev. Tool material : H.S.S.

Machine Time :

Operation 2. Spindle speed, $N = \dfrac{1000\,V}{\pi D}$

$$= \frac{1000 \times 40}{\pi D} \cong 500 \text{ rev./ min.}$$

$$\text{Time} = \frac{L}{fN} = \frac{50}{0.4 \times 500} = 0.25 \text{ min.}$$

Total time = 0.25 × 2 = 0.5 min., From Turret.

Operation 3. $N = \dfrac{1000 \times 40}{\pi \times 15} \cong 849$ rev./ min.

$\therefore$ $\text{Time} = \dfrac{30}{0.4 \times 849} = 0.09$ min., From Turret.

Operation 4. Let us take end forming time

= 0.15 min., From Turret.

Operation 5. $N = \dfrac{1000 \times V}{\pi \times 10} = 255$ rev./ min

feed = pitch = 1.5 min

$\therefore$ $\text{Time} = \dfrac{15}{1.5 \times 255} \cong 0.04$ min., From Turret

Operation 6. $N = \dfrac{1000 \times 8}{\pi \times 25} \cong 100$ rev./min.

$\text{Time} = \dfrac{10}{0.4 \times 100} = 0.25$ min., From Turret.

Operation 7. Let the time for chamfering be taken as 0.15 min., From front cross slide.

Operation 8. Dave = 25/2 = 12.5 mm

$\therefore$ $N = \dfrac{1000 \times 40}{\pi \times 12.5} \cong 1000$ r.p.m.

$\therefore$ $\text{Time} = \dfrac{12.5}{1000 \times 0.2} = 0.06$ min. (from rear cross slide)

Total machining time = 0.50 + 0.09 + 0.15 + 0.04 + 0.25 + 0.15 + 0.06

= 1.24 min.

Handling time = Feed bar to stop + Turret indexing (6)

+ Speed changes (4) + Feed changes (3)

= 0.06 + 6 × 0.06 + 4 × 0.05 + 3 × 0.05

= 0.06 + 0.36 + 0.20 + 0.15 = 0.77 min.

$\therefore$ Total operation time = 1.24 + 0.77 = 2.01 min.

Contingency time = 15% = 0.30 minute

$\therefore$ Total cycle time = 2.01 + 0.30 = 2.31 min.

Now set up time for turret lathe = 60 min.

$\therefore$ Set up time per piece = $\dfrac{60}{100} = 0.6$ min.

Total production time per piece = 2.31 + 0.60 = **2.91 min.**

PROBLEM

1. Define cost estimating.
2. What is the purpose of cost estimating?
3. Name the various constituents of cost.
4. Name the indirect material and indirect labour.
5. What are : direct expenses, indirect expenses, factory expenses, office and administrative expenses, selling and distribution expenses.
6. Define : Prime cost, Factory cost, Manufacturing cost. Total cost and Selling price.
7. How will you estimate : direct material cost and direct labour cost.
8. Define : Set up time, Handling time, Machining time. Tear down time and down or lost time.
9. What are the various ways of attributing indirect expenses towards the cost of a product?
10. Discuss the various methods of cost estimating.
11. What data is required for making a cost estimate?
12. List the various steps of cost estimating.
13. Discuss the chief factors in cost estimating.
14. A certain product is to be manufactured in batches of 100. The direct material cost if found to be Rs 320.00, direct labour cost Rs 560.00, and overhead chargeable to Rs 420.00. If the selling cost is 50% of the factory cost, what will be the selling price of each product to realize a profit of 15% of the selling price?
15. A company manufactures, 16,000 parts per month. The cost per piece is Rs 30 and the set up cost is Rs 65 per set up. Take carrying charge factor as 25%, determine

    1. the economic lot size

    2. the set up time per lot.
16. A company is to meet a production contract of 5000 pieces per month. The carrying charge factor is 25% and the set up cost is Rs 50. It takes 4.5 hours for each set up. Calculate :

    1. economic lot size

    2. the cost per piece.
17. A plant needs 500 components each month. Cost of set up is Rs 25 per set, cost per piece is Rs 40 and the carrying charge factor is 20%. Determine :

    1. lot size

    2. time per set up.
18. Define Performance factor. How it is related to 'Downtime'.
19. Define 'down-time'. Give examples.
20. Define 'economic lot size'. How it is related to set up time?
21. A company produces a component in 8 set ups at the rate of 200 pieces per set up. Calculate the cost of production from the following data :

Machining time	=	18 min. per piece
Non-Machining time	=	12 min. per piece
Set up time	=	80 min. per piece
Tool sharpening	=	2 min. per piece
Tool change	=	10 min.
Tool life	=	15 hours
Checking time	=	20 sec. and 6 checks per piece

Fatigue	=	15%
Personal needs	=	5%
Performance factor	=	1.3
Direct labour cost	=	Rs 5 per hour

22. A 20 mm hole is to be drilled in a cast iron block with a feed of 0.038 cm per rev. The thickness of the block is 6 cm and tool is of H.S.S.

Determine :

1. r.p.m.
2. the approach
3. machining time

24. For the data given in Problem 23 and applied to a Face cutter, Fig. 4.15, determine :

1. the r.p.m.
2. the approach
3. the time to take one cut.

25. For the data given in Problem 23, except that the work is 12.5 cm wide and the cutter is a face mill, determine :

1. approach
2. the r.p.m.
3. the time to take one cut

26. A medium-steel block is to be machined with a H.S.S. tool on a crank-shaper at a feed of 0.25 mm/stroke. The width of the block is 10 cm and the stroke length is 14 cm. Determine.

1. the number of strokes/min.
2. machining time.
3. the return stroke.

27. Define 'Tool approach' and Tool overtravel'. Explain their importance while determining the tool travel.

28. List the principal elements of metal machining. Give their definitions.

29. Discuss the various factors for the selection of : cutting speed, feed and depth of cut.

30. For a piece made of bar stock, how is the total amount of raw material calculated?

31. How is the amount of material needed for a casting estimated?

32. List the elements for typical operations on the following machines :

(*a*) milling machine (*b*) turret lathe (*c*) drilling machine.

33. Why is it common practice to apply a general performance factor to estimates? How is that factor obtained?

34. Define the following terms :

(*a*) Recurring costs.

(*b*) Non-recurring costs.

(*c*) Working capital.

(*d*) Turn over ratio.

35. Explain as to how the "Tool cost" is estimated?

36. Compare the various methods of tooling set ups.

5

ECONOMICS OF TOOLING

5.1. INTRODUCTION

The term "tool" as used in this chapter is a broad one. It includes machine tools such as engine lathe, turret lathe etc., equipment such as transfer mechanisms and small tools such as jigs, fixtures, dies, taps etc. The problems which are usually encountered in the economics of tooling are :

(*i*) Whether an old existing machine should replaced by a new machine or not.

(*ii*) For a given job which machine tool will be most economical.

(*iii*) Whether to go in for use of special tools such as jigs and fixtures for a given job or not.

(*iv*) Which is the most suitable process to produce a part.

(*v*) To determine "Break-even point" and "Economic batch size" and so on.

For all these purposes, the annual cost method is usually applied. The annual cost is the sum of fixed cost and variable cost.

Fixed costs. Fixed costs associated with a productive unit (machine, department, or plant) are those costs which are independent of the production rate. For a given machine tool, these fixed charges are made up of depreciation, interest, taxes, insurance and maintenance, that is the fixed overhead items.

Variable costs. Variable costs are those which increase or decrease in sympathy, though, not directly, with the sales or production volume. These costs include basic materials, direct labour and variable overheads.

In theory, if production were zero, fixed cost would not change, but variable costs would be zero.

Interest is included to represent the income which would have been derived from the investment if it would have been used otherwise or loaned in the market. It may also mean the case if the amount to purchase the machine tool has been borrowed and the interest is being paid to the loaner. To avoid accidents, the equipment must be insured and regularly inspected.

Depreciation. It is the most important item in the fixed costs. It represents the general decrease in the useful value of the machine with the passing of time, although the actual loss may or may not result from age. There are two types of depreciation :

(*a*) Physical depreciation (*b*) Functional depreciation.

Physical depreciation of mechanical equipment results from use and deterioration and is nearly always a function of time. Due to use, there is an impairment of strength, accuracy or efficiency of a machine tool. Even an idle machine will also lose some of its substance from the action of rust and rot.

Functional depreciation or obsolescence is largely independent of time. It is encountered in the case of comparatively new equipment which has ceased to be economical because other designs have come in the market which perform the same function better. It is also found in the case of special equipment for a product which is no longer in demand because of style changes or market saturation.

To account for the depreciation, a certain fixed amount of money is set aside every year so that the total amount accumulated at the end of the useful life of the machine, is the original cost of the machine less the expected salvage value. The determination of its magnitude in advance, as must be done in economy studies, is not easy because future unknown factors exist. For example, no one knows what the useful life of a machine or its salvage value will be until its end is reached. Consequently, resort is usually made to past experience as a basis for an enlightened guess. The useful life may vary from ten years or more for general purpose machine tools, to a year or less for specialized jigs, fixtures and dies for products subjected to frequent model or style changes. Hence, the magnitude of depreciation will only be an estimate and most likely it will not be entirely accurate.

Methods of distributing depreciation. Numerous methods for computing depreciation have been devised. Each is based on some hypothesis regarding loss of an asset's value versus time and is an attempt to solve the complex depreciation problem in a reasonably simple and satisfactory manner. The various methods are discussed below :

1. **Straight-line method.** This is the most widely favoured method because of its simplicity and adaptability. It assumes that the decrease in value during the life time is directly proportional to the time elapsed.

In this method, the same fixed amount is set aside every year throughout the useful life of the machine. The interest earned on the amount is not added to the depreciation reserve but is considered as the income of the firm or the company. That is,

$$\text{Depreciation per year, } d, \text{ is} = \frac{C_o - C_s}{n} \qquad \text{...(5.1)}$$

where C_o = original or initial value of the machine

C_s = salvage value of the machine

and n = useful life of machine, years

The drawbacks of this method are that very few articles depreciate actually in direct proportion to time and also that it does not take interest into account. However, these drawbacks are not very serious considering that even otherwise the estimate will not be very accurate as discussed above. Hence, in most cases, the simplicity of this method offsets its drawbacks. Also, if during the life of the machine, conditions develop to revise the original estimate, the allowance for depreciation can be adjusted most easily in this method.

$$\text{Total depreciation upto age } N \text{ year } (N \leq n),\ d_N = \frac{N(C_o - C_s)}{n}$$

$$\therefore BV_N = \text{Book value at the end of } N \text{ years} = C_o - \frac{N(C_o - C_s)}{n} \qquad \text{...(5.2)}$$

Book value. The book value of a machine at any time during its useful life, is its original cost less the amount that has been charged as depreciation expense. It goes to zero over the number of years of useful life. It thus represents the amount of capital that remains invested in the property and must be recovered in the future through the depreciation accounting process.

2. **Sinking fund method.** This method is based on the theory that interest earnings should be credited to the depreciation, reducing its charge to a certain extent.

In this method, the interest earned on the yearly amount is compounded periodically and it forms a part of the depreciation reserve. The total amount accumulated by this method is the sinking fund plus the interest earned and the sum should be equal to the original cost of the machine minus its salvage value.

Let d = annual deposit in the reserve fund, Rs

S = sum to be provided at the end of useful life, Rs

$= C_o - C_s$

n = expected useful life, years

I = annual interest rate on the investment d

Then, accumulation of first year's investment will be $= d(1+I)^{n-1}$

accumulation of second year's investment will be $= d(1+I)^{n-2}$

accumulation of $(n-2)$ year's investment will be $= d(1+I)^2$

accumulation of $(n-1)$ year's investment will be $= d(1+I)$

and accumulation of nth year's investment will be $= d$

$\therefore$ Total accumulation,

$$S = d\left[1+(1+I)...+(1+I)^2+...+(1+I)^{n-2}+(1+I)^{n-1}\right]$$

This is geometrical progression with a common ratio of the

$$r = (1+I)$$

$$\therefore \quad S = d\left[\frac{(1+I)^n - 1}{1+I-1}\right]$$

$$= d\left[\frac{(1+I)^n - 1}{I}\right]$$

$$\therefore \quad d = S\left[\frac{I}{(1+I)^n - 1}\right] = \frac{I(C_o - C_s)}{(1+I)^n - 1} \qquad ...(5.3)$$

The relation (d/S), *i.e.,* the ratio of annual depreciation to the sum accumulated is called as the depreciation rate.

The difference between the above two methods is made clear by the following examples.

Example 5.1. *The original value of a machine tool is Rs 250,000 and its salvage value at the end of its useful life of 20 years is Rs 25,000. Find the value of the machine tool at the end of 10 years of its use by : (i) Straight line depreciation method (ii) Sinking fund method when the fund is compounded annually at 8%.*

Solution C_o = Rs 250,000, C_s = Rs 25,000

(*i*) Straight line depreciation method :

$$\text{Depreciation per year} = \frac{C_o - C_s}{n} = \frac{250{,}000 - 25000}{20}$$

$$= \text{Rs } 11250$$

$\therefore$ Value of machine tool at the end of 10 years

$$= \text{Rs } (250{,}000 - 10 \times 11250)$$

$$= \text{Rs } 137500$$

(*ii*) **Sinking fund method :**

$$S = \text{Rs } (250{,}000 - 25000) = \text{Rs } 225{,}000;\ I = 8\%$$

∴ Annual deposit,

$$d = S\left[\frac{I}{(1+I)^n - 1}\right]$$

$$= \text{Rs } 225000\left[\frac{0.08}{(1.08)^{20} - 1}\right]$$

$$= \text{Rs } 4920$$

Now amount collected at the end of 10 years will be

$$= \text{Rs } d\left[\frac{(1+I)^{10} - 1}{I}\right]$$

$$= \text{Rs } 4920\left[\frac{(1.08)^{10} - 1}{0.08}\right] = \text{Rs} 71250$$

∴ Value of equipment at the end of 10 years

$= \text{Rs } (250{,}000 - 71{,}250) =$ **Rs 178,750.**

The sinking fund method is not generally used for accounting purposes because of its low depreciation charges in the early years of asset life.

3. **Declining balance method.** This method is also known as 'Reducing or Diminishing Fund Method', 'Constant Percentage Method' or "Matherson Formula'. In this method, heavier annual amounts are charged off for depreciation during the early years of life. This is in accordance with the actual depreciation trend of machines which depreciate rapidly in the early years and later on slowly. Therefore, it is better to depreciate more during the early year, when the repairs and renewals are not costly. Under this method, a constant fixed percentage is deducted each year from the balance (Book Value) outstanding at the beginning of every year. It should be remembered, however, that any such declining rate is based upon past experience, of which there is no guarantee of conformity in future.

Let F = fixed percentage for calculating the yearly depreciation

∴ At the end of one year depreciation fund $= C_o \times F$

" " " Two year " " $= FC_o + F(C_o - FC_o)$

$= C_o(2F - F^2)$

$= C_o[1 - (1 - F)^2]$

∴ Total fund at the end of n years $= C_o[1 - (1 - F)^n]$

∴ $C_o[1 - (1 - F)^n] = C_o - C_s$

From here,

$$F = 1 - \left(\frac{C_s}{C_0}\right)^{\frac{1}{n}}$$

Now the outstanding amount at the end of each year is nothing but the 'Book Value' of the machine. It means that a certain fixed percentage of the current book value is taken as the depreciation. Therefore, this method is also called "percentage on Book Value" method.

Now, $$F = 1-\left(\frac{C_s}{C_0}\right)^{\frac{1}{n}} = 1-\left(\frac{BV_N}{C_o}\right)^{\frac{1}{N}}$$

$$BV_N = C_o\,(1-F)^N = C_o\left(\frac{C_s}{C_o}\right)^{N/n} \quad ...(5.4)$$

This method is simple to apply. However, it has two drawbacks. One, the annual depreciation is different each year and from a calculation view point, this is inconvenient. Also, with this formula, an asset can never depreciate to zero value. This is not a serious difficulty and in actual practice, computation of the theoretical depreciation rate F is seldom made. Instead, a reasonable rate is assumed or taken from the *IRS* (Internal Revenue Service) guide lines, for example,

For all new depreciable property except real estate, $F = \frac{2}{n}$

For all used depreciable property and new real estate, $F = \frac{1.5}{n}$

For used rental residential property $F = \frac{1.25}{n}$

when $F = \frac{2}{n}$, it is known as "Double Declining Balance" depreciation.

4. **The annuity charging method.** This method differs from the above method in that the interest earned by the depreciation fund each year is also considered.

∴ Depreciation fund at the end of one year $= C_oF$

Depreciation fund at the end of two years

$$= C_oF\,(1+I) + F\,(C_o - C_oF)$$

$$= \frac{C_oF}{I+F} \times \left[(1+I)^2 - (1-F)^2\right]$$

∴ Depreciation fund at the end of n years

$$= \frac{C_oF}{I+F}\left[(1+I)^n - (1-F)^n\right] \quad ...(5.5)$$

$$= C_o - C_s$$

5. **The sum of the 'Years'-digits method : (SYD).** In this method, the digits corresponding to the number of each year of life are listed in reverse order. The sum of these digits is then determined, (if life is n years, sum of digits will be $\frac{1}{2}n(n+1)$). The depreciation factor for any year is the reverse digit for that year divided by the sum of the digits. For example, if $n = 5$, then,

Year	*No. of years in reverse order (digits)*	*SYD depreciation factor*
1	5	5/15
2	4	4/15
3	3	3/15
4	2	2/15
5	1	1/15
	15	

The depreciation for any year = *SYD* depreciation factor for that year × depreciable value, $C_o - C_s$.

The general expression for the annual cost of depreciation for any year N, when the total life is n year, is

$$d_N = (C_o - C_s) \times \frac{2(n - N + 1)}{n(n+1)} \qquad ...(5.6)$$

It is clear that *SYD* method, like the declining balance method, provides for very rapid depreciation during the early years of life. Further *SYD* method enables properties to be depreciated to zero value and is easier to use than the declining balance method. This method also tends to reduce chances that the book value of an asset will exceed actual or resale value at any time. However, use of *SYD*, or any other accelerated depreciation method, in effect, reduces the computed profits of a corporation during the early years of asset life and thus reduces income taxes in those early years.

6. **The insurance policy method.** In this method, the machine is insured and the premium paid on the policy is considered as the depreciation. Its added advantage is that risk of any damage is also covered and is given by the insurance company.

The various depreciation methods are compared in Fig. 5.1. below :

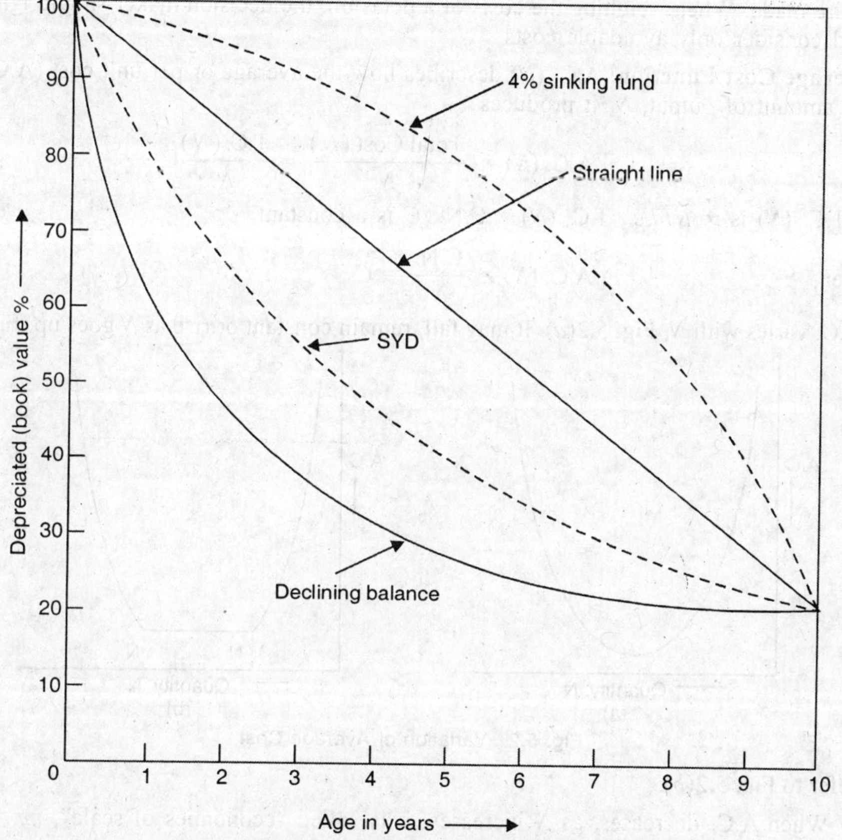

Fig. 5.1. Depreciation Methods.

SOME MORE DEFINITIONS OF COSTS

As noted above,

Variable Cost (V.C.): Direct labour cost, basic materials and commission to sales people.

Fixed Cost (F.C.): It includes salaries and wages of permanent employees (general and administrative expenses), rent, land, building, machinery (property taxes) etc. This cost pertains to the investment required to set up the facilities to produce a product (capital cost). It does not depend upon the quantity or volume of components produced. It tends to be fixed (non-variable) for a given period of time and for a given level of installed capacity. It is related more to the period rather than the quantity.

Line dividing V.C. and F.C. is often fuzzy. Some costs such as maintenance and advertising or promotional expenses, may have both fixed and variable components.

Semi-fixed Costs: These costs are fixed over certain ranges of output, but variable over other ranges. For example, a beer distributor may be able to deliver 5000 barrels of beer per week using a single truck. But, when it must deliver between 5000 and 10,000 barrels, it needs 2 trucks, between 10,000 and 15,000 barrels, three trucks and so on. The cost of trucks is fixed within the intervals (0, 5000); (5000, 10,000); (10,000, 15,000), and so forth, but is variable between these intervals.

Sunk Costs: When assessing the costs of a decision, the manager should consider only those costs that the decision actually affects. Some costs must be incurred, no matter what the decision is. These are the costs that have already been incurred and cannot be recovered. These are called "Sunk Costs".

Avoidable Costs: These are the opposite of sunk costs. These can be avoided if certain choices are made. When weighing the costs of a decision, the decision maker should ignore sunk costs and consider only avoidable costs.

Average Cost Function: A.C. (N) describes how the average or per unit cost (A.C.) varies with the amount of output, N, it produces

$$\text{A.C. }(N) = \frac{\text{Total Cost }(N)}{N} = \frac{\text{T.C. }(N)}{N}$$

If T.C. (N) is $\propto N$, *i.e.*, T.C. (N) = C.N., C is a constant.

Then,
$$\text{A.C. }(N) = \frac{\text{C.N.}}{N} = C$$

Often A.C. varies with N, Fig. 5.2(a). It may fall, remain constant or rise as N goes up, Fig. 5.2(b)

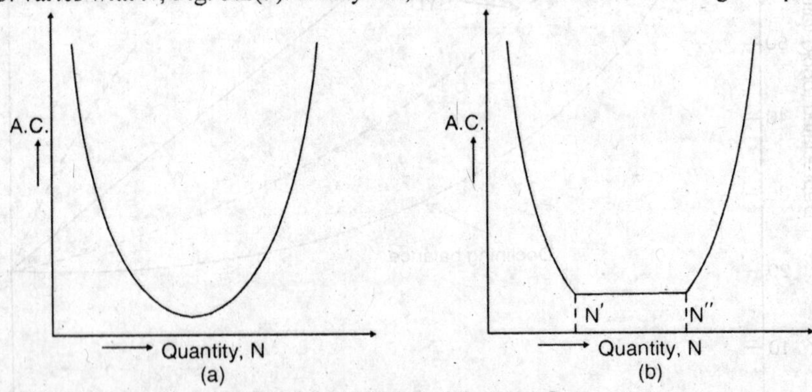

Fig. 5.2 Variation of Average Cost

Refer to Fig. 5.2(b):

(i) When A.C. decreases, as N increases, it is called "economies of scale", *i.e.*, at outpu levels upto N'.

(ii) When A.C. increases, as N increases, there are "diseconomies of scale", *i.e.*, at outpu levels of N'' and above.

(iii) When A.C. remains constant with N, we have constant returns to scale, *i.e.*, between N and N''.

The "economies of scale" and "diseconomies of scale" are defined ahead. N′ is the smallest output level at which economies of scale are exhausted. It is thus known as the minimum efficient scale.

Marginal Cost (M.C.): This cost refers to the rate of change of T.C. w.r.t. N. M.C. may be thought of as the incremental cost (I.C.) of producing exactly one more unit of output. This is also called as I.C., out-of-pocket cost or differential cost.

Let N = initial output

ΔN = change in output in units

Then, $$\text{M.C. } (N) = \frac{\text{T.C. } (N + \Delta N) - \text{T.C. } (N)}{\Delta N}, \qquad \text{Fig. 5.3}$$

Business often use information about A.C. to estimate the M.C. of a change in output. A.C. and M.C. are generally different. The exception is when T.C. varies in direct proportion to N, that is,

$$\text{T.C. } (N) = C \times N$$

So, $$\text{M.C. } (N) = \frac{C\,(N + \Delta N) - CN}{\Delta N} = C$$

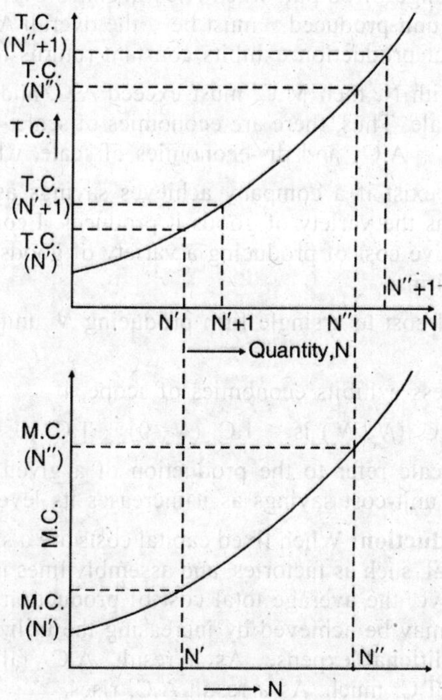

Fig. 5.3 Marginal Cost

which is also A.C.

(*i*) When A.C. $\propto \frac{1}{N}$, *i.e.*, decreasing function of N,

$$\text{M.C.} < \text{A.C.}$$

(*ii*) When A.C. is independent of N, or is at a minimum print,

M.C. = A.C.

(*iii*) When A.C. is an increasing function of N, then

M.C. > A.C., see Fig. 5.4

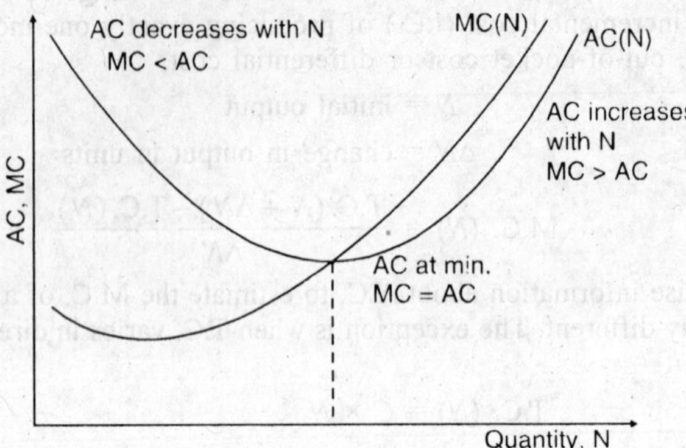

Fig. 5.4 Marginal Cost and Average Cost

Economies of Scale: The production process for a particular product or service exhibits economies of scale over a range of output, when A.C. declines over that range, Fig. 5.2(*b*). For this, M.C. – the cost of last unit produced – must be < the overall A.C. If A.C. is constant, then M.C. = A.C. and we say that production exhibits constant returns to scale.

If A.C. is increasing with N, then M.C. must exceed A.C. and we say that the production exhibits dis-economies of scale. Thus, there are economies of scale, when M.C. < A.C., constant returns to scale, when M.C. = A.C., and dis-economies of scale, when M.C. > A.C.

Economies of Scope: exist if a company achieves savings as it increases the variety of activities it performs, such as the variety of goods it produces. Economies of scope are usually defined in terms of the relative cost of producing a variety of goods together in one firm versus separately in two or more firms.

If T.C. (N_x, N_y) = Total cost to a single firm producing N_x units of good X and N_y units of good Y.

Then, production process exhibits economies of scope, if

$$\text{T.C. } (N_x, N_y) \text{ is } < \text{T.C. } (N_x, 0) + \text{T.C. } (0, N_y)$$

Note: Economies of scale refer to the production of a given good. Economies of scope exhibit if the firm achieves unit-cost savings as it increases its level of a given activity.

Capital Intensive Production: When fixed capital costs are a significant percentage of total cost. Much production capital, such as factories, and assembly lines is indivisible. Thus, when the production is capital intensive, the average total cost of production contains a substantial fixed amount. Increase in output may be achieved by increasing the utilization of existing production facilities often at little additional expense. As a result, A.C. falls.Conversely, cut backs in production may not reduce T.C. much. As a result A.C. rises.

Materials and Labour Intensive Production: Here, most production expenses go to raw materials and labour. Average T.C. of production depends mainly on the amount of materials and labour that goes into each unit of output. Because materials and labour are divisible, they can change in proportion to changes in output, with the result, A.C. does not vary with output.

5.2. MACHINE TOOL REPLACEMENT

Whether the existing machines and tools etc. should be replaced with new and more modern equipment is a problem frequently faced by the management of a company. The wearing out of the equipment as well as its being made obsolete by new developments and improved devices makes this an ever present problem. Some major reasons for machine replacement or purchase of new machine tool are :

1. To increase productivity, *i.e.*, "the production of an ever increasing amount of parts per unit of time, per unit of floor space, per unit of material, light, heat and power".
2. To improve product quality.
3. To accommodate changes in product size.
4. To increase flexibility of use of machine tools by addition of new features.
5. To extend the range of use of new product kind or style.
6. To eliminate safety hazards.
7. To reduce an indirect cost associated with the machine or the product it produces.
8. To obtain maximum rate of return.
9. To reduce physical strain or workers.
10. To cut cost of production.
11. To increase the capacity and revenue.
12. To obtain greater convenience and versatility.
13. To achieve a reduction of inventory and tool cost.

5.2.1. Return on Investment. The knowledge of percentage return on investment is essential to determine the most effective use of capital. Rate of return on investment refers to the annual interest on the unpaid balance of invested capital. On this basis, the annual capital charge against the replacement consists of two parts : (*a*) the annual amortization or reduction of the capital invested in the equipment, and (*b*) the annual interest on the unamortized portion of the investment.

The following elements should be considered for the calculations.

Annual gross savings. These include the savings in direct labour, overhead costs, maintenance and repair costs expected from the proposed replacement as compared with the corresponding costs for the existing equipment.

Direct labour savings. Direct labour savings refer to those savings in operator costs which can be utilized.

Savings in overhead costs. These include savings in only those incremental overhead cost factors affected by the replacement. Such cost factors include – inspection, serivce, power costs and similar overhead items. Depreciation rates should not be considered in analysing the operation cost factors.

Maintenance and repair costs. These costs include day-to-day equipment maintenance and normal repair costs exclusive of major overhaul.

5.2.2. Economy studies.

Economy studies are done to select between alternatives and to decide equipment replacement policy. Before we discuss the common methods used for economy studies, we discuss below the various terms used in the studies :

1. **Asset.** An asset of a company is a valuable like land, building, machinery and equipment, materials etc. owned by the company.

2. **Salvage value.** It is the actual or estimated value of an existing asset at which it can be disposed off now or on a future date.

3. **Receipt.** Receipt is any sum received by a company from sales of its products and services. It is also known as return, income, profit or revenue.

4. **Payment.** It is any sum paid by the company for buying materials, paying wages,

operating charges, depreciation etc. It is also known as cost, disbursement, outlay or expenditure.

5. **Equivalence.** Alternatives should be compared on "equivalent" basis, that is, when they produce similar results, serve the same purpose or accomplish the same function.

6. **Discounting.** To find the present value or worth of a future receipt is called discounting.

7. **Time value of money.** The value of money at any time will depend upon the criteria of interest we use in calculations :

(*a*) **Simple interest :**

P = Present worth (P.W.) of money or the principal amount borrowed or lent.

F = future worth (F.W.) of money

I = Nominal annual rate of interest, %

n = period in years.

Then we know that,

$$F = P(1 + nI) \quad \text{...(5.7)}$$

It means that the present sum, P, will become, F, in n years at simple interest rate of I%, which will be paid back in lump sum.

So, $$P = F(1 + nI)^{-1} \quad \text{...(5.8)}$$

Here P is the P.W. of F available in n years and interested at I%. Here we have discounted the future sum, F, back to the present time. 'Discounted' means bringing future value back in time to the present.

The Present Worth (P.W.) or Present Value (P.V.) of cash flow 'C' (see point 16 ahead) in n years is equal to the amount of money that must be invested to-day at the interest rate, I, so that after n years, the principal plus interest equals C.

Mathematically, $$\text{P.V.} = \frac{C}{(1+I)^n} \quad \text{(see eqn. 5.10)}$$

Here, C = Cash flow or future worth (F.W.) after n years.

The P.V. of a stream of cash flows received over a period of years is the sum of P.V.'s of the individual sums. Thus, the P.V. of Cash flows, C_1, C_2, ..., C_T, received one year from now, 2 years from now, ..., T years from now, is

$$\text{P.V.} = \frac{C_1}{1+I} + \frac{C_2}{(1+I)^2} + \ldots\ldots + \frac{C_T}{(1+I)^T}$$

$$= \sum_{n=1}^{T} \frac{C_n}{(1+I)^n}$$

(*b*) **Compound interest :**

$$F = P(1 + I)^n, \text{ (compounded annually)} \quad \text{...(5.9)}$$

or $$P = F(1 + I)^{-n} \quad \text{...(5.10)}$$

The factor $(1 + I)^n$ is called the "single payment compound amount factor" and the factor $(1 + I)^{-n}$ is called the "Single payment present worth factor".

Equations (5.9) and (5.10) can also be written as,

$$F = P(F/P,\ I\%,\ n) \quad \text{...(5.9a)}$$

and $$P = F(P/F,\ I\%,\ n) \quad \text{...(5.10a)}$$

Equation 5.9 (*a*) reads : Find F, given P, at I% interest per period for n interest periods and similarly equ. 5.10 (*a*).

If the interest is compounded more than once in a year, and

p = number of interest periods per year, that is,

= number of times the interest is compounded in a year.

$\therefore$ Interest rate per interest period, $i = \frac{I}{p}$ and number of interest periods in n years = np.

Then equ. (5.9) becomes :

$$F = P\left[\left(1 + \frac{I}{p}\right)^p\right]^n \qquad ...(5.11)$$

(*c*) **Effective rate of interest :**

From equ. (5.11), F_1 = future worth after 1 year

$$= P\left(1 + \frac{I}{p}\right)^p$$

$\therefore$ Amount of interest in one year $= F_1 - P$

$$= P\left(1 + \frac{I}{p}\right)^p - P$$

$\therefore$ Effective rate of interest, $I_e = \frac{F_1 - P}{P}$

$$= \left(1 + \frac{I}{p}\right)^p - 1 \qquad ...(5.12)$$

(*d*) **Continuous compounding :** In continuous compounding, p in equ. (5.11) tends to infinity. Let $\frac{p}{I} = k$, therefore, after n years,

$$F = P\left(1 + \frac{1}{k}\right)^{Ikn} = \left[\left(1 + \frac{1}{k}\right)^k\right]^{In} \times P$$

when $k \to \infty$ $\quad \left(1 + \frac{1}{k}\right)^k \to e$

$$\therefore \quad F = P \cdot e^{nI} \qquad ...(5.13)$$

8. **Rate of return.** The simplest measure of profitability of an enterprise is the rate of return on the investment (R.O.I.). While calculating this, the time value of money is not considered. It is a simple ratio of profit or cash income and the capital investment. There are many ways of writing this ratio. The net profit over a period (usually a year) can be before taxes or after taxes. Similarly, the annual cash income can be before taxes or after taxes. Thus,

$$\text{R.O.I.} = \frac{\text{Net profit/ Cash income}}{\text{Sum invested}}$$

9. **Minimum attractive rate of return.** Minimum attractive rate of return (M.A.R.R.) can be viewed as a rate at which an enterprise can always invest, since it has large number of opportunities that yield such a return. There is no completely satisfactory method for precisely determining this factor. Normally, MARR is substantially higher than the cost of capital. When an enterprise's cost of capital may be 12 percent, its MARR may be 20 per cent. This difference occurs because few firms are willing to invest in projects that are expected to earn slightly more than the cost of capital due to risk elements in most projects and because of uncertainty about the future.

10. **Life of an asset.** There are three types of life of an asset :

(*i*) **Technological life.** It is the life during which the asset or the equipment is able to fulfill its functions, purely from the technological considerations.

(*ii*) **Accounting life.** It is the period or the duration during which the investment made in acquiring the asset is to be recovered in the form of depreciation.

(*iii*) **Economic life.** Economic life of an asset can be defined as its service life during which it is financially viable in terms of its operating and maintenance costs and cash inflows.

11. **Retirement.** It is the disposal of an existing asset through sales or abandonment as scrap.

12. **Replacement.** It means acquiring a new asset to replace the existing asset which has been retired.

13. **Defender.** The existing old asset being considered to be replaced.

14. **Challenger.** The asset proposed to be the replacement.

15. **Uniform annual series.** The interest formulas discussed above are based on 'Single payment or receipt', that is, 'Lump sum' mode. However, in many situations, we come across uniform series of receipts (or income) and payments or disbursements (or costs) occurring equally at the end of each period. Such a uniform series is often called an 'Annuity'. The examples are :

(*a*) Paying back a loan on the installment basis.

(*b*) Acquiring an asset on uniform installment plan instead of paying for it in lump sum.

(*c*) Setting aside a sum at the end of each period, that will be available as an accumulated sum at a future date for replacement of an existing equipment or asset.

(*d*) Retirement annuity that consists of series of equal payments instead of a lump sum.

Let A = Uniform annual amount of payment or receipt. Let us suppose that this annual sum is invested at the end of each year for 4 years. Then, the total sum F available at the end of 4 years will be the sum of the compound amount of the individual investments A, that is,

$$F = A(1+I)^3 + A(1+I)^2 + A(1+I) + A$$

It is clear that F (future worth) occurs at the same time as the last A, and n periods (here 4 years) after the present time. If,

P = present worth of F

it occurs one period before the first A.

For general case of n years,

$$F = A(1+I)^{n-1} + A(1+I)^{n-2} + ... + A(1+I)^2 + A(1+I) + A$$
$$= A\,[(1+I)^{n-1} + (1+I)^{n-2} + ... + (1+I)^2 + (1+I)^1 + (1+I)^0]$$

The terms in the bracket form a G.P. whose common ratio is $(1+I)^{-1}$.

Hence, $$F = A\left[\frac{(1+I)^{n-1} - (1+I)^{-1}}{1-(1+I)^{-1}}\right]$$

which reduces to, $$F = A\left[\frac{(1+I)^n - 1}{I}\right] \qquad ...(5.14)$$

The factor $\left[\{(1+I)^n - 1\}/I\right]$ is called the 'uniform series compound amount factor'.

Equation (5.14) can be written symbolically as,

$$F = A\,(F/A,\ I\%,\ n) \qquad ...(5.14a)$$

From (5.14), $$A = F\left[\frac{I}{(1+I)^n - 1}\right] \qquad ...(5.15)$$

or $$A = F\,(A/F,\ I\%,\ n) \qquad ...(5.15a)$$

The factor within the bracket is called 'Sinking fund factor'.

Equation (5.9) is written again,

$$F = P(1 + I)^n \qquad ...(5.9)$$

Combining equation (5.9) and equation (5.14), we get

$$P = A\left[\frac{(1+I)^n - 1}{I(1+I)^n}\right] \qquad ...(5.16)$$

The term within the bracket is called 'Uniform series present worth factor'.

or $$P = A\,(P/A,\ I\%,\ n) \qquad ...(5.16a)$$

From (5.16), $$A = P\left[\frac{I(1+I)^n}{(1+I)^n - 1}\right] \qquad ...(5.17)$$

or $$A = P\,(A/P,\ I\%,\ n) \qquad ...(5.17a)$$

The term within the bracket is called 'Capital recovery factor'.

Note. The values of the above mentioned factors can be caluclated from the given data. Their values have been calculated and tabulated for different rates of interest and for different periods, and are available in any book on 'Engineering Economy'.

Example 5.2. *A Rs. 200,000 machine is to be replaced in 20 years. Find the annual investment that must be made at 10% to provide the sum to replace the machine, using (a) Sinking fund factor (b) Capital recovery factor.*

Solution. (*a*) Using Sinking fund factor, so from eqn. (5.15),

$$A = 200{,}000\left[\frac{0.10}{(1.10)^{20} - 1}\right]$$

$$= 200{,}000 \times 0.01746 = \textbf{Rs 3492/- per year.}$$

(*b*) Using Capital recovery factor, that is, eqn. (5.17),

$$A = P\left[\frac{I(1+I)^n}{(1+I)^n - 1}\right] = 200{,}000\left[\frac{0.10(1.10)^{20}}{(1.10)^{20} - 1}\right]$$

$$= 200{,}000 \times 0.11746 = \textbf{Rs 23492/– per year.}$$

We see from above that, Capital recovery factor (*CRF*) is 0.11746 and Sinking fund factor (SFF) is 0.01746 and interest is 0.10.

So, $$CRF = SFF + I$$

i.e., Annual cost (Capital recovery) = Annual cost (Sinking fund) + Annual interest cost

$$23492 = 3492 + 0.10\,(200{,}000).$$

So, in Sinking fund method, the sum set aside every year, over *n* years, together with accumulated compound interest equals the required future sum *F*. But in Capital recovery method, the annual payment *A* is used to return the initial investment *P* plust interest on that investment at a rate *I* over *n* years.

16. **Cash flow diagram.** We have seen that Engineering Economy deals with sums of money at different times in future. Cash flow measures the flow of funds into or out of an enterprise. It is a series of actual or estimated payments (or costs) and receipts (or income) of an enterprise over a period of time. The procedure of showing this graphically is called "Cash flow diagram", Fig. 5.5. Receipts (income) are shown above the time line and the payments (Disbursements, costs) below this line. The length of the arrow is proportional to the Rupee amount.

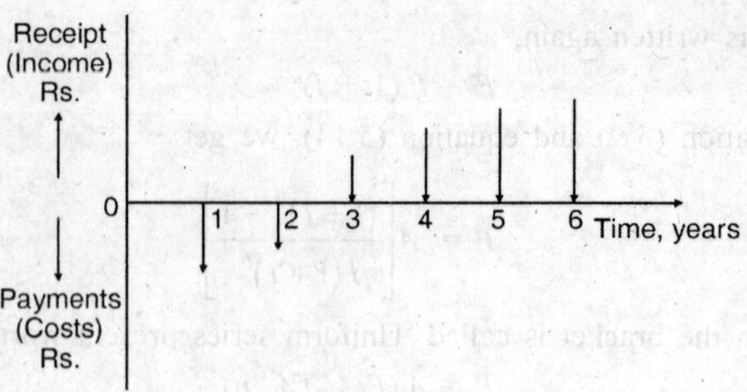

Fig. 5.5. Cash Flow Diagram.

17. **Beginning-of-period payment with uniform series.** In the uniform annual series above, the amount A occurs at the end of each period. However, if a beginning-of-period payment, A_b is required, it can be done as given below :

Discount A one year to the present, that is, in equation (5.7),

$$F = A,\ P = A_b \text{ and } n = 1$$

$$\therefore \quad A = A_b\ (1 + i) \qquad ...(5.18)$$

This can be substituted in equations (5.14 to 5.17)

18. **Deferred annuity.** In the analysis of uniform annual series discussed above, the first cash flow takes place at the end of first period. Such cash flows (payment or receipt) are called 'Ordinary annuities'. If the first cash flow does not take place at the end of first period but after some periods, the annuity is known as 'deferred annuity', Fig. 5.6 (*b*).

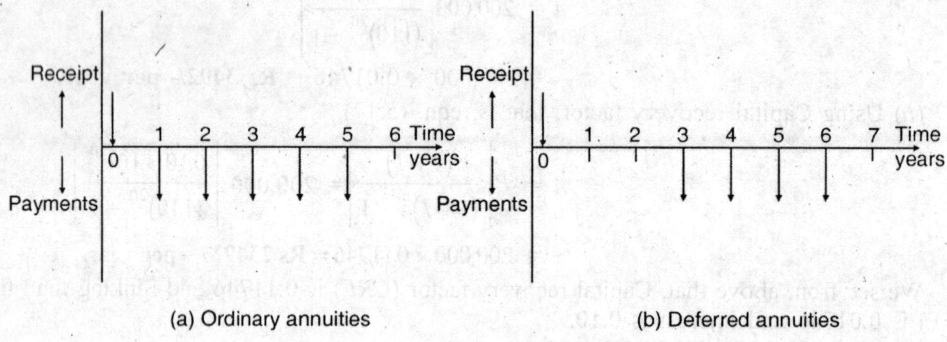

Fig. 5.6.

19. **Net present worth.** Net present worth (*NPW*) is defined as :

$$NPW = \text{PW of incomes} - PW \text{ of costs}$$

when comparing two assets, the asset with the largest positive value of *NPW* is preferred.

N.P.V. or N.P.W. of an investment = P.V. of cash flows the investment generates − cost of investment

Given the assumptions that the investment has an infinite life and that the price and revenues are expected to remain the same over the forceable future,

$$\text{N.P.V.} = \sum_{n=1}^{\infty} \frac{\text{Total costs/year}}{(1 + I)^n} - I \times \text{investment}$$

This looks intimidating, but fortunately the term in the summation is the P.V. of a "perpetuity". A perpetuity is the level cash flow *C* received each year for ever. The P.V. of a perpetuity has a convenient formula. It is equal to the cash flow divided by the interest rate, *I*, *i.e.*, *C*/*I*, so

$$\text{N.P.V.} = \frac{\text{Total costs/year}}{I} - \text{I} \times \text{investment}$$

If N.P.V. is +ve, the firm should undertake the investment.

With a constant annual cash flow and an infinitely lived investment,

Economic profit = N.P.V. × cost of capital

20. **Discounted cash flow.** Discounted cash flow, *DCF*, is the rate of return for which *NPW* is zero.

So, *PW* of income = *PW* of costs

'Discounted' means the time value of money is considered.

21. **Accounting profit.** It is defined as given below:

Accounting profit = Revenues – Expenses

22. **Economic profit.** It is the profit earned by investing resources in a particular activity. These are the profits that could have been earned by investing the same resources in the most lucrative alternative activity.

Basic methods of economy studies (Selection of alternatives). The following are the basic methods of economy studies :

1. Payback period method.
2. Present worth method.
3. Annual worth or Annual cost method.
4. Internal rate of return method.
5. Explicit reinvestment rate of return method.
6. Break-even point method.

The above methods can be used for : Replacement policy and for selection between alternatives. These methods are for "before-tax studies". However, because the capital is subject to income tax effects, the studies are not materially affected even for "after-tax" mode.

1. **Payback period method.** Payback period is the time during which the cash flow will fully recover the invested capital. The drawbacks of the method are : it does not consider the time value of money. Also it does not take into account the economic life of the asset. The emphasis is only on the rapid recovery of the investment. Again, no account is taken of cash flows which take place after the payback period. So, this method is normally not recommended for economy studies.

$$\text{Payback period} = \frac{\text{Initial capital investment}}{\text{Annual return expected (Net annual income)}}$$

2. **Present worth (*PW*) method.** In this method, the present worth of all cash flows for various alternatives is calculated. The alternative with lowest *PW* of costs is selected, or as has been mentioned above under point 19, the alternative with the greatest positive of NPW is selected.

Net *P.W.* = *P.W.* of cash inflows (receipts) – *P.W.* of cash out flows (payments).

Note. For a project or an equipment to be economically justified, $N.P.W > 0$.

Note. If the above analysis is based on the present worth of all costs (here costs will be taken as positive and income or receipts will be taken as negative) or if receipts are not known so that only the cash outflows (payments, costs) are relavent, the method is called the present worth-cost (P.W.C) method. The alternative with the smallest net present worth of costs will be selected.

Example 5.3. *An enterprise can make an investment of Rs 40,000 in a project. The uniform annual revenue of the project will be Rs 21,240 for 5 years. After that, its salvage value will be Rs 8000. Annual payments per year for operation and maintenance will be Rs. 12000. Will the project be economical by using P.W. method if the company is prepared to accept the project that will earn 10% or more before taxes ?*

Solution. (*a*) **Cash in flows :**

(*i*) *Annual revenue.* Using equ. 5.16 (*a*),

$$P = A\,(P/A, I\%, n)$$

Here, A = Rs 21,240, I = 10%, n = 5 years.

$$\therefore \quad P = 21240\left[\frac{(1.10)^5 - 1}{0.1 \times (1.10)^5}\right] \qquad \text{...(5.16)}$$

$$\therefore \quad P = \frac{21240 \times 0.61}{0.1 \times 1.61} = \text{Rs. } 80474.53$$

(*ii*) *Salvage value.* Using eqn. (5.10),

$$F = \text{Rs. } 8000$$

$$\therefore \quad P = \frac{8000}{(1.1)^5} = \text{Rs } 4967.40$$

$\therefore$ Total cash in flows = Rs. (80474.53 + 4967.40)

= Rs 85441.93

(*b*) **Cash out flows :**

(*i*) Investment = Rs 40,000 (It is the *P.W.*)

(*ii*) Annual payments. Using eqn. (5.16)

$$P = \frac{12000 \times 0.61}{0.1 \times 1.61} = \text{Rs } 45{,}465.84.$$

Total cash out flows = Rs 85,465.84

Since total cash out flows are more than total cash in flows on the P.W., therefore, the project is not economical.

Example 5.4. *The data for two machines A and B is given below :*

Item	*Machine A*	*Machine B*
Initial cost	*Rs 100,000*	*Rs 60,000*
Operating and maintenance cost per year	*Rs 8000*	*Rs 16000*
Reconditioning at the end of third year	—	*Rs 14000*
Salvage value	*Rs 12000*	
Annual benefit from better quality production	*Rs 2000*	

The interest rate is 10% and useful life of both the machines is 5 years. Which machine is more economical. Use P.W. method.

Solution. The Cash flow diagrams for the two machines are shown in Fig. 5.7.

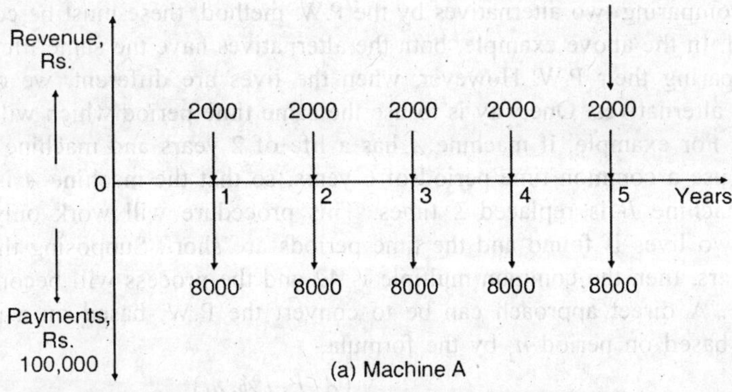

(a) Machine A

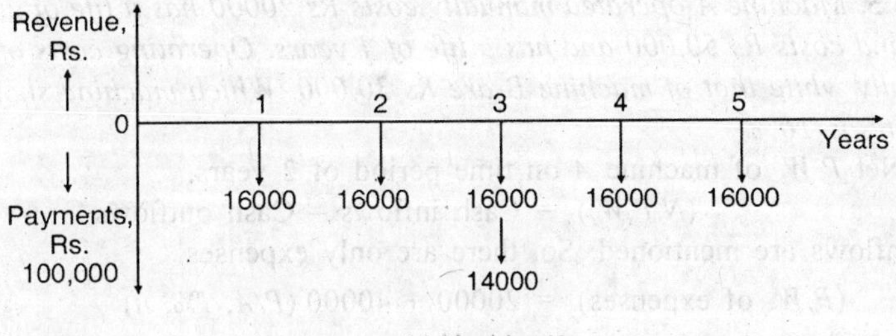

(b) Machine B

Fig. 5.7. Cash Flow Diagrams.

(*a*) **Machine *A* :** Net P.W. = Cash in flows – Cash out flows

$$\therefore \quad P_A = 2000\,(P/A, I\%, n) + 12000\,(1 + I)^{-n} - 8000\,(P/A, I\%, n) - 100{,}000$$

Eqn. (5.16*a*) Eqn. (5.10) Eqn. (5.16*a*)

$$= 2000\left[\frac{(1.1)^5 - 1}{0.1 \times (1.1)^5}\right] + 12000(1.1)^{-5} - 8000\left[\frac{(1.1)^5 - 1}{0.1 \times (1.1)^5}\right] - 100{,}000$$

$$= 7577.64 + 7453.42 - 30310.56 - 100{,}000 = -\text{ Rs } 1{,}15{,}279.50$$

That is cash outflows are more than cash inflows.

(*b*) **Machine *B* :**

$$P_B = -\,60{,}000 - 16000\,(P/A, I\%, n) - 14000\,(1 + I)^{-5}$$

$$= -\,60{,}000 - 60621.10 - 10518.41 = -\text{ Rs } 1{,}31{,}139.50$$

It is clear that N.P.W. of machine *A* is less negative as compared to that of machine *B* or N.P.W. of machine *A* is more positive as compared to that of machine *B*, therefore, machine *A* is economical. Or, since P_A and P_B are both negative above, they represent costs. The machine with least negative cost will be selected.

On the basis of N.P.W. of costs :

$$P_A = \text{Rs } 1{,}15{,}279.50$$

$$P_B = \text{Rs } 1{,}31{,}139.50$$

Since N.P.W. of costs of machine A is < N.P.W. of costs of machine B.

∴ Machine A is economical.

2*a*. When comparing two alternatives by the P.W. method, these must be compared for the same time period. In the above example, both the alternatives have the same life, so, there is no problem in comparing their P.W. However, when the lives are different, we can not directly compare the two alternatives. One way is to use the same time period which will be the L.C.M. of the two lives. For example, if machine A has a life of 2 years and machine B has a life of 3 years, we will use a common time period of 6 years, so that the machine A is replaced three times and the machine B is replaced 2 times. This procedure will work only if a common multiple of the two lives is found and the time periods are short. Supposing the two lives are 6 years and 7 years, then the common multiple is 42 and the process will become very lengthy and cumbersome. A direct approach can be to convert the P.W. based on a period n_1 to an equivalent P.W. based on period n_2 by the formula,

$$P_{n2} = P_{n1} \times \frac{(A/P, I\%, n_1)}{(A/P, I\%, n_2)} \quad ...(5.19)$$

Example 5.5. *Machine A operated manually costs Rs 20000 has a life of 2 years. Machine B is automatic and costs Rs 50,000 and has a life of 4 years. Operating costs of machine A are Rs 40000 annually while that of machine B are Rs 30,000. Which machine should be selected. Take interest rate as 10%.*

Solution. Net $P.W.$ of machine A on time period of 2 years,

$$(N.P.W.)_A = \text{Cash inflows} - \text{Cash outflows}$$

No Cash inflows are mentioned. So, there are only expenses.

So, $$(P.W. \text{ of expenses})_A = 20000 + 40000\,(P/A, I\%, n)$$

Now $$(P/A, I\%, n) = (P/A, 10\%, 2) = \frac{(1+I)^n - 1}{I(1+I)^n} = \frac{(1.1)^2 - 1}{0.1 \times (1.1)^2} = 1.735$$

∴ $$(P.W. \text{ of expenses})_A = 20000 + 40000 \times 1.735$$

$$= 20000 + 69400 = \text{Rs } 89{,}400.$$

$$(P.W. \text{ of expenses})_{B \text{ at } n=4} = 50000 + 30000\,(P/A, I\%, n)$$

Now $$(P/A, I\%, n) = (P/A, 10\%, 4)$$

$$= \frac{(1.1)^4 - 1}{0.1 \times (1.1)^4} = \frac{1.4641 - 1}{0.1 \times 1{,}4641} = 3.1698$$

$$(P.W. \text{ of expenses})_{B \text{ at } n=4} = 50000 + 95096$$

$$= 1{,}45{,}096$$

∴ $$(P.W. \text{ of expenses})_{B \text{ at } n=2} = 1{,}45{,}096 \times \frac{(A/P, I\%, 4)}{(A/P, I\%, 2)}$$

$$= \frac{1{,}45{,}096 \times 1.735}{3.1698}$$

$$= \text{Rs } 79{,}413.75$$

∴ $$(P/A, I\%, n) = 1/(A/P, I\%, n)$$

As the expenses of machine B are less, so this is economical.

3. Annual Worth (A.W.) or Annual Cost Method :

The net $A.W.$ of an asset or a project is given as,

$$A.W. = R - E - C.R.$$

where R = annual equivalent receipts

E = " " expenses or payments

$C.R.$ = " " capital recovery

R, E and $C.R.$ are calculated at $M.A.R.R.$

An asset or a project is economically justified if

$$N.A.W \geq 0.$$

This method is called as 'annual cost' ($A.C.$) method, if only costs are considered. The project with least negative of $A.C.$ will be selected.

There are many ways to determine $C.R$:

(*a*) $C.R.$ = annual equivalent of investment – annual equivalent of salvage value

$$= P\,(A/P,\ I\%,\ n) - S\,(A/F,\ I\%,\ n) \quad ...(5.20)$$

(*b*) Another method is,

$C.R.$ = annual sinking fund depreciation charge + interest on the original investment

$$= (P - S)(A/F,\ I\%,\ n) + P\,(I\%)$$

$$= (P - S).SFF + P\,(I\%) \quad ...(5.21)$$

(*c*) Another very popular way is,

$C.R.$ = Annual cost of the depreciable portion of investment + interest on the salvage value

$$= (P - S)(A/P,\ I\%,\ n) + S\,(I\%)$$

$$= (P - S) \times CRF + S\,(I\%) \quad ...(5.22)$$

Based on the last formula, the annual cost of a project will be :

$$A.C. = (P - S) \times CRF + S\,(I\%) + O \quad ...(5.23)$$

where O = operating costs.

(*d*) Annual cost can also be calculated as follows :

$A.C.$ = Annual taxes and insurance + Annual depreciation + Annual operating and maintenance expenses + Minimum annual profits required.

$\therefore$ Net $A.C.$ = Total $A.C.$ – Annual income or revenue

Note. If the total $A.C.$ is > the desired annual revenue, the project is not economical.

The annual depreciation is calculated by the Sinking Fund method, that is,

$$d = (C - S)SFF$$

where SFF = Sinking fund factor = $\dfrac{I}{(1+I)^n - 1}$.

Example 5.6. *Select the economical machine out of A and B for which the data is given below :*

Data	*A*	*B*
First cost, Rs	*46000*	*60,000*
Salvage value, Rs	*8000*	*10,000*
Operating charges, Rs	*10,000*	*9200*
Economic Life, years	*10*	*15*
Interest rate, %	*8*	*8*

Solution. For machine A, $A.C. = (46000 - 8000)CRF + 8000 \times \frac{8}{100} + 10{,}000$

now $$CRF = \frac{I(1+I)^n}{(1+I)^n - 1}$$

$$\therefore \quad CRF = \frac{0.08(1.08)^{10}}{(1.08)^{10} - 1} = \frac{0.08 \times 2.159}{1.159}$$

$$= 0.149$$

$$\therefore \quad A.C. \text{ of machine } A = 3800 \times 0.149 + 640 + 10000$$

$$= \text{Rs } 16302.00$$

For machine B : $$CRF = \frac{0.08(1.08)^{15}}{(1.08)^{15} - 1} = \frac{0.08 \times 3.172}{2.172}$$

$$= 0.1168$$

$$\therefore \quad A.C. \text{ of machine } B = 50000 \times 0.1168 + 1000 + 9200$$

$$= \text{Rs } 16040.00$$

∴ Machine B will be economical.

4. **The Internal Rate of Return (*I.R.R.*) method :**

This method is also known by the names : Discounted Cash Flow (*DCF*) method, receipts versus payments method, investors method and profitability index.

Discounted Cash Flow has already been defined according to which

$$N.P.W. \text{ of receipts} = N.P.W. \text{ of expenses}$$

The interest rate for which this equation holds is defined as *I.R.R.*, I'

$$\therefore \quad \sum_{j=0}^{n} R_j (P/F, I', j) = \sum_{j=0}^{n} E_j (P/F, I', j) \qquad ...(5.24)$$

where R_j = net receipts (income, savings) for the jth year

E_j = " expenses (payments, costs) for the jth year

n = project life

The above equation can also, be written as,

$$\sum_{j=0}^{n} R_j (P/F, I', j) - \sum_{j=0}^{n} E_j (P/F, I', j) = 0 \qquad ...(5.25)$$

The above equation is solved to find I'. The project or alternative is economical if $I' \geq$ *M.A.R.R.*

Eventhough, the *I.R.R.* method is widely used in economic studies, it has the following drawbacks :

(*i*) The method is based on the assumption that the capital is reinvested at I' rather than *M.A.R.R.* In many problems, this assumption does not mirror reality. Due to this, the method is not preferred for study of alternatives.

(*ii*) Computations to find I' are difficult, and sometimes we may get multiple values of *I.R.R.*

5. **The External Rate of Return (E.R.R.) method :**

This method eliminates the drawbacks of *I.R.R.* method. Here, an external interest rate, *e*, is considered at which net cash flows generated or required by a project over its life can be reinvested or borrowed outside the company.

While deriving the formula for *E.R.R.* method, all cash outflows are discounted to the present time at *e*% per period and all cash inflows are compounded to period *n* at *e* %, per period. Then, the *E.R.R, I'*, is determined for which the two become equal, that is,

$$\sum_{j=0}^{n} E_j \,(P/F, e\%, j)(F/P, I'\%, n) = \sum_{j=0}^{n} R_j \,(F/A, e\%, n-j) \qquad \text{...(5.26)}$$

A project is acceptable or economical, if I' found from the above equation is ≥ M.A.R.R. of the company.

Example 5.7. *A new equipment costing Rs 100,000 is being proposed to increase the productivity of an operation. Its useful life is 5 years and its salvage value is Rs 20,000. Annual net savings will be Rs 32000. If e = M.A.R.R. = 20%, Find E.R.R. and examine if the project is economical ?*

Solution. Equation (5.26) is,

$$\sum_{j=0}^{n} E_j \,(P/F, e\%, j)(F/P, I'\%, n) = \sum_{j=0}^{n} R_j \,(F/A, e\%, n-j)$$

Here, $e = 20\%,\ n = 5$

Now $E_j\ (P/F,\ e\%,\ j)$ = Rs 100,000 for $j = 0$.

There are no other expenses.

$\therefore$ 100,000 (*F/P*, *I'*%, 5) = 32000 (*F/A*, 20%, 5) + 20,000

$$= 32000\left[\frac{(1+I)^n - 1}{I}\right] + 20000$$

$$= 32000\left[\frac{(1.20)^5 - 1}{0.20}\right] + 20000 = \text{Rs } 25{,}8131.20$$

$$\therefore \quad (F/P, I'\%, 5) = \frac{258131.20}{100{,}000} = 2.5813$$

From here $(1 + I')^5 = 2.5813$

$\therefore$ $1 + I' = (2.5813)^{0.2} = 1.2088$

$\therefore$ $I' = 0.2088$, that is 20.88%.

Since I' is > *M.A.R.R.*, therefore, the project is just feasible.

Example 5.8. *The cash flow diagram of a project is shown in Fig. 5.8. Determine whether the project is feasible or not? Take e = 15% and M.A.R.R. = 20% in the E.R.R. method.*

Solution.

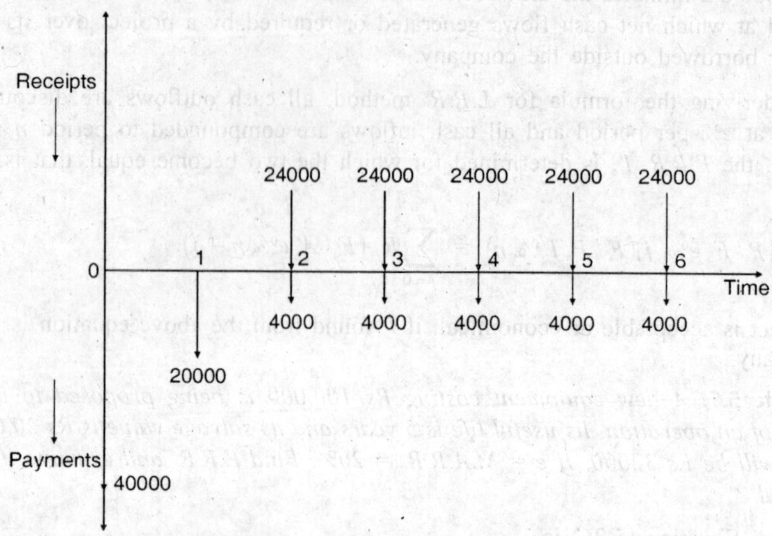

Fig. 5.8. Cash Flow Diagram.

The equation to be solved is eqn. (5.26), that is,

$$\sum_{j=0}^{n} E_j \left(P/F, e\%, j\right)\left(F/P, I'\%, n\right) = \sum_{j=0}^{n} R_j \left(F/A, e\%, n-j\right)$$

Here E_0 = Rs 40,000; E_1 = Rs 20,000

The expenses E for j = 2 to 6 can be subtracted from R for j = 2 to 6.

Therefore, net R_j = Rs 20,000 for j = 2 to 6

$$\therefore \left[40{,}000 + 20000\left(P/F, 15\%, 1\right)\right]\left(F/P, I'\%, 6\right) = 20000\ (F/A,\ 15\%,\ 5)$$

Now $$20000\ (P/F,\ 15\%,\ 1) = \frac{20000}{(1+0.15)^1} = \text{Rs } 17351.30$$

and $$20000\ (F/A,\ 15\%,\ 5) = 20000\left[\frac{(1.15)^5 - 1}{0.15}\right]$$

$$= \text{Rs } 134847.62$$

$$\therefore \quad 57351.30\ (F/P,\ I'\%,\ 6) = 134847.62$$

$$\therefore \quad (F/P,\ I'\%,\ 6) = \frac{134847.62}{57351.30} = 2.351$$

$$\therefore \quad (1 + I')^6 = 2.351$$

$$\therefore \quad 1 + I = (2.351)^{0.167} = 1.15345$$

$$\therefore \quad I = 0.15345, \text{ that is, } 15.345\%.$$

Since I is < *M.A.R.R.*, therefore, the project is not feasible.

6. The Explicit Re-investment Rate of Return (E.R.R.R.) Method :

This is another method to overcome the drawbacks of *I.R.R.* method. *E.R.R.R.* is defined as,

$$E.R.R.R. = \frac{\text{Net annual profit}}{\text{Initial investment}}$$

Now, Net annual profit = Receipts (incomes) – Payment (all expenses) – Depreciation calculated at e %.

$$= R - E - (P - S)(A/F, e\%, n)$$

The annual depreciation is calculated by the Sinking Fund method. Thus, the relation becomes,

$$E.R.R.R. = \frac{R - E - (P - S)(A/F, e\%, n)}{P} \quad ...(5.27)$$

The project is acceptable if *E.R.R.R.* is ≥ *M.A.R.R.* of the company.

Example 5.9. *For the data given in Ex. 5.5, calculate E.R.R.R. and see if the project is economical.*

Solution. Here e = *M.A.R.R.* = 20%, n = 5. Annual net savings = Rs 32000 = $R - E$, P = Rs 100,000; S = Rs 20,000.

$$\therefore \quad (P - S)(A/F, e\%, n) = (100{,}000 - 20{,}000)\left[\frac{I}{(1+I)^n - 1}\right]$$

Here, $I = e = 20\%$

$$\therefore \quad (P - S)(A/F, e\%, n) = 80000 \times \frac{0.20}{(1.20)^5 - 1} = \text{Rs } 10752.70$$

$$\therefore \quad E.R.R.R. = \frac{32000 - 10752.70}{100{,}000} = 0.2125 = 21.25\%$$

Since E.R.R. is > M.A.R.R.

∴ the project is feasible.

Note. When comparing two alternatives including replacing old equipment with a new equipment, then E.R.R.R. is found as follows :

$$\text{E.R.R.R.} = \frac{(\text{Total net annual expenses})_1 - (\text{Total net annual expenses})_2}{(\text{Investment})_1 - (\text{Investment})_2}$$

$$= \frac{\Delta\,(\text{Net annual expenses})}{\Delta\,(\text{Investment})} \quad ...(5.28)$$

If *E.R.R.R.* is > *M.A.R.R.*, then it is economical to go in for replacement or M_2 is economical then M_1.

or
$$E.R.R.R. = \frac{\Delta\,(\text{Net annual revenue})}{\Delta\,(\text{Investment})} \quad ...(5.29)$$

or calculate E.R.R.R. for each alternative and compare these.

Example 5.10. *The data for two machines is given below :*

Item	*A*	*B*
Initial cost, Rs	*46000*	*32000*
Annual net Cash flow, Rs	*9600*	*7200*
Life, years	*6*	*6*
Salvage value	*0*	*0*

Take M.A.R.R. = 8%. Determine whether the machines are acceptable with E.R.R.R. The investment rate = M.A.R.R.

Solution. Equation (5.27) is

$$E.R.R.R. = \frac{R - E - (P - S)(A/F, e\%, n)}{P}$$

(*a*) **Machine *A*:** $R - E =$ Rs 9600

$P =$ Rs 46,000; $S = 0$, $e = 8\%$, $n = 6$

$$(A/F, e\%, n) = \frac{I}{(1+I)^n - 1} = \frac{0.08}{(1.08)^6 - 1} = 0.1363$$

$$\therefore \quad E.R.R.R. = \frac{9600 - 46000 \times 0.1363}{46000} = 0.0724 = 7.24\%$$

(*b*) **Machine *B* :** $E.R.R.R. = \dfrac{7200 - 32000 \times 0.1363}{32000} = 0.0887 = 8.87\%$

∴ Only machine *B* is acceptable.

7. **Break Even Point (B.E.P.) method.**

Break-even point method is another very useful method for comparing alternatives. This has been discussed in detail ahead under Art. 5.5.

5.2.3. Equipment replacement. The replacemente of the existing asset or equipment may be warranted by the following reasons.

1. **Deterioration.** Determination or physical impairment means that the present asset has worn out due to normal use or accident and is no longer able to perform its intended function unless expensive overhandling is carried out. It means the end of the technological service life of the asset, resulting in its poor performance.

2. **Obsolescence.** Obsolescence is of two types : (*a*) Functional (*b*) Economical. Functional obsolescence means that the existing asset is not able to meet the quality demanded in the market. This results in decrease in the demand of the product and so loss of revenue. Economical obsolescence means that a new superior, more efficient asset has come in the market which can produce the demand at a lower cost.

3. **Inadequacy.** The existing asset is not capable of meeting the increased demand of the product or there is a change in the product design.

Other factors have been discussed under Art. 5.2.

Economic Studies of Replacement. Economic study of asset or equipment replacement is done on on the same lines as the economic study of alternatives as discussed under Art. 5.2.2., with the difference that one alternative (the existing asset) is called the "challenger". Replacement studies are very important for a company. The most important consideration is "when" to replace the existing asset. There are no set rules to decide this "when". If the replacement is done not at the appropriate time, there can be unnecessary drain of capital (early replacement) or the product may become non competitive (delayed replacement). So the problem is not so simple as it looks. A wrong replacement decision may shatter the economy of a company. A very systematic, scientific and realistic approach is necessary.

The economic studies of replacement should be done on "after tax" mode, because replacement of an asset can result in capital gains or loss which can result in tax benefits or penalties. This can change the decision, taken otherwise.

Factors affective replacement studies. The following factors should be considered when undertaking replacement studies :

1. Recognition and acceptance of past mistakes or errors.
2. Existence of sunk costs. Sunk costs are past payments or receipts, the uncovered balance of an investment etc. In replacement studies, the sunk costs are normally not considered.
3. Remaining life of the existing asset (Defender)
4. Economic life of the new asset (challenger)
5. Possible capital gains or losses.

Economic life. In the replacement studies, determining the economic life of the defender and the challenger is very important, since the two should be compared over their economic life. As already defined, the economic life of an asset is the time interval that minimizes' asset's total equivalent annual costs or maximizes its equivalent annual net income, Fig. 5.9. It is also known as minimum-cost life or the optimum replacement interval.

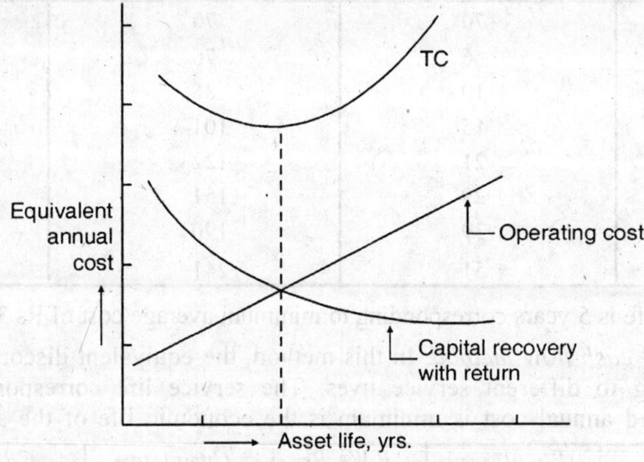

Fig. 5.9. Economic Life.

Example 5.11. *The initial cost of a machine purchased 5 years ago was Rs 60,000. The depreciation calculated by straight line formula is Rs 6000 per year. So, its present book value is Rs 30,000. The market price of the machine, if sold now, is Rs 15000. It would need Rs 6000 to overhaul it to make it serviceable for another 5 years. What is the investment cost and unamortized value of the machine ?*

Solution. Investment cost = Present market value + sum needed to make it serviceable

= 15000 + 6000 = Rs 21000.

Unamortized investment = Present book value – Salvage value

= 30000 – 15000 = Rs 15000.

Example 5.12. *A machine has initial cost of Rs 70,000. Its annual operating and maintenance charges for the first seven years are : Rs 6000, 10000, 150000, 21000, 29000, 39000 and 51000 respectively. Determine the economic life of the machine, using :*

(a) Undiscounted cash flow method.

(b) Discounted cash flow method at a discount rate of 10%.

Solution. The cash flow diagram is shown in Fig. 5.10.

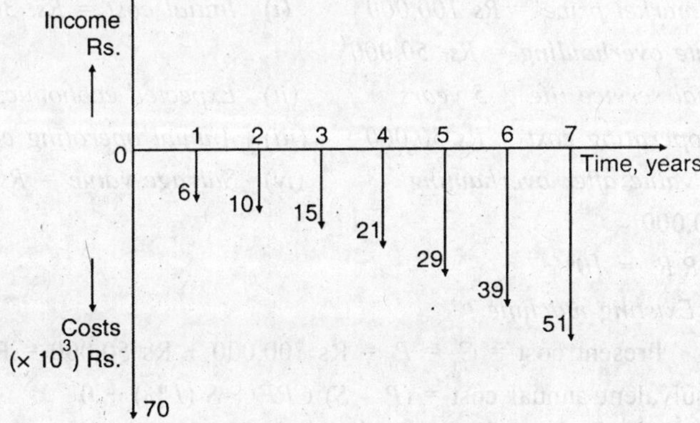

Fig. 5.10. Cash Flow Diagram.

(*a*) *Undiscounted cash flow method.* In this method, the average annual undiscounted cost is calculated. The service life corresponding to the minimum average undiscounted annual cost will give the economic life of the machine, as shown below :

Service life, year	*Cost, Rs (× 10³)*	*Cumulative Cost, Rs. (× 10³)*	*Average Cost, Rs. (× 10³)*
0	70	70	—
1	6	76	76
2	10	86	43
3	15	101	33.67
4	21	122	30.50
5	29	151	30.20
6	39	190	31.67
7	51	241	34.43

So, economic life is 5 years corresponding to minimum average cost of Rs 30.20×10^3 per year.

(*b*) *Discounted cash flow method.* In this method, the equivalent discounted annual cost is found corresponding to different service lives. The service life corresponding to whch the equivalent discounted annual cost is minimum is the economic life of the asset.

Service, life years	*Cost Rs (× 10³)*	*(P/F, I%, n)*	*P.W., Rs. (× 10³)*	*Cumulative P.W., Rs (× 10³)*	*CRF*	*Equivalent Annual Cost, Rs (× 10³), P.W CRF*
0	70	1.000	70	70	—	—
1	6	0.909	5.554	75.454	1.100	83.00
2	10	0.826	8.260	83.714	0.5762	48.23
3	15	0.751	11.27	94.984	0.4021	38.19
4	21	0.683	14.34	109.324	0.3155	34.49
5	29	0.621	18.01	127.334	0.2638	33.59
6	39	0.565	22.02	149.354	0.2296	34.29
7	51	0.513	26.17	175.524	0.2054	36.05

∴ Economic life is 5 years at minimum equivalent annual cost of Rs 33.59×10^3.

Example 5.13. *An enterprise is contemplating the replacement of the existing machine with a new machine. Make a decision on the basis of equivalent annual cost method with the following given data :*

Existing Machine	*Proposed Machine*
(*i*) *Present market price* = *Rs 100,000*	(*i*) *Initial cost* = *Rs. 300,000*
(*ii*) *Immediate overhauling* = *Rs. 50,000*	
(*iii*) *Additional service life* : *5 years*	(*ii*) *Expected economic life* = *10 years*
(*iv*) *Annual operating cost* = *Rs 50,000*	(*iii*) *Annual operating cost* = *Rs 30,000*
(*v*) *Salvage value after overhauling* = Rs. 10,000	(*iv*) *Salvage value* = *Rs. 100,000*

Take M.A.R.R. = *10%.*

Solution. (*a*) *Existing machine* :

$$\text{Present cost} = C_1 = P_1 = \text{Rs } 100{,}000 + \text{Rs } 50{,}000 = \text{Rs } 150{,}000$$

$$\text{Equivalent annual cost} = (P - S)\, CRF + S\,(I\%) + 0$$

Now CRF = Capital recovery factor

$$= \frac{I(1+I)^n}{(1+I)^n - 1} = \frac{0.10(1.10)^5}{(1.10)^5 - 1} = 0.2638$$

$$\therefore \quad A.C. = (150{,}000 - 10{,}000) \times 0.2638 + 10{,}000 \times \frac{10}{100} + 50{,}000$$

$$= 36.932 \times 10^3 + 1 \times 10^3 + 50 \times 10^3$$

$$= \text{Rs } 87.932 \times 10^3.$$

(*b*) *Proposed machine :*

$$CRF = \frac{0.10(1.10)^{10}}{(1.10)^{10} - 1} = 0.1628$$

$$\therefore \quad A.C. = (300{,}000 - 100{,}000) \times 0.1628 + 100{,}000 \times \frac{10}{100} + 30{,}000$$

$$= 32.56 \times 10^3 + 10 \times 10^3 + 30 \times 10^3 = \text{Rs } 72.56 \times 10^3.$$

Since the equivalent annual cost of proposed machine is less than that of the existing machine, therefore, the replacement is justified.

Dynamic equipment replacement policy.

We all know that every equipment is subjected to deterioration and obsolescence in varying degrees with passage of time. As a result, the old equipment does not operate as well as the best available new piece of equipment. As the time passes, this operating inferiority gets larger, when expressed in Rupees. So, a decision has to be taken to replace the old equipment with the new one. When the new equipment is purchased, its capital cost is maximum but its operating inferiority is at minimum [see Fig. 5.11]. As the time passes, capital charges decreases and operating inferiority increases. Because of these adverse costs, the management is in dilemma. The choice has to be made between more capital cost and less imperfection, on one hand, and less capital cost and more imperfection, on the other. Neither situation is good. However, when we have two such adverse costs, the best course is to find where the sum of the two adverse costs of the new machine by such an amount as to make the purchase desirable and economically feasible.

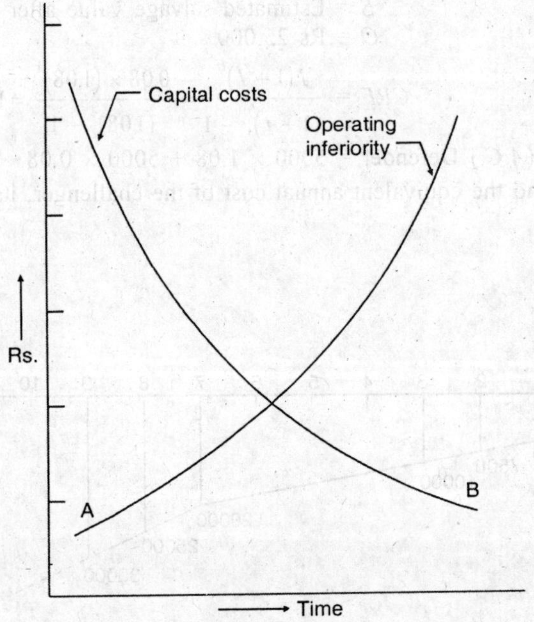

Fig. 5.11.

When we start comparing the costs of a proposed machine with cost of the existing machine, the problem arises as to whether we should compare only the next year's costs for two machines or for a longer period. Next year's costs can be estimated fairly accurately. The reliability of the estimates drops sharply as we extended the period. We have two models under the above mentioned replacement policy :

Turbogh's plan of equipment replacement

In this method, the so-called "Challenger's Adverse Minimum" is compared with the cost of keeping the existing equipment (Defender) by another year.

Challenger's adverse minimum

This is defined as the equivalent cost (annual) of the challenger (Proposed Replacement) over its economic life. To calculate this, the following assumptions are made :

(*i*) The challenger has zero salvage value.

(*ii*) The operating costs of the challenger increase at a constant rate during its economic life.

Cost of keeping the Defender by another year

This is determined by :

(*i*) Defender's estimated present salvage value and that at the end of one year.

(*ii*) Operating costs of the Defender over the next one year.

Turbogh's model or plan can be explained with the help of an example :

Example 5.14. *The present estimated salvage value of the existing equipment is Rs 10,000. It is estimated to decrease to Rs. 5000 next year. Its operating costs for the next one year are estimated to be Rs 25,000. The proposed replacement has an initial cost of Rs 80,000. The present estimated annual operating costs of the new equipment are Rs 2500 and they are estimated to increase by Rs 2500 every year of its service life till the end of 7 years. After that the operating costs are estimated to increase by Rs 5000 per year. If the discount rate is 8%, which alternative should be preferred ?*

Solution. Defender :

Equivalent cost (annual) = Cost of extending service by one year

From eqn. (5.23), $A.C. = (P - S).CRF + S(I\%) + O$

Now for the defender, P = Present estimated salvage value = Rs 10,000

S = Estimated salvage value after 1 year = Rs 5000

O = Rs 25,000

Now
$$CRF = \frac{I(1+I)^n}{(1+I)^n - 1} = \frac{0.08 \times (1.08)^1}{(1.08)^1 - 1} = 1.08$$

$\therefore$ (*A.C.*) Defender = 5000 × 1.08 + 5000 × 0.08 + 25000 = Rs 30800

Challenger. To find the equivalent annual cost of the challenger, its cash-flow diagram is shown in Fig. 5.12.

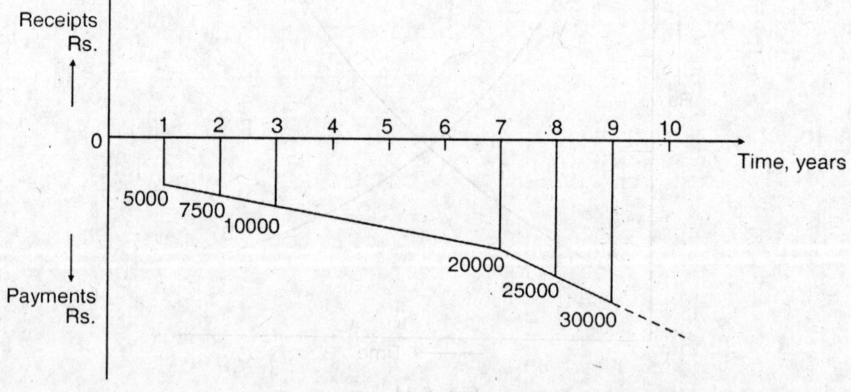

Fig. 5.12. Cash Flow Diagram.

Life, years	*Costs, (× 10³), Rs*	*(P/F, I%, n)*	*P.W. (× 10³), Rs*	*Cumulative, P.W. (× 10³),*	*CRF*	*Equi., A.C. (× 10³), Rs*
0	80	1.000	80	80	—	—
1	5	0.926	4.63	84.63	1.08	91.40
2	7.5	0.857	6.43	91.06	0.561	51.08
3	10.0	0.794	7.94	99.0	0.388	38.41
4	12.5	0.735	9.19	108.19	0.302	32.67
5	15.0	0.681	10.20	118.39	0.250	29.60
6	17.5	0.630	11.03	129.42	0.216	27.95
7	20.0	0.584	11.67	141.09	0.192	27.09
8	25.0	0.540	13.50	154.59	0.174	26.90
9	30.0	0.500	15.00	169.59	0.160	27.13
10	35.0	0.463	16.21	185.80	0.149	27.68

So, the minimum equivalent annual cost of the challenger is Rs 26900 and that of the Defender is Rs 30,800. Hence, Replacement is preferred.

2. **MAPI plan of equipment replacement.** The term MAPI stands for "Machinery and Allied Products Institute". It is based in Washington (U.S.A.). This plan is similar to the one given by Turbogh discussed above, with the following difference : while calculating the challenger's adverse minimum, its salvage value and also the Income tax are taken into consideration. The principle of the method is quite simple : A new equipment has maximum capital cost and minimum operating cost when installed and operated. With passage of time, the initial cost decreases and the operating cost rises. With passage of time, the initial cost decreases and the operating cost rises. The net effect of the two costs is expressed by a third curve of cummulative cost. These calculations are conveniently done with the help of charts known as MAPI charts. The full treatment of the topic is out of the scope of the book.

5.3. MATHEMATICAL ANALYSIS FOR ECONOMIC EQUIPMENT SELECTION :

Let n = number of years of economic repair life

C = first cost of machine (installed)

S = scrap value of machine

b = annual increase in cost of repairs

Then assuming straight line rising cost of repairs, we have

$$n = \sqrt{\frac{2(C-S)}{b}} \quad \text{...(5.30)}$$

Example 5.15. *A machine costs Rs 20,000 installed. The scrap value of the machine is Rs 1000 and the cost of repairs increases Rs 180 annually. Determine the economic repair life.*

Solution. From equation (5.1),

C = Rs 20,000, S = Rs 1000, b = Rs 180

$$\therefore \quad n = \sqrt{\frac{2(20{,}000-1000)}{180}}$$

$$= 14.53 \text{ years}$$

2. **Time required for new machine to pay for itself**

Let n = number of years required for new machine to pay for itself

C_n = Cost of new machine installed and tooled

C_o = Cost of old machine installed and tooled

S_n = probable scrap value of new machine at the end of its useful life.

S_o = scrap value of old machine (present)

K_o = present book value of old machine

$$= C_o - Y_o \frac{(C_o - S_e)}{Y_e}$$

Y_o = present age of old machine

Y_e = original estimated life of old machine

S_e = original estimated scrap value of old machine

P_o = labour and machine cost per unit on old machine

P_n = labour and machine cost per unit on new machine

N_n = estimated annual production on new machine

I = interest on investment, %

T = annual taxes, %

M = annual allowance for maintenance, %

D = annual allowance for depreciation on new machine, %

Annual saving on labour and machine cost = $N_n (P_o - P_n)$

Annual expenses on new machine = $C_n (I + T + M + D)$

$\therefore$ Net savings per year = $N_n (P_o - P_n) - C_n (I + T + M + D)$

Amount to be realised for the new machine = $C_n - S_n$.

Amount to be realised for the old machine which is to be replaced by the new machine

$$= K_o - S_o.$$

$\therefore$ Total amount to be realised from the saving due to the instalment of new machine

$$= (C_n - S_n) + (K_o - S_o)$$

$$\therefore \quad n = \frac{(C_n - S_n) + (K_o - S_o)}{N_n (P_o - P_n) - C_n (I + T + D + M)} \quad \text{...(5.30}a\text{)}$$

Example 5.16. *The first cost of the machine now being used was Rs 28000 and it was purchased five years ago, its original estimated life being ten years. It has a scrap value of Rs 6400. The present book value of the old machine is Rs 17200. The operator received Rs 2.50 per hour and the total machine cost per hour is Rs 48. The new machine will cost Rs 72000 fully installed and tooled. It is estimated that its scrap value at the end of its useful life will be Rs 8000. The hourly output is 16 pieces. The machine being more automatic, can be operated by a man who receives Rs 2 per hour. The total hourly machine rate is Rs 62. The shop works 2200 hours per year. The production rate of old machine is 10 pieces per hour. Find the number of years in which the new machine will pay for itself. I = 6%, T = 6%, D = 10%, M = .3%.*

Solution. C_n = Rs 72000, C_o = Rs 28000

$N_n = 2200 \times 16 = 35{,}200$ pieces, K_o = Rs 17200

S_o = Rs 6400, S_n = Rs 8000

Now $$P_o = \frac{2.50 + 48}{10} = \text{Rs } 5.05$$

$$P_n = \frac{2 + 62}{16} = \text{Rs } 4.0$$

From eqn. (5.30*a*), $n = \dfrac{(72000-8000)+(17200-6400)}{35200(5.05-4.00)-72000(0.06+0.06+0.10+0.03)}$

$$= \frac{64000+10800}{36960-18000} = \frac{74800}{18960} = \mathbf{3.945\ years}$$

3. **Which of the two suitable machines to select.** If for a job, the choice lies between two suitable machines, then the machine with the lower cost per unit of production will be selected. For this, the unit cost for each machine is calculated in the following manner :

Let C = first cost of first machine (installed)

c = first cost of second machine (installed)

N = Annual production of first machine

n = Annual production of second machine

L = annual labour cost on first machine

l = annual labour cost on second machine

B = annual labour burden on first machine, %

b = annual labour burden on second machine %

I = rate of interest, %

T = rate of taxes and insurance, %

D = annual allowance for depreciation of first machine, %

d = annual allowance for depreciation on second machine %

M = annual allowance for maintenance on first machine, %

m = annual allowance for maintenance on second machine %

P = annual cost of power for first machine

p = annual cost of power for second machine

X = unit production cost on first machine

x = unit production cost on second machine

F = saving in floor space per year, of first over second machine.

As already noted, annual cost of a machine is given as,

Total Cost = Fixed Cost + Variable Cost

For the first machine,

Fixed Cost = $C\,(I + T + D + M)$

Variable Cost = $L + BL + P$

$$\therefore \quad X = \frac{L + BL + P + C(I + T + D + M) - F}{N} \quad \text{...(5.31)}$$

Similarly
$$x = \frac{l + bl + p + c(I + T + d + m)}{n} \quad \text{...(5.32)}$$

If the use of first machine results in a loss rather than a saving in floor space, the sign of F must be changed to plus in eqn. (5.31).

Example 5.17. *The annual requirement of an article (about 4000 pieces) can be met by one semiautomatic turret lathe or by two engine lathes. The cost of turret lathe is Rs 80,000 (installed) and that of the engine lathe is Rs 32,000. The useful life of each type of machine is 10 years. The turret lathe takes up the same floor space as one engine, lathe, saving Rs 480 per year in floor space. The turret lathe can produce 4000 parts in 2256 hours including set up time.*

Its operator receives Rs 4 per hour. One operator is required on each engine lathe, the wage rate being Rs 5 per hour. Both engine lathes are operated for 2300 hours and produce a total of 3800 pieces $I = 6\%$, $T = 4\%$, $D = d = 10\%$, $M = m = 6\%$, $B = b = 55\%$. A 4-hp motor is used on turret lathe and a $2\frac{1}{2}$ hp motor on each of the engine lathe. The power rate is Rs 0.35 per kWh. Which type of machine should be selected for the job.

Solution. C = Rs 80,000, $c = 32000 \times 2$ = Rs 64,000

$N = 4000$, $n = 3800$

$L = 2256 \times 4$ = Rs 9024, $l = 2300 \times 2 \times 5$ = Rs 23,000

$$P = \frac{4 \times 746 \times 2300 \times 0.35}{1000} = \text{Rs } 2356$$

$$p = \frac{2 \times 2.5 \times 746 \times 2300 \times 0.35}{1000} = \text{Rs } 3002$$

F = Rs 480

$$\therefore \quad X = \frac{9024 + 0.55 \times 9024 + 2356 + 80000(0.06 + 0.04 + 0.10 + 0.06) - 480}{4000}$$

$$= \frac{9024(1.55) + 2356 + 80,000 \times 0.26 - 480}{4000}$$

= Rs 9.16 per piece

$$x = \frac{23000(1.55) + 3002 + 64000 \times 0.26}{3800}$$

= **Rs 14.55 per piece**

∴ One turret lathe will be more economical than two engine lathe.

Example 5.18. *For the problem given above, calculate the maximum investment on turret lathe that is justified by savings.*

Solution. Considering the equal number of pieces for engine lathes and turret lathe, say 4000. Now the cost to produce 4000 pieces on turret lathe

= 9.16 × 4000 = Rs 36,640

Cost to produce 4000 pieces on two engine lathe

$$= \frac{4000}{3800}(23000 \times 1.55 + 3002) + 64000 \times 0.26$$

= 40686 + 16640 = Rs 57326

∴ the annual saving with the turret lathe over the engine lathes,

= Rs (57326 – 36640) = Rs 20686

∴ the amount that could be invested in the turret lathe over the cost of engine lathes, is

$$= \frac{20686}{I + T + D + M} = \textbf{Rs 80,000 (approx.)}$$

Example 5.19. *An old machine which has become obsolete is to be replaced by a new machine costing Rs 60,000. It would fetch Rs 5000 on discard after ten years. Monthly labour charges for the new machine are Rs 500, whereas these are Rs 300 for the old machine. Yearly running charges and production rates are respectively Rs 10,000 and 200,000 pieces for the new machine, and Rs 15,000 and 50,000 pieces for the old machine. In how many years, the new machine will pay for itself ? $I = 10\%$, $M = 7\%$, $T = 6\%$, and scrap value of old machine is Rs 1000.*

Solution. For the cases, where the old machine has become obsolete, that is, has completed its useful life, relation (5.30*a*) gets modified as,

$$n = \frac{(C_n - S_n) - S_o}{N_n(P_o - P_n) - C_n(I + T + D + M)} \quad ...(5.33)$$

Now,

$$P_o = \frac{300 \times 12 + 15000}{50,000} = \text{Rs } 0.372$$

$$P_n = \frac{500 \times 12 + 10000}{200,000} = \text{Rs } 0.08$$

$$C_n = \text{Rs } 60,000,\ S_n = 5000,\ S_o = \text{Rs } 1000$$

$$N_n = 200,000$$

$$I = 10\%,\ M = 7\%,\ T = 6\%,\ D = \frac{1}{10} = 10\%$$

$$\therefore \quad n = \frac{(60,000 - 5000) - 1000}{200,000\,(0.372 - 0.08) - 60000(0.10 + 0.06 + 0.10 + 0.07)}$$

$$= \frac{54000}{58400 - 19800}$$

$$= 1.399 \text{ yrs.} = \mathbf{1.40\ Yrs.}$$

5.4. ECONOMICS OF SMALL TOOL SELECTION

The use of small tools such as jigs and fixtures facilitates the manufacture of a workpiece. Fast machining rates can be employed by the use of jigs and fixtures as they give adequate support to the work. This will result in saving in direct machining costs. So, a jig or fixture will be used for a job if the annual savings from their use are at least equal to the annual cost of jig or fixture. For this, the following mathematical analysis is very useful :

Let
N = number of pieces manufactured per year
C = first cost of small tool
I = annual allowance for interest on investment, %
M = annual allowance for maintenance and repairs, %
T = annual allowance for taxes, %
D = annual allowance for depreciation %
S = yearly cost of set up of the tool
a = saving in labour cost per unit
b = burden applied on labour saved, %
V = yearly operating profit over fixed charges
H = number of years to amortize the investment out of earnings.

Then, annual cost of the tool $= C(I + T + M + D) + S$

annual saving $= Na + Na.b$

$= N.a\,(1 + b)$

$\therefore$ If no profit is desired,

$$C(I + T + M + D) + S = N.a\,(1 + b) \quad ...(5.34)$$

(*i*) Number of pieces required per year to pay for the small tool,

$$N = \frac{C(I + T + M + D) + S}{a(1 + b)} \quad ...(5.35)$$

(*ii*) If an annual profit of V is also desired, then

$$C(I + T + M + D) + S + V = Na(1 + b)$$

$$\therefore \quad N = \frac{C(I+T+M+D)+S+V}{a(1+b)} \quad ...(5.36)$$

(*iii*) Economic investment in small tool for given production,

$$C = \frac{Na(1+b)-S}{I+T+M+D} \quad ...(5.37)$$

(*iv*) Profit from improved jig or fixture designs,

$$V = Na(1 + b) - C(I + T + M + D) - S \quad ...(5.38)$$

(*v*) To find the number of years required for amortization of investment out of earnings or the number of years required for a jig or fixture to pay for itself, we know that

$$D = \frac{1}{H}$$

Putting this in relation (5.34), we get

$$H = \frac{C}{Na(1+b)-C(I+T+M)-S} \quad ...(5.39)$$

Example 20. *The following data are given,*

a = Rs 1.50, b = 55%, T = 4%, M = 5%, I = 8%, D = 50%, H = 2 years, S = cost of each set up = Rs. 50.00.

Solution.

1. If a fixture costs Rs 3000, and one run is made per year then, from relation (5.34),

$$N = \frac{C(I+T+D+M)+S}{a(1+b)}$$

$$= \frac{3000(0.08+0.04+0.50+0.05)+50}{1.50 \times 1.55} = \textbf{886 pieces}$$

$\therefore$ 2 yearly runs of 886 pieces will pay for the fixture.

2. If 5 runs are made per year, then annual production,

$$N = \frac{3000 \times 0.67 + 5 \times 50}{1.5 \times 1.55} = \textbf{972 pieces}$$

3. If the fixture must pay for itself in a single run, then $H = 1$, and $D = 100\%$,

$$\text{Therefore,} \quad N = \frac{(3000 \times 1.17)+50}{1.5 \times 1.55} = \textbf{1531 pieces}$$

4. For a single run of 1530 pieces with an estimated savings of Rs 1.50 per piece, the economical investment on fixture will be (equ. 5.37),

$$C = \frac{1530 \times 1.5 \times 1.55 - 50}{1.17} = \textbf{Rs 3000.}$$

5. If 950 pieces are made per year in 6 (six) runs and the saving in labour cost is Rs 2 per piece, then the annual profit can be found as (eqn. 5.38),

$$V = 950 \times 2 \times 1.55 - 3000 \times 0.67 - 300$$
$$= 2945 - 2010 - 300 = \textbf{Rs 635 per year.}$$

Example 5.21. *How many components must be produced to warrant making a fixture costing Rs 2400 and Rs 80 to set up, if it saves $12\frac{1}{2}$ paise on each piece compared with previous method of manufacture and must pay for itself in two years. I = 6%, T = 4%, M = 10%, overhead applied on direct labour saved = 40%.*

Solution. $a = 0.125$, $b = 0.4$, $D = \frac{1}{2} = 50\%$

$$\therefore \quad N = \frac{C(I + T + D + M) + S}{a(1 + b)}$$

$$= \frac{2400(0.06 + 0.04 + 0.50 + 0.10) + 80}{0.125 \times 1.4}$$

$$= 10060 \text{ pieces per year.}$$

∴ Total number of pieces to be produced = 2 × 10060 = 20120.

Example 5.22. *If in the above example, the fixture is used to make pieces in six batches per year, what effect will this have on the number of pieces produced.*

Solution. The only change is the yearly cost of set up = 6 × 80 = Rs 480.

$$\therefore \quad N = \frac{2400 \times 0.7 + 480}{0.125 \times 1.4} = 12340 \text{ pieces per year}$$

∴ Total number of pieces to be produced = 2 × 12340 = **24680.**

Example 5.23. *How long it will take a fixture costing Rs 2000 to pay for itself, if by its use it saves 15 paise per piece over the previous method. The parts made per year are 5000 made in five batches. Cost of each set up is Rs 50. Yearly percentages are : I = 10%, T = 5%, M = 10%, burden applied on labour saved = 50%.*

(i) If later, the fixture could be made for Rs 1600, what profit would be made if the conditions remain as before.

Solution. C = Rs 2000, $N = 5000$, $S = 5 \times 50$ = Rs 250, a = Rs 0.15, $b = 50\%$, $I = 10\%$, $T = 5\%$, $M = 10\%$.

From eqn. (5.39).
$$H = \frac{C}{Na(1 + b) - C(1 + T + M) - S}$$

$$= \frac{2000}{5000 \times 0.15 \times 1.5 - 2000(0.10 + 0.05 + 0.10) - 250}$$

$$= \frac{2000}{1125 - 500 - 250} = 5.33 \text{ years}$$

Fixture will pay for itself in 5.33 years.

(*ii*)
$$C = \text{Rs } 1600, \; D = \frac{1}{H} = \frac{1}{5.33} = 18.76\%$$

∴ From eqn. (5.38),

$$V = Na(1 + b) - C(I + T + D + M) - S$$
$$= 5000 \times 0.15 \times 1.5 - 1600(0.10 + 0.05 + 0.10 + 0.1876) - 250$$
$$= 1125 - 700 - 250$$
$$= \textbf{Rs 175}$$

Example 5.24. *The cost of a fixture is Rs 1000. The machine cost per component using existing equipment is Rs 3 and that by using the fixture is Rs 1.00. Calculate the minimum number of components to be produced if the cost of the fixture is to be recovered in one setting only.*

Solution. Saving in machine cost per piece = 3 – 1 = Rs 2.

$$\therefore \quad N = \frac{1000}{2} = 500$$

5.4.1. Small tool replacement. *The replacement analysis for a small tool (jig or fixture) may be done as for any machine tool. That is, it is used to make a profit by paying for itself from savings which result because of its use. If,*

$$C = \text{cost of the new fixture}$$

and C_o and C_s are the original and scrap values of the old fixture and using the same notations as in Art. 5.4 we have,

$$N = \frac{C(I + T + D + M) + (C_o - C_s)\cdot I}{a(1 + b)} \qquad \text{...(5.40)}$$

For break even point.

Cost of the fixture will be,

$$C = \frac{NA(1 + b) - (C_o - C_s)I}{I + T + M + D} \qquad \text{...(5.41)}$$

The number of years required to amortize the fixture is,

$$H = \frac{1}{D} = \frac{C}{Na(1 + b) - (C_o - C_s)I - C(I + T + M)} \qquad \text{...(5.42)}$$

Net savings will be,

$$V = Na(1 + b) - C(I + T + M + D) - (C_o - C)\cdot I \qquad \text{...(5.43)}$$

Example 5.25. *A fixture is to cost Rs 1000. The old fixture which originally cost Rs 700, has a scrap value of Rs 250. The new fixture will save 10 paise per piece and the percentage of overhead charged to this fixture is 30%. Taking I = 8%, M = 3%, T = 12% and amortization = $1\frac{1}{2}$ years, calculate the number of pieces which must be produced to break even so that the fixture may pay for itself in one year.*

Solution. C = Rs 1000, C_o = Rs 700, C_s = Rs 250, a = Rs. 0.10, b = 0.30, I = 0.80, M = 0.03, T = 0.12, $D = \frac{1}{15}$.

$$N = \frac{C(I + T + D + M) + (C_o - C_s)I}{a(1 + b)}$$

$$= \frac{1000(0.08 + 0.12 + 0.67 + 0.03) + 450 \times 0.08}{0.10 \times 1.3}$$

$$= \textbf{7200 pieces per year.}$$

Example 5.26. *Using the data from example 5.25, and changing the number of pieces to be produced to 9000, how much can be spent for a new fixture ?*

Solution.
$$C = \frac{Na(1+b) - (C_0 - C_s)I}{(I + T + M + D)}$$

$$= \frac{9000 \times 0.10 \times 1.3 - 36}{0.90}$$

$$= \textbf{Rs 1260}$$

Example 5.27. *Assume the cost of the new fixture to be Rs 1350 when the yearly production is 6500 pieces. Using the same data as in example 5.25, how long will it take to amortize the fixture ?*

Solution.
$$H = \frac{1}{D} = \frac{C}{Na(1+b) - (C_0 - C_s)I - C(I + T + M)}$$

$$= \frac{1350}{6500 \times 0.10 \times 1.3 - 36 - 1350 \times 0.23}$$

$$= \textbf{2.7 years.}$$

Example 5.28. *In example 5.27, if the fixture results in a production of 9000 pieces per year, and the fixture costs Rs 1000, what is the profit ?*

Solution.
$$V = Na\,(1 + b) - C(I + T + M + D) - (C_o - C_s)I$$

$$= 9000 \times 0.01 \times 1.3 - 1000 \times 0.9 - 36$$

$$= \textbf{Rs 234.}$$

5.5. BREAK-EVEN POINT ANALYSIS

Break even point is very important in the economics of tools, machine tools and manufacturing processes. It is based on the principle that when the cost of two alternatives is affected by a common variable, there must exist a value of the variable for which the two alternatives will incur equal cost. The cost of each alternative can be expressed as function of the common independent variable and will be of the form,

$$(T.C.)_1 = f_1\,(x);\; (T.C.)_2 = f_2\,(x)$$

where $(T.C.)_1$ = Total cost per time period, per project or per piece for alternative 1 and similarly $(T.C.)_2$.

At Break-Even Point (*B.E.P.*),

$$(T.C.)_1 = (T.C.)_2$$

$$\therefore \quad f_1\,(x) = f_2\,(x)$$

which can be solved for *x*.

Break-Even Point analysis has many applications in Engineering and business.

(*a*) It usually refers to the number of pieces for which a business neither makes a profit nor incurs a loss. In other words, the selling price of the product is the total cost of production of the component. As already noted, the total cost of a product is the sum of fixed cost which is not dependent upon the production volume, and variable cost which depends upon the production volume.

If Q = quantity of production at Break-Even Point, then

$$F.C. + Q \times V.C = Q \times S.P. \quad ...(5.44)$$

where $F.C.$ = Fixed Cost

$V.C.$ = Variable Cost per unit

and $S.P.$ = Selling Price per unit

Break Even Point analysis is also known as Cost – Volume – Profit, CVP, analysis.

Example 5.29. *The following data refer to a manufacturing unit*

Fixed Cost = Rs 100,000

Variable Cost = Rs 100 per unit

Selling Price = Rs 200 per unit

(i) Calculate the Break-Even Point

(ii) If the fixed cost increases to Rs 125,000 and variable cost reduces to Rs 90 per unit, obtain the new Break-Even Point.

(iii) For (i) calculate the number of components needed to be produced to get a profit of Rs 20,000.

Solution. (*i*) Using eqn. (5.44),

$$F.C. + Q\,(V.C.) = Q\,(S.P.)$$

$$\therefore \quad 100{,}000 + Q \times 100 = Q \times 200$$

$$Q = 1{,}000 \text{ pieces}$$

(*ii*) Now $F.C.$ = Rs 125,000

$V.C.$ = Rs 90 per unit

$S.P.$ = Rs 200 per unit

Putting these values in eqn. (5.44),

$$125{,}000 + Q \times 90 = Q \times 200$$

$$\therefore \quad Q = 1136 \text{ pieces}$$

(*iii*) For this case, the relation will be

$$F.C. + Q \times V.C. + \text{profit} = Q \times S.P.$$

$$\therefore \quad 100{,}000 + Q \times 100 + 20000 = Q \times 200$$

$$Q = 1200 \text{ pieces.}$$

(*b*) Break-Even Point analysis is also used to make a choice between two machine tools to produce a given component.

To determine which of the two machines is most economical, the total cost of the two machines (fixed cost + variable cost) is plotted against the number of units. The point at which the two lines representing the total costs of the two machines meet each other, is termed as break even point. Towards the left of break-even point (Fig. 5.13), machine A is economical than machine B and if the quantity of production is more than that corresponding to break-even point machine B becomes economical than machine A.

Mathematically, the above discussion can be written as,

$$FC_A + Q \times VC_A = FC_B + Q \times VC_B$$

$$\therefore \quad Q = \frac{FC_B - FC_A}{VC_A - VC_B}$$

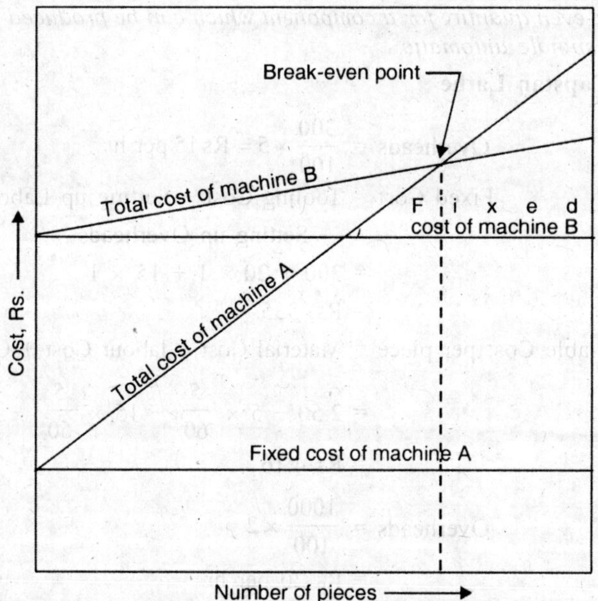

Fig. 5.13. Break-Even Point.

This would give a positive value when *FC* of a process is greater and *VC* less than those of the other process. If both *FC* and *VC* are lower than for the other process, then the latter process is always uneconomical whatever may be the production quantity.

Example 5.30. *The initial cost for machine A is Rs 12,000 and the unit production cost of the machine is Rs 6.00 each. For the other machine B, the initial cost is Rs 48,000 and the unit production cost is Rs 1.20 each. Do the break-even point analysis.*

Solution. At the Break-Even Point,

$$\text{Total Cost of machine } A = \text{Total Cost of machine } B$$

$$\text{Total Cost} = \text{Fixed Cost} + \text{Variable Cost}$$

$$\text{Variable Cost} = \text{unit production cost} \times \text{number of pieces}$$

∴ If 'Q' is the quantity of production at break-even point, then,

$$12000 + Q \times 6 = 48000 + 1.20 \times Q$$

$$Q = \frac{36000}{4.8} = 7500 \text{ pieces}$$

If production does not exceed 7500 pieces, it is more economical to purchase machine *A*. For higher quantity production, the economy lies with machine *B*.

Example 5.31. *The following information is available for two machines :*

	Item	*Capstan Lathe*	*Automatic (Single Spindle)*
(*i*)	*Tooling Cost*	*Rs 300*	*Rs 300*
(*ii*)	*Cost of Cams*	—	*Rs 1500*
(*iii*)	*Material cost per piece*	*Rs 2.50*	*Rs 2.50*
(*iv*)	*Operation Labour Cost*	*Rs 5 per hour*	*Rs 2 per hour*
(*v*)	*Cycle time per piece*	*5 min.*	*1 min.*
(*vi*)	*Setting up Labour Cost*	*Rs 20 per hr*	*Rs 20 per hr*
(*vii*)	*Setting up time*	*1 hr*	*8 hr*
(*viii*)	*Machine Overheads (Setting and operation)*	*300% of (iv)*	*1000% of (iv)*

Find the break-even quantity for a component which can be produced on either the capstan lathe or the single spindle automatic.

Solution. 1. Capstan Lathe

$$\text{Overheads} = \frac{300}{100} \times 5 = \text{Rs } 15 \text{ per hr}$$

$$\begin{aligned}\text{Fixed Cost} &= \text{Tooling Cost + Setting up Labour Cost} \\ &\quad + \text{Setting up Overheads} \\ &= 300 + 20 \times 1 + 15 \times 1 \\ &= \text{Rs. } 335\end{aligned}$$

$$\begin{aligned}\text{Variable Cost per piece} &= \text{Material Cost + labour Cost + Operation Overheads} \\ &= 2.50 + 5 \times \frac{5}{60} + 15 \times \frac{5}{60} \\ &= \text{Rs } 4.16\end{aligned}$$

2. **Automatic**

$$\begin{aligned}\text{Overheads} &= \frac{1000}{100} \times 2 \\ &= \text{Rs } 20 \text{ per hr}\end{aligned}$$

$$\begin{aligned}\text{Fixed Cost} &= (300 + 1500) + 20 \times 8 + 20 \times 8 \\ &= \text{Rs } 2120\end{aligned}$$

$$\begin{aligned}\text{Variable Cost per piece} &= 2.50 + 2 \times \frac{1}{60} + 20 \times \frac{1}{60} \\ &= \text{Rs } 2.863\end{aligned}$$

∴ If Q is the Break Even Quantity, then,

$$335 + Q \times 4.16 = 2120 + Q \times 2.863$$

$$\therefore \quad Q = \mathbf{1373 \text{ pieces.}}$$

(*c*) Another simple approach to do break even point analysis is to compare the cost of production by the two machines, by considering the following items of cost :

1. Time to produce a part
2. Set up
3. Direct Labour Cost
4. The Overhead

Let, for,

I[st] *Machine*	*2*[nd] *Machine*
t = time/piece, min	T = Time/piece, min
o = overhead cost/hr	O = Overhead Cost/hr
l = direct labour cost/hr	L = direct labour Cost
s = set up time, hr	S = set up time, hr
sr = set up rate/hr	Sr = set up rate/hr

For first machine

$$\text{Fixed Cost} = \text{set up Cost} = ss_r$$

$$\begin{aligned}\text{Variable Cost} &= \text{Direct Labour Cost + Overhead Cost} \\ &= \frac{t}{60}(l + o) \text{ per piece}\end{aligned}$$

$\therefore$ $$\text{Total Cost} = ss_r + \frac{t}{60}(l+o).Q$$

Similarly for second machine

$$\text{Total Cost} = ss_r + \frac{T}{60}(L+O).Q$$

$\therefore$ For Break-Even Point,

$$ss_r + \frac{t}{60}(l+o).Q = SS_r + \frac{T}{60}(L+O).Q$$

$$\therefore \quad Q = \frac{60(SS_r - ss_r)}{t(l+0) - T(L+O)} \quad \text{...(5.45)}$$

Example 5.32. *Do the break even analysis for engine lathe and turret lathe, from the following data :*

Engine lathe	*Turret lathe*
$t = 12$ *min*	$T = 5$ *min*
$l =$ *Rs 7 per hr*	$L =$ *Rs 5 per hr*
$o =$ *Rs 4 per hr*	$O =$ *Rs 8 per hr*
$s = 2$ *hrs*	$S = 8$ *hrs*
$sr =$ *Rs 8 per hr*	$Sr =$ *Rs 8 per hr*

Solution. From eqns. (5.45)

$$Q = \frac{60(8 \times 8 - 2 \times 8)}{12(7+4) - 5(5+8)} \cong \textbf{43 pieces}$$

Thus a job of 43 or more pieces should be done on turret lathe.

Example. 5.33. *A semiautomatic turret lathe costs Rs 80,000 and it produces 16 pieces per hour and its operator receives Rs 2 per hour. An engine lathe which costs Rs 32,000 produces 10 pieces per hour and its operator receives Rs 2.50 per hour. Calculate the minimum number of pieces which makes turret lathe more economical.*

Solution. Total Cost of turret lathe = Total Cost of engine lathe

(Fixed Cost + Variable Cost) of turret lathe

= (Fixed Cost + Variable Cost) of engine lathe

$\therefore$ If Q is the minimum number of pieces, then

$$80000 + \frac{1}{16} \times 2 \times Q = 32000 + \frac{2.5\,Q}{10}$$

$$\therefore \quad Q = \frac{48000}{0.125} = \textbf{384,000 pieces}$$

Example 5.34. *The following time data are available for engine lathe and automatic lathe :*

Type of Machine	*Set up time*	*Unit time*
Engine lathe	*15.0 min*	*15.0 min*
Automatic lathe	*90.00*	*1.5 min*

Determine the point at which the automatic lathe will be justified.

Solution. This problem can be solved by machine shop estimator criteria who is primarly interested in time. According to this, the best method is that which produces the parts in the shortest possible time, including the set up time. The unit time decreases as the number of parts increases. At a particular point, an increased set up time for a more complex machine that produces the units in less time will equal the set up time on a less complex machine having a greater unit-run time. Any job-lot size beyond this point will justify the increased set up time of more complex machine. So for comparision,

Set up time of machine A + (unit time of machine A) (pieces)

= Set up time of machine B + (unit time of machine B) (pieces)

For the given problem,

$$15.0 + 15.Q = 90.0 + 1.5\,Q$$

$$Q = 5.55$$

Therefore it will be more economical to use the automatic lathe on job lots of more than five pieces.

(*d*) Some companies use a formula to calculate the break-even point between two machines. The formula is based on known or estimated elements that make up the production costs. The formula is given as :

$$Q = \frac{pP(SL+SD-sl-sd)}{P(l+d)-p(L+D)} \qquad \text{...(5.46)}$$

where Q = quantity of pieces at break-even point

p = number of pieces produced per hour by the second machine

s = set up time required on the first machine, hrs

S = set up time required on the second machine, hrs

l = labour rate for the first machine, Rs

L = labour rate for the second machine, Rs

d = hourly depreciation rate for first machine (based on machine ours)

D = hourly depreciation rate for second machine (based on machine hours)

Example 5.35. *The following data is given for turret lathe and automatic lathe :*

Turret lathe	*Automatic lathe*
p = *10 pieces per hour*	P = *30 pieces per hour*
s = *2 hour*	S = *4 hours*
l = *Rs 4 per hour*	L = *Rs 4 per hour*
d = *Rs 1.50 per machine hour*	D = *Rs 4.50 per machine hour*

Solution. Putting these values in eqn. (5.46)

$$Q = \frac{pP(SL+SD-sl-sd)}{P(l+d)-p(L+D)}$$

$$= \frac{10\times 30(4\times 4+4\times 4.50-2\times 4-2\times 1.50)}{30(4+1.50)-10(4+4.50)}$$

= **86 pieces.**

(*e*) **Process-cost comparison.** For a given job, more than one manufacturing process may be used. The most economical process is that which gives the lowest total cost per part.

Let Nt = total number of parts to be produced in a single run

Q = number of parts for which the unit cost will be equal for each of the two compared methods A and B (break-even point), Fig. 5.10.

T_a = total tool cost for methods 'A'

T_b = total tool cost for methods 'B'

P_a = unit tool process cost for method 'A'

P_b = unit tool process cost for method 'B'

C_a = tool unit cost for method 'A'

C_b = tool unit cost for method 'B'

Equating the total cost of the two methods :

$$T_a + Q.P_a = T_b + Q.P_b$$

$$\therefore \qquad Q = \frac{T_a - T_b}{P_b - P_a} \qquad ...(5.47)$$

Now, Total Unit Cost = Fixed Cost + Variable Cost

$\therefore$ For method 'A',

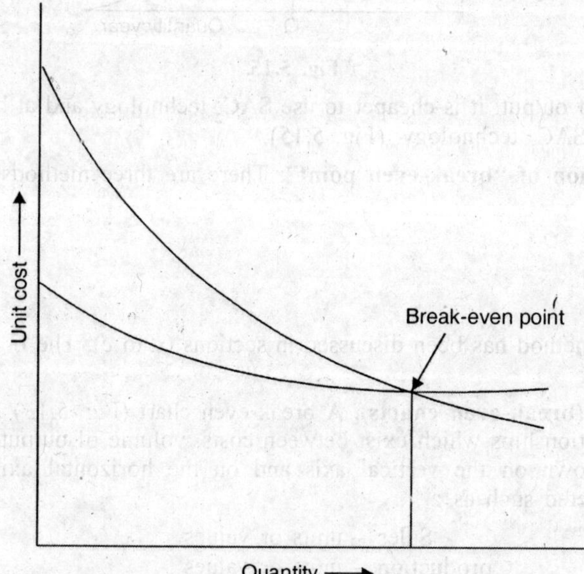

Fig. 5.14. Break-even Point.

$$C_a.N_t = T_a + P_a.N_t$$

$$\therefore \qquad C_a = \frac{T_a + P_a \cdot N_t}{N_t} \qquad ...(5.48)$$

Similarly,
$$C_b = \frac{T_b + P_b \cdot N_t}{N_t} \qquad ...(5.49)$$

Example 5.36. *The aircraft-flap nose rib can be produced either by hydropress or by steel draw die. The following data is available for Nt = 500.*

Pa = Rs 8.40, P_b = Rs 14.80, T_a = Rs 6480, T_b = Rs 1616, Ca = Rs 21.36, C_b = Rs 18.00

Determine the quantity of production at "break-even point"

Solution. From eqn. (5.47) $Q = \dfrac{6480 - 1616}{14.80 - 8.40} = \dfrac{4864}{6.40}$

$= \mathbf{760\ pieces.}$

If SAC = short-term Average Cost

$SAC_1 \rightarrow$ high fixed – low variable cost technology

$SAC_2 \rightarrow$ low F.C. – high V.C. technology

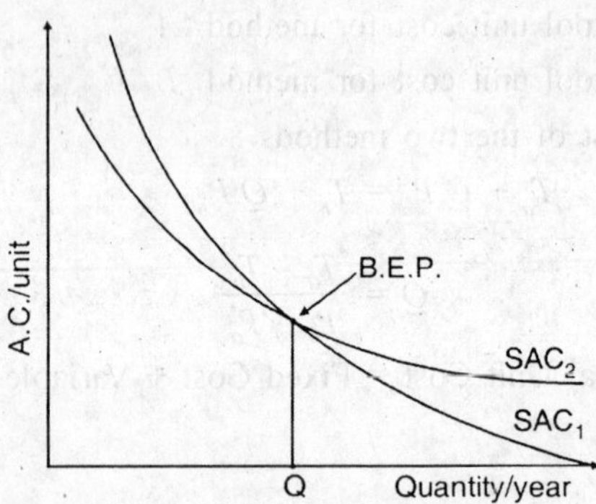

Fig. 5.15.

At low levels of output, it is cheaper to use SAC_2 technology and at high levels of output, it is cheaper to use SAC_1 technology, (Fig. 5.15).

(*f*) **Determination of "break-even point".** There are three methods for determining the break-even point.

(*i*) Algebraic
(*ii*) Graphical
(*iii*) Arithmetic

The algebraic method has been discussed in sections (*a* to *e*). The other two methods will be discussed below :

(*ii*) **Graphical (break-even charts).** A break-even chart (Fig. 5.13) is very useful device as it depicts the relationships which exist between costs, volume of output and profit. To draw charts, costs are shown on the vertical axis and on the horizontal axis a suitable unit of measurement is selected such as :

Sales – units or values
production – units or values
Capacity – production and/or sales shown in %

On the vertical axis, firstly the fixed costs will be entered. Since these remain fixed or constant irrespective of the volume of output or sales, they are shown as a straight, horizontal line parallel to the horizontal axis. Next, the variable costs are entered on the chart. They are put above the fixed cost line and slope upward from left to right. Because they are added to fixed cost, the line drawn becomes the total cost line. If break-even point for sales is to be found out, then the last step will be to insert sales revenue line. This starts from zero sales and goes to the point of maximum sales. Where the sales line and total cost line intersect is the break-even point. Below this point losses occur and above it profits are earned, Fig. 5.16. The vertical width between the sales and total cost lines above the break-even point (B.E.P.) shows how quickly the profit is earned.

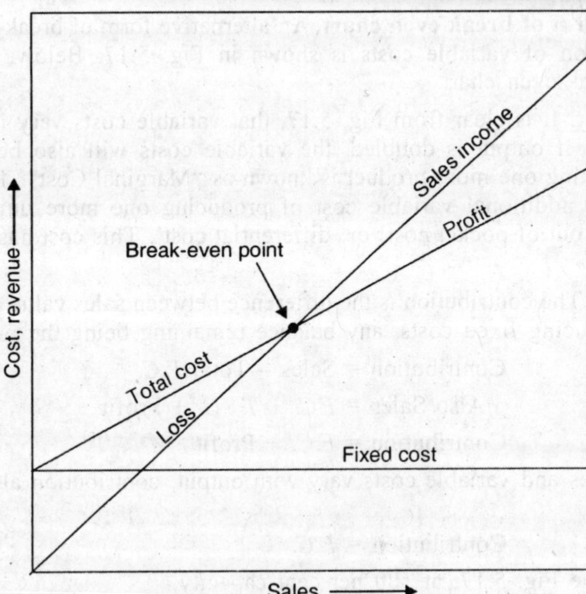

Fig. 5.16. Break-even Chart.

Margin of Safety = Actual Sales – *B.E.P.* Sales income

Greater the margin of safety, higher will be the profits. A company should have a reasonable margin of safety to avoid incurring losses during the period of reduced business activity.

Thus, for a healthy organization, B.E.P. should be as small as possible. There are three ways to achieve this: (see Fig. 5.16)

1. By reducing F.C.
2. By reducing V.C.
3. By increasing S.P.

Figs. 5.16 and 5.17 show the B.E.P. for a single product.

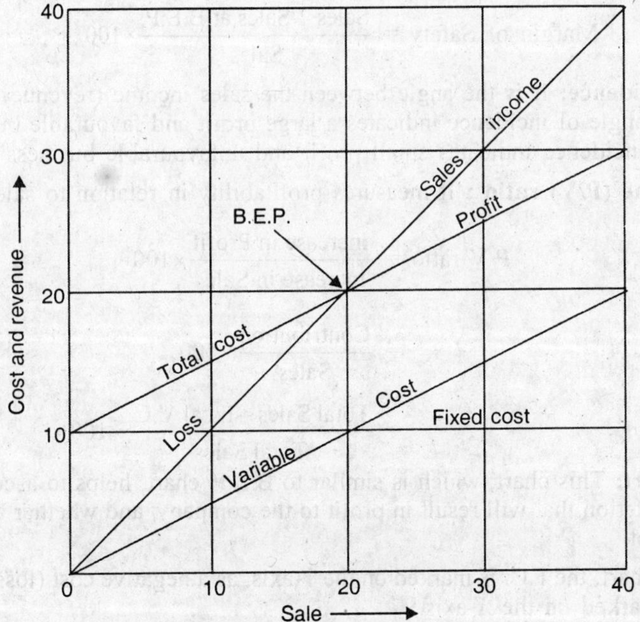

Fig. 5.17. Break Even Chart.

Alternative form of break even chart. An alternative form of break-even chart which also includes the variation of variable costs is shown in Fig. 5.17. Below, we define two terms connected with break-even chart.

Marginal Cost. It is clear from Fig. 5.17, that variable costs vary in direct proportion to the sales or output. If output is doubled, the variable costs will also be doubled. The added variable cost of making one more product is known as "Marginal Cost". Thus the marginal cost of a product is the additional variable cost of producing one more unit. This is also called 'Incremental cost', 'out of pocket cost' or 'differential cost'. This cost has already been defined under Art. 5.1.

Contribution. The contribution is the difference between sales value and the variable costs. It goes towards reducing fixed costs, any balance remaining being the profit. Thus,

$$\text{Contribution} = \text{Sales} - \text{Total } V.C.$$

$$\text{Also Sales} = F.C. + T.V.C. + \text{Profit}$$

or

$$\text{Contribution} = F.C. + \text{Profit}$$

Since both sales and variable costs vary with output, contribution also varies with output. At *B.E.P.*,

$$\text{Contribution} = F.C.$$

Referring to the Fig. 5.17, at 100 per cent capacity,

$$\text{Sales} = \text{Rs } 40{,}000$$

$$\text{Variable Cost} = \text{Rs } 20{,}000$$

$$\therefore \quad \text{Contribution} = \text{Rs } 20{,}000$$

This is used first, to eliminate the fixed overhead costs and then to build up a fund of profit. At total capacity, the profit earned is Rs 10,000.

Some more definitions regarding B.E.P. analysis are given below :

Margin of Safety: It is the distance between the break even point and the output volume. A large safety margin indicates that the company will still be able to earn profit even if there is a significant drop in production volume. With small margin of safety, there will be a considerable decrease in profit if there is a small drop in output volume.

$$\text{Margin of Safety} = \frac{\text{Sales} - \text{Sales at B.E.P.}}{\text{Sales}} \times 100$$

Angle of Incidence: It is the angle between the sales income (revenue) line and the total cost line. A large angle of incidence indicates a large profit and favourable business conditions. A small angle of incidence indicates small profit and unfavourable business conditions.

Profit Volume (P/V) ratio : It measures profitability in relation to sales.

It is given as,

$$\text{P/V ratio} = \frac{\text{Increase in Profit}}{\text{Increase in Sales}} \times 100$$

$$= \frac{\text{Contribution}}{\text{Sales}} \times 100$$

$$= \frac{\text{Total Sales} - \text{Total V.C.}}{\text{Total Sales}} \times 100$$

The P/V chart: This chart, which is similar to B.E.P. chart, helps to ascertain the quantity or volume of production that will result in profit to the company, and whether a new venture will be profitable or not.

To draw the chart, the F.C. is marked on the *Y*-axis, as a negative cost (loss) and the quantity of production is marked on the *X*-axis.

Now, maximum loss will be at zero production and will be F.C. and maximum profit will be when the establishment works at its full capacity. The line joining maximum loss and maximum profit points is known as "contribution line" or "profit line", Fig. 5.18.

The point where this line cuts the *X*-axis is the B.E.P. The slope of the line is called the *P*/*V* ratio or profitability of the product, ϕ. So,

$$\phi = \frac{DB}{AB} = \frac{DC + CB}{AB}$$

$$= \frac{P + FC}{N}$$

where P = Max. Profit and N = Max. quantity of production, (Capacity)

$\therefore \quad P = \text{Profit} = \phi \times N - \text{F.C.}$

Also, at any production q, between *BEP* and *N*, the profit

$P = (\text{S.P.} - \text{V.C.}) \times q - \text{F.C.}$

It is clear from the *P*/*V* chart, that it depicts profit or loss earned by the company at different levels of production.

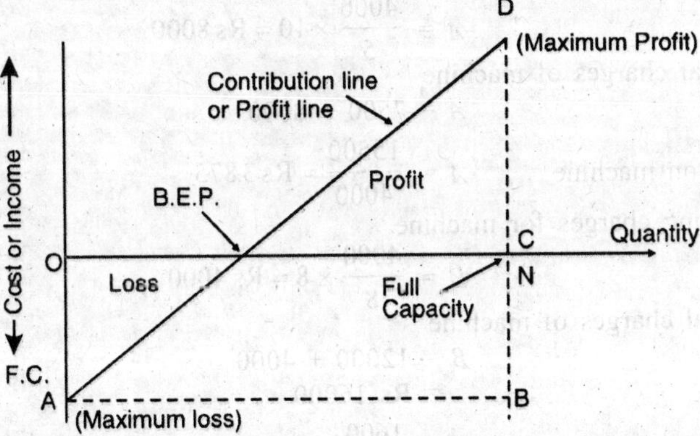

Fig. 5.18. P/V chart

Example 5.37. *A product can be produced by both the machines A and B, for which the data is given as below :*

	Machine A	*Machine B*
Initial cost, Rs	*50,000*	*80,000*
Hourly operating charges, Rs	*10*	*8*
Pieces produced per hour	*5*	*8*

Annual operating hours for both the machines = 2000.

Interest rate = 15%.

(i) Which machine will you prefer ?

(ii) If the annual output is 4000 pieces, which machine would give lower cost per piece ?

(iii) What should be the annual production to make the cost per piece equal for the two machines ? (*I.CAW, 1986*)

Solution. Machine *A*

Annual interest charges (fixed cost) $= \frac{50000 \times 15}{100} = \text{Rs } 7500$

Annual operating charge (variable cost) = 2000 × 10 = Rs 20000

Total Annual charges = *F.C.* + *V.C*

= 7500 + 20000 = Rs 27500

$$\text{Annual out put} = 2000 \times 5 = 10{,}000 \text{ pieces}$$

$$\therefore \quad \text{Cost per unit} = \frac{27500}{10000} = \text{Rs } 2.75$$

Machine *B*

$$\text{Annual interest charges} = \frac{8000 \times 15}{100} = \text{Rs } 12000$$

$$\text{" operating "} = 2000 \times 8 = \text{Rs } 16000$$

$$\text{Total annual charges} = \text{Rs } 28000$$

$$\text{Annual output} = 2000 \times 8 = 16000 \text{ pieces}$$

$$\therefore \quad \text{Cost per unit} = \frac{28000}{16000} = \textbf{Rs 1.75}$$

(*i*) Machine *B* will be preferred.

(*ii*) New annual output = 4000 pieces

∴ Operating (annual) charges for machine

$$A = \frac{4000}{5} \times 10 = \text{Rs } 8000$$

∴ Total annual charges of machine

$$A = 7500 + 8000$$

$$\therefore \text{Cost/piece on machine} \quad A = \frac{15500}{4000} = \text{Rs } 3.875$$

Annual operating charges for machine

$$B = \frac{4000}{8} \times 8 = \text{Rs } 4000$$

∴ Total annual charges of machine

$$B = 12000 + 4000$$

$$= \text{Rs } 16000$$

$$\therefore \quad \text{Cost/piece on machine } B = \frac{1600}{4000} = \text{Rs } 4$$

∴ Machine *A* will be preferred.

(*iii*) Operating cost/piece on Machine

$$A = \frac{10}{5} = \text{Re } 2$$

$$B = \frac{8}{8} = \text{Re } 1$$

$$\therefore \quad 7500 + 2Q = 12000 + 1Q$$

$$\therefore \quad Q = \textbf{4500 pieces.}$$

Example 5.38. *Following data relate to a manufacturing organisation :*

Annual sales (8000 units at the rate of Rs 10 per unit)

= Rs 80,000

Variable expenses = Rs 64,000

Contribution = Rs 16,000

Fixed expenses = Rs 24,000

Losses = Rs 8,000

(*a*) *What sales are needed to break even ?*

(*b*) *What sales are necessary to result in a net income of Rs 9,000, the corporate tax rate being 5.5%.*

(*c*) *What should be the selling price per unit if the B.E.P. is to be brought down to 10,000 units.*

Solution. (*a*) At *B.E.P.* Profit = Loss = 0

$\therefore$ Sales = *F.C.* + *V.C.*

= 24000 + 64000

= Rs 88,000

i.e., **8800 units**

(*b*) Now, Sales = *F.C.* + *V.C.* + Profit

= 24000 + 64000 + (9000 + 0.055 × Sales)

$\therefore$ Sales = $\dfrac{97,000}{0.945}$

= **Rs 10,2645.50**

(*c*) Q = 10,000 units

$\therefore$ $Q \times S.P.$ = 24000 + 64000

$\therefore$ $S.P.$ = **Rs. 8.80 per unit.**

(*iii*) **Arithmetic Method.** The following four steps are followed to determine the break-even point :

1. Separate fixed costs from variable costs.
2. Calculate the percentage of variable costs to sales.
3. Calculate the difference between 100 and the percentage of variable costs.
4. The break-even point is obtained by dividing the fixed costs by the percentage of marginal income.

Example 5.39. *Determine the break-even point from the following data :*

Fixed costs = *Rs 55,000*

Variable cost = *Rs 45 per piece*

Selling price = *Rs 100 per piece.*

Solution. Percentage of variable costs to selling price

$$= \frac{45}{100} = 45\%$$

Profit margin = 100 – 45 = 55%

$$\text{Break-even point} = \frac{55,000}{55\%}$$

= Rs 1,00,000 (minimum sales)

or **1000 pieces.**

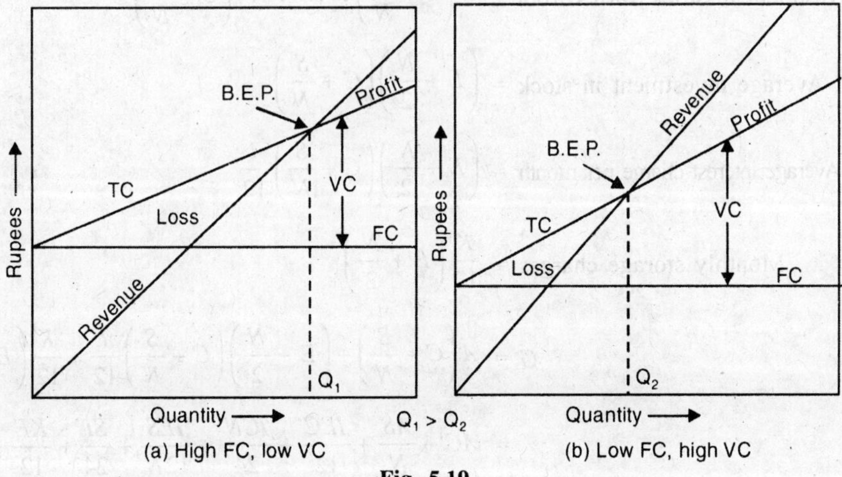

(a) High FC, low VC (b) Low FC, high VC

Fig. 5.19

Referring to Fig. 5.19, it is clear that the case (*a*) requires a large volume of output to reach break even, but once it has attained it, profitability increases rapidly. For case (*b*), profitability after BEP increases slowly.

5.6. ECONOMIC LOT SIZE

In some production units, a machine can be utilized continually throughout the year, to produce the same component. However, very often we are able to produce at higher rate than we can sell. Then the machine must be used to produce other parts for the rest of the year. In other words, the problem which arises is this : if the entire year's supply is produced as a batch, great deal of money will be tied up in inventory and storage costs. If, on the other hand, small lots are produced as needed, costs of setting up the machines will be large. Hence, there should be an optimum lot size. It is obtained so that the unit cost of a part is minimum. The minimum is reached, when the cost of operating the production unit is minimum. Depending upon the number of variables to consider, the formula to determine the economic lot size may range from a simple to a complex one.

Let N = Number of pieces in a batch for minimum costs

A = Average monthly requirements per schedule

C = Cost per piece, including labour, material, over heads, etc.

S = Setting costs, including tools etc.

I = yearly percentage for interest, insurance, maintenance, taxes and depreciation.

E = quantity in stores when new batch is started

K = annual storage charges per piece

G = Total cost of monthly supplies A.

$$\text{Now, cost per part} = \left(C+\frac{S}{N}\right)$$

The number of parts in stock varies from E to $E + N$, so,

$$\text{Average investment in stock} = \left(E+\frac{N}{2}\right)$$

$$\text{Investment in stock varies from} = E\left(C+\frac{S}{N}\right) \text{ to } (E+N)\left(C+\frac{S}{N}\right)$$

$$\text{Average investment in stock} = \left(E+\frac{N}{2}\right)\left(C+\frac{S}{N}\right)$$

$$\therefore \quad \text{Average interest charge per month} = \left(E+\frac{N}{2}\right)\left(C+\frac{S}{N}\right)\frac{I}{12}$$

$$\text{Monthly storage charges} = \frac{K}{12}\left(E+\frac{N}{2}\right)$$

$$\therefore \quad G = A\left(C+\frac{S}{N}\right)+\left(E+\frac{N}{2}\right)\left(C+\frac{S}{N}\right)\frac{I}{12}+\frac{K}{12}\left(E+\frac{N}{2}\right)$$

$$= AC+\frac{AS}{N}+\frac{IEC}{12}+\frac{ICN}{24}+\frac{IES}{12N}+\frac{SI}{24}+\frac{KE}{12}+\frac{KN}{24}$$

To find the minimum value of N,

$$dG/dN = 0$$

$$\therefore \quad \frac{dG}{dN} = -\frac{AS}{N^2} + \frac{IC}{24} - \frac{IES}{12\,N^2} + \frac{K}{24} = 0$$

From here,

$$N^2 = \frac{24\,AS + 2\,IES}{IC + K}$$

$$\therefore \quad N = \sqrt{\frac{24\,AS + 2\,IES}{IC + K}} \qquad \text{...(5.50)}$$

5.6.1. Minimum Cost Analysis. As already noted, the analysis of break-even point or economic lot size can be complex or simple depending upon the factors considered in the analysis. The economic lot size will be one for which the total cost of production per piece is minimum. The total cost will consist of :

(*i*) Fixed cost

(*ii*) Variable cost

(*iii*) Storage carrying cost which will include : the interest on the capital invested in creating the stock, stores overheads, salaries of stores personnel and the cost of deterioration of the stocked parts.

Considering the fixed cost and variable cost only, the total Cost Chart will be as shown in Fig. 5.14, which is reproduced in Fig. 5.20.

This is clear that the variable cost, V, rises uniformly, *i.e.*,

$$V = a.N \qquad \text{...(5.51)}$$

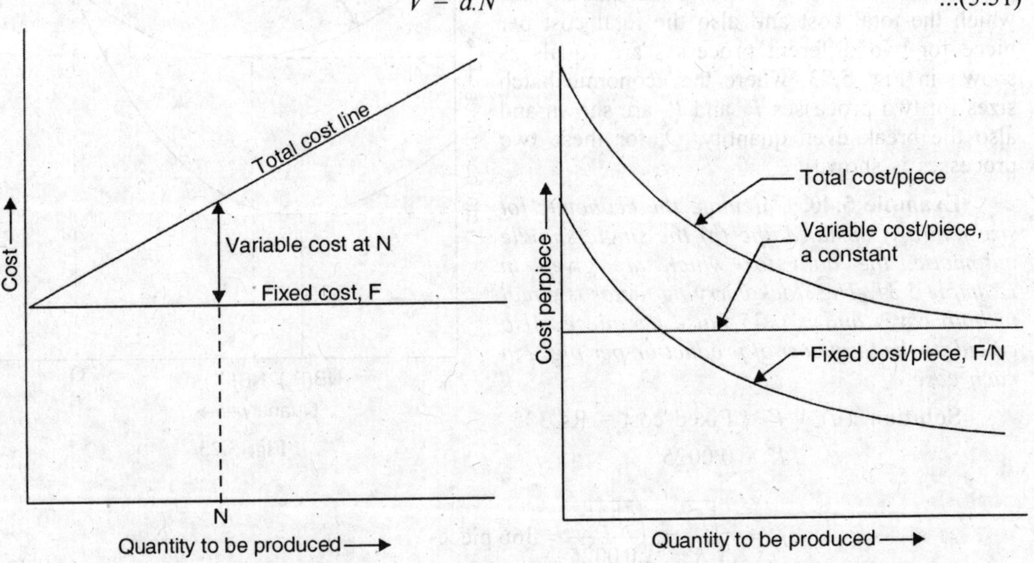

Fig. 5.20. Total Cost Chart.

Fig. 5.21. Total Cost/Piece Chart.

This chart can be converted into total cost per piece chart as shown in Fig. 5.21. Next, the total cost per piece chart has to be modified by including the storage carrying cost which is assumed to vary linearly with N, *i.e.*,

$$\text{Storage Carrying Cost} = KN$$

where K = carrying cost factor, being of different meaning than 'K' used in the last article.

The total cost per piece chart including the storage carrying cost is shown in Fig. 5.22.

The value of N at which the total cost per piece (considering fixed, variable and storage carrying costs) is minimum will give the economic lot size.

Mathematically, Total cost/piece = Fixed cost per piece + Variable cost per piece + Carrying cost per piece

or $$G = \frac{F}{N} + a + KN \qquad ...(5.52)$$

$N = Nb$ (economic batch size)

$$DG/DN = 0$$

$$\therefore \quad -\frac{F}{N^2} + K = 0$$

$$\therefore \quad N = N_b = \sqrt{\frac{F}{K}}$$

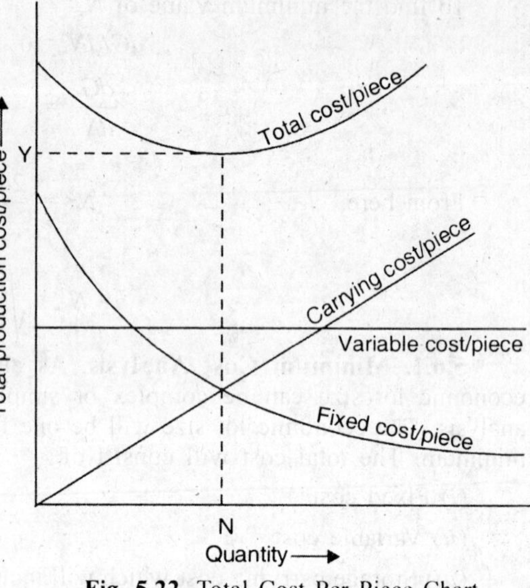

Fig. 5.22. Total Cost Per Piece Chart.

5.6.2. Difference between Economic Batch Quantity and Break-even Quantity. It should be noted that the economic batch quantity and break-even quantity are two different things. Whereas, the former is the batch quantity for which the total cost per piece is minimum for one particular process, the latter is the quantity for which the total cost and also the total cost per piece for two different processes are equals as shown in Fig. 5.23, where the economic batch sizes for two processes P_1 and P_2 are shown and also the break even quantity, Q, for these two processes is shown.

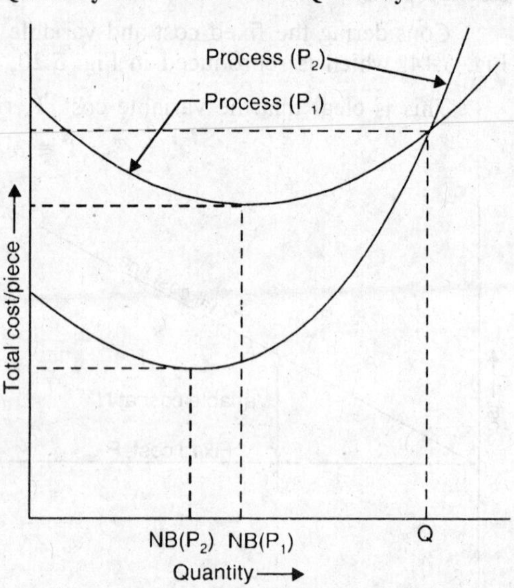

Fig. 5.23

Example 5.40. *Calculate the economic lot size for (a) Capstan lathe (b) the single spindle automatic, the data for which are given in Example 5.31. The stock carrying factor is equal in both cases and is 0.25 paise per piece. Also calculate the total cost of production per piece in each case.*

Solution. (*a*) F = Fixed cost = Rs 335

$K = 0.0025$

$$\therefore \quad N = \sqrt{\frac{F}{K}} = \sqrt{\frac{335}{0.0025}} = 366 \text{ pieces}$$

$$\text{Total cost per piece} = a + \frac{F}{N} + KN$$

$$= 4.16 + \frac{335}{336} + 0.0025 \times 306$$

$$= 4.16 + 0.915 + 0.915$$

$$= \textbf{Rs 5.99}$$

It should be noted that at quantity 'N', the fixed cost per piece is equal to the carrying cost per piece.

(*b*) F = Rs 2120

$$\therefore \quad N = \sqrt{\frac{F}{K}} = \sqrt{\frac{2120}{0.0025}} = 920 \text{ pieces}$$

$$\text{Total cost per piece} = a + \frac{F}{N} + KN$$

$$= 2.863 + \frac{2120}{920} + 0.0025 \times 920$$

$$= 2.863 + 2.30 + 2.30$$

$$= \text{Rs } 7.46$$

5.6.3. Other simple relations for economic batch quantity. (*a*) if the stock is zero, when the new batch is started then,

$$\text{Average stock inventory} = \frac{N}{2}$$

If A = Annual requirement of the parts, then,

$$\text{Total cost, } G = AC + S \times \frac{A}{N} + K \cdot \frac{N}{2}$$

$$\therefore \quad \frac{dG}{dN} = 0 = -\frac{SA}{N^2} + \frac{K}{2}$$

$$\therefore \quad N = \sqrt{\frac{2SA}{K}} \qquad \text{...(5.53)}$$

(*b*) When the quantities are ordered from outside, then,

$$G = C.A + \text{ordering cost} + \text{storage cost}$$

$$= C.A + R.\frac{A}{N} + K.\frac{N}{2}$$

$$dG/dN = 0$$

$$\therefore \quad N = \sqrt{\frac{2RA}{K}} \qquad \text{...(5.54)}$$

where R = ordering cost.

and A/N = number of orders.

Example 5.41. *For a certain inventory item, the following data are available :*

Annual consumption	=	*12,000 units,*
Cost of ordering	=	*Rs 500 per order*
Unit cost of item	=	*Rs 3.00*
Interest rate	=	*20%*
Unit storage cost	=	*Rs 1.50*

Calculate the economic order quantity and the cost. If the storage cost is zero, how would be inventory cost be affected? Discuss.

Solution. Average stock $= \frac{N}{2}$

Average investment in stock $= C \times \frac{N}{2}$

Average interest per year $= C \times \frac{N}{2} \times I$

Storage charges $= K \times \frac{N}{2}$

Number of orders $= \frac{A}{N}$

Ordering charges $= \frac{AR}{N}$

$\therefore$ Total cost per year, $G = CA + \frac{ICN}{2} + \frac{KN}{2} + \frac{AR}{N}$

$$\frac{dG}{dN} = 0$$

$$\therefore \quad \frac{IC}{2} + \frac{K}{2} - \frac{AR}{N^2} = 0$$

$$\therefore \quad N = \sqrt{\frac{\mathbf{2RA}}{\mathbf{IC+K}}}$$

Now R = Rs 500, A = Rs 12,000, C = Rs 3.00, K = Rs 1.50, I = 20% = 0.2

$$\therefore \quad N = \sqrt{\frac{2 \times 500 \times 12000}{2.1}}$$

$$= \mathbf{2390 \text{ units}}$$

$$\therefore \quad \text{Number of orders} = \frac{12000}{2390} = 5.021 \text{ say } 5$$

$$\therefore \quad N = \mathbf{2400 \text{ units}}$$

$$\text{Total cost} = 3 \times 12000 + \frac{3 \times 2400 \times 0.2}{2} + \frac{3 \times 2400}{2} + \frac{12000 \times 500}{2400}$$

$$= 36000 + 720 + 3600 + 2500$$

$$= \text{Rs } 42820 \text{ per year}$$

When $K = 0$,

Then $N = \sqrt{\frac{2RA}{IC}}$

$$\therefore \quad I = \frac{2 \times 500 \times 12000}{3 \times 2400 \times 2400} = \mathbf{0.6994}$$

It is clear that the inventory cost will get increased very greatly.

Example 5.42. *The forecast demand for a particular item is 40,000 units per year. If the ordering costs are Rs 250 per order and the inventory carrying costs are 25 per cent per year of the inventory value, determine the optimum lot size to be ordered. The manufacturer offers the item at the following rates :*

For $0 < N < 1000$*, Cost is Rs 8 per unit*

$1000 < N < 10000$*, Cost is Rs 7.50 per unit*

$N \geq 10{,}000$ *Cost is Rs 7.25 per unit*

Solution. To determine the optimal quantity, we begin on the lowest cost curve, for that

$$N = \sqrt{\frac{2RA}{IC}} = \sqrt{\frac{2 \times 250 \times 40000}{0.25 \times 7.25}} = 3322.$$

Since 10,000 or more units must be ordered to realize a price of Rs 7.25 per unit, $N = 3322$ is not feasible. Computing the total cost at the lowest feasible quantity 10,000, we get

$$\text{Total cost, } G = CA + \frac{AR}{N} + \frac{ICN}{2}$$

$$= 7.25 \times 40{,}000 + \frac{40{,}000 \times 250}{10000} + 0.25 \times 7.25 \times \frac{10000}{2}$$

$$= 290{,}000 + 1{,}000 + 9062.5 = \text{Rs } 300062.50$$

Moving to the next higher curve,

$$N = \sqrt{\frac{2RA}{IC}} = \frac{\sqrt{2 \times 250 \times 40000}}{0.25 \times 7.5} = 3266$$

This is a feasible quantity, since the cost of Rs 7.50 is for a volume of 1000 to 10,000 units. So, total cost for

$N = 3266$ will be

$$G = 7.5 \times 40{,}000 + \frac{40{,}000 \times 250}{3266} + 0.25 \times 7.50 \times \frac{3266}{2}$$

$$= 300{,}000 + 3061.85 + 3061.88 = \text{Rs } 306{,}123.73$$

Moving to highest cost curve,

$$N = \sqrt{\frac{2 \times 250 \times 40{,}000}{0.25 \times 8}} = 3162.28$$

The cost of Rs 8 per unit is for $N > 0 < 1000$, therefore, $N = 3162.28$ is not feasible. Computing the total cost at the first feasible quantity (I) in this range,

$$G = 3{,}20{,}000 + 100{,}00000 + 1 = \text{Rs } 10{,}320{,}001$$

Comparing all total costs, we see that the lowest total cost is Rs 300,062.50 for an order quantity of 10,000.

$\therefore$ $N = 10{,}000$ and No. of orders = **4.**

Example. 5.43. *A manufacture has to produce 50,000 components a year and wishes to select the most economical lot size to be equally spaced during the year. The annual cost on storage and interest on investment varies directly with the lot size and is Rs 1000 when the entire annual requirement is manufactured in one lot. The cost of setting up and dismantling the machine for each run is Rs 30. What is the most economical lot size ?*

Solution. If N is the most economical lot size, then the number of runs = 50000/N.

By proportion, we obtain the following :

Annual storage and interest expense

$$= \frac{1000}{50000} \times N = 0.02 \text{ N}$$

Setting up and dismantling cost

$$= \frac{50000}{N} \times 30 = \frac{1,500,000}{N}$$

$$\therefore \quad \text{Total cost } G = 0.02\,N + \frac{1,500,000}{N}$$

$\therefore$ For minimum value of N,

$$dG/dN = 0$$

$$0.02\,N - \frac{1,500,000}{N^2} = 0$$

$$\therefore \quad N = \sqrt{\frac{1,500,000}{0.02}} = 8660 \text{ components per lot.}$$

$$\therefore \quad \text{Number of lots required} = \frac{50,000}{8660} = 5.77$$

If 6 lots are used, the lot size is,

$$N = \frac{50,000}{6} = \textbf{8333 components.}$$

PROBLEMS

1. Differentiate between "fixed costs" and "variable costs".

2. What is meant by "Depreciation"? Explain the methods to account for depreciation.

3. Define "Book Value".

4. Enumerate the various reasons of machine tool replacement.

5. Define "Break-even Point". Explain its importance in studying the economics of tooling.

6. Discuss the various methods of obtaining the "Break-even Point".

7. What is the significance of "Economic lot size"? How it is obtained ?

8. Which would be the better fixture to employ for a component which is made in eight batches of 1000 pieces each per year :

(*a*) Fixture costing Rs 4000, saving 25 paise on each piece and costing Rs 30 to set up, or

(*b*) Fixture costing Rs 5000, saving 33 paise on each piece and costing Rs 35 to set up.

I = 20%, T = 20%, M = 10% and overhead on labour saved = 50%, are the same in each case.

9. A company is manufacturing brackets in batches of 2000 and usually makes six such batches per year. How much it can afford to be spent on a fixture for this component if it would save $12\frac{1}{2}$ paise per piece, 40% of overhead on labour saved and costs Rs $37\frac{1}{2}$ to set up. It must pay for itself in two years and I = 10%, T = 10% and M = 5%. (Rs 2500.00).

10. The following data were obtained for the production of a part with the help of a special fixture :

Cost of special fixture = Rs 500, I = 6%, M = 10%, D = 20%, T = 5%

Labour cost per piece without the use of fixture = 70 p

Labour cost per piece with of use of fixture = 50 p

Cost of set up of fixture = Rs 20 per year.

Overhead saving = 20% of labour saved

(*i*) Determine the quantity of production per year for "Break-even Point".

(*ii*) If the quantity of production per year is 3000 pieces, determine the profit on the investment of the special fixture. (Rs 495.00)

11. A fixture costs Rs 1000 and lasts for five years. It has to make four runs per year and it costs Rs 20 for each set up. $I = 12\%$, $T = 8\%$. Maintenance cost per year = Rs 50. The labour cost is reduced by 15 p per piece with the use of the fixture. At what rate of production can a profit of Rs 200 per year be made?

12. In considering the machining of component, two schemes are suggested :

(*a*) A turning fixture costing Rs 4000, saving 16 per piece on previous methods costing Rs 30 to set up.

(*b*) A milling fixture costing Rs 6000, saving 28 per piece and costing Rs 70 to set up.

Both save 50% overhead on labour saved and $I = 10\%$, $T = 10\%$, $M = 10\%$ are the same in each case, and 20,000 pieces are made per year. Which will be the better fixture to employ, assuming that the type of component is liable to change or become obsolete in four years.

13. Small steel screws can be made either on capstan lather or on automatic lathe. The data for the two machines is :

Capstan :

Setting time = 70 mins.
Setting cost = Rs 8
Tooling costs = Rs 200
Cost per 1000 pieces = Rs 150

Automatic :

Setting time = 130 mins.
Setting cost = Rs 16
Tooling costs = Rs 320
Cost per 1000 pieces = Rs 68.

Find the number of steel screws that will give equal total cost for both the machines.

14. A work piece can be held in a chuck on a turret lathe, but if a special fixture that costs Rs 400 is used, loading time is shortened by $\frac{1}{2}$ min/piece. The composite rate for interest, insurance, taxes and maintenance is 20%. Set up time is not changed. The rate for labour and overhead is Rs 12 per hour. For how many pieces is the fixture justified?

15. Following data were obtained for the production of a part with the help of a special fixture :

Cost of special fixture = Rs 500
Interest = 10%
Cost of repairs = Rs 50 per year
Depreciation = 20% per year
Insurance, taxes, etc. = 5% per year

Labour cost per piece without the use of fixture = 70 p

Labour cost per piece with the use of fixture = 40 p

Cost of set up of the fixture = Rs 20 per year

(*i*) Determine the quantity of production per year corresponding to the "break-even" point.

(*ii*) If the quantity of production per year is 4000 pieces, determine the profit on investment on the special fixture.

16. A new machine tooled and installed, costs Rs 50,000. It is expected to fetch Rs 2000 after its useful life of $12\frac{1}{2}$ years. The older machine which is proposed to be replaced, has scrap value of Rs 500. The older machine is used to produce 50 pieces per hour for labour charges of Rs 3 and machine cost of Rs 3. On the other hand, the new machine produced 100 pieces per hour for labour and running charges of Rs 4 and Rs 3 respectively. The machines run for 2200 hours per year. I = 15%, T = 5%, M = 6%. Find after how many years the new machine will start giving a net profit.

17. It is proposed to replace and old machine with a new one with installed cost of Rs 40,000. The scrap value of the old machine is Rs 500, and the scrap value of the new machine would be Rs 5000 after a useful life of 12 years. The old machine used to produce 40 pieces per hour and labour charges and running charges for this machine are Rs 3 and Rs 3.50 per hour, respectively. On the other hand, the new machine produces 100 pieces per hour and labour charges and running charges for this machine are Rs 4 and 3 per hour respectively. The machines run for 2200 hours per annum. Assuming interest rate 15%, taxes 5%, and maintenance charges 5%, determine the number of years the new machine will take to render itself profitable.

18. A component can be conveniently manufactured either on a milling or on a broaching machine. Each machine is worked 2000 hours per year and the useful life of each machine is 10 years. Assume straight line depreciation : Find the quantity at break even point for the following data :

	Item	*Broaching Machine*	*Milling Machine*
(*i*)	Initial cost	Rs 1,35,000	Rs 75,000
(*ii*)	Tooling cost	Rs 9000	Rs. 600
(*iii*)	Material cost/piece	Rs 1.50	Rs 1.50
(*iv*)	Labour cost/hour	Rs 10.00	Rs 16.00
(*v*)	Cycle time/piece	1 min	5 min
(*vi*)	Setting up labour cost	Rs 32	Rs 32
(*vii*)	Setting up time	2 hours	2 hours
(*viii*)	Machine overheads/hr.	Rs 120	Rs 75

19. The cost of new fixture which is to replace the old fixture is Rs 500. The percentage of overhead charged to this fixture is 20% and the expected saving per piece is 6 paise. The old fixture had original cost of Rs 300 and its probable scrap value is Rs 75. The fixed costs are : I = 6%, M = 2%, T = 15%, amortization 2 years. Calculate the number of pieces which should be produced to break even so that it may be paid off in one year.

20. If the number of pieces to be manufactured is 2000, how much may be spent for a new fixture ? Use data of problem 20.

21. If the cost of the new fixture is Rs 1500 and the yearly production is 1800 pieces, determine the length of time needed to amortize the cost. Use data of problem 20.

22. Calculate the saving if the fixture in problem 22 produces 2500 pieces per year.

23. What do you understand by Break-even point in production planning? How do you calculate the break-even point (indicate two approaches of finding this)? What is the practical significance of this analysis?

24. How many components would have to be made to warrant putting into use a jig costing Rs 1500, if by its use it saves 25 p per piece over previous methods of machining ? The cost of each set up is Rs 15 and the overhead on labour saved is 50 per cent. I = 10%, T = 4%, M = 6%. Assume that the jig must pay for itself in 18 months.

25. Determine the number of parts to be made in a continous run by a jig which by its use saves 16 p per piece, costs Rs 5000, Rs 40 to set up and the burden applied on labour saved is 50%. I = 10%, T = 12%, M = 8%. The jig must pay for itself out of earnings in two years. If the jig is used to machine the parts in 5 batches per years, find the increased number of parts to be made under the new conditions.

26. A company purchases all metal needed in bar stock form. The annual demand is 3000 units. The purchases ordering cost is Rs 60.00 and storage costs are 20 per cent of the unit cost, what is the optimal order quantity given these cost breaks :

(*i*) 0—299, Rs. 50.00 per unit

(*ii*) 300—499 Rs 40.00 per unit

(*iii*) 500 or more, Rs 30.00 per unit.

27. Make a selection of the machine tool as per data given below :

	Machine A	*Machine B*
No. of working hours/yr	1800	1800
No. of components/yr	10,000	10,000
Time each in min.	10	7
Machine hour rate, Rs	20	25
Tool cost, Rs	850	1750
Tool life in yrs	2	2

28. A technician is planning to open repair booth for refrigerators. The annual rent of the booth, the annual cost of the machinery and tools and overheads are estimated to be Rs 80,000. The average labour and material cost per unit of service are expected to be Rs 60. The average unit selling price is estimated to be Rs 80. Calculate the break even number of service units per month. If he gets 500 service units per month, estimate his profit. (334/month, Rs 40000/year).

29. A company is planning to manufacture a product.

Estimated fixed costs per year = Rs 3 106

" variable cost/unit = Rs 150

Selling price/unit = Rs 250

Expected Sales = 50000 units

Will there be profit or loss? Determine the *B.E.P.* Quantity.

(Profit = Rs 2 106; Q = 30,000 units)

30. A product can be manufactured by two machines A and B, for which the data is given as below :

	Machine A	*Machine B*
Initial cost, Rs	60,000	100,000
Operating cost per hour, Rs	12	10
Production per hour, pieces	6	10

Interest rate is 15%.

The overhead costs of the factory are Rs 1,20,000. It works for 4000 hours in a year.

(*i*) Which machine will you prefer ?

(*ii*) If annual output is 4000 pieces, which machine will give lower cost per unit.

(*iii*) What should be the output per year to make the cost of production per unit equal.

(*ICAW, 1988*) (**Ans.** Machine *B*; Machine *A*; 6000 pieces).

31. What is meant by the term "Discounting"?

32. Define : "Single payment compound amount factor" and "Single payment present worth factor".

33. What is effective rate of interest?

34. Define the following terms :

(*a*) Rate of return

(*b*) M.A.R.R.

(*c*) Life of an asset.

(*d*) Retirement

(*e*) Replacement

(*f*) Defender

(*g*) Challenger.

35. Define : Annuity, deferred annuity.

36. Define the following terms :

(*a*) Uniform Series Compound amount factor.

(*b*) Sinking fund factor.

(*c*) Uniform Series present worth factor.

(*d*) Capital recovery factor.

37. Define the following terms :

(*a*) Cash flow

(*b*) Discounted cash flow

(*c*) Cash flow diagram.

38. Discuss the following methods of Economy Studies :

(*a*) P.W. method.

(*b*) Annual worth method.

(*c*) Annual cost method.

(*d*) *I.R.R.* method.

(*e*) *E.R.R.* method.

(*f*) *E.R.R.R.* method.

39. Discuss the various ways of calculating capital recovery.

40. Why the existing equipment has to be replaced?

41. Define "Economic life" of an asset.

42. What is challenger's adverse minimum?

43. What is Turbogh's plan of equipment replacement?

44. What is MAPI plan of equipment replacement?

45. Machine *A* costs Rs 34000 and its annual operating costs are Rs 18000. The corresponding values for machine *B* are : Rs 28000 and Rs 19200 respectively.

Economic life of each machine = 10 years

M.A.R.R. = 10%

Make your selection by :

(*a*) P.W. method

(*b*) Annual cost method.

46. Determine the economic life of a machine that has an initial cost of Rs 40,000. The estimated operating charges for years 1 to 6 are respectively : Rs 6400, 7200, 8800, 8800, 10400, 11200 and the corresponding salvage values are : Rs 32000, 28000, 24000, 16000, 12000, 8000. The interest rate is 10%.

47. For a production process, the unit variable cost (V) of production increases with the number of units (N) produced but the unit selling price (S) decreases with the number of units sold and are related by the following empirical relations :

$$V = 0.001\ N + 5\ \text{Rs}; \qquad S = 15 - 0.001\ N,\ \text{Rs}.$$

The fixed over head cost of production is Rs. 10,000. Determine the break even output and the output for maximum profit. **(GATE 1992)**. (**Ans.:** 1382 units, 2500 units).

48. A machine is purchased for Rs. 32,000, and its assumed life is 20 years. The scrap value at the end of its life is Rs. 8000. If the depreciation is charged by the diminishing balancing method, then the percentage reduction in its value, at the end of its first year is

(*a*) 6.7% (*b*) 7.2% (*c*) 7.1% (*d*) 7.6 % **(GATE 1997)** (**Ans.** c)

49. Why is it necessary for a manufacturing engineer to have a detailed knowledge of economics of manufacturing ?

50. Discuss the factors which determine the selling price of a product.

51. Classify the following costs into fixed cost and variable cost :-

(*a*) Tooling cost (*b*) Tool grinding cost (*c*) Direct material cost

(*d*) Depreciation (*e*) Salaries and wages (*f*) Investments on machinery and building

(*g*) Electrical charges

52. What is the basic concept involved in break-even point analysis.

53. Write the limitation of break-even point chart.

54. How break-even point analysis may be used for process selection. Explain.

55. What are the different methods of lowering the break-even point. Explain graphically.

56. How is the profit - volume chart different from break-even point chart?

57. How is the P-V chart constructed?

58. A certain part can be either fabricated by welding or manufactured by forging process. The fixed cost and the variable cost for the two methods are given below :-

	Welding	Forging
Fixed cost	Rs 16,000	Rs 95,000
Variable cost	Rs 5	Rs 4

Select the economical process if the demand order is of 500,000 parts.

Solution : (*a*) Refer Art. 5.5 (b) and Fig. 5.13

No. of pieces at B.E.P.

$$Q = \frac{FC_B - FC_A}{VC_A - VC_B} = \frac{95000 - 16000}{5 - 4} = 79000 \text{ parts}$$

Since the order is of 500,000 parts, the forging process will be more economical. However, upto 79000 parts, welding will be the choice.

(*b*) What will be the loss of a wrong choice of the process.

For welding process,

$$\text{T.C.} = \text{F.C.} + 500,000 \times 5$$
$$= \text{Rs. } 25,16,000$$

For forging process

$$\text{T.C.} = 95000 + 500,000 \times 4$$
$$= \text{Rs. } 20,95,000$$

$$\therefore \quad \text{Loss} = \text{Rs. } 25,16,000 - 20,95,000$$
$$= \text{Rs } 4,21,000$$

6

PROCESS PLANNING

6.1. GENERAL

After the product has been designed, the engineer the most complex problem of developing and co-ordinating plans for manufacturing the product. The only information available to him is the part print. Using this information, he must create and follow through a properly sequenced series of operation to convert materials into useful product. He must select the types of tooling and equipment needed to carry it out. He must at the same time be concerned with product quality and manufacturing economy. This function is called "Product Engineering". Before, we study in detail, the various functions of "Prrocess Engineering", we must understand the significance of "Product Engineering" and the role of a "Product Engineer" in the manufacture of a product.

Product Engineering. The product to be manufactured is first conceived by the product engineer. He determines the need for a product. It may be an entirely new product or a new model of an old product. Experimental designs are made, scale models are made and tested. Finally, a production design is created after all the faults have been corrected. Part prints are drawn to illustrate the product graphically. All dimensions and specifications required are included on the print. The material to be used in the product is specified and the product name and number is included. The role of "Product Engineering" is illustrate in Fig. 6.1.

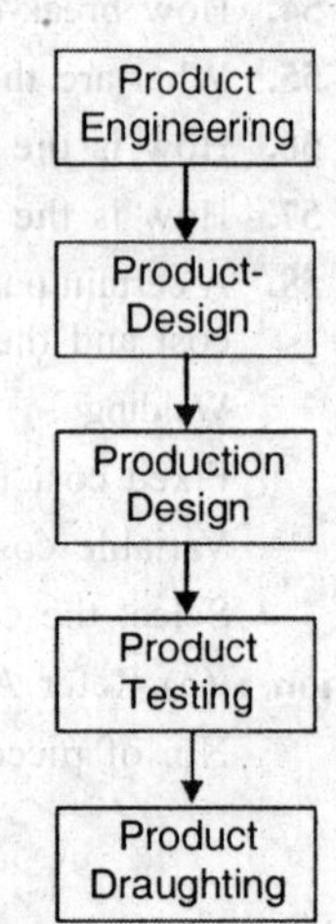

Fig. 6.1. Role of Product Engineering

Process Engineering. Process engineering takes place directly after product engineer has completed the design of a product. It takes the information received and then creates the plan for manufacture. Processing then, is the function of determining exactly how a product will be made. The process engineer will develop a set of plans or directions on part manufacture. He will initiate orders required to put the plan into effect. Functions of process engineering are :

1. To determine the basic manufacturing process to be used.

2. To determine the order or sequence of operations to manufacture the part.

(*i*) operating routing or line up.

(*ii*) process pictures.

3. To determine and order the tooling and gauges needed to manufacture the part.

(*i*) orders to design

(*ii*) orders to build, or

(*iii*) orders, to buy.

4. To determine, select and order the equipment needed to manufacture the part.

5. To determine the need and originate orders for all process revisions necessary when part print changes occur.

6. To follow up the tooling and equipment to determine if all functioning as planned and if not, make the necessary revisions.

7. To provide estimates of the cost of tooling and equipment needed to manufacture new products for the purpose of quotations or bids.

8. To determine part changes necessary to easy manufacture or reduce cost and request part print changes.

9. To take part in product study groups to assist the product engineer in the design of a product that will be feasible and economical to make.

From above, "Process Planning" can be defined in the manner given below :

Process planning can be defined as the systematic determination of the methods by which a product is to be manufactured, economically and competitively. It consists of devising, selecting and specifying processes, machine tools and other equipment to convert raw material into finished and assembled products. The purpose of process planning is to determine and describe the best process for each job so that

1. Specific requirements are established for which machines, tools and other equipments can be designed or selected.

2. The efforts of all engaged in manufacturing the product are co-ordinated.

3. A guide is furnished to show the best way to use the existing or the provided facilities.

6.2. CONTENTS OF A PROCESS PLAN

A process plan should contain those items which clearly show as to how it proposes to solve efficiently the problems at hand. The items to be included in a process plan are :

1. An identification of the purpose of the process. This includes . name and number of the component to be produced, the quantity and lot sizes to be made, a description of the rough material, the model or assembly for which the component is intended, the number of components for each assembly, the effective date of project, the name of the planner and the number of order authorizing the project.

2. List of the operations making up the process, an enumeration of the operations to show their sequence and designation of the place where each operation is performed.

3. Specifications needed to make each operation conform to the principles of interchangeable manufacture and quality control. These items include the locating surfaces and clamping areas on the work piece, dimensions, tolerances, geometric relationship, surface quality any material properties.

4. Specifications of the methods, machines, tools and equipment to produce the required quantity and quality of the component at the lowest cost. These include a description of what is to be done in each operation; the size, type and kine of each machine and its location, accessories and attachments, listing and identification of standard and special cutting and forming tools, fixtures, jigs, dies, gauges and so on, and instructions for the proper set up and operation of the equipment, optimum speeds, feeds and so on.

5. Specifications of performance expected from each operation, in the form of the estimated or standard cycle time per piece, set up time per lot, the out put expected in a certain length of time and the capacity of the equipment.

6.3. PROCESS OPERATIONS

During process planning, before the manufacturing sequence is established, the many types of manufacturing operations should be understood. Certain operations invariable have more influence on the manufacturing sequence than others. In addition, some operations, by their nature, must be performed ahead of others, *i.e.,* casting before machining, machining before hardening and drilling before tapping etc.

1. **Basic Process operations (Founding operations).** These are those operations which give the material initial shape or form prior to the process being planned, *e.g.*, sand castings, forging, bar stock and strip stock.

2. **Principal Process operations.** These operations include all the operations forming the backbone or nucleus of the type of manufacturing, *e.g.*,

(*i*) Cutting (machining)

(*ii*) Forming

(*a*) Hot: forging, rolling, drawing, extruding, spinning, pipe welding etc.

(*b*) Cold : bending, drawing, seqeezing, etc.

(*iii*) Casting and moulding.

(*iv*) Assembly : soldering, brizing, welding, cementing, press fitting, shrink fitting, mechanical fastening.

3. **Major operations.** Major operations are those operations, performed, within the principal process, that may be classified either by the manner in which they must be performed or their importance in the sequence.

Where cutting is the principal process, the major operations will be, turning, milling, shaping, drilling, broaching etc.

For cutting and forming, the major operations may be classified as :

(*a*) Critical operations

(*b*) Secondary operations

(*c*) Qualifying operations

(*d*) Requalifying operations

(*a*) **Critical operation.** Critical operations are those that must be given special consideration in order to accomplish some unique characteristic on or from surface of the workpiece. These areas or surfaces are called "critical areas" and so far as their processing is concerned, fall into two categories : (*i*) Product critical areas (*ii*) Process critical areas.

Critical areas are generally identified through close tolerances, surface conditions and their relationship to other areas as indicated by base line dimensioning. These are the surfaces on the workpiece which are generally best qualified for locating and measuring the workpiece on each of its operations.

(*i*) **Product Critical Areas.** These are the areas on the workpiece where control of the product specification is necessary to the functioning of the product, but may or may not have a direct influence on the dimensional control of other surfaces on the workpiece. Such areas are generally described, through specifications on surface finish, flatness, roundness, concentricity, close tolerance and so on, but are not necessarily used as a baseline for locating other areas of the workpiece for processing.

(*ii*) **Process Critical Areas.** These are the areas or the surface on the workpiece which have a critical relationship to other areas on the workpiece and as such as registering surfaces for the location system.

(*b*) **Secondary operation.** Secondary operations are those within the sequence which the normal in the normal sequence of processing the part, but which are less than critical in importance. These have a functional purpose on the workpiece but are generally performed to standard part print tolerance. No special effort must be made to accomplish them. Drilling and tapping, for example, incorporate two separate secondary operations in a sequence. In the normal sense, tapped holes are not held to usually close tolerances nor are they used for locating further detail on the workpiece. Thus, they are non-critical in nature and require no special treatment. They may occur either before or after critical operations in a sequence, depending upon their influence in the process.

(*c*) **Qualifying operations.** Where castings and forging are used, certain preliminary steps may be required in order to get the workpiece "out of rough". Operations thus performed on the workpiece to establish qualifying locating surfaces prior to accomplishing process critical areas the called "Qualifying operations".

(*d*) **Requalifying operations.** During the course of processing, certain operations performed on the workpiece may cause it to change its shape to the extent that original surfaces may have to be re-established before continuing the sequence. A requalifying operation is one that is performed the sequence. A requalifying operation is one that is performed on the workpiece in order to return it to its original machined geometry. Relief of casting stresses, sometimes causes surfaces to wrap. Handling and clamping may damage or destroy locating surfaces and heat treating operations frequently cause distortions in surfaces or natural centre lines. If such operational disturbances can be predicted, additional stock allowances should be made in order to ensure sufficient material to redefine them.

4. **Auxiliary Process operations.** In the succession of operations performed within a given principal process, the sequence of major operations is occasionally interrupted by the need to borrow from other processes. These borrowed processes may or may not qualify as principal processes and frequently are found accompanying other principal processes as a part of the sequence of manufacture. Auxiliary process operations are those necessary to ensure continuity and completion of the principal process operations. They generally change the physical characteristics or appearance of the workpiece. Some important, auxiliary process operations are, welding, heat treating, straightening, cleaning, finishing, shot peening etc.

5. **Supporting operations.** These are the operations which are necessary to the successful completion of the product. Such operations commonly accompany all principal process operations. The major supporting operations are : Shipping and receiving, Inspection and quality control, Handling and Packaging.

6.4. STEPS OF PROCESS PLANNING

The problems met in planning a process can be solved by proceeding along well-defined steps. The procedure may differ slightly with each individual, but in general, the steps of process planning are as follows :

1. **Requirements and conditions of the process.** Before any problem can be solved, its requirements and conditions must be defined. In process planning, these things include

(*a*) The specifications of the finished product.

(*b*) The size, shape and other properties of the raw material.

(*c*) The quantity or number of pieces of the product to be made and date of its delivery.

The specifications of the product can be obtained from : part and assembly prints, an engineering release for production, a manufacturing order and a list of parts or materials. If the quantity of the product to be manufactured is not given in the product specifications, it can be obtained from the sales department or the management. Similarly, if the form of the raw material is not given, the tool engineer may be required to study and choose from several possible forms of raw material. The process engineer should make a mental or written list of every item in the specifications to obtain a full grasp of the project. The process engineer must pay special attention to the form of the component; the raw material and its from, *e.g.*, bar stock, casting or forging; surfaces to be finished, quality of finished surface (surface finish), hardness and tolerance requirements, individual dimensions, geometric relationship among surfaces, special processes (surface painting or plating), assembling and other requirements.

2. **Improvements of the Specifications.** Merely noting down the various items of specifications is not enough. All the specifications must be clear and explicit. If these are confusing, ambiguous or incomplete, the tool engineer has a right to insist that they be corrected. Faulty specifications include : Overlapping dimensions, omitted tolerances or indefinite notations for surface finish quality. The process engineer consults with the product

designer to clarify all points which are not explicit. It is the responsibility of the process engineer to know the function of each component and specification and to look for and to investigate all possibilities for improving the part design to reduce production costs. A capable process engineer is often able to suggest changes in the design of a product for production ease and economy without changing the functional requirements, as he is in close touch with the problems of production. Such suggestions may range from a complete change to a minor alternation in the design of a component. The right to accept or reject changes belongs to the product designer (engineer) who is responsible for the performance of the product.

Tool Engineering. It is concerned with planning the process, supplying the tools and co-ordinating the facilities for economical manufacturing. Thus it contains two main activities : process planning and tool designing.

Process Engineering. It concerns determining the method of manufacturing a product, establishing the sequence and type of operations involved, selecting the tools and equipment required, and analyzing how the manufacturing of the product will fit into the facilities.

3. **List the basic operations.** Under this step, the basic operation required to satisfy the specific surface relationship are listed. This listing is in no particular sequence and is merely the first recognisation of the basic operations required to manufacturing the product.

4. **Determine the most practical and economical manufacturing method.** In this step, the basic operations listed under step (3) are examined to determine the most practical and economical methods to manufacture the product. Then the best selected method is that which can produce the part at the lowest overall cost. The overall cost includes materials, tooling, direct and indirect labour and overhead. The most economical method is decide by calculating and comparing the total costs for two or more feasible methods. In some cases, the past experience may be helpful but that approach is not dependable.

Just as the product engineer may have several materials to choose from when he designs the part, the process engineer may have many processes to choose from when planning its manufacture. Ordinarily, the principal process (the main process by which the part is to be produced) is not difficult to determine once the method by which the workpiece was originated is known. It is quite obvious that a forging must be machined to produce the final geometry required on the part print, a sheet metal part may require a series of stamping operations or a product made up from several fabricated parts must be assembled.

5. **Selection of Equipment.** The selection of the correct equipment is closely related to the selected process of manufacture. In fact, it is difficult to separate one from the other. However, there is a major difference between the selection of a process and the selection of a machine. Machines generally, represent long-term capital commitments, whereas, process may be designed for relatively short duration. The following factors must be taken into account while selecting a correct equipment :

(*i*) the size and shape of the workpiece,

(*ii*) the work material,

(*iii*) the accuracy and surface quality required,

(*iv*) the quantity of parts and the sizes of lots required, and

(*v*) personal preferences.

If number of surfaces are to be machined on a part, the choice is offered of machining them separately, all together or in various combinations. If surfaces on a part are similar in shape and size, they are better suited to being treated in one operation that if they are different from one another. More powerful machines may be needed to work hard material than soft material. Workpiece size and dimensions may dictate particular features that a machine tool must have. Small workpieces are handled on equipment different from that used for large parts. As an example, small and medium size parts are turned on horizontal lathes, but short pieces of large diameters are commonly machined on vertical lathes. Small tolerance call for certain types of equipment, whereas large tolerance are not so exacting.

Generally speaking, very large parts must be produced by slower methods, because, of difficulty in handling. As a result, they must be produced on the larger and slower machines, mainly on a tool room basis. Smaller parts whose shape is conducive to ease of handling can be produced on faster machines and are more readily adapted to mass production. There are exceptions, of course, for example, in the pressed metal industry, production techniques have been developed to the point where large sheet metal parts can be produced at relatively high speeds. Generally speaking, however, the size and shape of the workpiece associates itself closely with the size and type of machine required to produce it. Choice may be made between general purpose machine (centre lathes, planners, shapers, drill presses etc.) and special purpose machines. The general purpose machines have the following characteristics :

(*i*) Usually less initial investment in equipment

(*ii*) Greater machine flexibility.

(*iii*) Fewer machines may be required.

(*iv*) Less maintenance cost.

(*v*) Less set up and debugging time.

(*vi*) Less danger of obsolescence.

The special purpose machines have the following characteristics

(*i*) Uniform product flow.

(*ii*) Reduced in-process inventory.

(*iii*) Reduced manpower requirements.

(*iv*) Reduced factory floor space.

(*v*) Higher output.

(*vi*) Higher product quality.

(*vii*) Reduced inspection cost.

(*viii*) Reduced operator skill requirements.

The machines and equipments that will do a job at the lowest total cost are the ones that should be selected. Direct, overhead and fixed costs should all be considered. Generally, more items put into one operation, the less the handling time, the more the chance for simulation and the lower the direct costs. But, the operation is likely to become more complex, calling for expensive equipment. As a rule, a high rate of production justifies a large investment in equipment to reduce direct costs.

To select a machine tool, an investigation must be made to ascertain the aptitude, range and capacity required for the job. Each type of machine is best suited for certain kind of work : lathes for turning, drilling and boring machines for holes. A machine must have adequate range and capacity for the work it is to do, but not an excessive amount at unnecessary expense. The factors determining the range and capacity may be the size of the workpiece, the working area, length of stroke or other motions, speeds and feeds, forces and energy or power required.

Personal preference or specific conditions may influence the selection of a machine tool. A particular type or make of machine may be favoured because a person in the past found it dependable, easy to operate, safe and accurate. Often a new machine is not purchased if one almost as good is already in the plant and not fully loaded.

6. **Combine the operations and put them in proper sequence.** The purpose of this step is to combine the basic operations which have been already determined and put them in the best sequence. In general, as many basic operations should be combined as are practical and economical. Simple operations can usually be done singly with low cost tooling. For combined operations, more complex and expensive equipment is justified when it results in a saving for each piece produced. Operations can be combined in two ways : by simulation and by integration.

Simulation. This involves those combinations where two or more elements of an operation or two or more operations are performed at the same time, for example, drilling a series of holes simultaneously using a multiplc spindle drill head.

Integration. Where several individual elements of an operation or group of operations are combined in succession but not simultaneously, the performance is said to be integrated. It is obvious that simulation is desirable but is not always possible from practical point of view, *e.g.*, a hole must be drilled before it is tapped. However, operations can be combined to follow a sequence without requiring additional loading and `unloading time or additional set ups, *e.g.*, multispindle automatics.

Advantages of combining operations are :

(*i*) improved accuracy.

(*ii*) reduced labour cost.

(*iii*) reduced plant fixed cost.

(*iv*) less tooling required.

(*v*) less handling required.

(*vi*) fewer set ups.

(*vii*) smaller in-process inventory.

(*viii*) less scrap.

(*ix*) fewer inspection points required.

The disadvantages of combining the operations are :

(*i*) maintaining tool accuracy.

(*ii*) possible higher tool cost.

(*iii*) maintaining dimensions from several base lines.

(*iv*) combining tooling subjected to downtime.

(*v*) more costly set ups.

(*vi*) more costly scraps.

(*vii*) compromise on operation speed.

(*viii*) chip disposal.

The key to determining a good manufacturing sequence lies in the selection of suitable terminal points. The terminal points (operations) must be established before the balance of the operation sequence can be planned. Knowing the condition of the material as it is received (which is stated in the material specifications) is vital to setting up the initial machining operations and establishing control of workpiece.

The following steps lead to the proper sequence of operations :

(*i*) Determine the primary areas for locating and gauging.

(*ii*) Set up the primary manufacturing operations.

(*iii*) Set up the secondary operations, and arrange them in their best apparent order.

(*iv*) Set up and insert the necessary allied operations, such as burring, washing, and heat treating in the sequence.

Primary areas are those best suited for consistent locating and gauging throughout the process. Machining should begin from a surface which is to be employed as a locating surface in locating the workpiece in the process of its manufacture. This first surface is to be machined to an accuracy that will ensure the required accuracy of location for all the subsequent operations. The order in which the subsequent operations are performed should be the reverse of the accuracy required in them, that is, (*i*) First operation in sequence should be one in which largest layer of metal is removed. This has the following advantages :

(*a*) It reveals the internal defects in a casting.

(*b*) The work piece is relieved of internal stresses which eliminates the danger of warping in subsequent operations.

(*c*) Thick layer removal requires large cutting forces and so some machine tools can be assigned for such roughing operating only.

(*ii*) The next to be performed are the finishing operations and finally the fine finishing operations.

(*iii*) Roughing and finishing operations should be done on different machines.

(*iv*) Inspection stages should be introduced between

(*a*) roughing operations.

(*b*) before operations which are to be performed in other departments.

(*c*) before laborious and important operations.

(*d*) after the last machining operation.

(*v*) Surfaces whose machining will not greatly affect the rigidity of the work should be machined earlier in the sequence.

(*vi*) The sequence of machining operations should be coordinated with heat treating operations performed in the process of manufacture.

The division of the machining operation into roughing and finishing operations is done due to the following reasons :

1. There is always a certain amount of distortion of a machined surface due to redistribution of internal stresses in the workpiece as stock is removed and also as a result of clamping the workpiece for machining. If a surface requiring high accuracy is finish machined in a single operation only, it will invariably lose its accuracy due to redistribution of internal stresses caused by the subsequent machining of other surfaces. Moreover, this surface may get damaged while clamping the workpiece or in handling when the workpieces are delivered from one operation to another. The thinner the layer of metal removed in machining, the less the distortion due to redistribution of internal stresses. In a finishing operation, that is, in final machining operation, very small machining allowances are removed and no appreciable distortion of the workpiece is observed.

2. It is especially important to divide, the process into roughing and finishing operations in machining parts which are insufficiently rigid.

3. Such a division of machining operations, not only enables the equipment to be more efficiently employed but also allows the features of different machining methods to be utilised. For example, a large amount of stock is removed in rough machining, but high accuracy is required. Hence, roughing operations can be performed on machine tools capable of cutting a heavy chip of large section. Finish machining, whose purpose is to produce surfaces of specified accuracy, can be performed on other machine tools and by other methods which are capable of ensuring the required accuracy. For instance, roughing and semi-finishing of cylindrical surfaces can be done on lathes and the final finishing operations on a grinder. This will provide best results as a whole, both with respect to praduction capacity and to accuracy.

However, parts with small machining allowances and which are sufficiently rigid and have undergone stress relieving operation, can be finish machined without roughing it if no high accuracy is required. Also, in machining large housings, division into roughing and finishing operations is frequently undersirable because of the difficulty of setting up such parts in a machine tool. But, if machining accuracy requirements are high, a division into roughing and finishing operations is inevitable even for such parts.

When the machining operations are divided into roughing and finishing operations, each surface of the part acquires its final shape and size not at once, but is gradually changed, at the same time as other surfaces undergo similar changes. Each surface is machined several times in different operations, each preceding operation preparing the surface for machining in the next operation. In passing over from one operation to the next, the accuracy of the surface is gradually increased; the accuracy of its co-ordination to other surfaces is also increased.

The position of the secondary operations in the sequence is dictated by logic more than specific rule, for example, the sequence of drilling followed by reaming, drilling by spot facing, turning by threading or milling by grinding, are all determined by the only practical order in which they can occur. However, some operations may be independent from other and can be placed into the sequence where most convenient.

Thus, the best sequence of manufacturing operations is determined by both, the degree of control which can be maintained throughout the process and logical process order.

Figure 6.2 shows how the various types of operations fit together into a complete sequence Explanation of the diagram :

As the material is received from its basic process it becomes associated with one or more of the principal process operations. If for example, the material is received as a forging or casting, it must be completed by machining. Thus, machining becomes the principal process to follows. If the material from the basic process operation is in the form of pig iron, then casting

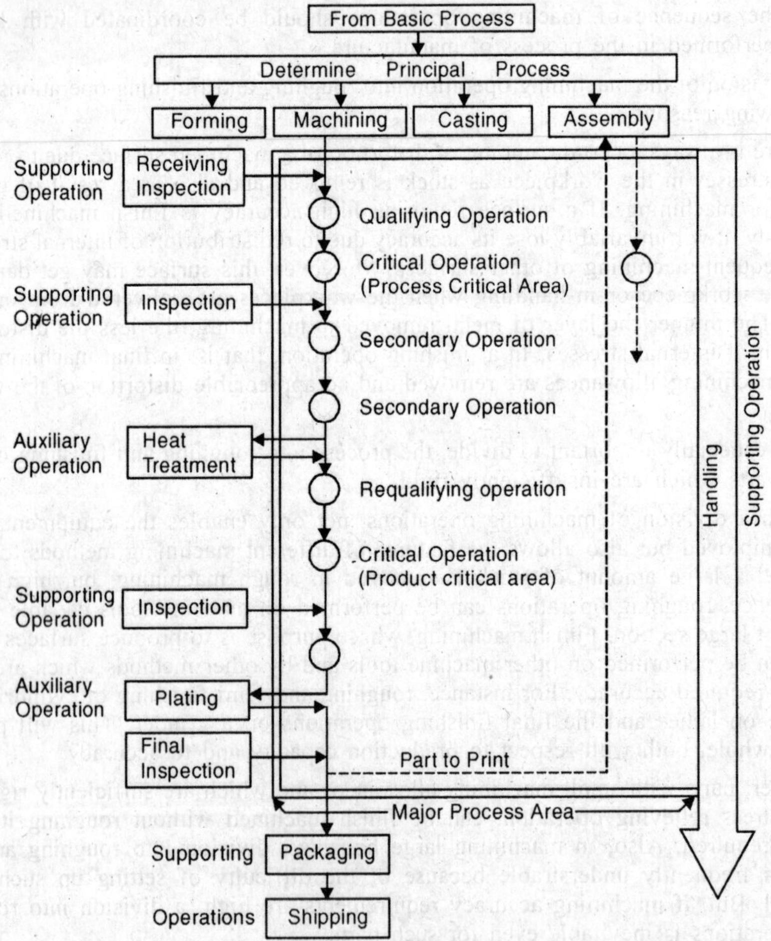

Fig. 6.2

becomes the principal process of manufacture. It is interesting to note that if casting is the principal operation, it would ordinarily be recycled through an other principal process operation, *i.e.,* machining. The diagram shows the operation sequence for machining as the principal process operation. The sequence is supplemented by various auxiliary and supporting operations. If the part is to become a part of an assembly as a continuation of the manufacturing process, then assembly becomes a principal operation (shown by dotted line).

After establishing the sequence of operations as above, it should be reviewed. There is always some scope of improvement. It is desirable to review the sequence of operations continuously. If a new plant is to be set up for a product, the process planner has a free choice. But if a product is to be manufactured in an existing plant, the operation sequence is to suit the equipment available and the present loading conditions. To establish an operation sequence, the process planner should have the following information :

(*i*) List of available machines

(*ii*) List of available general purpose tooling.

(*iii*) Capability of equipment.

(*iv*) Machine load charts.

(*v*) Plant layout.

(*vi*) Standard data.

7. **Specify the gauging required for the process.** The last step in the procedure of process planning is to specify the gauging required to maintain the quality of the product manufactured by the production process just completed. Gauging must be adequate to maintain the part to the print specifications and assure its proper function. The gauging equipment is selected and specified by the process engineer in collaboration with quality control department. However, the design of the special gauging equipment is within the province of tool design department.

6.5. HOW PROCESS PLANS ARE EXPRESSED

A process plan is a complete concept of a process. It is recorded and transmitted in a number of ways to suit various conditions. In a small plant or where skilled workers may be relied upon to perform without detailed instructions process plans may be recorded quite incompletely. In a large organisation with a complex product and highly refined procedures, process plans may be recorded in minute detail.

OPERATION PLANNING SHEET

<table>
<tr><td colspan="3" rowspan="2">PART NAME...........</td><td colspan="3">MATERIAL...........</td><td rowspan="2">PART NO........</td></tr>
<tr><td colspan="3">MATERIAL SPECIFICATION...........</td></tr>
<tr><td>Operation No.</td><td>Description of operation</td><td>Machine</td><td>Tools</td><td>Jigs</td><td>Gauges</td><td>Time Analysis</td></tr>
<tr><td></td><td></td><td></td><td></td><td></td><td></td><td></td></tr>
</table>

Fig. 6.3. Operation Planning Sheet.

A process planning medium almost universally used is the routing, also known as route sheet, process sheet, operation planning sheet and so forth, that lists and describes and operations of a process. Routings are written as briefly as possible to save time and they completely designate departments, machines, tools etc. The operation sheet form will vary for different companies. However, the description of the operation special instructions is usually similar. An operation planning sheet is shown in Fig. 6.3.

6.6. PLANNING AND TOOLING FOR LOW COST PROCESSING

Direct, indirect and fixed costs must all be taken into account in planning a process. A low direct cost may be realised with labour saving equipment involving high fixed cost. On the other hand, direct cost may be high when cheap equipment is used. In each case, that balance of the components must be found that results in the lowest total cost. In addition, a process must be planned and tooled so that all costs are as low as possible, whatever relationship among them.

1. Operation analysis for low direct costs. An operation is analysed by being broken down into elements. Each element is studied to find out how it can be done best. The tested element is studied to find out how it can be done best. The tested elements are put together in the most efficient way and facilities are provided to carry out the operations as planned. For convenience, the elements of an operation may be classified as handling elements and machine elements. The former are done by the operator, such as loading, unloading and clamping. The latter elements are those which are performed by the machine and tools. The major direct costs arise from these two activities, in most of the operations.

Direct handling costs

Direct handling costs can be reduced by reducing, eleminating or combining the operation elements. For this, the following principles are used :

1. Motion economy
2. Prevention of undue strain and fatigue.
3. Alleviation of heavy manual labour.
4. Conservation of skill.
5. Combination of operation below, in short :

A. Motion Economy. The principles of motion economy of most importance to tool engineering are :

(*a*) Eliminate all unnecessary motions.

(*b*) Shorten and simplify all necessary motions.

(*c*) Balance the work.

(*d*) Eliminate the use of eyes.

(*e*) Eliminate the use of hands as holding devices.

To eliminate the unecessary motions", the following rules are recommended :

(*i*) Replace hand motions by automatic machine motions, where practical. This is known as the principle of transfer of thought or human attention to machinery. The example can be of an automatic screw machine, for which constant observation and manipulation by the operator are not required to assure continuous performance.

(*ii*) Replace hand motions by foot or knee motions. This is commonly done on punch presses which are tripped by foot pedals.

(*iii*) Eliminate the necessity of passing work, tools and controls from one hand to another.

(*iv*) Combine motions by providing controls with multiple functions. Also combine two or more tools into one, wherever possible. A common example of first rule is a control with a dual function on a punch press. There, the lever which clamps the work holding jaws is interconnected with the tripping mechanism. As soon as the jaws lock the work in position, the press is tripped and the tool is forced downward. Common examples of two tools combined into one are : double end wrenches, combination tack hammer and puller and combined pliers and wire cutter. A dual-purpose cutting tool is the step drill, used for mass production of holes which must be drilled and counterdrilled or drilled and countersunk.

(*v*) Use mechanical ejectors (Fig. 1.34).

(*vi*) Use drop discharge chutes and the push through idea.

(*vii*) Aid locating by means of slides, guides, flanges, stops, bell mouthed holes and bullet-nosed pins, etc. These help direct parts into proper alignment, thereby saving a finite amount of time.

(*viii*) Preposition tools. Prepositioning a tool to bring it into proper position for use often involves a set of motions during each cycle of an operation. These can be totally or partially eliminated by an arrangement to hold the tool in a predetermined place so that when needed it may be grasped in the position it will be used.

(*ix*) Provide for fast feed of tools or carriage upto work and fast return to unloading point.

"To shorten and simplify necessary motions" the following rules apply :

(*i*) Make necessary motions as short as possible without crowding the operator. Provide for a small work space and keep movement within the workspace.

(*ii*) Arrange to get new work from tray, hopper or chute as close as possible to loading point or from location close to discharge point so as to overlap movements.

(*iii*) When possible keep hand motions within radius of forearm pivoted from elbow. In all events keep within radius of full arm without body bend or twist and without necessity for stepping to reach point desired.

(*iv*) Eliminate barriers so that movement can follow shortest path.

(*v*) Cluster and centralize all control liners and starting buttons within the normal working area.

(*vi*) Make controls "quick-acting" for high production. For example, instead of screw clamps, use hand operated toggles or cames for most case, or supplemented by air or hydraulic means when heavy pressures are required.

(*vii*) Design machines and tools as much as possible to shed oil and dirt by eliminating flat surfaces.

(*viii*) Make locating surfaces as readily cleanable as possible consistent with durability. Small locating areas are easier to keep clean (see Chapter 1).

(*ix*) Provide sufficient room for chips and trimmings and make these places easily accessible for cleanout (See chapter 1).

(*x*) Provide compound lines large enough to wash chips and turning off fixtures, or position air lines to blow off chips. This is to relieve the operator from having to brush them away.

"To balance the work" use the following rules :

(*i*) Avoid idleness of one hand.

(*ii*) Keep both hands busy with useful work.

(*iii*) Provide double station fixtures.

"To minimize use of eyes" the following apply :

(*i*) Eliminate hard to find controls. (like small buttons).

(*ii*) Aid positioning of work by means of slides, guides, flanges, stops, bell-mouthed holes and bullet pins etc. Preposition tools.

(*iii*) Keep necessary eye use within small space (about within a 15-cm circle, if possible).

(*iv*) Provide definite location for loose tools.

"To eliminate use of hands for handling".

(*i*) *Eliminate use of hands for holding machine parts* : For machine starting, feed or clamping levers, detents or cam locks are automatically applied to hold the levers in position after they are engaged so that the hand may release them. However attention has to be paid to rapid transverse levers on machine tools, otherwise the work may sun into the cutter at too rapid a rate with disastrous results.

(*ii*) *Eliminate use of hands for holding work* : For this, take the help of jigs and fixtures, which act holding as well as locating devices. The clamping devices should be self-locking.

B. Prevention of undue strain and fatigue. Strains hasten fatigue which slows down workers and brings about lags in production. The effects of undue strain and fatigue are mitigated by :

(*i*) Avoid necessity for worker to assume uncomfortable position. Most of the work should be at elbow level when seated and 15 cm below elbow when standing.

(*ii*) Provide sufficient space to allow worker to stand up to machine as he should. If seated, provide knee clearance so operator may sit close to working area.

(*iii*) Where feasible, arrange so that machine can be operated equally well from standing or sitting position. This can be accomplished by arranging the work at 15 cm below elbow height when standing, then by providing a high "posture chair" so that the operator may sit and work at proper levels. When foot controls are used, duplicate upper and lower pedals may be necessary.

(*iv*) Avoid circumstances which make it necessary for the worker to walk from one station to another.

(*v*) Provide means to position workpiece conveniently for the operator.

(*vi*) Illumination of proper quality and intensity should be provided.

(*vi*) Illumination of proper quality and intensity should be provided.

(*vii*) The colour of work place should aid visual perception and reduce eye fatigue.

(*viii*) Provide for proper ventilation, temperature and humidity control.

"To conform to physiological traits" use the follow :

(*i*) Build foot pedals so that they may be operated with comfort and by either foot.

(*ii*) Where possible levers such as feed handles on drill presses and arbor presses should be made reversible so that they can be operated by either the right or left hand.

(*iii*) The operator who is seated can use both feet to operate pedals.

(*iv*) Build controls of proper size, shape, and weight, and build to operate without undue effort.

(*v*) Levers, clamps and controls should be so placed that they can be moved by the operator with the most mechanical advantages and the least change in body position.

C. Alleviation of heavy manual labour. Heavy labour increases fatigue and narrows the field of suitable labour supply. Its requirements can be decreased by :

(*a*) Transfer of power, whereby means are provided for applying power from an external source to supplement and even replace human exertion. Transfer of power is found in :

(*i*) Power-driven machine tools.

(*ii*) Air and hydraulic claiming devices.

(*iii*) Auxiliaries and devices for moving and lifting heavy work and tools, for example cranes, general purpose lifting devices and conveyors of various types, etc.

(*b*) **Lightening of burden** : It includes means to reduce the weight or load an operator must bear. In addition to the shape, the material used and the method of fabrication are considerations in achieving desired properties of tools with minimum material. Welding of steel plates to form jigs, fixtures, gauges and other tools is a means of securing these attributes.

D. Conservation of skill. Skill may be conserved by reducing the level of skill necessary to do a job and by utilizing the skill available to the utmost. Two basic principles of tooling applicable to this consideration are :

(*a*) Transfer of skill

(*b*) Advantage of perception.

Skill is transferred when built into a machine or tool, making tools foolproof and providing means to compensate for variations in work pieces. Transfer of skill to machines and tools duplicate the craftman's performance and need relatively inexpert attention. The common drill jig is an example. It embodies the accuracy imparted by a skilled toolmaker in positioning the cutting tool so as to require less time and skill on the part of the operator in performing the drilling operation. An automatic machine which repeats each step of a cycle without continuous attention is another example. Foolproofing and providing means to compensate for variations in workpieces has been discussed in Chapter 1.

Perception means the ability of a workman to visualize the going state of conditions in an operation. It is improved by magnification of movements. It is illustrated by the dial and lead screw on a machine tool whereby minute increments of movement or distance are magnified so as to be readily visible to the operator. When the perception of a job is magnified, the skill necessary to do it is reduced.

E. combination of Operations. The combination of operations is one of the most profitable means of eliminating elements. As already discussed on p. 306, it can be achieved in two ways : by simulation and by integration. Simulation means the occurrence of two or more elements at the same time, such as when machine and operator are both working at the some time. An example of simulation is found in an operation arranged so that the operator can load one fixture, while the machine is working on a part in another fixture. In an operation, the shorter the machining time as compared with the non-machining time, the more important motion economy becomes.

Integration of operations means to incorporate several elements of an operation or to incorporate two or more cuts successively in an operation, eventhough, they are not down simultaneously. For example, if different drillings on a job are done in separate operations, elements of loading, positioning and unloading are needed for each operation. But if the several holes are drilled in one operation (with the help of a box jig), all but one loading and one unloading element is climinated. One set up instead of several is likewise incurred.

Machining as well as handling costs can be reduced by combining operations. Special machines are commonly built for high production to carry out these principles.

Direct Machining Costs. Machining time is influenced by the design of the work-piece, the machining method, the capacity of the machine, the design of the tools, the condition of the work material, and the conduct of the operations. Some machining processes are quicker than others under favourable conditions. For instance, broaching is faster than million but can be applied only to parts offering no obstruction to the broach. The maximum obtainable machining rate may depend upon the strength and power of the machine tool. Rugged and powerful machines are necessary to produce pieces in large quantities. With adequate machine capacity, carefully designed cutting tools are necessary to realise the most from an operation. Another factor for fast machining rates is the use of heavy jigs and fixtures to give adequate support to the work.

An equally important factor in the conduct of an operation is the distance the cutter or work must move while cutting and also while not cutting. A work-piece should usually be cut in the action requiring the shortest travel, Fig. 6.4. An exception to this rule is found in planning

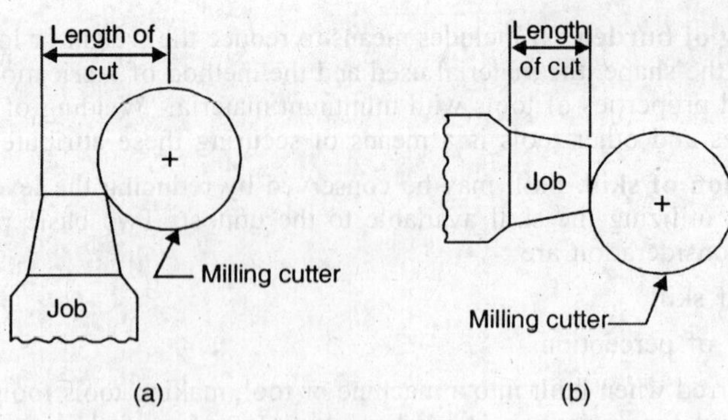

Fig. 6.4

or shaping, where the cut is taken in the longest direction to cover the surface in the fewest number of strokes. Tools should be moved to and from the work fast and over as short a distance as is safe.

2. **Economy of Indirect Costs.** All production effort is dependent upon and closely linked with a number of auxiliary functions. So, it is as important and profitable to minimize their costs as those of direct effort. The tool engineer can contribute to low overhead costs by planning and providing facilities to enable the auxiliary functions to serve the direct effort most efficiently, promote safety, and co-ordinate each operation with the material handling facilities between operations. Important auxiliary services are :

(*a*) Inspection, (*b*) Scrap disposal, (*c*) Set up and changeover (*d*) Cleaning, and (*e*) Maintenance, upkeep and, repairs.

For inspection, gauges should conform to the principles of motion economy as have been discussed in the case of tools. In addition, inspection can be co-ordinated with production. Inspection can often be facilitated by providing means to check work at the point of operation, even before it is removed from the machine, in order to uncover defective tendencies before they have gone too long.

Scrap disposal can be expedited by mechanical means, by segregating chips and scrap from the finish work, by readily cleanable equipment giving easy access to accumulation of chips and scrap, and by means to carry scrap from machines to collection centres, such as chip conveyors where warranted.

Set up and change over may be an important item in lot production. Simple and easily adjustable tooling (flexible and versatile) is desirable for such situations. Considerable savings can often be realized by using tools which have to be only partially changed or adjusted from job to job. For example, special vice jaws used with a standard vice make inexpensive and easily changed fixtures (See Milling Fixtures, Chapter 1). Also, quick change adapters for drill press and milling machine spindles are available commercially and help greatly in changing tools. The set up and changeover are also aided by preset tooling and by readily accessible arrangements, such as racks for reserve tools.

Under "Cleaning" equipment should be provided to have motions and labour. It is usually economical to provide baskets or racks to handle small-and medium-size parts in lots, with suitables lifting devices. For large parts, conveyors are helpful. This part of cleaning is an indirect function in the form of an enterprise protection operation. Cleaning may also be a part of direct effort with... ...ation.

Maintenance, upkeep and repair costs can be reduced by building and servicing. Repairs can be speeded and their costs reduced by the use of commercially obtainable details and tools whenever possible.

Promotion of Safety. Promotion of safety is another important responsibility of tool engineer. Safety should begin with the planning and designing of tools and equipment. For all this, due consideration should be given to the following aspects :

(*a*) Safeguarding the working zone.

(*b*) Safety of controls and machine machanism.

(*c*) Protection against moving parts.

The problem of 'safeguarding the working' will be different for operations where the stock is fed automatically or semi-automatically from that when the stock is fed manually and the worker is liable to come within the area of action of the tools. This aspect is more common and the protective arrangement for such operations can be listed as follows :

(*i*) Devices for feeding the jobs at a point outside the zone of tool action with mechanical means to pass the jobs into position. Such devices preferably include automatic ejection.

(*ii*) Automatic interlocking gate guards.

(*iii*) Sweep guards.

(*iv*) Electronic devices.

(*v*) Pull out devices.

(*vi*) Nonrepeating arrangements.

(*vii*) Tongs and other hand tools.

(*viii*) Two-hand tripping devices and similar interlocking devices for two persons.

Safety of controls and machine mechanisms is a consideration not only for actual operation but also for set up and machine repair. A common source of accident on power presses is the falling ram. To avoid this, an adequate brake maintained in good working order is necessary. On large presses, spring or compressed air cylinder counterbalances should be provided to support the weight of the ram. Foot pedals should be covered by guards so that they don't get pressed by falling objects or by being stepped on inadvertently. Care must also be taken that the pedal does not bind in the guard.

On some machine tools, speeds and feeds are changed by replacing pick-off gears manually. Accidents may occur if the worker does so before the machine has stopped or the machine being inadvertently started while the operator is making a change. A safety measure on some machines is to connect the door of the change gear compartment to a switch so that the power to the machine is cut off whenever the door is open.

"Protection against moving parts" of machines calls for guards completely enclosing all belts, gears, shafts, flywheels to prevent the possibility of anyone around the machine making contact with them. All exposed screws, nuts, lugs etc. whose moving might catch the worker, should be avoided, sharp coeners and edges should be broken.

Material handling costs are usually a sizable part of manufacturing cost. Even though, material handling is not ordinarily within the province of tool engineering, but planning and tooling must not overlook any contributions that might be made to keeping down the cost of moving materials.

A machine or tool that is readily available is usually preferable at least on a temporary basis, to a more efficient one that may take time to procure. Nothing is more expensive in industry than not producing at all.

3. **Economy of Fixed Costs.** The cost of equipment for an operation be kept at 'a minimum by making it as simple as feasible to produce the required quality and quantity of product.

Low initial costs of tools, machines and equipment can be promoted by :

(*a*) Observance of principles of economical construction.

(*b*) Use of versatile equipment.

(*c*) Adaptation of standard machine units for construction of special machines.

The principles of "economical construction" include the following requirements :

(*i*) With other factors constant, the amount of production determines how complex and refined a tool should be. A simple mechanism that will fulfill the requirements of a job is preferred to more elaborate devices.

(*ii*) Tools should have sufficient quality for the purposes intended, but should not be built too well.

(*iii*) Full advantage should be taken of the techniques of tool making to get low-cost construction. Low cost tooling is often devised by making minor additions to general purpose tools easily changed over to other work.

(*iv*) Construct tools as much as possible so that they have a wide range of usefulness. The cost of one tool may thus be spread over more than one job-and the expense of additional tooling can be reduced.

(*v*) The cost of a tool usuable for only one job must be realised entirely from that job.

"Standard tool details" are available in commercial form. Standard commercial machines, holding devices, cutting tools, gauges etc. are cheaper, more quickly obtainable, are based upon more comprehensive design effort, are produced in sufficient quantities to justify the most suitable equipment for their manufacture, and are subjected to much more extensive and severe testing than specially made tools and equipment. Therefore, if properly selected, they are more likely to possess desirable qualifications for any job for which they are adaptable and should be chosen whenever available and suitable. A representative commercial tooling item is the Universal jig. Usually, it is only necessary to add locations and bushings to an appropriate universal body to construct an efficient drill jig. A widely used commercial accessory for punch press work is the die-set. Substantial savings in tool costs can be realized by contriving to use as many standard tools as can be substituted for special ones. Examples of standard machine units can be : Drill presses, radial drills, knee-and Column-type milling machines and production type milling machines.

From the standpoint of economy in special machine tool construction, the principle of "unit construction" is adopted. In it, units (major components) such as head stock, base and ram etc of standard machines are arranged in a desirable manner and combined with supplementary adjuncts to make a special machine tool. For example, whether drilling is done on a simple drill press or on a complex automatic machine a spindle revolving in a drill head is necessary. This practice of unit construction has contributed markedly to reducing the extent and hazard of the investment requiring for special machinery.

6.7. SOLVED EXAMPLES

Example 6.1. *The hub shown in Fig. 6.5 is made of carbon steel. Discuss the steps of planning the process to manufacture this part. Draw the operation sheet also.*

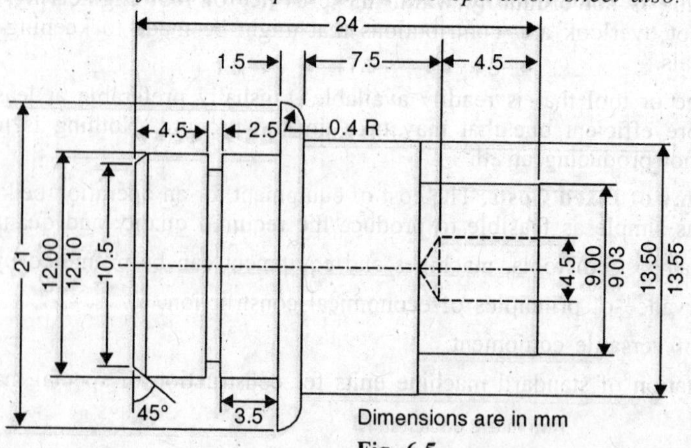

Fig. 6.5

Solution. The problem can be solved by following the steps given below :

1. **Study and analyze the part print.** From the study of the part print, the first factors to consider are : shape and size of the bar stock. From the part print, it is clear that some dimensions are given in fractions. It is important to clarify with the product engineer about the tolerances from fractional dimensions. The largest diameter given fractionally is 21 mm. Assuming a tolerance of 0.250 mm, it would be possible to finish this outside diameter to + 20.75 mm. For this, the part can be produced from 21 mm bar stock. However, if the final diameter had been specified as 21 mm, a larger diameter bar stock has to be used.

2. **Consult with product designer.** Under this step, the process planner will discuss with the product designer regarding the dimensions and tolerances. After getting everything clarified, he will proceed further.

3. **List the basic operations required to produce the part.** The following operations are required to produce the part :

(*i*) Stop drill.

(*ii*) Drill 4.5 mm-diameter hole.

(*iii*) Turn 13.525 mm dia.

(*iv*) Turn 9.15 mm dia.

(*v*) Form 12.00 mm dia.

(*vi*) Face outer end of part.

(*vii*) Generate 0.4 mm radius next to shoulder.

(*viii*) Cut off.

4. **Select the machine tool.** From the shape of the part, it is clear that a machine tool of lathe family will be suitable to produce this part. Depending upon the quantity to be produced, the machine tool can be : Engine lathe, Turret lathe or Automatic lathe. Assuming that the quantity to be produced is not large, turret lathe can be selected for the part.

5. **Combine the operations and put them in sequence.** In this step, the workpiece is distributed as evenly as possible among the tooling stations. After combining the operations, the operation sequence can be as follows :

(*i*) Feed out bar, position for length and close collect.

(*ii*) Spot drill end of bar and knee turn the 13.525 mm dia. slightly larger than the finish size using turret slide.

(*iii*) Drill 4.5 mm dia. hole and knee turn 9.15 mm dia. oversize using the turret slide. Rough form the area behind the shoulder with the cross slide.

(*iv*) Face the outer end using the end slide. Rough form the area behind the shoulder with the cross slide.

(*v*) Finish for all areas back of shoulder.

(*vi*) Rollers shave the area in front of the shoulder using the cross slide.

(*vii*) Cut off the part using the end slide.

6. **Specify the gauging.** The gauging requirements for this problem are quite simple. The following gauges are needed :

(*i*) micrometers for checking the various diameters.

(*ii*) a depth plug gauge for the 4.5 mm-dia, hole.

(*iii*) a template gauge to check the overall shoulder lengths and the width and position of grooves.

7. **Operation Sheet.** A simplified operation sheet for the problem is given below :

OPERATION SHEET

PART NAME : Hub Part No.			Material : Carbon Steel, Material Spec.
Oper. No	*Description of Operation*	*Machine*	*Tools, Gauge*
1.	Feed out bar, position for length and close collect.	Turret Lathe	
2.	Spot drill end of bar, knee turn the 13.525 mm dia.	"	Drill, knee turning tool, Micrometer
3.	Drill 4.5 mm dia. hole, knee turn 9.15 mm dia. Rough form the area behind the shoulder.	"	Drill, Depth Plug gauge, knee turning tool
4.	Face the outer end, Rough form the area behind the shoulder.		Facing tool, Form tool
5.	Finish form all area back of shoulder.		Forming tool Template gauge
6.	Roller shave the areas in front of the shoulder.		Roller shaving tool. Micrometer
7.	Cut off.		Cut off tool

Example 6.2. *The gear shown in Fig. 6.6. is to be made for a total quantity of 50,000 pieces. Prepare an operation sheet for this. The material of the gear is medium C-steel.*

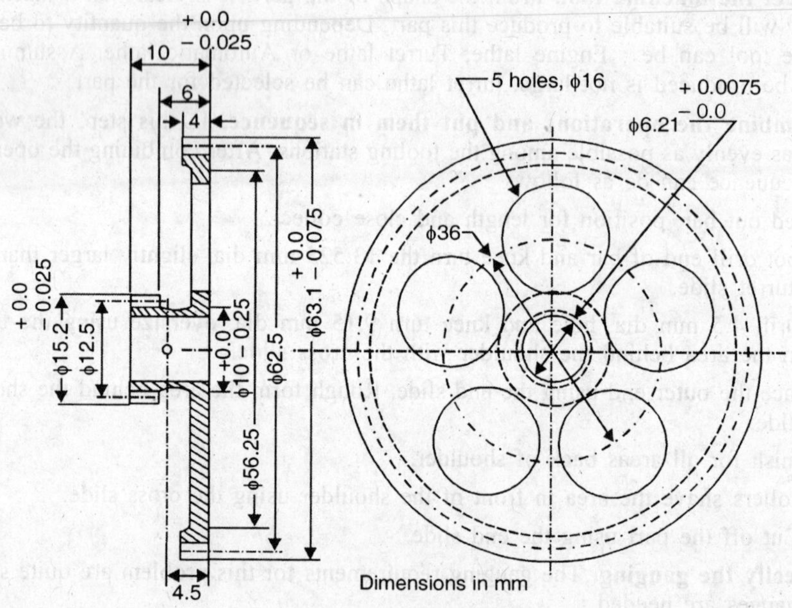

Fig. 6.6. Gear.

Solution. After the process engineer has satisfied himself with the part print, dimensions, specifications and tolerances, in consultation with the product engineer, the next step is to select the initial form of the material and the machine tools to manufacture the gear. The two convenient material sources from which the gear blanks can be made are : Bar Stock and forging.

The major drawback of starting with bar stock is that lot of stock will have to be removed which will go as waste. With forging, there is lot of material saving and we can start directly with the machining operations. So, we will start with blanks made from forgoing. It clear, then, that the principal process operation for manufacturing the gear will be : Machining.

Selection of Machine Tool. Considering the quantity to be made, the use of engine lathe for the turning and boring operations, is ruled out. Next, we have the choice from three machines : Turret lathe, automatic chucking machine and multiple-spindle automatic. The multiple-spindle automatic is ideal for rapid stock removal, but cannot be relied upon for holding the tolerances required. So, we have a choice between a turret lathe and an automatic chucking machine. The other machine tools needed are : Gear shaper for cutting the teeth, drilling machine for drilling the holes and machines for finishing the gear (say buffing machine).

Operation Sequence. The sequence of operations can be as follows :

1. Rough turning all over.
2. Drilling of 5 holes.
3. Heat treating.
4. Flattening.
5. Finish turning all over, drilling and rough-boring centre hole.
6. Precision-boring hole in hub.
7. Cutting teeth.
8. Shaving teeth.
9. Drilling holes in neck of the part.
10. Finishing.
11. Stamping part number.

The reasons for the above sequence of operations may be explained as follows :

Rough-turning. The purpose of rough turning the forged blank is to : remove surface scale and hardness and also to remove as much material as possible to enable finish-machining operations more precise. This will also provide reasonably accurate location points for drilling the five holes.

Drilling. Drilling of five holes is done before heat treatment so as to increase the tool life of drill.

Heat Treating. This operation has to be done before final finishing operations to ensure accuracy of the finished component.

Flattening. This operation may be necessary as the part may get warped after heat treatment.

Finish turning. Finish turning of external surfaces will provide accurate location points of precision boring and cutting teeth.

Finish boring. Finish boring should be done before cutting of teeth as this will provide best location for cutting teeth.

Cutting teeth. This is one of the major operations and should be done after all final finishing of external surfaces and precision boring has been done.

Shaving teeth. This operation is done to produce surface finish requirements of teeth.

Drilling holes in neck. This operation could have been done before teeth cutting but after precision boring.

Finishing. After the part is completely machined, its all surfaces must be covered as per plating specification.

Finally, the part number should be stamped on the finished part.

Now, we are ready to write the operation sheet.

OPERATION SHEET

Part Name : gear *Material : Medium C-steel*
Part No.

Oper.	*Description of operation*	*Machine*	*Tools, Gauge*
10	Forge blanks and trim flash	Drop hammer, punch press	Forging die, a trimming die
20	Stress-anneal	Electric furnace	
30	Inspect		Snap gauges
40	Rough turn hub; rough-face hub; Rough-recess hub and face; Roughface large diameter.	Turret lathe	3 Jaw chuck
50	Inspect		Snap gauges, depth pin gauges, profile gauges
60	Rough-turn outside diameter, Rough large face	Turret lathe	3-Jaw chuck
70	Inspect		Snap gauges
80	Drill 5, 15 mm diameter holes	Drill press	Drilling jig, 16 mm diameter drill
90	Heat-treat for hardness	Electric oven	
100	Inspection		Rockwell tester
110	Flatten, if necessary	Punch press	Flattening die
120	Inspection		Straight edge
130	Centre drill; drill; rough bore; finish large face; finish-turn outside diameter' burr sharp corner	Turret lathe	3-Jaw chuck, centre drill, drill, boring tool, mill file.
140	Inspection		Plug gauge, snap gauges
150	Finish turn small diameter; finish face small diameter' finish face relief' finish groove' finishface large diameter; burr sharp corners.	Turret lathe	3-Jaw chuck, 3 tool bits, form tool, mill file
160	Inspection		Snap gauges, depth pin gauges, profile gauge
170	Precision boring	Precision boring machine	Boring fixture, Boring bar and bit
180	Inspection		Plug gauge
190	Cut teeth-small gear	Gear shaper	Fixture, gear cutter
200	Cut teeth-large gear	Gear shaper	Fixture, gear cutter
210	Shave teeth-small gear	Gear shaving machine	Shaving cutter
220	Inspection		Master gear, adapter, pitch diameter gear gauge
230	Shave teeth-large gear	Gear shaving machine	Shaving cutter
240	Inspection		Master gear, adapter, pitch diameter gear gauge
			Drilling Jig, drills

250	Drill holes in the neck of gear	Multispindle drill press	
260	Remove all burrs	Hand operation	File, Scraper, emery cloth
270	Buff edges of gear teeth	Buffing machine	Buffing wheel
280	Finish (black oxide)		
290	Stamp		Hand stamp
300	Final Inspection		Various gauges already mentioned

PROBLEMS

1. Define Process Planning.
2. Explain the purpose of Processes Planning.
3. Discuss the contents of a Process Plan.
4. Discuss the various steps in the process planning.
5. Discuss the various considerations in determining the sequence of operations.
6. How does a process planner determine the most economical process for the manufacture of a product?
7. What is the significance of Process Planning in industry?
8. Prepare a process Planning Sheet to manufacture hollow piston pins.

   Note: Small ϕ piston pins are made from seamless tubes.
9. Construct a Process chart for the manufacture of small gears from gray iron castings.
10. A factory is engaged in the manufacture of bushes in moderate quantities of 50 per batch. Prepare a suitable operation plan for this production.
11. A factory is engaged in the manufacture of automobile pistons. Make an appropriate process layout, specifying the required gauges, tools and machines.
12. Prepare an operation sheet for a factory manufacturing crankshafts for a four cylinder petrol engine. The rate of production is 500 pieces a month. Specify the method of production and the machines used.
13. Construct a typical Process Chart for machining connecting rod forging.
14. Construct the suitable operation planning sheets for the manufacture of components shown in Fig. 6.7 to 6.8.

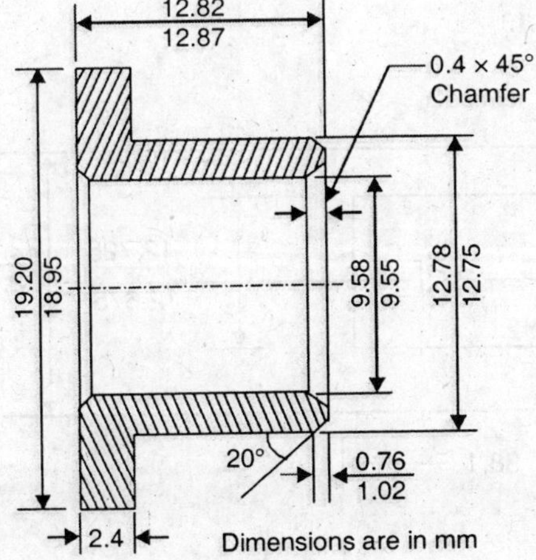

Fig. 6.7. Bush

Fig. 6.8. Starter Pinion-Gear Blank.

15. Prepare process planning sheets to manufacture components shown in Fig. 6.9 to Fig. 6.14.

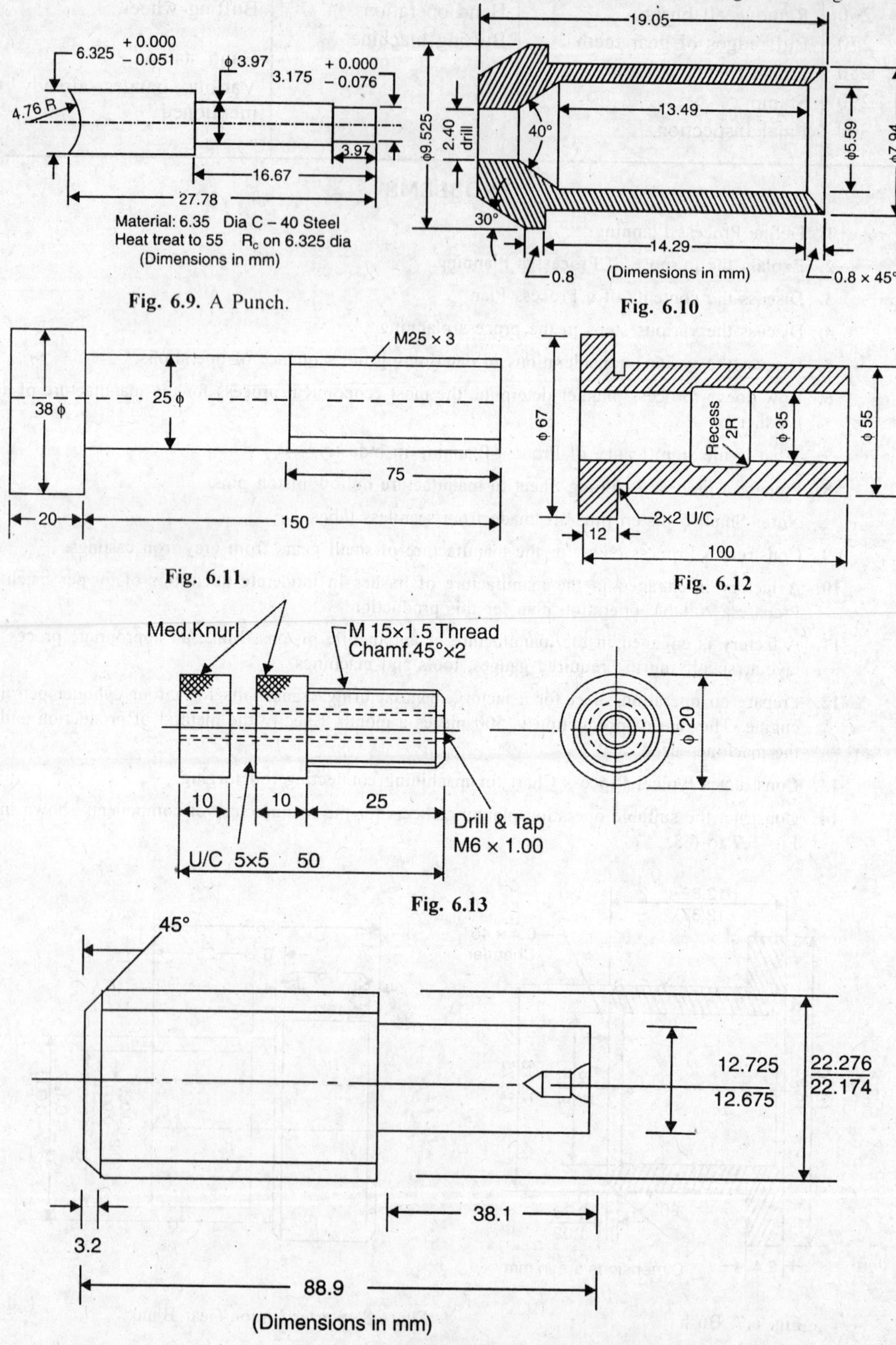

Fig. 6.9. A Punch.

Fig. 6.10

Fig. 6.11.

Fig. 6.12

Fig. 6.13

Fig. 6.14

7

TOOL LAYOUT FOR CAPSTANS AND TURRETS

7.1. GENERAL

Both Capstan and Turret lathes are sometimes called simply as "Turret lathes". A simple block diagram of a turret lathe is shown in Fig. 7.1. It differs from an engine lathe in that, the tail stock is replaced by a hexagonal turret which can be swivelled about a vertical axis and

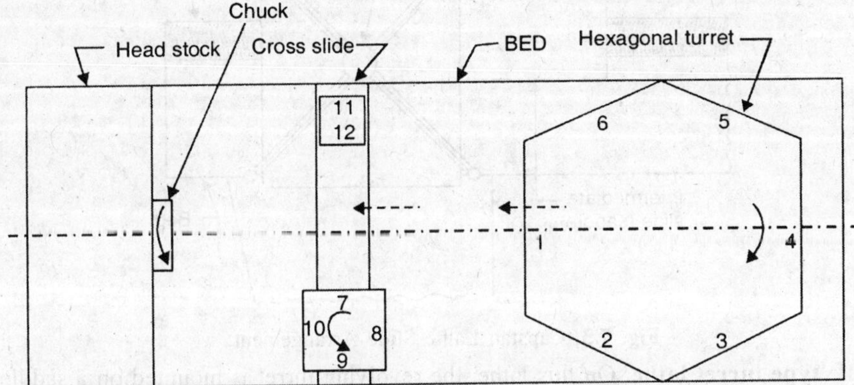

Fig. 7.1. Block Diagram of a Turret Lathe.

moved longitudinally along the bed to bring the tools into cutting positions, the regular tool post is replaced by a square cross-slide turret and the cross slide may support a square turret on the rear. In this way a minimum of 14 tools can be mounted at one time. As will be described later, this number can be increased by using multiple tooling at each position. The advantages of incorporating multiple tool holders is the chief distinguishing feature of a turret lathe. This enables the setting up of all tools for a job. Except for sharpening, they don't need further handling. The various elements of the Turret lathe are shown in Fig. 7.2.

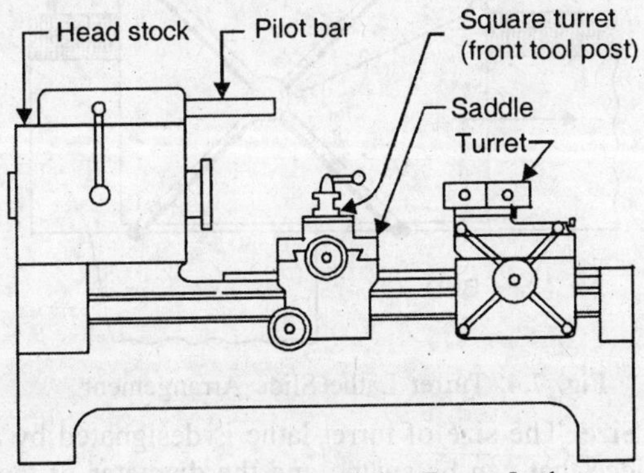

Fig. 7.2. Elements of the Turret Lathe.

Turrets take their place between centre lathes and automatics with regard to the volume production. They are essentially meant for medium batch production.

7.2. TYPES OF TURRET LATHES

The two main types of turret lathe are : (*a*) Ram type turret lathe, and (*b*) Saddle type turret lathe.

Ram type turret lathe. In this type of turret lathe, the revolving turret is mounted on a ram or a slide carried in a base which can be clamped in any position along the bed of the machine, as shown in Fig. 7.3. This type of lathe is somewhat lighter in construction and therefore can be quickly and easily operated. This lathe is also called as 'Capstan Lathe'.

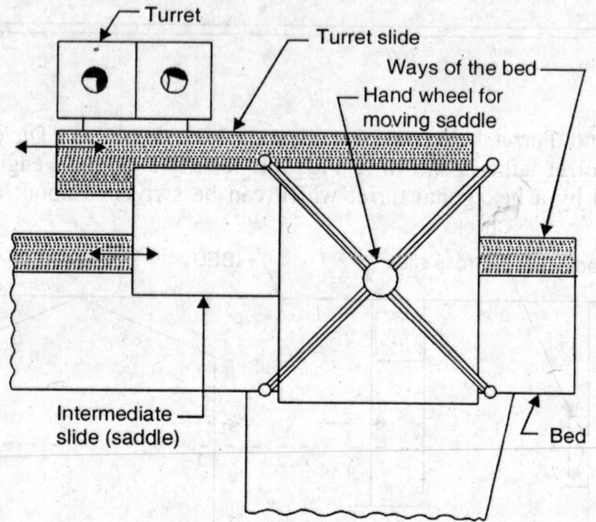

Fig. 7.3. Capstan Lathe Slide Arrangement.

Saddle type turret lathe. On this lathe, the revolving turret is mounted on a saddle which moves back and forward directly on the machine, Fig. 7.4. This design is also known as simply turret lathe. This type is heavier, has a longer stroke, and is more rigid.

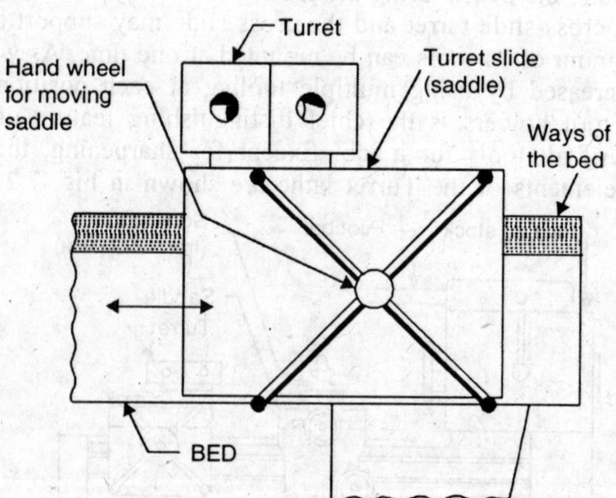

Fig. 7.4. Turret Lathe Slide Arrangement.

7.2.1. Turret Lathe Size. The size of turret lathe is designated by a number that indicates the diameter of the workpiece that can be swung and the diameter of the workpiece that can be swung and the diameter of the bar that can be passed through the hole spindle. For example, if the capacity of a turret lathe is designated as : 40-80-150, then it means that the bar diameter is 40 mm, swing over cross slide is 80 mm and swing over bed is 150 mm.

7.2.2. Applications of Capstan Lathes. The ram or slide is lighter and can be moved more quickly than a saddle but lacks some rigidity. Because of convenience and speed and limited travel of the turret, the ram type construction is ideally suited for the production of parts relatively short in length which are turned from the bar, and where the ram everhang can be kept short. Forgings, castings and stampings can also be machined by using jaw chucks.

The ram type turret lathe (capstan lathe) is made in the following size ranges :

Bar diameters = 12 mm to 60 mm

Swing over bed = 200 mm to 300 mm

Swing over cross-slide = 100 mm to 200 mm

Maximum turret movement = 75 mm to 350 mm

7.2.3. Applications of Turret Lathes. The saddle type construction, *i.e.,* the turret lathe is heavier, has a longer stroke and is more rigid. It provides good support for the tools and the means to move the tools a long way when necessary. It is more suitable for longer and heavier chucking work which requires longer and heavier cuts, *i.e.*, for the machining of forgings and castings.

The saddle type lathe (turret lathe) is made usually in the following size ranges :

Bar diameter = Small lathes upto 50 mm, and

large lathes upto 100 mm

Swing over bed = 350 mm to 900 mm and over

Swing over cross-slide = 250 mm to 650 mm

Maximum turret movement = 750 mm to 2000 mm

7.3. MAIN PARTS

The main parts of a turret lathe are as given below :

Head-Stock. The head-stock of a turret lathe is similar in construction to that of the engine lathe. However, it is larger and heavier in construction and with a wider choice of speeds and is fitted with a more powerful motor. Speed may range from 30 to 2000 r.p.m. There are two principal types of headstocks : (*a*) an electric head in which a variable speed motor is mounted directly in the head-stock (*b*) the all geared head which gives a wider range of speed and allows heavier cuts to be taken. For this, the electric motor is sometimes located in or on the motor leg below the head-stock and connected to the geared-head pulley by *V*-bets or a flange mounted motor may be directly connected to the drive shaft in the head.

The spindle is hollow and bar stock can be fed through it into a collet chuck. Direction of spindle rotation is reversible to enable the withdrawal of a solid tap from a tapped hole or the removal of a button die after the threading operation. The various spindle speeds, as well as forward and reverse movements of the spindle are controlled by levers extending from the head.

Turret. On all but very small capstan lathes, the turret is a hexagonal block mounted on vertical spindle. On each face, there is a hole to receive the shank, pilot or register of a flange mounting tool holder. To fasten the tool holder in position, four set pins are passed through the drilled holes in the flange into matching drilled and tapped holes in the turret face. On small capstan lathes, the turret is circular with six holes equally spaced around the circumference. The tool holders are provided with parallel circular shanks which fit into these holes. A clamp bolt is used to lock it in position.

A large star wheel at the front of the machine is used for moving the turret forward for each tool position. The forward movement is controlled by a preset stop and then the turret is returned to its starting position. As it returns, the turret is automatically indexed one-sixth of a resolution, bringing the next tool in the cutting position. There are six adjustable stops, one for each of the six turret faces and each stop is adjusted during the set up. These six preset stops are indexed automatically whenever the turret is indexed, so that the stop corresponding to a cutting tool is in use. The turret is moved to the workpiece by hand, and feeding may be done by hand or by power feed. For power feeding when the desired turret travel is reached, an automatic trip lever stops further movement of the tool by disengaging the drive clutch.

Cross Slide. There are two types of cross-slides which can be used for turret lathes : (*a*) side hung type, and (*b*) reach over type. In side hung type, the carriage does not extend across the entire top of the bed, but is supported entirely on the front way and another way located on the lower front portion of the bed. This design gives greater swing capacity across the cross-slide, but the rear tool post on the cross-slide is lost. This type of cross-slide is best suited for saddle type lathes which are used for large diameter jobs. However, this design has the disadvantage of lower rigidity, which reduces the number of tools that can be used simultaneously.

The reach over or bridge type of cross-slide is supported on the two bed ways in the conventional manner. In addition it may also be supported by a lower rail. This design is preferable for small capstan lathes (ram type machines) engaged in bar work where a large swing across cross-slide is not necessary, since this design restricts the swing across cross-slide. This type has advantage that a rear tool post may be provided which is frequently used for cut off operations. Also, this design is more rigid than the side hung type.

The tool post mounted on the cross-slide at the front of the machine, *i.e.*, the front tool post is a square turret with four faces for the mounting of the cutting tools. This turret can move in two directions; perpendicular and parallel to the rotational axis of the spindle. Stops are there for each motion of each cutting tool. These stops are indexed automatically when the square turret is indexed, through 90. This brings the next turret face with its tool or tools into position for the next machining operation. Feeding or movement in either direction may be by hand or by power feed. The rear tool post (when provided) usually carries the cut off or parting tool. Since the surface of work piece at the rear end is moving upward, the cut off tool is fitted upside down in the rear tool post.

7.3.1. Indexing of Turret and Cross Slides. As already noted, the saddle, in the case of Capstan Lathes, is mounted on the bed and the turret slide or ram moves back and forth in the guides of the saddle. In Turret Lathes, the turret slide (saddle) travels directly along the bed ways. The turret slide gets its movement from a rack and pinion drive. The shaft on which the large star-wheel is mounted (the starwheel is for providing handfeed to the turret slide) carries a pinion on its inner end. This pinion engages with a rack on the slide to provide longitudinal motion in either direction. For power feed, a shaft passes through the apron and is driven from the feed gearbox at the base of the headstock.

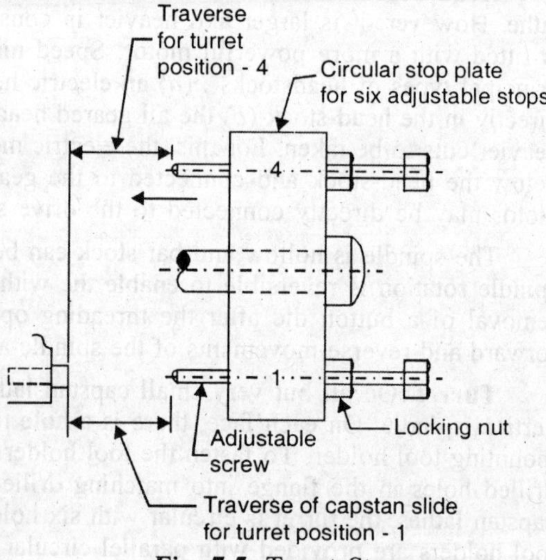

Fig. 7.5. Turret Stops.

For controlling the linear dimensions of length or diameter, the forward movement of the turret slide is limited by adjustable stops (which are six in number) mounted on a rotating circular stop plate or roll. The stop roll is linked to the turret through bevel gearing and rotates automatically as the turret indexes, thus bringing a separate stop screw into working position for each turret face. These stop screws act as automatic feed trips, and dead stops for hand feed. The movement of the turret slide (travel of the tool) is fixed by the adjustment of the stop screw which is locked in position by the locking

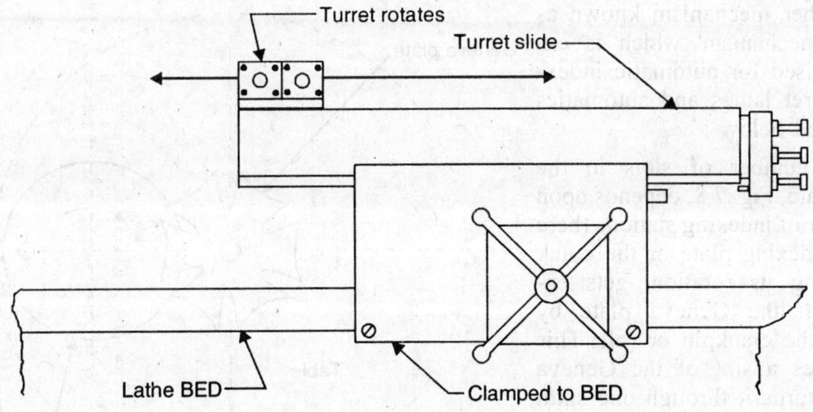

Fig. 7.6

nut, Fig. 7.5. When the turret slide reaches the end of its travel for a certain position of the turret, the corresponding stop screw runs against the fixed stop pin mounted in the saddle, Fig. 7.6. By means of a shaft, this stop pin actuates the mechanism located in the apron of the turret saddle for disengaging the working feed. This mechanism disconnects the feed gear train by means of a claw clutch and the slide (ram) stops in the preset position. When high precision work is required on high quality turret lathes, dial indicator type stops are used in place of stop screw, Fig. 7.7. This arrangement affords a visual indication to the operator, about exact end of the turret slide movement. At that time, the pointer of the dial indicator will read zero. This also avoids the variation of pressure against the fixed stop.

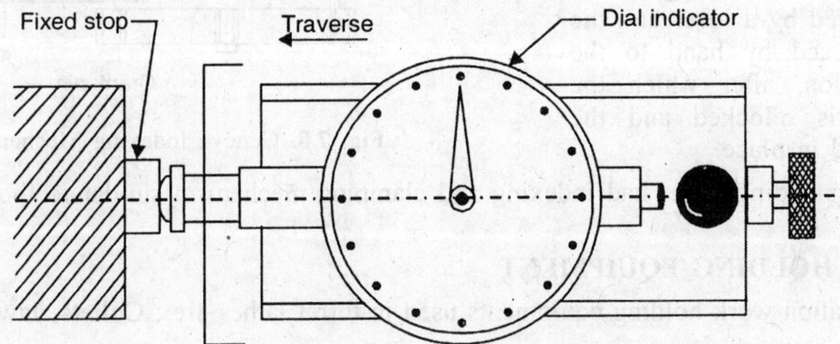

Fig. 7.7. Indicator Type Stop.

A tapered locking pin ensures that the turret holes are in strict alignment with the spindle axis after each indexing of the turret. The pin enters one of the six bushings fitted in the base of the turret. A clamping ring serves to clamp the turret rigidly after each indexing. It also relieves the locking pin of shearing stress due to the cutting force and the consequent torque developed in turret.

The turret of the capstan lathe being lighter, is indexed automatically, whereas the turrets of turret lathes do not index automatically, the saddles carrying them being heavy. The automatic indexing mechanism may vary from manufacturer to manufacturer. Here, we shall discuss a simple indexing mechanism. When the slide or ram is retracted at the end of its cutting travel, a withdraw finger encounters a cam on the turret rest and the lockpin is withdrawn from its slot. At the same time, a dog engages a cam and releases the turret clamp. Further withdrawal of the capstan slide causes the turret to rotate one-sixth of a revolution by a rotating finger which engages a six toothed ratchet. Once the turret has indexed one sixth of a turn the locking pin enters the next bush in the base of the turret and relocks the turret in its new position.

Another mechanism known as 'Geneva mechanism' which is extensively used for automatic indexing on turret lathes and automatics is explained below :

The number of slots in the Geneva plate, Fig. 7.8, depends upon the number of indexing stations (here 6). The indexing plate or the crank disc, during its rotation, gets engaged with the Geneva plate by means of the crankpin or roll. This pin engages a slot of the Geneva plate and turns it through one sixth of its revolution per revolution of the disc.

As mentioned earlier, the turrets of the turret lathes do not index automatically. The saddle carrying the turrets are heavy and they have long travel. To bring the saddle back to an automatic indexing position, will mean a considerable wasted movement. Due to this, indexing is not done automatically, but is done by hand. The turret clamping mechanism is released by a lever and the turret is indexed by hand to the desired position, after which the locking pin is relocked and the turret clamped in place.

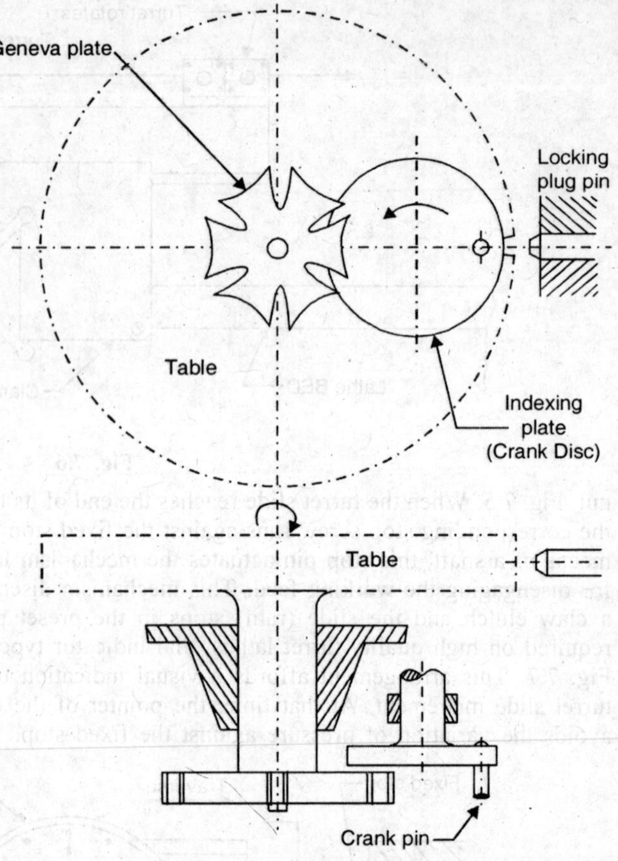

Fig. 7.8. Geneva Indexing Mechanism.

The cross-sliding turret had indexing and clamping mechanism similar to those explained above.

7.4. WORK HOLDING EQUIPMENT

The variation work holding equipments used in turret lathes are : Collets, Jaw Chucks and Fixtures.

Collets. Collets or collet chucks are used mainly to hold bar stock, especially in the smaller sizes. A collet is a circular steel shell having three or four equally spaced slits extending the greater part of its length, Fig. 7.9 (*a*). These slits impart springing action to the collet. That is why, collets are also known as "Spring Collets". The collet nose is made thicker to form the jaws. The inside of the collet is made according to the shape of the work to be held. The various types of collect chucks are shown in Fig. 7.9 (*b-d*).

1. **Drawback Collet.** In drawback collet, Fig. 7.9 (*b*), the spring collet is pulled by the thrust tubeor collet tube, to the left into the taper bore of the spindle nose. This action puts pressure on the tapered sections of the collet, forcing them inward and tightly clamping the bar stock. The thrust tube is placed behind the collect in the spindle. It can be moved axially a short distance by either a handwheel or a lever. The bar to be machined is passed through the tube and the collet jaws.

2. **Push-out Collet.** In push-out collet, Fig. 7.9 (*c*), the thurst tube pushes the spring collet to the right into the tapered seat in the spindle rose. This clamps the bar stock in the same manner as before.

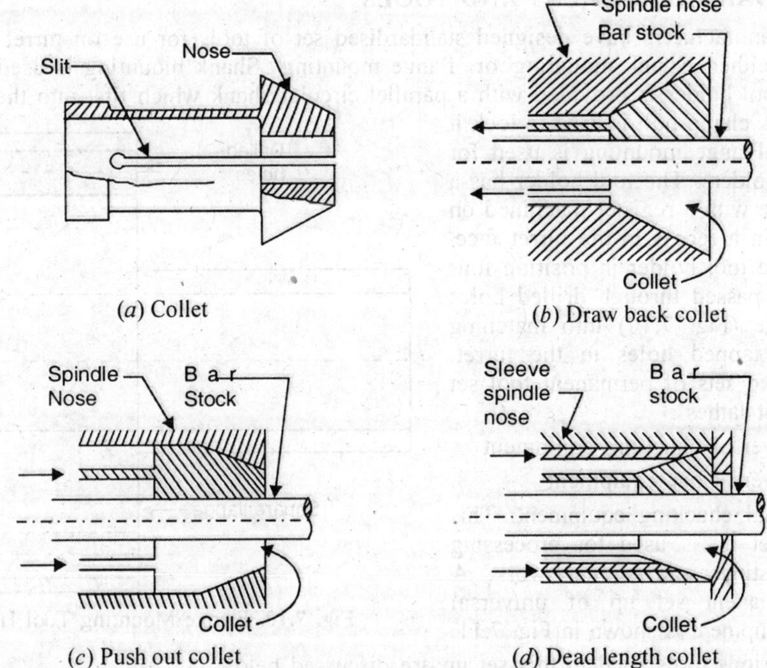

Fig. 7.9. Collet.

3. **Dead length Collet.** In this design (Fig. 7.9. *d*) the spring collet has no axial movement during operation. This chuck is closed when the thrust tube pushes a sleeve with an internal taper onto the taper of the collect forcing it inward to clamp the work.

Collets are designed to take only one size of work. Standard sizes are available for bars upto about 50 mm diameter.

Jaw Chucks. The types of jaw chucks commonly used on capstan lathes and turret lathes are :

(*a*) Two Jaw Chucks (*b*) Three Jaw Chucks (*c*) Four Jaw Chucks.

(*a*) **Two Jaw Chucks.** These chucks are of the self-centering type and are used for bar work. Many of these chucks may have blank jaws to which may be attached specially shaped holding devices or jaws for quickly locating and clamping irregular or odd-shaped pieces, for example, small castings and forgings. Two jaws hold the irregular work more readily since the clamping is at two points which are diametrically opposite. Two jaw chucks are available in size from about 125 mm to 250 mm outside diameter to hold bar stock diameter of about 20 mm to 45 mm diameter.

(*b*) **Three Jaw Chucks.** Three jaw universal chucks are used for holding round or hexagonal bar stock or other symmetrical work. They are of self-centering type and are used mostly for machining of forgings and castings. The size of these jaw chucks ranges from 100 mm to 750 mm diameter and they can hold work upto about 650 mm diameter.

(*c*) **Four Jaw Chucks.** These chucks are used for holding rough castings and square or octagonal work. The jaws move independent of one another. These chucks are available in sizes upto about 1000 mm diameter.

Two jaw chucks and three jaw chucks are generally power operated. They may be either pneumatically or hydraulically operated. Four jaw chucks are manually operated.

Fixtures. A special fixture may also be designed and used for holding the work, if the quantity of parts to be made is sufficient. The fixture is mounted on the machine spindle in place of a chuck or a collet. For details, readers may refer to chapter 1.

7.5. STANDARD EQUIPMENT AND TOOLS

The manufacturers have designed standardised set of tools for use on turret lathes. Tool holders are either `Shank mounting' or `flange mounting'. Shank mounting is used for smaller sizes. The tool holder is provided with a parallel circular shank which fits. into the bore in the turret face. A clamp bolt is used to lock it in position. Flange mounting is used for larger tool holders. The tool holder has a square flange with a register machined on it to locate in a recess in the turret face. To fasten the tool holder in position four set pins are passed through drilled holes in the flange (Fig. 7.10) into matching drilled and tapped holes in the turret. There are two sets of permanent tool set ups for turret lathes :

1. Universal chucking equipment
2. Universal bar equipment

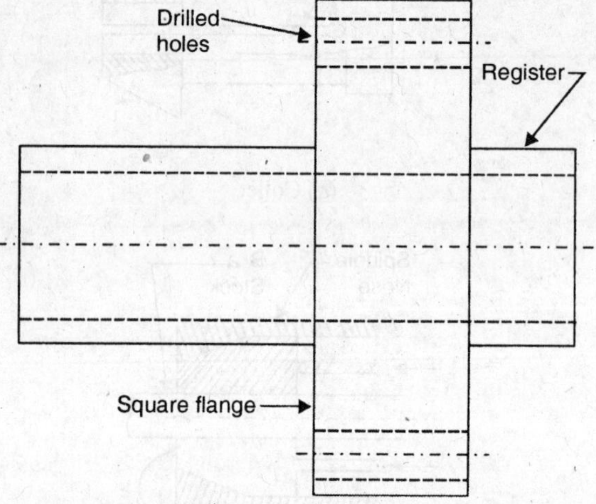

Fig. 7.10. Flange Mounting Tool Holders.

Universal chucking equipment. This permanent set up is used for processing forgings, castings and similar work. A typical permanent set up of universal chucking equipment is shown in Fig. 7.11.

The various tools used in this set up are discussed below.

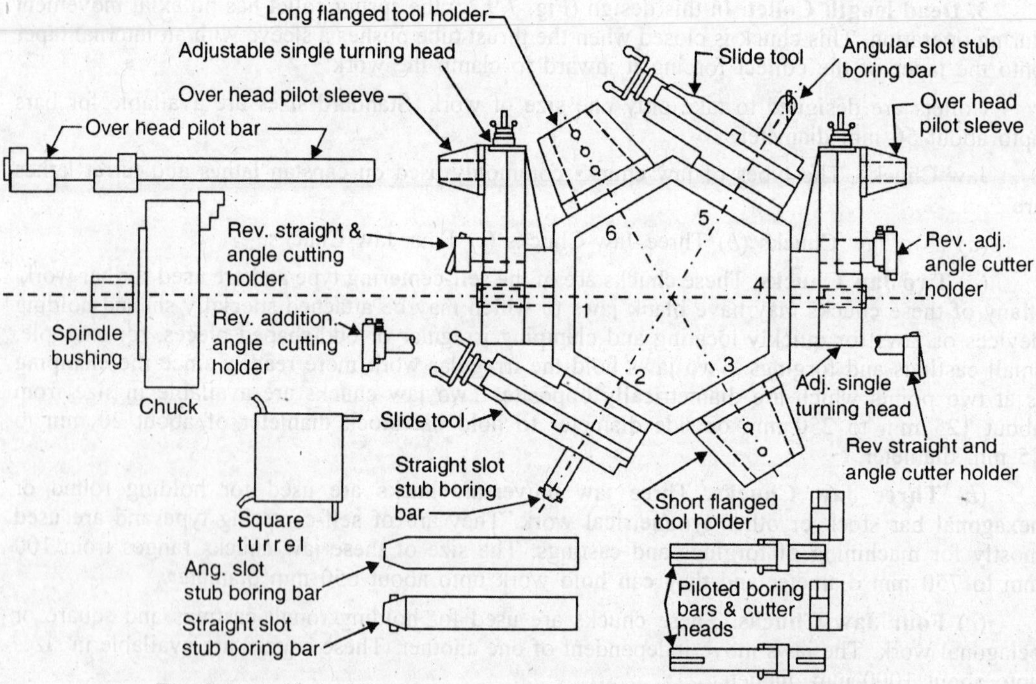

Fig. 7.11. A Typical Permanent set up of Universal Chucking Equipment.

1. **Adjustable single and multiple turning heads.** These are shown at stations 1 and 4 (Fig. 7.11). They are used for heavy roughing. A pilot bar is used in combination with this overhead turning head. It is generally mounted over the headstock, Fig. 7.12, but in some machines it may be mounted over the turret at each tool position requiring it. The pilot bar engages a sleeve as it travels (or the tool does) and provides support and rigidity as well as piloting the tool to

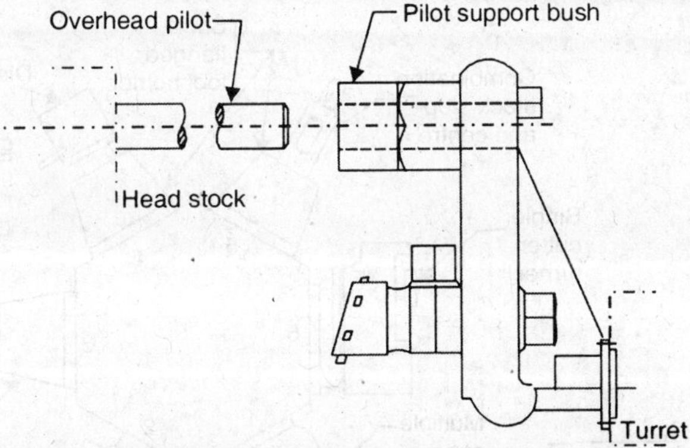

Fig. 7.12. Overhead Turning Tools.

increase accuracy. This is particularly important for heavy cuts which must also be accurate. The cutters for these overhead cuts are held in shank-type holders. It is also possible to use boring bars, drills or core drills in the centre holes of these heads working with the overhead turning cutters.

2. **Adjustable Slide Tool.** This is a tool which has a limited amount of travel in a direction usually perpendicular to the rotational axis of the spindle. The movement of this slide can be used for adjusting depth of cut or for feeding when facing, cutting grooves, or cutting recesses. The amount of movement can be accurately adjusted with the micrometer dial on the screw which moves the slide. Recessing with a slide tool is shown in Fig. 7.13.

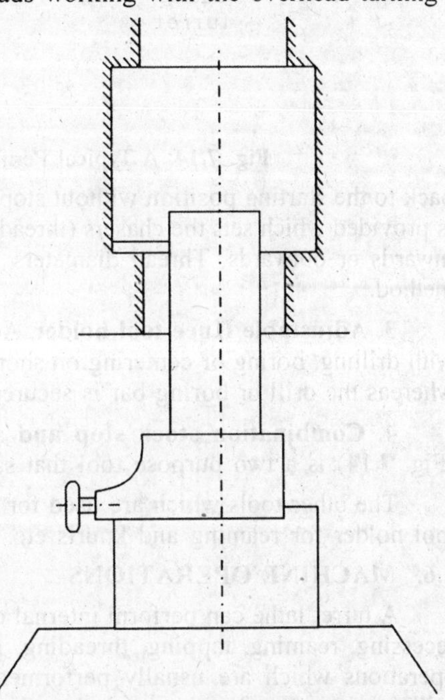

Fig. 7.13. Recessing with a Slide Tool.

The flanged tool holders (stations 3 and 6) provide positions for shank-type tools, for example, drills, taps, reamers etc. that can quickly be inserted during set up.

Universal Bar Equipment. A typical permanent set up of universal bar equipment is shown in Fig. 7.14. The various tool holders are discussed below :

1. **Box Tools.** Box tools are the predominant type of turning tools on turret lathes because of the lack of end support. The box tool supports the work with rollers or a crotch while the tool is cutting, thus enabling heavy cuts to be taken with accuracy and without springing. This tool has the limitation in that the length of cut taken is limited. The single cutter turner in position 6, is the simplest example of a box tool. The roll support the work at the point of cut. If the rollers are set behind the cutter, they will burnish the work to a fine finish and help in holding accurate sizes. The rolls are also used ahead of the cut if this surface has already been machined. Rolling ahead will give a turned diameter concentric to the rolling diameter but will also result in holding a tolerance only as good as that on the rolling diameter. When rolling ahead of the cut, lighter feeds must be used for finishing, as there is no burninshing. Box tool at station 1 is a multiple cutter turner.

2. **Die head.** On turret lathes, most external threads are cut using a self-releasing threading die head. As it cuts, it screws itself onto the threaded workpiece. The turret carrying the die head stops moving just prior to the full length of thread being reached, but a portion of the die head continues to be pulled ahead by the thread being cut. At a predetermined location, the cutters automatically open outward, clearing the thread, so that the die head and turret can be moved

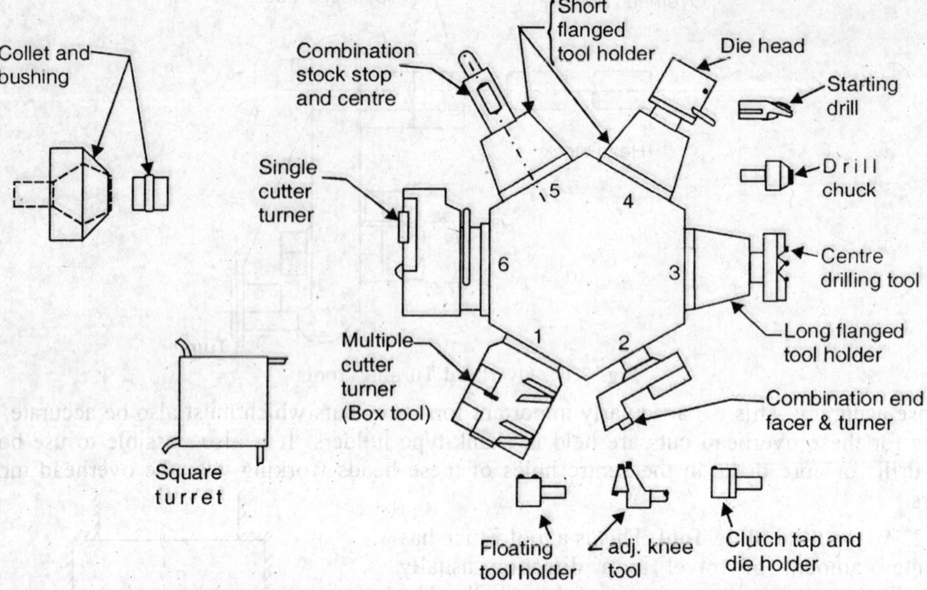

Fig. 7.14. A Typical Permanent Set up of Universal Bar Equipment.

back to the starting position without stopping and reversing the rotation of the workpiece. A lever is provided which sets the chasers (thread cutters) by operating the scroll to move the chasers either inwards or outwards. Thread diameters from about 6 mm to about 100 mm can be cut by this method.

3. **Adjustable Knee tool holder.** Adjustable knee tool holder is used for turning combined with drilling, boring or centering on short pieces. The turning tool is carried in an overhead arm, whereas the drill or boring bar is secured in a central bore that is concentric with the turret bore.

4. **Combination stock stop and starting drill.** This tool which is shown at position 5 (Fig. 7.14) is a two purpose tool that saves one turret face and one index.

The other tools which are used for bar work on turret lathes include : Drill chuck, Floating tool holder for reaming and knurls etc.

7.6. MACHINE OPERATIONS

A turret lathe can perform internal or external work such as turning, boring, drilling, facing, recessing, reaming, tapping, threading, forming, knurling, chamfering and parting off etc. The operations which are usually performed from the hexagonal turret include : turning, boring, drilling, reaming, tapping and threading. The operations which are done from cross slides include : facing, forming, knurling, chamfering, recessing and parting off. Internal cuts are almost always made by tools in hexagonal turret.

The cuts which can be taken on a turret lathe are of three types :

1. **Combined cut.** A combined cut is one where tools in both the hexagonal turret and square-slide turret are made to cut at the same time.

2. **Multiple cut.** A multiple cut is one where two or more tools are applied at the same time from the turret station.

3. **Successive cut.** A successive cut is one where cuts are made from successive faces of the turret consecutively.

Most threads are cut on a turret lathe from the hexagonal turret. Self-opening dies and collapsible taps allow rapid retraction of the tool after the thread is cut. The length of cut, however, is limited. Threading is done mostly on the end of the workpiece. For longer threads, a single point threading tool in the front cross slide must be used. This method is much slower when a good, clean and accurate thread is desired.

In the production of holes, provision must be made for the proper order of internal cuts. For example, a roughing operation using a drill is usually performed first to remove metal quickly. Then another roughing operation is carried out with a boring bar, after which a reamer is used to size the hole quickly and give a good finish.

There are three methods to obtain taper on a turret lathe. (1) with a form cutter : This method is suitable only for short, steep tapers, (2) with a taper turning attachment as on an engine lathe, and (3) with a roller-rest taper attachment. In this method, a tapered guide bar is mounted over the work in the same position as the pilot bar, replacing it. The cutting tool resembles the bar turner. It can be easily adjusted for size by means for a graduated dial. The tool and the rolls recede as they progress along the work, guided by the bar, thus producing the desired taper.

The parting off operation is always done from the rear cross-slide in the case of bridge type of cross-slide.

7.7. ADVANTAGES OF TURRET LATHE

The turret lathe is adapted to quantity production work. For this, the turret lathe has got the following advantages over a simple engine lathe :

1. A large number of tools can be permanently set up in the machine in the proper sequence of their use.

2. Each cutting position is provided with a feed stop or feed trip which ensures identical cuts on successive pieces.

3. Multiple cuts and combined cuts can be taken at the same time.

4. The turret lathe can be fitted with various attachement such as for taper turning, thread chasing, and duplicating. It can be also tape controlled.

5. A less skilled operator is required.

7.8. TOOL LAYOUT

Once a turret lathe is properly tooled, an experienced operator is not required to operate it. However, skill is required in the proper selection and mounting of tools. The tool layout for a job constitutes the predetermined plan for the order and method of machining operations necessary to produce it. Accuracy and cost are largely dependent on an efficient layout and the layout is itself influenced by the number of pieces to be made. As a general rule, standard tools should be used as much as possible and for small batches of work, the layout should be simple. For large quantities and long runs, it will be more economical to use special tools which minimise machining time and retain their cutting quantities for the maximum period.

For the preparation of layout, it will be necessary to have a finished drawing of the part to be produced and if it is a forging and casting, a rough blank will show how much machining has been left on the various faces. After the preliminary list giving the order of operations has been decided upon with details as to the tools required, it may be necessary to prepare a scale layout on the drawing sheet. Thus tool layout for a turret lathe consists of two steps :

1. Preparation of operation sheet.

2. Sketching the plan showing the various tools fitted into the turret faces and on the cross-slides, in proper sequence.

7.8.1. Planning Turret Lathe Operations. Time is the most important item of cost in a turret lathe operation. As already discussed in chapter 4, the total production time for a workpiece consists of : set up time, work-handling time, machine-handling time and actual cutting time.

A major portion of the set up time is used up in mounting and adjusting the cutting tools. This can be reduced considerably by using universal tooling and maintaining a permanent set up on a turret lathe.

The work handling time is that which is consumed in mounting or removing the work from the machine. It largely depends on the type of raw material and on the type of work holding devices used. For bar-stock, this time is minimised by using bar-stock collets. Also, for average work in small or moderate quantities, standard holding devices are best. If need be, special jaws, arbors and simple fixtures can be incorporated in these. Special fixtures are justified only if the job is hard to hold in standard devices and is made in large quantities.

Machine-handling time consists of : bringing the respective tools into cutting positions, changing the feeds and speeds and time for indexing. This time can be reduced by having the tools in proper position and sequence for convenient use. Machine handling time is also reduced by taking multiple and combined cuts to save indexing.

The actual cutting time for a given operation is largely dependent on the proper use of cutting tools, proper cutting tools, speeds and feeds. Multiple and combined cutting also reduce the cutting time.

The following main rules should be followed in laying out the sequence of operations necessary to produce a workpiece :

1. Overlap working operations whenever possible and try to increase the number of tools operating simultaneously.

2. Overlap handling operations with each other and with the working operations.

3. Overlap finishing and roughing cuts only in cases when this does not have a deterimental effect on the quality of the workpiece. Finishing operation should be done at the end whenever possible.

4. Arrange the cutting tools so as to counterbalance the cutting forces to a maximum extent.

5. Provide a finish cut over the full length of a surface if it was roughed in sections by several tools.

6. Do not permit substantial reduction in the rigidity of the workpiece (cutting deep grooves) until all the machining is completed. This is especially important in thread cutting.

7. Turn precise form surface in two cuts whenever possible.

8. Centre drill before drilling small diameter holes.

9. In drilling deep holes, pull out the drills several times to break up and remove chips and to facilitate drill cooling.

10. Cored holes should not be extended in diameter by using a drill but they should be bored only.

11. In drilling a stepped hole, first drill the largest hole and then the next small diameters in order. This will decrease the total travel required by all the drills.

12. Finish machine external and internal surfaces at a single position to obtain the exact alignment.

7.9. Bar Stock feeding mechanism. As is known, the rotating bar is fed through the hollow spindle of the lathe. The rotating bar beyond the back of the head stock of the lathe is protected and is usually enclosed in a tube which carries the feed mechanism. The bar feed can be; By hand, pneumatically, electrically or may function automatically upon release of the bar chuck (collet) or by a separate control. As soon as the first component is machined the mouth of the collet is opened. The required length of the bar stock is fed through the collet and then it is closed. Electric feed is less commonly used. A small separate motor provides the power and control is usually by foot operated switches.

Peneumatic feed is interconnected with the bar chuck so that depression of a pedal opens the chuck and feeds the bar forward. A rear pusher is unsuitable for the small sizes of the bar and the feed mechanism at the rear end of the spindle comprises a collet tube which clamps on the bar, pulls-it forward from the enclosing tube and pushes it into the spindle.

Automatic bar feeding takes less time and the bar can be fed without stopping the machine. A simple and effective method, whereby the bar can be readily fed forward with the minimum loss of production time, is shown in Fig. 7.15.

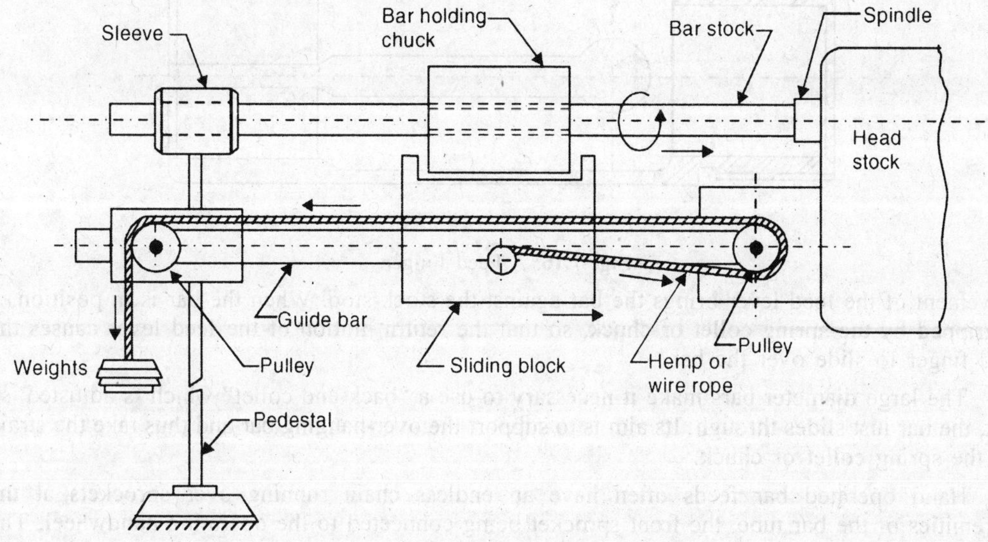

Fig.7.15. Automatic Bar Feeding Method.

The bar stock passes through the pedestal bushing or sleeve and is held in a bar holding chuck before it enters the lathe hollow spindle and is clamped in the collet (spring chuck) of the lathe. The bar holding chuck rotates within the sliding block, when the feeding bar is rotated by the lathe spindle. A heavy weight is attached to the sliding block, by means of a hemp or wire rope, that passes over a series of groove pulleys. As soon as the bar holding collect in the lathe spindle is released by a lever, on the completion of a component, the dead weight tends to move downward. This exerts an end thrust on the sliding block. This end thrust forces the sliding block to carry the bar holding chuck and the bar forward until the bar touches the stock-stop on the face of the turret slide. As the length of the bar decreases, the sliding block and the bar holding chuck near the machine spindle and then their position on the bar is adjusted. On placing a new bar in the feed mechanism, similar adjustment becomes necessary.

In a variation of the above method, the bar is enclosed in a tube which carries the feed mechanism. The bar tube, having a slot in one side throughout its length, is carried by supports and is accurately aligned with the lathe spindle. The bar pusher inside the tube has a lug projecting through the slot. The lug is externally attached to a wire which passes over a pulley at the headstock end of the tube and returns parallel to its initial direction, to the rear support. At this point, the wire runs round two series of pulleys, one of which supports a group of weights. The pulleys are so arranged that the available descent of the weights is sufficient to feed the bar through the length of the tube. The latter is pivoted at the rear support and the front end is supported by the transverse slide. For loading purposes, the tube assembly can be pivoted by pulling out the front end along the slide after which the bar is fed in from the front end. A screw pad is clamped on to the wire, retaining the bar inside the tube until the latter has been returned to the working position, where it is locked by a handle.

The above methods of automatic bar feeding are suitable only for small-diameter bars. With the larger and heavier bars, the weight of the meterial and the impact load created, as the mass is brought to rest by the stock stop, is such that the wire feed mechanism no longer-suffices. For such cases, it is essential to use the feed finger, Fig. 7.16, which also is the standard equipment for bar feeding on automatic lathes, as discussed in the next article. Upon heat-treatment, the jaws of the feed finger are closed, to form the spring grip. The action of the feed finger is to grip the bar with sufficient force, so that, when the spring collet is open, the

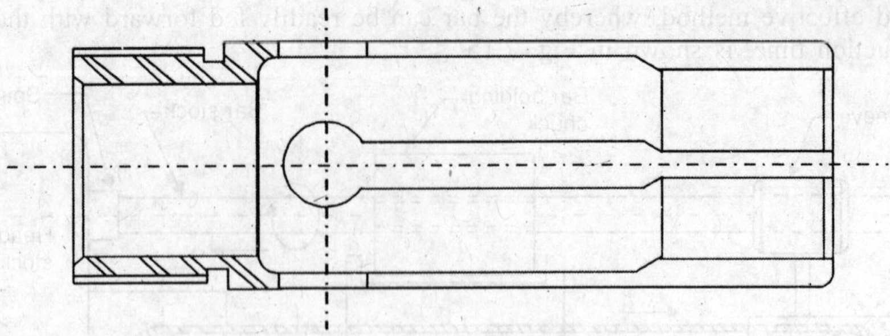

Fig. 7.16. A Feed Finger.

movement of the feed lever brings the bar against the stock stop. When the bar is in position, it is gripped by the spring collet or chuck, so that the return motion of the feed lever causes the feed finger to slide over the bar.

The large diameter bars make it necessary to use a "back end collet" which is adjusted, so that, the bar just slides through. Its aim is to support the over-hanging bar and thus take the strain off the spring collet or chuck.

Hand operated bar feeds often have an endless chain running over sprockets at the extremities of the bar tube, the front sprocket being connected to the operating handwheel. The tube is carried by two supports and pivots on the rear one. The tube is slotted along its full length and a lug on the bar pusher inside it projects through the slot and is connected to a roller chain. After the chuck is opened, operation of a hand wheel feeds out the desired length of bar. Loading is performed from the front end, the arrangement for pulling forward the tube assembly, clear of the head stock being similar to the one explained above. Hand feeding takes more time and the machine has to be stopped before feeding the bar.

Nearly all bar feeding mechanisms are equipped with extended pusher or Collet tubes, enabling the end of the bar to be fed right upto the back of the chuck to minimise wastage of stock.

7.10. SOLVED EXAMPLES

Example 1. *Draw the tool layout for the component shown in Fig. 7.17.*

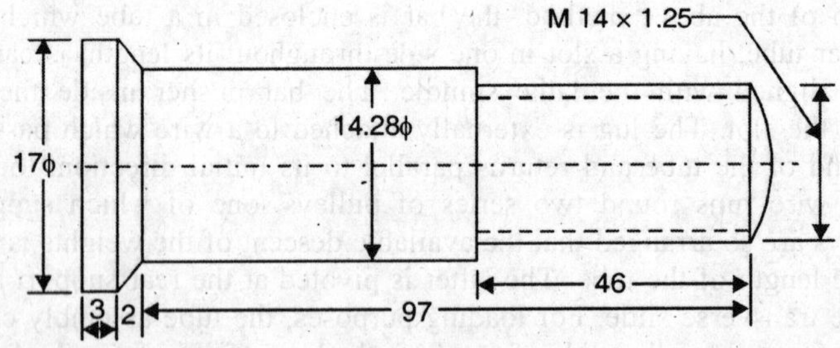

Fig. 7.17

Solution.

Material : 18-mm steel bar

Machine : Capstan lathe

Operations :

1. Feed the bar stock to stop.
2. Turn 14 mm diameter with box tool.

3. Turn 14.28 mm diameter with box tool.
4. Round end with roller steady ending tool.
5. Centre with centre drill.
6. Cut threads with die-head.
7. Form 17-mm diameter and chamfer with tool from front square slide.
8. Part off with cut of tool in rear tool post. The tool layout in shown in Fig. 7.18.

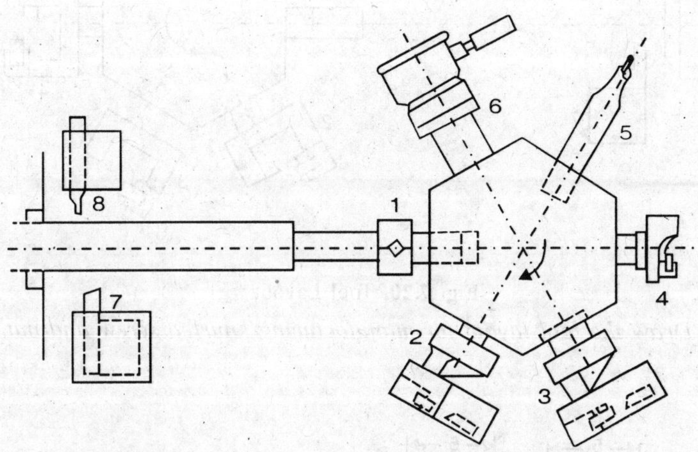

Fig. 7.18. Tool Layout.

Example 2. *Draw the tool layout for the component shown in Fig. 7.19, which is to be made on a capstan lathe.*

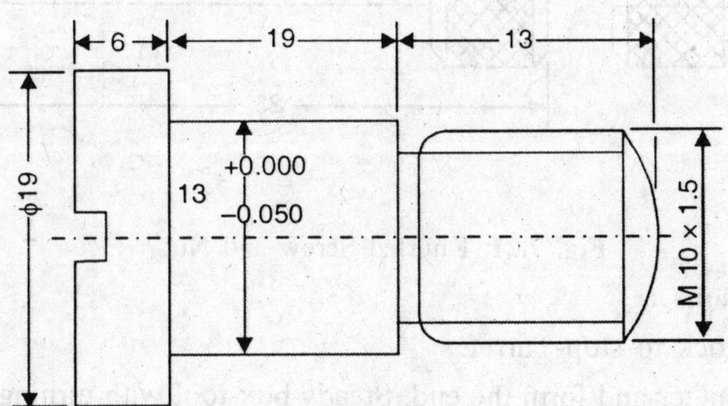

Fig. 7.19

Solution. Operations :

1. Feed to-Turret.
2. Turn 13/12.95 mm diameter, Roller box-Turret.
3. Turn 10 mm diameter, Roller box-Turret.
4. Turn end, Roller box-Turret.
5. Undercut and Chamfer-cross slide (Rear).
6. Threading, Self opening die head-Turret

7. Part off, Part-off tool-cross slide (Front). The tool layout is shown in Fig. 7.20.

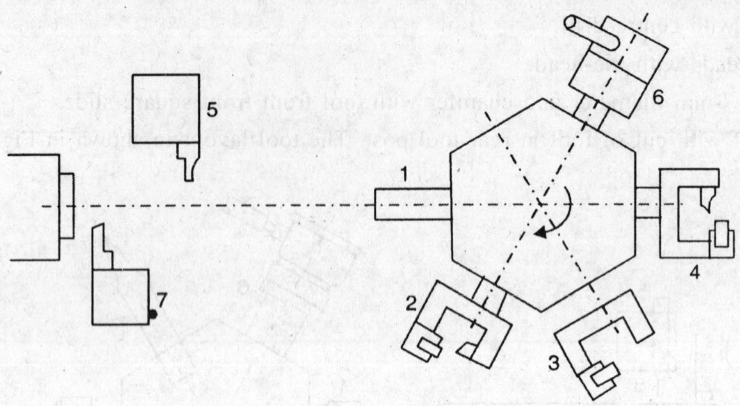

Fig. 7.20. Tool Layout

Example 3. *Draw the tool layout for manufacturing knurled screw and nut, Fig. 7.21, on capstan lathe.*

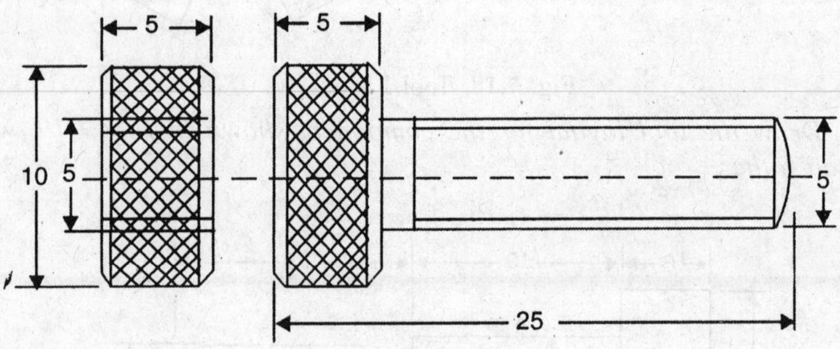

Fig. 7.21. Knurled Screw and Nut.

Solution. Operations :

1. Feed the bar stock to stop-Turret.
2. Turn 5 mm diameter and form the end, Steady box-tool with turning and radiusing tools-Turret.
3. Cut the threads on screw, die head-Turret.
4. Knurl, Knurl tool-Turret.
5. Chamfer the four diameters, chamfer tools-Front cross slide.
6. Part off the screw, parting tool-Rear cross slide.
7. Drill and face the nut, drill and facing tool-Turret.
8. Tap the nut, Tap-Turret.
9. Part off the nut, parting tool-Rear cross slide. The tool layout is shown in Fig. 7.22)

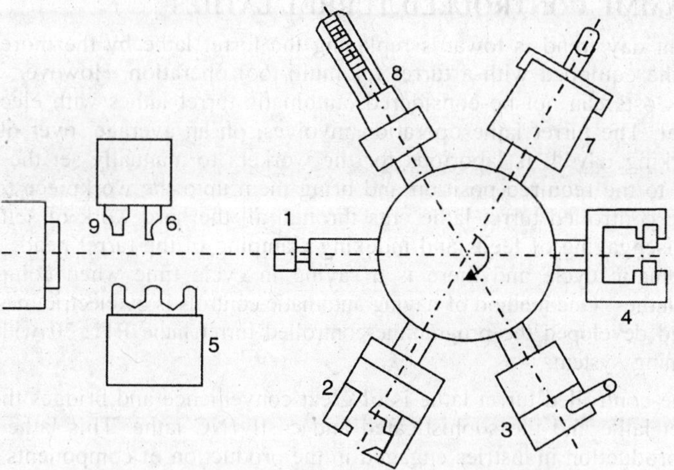

Fig. 7.22. Tool layout for knurled screw and Nut.

Example 4. *Draw the tool layout for the component shown in Fig. 7.23.*

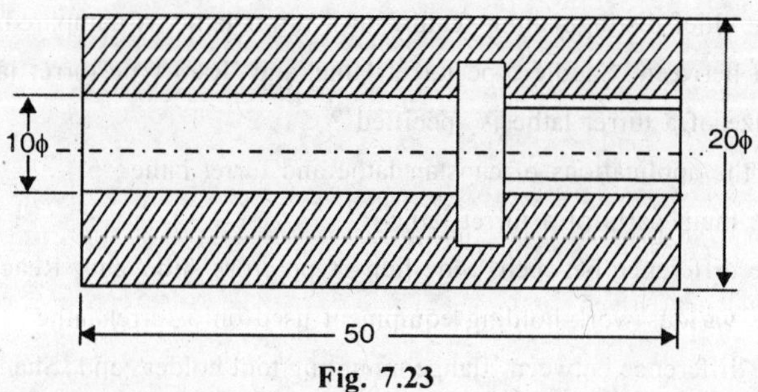

Fig. 7.23

Solution. Machine Capstan lathe

Operation :

1. Feed the bar stock to combined stock stop and start drill. Close the collect. The end of bar is then centred by advancing the start towards it—Turret.
2. Drill the internal diameter — Turret.
3. The thread diameter is bored to correct size, boring bar in a slide tool — Turret.
4. Ream the drilled hole, Reamer — Turret.
5. Make the recess, slide tool — Turret.
6. Cut the internal threads, Tap-Turret.
7. Cut-off, parting tool. — rear cross slide. The tool layout is shown in Fig. 7.24.

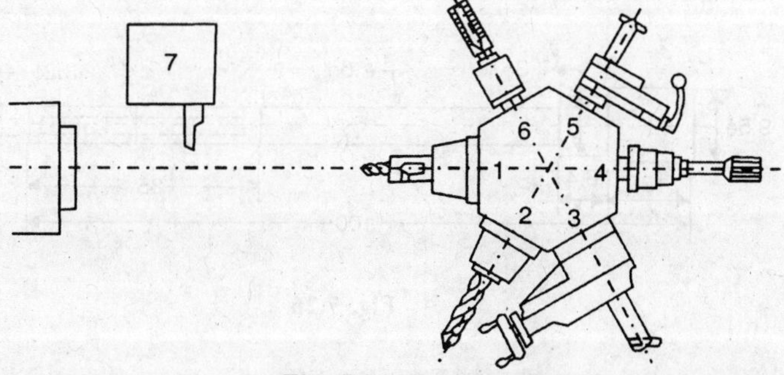

Fig. 7.24. Tool Layout.

7.11. PROGRAMME-CONTROLLED TURRET LATHES

The present day trend is towards replacing the turret lathe by the more sophisticated but costly N.C. Lathe equipped with a turret for multi-tool operation. However, where high initial and peripheral Costs can not be considered, automatic turret lathes with electronic programme control are ideal. The turret lathe operation involves, on an average, over 6000 manipulations during one working day. It is laborious for the workers to manually set the speeds and feeds, swing the tools to the required position and bring them upto the workpiece to engage the feed. The programme controlled turret lathe cuts through all the hard work of setting speeds, feeds, engaging and disengaging of feeds and indexing clamping of the turret head. The lathe operates on a fully automatic cycle and there is a saving in cycle time when compared to manually operated turret lathes. One method of having automatic controls is by electric master switches. HMT has designed and developed the programme-controlled turret lathe PTL 30 with a microprocessor based programming system.

Programme-controlled turret lathe is of great convenience and bridges the gap between the traditional turret lathe and the sophisticated and costly NC lathe. This lathe is mainly used in medium batch production industries engaged in the production of components like brake linings, bushes, bearing housings, gear blanks etc.

PROBLEMS

1. What is the chief distinguishing feature of a turret lathe as compared to an engine lathe ?
2. Distinguish between 'Saddle type turret lathe' and 'Ram type turret lathe.'
3. How the size of a turret lathe is specified ?
4. Enumerate the applications of capstan lathe and turret lathe.
5. Discuss the main parts of a turret lathe.
6. What is the difference between 'Side hung type cross slide' and 'Reach over type cross slide.
7. Discuss the various work-holding equipment used on a turret lathe.
8. What is the difference between 'flange mounting tool holder', and 'Shank mounting tool holder'
9. What is meant by 'multiple cut' and 'combined cut'.
10. Enumerate the advantages of a turret lathe.
11. What is meant by 'tool layout' of a turret lathe ?
12. How the total production time on a turret lathe be minimised ?
13. Enumerate the various rules which must be followed while laying out the sequence of operations for a turret lathe.

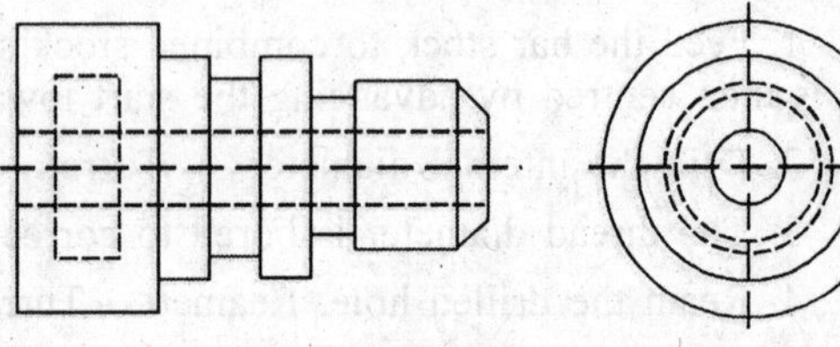

Fig. 7.25

14. Draw the tool layouts for the components shown in Fig. 7.25 to 7.27, which are to be produced on turret lathe.

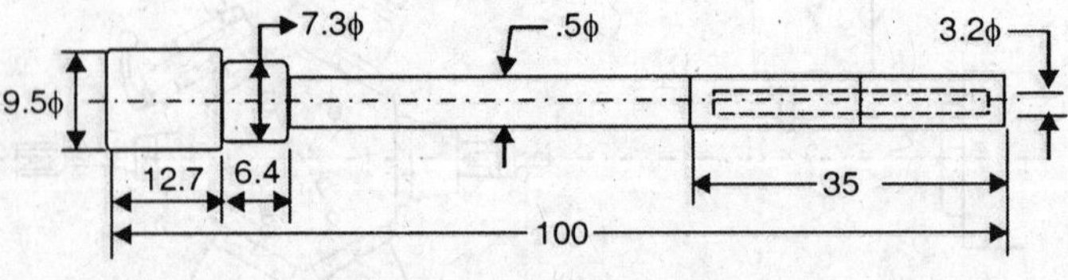

Fig. 7.26

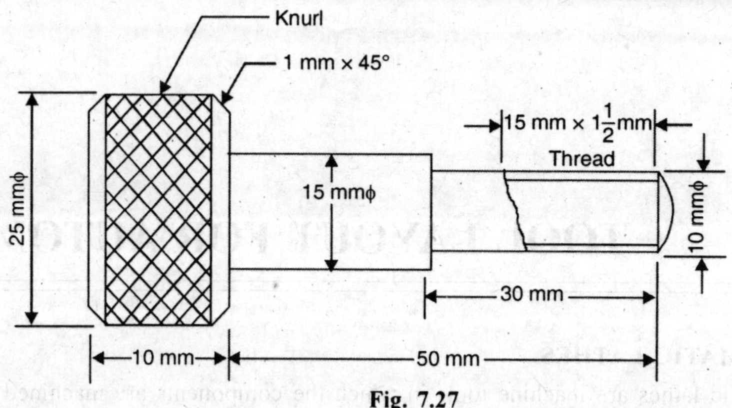

Fig. 7.27

15. Draw the tool layout for the component shown in Fig. 7.28, which is to be produced on a turret lathe.

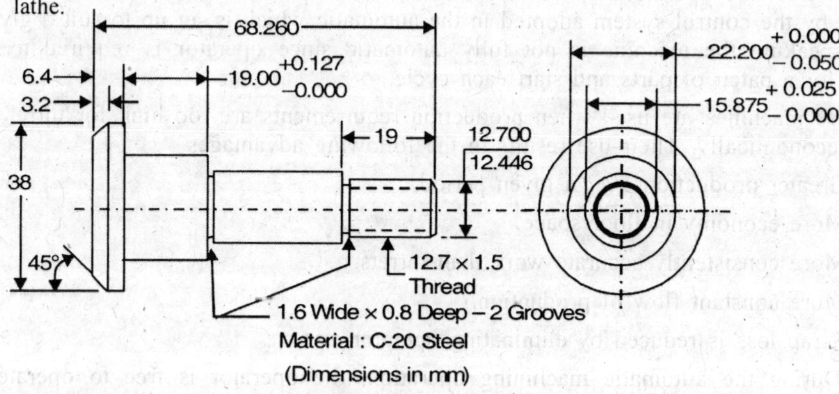

Fig. 7.28. A Stud

16. Draw the tool layout for capstan lathe to manufacture the components shown in Fig. 7.29 and Fig. 7.30.

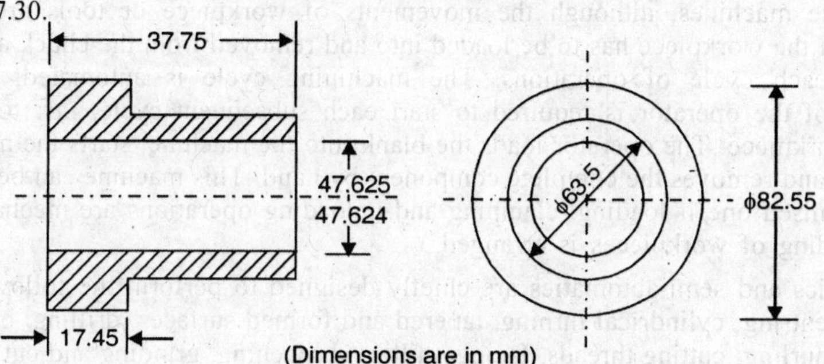

Fig. 7.29. A Flanged Bush.

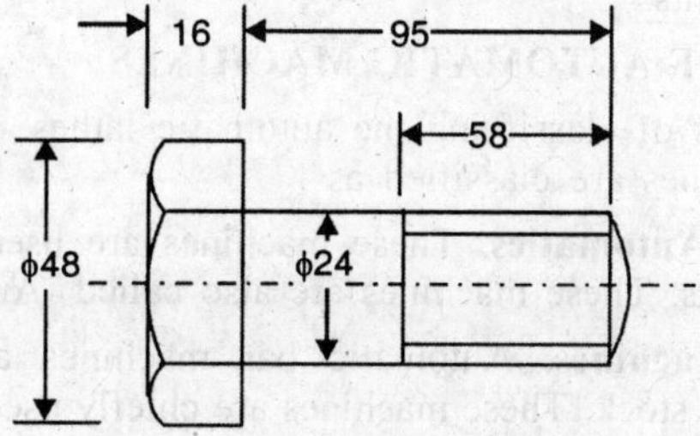

Dimensions are in mm

Fig. 7.30. A Hexagonal Bolt.

8

TOOL LAYOUT FOR AUTOMATICS

8.1. AUTOMATIC LATHES

Automatic lathes are machine tools in which the components are machined automatically. The working cycle is fully automatic that is repeated to produce duplicate parts without participation of the operator. All the working and idle operations are performed in a definite sequence by the control system adopted in the automatic which is set up to suit a given work. Strictly speaking the machine is not fully automatic since operator is required to load the machine for a batch of parts and start each cycle.

These machines are used when production requirements are too high for turret lathes to produce economically. Their use results in the following advantages :

1. Greater production over a given period.
2. More economy in floor space.
3. More consistently accurate work than turrets.
4. More constant flow of production.
5. Scrap loss is reduced by eliminating operator error.
6. During the automatic machining operation, the operator is free to operate another machine or inspect the completed parts.

Semi automatics. Semi-automatic machines are usually turning machines adapted to chuck work. In these machines, although the movements of workpiece or tools are automatically controlled, but the workpiece has to be loaded into and removed from the chuck at the beginning and end of each cycle of operations. The machining cycle is automated, but the direct participation of the operator is required to start each subsequent cycle, *i.e.*, to machine each subsequent workpiece. The operator loads the blank into the machine, starts the machine, checks the work size and removes the complete component by hand. This machine can be converted into a fully mechanised one if loading, clamping and unloading operations are mechanised and also magazine loading of workpieces is arranged.

Automatics and semi-automatics are chiefly designed to perform the following machining operations : centring, cylindrical turning, tapered and formed surfaces, drilling, boring, reaming spot facing, knurling, cutting threads, facing, milling, broaching, grinding and cut off. Additional operations such as slotting and cross drilling etc. can also be performed on these machines with the help of special attachments.

8.2. CLASSIFICATION OF AUTOMATIC MACHINES

These are various ways of classifying the automatic lathes. Depending upon the type of work machined, these machines are classified as :

1. **Magazine Loaded Automatics.** These machines are used for producing component from accurate separate blanks. These machines are also called `Automatic chucking machines.

2. **Automatic Bar Machines.** Automatic bar machines are designed for machining components from bar or pipe stock. These machines are chiefly used for the manufacture of high

quality fasteners (screws, nuts and studs), bushings, shafts, rings, rollers, handles and other parts, usually made of bar or pipe stock, and recently of separate blanks as well.

Depending upon the number of work spindles, the automatic lathes can be classified as : single spindle automatics and multispindle automatics.

Automatics can also be classified according to : purpose and arrangement of the spindle. According to purpose, automatics are classified into : general and single purpose machines. According to the arrangement of the spindle, automatics are classified into horizontal and vertical machines.

In this chapter, we shall deal with automatics and semiautomatics in which automaticity is achieved by mechanical means using cam drives. The operating cycle is automatically controlled with the help of a camshaft carrying cams which are linked to the operative mechanisms (slides, feeding and clamping mechanisms) through a system of levers, gears etc. when the camshaft rotates, the cams actuate the various operative units, Fig. 8.1. According to their forms the cams are classified as plate cams and cylinder cams.

In accordence with their cycle control, the automatics may be grouped into three categories :

1. In the first category, automatics have a single camshaft which rotates at a constant speed for the given set up and which controls both working and idle motions. The speed of the camshaft corresponds to slow working motion and since the idle motions should be completed in shortest possible time, considerable time will be wasted for these motions for the machines of this category. However, due to its simplicity, this category is justified for small automatics with only few idle motions.

2. Automatics of the second category have also only one camshaft but it has two speeds of rotation : slow for working motions and fast for idle motions. However, the camshaft has quite a large moment of inertia (due to whole set of working and idle cams) and due to this,

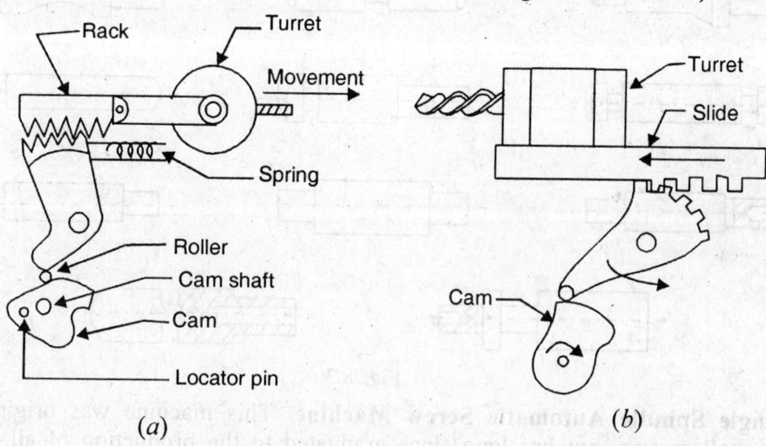

Fig. 8.1

the changeover mechanisms will experience impact loads when switching over from one speed to the other. Thus, it is rather impossible to obtain any significant difference between the two speeds of rotation.

3. The third group of automatics have one camshaft for working motion which rotates at slow speed and a high speed auxiliary camshaft for actuating the idle motions. This group finds applications for machining complex shaped work-pieces where many auxiliary motions are required.

Automaticity can also be obtained by Hydraulic, Pneumatic, Electric and combined systems and also by punch tapes, punched cards, magnetic tapes or by a drum.

8.2.1. Type of Single Spindle Automatics. The single spindle automatic lathes are of the following types :

1. **Automatic cutting off machines.** These machines produce short workpieces of simple form by means of cross-sliding tools. The machines are simple in design. The head stock with the spindle is mounted on the bed. Two cross slides are located on the bed at the front end of the spindle. Cams on a camshaft actuate the movements of the cross-slides through a system of levers.

The principle of operation of such a machine is explained below with the help of Fig. 8.2. The required length of work (stock) is fed out with a cam mechanism, upto the stock stop which is automatically advanced in line with the spindle axis, at the end of each cycle. The stock is held in the collect chuck of the rotating spindle. The machining is done by tools held in slides operating only in the cross-wise direction. Special attachments can be employed if holes or threads are required on the simple parts. Typical simple parts (from 3 to 20 mm in diameter) machined on such a machine are shown in Fig. 8.3.

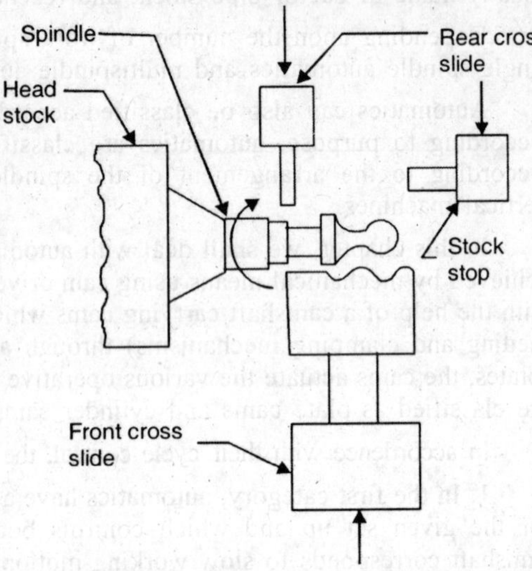

Fig. 8.2. Automatic Cutting-off Machine.

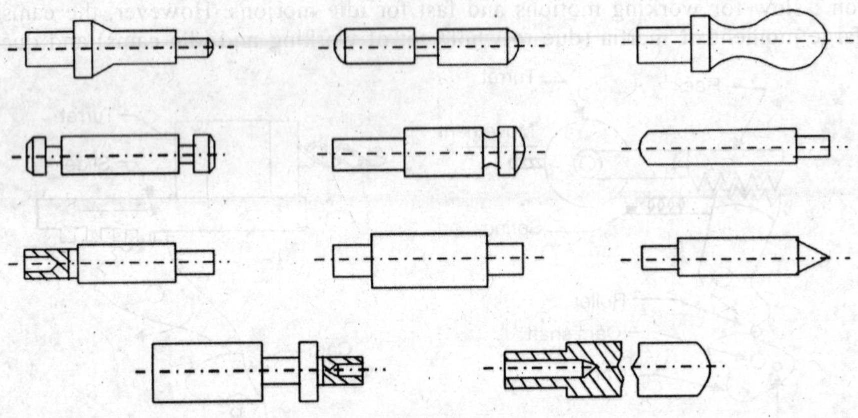

Fig. 8.3

2. **Single Spindle Automatic Screw Machine.** This machine was originally used for producing small screws, but has long since graduated to the production of all sorts of small turned parts. These machines are essentially wholly automatic bar type turret lathes. They differ from the horizontal turret lathe in that the turret revolves about a horizontal pivot pin instead of about a vertical axis. They are designed for machining complex internal and external surfaces on parts made of bar stock or of separate blanks. Upto ten different cutting tools may be employed at one time in the tooling of such a screw machine. The tools are fixed in indexing turret and in cross-slides. The turret carries six tools. Two cross-slides (front and rear) are employed for cross-feeding tools. A vertical slide for parting off operation may also be provided. It is installed above the work spindle. The head stock is stationary and houses the spindle which rotates in either direction. The bar stock is held in a collet chuck and advanced by a feed finger after each piece is finished and cut off. All movements of the machine units are actuated by cams mounted on the camshaft. These machines are made in several sizes for bar work from 12.7 mm to 60 mm diameter.

Fig. 8.4. shows a general layout of a single spindle automatic screw machine. The bar stock is pushed through stock tube in a bracket and its leading end is clamped in rotating spindle by means of a collet chuck. The bar is then fed out for the next part by stock feeding mechanism. Longitudinal turning and machining of the central hole are performed by tools mounted on turret slide. The cut off and form tools are mounted on the cross-slides. At the end of each cut, turret slide is withdrawn automatically and indexed to bring the next tool into position. The indexing motion is accompanied by an additional rapid retraction.

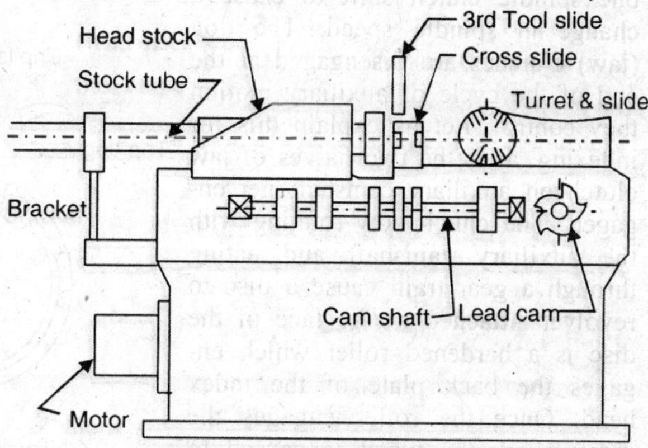

Fig. 8.4. Automatic Screw Machine.

Gearing mechanism of a single spindle automatic screw machine is shown in Fig. 8.5. The camshaft is at the front of the machine. It is driven from a shaft at the back of the machine (called back-shaft or auxiliary camshaft), through cycle time change gears. The backshaft rotates at a constant speed, usually 2 rev/s or 4 rev/s. One revolution of camshaft produces one component. The speed of the camshaft can be changed for each new component by means of cycle time change gears.

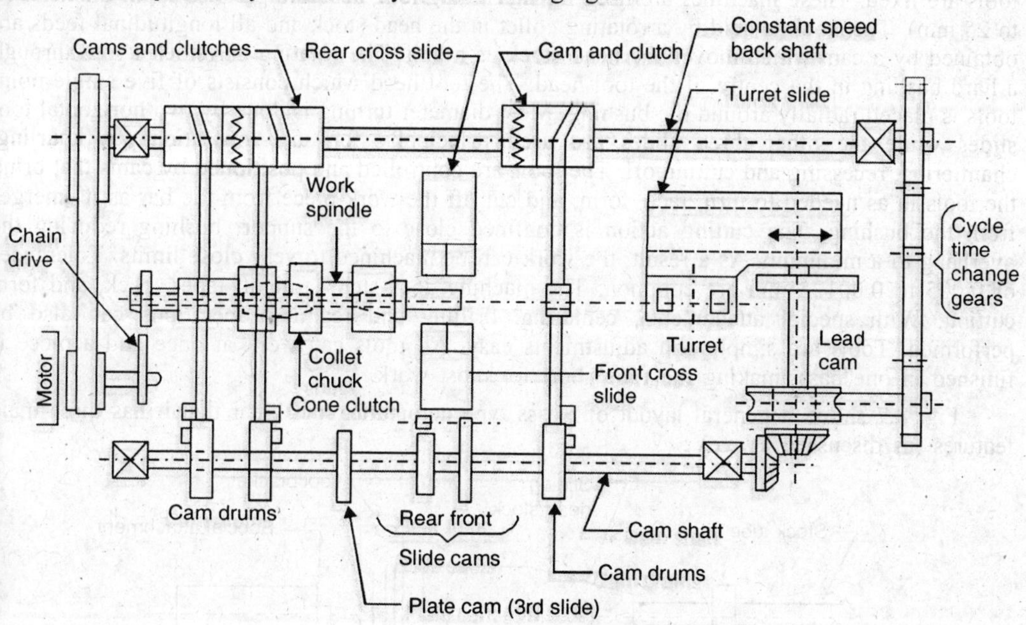

Fig. 8.5. Gearing diagram of Automatic Screw Machine

The auxiliary camshaft carries dog clutches in three positions. These are operated through drum cams, which are actuated through levers from cam drums on main camshaft. Trip dogs are bolted in T-slots, in the sides of cam drums or dog carriers and can be adjusted circumferentially in T-slots. The trip lever, Fig. 8.6, is actuated by a trip dog (on main camshaft). The lever releases a pin in one half of the clutch on the drive shaft (auxiliary camshaft). This clutch is then caused to slide over the auxiliary camshaft until it engages the other half of the clutch which is revolving, by a dowel pin. Depending upon which trip lever is raised, the respective clutch (which then acts through a gear train) stays engaged long enough to complete : one (or two) index, one stock feed (by causing the collet chuck to open, hence the bar feeds forward to the bar stop in the turret, and the collet closes against gripping the bar). or one spindle clutch shift to cause a change in spindle speed. The dog (jaw) clutches are disengaged at the end of the cycle of auxiliary motion they control. Let us explain this for indexing, after the two halves of jaw clutch on auxiliary camshaft get engaged. The clutch now rotating with the auxiliary camshaft and acting through a gear train cause a disc to revolve. Attached to the face of the disc is a hardened roller which engages the back plate of the index head. Once the roller engages the index head, the turret is caused to revolve in the manner of Geneva motion.

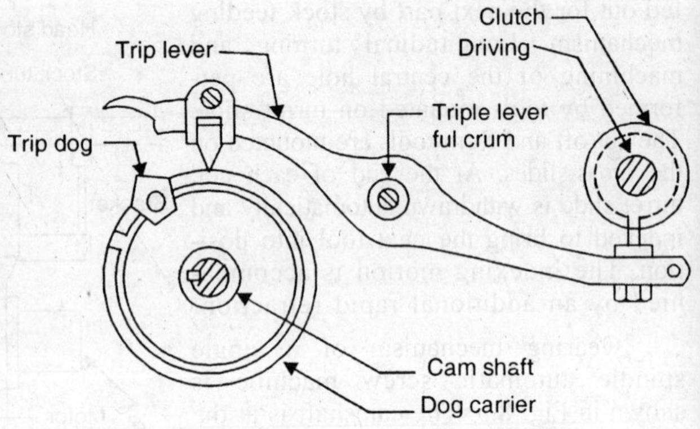

Fig. 8.6. Trip Dog and Lever Mechanism.

3. **Swiss type Automatic Screw Machine.** The machine is also known as `Sliding head screw machine', or `movable headstock machine', because the head stock is movable and the tools are fixed. These machines are used for machining long accurate parts of small diameter (2 to 25 mm). The stock is held by a rotating collet in the head stock and all longitudinal feeds are obtained by a cam which moves the head stock as a unit. The rotating bar stock is fed through a hard bushing in the centre of the tool head. The tool head which consists of five single point tools is placed radially around the bushing. Most diameter turning is done by two horizontal tool slides while the other three slides are used principally for such operations as knurling, chamfering, recessing and cutting off. The tools are controlled and positioned by cams that bring the tools in as needed to turn, face, form, and cut off the work-piece from the bar as it emerges from the bushing. The cutting action is confined close to the support bushing reducing the overhang to a minimum. As a result, the work can be machined to very close limits. Tolerances of 0.005 to 0.00125 mm are common. The machine does step, straight, taper, back, and form cutting. With special attachments, centering, drilling and reaming operations can also be performed. Tools are simple and adjustments easy. All tools can work at once and a piece is finished in one pass, making the time short for most work.

Fig. 8.7 shows a general layout of Swiss type automatic. It is clear that it has three main features (as discussed above) :

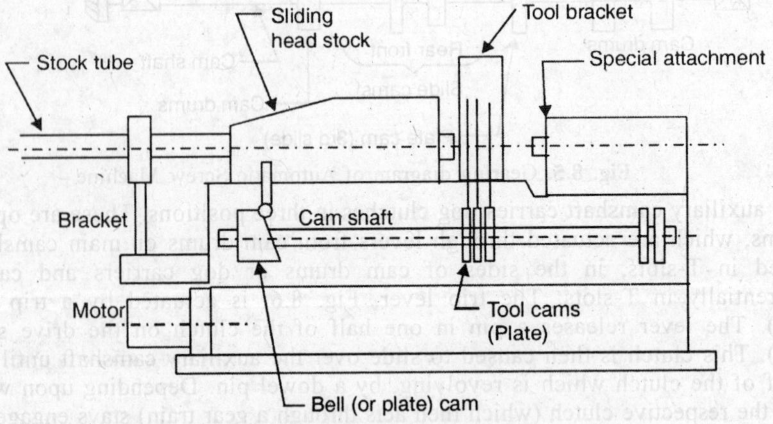

Fig. 8.7. Swiss Type Automatic

1. A sliding head stock through which the bar stock passes.

2. A tool bracket which supports five tool slides and also guide bush to support the bar stock.

3. A camshaft at the front of the machine. Cams on the camshaft control the movement of the tool slides and headstock.

In addition to the above three main features, the machine has a special attachment mounted on the right hand side of the bed, for machining of the central hole, i.e., drilling, boring, cutting threads with taps or dies.

Fig. 8.8 shows the arrangement of tool slides viewed in the direction normal to the horizontal axis of head stock.

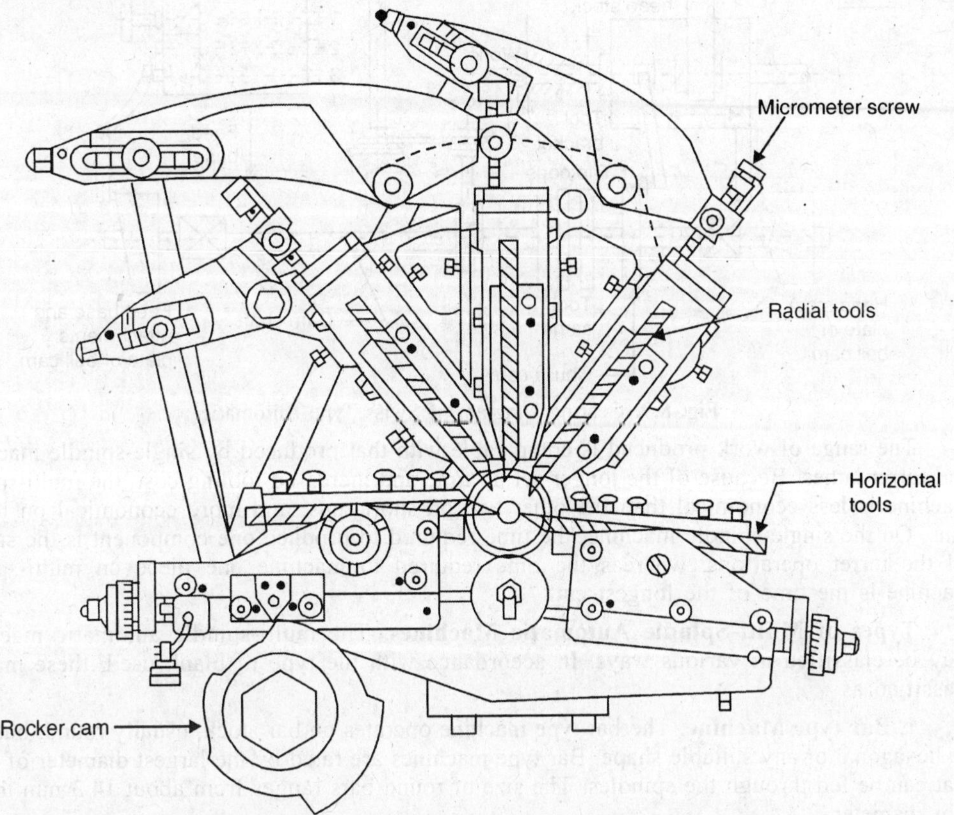

Fig. 8.8. Tool Slide Arrangement of Swiss Type Automatic

Fig. 8.9 shows the gearing diagram of a Swiss type automatic. The machine has a single camshaft which controls both the working and idle motions (first category of automatics). The camshaft is driven from the constant speed backshaft through cycle time change gears. One revolution of the camshaft produces one component. The cycle time gears can be changed for different parts.

8.2.2. Multi Spindle Automatic Machines. The multi-spindle automatic machines are the fastest type of production machines and are made in a variety of models with two, four, five, six or eight spindles. In contrast to the single spindle machine, where one turret face at a time is working on one spindle, the multi-spindle machine has all turret faces working on all spindles at the same time. Their production capacity is higher than that of single-spindle machines but their machining accuracy is somewhat lower. The rate of production of a multi-spindle machine, however, is less than that of the corresponding number of single-spindle machines. For example, the production capacity of a four spindle machine is not four times but only $2\frac{1}{2}$ to 3 times more than that of a single-spindle machine.

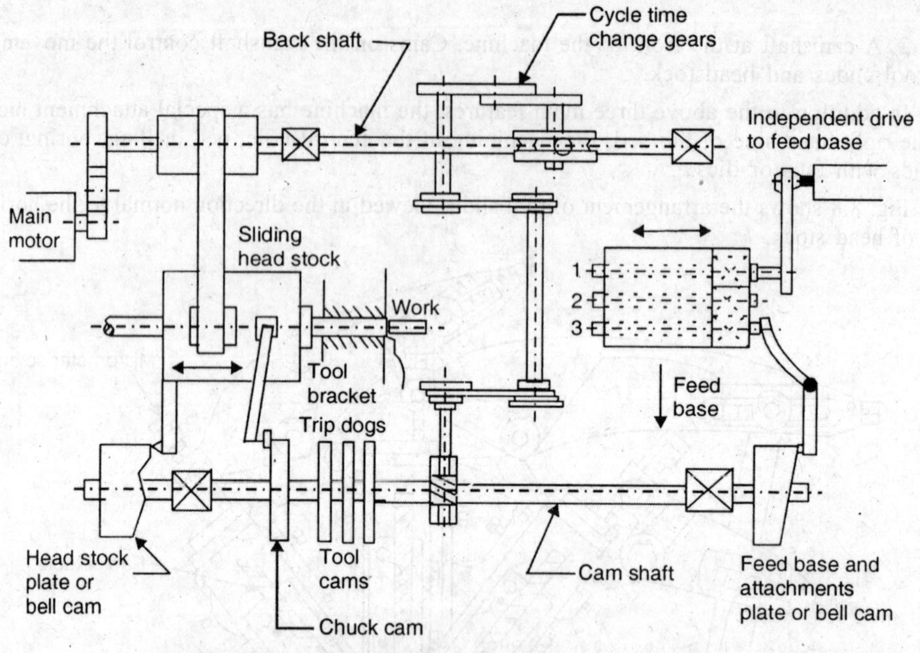

Fig. 8.9. Gearing Diagram of Swiss Type Automatic.

The range of work produced is comparable with that produced by single-spindle machines and turret lathes. Because of the longer set up time and increased tooling cost, the multi-spindle machine is less economical than the other two on short runs, and more economical on longer runs. On the single-spindle machine, the time required to produce one component is the sum of all the turret operations, whereas the time required to machine one piece on multi-spindle machine is the time of the longest cut.

Types of Multi-Spindle Automatic Machines. The multi-spindle automatic machines may be classified in various ways. In accordance with the type of blank used, these may be classified as :

1. **Bar type Machine.** The bar type machine operates on bar stock, usually round, but may be hexagonal or any suitable shape. Bar type machines are rated by the largest diameter of stock that can be fed through the spindles. The size of round bars ranges from about 14.3 mm to 197 mm diameter.

2. **Magazine-loading type or Chucking-type Machine.** The chucking machine is identical to bar type machine and operates in the same manner but handles several individual work-pieces, such as castings or forgings held in several chucks. The capacity of this machine is the diameter of the work that can be swung over the tool slides. The common sizes are upto about 250 mm diameter.

In accordance with their principle of operation, the multi-spindle automatic machines may be classified as :

(*i*) Parallel action automatics, and

(*ii*) Progressive action automatics.

(*i*) **Parallel action multi-spindle automatics.** Parallel action multi-spindle automatic machine is also called as `multiple-flow' machine. In this type of machine, the same operation is performed on each spindle and a work-piece is finished in each spindle in one working cycle. It means that the number of components being machined simultaneously is equal to the number of spindles in the machine. The rate of production is thus very high, but the machine can be employed to machine simple parts only since all the machining processes are done at one position.

These machines are usually automatic cutting-off bar-type machines. They are used to perform the same work as single-spindle automatic cutting off machines. Centering or a single drilling operation can also be performed on certain models. The machine consists of a frame with a head stock at the right end. The horizontal work spindles which are arranged in a line, one above the other, are housed in this headstock (Fig. 8.10). Cross-slides are located at the right and left-hand sides of the spindles and carry the cross-feeding tools. Tool slide on each side accommodates tools for all of the spindles. All the working and the auxiliary motions of the machine units are obtained from the cam mounted on the cam-shaft.

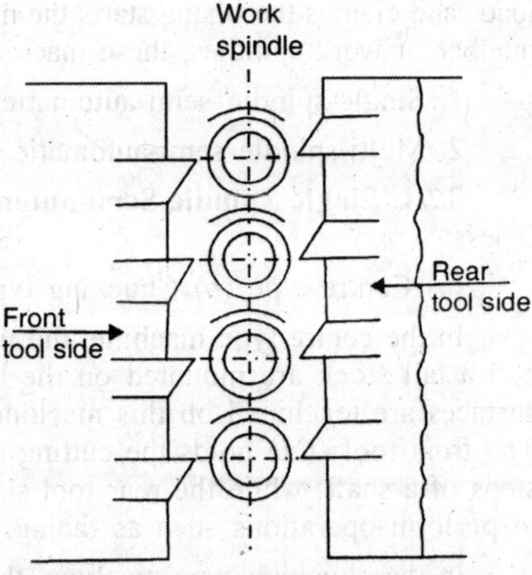

Fig. 8.10. Parallel-action multi-spindle Automatic.

(*ii*) **Progressive-action multi-spindle Automatics.** In this design of multi-spindle automatic machine, the workpiece is machined in states and progressively in station after station. The head stock is mounted at the left end of the base of the machine. It contains a spindle carrier which rotates about a horizontal axis through the centre of the machine. The working spindles are mounted in this spindle carrier. The spindles carry the collets and bars from which the workpieces are machined. The bar stock is fed through each spindle from the rear. On the face of the spindle-carrier support are mounted cross slides which carry tools for operations such as cut off,

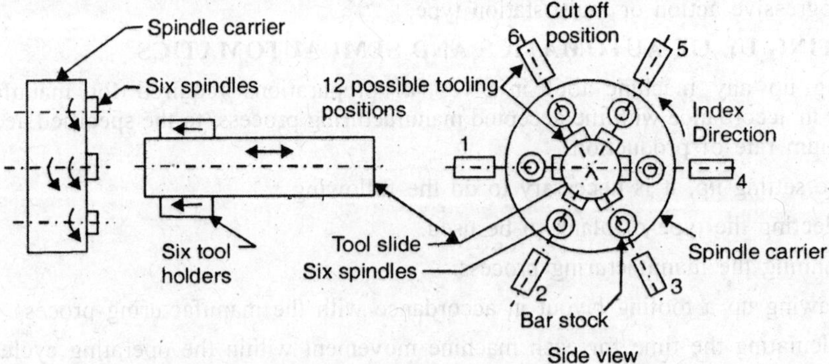

Fig. 8.11. Six Spindle Progressive Action Automatic.

turning, facing, forming and chamfering. The number of these slides equals the number of spindles. The main tool slide (end tool slide) extends from the middle of this support. Here also, the number of tool slides or faces is equal to the number of spindles. These tool slides move forward at the proper time and the proper amount and are operated by their own shaft and gears. The fed of each tool, both cross-slide tools and end-slide tools, is controlled by its own individual cam.

A six-spindle progressive action automatic is shown in Fig. 8.11. The spindle carrier indexes on its own axis by 60 (360 divided by the number of spindles) at each tool retraction. As the spindle carrier indexes, it carries the work from station to station, where various tools operate on it. The stock moves around the circle in counter clockwise direction and returns to the station number 6 for cutting off. A finished component is obtained each time the spindle carrier indexes.

8.3. CLASSIFICATION OF SEMI-AUTOMATICS

Semi-automatic lathes are employed for machining work from separate blanks. The operator loads and clamps the blank, starts the machine and unloads the finished work. Depending upon the number of work spindles, these machines are classified :

1. Single-spindle semi-automatic machine.
2. Multi-spindle semi-automatic machine.

8.3.1. Single Spindle Semi-automatics. Single-spindle semi-automatics are of the following types.

(*a*) Centre type (*b*) Chucking type.

In the centre type machine, the workpiece is held between centres, for which a head stock and a tail stock are mounted on the bed of the machine. Usually, external stepped or formed surfaces are machined on this machine. The work is machined by two groups of cutting tools. The front tool slide holds the cutting tools which require a longitudinal feed motion to turn the steps of a shaft, while the rear tool slide carries the tools that require a transverse feed motion to perform operations such as facing, shouldering, necking and chamfering etc.

In the chucking type machine, the workpiece is held in a chuck. Such a machine may be equipped with various tool slide arrangements. In addition to longitudinal and transverse feed tool slides, these machines may also be equipped with a central end-working tool slide or a turret if internal surfaces are also to be machined in addition to the external surfaces.

8.3.2. Multi-spindle Semi-automatics. This machine may also be built in two designs.

(*a*) Centre type, and (*b*) Chucking type

These semi-automatics, in the same way as automatics, are classified as :

1. Parallel action or single station type, and
2. Progressive action or multi station type.

8.4. SETTING UP OF AUTOMATICS AND SEMI AUTOMATICS

Setting up any machine tool involves all preparations required for manufacturing a component in accordance with the accepted manufacturing process, to the specified accuracy and at a maximum rate of production.

Before setting up, it is necessary to do the following :

1. Selecting the type of blank to be used.
2. Planning the manufacturing process.
3. Drawing up a tooling layout in accordance with the manufacturing process.
4. Calculating the time for each machine movement within the operating cycle.
5. Selecting the standard cutting tools, gauges and cams in accordance with the tooling layout.
6. Designing and manufacturing all special tools.

When the above preparatory work has been completed, the set up man proceeds to set up the machine tool. As already discussed in Chapter 7, the machining process is broken into separate operation elements, so that the work is completed at a maximum rate of production and at minimum cost. The processing breakdown between the various cutting tools must be checked several times to select the most suitable method. For this, the guidelines as discussed in Chapter 7 (Article 7.8.1) are also valid for automatics and semi automatics.

8.5. TOOLING LAYOUT AND OPERATION SHEET

The tooling layout for an automatic contains the sketches of the work in the various stages of machining, by operation elements. The required cutting tools are indicated and slown at the end of the feed motion. The operation sheet usually has a sketch of the finished workpiece. It indicates the material of which the workpiece is to be made and lists the operations in their proper sequence. To fill up the operation sheet, it will be necessary to select :

(*a*) cutting speeds (*b*) cutting feeds, and (*c*) to calculate the lengths of travel of all the tools in each operation element. Cutting speeds and feeds are given in Table 4.1. The following rules apply for the cutting speeds and feeds on automatics :

1. The spindle speed is the same for two tools working simultaneously and is determined by the maximum permissible cutting speed on the largest diameter machined in the given operation element.

2. Cutting tools mounted on the same slide and operating simultaneously, will always have the same rate of feed which is determined by the tool permitting the minimum feed.

The spindle speed is determined by substituting the selected cutting speed in the formula (4.9), that is,

$$N = \frac{1000\,V}{\pi D}$$

where V is the cutting speed in mpm of work or cutter, and D is the diameter of work or cutter in mm.

The nearest values of available spindle speeds are then selected from a table of speed steps and the corresponding change gears are installed in the main drive of the automatic.

The next step is to determine the travel of the cutting tools required in each operation element. This is obtained from the workpiece drawing. To this should be added about 0.5 to 3 mm (tool approach) to prevent the tool from cutting into the work during its rapid approach movement.

8.5.1. Time Required for each Operation Element. The number of spindle revolutions for each operation element is obtained by the relation.

$$n = \frac{L}{f}$$

where L is the tool travel in mm, and f is the rate of feed in mm per revolution.

The time required for each operation element is determined by the formula,

$$Tm = \frac{L}{fN}, \text{ min}$$

The total time is based on the total number of spindle revolutions to complete a piece if the spindle runs continuously at one speed. If different spindle speeds are used for different operation elements it is necessary to recalculate the number of spindle revolutions in each operation element to a single basis, by using the so called 'equivalent revolutions factor', which is the ratio of the higher to the lower speed.

8.5.2. Time Required for Idle Motions and Working Travel. The various idle motions in an automatic machine operation can be : feed stock to stop and index turret etc. The time needed for these idle motions depends on the design and type of automatic machine. In certain types, the idle motion time is constant and does not depend on the workpiece. The idle motion time is usually determined from the table given in the setting-up instructions of each automatic. The time required for the idle movements may be determined by estimating the approximate time for whole cycle, during which one workpiece is produced. This is obtained on the basis of the working travel. The total time is then increased by 25 or 30 per cent. Then, the time required for the idle movements is obtained from table corresponding to the estimated cycle time. After that the cycle time can be calculated exactly.

Cams are usually divided into either 100 or 360 parts. The time for idle movements is usually given in terms of divisions on the cam surface. For example, the following values are usually taken for the various idle movements :

Feed stock to stop = 2.5 hundredths of cam surface

Index turret = 1.5 hundredths of cam surface

Withdrawal of the cut-off tool = 3 hundredths of cam surface

To determine the time required for the working and idle movements, in terms of hundredths of cam surface, it is necessary to find the relationship between the spindle revolutions and hundredths of cam surface. If K_i is the number of hundredths required for the idle movements, then the number of hundredths of cam surface for the total working travel will be,

$$K_w = (100 - K_i)$$

Now, if 'n_w' is the number of spindle revolution for working travel, then the number of spindle revolutions for the whole cycle will be

$$n_{cycle} = \frac{n_w \times 100}{K_w}, \text{ revolutions per cycle}$$

∴ Number of spindle revolutions for each hundredths of cam surface = n cycle/100.

8.6. CAM DESIGN

For laying out the cam, the radii of the plate cam curves are determined in the following manner :

1. The main data on the cam and the drive of the operative unit are taken from the service manual. This includes (*a*) maximum radius of cam R_{max}, (*b*) minimum radius of cam R_{min}, (*c*) height of the cam lever fulcrum above the cam centre, (*d*) ratio of the lever system.

2. The cam radii are determined for each operation element from the final position of the operative unit (turret or cross-slide).

3. For hexagonal turret slide or end working slide, R_{max}, corresponds to the minimum, distance of the slide from the spindle nose. For cross-slides, R_{max} will correspond to the position where the tool has advanced slightly beyond the spindle axis.

4. The other radii for the final positions of the operative units are determined as follows :

For first operation, $R_1 = R_{max} - (L_1 - L_{min})$

where L_1 is the distance between the turret face and the spindle for first operation and L_{min} is the minimum distance between the turret face and spindle.

5. The cam radii of the beginning of the working travel for each operation is determined by subtracting the travel of the cutting tool from the radius at the end of each corresponding travel.

After knowing the various can radii, the cam can be drawn. In doing so, the following points should be taken care of :

(*i*) The movement of the operative unit (turret, cross-slides) usually consists of : rapid approach, working travel, and rapid return.

(*ii*) Rapid approach and rapid return being idle motions, should be performed with minimum time loss. Therefore, the cam curves should rise or drop sharply at these places. Curves for these two movements are drawn to a special template in accordance with cycle time.

(*iii*) The working travel curve should provide for motion at a constant speed, *i.e.*, at a uniform rate of feed. This uniform curve is called Archimedean spiral. This curve can be drawn to a special template or geometrically as explained below :

The cam is drawn full size and its surface is divided into hundredths. The lines of the hundredths are arcs drawn with a radius equal to that of the cam roll lever. The beginnings and ends of working travel lobes are marked off on the corresponding hundredths by drawing arcs of the required radii. Then the cam roll is drawn full size, tangent to the above points. The distance between the two roll positions, *i.e.*, the cam rise is divided into as many equal parts as there are hundredths in the given working travel. The points of intersection of the arcs with Arabic numerals and Roman numerals are marked, Fig. 8.12. A uniform curve passing through these points will ensure a uniform feed of the operative unit.

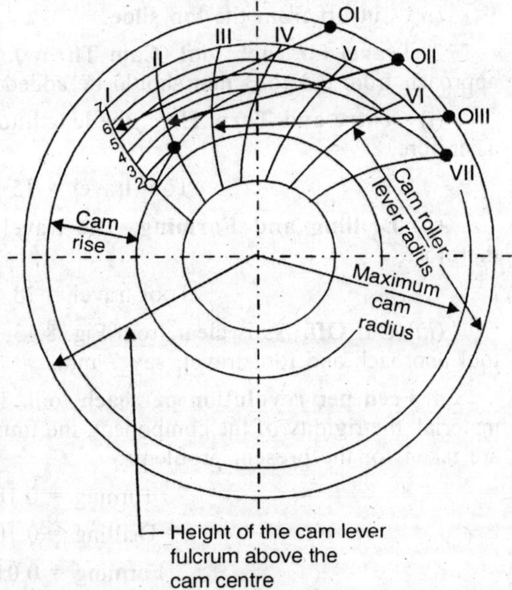

Fig. 8.12. Developing a uniform cam rise curve.

8.7. TOOL LAYOUT FOR AUTOMATIC SCREW MACHINE

In this article, the procedure for laying out the tools for an automatic screw machine to manufacture the component shown in Fig. 8.13, will be explained. The following order of procedure is recommended :

1. Determine the spindle speed for the material and the diameter of the component. Select the nearest speed in the range available.
2. Decide the proper sequence of operations.
3. Determine the travel for each tool.
4. Select the feed per revolutions for each tool.
5. Determine the number of spindle revolutions for each operation.
6. Determine the number of spindle revolutions for each of the idle movements.
7. Determine the hundredths of cam surface required for dwells and clearance.
8. Calculate the total spindle revolutions to machine one component.
9. Calculate the hundredth of cam surface for each operation.
10. Complete the operation sheet and draw the cams.

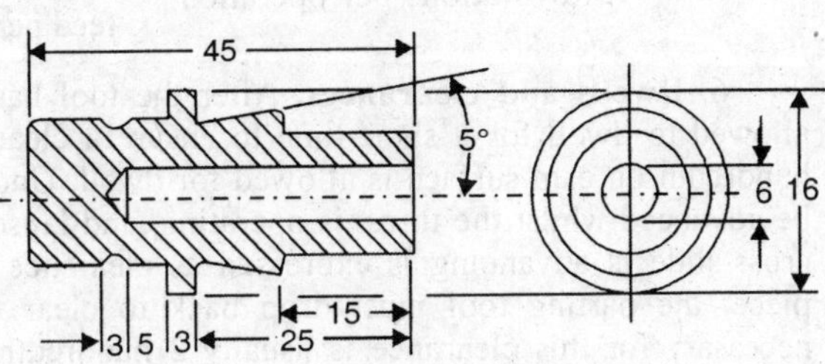

Fig. 8.13

1. **Spindle Speed.** The material of the component is Brass 16 mm diameter. The cutting speed for turning may be taken as 150 metre per min. Therefore the spindle speed is,

$$N = \frac{1000\,V}{\pi D} = \frac{1000 \times 150}{\pi \times 16}$$

$$= 2984 \text{ r.pm.}$$

The nearest standard spindle speed of 3000 r.p.m. is selected.

2. **Sequence of Operations.** This may be fixed as follows :

(*i*) Feed stock to turret stop.

(*ii*) Centre and turn-Turret slide.

(*iii*) Drill 6 mm diameter from turret slide. The forming from the front slide, the front portion of the component and the forming of the back portion of the component from the rear slide is overlapped with the drilling operations.

(*iv*) Cut off from the top slide.

3. **Travel for each tool (Cam Throw).** To calculate the total travel for each tool, a tool approach from 0.5 to 3 mm should be added to avoid damage to tools or component.

(*i*) **Centre and Turn.** Here, the length to be turned will be taken for calculation purposes. Therefore,

$$\text{Tool travel} = 15 + 1 = 16 \text{ mm}$$

(*ii*) **Drilling and Forming.** The travel of the drilling tool exceeds that for forming, therefore,

$$\text{Tool travel} = 33 + 2 = 35 \text{ min.}$$

(*iii*) **Cut Off.** As is clear from Fig. 8.13, the tool travel for cut off tool will be 6 mm plus tool approach and tool travel, say 7 mm.

4. **Feed per revolution per each tool.** The feed per revolution depends upon the type of material, the rigidity of the component, the finish required and the tool used. The following feeds are taken for the present problem :

Turning = 0.18 mm/rev.

Drilling = 0.10 to 0.12 mm/rev.

Forming = 0.018 to 0.050 mm/rev.

Cut off = 0.025 to 0.075 mm/rev.

The throw of the parting tool is divided into a low feed to cut-off the component, followed by an increased feed to clean up the end of the next component.

5. **Spindle revolutions for each working operations.** Having determined the tool travel for each operation and selecting the suitable feed, the revolutions of the spindle are calculated as :

$$\text{Revolutions per operation} = \frac{\text{Travel of tool}}{\text{feed per revolution}}$$

6. **Dwells and clearances.** After the tool has reached the end of its travel, it should be allowed to dwell for a short time to ensure a clean face and a uniform diameter. Usually, one hundredth on cam surface is allowed for dwell. Under certain conditions, the cross-slides cannot be advanced whilst the turret is operating, and lapse of time whilst the turret is receding and the cross-slide is advancing is expressed as clearance idle time. Also, after cutting off the work-piece, the parting tool must drop back to clear the bar on the next feed out. The interval necessary for this clearance is usually 2 hundredths of cam rotation.

7. **Tool spindle revolutions to machine one component.** In table 8.1, the spindle revolutions for working travels shown within parentheses are for overlapped operations and are not to be counted for calculating production. Similarly, cam hundredths for idle motions, shown within parenthesis are for overlapped operations and are not to be considered. From table 8.1.

Total number of spindle revolutions for working travel = 528

Number of hundredths for idle motions, $K_i = 30$

Number of hundredths for working travel,

$$K_w = 100 - 30 = 70$$

Number of spindle revolution to make one workpiece

Table 8.1. Operational Sheet for Automatic Screw Machine

Material : Brass 16 mm diameter **Production Time : 15 seconds**

Cutting Speed for turning : 150 m/min **Spindle R.P.M. : 3000**

Operative Unit	Operation element	Tool Travel, mm	Feed mm/rev	Spindle revolution for each operation element	Accepted for calculation		Number of hundredths on cam	Hundreths on cam for			
					Working travel	Idle motion		Calculating production		Cam Design	
					Spindle revolution	Hundredths on cam		Start	Finish	Start	Finish
1	2	3	4	5	6	7	8	9	10	11	12
Turret	1. Feed Stock to stop	-	-	-	-	4.0	4.0	0	4.0	0	4.0
	Double Index	-	-	-	-	4.0	4.0	4.0	8.0	4.0	8.0
Turret	2. Centre and turn	16	0.19	84	84	0	11.0	8	19	8	19
	Dwell	—	—	—	—	1.5	1.0	19	20	19	20
	Drop back	—	—	—	—	5.0	5.0	20	25	20	25

	Double Index	–	–	–	–	4.0	4.0	25	29	25	29
Turret	3. Drill 6 mm diameter	22	0.13	183	183	–	24	29	53	29	53
	Clear drill	–	–	–	–	8.0	8	53	61	53	61
	Drill	13.3	0.10	133	133	–	18	61	79	61	79
	Dwell		–	–	–	1.0	1.0	79	80	79	80
	Double Index	–	–	–	–	(4.0)	(4.0)	–	–	–	80
Cross Slide	3. Form	3.3	0.05	(66)	(66)	–	9	–	–	49	50
Front	Dwell	–	–	–	–	(1.0)	(1.0)	–	–	49	50
Rear Cross	3. Form	3.5	0.017	(203)	(203)	–	27	–	–	50	77
Slide	Dwell	–	–	–	–	(1.0)	(1.0)	–	–	77	78
	4. Cut off	1.5	0.027	(37)	(37)	–	5	–	–	75	80
	Cut off	5.5	0.043	128	128	–	17	80	97	80	97
	Dwell	–	–	–	–	1.0	(1.0)	97	98	97	98
	Clear	–	–	–	–	2.0	(2.0)	98	100	98	100

$$= \frac{528 \times 100}{70} = 750$$

Cycle time (seconds to make one piece)

$$= \frac{60 \times 750}{3000} = 15 \text{ secs}$$

$$\text{Gross production} = \frac{3600}{15} = 240 \text{ pcs/hr}$$

Actual production (assuming 15% losses)

$$= 240 \times 0.85$$

$$= 204 \text{ pcs/hr.}$$

8. **Number of hundredths on cam surface for each spindle revolution.** For drawing the cam, it is necessary to determine the portion on cam surface devoted to each operation. The cam surface is divided into 100 divisions round its circumference.

In the present example,

Total spindle revolution per piece = 750

$$\therefore \quad \text{Revolution per hundredth} = \frac{750}{100} = 7.5$$

Hence, by dividing the spindle revolutions for each working travel with 7.5 the equivalent hundredths on cam surface can be calculated.

9. **Drawing the cams.** To draw the cam layout, the following procedure is adopted. Let us take the case of turret slide cam :

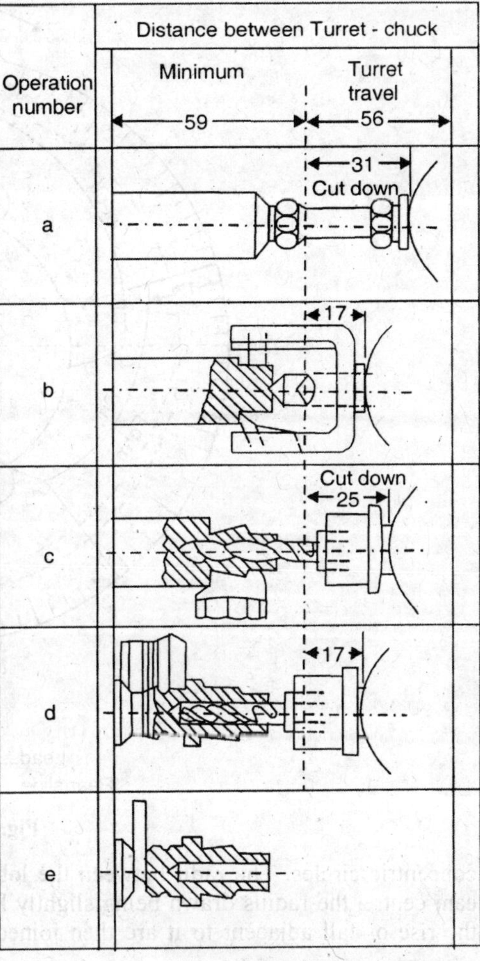

Fig. 8.14. Tool Layout for Cam Design.

(*a*) A circle is drawn for the maximum diameter of cam blank and this is divided off into 100 equal parts.

(*b*) To know the maximum radius of the cam lobes for each operation, we ought to determine the "cut down". For this, the accurate scale drawings for each operation (at the end of tool travel) have to be made. A vertical line is drawn touching the face of the collect chuck. Another line to the right is drawn which represents the forward-most position of the turret. A third line is drawn on the extreme right to represent the turret face at its maximum backward position Fig. 8.14. It will now be observed that the turret face is at varying distances behind the vertical line drawn for the turret forward position. These distances represent the "cut drawn", that is, the amounts by which the cam lobes must be reduced from the original outside radius of the cam blank.

(*c*) After determining the "cut down" for each operation (at the end of the tool travel), moving clockwise from the vertical line (zero division), Fig. 8.15, the hundredths of cam are divided off in the order and amounts set out in column 9 to 12 (Table 8.1). The radial lengths from the outer circle of the cam may now be set out by taking into account the "cut downs" at the end of the operation plus the cam throw for that operation. The cam surfaces joining the start and end of cutting operations may now be drawn as explained under article 8.6 or by means of a template. The cam surfaces for dwells and feeding stock to stop are drawn as

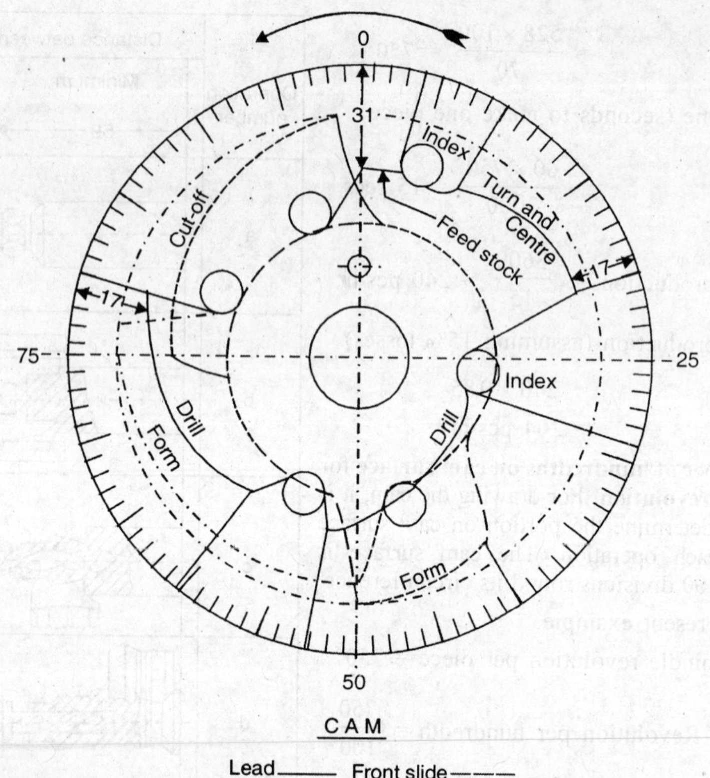

Fig. 8.15. Cam Layout.

concentric circles. The radii between the lobes, where the indexing occurs, are swung from the cam centre, the radius drawn being slightly less than the start of the next throw. This radius and the rise or fall adjacent to it are then joined by an arc equal to the cam roller radius.

The cams for the front, rear and top cross-sides may be set in the same manner as the turret cam.

8.8. PROGRAMMED AUTOMATIC LATHES

The standard automatic lathe is programmed to produced parts by means of Cams, stops or other mechanical methods. However, cams are suited to the production of a specific part and cam controlled automatic lathes are not referred to as programmed lathes. Complete versatility in programming is provided by Numerical Control. In a N.C. lathe, the programme is provided by the punched tape and no Cams are required.

8.9. BAR STOCK FEEDING

Bar stock may be fed by one of the following methods :

(*a*) By gravity. The devices are simple in design but are inconvenient in renewing long bars. Such devices are seldom used and then mainly in multi-spindle vertical automatic bar machines for small diameters and comparatively short bar stock.

(*b*) By driven rolls. This method is used for machining long components.

(*c*) By advancing the bar stock together with the head stock as in Swiss type machine.

(*d*) By the use of a feeding finger (stock pusher) and tube. This is the main method applied in general purpose single spindle automatic screw machines and multi spindle automatic bar machines.

The spring feeding finger is carried in the feeding tube which is arranged inside the collet tube. The stock pusher is screwed on the front end of the feeding tube. The feeding finger is a slitted spring bushing in which the jaws were compressed or closed before heat treatment. The jaws maintain a constant grip on the stock sufficient to feed the bar forward the moment it is released by the collet chuck.

The spring action of the jaws should not be excessive, however, so that the feeding finger may be drawn back over the stock to the initial position when the bar is clamped.

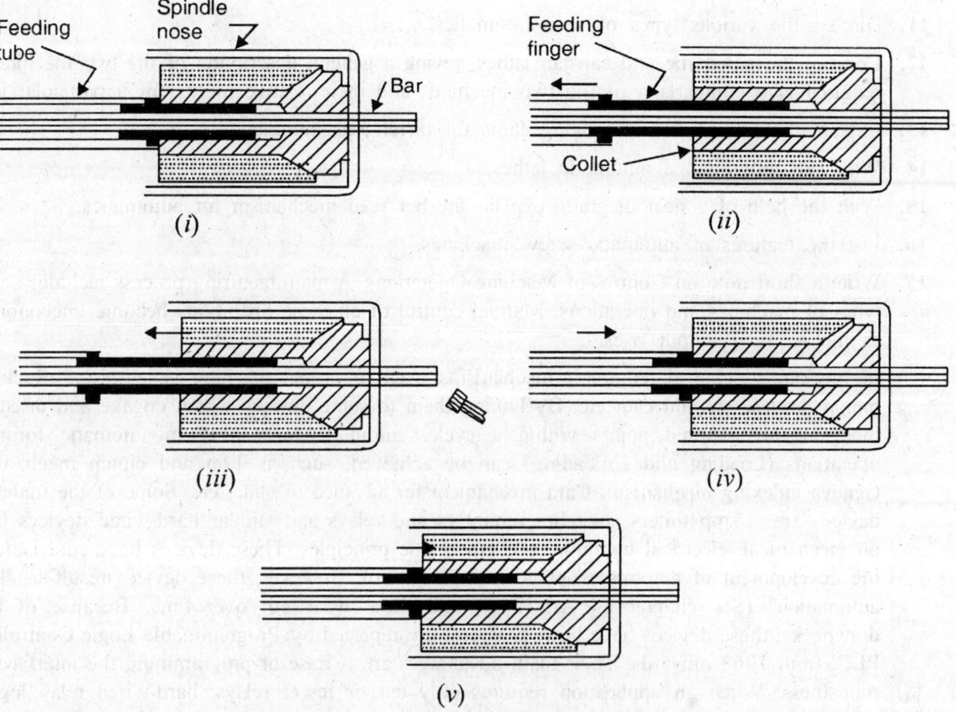

Fig. 8.16. Bar Stock Feeding for Automatics.

Bar stock feed and chucking proceed in the following order, Fig. 8.16.

(*i*) bar stock is clamped.

(*ii*) feeding finger is retracted

(*iii*) work is cut-off and collet opens

(*iv*) feeding tube and finger feed the bar out (to the right) upto the stock stop. After this, the stock is clamped by the movement of the collet tube. When the collet grips the bar, the feeding tube is returned to its initial left position and, the feeding finger slides along the clamped bar.

The feeding tube is operated by permanent cylinder cams through a lever system.

PROBLEMS

1. Define an automatic and a semi-automatic lathe.

2. Differentiate between an automatic and a semi-automatic lathe.

3. Discuss the various types of automatic lathes.

4. Describe a typical single-spindle automatic chucking machine?

5. Describe a typical single-spindle automatic bar machine?

6. Describe a Swiss-type automatic and tell how it differs from other single-spindle automatics?

7. Describe a single spindle cutting off automatic machine.

8. What are the relative advantages and disadvantages of single-spindle and multi-spindle automatics?

9. Discuss the various types of multi-spindle automatics.

10. Differentiate between parallel action and progressive action multi-spindle automatics.

11. Discuss the various types of semi-automatics.

12. Contrast the automatic and capstan lathes, giving a general description of the two machines as required in a comparison of the two machines and their relative merits for a particular job.

13. Explain the procedure for drawing cams for turret type automatic.

14. What are programmed automatic lathes.

15. With the help of a neat diagram, explain the bar feed mechanism for automatics.

16. List the features of automatic screw machines.

17. Write a short note on Control of Machine Operations. A manufacturing process includes several types of machines and operations. Manual control of all these operations become uneconomical as the required output rises.

    In mechanical control, common mechanisms used to propel machine slides are :- Rack and pinion, lead screw and cam etc. By linking them to some devices which engage and disengage them at the required points within a cycle, automatic form or Semi-automatic forms of operations (Loading and Unloading) can be achieved, such as Trip and clutch mechanisms, Geneva indexing mechanism, Cam mechanism for advance of slide, etc. Some of the traditional devices are :- stop timers, switches, counters and relays and similar hard-wired devices based on mechanical, electrical hydraulic and pneumatic principles. These devices have contributed to the development of automatic operations and control. However, these devices result in "Fixed automation" (See chapter 24 Art. 24.6) and these wear out over time. Because of these drawbacks, these devices have been progressively replaced by 'Programmable Logic Controllers", PLC, from 1968 onwards. Their main advantages are :- Ease of programming the interface and robustness. When an application requires only ten or fewer relays, hard-wired relay logic is probably more economical. When the applications require more computer control, then the PLC becomes more attractive alternative.

    Flexible manufacturing marks the latest trend in manufacturing. This has been made possible due to the development of Numerical Control of machines, See Chapter 16, A.T.B.O. Production Technology by the author, which have been further developed into CNC and DNC. PLC is not as flexible as CNC. However, since it is micro-processor based and specifically designed for industrial use, it is much convenient to use it in harsh workshop conditions, in which a computer will not survive for long.

9

LIMITS, TOLERANCES AND FITS

9.1. GENERAL

With all the advancement in the machine tool technology, it is not possible to achieve dimensional perfection due to the following reasons : Temperature changes, tool wear, deflections and vibrations of the machine and the work and human error. Even if the dimension is to be maintained within a very close degree of accuracy, lot of time will be consumed resulting in increased cost of manufacture, Fig. 9.1, and even then there would still be small variation. Other drawbacks of manufacturing a component exactly to a a nominal size are:

(*i*) Components will have to be produced individually which is slow and costly.

(*ii*) Highly intensive skilled labour is required, which is also costly.

(*iii*) Automation and its accompanying benefits can not be realised.

(*iv*) Replacement points have to be made the same way and so hold up other processes.

(*v*) Some fits will be better than other, and the variation will be unknown.

In mass production where the work has to be done in a set competitive time, greater variations will result. This fact is recognised and certain variations are allowed in the sizes of the machine elements or parts. This system of manufacture in which the dimensions of a part lie within some specified limits leads to "Interchangeable manufacture". Interchangeable part manufacture is a major feature of modern serial and mass production. The term "interchangeable manufacture" implies that the parts which go into assembly may be selected at random from large number of parts. These parts which have been produced with all their dimensions within their specified limits, need not be made in the same shop or even in the same company. This system of interchangeable part manufacture, that is, the system of limits and fits results in the following advantages.

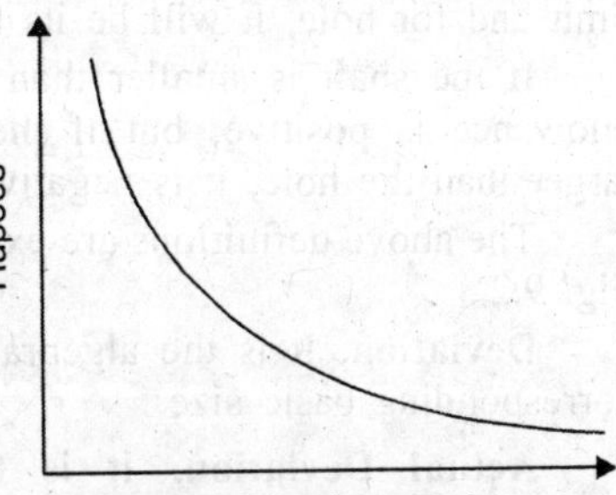

Fig. 9.1. Cost of Production vs Degree of Accuracy.

1. The components will assemble together at random. During assembly, individual fitting is not necessary. This results in reduction in cost of production because the elimination of fitting reduces the time required to build the product.

2. Components can be manufactured in large batches or lots and all treated alike.

3. Machine tools which have been developed for quantity production enable the components to be manufactured more rapidly using cheaper labour.

4. Repair of existing machines or products is simplified because component parts can be easily replaced.

5. Parts can be checked by gauging which considerably simplifies inspection procedure.

9.2. TERMINOLOGY FOR LIMITS AND FITS

1. The type of fit which is required between the mating parts.

2. The tolerance which is to be allowed upon each dimension.

Types of fits will be discussed in the next section. Below, we give terminology for limits and fits.

Shaft. The term shaft refers not only to the diameter of a circular shaft but also to any external dimension on a component.

Hole. The term hole refers not only to the diameter of a circular hole but also to any internal dimension on a component.

When an assembly is made of two parts, one is known as male (enveloped) surface and the other one as female (enveloping) surface. The male surface is referred to as 'Shaft' and the female surface as 'Hole'

Basic Size. Basic size or Nominal size is the standard size for the part and is the same both for the hole and its shaft. This is the size which is obtained by calculation for strength.

Actual Size. Actual size is the dimension as measured on a manufactured part. As already noted the actual size will never be equal to the basic size and it is sufficient if it is within predetermined limits.

Limits of Size. These are the maximum and minimum permissible sizes of the part.

Maximum Limit. Maximum limit or High limit is the maximum size permitted for the part.

Minimum Limits. Minimum limit or Low limit is the minimum size permitted for the part.

Tolerance. Tolerance is the difference between maximum limit of size and minimum limit of size.

Allowance. Allowance is an intentional difference between the maximum material limits of mating parts. For shaft, the maximum material limit will be its high limit and for hole, it will be its low limit.

If the shaft is smaller than hole, the allowance is positive, but if the shaft is larger than the hole, it is negative.

The above definitions are explained in Fig. 9.2.

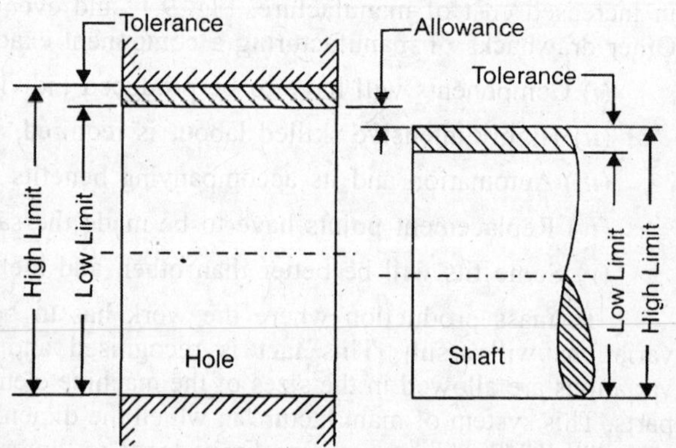

Fig. 9.2. Limits and Tolerance.

Deviation. It is the algebraic difference between a size (actual, maximum, etc.) and the corresponding basic size.

Actual Deviation. It is the algebraic difference between an actual size and the corresponding basic size.

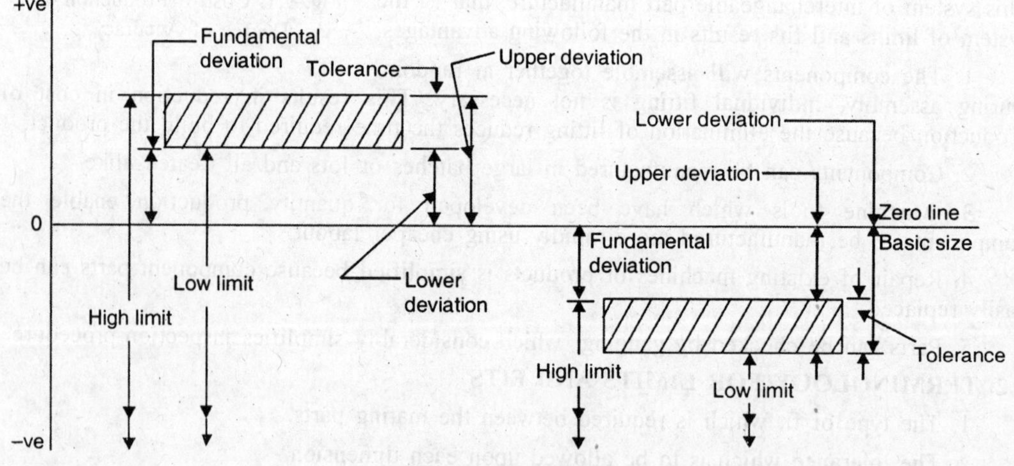

Fig. 9.3. Deviations.

Upper Deviation. It is the algebraic difference between the maximum limit of size and the corresponding basic size. It is a positive quantity when the maximum limit of size is greater than the basic size and a negative quantity when the maximum limit of size is less than the basic size.

Lower Deviation. It is the algebraic difference between the minimum limit of size and the corresponding basic size. It is a positive quantity when the minimum limit of size is greater than the basic size and a negative quantity when the minimum limit of size is less than the basic size.

Mean Deviation: - It is the arithmatic mean deviation between the upper deviation and Lower deviation.

Zero Line. It is a straight line to which the deviations are referred to in a graphical presentation of limits and fits. It is a line of zero deviation and represents the basic size. When the zero line is drawn horizontally, positive deviations are shown above and the negative deviations below this line.

Fundamental Deviation. This is the deviation, either the upper or the lower deviation, which is the nearest one to the zero line for either a hole or a shaft. It fixes the position of the tolerance zone in relation to the zero line.

Tolerance Zone. It is the zone bounded by the two limits of size of a part in the graphical presentation of tolerance. It is defined by its magnitude and by its position in relation to the zero line.

These terms are explained in Fig. 9.3.

Unilateral Limits. In the method of presenting the limits, both the limits of size are on the same side of the zero line. That is, the permitted tolerance is stated or indicated as wholly + ve or wholly –ve, e.g., $30 \text{ mm} \begin{matrix} +0.13 \\ +0.00 \end{matrix}$ or $100 \text{ mm} \begin{matrix} -0.12 \\ -0.26 \end{matrix}$. One of the limits of the size may be the basic size.

Bilateral Limits. Here, one of the limits of size is on one side of the zero line and the other limit of size is on the other side of the zero line, *i.e.*, the permitted tolerance is indicated partly + ve and partly negative, *e.g.*, $90 \text{ mm} \begin{matrix} +0.010 \\ -0.025 \end{matrix}$.

Both the above methods are illustrated in Fig. 9.4.

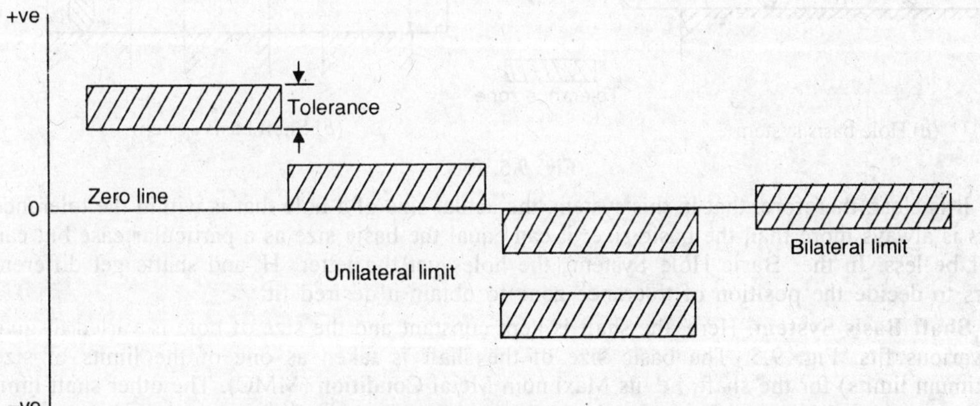

Fig. 9.4. Unilateral and Bilateral Limits.

9.2.1. Fits. The fit between two mating parts is the relationship existing between them with respect to the amount of play or interference which is present when they are assembled together. According the fit may result either in movable joint or a permanent joint. For example, a shaft running in a bush can move in relation to it and so forms a moving joint, whereas a pulley mounted on the shaft forms a fixed joint. The nature of joint or "fit" is characterised by the presence and size of 'clearance' (for movable joints) or 'interference' (for fixed joints). There are three basic types of fits.

1. **Clearance Fit.** In clearance fit or running fit, the shaft is always smaller than hole. A positive allowance exists between the largest possible shaft and the smallest possible hole, *i.e.*, when the shaft and hole are at their maximum metal conditions. The tolerance zone of the hole is entirely above that of the shaft.

Minimum Clearance. It is the difference between the maximum size of shaft and minimum size of hole.

Maximum Clearance. It is the difference between the minimum size of shaft and maximum size of hole.

2. **Interference, Press or Force Fit.** In this type of fit, the shaft is always larger than the hole. The tolerance zone of the shaft is entirely above that of the hole.

Minimum Interference. It is the difference between the maximum size of hole and the minimum size of shaft prior to assembly.

Maximum Interference. It is the difference between the minimum size of hole and the maximum size of shaft prior to assembly.

3. **Transition or Sliding Fit.** It occurs when the resulting fit due to the variations in size of male and female components due to their tolerance, varies between clearance and interference fits. The tolerance zones of shaft and hole overlap.

The various allowances for different fits may be obtained in two ways.

Hole Basis System. In this system, the hole is kept constant and the shaft diameter is varied to give the various types of Fits, Fig. 9.5. The basic size of the hole is taken as the low limit of size of the hole, i.e., the Maximum Metal Condition (MMC), of the hole. The high limit of size of the hole and the two limits of size for the shaft are then selected to give the desired fit.

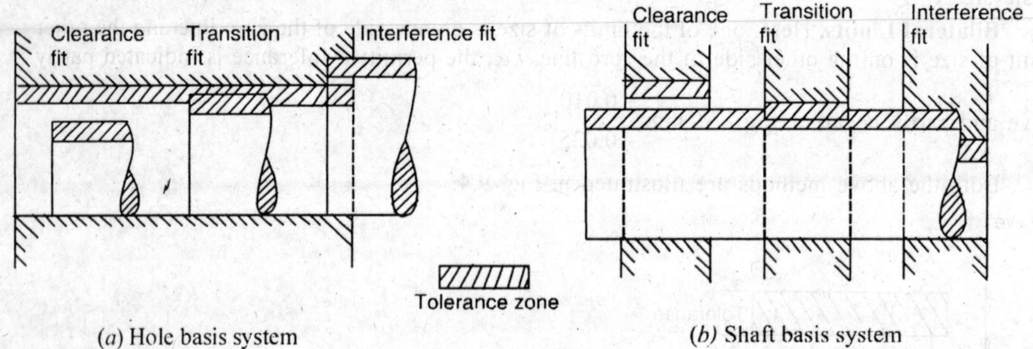

(*a*) Hole basis system (*b*) Shaft basis system

Fig. 9.5. Fits.

It is clear, therefore, that in this system, the actual size of a hole that is within the tolerance limits is always more than the basic size; it can equal the basic size as a particular case but can never be less. In the 'Basic Hole System', the holes get the letter 'H' and shafts get different letters to decide the position of tolerance zone to obtain a desired fit.

Shaft Basis System. Here, the shaft is kept constant and the size of hole is varied to give the various fits, Fig. 9.5. The basic size of the shaft is taken as one of the limits of size (maximum limits) for the shaft, i.e. its Maximum Metal Condition (MMC). The other shaft limit of size and the two limits of size for the hole are then selected to give the desired fit.

It is clear, therefore, that in this system, the actual size of a shaft that is within the tolerance limits is always less than the basic size. As a particular case, it can equal the basic size but can never be larger. In the 'Basic Shaft System', the shaft gets the letter '*h*' and holes get different letters to decide the position of tolerance zone to obtain a desired fit.

From a manufacturing point of view, it is preferable to use the "hole basis" system, because it is economical. This is because a great many holes are produced by standard fixed size tools, such as, twist drills, reamers, core drills, taps, broaches, etc. The advantages of using fixed size tools is that the machine need not be set up to obtain the proper size of the hole, setting up operations can consequently be made quicker and cheaper. Subsequently, the shaft sizes are more readily variable about the nominal size by means of turning or grinding operation. It is easier and more convenient to manufacture shafts of varying sizes than holes of varying sizes, as given above. The hole basis system is preferred, because it lessons the range of cutting and measuring tools for machining of holes, which are more expensive than tools to machine shafts. Also, the control of the size and shape of holes is more complicated and less accurate than the control of shafts. Applications : Machine and engine building, locomotive, construction.

The shaft basis system is more advantageous in certain cases, for example, this system can be efficiently applied for long shafts machined to the same size over their full lengths (smooth drawn shafts, shafts ground on centreless grinding machines etc.), if the shaft is to mate with at least two parts having holes that require different types of fit. Examples of "shaft basis" system are : the mating of a piston pin with both the piston and the connecting rod, and the outer rings of antifriction bearings with various bores in housings, electric motors, power transmission and products made from bright drawn bars.

It has been found in practice that a number of different fits of each basic type of fit are required which can provide different degrees of tightness or freedom between the mating parts.

The most commonly used fits of clearance type are : (1) Slide fit (2) Easy slide (3) running fit (4) slack running fit and (5) loose running fit.

1. **Slide fit.** Slide fits have a very small clearance, the minimum clearance being zero. Due to this, a sliding fit is close to the group of transition fits. They are employed when the mating parts move slowly relative to each other (for example, the tail stock spindle in a lathe, the feed movement of the spindle quill in a drilling machine etc.), or for mounting purposes (for example, stopper rings, keyed change gears etc.)

2. **Easy slide fit.** An easy slide fit provides for a small guaranteed clearance. It serves to ensure alignment between the shaft and hole and is supplied for slow and non-regular motion, for example, spindle of lathes and dividing heads, piston and slide valves, spigot or location fits.

3. **Running fits.** Running fits have appreciable clearance. It is employed in engineering for rotation at moderate speeds (shafts, gears, pulleys, couplings, crank shafts in their main bearings, throttles in the valve sleeves of a steam and air power forging hammer, bearings of small electric motors and pumps etc., gear box bearings). The clearance provides sufficient space for lubrication between mating friction surfaces.

4. **Slack running fits.** A slack running fit has considerable clearance which may be required as compensation for mounting errors, as in multi-support shafts (Cam shafts of I.C. engines) or shafts with widely spaced supports or if the bearings are very long (shafts of centrifugal pumps, shafts in the drives of cylindrical grinding machines etc.)

5. **Loose running fits.** These fits have the largest clearance and are employed for rotation at very high speeds and if misalignment of the mating parts may occur in assembly (shafts in specially long bearings (Plummer blocks), idle pulleys on their shafts.

Interference fits (shrink, heavy drive and light drive) are used for fixed permanent joints in which no additional fixing elements are needed. Elastic strains developed on the mating surfaces during the process of assembly prevent relative movement of the mating parts. Example of interference fits are : steel tyres on railway car wheels, gears on the intermediate shafts of trucks, bushing in the gear of a lathe head stock, pump impeller on shaft, drill bush in jig plate and cylinder linear in block etc.

Transition fits lie midway between clearance and interference fits. The use of transition fits does not guarantee either interference or clearances. Their main use is to ensure proper location of mating parts that are to be repeatedly disassembled and reassembled. In transition fits, torque is fitted between mating parts by means of additional fastening elements, such as keys, cotters, pins etc., which prevent relative motion between mating parts. The same purpose is served by spline shafts and bushings. The selection of a transition fit (force, tight, wringing or push) usually depends on how often the given joint is to be disassembled and reassembled during regular operation.

1. **Force fit.** Force fits are employed for mating parts that are not to be disassembled during their total service life, for example, gears on the shafts of a concrete mixer, forging machine etc.

2. **Tight fit.** Tight fits provide less interference than force fits. These fits are employed for mating parts that may be replaced while overhauling the machine, for example, stepped pulleys on the drive shafts of a conveyor, cylindrical grinding machine etc.

3. **Wringing fit.** A wringing fit provides, either, zero interference or a clearance. These find use where parts can be replaced without difficulty during minor repairs, for example, gears of machine tools.

4. **Push fit.** This fit is characterised by a clearance. It is employed for parts that must be disassembled during operation of a machine, for example, change gears slip bushing etc.

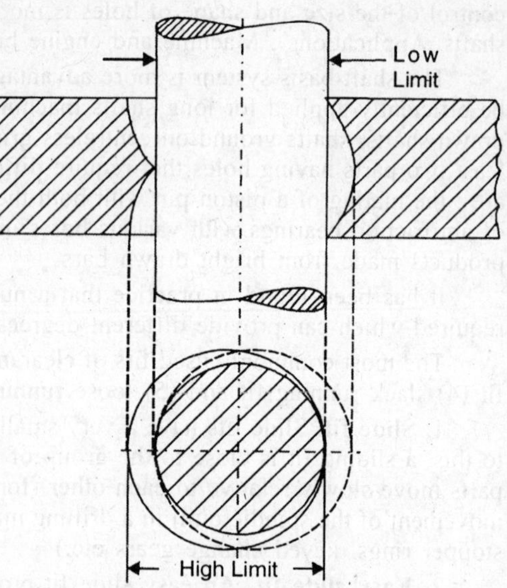

Fig. 9.6. Limits on Hole.

9.3. MEANING OF LIMITS

In the case of limits on hole, more is meant than simply the maximum and minimum size of the hole. The error of the hole may be as shown in Fig. 9.6 and yet a simple measurement of diameter would not indicate the lack of axial truth. The correct interpretation of a pair of limits can thus be stated as follows :

1. **Hole.** The upper limit refers to the greatest diameter at any point in hole, the lower limit refers to the diameter of the inscribed circle or cylinder which will just pass through the hole.

2. **Shaft.** The upper limit refers to the escribed ring which will just pass over the shaft, the lower limit refers to the minimum diameter at any point on the shaft, Fig. 9.7.

Although, normally limits refer to the extreme dimensions which may occur on a component, a more recent outlook on the matter is that the limits refer to the dimensions inside which all but a small percentage of parts must lie. Thus probability effect is taken into account. However, the definition given above still holds good.

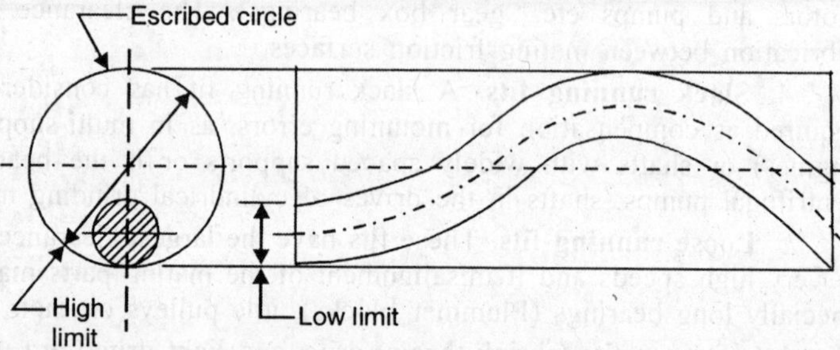

Fig. 9.7. Limits on Shaft.

The most important function of limits, is however, to enable interchangeable components to be produced. General limits are required to control general aspects of the specifications. Special limits are necessary with interchangeable manufacture.

(*i*) to control the fits between mating parts.

(*ii*) to maintain the desired clearance.

9.4. GENERAL LIMITS OF TOLERANCE

As mentioned before, tolerances are necessary to attain manufacturing control. The amount of tolerance to be given on a part depends upon :

(*i*) the function of the product, *i.e.,* the allowance desired in the fit.

(*ii*) manufacturing process available.

(*iii*) cost of production and assembly.

Cost of production and assembly increases as the tolerance is decreased. This increase in cost is due to :

(*a*) need for more sophisticated machinery.

(*b*) higher skills required.

(*c*) increased attention to inspection and handling.

The variation of relative cost of production as a function of tolerances will be similar to as shown in Fig. 9.1.

Therefore, tolerance that should be assigned to dimension is that which gives an economic balance between cost and quality. No part should be made with a greater of accuracy than is required by its use in a given mechanism or machine.

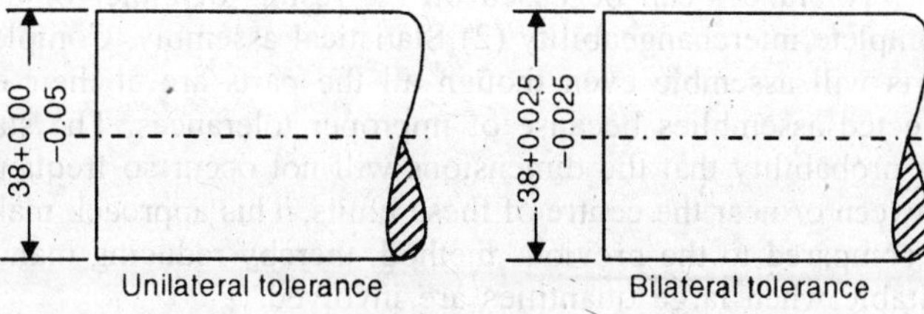

Fig. 9.8. Method of assigning tolerances

As already mentioned under article 9.2, manufacturing tolerance may be assigned to a dimension in two different manners :

1. Unilateral system
2. Bilateral system.

For unilateral tolerance, it is more critical for a certain dimension to deviate in one direction that in another. This system is also more satisfactorily and realistically applied to certain machining processes where it is common knowledge that dimensions will most likely deviate in one direction. For example, in drilling a hole with a standard size drill, the drill will most likely produce an oversize rather than an undersize hole. The operator machines to the lower limit of the hole (upper limit of a shaft), knowing fully well that he still has the whole tolerance left for machining before the parts are rejected. Also in this system the tolerance can be revised without affecting the allowance or clearance conditions between mating parts, *i.e.*, without altering the type of fit, Fig. 9.9. This system is most commonly used in industrial application. For interchangeable parts, unilateral system should always be used.

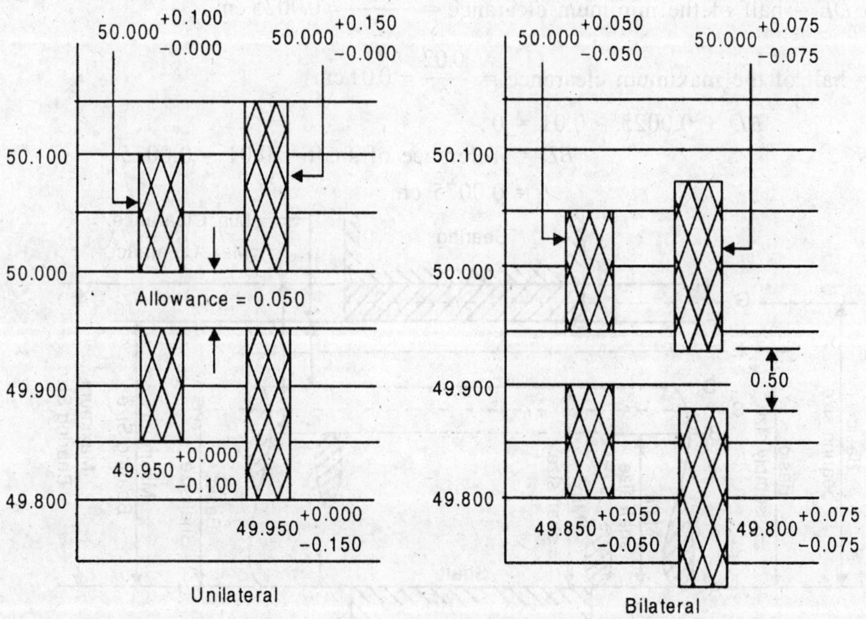

Fig. 9.9

In bilateral system, it is not possible to retain the same fit when tolerance is varied. The nominal size dimension of one or both of the mating parts will also have to be changed Fig. 9.9. This system clearly points out the theoretically desired size and indicates the possible and probable deviations that can be expected on each side of the nominal size. Moreover, it is easier to add bilateral tolerance together. This system permits operators to take full advantage of the limit system, especially in positioning of hole centre from base. On commencing machining, the position may be over an area of acceptance, the boundaries of which are governed by limits.

Tolerances can be based on the results of either of the two methods of analysis i.e., (1) Complete interchangeability (2) Statistical assembly. Complete interchangeability means that all parts will assemble even though all the parts are at their extreme limits. There should be no rejected assemblies because of improper tolerances. The statistical assembly method considers the probability that the dimensions will not occur so frequently at or near the extreme limits as between or near the centre of these limits. This approach makes it possible to increase tolerances as compared to the previous method, thereby reducing manufacturing costs. This method is not suitable when large quantities are involved.

9.4.1. Complete Interchangeability. In this method, the tolerances are obtained as explained in Fig. 9.10. It is necessary to choose one direction as positive and the other as negative, the choice is optional. The sum of the distances is zero if the starting and the end points coincide. This results in what is called the closed-path or loop equation. In order to determine the tolerances of the individual components, it is necessary to consider the tolerances of the entire assembly. Consider the shaft and bearing assembly shown in Fig. 9.11 The nominal assembly size is 5 cm with a minimum total clearance of 0.005 cm and a maximum total clearance of 0.02 cm. The tolerances have been shown bilaterally. The shaft has been shown to be resting on the bottom of the bearing for the bearing for the sake of analysis and clarity. To find the tolerance on the shaft, the loop equation will be starting at *B*.

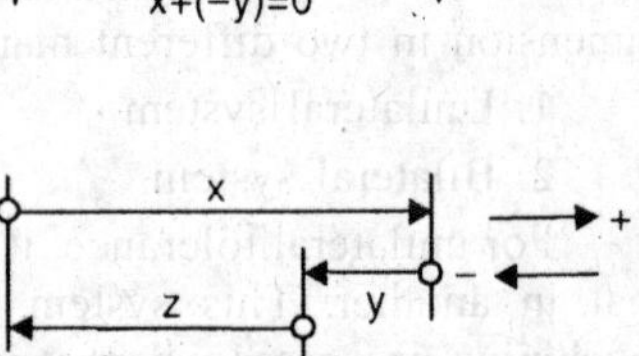

Fig. 9.10. Closed Path Method.

$$BD + DE + (-EB) = 0$$

Now DE = half of the minimum clearance $= \dfrac{0.005}{2} = 0.0025$ cm

BE = half of the maximum clearance $= \dfrac{0.02}{2} = 0.01$ cm

$\therefore \qquad BD + 0.0025 - 0.01 = 0$

Thus $\qquad BD$ = Tolerance of shaft = 0.01 – 0.0025

$= 0.0075$ cm

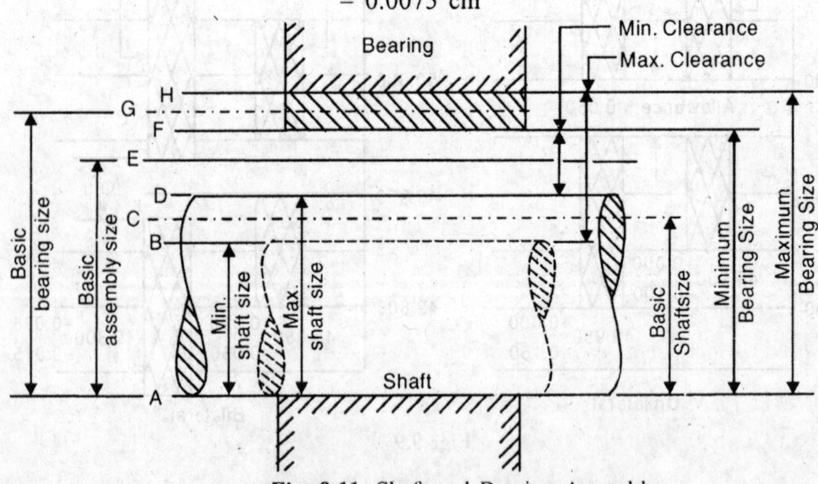

Fig. 9.11. Shaft and Bearing Assembly.

Similarly, $EF + FH + (-HE) = 0$

$\therefore$ $0.0025 + FH - 0.01 = 0$

$\therefore$ FH = Tolerance on bearing = 0.0075 cm

Max. Shaft Size = $5 - DE$ = 4.9975 cm

Min Shaft Size = 4.9975 – 0.0075 = 4.99 cm

Min. bearing Size = $5 + EF$ = 5.0025 cm

Max. bearing Size = $5.0025 + FH$

= 5.0025 + 0.0075 = 5.01 cm

9.4.2. Statistical Assembly. This method is based on the observation that a well-controlled machining operation will not produce many parts with dimensions exactly at the extreme limits. If the frequency of occurrence of a dimension is plotted against the dimensions, the distribution curve will or may take the form shown in Fig. 9.12. Such a curve is known as normal distribution curve.

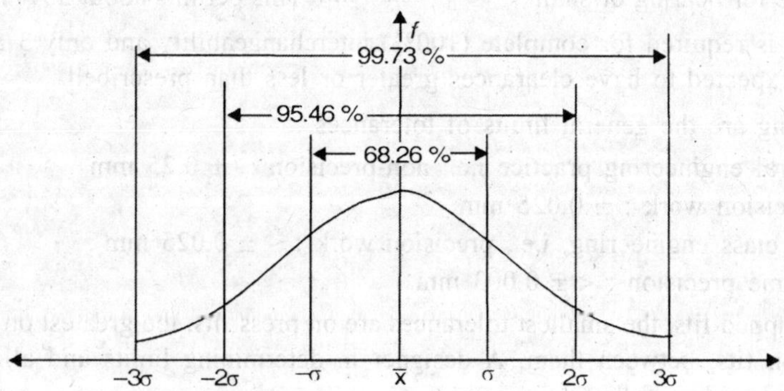

Fig. 9.12. Normal Distribution Curve.

This curve can be described by its arithmetic average, $\overline{X}$, and its standard deviation (measure of variation), σ. The average is given as,

$$\overline{X} = \frac{\Sigma X}{n} \qquad \text{...(9.1)}$$

where $\overline{X}$ = average

ΣX = sum of all dimensions considered

n = number of dimensions considered

The standard deviation is given as,

$$\sigma = \sqrt{\frac{\Sigma\left(X - \overline{X}\right)^2}{n}} \qquad \text{...(9.2)}$$

The curve (Fig. 9.12) is symmetrical about its vertical axis. The greater portion of the area the curve is included between the limits $\overline{X} \pm 3\sigma$. The curve actually continues out to plus and minus infinity, but the area under it beyond $\pm 3\sigma$ from the mean is practically negligible. 99.73% of the values will probably appear within the limits of $\pm 3\sigma$. Only 0.27% are expected to fall beyond these limits.

Applying this technique to the problem discussed above, the mean value of the clearance is

$$\overline{X}_{clearance} = \overline{X}_{bearings} - \overline{X}_{shafts} \qquad \text{...(9.3)}$$

and the standard deviation of the variation in the clearance is

$$\sigma_{clearance} = \sqrt{\sigma^2_{bearings} + \sigma^2_{shafts}} \quad ...(9.4)$$

Now $$6\sigma_{clearance} = 0.02 - 0.005 = 0.015 \text{ cm}$$

$$\therefore \quad \sigma_{clearance} = 0.0025 \text{ cm.}$$

Let $$\sigma_{bearing} = \sigma_{shafts} = \sigma_p$$

$$\sqrt{\sigma^2_p + \sigma^2_p} = 0.0025$$

$$\therefore \quad \sigma_p = 0.0017$$

∴ Tolerance for bearing or shaft = $6\sigma_p = 0.01$ cm. This permits about 33% more tolerance on the parts than is required for complete (100%) interchangeability and only 3 assemblies out of 1000 can be expected to have clearances greater or less than prescribed.

The following are the general limits of tolerances :

1. For general engineering practice i.e., non-precision : ± 0.25 mm
2. Semi-precision work : ± 0.025 mm
3. For high class engineering, i.e., precision work : < ± 0.025 mm
4. For extreme precision : < ± 0.003 mm.

Excluding lapped fits, the smallest tolerances are on press fits, the greatest on clearance fits, with the transition fits, between them. A designer in determining limits and allowances for a product should proceed as follows :

1. Settle the class of fit required by considering the minimum clearance or interference required taking into account the design and the question of lubrication etc.

2. Settle the manufacturing tolerances to be allowed, with particular reference to the available facilities.

3. Finally, consider the resultant variation of fit and determine if the mean fit is satisfactory from the functioning and the service life point of view.

9.5. LIMIT SYSTEMS

The primary aim of any limit system should be to give guidance to the user on many factors which cannot be analysed. In connection with mating parts between which there is relative motion, the conditions of loading, the speed of rotation, type and method of lubrication, the ambient conditions etc. are all conditions which are difficult to analyse completely and for which a complete analysis on every occasion would hardly be an economic proposition. The designer must, therefore, be largely dependent upon experience or upon guidance of others. The primary aims of any general system of standard fits and limits should be to provide guidance to the user in :

1. selecting basic functional clearances and interferences for a given application or type of fit.

2. establishing tolerances which provide a reasonable and economic balance between fit consistency and cost.

9.5.1. ISO (International Organization for Standardization) System of Limits and Fits. The ISO system covers holes and shafts from the smallest size upto 3150 mm. For any size over this range there is a wide choice of fits available and for each of the fits, there is a series of tolerance grades from very fine to wide tolerances. Either a hole based or a shaft based system can be obtained the standard.

For any basic size there are 28 different holes. These are obtained by providing a series of holes which are progressively oversize and a series of holes which are progressively undersize. The difference from basic size of the various holes is given by the fundamental deviation and it is these differences in size which give the fit required. The "28 holes" are designated by the capital letters : *A, B, C, CD, D, E, EF, F, FG, G, H, J, JS, K, M, N, P, R, S, T, U, V, X, Y, Z, ZA, ZB* and *ZC*. Each of the 28 holes has a choice of 18 grades of tolerances which are designated as ITO 1, ITO, IT 1 to IT 16. The tolerance grade decides the accuracy of manufacture. Similarly for the shafts, there are 28 shafts for a given basic size and these are designated by small letters '*a*' to '*zc*'. Also, each shaft has 18 grades of tolerance which are designated as for the holes. The general arrangement of holes and shafts is shown in Fig. 9.13. The upper deviation for shafts '*a*' to '*g*' is below the zero line and it is above the zero line for shafts '*j*' to '*zc*'. Similarly the lower deviations for holes '*A*' to '*G*' are above the zero line and it is below the zero line for holes '*J*' to '*ZC*'. The shaft '*h*' for which the upper deviation is zero is called the 'Basic Shaft'. Similarly the hole '*H*' for which the lower deviation is zero is called the 'Basic Hole'.

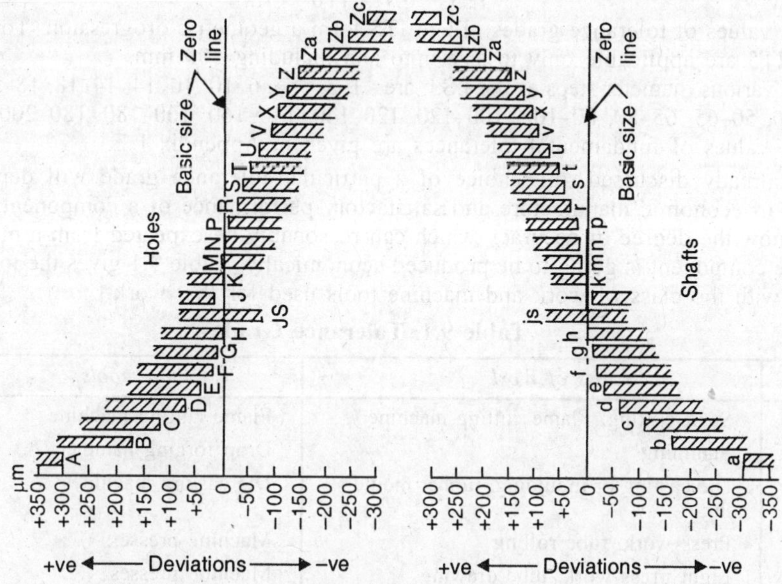

Fig. 9.13. General Arrangement of holes and Shafts.

The general trend for the shafts is that for the shafts '*a*' to '*g*', both the upper limit and the lower limit fall below the zero line. Therefore, these shafts tend to give clearance fits. For '*h*' shaft, the upper limit coincides with the basic size and this shafts will give close running fits when assembled with '*H*' holes. For '*j*' shafts, the tolerance zone is disposed above and below the zero line. These shafts will tend to give transition fits. For shafts '*k*' to '*zc*', the tolerance zones is entirely above the zero line (basic size). These shafts will always give interference fits.

Calculation of Tolerance Grades. The values for the tolerance grades are the same over a range of basic sizes, the minimum and maximum size in each range being known as the diameter steps involved. For all sizes the values of the tolerance grades IT16 are multiples of the tolerance unit '*i*' calculated using the R5 series of preferred numbers. For sizes upto and including 500 mm, '*i*' can be determined as :

$$i = 0.45\sqrt[3]{D} + 0.001\,D \qquad \text{...(9.5)}$$

Experience has shown that in manufacturing processes, the dimensional inaccuracies are proportional to the cube root of the absolute size. Therefore, in the ISO, 'i' is calculated in units of 10^{-3} mm or μm from eqn. 9.5.

For sizes above 500 mm,

$$i = 0.004\,D + 2.1 \qquad \text{...(9.6)}$$

'i' is given in units of 0.001 mm. D is in mm and is the geometric mean of diameter steps between which a particular basic size lies. The values of the tolerances for tolerance grades IT6 to IT16 are given below :

Grade	IT6	IT7	IT8	IT9	IT10	IT11	IT12	IT13	IT14	IT15	IT16
Tolerance	$10i$	$16i$	$25i$	$40i$	$64i$	$100i$	$160i$	$250i$	$400i$	$640i$	$1000i$

For tolerance grades IT01 to IT5, the tolerances are calculated as follows :

$$\begin{aligned} &\text{For IT01, tolerance} = 0.3 + 0.008\text{ D} \\ &\text{For IT0, tolerance} = 0.5 + 0.012\text{ D} \\ &\text{For IT1, tolerance} = 0.8 + 0.02\text{ D} \end{aligned} \qquad \text{...(9.7)}$$

Here tolerance is given in units of 0.002 mm.

Now $\quad IT5 = 0.7 \times IT6 = 7i$

The values of tolerance grades IT1 to IT5 form a geometric progression. Tolerance grades IT01 to IT5 are applicable only to size upto and including 500 mm.

The various diameter steps as per I.S.I. are : 1–3, 3–6, 6–10, 10–14, 14–18, 18–24, 24–30, 30–40, 40–50, 50–65, 65–80, 80–100, 100–120, 120–140, 140–160, 160–180, 180–200 mm etc.

The values of fundamental tolerances are given in Appendix I E.

As already discussed, the choice of a particular tolerance grade will depend upon the principle of economic manufacture and satisfactory performance of a component. For this, one should know the degree of accuracy which can reasonably be expected from a machine tool on which the component is going to be produced economically. Table 9.1 gives the tolerance grade together with the class of work and machine tools used for that work.

Table 9.1. Tolerance Grades.

IT	*Class of Work*	*Machine tools*
16	Sand casting, Flame cutting machine	Flame cutting machine
15	Stamping	Drop forging hammer
14	Die casting or moulding, rubber moulding	Die casting machines
13	Press work, tube rolling	Machine presses
12	Light press work, tube drawing	Machine presses
11	Drilling, rough turning and boring, precision tube drawing	Drilling machines, lathes
10	Milling, slotting, planning, metal rolling and extrusion	Milling, slotting and planning machine
9	Worn capstan or automatic lathe horizontal and vertical boring machines	Capstan and automatic lathes, borers
8	Centre lathe turning and boring reaming, capstan or automatic in good condition	Lathes, capstan and automatic
7	High quality turning, broaching	Lathes, honing and broaching machines
6	Grinding, fine honing	lapping, boring and
5	Machine lapping fine or diamond boring, fine grinding	grinding machines

4	Gauges, precision lapping	Precision lapping machines
3	Good quality gauges	–
2	High quality gauges	–
1	Workshop standards and gauges	–
0	Inspection standards and gauges	–
01	Work of the highest quality	–

Calculation of upper and lower deviation. For shafts, the upper deviation is denoted by 'es' and the lower deviation by '*ei*'. The corresponding deviations for holes are denoted by 'ES' and 'EI' respectively. The fundamental deviations can be calculated with the help of formulas. The other deviations can be evaluated by using the absolute value of the tolerance IT. Thus, for shafts,

$$ei = es - \text{IT} \quad \text{...(9.8)}$$

or $$es = ei + \text{IT}$$

Similarly for hole; $$\text{ES} = \text{EI} + \text{IT}$$

The values of the deviations for the shafts for sizes upto 500 mm are given in Table 9.2.

Table 9.2. Fundamental Deviations, Microns (0.001 mm).

Upper Deviations es		*Lower Deviation ei*	
Shaft Designation	*es*	*Shaft Designation*	*ei*
a	$-(265 + 1.3D)$; $D \leq 120$	j_5 *to* j_8	
	$-3.5\,D$; $D \geq 120$	k_4 to k_7	$+\ 0.6\sqrt[3]{D}$
b	$-(140 + 0.85D)$; $D \leq 160$	*k* for grades	
	$-1.8\ D \geq 160$	≤ 3 and ≥ 8	0
c	$-52D^{0.2}$; $D \leq 40$	*m*	$+(\text{IT7} - \text{IT6})$
	$-(95 + 0.8D)$; $D \geq 40$	*n*	$+\ 5D^{0.34}$
		p	+ (IT7 + 0 to 5)
d	$-16D^{0.44}$		
e	$-11D^{0.11}$	*r*	geometric mean of values of *ei* for shafts *p* and *s*
f	$-5.5D^{0.41}$	*s*	+ (IT8 + 1 to 4);
g	$-2.5D^{0.34}$		$D < 50$, + (IT7
h	O		+ 0.4D); $D > 50$
		t	+ (IT7 + 0.63 D)
		u	+ (IT7 + D)
		v	+ (IT7 + 1.25 D)
		x	+ (IT7 + 1.6D)
		y	+ (IT7 + 2D)
		z	+ (IT7 + 2.5D)
		za	+ (IT8 + 3.15D)
		zb	+ (IT9 + 4D)
		zc	+ (IT10 + 5D)

For holes, the deviations are derived from the corresponding values of shafts. For holes, the limits are the same as the shaft limits of the same symbol *i.e.,* letter and grade, but disposed on the other side of the zero line. Thus,

EI = Upper Deviation, *es* of shaft of the same letter symbol but of opposite sign.

and ES = Lower deviation, *ei*, of shaft of the same letter symbol but of opposite sign.

Designation of Holes, Shafts and Fits. A hole or shaft is completely described if the basic size, followed by the appropriate letter and by the number of tolerance grade, is given. For example 60 mm H-hole with the tolerance grade IT7 is given as

60 mm H7

Similarly a 60 mm *f*-shaft with the tolerance grade IT8 is given as,

60 mm *f*8

A fit is designated by the basic size common to both hole and shaft followed by symbols corresponding to each element, the hole is quoted first. Thus, if basic size is 60 mm, the hole is H7 and the shaft is *f*8, then the fit can be indicated as,

60 mm H7-f8

or 60 mm H7/f8

Commonly used Holes and Shafts :

Commonly used holes : H7, H8, H9, and H11

Commonly used shafts : *c*11, *d*10, *e*9, *f*7, *g*6, *h*6, *k*6, *n*6, *p*6, *s*6.

The recommended fits, when the above mentioned shafts are associated with 'H' holes are given in Table 9.3 and Fig. 9.14.

Table 9.3. Recommended Selection of Fits.

TYPE OF FIT	SHAFT AND TOLERANCE	HOLE AND TOLERANCE			
		H7	H8	H9	H11
CLEARANCE	*c* 11				■
	d 10			■	
	e 9			■	
	f 7		■		
	g 6	■			
	h 6	■			
TRANSITION	*k* 6	■			
	n 6	■			
INTERFERENCE	*p* 6	■			
	s 6	■			

The various fits given above can be explained below :

Loose clearance, H11/c11 for example, some farm equipment.

Loose running H9/d10, for example, large bearings, Gland seals.

Easy running H9/*e*6, for example, Cam-shaft bearings, several bearings in line.

Normal running H8/*f*7, for example, small shaft bearings, pump or gear box bearings.

Precision running or location H7/g6, for example, precision slide ways and bearings.

Average location H7/*h*6, for example, location purposes (non running assemblies)

Easy keying H7/*k*6, for example, couplings keyed to shaft,

Push fit H7/*n*6, for example, transition fit

Press fit H7/*p*6, for example, small amounts of interference (gears, nuts)

Heavy press fit H7/*s*6, for example, permanent assemblies.

Selected I.S.O. fits are given in Table 9.4 :

Table 9.4. Selected I.S.O. fits

Fit	Hole Basis System	Shaft Basis System	Type of fit
Clearance fits	$H_{11} - c_{11}$	$C_{11} - h_{11}$	Loose running fit
	$H_9 - d_9$	$D_9 - h_9$	Free running fit
	$H_8 - f_7$	$F_8 - h_7$	Close running fit
	$H_7 - g_6$	$G_7 - h_6$	Sliding fit-intended to more freely
	$H_7 - h_6$	$H_7 - h_6$	Locational Clearance fit
Transition fits	$H_7 - k_6$	$K_7 - h_6$	Locational Transition fit
	$H_7 - n_6$	$N_7 - h_6$	— do —
Interference fits	$H_7 - p_6$	$P_7 - h_6$	Locational interference fit-Press fit
	$H_7 - s_6$	$S_7 - h_6$	Medium Drive Fit, or Shrink fit
	$H_7 - u_6$	$U_7 - h_6$	Force fit

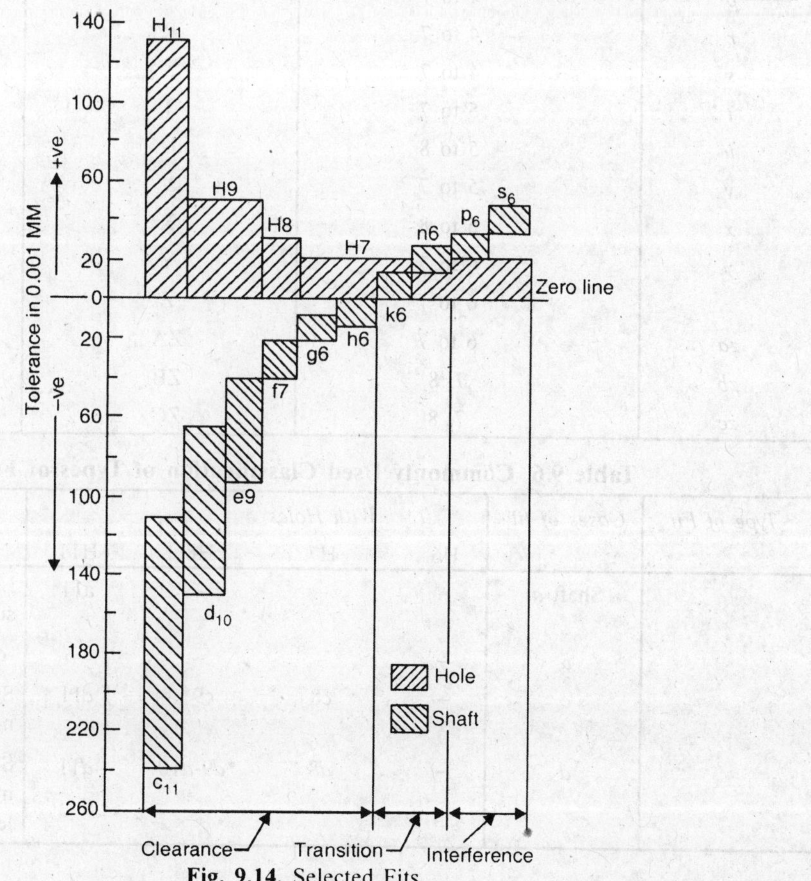

Fig. 9.14. Selected Fits.

The most commonly used shafts and holes (as per IS : 919–1963) are given in Table 9.5. Table 9.6 gives the commonly used classification of types of fits. To find the limits of the Basic size, the first step is to go through the Table 9.6 and select a proper fit on the basis of product application given in the last column under "Remarks".

Table 9.5. Most Commonly Used Shafts and Holes.

Shafts		*Holes*	
a	9, 11	A	9, 11
b	8, 9, 11	B	8, 9, 11
c	8, 9, 11	C	8, 9, 11
d	5 to 11	D	6 to 11
e	5 to 9	E	5 to 10
f	4 to 9	F	5 to 9
g	4 to 7	G	5 to 7
h	1 to 16	H	1 to 16
j	5 to 7	J	6 to 8
js	1 to 16	JS	1 to 16
k	4 to 7	K	5 to 8
m	4 to 7	M	5 to 8
n	4 to 7	N	5 to 11
p	4 to 7	P	5 to 9
r	4 to 7	R	5 to 8
s	4 to 7	S	5 to 7
t	5 to 7	T	5 to 7
u	5 to 8	U	6, 7
v	5 to 7	V	6, 7
x	5 to 8	X	6, 7
y	6 to 7	Y	7
z	6 to 7	Z	7, 8
za	6 to 7	ZA	7, 8
zb	7, 8	ZB	8, 9
zc	7, 8	ZC	8, 9

Table 9.6. Commonly Used Classification of Types of Fits

Type of Fit	*Cases of fit*	*With Holes*				*Remarks*
		H6	H7	H8	H11	
	Shaft *a*	–	–	–	a11	Large clearance fits result, not widely used
	b	–	–	–	–	-do-
	c	–	c8	c9*	c11	Suitable for slack running fit.
	d	–	*d*8	**d*9-*d*10	*d*11	Suitable for loose running fits *e.g.* loose pulleys, plummer blocks

CLEARANCE	*e*	*e7*	*e8*	*e8-e9	*e9*	For loose running or clearances fits *e.g.* properly lubricated bearings.
	f	$*f_6$	$*f_7$	$*f_8$	–	suitable for normal running fits, *e.g.*, bearings of small electric motors
	g	$*g_5$	$*g_6$	g_7	–	suitable for close running fit or sliding fit, *e.g.*, piston and slide valves
	h	$*h_5$	$*h_6$	$*h_7$-h_8	h_{11}	For precision sliding fit. Used widely for non-running parts.
TRANSITION	*j*	$*j_5$	$*j_6$	j_7	–	Push fit for very accurate location with easy assembly and dismantling, *e.g.*, coupling spigots and recesses.
	k	$*k_5$	$*k_6$	k_7	–	True transition fit, *e.g.*, Keyed shaft, non-running locked pins etc.
	m	$*m_5$	$*m_6$	m_7	–	Medium keying fit, *e.g.*, ball bearing races of medium duty.
	n	$*n_5$	$*n_6$	n_7	–	Heavy keying fit, for tight assembly of mating parts.
	p	p_5	$*p_6$	–	–	Light press fit with easy dismantling for non-ferrous parts. Standard press fit with easy dismanting for non-ferrous parts assembly.
	r	r_5	$*r_6$	–	–	Medium drive fit with easy dismantling for ferrous parts assembly. Light drive fit with easy dismantling non-ferrous parts assembly.
	s	s_5	$*s_6$	s_7	–	Heavy drive fit ferrous parts permanent of semi-permanent assembly. Standard press fit for non.
	t	t_5	$*t_6$	t_7	–	Force fit on ferrous parts for permanent assembly.
	u	u_5	u_6	$*u_7$	–	Heavy force fit or shrink fit.
	v, x, y, and *z*	–	–	–	–	Very large interference fit. Not recommended for use.

* Fits recommended for common use.

9.6. SELECTIVE ASSEMBLY

The discussion so far has been in connection with full interchangeability or random assembly in which any component assembles with any other component. Often special cases of accuracy and uniformity arise which might not be satisfied by certain of the fits given under a fully interchangeable system. For example if a part at its low limit is assembled with the mating part at high limit, the fit so obtained may not fully satisfy the functional requirements of the assembly. Also machine capabilities are sometimes not compatible with the requirements of interchangeable assembly. Complete interchangeability in the above two cases can be obtained at some extra cost in inspection and materials handling by using the selective assembly method. In this system, the parts are graded according according to size and only matched grades of mating parts are assembled. This method is especially useful where close fits of two component assemblies are required. It results in complete protection against defective assemblies and reduces machining costs since the tolerances can be increased.

Taking the case of mating of pistons in motor car cylinder bores. Let the bore size be 63.5 mm and the best skirt clearance for a certain type of piston is 0.13 mm on the diameter. Let the tolerance on bore and on piston skirt each be 0.04 mm.

Dimension of bore diameter is $63.5 \left.\begin{matrix} +0.02 \\ -0.02 \end{matrix}\right\}$ mm

Dimension of piston skirt is $63.37 \left.\begin{matrix} +0.02 \\ -0.02 \end{matrix}\right\}$ mm

Now minimum clearance = smallest bore – largest piston

= 0.09 mm

Maximum clearance = largest bore-smallest piston

= 0.17 mm

By grading and marking the bores and the pistons as shown below, they may be selectively assembled to give the conditions required :

Cylinder bore	:	63.48	63.50	63.52
Piston	:	63.35	63.37	63.39

Another application of manufacturing by selective assembly is that of ball bearings. The variation in sizes of the inner race, balls and outer race that make the difference between a snug and loose bearing are very small, of the order of 2.5 thou of a mm. To reduce their cost, these components are manufactured more liniently and graded into groups. The bearings are assembled from comparable groups. Only the inside and outside diameters of each bearing need be finished to interchangeable dimensions. An incidental advantage of this system is that the manufacturer is able to arrange bearings suitable for various purposes by combining components from different groups. By that means, three desirable classes are drived : a free running and loose bearing, a snug but free bearing, and a preloaded bearing for close constraint of a shaft or spindle.

9.7. SOLVED EXAMPLES

Example 1. *Find the values of allowance, hole tolerance and shaft tolerance for the following dimensions of mated parts according to basic hole system.*

Hole : *37.50 mm* *Shaft :* *37.47 mm*

37.52 mm *37.45 mm*

Solution.

(*i*) Hole tolerance = High limit – Low limit

= 37.52 – 37.50 mm

= 0.02 mm

(*ii*) Shaft tolerance = High limit – Low limit

= 37.47 – 37.45

= 0.02 mm

(*iii*) Allowance = Maximum metal condition of Hole

– Maximum metal condition of shaft

= Low limit of hole – High limit of shaft

= 37.50 – 37.47

= 0.03 mm

Example 2. *A 75 mm shaft rotates in a bearing. The tolerance for both shaft and bearing is 0.075 mm and the required allowance is 0.10 mm. Determine the dimensions of the shaft, and the bearing bore with the basic hole standard.*

Solution. Refer to Fig. 9.15

It is clear with the basic hole standard that,

Low limit of hole = 75 mm

High limit of hole = Low limit + tolerance

= 75 + 0.075 = 75.075 mm

High limit of shaft = Low limit of hole – allowance

= 75 – 0.10 = 74.90 mm

Low limit of shaft = High limit – tolerance

= 74.90 – 0.075 = 74.825 mm

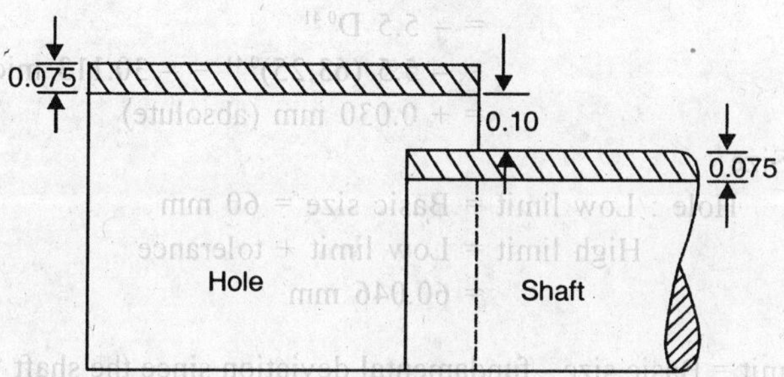

Fig. 9.15

Example 3. *A medium force fit on a 75 mm shaft requires a hole tolerance and shaft tolerance each equal to 0.225 mm and an average interference of 0.0375 mm. Determine the proper hole and shaft dimensions with the basic hole standard.*

Solution. Refer Fig. 9.16.

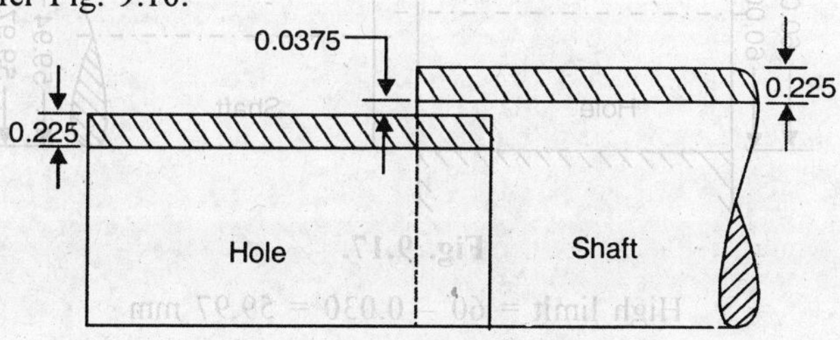

Fig. 9.16.

It is clear with the basic hole standard that,

Low limit of hole = 75 mm

High limit of hole = Low limit of hole + Tolerance

= 75 + 0.225 mm = 75.222 mm

Low limit of shaft = 75.225 + 0.0375

= 75.2625 mm

High Limit of Shaft = 75.2625 + 0.225

= 75.4875 mm

Example 4. *Calculate the fundamental deviation and tolerances and hence the limits of size for the shaft and hole for the following fit : 60 mm H8–f7. The diameter steps are 50 mm and 80 mm.*

Solution. (*a*) Now the tolerance unit is given as,

$$i = 0.45 \sqrt[3]{D} + 0.001\,D, \text{ microns}$$

where $D = \sqrt{50 \times 80} = 63.25$ mm

$$\therefore \quad i = 0.45 \times \sqrt[3]{63.25} + 0.001 \times 63.25$$

= 1.856 microns = 0.001856 mm.

Now for hole H8, the tolerance = $25i$ = 0.0464 mm

= 0.046 mm (rationalized)

For shaft *f*7, tolerance = $16i$ = 0.0297 mm

= 0.030 mm (rationalized)

(*b*) We know that for hole 'H', fundamental deviation is zero. The fundamental deviation for shaft '*f*' (Table 9.1) is,

$= -\,5.5\ D^{0.41}$

$= -\,5.5\,(63.25)^{0.41} = -\,30.113$ microns

= + 0.030 mm (absolute)

(*c*) Limits of Size

Hole : Low limit = Basic size = 60 mm

High limit = Low limit + tolerance

= 60.046 mm

Shaft : High limit = Basic size - fundamental deviation since the shaft '*f*' lies below the zero line.

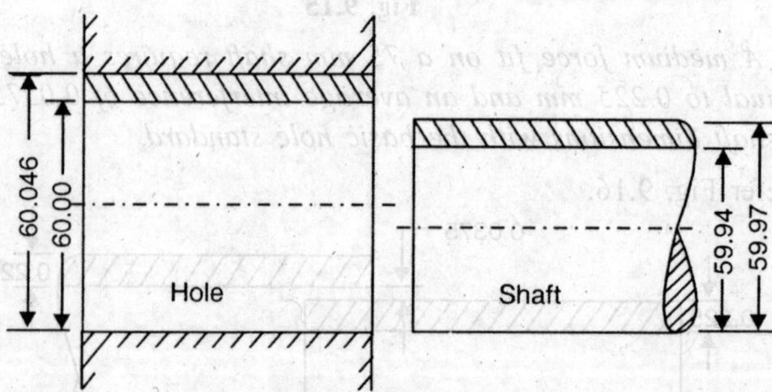

Fig. 9.17.

$\therefore$ High limit = 60 – 0.030 = 59.97 mm

Low limit = High limit – tolerance

$= 59.97 - 0.03 = 59.94$ mm

The fit is shown in Fig. 9.17. It is a clearance fit with 0.03 to 0.106 mm clearance.

Example 5. *In a limit system, the following limits are specified to give a clearance fit between a shaft and a hole :*

$$\text{shaft } 30 \begin{matrix} -0.005 \\ -0.018 \end{matrix} \text{ mm}\phi$$

$$\text{hole } 30 \begin{matrix} +0.020 \\ -0.000 \end{matrix} \text{ mm}\phi$$

Determine : (a) Basic size (b) shaft and hole tolerances (c) the shaft and hole limits (d) the maximum and minimum clearance.

Solution.

(*a*) Basic Size = **30 mm**

(*b*) Shaft tolerance = 0.018 – 0.005 = **0.013 mm**

Hole tolerance = 0.020 mm

(*c*) High limit of shaft = 30 – 0.005

= **29.995 mm**

Low limit of shaft = 30 – 0.018

= **29.982 mm**

High limit of hole = 30 + 0.020

= **30.020 mm**

Low limit of hole = 30 mm

(*d*) Maximum clearance = High limit of hole – low limit of shaft

= 30.020 – 29.982

= **0.038 mm**

Minimum clearance = Low limit of hole – High limit of shaft

= 30.000 – 29.995

= **0.005 mm**

Example 6. *A hole and shaft have a basic size of 25 mm, and are to have a clearance fit with maximum clearance of 0.02 mm and a minimum clearance of 0.01 mm. The hole tolerance is to be 1.5 times the shaft tolerance. Determine : limits for both hole and shaft* (*a*) *using a hole basis system* (*b*) *using a shaft basis system.*

Solution. Referring to Fig. 9.18.

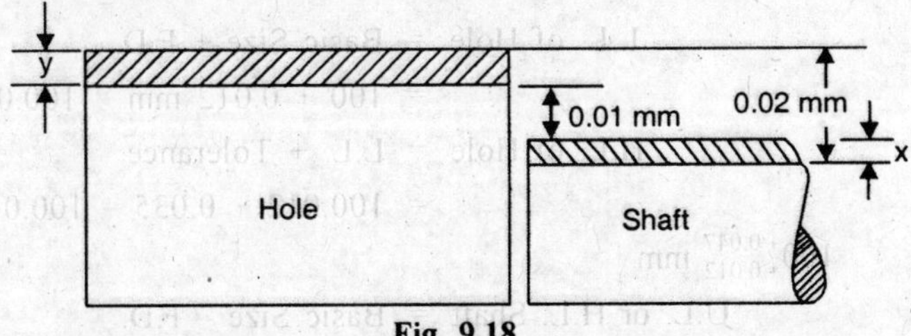

Fig. 9.18

If x is the shaft tolerance and y is the hole tolerance, then

$$y + 0.01 + x = 0.02$$

$$\therefore \quad 1.5x + 0.01 + x = 0.02$$

$\therefore$ $x = 0.004$ mm and $y = 0.006$ mm

(*a*) **Hole basis system.** The lower deviation is zero,

$\therefore$ Low limit of hole = 25 mm (Basic size)

High limit of hole = 25 + tolerance

= 25.006 mm

Upper (High) limit of shaft = low limit of hole – minimum clearance

= 25 – 0.01 = 24.99 mm

Low limit of shaft = 24.99 – 0.004 = 24.986 mm

(*b*) **Shaft basis system.** The upper deviation is zero.

$\therefore$ High limit of shaft = Basic size = 25 mm

Low limit of shaft = 25 – 0.004 = 24.996 mm

Low limit of hole = 25 + 0.01 = 25.01 mm

High limit of hole = 25.01 + 0.006 = 25.016 mm

Example 7. *A fit is designated as 100 G_7 / e_8. Find the dimensions of the hole and the shaft. The diameter steps are : 80 mm and 120 mm.*

Solution. Geometric mean diameter, $D = \sqrt{80 \times 120}$

= 98 mm

$\therefore$ Tolerance grade, $i = 0.45\,(D)^{\frac{1}{3}} + 0.001\,D$, microns

$= 0.45\,(98)^{\frac{1}{3}} + 0.001 \times 98$

= 2.079 + 0.098 = 2.177 microns

Now, for hole G_7, tolerance = 16 i = 34.832 microns

= 0.035 mm (rationalized)

For shaft e_8, tolerance = 25 i = 54.425 microns

= 0.054 mm (rationalized)

We can get these values of tolerances directly from Table : Appendix IE.

Now F.D. for hole G = $+ 2.5\,D^{0.34}$, microns (T 9.2)

= $25\,(98)^{0.34}$ = 0.012 mm (rationalized)

F.D. for shaft $e = -11\,D^{0.11}$, microns (T 9.2)

= $-11\,(98)^{0.11}$, microns

= 0.072 mm (rationalized)

Dimension See F 9.13.

(*i*) **Hole** L.L. of Hole = Basic Size + F.D.

= 100 + 0.012 mm = 100.012 mm

H.L. of Hole = L.L. + Tolerance

= 100.012 + 0.035 = 100.047 m

or $100^{+0.047}_{+0.012}$ mm

(*ii*) **Shaft** U.L. or H.L. Shaft = Basic Size – F.D.

= 100 – 0.072 = 99.928 mm

L.L. of Shaft = H.L. – Tolerance

= 99.928 – 0.054

$= 99.874$ mm

or $100^{-0.072}_{-0.126}$ mm

Example 8. *A 100 mm diameter journal and bearing assembly has a clearance fit, with the following specifications :*

Tolerance on bearing = 0.005 mm

Tolerance on Journal = 0.004 mm

Allowance = 0.002 mm

Determine the sizes of the bearing and the journal on (i) Hole Basis System (ii) Shaft Basis System. Take Unilateral System of tolerances.

Solution.

(*a*) **Hole-Basis System :** Refer Fig. 9.13,

Lower limit of Bearing = Basis Size = 100 mm

Highest limit of Bearing = L.L. of Bearing + Tolerance

= 100 + 0.005 = 100.005 mm

Refer to Fig. 9.2,

Higher limit of Journal = Lower limit of bearing – allowance

= 100 – 0.002

= 99.998 mm

Lower limit of Journal = Higher limit – Tolerance

= 99.998 – 0.004

= **99.994 mm**

(*b*) **Shaft-Basis System :** Refer Fig. 9.13

Upper limit of Journal = Basic Size = 100 mm

Lower limit of Journal = Upper limit – Tolerance

= 100 – 0.004

= **99.996 mm**

Refer to Fig. 9.2,

Lower limit of Bearing = Upper limit of Journal + allowance

= 100 + 0.002

= 100.002 mm

Upper limit of Bearing = Lower limit + tolerance

= 100.002 + 0.005

= **100.007 mm.**

Example 9. *In an assembly of two mating parts of 100 mm basic size, the fit is Interference and the interference varies from 0.05 mm to 0.12 mm. The tolerance on the two mating parts is equal. Determine the sizes of the two mating parts on (a) Hole Basis System (b) Shaft Basis System.*

Solution.

(*a*) **Hole Basis System :** Refer Fig. 9.13,

Lower limit of hole = Basic size = 100 mm

Now Refer to Fig. 9.5 (*a*),

Maximum interference will be when the hole is at its lower limit and the shaft is at its upper limit,

∴ Upper limit of shaft = Lower limit of hole + Maximum interference

= 100 + 0.12 = 100.12 mm

Now it is clear that,

Maximum interference – Minimum interference = Upper limit of shaft – Lower limit of hole – (Lower limit of shaft – Upper limit of hole)

= (Upper limit of shaft – Lower limit of shaft)

+ (Upper limit of hole – Lower limit of hole

= Tolerance on shaft + Tolerance on hole

= 2T

∴ 2T = 0.12 – 0.05 = 0.07 mm

∴ Tolerance on shaft = Tolerance on hole = 0.035 mm

∴ Higher limit of hole = Lower limit + Tolerance

= 100.035 mm

Lower limit of shaft = Upper limit – Tolerance

= 100.12 – 0.035

= **100.085 mm.**

(*b*) **Shaft Basis System :**

Upper limit of shaft = Basic zize = 100 mm

Lower limit of shaft = Upper limit – Tolerance

= 100 – 0.035 = 99.965 mm

Lower limit of hole = Upper limit of shaft – Maximum interference

= 100 – 0.12 = 99.88 mm.

∴ Upper limit of hole = Lower limit + Tolerance

= 99.88 + 0.035 = **99.915 mm.**

Example 10. *For a number of interchangeable mating parts (holes and shafts), the average allowance is 0.04 mm and the allowance must not exceed ± 0.012 mm from the average value. The basic size is 100 mm. Tolerance on hole = 2 × tolerance on the shaft. Determine the sizes of holes and Shafts using Hole basis system and Unilateral system of tolerances.*

Solution. Refer to Fig. 9.2

Maximum allowance – Minimum allowance = Tolerance on shaft + Tolerance on hole

(0.04 + 0.012) – (0.04 – 0.012) = T + 2T

∴ 3T = 0.024, T = 0.008 mm

∴ Tolerance on shaft = 0.008 mm

or Tolerance on Hole = 0.016 mm

Lower limit of Hole = Basic Size = **100 mm**

Upper limit of Hole = Lower limit + tolerance

= 100 + 0.016 = **100.016 mm**

Now Min. allowance = Lower limit of hole – Upper limit of shaft

∴ Upper limit of shaft = 100 – 0.028 = **99.972 mm.**

Lower limit of shaft = Upper limit – Tolerance
= 99.972 – 0.008
= **99.964 mm.**

PROBLEMS

1. Define "Interchangeability" and discuss its importance.
2. Define tolerance.
3. Why is it impossible to obtain an exact dimension?
4. What is meant by the term "fit". Explain the various types of fits.
5. Define : allowance, clearance and interference.
6. What is zero line?
7. Define : Upper deviation, Lower deviation and Fundamental deviation.
8. Explain unilateral system and bilateral system of tolerances.
9. What is meant by "Basic Hole" and "Basic Shaft"?
10. Explain and compare "Hole basis system" and "Shaft basis system" of fits.
11. A hole and mating shaft are to have a nominal assembly size of 38 mm. The assembly is to have a maximum total clearance of 0.15 mm and a minimum total clearance of 0.05 mm. Determine the specifications of the parts (*a*) for 100% interchangeability and (*b*) for statistical average interchangeability.
12. How will you write the fit : Shaft '*m*' of grade 5 and hole '*H*' of grade 11 and the basic size is 40 mm?
13. Give the designation for. :

    (*i*) 30 mm H-hole to the tolerance grade IT6.

    (*ii*) 30 mm *f*-shaft to the tolerance grade IT7.

    (*iii*) the above fit.
14. Calculate the fundamental deviation and tolerances and hence obtain the limits of size for hole and shaft in the fit : 25 mm H_8-d_9. The diameter steps are 18 mm and 30 mm. The fundamental deviation for '*d*' shaft is given as $-\ 16\ D^{0.44}$. The tolerance unit is,

    $$i = 0.45\ \sqrt[3]{D} + 0.001\ D$$

    The tolerance grade for number 8 quality is 25i and for number 9 quality is 40*i*.
15. A gear ring of 85 mm diameter bore is fitted on to a hub resulting in a H_7/j_6 fit. Calculate the tolerances and hence the limits of size for the hub and the gear bore. Specify the type of fit. The diameter steps are 80 mm and 100 mm. The fundamental deviation for *j* shaft is 0.009 mm.
16. Calculate the fundamental deviations and tolerances and hence the limits of size for shaft and hole pair designated as 60 mm H_7/m_6. The tolerance unit is given as

    $$i = 0.45\ \sqrt[3]{D} + 0.001\ D \text{ microns.}$$

    The diameter steps are 50 mm and 80 mm. The fundamental deviation for *m* shaft is = + (IT7 – IT6). For quality 7, the multiplier is 16 and that for quality 6 it is 10.
17. Calculate the tolerances, limits and allowances for a 25 mm shaft and hole pair designated as H_8/d_8. The diameter steps are 24 mm and 30 mm. The multiplier for quality 8 is 25. The FD for '*d*' shaft is $-\ 16D^{0.44}$ microns. Name the type of fit.
18. Determine the types of fits produced by the following mating of holes and shafts.

    (*a*) 30 mm H_7/f_6

    (*b*) 30 mm H_6/f_6

    (*c*) 50 mm P_8/h_5

19. A fit is designated as : 60 mm $H_7 - h_8$. Determine the minimum clearance and maximum clearance of the fit.

20. A turned shaft is to rotate in a reamed hole. The selected fit is H_7/c_7. Determine the actual dimensions of the hole and the shaft. The basic size is 60 mm. The diameter steps are 50 mm and 80 mm. The fundamental deviation for the shaft c is,

$$-(95 + 0.8\ D) \text{ microns.}$$

For quality 7, the multiplier is 16.

21. An idler gear is to rotate over a 30 mm shaft. The chosen fit is H_7/c_8. Determine the actual dimensions of the shaft and the bore of the idler. The diameter steps are 30 and 40 mm. The fundamental deviation for the hole H is zero and that for the shaft c is

$$-(95 + 0.8\ D) \text{ microns.}$$

The multiplier for quality 7 is 16 and that for quality 8 is 25.

22. Define and compare : Random assembly, statistical assembly and selective assembly.

23. Give some common applications of selective assembly.

24. What are the main aims of any Limit System?

25. Two shafts A and B have their diameters specified as 100 ± 0.1 mm and 0.1 ± 0.0001 mm, respectively. Which of the following statements is/are true? **(GATE 1992)**

(*a*) Tolerance in the dimension is greater in shaft A.

(*b*) The relative error in the dimension is greater in shaft A.

(*c*) Tolerance in the dimension is greater in shaft B.

(*d*) The relative error in the dimension is greater in shaft B. **(Ans.: a)**

26. A shaft (diameter $20^{+0.05}_{-0.15}$ mm) and a hole (diameter $20^{+0.20}_{+0.10}$ mm) when assembled would yield :-

(*a*) Transition Fit (*b*) Interference F it (*c*) clearance fit (*d*) None **(GATE 1993) (Ans. : c)**

27. The fit on a hole-shaft system is specified as $H_7 - s_6$. The type of fit is :

(*a*) Clearance fit (*b*) running fit (sliding fit)

(*c*) push fit (transition fit) (*d*) force fit (interference fit) **(GATE 1996) Ans. : d)**

28. In the specification of dimensions and fits.

(*a*) allowance is equal to bilateral tolerance.

(*b*) alowance is equal to unilateral tolerance.

(*c*) allowance is independent of tolerance.

(*d*) allowance is equal to the difference between maximum and minimum dimension specified by the tolerance. **(GATE 1998) [Ans. : c]**

29. A journal and bearing assembly has the following sizes : -

Journal : $50^{-0.001}_{-0.002}$ mm

Bearing : $50^{+0.001}_{-0.002}$ mm

Determine : Tolerance on Journal, Tolerance on Bearing, Minimum clearance, Maximum clearance and type of fit.

(Ans : 0.001 mm, 0.003 mm. –0.001 mm, 0.003 mm, Transition fit).

30. In an assembly of journal and bearing, the basis size is 55 mm. The lower deviation, and upper deviation for bearing is 0 micron and 4 microns respectively. the corresponding values for journal are – 3 and –7 microns. Determine the sizes of journal and bearing and the type of fit.

(**Ans** : - Journal : $\begin{matrix}\text{H.L.} = 54.997 \text{ mm} \\ \text{L.L.} = 54.993 \text{ mm}\end{matrix}$; Bearing $\begin{matrix}\text{H.L.} = 55.004 \text{ mm} \\ \text{L.L.} = 55.000 \text{ mm}\end{matrix}$, clearance fit).

31. Explain the difference between tolerance and allowance.

32. Why do manufacturing processes produce parts with such a wide range of tolerances?

33. The journal and bearing assembly has a basic size of 200 mm. For the bearing : F.D.; (Here, L.D.) = 0 micron, and Tolerance = 46 micron.

For the journal, the values are – 820 and 115 microns respectively. Find the dimensions of journal and bearing, assuming unilateral system of tolerances. Also, determining the allowance and type of fit. Which systems of fit has been adopted?

(**Ans.** Hole: 200, 200.46 mm; Journal: 199.180, 199.065 mm, Clearance fit, Hole – basis system).

34. The following are the sizes of a shaft and a hole :-

$$\text{Shaft} \ : \ 30^{+0.000}_{-0.013} \text{ mm}$$

$$\text{Hole} \ : \ 30^{+0.000}_{+0.013} \text{ mm}$$

Determine the type of fit obtained when these are assembled.

Sol. Maximum size of the shaft = 30 mm

Minimum size of the shaft = 30 – 0.013 = 29.987 mm.

Maximum size of the hole = 30 + 0.013 mm = 30.013 mm

Minimum size of the hole = 30 mm

∴ Maximum clearance = Maximum size of the hole – Minimum size of the shaft.

= 30.013 – 29.987 = 0.026 mm

Minimum clearance = Minimum size of Hole – Maximum size of shaft

= 30 – 30 = 0 mm

Since both the maximum and minimum clearances are greater than or equal to zero, the fit will be "Clearance Fit".

35. The following are the sizes of the shaft and the hole :-

$$\text{Shaft} \ : \ 50.000^{+0.000}_{-0.011} \text{ mm} ; \quad \text{Hole} : 50.000^{-0.026}_{-0.065} \text{ mm}$$

Determine the fit obtained

Sol. Maximum size of the shaft = 50 mm.

Minimum size of the shaft = 50 – 0.011 = 49.989 mm

Maximum size of the hole = 50 – 0.026 = 49.974 mm

Minimum size of the hole = 50 – 0.065 = 49.935 mm

Maximum Clearance = Maximum size of hole - Minimum size of the shaft

= 49.974 – 49.989 = – 0.015 mm

Minimum Clearance = Minimum size of hole - Maximum size of the shaft

= 49.935 – 50.000 = – 0.065 mm

Since both the maximum and minimum clearances are negative, the fit obtained will be "interference fit"

36. Determine the type of fit :

$$\text{Hole size} \ : \ 20^{+0.05}_{-0.05} \text{ mm} ; \quad \text{Shaft} : 20^{+0.05}_{-0.05} \text{ mm}$$

Sol. Maximum size of the hole = 20.05 mm

Minimum size of the hole = 19.95 mm

Maximum size of the shaft = 20.05 mm

Minimum size of the shaft = 19.95 mm

∴ Maximum Clearance = Maximum size of hole – Minimum size of the shaft = 20.05 – 19.95 = 0.10 mm

Minimum Clearance = Minimum size of hole – Maximum size of shaft = 19.95 – 20.05 = – 0.10 mm

Since the Maximum Clearance is positive and the Minimum Clearance is negative, the fit obtained will be the 'Transition fit'.

10

GAUGES AND GAUGE DESIGN

10.1. INTRODUCTION

There are several methods available for the control of dimensions of components in a system of limits and fits. Each component, for example, be measured with an instrument giving a suitable accuracy and this method is often adopted, particularly for closely limited work. The method used for the majority of the work in quantity production is the system of limit gauging. This has the advantages that it can be operated in many cases by quite unskilled persons.

Gauges are inspection tools of rigid design, without a scale, which serve to check the dimensions of manufactured parts. Gauges do not indicate the actual value of the inspected dimensions of the component. They are only used for determining whether the inspected part has been made within the specified limits. A workman checking a component with a gauge does not have to make any calculations or to determine the actual dimensions of the part. Gauges are easy to employ. This is one reason for their wide application in engineering. Gauges differ from measuring instruments in the following respects :

(*i*) no adjustment is necessary in their use.

(*ii*) they usually are not general-purpose instruments but are specially made for some particular part, which is to be produced in sufficiently large quantities.

Gauging is used in preference to measuring when quantities are sufficiently high, because it is faster and easier with resulting lower costs.

10.2. PLAIN GAUGES

Plain gauges are used in checking plain, that is, unthreaded holes and shafts.

10.2.1. Classification of Plain Gauges. Plain gauges are classified in the following ways.

1. According to type.
2. According to purpose.
3. According to form of tested surface, and
4. According to design.

1. **According to type.** (*a*) Standard Gauges (*b*) Limit Gauges.

(*a*) **Standard Gauges.** Every gauge is almost a copy of the part example, a bushing is to be made which is to mate with a shaft. In this case, shaft is the mating part. The bushing is checked by a gauge which in so far as the form of its surface and its size is concerned is a copy of the mating part, that is, the shaft.

If a guage is made as an exact copy of the opposed (mating) part, in so far as the dimensions to be checked are concerned, it is called a 'standard gauge'. The first gauges to be developed were the standard gauges. The first standard gauges were the opposed (mating) parts themselves. When a component is assembled with its mating part, a (mating) part itself. However, such individual fittings are not convenient or even possible in mass production conditions. Moreover, the two parts to be assembled might be in production in two different shops or even at two different plants. Therefore, it is more proper to use, as a checking tool, not the mating part, but its exact copy as far as the tested dimension is concerned.

Such a standard gauge has two drawbacks :

(*i*) The quality of the manufactured part will depend upon the freedom with which it mates with the standard gauge. The judgement of this freedom is a relative thing and it usually creates misunderstanding between the purchaser and the manufacture.

(*ii*) A standard gauge cannot be used to check an interference fit. For example, if a bushing (Fig. 10.1) of 50 mm diameter is to be made for assembly with a shaft of 50.1 mm diameter, then the standard gauge diameter will be 50.1 mm of opposed part. Such a gauge will not pass into a properly produced bushing of diameter 50 mm.

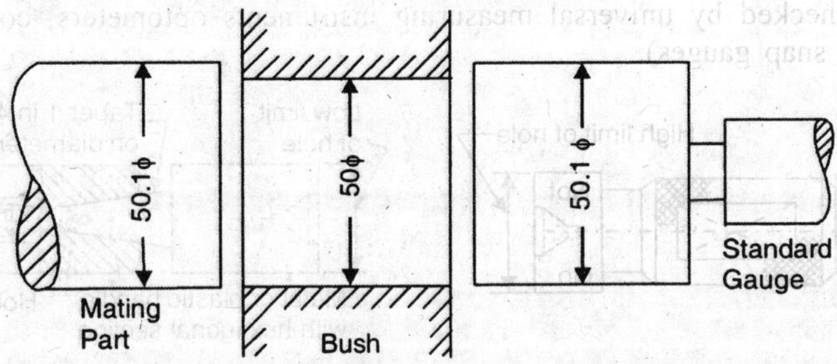

Fig. 10.1. Application of a Standard Gauge.

(*b*) **Limit Gauges.** The system of limit gauges is very widely used in industries. Limit gauges are made to the limits of the dimensions of the part to be tested. As there are two limits of the dimensions of a part, high and low, two gauges are needed to check each dimension of the part. The part is checked by successively assembling each of the gauges with it. Since the dimensions of a properly manufactured part must be within the prescribed limits, one of the gauges called a "Go Gauge" should pass through or over the part, while the other gauge called a "Not Go Gauge" should not pass through or over a part, Fig. 10.2. Gauges should pass through or over a part under their own weight and the part and the gauge must be at the same temperature.

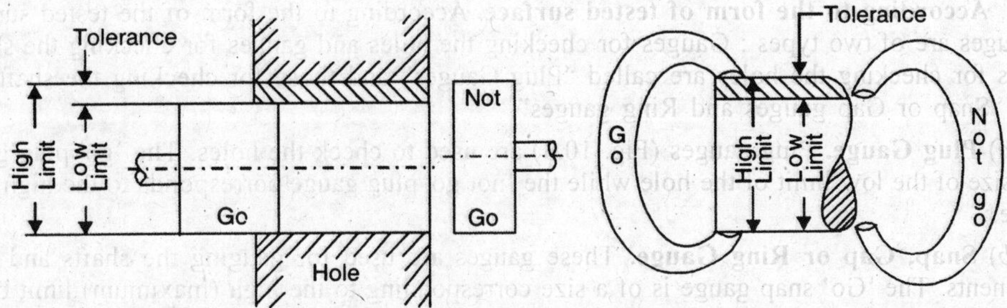

Fig. 10.2. Limit Gauges.

2. **According to Purpose.** According to purpose, the gauges may be classified as : (*a*) Workshop gauge or Working gauge (*b*) Inspection gauge (*c*) Purchase inspection gauge, and (*d*) Reference or Master gauge.

(*a*) **Workshop Gauge.** Workshop gauge or the manufacturing gauge is used by the machine operator to check the dimensions of the parts as they are being produced. These gauges usually have limits within those of the component being inspected. They are designed so as to keep the size of the part near the centre of the limit tolerance.

(*b*) **Inspection Gauge.** Inspection gauges are those used by inspectors in the final acceptance of manufactured parts when finished. These gauges are made to slightly larger tolerances than the workshop gauges so as to accept work slightly nearer the tolerance limit than the workshop gauges. This is to ensure that work which passes the working gauge will be accepted by the inspection gauge also.

(*c*) **Purchase Inspection Gauge.** The need of such gauges arises when the products of other plants are to be accepted. The purchaser must remember that the parts may have been made and checked by working gauges worn to the maximum permissible degree. Therefore the 'Go' side of the purchase inspection gauge must be designed accordingly. Thus, nominal size of "Go" purchase inspection gauge will be equal to the lower limit of the hole. 'No-Go' purchase inspection gauge design is similar to "No-Go" working gauge.

(*d*) **Reference or Master Gauges.** Reference or master gauges are used only for checking the size and condition of other gauges. Reference gauges are the reverse or opposite in form to working or inspection gauges. Due to the expenditure involved, reference gauges are seldom used and gauges are checked by universal measuring instruments optometers, comparators etc. or gauge blocks (for snap gauges).

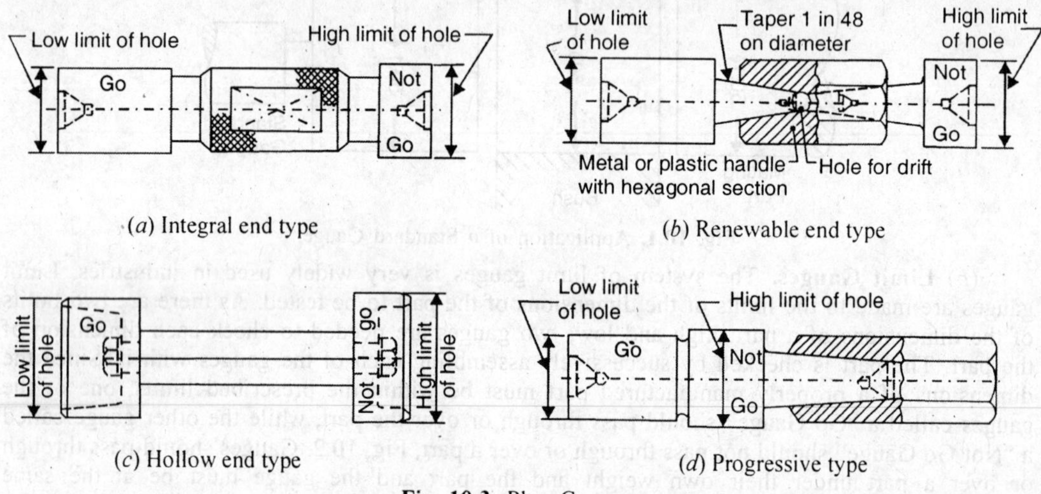

(*a*) Integral end type

(*b*) Renewable end type

(*c*) Hollow end type

(*d*) Progressive type

Fig. 10.3. Plug Gauges.

3. **According to the form of tested surface.** According to the form of the tested surface, the gauges are of two types : Gauges for checking the holes and gauges for checking the shafts. Gauges for checking the holes are called "Plug Gauges" and those for checking the shafts are called "Snap or Gap gauges and Ring gauges".

(*a*) **Plug Gauge.** Plug gauges (Fig. 10.3) are used to check the holes. The 'go' plug gauge is the size of the low limit of the hole while the 'not go' plug gauge corresponds to the high limit of hole.

(*b*) **Snap, Gap or Ring Gauge.** These gauges are used for gauging the shafts and male components. The 'Go' snap gauge is of a size corresponding to the high (maximum) limit of the shaft, while the 'Not Go' gauge corresponds to the low (minimum limit). The various snap, gap and ring gauges are shown in Fig. 10.4. Gap gauges can also check lengths or widths flangs thickness etc.

4. **According to Design.** According to design, the gauges may be classified as :

(*i*) (*a*) Single limit (*b*) Double limit
(*ii*) (*a*) Single ended (*b*) Double ended
(*iii*) (*a*) Fixed (*b*) Adjustable
(*iv*) (*a*) Integral end (*b*) Renewable end
(*v*) (*a*) Solid end (*b*) Hollow end

(*i*) **Single limit and Double limit Gauge.** The gauge cannot be made exactly according to the dimensions it has to check.

The dimensions of 'Go' and 'Not Go' sides of the gauge will vary over a small range. Depending upon the manner of this variation, the gauge can be single or unilateral limit gauge and double or bilateral limit gauge. For example, if the low limit of hole is 6.25 cm and the high limit is 6.27 cm and the tolerance on gauge manufacture is 0.002 cm, then

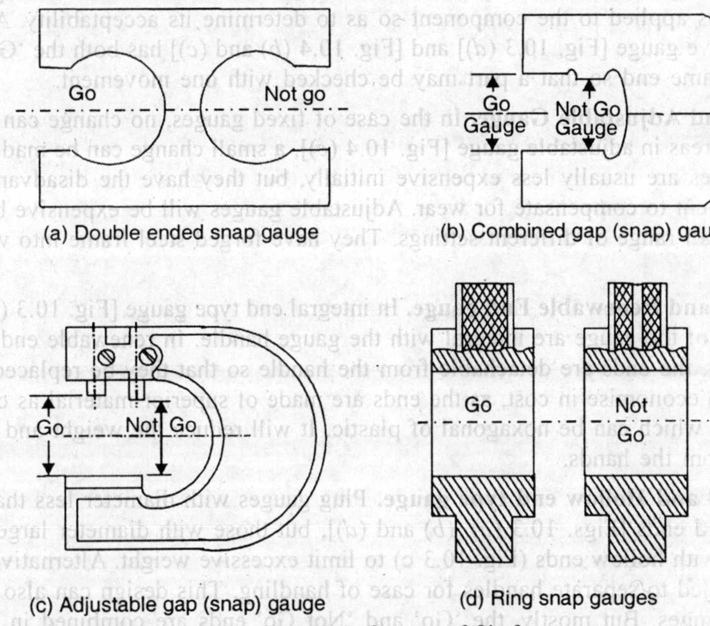

Fig. 10.4. Gauges of Shaft.

In Single limit (unilateral) system :

Size of 'Go' Plug gauge (For maximum metal condition of hole)

$$= 6.25 \begin{array}{l} + 0.002 \text{ cm} \\ - 0.000 \end{array}$$

Size of 'Not Go' Plug gauge $= 6.27 \begin{array}{l} + 0.000 \text{ cm} \\ - 0.002 \end{array}$

In Double limit (Bilateral) system

Size of 'Go' Plug gauge $= 6.25 \begin{array}{l} + 0.001 \text{ cm} \\ - 0.001 \end{array}$

Size of 'Not Go' Plug gauge $= 6.27 \begin{array}{l} + 0.001 \text{ cm} \\ - 0.001 \end{array}$

This has been made clear in Fig. 10.5.

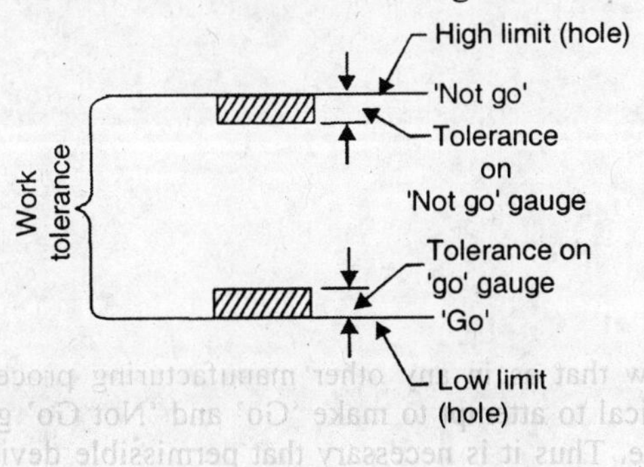

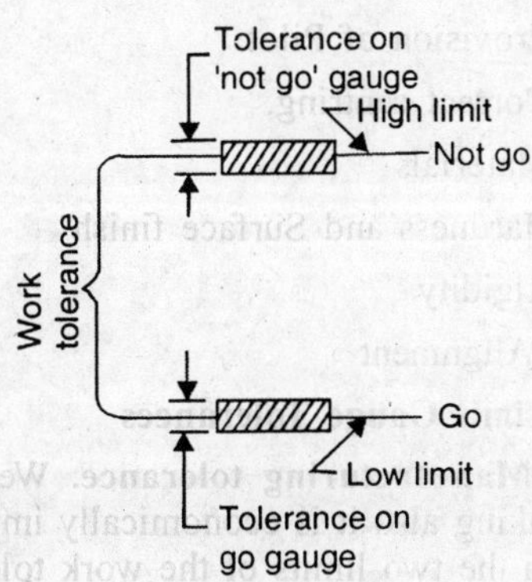

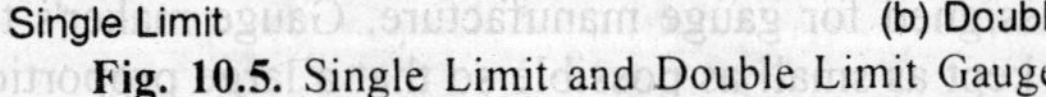

Fig. 10.5. Single Limit and Double Limit Gauge.

(*ii*) **Single end and double end gauge.** A double end limit gauge [Fig. 10.3 (*a*) to (*c*)], [Fig. 10.4 (*a*)] has the 'Go' side at one end and the 'Not Go' side at the other end of the gauge. Each end of the gauge is applied to the component so as to determine its acceptability. A single end gauge or progressive gauge [Fig. 10.3 (*d*)] and [Fig. 10.4 (*b*) and (*c*)] has both the 'Go' and 'Not Go' sides at the same end so that a part may be checked with one movement.

(*iii*) **Fixed and Adjustable Gauge.** In the case of fixed gauges, no change can be made in the size range whereas in adjustable gauge [Fig. 10.4 (*c*)], a small change can be made in the size range. Fixed gauges are usually less expensive initially, but they have the disadvantage of not permitting adjustment to compensate for wear. Adjustable gauges will be expensive but they can be used over a small range of different settings. They have forged steel frame into which anvils are fitted.

(*iv*) **Integral and Renewable End gauge.** In integral end type gauge [Fig. 10.3 (*a*), the 'Go' and 'Not Go' end of the gauge are integral with the gauge handle. In renewable end type gauge [Fig. 10.3 (*b*), (*c*)], the ends are detachable from the handle so that they be replaced separately when worn, and to economise in cost, as the ends are made of superior material as compared to that of the handle, which can be hoxagonal of plastic. It will reduce the weight and prevent the transfer of heat from the hands.

(*v*) **Solid end and Hollow end type gauge.** Plug gauges with diameter less than 63.5 mm are made with solid ends [Figs. 10.3 (*a*), (*b*) and (*d*)], but those with diameter larger than 63.5 mm are designed with hollow ends (Fig. 10.3 c) to limit excessive weight. Alternatively, the two ends may be attached to separate handles for case of handling. This design can also be adopted for lighter plug gauges. But mostly, the 'Go' and 'Not Go' ends are combined in one unit as shown in Fig. 10.3. This makes the use of the gauge convenient and also it ensures that both ends of the gauge are kept together with no risk of misplacing one or the other.

10.3. DESIGN OF LIMIT GAUGES

The design of a limit gauge must ensure proper inspection of the part for which it is intended. The following points and factors must be kept in mind while designing the limit gauges :

1. Limit gauge tolerance :

(*a*) Manufacturing tolerance

(*b*) Wear allowance

and the disposition of these tolerances with respect to the work tolerance.

2. Taylor's principle of gauge design.
3. Fixing of gauging elements (ends) with handle.
4. Provision of Pilot.
5. Provision of Pilot.
6. Correct centring.
7. Materials
8. Hardness and Surface finish
9. Rigidity
10. Alignment

1. **Limit Gauge Tolerances**

(*a*) **Manufacturing tolerance.** We know that as in any other manufacturing process, in gauge making also it is economically impractical to attempt to make 'Go' and 'Not Go' gauges exactly to the two limits of the work tolerance. Thus it is necessary that permissible deviations in accuracies must be assigned for gauge manufacture. Gauge maker's tolerance or manufacturing tolerance should be kept as small as possible so that a large proportion of the work tolerance is still available for the manufacturing process. However the small the gauge tolerance, the more the gauge will cost.

There is no universally accepted policy for the amount of gauge tolerance. However, the following norms are generally accepted : Limit gauges are made 10 times more accurate than the tolerances they are going to control. That is, the tolerance on each gauge whether 'Go' or 'Not Go', is 1/10th of the work tolerance. For example, if the work tolerance is 10 units, then the manufacturing tolerance for `Go' and `Not Go' gauge each will be 1 unit. This makes it possible, although the probability is small, for the work tolerance available in the shop to be cut down to 80% of the specified tolerance. The amount of tolerance on inspection gauges is generally 5% of the work tolerance. Tolerance on reference or master gauges is generally 10% of the gauge tolerance.

Allocation of Manufacturing Tolerance. As already discussed, there are two systems for the allocation of manufacturing tolerance, unilateral system and bilateral system, Fig. 10.6.

(*i*) **Unilateral System.** In this system, the gauge tolerance zone lies entirely within the work tolerance zone. Due to this, the work tolerance zone becomes smaller by the sum of the gauge tolerances. But this ensures that every component passed by such a gauge regardless of the amount of gauge size variation will be within the work tolerance zone.

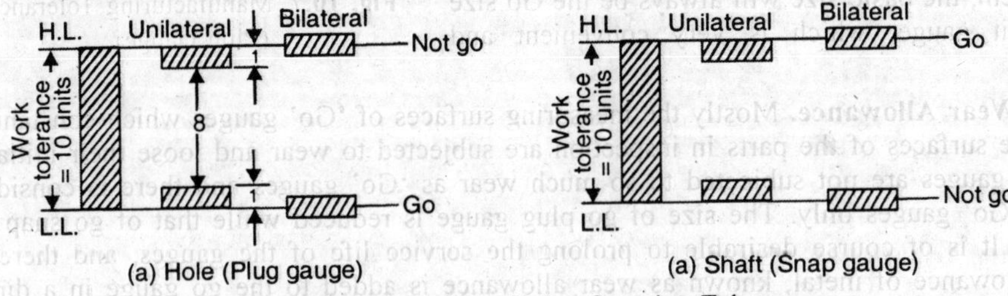

Fig. 10.6. Allocation of Manufacturing Tolerance.

For example, let the size of the hole to be tested be 25 ± 0.02 mm.

Therefore,

High limit of hole = 25.02 mm

Low limit of hole = 24.98

Work tolerance = 0.04

∴ Gauge tolerance = 10% of tolerance

= 0.004 mm

Dimension of 'Go' Plug/gauge

$$= 24.98 \text{ mm} \begin{array}{l} +0.004 \\ -0.000 \end{array}$$

Dimension of 'Not Go' Plug gauge

$$= 25.02 \text{ mm} \begin{array}{l} +0.000 \\ -0.004 \end{array}$$

The disadvantage of this system is that certain components may be rejected as being outside the working limits when they are not. However, the unilateral system has found wider use in industry than the bilateral system.

(*ii*) **Bilateral System.** In this system, the 'Go' and 'Not Go' gauge tolerance zones are bisected by the high and low limits of the work tolerance zone, Fig. 10.6. Taking the example as above,

$$\text{Dimension of 'Go' Plug gauge} = 24.98^{+0.002}_{-0.002} \text{ mm}$$

$$\text{Dimension of 'Not Go' Plug gauge} = 25.02^{+0.002}_{-0.002} \text{ mm}$$

The disadvantages of this system are that components which are within working limits can be rejected and parts which are outside the working limits can be accepted. But we already know (Chapter 9) that the percentage of such parts is very small, if the process is under control.

Another way of providing manufacturing tolerance in the unilateral system is shown in Fig. 10.7 in which the manufacturing tolerance is disposed opposite to the direction of wear, both for Go gauge and No Go gauge.

In the modern limit systems, unilateral system of providing tolerances is preferred, because in a hole basis system, the basic size will always be the Go size of a limit gauge, which is very convenient and practical.

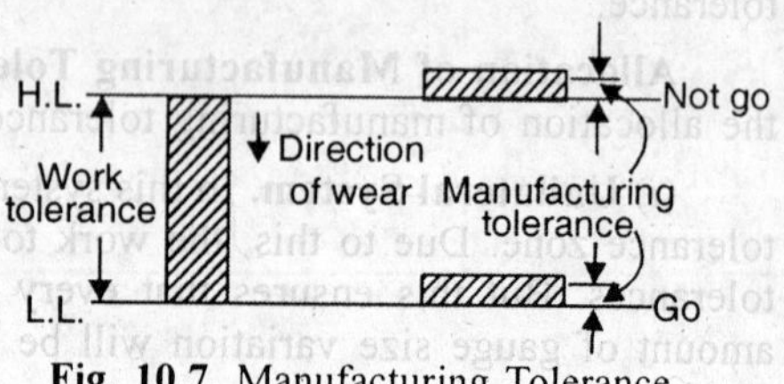

Fig. 10.7. Manufacturing Tolerance (Plug Gauge)

(*b*) **Wear Allowance.** Mostly the measuring surfaces of 'Go' gauges which constantly rub against the surfaces of the parts in inspection are subjected to wear and loose their initial size. 'Not Go' gauges are not subjected to so much wear as 'Go' gauges and there is considerable wear on 'Go' gauges only. The size of go plug gauge is reduced while that of go snap gauge increases. It is of course desirable to prolong the service life of the gauges, and therefore a special allowance of metal, known as wear allowance is added to the go gauge in a direction opposite to wear. Wear allowance is usually taken as 5% of work tolerance. Wear allowance is applied to a nominal go gauge diameter before gauge tolerance is applied. Taking the example discussed above,

$$\text{Wear allowance} = 5\% \text{ of work tolerance}$$
$$= 0.002 \text{ mm}$$

$$\text{Nominal size of go plug gauge} = 24.98 + 0.002 = 24.982 \text{ mm}$$

$$\left.\begin{aligned} \therefore \text{Dimensions of Go Plug gauge} &= 24.982^{+0.004}_{-0.000} \text{ mm} \\ \text{Dimension of Not Go Plug gauge} &= 25.02^{+0.000}_{-0.004} \text{ mm} \end{aligned}\right\} \text{Unilateral System}$$

This is shown in Fig. 10.8

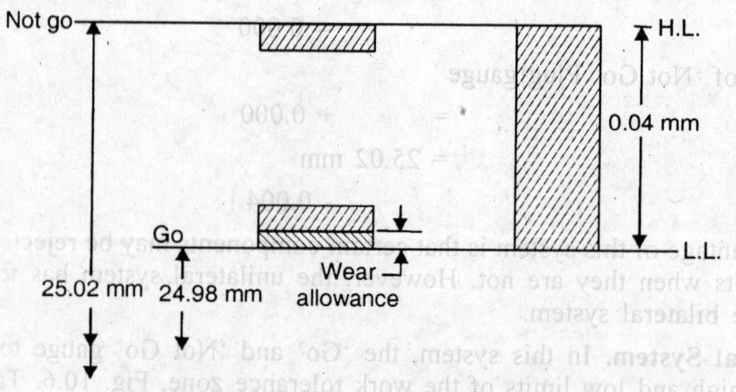

Fig. 10.8. Application of Wear Allowance.

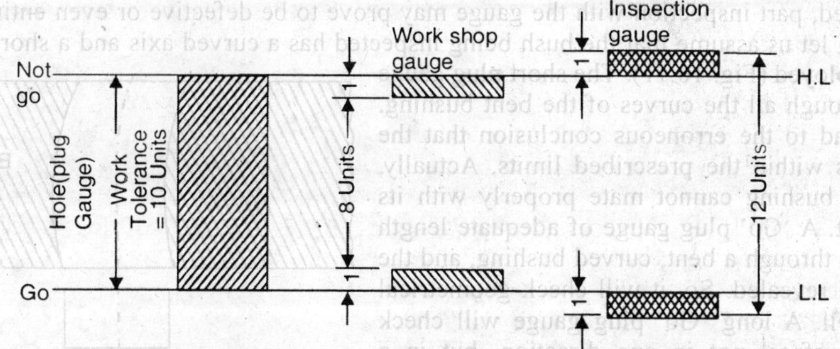

Fig. 10.9. Tolerances of Workshop and Inspection Gauges.

According to British Standards, wear allowance is provided when the work tolerance is greater than 0.09 mm. When the work tolerance is small and no wear allowance is provided, gauges should be of specially has working material in order to give a reasonable life. The method of providing tolerances on the workshop gauge and the inspection gauge according to British system is shown in Fig. 10.9. The tolerance on the workship gauge is arranged to fall inside the work tolerance (unilateral system) while on the inspection gauge it falls outside the work tolerance. The effect of these two types of gauges is that although the workshop gauge tends to cut down the tolerance available in the shop, it also ensures that all work passed by the working gauge will automatically pass the inspection gauge. It is theoretically possible for work which is slightly limits to be passed by the inspection gauge, although in practice very little trouble is experienced in this way. In this system, it may also happen that after wear a shop gauge may become inspection gauge (not a very desirable feature). Fig. 10.10 shows a revised system. In this system, the disadvantages of the inspection gauge are reduced by reducing the tolerance zone of the inspection gauge while the workshop gauge tolerance remains the same.

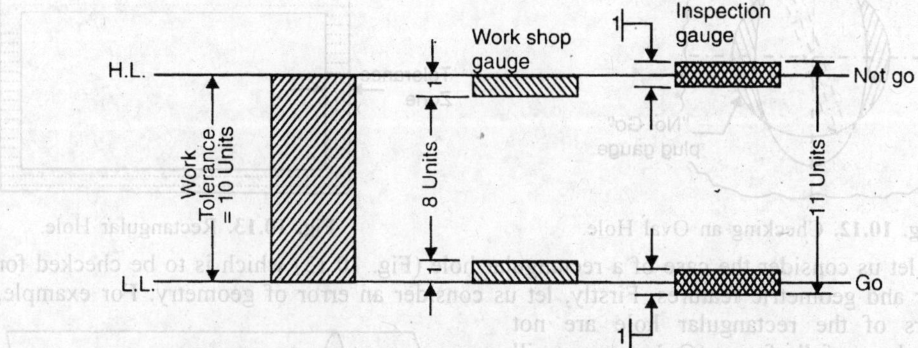

Fig. 10.10. Tolerances of Workshop and Inspection Gauges (Revised).

2. **Taylor's Principle.** This principle states that the Go gauge should always be so designed that it will cover the maximum metal condition (MMC) of as many dimensions as possible in the same limit gauge, whereas a Not Go gauge to cover the minimum metal condition of one dimension only, as shown in Fig. 10.2. According to this rule, a Go plug gauge should have a full circular section and be of full length of the hole, it has to check. In addition to control the diameter at any one point this ensures that any lack of straightness or parallelism of the hole will prevent the entry of full length Go plug gauge. For example, let us assume that a bushing is to be inspected. The bush is to mate with a shaft. The shaft is, therefore, the opposed part in relation to the bushing. Therefore the form of 'Go' plug gauge should exactly coincide with the form of the shaft. For this purpose, the "Go" plug gauge must be of adequate length, not less than the length of the future association of bushing and shaft. If this condition

is not satisfied, part inspection with the gauge may prove to be defective or even entirely wrong. For instance, let us assume that the bush being inspected has a curved axis and a short 'Go' plug gauge is employed (Fig. 10.11). The short plug gauge will pass through all the curves of the bent bushing. This will lead to the erroneous conclusion that the workpiece is within the prescribed limits. Actually, such a bent bushing cannot mate properly with its opposed part. A 'Go' plug gauge of adequate length will not pass through a bent, curved bushing, and the error will be revealed. So, it will check geometrical shape as well. A long 'Go' plug gauge will check cylindrical surface, not in one direction, but in a number of sections simultaneously. Generally, the length of 'Go' plug gauge will check cylindrical surface, not in one direction, but in a number of sections simultaneously. Generally, the length of 'Go' plug gauge should not be less than 1.5 times the diameter of the hole. Length of 'Not Go' gauge is kept smaller than 'Go' gauge.

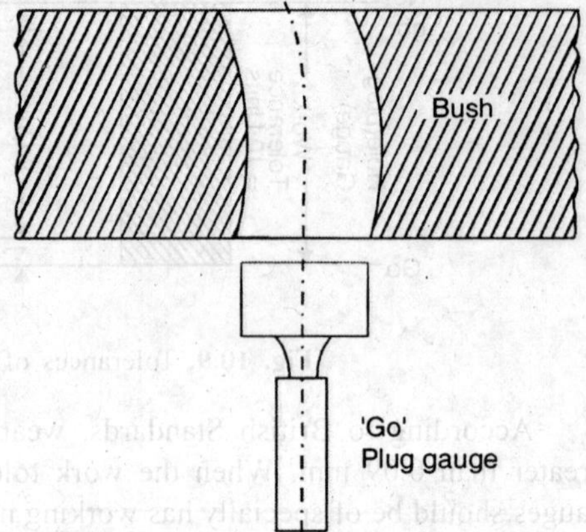

Fig. 10.11. Checking a Bush with a Curved Axis.

Now let us take the case of checking an oval hole by a cylindrical 'Not Go' gauge, Fig. 10.12. As the faces of the plug gauge and the hole under inspection overlap (hatched portion) the plug will obviously not enter the hole. This will again lead to erroneous conclusion that the part is within the prescribed limits. It will be more appropriate to make the 'Not go' gauge in the form of a pin or bar, shown with dashed lines. Turning such a pin gauge about the axis of the bushing will reveal the improper form of the hole.

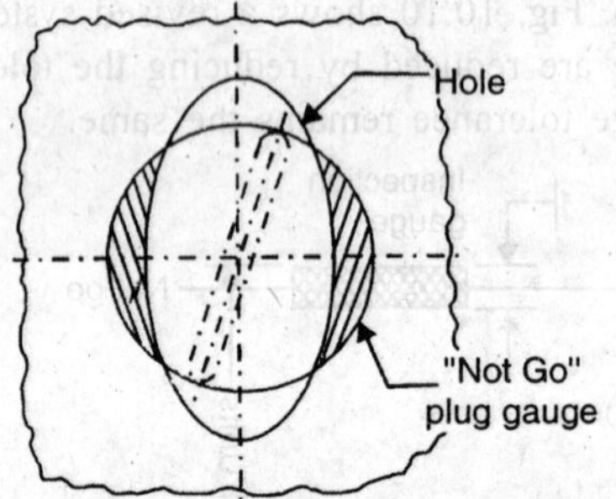

Fig. 10.12. Checking an Oval Hole.

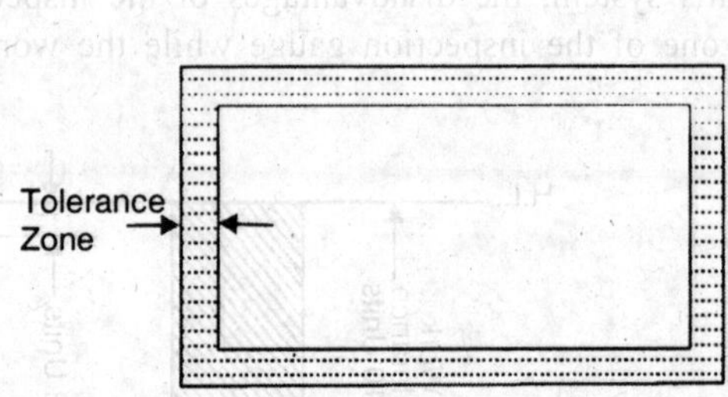

Fig. 10.13. Rectangular Hole.

Now let us consider the case of a rectangular hole (Fig. 10.13) which is to be checked for both linear and geometric features. Firstly, let us consider an error of geometry. For example, the corners of the rectangular hole are not square. Only a full form 'Go' gauge will indicate that the part is wrong. If pin gauges, made to the low limit of the hole (MMC), are used to check the hole they will enter the hole (Fig. 10.14) and the error will remain undetected. This will lead to wrong conclusion that the hole is satisfactory, when actually it is not.

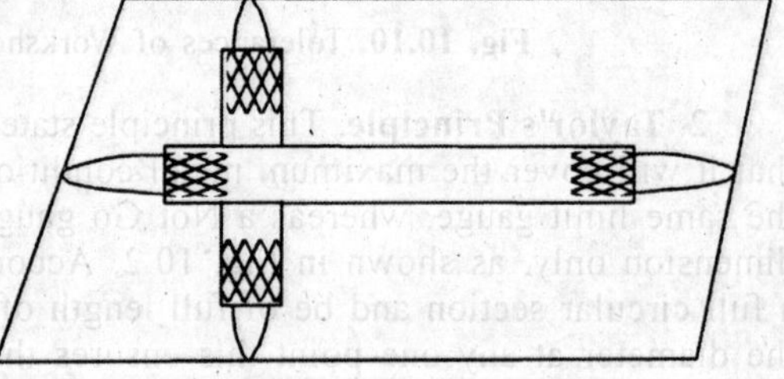

Fig. 10.14. Pin Gauges.

Next, let us consider an error of size, say for example, the length of hole is outside the high limit. A full form 'Not Go' gauge made to high limit of hole will not detect this error. As the width of hole is within limits, the gauge will not enter the hole, even though the length is outside limits. Again, this gauge will indicate that the hole is satisfactory when it is not. However, a 'Not Go' gauges of pin type to check the width and length of the hole.

From above, the design of various types of gauges based on Taylor's principle can be summarised as follows :

(*i*) **Circular Holes.** For checking circular holes 'Go Plug Gauge' should have a minimum length equal to the length of the hole or the length of engagement of the gauge with associated component, which ever is smaller. The 'Not Go Plug Gauge' will not be of full form. It should be pin gauge which would check the upper limit of hole (minimum metal condition) across any diameter at any position along the length of the hole. As already explained, turning such a gauge about the axis of hole will reveal the defect if there is any, because, even if the pin gauge accepts the hole along one axis, it will reject the hole when used along the other axis, as shown in Fig. 10.15.

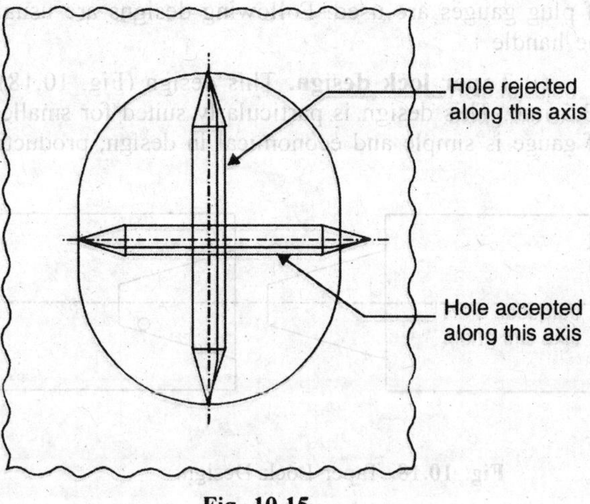

Fig. 10.15

(*ii*) **Circular Shafts.** To check the circular shafts, the ring gauge should be used for the 'Go Gauge' whose minimum length should be equal to the length of shaft or the length of engagement of the 'Ring Go Gauge' with the associated component, whichever is shorter. The 'Not Go Gauge' will not be ring gauge, but it should be in the form of a Snap Gauge or Gap Gauge, so that it is able to reject the shaft which is not circular, as shown in Fig. 10.16

Note. In many cases, shaft inspection with ring gauges would be difficult, for example, in checking stepped shafts or crank shafts. For this reason, "Go" snap gauges are much used for shafts, though their design is not according to rules.

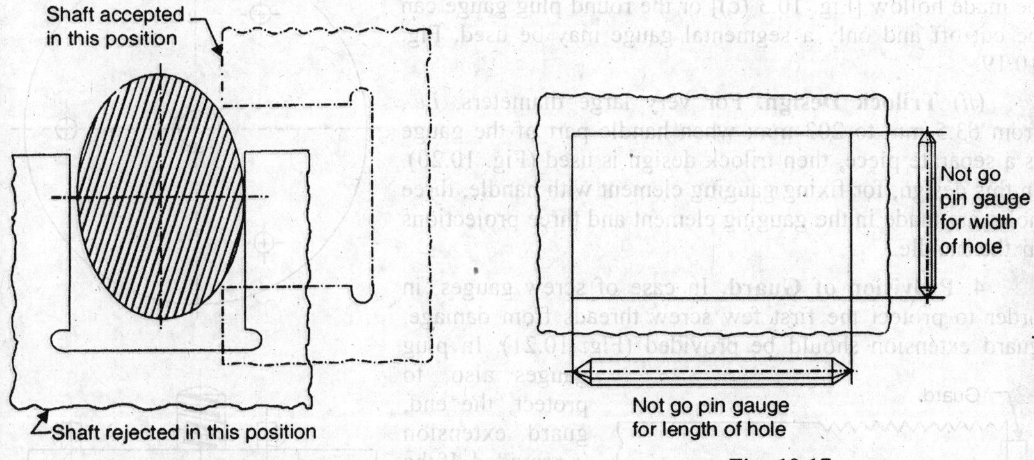

Fig. 10.16 **Fig. 10.17**

(*iii*) **Non-circular Holes and Shafts.** For checking non-circular holes and shafts, the above principles will apply, but for each dimension there will be a separate 'Not Go Gauge', which will correspond to the minimum metal condition of the component, Fig. 10.17. The 'Go Gauge' would of course be of full form and would correspond to the maximum metal condition of the component.

3. **Fixing of guage elements with handle.** Plug gauges can be of solid type in which the gauging members are integral with the handle or the gauging elements can be separated from

the handle and suitable fixed together. These are known as "Renewable"type of gauges. Below 050 mm diameter, solid type gauges are mostly used. For larger diameters, the renewable end type of plug gauges are used. Following designs are usually used for fixing the gauging element to the handle :

(*i*) **Taper lock design.** This design (Fig. 10.18) is used for diameters upto and including 63.5 mm. This design is particularly suited for smaller size of plain and screw gauges. This type of gauge is simple and economical in design, production and maintenance. The gauging member

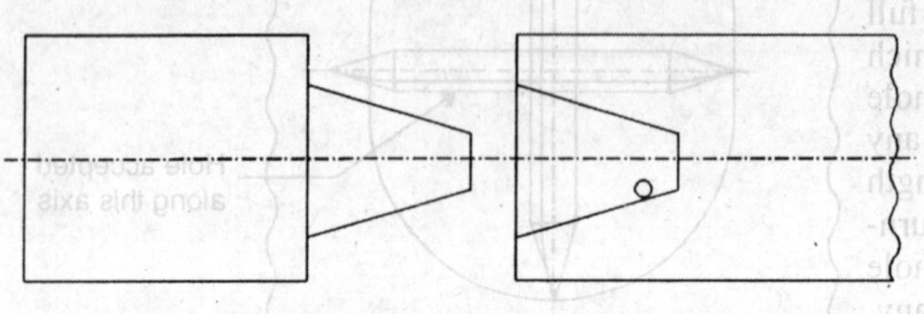

Fig. 10.18. Taper Lock Design.

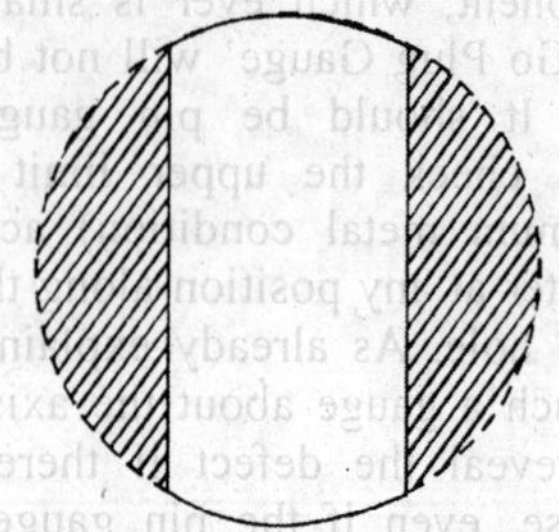

Fig. 10.19. Segmental Plug Gauge.

has a taper shank which is pushed into a taper hole in the handle. When properly made and assembled, the taper lock gauge has been found to possess the rigidity of a solid gauge and is entirely free from shock or vibrations. A drift hole or slot is provided near one end of the handle to enable the gauging element to be removed when replacement is necessary. In case of double ended gauges, the second member is removed by running a rod through the hollow handle.

For very large size holes where the question of a very heavy weight of plugs comes in, the ends of the gauge can be made hollow [Fig. 10.3 (c)] or the round plug gauge can be cut off and only a segmental gauge may be used, Fig. 10.19.

(*ii*) **Trilock Design.** For very large diameters, *i.e.*, from 63.5 mm to 203 mm; when handle part of the gauge is a separate piece, then trilock design is used (Fig. 10.20). In this design, for fixing gauging element with handle, three holes are made in the gauging element and three projections in the handle.

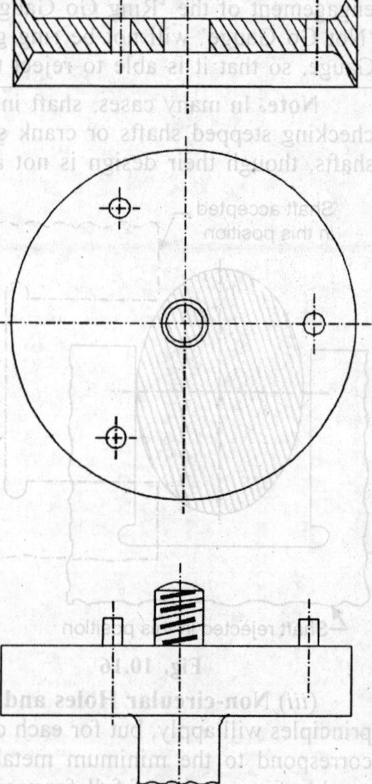

Fig. 10.20. Trilock Design.

4. **Provision of Guard.** In case of screw gauges, in order to protect the first few screw threads from damage, guard extension should be provided (Fig. 10.21). In plug gauges also, to protect, the end, guard extension is provided. If the plug gauge is used for a blind hole, then guard extension will not be provided.

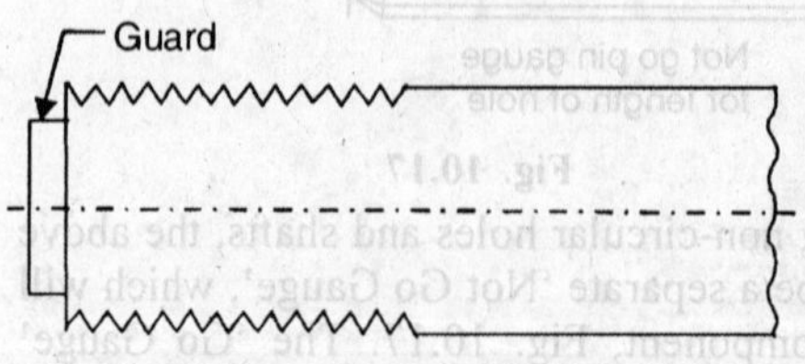

Fig. 10.21. Provision of Guard.

5. **Provision of Pilot.** In case of very closely toleranced parts, sometimes it happens that the plug gauge does not easily enter the hole. This requires skill and practice. Since the gauges are generally used by semi-skilled workers in industry, therefore,

to solve this problem, we use what is known as piloting of plug gauge. (Fig. 10.22). The diameter of land is the same as that of the plug gauging portion.

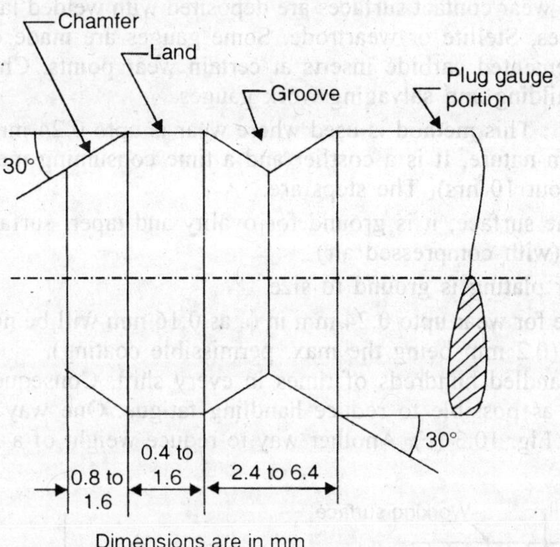

Fig. 10.22. Provision of Pilot.

6. **Correct Centring.** For the purpose of fabrication of plug gauges, grinding is very extensively used, and sometimes lapping may be used. In grinding, the centre of job and centre of machine plays an important role. Centres for high grade job should be very very perfect. Sometimes, before grinding, the centres are even honed after heat treatment. After making the centre in the plug gauges, for heat treatment purposes these centres should be protected so that the same maynot get spoiled due to heat and burn. A typical centre in the gauge is shown in Fig. 10.23. The recess and groove in the centre protect the centre from external contacts and burns etc.

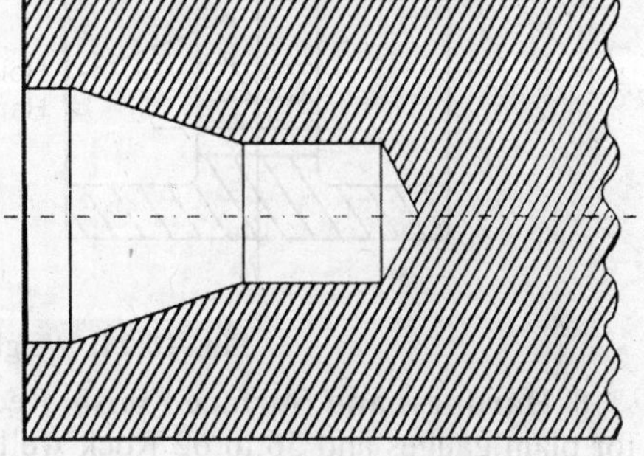

Fig. 10.23. Centre

7. **Materials.** Most gauges are subjected to considerable abrasion during their use and must, therefore, be made of wear resistant materials. Hence, the materials for limit gauges should meet the following requirements :

(*i*) Uniformity of structure and required co-efficient of linear expansion.

(*ii*) Proper workability, especially in grinding and polishing.

(*iii*) Stability of dimensions and forms of parts in the process of operation and possibly lower deformation in heat treatment and during manufacture.

(*iv*) High resistance to mechanical wear and corrosion.

(*v*) Optimal hardness which is a property characterising a high durability and resistance to damage in use.

High Carbon and alloy steel have been used as gauge materials because of their relatively high hardenability and abrasion resistance. For high volume production runs, gauge wear surfaces are often chrome plated. The durability of steel gauges coated with a layer of chrome of 5 to 8 μm is 10 to 12 times that of uncoated ones. For economy in material as well as hardening costs, gauges are designed in such a way that only the parts subjected to wear are made of hardened steel. Handles are made of cheaper MS. For high degree of accuracy, long production runs and excessive wear conditions and particularly in bigger gauges, the entire body is made of M.S. and only wear contact surfaces are deposited with welded layer of hard materials such as cemented carbides, Stellite or weartrode. Some gauges are made entirely of cemented carbides or they have cemented carbide inserts at certain wear points. Chrome plating is also used as a means of rebuilding and salvaging work gauges.

Chromium Plating : This method is used where wear is upto 0.24 mm. Chromium is very hard and anti-corrosive in nature. It is a costlier and a time consuming method (about 0.1 mm coating is built up in about 10 hrs). The steps are:

1. Before plating the surface, it is ground for ovality and taper, surface polished, cleaned with benzene and dried (with compressed air)

2. The surface after plating is ground to size.

3. Cr plating is done for wear upto 0.24 mm in ϕ, as 0.16 mm will be necessary for grinding before and after plating (0.2 mm being the max. permissible coating).

Gauges are often handled hundreds of times in every shift. Consequently they should be made as light in weight as possible to reduce handling fatigue. One way do so is to provide hollow ends as shown in Fig. 10.3 (*c*). Another way to reduce weight of a Plug gauge is shown below in Fig. 10.24.

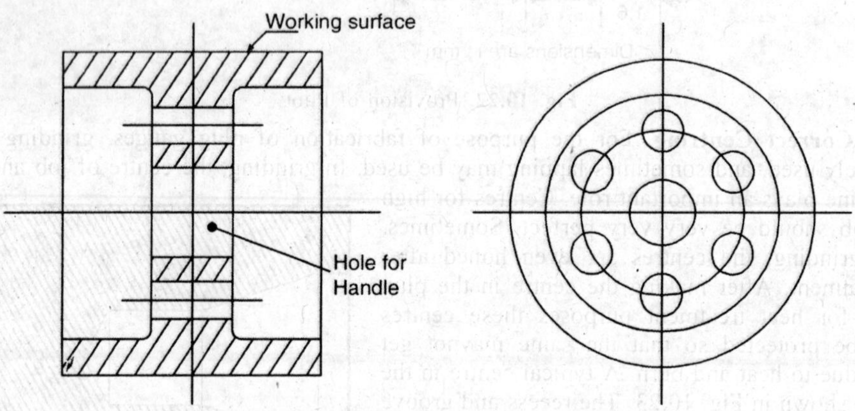

Fig. 10.24. Weight reduction of a Plug gauge.

8. **Hardness and Surface Finish.** Recommended hardness for gauges is 62 to 64 Rock well-C for plain gauges and 56 to 62 Rock well-C for screw gauges. The recommended surface finish is : 0.127 to 0.254 μm, Ra for ground gauges and 0.05 to 0.20 μm, Ra for lapped gauges.

9. **Rigidity.** Rigidity is one of the most important features of gauge design. Gauges such as gap gauges must always be designed with adequate rigidity as well as with robustness suitable for use in shop where they seldom meet with the treatment they deserve.

10. **Alignment of Gauge Faces.** In a normal gap gauge, the faces must be parallel and opposite to each other and the points of contact with the work at each face must lie on a line normal to the gauging face.

10.4. MANUFACTURING OF LIMIT GAUGES

10.4.1. Limit Plug Gauges. Shank type plug gauging elements are produced in the tool room of any engineering plant in small quantities (10 to 12 pieces) to satisfy the plant needs. Here, a bar sized for several gauges is the most suitable blank. The gauging elements are manufactured in the following steps :

1. Rough machining of the plug on a turret lathe which includes the operations : rough turning of shank, rough turning of body, facing the shank and cutting off the blank.

2. The next operation is also done on the turret lathe. The blank is located from the machined body and its correct centring is achieved by mounting it in a collect chuck. The operation comprises : facing of body, boring out a recess in the face, centre drilling and countersinking tapered centres hole to 60° to 120° with a combined drill and counter sink.

3. In the third step, the shank centre hole is made on the same machine and fixtures.

4. The subsequent machining is done between centres and followed by heat treatment : hardening, tempering and aging.

5. Thereafter, the centre holes are restored by grinding on a centre grinding machine or by machining with a carbide counter sink on a drill press and subsequent lapping on the same machine. As a result, accurate centre holes defining a correction location for subsequent machining are obtained with an accuracy to microns.

6. After that, grinding between centres follows.

7. Finally, all the measuring surfaces are lapped manually with a C.I. lap. For rough lapping, Al_2O_3 powder and for fine lapping, (5 to 2 μm) abrasive paste are used. More accurate geometric shape is obtained with the use of annular expendable laps.

10.4.2. Snap Gauges. The manufacture of snap gauges involves the following steps :

1. Stamping of blanks from a strip, annealing and normalising.
2. Preparatory grinding of wide surfaces with an allowance of 0.2 to 0.3 mm for grinding on hardening.
3. Milling of working surfaces with a set of milling cutters in a multipoint fixture, milling of gauge face to size and of a goove which divides the Go and No-Go parts of one working surface of gauge.
4. Deburring and filing the chamfers on the gauge blocks.
5. Temper hardening of gauges to 58-64 Rc.
6. Finish grinding of the wide surfaces of gauge on the magnet plate of a surface grinding machine.
7. Finish grinding of working surfaces to size with a lapping allowance. This operation is performed on a cup wheel or by the face of a recessed wheel.
8. Artificial aging of snap gauges.
9. Trimming and rounding of sharp edges on a hand grinding machine.
10. Preliminary lapping of working surfaces on machines with flat C.I. laps.
11. Finish hand lapping of working surfaces on glass laps with the use of abrasive paste.
12. Marking of gauges by electrographing.
13. Inspection of gauges and slushing.

10.5. CHOICE OF LIMIT GAUGES

The choice of a particular type of limit gauge mainly depends upon the size of the parts to be checked.

10.5.1. Plug gauges. (*a*) To check holes from 1 to 6 mm in diameter, double-end gauges with moulded members are used (Fig. 10.25 *a*). These members are cylindrical pins, 15-18 mm long, with a longitudinal milled groove or with two recesses over circumference, which ensures

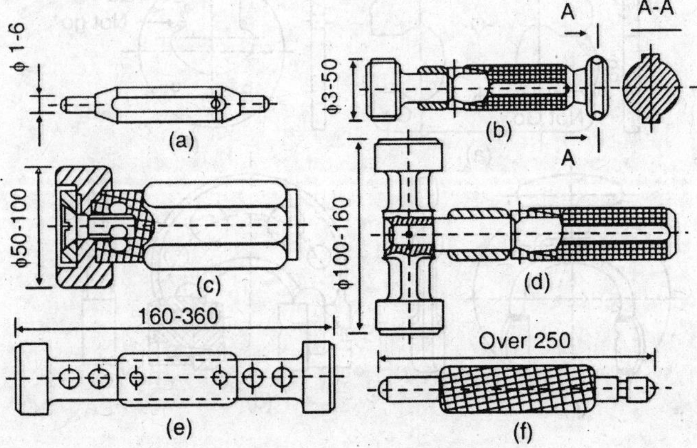

Fig. 10.25. Plug Gauges

a reliable connection of the pin with the plastic handle. There is a round mark on the handle from the Not-Go end.

(*b*) For gauging holes from 3 to 50 mm in diameter, double-and single-end plug gauges are used. The plugs have tapered shanks with a 1 : 50 taper (Fig. 10.25 *b*). The Not-Go plug has the working length 1/3 shorter than the go plug working length. Sometimes the plug Not-Go portion is made of the shape shown in the Figure (Section A-A). Such segmental-type plugs are more reliable in checking the hole diameter, especially of oval holes.

(*c*) For checking holes from 50 to 100 mm in diameter, single-end Go and Not-Go plug gauges lighter in weight are used. The plug gauge may be of full-form or segmental-type shape over the circumference; the plastic handle is fastened by a screw as shown in Fig. 10.25 *c*. When one side of a detachable plug becomes worn-out, the plug is reversed, and the opposite side is used for gauging.

(*d*) Segmental-type stamped or milled plug gauges (Fig. 10.25 *d*) are used for checking holes 100 to 160 mm in diameter. A tapered steel shank with a steel or plastic handle is pressed into the plug.

(*e*) Segmental-type Go and Not-Go plug gauges with plastic grip pads as shown in Fig. 10.25 (*e*) are used for checking holes from 160 to 360 mm in diameter.

Internal caliper gauges with spherical precise surfaces on the ends (Fig. 10.25 *f*) are produced for checking deviations from the sizes of large-diameter holes (250-1000 mm). They are made separately for the Go and Not-Go sizes. The sphere radius is 30-50 mm; the steel rod diameter is 10-12 mm. A heat-protection handle in the form of a sleeve is fitted on the rod, the handle being 25-30 mm in external diameter.

Many plants manufacture for in-plant use plug gauges of a simplified design when the Go and Not-Go working portions are on one side. Single- and double-end plug gauges in the form of plates made of sheet steel, 5 to 10 mm thick, are also often used.

10.5.2. Snap Gauges. Sheet steel double end gauges (Fig. 10.26a) are usually made in tool rooms of engineering plants for in-plant use and applied for checking sizes 1 to 70 mm. These are easy in manufacture, but less convenient in use and take more time to check than two-size single-end gauges. The Go portion of gauging surface of single-end gauges is made considerably longer than the Not-Go one, and both sourfaces are divided by a groove, (Fig. 10.26 *b*).

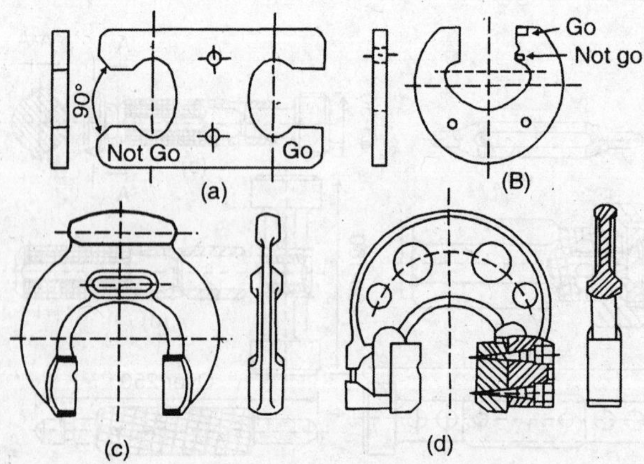

Fig. 10.26. Snap Gauges.

Single-end stamped snap gauges are usually used. Moulded snap gauges, Fig. 10.26 (*c*), are provided with stiffening ribs and wider measuring faces, which make them more rigid and wear resistant. There are two designs of these snap gauges used for checking sizes from 3 to 50 mm and from 50 to 170 mm (with heat-protection plastic handles).

To check sizes from 100 to 325 mm, single-end moulded snap gauges with detachable anvils are available (Fig. 10.26 *d*). To reduce the weight, these gauges are made with holes. In recent years single-end stamped snap gauges with adjustable anvils of hard metals have been widely employed. They are intended to check the sizes up to 340 mm with adjustment within 6 to 15 mm. The durability of snap gauges with anvils made from hard metal is 40 times that of snap gauges made from steel.

10.6. THREAD OR SCREW GAUGES

These gauges are designed along the same general principles as all other types of gauges. These gauges are of Plug and Ring type and are made in the 'Go' and 'Not Go' models. These are of special quality gauge steel, hardened and seasoned before the threads are ground and finally lapped to dimensions. Nuts or internal threads are checked with Plug thread gauges and screws or external threads, with ring-thread gauges.

Thread gauges similar to plain gauges are designed with manufacturing tolerance for new gauges and wear allowance. Both tapered and trilock methods securing the plug gauges into their holders are employed. In case of 'Go' and 'Not Go' plug gauges, it is common practice to make the wearing side, *i.e.*, the 'Go End' at least twice as long as the 'Not Go End'. The maximum wear in the latter case occurs on the end thread or threads.

In screws threads, there are three classes of fit : (*a*) Close fit (*b*) Medium fit (*c*) Free fit. Tolerances for major, minor and effective or pitch diameters for all these three types of fits are given in national standards. Screw or thread gauges take the form of the mating thread and are assembled with the thread to be checked. By suitable designing the gauges, it is possible to provide a limit gauging system which will control the complex dimensions of the threads within the tolerance limit.

1. **Plug Screw Gauges.** Four types of plug screw gauges are used for checking the internal threads :

(*a*) Full form 'Go' gauge for controlling maximum metal condition of major and effective diameters simultaneously.

(*b*) 'Not Go' effective gauge with crests and roots removed.

(*c*) 'Not Go' gauge for major diameter.

(*d*) 'Not Go' gauge for minor diameter.

(*a*) **Full form 'Go' Gauge.** This is made accurately to the minimum dimensions of the internal thread and will ensure that all, dimensions are not less that the minimum, if it assembles with the thread. Major, minor and effective diameters are checked simultaneously.

(*b*) **'Not Go' effective diameter gauge.** It is a thread gauge whose diameter is truncated and whose minor diameter is cut away to be clear of the minor diameter of the work to be checked. The effective diameter of this gauge is made on the upper limit of the effective diameter of thread to be checked. This gauge (Fig. 10.27) is very extensively used in production work.

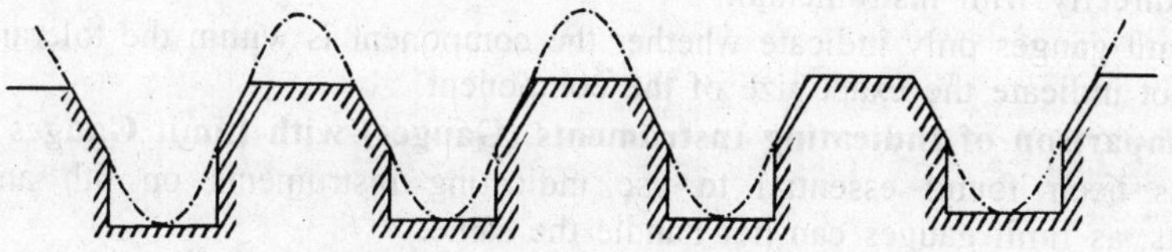

Fig. 10.27. 'Not Go' Effective Diameter Gauge.

(*c*) 'Not Go' major diameter gauge. This gauge has its major diameter on the corresponding upper limit of the work. Its flanks and minor diameter are well clear of the work dimension. This gauge is used to ensure that the threads of the work are not thin even though major or minor diameter is within its limits. This is not much used in production work, (Fig. 10.28).

Fig. 10.28. 'Not Go' Major Diameter Gauge.

(*d*) **'Not Go' Minor Diameter Gauge.** This gauge is used to check the core diameter only. It has its minor diameter on the corresponding upper limit of the work (Fig. 10.29). This is also not in much use.

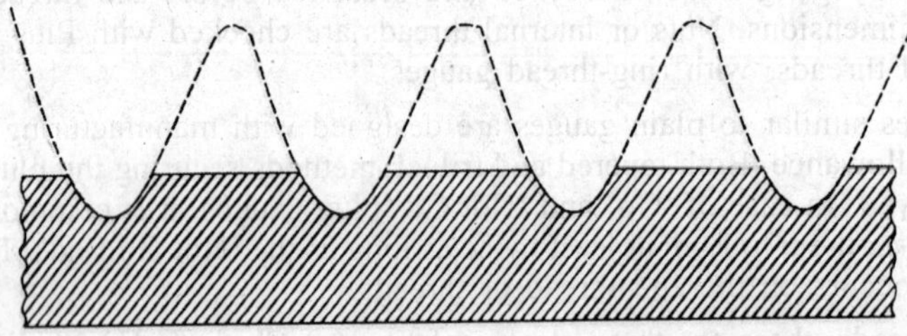

Fig. 10.29. 'Not Go' Minor Diameter Gauge.

Plug screw gauges are usually made double ended, either solid with the 'Go' and 'Not Go' threads ground on it or as a built up gauge with a metal or plastic handle having a grooved gauging unit at each end fitting into a taper socket or otherwise attached. In use, a 'Not Go' gauge may enter the work by one or two turns and still be regarded as a 'Not Go' gauge.

2. **Ring Screw Gauges.** These gauges are used for checking external threads. Here also, the same four types of gauges are used. But the first two are very common. The 'Not Go' effective diameter ring gauge is truncated on its minor diameter and cleared on its major diameter. Other features are the same.

Thread Calliper Gauges. These days, instead of popular ring gauges, thread calliper gauges are much favoured. In principle, these gauges are the equivalent of gap gauges with thread forms on the anvils.

10.7. ADVANTAGES OF LIMIT GAUGES

1. These are conveniently used in mass production for controlling various dimensions and thus ensure interchangeability.

2. These can easily be used by semi-skilled labour.

3. These are economical in their own cost as well as engaging cost.

10.8. LIMITATIONS OF LIMIT GAUGES

1. It is generally uneconomical even with excellent equipment and cost control to manufacture limit plug gauge to a tolerance on the gauge finer than 0-.0013 mm corresponding to a work tolerance of about 0.013 mm. There is also difficulty in the use of such fine limit gauges in the production shops. Finer tolerances than those specified above should preferably be measured directly with instrumental.

2. Limit gauges only indicate whether the component is within the tolerance limit or not. They do not indicate the exact size of the component.

10.8.1 Comparison of Indicating Instruments (Gauges) with Limit Gauges

It has been found essential to use indicating instruments on 6th and finer quality components, as limit gauges can not handle the same.

The cost of investment on indicating gauges will be recovered in the form of savings in maintenance and reconditioning costs. The indicating instruments have the following additional benefits over limit gauges:

Indicating Gauges	Limit Gauges
1. Indicate the actual size of the component.	1. Do not indicate the actual gauge.
2. Free from wear, expansion or collapse (as it in calibrated before use)	2. Susceptible for collapse, expansion and wear
3. Require less space as a small number of gauges is sufficient	3. Large number of gauges is required necessitating larger space.
4. Can handle 6th and finer quality components.	4. Can not handle finer quality jobs.
5. Require occasional checking	5. Require frequent checking.

10.9. CARE OF GAUGES

Gauges should be used and cared for properly to ensure their maximum useful service life. Some suggestions for their use and care are :

1. Master, inspection and working gauges should be applied only to the uses for which they are intended, *i.e.,* a master gauge shall be used to check inspection and working gauges, an inspection gauges will be used to check the finished product and a working gauge should be used to check the product as it is being manufactured.

2. A plain cylindrical gauge should be cleaned and a thin film of light oil should be applied to the

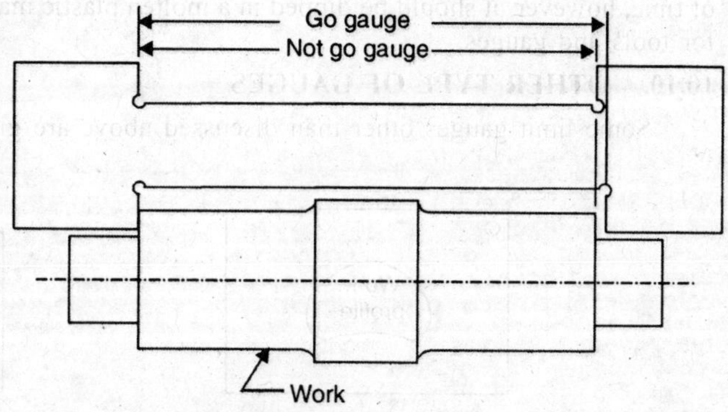

Fig. 10.30. Length Gauge.

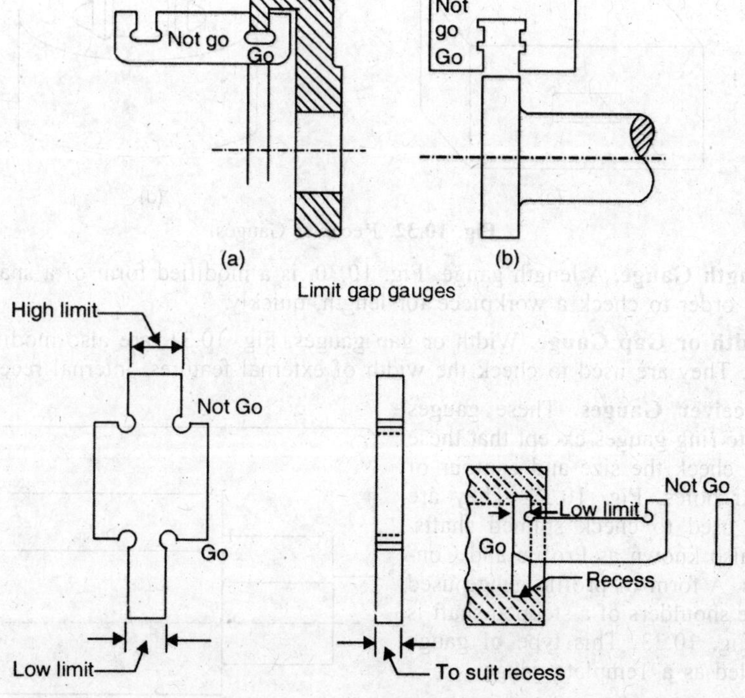

Fig. 10.31. Width or gap gauges.

gauging surface before it is used. The work should also be cleaned. Then the gauge should be aligned with the hole to be measured and given a forward motion combined with a slight rotation. The 'go plug' gauge will enter the hole if the latter is of correct size but if not so, then the gauge will not enter it. Keep the gauge moving into and out of the hole when the fit is closed, to prevent seizure of the parts. The same suggestions are applicable to the use of plain cylindrical ring gauges.

3. Don't force a snap gauge over work, because forcing will cause the gauge to pass oversized parts and it may also spring the frame of the gauge. In fact force should be avoided in any gauging operation as it tends to harm the gauge, the work or both.

4. A gauge should be cleaned after use and prepared for storage. If it is to be stored for a short time only, it should be coated with a rust preventive oil. If it is to be stored for a long period of time, however, it should be dipped in a molten plastic material designed as a protection coating for tools and gauges.

10.10. OTHER TYPE OF GAUGES

Some limit gauges other than discussed above are given below :

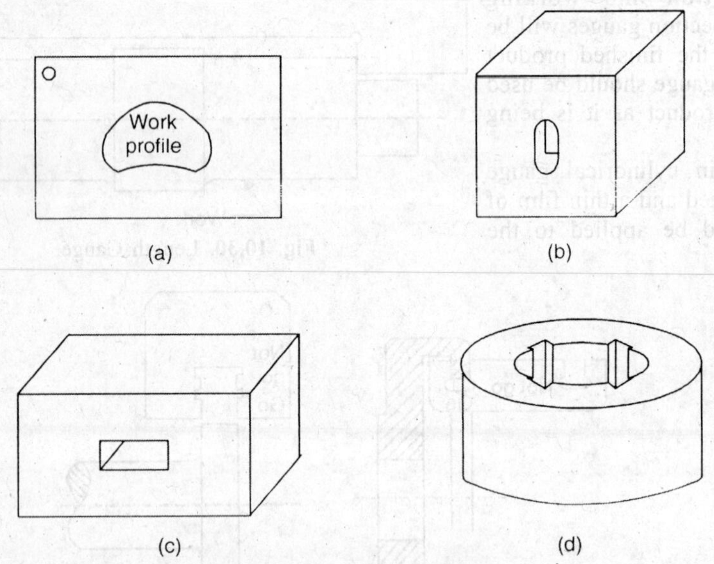

Fig. 10.32. Receiver Gauges.

1. **Length Gauge.** A length gauge, Fig. 10.30, is a modified form of a snap gauge and is designed in order to check a workpiece for length, quickly.

2. **Width or Gap Gauge.** Width or gap gauges, Fig. 10.31, are also modified forms of a snap gauge. They are used to check the width of external features, internal recesses.

3. **Receiver Gauges.** These gauges are similar to ring gauges except that these are used to check the size and contour of non circular holes, Fig. 10.32. They are extensively used to check splined shafts. These are also known as Profile and Contour gauges. A form of profile gauge used to check the shoulders of a stepped shaft is shown in Fig. 10.33. This type of gauge may be called as a Template gauge.

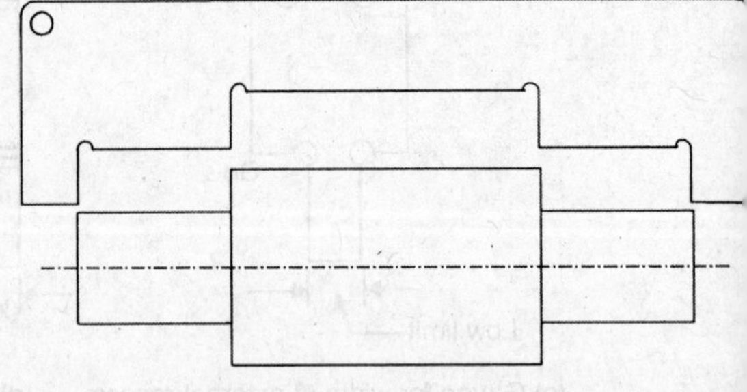

Fig. 10.33. Template Gauge.

4. **Flush Pin Gauges.** These types of gauges are mainly used to check the depths

of slots, but can also be used for gauging lengths and tapers. A flush pin is defined as : a gauge for checking the distance between two surfaces, comprising a body having a through hole, and a pin in the hole which projects from a face of the body, a distance equal to the dimension to be gauged when the opposite or indicating end of the pin is flush with the opposite face of the body. The indicating end of the pin, or the adjacent face of the body, has a step of a depth equal to the tolerance on the dimension gauged.

As is given in the definition such a gauge has two designs. In the first, a sliding pin is used in a female member (a bush) and in the second, the work to be checked itself acts as a pin. The essential and common feature in both designs is a step provided on the member in which the pin fits. In Fig. 10.34 (*a, b*), a pin is fitted in the bush to check the depth and length of work. The step is provided on the bush in which the pin slides. In Fig. 10.34 (*c*), the work (whose length is to be checked) itself acts as the pin and the step is provided on the bush as usual. In Fig.

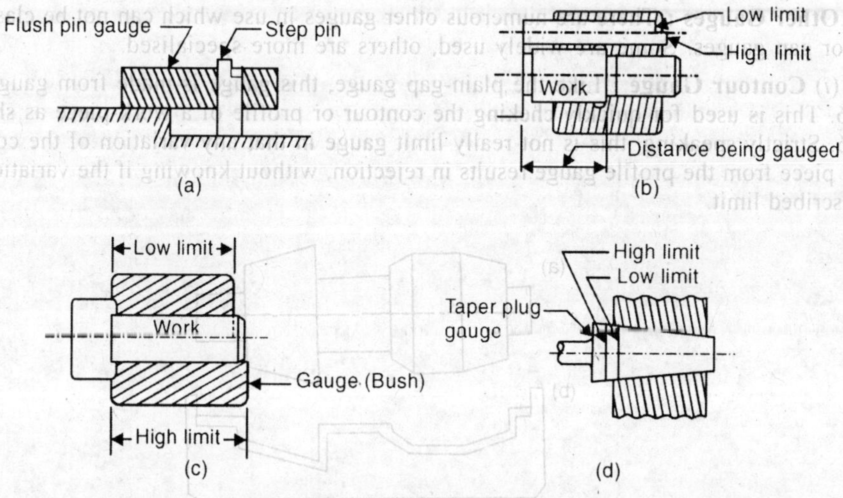

Fig. 10.34. Flush Pin Gauges.

10.34 (*d*), the principle of flush pin gauge is used to check the taper of a hole in a component. In all these cases, checking may be done by sight or feel (by using feeler gauges of 0.01 mm). Due to this, these gauges are also known as 'Feeler-pin Gauges'.

To check depth, Fig. 10.34 (*a*), the main section of the gauge is placed on the higher side of the two surfaces, with the movable step pin resting on the lower surface. If the depth between the two surfaces is sufficient but not too great, the top of the pin, but not the lower step, will be slightly above the top face of the gauge body. If the depth is too great, the top of the pin will be below the surface. Similarly, if the depth is not great enough, the lower step on the top of the pin will be above the surface of gauge body. By running a finger or a fingernail across the top of the pin, its position with respect to the surface of the gauge body can be easily found.

Taper Plug and Ring Gauges : These are used for checking tapered holes and shafts (Also see Art. 12.13.7). They are usually made in one piece, with a knurled portion for handling. The diameter and angle of the taper can be checked by means of a step in the end of the gauging portion, Fig. 10.35, although a line is sometimes engraved on the periphery of the taper plug gauge. Any slight variation is the angle of the taper, or if the permissible maximum or minimum diameters are exceeded, will result in the end face of the component standing proud or lying below the vertical face of the step, Fig. 10.35.

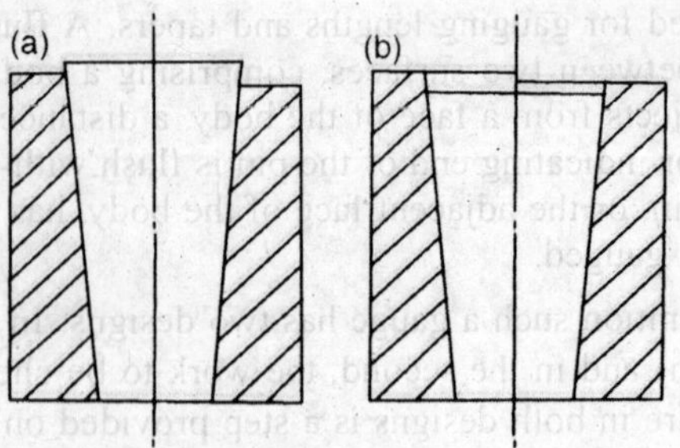

Fig. 10.35. The rejection of a component by a taper ring gauge
(*a*) the top face is above the step, indicating an oversized component
(*b*) the top face is below the step indicating an undersized component.

Other Gauges : There are numerous other gauges in use which can not be classed as plug, ring or gap gauges. Some are widely used, others are more specialised.

(*i*) **Contour Gauge :** Like the plain-gap gauge, this gauge is made from gauge plate, Fig. 10.36. This is used for quickly cheking the contour or profile of a work-piece as shown in Fig. 10.36. Strictly speaking, this is not really limit gauge in that any variation of the contour of the work piece from the profile gauge results in rejection, without knowing if the variation is beyond a prescribed limit.

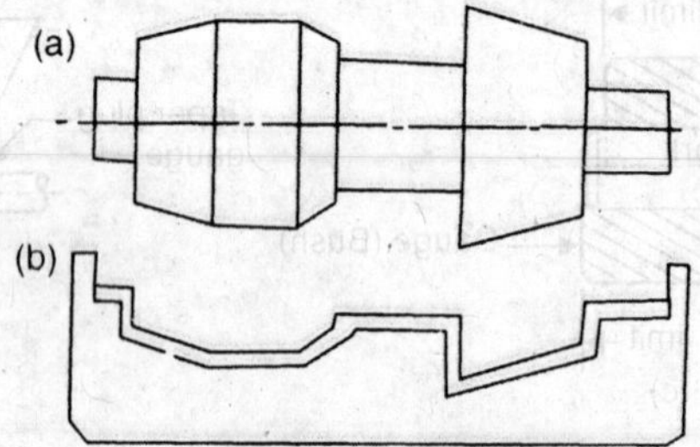

Fig. 10.36. Contour gauging (*a*) the component (*b*) the gauge.

(*ii*) **Flush pin Depth Gauge :** This gauge works on the same principle as the taper-plug gauge to chek the depth of a hole. The sliding pin is retained in the collar by means of a small screw located in a groove on the side, which allows sufficient vertical movement for gauging. The step on the top of the pin is equal to the depth tolerance of the hole, and the top face of the collar provides the datum.

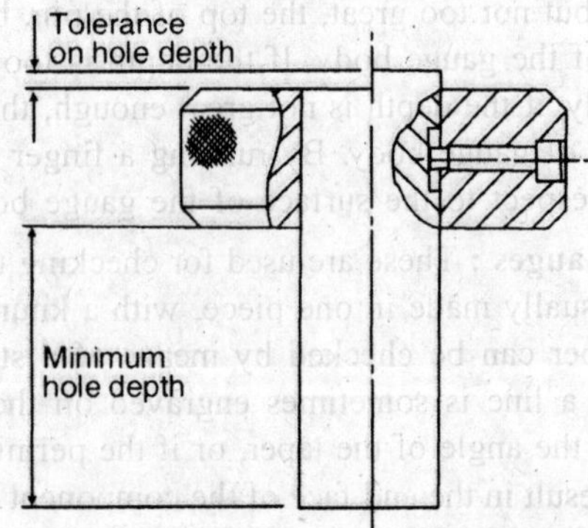

Fig. 10.37. A flush pin depth gauge.

(*iii*) **Combined Bore / Face Gauge :** The position and parallelism of a bore in relation to a datum face can be checked by means of a combined bore/face gauge, Fig. 10.38. The pin which locates in the bore is in effect, 'Go' plug gauge, and the steps ground on the other pin are the 'Go' and 'Not Go' limits for the datum face to hole axis dimension. The length of the plug gauge needs to be sufficient to enable the length of the 'Go' step on the pin to check for parallelism. The tolerance on the hole must be less than the tolerance on the dimension to the face for the gauge to operate satisfactorily.

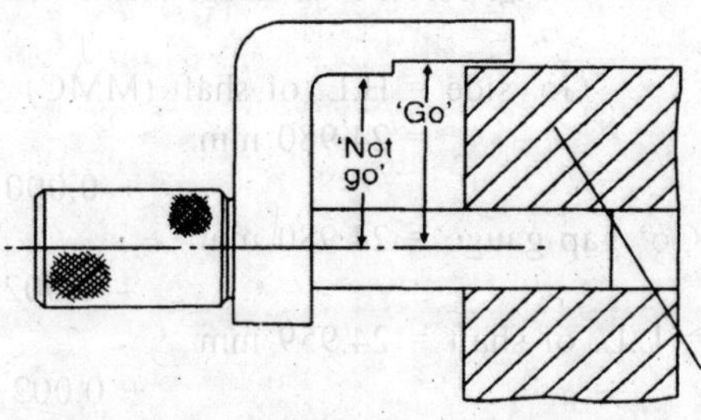

Fig. 10.38. A combined bore/face gauge.

10.11. SOLVED EXAMPLES

Example 1. *A 25 mm H8--f7 fit is to be checked. The limits of size for H8 hole are : High limit 25.033 mm, low limit 25.000 mm. The limits of size for f7 shafts are : High limit 24.980 mm, low limit 24.959 mm. Taking gauge maker's tolerance to be 10% of the work tolerance, design plug gauge and gap gauge to check the fit.*

Solution. Tolerance for hole = H.L – L.L.

$= 25.033 - 25.000 = 0.033$ mm

$\therefore$ Gauge makers tolerance for plug gauge $= 0.1 \times 0.033$ mm $= 0.0033$ mm

$= 0.003$ mm (rationalised)

Gauge makers tolerance for gap gauge $= 0.0021$ mm $= 0.002$ mm (rationalised)

As the work tolerances are less than 0.09 mm, wear allowance may not be provided.

(*i*) **Plug Gauge**

Basic size of 'Go' plug gauge = L.L. of the hole (MMC) = 25.000 mm

$\therefore$ In unilateral system,

$$\text{Dimensions of 'Go' plug gauge} = 25.00^{+0.003}_{-0.000} \text{ mm}$$

That is,

High limit of 'Go' plug gauge = 25.000 + 0.003

= 25.003 mm

Low limit of 'Go' plug gauge = 25.000 mm

Now,

Basic size of 'Not Go' plug gauge = 25.033 mm

$$\therefore \text{Dimensions of 'Not Go' plug gauge} = 25.033^{+0.000}_{-0.003} \text{ mm}$$

(Fig. 10.40 shows a sketch of combined 'Go' and 'Not Go' plug gauge.)

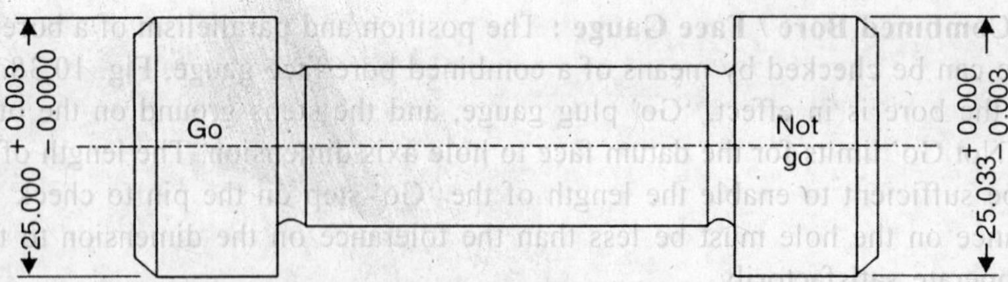

Fig. 10.40. Plug Gauge (combined type)

(*ii*) **Gap Gauge**

'Go' side = H.L. of shaft (MMC)
= 24.980 mm.

$\therefore$ Dimensions of 'Go' gap gauge = $24.980^{+0.000}_{-0.002}$ mm

'Not Go' side = L.L. of shaft = 24.959 mm

$\therefore$ Dimensions of 'Not Go' gap gauge = $24.959^{+0.002}_{-0.000}$ mm

(Fig. 10.41 shows a sketch of combined 'Go' and 'Not Go' gap gauge)

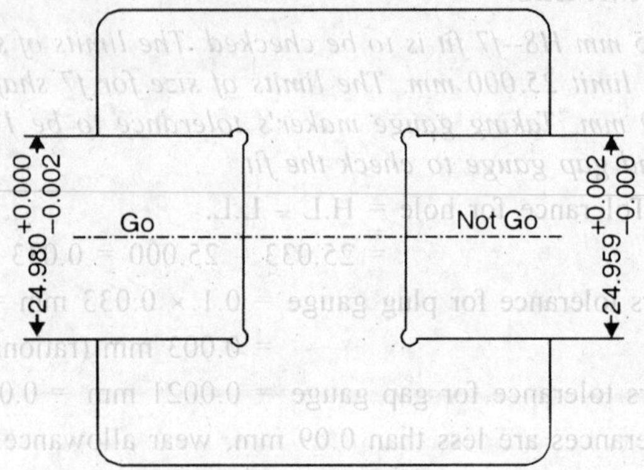

Fig. 10.41. Gap Gauge (combined type).

Example 2. *Shafts of 75 ± 0.02 mm diameter are to be checked by the help of a Go, No Go snap gauges. Design the gauge, sketch it and show its Go size and Not go size dimensions Assume normal wear allowance and gauge maker's tolerance.*

Solution. High limit of shaft = 75.02 mm

Low limit of shaft = 74.98 mm

Work tolerance = 75.02 – 74.98 = 0.04 mm

$\therefore$ Gauge makers tolerance (10%) = 0.004 mm

Wear tolerance = 0.002 mm

'Go side' of snap gauge = H.L. of shaft, (MMC) = 75.02 mm

'Not Go' side of snap gauge = 74.98 mm

Wear allowance is to be applied first to 'Go' side, before gauge maker's tolerance is applied (Refer to Fig. 10.8).

'Go' side of snap gauge after considering the wear allowance

= 75.02 – 0.002 = 75.018 mm

$\therefore$ Dimensions of snap gauge are given as :

Unilateral System

+ 0.000	+ 0.004
'Go' 75.018 mm	'Not Go' 74.98 mm
− 0.004	− 0.000

Bilateral System

+ 0.002	+ 0.002
'go' 75.018 mm	'Not Go' 74.98 mm
− 0.002	− 0.002

Example 3. *Find the 'Go' and 'Not Go' gauge dimensions of a plug gauge using Bilateral and Unilateral Systems and including wear allowance for gauging 75 ± 0.05 mm diameter holes.*

Solution. High limit of hole = 75.05 mm

Low limit of hole = 74.95 mm

work tolerance = 75.05 - 74.95 = 0.1 mm

Gauge maker's tolerance = 0.01 mm

Wear allowance = 0.005 mm

'Go' side of plug gauge = L.L. of hole (M.M.C.)

= 74.95 mm

'Go' side of plug gauge after the application of wear allowance

= 74.95 + 0.005 (Fig. 10.8)

= 75.955 mm

Dimension of plug gauge are given as :

Unilateral System

+ 0.010	+ 0.000
'Go' 74.955 mm	'Not Go' 75.05 mm
− 0.000	− 0.010

Bilateral System

+ 0.005	+ 0.005
'Go' 74.955 mm	'Not go' 75.05 mm
− 0.005	−0.005

Example 4. *The rectangular hole shown in Fig. 10.42 is to be checked.*

The limits of size for the width of hole are :

+ 0.04
60 mm
− 0.00

The limits of size for the breadth of hole are :

+ 0.05
80 mm
− 0.00

Design the suitable gauges based on Taylor's principle.

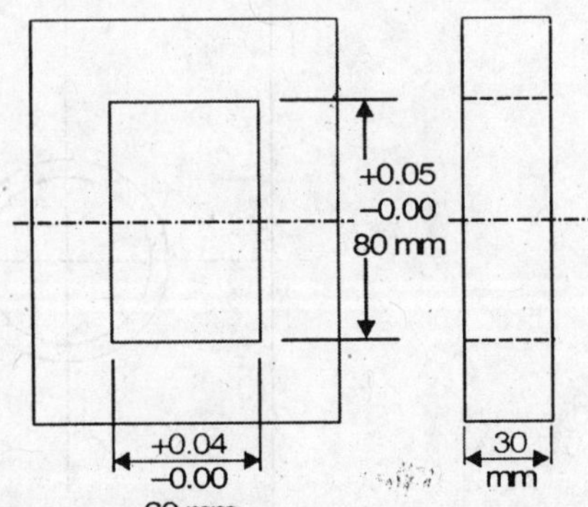

Fig. 10.42

Solution. According to Taylor's principle, there will be one 'Go Gauge' of full form and of length equal to the length of the hole. There will be two 'Not Go Gauges;' of pin form to check the width and breadth of the hole.

High limit of width of hole = 60.04 mm

Low limit of width of hole = 60.00 mm

High limit of breadth of hole = 80.05 mm

Low limit of breadth of hole = 80.00 mm

Tolerance on width of hole = 0.04 mm

Tolerance on breadth of hole = 0.05 mm

Gauge marker's tolerance (10%) :

For breadth of hole = 0.005 mm

For width of hole = 0.004 mm

Go Gauge

(*i*) *60 mm dimension :*

Go gauge will correspond to low limit of width of hole (MMC), *i.e.*, 60 mm.

∴ Basic size of 'Go Gauge' for 60 mm dimension

= 60.00 mm

Using unilateral system, the limits of size for 'Go Gauge' for 60 mm dimension are :

$$60.00\ \text{mm}\ {}^{+0.004}_{-0.000}$$

(*ii*) 80 *mm dimension :*

Basic size of 'Go Gauge' for 80 mm dimension

= 80.00 mm

∴ Limits of size of 'Go Gauge' for 80 mm dimension are :

$$80.00\ \text{mm}\ {}^{+0.005}_{-0.000}$$ (unilateral system)

The 'Go Gauge' is shown in Fig. 10.43.

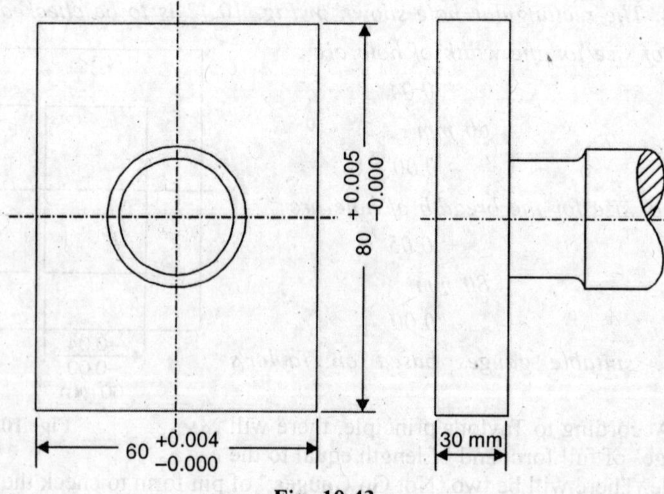

Fig. 10.43.

Not Go Gauges (Refer Fig. 10.7)

The 'Not Go Gauges' will correspond to the minimum metal condition of hole i.e. high limits of dimensions.

∴ Basic size of 'Not Go Gauge' for 60 mm dimension

$$= 60.04 \text{ mm}$$

Basic size of 'Not Go Gauge' for 80 mm dimension

$$= 80.05 \text{ mm}$$

∴ Limits of size of 'Not Go Gauge' for 60 mm dimension are :

$$60.04^{+0.004}_{-0.000} \text{ mm}$$

Limits of size of 'No Go Gauge' for 80 mm dimension are :

$$80.05^{+0.005}_{-0.000} \text{ mm}$$

The two pin gauges are shown in Fig. 10.44.

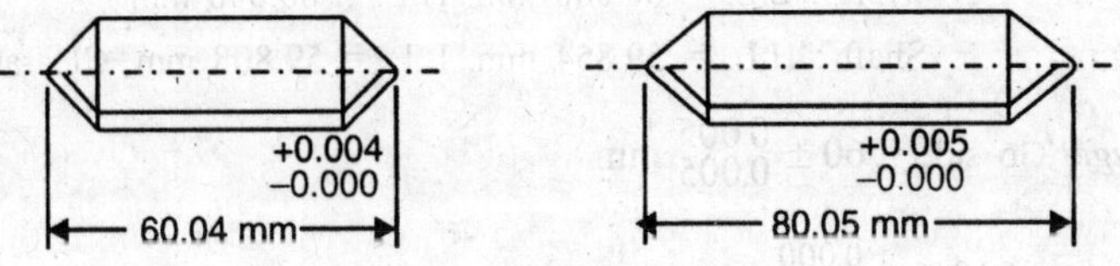

Fig. 10.44

PROBLEMS

1. What is a gauge?
2. How does a gauge differ from a measuring instrument?
3. How the plain gauges are classified?
4. What is the difference between standard gauge (non-limit gauge) and a limit gauge?
5. Differentiate between : Workshop gauge, inspection gauge and master gauge.
6. What is gauge maker's tolerance? How is it applied in the design of gauges?
7. Give the advantages and disadvantages of unilateral and bilateral system of gauging.
8. What is wear allowance? How is it applied in the design of gauges?
9. State and explain Taylor's principle of limit gauging.
10. Discuss the various materials used for gauge manufacture.
11. What are the advantages of limit gauges?
12. What are the limitations of limit gauges?
13. How the gauges should be cared for before and after use?
14. What is MMC?
15. Determine the specification for the 'go' and 'not go' ends of a set of manufacturing and inspection plug gauges to be used in checking a hole with

    diameter specification of $25^{+0.075}_{-0.000}$ mm

16. A shaft $100 \pm \frac{0.02}{0.02}$ mm diameter is to be checked using a 'go-not go' snap gauge. Give its dimensions. Allow for wear and gauge maker's tolerance.

17. A limit gauge is required to check the hole $50 \begin{smallmatrix} +0.339 \\ +0.00 \end{smallmatrix}$ mm (50 H_8). The depth of hole is 200 mm. Design the gauge and sketch it with dimensions.

18. Discuss briefly various aspects for deciding the tolerance on limit gauges.

19. A hole and shaft system has the following dimensions :

60 mm H8/c8

The standard tolerance is given by

$$i = 0.45\,(D)^{1/3} + 0.001\,D$$

Where D = Diameter of geometric mean of steps. mm

i = standard tolerance, micron

The multiplier for grade 8 is 25. The fundamental deviation for shaft c for $D > 40$, is given by

$$-(95 + 0.8\,D)$$

The diameter range lies between 50 to 80 mm. Sketch the fit and show on it the actual dimensions of hole and shaft. Name the class of fit. Also, design the suitable gauges to check the hole and the shaft. (AMIE 1974 S)

[Ans. Hole : L.L. = 60.000 mm, H.L. = 60.046 mm

Shaft : H.L. = 59.854 mm, L.L. = 59.808 mm, Clearance fit.

Plug Gauge, Go side : $60 \pm \frac{0.005}{0.005}$ mm

Not Go side : $60.046 \pm \frac{0.000}{0.005}$ mm

Snap Gauge, Go side $59.854 \pm \frac{0.000}{0.005}$ mm

Not Go side : $59.808 \begin{smallmatrix} +0.000 \\ -0.005 \end{smallmatrix}$ mm

Note. Unilateral system has been used. Wear allowance has been neglected; work tolerance being less than 0.09 mm.

20. The minimum size of a hole is 25.00 mm. Its maximum size is 25.002 mm. When matching shaft of this hole is measured, the fundamental deviation is found to be – 0.02 mm. Shaft tolerance is 0.003 mm. Design gauges for hole and shaft. Take the usual values of gauge maker's tolerance and wear allowance.

21. Design 'GO' and 'NO GO' ends of a plug gauge to measure a hole of size 28.000 ± 0.014 mm adopting (*a*) Unilateral system (*b*) Bilateral system.

22. A bore of $\frac{27.50}{27.52}$ mm dia × 45 mm long is to be checked. Design, draw and dimension a plug gauge for this, based on Taylor's principle of gauge design.

23. A square peg having limits of 25.00 mm and 24.97 mm is to be checked. Design a gauge (gauges) for checking this, based on Taylor's principle of gauge design.

24. Discuss the principle of Taylor's for the design of gauges for checking :

(*a*) of an oval hole with a cylindrical 'Not Go' gauge.

(*b*) of rectangular hole

(*c*) circular holes

(*d*) circular shafts.

(*e*) Non-circular holes and shafts.

25. Sketch and discuss various types of Plug gauges.

26. Sketch and discuss various types of snap gauges.

27. Discuss the procedure of manufacturing Limit plug gauges.

28. Discuss the procedure of manufacturing Limit snap gauges.

29. What role do gauges play in the mass production system?

30. Where are plug gauges used?

31. Where are Snap gauges used?

32. Sketch and discuss the use of following gauges :

(*a*) Length gauges

(*b*) width or Gap gauge

(*c*) Receiver gauge

(*d*) Flush pin gauge.

33. What are screw gauges? How are they used to control the complex dimensions of threads?

34. Sketch and discuss the use of :

(*a*) Plug screw gauges

(*b*) Ring screw gauges

(*c*) Thread caliper gauges.

35. Exlain the meaning of the expression 30 mm H8–f7.

36. Derive the formula to give the depth of the step '*h*' of the taper plug gauge in terms of max. and min. diameters of the large end. Calculate '*h*' given $\theta = 1°–58'$. Max. dia, = 23.42 mm, min. dia = 23.40 mm **[Ans:** $h = (D - d)/2 \tan \theta/2$; 0.588 mm**]**

37. Write on materials resisting wear for gauges.

Sol. The various materials for gauges have been discussed under Art. 10.3(7). Special wear resisting materials for applications where the plug will lose its size very rapidly due to abrasive wear (for example, a small diameter screw plug gauge used on a cast iron component), can be :-

(*a*) H.S.S. (Better wear resistance than C-steel).

(*b*) 'Hard' Chromium plating on hardened steel (See Art. 10.3(7)), which is finally ground to size.

(*c*) WC, sintered metal-powder mouldings ground with diamond grit abrasive, though very expensive, have a greatly improved wear life.

38. Write the disadvantages of limit Gauging.

Sol. Two limitations of limit Gauges have been discussed in Art. 10.8. The other drawbacks are:

(*a*) Suitable for Mass production of components only. Their cost may not be recovered from a small quantity of work.

(*b*) The use of limit gauging is restricted to certain gauge sizes and types because of practical considerations such as weight, flexture and manufacturing complexity.

(*c*) As with any components, the limit gauges can not be manufactured entirely without error. Also, they lose their accuracy due to wear during use.

(*d*) Particular sources of error in the component are less easily revealed. Refinements in machine tool control and better instrumentation (mainly due to electronics) have allowed the dimensional control to be easily integrated with the machining process. Thus inspection can be limited to "first off" checks and subsequently by sampling methods for important dimensions. However, limit gauging is still the cheapest and most convenient method of checking plain bores, both internal and external screw threads, splines and serrations.

11

SURFACE FINISH

11.1. INTRODUCTION

Whatever may be the manufacturing process, an absolutely smooth and flat surface (or any other perfect geometrical form) can not be obtained. The machine elements or parts retain the surface irregularities left after manufacturing. The surface of a part is its exterior boundary and the surface irregularities consist of numerous small wedges and valleys that deviate from a hypothetical nominal surface (Fig. 11.1, which shows a surface on a highly magnified scale). These irregularities are responsible to a great extent for the appearance of a surface and its suitability for an intended application of the component. These surface irregularities are usually understood in terms of surface finish, surface roughness, surface texture or surface quality. Heat exchanger tubes transfer heat better when their surfaces are slightly rough rather than highly finished. Brake drums and clutch plates etc. work best with some degree of surface roughness. However, if a film of lubrication must be maintained between two moving parts, the surface irregularities must be small enough (smooth surface) so that they do not penetrate the oil film under the most severe operating conditions. The examples are : Bearings, journals, cylinder bores, piston pins, bushing, helical and worm gears, seal surfaces and machine ways etc. In gears, smooth surfaces are also necessary to ensure quiet operations. For components which are subjected to load reversals, sharp irregularities act as stress raisers constituting the greatest potential source of fatigue cracks. Therefore, the surfaces of components which are subjected to high stresses and load reversals are finished highly smooth.

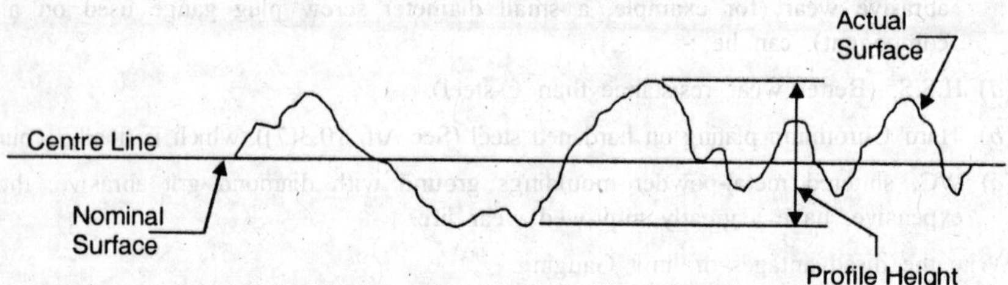

Fig. 11.1. A Profile of Surface Irregularities.

A certain magnitude of surface roughness is also needed to accommodate wear-in of certain parts. Most new moving components do not attain a condition of complete lubrication because of imperfect geometry, running clearances and thermal distortions. Therefore, the surfaces must wear in by a process of actual removal of metal during operation. The surface finish has to be a compromise between sufficient roughness for proper wear-in and sufficient smoothness for expected service life. If the surface is too rough, the initial wear particles are large which act as abrasives and wear continues at a high rate. On the other hand, if the surface is too smooth, the initial wear will be very slow. May be, the surfaces will never wear-in and improper clearances may result in local hot spots and high oil consumption.

The factors effecting surface roughness are : 1. Vibrations 2. material of work piece 3. type of machining 4. rigidity of the system consisting of machine tool, fixture, cutting tool and work 5. type, form, material and sharpness of the cutting tool 6. cutting conditions, *i.e.*, feed, speed and depth of cut, and 7. type of coolant used.

11.2. ELEMENTS OF SURFACE ROUGHNESS

The various elements of surface roughness can be defined and explained with the help of Fig. 11.2 which shows a typical surface highly magnified.

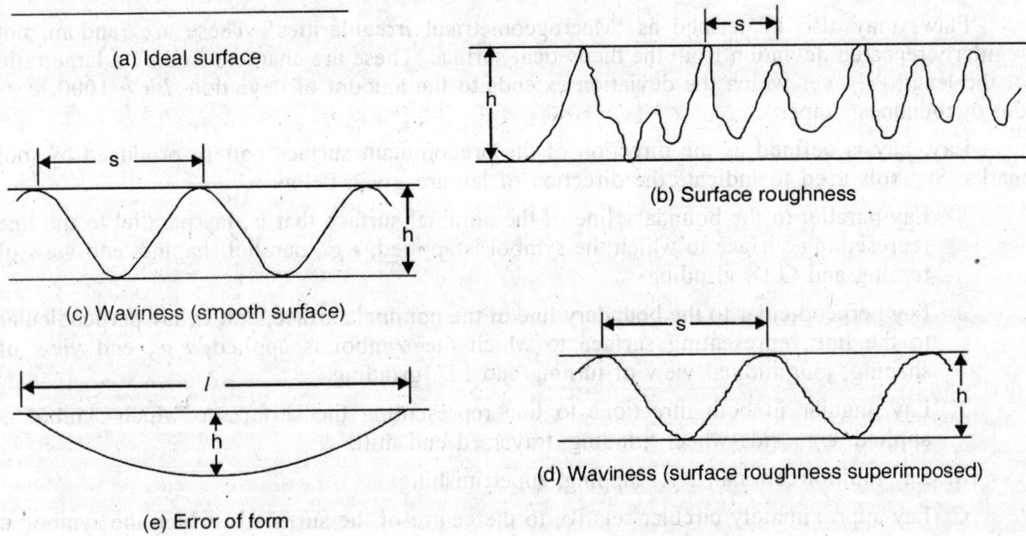

Fig. 11.2. A Typical Surface Highly Magnified

Surface. The surface of a part is confined by the boundary which separates that part from another part, substance or space.

Actual Surface. This refers to the surface of a part which is actually obtained after a manufacturing process.

Nominal Surface. A nominal surface is a theoretical, geometrically perfect surface which does not exist in practice, but is an average of the irregularities that are superimposed on it.

Profile. Profile is defined as the contour of any section through a surface.

Roughness. Roughness refers to relatively finely spaced irregularities such as might be produced by the action of a cutting tool. Roughness is sometimes referred to as a 'primary' texture. This is also known as 'Microgeometrical deviation' or 'Microgeometrical irregularities. These are characterised by a low ratio of their pitch '*s*' to height '*h*' of irregularity (peak to valley distance). s/h is < 50. Lower values of s/h are the irregularities of great height.

Roughness Height. This is rated as the arithmetical average deviation expressed in microinches or micro-metres normal to an imaginary centre line, running through the roughness profile (Fig. 11.1).

Roughness Width. Roughness width is the distance parallel to the normal surface between successive peaks or ridges that constitute the predominant pattern of the roughness.

Roughness Width Cut Off. (Sampling Length). This is the maximum width of surface irregularities that is included in the measurement of roughness height. This is always greater than roughness width and is rated in inches or centimetres.

Waviness. Waviness consists of those surface irregularities which are of greater spacing than roughness and it occurs in the form of waves. It may be caused by vibrations, machine or work deflections, warping etc. It is also referred as 'secondary texture'. Waviness is characterised by the ratio of their pitch to the peak to valley distance, s/h, being equal to 50 to 1000. The wavelength is greater than about 1 mm. The pitch or spacing exceeds the sampling length chosen for measurement of roughness.

Flaws. Flaws are surface irregularities or imperfections which occur at infrequent intervals and at random intervals. Such imperfections are : scratches, holes, cracks, pits, checks, porosity etc. These may be observed directly or with the aid of a penetrating dye or other materials that make them visible for examination and evaluation.

Flaws may also be termed as 'Macrogeometrical irregularities'. These are random, not regularly repeated deviation from the theoretical surface. These are characterised by a large ratio of the length '*l*' over which the deviation extends to the amount of deviation, $l/h > 1000$, *e.g.*, out of roundness, taper etc.

Lay. Lay is defined as the direction of the predominant surface pattern produced by tool marks. Symbols used to indicate the direction of lay are given below :

|| Lay parallel to the boundary line of the nominal surface that is, lay parallel to the line representing surface to which the symbol is applied, *e.g.*, parallel shaping, end view of turning and O.D. grinding.

⊥ Lay perpendicular to the boundary line of the nominal surface, that is, lay perpendicular to the line representing surface to which the symbol is applied, *e.g.*, end view of shaping, longitudinal view of turning and I.D. grinding.

X Lay angular in both directions to line representing the surface to which symbol is applied, *e.g.*, side wheel grinding, traversed end mill.

M Lay multidirectional, *e.g.*, lapping, superfinishing.

C Lay approximately circular relative to the centre of the surface to which the symbol is applied, *e.g.*, facing on a lathe.

R Lay approximately radial relative to the centre of the surface to which the symbol is applied, *e.g.*, surface ground on a turntable, fly cut and indexed on a mill.

The various types of 'Lay' are shown diagrammatically in Fig. 11.3. It should be noted that surface roughness is measured at 90° to the direction of lay.

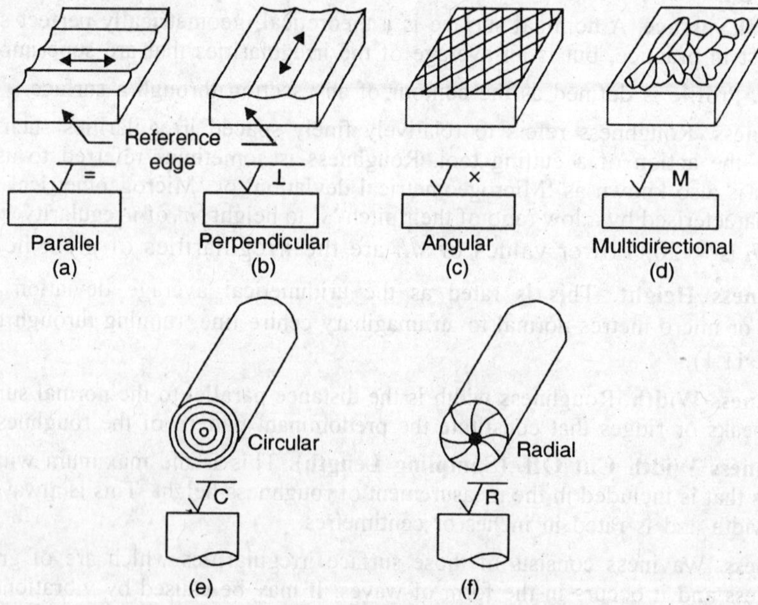

Fig. 11.3. Types of Lay.

Traversing Length. It is the length of the profile (measured in a direction parallel to the general direction of the profile) necessary for the evaluation of the surface roughness parameters. The traversing length includes from 3 to 10 roughness widths cut off, which is also called as sampling length, cut off wavelength or meter (instrument) cut-off. It is apparent that the sampling length is also measured in a direction parallel to the general direction of the profile.

It is clear from above that any finished surface could contain both types of irregularities (waviness and surface roughness) superimposed on each other. When measuring surface roughness, the problem of separating waviness from it is usually encountered. However, conventionally the surface roughness is defined within the area where waviness and deviations of form are eliminated.

11.3. EVALUATION OF SURFACE ROUGHNESS

1. Root mean square value : r.m.s. value.
2. Centre line average (CLA) or arithmetic mean deviation denoted as R_a.
3. Maximum peak to valley height, denoted as R_t or R_{max}.
4. The average of the five highest peaks and five deepest valleys in the sample, denoted as R_z
5. The average or levelling depth of the profile, denoted as R_p.

The two most accepted methods of assessing the surface roughness are : Root mean square value and the arithmetic average or centre line average value. In both the methods, the surface roughness is measured as the average deviation from a nominal surface. The value of surface roughness is expressed in micro-inches (μ in) or in micro-metres (μm). Referring to Fig. 11.4, the centre line AB is located such that the sum of the areas above the line is equal to the areas below the line. If n measurements are made (plus or minus) from the centre line vertically to points on the profile and are called y_i, then Root means square value is the positive square root of the arithmetic mean of the value of the squares of the values in the set. The distances above the centre line are taken as positive and below is as negative. Therefore,

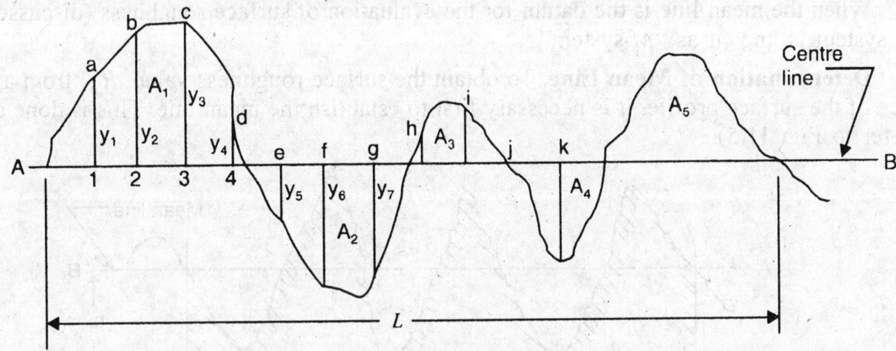

Fig. 11.4. Surface Roughness.

$$rms \text{ average} = \sqrt{\frac{y_1^2 + y_2^2 + y_3^2 + \ldots}{n}}$$

$$= \left[\left(\sum y_i^2\right)/n\right]^{\frac{1}{2}}$$

where (y_i) is the profile deviation.

Arithmetic average or centre line average is defined as the average value of the ordinates from the centre line, the algebraic sign of the ordinates is not considered, that is,

Arithmetic average $$AA = Ra = \frac{1}{L}\int_0^L |y(x)|\, dx$$

$$\text{or, approximately, } R_a = \frac{i = \sum_{1}^{n} |y_i|}{n}$$

$$= \frac{y_1 + y_2 + y_3 + \ldots}{n}$$

or
$$= \frac{A_1 + A_2 + A_3 + \ldots}{L}$$

$$= \sum A / L$$

where L is the roughness-width cut off (sampling length).

RMS values for a given profile are about 11% higher than AA figures,

1 microinch = 10^{-6} inch

1 micron = 10^{-3} mm.

According to the Indian standards, the surface roughness is assessed in terms of 'CLA' value in µm denoted as R_a.

Mean Line of the Profile. Line having the form of the geometrical profile and dividing the effective profile so that within the sampling length, the sum of the squares of the distances (y_1, y_2, ..., y_n) between effective profile points and the mean line is a minimum. When the wave form is repetitive, the mean line and the centre line are equivalent.

When the mean line is the datum for the evaluation of surface roughness (discussed above), the system is known as "*M*-system".

Determination of Mean Line. To obtain the surface roughness value, R_a, from a graphical trace of the surface profile, it is necessary first to establish the mean line. This is done as follows (Refer to Fig. 11.5).

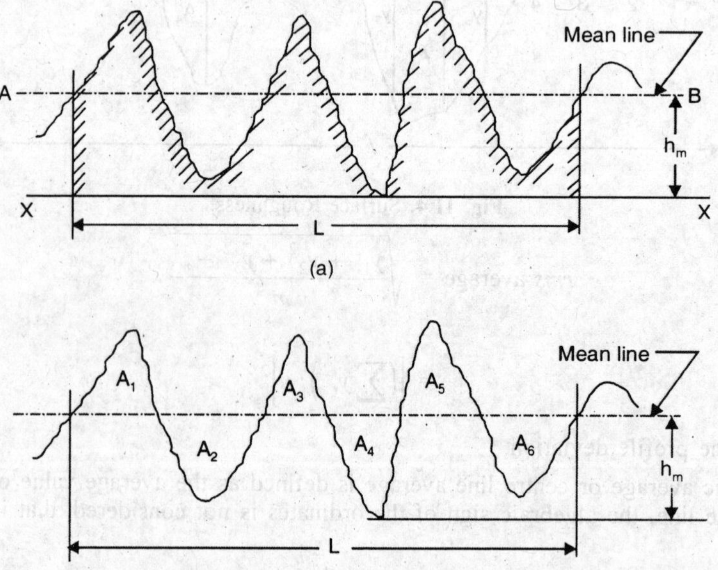

Fig. 11.5. Determination of Mean Line.

(*i*) Select a suitable sampling length, *L*.

(*ii*) Draw a line *X–X* parallel to the general direction of the profile and touching the deepest valley, Fig. 11.5 (*a*).

(*iii*) Find the area under the curve using a planimeter or ordinate method.

(*iv*) Then the height of the centre line or mean line *A-B*, from the line *X-X*, will be,

$$h_m = \frac{\text{Area under the curve}}{L}$$

Draw the mean line.

(*v*) The profile has now been divided so that the sum of areas above the mean line is equal to the sum of areas below the line. Fig. 11.5 (*b*).

(*vi*) Then $$R_a\ (\mu m) = \frac{\sum A}{L}$$

The profile for study is greatly magnified. The vertical magnification is upto 50,000 times or even greater, whereas, the horizontal magnification is from 50 to 300 times. The horizontal magnification appears in both the areas and the sampling length and thus does not appear in the calculations. Thus,

$$R_a(\mu m) = \frac{\sum A}{L} \times \frac{1000}{V}$$

where *A* is in mm², *L* is in mm and *V* is the vertical magnification.

The average of a number of samples is taken as the R_a value for the measuring traversing length.

Determination of Mean Line as per ISI. (Fig. 11.6)

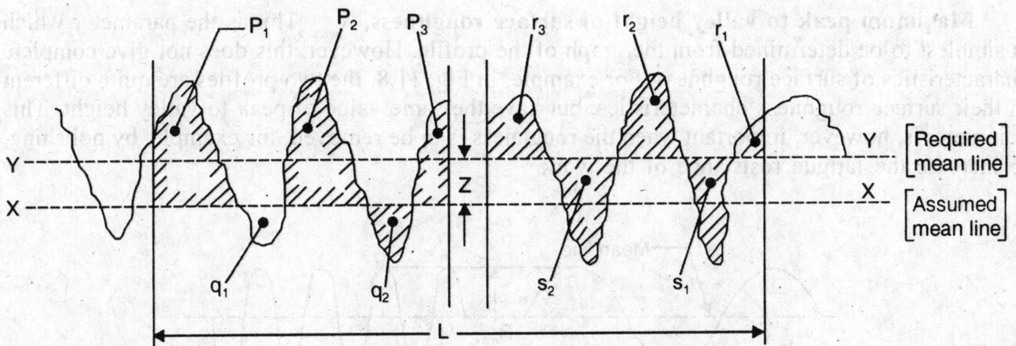

Fig. 11.6. Determination of Mean Line.

1. By eye judgement draw assumed mean line X-X.
2. Measure the total areas above and below the line X-X.
3. Then correction is found as,

$$Z = \frac{\text{Sum of area } p - \text{Sum of areas } q}{L}$$

4. Apply the correction to the line X-X and determine the required mean line Y-Y. Now the sum of areas above and below this line will be equal.

5. Then,

$$R_a\ (\mu m) = \frac{\text{Sum of area } r + \text{Sum of area } s}{L} \times \frac{1000}{V}$$

'E' System (Envelope System) of Assessment of Surface Roughness. This system (developed by Germany) expresses the arithmetic average departure of a surface both above and below a "mean curve". The mean curve is developed from what is known as a "contacting envelope" by displacing it inwards to a position where the areas enclosed by the profile above and below the mean curve are equal. The contacting envelope, in turn, is obtained by rolling across the surface, a sphere of a radius '*r*' which is normally 25 mm. The locus of centre of the sphere is displaced towards the surface by an amount equal to '*r*'. This curve now obtained, touching the peaks of the profile, is called the "contacting envelope", Fig. 11.7.

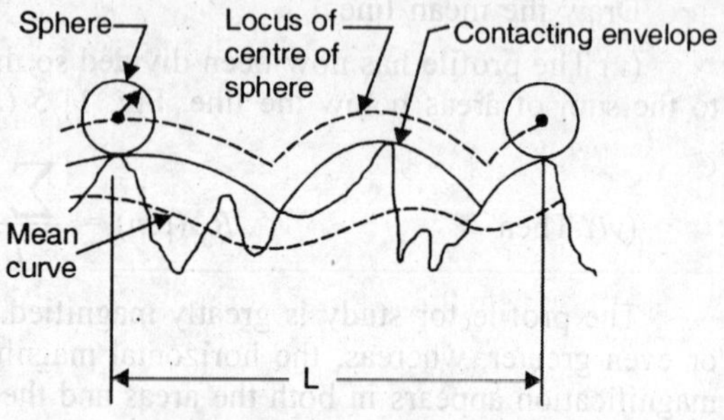

Fig. 11.7. E-System.

***M*-System.** It is more useful and a satisfactory means of controlling at the point of production, the consistency of results from a process when the production parameters have been established. The instruments based on this system are easily available. However, this system has the limitation that it is unable to control the functional qualification of a surface when associated with a machine process. The surface roughness parameter, in this system, R_a, is very useful for comparison of various surfaces produced by similar processes. However, it does not indicate the limits of irregularity and, also the same numerical value may be obtained on surfaces of greatly different profile. Thus, R_a value is often inadequate to describe the surfaces for specific applications.

***E*-System.** This system is more easily applied to the surface finish instruments based on the interference type. Also, the definition of the *E*-system peak to peak measure is superior to the *M*-system. However, in this country, the instruments based on this system are not easily available.

Maximum peak to valley height of surface roughness, R_{max}. This is the parameter which is simplest to be determined from the graph of the profile. However, this does not give complete characteristics of surface roughness. For example, in Fig. 11.8, the two profiles are quite different in their surface roughness characteristics, but have the same value of peak to valley height. This parameter is, however, important when the roughness is to be removed, for example, by polishing, to increase the fatigue resistance of the work.

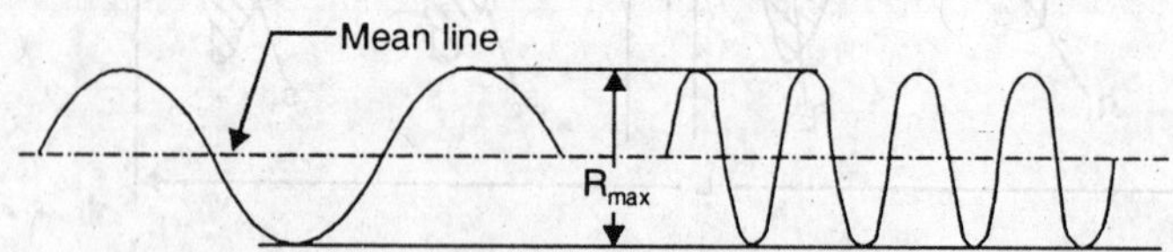

Fig. 11.8

The averages peak to valley height, R_z. This parameter is easier to obtain from a profile graph than R_a. This is the average difference between the five highest peaks and the five deepest valleys within sampling length measured from a line drawn parallel to the general direction of the profile. This line does not cross the profile. From Fig. 11.9, this parameter is obtained as,

$$R_z = \frac{1}{5}\left[(R_1 + R_3 + R_5 + R_7 + R_9) - (R_2 + R_4 + R_6 + R_8 + R_{10})\right] \times \frac{1000}{V}$$

where V = vertical magnification.

As per Indian Standard on surface roughness (IS 3073-1967), the sampling lengths in mm are given below :

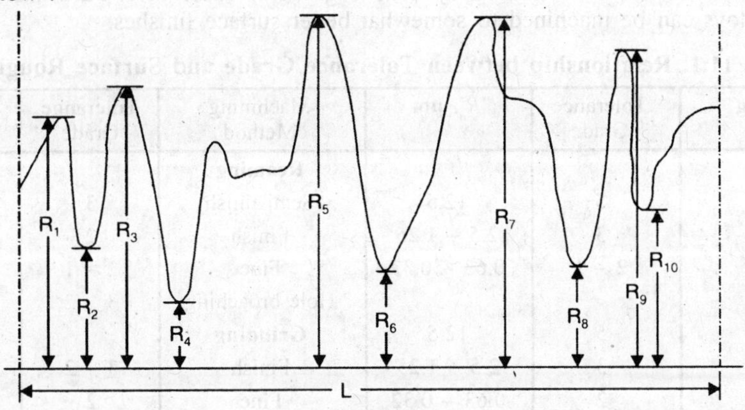

Fig. 11.9.

1. **For Machined Surface :**

Milling	:	0.8, 2.5, 8, 10
Boring	:	0.8, 2.5, 8, 10
Turning	:	0.8, 2.5
Grinding	:	0.25, 0.8, 2.5
Planning	:	25, 8, 10, 25
Reaming	:	0.8, 2.5
Broaching	:	0.8
Diamond boring	:	0.25, 0.8
Diamond turning	:	0.25, 0.8
Honing	:	0.25, 0.8
Lapping	:	0.25, 0.8
Super-finishing	:	0.25, 0.8
Buffing	:	0.8, 2.5
Polishing	:	0.8, 2.5
Shaping	:	0.8, 2.5, 8, 10
Spark machining	:	0.8

2. **For non-Machined Surfaces :**

Burnishing, Drawing, Extrusion, Moulding, Electro-polishing : 0.8, 2.5.

Preferred Values for R_a (μm) : 0.025, 0.05, 0.1, 0.2, 0.4, 0.8, 1.6, 3.2, 6.3, 12.5, 25.

There is a close relationship between surface roughness and tolerance. A high machining accuracy always implies good surface finish. In general tolerances must be greater than R_{max} (and waviness if any), unless the fit is a force fit and the surface roughness can be at least partially smoothed out in the fitting process. R_{max} can be taken approximately as 10 times R_a. The ratio of R_z and R_a ranges from 4 to 5. It has been established that the mean height of irregularities should not exceed 10 to 25 per cent of the machining tolerance. With *a* surface roughness of 5 to 0.2 μm R_a, this ratio must be 0.10 to 0.12; 0.08 to 0.10; 0.05 to 0.07, for force fits, transition fits and running fits respectively.

The average values are given below:

Dimensional Tolerance, Surface Roughness, *Ra*.

m m	μ m
< 0.005	< 0.2
0.005 to 0.012	0.2 to 0.4
0.012 to 0.025	0.4 to 0.8
0.025 to 0.05	0. 8 to 1.6
0.05 to 0.25	1.6 to 6.3

The relationship between tolerance grades and the surface roughness for parts machined by different methods from C-steels and Grey C.I., is given in Table 11.1. Parts from non-ferrous metals and alloys can be machined to somewhat better surface finishes.

Table 11.1. Relationship between Tolerance Grade and Surface Roughness

Machining Method	Tolerance Grade	R_a, µm	Machining Method	Tolerange Grade	R_a, µm
Turning			**Reaming**		
Rough	5	12.5	Semi-finish	3	2.5
Finish	3	2.5 – 1.25	Finish	2	1.25 – 0.63
Fine	2 – 1	0.63 – 0.32	Fine	2 – 1	0.32
Milling			Hole broaching	3 – 2	1.25 – 0.63
Rough	5	12.5	**Grinding**		
Finish	3	2.5 – 1.25	Finish	3 – 2	0.63 – 0.32
Fine	2	0.63 – 0.32	Fine	2	0.32 – 0.08
Drilling	4 – 5	2.5 – to 6.3	Lapping	1	0.16 – 0.04
Finish core drilling	4	6.3 – 2.5			

11.4. REPRESENTATION OF SURFACE ROUGHNESS

The surface roughness is represented in Fig. 11.10 (*a*). If the machining method is milling, sampling length is 2.5 mm, direction of lay is perpendicular to the surface, machining allowance is 2 mm and the representation will be as shown in Fig. 11.10. (*b*).

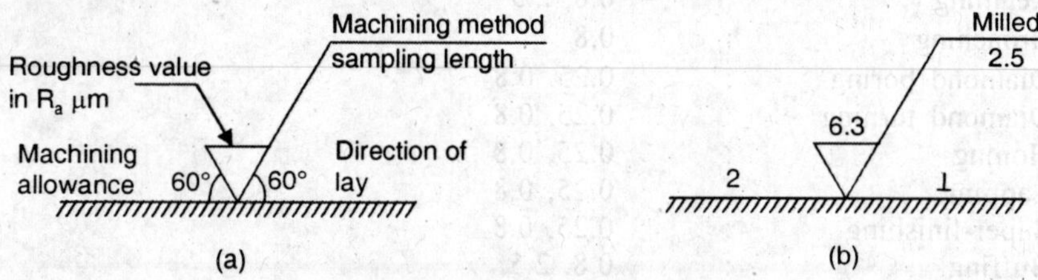

Fig. 11.10. Representation of Surface Roughness.

(*i*) The limits of surface roughness can be represented as :

$$R_{a\,16.0}^{\;8.0} \quad \text{or} \quad R_a^{\,8.0\,-\,16.0}$$

(*ii*) The surface roughness and sampling length can be represented as :

R_a 8.0 (2.5)

Here surface sampling length is 2.5. mm.

(*iii*) The surface roughness and Lay can be stated as :

R_a 1.6 Lay Circular

However, in most cases, one single piece of information is sufficient which is indicated as follows :

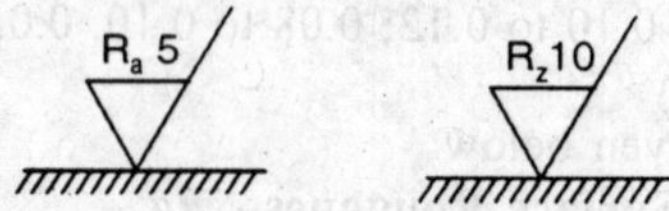

The I.S.O. has recommended a series of preferred roughness values and corresponding roughness grade numbers to be used when specifying surface roughness on drawings. These values and grade numbers are listed in Table 11.2.

The roughness symbols indicate the practice followed in the industry.

Another method adopted is : N8 and so on.

Table 11.2. Roughness Values

Roughness Values R_a, (μm),	*Roughness grade number*	*Roughness Symbol*
50	N 12	~
25 12.5	N 11 N 10	∇
6.3 3.2 1.6	N 9 N 8 N 7	∇∇
0.8 0.4 0.2	N 6 N 5 N 4	∇∇∇
0.1 0.05 0.025	N 3 N 2 N 1	∇∇∇∇

11.5. RELATIONSHIP OF SURFACE ROUGHNESS TO PRODUCTION METHODS

Each manufacturing process ordinarily produces surfaces in a certain range. This variation is due partly to the obvious control over roughness, i.e., at the operator's disposal such as the shape of the tool and speed of the cut, partly due to factors beyond the operator's control, variation in hardness of material or changes in grinding wheels and largely due to differences in shop practice of the individual plants. Table 11.3 gives the range of surface roughness found for the various production methods.

Table 11.3. Typical Ranges of Surfaces Roughness, R_a μm

Process	*Average Application*	*Less Frequent Application*
Flame Cutting	12.5 to 25	25 to 50; 6.3 to 12.5
Snagging	6.3 to 25	25 to 50; 3.2 to 6.3
Sawing	1.6 to 25	25 to 50; 0.8 to 1.6
Planning, Shaping	1.6 to 12.5	12.5 to 25; 0.4 to 1.6
Drilling	1.6 to 6.3	6.3 to 12.5; 0.8 to 1.6
Chemical Milling	1.6 to 6.3	6.3 to 12.5; 0.8 to 1.6
EDM	1.6 to 63	6.3 to 12.5; 0.8 to 1.6
Milling	0.8 to 6.3	6.3 to 25; 0.2 to 0.8
Broaching	0.8 to 3.2	3.2 to 6.3; 0.4 to 0.8
Reaming	0.8 to 3.2	3.2 to 6.3; 0.4 to 0.8
EBM	0.8 to 6.3	0.2 to 0.8
LBM	0.8 to 6.3	0.2 to 0.8
ECM	0.2 to 3.2	3.2 to 12.5; 0.05 to 0.2
Boring, turning	0.4 to 6.3	6.3 to 25; 0.05 to 0.4
Barrel finishing	0.2 to 0.8	0.8 to 3.2; 0.05 to 0.2
Electrolytic grinding	0.2 to 0.6	0.6 to 0.8; 0.1 to 0.2
Roller Burnishing	0.2 to 0.8	0.1 to 0.2
Grinding	0.1 to 1.6	1.6 to 6.3; 0.025 to 0.8

Honing	0.1 to 0.8	0.8 to 1.6; 0.025 to 0.1
Electro-polish	0.1 to 0.8	0.8 to 1.6; 0.012 to 0.1
Polish	0.1 to 0.4	0.4 to 0.8; 0.012 to 0.1
Lapping	0.05 to 0.4	0.4 to 0.8; 0.012 to 0.05
Superfinishing	0.05 to 0.2	0.2 to 0.8; 0.012 to 0.05
Sand Casting	12.5 to 25	25 to 50; 6.3 to 12.5
Hot Rolling	12.5 to 25	25 to 50; 6.3 to 12.5
Forging	3.2 to 12.5	12.5 to 25; 1.6 to 3.2
Permanent Mould Casting	1.6 to 3.2	3.2 to 6.3; 0.8 to 1.6
Investment Casting	1.6 to 3.2	3.2 to 6.3; 0.4 to 0.8
Extruding	0.8 to 3.2	3.2 to 12.5; 0.2 to 0.8
Cold Rolling, Drawing	0.8 to 3.2	3.2 to 6.3; 0.2 to 0.8
Die Casting	0.8 to 1.6	1.6 to 3.2; 0.4 to 0.8

The ranges given above are typical of the processes listed. Higher or lower values may be obtained under special conditions. It is clear from the table that the values for surface roughness overlap for different manufacturing processes but it has its own surface pattern. Turning and shaping have parallel feed lines, and face milling produces curved or crossed lines on a surface. The large number of small cutting edges acting at random in grinding give a directional pattern of small scratches that vary in length and often overlap. Honing and lapping may produce multidirectional or crisscross pattern.

The selection of the method by which a surface is to be produced will depend on a large number of variables ranging from the physical limitations imposed by the material and the duties to be imposed on the surface, to the commercial considerations of output. The problems and responsibility facing an engineer charged with the production of suitable surfaces will be found to come under the following six headings :

1. To determine and produce consistently the class of surface necessary for the functioning of the component whether it is moving or stationary.

2. to select method of finishing which will give the necessary dimensional control.

3. to produce the required surface mechanically in such a manner that it needs the least amount of hand fitting and the shortest running in period.

4. to develop a surface which will give maximum life and minimum wear.

5. to select classes of surfaces and methods of producing them which do not create the risk of fatigue failure or lower the physical properties of the material in highly stressed parts.

6. to select surface treatments or coatings necessary to protect the part from the effect of corrosion which may be harmful to its function or life.

Cost of producing a surface increases as its quality increases, i.e., as the value of surface roughness decreases. It is important, therefore, that no better finish than really needed be specified for a surface.

The following are the examples of various levels of surface finish :

(*a*) *0.10 μm.* It is a mirror like surface which is free from any kind of visible marks. It is a very high cost finish which is produced by processes such as lapping, honing and precision grinding (see Table 11.2). Such a finish is used typically in high quality bearings.

(*b*) *0.20 μm.* It is scratch-free, close tolerance finish that is obtained by diamond turning or boring and precision grinding. This type of finish is used for parts like the internal diameter of hydraulic cylinders, pistons, journal bearings and cam faces.

(*c*) *0.40 μm.* Such a surface is also produced by precision grinding and processes using diamond tools. Its typical uses are in : hydraulic applications, static sealing rings, heavily loaded bearings and rapidly rotating shaft bearings.

(*d*) *0.80 μm.* It is a fine surface finish produced by various machining operations (turning, drilling, milling) performed very carefully, followed by grinding or broaching. Such a finish is used for parts subjected to stress concentration and vibrations such as gear teeth and brake drums.

(*e*) *1.60 μm.* This is high quality surface finish produced by turning, milling etc. and without subsequent operations. It is used for parts with fairly close tolerances but where fatigue is not a likely problem, such as ordinary bearings and ordinary machine parts.

(*f*) *3.2 μm.* Such a surface is produced by high quality machining using light cuts, fine feeds and sharp tools in the finishing pass. It finds applications in lightly loaded bearing surfaces and moderately stressed parts but not for sliding surfaces.

(*g*) *6.25 μm.* This surface finish results from ordinary machining operations using medium feeds. It finds uses for the surfaces of noncritical parts.

Typical cases can be, clutch disk faces : 3.2 *μm*; Brake drums : 1.6 *μm*; crankshaft bearings : 0.40 *μm*; Bearing Balls : 0.025 *μm*.

11.6. EFFECT OF SURFACE ROUGHNESS ON THE PERFORMANCE OF MACHINE PARTS.

The surface roughness of machined surfaces has an important effect on the functional properties and performance of machine parts. It greatly affects the wear resistance of the surfaces of the part, its strength, corrosion resistance and the reliability of fixed joints of mating parts.

1. **Wear Resistance.** The wear resistance of two rubbing surfaces depends upon the stability of the surface layers against failure, which in turn, depends to a great extent on the specific pressure between the two rubbing surfaces. Micro-irregularities and waviness on the surfaces reduce the area of contact between them, since the surface make contact only at the crests. This increases the specific pressure and temperature at the places of contact. This leads to more intensive deformation, crushing, shearing and chipping of the ridges on both surfaces, resulting in increased wear of surfaces. The peaks on the two surfaces also break the lubricating oil film, resulting in dry friction conditions which lead to increased wear. However, some sort of surface roughness is desirable to retain the lubricating oil, which is not possible with perfectly smooth surfaces. The retaining of the lubricating oil on the surfaces, under various conditions of friction (in accordance with the load, velocity and material of the mating parts etc.) depends largely on the size of the micro-irregularities. A smooth surface is not the most wear resistant one in all cases. If the height of the micro-irregularities is extremely small, there may also be intensive wear between two rubbing surfaces since the co-efficient of friction is first reduced with a decrease in the height of the micro-irregularities, reaches a minimum value and then increases. Thus, an optimum value of surface roughness can be established to suit definite friction conditions. Macro-geometric deviations may also lead to uneven wear of certain sections of the surface.

2. **Fatigue Strength.** It is known that lines and deep and sharp scratches across a surface become points of internal stress concentration. The bottom of the valleys between the ridges of surface micro-irregularities may be such points of stress concentration. This, in turn, may lead to the formation of cracks which substantially reduce the strength of the part and ultimately may result in failure of the part (especially if it is subjected to fatigue loading).

3. **Corrosion Resistance.** The valleys are the places where the corrosive agents may accumulate. Therefore, the rougher the surface, that is, the deeper and the more sharply defined the valleys between the ridges of micro-irregularities, the more favourable the conditions for corrosive action and its penetration into the depth of the metal. The smoother the surface, the less the area of contact of the surface with the corrosive medium will be, and consequently, the less the medium will attack the surface.

4. **Strength of Inference Fits.** This is also affected by the surface quality of the mating

parts, especially on the height of the microirregularities. When one part is pressed into the other, the ridges are crushed, that is, smoothed out. Thus the actual interference obtained may be less than the designed interference resulting in reduction of the strength of the interference fits. Also, the actual interference obtained for rough surfaces will differ from that obtained for smooth surfaces for the same measured diameter of the pressfitted surfaces.

It should be noted that the serviceability and performance of machine parts depend not only on the surface quality (as discussed above), but also on many other factors, such as, the degree and depth of workhardening and the magnitude of the residual stresses in the surface layer. It has been investigated that the smoother the machined surface, the greater the depth and degree of workhardening and the residual compressive stresses in the surface layer, the higher the fatigue strength of the part. The residual compressive stresses also reduce the effect of corrosion and so does the work hardening of the surface. The residual tensile stresses, however, reduce the fatigue strength. Residual stresses (both tensile and compressive) lower the mobility of the atoms in the metal and thus increase the resistance to wear. Workhardening of the surface layer also promotes an increase in wear resistance.

11.7. MEASUREMENT OF SURFACE ROUGHNESS

Shop and inspection departments in industry employ two principal methods to evaluate surface roughness :

(*a*) The surface is compared visually and by a feel with a standard.

(*b*) Roughness is measured with indicating, recording or optical instruments.

11.7.1. Visual Inspection. A visual examination is the simplest means of judging the finish on surfaces of any shape, size or material, but this method alone is unreliable because appearance is not a true indication of roughness. The reliability of this method may, however, be improved by comparing the work with graded standards which are copies of master surfaces finished by shaping, grinding and milling etc. and the roughness is graded in micrometers AA to agree with standard numbers. The machining method is indicated on the standards. Comparison of work and specimen of this kind by sight and feel is adequate for many requirements. Comparison by feel is accomplished by running the edge of the finger nail or the tip of the finger over the work and specimen to estimate and compare their roughness. A number of manufacturing companies have produced surface finish samples for use in production and inspection. Although visual comparison may be accomplished with the naked eye, it may be desirable to use a microscope to bring out more details. A comparison microscope designed specially for surface inspection has two stages, so that a workpiece and standard specimen may be mounted in the instrument simultaneously and viewed as a dual image in the eye piece. The method is quite simple but has its drawbacks. Only a subjective evaluation can be made. Also, no quantitative evaluation can be obtained by this method.

11.7.2. Instrument Inspection of Surface Roughness. According to the method employed for evaluating the surface roughness, instruments are classified as : (*a*) Profilometers (*b*) Profilographs. Profilometers are used for a quantitative evaluation of surface roughness by means of a single numerical rating. It may measure either the maximum, root mean square, or arithmetical mean height of the irregularities. A profilograph, on the other hand, reproduces the microirregularities in a magnified scale for subsequent analysis of the profilogram, so obtained, to determine the value of the microirregularities.

Most instruments in general used for the measurement of surface roughness are designed to respond to the irregularities of the surface through the agency of a stylus, which rests on the surface and is traversed across it. It is necessary to have a datum level to which the vertical movements of the stylus can be referred. The datum most commonly used is provided by a skid which has a relatively large radius of curvature in the direction of the traverse. This also rests on the surface but follows its general contour, riding over the crests of smaller irregularities without responding to them individually. These movements of the stylus normal to the surface, measured relative to a datum corresponding to the path followed by the skid, are recorded by an instrument.

Electrical Stylus Instruments. In these instruments, the stylus movements (vertical) are transduced to electrical signals. These instruments provide either a numerical assessment of the surface roughness displayed on a meter or a graphical trace. These instruments are calibrated against a standard etched or electroformed surfaces of known dimensions. A scratch-free optical flat may be used as a test piece to check the minimum surface variation obtainable by the instrument. It should be noted that during measurement, the waviness is usually filtered out by electronic processing of the stylus signal.

1. **Moving Coil Type Profilometer.** It is an indicating and recording instrument that measures roughness in μm, rms, or AA. It is used to check comparison standards for accuracy and to measure work directly as a means of controlling surface finish. This measure work directly as a means of controlling surface finish. This measurement is accomplished by moving a diamond tracer point, stylus or a pick-up similar to gramophone, over the work. During this movement, the tracer point is displaced by the minute ridges and valleys on the surface and its displacement is magnified and registered on a meter or a chart for reading and analysis. The profilometer is a tracer element and consists of two principal units; a tracer and an amplifier. The tracer is a conically shaped diamond point and is used to explore the surfaces which are to be inspected. The stylus or tracer point has a diamond tip with a point radius of about 12 microns and is suspended on the flat spring (Fig. 11.11). The upper end of the stylus is linked to an induction coil which is located in the field of a permanent magnet, so that the movement of the coil will induce a current. As the tracer is moved over the surface, irregularities cause the tracer point to vibrate. The vibration is transmitted to the coil in the magnetic field and the induced current is amplified and measured with an amplimeter (a galvanometer). The amplimeter amplifies the current produced by the tracer, so that it actuates the meter on the panel. The meter reading represents the surface roughness in μm (rms or AA) of the part being inspected. If the surface is uniform in character a steady reading will be inspected. If the surface is uniform in character a steady reading will be obtained but if it is not, average variations may be estimated from the meter fluctuations. Some tracers may be operated manually, and all can be operated mechanically or with motor. Manual tracing is desirable in many cases because of its speed, convenience and ability to inspect hard to reach and large surfaces.

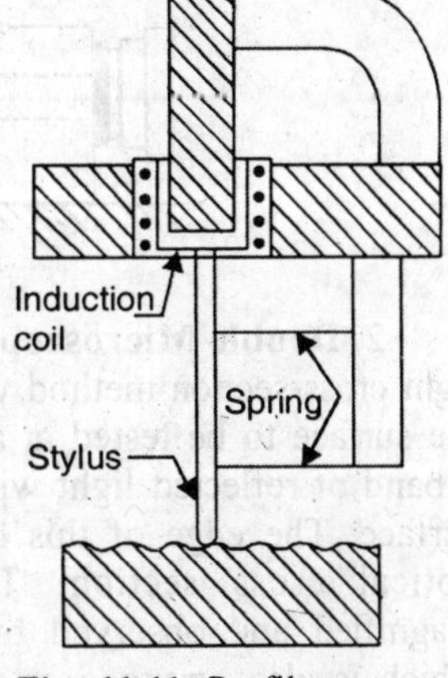

Fig. 11.11. Profilometer.

The motor drive makes the tracer head to traverse over the tested surface at a speed from 10 to 20 mm/s. The pressure of the tracer point on the surface is from 0.5 to 2.5 gms.

2. **Piezo-Electric Instrument.** This instrument is shown in Fig. 11.12. This is a voltage generating instrument. The stylus arm is connected to a crystal which is pivoted to the body of the instrument. The arm or the body of the instrument with a reference shoe (skid) is drawn across the surface; by means of a drive motor and a gear box. The stylus following the finer surface details. The stylus movements are transmitted through the stylus arm to the crystal. Due to this, mechanical vibrations are caused at the crystal. These vibrations are transformed into proportional voltages. From the amplified voltage signals, either a meter reading or a graphical trace can be obtained. The radius of curvature of the skid ranges from 5 to 50 mm. The skid may consist of two parts arrangement on either side of the stylus. For wavy surface, the skid length should be greater than twice the wavelength. This method of magnifying stylus movement employs an inductive transducer (crystal), to convert stylus movements into an electrical signal, which is amplified and then to a graphical recorder or direct reading meter, as noted above.

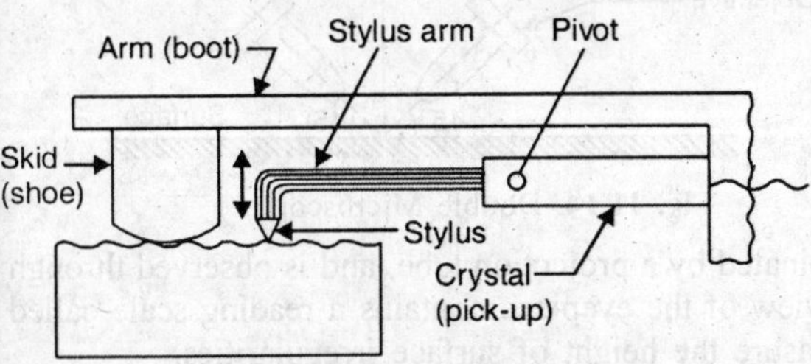

Fig. 11.12. Piezo-electric Instrument.

Optical Instruments

1. **Profilograph.** The principle of working of a tracer-type profilograph is shown in Fig. 11.13. The work to be tested is placed on the table of the instrument. The work and the table are traversed with the help of a lead screw. The stylus which is pivoted to a mirror moves over the tested surface. Oscillations of the tracer point are transmitted to the mirror. A light source sends a beam of light through lens and a precision slit to the oscillating mirror. The reflected beam is directed to a revolving drum, upon which a sensitised film is arranged. This drum is rotated through two bevel gears from the same lead screw that moves the table of the instrument. A profilogram will be obtained from the sensitised film, that may be subsequently analysed to determine the value of the surface. Another optical transducer uses a beam splitter to feed the light source to two photocells through a slit in a plate at the end of the stylus arm. When the stylus is at its balanced mid position, the slit allows equal amounts of light to fall on the photocells, but any vertical movement increases the light to one cell and reduces it to the other. This imbalance is amplified and fed to a graphical recorder or a meter.

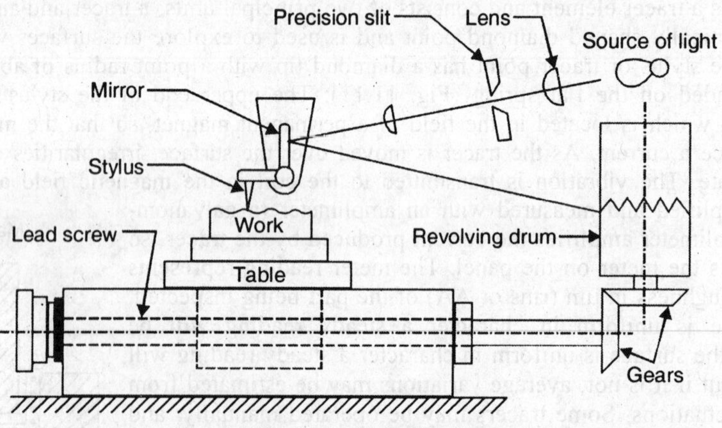

Fig. 11.13. Tracer-Type Profilograph

2. **Double Microscope.** This is an optical method. The surface roughness is evaluated by the light cross-section method which is based on the following principle. If a thin film of light strikes the surface to be tested at an angle of 45°, a band of reflected light will appear on the surface. The edge of this band will be an optical cross section. This profile is magnified and observed by a microscope which is also arranged at an angle of 45° to the surface and hence normal to the light source. This principle is explained in Fig. 11.14. From a source of light, a beam of light, passes through the condenser and a precision slit and is directed at an angle of 45° to the surface being tested. The observing microscope is inclined at an angle of 45° to the tested surface. The microscope has the objective and an eyepiece. The surface is illuminated by a projection tube, and is observed through the eyepiece of the microscope. The field of view of the eyepiece contains a reading scale called an eyepiece micrometer, which is used to measure the height of surface irregularities.

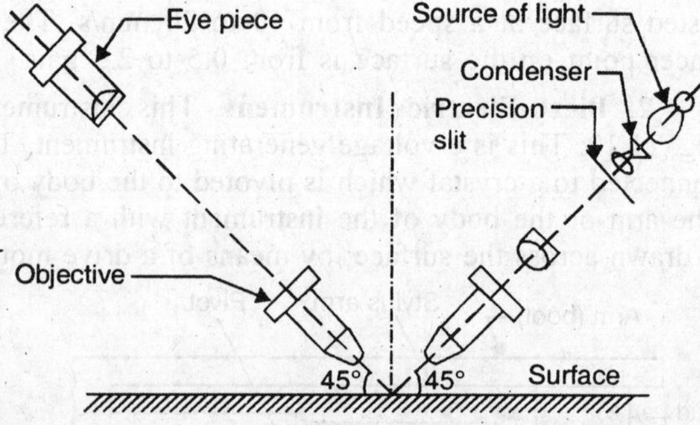

Fig. 11.14. Double Microscope.

11.8. SURFACE FINISHING PROCESSES

The various methods used for finishing the surfaces of the parts are discussed below :

1. **Diamond Turning and Boring.** It is customary to find, where the application demands it, light alloys, bronzes and tin alloys, bearing metal, being turned or bored using diamond tools

with a geometric control of about 0.0125 mm or below and with surface roughness measurement of between 0.075 and 0.125 μm (AA). Fine finishing by the diamond tools is more or less confined to those materials which do not include hard or abrasive particles in their make up which would chip or damage the stone, and which cut cleanly with a definite chip. Materials which come away in an abrasive powder rapidly erode the cutting edge and thus make the use of diamond tools expensive and uncertain. The machines employed for fine turning and boring must obviously be in perfect conditions, with all ways and guides straight and true and all bearing and spindles running in perfect truth with minimum clearance and no vibration.

2. **Grinding.** Grinding is the general method of finishing steel but it requires a high degree of skill to repeat continuously a restricted grade of surface finish to fine geometric tolerances. In works concerned with the production of hardened components with tolerances of between 0.05 mm and 0.025 mm, finishes in the range of 0.2 to 0.3 μm (AA) are commonly produced and it has to be noted that in producing finishes finer than the above figures, the danger of surface burning and cracking is greatly increased. Machine employed for fine grinding must be in first class condition with all ways and guides straight and true and particular attention must be paid to the balancing of wheels.

Wheel surface preparation is of utmost importance. Adequate flow of chip and granule free coolant is essential as it has been found that the cleanliness of the coolant has a direct bearing on the quality of the surface produced. The choice of wheel grit, speeds, coolants will vary with the material being cut and the finish demanded. Whilst grinding is convenient and a reasonable economic method of producing fine surface, it brings in its train a number of pitfalls. An inevitable result of grinding is the formation by the high temperature generated at the work and wheel contact line, of an extremely thin layer of decarburised material which has a considerable effect on the initial efficiency of a surface as a bearing and has to be removed in many cases before really satisfactory bearing conditions are attained. A further effect of high temperature is burning of the surface and the formation of grinding cracks which are large enough to affect the fatigue strength of the component. Chatter is also evident on ground cylindrical components, particularly on those which have been centreless ground.

3. **Lapping.** The purpose of lapping process is : (*a*) to produce geometrically true surface (*b*) to correct minor surface imperfections and (*c*) improve dimensional accuracy or provide a very close fit between the contact surfaces. This is called lapping in. It is used to get snug and tight (hermetic) movable and detachable joints such as fluid valves and fitting components.

Lapping is the abrading of a metal or other surface by means of a lap which is made of a second softer material, which has been charged with fine abrasive particles. The lap which may be made of fine grain cast iron, copper or cloth, always is softer than the material which is to be lapped. When the lap and the harder surface are rubbed together with the fine abrasive particles between them, these particles become embedded in the softer lap. It then becomes a holder for the hard abrasive. As a charged lap is rubbed against a hard surface, the hard particles in the surface of the lap remove small amounts of materials from the harder surface. Thus it is the abrasive which does the cutting and the soft lap is not worn away, because the abrasive particles become embedded in its surface, instead of moving across it.

In lapping, the abrasive is usually carried between the lap and the work in some sort of a vehicle. The vehicle or lubricant controls to some extent the cutting action and prevents scoring the work and caking of the abrasive. Some of the vehicles used include : kerosene plus a small amount of machine oil, greases, fine sperm oil for fine job, olive oil, lard oil, spindle oil, benzene, and soapy water. Naphtha is used to clean the laps.

In lapping, the material removal is usually less than 0.025 mm, although rough lapping may remove as much as 0.075 mm and finish lapping as little as 0.0025 mm. Commercial lapping operations can produce parts to limits of 0.000625 mm. Since it is such a slow metal removal process, it is used only to remove scratch marks left by grinding or honing or to obtain very flat or smooth surfaces such as required on gauge blocks or liquid tight seals where high pressures are involved. Materials of almost any hardness may be lapped. However, it is difficult to lap soft materials since the abrasives tend to become embedded.

Laps may be made of almost any material soft enough to receive and retain the abrasive grains. They may be made of soft cast iron brass, copper, lead or soft steel. The most common lap is fine grain cast iron. Copper is used rather often and is the common material for lapping diamonds. For lapping hardened metals for metrolographic examination, cloth laps are used.

For steel surfaces, artificial corrundum is used as an abrasive for preparatory lapping and again used in a final state for finishing. Silicon carbide gives good results on cast iron and alumina for finest lapping. For lapping small components, diamond dust or boron carbide in the finest grain size give good results. The other abrasives used are : rouge (Ferric oxide Fe_2O_3), green rouge (chromium oxide Cr_2O_3) and crocus powder. The abrasive particles are from 120 grit up to the finest powdered sizes. In nearly all cases, a paraffin lubricant is used. The exception being for soft materials when a soluble oil or water lubricant is used. The use of lubricants/cutting fluids speeds up lapping process, prolongs the sharpness of the abrasive grains, enhances the product accuracy and surface finish and cool the work surface.

Lapping may be done by hand or by special lapping machines. In hand lapping, the lap is flat similar to a surface plate. Grooves are usually cut across the surface of a lap to collect the excessive abrasive and chips, using an irregular rotary motion and turning the work frequently to obtain uniform cutting action.

An external lap for external pieces (round) is shown in Fig. 11.15. It is split by a saw cut and can be closed in by tightening one or more screws. The diameter of the hole is made the same as that of the piece to be lapped and the hole is, of course bored before the saw cut is made. Internal laps are made to expand.

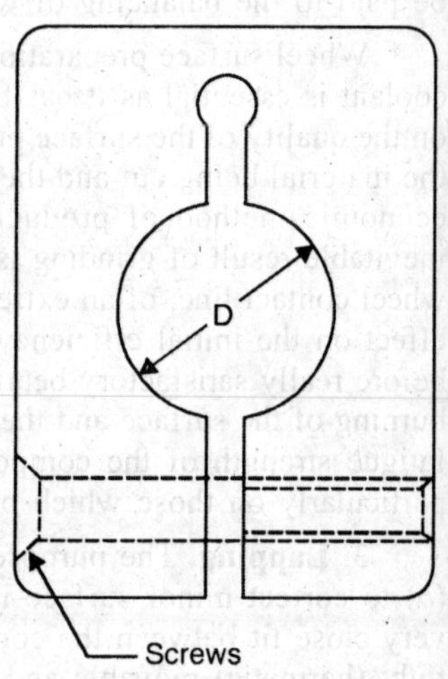

Fig. 11.15. External Lap.

Mechanical lapping is a high production process, for example, gudgeon pins 25 mm diameter and 75 mm long are lapped at the rate of 500 pieces per hour, removing 0.05 to 0.075 mm of material with a limit of accuracy of roundness, straightness and size within 0.025 mm. Mechanical lapping machines are of vertical construction with the work holder mounted on the lower table which is given an oscillating motion. The upper lap is stationary and floating, while the lower one revolves at about 60 rev/min. Several types of lapping machines are available for lapping round surfaces. A special type of centreless lapping machine is made for lapping small parts such as piston pins, ball bearing races etc.

The following are examples of work done by lapping : aircraft piston pins, automotive wrist pins, diesel engine injector-pump parts and spray nozzles, plug gauges, certain dies and moulds, gauge blocks, refrigerator-compressor parts, oil-burner parts, micro-meter spindles, roller bearings, taper rollers, worm and worm gears, crankshafts, camshafts, ball bearing raceways etc.

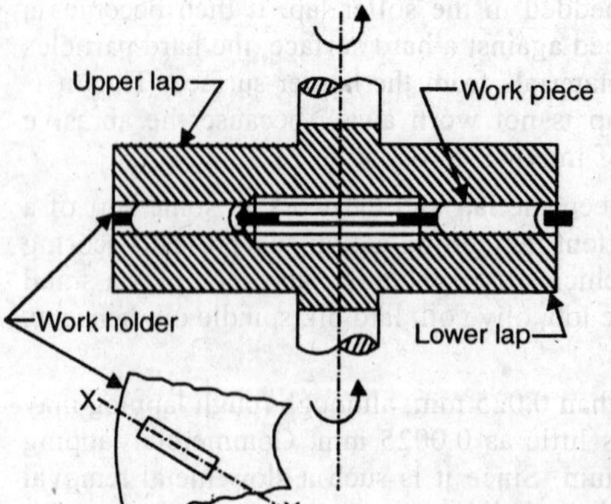

During machine lapping, a pressure of 0.007 to 0.02 N/mm² for soft materials and upto 0.07 N/mm² for hard materials is satisfactory. Mechanical lapping machines can be used for lapping : (*i*) External cylindrical surfaces, and (*ii*) flat surfaces.

A general purpose machine for lapping both cylindrical and flat surfaces is shown in Fig. 11.16. A number of workpieces are placed between the upper and the lower lap, whose surfaces have previously been lapped flat. The workpieces are placed in slots in a workholder so that their axes X-X are not quite radial. The shape of the slots will depend upon that of the workpieces. The two laps are rotated and the workholder is

Fig. 11.16. Machine Lapping.

given as oscillation of about 25 mm amplitude. A stream of vehicle in which fine abrasive flour is suspended is fed to the centre of the laps and flows outwards and the workpieces are thus gradually lapped to size.

Lapping has become a common production process with the demand for hardened surfaces having only a few micrometers of surface finish. However, because, it is such a slow method of metal removal, it is obviously relatively expensive and is not economically justified unless operating requirements make such surface finishes absolutely necessary. Lapped surfaces are well resistant to corrosion and wear, which is vital for proper functioning of measuring tools, instruments and various high precision parts.

Lapping and polishing differ in the following manner-Polishing is meant to produce a shiny surface whereas a lapped surface does not usually have a bright shiny appearance. Lapping definitely removes metal from lapped surface, whereas, polishing as a rule does not remove any appreciable amount of metal. Lapping improves the geometrical shape of the body, whereas, polishing does not. Lapping is essentially a cutting process, while polishing consists of producing a kind of plastic flow of the surface crystals so that the high spots are made to fill the low spots.

4. **Honing.** Honing is a grinding or abrading process. In it, a very little material is removed. This process is used primarily to remove the grinding or tool marks left on the surface by previous operations. The cutting action is obtained from abrasive sticks (aluminium oxide or silicon carbide) mounted in a mandrel or fixture. A floating action between the work and the tool prevails so that any pressure exerted on the tool is exerted and transmitted equally on all sides. The honing tool is given a slow reciprocating motion as it rotates, having resultant honing speeds from 0.25 to 1.0 m/s. This action results in rapid removal of stock and at the same time, the generation of a straight and round surface. Defects such as slight eccentricity, a wavy surface, or a slight taper caused by previous operation can be corrected by this process. Parts honed for finish remove only 0.025 mm or less. However, when certain inaccuracies must be corrected, amounts upto 0.50 mm

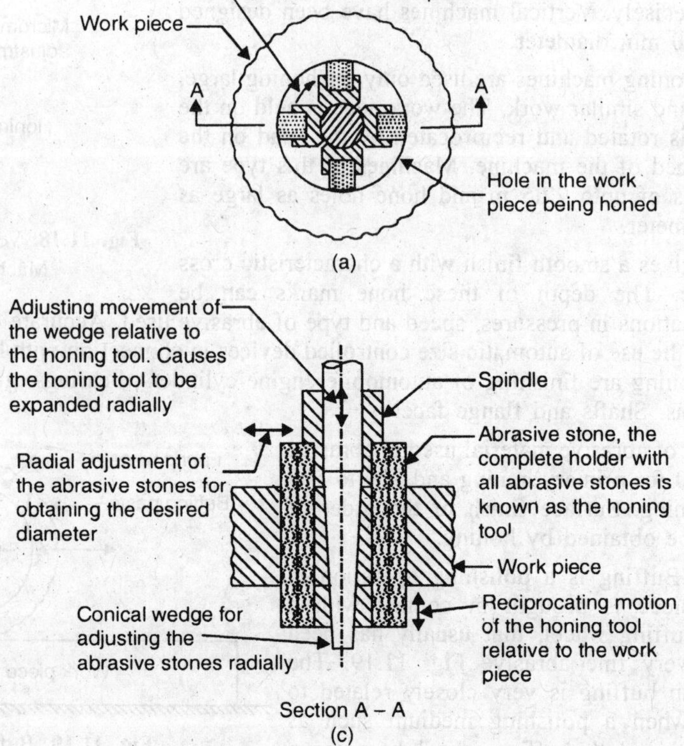

Fig. 11.17. Honing Tools

represent usual practice. Coolants are essential to the operation of this process to flush away small chips and keep temperatures uniform. Sulphurised mineral base or lard oil mixed with kerosene is generally used. Paraffin is also used.

Most honing is done on internal surfaces, or holes, such as automobile cylinders. There are a few applications of honing to external surfaces. Parts can be of any shape, but the surface must be cylindrical. Practically any material can be honed. Soft materials which cannot be lapped, can be honed because of the use of bonded abrasive. Hard and soft cast iron, steel, carbides, bronze, aluminium, brass and silver, as well as glass, ceramics and some plastics can be honed.

Honing machines are similar in general construction to vertical drilling machines, but the spindle reciprocation is usually by hydraulic means. The rotary motion may be from a hydraulic motor or by gearing. The speed ratio of two motions affects the work finish and may be varied throughout the operation or for different materials. For cast iron, the speed ranges from 1.0 to 2.5 m/s for rotation; with 0.25 to 0.35 m/s for reciprocation. The corresponding speeds for steel are 0.75 to 1.0 m/s for rotation and 0.2 m/s for reciprocation. The reciprocation motion distributes the wear over the whole length of the sticks and keeps the bore cylindrical. Semi-automatic honing machines used in the finishing of automobile cylinder bores are of vertical type. Both single and multispindle machines are used for this operation. The abrasive stones are mounted on a honing tool. In order to expand the abrasive stones outward to fit the hole to be honed, the conical wedge is moved relative to the honing tool, Fig. 11.17.

A general arrangement of a vertical honing machine is shown in Fig. 11.18. The abrasive sticks (upto 8 in number) are expanded while honing takes place, if required, by micrometer controlled, mechanical or hydraulic, means. The honing tool will follow the axis of the original hole, therefore, the honing tool or fixture must be free to float. This is done by using universal joints as shown in the figure. Due to this, the honing tool becomes self centring and it is not necessary to line up the hole and hone axes precisely. Vertical machines have been designed for work upto 500 mm diameter.

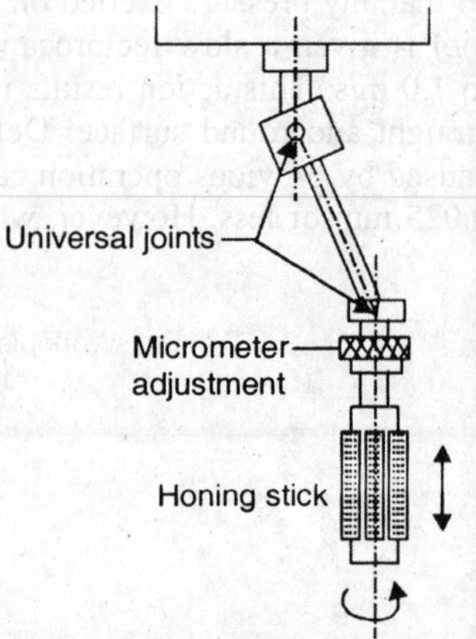

Fig. 11.18. Vertical Honing Machine.

Horizontal honing machines are used only for honing large, long gun barrels and similar work. The workpiece is held on the left and the tool is rotated and reciprocated by the head on the right end of the bed of the machine. Machines of this type are made with strokes of upto 22.5 m and hone holes as large as about 1 m in diameter.

All honing gives a smooth finish with a characteristic cross hatch appearance. The depth of these hone marks can be controlled by variations in pressures, speed and type of abrasive used. Accurate dimensions can be maintained by the use of automatic size controlled devices in connection with honing. Typical applications of honing are finishing of automobile engine cylinders, bearings, gun barrels, ring gauges, piston pins. Shafts and flange faces.

The grit size of abrasive material used in abrasive sticks is 80 to 180 for primary honing and 300 to 500 for secondary honing. Surface finish of the order of 0.05 μm R_a can be obtained by honing.

5. **Buffing.** Buffing is a polishing operation in which the workpiece is brought in contact with a revolving cloth buffing wheel, that usually has been charged with a very fine abrasive Fig. 11.19. The polishing action in buffing is very closely related to lapping in that when a polishing medium such as 'rouge' is used, the cloth buffing wheel becomes a

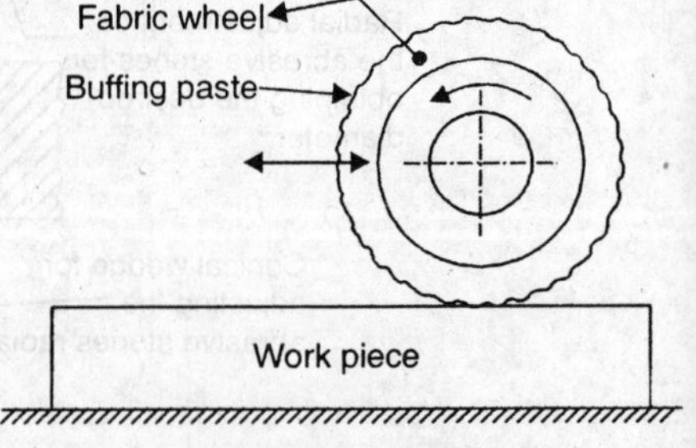

Fig. 11.19. Buffing.

carrying vehicle for the fine abrasives. In (this action the abrasive removes amounts of metal from the workpiece, thus eliminating the scratch marks and producing a very smooth surface. When softer metals are buffed, particularly without the use of an abrasive, there is some indication that a small amount of metal flow may occur which helps to reduce the high spots and produce a high polish.

Buffing wheels are made of discs of linen, cotton, broad cloth and canvass. They are made more or less firm by the amount of stitching used to fasten the layers of the cloth together. Buffing wheels for very soft polishing or which can be used to polish into interior corners may have no stitching, the cloth layers being kept in position by the centrifugal force resulting from the rotation of the wheel. Buffing wheel speeds are in the range of 32.5 to 40 m/s.

Various types of buffing rouges are available. Most of them being primarily ferric oxide in some soft type of binder. Buffing should be used only to remove very fine scratches or to remove oxide or similar coatings which may be on the work surface. It ordinarily, is done manually, the work being held against the rotating wheel. This procedure is apt to be relative expensive because of the labour cost. There are semi-automatic buffing machines available consisting of a series of individually driven buffing wheel which can be adjusted to the desired position so as to buff different portions of the workpiece. The workpieces are held in fixtures on a rotating circular worktable so as to move past the buffing wheels. If the workpieces are not too complex in shape, very satisfactory results can be achieved with such equipment and the buffing cost will be low.

Product applications of buffing process which produces mirror-like finish are : objects used on mobile homes, automobiles, motor-cycles, boats, bicycles, sporting items, tools, store fixtures, commercial and residential hardware and household utensils and appliances.

6. **Barrel Tumbling.** Barrel tumbling is the process of revolving workpiece in a barrel with abrasive and water for the purpose of producing a high lustre or for the purpose of removing burrs.

A typical tumbling barrel is eight sided, lined with wood and about 1.8 m long and 1.2 m in diameter. It rotates at about 24 rpm. The barrel may have one to six compartments. The time involved may be from 1 to 4 hours depending upon the job.

Parts are packed into the barrel or drum until it is nearly full, together with slugs, stars or jacks or some abrassive such as sand, granite chips or aluminium oxide pallets. The barrel is then rotated. The movement of the parts as they tumble and roll over one another and the accompanying impingement of the slug or abrassive against the parts produce a fine cutting action, Fig. 11.20. Delicate parts should not shift loosely during tumbling and in some cases the parts must be attached to racks within the barrel so that they will not strike against one another. Tumbling is an inexpensive cleaning method. Various shapes of slug materials may be used. Several shapes may be mixed in a given load since shapes must be provided which will reach into all sections and corners that must be cleaned. Tumbling usually is done dry, but sometimes it may also be wet.

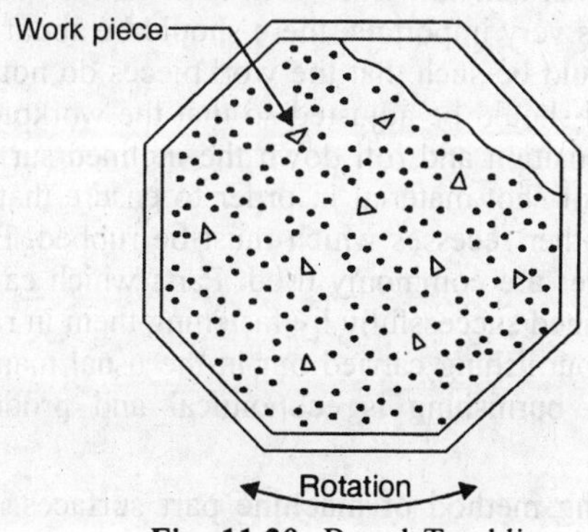

Fig. 11.20. Barrel Tumbling

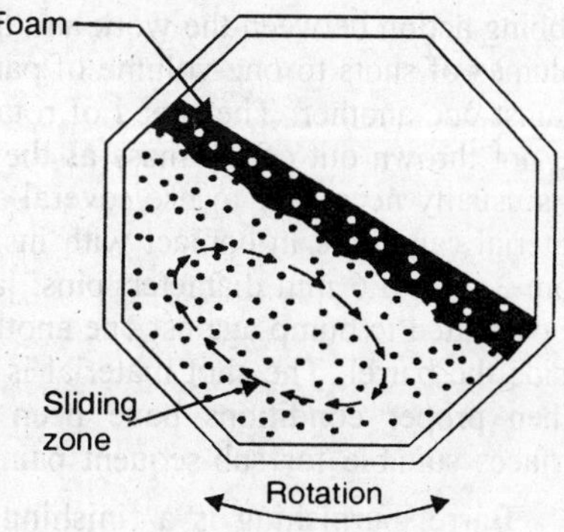

Fig. 11.21. Barrel Rolling.

Tumbling may be employed for any of the following purposes :

(*a*) Removing fins, flashes and scales from parts.

(*b*) Cleaning of forgings, stampings and castings.

(*c*) Deburring.

(*d*) Improving micrometer finish.

(*e*) Finishing high precision work to a high lustre.

(*f*) Forming uniform radii.

(*g*) Finishing gears and threaded parts without damage.

(*h*) Removing paint or plating.

After tumbling, the parts must be thoroughly washed and dried by sawdust or infrared lamps and then oiled to prevent the formation of rust.

7. **Barrel Rolling.** Barrel Rolling is similar to tumbling except that the barrel is loaded only about 40 to 60% full and its purpose is not to clean but no cut down the surfaces through the use of suitable cutting abrasive. Rolling is done either in open tilted barrels or in close horizontal barrels. The workpieces, abrasive and water or dilute acid solutions are loaded into the barrel to such a height that as barrel turns, the mass will be carried about 3/4 of the distance up the side and then roll over and fall to the bottom. The abrasives most commonly used are : slag, cinders, sharp sand, granite chips, broken chips of glass or carborundum. The abrasives vary in shape from round to triangular and in size from about 5 mm to 25 mm. The rolling action must be such that there is a relative motion between the work and the abrasive particles, since no cutting action will occur unless the sharp edges of the abrasive pass across the surfaces to be finished. Rolling usually is done wet since this gives a faster cutting action. However dry rolling is also done usually with an abrasive mixed with saw dust to brighten small parts.

By using the proper abrasive, a wide range of finishes can be obtained. Rolling time varies from 10 minutes for non-ferrous parts to 2 or 3 hours for steel. Since it usually is done as a batch process, it is simple and very economical. The resulting surface is remarkable uniform, but deep scratches must be removed prior to rolling.

8. **Barrel Burnishing.** Results very nearly comparable with those obtained by buffing may often be obtained by barrel burnishing. It is similar to barrel rolling except that instead of using an abrasive medium, medium balls, shots or round pins are added to the work in the barrel. There is no cutting action in burnishing. Instead, the slug material producing peening and rubbing action on the work rough surface, spreading the minute surface irregularities to an even surface. Burnishing will not ordinarily remove visible scratches or pits, but will produce a smooth, uniform surface and reduce the porosity in surfaces which are to be or have been plated. Parts which are to be barrel burnished, usually, first should be rolled with a fine abrasive. Barrel burnishing normally is done wet, using water to which has been added some lubricating or cleaning agent such as soap. The barrel should not be loaded more than half full with work and shot. Since the rubbing action between the work and the shot material is very important, there should be about two volumes of shots to one volume of parts. The ratio should be such that the workpieces do not rub against one another. The speed of rotation of the barrel should be adjusted so that the workpiece are not thrown out of the mass as they reach the top position and roll down the inclined surface. It is usually necessary to use several sizes and shapes of shot material in order to ensure that the material can come in contact with inside corners and other recesses which must be rubbed. Balls from 3 mm to 6 mm diameter, pins, jacks and ball cones are commonly used. Parts which cannot be permitted to bump against one another may be burnished successfully by fastening them in racks inside the barrel. The shot material is then added and burnishing carried out in the usual manner. When proper conditions have been achieved, barrel burnishing is economical and produce surfaces suitable for sub-sequent painting or plating.

Barrel burnishing is a finishing and strengthening method of machine part surfaces. The process is based upon plastic deformation of the surface layer (non-cutting methods). When ball and shots fall on the machine surface, it is subjected to impact loads, which results in the setting up of residual compressive stresses (400 to 700 N/mm^2) in the layer due to its cold working. The method also results in higher work hardness of the layer. A similar method is "Shot Peening" which is discussed under Art. 4.10 in the book "A Text Book of Production Technology" by the author.

An effective method for finishing and strengthening machine parts is "Roller and Ball Burnishing" (see Art. 17.2.9 of the above mentioned book). Cylindrical surfaces are burnished by hardened steel or cemented carbide rollers (and less frequently with steel balls/rollers) mounted in a holder. Fillets and grooves are burnished by rollers rounded to a radius and the non-rigid parts held at one end (*e.g.*, on automatic lathes), are burnished with three-roller burnishing heads.

Treatment with power brushes is an effective method of strengthening surfaces to a depth of 0.04 to 0.06 mm. Brushes made from steel wires (0.3 to 1 mm ϕ) are rotated at 30 to 45 m/s. The initial surface roughness reduces by 50 to 80%. However, after about 4 to 6 s, it starts to increase. The surface hardness can reach 3 to 4 times the initial hardness. The process can be automated for treating workpieces of different types and sizes.

11.9. SUPER-FINISHING

Super-finishing is a microfinishing process that produces a controlled surface condition on parts which is not obtainable by any other method. The operation which is also called 'microstoning' consists in scrubbing a stone against a surface to produce a fine quality metal finish. The process consists of removing chatter marks and fragmented or smear metal from the surface of a dimensionally finished parts. As much as 0.03 to 0.05 mm of stock can be efficiently removed with some production applications, the process becomes most economical if the metal removal is limited to 0.005 mm.

The method is performed by rapidly reciprocating a fine grit stone with a soft bond and pressing it against a revolving round work-piece, Fig. 11.22. The stone quickly wears to conform to the contour of the work-piece. The work-piece and tool are flooded with a cutting fluid to carry away heat and particles of metal and abrasive. The time needed for superfinishing is quite small. Parts may be superfinished to a smoothness of 0.075 μm as rapidly as 15 to 50 seconds. However, to obtain a better finish 2 or 3 minutes may be required. Product applications of super-finishing are : computer memory drums, sewing machine parts, automotive cylinders, brake drums, bearings, pistons, piston rods and pins, axles, shafts, clutch plates, tappet bodies, guide pins. etc.

11.10. POLISHING

Polishing is a surface finishing process, used to obtain a smooth and lustrous surface on a part. It removes scratches, tool marks and similar irregularities from the job surfaces produced through other operations, like machining, casting or forging. The process is carried out with the help of soft/resilient wheels or belts made of felt, leather, wood, or coarse calico cloths etc. and coated with abrasive particles (of Al_2O_3 or diamond) paste or used with fluid carrying abrasives. The process is based on the simultaneous action of the tool (polishing wheel or belt) and the surface active agents of the pastes. Two basic mechanisms are involved in the polishing process: (*i*) fine-scale abrasive removal, and (*ii*) softening and smearing of surface layers by frictional heating during the process. The fine lustrous (shining) surfaces result from the smearing action. The process reduces surface roughness to 0.032 – 0.012 μm R_a. The polishing can be done by hand, but for production work, specially designed semi-automatic and automatic polishing machines are available. Polished surfaces may be buffed to obtain an even finer surface roughness/finish. However, unlike lapping, polishing does not improve dimensional accuracy.

There are three variations of the process: roughing, dry fining and fine finishing. Roughing and dry fining are carried out with dry wheels, whereas in fine finishing process, oil, tallow or beeswax is used. The grit size of the abrasive is: 20 to 80 for roughing, 90 to 120 for dry fining and 150 for fine finishing.

The limitation of the process is that parts with irregular shapes, sharp corners, deep recesses or sharp projections are difficult to polish.

Product applications: Cutlery and small hand tools, internal work on tools and dies, bicycle parts, golf club heads, parts for fountain pens, jet engine turbine blades, sole plates for electric irons etc.

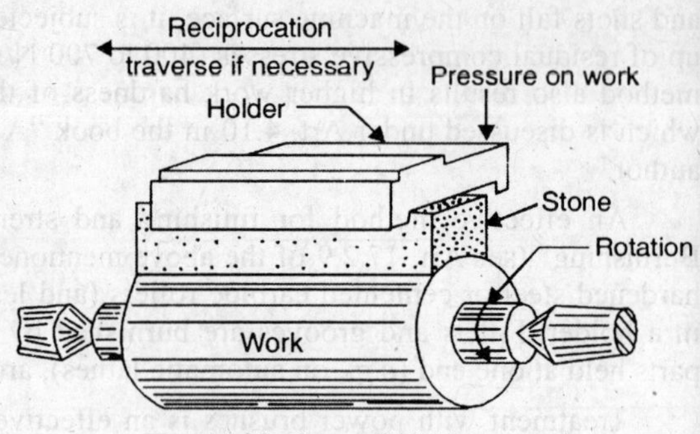

Fig. 11.22. Super-finishing.

11.11. NUMERICAL EXAMPLES

Example 11.1. *Calculate the C.L.A. value of a surface for the following data :*

The sampling length is 0.8 mm and the graph drawn to a vertical magnification of 15,000 and horizontal magnification of 100 and the areas above and below the datum line are 160, 90, 180, 50 mm² and 95, 65, 170, 150 mm² respectively.

Solution.
$$\sum A = \frac{\text{Sum of all the areas}}{\text{vertical magnification} \times \text{horizontal magnification}}$$

$$= \frac{(160 + 95 + 90 + 65 + 180 + 170 + 50 + 150)}{15000 \times 100}$$

$$= \frac{960}{15000 \times 100} = 0.00064 \text{ mm}^2$$

$$\text{C.L.A. value} = \frac{\sum A}{L}$$

$$= \frac{0.00064}{0.8} = 0.0008 \text{ mm} = 0.0008 \times 1000 \text{ µm}$$

$$= \mathbf{0.8\ µm}$$

Example 11.2. *Calculate the arithmetic average and the rms values for surface irregularities shown in Fig. 11.23.*

Solution. The values of y and y^2 are

Ordinate	y	y^2
1	0.15	0.0225
2	0.25	0.0625
3	0.35	0.1225
4	0.25	0.0625
5	0.30	0.0900
6	0.15	0.0225
7	0.10	0.0100
8	0.30	0.0900
9	0.35	0.1225
10	0.10	0.0100

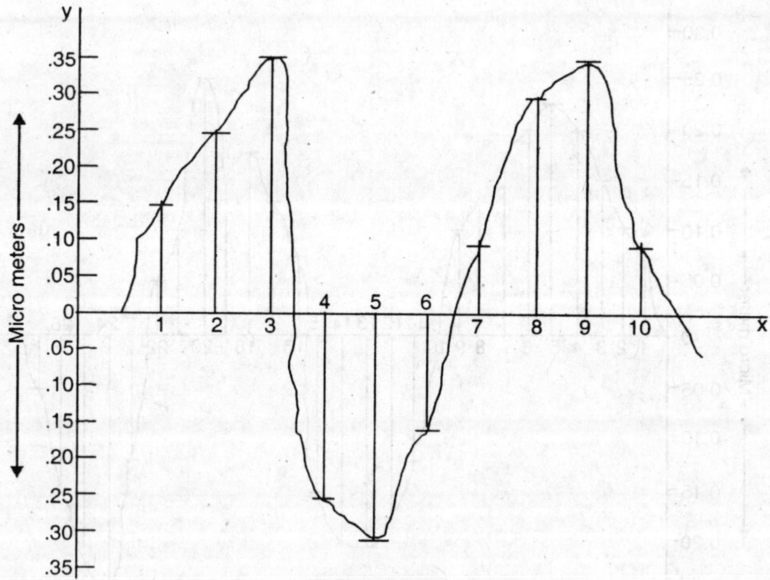

Fig. 11.23

$$\Sigma Y_n = 2.30; \qquad \Sigma Y_n^2 = 0.615$$

$$R_a = \frac{\Sigma Y_n}{n} = \frac{2.30}{10} = 0.23 \ \mu\text{m}$$

The r.m.s. value is, $\sqrt{\Sigma Y_n^2 / n} = \sqrt{\dfrac{0.615}{10}} = 0.248 \ \mu\text{m}$

PROBLEMS

1. Explain the importance of surface finish.
2. Name and explain the various elements of surface roughness.
3. What is meant by 'Primary texture' and 'Secondary texture'.
4. What is meant by 'Lay' ? Enumerate the various types of Lay and give the symbol for each.
5. Discuss the various methods of evaluating surface roughness.
6. Give the units of surface roughness.
7. How surface roughness is represented?
8. Discuss the various methods of measurement of surface roughness.
9. What is the difference between a 'Profilometer' and a 'Profilograph'?
10. What are the purposes of lapping and honing?
11. Briefly explain the process of lapping?
12. Briefly explain the process of honing.
13. Discuss two general types of honing machines.
14. What is the difference between lapping and honing?
15. Briefly express the process of buffing.
16. Give the product applications of lapping, honing and buffing process.
17. Differentiate between 'Barrel Tumbling', 'Barrel Rolling' and 'Barrel Burnishing'.
18. Briefly explain the processes of 'Barrel Tumbling', 'Barrel Rolling' and 'Barrel Burnishing'.
19. How is super-finishing done?
20. Discuss the effects of surface quality on functional properties. Give the factors affecting surface finish.
21. Explain with a sketch, one possible method for measuring roughness for a lapped shaft, and state why this particular method is preferable.
22. Name the common abrasives used in 'lapping' and 'honing' process.

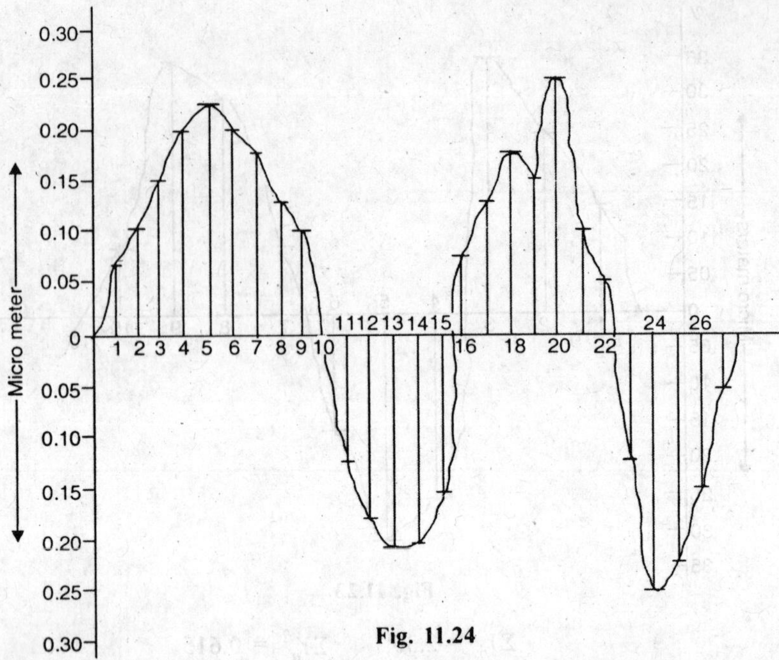

Fig. 11.24

23. Find (*a*) the arithmetic average and (*b*) the rms value for surface irregularities shown in Figures 11.24 and 11.25. (0.145 μm, 0.1515 μm; 0.3425 μm 0.3725 μm)

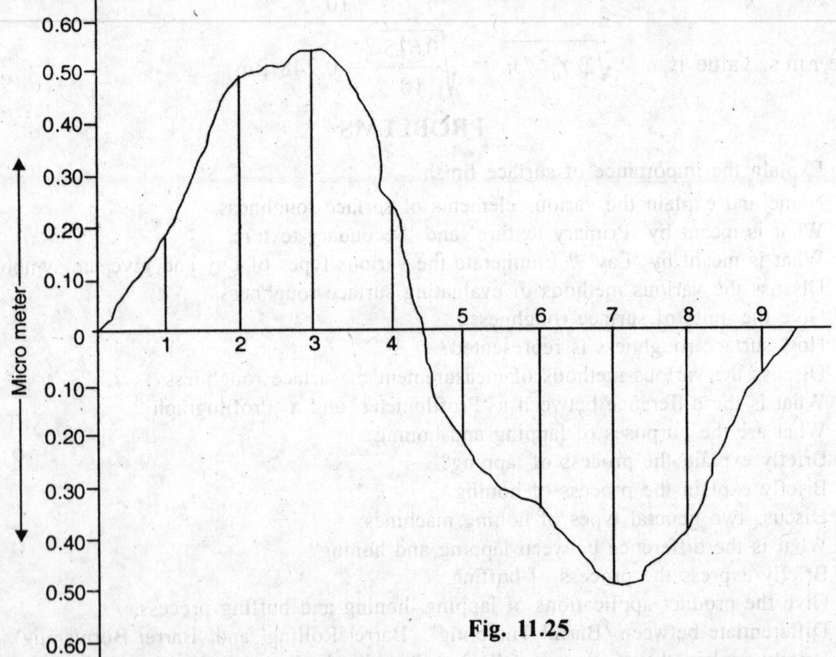

Fig. 11.25

24. How does the production methods affect the surface finish ? Explain with examples.

25. Specification of Surface Finishing Machines:

i) Lapping Machines, Flat Surface	:	Job Size
ii) Lapping Machines, cylindrical Surface	:	Job ϕ
iii) Honing Machines, Internal, Manually operated	:	Bore ϕ, stroke
iv) Honing Machines, Internal, Power Operated	:	Bore ϕ, stroke
v) Honing Machines, External (including superfinishing)	:	Job ϕ
vi) Polishing and Buffing Machines	:	Wheel ϕ

26. State the magnitude of the stylus-tip radius used for assessing surface finish.
[**Ans.** Maximum of 0.0025 mm]

27. State the range of vertical magnification used for graphical displays.
[**Ans.** X 500 to X 10,000]

28. What equipment is available for carrying out visual examination of surface finish?
[**Ans.** Micro-scopes, Optical flats]

29. To what degree of accuracy can visual assessment be made?
[**Ans.** 0.00015 mm]

30. What type of light is necessary when using optical flats?
[**Ans.** Monochromatic light]

31. The graphical trace of a surface roughness profile was measured by means of a planimeter over a length of 120 mm. The following areas in mm² were recorded: 2260, 2240, 2380, 2315, 2275, 2280, 2290 and 2330. If the vertical magnification was X 5000, determine the R_a value of the profile.

Sol. $R_a = \dfrac{\Sigma A}{120} \times \dfrac{10^3}{5000}$ = **30.6 μm.**

32. The five highest peaks and five deepest valleys were measured in mm from a line drawn on a surface roughness graphical trace as follows: 15, 44, 21, 38, 16, 46, 17, 42, 18, 49. If the vertical magnification was × 5000, calculate the R_z value of the profile.

Sol. $R_z = \left[\dfrac{44+38+46+42+49}{5} - \dfrac{15+21+16+17+18}{5}\right] \times \dfrac{10^3}{5000}$ = **5.28 μm.**

33. What is the meaning of the following terms :- Surface texture; Surface waviness; Surface flaw; Micro-geometric irregularities;
Macro-geometric irregularities; Arithmetical Mean Deviation; Root Mean square; Peak to Valley; 0.8 and 2.4 mm wavelength.

34. Enumerate the relative merits of stylus type and an light interferometer type of surface measuring instrument.

35. TRUE OR FALSE

(*a*) Lap is harder than the work piece material.
(*b*) Lapping is a material removal process.
(*c*) Lapping is carried out before grinding operation.
(*d*) Polishing wheels are made of very hard abrasive materials.

36. Multiple Choice Questions :

(*a*) In lapping process :-
 (i) Form tool is used.
 (ii) The shape of the lap (tool) is imparted to the component
 (iii) There is an improvement in the surface quality of the part.
 (iv) None of the above

(*b*) Honing process is used for :-
 (*i*) Cutting Castings
 (*ii*) Finishing the surface of a part
 (*iii*) Generating roundness
 (*iv*) All of the above.

(*c*) The use of Polishing operation is to :-
 (*i*) Remove minor imperfections
 (*ii*) Remove scratches
 (*ii*) Improve the appearance of the job.
 (*iv*) All of the above.

37. Define the surface finish terms denoted by the symbols Ra and Rz.

38. What factors contribute to the formation of Waviness.

39. What is the limitation of a Surface roughness trace.

40. Explain why Polishing can not eliminate surface roughness completely.

12

MEASUREMENT

12.1. GENERAL

Gauges (discussed in chapter 10) are used mainly to check the components produced on mass scale, where the job is usually handled by semi-skilled workers. This type of measurement cannot be relied upon where accuracy is most important, for example, the tool room where gauges, tools etc. are made for rest of the factory; the machine repair shop; the plant making machines and assemblies in small quantities and so on. For such work, one must rely for accuracy on the fundamental methods of measurement, which will be discussed in this chapter.

The basis of the metric system of length is the international prototype standard metre, which is defined as the distance between two scratches on a platinum-iridium bar. The prototype metre is kept under special conditions in the Achieves of the International Bureau of Weights and Measurements near Paris in France.

The measuring instruments/methods can be classified in various manners as given below :

1. (*a*) **Direct Measuring.** In which the measured value is determined directly, *e.g.*, micrometer, vernier caliper, vernier height gauge, Bevel protractors etc. Such instruments are simple and most widely used in production.

(*b*) **Indirect Measuring.** In which the dimension is determined by measuring other values functionally related to the required value, *e.g.*, divider, caliper, sine bar etc.

2. (*a*) **Absolute.** Here, the zero division of the instrument corresponds with the zero value of the measured dimension, *e.g.*, steel rules, vernier calipers etc.

(*b*) **Comparative.** Here, only the deviations of the measured dimensions from a master gauge are determined, *e.g.*, dial indicators, optimeters etc.

3. (*a*) **Contact.** Here, the measuring tip of the instrument actually touches the surface to be measured, *e.g.*, micrometers, calipers, dial indicators etc.

(*b*) **Contactless.** Here, no contact is required for measurement, *e.g.*, tool maker's microscope, projection comparator etc.

4. According to their functions, the measuring instruments are classified as :

(*a*) **Linear or length measuring instruments.** For example, steel rule, caliper, divider, micrometer, vernier caliper etc.

(*b*) **Angular or Angle measuring instruments.** For example, combination set, bevel protractor, sine bar, square, dividing head etc.

5. Depending on their accuracy, the measuring instruments can be grouped into three categories :

(*i*) Most accurate group includes light interference instruments.

(*ii*) Second group includes : Optimeters, tool maker's microscope and dial comparators.

(*iii*) Third group includes : Dial indicators, vernier calipers, micrometers and rules with vernier scales.

6. Measuring instruments can also be classified in accordance with their meterological properties, such as range of measurement scale, graduation value, scale spacing, sensitivity and reading accuracy.

The various terms connected with the measuring instruments can be defined as given below :

(*i*) **Range of measurement.** It indicates the size values between which measurements can be made on the given instrument, *e.g.,* micrometers are usually available for the following size ranges : 0 to 25, 25 to 50, 50 to 75, 75 to 100, 100 to 125 and 125 to 150 mm.

(*ii*) **Scale division value.** It is the measured value corresponding to one division of the instrument scale, *e.g.*, the scale division value of a micrometer is 0.01 mm etc.

(*iii*) **Scale spacing.** It is merely the distance between the axes of two adjacent graduations on the scale.

(*iv*) **Sensitivity.** The sensitivity of an instrument is the ratio of the scale spacing to the scale division value. If on a dial indicator, the scale spacing is 1.5 mm and the scale division value is 0.01 mm, then sensitivity is 150. Sensitivity is also called 'amplification factor' or 'gearing ratio.'

The reading accuracy of an instrument depends to a large extent on the contact pressure. To prevent error due to variation of contact pressure, special devices, which automatically adjust the contact pressure are incorporated in the design of measuring instrument, e.g., the ratchet stop of a micrometer.

The measuring accuracy also depends upon variations in temperature during measuring. Due to this, when taking measurements, the job and the instrument must be at the same temperature.

12.1.1. Types of Length Standard. A length may be expressed as the distance between two lines or as the distance between two faces. So, the instruments used for the direct measurement of linear dimensions fall into two categories :

(*i*) Line standards

(*ii*) End standards.

(*i*) **Line Standards.** Here, the measurement is made between two parallel lines engraved across the standard. The most common example of line standard or line measurement is the 'rule' with its divisions shown as lines marked on it.

(*ii*) **End Standards.** Here, the measurement is made between two flat parallel faces. Examples are : slip gauges, end bars, micrometers, vernier calipers etc.

Line standards have the drawback that the engraved lines themselves possess thickness. Also, the assistance of a magnifying glass or a microscope is required if sufficient accuracy is to be obtained. Greater accuracy can be obtained with the use of end standards as compared to line standards. Due to this, end standards are employed in the shop as far as possible.

12.2. CALIPERS

Calipers are used to pick off diameters or distance from a piece of work. This setting is then measured with a scale, vernier caliper or micrometer. There is no provision for reading the calipers directly. That is why they are known as 'transfer measuring instruments'. Calipers can also be used in the reverse order. The dimensions may be set with a scale or a micrometer and then checked against the work piece. The technique of measuring with calipers depends upon the skill of the user.

Calipers are designated as : outside, inside, and hermaphrodite. They may also be classified as : spring, firm joint, lock joint, and transfer calipers. The various calipers are shown in Fig. 12.1.

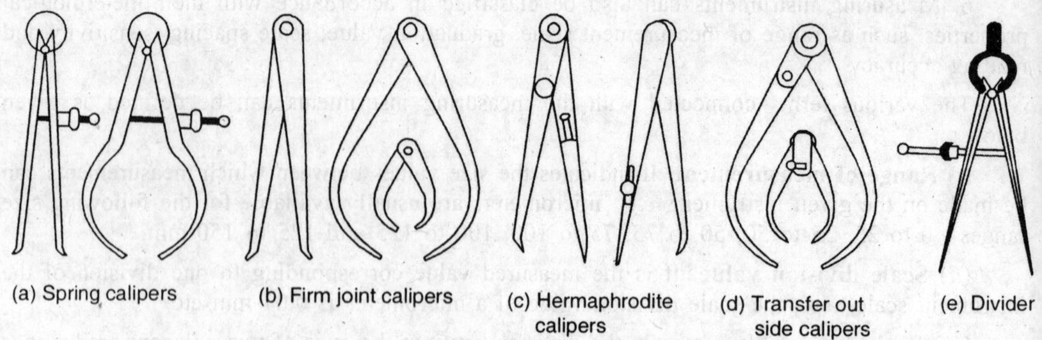

Fig. 12.1. Calipers

Outside caliper are used for measuring or comparing thickness, outside diameter and other outside dimensions. Inside calipers are used for measuring or comparing hole diameters, distance between shoulders or other parallel surfaces or any inside dimension. Hermaphrodite calipers have one leg bent and the other leg equipped with scriber. Distances from the edges of a work piece may be scribed or measured with this caliper. In the spring joint caliper, the spring holds the legs apart while adjustments are made with the adjusting nut. The firm joint caliper is set by tapping the outside of one leg while closing or opening the inside of the V formed by the two legs. The lock joint permits adjustment by turning the small knurled nut after the lock nut is secured. The transfer joint caliper permits setting the caliper to size, destroying the setting and resetting by opening the legs until the grooved nut enters the slot of the transfer arm. Dividers are used for taking accurate measurements or scribing arcs. They are used mainly for the layout of a work piece. They are equipped with two scribers.

12.3. VERNIER CALIPERS

When a component has to be measured to the second place of decimal, Vernier Calipers and micrometers are employed. A Vernier Scale is a scale in which use is made of the difference between two scales which are nearly alike for obtaining small differences. One scale (main scale) is fixed and the other scale (Vernier Scale) is sliding on the main scale. When the zero on the Vernier Scale coincides with the zero on the main scale, the number of divisions on the Vernier Scale is one more or one less than the number of divisions on main scale with which it coincides exactly.

In metric systems, there are two variations of the Vernier Scale :

1. Where the main scale is graduated into mm and 0.5 mm.
2. Where the main scale is graduated in whole mm only.

A relation can be derived between the size of division on main scale and the size of a division on Vernier Scale.

Let, C = size of division on main scale

C_v = size of a division on Vernier Scale

n = number of total divisions on the Vernier

Now, when the zeros on the two scales coincide, we have

$$(n-1)\,C = n.C_v$$

$$\therefore \qquad C_v = \frac{n-1}{n}.C$$

and

$$C - C_v = \frac{C}{n}.$$

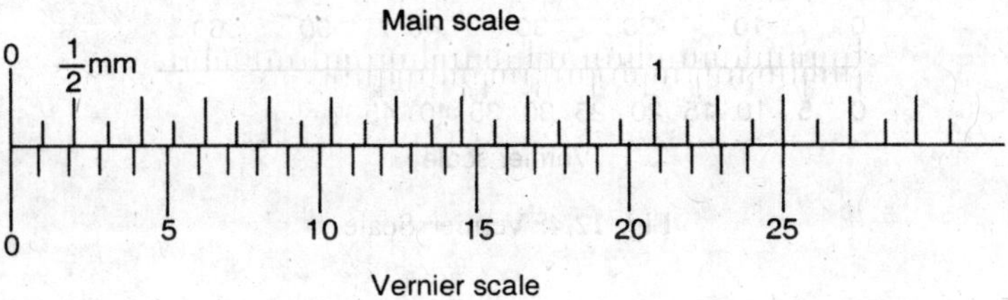

Fig. 12.2. Vernier Scale.

Let us take the first case, where the main scale is graduated into mm and 0.5 mm, Fig. 12.2. It is clear that 25 divisions on Vernier Scale coincide with the 12 mm graduation (24 dimensions) on the main scale. Therefore, a division on Vernier Scale is,

$$C_v = \frac{24}{25} \times 0.05 = 0.48 \text{ mm}$$

$$\therefore \qquad C - C_v = 0.02 \text{ mm}$$

which represents the accuracy to which the reading may be taken.

A dimension on a vernier caliper is read as follows : Note the reading on main scale against zero mark on vernier scale. Then count the number of divisions on vernier scale which coincides with a division (mark or graduation) on main scale. Then

Total reading = opening of jaws of vernier caliper
= Reading on main scale against zero mark on vernier scale + 0.02 × mark number on vernier scale which coincides with a mark on main scale.

For example, see Fig. 12.3.

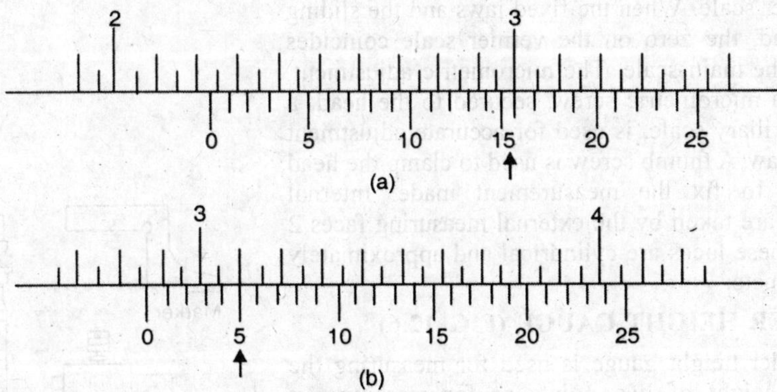

Fig. 12.3. Vernier Scale.

$$\text{Reading } (a) = 22 + 0.02 \times 16 = 22.32 \text{ mm}$$

$$\text{Reading } (b) = 28.5 + 0.02 \times 5 = 28.60 \text{ mm}$$

when the main scale is graduated into whole mm, then to get an accuracy of reading of 0.02 mm, the vernier is obtained by dividing 49 mm on the main scale into 50 divisions on the vernier scale, Fig. 12.4.

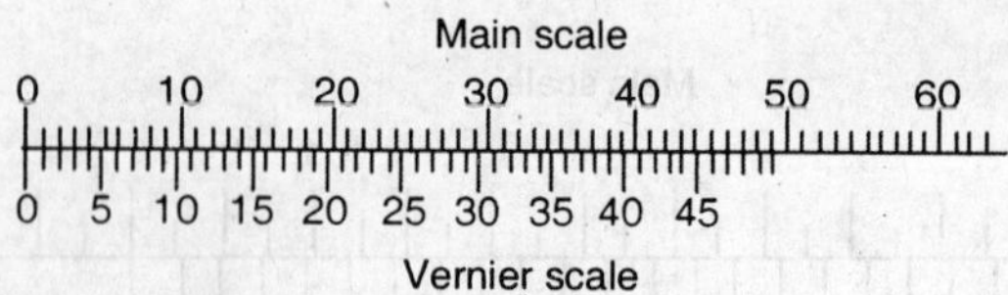

Fig. 12.4. Vernier Scale

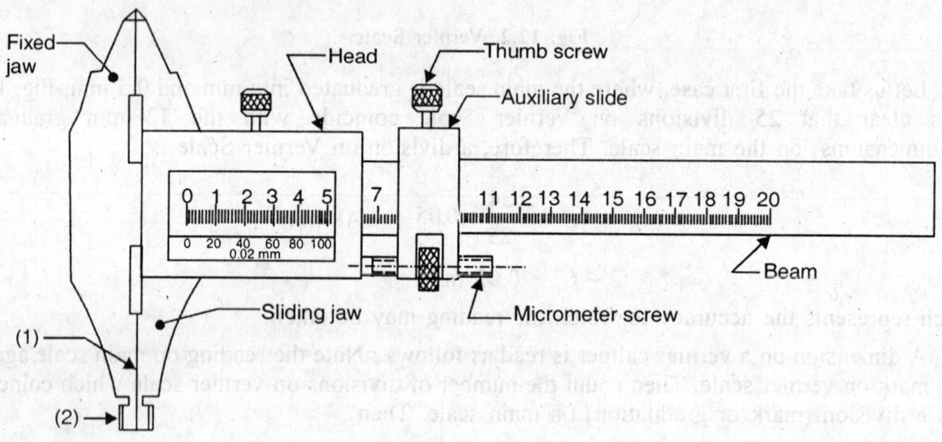

Fig. 12.5. Vernier Caliper.

A vernier caliper is shown in Fig. 12.5. It consists of a steel beam ending in the double fixed jaw with the measuring face 1. It is graduated in the main scale divisions. The fixed jaw is either integral with or welded to the beam. The head slides along the beam and carries the integral double sliding jaw whose measuring face is parallel to that of the fixed jaw. The sliding head carries a vernier plate with the vernier scale. When the fixed jaws and the sliding jaws are closed, the zero on the vernier scale coincides with zero on the main scale. The micrometric adjustment, consisting of a micrometric screw, secured to the head, a nut and an auxiliary scale, is used for accurate adjustment of the sliding jaw. A thumb screw is used to clamp the head on the beam to fix the measurement made. Internal measurements are taken by the external measuring faces 2 of the jaws. These faces are cylindrical and approximately 10 mm in diameter.

12.4. VERNIER HEIGHT GAUGE (FIG. 12.6)

The vernier height gauge is used for measuring the difference in height of two points or for marking out purposes. It is practically similar to a vernier caliper with its fixed jaw set rigidly into a base. Referring to Fig. 12.6, the base is a flat support. The main scale is carried on the vertical beam. A sliding arm which carries the vernier scale, has a clamp for holding either a jaw or a marker. The jaw has two measuring faces for measuring internal and external height dimensions. The distance '*d*' between the two faces must be taken into account when taking measurements.

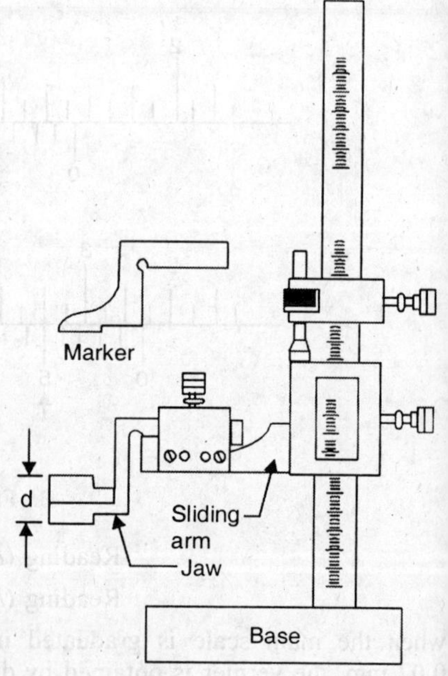

Fig. 12.6. Vernier Height Gauge

12.5. VERNIER DEPTH GAUGE

A vernier depth gauge, Fig. 12.7, is used to measure the depths of holes and the heights of shoulders in holes. It consists of a sliding head that slides on a beam or rule which carries the main scale. The vernier scale also slides on the rule, along with the head. The sliding head can be clamped at any desired position on the rule. In use, the flat end face of the rule or beam is made to contact the bottom of a blind hole whose depth is to be measured. Then, the base surface of the sliding head is brought into contact with the end surface from which the hole depth is to be measured. The depth is then determined with the help of main scale and the vernier scale.

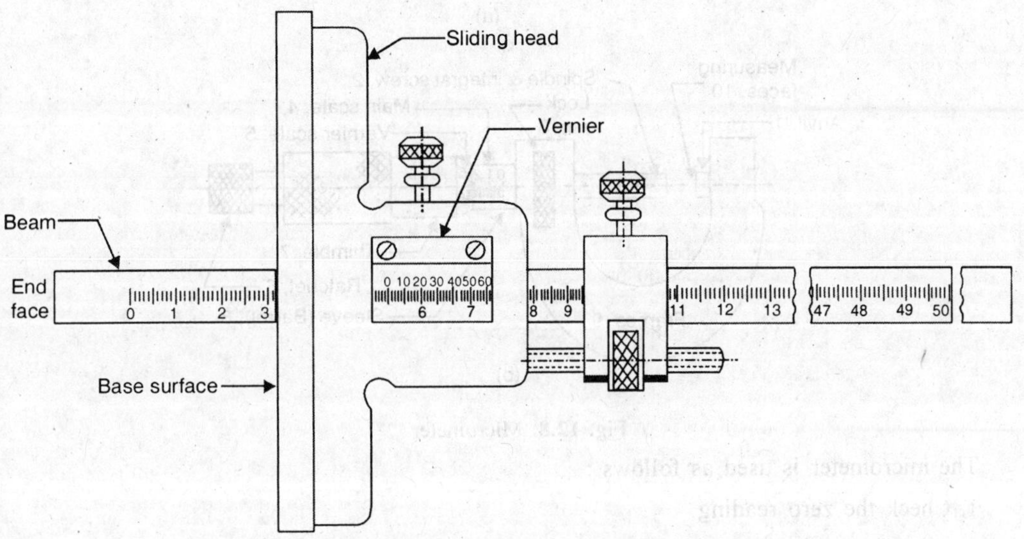

Fig. 12.7. Vernier Depth Gauge.

12.6. MICRO-METER CALIPER

The micrometer type tools are used to measure external and internal dimensions, as well as depths and heights. The principle of such tools consists of the employment of a screw and a nut, both having accurately cut threads. One complete revolution of the screw will advance it, in relation to the nut threads, a distance equal to the lead of the thread.

Fig. 12.8 (*a*) illustrates the principle of the micrometer caliper commonly known as 'Micrometer'. The micrometer screw 2 rotates in the fixed nut 11 and moves in an axial direction relative to the nut. The screw is rigidly attached to the thimble 7 while the fixed nut is fastened to the sleeve or barrel 8. Thus, the thimble rotates and moves axially along the barrel when the screw is rotated. The other main parts of the micrometer are shown in Fig. 12.8 (*b*). It has a U-shaped frame, to the left end of which an anvil is attached and to the right end. a barrel or sleeve is attached. A spindle 2-(integral with screw) passes through the barrel and a thimble which surrounds the barrel or sleeve is attached to the right end of measuring faces of the micrometer. A linear scale with graduations in mm and 0.5 mm is engraved on the barrel. This is the main scale. A circular scale with 50 divisions is engraved on the bevelled surface of the thimble. This is the vernier scale. If the micrometer screw has a pitch of 0.5 mm, then one turn of thimble is equal to 0.5 mm. As, there are 50 equal divisions on the bevelled edge of the thimble, the axial movement of the spindle and screw per division of the thimble scale will be equal to $\frac{0.5}{50} = 0.01$ mm. This is the scale division value and is known as the least count of the micrometer.

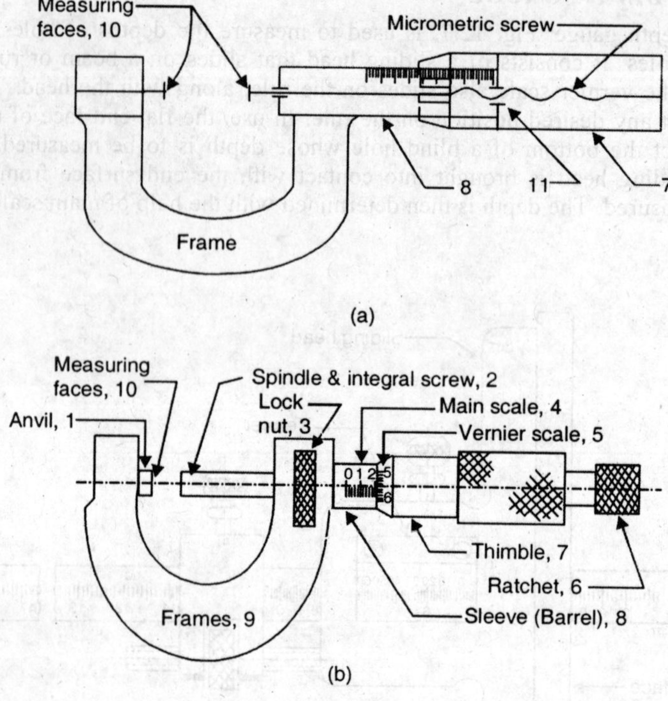

Fig. 12.8. Micrometer

The micrometer is used as follows :

1. Check the zero reading.

2. Place the part to be measured in between the measuring faces.

3. Advance the spindle by rotating the ratchet until it begins to slip and clicks are heard. This indicates that there is no further movement of the spindle.

4. Note the reading both on barrel scale and on the circular scale of the thimble. The barrel scale (main scale) indicates whole millimetres while the thimble scale (vernier scale) indicates tenths and hundredth of a millimeter.

Then

Reading on micrometer = Reading uncovered on the barrel + 0.01 × number of divisions on thimble scale which coincides with horizontal line on the barrel.

Thus reading in Fig. 12.9 is = 11.50 + 0.01 × 47 = 11.97 mm

A micrometer caliper is checked and adjusted to the zero position by means of gauge blocks, standards, or end bars. Micrometers are usually available in the ranges of 0 to 25, 25 to 50, 50 to 75, 75 to 100, 100 to 125 and 125 to 150 mm ranges.

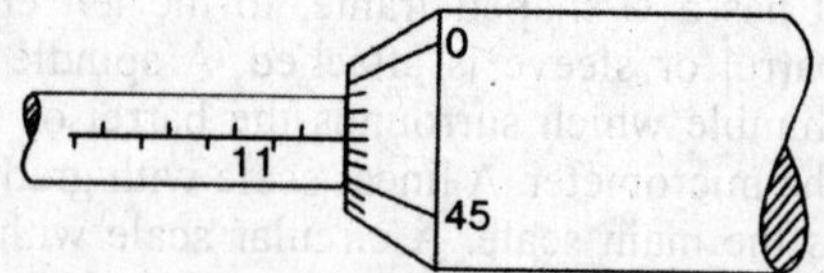

Fig. 12.9

12.6.1. Vernier Micrometer. The measurement to third degree of decimal can be made with a micrometer by applying Vernier Scale to it in addition to usual thimble and barrel graduations. The Vernier Scale is provided on the barrel or sleeve, Fig. 12.10.

Reading in Fig. 12.10

= 11.50 (reading of barrel) + 16 × 0.01 (reading on sleeve) + 6 × 0.001 (reading on Vernier)

= 10.50 + 0.16 + 0.006

= 10.666 mm.

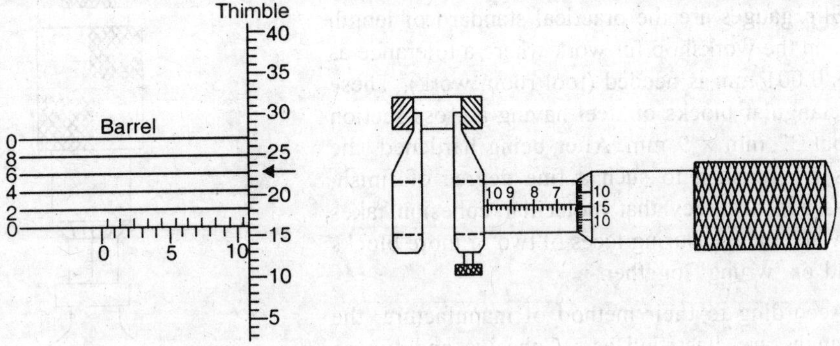

Fig. 12.10. Vernier Micrometer. **Fig. 12.11.** Inside Micrometer.

12.6.2. Inside Micrometer. This micrometer is used for measuring the internal dimensions and is graduated in hundredths of a mm. This micrometer is available in two designs :

1. One with jaws similar to Vernier Caliper and with the scale reading backwards, Fig. 12.11, and

2. The second is a straight bar with a micrometer barrel, Fig. 12.12. This type can be obtained with several interchangeable rods which allow a wide range of measurements.

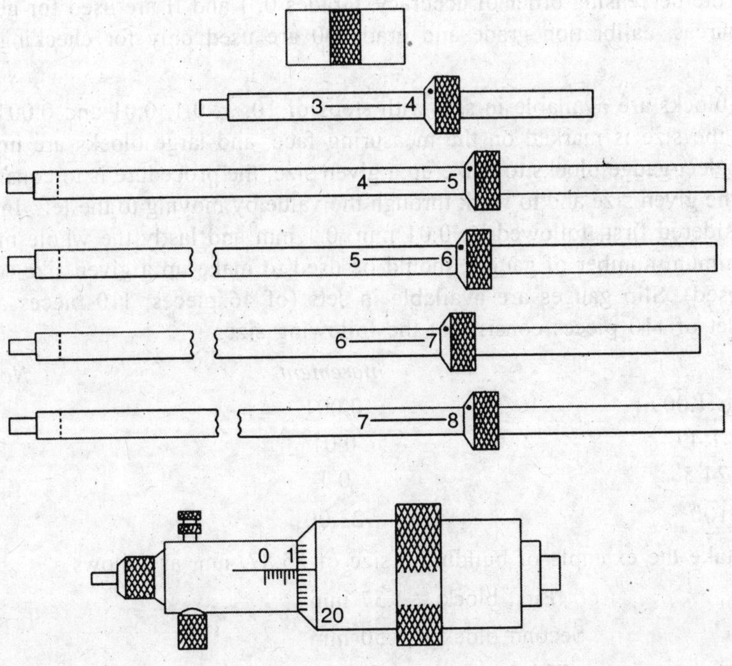

Fig. 12.12. Inside Micrometer.

12.6.3. Depth Micrometer. A depth micrometer, Fig. 12.13, has the same function as that of a Vernier depth gauge, that is to measure the depth of a slot or hole. They give the dimensions (depth) from some surface to the bottom of a slot or hole. Depth micrometers with the use of extension rods are available from 0 to 23 cm.

12.7. SLIP GAUGES

Slip gauges are the practical standard of length for use in the workshop for work where a tolerance as low as 0.001 mm is needed (tool room work). These are rectangular blocks of steel having a cross-section of about 32 mm × 9 mm. After being hardened, the blocks are finished to such a fine degree of finish, flatness and accuracy that molecular cohesion takes place when the measuring faces of two or more blocks are slid or 'wrung' together.

According to their method of manufacture, the slip gauges are classified as : Cohesive and 'wring together' types. The cohesive type is machine lapped with high precision so as to obtain a mirror like polished surface. The second type has a surface with a scratch pattern are more accurate than the "wring-together' type, but their surfaces wear rapidly and they become undersized. Therefore, these should be used only as reference measures.

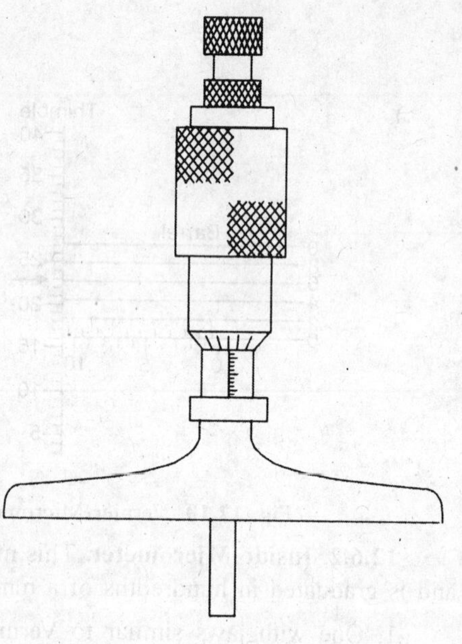

Fig. 12.13. Depth Micrometer.

Slip gauges are made in five grades of accuracy. Calibration grade, Grade 00, Grade I and Grade II, in the decreasing order of accuracy. Grades 0, I and II are used for general workshop purpose, whereas, calibration grade and grade 00 are used only for checking other types of blocks.

Gauge blocks are available in sets with steps of 10, 1, 0.1, 0.01 and 0.001 mm. On small size blocks, the size is marked on the measuring face, and large blocks are marked on a side surface. To select gauge blocks to make up a given size, the procedure is to consider the smallest unit first in the given size and to work through the value by moving to the left. In this way, 0.001 mm are considered first, followed by 0.01 mm, 0.1 mm and lastly the whole millimeters. As a rule, the minimum number of gauges should be used to make up a given size (generally 3 to 4 blocks are used). Slip gauges are available in sets (of 46 pieces, 110 pieces and so on). For example, a set of 110 pieces consists of the following sizes :

Size	*Increment*	*No. of pieces*
1.001 to 1.009	0.001	9
1.01 to 1.49	0.01	49
1.0 to 24.5	0.1	48
25 to 100	25.00	4

Let us take the example of building a size of 55.87 mm, as follows :

First block = 1.37 mm

Second block = 4.50 mm

Third block = 50.00 mm

Total = 55.87 mm

This is shown in Fig. 12.14.

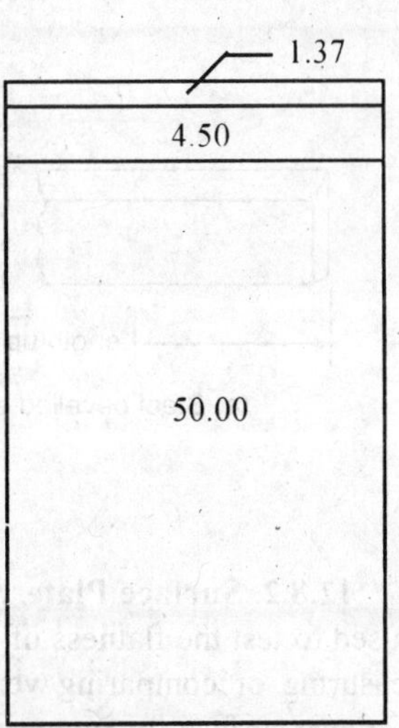

Fig. 12.14. Gauge Blocks.

Care of slip gauges. Because of their initial high cost and in order to preserve their accuracy, the slip gauges should receive great care during their use and also when not in use. Following points should be kept in mind regarding the care of slip gauges :

1. When not in use, the slip gauges should be kept in their case.

2. Before wringing the blocks together, ensure that their faces are perfectly clean.

3. After use, do not break the wring but slide one gauge over the other to separate them.

4. Slip gauges should be used in an atmosphere free from dust.

5. If, after use, the blocks are not to be used again for some time, a thin layer of good quality grease should be applied on their faces, before they are kept in their case.

Gauge blocks are used to check snap gauges, electrical and other comparators, optical inspection devices etc. The gauge blocks are checked with optical flats which tell the flatness of the gauge blocks as close as the wavelength of light. The finish of the blocks is about 0.05 μm Ra.

12.7.1. Length Bars. Slip gauges are usually used for measuring a linear dimension of about 250 mm. For longer dimensions, length bars or end measuring bars are used. The length bars are made from high grade carbon steel with their ends hardened and finished to a very fine accuracy. They are circular in section, about 22 mm in diameter. They have the same wringing property as the slip gauges. In addition, the ends of the bars are drilled and tapped for studs to connect them. Length bars are made in three grades : Reference grade, Inspection grade, and Workshop grade. The length bars are available in following lengths : 25, 50, 75, 100, 125, 150, 175, 200, 375, 575 and 775 mm.

12.8. CHECKING STRAIGHTNESS AND FLATNESS

Straightness and flatness of parts may be checked by means of (*a*) Straight Edges (*b*) Surface plates (*c*) Spirit levels, and (*d*) Optical flats.

12.8.1. Straight Edges. Straight edges may be checked by means of (*a*) Straight Edges (*b*) Surface plates (*c*) Spirit levels, and (*d*) Optical flats.

12.8.1. Straight Edges. Straight edges may be classified as : tool maker's straight edges and wide edge straight edges, Fig. 12.15. Tool maker's straight edges have the highest accuracy. They are available in lengths from 75 to 175 mm and with one to four working edges. With these edges, the straightness and flatness are checked by sight test. Straight edges with a single edge are mainly for checking straightness. For checking the straightness of an element of a cylindrical or taper surface, the straight edge is applied along the full length of the surface and is held before a bright background. The absence of light between the straight edge and work surface indicates the straightness of the element and vice versa. For checking the flatness, single edged as well as three and four-edged straight edges may be used. With single edge straight edge, it is applied in different directions at different places on the surface to be tested. The flatness of the surface is judged by the light showing through.

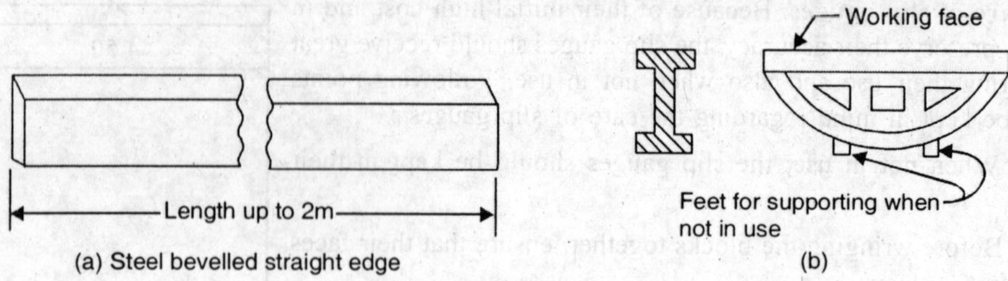

Fig. 12.15. Straight Edges.

12.8.2. Surface Plate. A surface plate or a flat plate has a highly accurate flat surface which is used to test the flatness of a surface. In addition, it supplies a datum plane from which scribing, measuring or comparing vertical distances can be done. For this the surface plate is used as reference surface for mounting mechanical comparators, optimeters, sine bars, slip gauge blocks, vernier height gauges and so on. The plate is usually a good quality casting, heavily ribbed for rigidity and with a thick top, Fig. 12.16. The surface plate is then made by planning, perhaps grinding the finally is finished by a hand operation called "scraping".

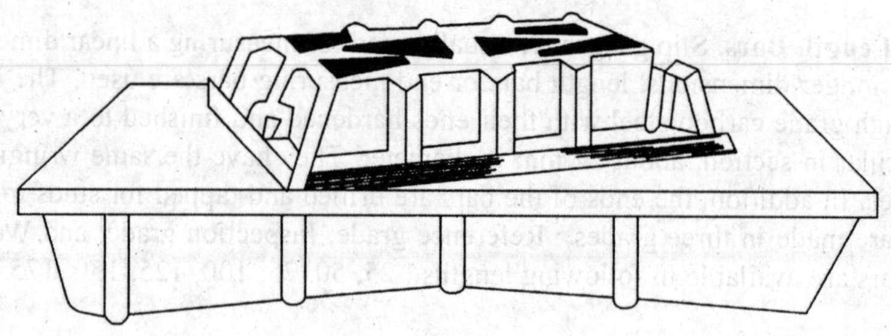

Fig. 12.16. Surface Plate.

One method of testing the flatness of a surface is to first rub the top of the plate with a thin smear of a paste made up of red lead and oil. The surface to be checked is then wiped clean and is placed on the top of the surface plate and moved about. If the surface is reasonable flat, the red spots will be visible all over it.

12.8.3. Optical Flats. Optical flats are used to check the flatness of work-pieces, gauge blocks, micrometer anvils etc. They are cylindrical pieces (usually), from 25 by 38 mm to 300 by 70 mm. These are made of a clear material, such as fused quartz (it wears considerably better than glass) and have two faces perfectly parallel to each other, which are accurate within a few millionth of a centimeter. Some optical flats have both the parallel faces ground and lapped, others are truly only on one side.

The principle of an optical flat is illustrated in Fig. 12.17. The optical flat is placed over the work piece to be checked so that a thin workpiece in turn is resting on a flat surface plate. Monochromatic workpiece in turn is resting on a flat surface plate. Monochromatic light (single wavelength) is now thrown on the optical flat and let a ray enters the flat at *A*. It is refracted along *AB* through the flat. At *B*, a part of this light is reflected from the bottom face of the flat along *BC* and it leaves the flat at *C*. The other part of light continues along *BD* (in air gap)

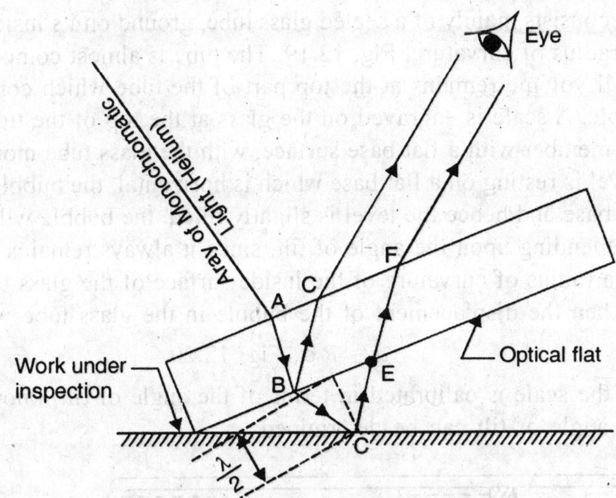

Fig. 12.17. Principle of Optical Flats.

and gets reflected from the work surface at *D*, along *DE*. After refracting through the flat (along *EF*) it leaves it at *F* parallel to the first part of the light. The second part of light travels an extra distance of (*BD* + *DE*) than the first part of light. Also, it is thrown out of phase by π radians when reflecting from the work surface at *D* (when light travelling in air along *BD* is reflected at the work surface, which is optically more dense than air, there is a change in phase of π radians). Now, if the air gap at *D* is half the wavelength of light $\left(\frac{\lambda}{2}\right)$, the second part of light travels $\frac{\lambda}{2}$ down (*BD*) and $\frac{\lambda}{2}$ up (*DE*) in air. Thus, this part of light will be one full wavelength behind and out of phase by π radians relative to the first part of light. Thus, the two portions of light leaving the optical flat at *C* and *F* will cancel each other and a dark band will be seen by the observer. This will happen wherever the air gap is zero or product of any whole number times the half wave-length. On the other hand, if air gap is $\frac{1}{4}$ wavelength, $\frac{3}{4}$ wavelength and so on, the two parts of light will reinforce each other and we will get a right band.

A trained inspector can tell the condition of the face of the work piece from the pattern of bright and dark bands. When these bands are straight, parallel and equally spaced, the surface is flat, [Fig. 12.18 (*a*)]. If the bands curve towards the contact edge, the surface is concave [Fig. 12.18 (*b*)] and if the bands curve away from the contact edge, it is convex [Fig. 12.18 (*c*)].

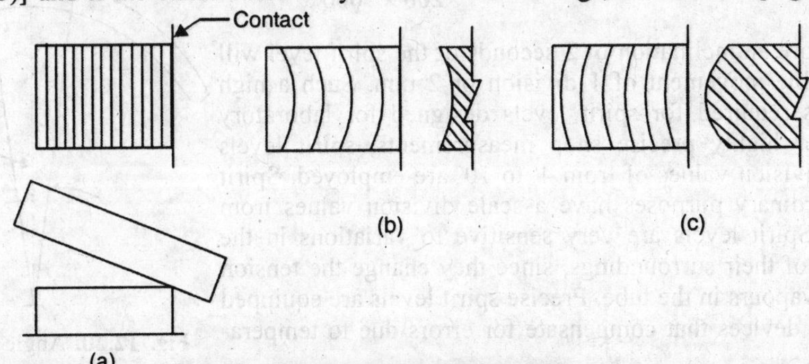

Fig. 12.18. Band Patterns.

12.8.4. Spirit Levels. Spirit levels are measuring devices which enable the position of a surface to be determined with respect to the horizontal. They are widely used for the static levelling of the machinery and other equipment. However, they can also be used for measuring small inclinations or angles.

A spirit level consists mainly of a sealed glass tube, ground on its inside surface to a concave form with a large radius of curvature, Fig. 12.19. The tube is almost completely filled with ether, so that only a small volume remains at the top part of the tube which contains ether vapours in the form of a bubble. A scale is engraved on the glass at the top of the tube. A machinists' level consists of a body member with a flat base surface, with the glass tube mounted in its upper part. When the spirit level is resting on a flat base which is horizontal, the bubble will rest at the centre of the scale. If the base and hence the level is slightly tilted, the bubble will move along the scale a small distance depending upon the angle of tilt, since it always remains at the highest point of the tube. If R is the radius of curvature of the inside surface of the glass tube and α is the angle of tilt in radians, then the displacement of the bubble in the glass tube will be,

$$d = R\,\alpha, \text{ Fig. } 12.20$$

Therefore, if the scale is calibrated in terms of the angle of tilt, knowing the displacement of the bubble, the angle of tilt can be determined.

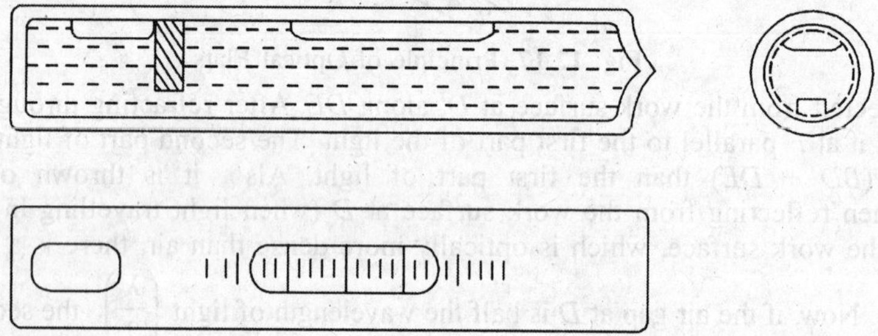

Fig. 12.19. Glass Tube of a Spirit Level.

The sensitivity of a spirit level is defined as the angle of tilt which will move the bubble through one division on the scale and is expressed in seconds per division. It increases as R increases. If α is expressed in seconds, then

$$d = \frac{R\alpha''}{206{,}265} \qquad \text{(1 radians = 206,265 seconds of arc)}$$

The scale spacing of a spirit level scale is usually 2 mm. So for a radius R = 206 m,

$$\alpha'' = \frac{206{,}265 \times 2}{206 \times 1000} \approx 2''$$

Therefore an inclination of 2 seconds of the spirit level will cause a bubble movement of 1 division or 2 mm. Such a high sensitivity is required for spirit levels designed for laboratory research. For highly precise shop measurements, spirit levels with scale division values of from 4′ to 10′ are employed. Spirit levels for ordinary purposes have a scale division values from 10′ to 40′. Spirit levels are very sensitive to variations in the temperature of their surroundings, since they change the tension of the ether vapours in the tube. Precise spirit levels are equipped with various devices that compensate for errors due to temperature changes.

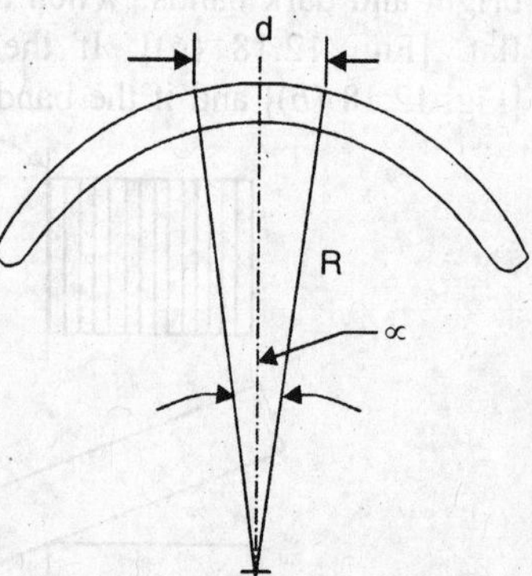

Fig. 12.20. Angle of tilt of a spirit level.

12.9. CHECKING SQUARENESS

The instruments commonly used to check the squareness of work-pieces are : (*i*) Try square (*i*) Combination square of combination set.

12.9.1. Try Square. The try square is the most common instrument for testing the squareness of a work piece. When using it, it should be ensured that its blade is held perpendicular to the surface being tested, Fig. 12.21.

12.9.2. Combination Square. The squareness can also be tested with the help of combination square of a combination set. The main part of a combination set is a straight steel blade to which different types of heads may be clamped, to enable the combination set to be used for different purposes. When a square head is attached to the steel blade, the combination set can be used as a square for checking squareness or for scribing 90° and 45° angles. Similarly, it can be used as a protractor for measuring and scribing other angles and with the help of a centre head, it can be used for locating and scribing the centre of a round bar. The combination square of the combination set is shown in Fig. 12.22. The square head also contains a bubble for levelling purposes. For checking the squareness to a high degree of accuracy, a try square is preferred to a combination square.

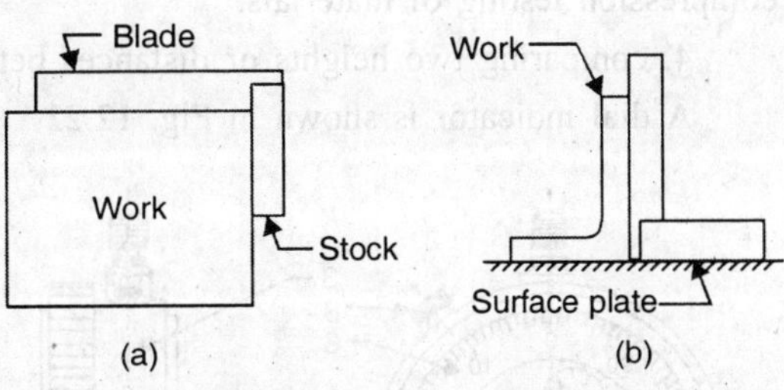

Fig. 12.21. Use of Try Square.

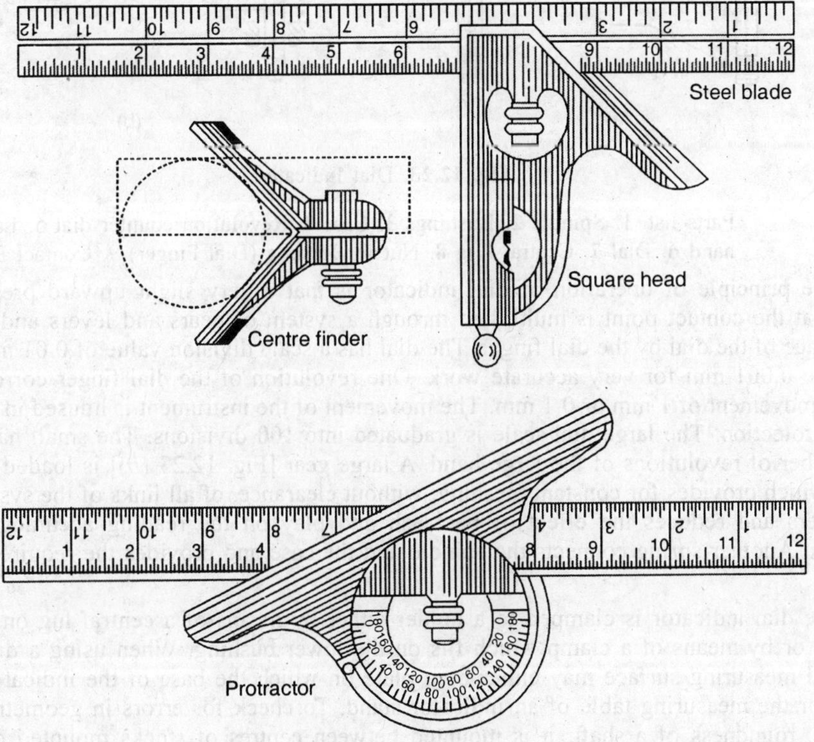

Fig. 12.22. Combination Square.

12.10. DIAL. GAUGE OR INDICATOR

A dial indicator (gauge) or clock indicator is a very versatile and sensitive instrument. It is used for :

1. determining errors in geometrical form, for example, ovality, out-of-roundness, taper etc.

2. determining positional errors of surfaces, for example, in parallelism, squareness, alignment etc.

3. taking accurate measurements of deformation (extension or compression) in tension and compression testing of materials.

4. comparing two heights or distances between narrow limits.

A dial indicator is shown in Fig. 12.23.

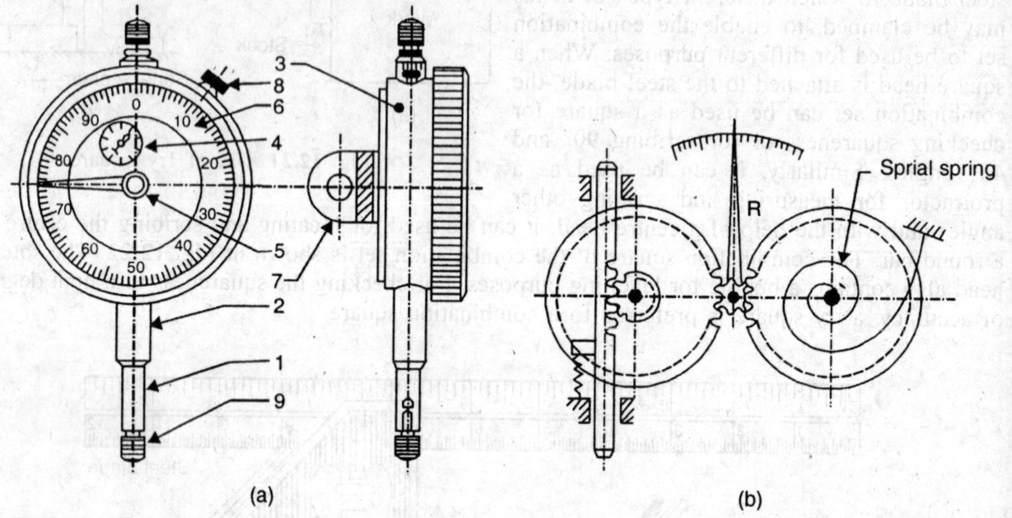

Fig. 12.23. Dial Indicator.

Parts list. 1. Spindle 2. Bushings 3. Case 4. Revolution counter dial 5. Large hand 6. Dial 7. Central Lug 8. Nut for Resting (Dial Finger) 9. Contact Point.

The principle of operation of dial indicator is that a very slight upward pressure on the spindle at the contact point is multiplied through a system of gears and levers and is indicated on the face of the dial by the dial finger. The dial has a scale division value of 0.01 mm (usually). It can be 0.001 mm for very accurate work. One revolution of the dial finger corresponds to a spindle movement of 1 mm or 0.1 mm. The movement of the instrument is housed in a metal case for its protection. The large dial scale is graduated into 100 divisions. The small hand indicates the number of revolutions of the large hand. A large gear [Fig. 12.23 (*b*)] is loaded by a helical spring which provides for constant meshing, without clearance, of all links of the system of gears and levers and reduces the effect of backlash and play on the reading accuracy of the dial indicator. Another spring connects the spindle and the case and provides the required measuring pressure.

The dial indicator is clamped to a holder either by means of a central lug on the back of the case or by means of a clamp which fits on the lower bushing. When using a dial indicator, the fixed measuring surface may either be a plate on which the base of the indicator test set is resting or the measuring table of an indicator stand. To check for errors in geometric form, for example, roundness of a shaft, it is mounted between centres of stocks mounted on a surface plate. Fig. 12.24. The indicator test set is also rested on surface plate and the contact point of the dial indicator is made to touch the shaft surface by moving the indicator up or down along the upright of the set. The shaft is then rotated by hand and any radial run out may be read on the dial. If the contact point is moved along the axis of the shaft, the dial band will indicate straightness errors.

A dial indicator may also be used for checking a dimension. For this, it is first set up with slip gauge blocks to a dimension near to that of the part being checked. For setting up, the hand is set to zero by rotating the scale. The slip gauge blocks are then removed from underneath the contact point and the part is placed there on the table of indicator stand. The dial hand will show the deviation of the size from the set up dimension.

The practical applications of the use of dial indicator are :

1. To check alignment of lathe centres by using a suitable accurate bar between centres.

2. To check trueness of milling machine arbors.

3. To check parallelism of the shaper ram with table surface or vice.

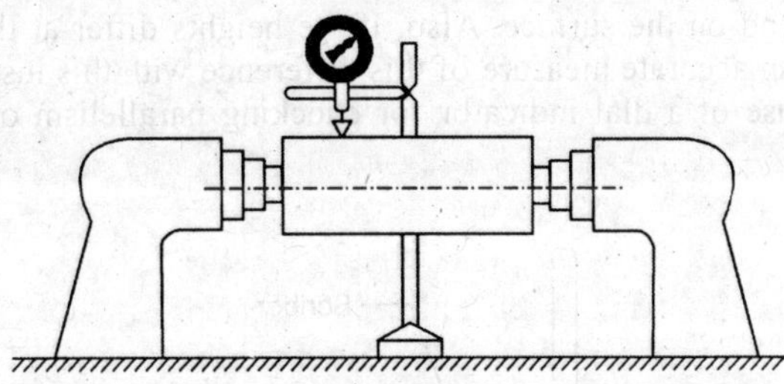

Fig. 12.24

12.11. SURFACE GAUGE

A surface gauge, Fig. 12.25, or scribing block is an indirect measuring instrument. It is not very accurate as it depends upon a separate steel scale for actual values. It is always used in connection with a surface plate. The scriber of the gauge has one end bent at right angles. The scriber is clamped in a vertical rod (upright) which is supported on the base of the gauge. The scribe can be moved up or down along the vertical rod. The surface gauge is used for the following purposes :

1. For scribing vertical distances from a datum plane (usually a surface plate).
2. For checking parallelism of surfaces (discussed in the next article).

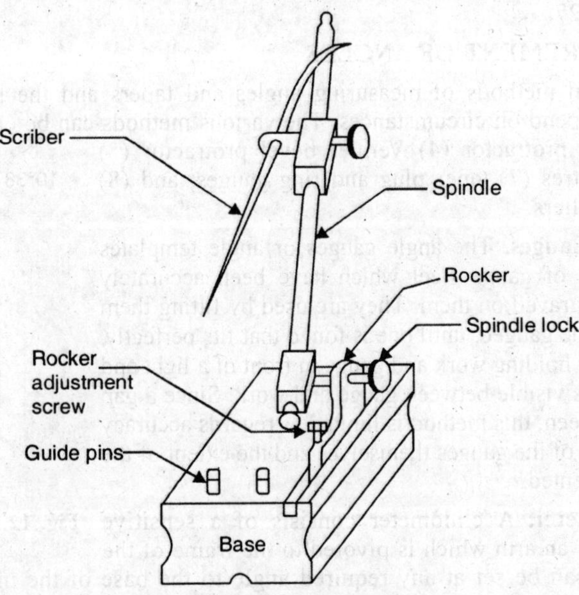

Fig. 12.25. Surface Gauge.

12.12. CHECKING OF PARALLELISM

The two commonly used instruments for checking parallelism of surfaces are :

1. Surface Gauge, and
2. Dial Indicator.

The use of surface gauge for checking the parallelism of a surface is shown in Fig. 12.26. The bent end of the scriber is moved along the surface to be checked. The draw-back of surface gauge is that accuracy of measurement depends upon the sensitiveness of our feel with the heat end on the surface. Also, if the heights differ at the ends of the faces being tested, we can't get an accurate measure of this difference with this instrument. These drawbacks are overcome by the use of a dial indicator for checking parallelism of surfaces.

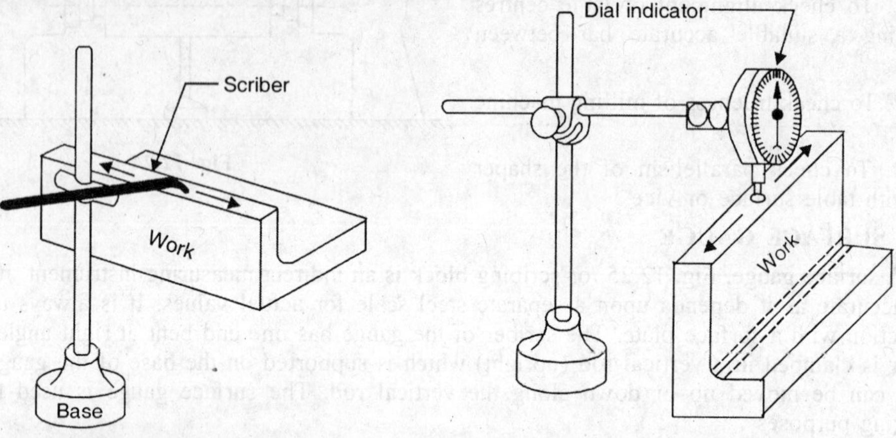

Fig. 12.26. Checking of parallelism with surface gauge.

Fig. 12.27. Checking parallelism with dial indicator.

A slight upward pressure on the plunger moves it upward and the movement is indicated on a dial gauge, Fig. 12.27.

12.13. THE MEASUREMENT OF ANGLES

There are several methods of measuring angles and tapers and the method used in any particular case will depend on circumstances. The various methods can be (1) Angles gauges (2) Clinometer (3) Bevel protractor (4) Vernier bevel protractor (5) Sine bar (6) Sine centres (7) taper plug and ring gauges, and (8) Precision balls and rollers.

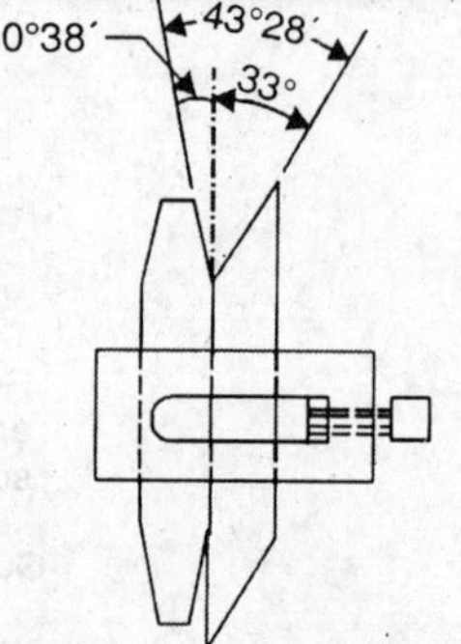

Fig. 12.28. Angle Gauges.

12.13.1. Angle Gauges. The angle gauges or angle templates Fig. 12.28, are pieces of gauge steel which have been accurately formed to the angle engraved on them. They are used by fitting them in turn on the piece to be gauged, until one is found that fits perfectly. The fitting is judged by holding work and gauge in front of a light and observing if any light is visible between gauge and work. Since a gap of 0.025 mm is easily seen, this method is limited as regards accuracy chiefly by the accuracy of the gauges themselves and the extent of the range of angles represented.

12.13.2. Clinometer. A clinometer consists of a sensitive bubble tube carried by an arm which is pivoted to the frame of the instrument. The arm can be set at any required angle to the base of the instrument by means of a worm which engages a toothed arc attached to the arm. The worm spindle is carried in the frame. The angle may be read off from a scale attached to the arm in conjunction with a graduated micrometer thimble carried by the worm spindle. Clinometers are much used to enable the surface of a piece of work to be set at a required angle to either a marking-out table or a machine work table as follows. The clino meter is placed on the table and is adjusted until the bubble is central, the scale reading is then observed. The clinometer is then adjusted until it gives the same reading plus the required angle. It is then placed on the surface of the work, which is being set, and the work is adjusted on the table until the bubble is again central. Precision clinometers can read direct to 3 seconds of arc. In some clinometers, the inclination of the bubble tube is read directly off a glass tube by means of a microscope eyepiece.

12.3.3. Bevel protractor. The protractor of the combination set, Fig. 12.22 can be used to measure angles. It has a scale of degrees marked on it. This type of instrument is not suitable for accurate work, since the scale is very fine and the degrees are not subdivided.

12.3.4. Vernier Bevel Protractor. For angular measurements, vernier bevel protractor is also commonly used, Fig. 12.29. The main scale is circular and is graduated in degree. The vernier scale has a least count of usually 5 minutes of a degree (5′).

The graduations are read to the right and left on both the main scale and the vernier scale. When the main scale is read to the right of zero of the main scale, the vernier is also read to the right of zero of the vernier scale. One straight edge is rotated relative to another straight edge and the angular relationship is read with the help of main and vernier scale. These adjustable angle gauges are not so accurate, being subjected to errors (*a*) in the graduated scale (*b*) in the reading of the scales (*c*) due to the eccentricity of scale and pivot.

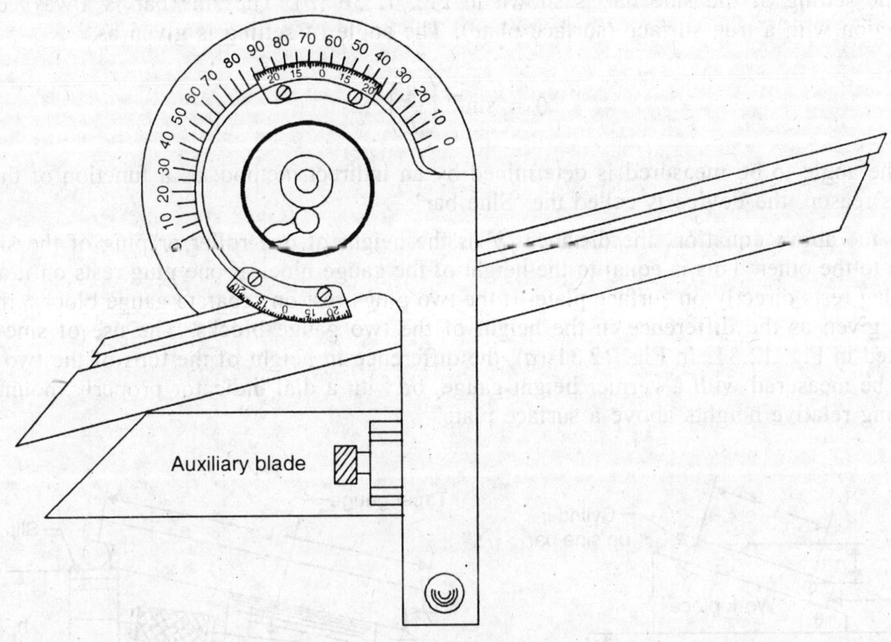

Fig. 12.29. Vernier Bevel Protractor.

12.13.5. Sine bar. Under shop conditions, the sine bar is the most accurate instrument for measuring angles. It consists of an accurate straight bar, Fig. 12.30 (*a*), in which two accurately lapped cylindrical plugs are located with extreme precision. In design (I), the plugs project about 12 mm from the front face of the bar and in design (II), the ends of the bar are stepped and the plugs are secured into each step by a screw. The centre to centre distance of the plugs is 100, 200, 250 or 300 mm. The following conditions must be satisfied to close tolerances :

1. The plugs must be of the same diameter.

2. The centre distance between the plugs must be absolutely correct, and the centre lines of two plugs must be parallel.

3. One surface (the measuring surface) of the sine bar must be absolutely parallel to the centre line of the plugs.

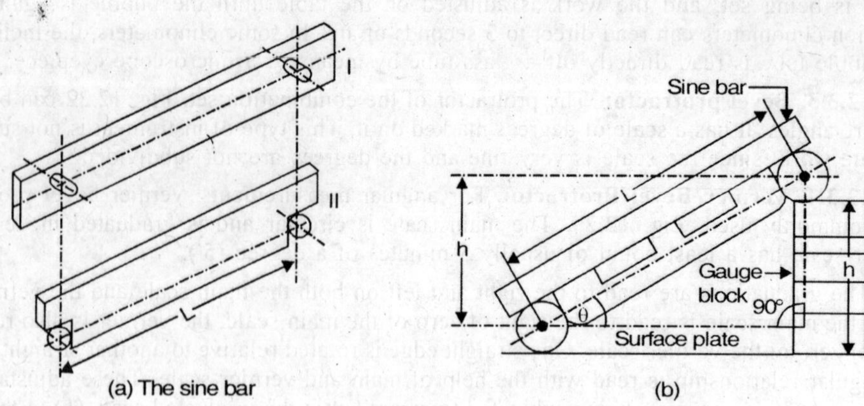

Fig. 12.30. Sine Bar.

The setting of the sine bar is shown in Fig. 12.30 (*b*). The sine bar is always used in conjunction with a true surface (surface plate). The angle of setting is given as,

$$\theta = \sin^{-1}\left(\frac{h}{L}\right).$$

The angle to be measured is determined by an indirect method, as a function of the sine. For this reason, the device is called the 'Sine bar'.

In the above equation, the distance '*h*' is the height of one roller or plug of the sine bar relative to the other. This is equal to the height of the gauge block if one plug rests on it and the other plug rests directly on surface plate. If the two plugs rest on separate gauge blocks, then '*h*' will be given as the difference in the height of the two gauge blocks. The use of sine bar is illustrated in Fig. 12.31. In Fig. 12.31 (*a*), the difference in height of the tops of the two plugs, *h*, may be measured with a vernier height gauge, or with a dial indicator properly mounted for measuring relative heights above a surface plate.

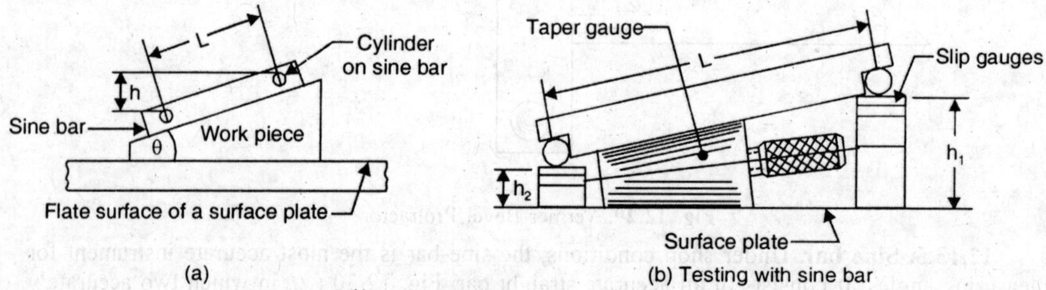

Fig. 12.31. Use of Sine Bar.

In Fig. 12.31 (*b*), the taper angle of the taper plug gauge will be given as,

$$2\theta = \sin^{-1}\left(\frac{h_1 - h_2}{L}\right).$$

12.13.6. Sine Centre. Sine centres are extremely useful for the testing of conical work, since the centres ensure correct alignment of the work piece. The procedure of its setting is the same as for sine bar.

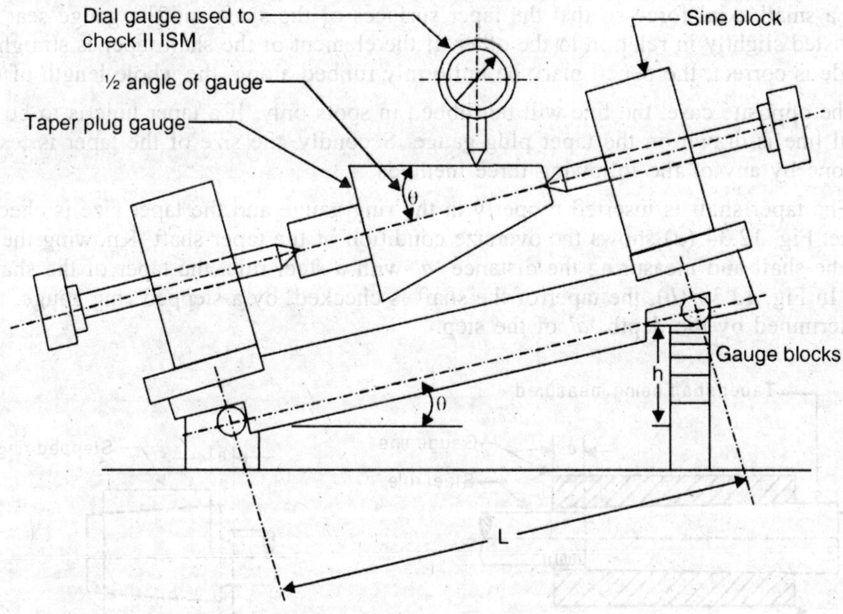

Fig. 12.32. Sine Centres.

In Fig. 12.32, the included angle of a taper plug gauge is to be checked. The sine block is set to half the included angle of the gauge. When the parallelism has been established using a dial test indicator and adjusting the height of the gauge blocks, the half included angle will be,

$$\theta = \sin^{-1}\frac{h}{L}$$

Total inclusive angle = 2θ

12.13.7. Taper Plug and Ring Gauges. A taper is difficult to measure. Various taper gauges are shown in Fig. 12.33. To check a taper, two tests are made. Firstly (let us take the case of a taper shaft), a straight line is drawn with a chalk, red lead, prussian blue, or pencil

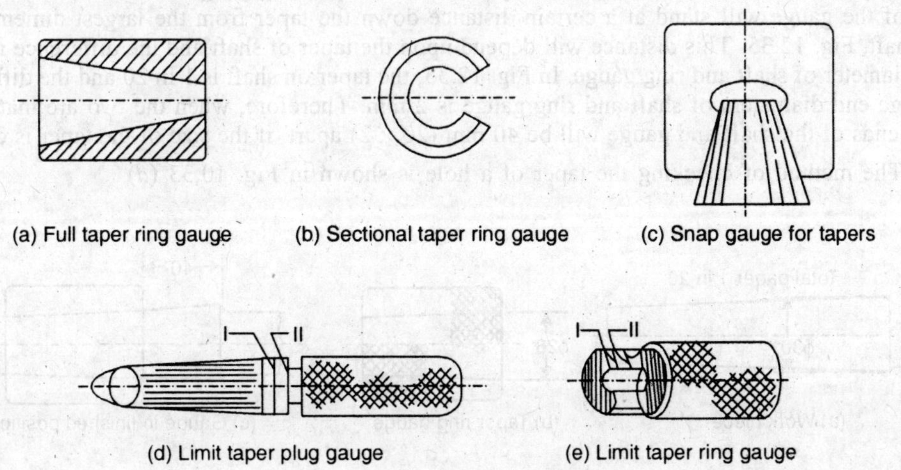

Fig. 12.33. Taper Gauges.

along an element of the taper. Then the shaft is carefully inserted into a taper ring gauge. After applying a small axial force so that the taper surfaces of the shaft and the gauge seat properly, one is twisted slightly in relation to the other. If the element of the shaft taper is straight and the taper angle is correct, the pencil mark be uniformly rubbed along, the whole length of the taper.

In the opposite case, the line will be rubbed in spots only. If a taper hole is to be checked, the pencil line is drawn on the taper plug gauge. Secondly, the size of the taper is tested. This can be done by any of the following three methods :

1. The taper shaft is inserted properly in the ring gauge and the taper size is checked with gauge line. Fig. 12.34 (*a*) shows the oversize condition of the taper shaft. Knowing the required taper of the shaft and measuring the distance '*a*' with a steel rule, the taper of the shaft can be checked. In Fig. 12.34 (*b*), the taper of the shaft is checked, by a stepped ring gauge, the limits being determined by the depth '*a*' of the step.

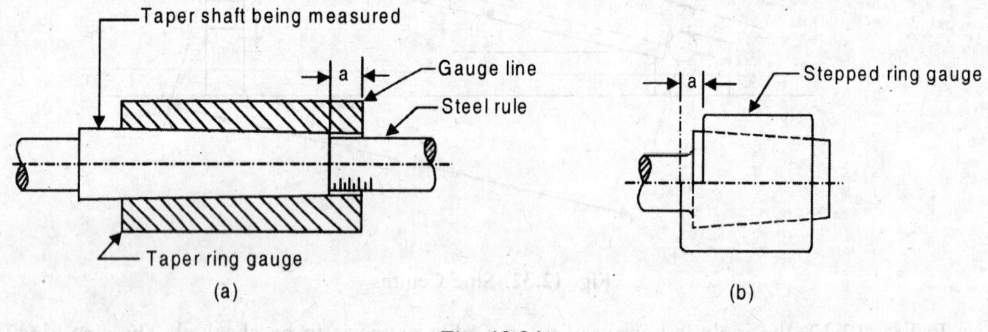

Fig. 12.34

2. In another method, two lines I and II [Fig. 12.33 (*d*) and 12.33 (*e*)] are provided on the gauge. A properly machined taper surface will seat so that its end surface is located between the lines of the limits taper gauge.

3. In the third method, a line is marked around taper plug gauge (say) at the larger diameter. The size of the gauge at this line is equal to the size of the hole (it is going to check) at its larger end face. When the taper plug gauge is inserted in the hole and is pushed forward, the line on the gauge will be flush with the large end face of the hole if its size is correct. When checking taper shafts, the large end diameter of the ring gauge is made smaller than the large end diameter of the taper shaft. Due to this, when the taper shaft and taper ring gauge are mated, the large end face of the gauge will stand at a certain distance down the taper from the largest dimension of the shaft, Fig. 12.35. This distance will depend upon the taper of shaft and the difference in large end diameter of shaft and ring gauge. In Fig. 12.35, the taper on shaft is 1 in 20 and the difference in large end diameters of shaft and ring gauge is 2 mm. Therefore, when the two are mated, the large ends of the shaft and gauge will be 40 mm (20 × 2) apart, if the size of the taper is correct.

The method of checking the taper of a hole is shown in Fig. 10.33 (*d*).

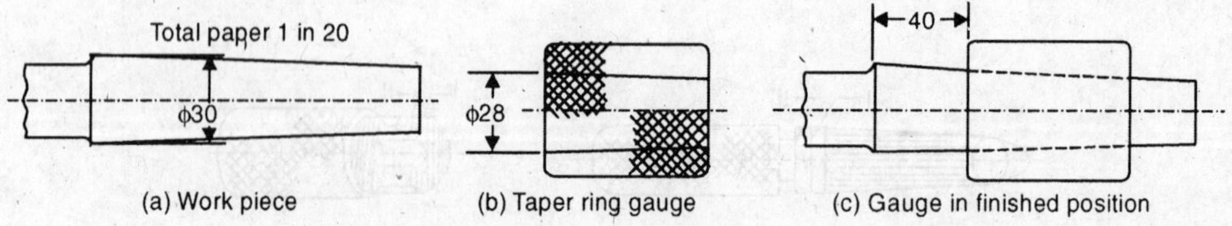

Fig. 12.35. Checking a Taper Shaft.

12.13.8. Precision Balls and Rollers. Precision balls and rollers are used to determine both linear and angular dimensions in conjunction with slip gauges. These are made of good quality steel and are hardened and tempered. The length of the roller is equal to its diameter and both the dimensions are within two micrometers of the stated size. The balls and rollers are available in sizes ranging from 1 to 25 mm diameter. The use of precision balls and rollers for determining both linear and angular dimensions is explained with the help of the following examples :

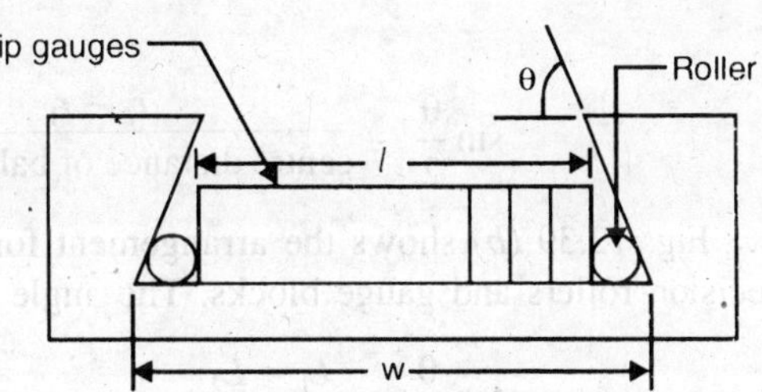

Fig. 12.36. External Slide.

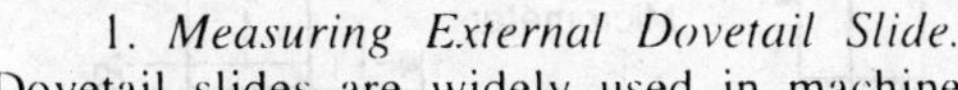

1. *Measuring External Dovetail Slide.* Dovetail slides are widely used in machine tool construction. Fig. 12.36 shows an external dovetail slide with angle of dovetail θ. It is required to check the width of the opening, *w.* Firstly, two rollers of equal diameter, *d*, are placed one each in the two corners. Then, the length, *l*, is obtained by hit and trial with the help of slip gauges (or end bars if *l* is greater than 250 mm). Then, the width, *w*, can be calculated by the relation:

$$w = l + d + d \cot \frac{\theta}{2}$$

2. *Measuring Internal Dovetail Slide.* Fig. 12.37 shows an internal dovetail slide. It is required to check the width of the opening. The same procedure is adopted as for external slide. Then, the width of the opening will be given as :

$$w = l - d - d \cot \frac{\theta}{2}$$

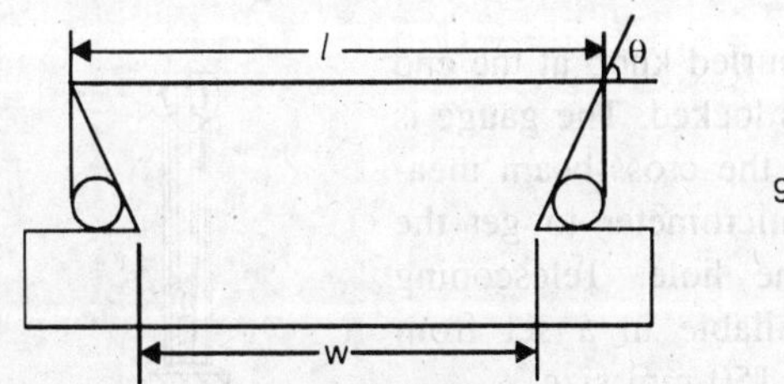

Fig. 12.37. Internal dove-tail

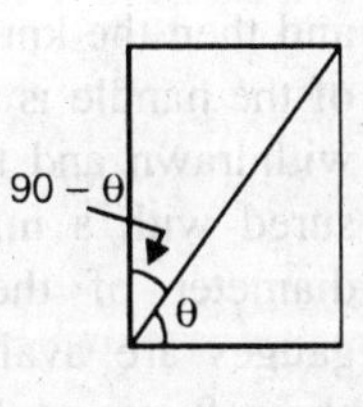

Here, the slip gauges will be held in a cage to get the length '*l*'.

3. *Checking the angle of Dovetail (Fig. 12.38).* In the above two examples, the dovetail angle has been assumed to be within limits. Here, we will see how to check the dovetail angle. First, two rollers of equal diameters are placed, one each at the two corners and distance 'l_1' determined with slip gauges. Then the rollers are placed on equal size slip gauges and distance l_2 determined. The dovetail angle, θ, will be give as :

$$\theta = \tan^{-1}\left[h / \frac{1}{2}(l_2 - l_1)\right]$$

4. *Checking the taper angle of gauges.* Fig. 12.39 (*a*) shows the arrangement for checking the internal taper of a taper ring-gauge using precision balls. The taper ring-gauge is placed on a surface plate and a small ball is positioned close to the small end of the taper. Two piles of slip gauges of equal height are then placed on either side. A depth micrometer is then used to determine the distance from the

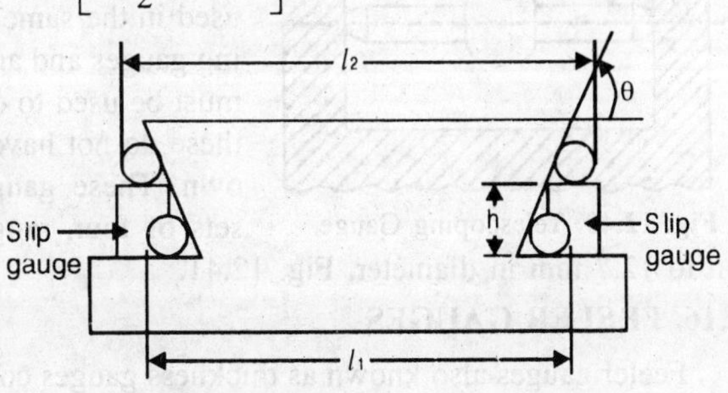

Fig. 12.38. Checking Dovetail Angle.

top face of the gauge blocks to the surface of the precision ball. Next, a bigger ball is positioned near the big end of the taper and the above procedure repeated. The angle of taper is then given as,

$$\sin\frac{\theta}{2} = \frac{r_2 - r_1}{\text{centre distance of balls } (0_1\ 0_2)} = \frac{r_2 - r_1}{h_2 - h_1 - r_2 + r_1}$$

Fig. 12.39 (*b*) shows the arrangement for checking the taper of a taper plug gauge using precision rollers and gauge blocks. The angle of taper will be given as,

$$\tan\frac{\theta}{2} = \frac{L_1 - L_2}{2h}$$

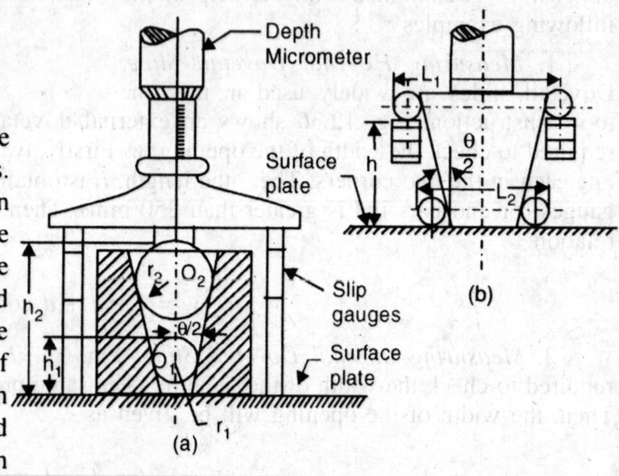

Fig. 12.39. Checking Taper of Gauges.

12.14. TELESCOPING GAUGE

These are T-shaped adjustable gauges used to measure holes or slots. They don't have any graduations, so an outside micro-meter is used to get the reading. A telescoping gauge is made with an adjustable cross beam fastened to a handle with a locking device at the end of the handle. When the diameter of a hole is to be measured, the cross beam of the gauge is compressed and inserted into the hole, Fig. 12.40. It is then released so that the cross-beam contacts the largest diameter of the hole and then the knurled knob at the end of the handle is locked. The gauge is withdrawn and the cross-beam measured with a micrometer to get the diameter of the hole. Telescoping gauges are available in a set from about 8 mm to 150 mm size.

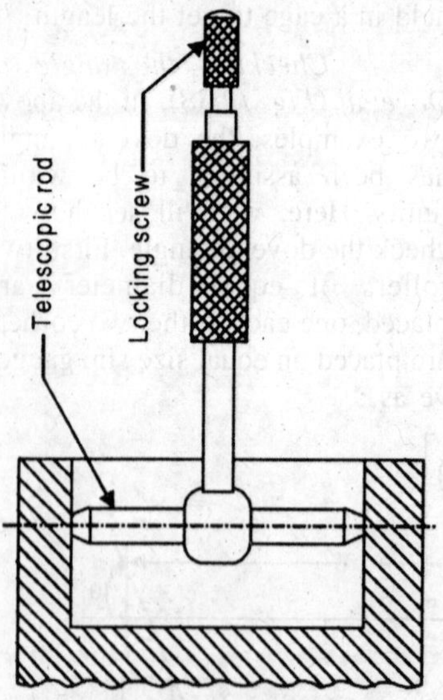

Fig. 12.40. Telescoping Gauge.

12.15. SMALL HOLE GAUGES

Small hole gauges are used for measuring holes too small in diameter or too deep to admit the inside micrometer or Vernier Caliper. They are used in the same manner as telescoping gauges and an outside micrometer must be used to obtain the reading as these do not have graduation of their own. These gauges are available in sets of four, ranging from about 3.2 mm to 12.7 mm in diameter, Fig. 12.41.

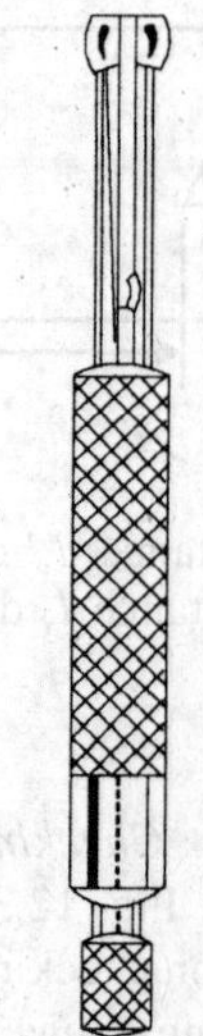

Fig. 12.41. Small hole Gauge.

12.16. FEELER GAUGES

Feeler gauges also known as thickness gauges consist of several thin hardened strips, blades or leaves. The blades vary in thickness from about 2 to 60 hundredths. The blades are held together at one end with a screw so that any blade may be selected for use, Fig. 12.42. Feeler gauges are used :

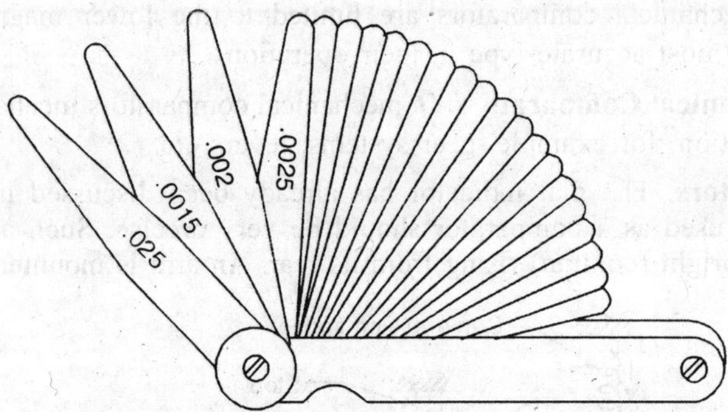

Fig. 12.42. Feeler Gauges.

1. for measuring and testing clearance such as tool clearance to work.
2. for setting milling or planning tool to gauges or setting pins.
3. for adjusting spark plugs, and so on.

12.17. RADIUS GAUGES

These gauges are a type of form gauges used to check the profile of various objects, Fig. 12.43.

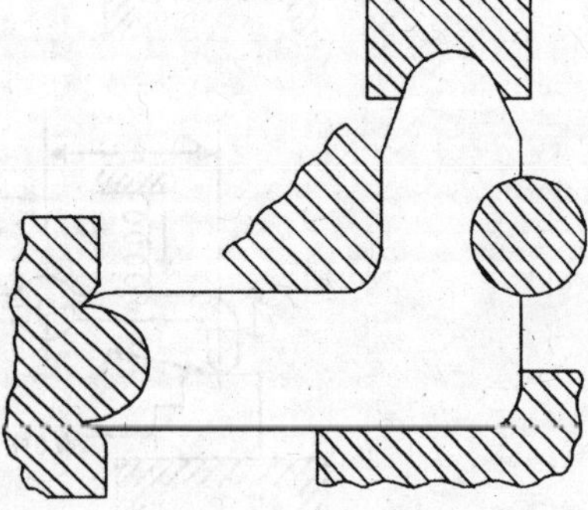

Fig. 12.43. Radius Gauges.

12.18. COMPARATORS

With all the measuring instruments discussed so far, the troublesome question of "feel" exists because of the involvement of the worker or inspector for taking a measurement. For example, if four persons take a measurement with the same instrument, it is certain that there will be discrepancies in their readings. Hence for precision measurements with consistent accuracy, the human element should be completely eliminated. This is achieved by using instruments called "Comparators". A comparator is an indirect type of instrument with the help of which an unknown dimension of a work piece is compared with a working standard (usually slip gauges). The basic principle of operation of a comparator is : The comparator is first adjusted to zero on its dial or recording device with a guage block in position. The gauge block is of dimension which the workpiece should have. Then the workpiece to be checked is placed in position and the comparator is used to check its dimension. The dimension of the work piece may be less than, equal to or greater than the standard dimension. If the dimension is less or greater than the standard, the difference will be shown on the dial or the recording device of the comparator. Thus, a comparator does not give the dimension of a workpiece, but only gives the difference between the standard and the actual dimension of the work piece. In comparators, this difference is shown as magnified on the dial or the recording device. Magnifications of 2000 times are quite common, and some may have as high as 40,000. If a comparator has a magnification of 1000 and if the difference between the standard and actual dimensions of a work piece is 0.02 mm, it will result in pointer movement of 20 mm on the dial or the recording device of the comparator.

Depending on the method by which the difference in dimensions is magnified, the comparators are classified as :

1. Mechanical
2. Electrical
3. Optical
4. Pneumatic

In general, mechanical comparators are limited to the lower magnification. Pneumatic comparators are the most accurate type in their operation.

12.18.1. Mechanical Comparators. In mechanical comparators mechanical means are used to get the magnification, for example, gear systems, levers etc.

1. **Dial Indicators.** The dial indicator has already been discussed in Article 12-10. The dial indicator to be used as a comparator should be very precise. Such a comparator consists of a base with an upright (column) rising from its rear. An arm is mounted on this column and

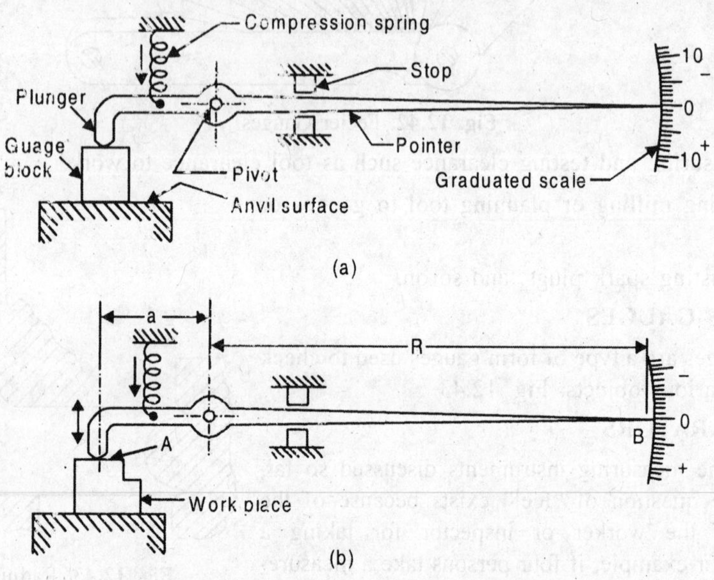

Fig. 12.44. Principle of Lever Type Comparator.

it carries a dial indicator at its outer end. The arm can be adjusted vertically up and down along the column. An anvil or a worktable is mounted on the base, which provides a reference on which work pieces are placed during measuring operations.

2. **Lever Comparators.** The principle of the lever type mechanical comparator is shown in Fig. 12.44. Firstly, a gauge block of standard dimension is wrung to anvil surface below the plunger and the pointer set to zero. Then the gauge block is removed and in its place, the work piece to be measured is placed on the anvil surface. Depending upon the difference in dimension of gauge block and work piece, the plunger moves up and down. This movement of plunger gets magnified and shown on a graduated scale. A compression spring controls measuring pressure. The main drawback of this comparator is that the linear movement of point *A* on plunger gets magnified into the movement of point *B* along an arc. Angular movement of point *B* must be small, otherwise, equal increments of plunger do not produce equal increments of the pointer. In this figure, magnification is equal to *R/a*.

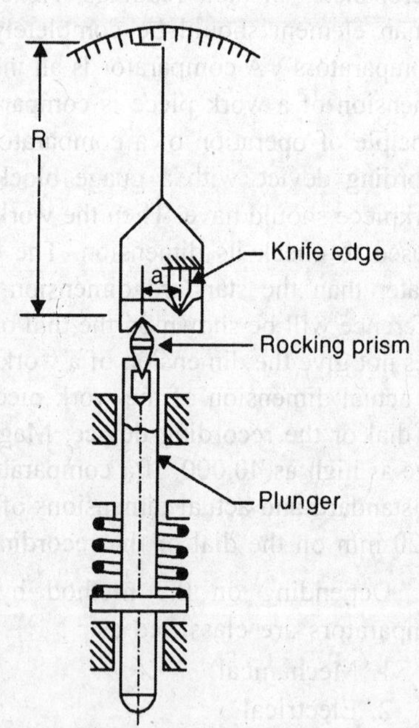

Fig. 12.45. Mechanical Comparator.

Fig. 12.45 shows the construction of a lever type mechanical comparator. At its upper end, the plunger bears against a rocking prism of knife type. The second edge of the prism enters the *V*-slot of a frame member. Another *V*-slot on the upper part of the frame member is offset from the first V-slot by a distance '*a*'. A stationary knife edge enters the upper *V*-slot. The apex of this knife edge is the centre for the movement of the needle on the dial. *R* is the lever arm. The amplification of this instrument is *R/a*. The contact pressure of the mechanical comparator is provided by a spring. The use of knife edge pivots in the comparator movement excludes the influence of possible clearance in the pivots on the accuracy of this instrument.

Fig. 12.46 shows another mechanical comparator known as 'twisted-strip comparator'. Its principle is explain in Fig. 12.46 (*a*). A disc is carried on a twisted strip. A small pull of the strip in the direction of arrows will make the disc to rotate at a very high speed and a point 'A' on the face of the disc will move through a very large distance. This principle is made use of in a mechanical comparator in which is a pointer takes the place of disc. The construction of such a comparator is shown in Fig. 12.46 (*b*). A slight upward movement of plunger will make the bell crank lever to rotate. Due to this, a tension will be applied to the twisted strip in the direction of the arrow. This causes the strip to untwist resulting in the movement of the pointer.The spring will ensure that the plunger returns when the contact pressure between the bottom tip of the plunger and the work piece is not there, that is, when the work piece is removed from underneath the plunger.

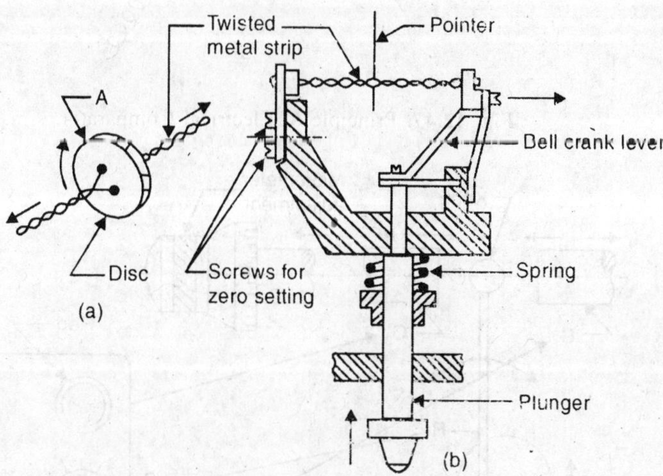

Fig. 12.46. Twisted-strip Comparator.

In mechanical comparators, the magnification ranges from about 250 to 5000. These comparators are relatively cheap and are used mainly for comparative measurement of external surface. Their main drawback is that any wear, play, backlash or dimensional faults in the mechanical devices used will also get magnified.

12.18.2. Electrical Comparators. In electrical comparators, the movement of the measuring contact is converted into an electrical signal. This electrical signal is recorded by an instrument which can be calibrated in terms of plunger movement. For this, an *A-C* wheat stone bridge circuit incorporating a galvanometer is used. The principle of an electrical comparator is shown in Fig. 12.47. An armature supported on thin steel strips is suspended between two coils *A* and *B*. When the distance of the armature surface from the two coils *A* and *B*. When the distance of the armature surface from the two coils is equal, the wheat stone bridge is balanced and no current flows through its galvanometer. Slight movement of the measuring plunger unbalances the bridge resulting in the flow of current through the galvanometer. The

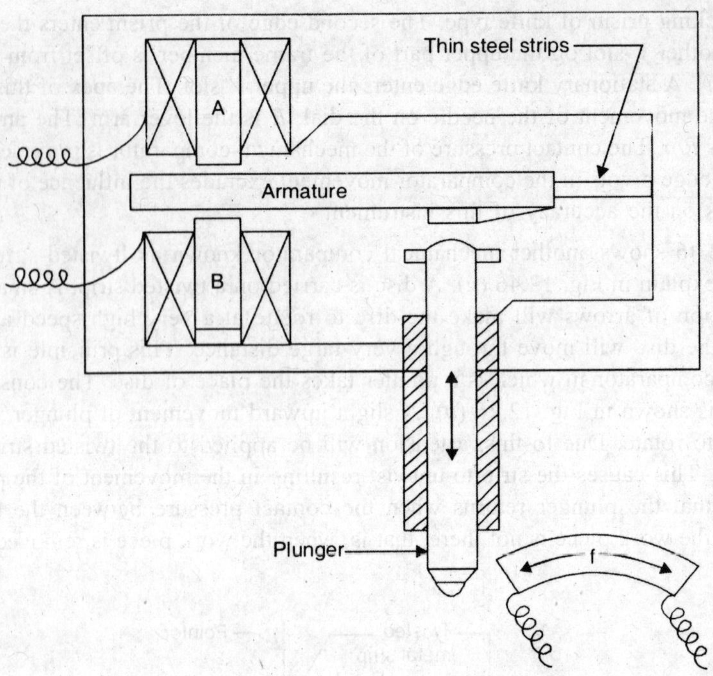

Fig. 12.47. Principle of Electrical Comparator.

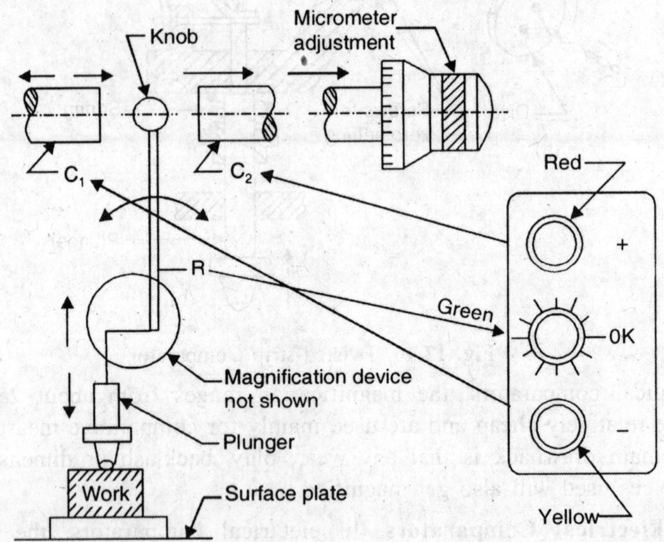

Fig. 12.48. Principle of Visual Gauging Head.

scale of the galvanometer is calibrated to give the movements of the plunger. Electrica comparators have a minimum of moving parts and so give a high degree of reliability Magnifications of the order of 40,000 or higher are obtainable with these comparators.

Another instrument based on electrical principle is known as "Visual gauging head." Thi instrument does not give the difference in dimension but gives only a visual indication with the help of coloured lamps whether the component is correct or not when it is compared with a preset dimension. The principle of a "Visual gauging head" is explained in Fig 12.48. A slight movement of the measuring plunger is transferred to a rod, R. through a magnification device (not shown in the figure). The rod sways to the left or right and a knob at its end makes contact with either of the electrical contacts C_1 and C_2. Precise adjustments of the electrical contacts in the direction of arrows can be done by a micrometer. If the knob is in mid-position between contacts C_1 and C_2, the dimensions of the work under test and correct and the green bulb will be on. If the work is oversize, the rod will move towards right and makes contact with C_2 and red bulb will be on. Conversely, if the dimension is undersize, the rod moves towards the left and the knob will make contact with C_1 and yellow bulb will be on.

12.18.3. Optical Comparators. In these comparators, use is made of a fundamental optical law and instead of a pointer, the edge of a shadow is projected on to a curved graduated scale to indicate the comparison measurement. The optical principle adopted is that of "optical lever" which is illustrated in Fig. 12.49. If a ray of light *OA* strikes a mirror, it is reflected as ray *AB* such that,

$$\angle OAN = \angle NAB$$

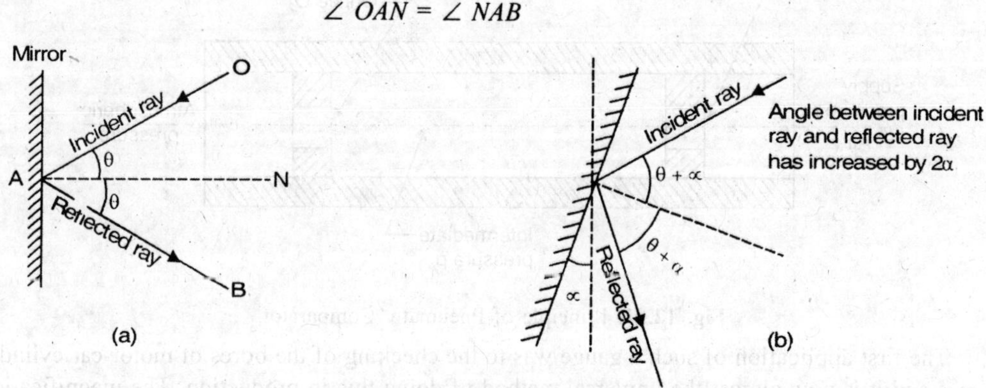

Fig. 12.49. Optical Lever.

Now, if the mirror is tilted through an angle α, the reflected ray of light has moved through an angle 2 α. In optical comparators, the mirror is tilted by the measuring plunger movement and the movement of the reflected light is recorded as an image on a screen. Fig. 12.50 shows the working principle of an optical comparator in which both mechanical and optical levers are used.

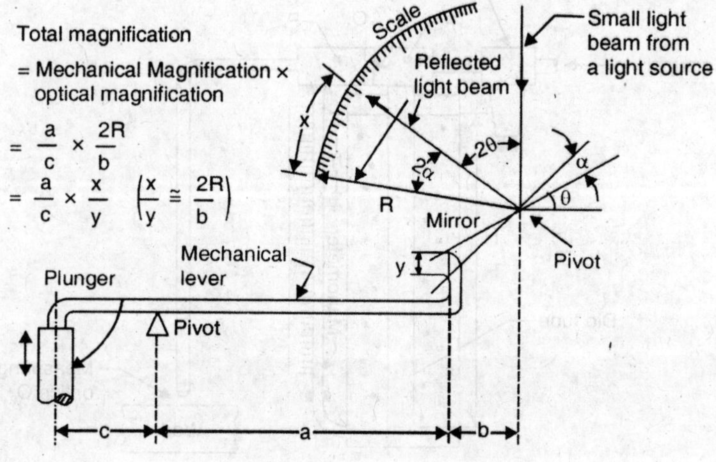

Fig. 12.50. Optical Comparator.

12.18.4. Pneumatic Comparator. In pneumatic comparators, either air flow or air pressure is measured to give measurement deviations from a standard. The response of the comparators working on air flow is quicker than those working on air pressure, but the latter are more versatile than the former. The pneumatic gauging is based on Bernouli's theory. The basic principle of a pneumatic comparator is explained in Fig. 12.51. If air at a low but constant pressure, p_s, is allowed to flow through a small jet or orifice O_1 into an intermediate chamber and then through a second orifice O_2 to atmosphere, then the intermediate pressure *pi* in the intermediate chamber will depend on the relative sizes of the two orifices. Since the flow of air through the two orifices is the same, the pressure drop across them will depend on the resistances they offer to the flow, that is, on their sizes. If the size of the orifice O_1 is kept constant, then the change in *pi* will depend upon the variation in the size of the orifice O_2. The change in the intermediate pressure, *pi,* can be calibrated interms of the change in the size of the orifice O_2 which can be the measurement deviation from a standard.

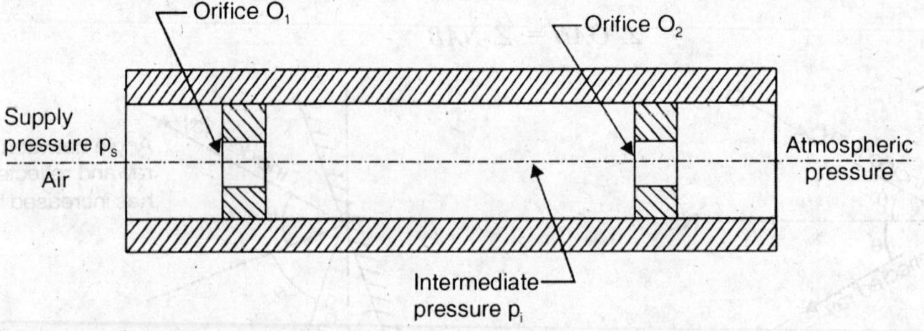

Fig. 12.51. Principle of Pneumatic Comparator.

The first application of such a gauge was to the checking of the bores of motor-car cylinder blocks and it is now almost the universal method of doing this in production. The magnification that can be provided depends on the relative size of the orifices O_1 and O_2 and may be as high as 20,000 : 1, but are usually about 1000-5000 : 1.

Fig. 12.52 shows the working principle of a pneumatic comparator. A pressure reducer and regulator reduces the high-pressure air supply to a value of about 0.007 N/mm² and this

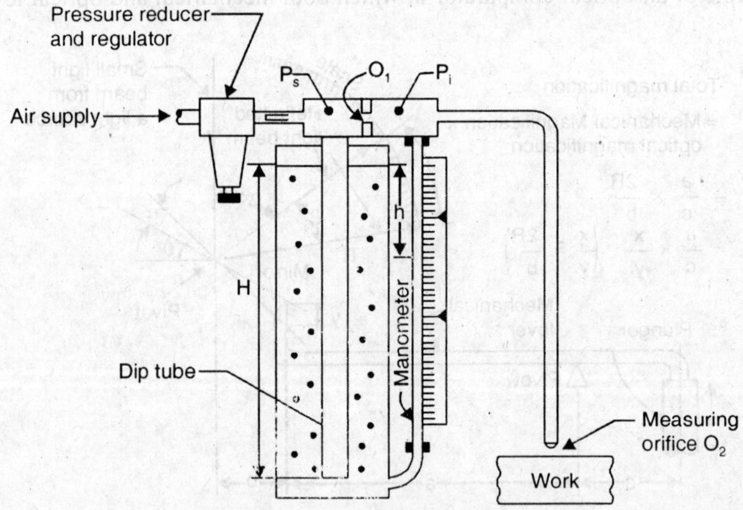

Fig. 12.52. Principle of Pneumatic Comparator.

pressure is maintained closely by including a restriction between the regulator and dip tube *A*. Initially, the levels of liquid in the tank *B* and the dip tube are the same. When the air under pressure is supplied to the system, the liquid in the dip tube will be pushed out into the tank. Any pressure greater than needed to just clear-the dip tube will escape through it into the tank as air bubbles. Thus, the pressure between the regulator and the control orifice O_1 will always be exactly the same, irrespectively of any variation in air pressure. Air after passing through the control orifice O_1, will reach the mouth of measuring orifice O_2. Back pressure will build up behind the measuring orifice if it is not able to pass the coming air flow. Due to this back pressure the level of the liquid in the manometer tube will change. Back pressure is created due to the restriction at the mouth of the orifice O2 caused by the variations in the dimensions of the work being checked. Thus, the manometer tube can be calibrated to give the measurement deviation from a standard.

Another design of pneumatic comparator is shown in Fig. 12.53. Air pressure from about 0.42 to 0.875 N/mm² is reduced to about 0.07 to 0.14 N/mm² as it enters a calibration valve. This air is set wih the help of a standard test piece and all other workpieces are checked against this standard. Any difference in size from that of the standard test piece will result in a difference in air pressure. These air differences will be indicated by a float in a calibrated glass

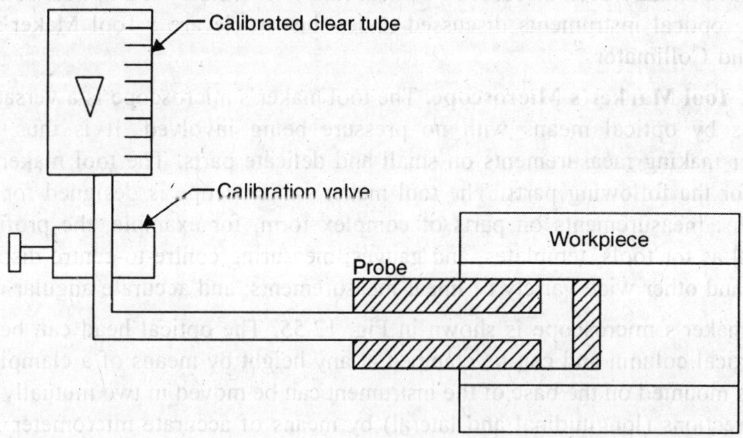

Fig. 12.53. Pneumatic Comparator.

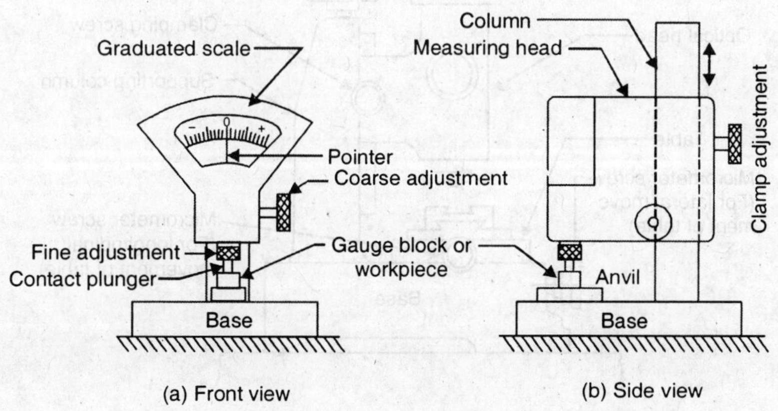

Fig. 12.54. A Typical Comparator.

tube, that is sealed in the low pressure side of the comparator. If a given workpieces is undersize, the float position will be below the standard position and a high float position indicates oversized work piece. For gauging external surfaces, the reverse will be true. This comparator can be used to check : true dimensions of out of round, bell mouthing, parallelism, flatness, camber and concentricity.

Important features of a typical comparator are shown in Fig. 12.54. Measuring head moves up and down the vertical column. For moving the head vertically, the clamp for the head is loosened. There are two devices for coarse adjustment and fine adjustment of the measuring head. Coarse adjustment causes the measuring head to move vertically along the column. Fine adjustment is used for bringing the pointer to zero. When the required movement of the measuring head is achieved, it is champed with the help of head clamp.

12.19. OPTICAL MEASURING INSTRUMENTS

Optical measuring instruments are used for highly precise measurements. They are known as optical instruments because light beams are used as the amplifying levers. The advantages of a light beam are its straightness and its weightlessness. Two optical instruments have already been discussed : Optical flats (Art. 12.8.3) and optical comparators (Art. 12.18.3). The optical instruments have already been discussed : Optical flats (Art. 12.8.3) and optical comparators (Art. 12.18.3). The optical instruments discussed under this article are : Tool Maker's Microscope, Optimeters and Collimators.

12.19.1. Tool Marker's Microscope. The tool maker's microscope is a versatile instrument that measures by optical means with no pressure being involved. It is thus a very useful instrument for making measurements on small and delicate parts. The tool maker's microscope is designed for the following parts. The tool maker's microscope is designed for the following measurements : measurements on parts of complex form, for example, the profile of external thread as well as for tools, templates, and gauges; measuring contre-to-centre distances of holes in any plane and other wide variety of linear measurements; and accurate angular measurements.

A tool maker's microscope is shown in Fig. 12.55. The optical head can be moved up or down the vertical column and can be clamped at any height by means of a clamping screw. The table which is mounted on the base of the instrument can be moved in two mutually perpendicular horizontal directions (longitudinal and lateral) by means of accurate micrometer screws having thimble scale and verniers.

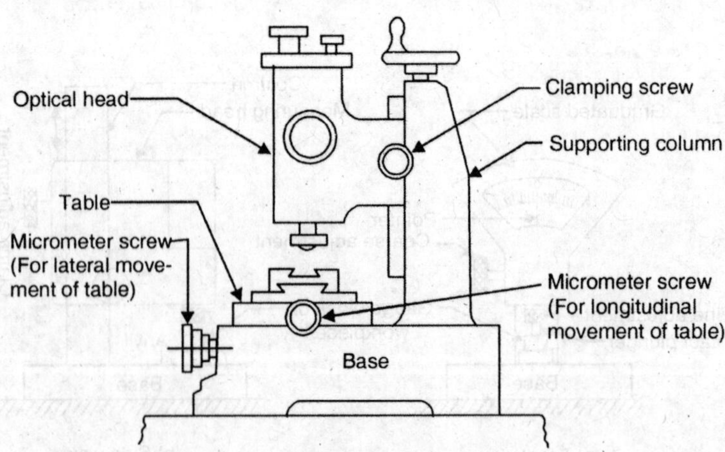

Fig. 12.55. Tool Maker's Microscope.

A ray of light from a light source (Fig. 12.56) is reflected by a mirror through 90°. It then passes through a transparent glass plate (on which flat parts may be placed). A shadow image of the outline or contour of the workpiece passes through the objective of the optical head, and is projected by a system of three prisms to a ground glass screen. Observations are made through an eyepiece. Measurements are made by means of cross-lines engraved on the ground-glass screen. The screen can be rotated through 360°, the angle of rotation is read through an auxiliary eyepiece.

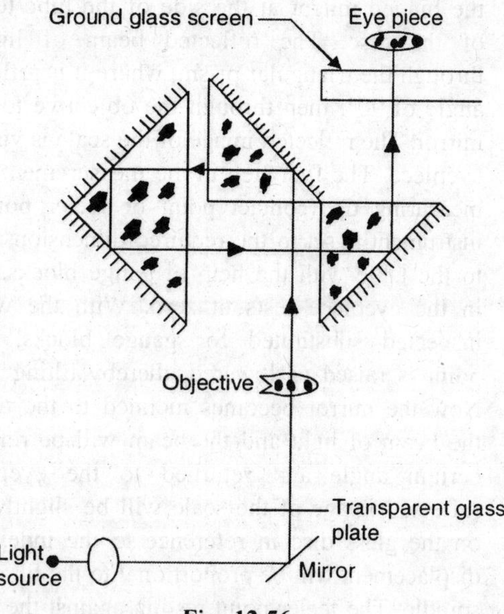

Fig. 12.56

The use of tool maker's microscope for taking the various measurements is explained below :

1. For taking linear measurements, the work piece is placed over the table. The microscope is focused and one end of the workpiece is made to coincide with the cross-line in the microscope (by operating micrometer screws). The table is again moved until the other end of the work piece coincides with the cross-line on the screen, and the final reading taken. From the final reading, the desired measurement can be obtained.

2. To measure the screw pitch, the screw is mounted on the table. The microscope is focused (by adjusting the height of the optical head) until a sharp image of the projected contour of the screw is seen on the ground glass screen. The contour is set so that some point on the contour coincides with the cross-line on the screen. The reading on the thimble of longitudinal micrometer screw is noted. Then the table is moved by the same screw until a corresponding point on the contour (profile) of the next thread coincides with the cross-line. The reading is again noted and the difference in the two readings gives the screw pitch.

3. To determine the pitch diameter, the lateral movement to the table is given.

4. To determine the thread angle, the screen is rotated until a line on the angle of screen rotation is noted. The screen **is further rotated until** the same line coincides with the other flank of the thread. The angle of thread on the screw will be difference in two angular readings.

Different types of graduated and engraved screens and corresponding eye-pieces are used for measuring different elements.

12.19.2. **Optimeters.** Optimeters known as optical comparators, because measurements are made by the comparative method, are gauges used for comparing the size of various parts with that of master gauges. In these instruments the basic principle of reflection is utilised. That is, if a beam of light strikes a flat reflecting surface which is perpendicular to the direction of ray, it is reflected back along its original path. However, if the mirror is tilted through a small angle, the reflected ray will get deflected by twice the angle of tilt of mirror. This principle is already explained under Art. 12.18.3.

The instrument mainly consists of an optimeter tube, a vertical column mounted rigidly in a wide base. The tube which is of right angular design is held in a bracket which can move up and down the column and can be clamπed at any position along the column. The optimeter table on which the parts τo be checked will be placed, is under the measuring tip of the tube and is well supported on the base.

The optical system of the instrument is shown in Fig. 12.57. The optimeter tube consists of the eyepiece through which scale deviations are observed, the scale being engraved on a flat circular disc in front of the eye piece. From the light source-an electric lamp or simply day light outside the instrument - a beam of light is directed by the hinged mirror at the side of the tube to the scale of the disc. The reflected beam of light passes through the triangular prism, where it is reflected at an angle of 90°, then through the objective to the tilting mirror. The reflected image of the scale is visible in the eyepiece. The first step in the measurement is that the measuring tip (contact point or feeler point) of the instrument is set to the required dimensions in relation to the table with the help of gauge blocks. The scale in the eyepiece rests at zero. With the work to be inspected substituted for gauge blocks, the feeler point is raised or lowered, thereby tilting the mirror. Now the mirror becomes inclined to the direction of the beam of light and the beam will be reflected at a certain angle and returned to the eyepiece. The reflected image of the scale will be slightly displaced on the glass disc in reference to the index line. The displacement will be proportional to the tilt of the mirror, that is, to the movement of the measuring spindle. The feeler point resting against the work may be raised slightly by means of a small lever which placing and removing the work. The pressure exerted by the feeler point on the work is very small and is kept uniform by the action of small springs.

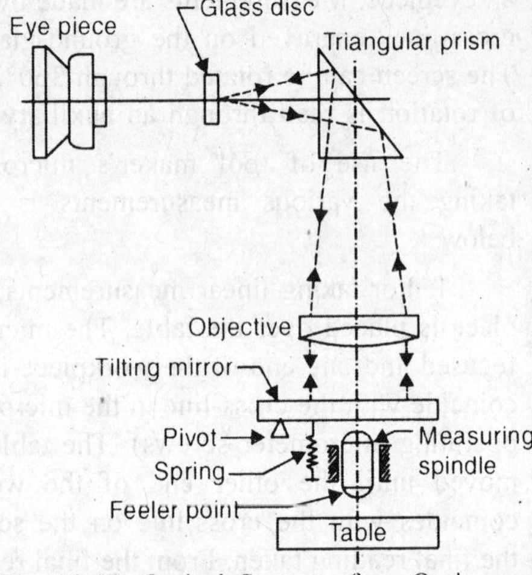

Fig. 12.57. Optical System of an Optimeter.

The vertical optimeter is designed mainly for external measurement of shafts and plates. The range of measurement is from 0 to 180 mm. It can be used expediently for checking plug gauges as well as verifying gauges blocks of certain grades. Optimeters may be fitted with a number of standard and extra attachments to make them adaptable to a wide range of work. The optimeter can also be horizontal with the axis of the measuring spindle horizontal.

12.19.3. Auto-Collimator. This instrument is used to measure small angular inclinations. It is also used to check : straightness, flatness and alignment.

A collimator is a tube for projecting parallel rays of light. It consists simply of a light source and a collimating lens (converging lens). This principle has been in use since long in the telescope. The optical instrument known as Auto-Collimator consists of a collimater and telescope (in one unit) with an eyepiece having either micrometer controlled setting lines or a scale. A source of light is also incorporated in the instrument.

The principle of working of this instrument is the basic principle of reflection of light discussed in the last article. Let a point source of light be placed at the pr nciple focus of a converging lens. After passing through the lens, the beam of light becomes parallel. If this beam of light strikes a flat reflecting surface (mirror) which is perpendicular to the optical axis, it will be reflected back along its original path and re-focused at the same point, Fig. 12.58 (*a*). Now, if the mirror is tilted by an angle 'θ' the beam of light will be reflected back but deflected through an angle '2θ' and will be focused in the same plane as the light source but displaced to one side of it. Fig. 12.58 (*b*). The displacement will obviously be,

$$Oo' = d = 2f\theta$$

where '*f*' is the focal length of the lens.

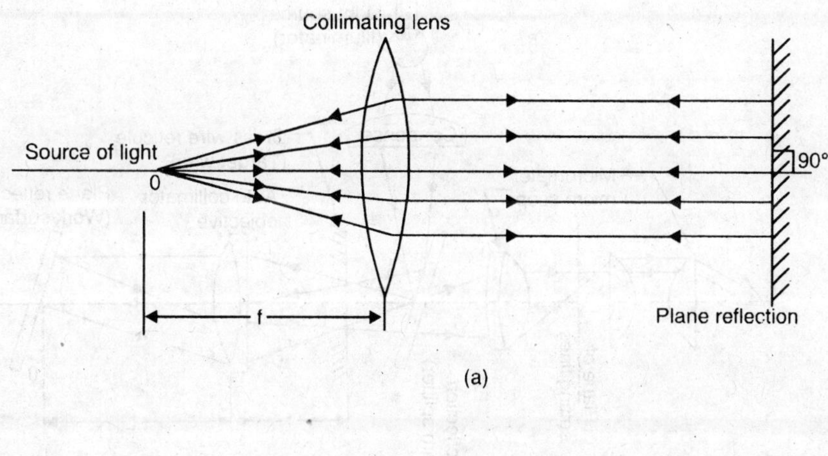

(a)

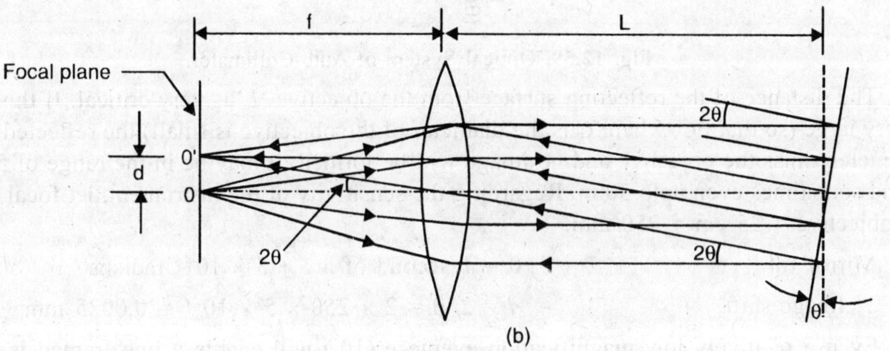

(b)

Fig. 12.58. Principle of Collimation.

So, by measuring the displacement of the image, the inclination of the reflecting surface can be estimated.

In the actual instrument, a pair of target wires is used in place of the point source of light, since it is not convenient to visualise the reflected image of a point and then to measure the displacement '*d*' accurately. Also, the reflecting surface is the work surface whose inclination is to be found. The displacement '*d*' is measured with the help of a precise microscope.

The optical system of the auto-collimator is shown in Fig. 12.59. The pair of target wires (cross wire) is placed at the focal point of the objective lens (collimating lens). Behind this is a reflector. A beam of light from the light source strikes the reflector and gets reflected at 90°. The image of the illuminated cross-wire travels to the right and after passing through the objective moves as a para lel beam of light. After striking the plane reflecting surface (work surface) it is reflected back and gets focused in the focal plane and is focused in the eye piece of eyepiece and if its reflected image has travelled back along its original path, it will coincide with the original cross wire. However, if the reflecting surface is inclined, the reflected image of the cross-wire will get displaced from the position of the original cross-wire. So, in the microscope, two sets of cross-wires will be seen. A pair of setting lines in the micrometer microscope mechanism is used to measure the displacement of the target wires, by first setting these to the original cross-wire and then moving over to those of the image. Normally, a calibration is provided with the instrument relating the displacement of the image and the angle of tilt of the reflecting surface.

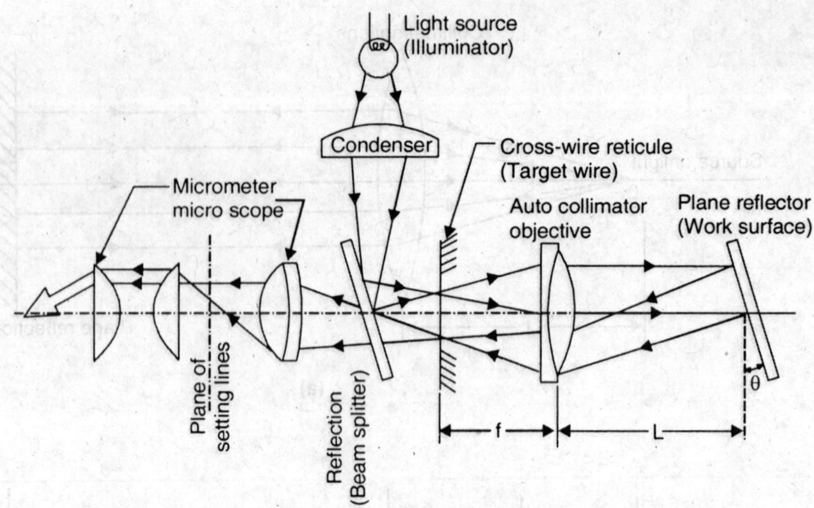

Fig. 12.59. Optical System of Auto-Collimator.

The distance of the reflecting surface from the objective 'L' is very critical. If this distance is very large (so that θC, where C is the diameter of the objective is small) the reflected rays will completely miss the objective and no image will be formed. It may be in the range of 5 m to 10 m and sometimes even upto 30 m. Regarding the sensitivity of the instrument, let focal length of the objective = 25 cm = 250 mm.

Mirror tilt, $\theta = 1$ second of arc $\approx 5 \times 10^{-6}$ radians

$\therefore$ Image shift, $d = 2f\theta = 2 \times 250 \times 5 \times 10^{-6} = 0.0025$ mm

Even a relatively low magnification eyepiece (10′) will enable a line framed is a double index to be judged well within 0.0025 mm.

These instruments are very popular due to their ability to sense remotely, to high accuracy, the angular rotation of a flat mirror around axes in the plane of the mirror. Because of their sensitivities (which is as less as $\frac{1}{10}$ s of an arc or even smaller under special conditions) it has become common practice to use auto collinators not only to monitor angular tilts as such, but to convert linear displacements into angular ones so that they can be monitored by this versatile instrument.

(*a*) *Examination of plane surfaces.* An auto-collimator is commonly used for checking the plane surfaces such as beds of machine tools. For this a test block is made. The base of the block is flat with its working surface square, flat and polished to form a reflecting surface.

The test block is placed at one end of the surface to the checked with its reflecting surface normal to the collimator axis which is mounted on a rigid stand. The collimator is adjusted until the cross wire and its image are coincident as seen through the eyepiece. The test block is then moved to another position on the surface keeping its working surface square with direction of movement. If the surface at this position is not flat, the working surface of the test block will become inclined to the optical axis of the collimator. Due to this, the image of the cross-wire in the eye piece will get displaced and the displacement measured with the micrometer. Continuing this procedure along the entire length of the surface, the flatness or straightness of the surface can be checked and its profile plotted.

(*b*) *Checking squareness of two surfaces :* Supposing there are two machined surfaces at right angle to each other. Their squareness can be checked with the help of auto-collimator to a high degree of accuracy $\left(\pm\frac{1}{2}s\right)$ in the same manner as explained above. The test block is first placed on the horizontal surface with its reflecting surface normal to the optical axis of the collimator, and reading in the micrometer noted. Next, the test block is placed on the vertical surface and an optical square (penta prism) is placed at the corner of the two surfaces to view the test block in the collimator in the new position. Another reading is taken in the eyepiece. The two readings are compared to know the squareness of the two surfaces.

(*c*) *Checking alignment or parallelism.* Parallelism or alignment of two holes in a part can also be tested by the auto-collimator. In this case, a test bar will be used, placed alternately in each hole and collimator readings taken.

PROBLEMS

1. In what ways, the measuring instruments are classified?
2. Differentiate between direct measuring and indirect measuring instruments.
3. Differentiate between absolute and comparative measuring instruments.
4. What is meant by 'sensitivity' of a measuring instrument?
5. Discuss the two length standards.
6. Why calipers are called 'transfer measuring instruments'?
7. Write about the various types of calipers.
8. What is the difference between measurement and gauging?
9. Explain the principle of Vernier Calipar.
10. Where are Vernier height gauge' and Vernier depth gauge' used?
11. Explain the principle of a micrometer.
12. Explain the use of slip gauges.
13. State about the care of slip gauges.
14. What is the use of end bars?
15. Explain principle of optical flats.
16. What is the principle of a dial indicator?
17. Explain the practical applications of a dial indicator.
18. Upon what principle is a sine bar based?
19. What is the use of telescoping gauges?
20. What is the use of precision balls and rollers?
21. How will you measure external dovetail, internal dovetail and angle of dovetail with the help of precision balls and rollers?
22. What is the main disadvantage of other instruments as compared to comparators?
23. State the principle of operation of a comparator.
24. Explain the principle of lever type of mechanical comparator.
25. Explain the principle of a twisted-strip comparator'.
26. Explain the principle of electrical comparator.
27. What is 'optical lever'?
28. Explain the principle of pneumatic comparator.
29. What is the use of toolmaker's microscope.
30. Explain the principle of collimation.
31. Explain the optical system of an optimeter.
32. Explain the optical system of an auto-collimator.
33. (*a*) Show a set up for the measurement of the cone angle of a taper plug gauge by a sine bar.
    (*b*) Calculate the cone angle of the taper plug guage for such a set up if the slip gauge heights were 50.667 mm and 38.667 mm, the length of the since bar being 125 mm.
34. How would you check the taper of the dead centre of a lathe by using a sine-bar in the metrology laboratory. Sketch the set up.
35. A set of 78 gauge blocks is made of the following :-

49 blocks 1.01 to 1.49 mm with step size = 0.01 mm
19 blocks 0.50 to 9.50 mm with step size = 0.50 mm
4 blocks 10 mm to 40 mm with step size = 10.00 mm
3 blocks 50 mm to 100 mm with step size = 25 mm
3 blocks 1.0025 mm to 1-0075 mm with step size = 0.0025 mm
2 protection blocks of 2-0 mm
How a length of 146.9925 mm will be built up ?

Sol. 2 × 2 mm protection blocks = 4
1 × 1.0025 mm block = 1.0025
1 × 1.49 mm blocks = 1.4900
1 × 0.50 mm block = 0.5000
4 × 10 mm block = 40.0000
1 × 100 mm block = 100.0000
146.9925 mm

36. A wedge shaped job was measured using a 250 mm sine bar and the top edge found to be level when gauge blocks totalling 46.376 mm were placed under one of the rollers. What is the angle of the wedge ?

Sol. Refer Art 12.13.5

$$\sin\theta \simeq \frac{h}{L} = \frac{46.376}{250} = 0.1855$$

$$\therefore \theta = \sin^{-1} 0.1855 = 10.69°$$

37. A taper plug gauge was measured using the method illustrated in Fig. 12.39 (b). The measurements over the rollers when on the base and when resting on a 70 mm gauge block was found to be 64.67 mm and 68.30 mm respectively. Find the rate of taper of the gauge.

Sol. $$\tan\theta/2 = \frac{L_1 - L_2}{2h}$$

$$= \frac{68.30 - 64.67}{2 \times 70}$$

Now rate of taper, i.e., increase in diameter per unit length,

$$= 2 \times \tan\theta/2 = \frac{3.63}{70} = 0.0519$$

38. A 250 mm sine bar was used to check the angle of a taper plug gauge by placing the sine bar on the plug gauge and using two piles of gauge blocks. If the two piles are 28.732 mm and 40.174 mm when the sine bar, plug gauge and surface table are in perfect contact, determine the rate of taper of the plug gauge.

Sol. Refer Art. 12.13.5, Fig. 12.31(b)

Rate of Taper of plug gauge = 2 × tan 2 θ

$$\cong 2 \times \tan x \sin^{-1}\left(\frac{h_1 \sim h_2}{L}\right)$$

when the angle is very small, tan ≅ sin

$$\therefore \text{Rate of Taper of Plug Gauge} \cong 2 \times \frac{40.174 - 28.732}{250}$$

$$\cong 0.0916$$

39. State the basic means of obtaining magnification of the measuring probe in comparator type measuring instruments.

40. What single feature gives the electrical comparator a unique advantage over all other comparators.

41. State the type of pivot which because of its efficiency, is used in the mechanical comparator.

42. State the size of the screen built into the tool maker's microscope.

43. What is the smallest angular displacement that can be resolved by the auto-collimator?

44. What is the maximum angular displacement that can be accommodated by the auto-collimator?

13

ANALYSIS OF METAL FORMING PROCESSES

13.1. THEORETICAL BASIS FOR METAL FORMING

In metal forming, large forces are applied to the work material to deform it plastically, to get the desired product. The most important item in the analysis of a metal forming processes is the determination of the magnitude of the applied force, since it is an item necessary for the design of processing equipment. Another important factor is to know the extent of deformation to which a workpiece can be subjected before it fails. So, we ought to know the relationship between a force and the deformation that it produces.

The most common method of determining the relationship between force (load) and deformation is the tension test. In this test, a suitable work specimen is subjected to an increasing axial load until it fractures. The measurements of force and deformation (elongation) are taken at frequent intervals during the test. The engineering load-deformation curve or stress-strain curve for a ductile material (based on original dimensions of the specimen) is shown in Fig. 13.1. The significant points on the curve have been shown. Upto the proportional limit, the stress is directly proportional to strain and the slope of the curve is the modulus of elasticity, E. Point A is the elastic limit upto which the workpiece will come back to its original dimensions when unloaded. The stress at this point is called the yield stress, σ_0. For most of the materials, it is quite difficult to locate the elastic limit and the yield stress is usually taken as the stress which will produce a small amount of permanent deformation, usually equal to a strain of a 0.002. This is known as "offset yield stress". Beyond elastic limit, plastic deformation begins. With the increase in plastic deformation, the specimen becomes stronger until the point 'C' is reached on the curve. The curve AC is called as "strain-hardening" or "work-hardening" curve, since the force required to deform the specimen increases with further straining. In this portion of the curve the specimen continues to become longer and thinner, the deformation being uniform

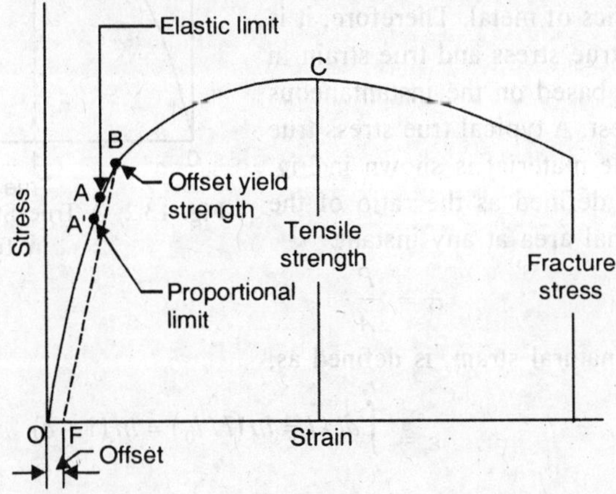

Fig. 13.1. Engineering Stress-strain Curve.

along the length of the specimen. The stress at point C (where load is maximum) is called "tensile strength" or "ultimate tensile strength" of the material. Beyond maximum load, the specimen starts to become thinner rapidly at some point between the two ends, until the specimen breaks or fractures. This phenomenon is called as "Necking". At the point of failure, the load is lower than the maximum load, and the stress at this point is called as "breaking stress" or "fracture stress".

The engineering stress in the specimen is defined as the ratio of the applied load to the original cross-sectional area of the specimen,

$$s = \frac{P}{A_0} \qquad \text{...(13.1)}$$

The engineering strain or the conventional strain is defined as the ratio of the elongation of the gauge length of the specimen by its original length.

$$e = \frac{l - l_0}{l_0} \qquad \text{...(13.2)}$$

where 'l' is the length at any stage during the test.

True Stress—True Strain Curve. The engineering stress-strain curve is based on the original dimensions of the specimen, and these dimensions change continuously during the test. Therefore, such a curve does not give a true indication of the deformation characteristics of metal. Therefore, it is necessary to obtain the true stress and true strain in the specimen which are based on the instantaneous dimensions during the test. A typical true stress true strain curve for a ductile material is shown in Fig. 13.2. The true stress is defined as the ratio of the load to the cross-sectional area at any instant.

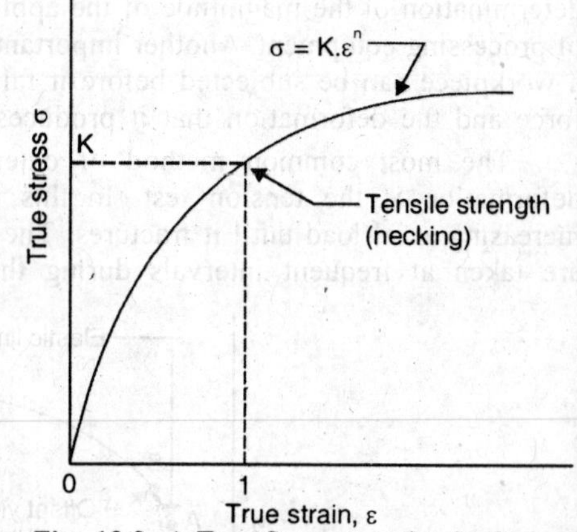

Fig. 13.2. A True Stress-true Strain Curve in Tension.

$$\sigma = \frac{P}{A}. \qquad \text{..(13.3)}$$

The true strain, or natural strain, is defined as,

$$\varepsilon = \int_{l_0}^{l} dl/l = ln\,(l/l_0) = ln\,(1+e) \qquad \text{...(13.4)}$$

For small values of strains, $ln\ (1 + e) \cong e$, so $\varepsilon \cong e$, but at larger values, the engineering and true strains diverge rapidly.

The volume of the specimen is assumed to be constant during plastic deformation, therefore,

$$\varepsilon = ln\,(l/l_0) = ln\,(A_0/A)$$

$$= ln\,(D_0^2/D^2) = 2ln\,(D_0/D)$$

Now $$\sigma = \frac{P}{A} = \frac{P}{A_0}\cdot\frac{A_0}{A}$$

$$= \frac{P}{A_0}\cdot\left(\frac{l}{l_0}\right) = \frac{P}{A_0}\,(1.+e)$$

$$\therefore \quad \sigma = s\,(l+e) \qquad \text{...(13.5)}$$

Note. In plastic deformation of metals, it is assumed that volume remains constant. Therefore,

$$A_0 l_0 = Al$$

$$\therefore \quad \frac{A_0}{A} = \frac{l}{l_0}$$

It is clear from Fig. 13.2, that, the true stress, true strain curve rises continuously up to fracture. This curve is also called as the "flow curve", since it represents the basic plastic-flow characteristics of the material. The most common mathematical equation to fit this curve is a power expression of the form

$$\sigma = K.\varepsilon^n \qquad ...(13.6)$$

where K is stress at $\varepsilon = 1.0$ and is called the 'strength coefficient', and n is the strain-hardening exponent and is the slope of a log-log plot of eqn. (13.6).

Ideal or Hypothetical stress-strain curves. The solution of the plasticity problems is quite involved. To make the analysis of plasticity problems simple, ideal or hypothetical stress-strain curves for ductile materials can be used, which are shown in Fig. 13.3. In Fig. 13.3 (*a*), the elastic strains are considered to be zero and once σ_0 is reached (σ_0 is the yield stress in simple tension), deformation proceeds at a constant stress value. Such a material is called 'rigid perfectly plastic'. Fig. 13.3 (*b*) is for elastic-perfectly plastic material. Fig. 13.3 (*c*) is for elastic strain-hardening material and Fig. 13.3 (*d*) is for a typical real material (elastic, non-linear strain hardening material).

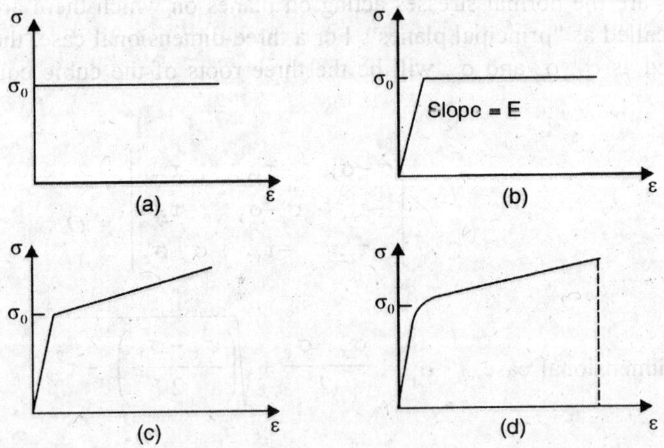

Fig. 13.3. Stress-Strain Curves.

The General State of Stress at a point in material. Above, the material has been subjected to only uniaxial simple stress. In metal forming process, the material is usually subjected to a complex state of stress. The general state of stress at a point in a material is shown in Fig. 13.4. The sign of the normal stresses, σ are taken as : Tensile stress are taken as positive and compressive stresses are negative. A shear stress (τ) is taken as positive if it points in the positive direction on the positive face of a unit cube, or if it points in the negative direction on the negative face of the unit cube. In Fig. 13.4. the shear-stresses shown are positive (the shear stresses on faces normal to Z-axis are not shown). It is well known that,

$$\tau_{xy} = \tau_{yx}, \ \tau_{xz} = \tau_{zx} \text{ and } \tau_{yz} = \tau_{zy}$$

so that the state of stress at a point is given by six components

$$\begin{matrix} \sigma_x & \tau_{xy} \\ \sigma_y & \tau_{yz} \\ \sigma_z & \tau_{zx} \end{matrix}$$

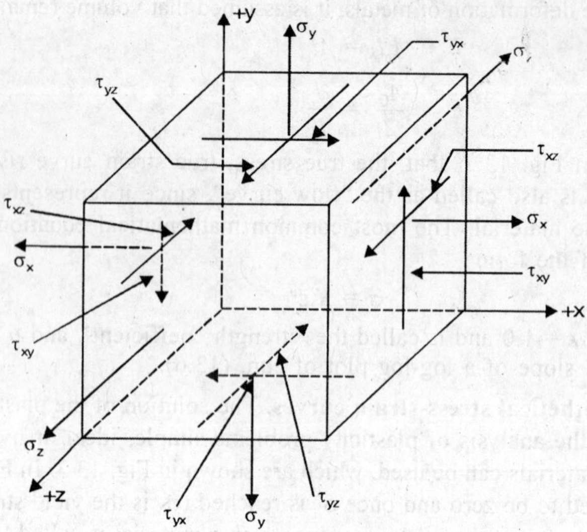

Fig. 13.4. General State of Stress at a Point.

In metal working processes, it is very useful and convenient to use "principal stress". Principal stresses are the normal stresses acting on planes on which there are no shear stresses. Such planes are called as "principal planes". For a three-dimensional case, the principal stresses, which are denoted as σ_1, σ_2 and σ_3, will be the three roots of the cubic equation given by the determinant,

$$\begin{vmatrix} \sigma-\sigma_x & -\tau_{yx} & -\tau_{zx} \\ -\tau_{xy} & \sigma-\sigma_y & -\tau_{zy} \\ -\tau_{xz} & -\tau_{yz} & \sigma-\sigma_z \end{vmatrix} = O \qquad ...(13.7)$$

For a two-dimensional case,
$$\sigma_1 = \frac{\sigma_x+\sigma_y}{2} + \sqrt{\left(\frac{\sigma_x-\sigma_y}{2}\right)^2 + \tau_{xy}^2} \qquad ...(13.8)$$

$$\sigma_2 = \frac{\sigma_x+\sigma_y}{2} - \sqrt{\left(\frac{\sigma_x-\sigma_y}{2}\right)^2 + \tau_{xy}^2}$$

The maximum shear stress is given as,

$$\tau_{max} = \frac{\sigma_{max}-\sigma_{min}}{2} \qquad ...(13.9)$$

For each pair of principal stresses, there are two planes of principal shear stresses, which bisect the directions of the principal stresses.

Yielding Under Complex Stresses. In simple uniaxial tension test, the material starts yielding when the normal stress (principal stress, there being no shear stress) equals the yield stress, that is, $\sigma_1 = \sigma_0$. The purpose of yield theories is to predict the onset of yielding under complex state of stress σ_1, σ_2, σ_3). For ductile materials there are two common theories : Maximum shear stress theory (Tresea theory) and the Distortion Energy theory (Von-Mise's theory).

(*i*) **Maximum Shear Stress Theory.** This yield criterion assumes that yielding occurs when the maximum shear stress within the material reaches the value of yield shear stress in uniaxial tension test.

Considering principal stresses (σ_1, σ_2, σ_3), ($\sigma_1 > \sigma_2 > \sigma_3$).

$$\tau_{max} = \frac{\sigma_{max} - \sigma_{min}}{2} = \frac{\sigma_1 - \sigma_3}{2}$$

For uniaxial tension, $\sigma_1 = \sigma_0$ and $\sigma_2 = \sigma_3 = 0$, and the shear yield stress K, is equal to $\frac{\sigma_0}{2}$.

$$\therefore \quad \tau_{max} = \frac{\sigma_1 - \sigma_3}{2} = \frac{\sigma_0 - 0}{2} = \frac{\sigma_0}{2} = K$$

∴ Maximum shear stress criterion is,

$$\sigma_1 - \sigma_3 = \sigma_0 \qquad \text{...(13.10)}$$

(*a*) **Plane Stress Case :**

$\sigma_2 = 0 \therefore \sigma_{min} = \sigma_2$ (If all the stresses are tensile).

∴ The criterion is,

$\sigma_1 = \sigma_0$, If σ_3 is +

and $\sigma_1 + |\sigma_3| = \sigma_0$, if σ_3 is –

(*b*) **Plane Strain Case :** If $\varepsilon_2 = 0$, it can be shown that

$$\sigma_2 = \frac{\sigma_1 + \sigma_3}{2}$$

Whatever may be the sign of σ_1 and σ_3, σ_2 can never be minimum, therefore, for plane strain case,

$$\sigma_1 = \sigma_0 + \sigma_3$$

(*c*) **State of Pure Shear :**

$$\sigma_1 = -\sigma_3 = K, \sigma_2 = 0$$

where K is the yield stress in pure shear. The criterion is,

$$\sigma_1 - \sigma_3 = 2K = \sigma_0$$

or

$$K = \frac{\sigma_0}{2}$$

(*ii*) **Von Mise's theory.** This theory gives the criterion as,

$$(\sigma_1 - \sigma_2)^2 + (\sigma_2 - \sigma_3)^2 + (\sigma_3 - \sigma_1)^2 = 2\sigma_0^2 \qquad \text{...(13.11)}$$

(*a*) **Plane Stress Case :**

Putting $\sigma_2 = 0$, we get

$$\sigma_1^2 + \sigma_3^2 - \sigma_1\sigma_3 = \sigma_0^2$$

(*b*) **Plane Strain Case :**

Putting $\sigma_2 = \frac{\sigma_1 + \sigma_3}{2}$, we get

$$\sigma_1 = \frac{2}{\sqrt{3}}\sigma_0 + \sigma_3$$

(*c*) **State of Pure Shear :**

Putting $\sigma_1 = -\sigma_3 = K, \sigma_2 = 0$, we get,

$$\sigma_1^2 + \sigma_3^2 + 4\sigma_1^2 = 2\sigma_0^2$$

$$\therefore \quad 6K^2 = 2\sigma_0^2$$

$$\therefore \quad K = \frac{1}{\sqrt{3}} \cdot \sigma_0 = 0.577\, \sigma_0$$

Stress-Strain Relations. Within the elastic region of the stress-strain curve, the strains are uniquely determined by the stresses with the help of Hooke's law, without regard to how the stress state was achieved. However in the plastic region, the strains depend upon the entire previous history of loading. Therefore, in plasticity problems, the strains are determined as increments throughout the loading path and the total strain is obtained by integration or summation. This is known as the "Incremental or flow theory of plasticity". It relates the stresses to the plastic strain increments.

In metal working processes, large deformations of the material take place. The elastic strains are small as compared to plastic strains and so can be neglected. Then the total strains produced in the material can be taken as only the plastic strains. For such a material (ideal plastic solid), the stress-strain relations are those given by Levy and Von Mises.

Levy-Mises Equations. The incremental plastic strains (which will be the total strains) are given as :

$$d\varepsilon_x = \frac{d\bar{\varepsilon}}{\bar{\sigma}} \left[\sigma x - \frac{1}{2} (\sigma_y + \sigma_z) \right]$$

$$d\varepsilon_y = \frac{d\bar{\varepsilon}}{\bar{\sigma}} \left[\sigma_y - \frac{1}{2} (\sigma_z + \sigma_x) \right]$$

$$d\varepsilon_z = \frac{d\bar{\varepsilon}}{\bar{\sigma}} \left[\sigma_z - \frac{1}{2} (\sigma_x + \sigma_y) \right] \quad \text{...(13.12)}$$

$$d\varepsilon_{xy} = \frac{3}{2} \cdot \frac{d\bar{\varepsilon}}{\bar{\sigma}} \cdot \tau_{xy}$$

$$d\varepsilon_{yz} = \frac{3}{2} \cdot \frac{d\bar{\varepsilon}}{\bar{\sigma}} \cdot \tau_{yz}$$

$$d\varepsilon_{zx} = \frac{3}{2} \cdot \frac{d\bar{\varepsilon}}{\bar{\sigma}} \cdot \tau_{zx}$$

Above, $\bar{\sigma}$ is the effective stress and $d\bar{\varepsilon}$ is the effective incremental strains and are given as below :

$$\bar{\sigma} = \frac{1}{\sqrt{2}} \left[(\sigma_x - \sigma_y)^2 + (\sigma_y - \sigma_z)^2 + (\sigma_z - \sigma_x)^2 + 6\left(\tau_{xy}^2 + \tau_{yz}^2 + \tau_{zx}^2\right) \right]^{1/2}$$

and $$d\bar{\varepsilon} = \frac{\sqrt{2}}{3} \left[(d\varepsilon_x - d\varepsilon_y)^2 + (d\varepsilon_y - d\varepsilon_z)^2 + (d\varepsilon_z - d\varepsilon_x)^2 + 6\left(d\varepsilon_{xy}^2 + d\varepsilon_{yz}^2 + d\varepsilon_{zx}^2\right) \right]^{1/2}$$

In equation (13.12), $d\bar{\varepsilon}/\bar{\sigma}$ can be determined from the effective stress-effective strain curve.

Prandtl-Reuss Equations. Levy-Mises equations are valid for large plastic deformations, where, elastic strains are negligible. But in the elastic-plastic region or for a real elastic-plastic solid, it is necessary to consider both elastic strains and plastic strains. For this, the equations have been proposed by Prandtl and Reuss. Here,

Total strain increment = elastic strain increment + plastic strain increment.

$$d\varepsilon_{ij} = d\varepsilon_{ij}^{e} + d\varepsilon_{ij}^{p}$$

The elastic strain increments are given by Hooke's law :

$$d\varepsilon_{ij}^{e} = \frac{1+v}{E}\, d\sigma_{ij} - \frac{v}{E}\sigma_{kk}\,\delta_{ij}$$

where 'v' is Poisson's ratio and δ_{ij} is Kronecker's delta, given as,

$$\delta_{ij} = 1, \text{ when } i = j$$
$$= 0, \text{ when } i \neq j$$

The plastic strain increments will be given by eqn. (13.12), which can be written as,

$$d\varepsilon_{ij}^{p} = \frac{3}{2}\cdot\frac{d\bar{\varepsilon}}{\bar{\sigma}}\cdot\sigma'_{ij}$$

where σ'_{ij} is the deviatoric stress tensor and is given as,

$$\sigma'_{ij} = \sigma_{ij} - \frac{1}{3}\delta_{ij}\,\sigma_{kk}$$

that is, $$d\varepsilon_{x}^{p} = \frac{d\bar{\varepsilon}}{\bar{\sigma}}\left[\sigma_x - \frac{1}{2}\left(\sigma_y + \sigma_z\right)\right]$$

and so on.

Thus, $$d\varepsilon_{ij} = \frac{1+v}{E}\, d\sigma_{ij} - \frac{v}{E}\sigma_{kk}\,\delta_{ij} + \frac{3}{2}\cdot\frac{d\bar{\varepsilon}}{\bar{\sigma}}\cdot\sigma'_{ij} \qquad ...(13.13)$$

13.2. CLASSIFICATION OF METAL FORMING PROCESSES

Depending upon the temperature at which a material is mechanically worked or formed, the metal forming processes can be classified as : Cold forming, Hot forming and Warm or Semi-hot forming.

13.2.1. Cold Forming. Cold forming or cold working can be defined as the plstic deforming of metals and alloys under conditions of temperature and strain rate such that the work hardening or strain hardening is not relieved. Theoretically, the working temperature for cold working is below the recrystallization temperature of the metal/alloy (which is about one-half the absolute melting temperature). However, in practice, cold working is carried out at room temperature.

As already discussed, cold working results in increase in strength and hardness and a decrease in ductility of the material, Fig. 13.5 (*a*). So, if the material is excessively deformed, it may fracture before it is formed. To avoid this, large deformations in cold working are obtained in several stages, with intermediate annealing. This will soften the cold worked material and restore its ductility, Fig. 13.5 (*b*).

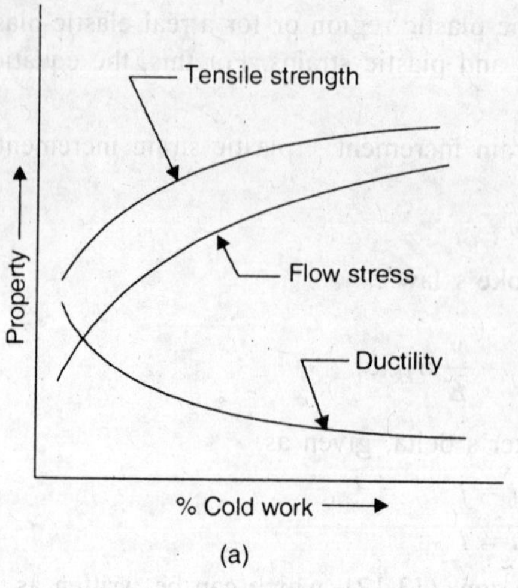

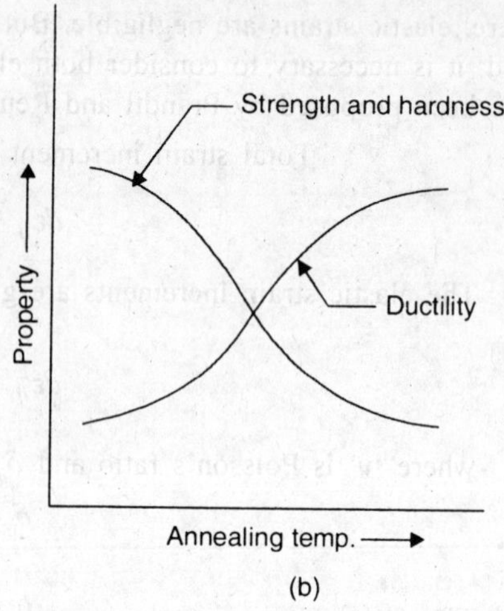

Fig. 13.5. Effects of Cold Working and Annealing.

In addition to the characteristics of cold working discussed in chapter 3 the cold working process has got the following characteristics :

1. Since cold working is done at room temperature or low temperatures, no oxidation and scaling of the work-material occurs. This results in reduced material loss.

2. Thin gauge sheets can be made by cold working.

3. Since higher forces are required, high capacity and costly machines are needed for cold working.

4. Severe stresses are set up in the material during cold working. This requires stress relieving or annealing treatment, which increases the cost of cold working.

Materials for cold-working. In principle, any material can be cold worked. In practice, however, the choice is limited by the following two factors :

1. The ability of the tool material to withstand the required pressures for cold working of a material. Obviously, the tool material must have a mechanical strength greater than that of the material to be cold worked. Also, from the point of view of economy, the tool (die etc). must have a reasonable working life, that is, it must be able to withstand the developed working stresses for a reasonable length of time.

2. The economic requirement that the maximum possible deformation of the material should be obtained in a single working operation. This will depend upon the cold ductility and cold flowability of the material.

Thus, the two principal limitations to cold working of a material are : the permissible stress placed on the tool material and the ductility of the material to be cold worked.

Both cold ductility and cold flowability of a material depend closely on its chemical composition. As for steel, with an increase in the percentage of carbon or alloying constituents, its deformability decreases and the resistance to deformation increases. The maximum limit is usually 0.45% carbon for steels used in cold extrusion and 1.6% carbon for other cold forging operations. Impurities such as S, P, O_2 and N_2, also impair the cold workability of the steel. For cold working, the micro-structure of the material, also plays an important role. Soft annealing known as spheroidize annealing of steel before cold working, improves its cold workability. The grain size is also an important factor. Large grain is easier to cold work, while the parts made from fine grained material are stronger. A good guiding rule for forging steels used to produce forgings is that the stress on the die should not exceed 2500 N/mm^2 and the material must allow at least a 25% deformation in a single step.

As a general rule, the requirements for a material to be cold worked are :

1. Yield stress curve of gentle slope.
2. Early yield point.
3. Then great elongation with pronounced necking before fracture.

The material commonly used for cold working include : low and medium carbon steel (0.25 to 0.45% C), low alloy steels, copper and light alloys such as Aluminium, Magnesium, Titanium, and Berylium.

13.2.2. Hot-Working. Hot working or hot forming can be defined as plastic deformation of metals and alloys under conditions of temperature and strain rate, such that recovery and recrystalization take place simultaneously with the deformation. The hot working is carried out above the recrystallization temperature of the material, and after hot working, a fine grained recrystallized structure is obtained. Hot working occurs at an essentially constant flow stress. Fig. 13.3 (*b*). In addition to the characteristics discussed in chapter 3, hot working of metals has got the following characteristics :

1. Blow holes and porosities in the work material are eliminated during hot working, as a result of internal welding.
2. Stress annealing is not required after hot working.
3. Because of large deformations possible in one stage, the number of stages needed to form a part is very much less compared to cold working. Also, interstage annealing can be avoided.
4. Because of higher temperatures, there is oxidation and scaling of the material's surface. This results in heavy loss of work material. Also, there oxide flakes may get embedded in the surface resulting in poor surface finish.
5. Due to the occurrence of surface decarbonization in steels, the surface strength and hardness of the component is reduced.
6. Thin gauge sheets cannot be made by hot working.
7. Because of high working temperatures, there is a serious problem of surface reactions between the metal and the furnace atmosphere, more so in the case of reactive metals like Titanium, calling for inert atmosphere.

Note. It should be noted that the difference between cold working and hot working depends only upon the temperature of recrystallization and not on any arbitrary temperature of deformation. Lead, tin and zinc recrystallize rapidly at room temperature after large deformations. Hence, the working of these metals at room temperature will constitute their hot working. Similarly working of Tungsten at about 1090°C will be termed its cold working, because it has recrystallization temperature above this value.

13.2.3. Warm working or Semihot working. It can be defined as plastic deforming of a metal or alloy under conditions of temperature and strain rate, such that the drawbacks of both cold working and hot working are eliminated and their advantages are combined together. For this, the selection of proper temperature for warm working is very important. This depends upon the following factors :

1. Yield or flow strength of the metal or alloy.
2. Ductility of the material.

3. Dimensional tolerance on the component.
4. Oxidation and scaling losses.

The variations of the above properties relative to the working temperature can be studied to arrive at the proper working temperature for warm working. For example, for 0.13% C steel; the following observations are made :

Yield-Strength. In general, yield strength decreases with increase in temperature. However, in the temperature range of 150°C to 350°C and 800°C to 900°C, it increases with, increases in temperature. The first temperature range is called blue brittleness range of steel and in the second range, structural changes occur in steel. Both these ranges are brittle ranges and if steel is worked in these temperature ranges, it will fracture. So, the best temperature ranges from the yield strength point of view are 400°C to 750°C and above 900°C.

Ductility. In general, the ductility or formability of a material increases with increase in temperature. However, in temperature ranges of 250°C to 350°C and 800°C to 900°C, it decreases with increase in temperature. Thus, from the ductility point of view, the best temperature ranges for the above mentioned steel are : 400°C to 750°C and above 900°.

Dimensional Tolerance and Scaling and oxidation losses. The dimensional tolerances increase rapidly above 700°C. Similarly, scaling and oxidation losses, which are negligible upto 700°C, increases very rapidly above this temperature.

From the above discussion, it is clear that the best temperature range for working the above-mentioned steel is 400°C to 700°C. Working of steel within this temperature range is called warm working or Semi-hot working.

13.3. EFFECT OF VARIABLES ON METAL FORM ING PROCESSES

The various variables affecting a metal forming process can be : temperature, strain rate or speed of deformation, friction etc.

1. **Temperature.** Temperature is an important variable in metal working process. Its effect has been studied in detail in the last article. In general, it can be said that with increase in temperature, the strength and hardness of the work-material decreases and its ductility increases, Fig. 13.6.

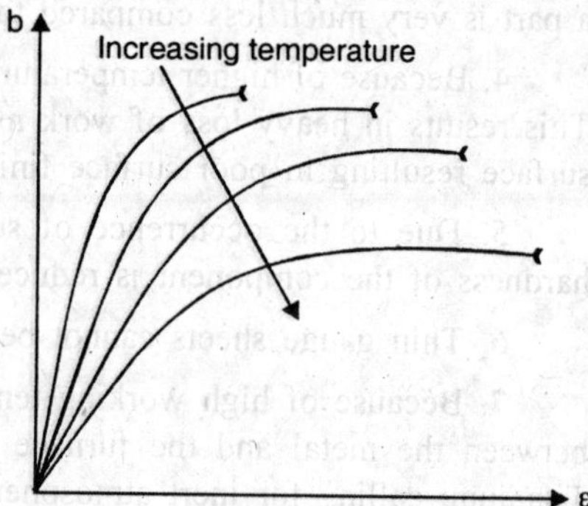

Fig. 13.6. Effect of Temperature on Stress-Strain Curve.

2. **Strain-rate.** Strain rate or deformation velocity has also great effect on metal working processes. Strain rate is defined as.

$$\dot{\varepsilon} = \frac{v}{h}$$

where v is the deformation velocity and h is the instantaneous height or length of the workpiece. It has been experimentally observed that the effect of strain rate is generally small at room temperature or low temperatures (cold working), but at high temperatures, i.e. hot working, there is substantial strain-rate sensitivity. In general, we can write the effects of strain rate as follows:

(*a*) The flow stress of the material increases with strain rate.

(*b*) The temperature of the work-material increases with strain rate, due to adiabatic heating.

(*c*) With temperature as a parameter, the tensile strength of the material increases with strain rate.

(*d*) Lubrication at the tool-workpiece interface improves, provided the film can be maintained.

It is clear from above that the strength of a material which is available at low temperatures when the material is worked at low strain rate, can be achieved at high working temperature by working the material at high strain rates.

Table 13.1 gives the typical values of deformation velocities for various tests and metal working processes.

Table 13.1. Typical Deformation Velocities.

Process	*Deformation velocity, m/s*
Tensile test	0.6×10^{-6} to 0.6×10^{-2}
Hydraulic press	0.025 to 0.35
Mechanical press	0.15 to 1.5
Tube drawing	0.05 to 0.5
Deep drawing	0.05 to 1.0
Hammer forging	2.5 to 10.0
Explosive forming	30 to 200

3. Friction. In metal forming processes where there is contact between the tool and the workpiece, there is always friction. The friction is a very important variable. It affects not only the forming forces but also the pattern of material deformation, which becomes increasingly non-uniform with friction. It is also a source of heat. In cold forming, where lubrication can be achieved efficiently, the coefficient of friction can be of the order of 0.1, but in hot working it may be of the order of 0.6 or greater. Under such conditions, there is sticking of the work material to the tool face and subsurface plastic flow occurs rather than sliding.

13.4. METHODS OF ANALYSIS OF MANUFACTURING PROCESSES

When a metal is deformed by a manufacturing process, the total work per unit volume done on the metal is given by,

$$W_T = W_p + W_f + W_r$$

where W_p = ideal work of deformation

W_f = work to overcome friction at the metal-tool interface

W_r = the redundant work

The redundant work is the work involved in the internal shearing process due to non-uniform deformation. It does not contribute to the change in shape of the body. This concept is explained in Fig. 13.7, where a billet is shown before compression and after compression. If there is perfect lubrication at the metal-tool interface, the grid will remain undistorted. The square grid (before compression) will become rectangular (after compression). The height of each grid block will get shortened and its length will get increased. Such a deformation is called uniform deformation'. However if friction is present at the metal-tool interface, then the grid

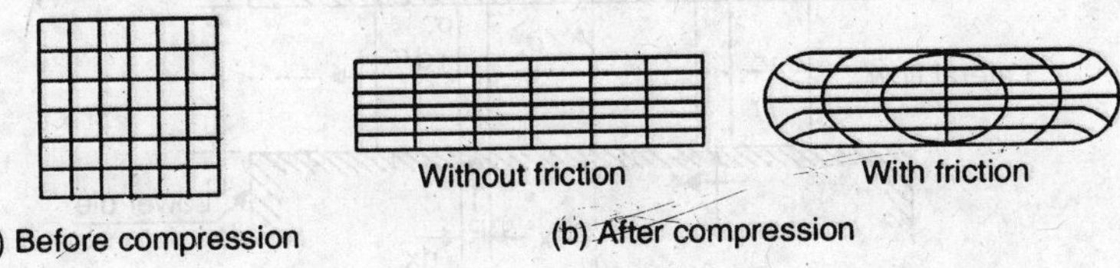

Fig. 13.7. Mode of Deformation.

will become distorted accompanied by barrelling or bulging on the sides. The extra work which goes to distort the grid and also to create bulge is a waste and is called the redundant work'. Such a deformation is called "non-uniform deformation."

The commonly used methods for the analysis of metal forming processes are :

1. Slab Method.
2. Upper-Bound Method
3. Slip-Line Field Method.

The analysis is basically same for cold forming and hot forming, with difference only in flow and friction characteristics. The analysis of the metal forming processes by the above methods is only approximate due to the many assumptions made regarding the behaviour of the work material and the mode of deformation. The approximation is better at low co-efficients of friction. Thus, the correlation is better with cold working, low friction processes than with hot working, high friction processes.

In this chapter, we shall deal with only the Slab Method.

Slab-analysis technique' or 'Elementary theory' makes the following assumptions :

1. The material is isotropic, incompressible and the elastic strains are neglected.
2. Deformation is homogeneous throughout the deforming materials under study.
3. Stresses on a plane normal to the flow direction are principal stresses.

In this method a 'slab' of infinitesimal thickness is considered and a force balance (equilibrium equation) is made on it. The resulting differential equation of static equilibrium is solved with the help of appropriate boundary conditions and yield criteria.

Here we shall not consider high-energy rate forming. Now, we shall discuss the application of slab method for the analysis of common metal forming processes.

13.5. OPEN DIE FORGING

In open die forging, the work piece is upset, compressed or forged between two flat over-hanging dies. If the co-efficient of friction at the die-workpeice interface is zero and if the material is assumed to be rigid perfectly plastic, then the force-required to forge a specimen is equal to the product of yield stress σ_0 and the projected area at any instant. Here, we shall take the case when friction is there in the interface.

1. **Plane Strain Forging.** Fig. 13.8 shows the schematic view of open die plane strain forging. In plane strain, the specimen is not free to flow in the direction perpendicular to the plane of the page. As the dies come closer to each other, there is lateral flow of the work

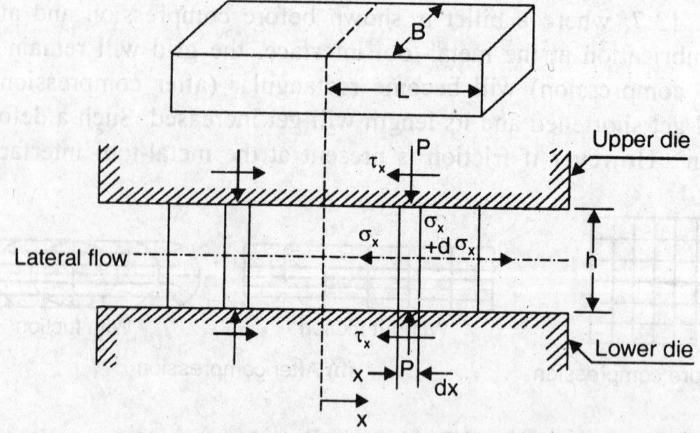

Fig. 13.8. Plane Strain Open Die Forging.

material. Due to this, frictional shear stresses are set up at die contact surface, which are directed towards the centre line, opposing the metal flow. Due to the presence of interface friction, the horizontal stress σ_x varies along the length of the rectangular workpiece. In the following analysis, it is assumed that the workpiece material behaves like an ideal plastic material, the entire workpiece is in a plastic state and that the stresses do not vary with height. At any instant of forging, the equilibrium equation of a small element of width dx, in the x-direction gives,

$$(\sigma_x + d\sigma_x)\, Bh - \sigma_x\, Bh - 2\tau_x\, dx\, B = 0$$

$$\frac{d\sigma_x}{dx} - \frac{2\tau_x}{h} = 0$$

(*i*) For sliding friction; assuming coulomb friction with constant co-efficient of friction μ, we have

$$\tau_x = \mu p$$

$$\therefore \quad \frac{d\sigma_x}{dx} - \frac{2\mu p}{h} = 0 \qquad \text{...(13.14)}$$

Assuming σ_x and p as principal stresses, and taking Tresca's yield theory, we have for plane strain,

$$\sigma_1 - \sigma_3 = \sigma_0 = 2k$$

$$\therefore \quad \sigma_x + p = \sigma_0 \qquad \text{...}p \text{ being compressive.}$$

$$\therefore \quad \frac{d\sigma_x}{dx} = -\frac{dp}{dx}$$

$\therefore$ Equation (13.14) becomes,

$$\frac{dp}{p} = -\frac{2\mu}{h}\cdot dx$$

Integrating,

$$\log_e p = -\frac{2\mu}{h}\cdot x + C$$

Now at $x = L$, $\sigma_x = 0$ (stress free surface) and we have

$$p = \sigma_0$$

$$\therefore \quad \log_e \sigma_0 = -\frac{2\mu}{h}\cdot L + C$$

$$\therefore \quad C = \log_e \sigma_0 + \frac{2\mu}{h}\cdot L$$

$$\therefore \quad \log_e \frac{p}{\sigma_0} = \frac{2\mu}{h}(L - x)$$

$$\therefore \quad p = \sigma_0 \cdot e^{[2\mu(L-x)/h]}$$

$$\text{or} \quad p/2k = p/\sigma_0 = e^{2\mu(L-x)/h} \qquad \text{..(13.15)}$$

$$\therefore \quad \sigma_x = \sigma_0 - p = \sigma_0\left[1 - e^{2\mu(L-x)/h}\right] \qquad \text{...(13.16)}$$

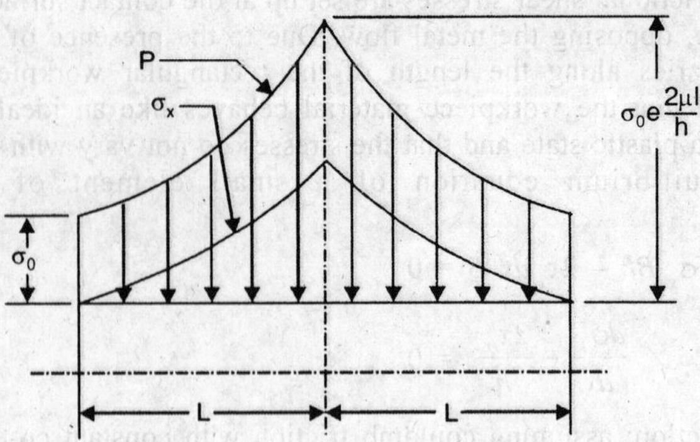

Fig. 13.9. Distribution of p and σ_x.

The distribution of p and σ_x is shown in Fig. 13.9. p and σ_x will be maximum at the centre, illustrating the "friction hill". So, putting $x = 0$.

$$p_{max} = \sigma_0 \cdot e^{2\mu L/h}$$

If μ is small, this can be approximated as,

$$(p/2k)_{max} \cong 1 + 2\mu \frac{L}{h} \qquad ...(13.16\ a)$$

$$(\sigma_x)_{max} = \sigma_0 \left[1 - e^{2\mu L/h}\right]$$

As h decreases, p will increase for the same value of μp is directly proportional to μ.

Now total forging load,

$$Pt = 2B \int_0^L p\, dx$$

Average pressure is,
$$p_a = \frac{P_t}{2BL} = \frac{1}{L} \int_0^L p\, dx$$

$$= \frac{\sigma_0}{L} \left| e^{\frac{2\mu}{h}(L-x)} \cdot \frac{1}{-\frac{2\mu}{h}} \right|_0^L$$

$$= \frac{\sigma_0}{L} \left[-\frac{1}{2\mu/h} + e^{\frac{2\mu L}{h}} \cdot \frac{1}{2\mu/h} \right]$$

$$\therefore \quad p_a = \frac{\sigma_0}{L} \cdot \frac{h}{2\mu} \left[-1 + e^{\frac{2\mu L}{h}} \right] \qquad ...(13.17)$$

If $2\mu L/h$ is small, then,

$$p_a/\sigma_0 = \frac{p_a}{2k} \cong \frac{h}{2\mu L}\left[1 + \frac{2\mu L}{h} + \frac{4\mu^2 L^2}{2h^2} - 1\right] \quad ...(13.17a)$$

$$\approx 1 + \frac{\mu L}{h} \quad ...(13.17b)$$

(*ii*) **Sticking friction.** When there is a condition of sticking friction, the workpiece material does not slide along the die face and actually becomes a part of the die face and there is subsurface flow of metal. The frictional shear stress at the die interface cannot be more than the yield shear stress of the material.

So, for sticking friction,

$$\tau_x = K = \frac{\sigma_0}{2} \quad \text{(Tresca condition)}$$

∴ Equilibrium equation is,

$$\frac{d\sigma_x}{dx} - \frac{\sigma_0}{h} = 0$$

or

$$-\frac{dp}{dx} - \frac{\sigma_0}{h} = 0$$

∴

$$p = -\frac{\sigma_0}{h}\cdot x + c \quad ...(13.18)$$

Now usually the sliding friction exists near the edges of the workpiece ($x = L$), where the pressure is low, but at some distance nearer to the centre line, sticking friction may exist.

Let sticking occurs at $x = x_s$, where

$$\tau_s = \mu p = \frac{\sigma_0}{2} \quad [\text{here } p = p_s \text{ (sticking)}]$$

∴ From equation (13.15)

$$\frac{\sigma_0}{2} = \mu\sigma_0 \cdot e^{2\mu(L - x_s)/h}$$

∴

$$e^{2\mu(L-x_s)/h} = \frac{1}{2\mu}$$

∴

$$x_s = L - \frac{h}{2\mu}\log_e\left(\frac{1}{2\mu}\right) \quad ...(13.18a)$$

∴ at $x = x_s$ $p = p_s$

∴ From equ. (13.18),

$$p_s = -\frac{\sigma_0}{h}\cdot x_s + C$$

∴

$$C = p_s + \frac{\sigma_0}{h}\cdot x_s$$

$\therefore$ In the sticking region, $$p = p_s + \frac{\sigma_0}{h}(x_s - x)$$

where $$p_s = \sigma_0 \cdot e^{[2\mu(L - x_s)/h]} \quad \text{...(13.18}b\text{)}$$

(*iii*) If the friction is high, for example, in hot forging it may reach sticking friction, then sticking regime extends over the whole interface. For this situation,

$$\tau_x = k = \frac{\sigma_0}{2}$$

$\therefore$ Equation (13.14) becomes,

$$hd\sigma_x - 2kdx = 0$$

Now $$\sigma_x + p = -\sigma_0 \quad \text{(yield condition)}$$

$$\therefore \quad d\sigma_x = -dp$$

$$\therefore \quad -h\,dp - 2k\,dx = 0$$

$$\therefore \quad \frac{dp}{2k} = -\frac{dx}{h}$$

$\therefore$ Integrating, $$\frac{p}{2k} = -\frac{x}{h} + c$$

Now at $x = L$, $p = 2k = \sigma_0$ (since $\sigma_x = 0$)

$$\therefore \quad c = 1 + \frac{L}{h}$$

$$\therefore \quad \frac{p}{2k} = 1 + \frac{1}{h}(L - x)$$

$$\therefore \quad \left(\frac{p}{2k}\right)_{max} = 1 + \frac{L}{h} \quad \text{(at centre, } x = 0\text{).} \quad \text{...(13.18}c\text{)}$$

Average pressure will be, $$p_a = 2k\left(1 + \frac{L}{2h}\right). \quad \text{...(13.18}d\text{)}$$

Note. The value of co-efficient of friction at sticking conditions :

μ = 0.5, Tresca yield condition

= 0.577, Von-Mises yield condition

2. **Axisymmetric forging.** Fig 13.10 shows the stresses acting on a sector of circular disc. Taking the equilibrium of forces in the radial direction,

$$\sigma_r \cdot h \cdot rd\theta - (\sigma_r + d\sigma_r)\,h\,(r + dr)d\theta + 2\sigma_\theta\, h\,dr \sin\frac{d\theta}{2} - 2\tau r d\theta dr = 0$$

Taking $\sin\frac{d\theta}{2} \approx \frac{d\theta}{2}$, the above equation reduces to,

$$\sigma_r\, h\, d_r + d\sigma_r\, rh - \sigma_\theta\, hdr - 2\tau r\, dr = 0$$

The disc being axi-symmetric,

$$d\varepsilon_\theta = d\varepsilon_r$$

∴ From equn. (13.13), $\sigma_\theta = \sigma_r$

(The material being rigid perfectly plastic, the incremental plastic strains will be the total incremental strains).

Making this substitution, we finally get

$$\frac{d\sigma_r}{dr} - \frac{2\tau}{h} = 0$$

Now $\tau = \mu p = -\mu\sigma_z$ $(p = -\sigma_z)$

$$\therefore \quad \frac{d\sigma_r}{dr} + \frac{2\mu\sigma_z}{h} = 0$$

Now using Von-Mise's yield condition,

$$2\sigma_0^2 = (\sigma_1 - \sigma_2)^2 + (\sigma_2 - \sigma_3)^2 + (\sigma_3 - \sigma_1)^2$$

Taking σ_r, σ_θ and σ_z as principal stresses, we get

$$2\sigma_0^2 = (\sigma_r - \sigma_z)^2 + (\sigma_z - \sigma_\theta)^2 + (\sigma_\theta - \sigma_r)^2$$

We have $\sigma_r = \sigma_\theta$

$$2\sigma_0^2 = (\sigma_r - \sigma_z)^2 + (\sigma_z - \sigma_r)^2 + (\sigma_r - \sigma_r)^2$$

$$\therefore \quad 2\sigma_0^2 = 2(\sigma_r - \sigma_z)^2$$

$$\therefore \quad \sigma_0 = \sigma_r - \sigma_z = \sigma_r + p$$

$$\therefore \quad d\sigma_r = -\,dp$$

$$\therefore \quad \frac{dp}{p} = -\frac{2\mu r}{h}$$

$$\therefore \quad \log_e p = -\frac{2\mu r}{h} + C$$

At the outer surface, $r = R,\ \sigma_r = 0$

$$\therefore \quad p = -\sigma_z = \sigma_0$$

$$\therefore \quad C = \log_e \sigma_0 + \frac{2\mu R}{h}$$

$$\therefore \quad \log_e \frac{p}{\sigma_0} = \frac{2\mu}{h}(R - r)$$

or

$$p = \sigma_0 \cdot e^{\frac{2\mu}{h}(R-r)} \qquad \text{...(13.19)}$$

Average pressure will be

$$p_a = \frac{\int_0^R 2\pi pr\,dr}{\pi R^2}$$

$$= \frac{\sigma_0}{2}\left(\frac{h}{\mu R}\right)^2\left[e^{2\mu R/h} - \frac{2\mu R}{h} - 1\right] \quad ...(13.19a)$$

Now, the pressure will be maximum when

$$r = 0$$

$$\therefore \quad \left(\frac{p}{\sigma_0}\right)_{max} = e^{(2R\mu/h)} \quad ...(13.19b)$$

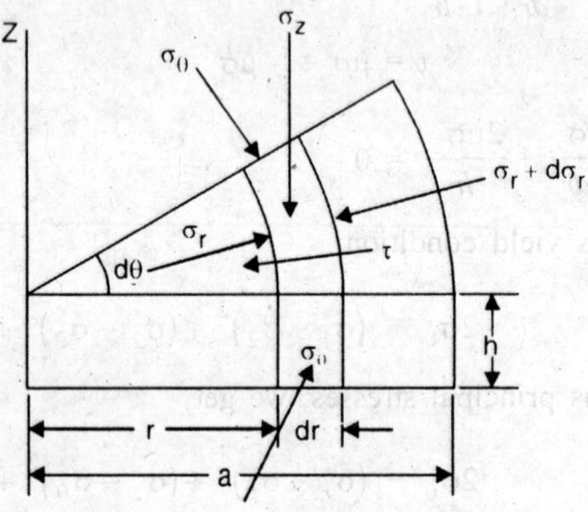

Fig. 13.10. Axisymmetric Forging.

(*i*) **Sticking Friction.** When sticking friction occurs over a portion of the disc, the problem can be analysed on the same lines as for plane strain forging. Thus, sticking radius will be given as, see equ. (13.18*a*).

$$R_s = R - \frac{h}{2\mu}\log_e\left(\frac{1}{2\mu}\right) \quad ...(13.19c)$$

For Von-Mises yield condition. $R_s = R - \frac{h}{2\mu}\log_e\left(\frac{1}{\sqrt{3}\,\mu}\right)$ ...(13.19*d*)

Pressure at the sticking radius, see equ. (13.18*b*)

$$p_s = \sigma_0\, e^{[2\mu(R-R_s)/h]} \quad ...(13.19e)$$

In the sticking region, $p = p_s + \frac{\sigma_0}{h}(R_s - R)$ ...(13.19*f*)

Pressure at the centre, that is, $R = 0$,

$$p_c = p_s + \frac{\sigma_0}{h}\cdot R_s \quad ...(13.19g)$$

(*ii*) For hot forging, the interface friction often reaches the sticking value. Then

$$\tau = k$$

$\therefore$ From $\quad \frac{d\sigma_r}{dr} - \frac{2\tau}{h} = 0$

$$\frac{d\sigma_r}{dr} - \frac{2k}{h} = 0$$

Now $\qquad d\sigma_r = -\,dp$

$\therefore \qquad -\,hdp - 2k\,dr = 0$

Integrating, $\qquad p = -\dfrac{2k}{h}\cdot r + \text{constant}$

At $\qquad r = R,\ \sigma_r = 0$ and $p = \sigma_0$ (From yield condition)

$\therefore \qquad \text{Constant} = \sigma_0 + \dfrac{\sigma_0}{h}\cdot R$

$\therefore \qquad p = \sigma_0 + \dfrac{\sigma_0}{h}\cdot(R - r)$

or $\qquad \dfrac{p}{2k} = 1 + \dfrac{1}{h}(R - r)$

$$p_{max} = 2k\left(1 + \frac{R}{h}\right) \qquad \text{...(13.19}h\text{)}$$

$$p_a = 2k\left(1 + \frac{R}{2h}\right)$$

Empirical Methods to Compute Forging Loads

1. **Open Die Forging.** The load required to forge a flat section in open dies may be estimated by,

$$P = \bar{\sigma}\, A.C., N$$

where $\bar{\sigma}$ = mean flow, stress N/mm²

A = forging projection area, mm²

C = constant (constraint factor) to allow for inhomogeneous deformation.

The deformation resistance increases with Δ which is defined as,

$$\Delta = \frac{\text{mean thickness deforming zone}}{\text{length of deforming zone}}$$

$$= \frac{h}{2L} \qquad \text{(see Fig. 13.7)}$$

Then C is given as, $\qquad C = 0.8 + 0.2\,\Delta$

2. **Closed-die Forging.** The prediction of forging load in a closed-die forging process is quite difficult. Fig. 13.11 shows a typical curve of forging load *vs.* press stroke. The flash land (Fig. 3.19) is so designed that the extrusion of metal through the narrow flash opening is more difficult than the filling of the most intricate detail in the die. However, excessive restriction to metal flow in flash land will result in very high forging loads accompanied by die wear and breakage. The ideal design is minimum flash to do a job.

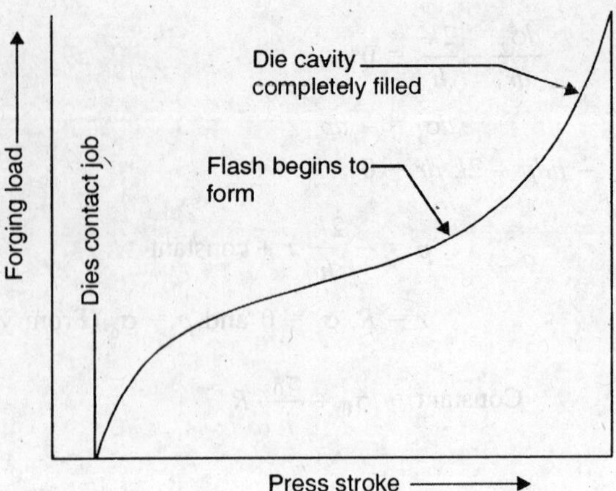

Fig. 13.11. Variation of Forging Load with Press Stroke.

For closed-die forging also, we can use the empirical relation,

$$P = \overline{\sigma} A.C.$$

where A = cross sectional area of forging at the parting line, including the flash.

C = a constant factor depending upon the complexity of the forging
= 1.2 to 2.5 for upsetting a cylinder between flat dies
= 3 to 8 for closed die forging of simple shapes with flash.
= 8 to 12 for more complex shapes

It may be of interest that every 10 MN of force corresponds approximately to 1000 kg of falling parts of hammer.

13.6. ROLLING

In rolling, the metal is plastically deformed by passing it between rolls. Rolling is done both hot and cold. The starting material is cast ingot, which is broken down by hot rolling into blooms, billets and slabs, which are further hot rolled into plate, sheet, rod, bar, pipe, rails or structural shapes. Cold rolling is usually a finishing process in which products made by hot rolling are given a good surface finish with increased mechanical strength of the material. The main objective in rolling is to decrease the thickness of the metal. Ordinarily, there is negligible increase in width, so that the decrease in thickness results in an increase in length.

Fig. 13.12 shows the typical geometry for rolling. A metal sheet with a thickness h_0 enters the rolls, passes through the roll gap and leaves with a reduced thickness h_1. Since the volume rate of metal flow has to remain constant, therefore, velocity at exit V_1 will be more than velocity at entrance V_0. The roll has a constant surface velocity V_r. Thus, there is relative sliding between the roll and the workpiece. The direction of this relative velocity changes at a point along the contact area, this point is known as the 'no slip point' or neutral point, N, where the velocities of roll and workpiece will be equal.

$$V_0 h_0 b_0 = V_1 h_1 b_1 \qquad \text{(volume constancy)}$$

For plane strain rolling, $b_0 \cong b_1 \cong b$

$$\therefore \quad V_1 = \frac{V_0 h_0}{h_1}$$

As $h_0 > h_1$

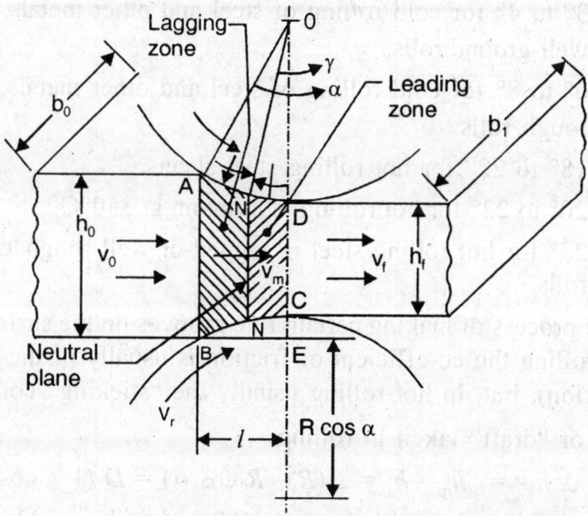

Fig. 13.12. Geometry of Rolling Process.

$\therefore$ $V_1 > V_0$, and we have

$$V_0 < V_r < V_1$$

At neutral plane $V_r = V_m$ (actually V_m is the component of V_r in horizontal direction, but as the angle of inclination is very small, we can take $V_r \cong V_m$).

Because of the relative velocities involved, we have two definitions :

$$\text{Backward slip} = \frac{V_r - V_0}{V_r}$$

$$\text{Forward slip} = \frac{V_1 - V_r}{V_r}$$

Zone *ANNB* is lagging zone ($V < V_r$) and zone *NDCN* is leading zone ($V > V_r$).

The angle α between the entrance plane and the centre-line of rolls is called the 'angle of contact' or 'angle of bite'.

Let us consider the moment when the rolling process is just going to start. The roll contacts the entering material (may be the strip) at point A, Fig. 13.13 (*a*). This contact results in a normal face P_r between the roll and workpiece, and the tangential frictional force *F*. If the resultant of P_r and *F*, that is *T*, is sloped to the right, its component along the *X*-axis $T_x > 0$ which tends to push the workpiece into the roll opening and thus ensures biting. Thus the condition of biting, that is the condition for unaided entry of the workpiece in the rolls is,

$$T_x = F \cos \alpha - P_r \sin \alpha > 0$$

or
$$\frac{F}{P_r} > \tan \alpha$$

but
$$F = \mu P_r \text{ (Coulomb friction)}$$

$\therefore$
$$\mu > \tan \alpha$$

Let μ = tan β, where β is the angle of friction, therefore, the limiting condition becomes,

$$\tan \beta > \tan \alpha \text{ or } \beta > \alpha, \text{ or } \alpha < \beta$$

which means that rolls 'bite' a workpiece, if the angle of friction is greater than the angle of tangency (biting). If tan α exceeds μ, then the workpiece cannot be drawn into the rolls. The values of biting angles are usually.

α = 3° to 4° for cold rolling of steel and other metals,with lubrication or well-ground rolls.

= 6° to 8° for cold rolling of steel and other metals, with lubrication on rough rolls.

= 18° to 22° for hot rolling steel sheets.

= 20° to 22° for hot rolling aluminium at 350°C

= 28° for hot rolling steel in ragged or well-roughed rolls.

"Ragging" is the process of making certain fine grooves on the surface of the roll to increase the friction. In cold rolling the co-efficient of friction is usually of the order of 0.1 (due to the possibility of lubrication), but, in hot rolling usually the "sticking" conditions exist.

Total reduction or "draft" taken in rolling,

$$\Delta h = h_0 - h_1 = 2(R - R\cos\alpha) = D(1 - \cos\alpha)$$

Usually, the reduction in blooming mills is about 100 mm and in slabbing mills, about 50 to 60 mm. The projected length of the arc of contact is,

$$l = R.\sin\alpha$$

or $$l = \sqrt{BC^2 - CE^2}$$

Now $$BC = \sqrt{R.\Delta h} \text{ and } CE = R(1 - \cos\alpha) = 0.5\,\Delta h$$

$\therefore$ $$l = \sqrt{R.\Delta h - (0.5\,\Delta h)^2}$$

$$P = \sigma$$

Usually,$(0.5\,\Delta h)^2$ is $< R\,\Delta h$

$\therefore$ $$l \cong (R\,\Delta h)^{1/2}$$

The vertical component of P_r is known as the "rolling load", P. This is the force with which the rolls press against the metal. Due to reaction, the metal tends to separate the rolls apart therefore, this force is also called the "Separating force".

The specific roll pressure, $p = \dfrac{P}{\text{contact area}} = \dfrac{P}{b.l}$

where b is the width of the sheet. For uniform deformation and no friction, the specific roo pressure may be taken equal to mean yield stress (flow stess), $\bar{\sigma}$, of the material. Thus, the rolling load is given as, $P = \bar{\sigma}.l.b$ The various methods to reduce the separating force are :

1. Smaller roll diameter (which reduces contact area).
2. Lower friction.
3. Higher workpiece temperature (even though friction will increase, but 'p' will be smaller)
4. Take smaller bites, thereby reducing the contact area.
5. From equation (13.10), it is clear that yield stress of the material in one direction is a function of the stress in the other principal stress. Therefore, if we apply tensile force to the workpiece in the horizontal direction, the compressive yield strength of the material in the vertical direction will be lower, hence, the separating force will be smaller. Both "back tension" and "front tension" can be applied.

(*a*) at the start of rolling

(*b*) after roll gap is filled by metal

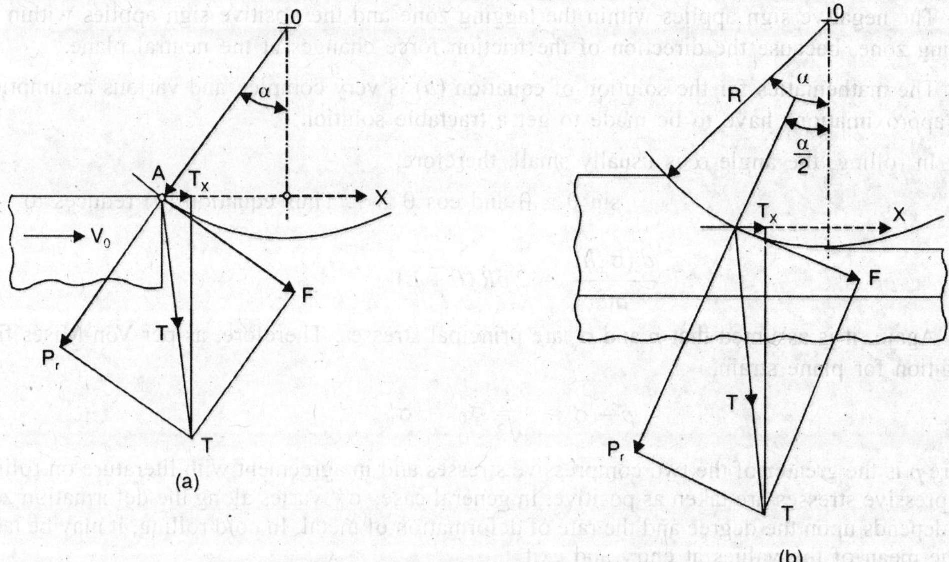

Fig. 13.13. Biting of Workpiece by Rolls.

Analysis. The stress equilibrium of an element in rolling is shown in Fig. 13.14. The following assumptions are made :

1. Rolls are straight, rigid cylinders.
2. Strip is wide compared with its thickness, so that no widening of strip occurs (plane strain conditions).
3. The material is rigid perfectly plastic (constant yield strength).
4. The co-efficient of friction is constant over the tool-work interface.

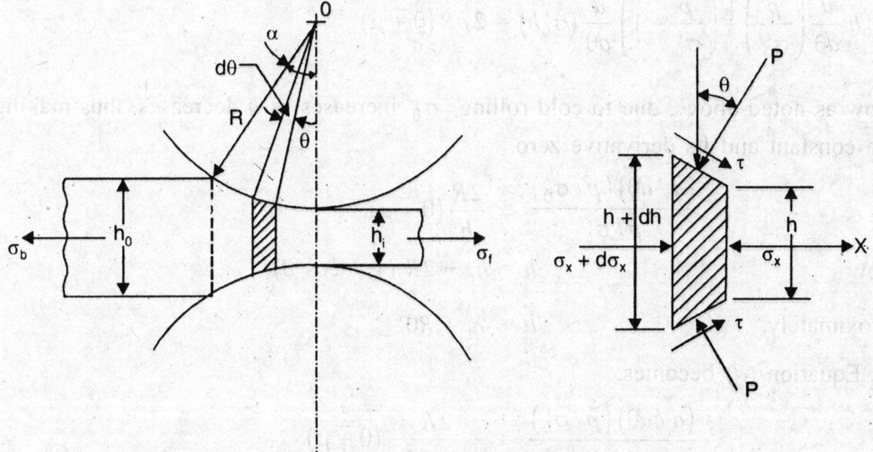

Fig. 13.14. Stress Equilibrium of an Element in Rolling

Considering the thickness of the element perpendicular to the plane of paper to be unity, we get equilibrium equation in x-direction as,

$$-\sigma_x h + (\sigma_x + d\sigma_x)(h + dh) - 2pR\,d\theta \sin\theta + 2\,\tau_x R\,d\theta \cos\theta = 0$$

For sliding friction, $\tau_x = \mu p$. Simplifying and neglecting second order terms, we get

$$\frac{d(\sigma_x h)}{d\theta} = 2pR(\sin\theta \mp \mu \cos\theta) \qquad ...(a)$$

The negative sign applies within the lagging zone and the positive sign applies within the leading zone, because the direction of the friction force changes at the neutral plane.

The mathematics for the solution of equation (*a*) is very complex and various assumptions and approximations have to be made to get a tractable solution.

In rolling, the angle α is usually small, therefore,

$\sin\theta \cong \theta$ and $\cos\theta \cong 1$. Thus equation (*a*) reduces to

$$\frac{d(\sigma_x h)}{d\theta} = 2pR(\theta \mp \mu) \quad ...(b)$$

Again, it is assumed that p and σ_x are principal stresses. Therefore, as per Von-Mises field condition for plane strain,

$$p - \sigma_x = \frac{2}{\sqrt{3}}\sigma_0 = \sigma_0' \quad ...(c)$$

where p is the greater of the two compressive stresses and in agreement with literature on rolling, compressive stresses are taken as positive. In general case, σ_0' varies along the deformation zone and depends upon the degree and the rate of deformation of metal. In cold rolling, it may be taken as the mean of the values at entry and exit.

With the help of (*c*), eqn. (*b*) can be written as

$$\frac{d}{d\theta}\left[h(p-\sigma_0')\right] = 2pR(\theta \mp \mu)$$

or
$$\frac{d}{d\theta}\left[\sigma_0' h\left(\frac{p}{\sigma_0'}-1\right)\right] = 2pR(\theta \mp \mu)$$

$$\therefore\ \sigma_0' h\frac{d}{d\theta}\left(\frac{p}{\sigma_0'}\right) + \left(\frac{p}{\sigma_0'}-1\right)\frac{d}{d\theta}(\sigma_0' h) = 2pR(\theta \mp \mu)$$

Now as noted above, due to cold rolling, σ_0' increases as h decreases, thus making $\sigma_0' h$ nearly a constant and its derivative zero.

$$\therefore\ \frac{(d/d\theta)(p/\sigma_0')}{p/\sigma_0'} = \frac{2R}{h}(\theta \mp \mu) \quad ...(d)$$

now
$$h = h_1 + 2R(1-\cos\theta)$$

or approximately,
$$h = h_1 + R\theta^2$$

$\therefore$ Equation (*d*) becomes,

$$\frac{(d/d\theta)(p/\sigma_0')}{p/\sigma_0'} = \frac{2R}{h_1 + R\theta^2}(\theta \mp \mu)$$

or
$$\frac{d(p/\sigma_0')}{p/\sigma_0'} = \frac{2R}{h_1 + R\theta^2}(\theta \mp \mu)\,d\theta$$

Integrating this equation,

$$\log_e(p/\sigma_0') = \int \frac{2R\theta\,d\theta}{h_1 + R\theta^2} \mp \int \frac{2R\mu}{h_1 + R\theta^2}\cdot d\theta \quad ...(e)$$

Now, the first term on the R.H.S.,

$$\int \frac{2R\theta\, d\theta}{h_1 + R\theta^2} = \int \frac{2R\theta\, d\theta}{h}$$

$$= \int \frac{2\theta\, d\theta}{h/R}$$

Now $$h/R = \frac{h_1}{R} + \theta^2$$

$\therefore$ $$\frac{d}{d\theta}(h/R) = 2\theta \text{ } (h_1 \text{ and } R \text{ being constants})$$

$\therefore$ $$\int \frac{2\theta\, d\theta}{h/R} = \log_e \frac{h}{R}$$

Second term on R.H.S.

$$\int \frac{2R\mu}{h_1 + R\theta^2} \cdot d\theta = \int \frac{2\mu}{h_1/R + \theta^2}\, d\theta$$

$$= 2\mu \sqrt{\frac{R}{h_1}} \tan^{-1} \sqrt{\frac{R}{h_1}}\, \theta$$

$\therefore$ Equation (*e*) becomes,

$$\log_e (p/\sigma_0') = \log_e (h/R) \mp 2\mu \sqrt{\frac{R}{h_1}} \tan^{-1} \sqrt{\frac{R}{h_1}} \cdot \theta + \log_e C$$

or $$p = C\sigma_0' (h/R)\, e^{\mp \mu H} \quad ...(f)$$

where $$H = 2\sqrt{\frac{R}{h_1}} \tan^{-1} \sqrt{\frac{R}{h_1}} \cdot \theta \quad ..(g)$$

Now at entry, $\theta = \alpha$

Hence $H = H_0$ with θ replaced by α in equ. (*g*).

At exit $\theta = 0$, $\therefore$ $H = H_1 = 0$

Also at entry and exit (stress free points), $\sigma_x = 0$

$\therefore$ from equ. (*c*), $p = \sigma_0'$.

$\therefore$ In the lagging zone (entry zone),

$$\sigma_0' = C\sigma_0' \left(\frac{h_0}{R}\right) e^{-\mu H_0}$$

$\therefore$ $$C = \frac{R}{h_0} e^{\mu H_0}$$

$\therefore$ $$p = \sigma_0' \cdot \frac{h}{h_0} \cdot e^{\mu (H_0 - H)} \quad ...(13.20)$$

In the leading zone (exit zone), $C = R/h_1$

$$\therefore \qquad p = \sigma_0' \cdot \frac{h}{h_1} \cdot e^{\mu H} \qquad \text{... (13.21)}$$

The pressure rises from both entry and exit (where it is equal to σ_0') to a maximum at an intermediate point (similar to Fig. 13.9), forming a friction hill, where the direction of friction force changes, that is, the neutral point which can be determined by equating equations (13.20 and (13.21),

$$\frac{h_n}{h_0} \cdot e^{\mu\left(H_{(0)} - H_n\right)} = \frac{h_n}{h_1} \cdot e^{\mu H_n}$$

or

$$\frac{h_0}{h_1} = e^{\mu(H_0 - 2H_n)}$$

or

$$H_n = \frac{1}{2}\left(H_0 - \frac{1}{\mu} \log_e \frac{h_0}{h_1}\right) \qquad \text{...(13.21a)}$$

$\therefore$ From equation (*g*), above,

$$\theta_n = \sqrt{\frac{h_1}{R}} \tan\left(\sqrt{\frac{h_1}{R}} \cdot \frac{H_n}{2}\right) \qquad \text{...(13.21b)}$$

or

$$h_n = h_1 + 2R\left(1 - \cos\theta_n\right) \qquad \text{...(13.21c)}$$

As discussed above, the rolling load can be reduced by applying both "back tension" and "front tension", either individually or together. Back tension is applied by controlling the speed of the uncoiler relative to the roll speed and front tension may be created by controlling the coiler. The application of back tension moves the neutral plane towards the exit. If its value is continuously increased, the neutral point will eventually reach the roll exit and the rolls will start slipping over the metal surface. The application of front tension shifts the neutral plane towards the roll entry. The possible height reduction decreases with back tension since it will be difficult to roll the metal continuously. With front tension, the maximum possible height reduction increases since the pulling force increases. A study of the effect of sheet tensions has shown that back tension is about twice as effective as front tension, in reducing the rolling load.

At entry, $\sigma_x = -\sigma_b$ (back tension)

and at exit, $\sigma_x = -\sigma_f$ (front tension)

$\therefore$ at entry, $p = \sigma_0' - \sigma_b$ From yield condition

and at exit, $p = \sigma_0' - \sigma_f$

The equations (13.20) and (13.21) will become,

$$p = \left(\sigma_0' - \sigma_b\right) \cdot \frac{h}{h_0} \cdot e^{\mu(H_0 - H)} \qquad \text{...(13.21d)}$$

and

$$p = \left(\sigma_0' - \sigma_f\right) \cdot \frac{h}{h_1} \cdot e^{\mu H} \qquad \text{...(13.21e)}$$

As already written, the rolling load also decreases with the decrease in roll diameter, as

$$l = \sqrt{R \cdot \Delta h}$$

Therefore, for same Δh, if R is less, l will decrease and so load will decreases since the contact area is reduced. As the thickness of the sheet to be rolled goes on decreasing, the roll diameters go on decreasing. The 'spread' which is defined as,

$$\Delta b = (b_1 - b_0)$$

increases with an increase in roll diameter and the co-efficient of friction and fall in temperature of the metal during hot rolling. The forward slip which is about 3 to 10%, increases with radius of the roll, co-efficient of friction and with a decreases in the thickness of the bar. The above analysis is only an approximate one and it tends to correlate better with cold-working, low friction processes.

Maximum Draft. It has already been proved that if the strip is to enter the rolls unaided then, the following relation has to be satisfied between the angle of bite and co-efficient of friction between the roll and material surfaces.

$$\mu > \tan \alpha$$

Now, from Fig. 13.12, the projected length of arc of contact,

$$l = \sqrt{R \cdot \Delta h}, \text{ and}$$

$$\tan \alpha = \frac{l}{R - \frac{\Delta h}{2}} = \frac{\sqrt{R \, \Delta h}}{R - 0.5 \, \Delta h}$$

Since $R >> 0.5 \, \Delta h$, it can be written that

$$\tan \alpha \approx \sqrt{\frac{\Delta h}{R}}$$

Since $\mu \geq \tan \alpha$

∴ The maximum draft is given by

$$\mu \geq \sqrt{\frac{\Delta h}{R}}$$

or, $$(\Delta h)_{max} = \mu^2 R \qquad \text{...(13.21 f)}$$

It is clear that large rolls and high friction allow heavy draft. It is, however, possible to roll with greater draft if the strip is pushed into the rolls or strip is accelerated prior to biting in order to make use of force of inertia. If the friction is too high, the load becomes excessive. Again a compromise can be obtained by the help of theory.

Also, after the gap between rolls is filled will metal, rolling may proceed with angle of tangency α greater than β. The limit for both these conditions may be found by considering the forces on the rolls. Let us assume that the resultant of pressure force P_r is located in the middle of the arc of contact of the strip with the roll, Fig. 13.13 (*b*). The rolls will start slipping when the horizontal component of the average roll pressure P_r equals that of the surface friction. For small angles, the condition for no slipping of rolls or for continuous rolling will be,

$$F \cos \frac{\alpha}{2} > P_r \sin \frac{\alpha}{2}$$

or $$F/P_r > \tan \frac{\alpha}{2}$$

or $\tan \frac{\alpha}{2} \approx \mu$

or $\alpha \approx 2 \tan^{-1} \mu$

or $\alpha \approx 2\beta$

It is thus clear that the angle of bite can be increased by a factor of 2 after the strip enters the rolls. This angle is called "angle of nip". Thus, the "angle of nip" is twice the "angle of bite". This angle is of importance when very large drafts are required. The rolling can then be started by tapering the front end of the strip.

Rolling Load. It is apparent from above that the roll pressure varies along the arc of contact and is maximum at the neutral point. Thus the total rolling load (vertical), is given as,

$$P = R.b \int_0^{\alpha} p \cdot d\theta$$

This is best evaluated by graphical integration.

Note. Here, the flattening of rolls and effect of inclination have been neglected.

Roll Torque and Power. The total rolling load is distributed over the arc of contact. However, it can be assumed to be concentrated at a point along the arc of contact at a distance 'a' from the centre line of the rolls. So, torque on each roll is,

$$T = P \cdot a$$

The moment arm 'a' is generally expressed through the arm factor, $\lambda = \frac{a}{l}$.

The values of λ can be determined experimentally. Its values can be taken as,

$$\lambda = 0.43 \text{ for cold rolling with matte finished rolls.}$$
$$= 0.48 \text{ for cold rolling with a smooth surface.}$$

These values of λ neglect the effect of elastic deformation of rolls. Considering this, however, the value of λ can be taken as,

$$\lambda \approx 0.3 \text{ to } 0.4$$

For hot rolling, for $l/h_m \geq 1.5$, its value can be taken as from about 0.35 to 0.45, the values increasing with temperature. For $l/h_m < 1.5$, λ depends heavily on the l/h_m ratio and its value should be taken from appropriate curve. Here,

$$h_m = \text{mean height} = 0.5\,(h_0 + h_1)$$

Assuming that the power consumed is equal on both rolls,

$$\text{Power} = \frac{2T\omega}{100}, \text{ kW}$$

where T is in Nm and ω in rad/s.

13.6.1. Powder Rolling. In conventional rolling, the starting material may be either a cast ingot or an electroplated slab. However, we can start with powder also. The powder is introduced between the rolls and compacted to obtain a "green strip". This is sintered to get the desired strength. This is subsequently hot worked and/or cold worked and annealed. This process is known as "Powder Rolling" and dense sheets may be produced by this method. The advantages of powder rolling are :

1. The initial hot-ingot breakdown is eliminated. This results in large saving in capital investment.

2. Contamination present in hot rolling is minimized.

3. Sheets with very fine grain size or with a minimum of preferred orientation can be obtained.

13.7. DRAWING (WIRE, ROD, TUBE)

In wire, rod or tube drawing, the workpiece is pulled through a die resulting in reduction in the cross-section. A typical drawing die is shown in Fig. 13.15. β is the entrance angle and it provides the entry zone to allow the introduction of lubricant into the working zone and also to protect the work material against scoring by die edges. The angle β is usually about 40°. α is half die angle (usually 6° to 24°) and here the actual reduction is section occurs. The die bearing surface or land (a few mm long) serves to guide the wire or rod as it comes out from the working zone of the die. It ensures accuracy of dimensions from the working zone of the die. It ensures accuracy of dimensions and of the section's shape. The exit zone prevents damage to die bearing and scoring of finished product. There are a number of variables in the drawing process : work-material properties, reduction in cross-sectional area, die angle, drawing speed, and lubrication. The performance of the operation is affected by a change in one or more of these variables. The degree of drawing is measured in terms of "reduction of area" (*RA*) :

$$RA = \frac{r_0^2 - r_1^2}{r_0^2} = 1 - \left(\frac{r_1}{r_0}\right)^2 = 1 - \left(\frac{D_1}{D_0}\right)^2$$

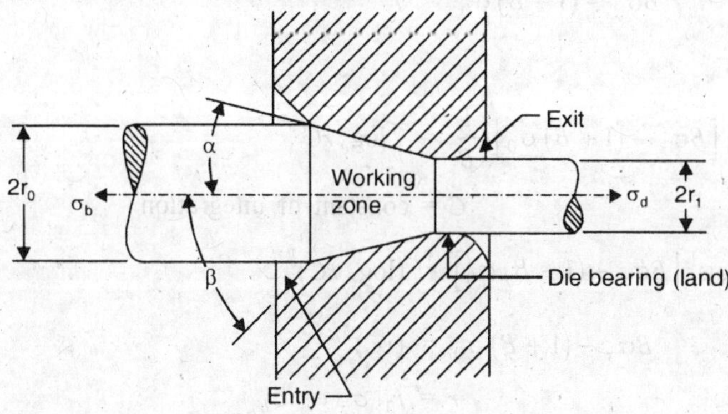

Fig. 13.15. A Drawing Die.

For fine wires, the reductions per pass of 15 to 25 percent are used, while for coarse wires this value per pass may be 20 to 50 per cent.

Die life is an important consideration. To increase die life, pressure on die should be reduced. This is accomplished by back tension (similar to rolling). This reduces the pressure but increases the drawing tension, σ_d. Die materials can be alloy steels, carbides and diamonds. Drawing speeds range from about 9 mpm for largest diameter rods, and 90 mpm for small rods and coils to 1500 mpm for very fine wires.

Analysis of Wire/Rod Drawing. Fig. 13.16 shows the stress acting on an element in drawing of a wire or rod. The equilibrium equation in *x*-direction will be,

$$(\sigma_x + d\sigma_x)\,\pi\,(r + dr)^2 - \sigma_x \cdot \pi r^2 + \tau_x \cdot \cos\alpha \left(2\pi r \frac{dx}{\cos\alpha}\right) + P_x \cdot \sin\alpha \left(2\pi r \frac{dx}{\cos\alpha}\right) = 0$$

From here, we get

$$\sigma_x \cdot 2r\,dr + d\sigma_x \cdot r^2 + 2r\tau_x\,dx + P_x \cdot 2r\,dx \tan\alpha = 0$$

Dividing by r^2dr and taking $dx/dr = \cot\alpha$, we get

$$\frac{d\sigma_x}{dr} + \frac{2}{r}(\sigma_x + P_x) + \frac{2\tau_x}{r}\cot\alpha = 0$$

Vertical component of $P_x \cong P_x$ and that of τ_x can be neglected due to small half die angles. Therefore, there are two principal stresses, σ_x and P_x.

$\therefore$ yield condition is (Tresca's),

$$\sigma_x + P_x = \sigma_0$$

Taking $\tau_x = \mu P_x = \mu(\sigma_0 - \sigma_x)$, we get

$$\frac{d\sigma_x}{dr} + \frac{2\sigma_0}{r} + \frac{2\mu}{r}(\sigma_0 - \sigma_x)\cot\alpha = 0$$

Taking $\mu\cot\alpha = B$,

$$\frac{d\sigma_x}{dr} = \frac{2}{r}\left[B\sigma_x - (1+B)\sigma_0\right]$$

$$\therefore \quad \frac{d\sigma_x}{B\sigma_x - (1+B)\sigma_0} = \frac{2}{r}dr$$

Integrating,

$$\log_e\left[B\sigma_x - (1+B)\sigma_0\right]\cdot\frac{1}{B} = 2\log_e rC$$

where C = constant of integration

$$\therefore \quad \log_e\left[B\sigma_x - (1+B)\sigma_0\right] = (\log_e rC)^{2B}$$

$$\therefore \quad B\sigma_x - (1+B)\sigma_0 = (rC)^{2B}$$

at $r = r_0, \sigma_x = \sigma_b$

$$\therefore \quad B\sigma_b - (1+B)\sigma_0 = (r_0C)^{2B}$$

$$C = \frac{\left[B\sigma_b - (1+B)\sigma_0\right]^{\frac{1}{2B}}}{r_0}$$

$$\therefore \quad B\sigma_x - (1+B)\sigma_0 = \left(\frac{r}{r_0}\right)^{2B}\cdot\left[B\sigma_b - (1+B)\sigma_0\right]$$

$$\therefore \quad \sigma_x = \frac{\sigma_0(1+B)}{B}\left[1 - \left(\frac{r}{r_0}\right)^{2B}\right] + \left(\frac{r}{r_0}\right)^{2B}\sigma_b \qquad \text{...(13.22)}$$

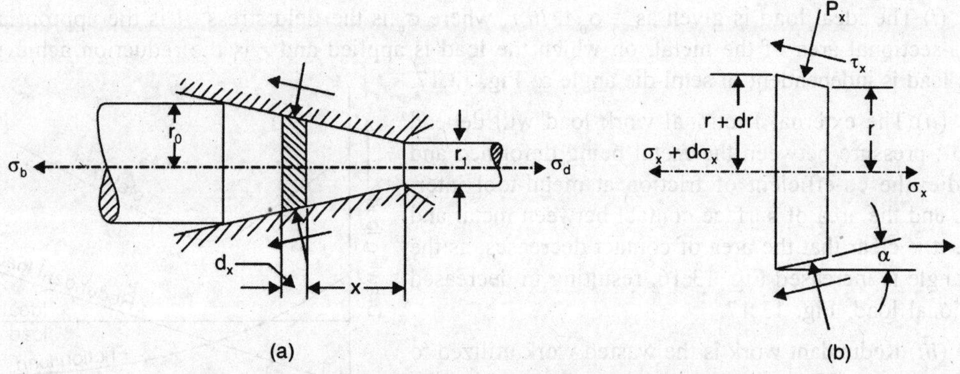

Fig. 13.16. Stress Equilibrium in Wire (Rod) Drawing.

Drawing stress,
$$s_d = \frac{\sigma_0 (1+B)}{B}\left[1-\left(\frac{r_1}{r_0}\right)^{2B}\right]+\left(\frac{r_1}{r_0}\right)^{2B}\cdot\sigma_b \quad \text{...(13.23)}$$

Now $\sigma_d \not> \sigma_0$ (In ideal case), therefore, maximum reduction can be found out.

Die Pressure : $P_x = \sigma_0 - \sigma_x$

By Calculating σ_x (eqn. 13.22) at different points along x-axis, the variation of P_x can be plotted from entry to exit of die.

For strip drawing, the expression will be,
$$\sigma_d = \frac{\sigma_0 (1+B)}{B}\left[1-\left(\frac{h_1}{h_0}\right)^{B}\right]+\left(\frac{h_1}{h_0}\right)^{B}\sigma_b \quad \text{...(13.24)}$$

The drawing load can be approximately determined by the "Work formula" (uniform deformation and no friction) as below :-
$$P = A_1.\bar{\sigma}.\ln\frac{A_0}{A_1}$$

Maximum Reduction or Draft per Pass. The maximum reduction taken per pass in wire or rod drawing, is limited by the strength of the deformed product. The exit end of the drawn rod will fracture at the die exit, when,

$\sigma_d / \sigma_0 = 1$, if there is no strain-hardening.

For Zero back stress, the condition will be,
$$\frac{1+B}{B}\left[1-(1-RA)^{B}\right] = 1$$

In wire and rod drawing, co-efficients of friction of the order 0.1 are usually obtained (by the use of proper lubricants).

Now, $B = \mu \cot \alpha$

Taking $\mu = 0.1$, and $\alpha = 6°$

$B = 0.1 \times 9.515 = 0.9515$

∴ From here, we will get the limiting or maximum reduction per pass, $RA = 50.5\%$.

Optimum die-angle for Wire drawing. As already discussed under Art. 13.4, the total load in any deformation process consists of three parts : ideal load, load to overcome external friction and the load to overcome redundant work.

(*i*) The ideal load is given as = $\sigma_0 A. \ln r$, where σ_0 is the field stress, A is the appropriate, cross-sectional area of the metal, on which the load is applied and r is the reduction achieved. This load is independent of semi-die angle α, Fig. 13.17.

(*ii*) The external frictional work load will depend upon : pressure between the metal being deformed and the die, the co-efficient of friction at metal-tool interface, and the area of surface contact between metal and tool. It is clear that the area of contact decreases, as the die angle is increased Fig. 13.16. resulting in decreased frictional load, Fig. 13.17.

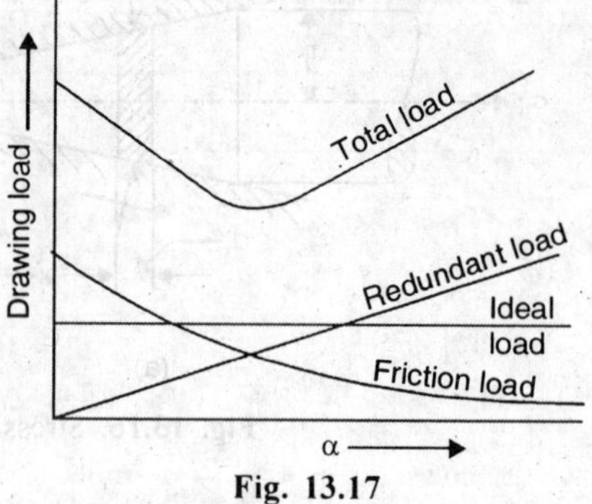

Fig. 13.17

(*iii*) Redundant work is the wasted work utilized to bend the metal fibres first one way and then back to the original direction of flow (in wire drawing). It is again clear from Fig. 13.16 that the angle of bend and hence the redundant work will increase as the die angle increases. This is made clear in Fig. 13.17.

By combining the above three types of load, the variation of total wire drawing load with semi-die angle, can be drawn, Fig. 13.17. It is clear that at a certain value of α (optimum semi-die angle), the total load is minimum. It is found in practice that the harder the metal, the smaller the optimum die angle, as given below :

$$\alpha = 24° \text{ for Aluminium} = 12° \text{ for Copper} = 6° \text{ for steel}$$

Tube or Pipe drawing. Tubes which are made by hot metal working processes (piercing, extrusion and rolling) are finally cold drawn to obtain better surface finish and dimensional tolerances, to enhance the mechanical properties of the pipe (by work hardening) and to produce tubes of reduced wall thickness or smaller diameter.

The three common methods of tube drawing are : Tube sinking. Tube drawing with a plug and Tube drawing with a moving mandrel, the last two methods being more widely used because in tube sinking, Fig. 13.18 (*a*), the inside of the tube is not supported and so during drawing operation, the inner surface becomes uneven and there will be tendency for the wall

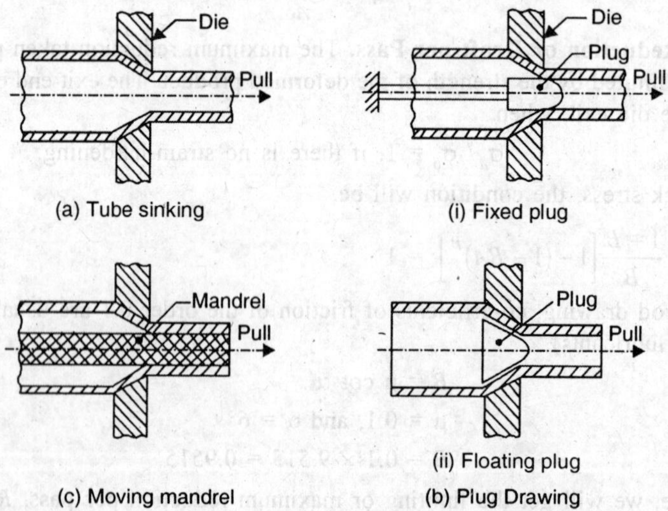

Fig. 13.18. Methods of Tube Drawing.

thickness to increase slightly. In plug drawing and movable mandrel drawing, both the inner and outer surfaces of the tube are controlled and we get tubes of better dimensional accuracy as compared to Tube sinking. In plug drawing, the plug (which may be either cylindrical or conical) can either be fixed, [Fig. 13.18 *b*(*i*)], or floating, [Fig. 13.18 *b*(*ii*)]. The friction with a fixed plug will be more than with a floating plug, so the reduction in area seldom exceeds 30% in this method. With a floating plug, this figure can be approximately 45% or with the same reduction, the drawing loads will be less with floating plug than with a fixed plug. The friction is minimized in Tube drawing with a movable mandrel. However, after tube drawing, the mandrel has to be removed by rolling which results in slightly increased tube diameter and reduced dimensional tolerances.

Analysis of Tube drawing. In tube or pipe drawing with a plug or a movable mandrel, most of the deformation occurs as a reduction in wall thickness. There is only a small reduction in inside diameter, needed only to insert plug or mandrel before drawing. Thus, there is no hoop strain and the analysis can be based on plane strain conditions.

The stress system for tube drawing is shown in Fig. 13.19.

The equilibrium equation can be written and solved as explained below :

If wall thickness h is small in comparison with pipe diameter D and the variation of D per pass is small, then pressure p varies little in the deformation zone across the thickness of pipe. The equilibrium equation can be written by considering the force acting on the annular element, with respect to the axis of drawing. Refer Fig. 13.19 (*a*) for plug drawing.

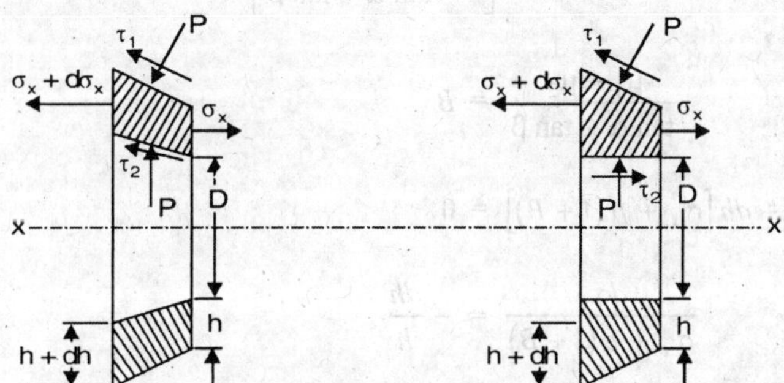

Fig. 13.19. Tube Drawing.

(*i*) Projection of the pressure forces on the die surface

$$\int_0^{2\pi} p \sin\alpha \frac{dx}{\cos\alpha} \cdot \frac{D}{2} \, d\theta = p\pi D \tan\alpha \, dx$$

where α = semi-die angle.

(*ii*) Projection of the pressure forces acting upon the plug

$$-\int_0^{2\pi} p \sin\beta \frac{dx}{\cos\beta} \cdot \frac{D}{2} \, d\theta = -p\pi D \tan\beta \, dx$$

where β = semi-cone angle of mandrel; for a cylindrical plug $\beta = 0$.

(*iii*) Projection of the friction forces $\tau_1 = \mu_1 p$ against the die plate

$$\int_0^{2\pi} \mu_1 p \cos\alpha \frac{dx}{\cos\alpha} \cdot \frac{D}{2} \, d\theta = \mu_1 p \pi D dx$$

where μ_1 = co-efficient of friction between tube and die wall (*iv*) Projection of the friction force $\tau_2 = \mu_2 p$ against the plug,

$$\int_0^{2\pi} \mu_2\, p \cos\beta \frac{dx}{\cos\beta} \cdot \frac{D}{2}\, d\theta = \mu_2 p \pi D dx$$

where μ_2 = co-efficient of friction between tube and plug

(*v*) Due to longitudinal stress σ_x,

$$(\sigma_x + d\sigma_x)(h + dh)\pi D - \sigma_x h \pi D$$

$$= (h d\sigma_x + \sigma_x dh)\pi D \text{, neglecting second order terms}$$

∴ Equilibrium equation will be,

$$(\sigma_x dh + h d\sigma_x)\pi D + p\pi D(\tan\alpha - \tan\beta)\,dx + p\pi D(\mu_1 + \mu_2)\,dx = 0$$

Now, the decrease in thickness dh as the element moves a distance dx is given by

$$dh = dx(\tan\alpha - \tan\beta)$$

$$\therefore \quad (\sigma_x dh + h d\sigma_x) + p.dh\left[1 + \frac{\mu_1 + \mu_2}{\tan\alpha - \tan\beta}\right] = 0$$

Let $$\frac{\mu_1 + \mu_2}{\tan\alpha - \tan\beta} = B$$

$$\therefore \quad h d\sigma_x + dh[\sigma_x + p(1 + B)] = 0$$

or $$\frac{d\sigma_x}{\sigma_x + p(1 + B)} = -\frac{dh}{h} \qquad \text{...(13.25)}$$

Principal stresses are approximately,

$$\sigma_1 = \sigma_x,\ \sigma_2 = \sigma_3 = -p$$

∴ Yield condition is (Tresca's).

$$\sigma_x + p = \sigma_0$$

Equation (13.25), then becomes,

$$\frac{d\sigma_x}{B\sigma_x - \sigma_0(1 + B)} = \frac{dh}{h}$$

Assuming that μ_1, μ_2, α and β are constant and there is no work hardening (σ_0 is constant), integrating the above equation,

$$\frac{1}{B}\log_e[B\sigma_x - \sigma_0(1 + B)] = \log_e h + \log_e C = \log_e Ch$$

$$\therefore \quad B\sigma_x - \sigma_0(1 + B) = Ch^B$$

Now, at entry, $h = h_0$, $\sigma_x = \sigma_{x_0} = 0$ (if there is no back tension)

$$\therefore \quad C = -\sigma_0(1 + B)\cdot h_0^{-B}$$

$$\therefore \quad \sigma_x = \sigma_0 \cdot \frac{1+B}{B}\left[1-\left(\frac{h}{h_0}\right)^B\right] \quad ...(13.26)$$

Now, the drawing stress at the exit σ_d, that is at $h = h_1$

$$\sigma_d = \sigma_0 \cdot \frac{1+B}{B}\left[1-\left(\frac{h_1}{h_0}\right)^B\right] \quad ...(13.27)$$

This equation is for conical plug. For cylindrical plug, $\beta = 0$,

therefore
$$B = \frac{\mu_1 + \mu_2}{\tan \alpha} \quad ...(13.27\ a)$$

Tube drawing with a moving mandrel. In plug drawing, the friction drag acts in the backward direction (that is towards the die entrance) both at the die tube interface and tube plug interface, Fig. 13.19 (*a*). But in tube drawing with a moving mandrel, the friction conditions at the tube mandrel interface get reversed because while the tube is getting elongated, the mandrel remains underformed. Thus, the friction force at the tube mandrel interface is directed toward the exit of the die, that is opposite to the direction of friction force at the die tube interface. Due to this, the relation for '*B*' will get changed to,

$$B = \frac{\mu_1 - \mu_2}{\tan \alpha - \tan \beta} \quad ...(13.27\ b)$$

The mandrels will be usually cylindrical, that is $\beta = 0$, therefore,

$$B = \frac{\mu_1 - \mu_2}{\tan \alpha} \quad ...(13.27\ c)$$

If $\mu_1 = \mu_2$, then $B = 0$ and the equilibrium equation will get reduced to

$$h d\sigma_x + dh(\sigma_x + p) = 0$$

From here, the equation can be obtained for draw stress, as

$$\sigma_d = \log_e \frac{1}{(h_0 / h_1)} \cdot \sigma_0 \quad ...(13.28)$$

Die Pressure. Now die pressure is given as,

$$p = \sigma_0 - \sigma_x, \text{ from yield condition}$$

Therefore calculating σ_x along the *x*-axis at different points, by equation 13.26, the variation of die pressure can be plotted from the point of entry to the die exit.

Maximum Reduction possible in Tube Drawing

As discussed under wire drawing, the maximum reduction is limited by the mechanical strength of the exit end of the tube, that is, when

$$\sigma_d / \sigma_0 = 1 \text{ (non-work hardening condition)}$$

∴ From eqn. (13.27), the limiting condition is,

$$\frac{1+B}{B}\left[1-\left(\frac{h_1}{h_0}\right)^B\right]_{max} = 1$$

Let us take, $\mu_1 = \mu_2 = 0.05$, $\alpha = 15°$, $\beta = 0$, then

$$B = \frac{\mu_1 + \mu_2}{\tan \alpha} = \frac{0.10}{0.268} = 0.373$$

$$\therefore \quad \frac{1+B}{B} = 3.68$$

$$\therefore \quad \left(\frac{h_1}{h_0}\right)_{max} = (0.7283)^{2.68} = 0.4275$$

∴ Maximum possible reduction is about 43%.

13.8. EXTRUSION

In general, extrusion may be defined as a process by which a block of metal (placed in a container) is reduced in cross-section, by forcing it to flow through a die (mounted in the container) under high pressure. The main advantage of extrusion is that high compressive stresses are set up in the billet due to its reaction with the container and the die. These stresses are effective in reducing the cracking of materials during primary breakdown from the billet. Due to this, large reductions are possible and also the difficult metals can be extruded, for example, stainless steels, nickel-based alloys and other high temperature materials. Extrusion can be used both for hot and cold working. Also bars, hollow tubes and shapes of irregular cross-sections can be extruded. Extrusion could well be considered as adaption of closed die forging, the difference being that in a forging, the main body of the metal is the product and flash is cut away and discarded, in extrusion the flash in the product and the slug remaining in the die is not used.

The various types of extrusion processes are shown in Fig. 13.20. The direct extrusion, [Fig. 13.20 (*a*)], is the simplest, but it is limited by the fact that as the ram moves, the billet must slide or shear at the interface between billet and container. These large friction forces must be overcome by very high ram forces, which produce very high residual stresses on the container. In the indirect method (also called reverse, back or inverted), the billet proper does not move relative to the container, instead the die moves. The friction involved is only between the die and container and this is independent of the billet length. The friction forces are lower and the power required for extrusion is less than for direct extrusion. Extruding force is 25 to 30% less than in direct extrusion. However, a long hollow ram is required and this limits the loads which can be applied. As is clear from the figure, in direct (forward) extrusion, the product emerges in the same direction as the movement of the ram, whereas in indirect extrusion, the product travels

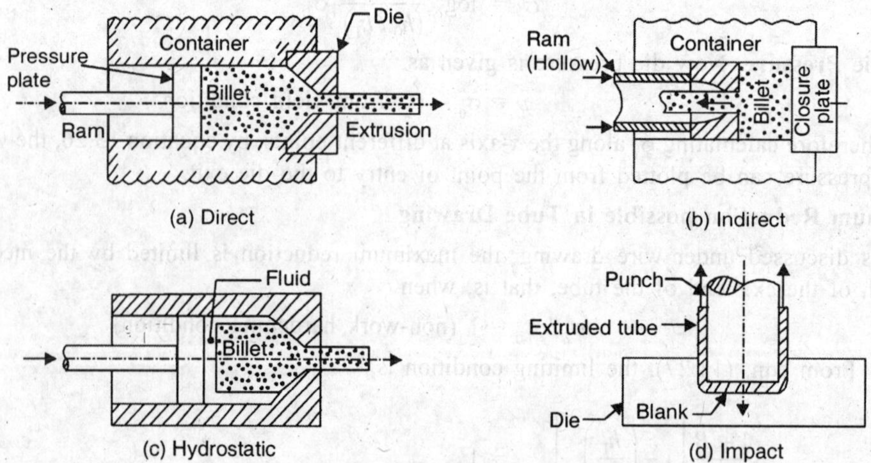

Fig. 13.20. Types of Extrusion Process.

against the direction of the ram. The container friction can be avoided by hydrostatic extrusion, [Fig. 13.20 (*c*)], where a pressurised liquid medium is used for the transmission of the force to the billet. Due to the hydrostatic pressure, the ductility of the material is increased. Even brittle materials like tungsten, cast iron and stainless steel etc. can be extruded. This also permits the extrusion of very long billets or even wires, accompanied by large reductions. The pressurised fluid also acts as a lubricant and because of this, the extruded product has a good surface finish and dimensional accuracy. However, the absence of container friction combined with reduced die friction can increase the tendency to internal crack formation. The pressure transmitting fluids commonly used for hydrostatic extrusion are : Glycerine, Ethylglycol, SAE 30 mineral lubricating oil, Castor oil with 10 percent alcohol and isopentane. The hydrostatic pressure ranges from 1100 to 3150 N/mm^2. The main commercial applications of this process are : Cladding of metals, making wires of less ductile materials and extrusion of nuclear reactor fuel rods. Impact extrusion (cold), [Fig. 13.20 (*d*)], is used to produce short lengths of hollow shapes such as collapsible tooth paste tubes. However the process is limited to soft ductile materials such as Aluminium, Tin and Lead. Hollow workpieces (Tubes and Pipes) can be extruded by attaching a mandrel to the end of the ram in forward extrusion. For this hollow forward extrusion known as Hooker extrusion, either a solid or a hollow billet can be used.

The "extrusion ratio", *R*, defined as the ratio of the cross-sectional area of the billet to the cross-sectional area of the product, reaches about 40 : 1 for hot extrusion of steel and may be as high as 400 : 1 for aluminium. The other advantages of extrusion are :

1. Cross-sectional shapes not possible by rolling can be extruded.

2. No time is lost when changing shapes since the dies may be readily removed and replaced.

3. Dimensional accuracy of extruded parts is generally superior to that of rolled ones.

4. The range of extruded items is very wide : rods from 3 to 250 mm in dia, pipes of 20 to 400 mm in dia and wall thickness of 1 mm and above, and more complicated shape which cannot be obtained by other mechanical working methods. Materials most commonly being extruded are light alloys, *e.g.*, aluminium alloys, copper, brass and bronze.

5. Very large reductions are possible as compared to rolling, for which, the reduction per pass (initial cross-sectional area divided by the final cross-sectional of the product) is generally ≤ 2.

6. Automation of extrusion is simpler as items are produced in a single passing.

However, extrusion has the following drawbacks :

1. Process waste in extrusion which consists of a poorly shaped leading end of the section and of an unextruded butt, is higher than in rolling. It is 18 to 20% of the billet weight in direct extrusion 5 to 6% of the billet weight in indirect extrusion. Whereas, the waste in rolling is only 1 to 3%.

2. Inhomogeneity in structure and properties of an extruded product is greater due to different flows of the axial and the outer layers of blanks.

3. Service life of extrusion tooling is shorter because of high contact stresses and slip rates.

4. Relatively high tooling costs, being made from costly alloy steels.

5. In productivity extrusion is much inferior to rolling.

6. Costs of extrusion are generally greater as compared to other methods.

Due to the above factors; the main fields of application of extrusion process are :

1. working of poorly plastic and nonferrous metals and alloys.

2. manufacture of shapes and pipes of complex configuration.

3. medium and small batch production.

4. manufacture of parts of high dimensional accuracy.

Analysis of Extrusion Process. The analysis of extrusion process for round-bar extrusion through a conical die is very similar to that for drawing of a round bar discussed in Art. 13.7. In the following analysis, container friction is not considered. We can proceed exactly in the same way, upto the following equation :

$$\log_e \left[B\sigma_x - (1+B)\,\sigma_0\right] = (\log_e rC)^{2B}$$

or
$$B\sigma_x - (1+B)\,\sigma_0 = (rC)^{2B}$$

Now at the die exit, the longitudinal stress is zero,

$\therefore$ at $r = r_f$, $\sigma_x = 0$ (r_f is the final radius)

$$\therefore \quad -(1+B)\,\sigma_0 = (r_f C)^{2B}$$

$$\therefore \quad C = \frac{\left[-(1+B)\,\sigma_0\right]^{\frac{1}{2B}}}{r_f}$$

$$\sigma_x = \frac{\sigma_0\,(1+B)}{B}\left[1-\left(\frac{r}{r_f}\right)^{2B}\right]$$

Now at $r = r_0$ (original radius)

σ_x = extrusion pressure, σ_{x0}

$$\therefore \quad \sigma_{x0} = \frac{\sigma_0\,(1+B)}{B}\left[1-\left(\frac{r_0}{r_f}\right)^{2B}\right]$$

Now extrusion ratio, $R = \dfrac{A_0}{A_f} = \left(\dfrac{r_0}{r_f}\right)^2$ for round bars

$= \dfrac{h_0}{h_f}$ for flat strip

$$\therefore \quad \sigma_{x0} = \frac{\sigma_0\,(1+B)}{B}\left[1-R^B\right] \qquad ...(13.29)$$

Note. For a 180° flat die it is assumed that the material shears internally to form an effective cone angle of $2\alpha = 90°$. The metal in the corners of the square container at the exit end acts as dead metal and the flow of metal past this dead metal takes place as through a conical die. Due to this reason, the dies at the exit end of the container are usually conical instead of square as shown in [Fig. 13.20 (*a*) and (*c*)].

The effect of container friction can be considered as given below :

Let p_f = ram pressure required by container friction

and τ_i = uniform interface shear stress between billet and container wall

Then, $$p_f \times \frac{\pi}{4} D^2 = \pi D \tau_i L$$

$$\therefore \quad p_f = \frac{4\tau_i L}{D}$$

where L = length of billet in the container and D = Internal diameter of the container

$\therefore$ Total extrusion pressure $\quad p_t = \sigma_{x0} + p_f$

$\therefore \quad$ Extrusion load $= p_t \times \pi r_0^2$

The extrusion load can be approximately determind by the "Work-formula" (Uniform deformation, no friction) as below :

$$P = A_0 . \bar{\sigma} . \ln . A_0 / A_f$$

This formula is modified for real conditions, as $P = k. A_0 .\ln. A_0/A_f$, where

k = the "extrusion constant". It accounts for the flow stress, friction and inhomogeneous deformation.

Variation of Extrusion Pressure. The extrusion pressure (extrusion force divided by the cross-sectional area of the billet) depends upon the following factors : (1) the type of extrusion, that is, direct or indirect (2) the extrusion ratio (3) the working temperature (4) the speed of deformation, and (5) the frictional conditions at the die and the container wall.

The variation of extrusion pressure with ram travel is shown in Fig. 13.21. Initially, there is steep rise in extrusion pressure with ram travel. This occurs due to the initial upsetting of the billet to fill the container. Once the extrusion pressure reaches the maximum value called the "breakthrough pressure", the material starts flowing through the die. After this, the pressure starts decreasing with decrease in the billet length. In the case of indirect extrusion, since there is no relative motion between the billet and the container wall, the extrusion pressure almost remains constant with increasing ram travel once it reaches the maximum value. This extrusion pressure represents the stress needed to deform the material through the die. This variation of extrusion pressure with ram travel is better than that for direct extrusion, but, indirect extrusion has its own limitations as discussed above. At the end of the ram stroke, the pressure builds up rapidly and it is a usual practice to stop the process at this point and leave a small portion of the billet in the container as discord. In extrusion, hydraulic presses with either vertical or horizontal plunger are most commonly used. The pressure of the working fluid (water or emulsion) may reach 400 atm and the force 30,000 tf. Mechanical presses are used to a lesser extent. Lubricants used are : mixture of machine oil and graphite, molten glass.

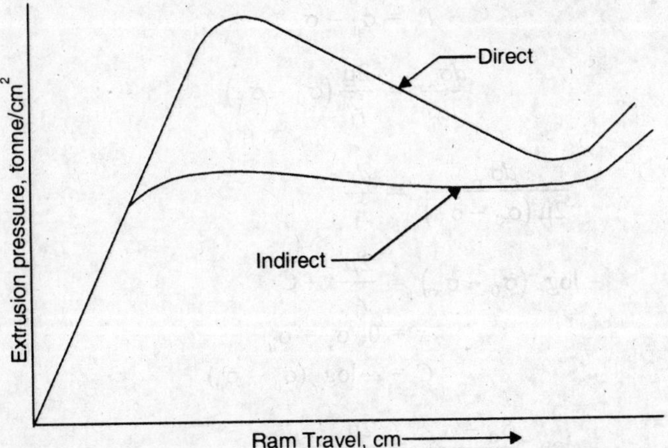

Fig. 13.21. Variation of Extrusion Pressure with Ram Travel.

13.9 Solved Examples

Example 13.1. *In a wire drawing operation, initial wire diameter is 5.5 mm and the final wire diameter is 5 mm. Die angle is 16°, die land is 3 mm and co-efficient of friction is 0.1. Find the drawing load.* σ_0 *= 240 N/mm².*

Solution. The drawing stress is given as

$$\sigma_d = \frac{\sigma_0\,(1+B)}{B}\left[1-\left(\frac{r_1}{r_0}\right)^{2B}\right],\ \sigma_b \text{ being not mentioned}$$

$$B = \mu \cot \alpha = 0.1 \times \cot 8° = 0.1 \times 7.115 = 0.7115$$

$$\frac{r_1}{r_0} = \frac{5}{5.5} = 0.91$$

$$\left(\frac{r_1}{r_0}\right)^{2B} = (0.91)^{2 \times 0.7115} = 0.873$$

$$\therefore \quad \sigma_d = \frac{240\,(1+0.7115)}{0.7115}(1-0.973) = 73.41\ \text{N/mm}^2$$

Stress equilibrium at die land. See, Fig. 13.22.

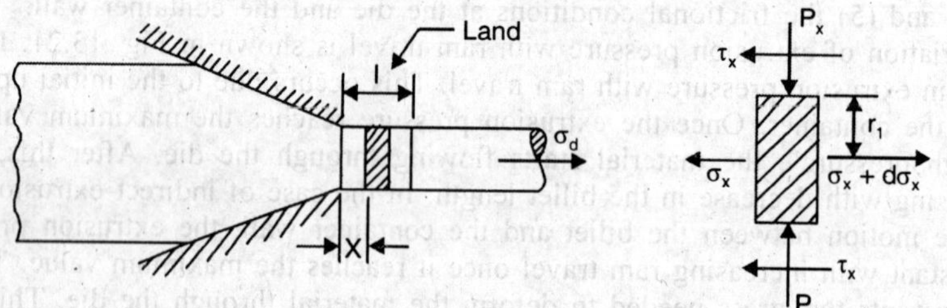

Fig. 13.22

$$(\sigma_x + d\sigma_x)\,\pi r_1^{\,2} - \sigma_x \cdot \pi r_1^{\,2} - 2\pi r_1 dx \tau_x = 0$$

$$\therefore \quad \frac{d\sigma_x}{dx} = \frac{2\tau_x}{r_1} = \frac{2\mu P_x}{r_1}$$

Now $\sigma_x + P_x = \sigma_0$ (Tresca yield condition)

$$\therefore \quad P_x = \sigma_0 - \sigma_x$$

$$\therefore \quad \frac{d\sigma_x}{dx} = \frac{2\mu}{r_1}(\sigma_0 - \sigma_x)$$

$$\therefore \quad \frac{d\sigma_x}{2\mu\,(\sigma_0 - \sigma_x)} = \frac{dx}{r_1}$$

Integrating, $$-\log_e(\sigma_0 - \sigma_x) = \frac{2\mu}{r_1}x + C$$

at $$x = 0,\ \sigma_x = \sigma_d$$

$$\therefore \quad C = -\log_e(\sigma_0 - \sigma_d)$$

$$\therefore \quad \log_e \frac{\sigma_0 - \sigma_d}{\sigma_0 - \sigma_x} = \frac{2\mu}{r_1}\cdot x$$

$$\therefore \quad \sigma_0 - \sigma_d = (\sigma_0 - \sigma_x)\cdot e^{2\mu x/r_1}$$

or $$\sigma_x = \sigma_0 - (\sigma_0 - \sigma_d)\cdot e^{-2\mu x/r_1}$$

Now σ_x will be maximum, at $x = l$

i.e., $\sigma_x = \sigma_t$ = total drawing stress

$$\therefore \quad \sigma_t = \sigma_0 - (\sigma_0 - \sigma_d)e^{-2\mu l/r_1}$$

$$= 240 - (240 - 73.41) \cdot e^{-2 \times 0.1 \times 3/2.5}$$

$$= 240 - 166.59 \times 0.786$$

$$= 109.0 \text{ N/mm}^2$$

$$\therefore \quad \text{Total drawing load,} = \sigma_t \times \pi r_1^2$$

$$= 109 \times \pi \times 6.25 = \mathbf{2140.5\ N}$$

Example 13.2. *A pipe of annealed steel, inside diameter of 50 mm and wall thickness of 2.5 mm is to be reduced down to 48.7 mm × 1.75 mm. Die-angle is 30°, $\mu = 0.1$ and draft = 3.12. Compare the pipe drawing force on plug and movable mandrels.*

Solution. From equation (13.27),

$$\sigma_d = \sigma_0 \frac{1+B}{B}\left[1 - \left(\frac{h_1}{h_0}\right)^B\right]$$

Here, $$B = \frac{\mu_1 + \mu_2}{\tan\alpha - \tan\beta}, \alpha = 15°, \beta = 0$$

$$\therefore \quad B = 0.2/\tan 15° = 0.747$$

$$\therefore \quad \sigma_d/\sigma_0 = \frac{1.747}{0.747}\left[1 - \left(\frac{1.75}{2.5}\right)^{0.747}\right] = 0.547$$

For movable mandrel, μ_1 being $= \mu_2$

$$\therefore \quad \sigma_d/\sigma_0 = \log_e \frac{1}{\left(\frac{h_0}{h_1} - 1\right)} = \log_e 2.333 = 0.368$$

∴ Use of movable mandrel substantially reduces the drawing force.

Example 13.3. *In rolling process, 25 mm thick plate is rolled to 20 mm in a four high mill. Determine the co-efficient of friction if this is the maximum reduction possible. Roll diameter is 500 mm. Find neutral section, Backward and forward slips and maximum pressure. $\sigma_0 = 100$ N/mm² for hot rolls of mild steel at about 1100°C.*

Solution. $$\Delta h = 2R\,(1 - \cos\alpha)$$

$$5 = 500\,(1 - \cos\alpha)$$

$$\therefore \quad \cos\alpha = \frac{99}{100}$$

$$\therefore \quad \alpha = 8.11°$$

(*i*) $$\mu = \tan\alpha = 0.142$$

(*ii*)

$$H_0 = 2\sqrt{\frac{R}{h_1}} \tan^{-1}\left(\sqrt{\frac{R}{h_1}} \cdot \alpha\right)$$

$$= 2\sqrt{\frac{250}{20}} \tan^{-1}\left(\sqrt{\frac{250}{20}} \times 0.1429\right) = 3.306$$

$$H_n = \frac{1}{2}\left(H_0 - \frac{1}{\mu}\log_e \frac{h_0}{h_1}\right)$$

$$= \frac{1}{2}\left(3.306 - \frac{1}{0.142}\log_e \frac{25}{20}\right) = 0.8678$$

$$\theta_n = \sqrt{\frac{h_1}{R}} \tan\left(\sqrt{\frac{h_1}{R}} \cdot \frac{H_n}{2}\right)$$

$$= \sqrt{\frac{20}{250}} \tan\left(\sqrt{\frac{20}{250}} \times 0.4339\right) = 0.0349 \text{ radians}$$

$$h_n = h_1 + 2R(1 - \cos\theta_n)$$

$$= h_1 + R\theta_n^2$$

$$= 20 + 250 \times (0.0349)^2 = 20.3 \text{ mm}$$

(*iii*)

$$\text{Backward slip} = \frac{V_r - V_0}{V_r} = 1 - \frac{V_0}{V_r}$$

now

$$\frac{V_0}{V_r} = \frac{h_n}{h_0} = \frac{20.3}{25}$$

$\therefore$

$$\text{Backward slip} = 1 - \frac{20.3}{25} = 18.8\%$$

$$\text{Forward slip} = \frac{V_1 - V_r}{V_r} = \frac{V_1}{V_r} - 1$$

now

$$\frac{V_1}{V_r} = \frac{h_n}{h_1} = \frac{20.3}{20}$$

$\therefore$

$$\text{Forward slip} = \frac{0.3}{20} = \mathbf{1.5\%}$$

(*iv*)

$$p_{max} = p_n = \sigma_0' \times \frac{h_n}{h_1} \times e^{\mu H_n}$$

$$= \frac{2}{\sqrt{3}} 100 \times \frac{20.3}{20} \times e^{0.142 \times 0.8678}$$

$$= \mathbf{132.4\ N/mm^2}$$

Example 13.4. *Determine the maximum force of a hydraulic press required to upset a low carbon steel blank of diameter 250 mm and height 250 mm at a reduction of 100 mm at 1000°C. σ_0 = 55 N/mm².*

Solution. $d_0 = 250$ mm, $h_0 = 250$ mm

$\Delta h = 100$ mm, $h = 150$ mm

Diameter d ($2R$), after reduction is,

$$d = d_0 \sqrt{\frac{h_0}{h_0 - \Delta h}}$$

$$= 250\sqrt{\frac{250}{150}} \approx 325 \text{ mm}$$

$\mu = 0.4$ to 0.45, say 0.42

Now
$$p_a = \frac{\sigma_0}{2}\left(\frac{h}{\mu R}\right)^2 \left[e^{2\mu R/h} - \frac{2\mu R}{h} - 1\right],$$

$$R = 162.5 \text{ mm} \qquad \text{...(13.19}a\text{)}$$

$\therefore$
$$p_a = \frac{55}{2}\left(\frac{150}{0.42 \times 162.5}\right)^2 \left[e^{\frac{2 \times 0.42 \times 162.5}{150}} - \frac{2 \times 0.42 \times 162.5}{150} - 1\right]$$

$$= 55 \times 0.5 \times 4.83\,(2.48 - 0.91 - 1)$$

$$= 76.285 \text{ N/mm}^2$$

$\therefore$ Force,
$$P = p_a \times \pi R^2$$

$$= 76.285 \times \pi \times (162.5)^2$$

$$= \mathbf{6328 \text{ kN}}$$

Example 13.5. *A circular disc of 150 mm diameter and 10 mm thickness is compressed between two flat dies. μ = 0.2. Determine the sticking radius, total load on the dies. σ_0 = 200 N/mm².*

Solution. From eqn. (13.18 *a*), the sticking radius is,

$$R_s = R - \frac{h}{2\mu}\log_e \frac{1}{\sqrt{3}\cdot\mu} \quad \text{(taking Von-Mise's yield condition)}$$

$$= 75 - \frac{10}{2 \times 0.2}\log_e \frac{1}{\sqrt{3} \times 0.2}$$

$$= 75 - 26.5 = 48.5 \text{ mm}$$

Pressure at sticking radius,
$$p_s = \sigma_0 . e^{[2\mu(R-R_s)/h]} \qquad \text{...(eqn. 13.18}h\text{)}$$

$$= 200 \cdot e^{[2 \times 0.2 \times 26.5/10]} = 200 \times 2.88 = \mathbf{576 \text{ N/mm}^2}$$

Total load = load on sliding portion + load on sticking portion

$$\text{Load on sliding portion} = \int_{48.5}^{75} 2\pi r dr \sigma_0 \cdot e^{2\pi/h(R-r)} \quad \text{...(eqn. 13.15)}$$

$$= 2\pi\sigma_0 \left[r \cdot e^{\frac{2\pi}{h}(R-r)} \cdot -\frac{h}{2\pi} + \frac{h}{2\mu} \int e^{\frac{2\mu}{h}(R-r)} \cdot dr \right]$$

$$= 2\pi\sigma_0 \left[-\frac{hr}{2\mu} \cdot e^{\frac{2\mu}{h}(R-r)} - \left(\frac{h}{2\mu}\right)^2 \cdot e^{\frac{2\mu}{h}(R-r)} \right]_{48.5}^{75}$$

$$= 2\pi\sigma_0 \left[-\frac{10 \times 75}{2 \times 0.2} \cdot e^0 - \left(\frac{10}{2 \times 0.2}\right)^2 \cdot e^0 \right.$$

$$\left. + \frac{10 \times 48.5}{0.4} \cdot e^{\frac{0.4}{10} \times 26.5} + \left(\frac{10}{0.4}\right)^2 \cdot e^{\frac{0.4}{10} \times 26.5} \right]$$

$$= 2\pi\sigma_0 [-1875 - 625 + 1212.5 \times 2.88 + 625 \times 2.88]$$

$$= 2\pi \times 200 (-1875 - 625 + 3499.72 + 1800)$$

$$= 3518.2 \text{ kN}$$

Load on sticking portion may be taken as

$$= \frac{p_s + p_c}{2} \cdot \pi R_s^2$$

$$p_s = 576 \text{ N/mm}^2$$

$$p_c = \text{pressure at centre} = p_s + \frac{2K}{h} \cdot R_s = p_s + \frac{2}{20} \cdot \frac{\sigma_0}{\sqrt{3}} \cdot 48.5$$

$$= 576 + 1120 = 1696 \text{ N/mm}^2$$

$$\therefore \quad P_s = \frac{1}{2}(576 + 1696) \times \pi \times 48.5^2$$

$$= 8396 \text{ kN}$$

$$\text{Total forging load} = 3518.2 + 8396$$

$$= 1194.2 \text{ kN} = \mathbf{11.9142 \text{ MN}}$$

Example 13.6. *A piece of lead 25 mm × 25 mm × 150 mm having a yield stress of 7 N/mm² is to be pressed between flat dies a size of approximately 6.25 mm × 100 mm × 150 mm. μ = 0.25. Determine pressure distribution and the total forging load.*

Solution. Final dimensions : $h = 6.25$ mm, $2R = 100$ mm

Sticking radius.

$$R_s = R - \frac{h}{2\mu} \log_e \frac{1}{2\mu}$$

$$= 50 - \frac{6.25}{0.5} \log_e \frac{1}{2 \times 0.25}$$

$$= 50 - 8.66 = 41.34 \text{ mm}$$

It is clear that on the major portion of piece, sticking conditions exist. In the sliding portion,

$$p = \sigma_0 \cdot e^{2\mu(L-x)/h} \qquad \text{for } 41.34 < x < 50$$

$$= \sigma_0 \cdot e^{2\mu(R-r)/h}$$

at $r = R$, *i.e.*, outer surface, $p = \sigma_0$

at $r = R_s$

$$p = \sigma_0 \cdot e^{0.5 \times 8.66 \times \frac{1}{6.25}} = 2\sigma_0$$

In the sticking region, $$p = p_s + \frac{\sigma_0}{h}(R_s - R)$$

$$= 2\sigma_0 + \frac{\sigma_0}{6.25}(41.34 - R)$$

$$= 2\sigma_0 + 0.16\sigma_0 (41.34 - R)$$

At the centre, $R = 0$

$$\therefore \quad p = 2\sigma_0 + 0.16\sigma_0 \times 41.34 = 8.61\sigma_0$$

The pressure distribution is shown in Fig. 13.23. Assuming that the entire distribution is linear, the forging load is the area of the curve.

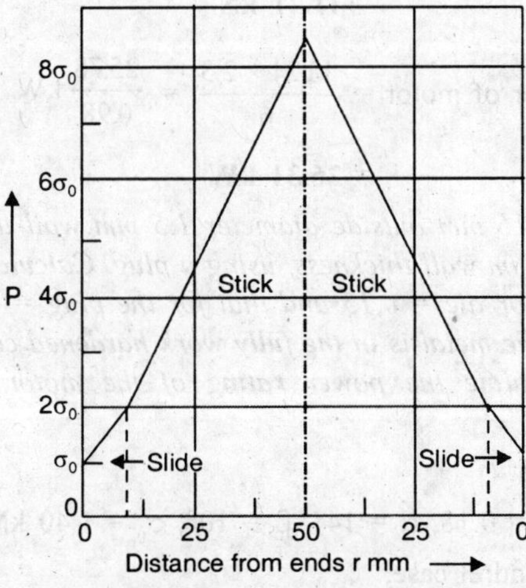

Fig. 13.23. Pressure Distribution.

$$\therefore \quad \text{Forging load per unit width} = \frac{1}{2} \times 100 \times 7.61\sigma_0$$

$$= 380.5\ \sigma_0$$

$\therefore$ Total forging load $= 380.5\sigma_0 \times 150$

$= 380.5 \times 7 \times 150$

$= \mathbf{399.5\ kN}$

Example 13.7. *Calculate the drawing load required to obtain 30% reduction in area on a 12 mm diameter copper wire. The following data is given :*

$$\sigma_0 = 240\ N/mm^2,\ 2\alpha = 12°,\ \mu = 0.10$$

(*b*) *Calculate the power of the electric motor if the drawing speed is 2.3 m/s. Take efficiency of the motor as 98%.*

Solution. (*a*) $RA = 0.30$

$$B = \mu \cot \alpha = 0.10 \times \cot 6° = 0.95$$

now
$$\sigma_d = \frac{\sigma_0 (1 + B)}{B}\left[1 - \left(\frac{r_1}{r_0}\right)^{2B}\right]$$

$$RA = 1 - \left(\frac{r_1}{r_0}\right)^2 \quad \therefore \quad \left(\frac{r_1}{r_0}\right)^2 = 0.7$$

$\therefore$
$$\sigma_d = \sigma_0 \times \frac{1.95}{0.95}\left[1 - (0.7)^{0.95}\right] = 141.60\ \text{N/mm}^2$$

now
$$r_1 = \sqrt{0.7} \times 6 = 5.02\ \text{mm}$$

$\therefore$ Drawing load $= 141.60 \times \pi \times (5.02)^2$

$= \mathbf{11.21\ kN}$

(*b*) Power of motor $= \dfrac{11.21 \times 2.3}{\eta} = \dfrac{25.78}{0.98}$ kW

$= \mathbf{26.31\ kW}$

Example 13.8. *A tube 16 mm outside diameter 1.5 mm wall thickness is to be drawn to 11 mm outside diameter 1.0 mm wall thickness, using a plug. Calculate the drawing load, given that, co-efficient of friction for die = 0.15 and that for the plug = 0.18. The die angle is 28° and the plug angle is 20°. The metal is in the fully work hardened condition with a yield stress of 1.40 kN/mm². Also calculate the power rating of the motor if the drawing speed is 0.65 m/s.*

Solution. It is given that

$$\mu_1 = 0.15,\ \mu_2 = 0.18,\ \alpha = 14°,\ \beta = 10°,\ \sigma_0 = 1.40\ \text{kN/mm}^2$$

Now it is a floating mandrel case,

$\therefore$
$$B = \frac{\mu_1 + \mu_2}{\tan \alpha + \tan \beta} = \frac{0.15 + 0.18}{\tan 14° + \tan 10°} = \frac{0.33}{0.425}$$

$$= 0.7765$$

$$h_0 = 1.5\ \text{mm},\ h_1 = 1\ \text{mm}$$

$\therefore$ Drawing stress is $\sigma_d = \sigma_0 \cdot \dfrac{1+B}{B}\left[1-\left(\dfrac{h_1}{h_0}\right)^B\right]$

$$= 1.40 \times \frac{1.7765}{0.7765}\left[1-\left(\frac{1}{1.5}\right)^{0.7765}\right]$$

$$= 0.865 \text{ kN/mm}^2$$

Area of cross-section of tube at exit $= \dfrac{\pi}{4}\left(11^2 - 9^2\right) = 31.416 \text{ mm}^2$

$\therefore$ Drawing load $= 0.865 \times 31.416 = 27.175$ kN

Now work done per second $= 27.175 \times$ drawing speed

$= 27.175 \times 0.65 = 17.66$ kJ/s

$\therefore$ Power rating of the motor = **17.66 kW**

Example 13.9. *For plane strain and sticking friction conditions, derive an expression for the forging load in terms of tool bite, material width, material thickness and yield stress. Calculate the forging loads at the start and the completion of hot forging of a steel billet, for the following data : Length of billet = 2m, width = 0.9 m, thickness = 0.2 m, tool bite = 0.3 m, σ_0 = 50 MPa at start and σ_0 = 150 MPa at completion of the forging, reduction in forging = 50%.*

Solution. Now for sticking friction and plane strain conditions, the average pressure is given as (eqn. 13.18 *d*),

$$p_a = 2k\left(1+\frac{L}{2h}\right) = \sigma_0\left(1+\frac{L}{2h}\right)$$

Now bite of tool, $b = 2L \therefore L = b/2$

$$\therefore \quad p_a = \sigma_0\left(1+\frac{b}{4h}\right)$$

Now Forging area $= B \times 2L = B \times b$

$$\therefore \quad \text{Forging load} = \sigma_0 Bb\left(1+\frac{b}{4h}\right)$$

(*i*) At the commencement of forging :

$\sigma_0 = 50$ MPa, $B = 0.9$ m, $h = 0.2$ m, $b = 0.3$ m

$$\text{Forging load} = 50 \times 0.9 \times 0.3\left(1+\frac{0.3}{4 \times 0.2}\right)$$

$$= 50 \times 0.9 \times 0.3 \times 1.375$$

$$= \mathbf{18.5625 \text{ MN}}$$

(*ii*) At the completion of the forging,

$h = 0.1$ m, $\sigma_0 = 150$ MPa

$$\therefore \quad \text{Forging load} = 150 \times 0.9 \times 0.3\left(1 + \frac{0.3}{4 \times 0.1}\right)$$

$$= 150 \times 0.9 \times 0.3 \times 1.75$$

$$= \mathbf{70.875\ MN}$$

Example 13.10. *An aluminium alloy is hot extruded at 400°C through square dies without lubrication, from 15 cm diameter to 5 cm diameter. The extrusion speed is 5 cm/s. The flow stress of the material at the above temperature is 250 MPa. The length of the billet is 37.5 cm. Determine the extrusion load.*

Solution. As already discussed under Art. 13.8 that a dead-metal zone forms in the corners against the die (square). This can be assumed as equivalent to die angle $\alpha = 45°$.

∴ extrusion pressure due to flow through the die,

$$\sigma_{x_0} = \frac{\sigma_0(1+B)}{B}\left[1 - R^B\right]$$

Now $$R = \left(\frac{r_0}{r_f}\right)^2 = \left(\frac{7.5}{2.5}\right)^2 = 9$$

$$B = \mu \cot \alpha = 0.1 \times \cot 45° \quad (\text{Assuming } \mu = 0.1)$$

$$= 0.1$$

$$\therefore \quad \sigma_{x_0} = 250 \times \frac{1.1}{0.1}\left(1 - 9^{0.1}\right)$$

$$= 675.8 \text{ MPa (Compressive)}$$

The maximum pressure due to container wall friction will occur at break through (as already discussed), when

$L = 37.5$ cm. Taking Tresca's condition,

$$\tau_1 \approx k = \frac{\sigma_0}{2} = 125 \text{ MPa (Sticking friction)}$$

∴ Total extrusion pressure, $$p_e = \sigma_{x0} + \frac{4\tau_1 L}{D}$$

$$= 675.8 + \frac{4 \times 125 \times 37.5}{15}$$

$$= 1925.8 \text{ MPa}$$

$$\therefore \quad \text{Extrusion load} = p_e \times \pi r_0^2$$

$$= 1925.8 \times 10^6 \times \pi \times \left(\frac{7.5}{100}\right)^2$$

$$= \mathbf{34\ MN}$$

Example 13.11. *Sheet steel is reduced from 4.05 mm to 3.55 mm with 500 mm diameter rolls having a co-efficient of friction of 0.04. The mean flow stress in tension is 210 N/mm². Neglect work hardening and roll flattening.*

(*a*) *Calculate the roll pressure at the entrance to the rolls, the neutral plane, and the roll exit.*

(*b*) *If the co-efficient of friction is 0.40, determine the roll pressure at the neutral point.*

(*c*) *If 35 N/mm² front tension is applied in problem (a), find the roll pressure at the neutral point.*

Solution. (*a*) $h_0 = 4.05$ mm, $h_1 = 3.55$ mm, $R = 250$ mm

$\mu = 0.04$, $\sigma_0 = 210$ N/mm²

The roll pressure at entry and exit,

$$p = \sigma_0' = \frac{2}{\sqrt{3}}\sigma_0 = 242.5 \text{N/mm}^2.$$

Now $$H_0 = 2\sqrt{\frac{R}{h_1}} \tan^{-1}\left(\sqrt{\frac{R}{h_1}}\alpha\right)$$

Now $$\Delta h = 2h(1 - \cos\alpha)$$

$\therefore$ $$0.50 = 500(1 - \cos\alpha)$$

From here, $$\alpha = 2.56°$$

$\therefore$ $$H_0 = 2\sqrt{\frac{250}{3.55}} \tan^{-1}\left(\sqrt{\frac{250}{2.55}} \times 0.0447\right)$$

$$= 6.02$$

$$H_n = \frac{1}{2}\left(H_0 - \frac{1}{\mu}\log_e \frac{h_0}{h_1}\right)$$

$$= \frac{1}{2}\left(6.02 - \frac{1}{0.04}\log_e \frac{4.05}{3.55}\right) = 1.363$$

$\therefore$ $$p_n = \sigma_0' \cdot \frac{h_n}{h_1} \cdot e^{\mu H_n}$$

Now $$\theta_n = \sqrt{\frac{h_1}{R}} \tan\left(\sqrt{\frac{h_1}{R}} \cdot \frac{H_n}{2}\right)$$

$$= \sqrt{\frac{3.55}{250}} \tan\left(\sqrt{\frac{3.55}{250}} \times 0.6815\right)$$

$$= 0.009672 \text{ radians}$$

$$h_n = h_1 + 2R(1 - \cos\theta_n)$$

$$= 3.55 + 500(1 - \cos 0.554°)$$

$$= 3.5734 \text{ mm}$$

$$\therefore \quad p_n = 242.5 \times \frac{3.5734}{3.55} \times e^{0.04 \times 1.363}$$

$$= 242.5 \times 1.0066 \times 1.056 = \mathbf{257.78\ N/mm^2}$$

(*c*)

$$p_n = \left(\sigma_0' - \sigma_f\right) \cdot \frac{h_n}{h_1} \cdot e^{\mu H_n}$$

$$= 207.5 \times 1.0066 \times 1.056 = \mathbf{220.57\ N/mm^2}$$

(*b*)

$$H_0 = 6.02;\ \mu = 0.40$$

$$H_n = \frac{1}{2}\left(6.02 - \frac{1}{0.40} \log_e \frac{4.05}{3.55}\right) = 2.845$$

$$\theta_n = \sqrt{\frac{3.55}{250}} \tan\left(\sqrt{\frac{3.55}{250}} \times 1.4225\right)$$

$$= 0.02 \text{ radians}$$

$$h_n = h_1 + R\theta_n^2 \text{ (approximately)}$$

$$= 3.55 + 250 \times (0.02)^2 = 3.65 \text{ mm}$$

$$p_n = 242.5 \times \frac{3.65}{3.55} \times e^{0.40 \times 2.845}$$

$$= \mathbf{777.9\ N/mm^2}$$

Example 13.12. *A wide-strip is rolled to a final thickness of 6.35 mm with a reduction of 30 per cent. The roll radius is 50 cm and the co-efficient of friction is 0.2. Determine the neutral plane.*

Solution. $h_1 = 6.35$ mm, $R = 50$ cm, $\mu = 0.20$

$$h_0 = h_1 \times \frac{100}{70} = 6.35 \times \frac{100}{70} = 9.07 \text{ mm}$$

$$\therefore \quad \Delta h = 9.07 - 6.35 = 2.72 \text{ mm}$$

Now $\Delta h = 2R\,(1 - \cos \alpha)$

$$\therefore \quad \cos \alpha = \frac{500 - 1.36}{500} = 0.9973$$

$$\therefore \quad \alpha = 4.23° = 0.0738 \text{ radians}$$

Now

$$H_0 = 2\sqrt{\frac{R}{h_1}} \tan^{-1}\left(\sqrt{\frac{R}{h_1}} \cdot \alpha\right)$$

$$= 2\sqrt{\frac{500}{6.35}} \tan^{-1}\left(\sqrt{\frac{500}{6.35}} \times 0.0738\right)$$

$$= 10.29$$

Now $$H_n = \frac{1}{2}\left(H_0 - \frac{1}{\mu}\log_e \frac{h_0}{h_1}\right)$$

$$= \frac{1}{2}\left(10.29 - \frac{1}{0.2}\log_e \frac{9.07}{6.35}\right) = 4.26$$

$\therefore$ $$\theta_n = \sqrt{\frac{h_1}{R}}\tan\left(\sqrt{\frac{h_1}{R}}\cdot\frac{H_n}{2}\right)$$

$$= \sqrt{\frac{6.35}{500}}\tan\left(\sqrt{\frac{6.35}{500}}\times 2.13\right)$$

$$= \mathbf{0.0273 \text{ radians} = 1.55°}$$

PROBLEMS

1. Differentiate between nominal stress and true stress.
2. Differentiate between conventional strain and true-strain.
3. What is offset yield strength?
4. What is flow curve?
5. Define : strain hardening co-efficient and strength co-efficient.
6. Define : principal stresses and principal planes.
7. Explain the two most common theories of yielding.
8. Write the two theories for (*a*) plane stress case (*b*) plane strain case.
9. Define : effective stress and effective strain.
10. List the assumptions used in "slab-method".
11. What is sticking friction?
12. What is meant by plane strain condition?
13. Define : back ward slip, forward slip and draft in the case of rolling.
14. List and explain the methods used to reduce the separating force in strip rolling.
15. How will you reduce the pressure in wire drawing?
16. Differentiate between hot working and cold working in metal forming. Bring out the advantages and disadvantages of each of these techniques.
17. Discuss "Warm working" of metals.
18. Discuss the effects of temperature, strain rate and friction on metal forming process.
19. For a thin strip being cold forged between two parallel overhanging platens, show that the maximum forging pressure is given by

$$(p/k)_{max} = \left[1 + \mu\frac{b}{h}\right]$$

20. What do you understand by the Term Limiting. Draw Ratio (LDR)? Explain why such a limit exists, and give the principles underlying the estimation of LDR in a given situation.
21. Analyse the reasons for the limits imposed on reduction achievable in a single pass in rolling of slabs and plates, and derive expression for the limiting value.

22. Determine the power required to draw hot drawn steel wire from 12.5 mm to 10 mm in diameter at 100 m/min. The co-efficient of friction between die and wire can be assumed to be equal to 0.1 and the die angle in 4°. Average flow stress for hot drawn steel can be assumed to be equal to 300 N/mm^2. Also determine maximum reduction possible. If the wire is subjected to a back pull of 50 N/mm^2, determine the draw stress and maximum reduction possible.
(σ_d = 292 N/mm^2, Draw load = 22.034 kN, Power = 50.96, *RA* = 37.6%; σ_d = 306 N/mm^2, *RA* = 34.67%)
23. Determine the drawing stress to produce a 25% reduction in a 10 mm stainless steel wire. Take semi die angle as 12°, Co-efficient of friction as 0.10 and flow stress as 640 MPa. The draw speed is 3m/s, determine the power needed for drawing.
24. A shaped wire is drawn from annealed, 3 mm diameter stainless steel wire. The cross-sectional area of the shape is 5.0 mm^2. Oil based lubricant is used that gives a co-efficient of friction of 0.05. The included angle of the dies is 12°. Draw speed is 2 m/s. Calculate the draw force and power requirement. (2.2 kN, 4.4 kW)
25. A strip with a cross-section 150 mm × 4.5 mm is being rolled with 20% reduction of area using 450 mm diameter rolls. The angle subtended by the deformation zone at the roll centre is (in radian)
(*a*) 0.01 (*b*) 0.02 (*c*) 0.03 (*d*) 0.06 (Gate 1998) (**Ans.** : D)
26. A metal strip is to be rolled from an initial wrought thickness of 3.5 mm to a final rolled thickness of 2.5 mm in a single pass rolling mill having rolls of 250 mm diameter. The strip is 450 mm wide. The average co-efficient of friction in the roll gap is 0.08. Taking plain strain flow stress of 140 MPa, for the metal, and assuming negligible spreading, estimate the roll separating force. (Gate 1997)
Hint. P can be found out by the relations :
P =*l*. *bm.Pm*
In rolling of thin strips ($l/hm \geq 1$), the mean specific pressure *pm* is found from equations describing the distribution of *p* along the area of contact. First find *hn* by equating expressions for *p* in the backward and forward slip zones. Then *Pm* can be found out as,

$$p_m = \frac{1}{\Delta h}\left[\int_{h1}^{hn} p.dh + \int_{hn}^{h0} p.dh + \int_{hb}^{ho} p.dh\right]$$

27. (*a*) A 400 mm-thick slab is to be cold rolled. The roll diameter is 800 mm and the co-efficient of friction is 0.08. Determine the maximum possible reduction.
(*b*) What is the maximum reduction on the same mill for hot rolling when co-efficient of friction is 0.5.

Solution : (*a*) $(\Delta h)_{max} = \mu^2 R = (0.08)^2 \times 400 = \mathbf{2.56\ mm}$

(*b*) For hot rolling, $(\Delta h)_{max} = (0.5)^2 \times 400 = \mathbf{100\ mm}$

28. A steel wire is drawn at speed of 90m/min. from 12.7 mm to 10.2 mm diameter. Tensile yield stress of the original steel specimen is 207 MPa, and is 414 MPa at a strain of 0.5. Assume linear stress-string relation for the material.
Take $\mu = 0.1$, $\alpha = 6°$, σ_b (back-tension) = 0. Determine the Drawing load, Drawing power and the maximum possible reduction. (17 kN, 25.5 kW, 50.5%)

Hint : Strain, $\epsilon = \ln \frac{A_0}{A_1} = 0.438$

Tensile yield stress at end of operation $= \left(207 + \frac{414-207}{0.5} \times 0.438\right) = 388$ MPa

Average yield stress $\sigma_0 = \frac{207+388}{2} = 297.5$ MPa

Find Drawing stress, σ_d, using equ. 13.23 ($\sigma_d = 0$)

Drawing Load $P = \sigma_d \cdot \pi r_1^2$

Drawing Power = P × Velocity of drawing in m/s.

14

THEORY OF METAL CUTTING

14.1. INTRODUCTION

Metal cutting or Machining is the process of producing workpiece by removing unwanted material from a block of metal, in the form of chips. This process is most important since almost all the products get their final shape and size by metal removal, either directly or indirectly. The major drawback of the process is loss of material in the form of chips. In this chapter, we shall have a fundamental understanding of the basic metal process.

14.2. THE MECHANICS OF CHIP FORMATION

A typical metal cutting process can be schematically represented as in Fig. 14.1. A wedge-shaped tool is made to move relative to the workpiece. As the tool makes contact with the metal,

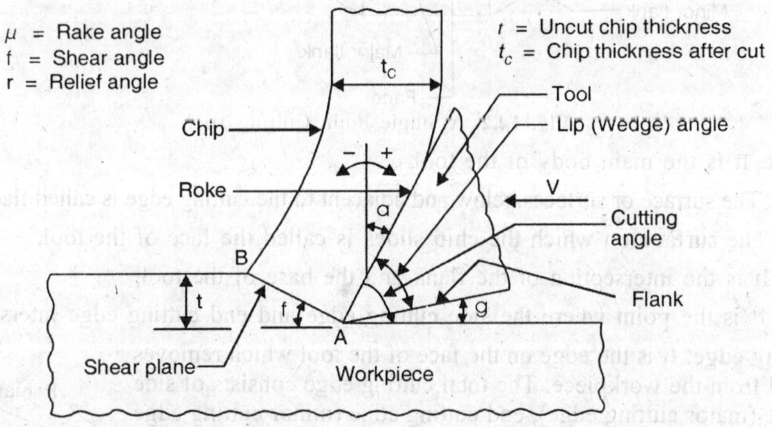

Fig. 14.1. Schematic Representation of Machining.

it exerts a pressure on it resulting in the compression of the metal near the tool tip. This induces shear-type deformation within the metal and it starts moving upward along the top face of the tool. As the tool advances, the material ahead of it is sheared continuously along a plane called the Shear plane. This shear plane is actually a narrow zone (of the order of about 0.025 mm) and extends from the cutting edge of the tool to the surface of the workpiece. The cutting edge of the tool is formed by two intersecting surfaces. The surface along which the chip moves upwards is called "Rake surface" and the other surface which is relieved to avoid rubbing with the machined surface, is called "Flank". The angle between the rake surface and the normal is known as "Rake angle" (which may be positive or negative), and the angle between the flank and the horizontal machined surface is known as the "relief or clearance angle". Most cutting processes have the same basic features as in Fig. 14.1, where a single point cutting tool is used (a milling cutter, a drill, and a broach can be regarded as several single-point tools joined together and are known as multi-point tools).

14.3. SINGLE POINT CUTTING TOOL

A single point cutting tool consists of a sharpened cutting part called its point and the shank, Fig. 14.2. The point of the tool is bounded by the face (along which the chips slide as they are cut by the tool), the side flank or major flank, the end flank, or minor flank and the base. The side cutting edge, a-b, is formed by the intersection of the face and side flank. The end cutting edge a-c is formed by the intersection of the face and the end flank. The chips are cut from the work the piece by the side-cutting edge. The point 'a' where the end and side-cutting edges meet is called the nose of the tool. Fig. 14.2 is for a right hand tool. Below, we give the definitions of the various tool elements tool elements and tool angles.

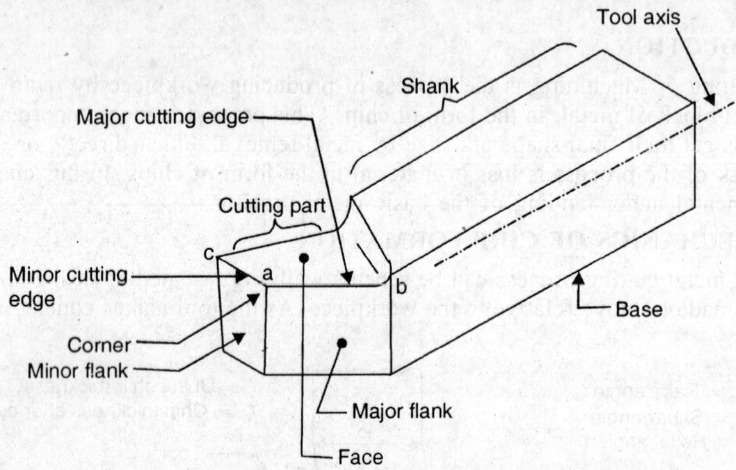

Fig. 14.2. A Single Point Cutting Tool.

Shank. It is the main body of the tool.

Flank. The surface or surfaces below and adjacent to the cutting edge is called flank of the tool.

Face. The surface on which the chip slides is called the face of the tool.

Heel. It is the intersection of the flank and the base of the tool.

Nose. It is the point where the side cutting edge and end cutting edge intersect.

Cutting edge. It is the edge on the face of the tool which removes the material from the workpiece. The total cutting edge consists of side cutting edge (major cutting edge), end cutting edge (minor cutting edge and the nose).

A single point cutting tool may be either right–or left hand cut tool depending on the direction of feed. In a right cut tool, the side cutting edge is on the side of the thumb when the right hand is placed on the tool with the palm downward and the fingers pointed towards the tool nose [Fig. 14.3 (*b*)]. Such a tool will cut when fed from right to left as in a lathe in which the tool moves from tailstock to headstock. A left-cut tool is one in which the side cutting edge is on the thumb side when the left hand is applied [Fig. 14.3 (*a*)]. Such a tool will cut when fed from left to right.

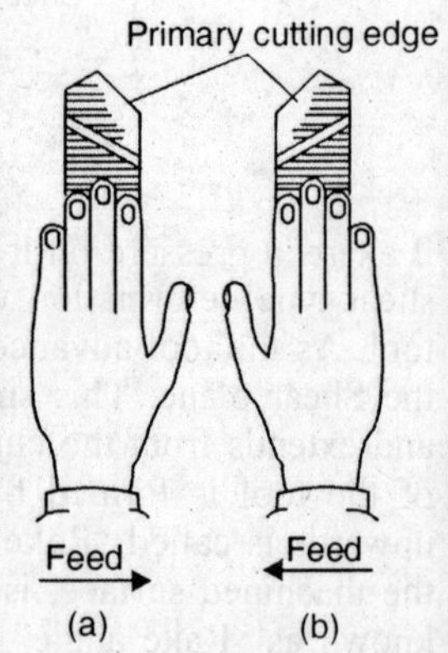

Fig. 14.3. Left and Right Cut Tools.

The various types of surfaces and planes in metal cutting are explained below with the help of Fig. 14.4, in which the basic turning process is shown. The three types of surfaces are :

1. the work surface, from which the material is cut.

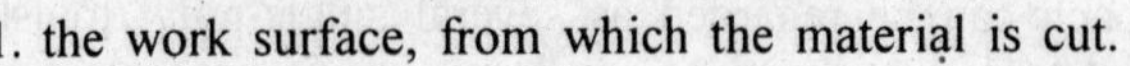

2. the machined surface which is formed or generated after removing the chip.
3. the cutting surface which is formed by the side cutting edge of the tool.

The references from which the tool angles are specified are the 'cutting plane' and the 'basic plane' or the 'principal plane'. The cutting plane is the plane tangent to the cutting surface and passing through and containing the side cutting edge. The basic plane is the plane parallel to the longitudinal and cross feeds, that is, this plane lies along and normal to the longitudinal axis of the workpiece. In a lathe tool, the basic plane concides with the base of the tool.

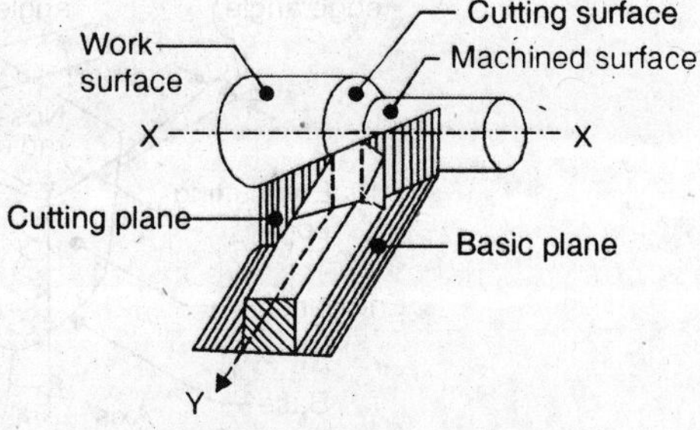

Fig. 14.4. Principal Surfaces and Planes in Metal Cutting.

14.3.1. Designation of Cutting Tools. By designation or nomenclature of a cutting tool is meant the designation of the shape of the cutting part of the tool. The two systems to designate the tool shape, which are widely used, are :

1. American Standards Association System (ASA) or American National Standards Institute (ANSI).
2. Orthogonal Rake System (ORS).

ASA System. In the ASA system, the angles of tool face, that, is its slope, are defined in two orthogonal planes, one parallel to and the other perpendicular to, the axis of the cutting tool, both planes being perpendicular to the base of the tool. For simple turning operation, this system is illustrated in Fig. 14.5.

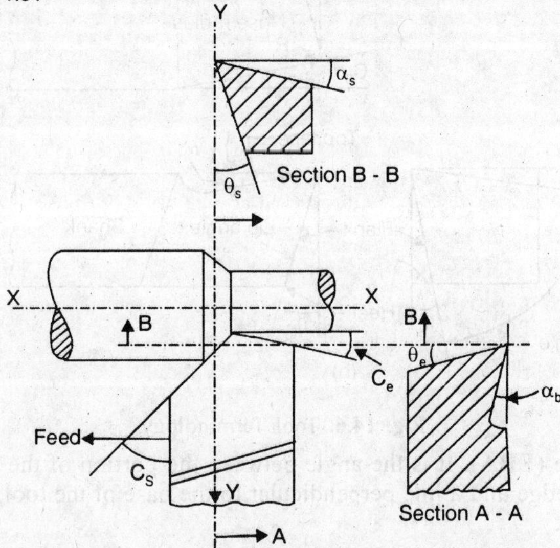

Fig. 14.5. ASA System.

The typical right hand single point cutting tool terminology is given in [Fig. 14.6 (*a*)]. Fig. 14.6 (*b*) gives the three views of the single point cutting tool, with all the details marked on it.

The various tool angles are defined and explained below :

Side Cutting Edge Angle (SCEA). Side cutting edge angle, Cs, also known as lead angle, is the angle between the side cutting edge and the side of the tool shank.

The complimentary angle of SCEA is called the "Approach angle".

End Cutting Edge Angle (ECEA). This is the angle between the end cutting edge and a line normal to the tool shank.

Side Relief Angle (SRA). It is the angle between the portion of the side flank immediately below the side cutting edge and a line perpendicular to the base of the tool, and measured at right angle to the side flank.

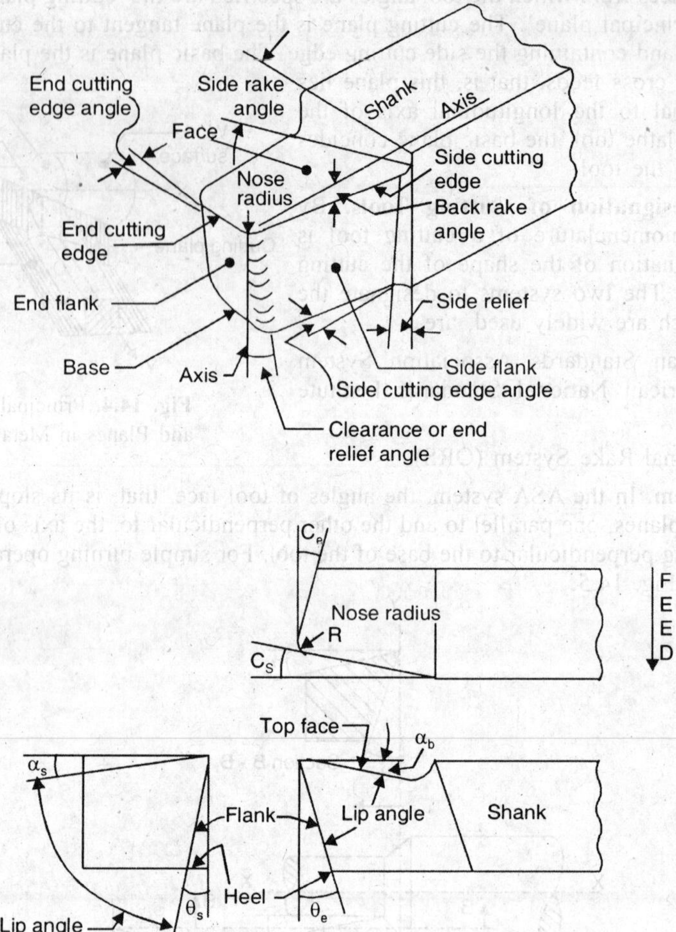

Fig. 14.6. Tool Terminology.

End Relief Angle (ERA). It is the angle between the portion of the end flank immediately below the end cutting edge and a line perpendicular to the base of the tool, and measured at right angle to the end flank.

Back-Ranke Angle (BRA). It is the angle between the face of the tool and a line parallel to the base of the tool and measured in a plane (perpendicular) through the side cutting edge. This angle is positive, if the side cutting edge slopes downwards from the point towards the shank and is negative if the slope of the side cutting edge is reverse. So this angle gives the slope of the face of the tool from the nose towards the shank.

Side-Rake Angle (SRA). It is the angle between the tool face and a line parallel to the base of the tool and measured in a plane perpendicular to the base and the side cutting edge. This angle gives the slope of the face of the tool from the cutting edge. The side rake is negative if the slope is towards the cutting edge and is positive if the slope is away from the cutting edge.

Importance of Tool Angles

1. **Side Cutting-Edge Angle, Cs.** It is the angle which prevents interference as the tool enters the work materials. The tip of the tool is protected at the start of the cut, Fig. 14.7, as it enables the tool to contact the work first behind the tip. This angle affects tool life and surface finish. This angle can vary from 0° to 90°. The side cutting edge at increased value of SCEA will have

more of its length in action for a given depth of cut and the edge lasts longer. Also, the chip produced will be thinner and wider which will distribute the cutting and heat produced over more of the cutting edge. On the other hand, the larger this angle, the greater the component of force tending to separate the work and the tool. This promotes chatter. Satisfactory values of SCEA vary from 15° to 30°, for general machining. The shape of the workpiece will also determine the SCEA. To produce a 90° shoulder, zero degree SCEA is needed. No SCEA is desirable when machining castings and forgings with hard and scaly skins, because the least amount of tool edge should be exposed to the destructive action of the skin.

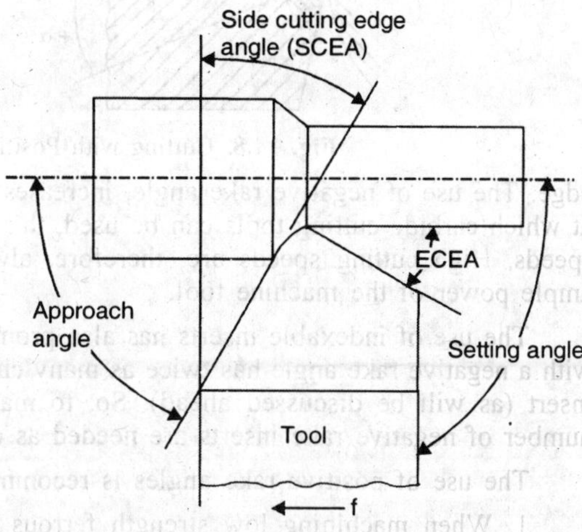

Fig. 14.7. SCEA and ECEA.

2. **End Cutting Edge Angle, Ce.** The ECEA provides a clearance or relief to the trailing end of the cutting edge to prevent rubbing or drag between the machined surface and the trailing (non-cutting) part of the cutting edge. Only a small angle is sufficient for this purpose. Too large an ECEA takes away material that supports the point and conducts away the heat. An angle of 8° to 15° has been found satisfactory in most cases on side cutting tools, like boring and turning tools. Sometimes, on finishing tools, a small flat (1.6 to 8 mm long) is ground on the front portion of the edge next to the nose radius, to level the irregular surface produced by a roughing tool. End cutting tools, like cut off and necking tools often have no end cutting angle.

3. **Side Relief Angle, (SRA) and End Relief Angle (ERA).** These angles (denoted as θ_s and θ_e in the figure) are provided so that the flank of the tool clears the workpiece surface and there is no rubbing action between the two. Relief angles range from 5° to 15° for general turning. Small relief angles are necessary to give strength to the cutting edge when machining hard and strong materials. Tools with increased values of relief angles penetrate and cut the workpiece material more efficiently and this reduces the cutting forces. Too large relief angles weaken the cutting edge and there is less mass to absorb and conduct the heat away from the cutting edge.

4. **Back and Side Rake Angle** (α_b, α_s). The top face of the tool over which the chip flows is known as the rake face. The angle which this face makes with the normal to the machined surface at the cutting edge is known as "Back-rake angle, α_b", and the angle between the face and a plane parallel to the tool base and measured in a plane perpendicular to both the base of the tool holder and the side cutting edge, is known as "Side-rake angle, α_s". The rake angles may be positive, zero, or negative. Cutting angle and the angle of shear are affected by the values for rake angles. Larger the rake angle, smaller the cutting angle (and larger the shear angle) and the lower the cutting force and power. However, since increasing the rake angle decreases the cutting angle, this leaves less metal at the point of the tool to support the cutting edge and conduct away the heat. A practical rake angle represents a compromise between a large angle for easier cutting and a small angle for tool strength. In general, the rake angle is small for cutting hard materials and large for cutting soft ductile materials. An exception is brass which is machined with a small or negative rake angle to prevent the tool form digging into the work.

The use of negative rake angles started with the employment of carbide cutting tools. When we use positive rake angle, the force on the tool is directed towards the cutting edge, tending to chip or break it, [Fig. 14.8 (*a*)]. Carbide being brittle lacks shock resistance and will fail if positive rake angles are used with it. Using negative rake angles, directs the force back into the body of the tool away from the cutting edge, [Fig. 14.8 (*b*)], which gives protection to the cutting

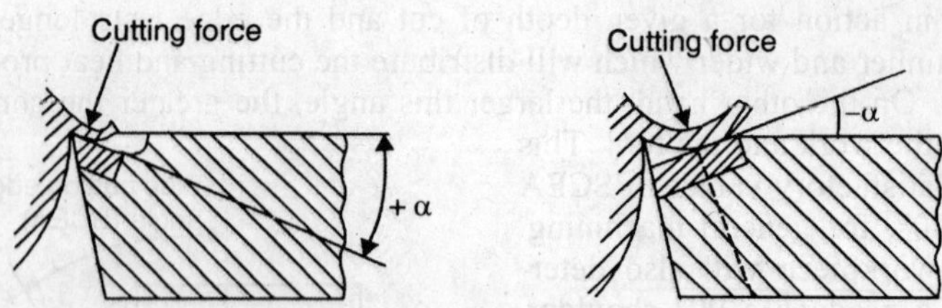

Fig. 14.8. Cutting with Positive and Negative Rake Tools.

edge. The use of negative rake angle, increases the cutting force. But at higher cutting speeds, at which carbide cutting tools can be used, this increase in force is less than at normal cutting speeds. High cutting speeds are, therefore, always used with negative rakes, which requires ample power of the machine tool.

The use of indexable inserts has also promoted the use of negative rake angles. An insert with a negative rake angle has twice as many cutting edges as an equivalent positive rake angle insert (as will be discussed ahead). So, to machine a given number of components, smaller number of negative rake inserts are needed as compared to positive rake inserts.

The use of positive rake angles is recommended under the following conditions :

1. When machining low strength ferrous and non-ferrous materials and work-hardening materials.
2. When using low power machines.
3. When machining long shafts of small diameters.
4. When the set up lacks strength and rigidity.
5. When cutting at low cutting speeds.

The use of negative rake angles is recommended under the following conditions :

1. When machining high strength alloys.
2. When there are heavy impact loads such as in interrupted machining.
3. For rigid set ups and when cutting at high speeds.

Recommended rake angles are given in Table 14.1.

Table 14.1. Recommended Rake angles

Work Material	*Tool Material*					
	H.S.S. and Cast Alloys		*Cemented Carbide*			
			Brazed		*Throw away*	
	Back	*Side*	*Back*	*Side*	*Back*	*Side*
Free Machining Steels	10	12	0	6	− 5	− 5
Mild Steel	8	10	0	6	− 5	− 5
Med. Carbon Steels	0	10	0	6	− 5	− 5
Alloy Tool Steels	0	10	− 5	− 5	− 5	− 5
Stainless Steel	0	10	0	6	− 5	− 5
Cast Iron	5	5 to 10	0	6	− 5	− 5
Aluminium Alloys	20	15	3	15	0	5
Copper Alloys	5	10	0	8	0	5
Magnesium Alloys	20	15	3	15	0	5
Titanium Alloys	0	5	0	6	− 5	− 5

5. **Nose Radius.** Nose radius is favourable to long tool life and good surface finish. A sharp point on the end of a tool is highly stressed, short lived and leaves a groove in the path of cut. There is an improvement in surface finish and permissible cutting speed as nose radius is increased from zero value. Too large a nose radius will induce chatter. The use of following values for nose radius is recommended :

R = 0.4 mm, for delicate components.

≥ 1.5 mm for heavy depths of cut, interrupted cuts and heavy feeds.

= 0.4 mm to 1.2 mm for disposable carbide inserts for common use.

= 1.2 to 1.6 mm for heavy duty inserts.

Tool Designation. The tool designation or tool signature, under ASA system is given in the order given next :

Back rake, Side rake, End relief, Side relief, End cutting edge angle, Side cutting edge angle, and nose radius that is,

$$\alpha_b - \alpha_s - \theta_e - \theta_s - C_e - C_s - R$$

If tool designation is :

$$8 - 14 - 6 - 6 - 6 - 15 - \frac{1}{8}, \text{ it means that,}$$

$$\alpha_b = 8° \qquad \alpha_s = 14°$$

$$\theta_e = 6° \qquad \theta_s = 6°$$

$$C_e = 6° \qquad C_s = 15° \qquad R = \frac{1''}{8}.$$

In ASA system of tool angles, the angles are specified independently of the position of the cutting edge. It, therefore, does not give any indication of the behaviour of the tool in practice. Therefore, in actual cutting operation, we should include the side cutting edge (principal cutting angle) in the scheme of reference planes. Such a system is known as Orthogonal rake system (ORS).

Orthogonal Rake System (ORS). As mentioned above, in this system the planes for designating tools are the planes containing the principal or side cutting edge and the plane normal to it. In the plane *NN* which is normal to the principal cutting edge and is known as Orthogonal plane or the chief plane, we have the following angles : side relief angle γ, the side rake angle (known as Orthogonal rake angle) α, wedge (lip angle) and the cutting angle (see Fig. 14.9).

The side relief angle is the angle between the side (main) flank and the cutting plane. The side rake angle, α, is the angle between the toolface and a plane normal to the cutting plane and passing through the main cutting edge. This angle is positive when the face slopes downward from the plane perpendicular to the cutting plane (as shown in Fig. 14.9), equal to zero when the face is perpendicular to the cutting plane and negative when the face slopes upwards. The "wedge angle, β" is the angle between the tool face and the main flank. The "cutting angle, δ" is the angle between the tool face and the cutting plane. When α is positive, we have,

$$\alpha + \gamma + \text{wedge angle} = 90°$$

$$\gamma + \beta = \delta$$

The usual values of α and γ are :

$$\alpha = -10° \text{ to } +15°, \ \gamma = 6° \text{ to } 12°$$

In the ORS, the back rake angle is the inclination angle (i) between the principal cutting edge and a line passing through the point of the tool parallel to the principal plane. This angle is measured in a plane passing through the main cutting edge and perpendicular to the basic plane. In Fig. (14.9), the angle i is negative with tool nose being the highest point of the cutting edge. It will be zero when the cutting edge is parallel to the basic plane and positive if the cutting edge is towards the right (Fig. 14.9) of the line passing through the point of the tool and parallel to the principal (basic) plane, that is, the tool nose is the lowest point of the cutting edge.

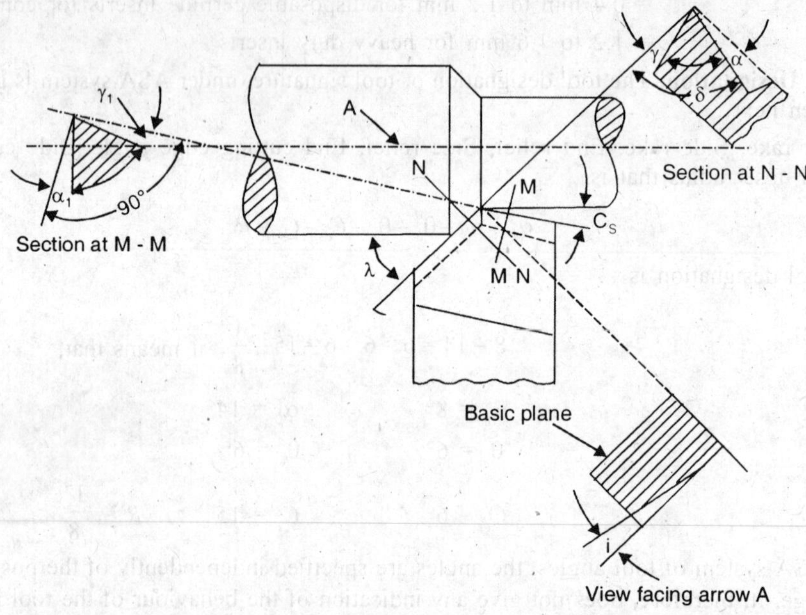

Fig. 14.9. ORS of Tool Angles.

In addition to the angles discussed above, angles are also measured in the plane *MM* (known as Auxiliary reference plane) which is normal to the projection of the end cutting edge on the basic plane. These angles are the end relief angle γ_1 and the back rake angle α_1 (also called auxiliary rake angle). The plan angles are the approach angle or entering angle λ which is equal to $(90° - C_s)$ and the end cutting edge angle, C_e.

$$\gamma_1 = 8° \text{ to } 10°, \lambda = 30° \text{ to } 70°, C_e = 10° \text{ to } 15°$$

The tool designation under ORS is :

$$i - \alpha - \gamma - \gamma_1 - C_e - \lambda - R$$

A typical tool designation (signature) is :

$$0 - 10 - 6 - 6 - 8 - 90 - 1 \text{ mm}$$

Interconversion between ASA system and ORS

$$\tan \alpha = \tan \alpha_s \sin \lambda + \tan \alpha_b \cos \lambda$$

$$\tan \alpha_b = \cos \lambda \tan \alpha + \sin \lambda \tan i$$

$$\tan \alpha_s = \sin \lambda \tan \alpha - \cos \lambda \tan i$$

$$\tan i = - \tan \alpha_s \cos \lambda + \tan \alpha_b \sin \lambda$$

In the second and third equations above, the values of angles α and i are taken with their signs.

14.4. METHODS OF MACHINING

In the metal cutting operation, Fig. 14.1, the tool is wedge-shaped and has a straight cutting edge. Basically, there are two methods of metal cutting, depending upon the arrangement of the cutting edge with respect to the direction of relative work-tool motion :

1. Orthogonal cutting or two dimensional cutting.
2. Oblique cutting or three dimensioning cutting.

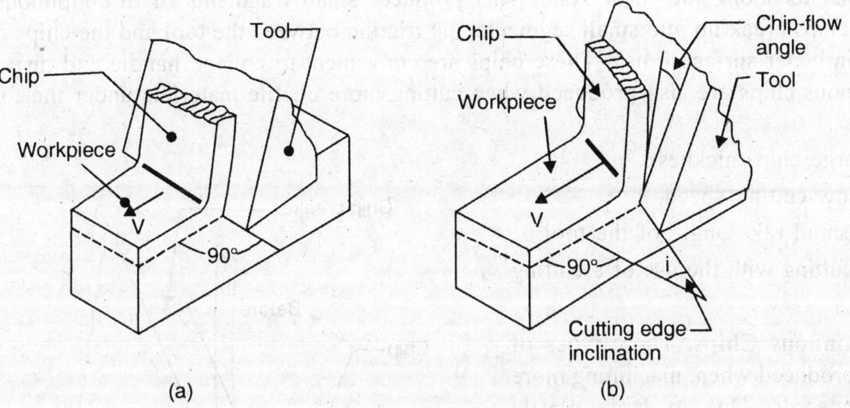

Fig. 14.10. Methods of Machining.

In orthogonal cutting, Fig. 14.10, the cutting edge of the tool is arranged perpendicular to the cutting velocity vector, V, whereas in oblique cutting, it is set at some angle other than 90° to the cutting velocity vector, which gives an "inclination angle i". The analysis of oblique cutting being very complex, the relatively simple arrangement of orthogonal cutting is, therefore, widely used in theoretical and experimental work.

In pure orthogonal cutting, $i = 0°$, $C_e = 0°$, and $\lambda = 90°$. This is also known as orthogonal system of second kind. When $i = 0$, and $0 < \lambda < 90°$, it is called as orthogonal system of first kind. A common example of pure orthogonal cutting process is the turning of a thin pipe with a straight edged tool set normal to the longitudinal axes.

In almost all other machining operations, two cutting edges take part in the cutting operation, primary (principal) cutting edge and the secondary cutting edge. The situation where the primary cutting edge is perpendicular to the cutting velocity vector is called" Restricted Orthogonal Machining". However, in practice, it is called as "Orthogonal machining".

14.5. TYPES OF CHIPS

Whatever the cutting conditions can be, the chips produced may belong to one of the following three types, Fig. 14.11 :

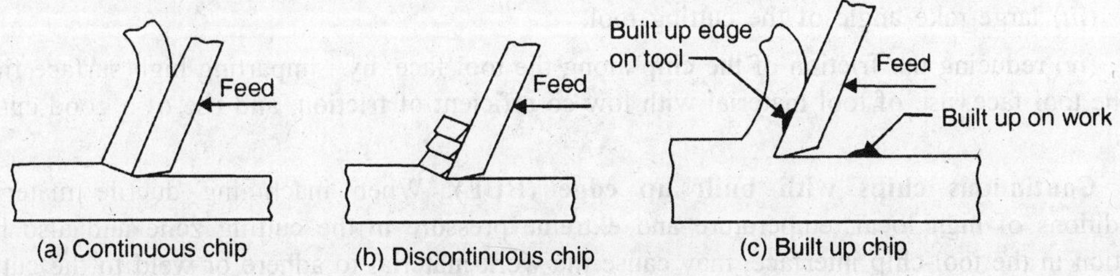

Fig. 14.11. Types of Chips.

1. Discontinuous chips.
2. Continuous chips.
3. Continuous chips with build up edge (BUE).

Discontinuous Chips. These types of chips are usually produced when cutting more brittle materials like grey cast iron, bronze and hard brass. These materials lack the ductility necessary for appreciable plastic chips formation. The material ahead of the tool edge fails in a brittle fracture manner along the shear zone. This produces small fragments of discontinuous chips. Since the chips break up into small segments, the friction between the tool and the chips reduces, resulting in better surface finish. These chips are convenient to collect, handle and dispose of *f*. Discontinuous chips are also produced when cutting more ductile materials under the following conditions :

(*i*) large chip thickness.

(*ii*) low cutting speed.

(*iii*) small rake angle of the tool.

(*iv*) cutting with the use of a cutting fluid.

Continuous Chips. These types of chips are produced when, machining more ductile materials. Due to large plastic deformations possible with ductile materials, longer continuous chips are produced. This type of chip is the most desirable, since it is stable cutting, resulting in generally good surface finish. On the other hand, these chips are difficult to handle and dispose off. The chips coil in a helix (chip curl) and curl around the work and the tool and may injure the operator when break loose. Also, this type of chip remains in contact with the tool face for a longer period, resulting in more frictional heat. These difficulties are usually avoided by attaching to the tool face or machine on the tool face, a 'chip breaker', Fig. 14.12. The function of chip breaker is to reduce the radius of curvature of the chip and thus break it. The following cutting conditions also help in the production of continuous chips :

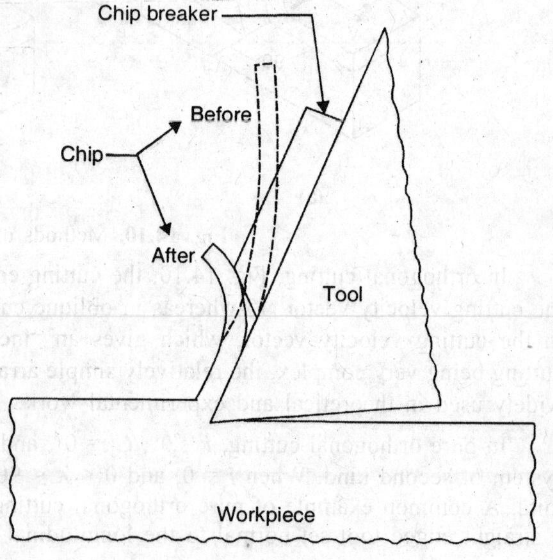

Fig. 14.12. Chip Breaker.

(*i*) small chip thickness.

(*ii*) high cutting speed.

(*iii*) large rake angle of the cutting tool.

(*iv*) reducing the friction of the chip along the tool face, by : imparting high surface finish to the tool face, use of tool material with low co-efficient of friction, and use of a good cutting fluid.

Continuous chips with built up edge (BUE). When machining ductile materials, conditions of high local temperature and extreme pressure in the cutting zone and also high friction in the tool-chip interface, may cause the work material to adhere or weld to the cutting edge of the tool forming the built-up edge. Successive layers of work material are then added to the built-up edge. When this edge becomes larger and unstable, it breaks up and part of it is carried up the face of the tool alongwith the chip while the remaining is left over the surface being machined, which contributes to the roughness of the surface. The built-up edge changes its size during the cutting operation. It first increases, then decreases, then again increases etc. This cycle is a source of vibration and poor surface finish. Although, the built-up edge protects the cutting edge of the tool, it changes the geometry of the cutting tool. Low cutting speed also contributes to the formation of the built-up edge. Increasing the cutting speed, increasing the rake angle and using a cutting fluid contribute to the reduction or elimination of the built-up edge.

14.6. DETERMINATION OF SHEAR ANGLE

In the simplified model of two dimensional cutting operation, the cutting tool is completely defined by the rake angle α and clearance angle γ. In addition, the following assumptions are made :

1. The tool is perfectly sharp and contacts the chip on its front, or rake face.
2. The primary deformation takes place in a very thin zone adjacent to the shear plane *AB*, Fig. 14.1.
3. There is no side flow of the chip, that is, plain strain condition.

Shear angle, φ, is defined as the angle made by the shear plane, with the direction of the tool travel. If,

t = uncut or underformed chip thickness

and t_c = chip thickness after the metal is cut.

Then
$$r = \frac{t}{t_v} \qquad \text{...(14.1)}$$

is called the "cutting ratio" or "chip-thickness ratio", or "chip-compression factor".

$$\zeta = \frac{1}{r} = \frac{t_c}{t} \qquad \text{...(14.2)}$$

is called the "chip reduction factor".

't' is the distance between two consecutive positions of the cutting edge, measured in a direction normal to the cutting edge along the plane of the tool face. In case of turning with λ = 90°, t = feed (see Art. 14.7).

The shear angle can be determined in the following ways :

(*a*) From Fig. 14.13, simple geometry gives,

$$r = \frac{t}{t_c} = \frac{AB \sin \varphi}{AB \cos(\varphi - \alpha)}$$

or
$$r = \frac{t}{t_c} = \frac{\sin \varphi}{\cos(\varphi - \alpha)} \qquad \text{...(14.3)}$$

Solving this equation for φ, we get,

$$\tan \varphi = \frac{r \cos \alpha}{1 - r \sin \alpha} \qquad \text{...(14.4)}$$

After the cut has been taken, the measurements of 't' and 't_c' can be taken, and so the cutting ratio can be determined. Then the shear angle existing during the cut can be obtained with the help of Eqn. 14.4.

(*b*) If the length of cut, *l*, is known then the continuity equation gives,

$$\rho\, lbt = \rho\, l_c btc$$

where *r* is the density of the material, assumed incompressible and l_c is the length of chip.

$$\therefore \quad \frac{l_c}{l} = \frac{t}{t_c} = r \qquad ...(14.5)$$

Then, with the help of equation (14.4), φ can be estimated.

(*c*) When the length of the cut is not directly known, then it can be estimated by weighing a known length of chip. Then, weight of the chip,

$$W_c = \rho l_c bt_c = \rho lbt$$

or

$$l = \frac{W_c}{\rho bt} \qquad ...(14.6)$$

Knowing ρ, *b* and *t*, the length of cut '*l*' can be determined. Then, φ can be estimated from equations (14.5) and (14.4).

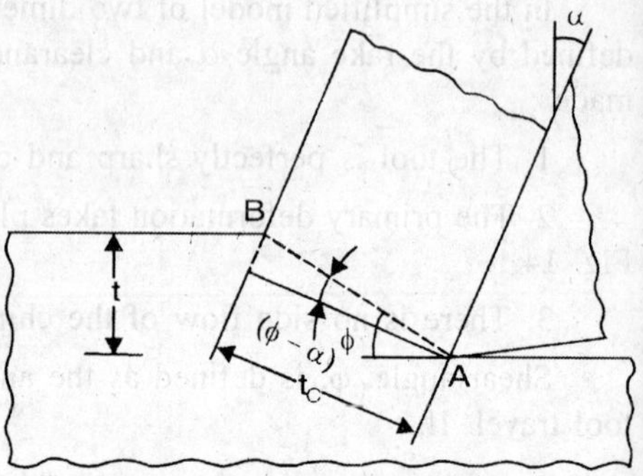

Fig. 14.13

(*d*) If the chip and workpiece velocities V_c and *V* are known, then from the continuity equation :

$$V\rho\, bt = V_c \rho bt_c$$

or

$$\frac{t}{t_c} = \frac{V_c}{V} = r \qquad ...(14.7)$$

Then φ can be determined from eqn. (14.4).

The shearing process in metal cutting can be represented by a simple model shown in Fig. 14.14.

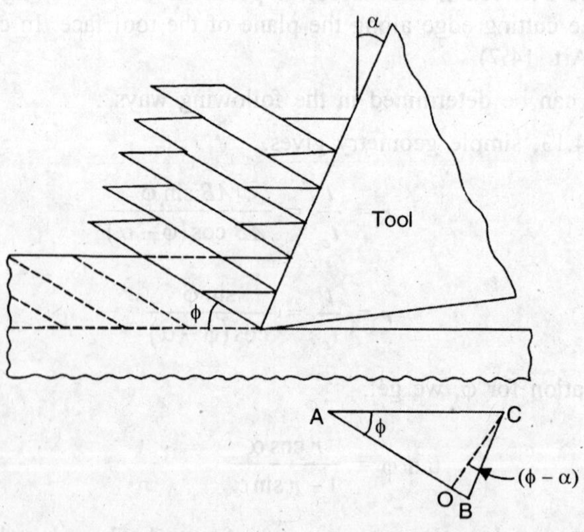

Fig. 14.14. Model of Chip formation.

The shearing process is analogous to the displacement of a stack of cards, with each card sliding slightly over the adjacent card.

From this model, the shear strain that the material is undergoing can be determined. Based on its definition, the shear strain will be given as, (View at tool tip)

$$s = \frac{\text{distance sheared}}{\text{thickness of zone}}$$

$$s = \frac{AB}{OC} = \frac{OA + OB}{OC} = \frac{AO}{OC} + \frac{OB}{OC}$$

or, $$s = \cot \varphi + \tan (\varphi - \alpha) \quad ...(14.8)$$

Velocities. There are three types of velocities involved in any metal cutting process :

(*i*) The velocity of the tool relative to the workpiece, V, usually called the cutting speed.

(*ii*) The velocity of the chip relative to the work, V_s, called the shear velocity, that is, the velocity at which shearing takes place.

(*iii*) The velocity of the chip up the face of the tool, V_c, called the chip velocity.

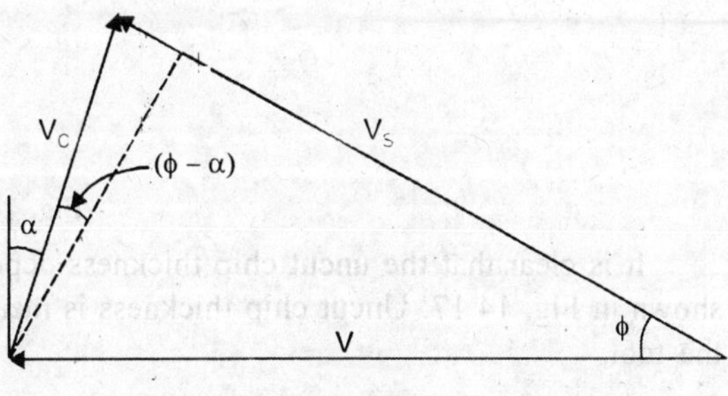

Fig. 14.15

We have from constancy of volume,

$$V.t = V_c t_c, \text{ width remaining constant}$$

$$\therefore \quad V_c/V = \frac{t}{t_c} = r \quad ...(14.9)$$

The relationship in velocities can be determined from Fig. 14.15. It can be shown that,

$$V_s/V = \frac{\cos \alpha}{\cos (\varphi - \alpha)} \text{ and } V_c = \frac{\sin \phi}{\cos(\phi - \alpha)}.V \quad ...(14.10)$$

From here, V_s, can be determined, and so the shear strain rate during cutting can be estimated, which is given as

$$\dot{s} = \frac{V_s}{\text{thickness of shear zone}} = \frac{V_s}{t_s} \quad ...(14.11)$$

The maximum value of 't_s' is approximately 25×10^{-3} mm.

14.7. DETERMINATION OF UNDERFORMED CHIP THICKNESS

To estimate the underformed or uncut chip thickness, 't', refer to Fig. 14.16, where the two consecutive cuts have been shown and the various parameters, that is, feed f, depth of cut d, width of cut b, t and t_c have been marked. It can be easily seen that the following relations exist :

$$t = f \sin \lambda$$

and $$b = \frac{d}{\sin \lambda} \quad ...(14.12)$$

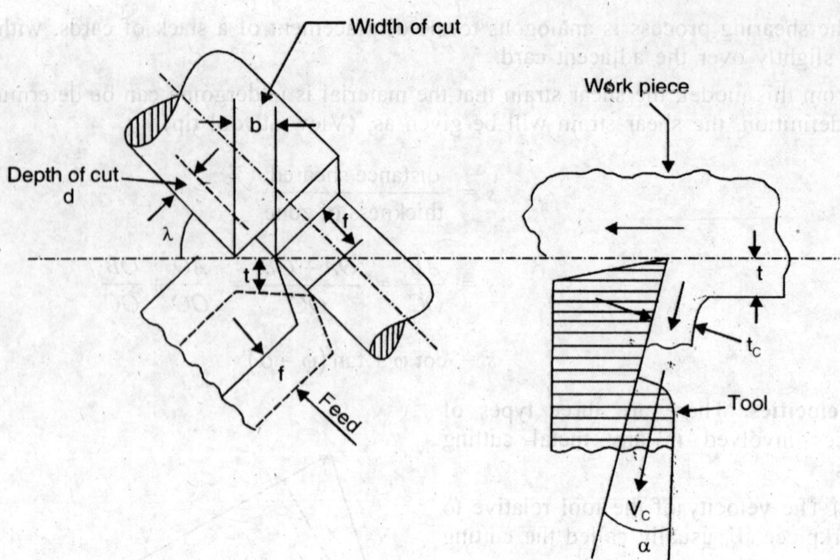

Fig. 14.16

It is clear that the uncut chip thickness depends upon the primary cutting edge angle λ, as shown in Fig. 14.17. Uncut chip thickness is measured perpendicular to the side cutting edge of the tool.

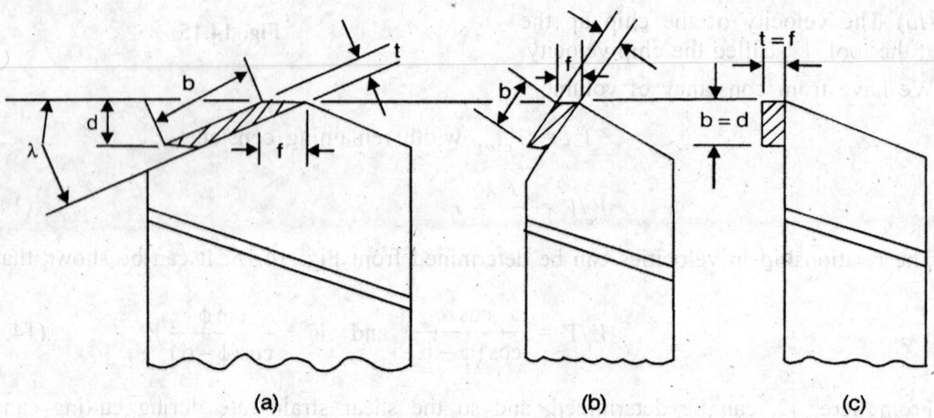

Fig. 14.17

For Fig. 14.17 (*c*) where $\lambda = 90°$ and $C_e \neq 0$.

$$t = f; \text{ and } b = d. \qquad ...(14.13)$$

i.e., uncut chip thickness, t = (feed, mm/rev)

and width of cut = depth of cut

Such a case is called " Restricted Orthogonal cutting".

These relations hold good for metal cutting with single point tools (turning, shaping, planing etc.).

For drilling operation, since, tool has two cutting edges, thus, the feed per tooth will be one-half of the feed. Thus

$$t = \frac{f}{2}\sin\lambda$$

14.8. FORCE RELATIONS

Here the analysis is limited to two dimensional or orthogonal cutting which is simpler to understand as compared to the complicated three dimensional cutting process. When a cut is made, Fig. 14.18 (*a*), the forces acting on the metal chip are :

1. Force F_s, which is the resistance to shear of the metal in forming the chip. It acts along the shear plane.

2. Force F_n normal to the shear plane. This is the "backing up" force on the chip provided by the workpiece.

3. Force N at the tool chip interface acting normal to the cutting face of the tool and is provided by the tool.

(4) Force F is the frictional resistance of the tool acting on the chip. It acts downward against the motion of the chip as it glides upwards along the tool face.

Fig. 14.18 (*b*) is a free body diagram showing the forces acting on the chip. R is the resultant force of F and N and R' that of F_s and F_n. Neglecting the couples which curl the chip, considering the equilibrium of the chip, R and R' are equal in magnitudes opposite in direction and collinear as shown.

All these forces can be represented with the help of a circle known as the 'Merchant force circle', Fig. 14.19. Here the two force triangles have been superimposed by placing the two equal forces R and R' together. In the figure, β is the angle of friction. In this diagram, for convenience, the resultant forces have been moved to the point of the tool. Since the forces F_s and F_n are at right angle to each other, their intersection lies on a circle with diameter R'. The forces F and N may be placed in the diagram as shown to form the circle diagram.

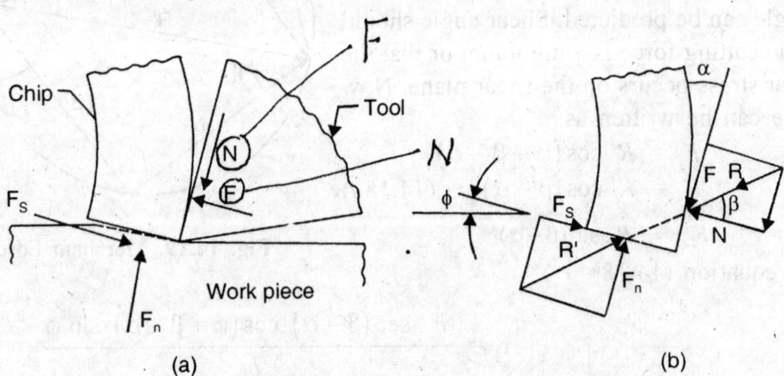

Fig. 14.18. Forces in Metal Cutting.

The two orthogonal components (horizontal and vertical) F_c and F_t of the resultant force R' can be measured by using a dynamometer. The horizontal component is the cutting force F_c and the vertical component is the thrust force, F_t. The power consumed during the process is the product of F_c and V. The thrust force does not contribute to the work done. It holds the tool against the workpiece. After F_c and F_t are known, they can be laid as shown. The rake angle α can be laid off and the force F and N determined. The shear angle can be determined as explained earlier. Knowing, F_c, F_t, α and φ, all the component forces acting on the chip can be determined. It is easily shown that

$$F = F_c \sin \alpha + F_t \cos \alpha \qquad \text{...(14.14)}$$

$$N = F_c \cos \alpha - F_t \sin \alpha$$

The co-efficient of friction will then be given as

$$\mu = \frac{F}{N} = \frac{F_c \tan \alpha + F_t}{F_c - F_t \tan \alpha} \quad ...(14.15)$$

$$\beta = \tan^{-1} \mu$$

On the shear plane.

$$F_s = F_c \cos \varphi - F_t \sin \varphi$$

$$F_n = F_c \sin \varphi + F_t \cos \varphi \quad ...(14.16)$$

From here,

$$F_t = F_n.\cos\phi - F_s.\sin\phi$$
$$F_c = F_n \sin\phi + F_s.\cos\phi$$

The shear angle, φ, can be determined from equation (14.4).

Now shear plane area,

$$A_s = bt/\sin \varphi \quad ...(14.17)$$

The average stresses on the shear plane area are then

$$\tau_s = F_s/A_s,$$
$$\sigma_s = F_n/A_s \quad ...(14.18)$$

The shear angle has great influence on the overall cutting geometry. So, its relationship to material and process variables should be determined so that this angle can be predicted. Shear angle should be such that the cutting force is a minimum or that the maximum shear stress occurs on the shear plane. Now the shear force can be written as

$$F_s = R' \cos(\varphi + \beta - \alpha)$$

and

$$F_c = R' \cos(\beta - \alpha) \quad ...(14.18a)$$

$$F_t = R'.\sin(\beta - \alpha)$$

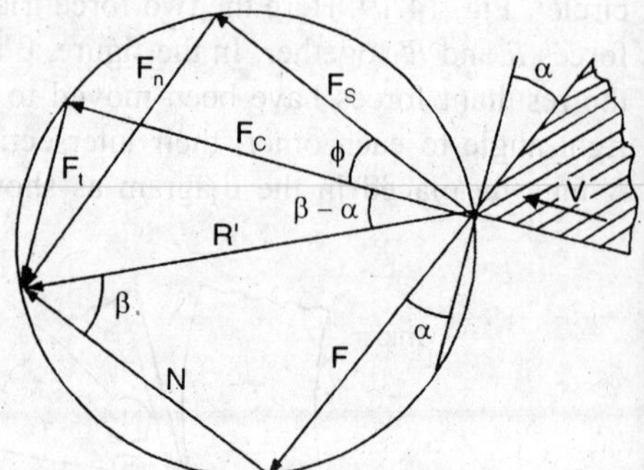

Fig. 14.19. Merchant Force Circle.

∴ From equation (14.18)

$$\tau_s = \frac{F_c \cdot \sec(\beta - \alpha) \cdot \cos(\varphi + \beta - \alpha) \cdot \sin \varphi}{bt}$$

Assuming that β is independent of φ, therefore, for maximum shear stress,

$$\frac{\partial \tau_s}{\partial \varphi} = 0$$

$$\therefore \quad \cos(\varphi + \beta - \alpha) \cos \varphi - \sin(\varphi + \beta - \alpha) \sin \varphi = 0$$

$$\therefore \quad \tan(\varphi + \beta - \alpha) = \cot \varphi = \tan(90 - \varphi)$$

$$\therefore \quad \varphi = 45° + \alpha/2 - \beta/2 \quad ...(14.19)$$

For practical purposes, the following approximate values for φ has been suggested,

$$\varphi = \alpha, \quad \text{for } \alpha > 15°$$

$$\varphi = 15° \quad \text{for } \alpha < 15° \quad ...(14.20)$$

It is clear from equation (14.19), that for the same rake angle, a low friction angle, *i.e.*, less friction between chip and tool results in a higher shear angle, which reduces the cutting force, which in turn results in less friction. A higher shear angle also indicates a thin chip and less severe deformation of the chip, since the shear strain is also lower, equation (14.8). The above results can also be obtained by increasing the rake angle.

In making the shear angle predictions (Eqn. 14.19), Merchant made the following assumptions :

1. The work material behaves like an ideal plastic.
2. The theory involves the minimum energy principle.
3. A_s, τ_s and β are assumed to be constant, independent of φ.

The second assumption is not supported by evidence and also, it is an experimental fact that β (friction angle) varies greatly with α, and also is not independent of φ. On reconsidering these assumptions, Merchant included the following relation into his theory :

$$\tau_s = \tau_{so} + k\sigma_s \qquad ...(14.21)$$

where k is a constant and σ_s is the normal stress on the shear plane.

At $\sigma_s = 0$

$$\tau_s = \tau_{so}$$

He, then obtained the following relation :

$$2\varphi + \beta - \alpha = c$$

where $c = \text{arcot}\, k$. ...(14.22)

Theory of Lee and Shaffer. Lee and Shaffer applied the theory of plasticity for an ideal-rigid-plastic material, for analysing the problem of orthogonal metal cutting. They also assumed that deformation occurred on a thin-shear plane. They considered that there must be a stress field within the metal chip to transmit the cutting forces from the shear plane to the tool face. For this, they constructed a slip-line field for stress zone and derived the following relation for shear angle,

$$\varphi = \frac{\pi}{4} - (\beta - \alpha) \qquad ...(14.23)$$

Many authors have given shear-angle relations and the problem is treated in detail in books on metal cutting.

Cutting Power. The cutting power or the rate of energy consumption, is the product of the cutting speed, V, and the cutting force, F_c. Thus,

$$E = F_c V \qquad ...(14.24)$$

now

$$V = \frac{\pi DN}{1000 \times 60} \text{ m/s}$$

where

D = Diameter of job/tool in mm

N = Velocity of job or tool, rev/min

Power consumed in cutting, $P_c = \dfrac{F_C \times V}{60 \times 102}$ kW, if F_c is in kgf., and V is in m/min.

If F_C is in newtons and V in m/s, then

$$P_C = \frac{F_C \times V}{1000}, \text{kW}$$

Design power rating of the electric motor of the main drive,

$$P_m = \frac{P_c}{\eta_{mt}}$$

where 'η_{mt}' is the efficiency of the machine tool.

The values for mean efficiency at full load for various machine tools as determined by experimental methods, are given below :

Lathes, Milling machines = 0.8 to 0.90

Drilling machines = 0.85 to 0.90

Shapers, Planers = 0.65 to 0.75

Grinding machines = 0.8 to 0.85

To find the cutting power, the cutting force should be known. The cutting force depends upon depth of cut, feed, cutting velocity and many other factors. The following relations can be used to determined the cutting force.

1. For turning operations with carbide tipped tool on which $\alpha = +10°$, $\lambda = 45°$, $R = 2$mm, $i = 0°$, $C_e = 10°$, and using no lubrication, the relation is :

$$F_c = Cd^x f^y V^n, kgf$$

where d = depth of cut, mm

f = feed, mm/rev.

V = cutting speed, m/min

The values of C, x, y and n for work material steel and steel castings with tensile strength of 75 kgf/mm² and for feed less than or equal to 0.75 mm/rev., are :

$$C = 300, x = 1.0, y = 0.75, n = -0.15$$

For the other machining conditions one can refer to hand books.

2. For H.S.S. cutting tools with tool designation (ASA) of 18 – 14 – 6 – 6 – 6 – 15 – 1.8 mm, the following expressions may be used to determine the cutting force (in absence of lubrication) :

(*a*) **Hot rolled 0.2% *C* – Steel**

$$F_c = 162.4 f^{0.85} d^{0.98} \text{ kgf}$$

(*b*) **18 – 8 Stainless Steel**

$$F_c = 196.7 f^{0.85} d^{0.96} \text{ kgf}$$

(*c*) **Yellow Brass**

$$F_c = 123.6 f^{0.81} d^{0.96} \text{ kgf}$$

(*d*) **C.I.** $F = Cf^{0.78} d^{0.89}$ kgf

The constant C is given as, C = 60.2 for BHN 126

= 111.6 for BHN 181

= 131.1 for BHN 241

Above, d and f are in mm

The cutting force may also be determined by the following simple relation:

$$F_c = K_s \cdot f \cdot d$$

where K_s = Specific cutting resistance, N/mm²

= (2.5 to 4.5) × σ_{ut} for ductile materials or continuous chips

= (0.5 to 1) × HB for brittle material or segmental chip,

HB = Brinell hardness number of the work material.

Note : Power needed to make a particular cut can be approximately estimated from the following formula :

$$\text{Power} = d \times f \times V \times C, \text{ watts}$$

where d and f are in mm and mm/rev respectively and V is in m/min. C is a power constant whose values are given below, for positive rake cutting :

Material →	Low C-Steel	Medium C-Steel	Ni-Cr Steel	Grey C.I.	Cu Brass	Bronze	Al Alloys
Values of C	24	32	56	12	16	28	16

For negative rake cutting, increase the value of C by 50 per cent.

Metal removal rate. Metal removal rate is expressed as mm³/min. A process which removes metal at a faster rate may not be the most economical process, since power consumed and cost factors must be taken into account. Due to this, to compare two processes, the amount of metal removed per unit of power consumed is determined. This is called "specific metal removal rate" and is expressed as, mm³/W/min, if the power is measured in watts For a single point tool, the rate of metal removal

$$= 1000\ A_C \times V,\ \text{mm}^3/\text{min. } (V \text{ is in m/min})$$

where $\quad A_C$ = cross-sectional area of the uncut chip, mm²

$$= b \times t$$

Now, $\quad b \times t = d \times f$ (see eqn. 14.12)

∴ $\quad \text{MRR} = 1000\ \text{dfV, mm}^3/\text{min}$

or $\quad = \pi DdfN$ mm³/min; N is in rev/min

and D is mean diameter in mm.

∴ For a cylindrical workpiece turning on a lathe,

$$\text{MRR} = \frac{\pi}{4}(D_1^2 - D_2^2) \cdot f \cdot N \cdot \text{mm}^3/\text{min}$$

where D_1 and D_2 are initial and find diameters.

Forces acting at the point of the single point tool: In orthogonal cutting (Fig. 14.17*c*), where the tool cutting edge is perpendicular to the cutting direction, only two forces will act at the cutting tool point: Cutting F_c and the feed force, F_f (or the thrust force F_t), Fig. 14.20*a*. F_c acts in the vertical plane and is tangential to the job (Tangential force). It will be downward for c.c.w. rotation of the job. F_f or F_t acts in the horizontal plane, parallel to the axis of the job and in opposite direction of the feed. It helps in holding the tool in position. However, in oblique cutting, where the cutting edge is inclined to the cutting direction (Fig. 14.17*a* and *b*), another force, normal force F_n or radial force F_r, acts in the horizontal plane along the axis of the tool and normal to the job surface Fig. 14.20*b*. This force will tend to push away the tool from the workpiece which may cause chatter. Now, $F_f = F_t$ = 0.3 to 0.6 F_c and $F_n = F_r$ = 0.2 to 0.4 F_c. The three forces: F_c, F_f and F_n are mutually perpendicular and can be represented vectorially as in Fig. 14.20*c*. The resultant force, F_e, which is the single equivalent force acting on the tool point, can be determined as,

$$F_e^2 = F_c^2 + R^2$$

But, $\quad R^2 = F_f^2 + F_n^2$; R is the diagonal of the horizontal plane

$$\therefore \qquad F_e = \sqrt{F_c^2 + F_f^2 + F_n^2}$$

The inclination of the resultant equivalent force, to the horizontal and vertical planes can easily be found. The magnitude of F_e and its direction are coincident with those of the chip as it is sheared from the surface. Thus, if the force F_e is used in power calculations, we would need to know the velocity of the chip. This can be simplified by using the component forces of the equivalent force, F

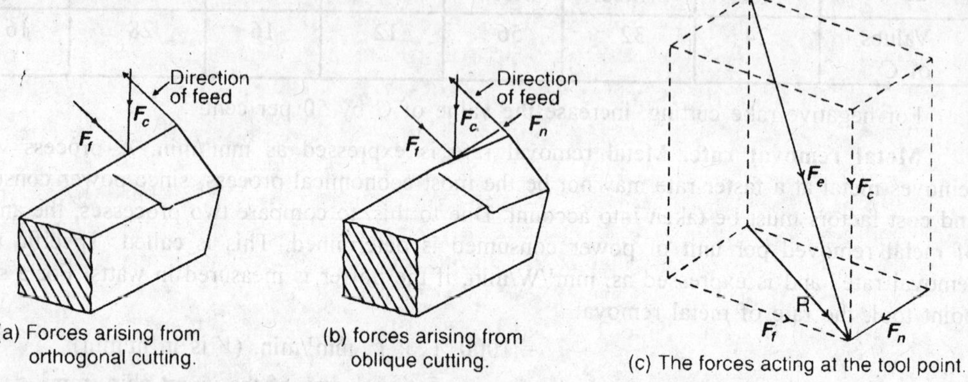

(a) Forces arising from orthogonal cutting.
(b) forces arising from oblique cutting.
(c) The forces acting at the tool point.

Fig. 14.20. Forces acting at the tip of the single point cutting tool

14.9. ENERGY CONSIDERATION IN METAL CUTTING

The total energy required per unit time (power) during metal cutting, E, is given (as already noted),

$$E = F_C \times V, \text{ Nm/s} \qquad (V \text{ being in m/s})$$

This total energy consists of two main parts : the shear energy, E_s, required to produce the plastic deformation in the shear zone, and the friction energy, E_f, used as the chip slides along the tool. Other small contributors to the total energy are ; energy required to curl the otherwise straight chip, kinetic energy required to accelerate the chip, and surface energy required to produce the new surface area. These three types of energies are negligible, so, as a first approximation,

$$E \cong E_s + E_f$$

The energy per unit time divided by the volume removed per unit time, is known as the specific energy,

Total specific energy, $e = \dfrac{E}{b.t.V} = \dfrac{F_C}{b.t} \text{ N/mm}^2$

Specific shear energy, $e_s = \dfrac{E_s}{b.t.V} = \dfrac{F_s.V_s}{b.t.V} \text{ N/mm}^2$

Specific friction energy, $e_f = \dfrac{E_f}{b.t.V} = \dfrac{F.V_c}{b.t.V} \text{ N/mm}^2$

Now $$\frac{V_s}{V} = \frac{\cos\alpha}{\cos(\varphi - \alpha)}$$

and $$\frac{V_c}{V} = \frac{t}{t_c} = r$$

$$\therefore \qquad e_s = \frac{F_s}{b.t} \cdot \frac{\cos \alpha}{\cos(\varphi - \alpha)}$$

$$e_f = \frac{F}{b.t_c}$$

Because of the many factors involved, it is not possible to exactly determine the cutting force and the cutting power. The reliable predictions are based on the experimental data as given on the last page or in Table 14.2. The wide range in values is due to the cutting variables, cutting fluids, friction conditions and the variation in the material properties under the same group. While arriving at these values, the efficiency of the drive motor is taken as 0.8. These values should be multiplied by 1.25 for dull tools.

Now, energy supplied by electric motor is partly used in cutting and partly in feeding the tool, so

$$E_m = F_c \times V + F_f \times \text{feed velocity}$$

Now, feed velocity is very nominal. Also F_f (or F_t) is very small as compared to F_c, so the term $F_f \times$ feed velocity can be neglected.

$$\therefore \qquad E_m \cong F_c \times V$$

Now, under ideal conditions, *i.e.*, no loss of work

$$E \cong E_m$$

$$\therefore \qquad E_s + E_f = E_m \text{ (as above)}$$

$$\text{or} \qquad F_s \times V_s + F \times V_c = F_c \times V$$

$$\therefore \qquad F_c \times V = F_s \times V_s + F \times V_c$$

where F_s = Shear force

V_s = Shear velocity

F = Frictional force

V_c = Chip velocity

Table 14.2. Approximate Cutting Power Requirements

Material	*Specific energy, W.s/mm³*
Aluminium Alloys	0.4 to 1.1
Magnesium Alloys	0.4 to 0.6
Copper Alloys	1.4 to 3.3
Cast Irons	1.6 to 5.5
Steels	2.7 to 9.3
Stainless Steels	3.0 to 5.2
High Temperature Alloys	3.3 to 8.5
Refractory Alloys	3.8 to 9.6
Titanium Alloys	3.0 to 4.1
Nickle Alloys	4.9 to 6.8

14.10. OBLIQUE CUTTING

A simplified model of oblique cutting or three dimensional cutting is shown in Fig. 14.21.

The cut chip flows up the face of the tool at an angle η_c, called the "chip flow angle". As shown in Fig. 14.20, the chip does not slide up the tool along the line *OA* which is perpendicular to the cutting edge, but flow up at an angle η_c to the line *OA*. A plane can be passed through *V* and V_e. The cutting process in this plane will appear quasi-two dimensional. Thus, a three dimensional cut referred to this plane can be treated as two dimensional,

α_n = normal rake angle, is between *OA* and *z*-axis.

α_e = effective rake angle, is between V_c and a line perpendicular to *V* in the plane passed through *V* and V_c.

when $i = 0$, α_n will be equal to α.

α_e is the rake angle which the material sees as it is cut. Thus α_e determines the strain etc. From geometry :

$$\sin \alpha_e = \sin \eta_c \sin i + \cos \eta_c \cos i \sin \alpha_n \quad ...(14.25)$$

Many authors have shown that for unrestricted cutting,

$$\eta_c = i \text{ (very nearly)}$$

$$\therefore \sin \alpha_e = \sin^2 i + \cos^2 i \sin \alpha_n \quad ...(14.26)$$

Thus, if α_n and *i* can be determined for any tool, the approximate value of α_c can be determined and the cutting mechanism can be treated as two dimensional.

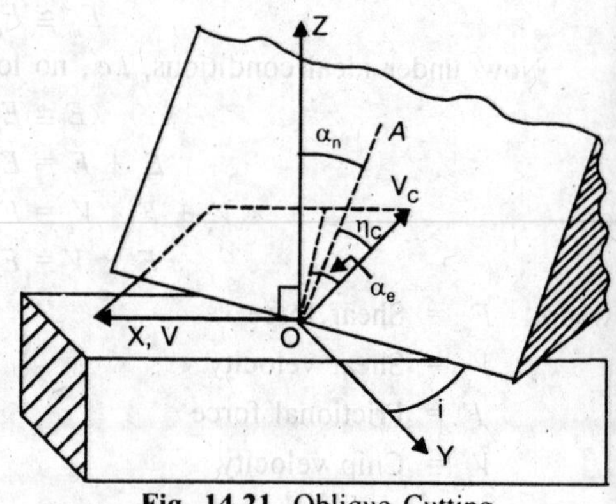

Fig. 14.21. Oblique Cutting.

For a given α_n, if *i* increases, α_e increases (Eqn 14.26). This reduces shear strain (Eqn. 14.8) and a thinner and longer chip results. This results in reduced cutting forces, reduced energy per unit volume and a reduced tendency to form a built-up edge. To get the same cutting characteristics in orthogonal cutting the increase of rake angle reduces the tool mass near the cutting edge. This results in increased tool breakage and reduced heat dissipation. Thus, in oblique cutting, the cut is improved by introducing inclination and without harming the tool.

14.11. TOOL WEAR AND TOOL LIFE

During any machining process the tool is subjected to three distinct factors : forces, temperature and sliding action due to relative motion between tool and the workpiece. Due to these factors, the cutting tool will start giving unsatisfactory performance after some time. The unsatisfactory performance may involve : loss of dimensional accuracy, increased surface roughness, and increased power requirements etc. The unsatisfactory performance results from tool wear due to its continued use. When the tool wears out, it is either replaced or reconditioned, usually by grinding. This will result in loss of production due to machine down-

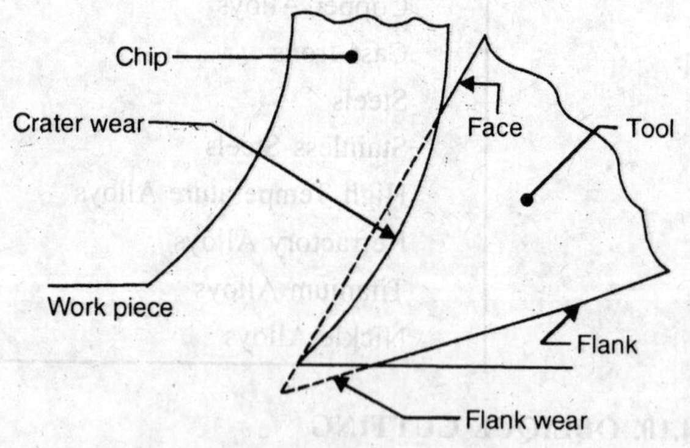

Fig. 14.22. Tool Wear.

time, in addition to the cost of replacing or reconditioning the tool. Thus, the study of tool wear is very important from the stand point of performance and economics. Due to a large number of factors over which the tool wear depends (hardness and type of tool material, type and condition of workpiece, dimensions of cut, *i.e.,* feed and depth of cut, tool geometry, tool temperature, which, in turn, is a function of cutting speed, surface finish of tool temperature and cutting fluid), the majority of studies in tool wear are based on experimental observations, since the analytical study will be very difficult.

Tool wear or tool failure may be classified as follows :

(*a*) Flank wear.

(*b*) Crater wear on tool face.

(*c*) Localized wear such as the rounding of the cutting edge, and

(*d*) Chipping off of the cutting edge.

Flank wear and crater wear are shown in Fig. 14.22. Flank wear is attributed usually to the following reasons :

1. Abrasion by hard particles and inclusions in the workpiece.

2. Shearing of the micro welds between tool and work-material.

3. Abrasion by fragments of built-up edge plowing against the clearance face of the tool.

Crater wear usually occurs due to :

1. Severe abrasion between the chip and tool face.

2. High temperatures in the tool-chip interface reaching the softening or melting temperature of tool resulting in increased rate of wear. The sharp increase in wear rate after the interface temperature reaches a certain temperature is attributed to 'diffusion'. Diffusion is the movement of atoms between tool and chip materials resulting in loss of material from the face of the tool. It depends upon the chemical composition and microstructure of tool and workpiece materials, in addition to temperature. So, unless these conditions are favourable, crater wear due to diffusion may be absent.

Crater wear is more common in cutting ductile materials which produce continuous chips. Also, it is more common in HSS (high speed steel) tools than ceramic or carbide tools which have much higher hot hardness.

The reasons for 'Nose wear' may be one or more of the reasons discussed above. Chipping of the tool may occur due to the following factors :

1. Tool material is too brittle.

2. As a result of crack that is already in the tool.

3. Excessive static or shock loading of the tool.

4. Weak design of the tool, such as a high positive rake angle.

14.11.1. Tool life. The total cutting time accumulated before tool failure occurs is termed as 'tool life'. There is no exact or simple definition of tool life. However, in general, the tool life can be defined as tool's useful life which has been expended when it can no longer produce satisfactory parts. The two most commonly used criteria for measuring the tool life are :

1. Total destruction of the tool when it ceases to cut.

2. A fixed size of wear land on tool flank. On carbide and ceramic tools where crater wear is almost absent, tool life is taken as corresponding to 0.038 or 0.076 mm of wear land on the flank for finishing respectively.

As discussed above, tool wear and hence tool life depends on many factors. The greatest variation of tool life is with the cutting speed and tool temperature which is closely related to cutting speed. Tool temperature is seldom measured and much study has been done on the effect of cutting speed on tool life. Tool life decreases with increased V, the decrease being parabolic. To draw these curves, the cutting tools are operated to failure at different cutting speeds. In 1907, Taylor gave the following relationship between cutting speed and tool life,

$$VT^n = C \quad \text{...(14.27)}$$

where V is the cutting speed (m/min), T is the time (min) for the flank wear to reach a certain dimension, *i.e.,* tool life, C is constant and n is an exponent which depends upon the cutting conditions. If cutting speed-tool life curves are plotted on a log-log graph, straight lines are obtained, Fig. 14.23, n is the negative inverse slope of the curve and C is the intercept velocity at $T = 1$. The results are valid only for the particular test conditions employed. Thus 'C' is the cutting speed for tool life of 1 min.

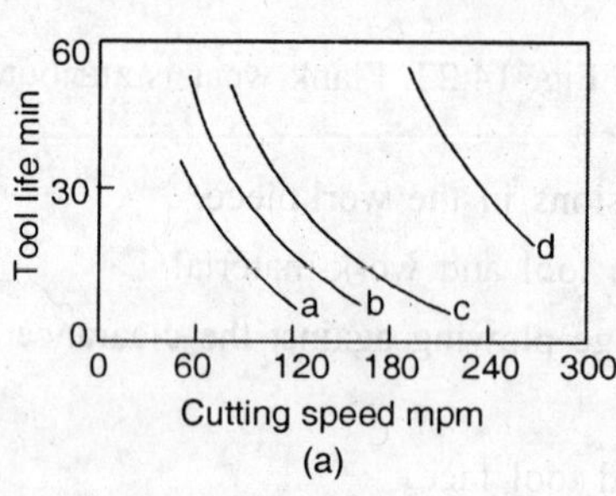

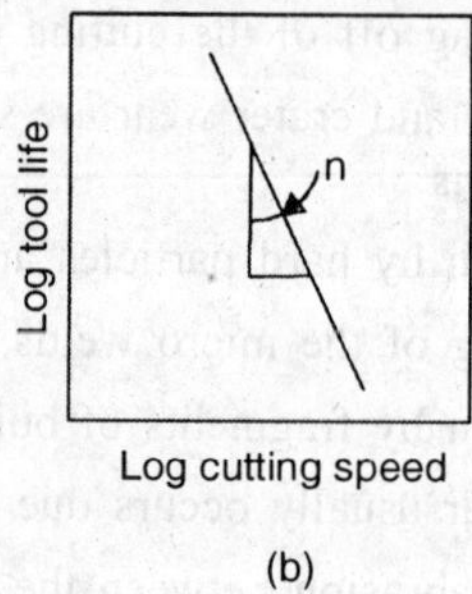

Fig. 14.23. Cutting Speed-Tool Life Curves.

The following values may be taken for 'n':

n = 0.1 to 0.15 for HSS tools
= 0.2 to 0.4 for carbide tools
= 0.4 to 0.6 for ceramic tools

The tool life also depends to a great extent on the depth of cut d and feed rate per revolution, f. Assuming a logarithmic variation of C with d, the equation (14.27) can be written as,

$$VT^n \cdot d^m = C \quad \text{...(14.28)}$$

It has been seen that decrease of life with increased speed is twice as great (exponentially) as the decrease of life with increased feed.

Considering feed rate also, the general equation can be :

$$VT^n . d^m . f^x = C \quad \text{...(14.29)}$$

14.11.2. Tool-life criteria. As pointed out above, there is no exact and simple definition of tool life. The tool life between reconditioning or replacement can be defined in a number of ways, namely

1. Actual cutting time to failure. In the case of interrupted cutting process, such as milling, it will be the total time to failure.

2. Volume of metal removed to failure.

3. Number of parts produced to failure.

4. Cutting speed for a given time to failure.

5. Length of work machined to failure.

Each one of the above methods may be useful at one time or other. A tool fails when it no longer performs its function properly. This will have different meaning under different circumstances. In a roughing operation, where, surface finish and dimensional accuracy are of little importance, a tool failure can mean an excessive rise in cutting forces and power requirements. In the case of finishing operation, where surface finish and dimensional accuracy are most important, a tool failure will mean that the specified conditions of surface finish and dimensional accuracy can no longer be achieved. All these failures are principally related to the wear on the clearance face of the cutting tool. The method of complete tool failure or total tool destruction which occurs due to high cutting forces or shock load, is usually not considered because of the total loss of cutting tool and possible damage to the component. It is clear that the tool life/tool failure is related to tool wear and condition of the finished parts. The various tool life criteria can be listed as given below :

1. Chipping or fine cracks developing at the cutting edge.
2. Total destruction of the cutting tool.
3. Wear land size on the flank of the tool.
4. Crater depth, width or other parameters of the crater wear on the rake face of the tool.
5. A combination of (3) and (4).
6. Volume of weight of material worn off the tool.
7. Limiting value of surface finish produced on the component.
8. Limiting value of change in component size.
9. Fixed increase in cutting forces or power required to perform a function.

'Tool Life' can now be defined as the cutting time required to reach a tool-life criterion. Thus when tool-life values are quoted or compared it is essential to clearly state the tool-life criterion. A 'tool-life criterion' can be defined as a predetermined threshold value of a tool wear measure or the occurrence of a phenomenon.

The first type of failure can occur due to : faults in tool design, poor tool grinding technique, wrong selection of tool material and non-steady cutting conditions. This type of failure can be prevented by improving the cutting tool design and production of the cutting edges. Thus, on a lathe tool, the SCEA is made as large as possible, so that the initial contact between the workpiece and the tool occurs away from the tool tip and the full depth of cut and the cutting forces are obtained gradually. In the case of tools made of cemented carbide and other brittle tool materials, the rake angle is small or even negative to strengthen the tool and allow initial contact to occur on the rake face away from the cutting edge.

The tool wear volume or weight criterion is applicable when radioactive tracer methods are used to study the tool wear. Such tests are expensive and are used only for laboratory experiments. Similarly, the limiting surface finish and component size criteria are expensive and cumbersome. The forces on a worn tool increase and a selected increase in force can be used as a failure criterion. This method is useful when the wear land is the major type of wear. When crater wear is predominant, the tool will act as a restricted contact tool and the cutting forces decrease, with increase in wear. As the wear further increases, the crater merges with the flank, creating a new cutting edge. This results in increased cutting forces followed by total destruction of tool.

The most important tool-life criteria are; wear land size and the crater width or depth. The wear of the face and flank is not uniform along the active cutting edge, therefore, it is necessary to specify the locations and degree of wear, when deciding on the amount of wear allowable before regrinding the tool. Common values of wear-land size, w, Fig. 14.24 (c), are given below, as a guide.

Table 14.3. Wear Land Size.

w mm	*Tool*	*Remarks*
0.76	Cemented carbide	Roughing passes
0.25 to 0.38	"	Finishing passes
1.50 or total destruction	H.S.S.	Roughing passes
0.25 to 0.38	"	Finishing passes
0.25 to 0.38	Cemented oxides	Roughing and Finishing passes

The crater depth *h*, Fig. 14.24 (*b*) is measured at the deepest point of the crater.

The 'Tool-life' criteria as per I.S.O., are given below :

(*a*) **For H.S.S. and ceramic tools :**

1. Catastropic failure, or
2. w = 0.3 mm, if the flank is regularly worn, or
3. w_{max} = 0.6 mm, if the flank is irregularly worn

(*b*) **Sintered carbide tools :**

1. w = 0.3 mm, or
2. w = 0.6 mm if the flank is irregularly worn, or
3. $h = 0.06 + 0.3 f$,

14.11.3. Variables affecting tool life. The various variables that play a role in tool life are :

1. Process variables–speed, feed and depth of cut.
2. Tool material.
3. Tool geometry.
4. Workpiece material, its hardness and micro-structure.
5. Surface condition of the work-piece.
6. Cutting fluid.

Process variables. The process variables, speed, feed and depth of cut play an important role (Eqn 14.29) since they control the rate of metal removal and the production rate. Equation 14.29 is reproduced below :

$$VT^n d^m f^x = C$$

or
$$T = \frac{C^{\frac{1}{n}}}{V^{\frac{1}{n}} \cdot d^{\frac{m}{n}} \cdot f^{\frac{x}{n}}}$$

$$= \frac{K}{V^{\frac{1}{n}} \cdot f^{\frac{1}{n_1}} \cdot d^{\frac{1}{n_2}}} \quad ...(14.30)$$

The values of the exponents $\frac{1}{n}, \frac{1}{n_1}, \frac{1}{n_2}$ and the constant *K* will depend upon the tool failure criterion. The values of the exponents which describe the effect of the variables on tool life, vary with different tool and workmaterial. It has been found that,

$$\frac{1}{n} > \frac{1}{n_1} > \frac{1}{n_2}$$

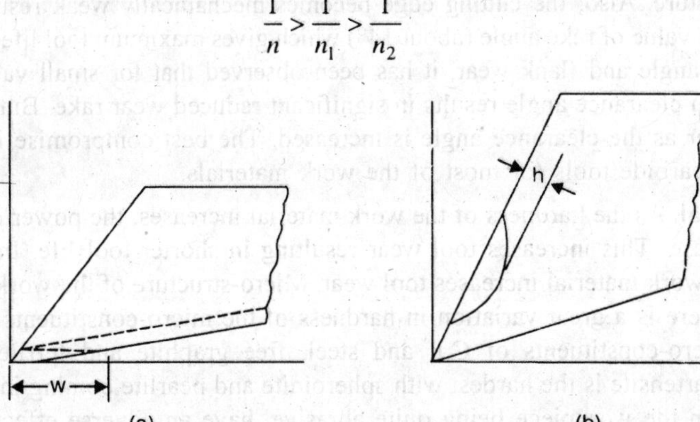

Fig. 14.24. Wear Land and Crater Depth.

It is clear that the cutting speed has the greatest effect on tool life followed by feed and depth of cut, respectively. Their effect on tool life can be explained in terms of tool-workpiece interface temperature. It has been seen that the tool life is a direct function of cutting temperature, irrespective of the cutting condition. The cutting temperature increases with increases in the three process variables.

Tool material. The major requirements of cutting tool materials (discussed in next chapter) are : Hot hardness, impact toughness, and wear and abrasion resistance. High hardness gives the tool good wear resistance but its toughness decreases. For thermal shock resistance, the tool material should have high thermal conductivity and specific heat, a low co-efficient of thermal expansion and high tensile strength.

The variations in the above tool material requirements with cutting temperature are of considerable importance to tool life. Taking hardness as a guide to wear resistanc and hence tool life, the effect of cutting temperature on tool material hardness is of major concern. Carbon steels are very sensitive to temperature and they rapidly lose their hardness at low temperatures. Thus, they are used for slow cutting of soft non-ferrous materials which give low cutting temperatures. The hardness of H.S.S. is affected only slightly till about 600°C, after which its hardness starts falling rapidly with temperature. So, H.S.S. will give good performance below 600°C, even better than cast alloys. Cast alloys are superior to H.S.S. above 600° C. These can be used at slightly higher speeds than H.S.S. and have less tendency to form a B.U.E. Cast alloys are widely used for the machining of C.I., malleable iron and hard bronzes. Cemented carbides can retain their hardness at temperatures as high as 1200°C and can therefore be used at much higher cutting speeds than H.S.S. or cast alloys. However, because of their reduced toughness, they have a greater tendency to chip and fracture under heavy loads and interrupted cutting conditions. Sintered oxides or ceramics can be used at cutting speeds of 2 or 3 times those employed with carbides. But these being very hard are extremely brittle and can only be used where shock and vibrations do not occur.

Tool geometry. The tool life is greatly affected by tool geometry. It has already been noted that larger the rake angle, smaller will be the cutting angle and larger will be the shear angle. This will reduce the cutting force and power and hence the heat generated during cutting. This means reduced cutting temperature resulting in longer tool life. But increasing the rake angle will reduce the mass of metal behind the cutting edge resulting in poor transfer of heat. This can tend to increase the cutting temperature. Also, the cutting edge becomes mechanically weak resulting as a guide, there is an optimum value of rake angle (about 14°) which gives maximum tool life. From the curves between clearance angle and flank wear, it has been observed that for small values of clearance angle, an increase in clearance angle results in significant reduced wear rake. But the cutting edge will become weaker as the clearance angle is increased. The best compromise is 8° with H.S.S. tools and 5° with carbide tools for most of the work materials.

Work material. As the hardness of the work material increases, the power consumption and temperatures increase. This increases tool wear resulting in shorter tool life. Impurities or hard constituents in the work material increases tool wear. Micro-structure of the work material is very significant since there is a great variation in hardness of the micro-constituents of an alloy. Out of the various micro-constituents of C.I. and steel, free graphite and ferrite are the softest constituents and martensite is the hardest with spheroidite and pearlite coming in between. Scales and oxide layers on the workpiece being quite abrasive, have an adverse effect on tool life. In such cases, the material should be cut below the oxide layer or below the work hardened zone. Pure metals tend to adhere to the tool surface and give high friction and high wear rates resulting in shorter tool life.

Cutting fluid. The function of a cutting fluid is to reduce the friction and interface temperature which have an adverse effect on tool life. So, tool life can be increased by the effective use of a cutting fluid.

14.12. ECONOMICS OF METAL CUTTING

While considering economics of metal cutting, the basic consideration should be to produce a satisfactory part at the best possible cost. At low cutting speeds, tools last long and tooling cost is low. But, the metal removal rate will be low and hence the cutting cost and the total cost are high. On the other hand, at high cutting speeds the metal removal rate will be high giving low cutting cost, but the tooling life will be shorter giving the high tooling cost, making the total cost high. At some intermediate cutting speed, the total cost is at a minimum, (Fig. 14.25). The tool life corresponding to this cutting speed is economical tool life. To find the optimum cutting speed, the total cost of manufacturing a batch of components is found.

Total cost of cutting a unit volume of metal = cost of machining metal per unit volume of metal cut + Cost of replacing or servicing tools per unit volume of metal cut.

Let C_m = machining cost per minute (labour cost per min + overheads per min)

C_t = tooling cost

Now, time to machine a unit volume of metal in minutes,

$$= \frac{1}{d.f.V} = \frac{C_1}{V}$$

where C_1 is a constant.

∴ Cost of machining metal per unit volume of metal cut

$$= \frac{C_m \cdot C_1}{V}$$

Now number of tool replacements or servicing in $\frac{C_1}{V}$ minutes $= \frac{C_1}{TV}$ minutes

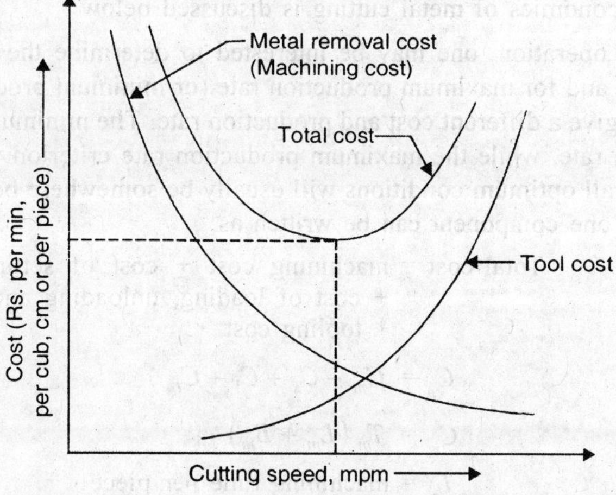

Fig. 14.25

$\therefore$ Tooling cost per unit volume of metal cut

$$= \frac{C_t \cdot C_1}{TV}$$

$$= \frac{C_t \cdot C_1}{\left(\frac{C}{V}\right)^{1/n} . V}, \; (VT^n = C)$$

$$= \frac{C_t \cdot C_1 \, (V)^{\frac{1-n}{n}}}{C^{1/n}}$$

$\therefore$ Total cost per unit volume of metal cut,

$$Y = \frac{C_m . C_1}{V} + \frac{C_t \cdot C_1 \cdot (V)^{\frac{1-n}{n}}}{C^{1/n}}$$

For optimum value of V, $\frac{dY}{dV} = 0$

$$\therefore \quad -\frac{C_m \cdot C_1}{V^2} + \left(\frac{1-n}{n}\right) \cdot \frac{C_t \cdot C_1}{C^{1/n}} \cdot (V)^{\frac{1-2n}{n}} = 0$$

From here, $C_m = \left(\frac{1-n}{n}\right) \frac{C_t}{C^{1/n}} \cdot (V)^{1/n}$

$\therefore$ Optimum value of (minimum total cost),

$$V = C. \left[\frac{C_m}{C_t} \frac{n}{1-n}\right]^n \qquad (14.31)$$

A more detailed economics of metal cutting is discussed below :

In a metal cutting operation, one may be interested to determine the cutting speed for a minimum cost per piece and for maximum production rate (or minimum production time). These two criteria will always give a different cost and production rate. The minimum cost criterion will give a lower production rate, while the maximum production rate criterion will result in higher cost per piece. The overall optimum conditions will usually be somewhere between the two. The total cost of machining one component can be written as,

Total cost = machining cost + cost of setting up the machine + cost of loading, unloading and machine handling + tooling cost

or
$$C_p = C_m + C_s + C_l + C_t \quad ...(14.32)$$

Now,
$$C_m = T_m (L_m + B_m) \quad ...(14.33)$$

where T_m = machining time per piece

$$= \frac{L}{fN} = \frac{\pi LD}{100 fV} \quad ...(14.34)$$

L_m = labour rate per hour

B_m = overhead charge or burden rate of the machine.

The set up includes mounting the cutter and fixture and preparing the machine for the cutting operation. This cost is a fixed value per piece.

Now,
$$C_l = T_l (L_m + B_m) \quad ...(14.35)$$

where T_l = time used in loading and unloading the component changing speed, feed rate etc.

Now,
$$C_t = \frac{1}{Z}\left[T_c (L_m + B_m) + T_g (L_g + B_g) + D_c\right] \quad ...(14.36)$$

where Z = number of pieces machined per tool grind

T_c = time needed to change the tool

T_g = time needed to grind the tool

L_g = labour rate per hour of tool grinder operator

B_g = burden rate per hour of tool grinder

D_c = depreciation of tool per grind, Rs

Now, the total time to produce one piece is,

$$T_p = T_l + T_m + \frac{T_c}{Z} \quad ...(14.37)$$

From Taylor's tool-life equation,

$$VT^n = C$$

We get
$$T = \left(\frac{C}{V}\right)^{1/n} \quad ...(14.38)$$

The number of pieces produced per tool grind is,

$$Z = \frac{T}{T_m} \quad ...(14.39)$$

From (14.34), (14.38) and (14.39), we get,

$$Z = \frac{100\, fC^{\frac{1}{n}}}{\pi LD.V^{\frac{1}{n-1}}} \qquad ...(14.40)$$

We can get equation (14.33) in terms of many variables. To find the optimum cutting speed for minimum cost,

$$\frac{\partial C_p}{\partial V} = 0$$

From here, we shall get.

$$V_0 = \frac{C(L_m + B_m)^n}{\left[\left\{\left(\frac{1}{n}\right) - 1\right\}\left\{T_c\left(L_m + B_m\right) + T_g\left(L_g + B_g\right) + D_c\right\}\right]^n} \qquad ...(14.41)$$

The corresponding optimum tool life will be, from (14.38),

$$T_0 = \frac{(1-n)\left[T_c\left(L_m + B_m\right) + T_g\left(L_g + B_g\right) + D_c\right]}{n\left(L_m + B_m\right)} \qquad ...(14.42)$$

For maximum production rate, $\frac{\partial T_p}{\partial v} = 0$

From eqn. (14.37), we will get

$$V_0 = \frac{C}{\left[T_c\left(\frac{1-n}{n}\right)\right]^n} \qquad ...(14.43)$$

and $$T_0 = T_c\left(\frac{1-n}{n}\right) \qquad ...(14.44)$$

14.13. MACHINEABILITY

In spite of efforts by a number of investigators, so far, there has been no exact quantitative definition of machineability. This is because of a large number of variables involved and their complexity. However the major factors involved in metal cutting are : forces and power absorbed, tool wear and tool life, surface finish, dimensional accuracy and machining cost. These factors depend upon a large number of variables such as properties of work materials, tool geometry, cutting conditions, machine tool rigidity etc. Due to this, it is impossible to combine these factors so as to give a suitable definition for machineability. Many authors give a qualitative measure of machineability of a material as :

1. the ease with which it could be machined,
2. the life of tool before tool failure or resharpening,
3. the quality of the machined surface, and
4. the power consumption per unit volume of material removed.

However, in production, tool life is generally considered the most important factor and, so, most of the investigators have related machineability with tool life. Higher the tool life, the better is the machineability of a work material. The various materials have been given machineability ratings, which are relative. Supposing a material is given the rating of 100. Those materials which have a better machineability will have higher ratings and those materials with lower machineability have a lower one.

According to one investigator, the machineability may be evaluated as given below :

1. Long tool life at a given cutting speed.
2. Lower power consumption per unit volume of metal removed.
3. Maximum metal removal per tool resharpening.
4. High quality of surface finish.
5. Good and uniform dimensional accuracy of successive parts.
6. Easily disposable chips.

The machineability rating or index of different materials is taken relative to the index which is standardised. The machineability index of free cutting steel is arbitrarily fixed at 100 per cent. For the other materials, the index is found as below :

$$\text{Machineability index, \%} = \frac{\text{Cutting Speed of Material for 20 min tool life}}{\text{Cutting speed of free cutting steel for 20 min tool life}} \times 100$$

The machineability indexes for some common materials are given below :

C-20 steel = 65
C-45 steel = 60
Stainless steel = 25
Copper = 70
Brass (red) = 180
Aluminium alloys = 300 – 1500
Magnesium alloys, = 600 – 2000

The above method is only one of the methods of machineability ratings of various materials. Another test is for the rate of tool wear under standard conditions. A third test is to measure cutting forces and compare these with those for a base material. These methods, however, do not give a uniform rating. For example, ratings obtained for a group of materials on a tool wear basis may get altered if surface finish or power consumption criteria are applied. Thus machineability will have some meanings depending upon the cutting process under consideration. With finishing operations, tool wear and surface finish are the most important considerations and with roughing operations, tool wear and power consumption are important. A more rational test is the machining of long runs of one component from different materials, but satisfying the same conditions of surface finish and size. Speeds, feeds and tool angles are varied to get the fastest rate of production with 6 to 8 hour tool life for each material. Then the machineability rating or index is the ratio of hourly output with one material to that of the standard. These production tests are based on maintenance of a good surface finish at a rapid rate of production.

Machineability ratings from cutting forces or tool life criteria generally decrease with increase in material hardness and strength. However with production criteria, the ratings of steel increase with tensile strength to a broad maximum around 595 MPa, but decrease for stronger and harder steels. The reason is that steel cuts cleaner and finishes more easily at higher rates as it is made harder. However, above a certain limit, the tool starts getting damaged resulting in reduced production rate and machineability rating.

Micro-constituents of the work material have a strong influence on its machineability. In C.I. and steels, ferrite is the softest constituent and next comes pearlite and martensite. As the percentage of ferrite gets reduced in relation to other constituents, the machinability of the work material will go on reducing.

The addition of certain materials like sulphur, lead and tellurium to non-ferrous metals, as well as steels, improves their machineability, resulting in increased production rates and improved surface finish. Sulphur is added to steel only if there is sufficient manganese in it, otherwise, a low melting iron sulphide net work is formed around the austenite grain boundaries. Such a steel is very weak and brittle. But with manganese, sulphur forms manganese sulphide which exists as an isolated phase and acts as an internal lubricant and chip breaker. Lead is added to steel ingots in the form of shot. Since it is insoluble in steel, it is distributed throughout the solidified steel in the free state. It also acts as internal lubricant and chip breaker. Selenium is used for the same purpose in some stainless steel compositions. Similarly, tellurium and bismuth are added in small amounts to certain steels. Nickel and molybdenum are added to C.I., lead to brass and bronze and copper lead and bismuth to aluminium to improve machineability.

Such compositions are said to be 'Free machining' or 'Free cutting'. The additives reduce the metallic contact between the tool and work material, hence, they reduce friction and tool wear rates.

The effect of tool angles on machinability is the same as on tool wear and tool life, already discussed.

14.14. SURFACE ROUGHNESS

The quality of surface produced during a machining process will depend upon :

(*i*) the tool geometry and the feed rate, and

(*ii*) the irregularities in the machining process. These irregularities include :

1. Occurrence of B.U.E.
2. Occurrence of vibration and chatter.
3. Deflections of the tool or workpiece.
4. Inaccuracies and irregularities in the movements of spindles, slides and feed mechanism of the machine tools.
5. Formation of discontinuous chips when cutting brittle materials.
6. Tearing of the ductile work material when machining at low cutting speeds.
7. Defects in the structure of the work material.
8. Flow of chips on the machined surface.

If the above contributing factors can be completely eliminated, then the surface roughness produced will depend entirely on tool geometry. Such a surface roughness is called "Ideal surface roughness". However in actual practice, the surface roughness obtained will depend upon the combined effect of tool geometry and the contributing factors. Such a surface roughness is called "natural surface roughness."

Ideal surface roughness. As indicated above, the ideal surface roughness will depend entirely on tool geometry. For a sharp cornered single point tool used for turning operation, the surface roughness obtained will be like that shown in Fig. 14.26. As, already discussed in Chapter 11, the arithmatic mean value is given as,

$$R_a = \frac{\text{Sum of absolute values of all the areas above and below the mean line}}{\text{Sampling length}}$$

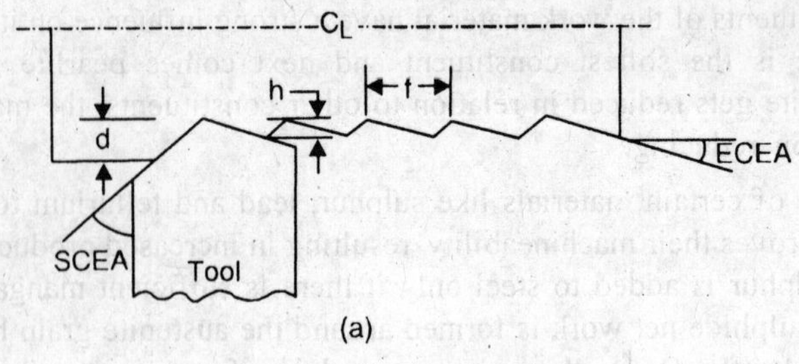

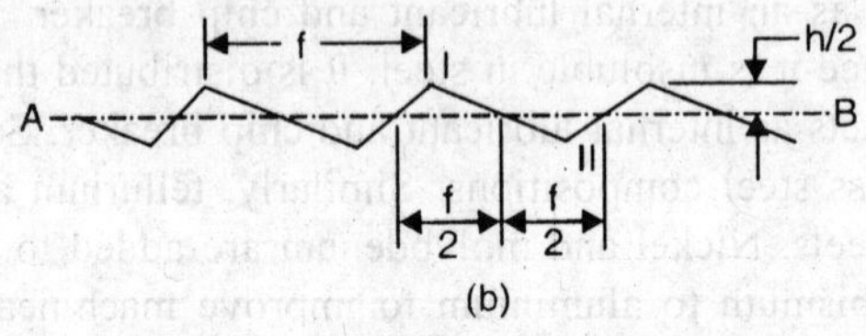

Fig. 14.26. Surface Roughness.

The mean or datum line *AB* is located such that the sum of the areas above the line is equal to the sum of the areas below the line. For Fig. 14.26 (*b*), if the sampling length is taken as one feed, *f*, then the surface roughness will be given as,

$$R_a = \frac{|\text{area I}| + |\text{area II}|}{f}$$

But areas I and II are equal, therefore.

$$R_a = \frac{2}{f}(\text{area I}) = \frac{h}{4}$$

where *h* is the height of the geometry of the surface roughness.

Now from geometry, $$h = \frac{f}{\tan SCEA + \cot ECEA}$$

$$\therefore \quad R_a = \frac{f}{4(\tan SCEA + \cot ECEA)} \quad ...(14.45)$$

(a)

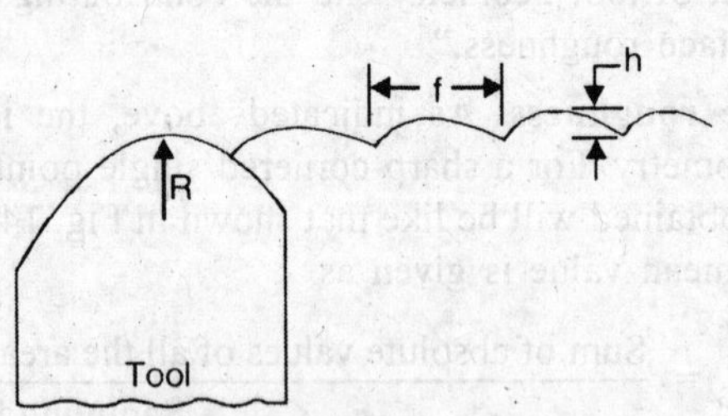

Fig. 14.27. Surface Roughness.

Practical cutting tools have their corners rounded off and surface geometry produced will be as shown in Fig. 14.27. For such a tool, the height of the geometry will be given as,

For Fig. 14.27 (*a*),

$$h = \left(1 - \cos C_e\right) R + f \sin C_e \cos C_e - \sqrt{\left(2 fR \sin^3 C_e - f^2 \sin^4 C_e\right)} \qquad \text{...(14.46)}$$

For small values of feed, so that the tool is cutting entirely on its nose radius, Fig. 14.27 (*b*), that is,

$$f \leq 2R \sin Ce$$

and h gets reduced to

$$h = R - \frac{1}{2}\sqrt{\left(4R^2 - f^2\right)}$$

By expanding the second term in binominal series, h, can be approximated as,

$$h \approx \frac{f^2}{8R} \qquad \text{...(14.47)}$$

Thus, a change in the rate of feed is more important than a change in the nose radius. The surface roughness will be given as,

$$R_a = \frac{f^2}{18\sqrt{3}.R} \qquad \text{...(14.48)}$$

It is clear that the depth of cut has no effect on the surface geometry.

Natural surface roughness. In actual practice, it is not possible to entirely eliminate the effects of the contributing factors. To minimise the effect of vibration and chatter, it is recommended that both tool and workpiece should be held rigidly with as little overhang as possible. The damping characteristics of the tool and the machine tool are also important. A major factor which contributes to natural surface roughness is the occurrence of B.U.E. The B.U.E. may be continually building up and breaking down and its broken portions get deposited on the machined surface. Larger the B.U.E. the rougher would be the surface produced. So, the factors which 'eliminate or reduce the formation of B.U.E. will give better surface furnish. A reduction in chip-tool friction will also give better surface finish. Such factors are : an increase in cutting speed, use of cutting tool materials with low co-efficient of friction, use of correct cutting fluid and the use of free machining materials.

At very low cutting speeds, the chip is usually discontinuous which results in a poor surface finish due to the interrupted cutting action. As the cutting speed increases, the chip becomes continuous, but B.U.E. starts forming. But at higher speeds the B.U.E. begins to disappear resulting in better surface finish.. The formation of B.U.E. is greatly dependent upon the temperature and friction at the tool-chip interface. An effective cutting fluid reduces the possibility of formation of B.U.E. and thus improves surface finish. Cemented carbide, ceramics and diamond have low co-efficient of friction as compared to H.S.S. tool material. Their use reduces the tendency of formation of B.U.E. and produces a better surface finish.

14. 15. Solved Examples

Example 1. *The useful tool life of a HSS tool machining mild steel at 18 m/min. is 3 hours. Calculate the tool life when the tool operates at 24 m/min.*

Solution. $VT^n = C$

$$V = 18 \text{ m/min}$$

$$T = 3 \times 60 = 180 \text{ min.}$$

$$\therefore \quad C = 18 \times (180)^n$$

Let $n = 0.125$

$$\therefore \quad C = 18 \times (180)^{0.125} = 34.45$$

Now $V = 24$ m/min.

$$\therefore \quad VT^n = C$$

$$T = \left(\frac{34.45}{24}\right)^{\frac{1}{0.125}} = \mathbf{18 \text{ min.}}$$

Example 2. *For a metal machining, the following information is available :*

Tool change time, = 8 min

Tool re-grind time = 5 min

Machine running cost, = Rs 5 per hour

Tool depreciation per re-grind, = 30 p

n = 0.25, C = 150

Calculating the optimum cutting speed.

Solution. Tooling cost C_t = Tool change cost + tool regrind cost + tool depreciation

$$= \frac{5}{60} \times 8 + \frac{5}{60} \times 5 + 0.30$$

$$= \text{Rs } 1.38$$

$$\text{Machining cost } C_m = \text{Rs } \frac{5}{60}$$

$$\therefore \quad V_T = C\left[\frac{C_m}{C_t} \cdot \frac{n}{1-n}\right]^n$$

$$= 150\left(\frac{5}{60 \times 1.38} \cdot \frac{0.25}{0.75}\right)^{0.25}$$

$$= \mathbf{56.5 \text{ m/min.}}$$

Example 3. *In an orthogonal cutting operation, the following data have been observed :*

Uncut chip thickness. $t = 0.127$ *mm*

Width of cut, $b = 6.35$ *mm*

Cutting speed, $V = 2$ *m/s*

Rake angle, $\alpha = 10°$

Cutting force, $F_c = 567$ *N*

Thrust force, $F_t = 227$ *N*

Chip thickness, $t_c = 0.228$ *mm*

Determine : Shear angle, the friction angle, shear stress along the shear plane and the power for the cutting operation. Also find the chip velocity, shear strain in chip and shear strain rate.

Solution. (*i*) Shear angle, $\varphi = \tan^{-1}\left(\frac{r\cos\alpha}{1-r\sin\alpha}\right)$

$$r = \frac{t}{t_c} = \frac{0.127}{0.228} = 0.557$$

$$\alpha = 10°$$

$$\therefore \quad \varphi = \tan^{-1}\left(\frac{0.557 \times 0.985}{1-0.557 \times 0.1736}\right)$$

$$= \tan^{-1}(0.607) = \mathbf{31.25°}$$

(*ii*)

$$\mu = \tan\beta = \frac{F_c \sin\alpha + F_t \cos\alpha}{F_c \cos\alpha - F_t \sin\alpha}$$

$$= \frac{567 \times 0.1736 \times + 227 \times 0.985}{567 \times 0.985 - 227 \times 0.1736}$$

$$= \frac{98.4 + 223.6}{558.5 - 39.4} = \frac{332}{519.1} = 0.64$$

or

$$\beta = \tan^{-1}(0.64) = \mathbf{32.62°}$$

(*iii*) Shear force, $F_s = F_c \cos\varphi - F_t \sin\varphi$

$$= 567 \times 0.855 - 227 \times 0.519$$

$$= 484.8 - 117.8 = 367 \text{ N}$$

$$\therefore \quad \tau_s = \frac{F_s}{A_s} = \frac{367 \times \sin\varphi}{bt}$$

$$= \frac{367 \times 0.519}{6.35 \times 0.127}$$

$$= \frac{190.5}{0.806} = \mathbf{236.5\ N/mm^2}$$

(*iv*) Cutting power, $= \frac{F_c V}{1000} = \frac{567 \times 2}{1000}$

$$= 1.134 \text{ kW}$$

(*v*) Now $\frac{V_c}{V} = r$

$\therefore$ Chip velocity, $V_c = 2 \times 0.557 = 1.114$ m/s

(*vi*) Shear strain, $s = \cot\varphi + \tan(\varphi - \alpha)$

$$= \cot 31.25 + \tan(31.25 - 10)$$

$$= \mathbf{2.037}$$

(*vii*) Shear plane length $= \dfrac{t}{\sin \varphi} = \dfrac{0.127}{0.519}$

$= 0.245$ mm.

Taking the thickness of deformation zone equal to one-tenth of shear plane length.

t_s, (eq. 14.11) $= 0.0245$ mm

$\therefore$ Shear strain rate, $\dot{s} = \dfrac{V_s}{t_s}$

Now $V_s = \dfrac{V \cos \alpha}{\cos(\varphi - \alpha)}$

$= \dfrac{2 \times 0.985}{\cos 21.25} = 2.11$ m/s

$\therefore \quad \dot{s} = \dfrac{2.11 \times 1000}{0.0245}$

$= 86.12 \times 10^3\ s^{-1}$

Example 4. *The following equation for tool life is given for a turning operation :*

$$VT^{0.13} f^{0.77} d^{0.37} = C$$

A 60 minute tool life was obtained while cutting at V = 30 m/min, f = 0.3 mm/rev. and d = 2.5 mm.

Determine the change in tool life if the cutting speed, feed and depth of cut are increased by 20% individually and also taken together.

Solution.

$$VT^{0.13} f^{0.77} d^{0.37} = C$$

$\therefore \quad C = 30 \times 60^{0.13} \times 0.3^{0.77} \times 2.5^{0.37}$

$= 30 \times 1.702 \times 0.396 \times 1.403$

$= 28.38$

(*i*) Now $V = 30 \times 1.2 = 36$ m/min.

$\therefore \quad T^{0.13} = \dfrac{28.38}{0.396 \times 1.403 \times 36}$

$= 1.419$

$\therefore \quad T = (1.419)^{\frac{1}{0.13}} = (1.419)^{7.69}$

$= \mathbf{14.75\ min.}$

(*ii*) Now $f = 0.3 \times 1.2 = 0.36$ mm/rev.

$$\therefore \quad T^{0.13} = \frac{28.38}{(0.36)^{0.77} \times 1.403 \times 30} = 1.48$$

$$\therefore \quad T = (1.48)^{7.69} = 20.39 \text{ min}$$

(*iii*) Now $d = 2.5 \times 1.2 = 3$ mm.

$$\therefore \quad T^{0.13} = \frac{28.38}{30 \times 0.396 \times (3)^{0.37}} = 1.591$$

$$\therefore \quad T = (1.591)^{7.69} = \mathbf{35.55 \text{ min}}$$

∴ The maximum effect on tool life is of cutting speed, and the least effect is of depth of cut.

(*iv*) Now $V = 36$ m/min, $f = 0.36$ mm/rev. $d = 3$ mm

$$\therefore \quad T^{0.13} = \frac{28.28}{38 \times 0.455 \times 150} = 1.154$$

$$\therefore \quad T = (1.154)^{7.69} = \mathbf{3.011 \text{ min.}}$$

Example 5. *During an orthogonal machining (turning) operation of C—40 steel, the following data were obtained :*

Chip thickness = 0.45 mm
Width of cut = 2.5 mm.
Feed = 0.25 mm/rev
Tangential cut force = 1130 N
Feed thrust force = 295 N
Cutting speed = 2.5 m/s
Rake angle = + 10°

Calculate : (a) Force of shear at the shear plane. (b) Kinetic co-efficient of friction at the chip-tool interface.

Solution. Give data are : $t_c = 0.45$ mm, $b = 2.5$ mm, $\alpha = 10°$, $t = f = 0.25$ mm, (Eqn. 14.13) $F_c = 1130$ N, $F_t = 295$ N, $V = 2.5$ m/s.

(*i*) Shear force, F_s, is given as :

$$F_s = F_c \cos \varphi - F_t \sin \varphi, \text{ where } \varphi \text{ is shear angle.}$$

$$\tan \varphi = \frac{r \cos \alpha}{1 - r \sin \alpha}$$

$$r = \frac{t}{t_c} = \frac{0.25}{0.45} = 0.555$$

$$\therefore \quad \tan \varphi = \frac{0.555 \times 0.985}{1 - 0.555 \times 0.1736} = 0.605$$

$$\therefore \quad \varphi = 31.2°$$

$$\therefore \quad F_s = 1130 \times 0.855 - 295 \times 0.518$$

$$= 966.2 - 152.8 - 813.4 \text{ N}$$

(ii)
$$\mu = \frac{F_c \tan \alpha + F_t}{F_c - F_t \tan \alpha} \quad \text{(eqn. 14.9)}$$

$$= \frac{1130 \times 0.1763 + 295}{1130 - 295 \times 0.1763}$$

$$\therefore \quad \mu = \frac{199.25 + 295}{1130 - 52} = 0.458$$

Example 6. *The following data relate to an orthogonal turning process :*

Chip thickness = 0.62 mm

Feed = 0.2 mm/rev.

Rake angle, = 15°

(i) Calculate cutting ratio and chip reduction co-efficient.

(ii) Calculate shear angle,

(iii) Calculate the dynamic shear strain involved in the deformation process.

Solution. $t = f = 0.2$ mm (equ. 14.13), $t_c = 0.62$ mm, $\alpha = 15°$

(i)
$$\text{Cutting ratio, } r = \frac{t}{t_c} = 0.322$$

$$\text{Chip-reduction co-efficient} = \frac{1}{r} = 3.1$$

(ii)
$$\text{Shear angle, } \tan \varphi = \frac{r \cos \alpha}{1 - r \sin \alpha}$$

$$= \frac{0.322 \times 0.966}{1 - 0.322 \times 0.259} = 0.33933$$

$$\therefore \quad \phi \approx 18.74°$$

(iii)
$$\text{Shear strain, } s = \cot \varphi + \tan (\varphi - \alpha) \quad \text{(Eqn. 14.8)}$$

$$= 2.947 + 0.065 = \mathbf{3.012}$$

Example 7. *The Taylorian tool-life equation for machining C-40 steel with a 18 : 4 : 1 H.S.S. cutting tool at a feed of 0.2 mm/min and a depth of cut of 2 mm is given by $VT^n = C$, where n and C are constants. The following V and T observations have been noted :*

V, m/min	*25*	*35*
T, min	*90*	*20*

Calculate (i) n and C.

(ii) Hence recommend the cutting speed for a desired tool life of 60 minutes.

Solution.
$$VT^n = C$$

$$\therefore \quad 25 \times 90^n = C \text{ and } 35 \times 20^n = C$$

$$\therefore \quad 25 \times 90^n = 35 \times 20^n$$

$$\therefore \quad \left(\frac{90}{20}\right)^n = \frac{35}{25} = 1.4$$

$$\therefore \quad n = 0.225$$

$$\therefore \quad C = 25 \times 90^{0.225} = 68.8$$

(*ii*) $$V \times 60^{0.225} = 68.8$$

$$\therefore \quad V \times 2.512 = 68.8$$

$$\therefore \quad V = \mathbf{27.39\ m/min.}$$

Example 8. *The following data from an orthogonal cutting test is available :*

Rake angle, $= 15°$

Chip-thickness ratio, $= 0.383$

Uncut chip thickness, $t = 0.5$ *mm*

Width of cut, $b = 3$ *mm*

Yield stress of material in shear, $= 280$ *N/mm*^{2}

Average co-efficient of friction on the tool face $= 0.7$

Determine the normal and tangential forces on the tool face.

Solution. Now,
$$\tan \varphi = \frac{r \cos \alpha}{1 - r \sin \alpha}$$

$$r = 0.383 \text{ and } \alpha = 15°$$

$$\therefore \quad \tan \varphi = \frac{0.383 \times 0.966}{1 - 0.383 \times 0.259} = \frac{0.37}{0.9}$$

$$\therefore \quad \varphi = \text{Shear angle} = 22.35°$$

Now $$\tan \beta = 0.7$$

$$\therefore \quad \text{Friction angle, } \beta = 35°$$

Now
$$F_c = \frac{\tau_s bt}{\sec(\beta - \alpha) \cdot \cos(\varphi + \beta - \alpha) \sin \varphi}$$

$$= \frac{280 \times 3 \times 0.5}{1.064 \times 0.739 \times 0.38} = 1405.7 \text{ N}$$

Now
$$\mu = \frac{F_c \tan \alpha + F_t}{F_c - F_t \tan \alpha}$$

$$0.7 = \frac{1405.7 \times 0.268 + F_t}{1405.7 - 0.268 \times F_t}$$

$$984 - 0.188 F_t = 376.6 + F_t$$

$$\therefore \quad F_t = 511.3 \text{ N}$$

$$\therefore \quad F = \text{Tangential force on tool face}$$

$$= F_c . \sin \alpha + F_t . \cos \alpha$$

$$\therefore \quad F = 1405.7 \times 0.259 + 511.3 \times 0.966$$
$$= 364 + 494 = 858 \text{ N}$$
$$N = \text{Normal force on tool face}$$
$$= F_c.\cos\alpha - F_t \sin\alpha$$
$$= 1405.7 \times 0.966 - 511.3 \times 0.259$$
$$= 1357.9 - 132.4 = \mathbf{1225.4\ N}$$

Example 9. *The following observations were made during orthogonal cutting of steel tube on a lathe :*

Width of cut, $b = 0.5$ cm

Cutting speed, $V = 8.2$ m/min

Rake angle, $\alpha = 20°$

$t = 0.25$ mm

$r = 0.351$

$\varphi + \beta - \alpha = 35°$

Find F_c *and* F_t *given tensile property of material as*

$$\bar{\sigma} = 784\,(\bar{\varepsilon})^{0.15} \text{ N/mm}^2$$

Solution. Shear angle, $\tan\varphi = \dfrac{r\cos\alpha}{1 - r\sin\alpha} = \dfrac{0.351 \times \cos 20°}{1 - 0.351 \times \sin 20°}$

From here, $\varphi = 20.5°$

$\therefore \quad \beta = 34.5°$

Now shear strain $s = \cot\varphi + \tan(\varphi - \alpha)$

$$= \cot 20.5° + \tan(20.5° - 20°)$$
$$= 2.774$$

Now from the relations of chapter 13, (Eqn. 13.6), for simple tension test,

$$\sigma = k \cdot (\varepsilon)^n$$

For generalised state of stress, $\bar{\sigma} = K.(\bar{\varepsilon})^n$

and $\bar{\varepsilon} = \dfrac{s}{\sqrt{3}} = 1.605$

(Considering Von Mise's yield condition)

$$\therefore \quad \bar{\sigma} = 784\,(1.605)^{0.15}$$
$$= 784 \times 1.0735 = 841.66 \text{ N/mm}^2$$

$\therefore$ Yield shear stress, $\tau_s = \dfrac{\bar{\sigma}}{\sqrt{3}} = \dfrac{841.66}{\sqrt{3}} = 485.95 \text{ N/mm}^2$

Now shear plane area, $A_s = \dfrac{bt}{\sin\varphi} = \dfrac{5 \times 0.25}{\sin 20.5} = 3.65 \text{ mm}^2$

$$\therefore \quad F_s = \tau_s . A_s = 485.95 \times 3.65 = 1773.7 \text{ N}$$

Now $$R' = \frac{F_s}{\cos(\varphi + \beta - \alpha)} = \frac{1773.7}{\cos 35°} = 2165.7 \text{ N}$$

Now $$F_c = R' \cos(\beta - \alpha) = 2165.7 \cos(34.5 - 20)$$
$$= 2096.4 \text{ N}$$

and $$F_t = R' \sin(\beta - \alpha) = 2165.7 \sin 14.5$$
$$= \mathbf{541.5\ N}$$

Example 10. *During machining of C-25 steel with 0 – 10 – 6 – 6 – 8 – 90 – 1 mm (ORS) shaped tripple carbide cutting tool, the following observations have been made :*

Depth of cut = *2 mm*
Feed = *0.2 mm/rev*
Speed = *200 m/min*
Tangential cutting force = *1600 N*
Feed thrust force = *850 N*
Chip thickness = *0.39 mm*

Calculate :

(*i*) *Shear force*
(*ii*) *Normal force at shear plane*
(*iii*) Friction force
(*iv*) *Kinetic co-efficient of friction*
(*v*) *Specific cutting energy.*

Solution. From tool designation,

$$\alpha = 10°, \lambda = 90°$$

Other given data are : $d = 2$ mm, $t = f = 0.2$ mm (Since $\lambda = 90°$, Equ. 14.13), $V = 200$ m/min, $t_c = 0.39$ mm, $F_c = 1600$ N, $F_t = 850$ N.

(*i*) Shear force, $$F_s = F_c \cos\varphi - F_t \sin\varphi$$

Now $$\tan\varphi = \frac{r\cos\alpha}{1 - r\sin\alpha}$$
$$r = t/t_c = 0.2/0.39 = 0.513$$

$\therefore$ $$\tan\varphi = \frac{0.513 \times \cos 10°}{1 - 0.513 \times \sin 10°} = 0.551$$

$\therefore$ $$\varphi = 32°$$

$\therefore$ $$F_s = 1600 \times \cos 32° - 850 \times \sin 32°$$
$$= 1600 \times 0.876 - 850 \times 0.482 = 992 \text{ N}$$

(*ii*) Normal force at shear plane,

$$F_n = F_c \sin\varphi + F_t \cos\varphi$$
$$= 1600 \times 0.482 + 850 \times 0.76 = 1515.8 \text{ N}$$

(*iii*) Friction force, $$F = F_c \sin\alpha + F_t \cos\alpha$$
$$= 1600 \times \sin 10° + 850 \times \cos 10° = 1089.8 \text{ N}$$

(iv) $$\mu = \frac{F_c \tan\alpha + F_t}{F_c - F_t \tan\alpha}$$

$$= \frac{1600 \times \tan 10° + 850}{1600 - 850 \times \tan 10°} = 0.753$$

(v) $$\text{Specific cutting energy} = \frac{F_c}{b.t}$$

Now $b = d = 2$ mm (See equ. 14.13)

$$\therefore \quad \text{Specific cutting energy} = \frac{1600}{2 \times 0.2} = \mathbf{4000\ N/mm^2}$$

Example 11. *A turning tool with side and end cutting edge of 20° and 30° respectively, operates at a feed of 0.1 mm/rev. Calculate the CLA of the surface produced if the tool nose radius is 3.00 mm.*

Solution. The given data are : $C_s = 20°$, $C_e = 30°$, $f = 0.1$ mm/rev, $R = 3.00$ mm

Refer to Fig. 14.26 (*a*), the peak to valley roughness is given as, eqn. 14.46,

$$h = \left(1 - \cos C_e\right) R + f \sin C_e \cos C_e - \sqrt{\left(2 fR \sin^3 C_e - f^2 \sin^4 C_e\right)}$$

$$= 0.402 + 0.0433 - 0.2727 = 0.1726 \text{ mm}$$

The centre-line average roughness can be taken roughly as,

$$R_a \approx \frac{h}{4} \approx 0.04315 \text{ mm}$$

$$= \mathbf{43.15\ \mu m.}$$

Example 12. *In 'ORS', the tool angles are :*

Inclination angle (i) = 0°

Orthogonal rake (α) = 10°

Principal cutting edge angle (λ) = 75°

Calculate : (i) Back rake (ii) Side rake. (*AMIE 1974 W*)

Solution. We know $$\tan a_b = \cos\lambda \tan\alpha + \sin\lambda \tan i$$

$$= \cos 75° \tan 10° + \sin 75° \tan 0°$$

$$= 0.259 \times 0.176 = 0.0456$$

$$\therefore \quad a_b = \text{Back rake} = \tan^{-1}(0.0456) = 2°37'$$

Also, $$\tan\alpha_s = \sin\lambda \tan\alpha - \cos\lambda \tan i$$

$$= \sin 75° \tan 10°$$

$$= 0.966 \times 0.176 = 0.17$$

$$\therefore \quad (\text{side rake})\ \alpha_s = \tan^{-1}(0.17) = \mathbf{9°, 40'}$$

Example 13. *In a single point cutting tool used for turning, the geometry as per ASA is :*

Back rake = 8°

Side rake = 4°

Side cutting edge angle = 15°

Find the values of inclination angle and rake angle in ORS of tool nomenclature.

(AMIE 1975 S)

Solution. As per ASA system, $\alpha_b = 8°, \alpha_s = 4°, C_s = 15°$

In ORS of nomenclature, λ = approach angle

$= 90° - Cs$

$= 90° - 15° = 75°$

Now we know $\tan \alpha = \tan \alpha_s \sin \lambda + \tan \alpha_b \cos \lambda$

$= \tan 4°.\sin 75° + \tan 8°.\cos 75°$

$= 0.07 \times 0.966 + 0.1405 \times 0.259$

$= 0.0676 + 0.0364$

$= 0.104$

$\therefore$ Orthogonal rake angle, $\alpha = \tan^{-1} 0.104$

$= \mathbf{5°, 56'}$

Also, $\tan i = \sin \lambda \tan \alpha_b - \cos \lambda \tan \alpha_s$

$= 0.966 \times 0.1405 - 0.259 \times 0.07$

$= 0.1357 - 0.0181$

$= 0.1176$

$\therefore$ Inclination angle, $i = \tan^{-1}(0.1176)$

$= \mathbf{6.7°}$

Example 14. *For a turning operation with H.S.S. tool for hot rolled 0.2% C-Steel, the following data is given :*

Cutting speed = *0.2 m/s*

Depth of cut = *3.2 mm*

Feed = *0.5 mm/rev.*

C_s = *15°*

Determine : Cutting power, motor power, specific cutting resistance and unit power.

Solution. The cutting force is, $F_c = 162.4\, f^{0.85}\, d^{0.98}$ kgf

$= 1593 \times f^{0.85} \times d^{0.98}$ newtons

$= 1593 \times (0.5)^{0.85} \times (3.2)^{0.98}$

$= 2763.74$ N

Cutting Power, $P_c = \dfrac{F_c \times V}{1000}$ kW

$= \dfrac{2763.74 \times 0.2}{1000} = 0.553$ kW

$$\text{Motor Power, } P_m = P_c / \eta_{mt}$$

$$\text{Let } \eta_{mt} \text{ for lathe} = 0.85$$

$$\therefore \quad P_m = \frac{0.553}{0.85} = 0.65 \text{ kW}$$

$$\text{Area of uncut chip, } A_c = t \times b = f \times d = 0.5 \times 3.2 = 1.6 \text{ mm}^2 \quad \text{(Eqn 14.12)}$$

$$\therefore \quad \text{Specific cutting resistance} = \frac{F_c}{f \times d} = 1727.34 \text{ N/mm}^2$$

$$\text{Unit power} = \frac{P_c}{A_c V} = \frac{0.553 \times 1000}{1.6 \times 0.2 \times 1000}$$

$$= \mathbf{1.728\ W/mm^3/s.}$$

Example 15. *Using Taylor equation and using $n = 0.5$, $C = 400$. Calculate the percentage increase in tool life when cutting speed is reduced by 50%.*

Solution.

$$VT^n = C$$

$$\therefore \quad V_1 T_1^n = V_2 T_2^n$$

$$= 0.5 V_1 \cdot T_2^n$$

$$\therefore \quad \left(\frac{T_2}{T_1}\right)^n = 2$$

$$\therefore \quad T_2 = 4T_1$$

$$\therefore \quad \text{Percentage increase} = \frac{T_2 - T_1}{T_1} \times 100 = \mathbf{300\%}$$

Example 16. *For an Orthogonal cutting process :*

Uncut chip thickness = 0.127 mm

V = 120 m/min

Rake angle = 10°

Width of cut = 6.35 mm

Chip thickness = 0.229 mm

Cutting force = 556.25 N

Thrust force = 222.50 N

Calculate the percentage of total energy that goes into overcoming friction at the tool-chip interface.

Solution.

$$\frac{\text{Friction energy}}{\text{Total energy}} = \frac{F.V_C}{F_C.V}$$

Now

$$\frac{V_C}{V} = \frac{t}{t_C} = r \quad \text{(Equ. 14.9)}$$

$$\therefore \quad \frac{\text{Friction energy}}{\text{Total energy}} = \frac{F.r}{F_C}$$

$$r = \frac{t}{t_C} = \frac{0.127}{0.229} = 0.555$$

It is clear from Fig. 14.19,

$$F = R' \sin \beta$$

$$F_C = R'.\cos(\beta - \alpha)$$

and $$R' = \sqrt{F_c^2 + F_t^2} = 600.725 \text{ N}$$

$\therefore$ $$F_c = 556.25 = 600.725 \times \cos(\beta - 10)$$

From here, $$\beta = 32°$$

$\therefore$ $$F = 600.725 \times \sin(32°)$$

$$= 318.34 \text{ N}$$

$\therefore$ $$\frac{\text{Friction energy}}{\text{Total energy}} = \frac{318.34 \times 0.555}{556.25} \times 100$$

$$= \mathbf{31.75\%}$$

Example 17. *A 300 mm diameter bar is turned at 45 rev/min with depth of cut of 2 mm and feed of 0.3 mm/rev. The forces measured at the cutting tool point are :*

Cutting force = 1850 N

Feed force = 450 N

Calculate :

(i) Power consumption

(ii) Specific cutting energy

(iii) Energy consumed if the total metal removed during the turning operation is 2.5×10^6 mm^3.

Solution. Cutting velocity,

$$V = \frac{\pi \times 300 \times 45}{60 \times 1000} = 0.707 \text{ m/s}$$

(*i*) $\therefore$ Cutting power, $$P_C = \frac{F_C.V}{1000}, \text{kW}$$

$$= \frac{1850 \times 0.707}{1000}$$

$$= \mathbf{1.308 \text{ kW}}$$

Now feed velocity $$= \frac{0.3 \times 45}{60 \times 1000}, \text{m/s}$$

$\therefore$ Feed power $$= \frac{0.3 \times 45}{60 \times 1000} \times 450$$

$$= \mathbf{0.10 \text{ W}} \text{ (negligible)}$$

(*ii*) Now,

$$MRR = A_c \cdot V$$
$$= b \times t \times V$$
$$= d \times f \times V \quad \text{(See Equ. 14.12)}$$
$$= 2 \times 0.3 \times \pi \times 300 \times 45$$
$$= 2.545 \times 10^4 \text{ mm}^3/\text{min.}$$

$\therefore$ Specific cutting energy $= \dfrac{1308 \times 60}{2.545 \times 10^4} = \mathbf{3.08\ W.s/mm^3}$

(*iii*) Energy consumed $= 3.08 \times 2.5 \times 10^6$, W.s

$$= \frac{3.08 \times 2.5 \times 10^6}{1000 \times 3600} = \mathbf{2.139\ kWh}$$

Example 18. *A M.S. bar of 100 mm is being turned with a tool having ASA tool significant as: 6° – 10° – 5° – 7° – 10° – 30° – 0.5 mm. Determine the various components of the machining force and the power consumption. Take : depth of cut = 2.5 mm, feed = 0.125 mm/rev, turning speed of job = 300 rev./min., co-efficient of friction at the tool-work interface = 0.6, ultimate shear stress of the work material = 400 MPa.*

Solution. It is clear from the tool designation, that the side cutting edge angle is 30°. Therefore, the tool approach angle, $\lambda = 90° - 30° = 60°$

Friction angle, $\beta = \tan^{-1}\mu = \tan^{-1} 0.6 = 30.96° \cong 31°$

Now, orthogonal rake angle is given as,

$$\tan a = \tan a_s \cdot \sin\lambda + \tan a_b \cdot \cos\lambda$$

Now, $a_s = 10°$ and $a_b = 60°$, $\lambda = 60°$

From here, $a = 11.6°$

Now, from Merchant's relation, the shear angle is,

$$\phi = 45° + \alpha/2 - \beta/2$$
$$= 45° + 5.8° - 15.5° = 35.3°$$

Merchant's theory is more accurate for plastics but agrees poorly for machining metals. With Lee and Shaffer relation,

$$\phi = \frac{\pi}{4} - (\beta - \alpha)$$
$$= 45° - 31° + 11.6° = 25.6°$$

Now, the cutting force is given as,

$$F_c = \frac{\tau_s \cdot b \cdot t}{\sec(\beta - \alpha) \cdot \cos(\phi + \beta - \alpha) \cdot \sin\phi}$$

Now, $b = \text{width of cut} = \dfrac{\text{depth of cut}}{\sin\lambda} = \dfrac{d}{\sin\lambda}$

and uncut chip thickness, $t = f \cdot \sin\lambda$

$\therefore$ $b \cdot t = d \cdot f$

$\therefore$ $F_c = \dfrac{400 \times 2.5 \times 0.125}{\sec(19.4°) \times \cos 45° \times \sin 25.6°} = 386$ N.

Now, from equation (14.18*a*), the thrust component is,

$$F_t = F_c \cdot \tan(\beta - \alpha)$$

$$\therefore \quad F_t = 386 \times \tan 19.4° = 135.932 \text{ N}$$

The thrust force is normal to the tool-job interface, that is, normal to the principal cutting edge of the tool, see Fig. 14.17.

∴ Feed force (along the axis of the job),

$$F_f = F_t \cdot \sin \lambda = 135.932 \times \sin 60° = 117.7 \text{ N}$$

Radial force (Normal to the axis of the job),

$$F_r = F_t \cdot \cos \lambda = 67.961 \text{ N}$$

Now, cutting power = $F_c \cdot V$ watts

and
$$V = \frac{\pi DN}{1000 \times 60}, \text{ m/s}$$

$$= \frac{\pi \times 100 \times 300}{1000 \times 60} = 1.57 \text{ m/s}$$

$$\therefore \quad \text{Power} = 386 \times 1.57 = \mathbf{606 \text{ watts}}$$

PROBLEMS

1. Define Machining Process.
2. Explain a basic machining operation with the help of a neat diagram.
3. Explain the various elements of a single-point cutting tool with the help of a neat diagram.
4. What is meant by 'hand' of a single point cutting tool?
5. With the help of a neat sketch, discuss the principal surfaces and planes is metal cutting.
6. Name the two systems of tool designation.
7. Why a negative rake angle is normally employed for cutting hard and strong materials?
8. Show the ORS of tool angles with the help of a sketch.
9. Write the relations between ASA nd ORS systems of tool angles.
10. What is meant by Orthogonal cutting and Oblique cutting?
11. How does rake angle affect the life of the cutting tool?
12. Differentiate between positive and negative rake angles.
13. How is the nose radius of a cutting tool selected?
14. Discuss the various types of chips produced during metal cutting.
15. Why are discontinuous type chips preferred over the continuous type?
16. Explain, why built up edge on a cutting tool is undesirable?
17. Name the factors that contribute to the formation of discontinuous chips.
18. Name the factors that contribute to the formation of B.U.E.
19. What is the use of 'chip breaker'?
20. Discuss the two methods of metal cutting.
21. Discuss the various forces encountered in metal cutting.
22. Explain 'Merchant force circle'.
23. Define Tool Life.

24. Why tool wear is important in metal cutting?

25. Discuss various types of tool wears.

26. Enumerate the factors on which tool wear and tool life depend.

27. Name the factors that contribute to flank wear.

28. Name the factors that contribute to crater wear.

29. Which two pressure areas of the cutting tools are subjected to wear?

30. With the help of a sketch, show crater wear and flank wear on a cutting tool.

31. Discuss Taylor's relationship for cutting speed-tool life.

32. Derive an expression for optimum value of cutting speed.

33. In an orthogonal cutting operation, the following data have been observed :

Cutting speed	=	0.223 m/s
Uncut chip thickness	=	0.06 mm
Width of cut	=	3.83 mm
Chip thickness ratio	=	0.51
Rake angle	=	20°
Cutting force	=	363 N
Thrust force	=	127 N

Determine : Shear angle, friction angle, shear stress along the shear plane, chip velocity, shear strain in chip, shear strain rate and the power for the cutting operation.

34. The following equation for tool life has been obtained for H.S.S. tool :

$$VT^{0.13} \cdot f^{0.6} \cdot d^{0.3} = C$$

A 60 min. tool life was obtained using the following cutting conditions :

$$V = 40 \text{ m/min}, f = 0.25 \text{ mm}; d = 2.0 \text{ mm}$$

Calculate the effect on tool life if speed, feed and depth of cut are together increased by 25% and also if they are increased individually by 25%.

35. During machining of C–20 steel with an orthogonal tool having a rake of 10° at a feed of 0.2 mm/rev., the value of the shear angle has been observed to be 30° under a shear angle microscope. If the principal cutting edge angle is 90°, what is the value of cutting ratio or the chip reduction co-efficient.

36. In a turning operation, it was observed that the tool life was 100 minutes and 50 minutes at cutting speeds of 25 m/min. and 100 m/min respectively. Find out the tool life at 200 m/min under the same cutting conditions.

37. The end of a tube is being turned on a lathe at 200 rev/min and at a feed rate of 25.4 mm/min.

The tube is 15 cm in diameter and 2.5 mm thick. The tube material obeys the equation,

$$\bar{\sigma} = K(\bar{\varepsilon})^n$$

where n = 0.26 and K = 540 N/mm²

V_c/V is $\frac{1}{3}$ and α = 5°. The friction energy is 25 per cent of the total energy consumed. Determine the power in operation.

38. In an orthogonal cutting operation, the orghogonal rake is 12°, while the principal cutting edge angle is 75°. Calculate,

(*i*) back rake (*ii*) side rake

If the principal cutting edge angle is 90°, how much are

(i) back rake (ii) side rake.

39. During machining of C–20 steel, a double carbide cutting tool of 0 – 10 – 6 – 6 – 8 – 75 – 1 mm (ORS) shape has been used. Feed is 0.15 mm/rev., depth of cut of 1.5 mm at a cutting speed of 120 m/min., a chip thickness of 0.30 mm have been obtained. Calculate :

(i) the chip reduction co-efficient (ii) the shear angle.

40. Sketch a single-point cutting tool and show on it the various tool elements and tool angles.

41. Give the function of each tool element. List the various tool angles and discuss their significance.

42. Discuss the two construction of tipped tools.

43. Why indexable inserts are better than brazed tool tips.

44. Why do carbide tools employ negative rake angles more often than H.S.S. tools.

45. When the use of positive rake angles and negative rake angles is recommended?

46. Give the significance of providing nose radius on tool tip.

47. What do you understand by the term. 'Tool Designation' or 'Tool Signature'.

48. Describe the tool represented by 10, 10, 6, 6, 8, 8, 1 mm in ASA system.

49. What is orthogonal rake angle?

50. Define cutting ratio.

51. What is the approximate thickness of shear zone in metal cutting?

52. In orthogonal cutting operation, the feed is 0.10 mm and the chip thickness is 0.25 mm. The cutting force is 1360 N and the feed thrust force is 770 N. The rake angle of the tool is + 10°. Find :

(a) The shear angle.

(b) The size of the force exerted by the tool on the chip.

(c) The coefficient of friction on the face of the tool.

(d) The sizes of the friction force and the normal force, on the tool face.

(e) The sizes of the shearing force and the normal force, on the shear plane.

53. An orthogonal cut 2.5 mm wide is made at a speed of 0.5 m/s and feed of 0.26 mm with a H.S.S. tool having a 20° rake angle. The chip thickness ratio is found to be 0.58, the cutting force is 1400 N and the feed thrust force is 360 N. Find :

(a) Chip thickness.

(b) Shear plane angle.

(c) Resultant force.

(d) Co-efficient of friction on the face of the tool.

(e) Friction force and normal force on the chip.

(f) Shearing force and normal force on the shear plane.

(g) Specific energy.

54. In orthogonal cutting, the feed is 0.127 mm and the depth of cut normal to the plane of the paper is 2.54 mm. The cutting speed is 4 m/s. The cutting force is found to be 1800 N and the feed thrust force 900 N. The rake angle of the tool is + 8°. Find :

(a) The power required for the cut in kW.

(b) The rate of metal removal, mm^3/s.

(c) The unit power in $W/mm^3/s$.

55. A workpiece is being cut a 1.25 m/s and the power is found to be 2.05 kW. The feed is 0.25 mm/rev. and the depth of cut is 5 mm. Estimate :

(*a*) Cutting force in *N*.

(*b*) Unit power consumption W/mm^3/s.

56. A tool making an orthogonal cut has a rake angle of – 10°. The feed is 0.10 mm, the width of cut 6.35 mm, the speed 2.7 m/s, and a dynamometer measures the cutting force to be 1800 N and the normal thrust force to be 1540 N. A high-speed photograph shows a shear plane angle of 20°. Estimate :

(*a*) Chip thickness.

(*b*) Co-efficient of friction.

(*c*) Shear and normal stresses on the shear plane.

(*d*) Shearing strain.

(*e*) Power to shear the metal.

57. What is understood by tool life? What is the significance to an engineer who is interested in productivity? What different criteria are used to identify that the tool has reached its limiting life?

58. Establish the simple tool-life equation with magnitudes of constants from the following data (Taylor's tool-life equation) :

A tool life of 100 min is obtained from a cutting tool at a cutting speed of 25 m/min, and 10 min, at 33.3 m/min. What is the cutting speed for a 60 min, tool life.

59. (*a*) State the general form of the Taylor's equation for tool life.

(*b*) Discuss the usefulness of Taylor's equation in selection of proper cutting speed for machining operations.

(*c*) Derive an expression for the most economic cutting speed.

(*d*) Is the most productive cutting speed the same as the most economic speed. Give reasons for your answer.

(*e*) Make a rough plot of the variation of the cost of machining per piece, with the cutting speed and discuss how this could be useful in cost reduction.

60. Discuss in brief the relationship between the mechanical properties of the work material and machineability.

61. Define the term machineability. Explain, how it is influenced by following :

(*i*) Work material micro-structure.

(*ii*) Type of cut.

(*iii*) Tool rake angle.

62. Following data were collected from an orthogonal machine test on steel :

Cutting speed	=	18 m/min.
Rake angle	=	20°
Clearance angle	=	10°
Width of cut	=	3.2 mm
Underformed chip thickness	=	0.10 mm
Deformed chip thickness	=	0.25 mm
Cutting force in the cutting velocity direction	=	800 N

Normal force in a direction normal to cutting velocity = 500 N. Draw the circle diagram of forces and evaluate : shear angle, shear strain, friction co-efficient against chip flow, friction force on the rake face.

63. Show that during orthogonal cutting with a zero degree of rake angle, the ratio of the shear strength, τ_s, of the work material to specific cutting energy, e, is given by

$$\tau_{s/e} = (1 - \mu r) \cdot \frac{r}{1 + r^2}$$

64. During orthogonal turning of M.S. tube of 5 mm wall thickness, the following data were recorded :

Speed	25 m/min.
Tool	H.S.S. 15° rake
Feed	0.15 mm/rev.
Chip thickness ratio	0.35
Co-efficient of friction	0.60

The friction force on the tool-chip interface was measured by means of a special device and was equal to 480 N. Determine the components of the cutting forces, shear angle and the specific energy of deformation.

65. In an orthogonal cutting operation, the cutting speed is 2.5 m/s, rake angle is 6° and the width of the cut is 10 mm. The underformed chip thickness is 0.2 mm. 13.36 gms of steel chips with a total length of 50 cm are obtained. The tool post dynamometer gives cutting and thrust forces 1134 N and 453.6 N. Determine :

(*a*) slip plane angle.

(*b*) friction energy at the tool-chip interface, as the percentage of total energy.

(*c*) shear energy as the percentage of total energy. (33°, 30.9%, 69.3%)

66. In orthogonal cutting of a low carbon steel, the specific cutting energy is 4080 N/mm². The uncut chip thickness is 0.2 mm and the chip width is 5 mm. The cutting speed is 1.1 m/s, and the rake angle of the tool is 10°. Assuming co-efficient of friction at the tool-chip interface as 0.7, determine :

(*a*) the cutting force

(*b*) the average shear stress in the shear plane

(*c*) the normal stress on the shear plane

(*d*) the average shear strain in cutting

(*e*) the average shear strain rate.

Use the relation of Lee and Shaffer to find the shear angle.

(4080 N, 861.8 N/mm², 861.8 N/mm², 3.906, 4.44 × 105 s⁻¹)

67. A 0.2% C–Steel is machined with a tripple carbide cutting tool having 0 – 10 – 6 – 6 – 8 – 75 – 1 mm ORS shape. A feed of 0.2 mm/rev and depth of cut of 2 mm at the cutting speed of 150 m/min have been employed. A chip thickness of 0.36 mm has been obtained. Calculate the chip reduction co-efficient and shear angle.

68. Select the speed in rev/min for turning a round steel bar of diameter 320 mm with a H.S.S. T–1 tool, having a tool life of 60 minutes. A feed of 0.2 mm/rev. and depth of cut 1 mm have been chosen. The cutting speed equation is given in the following Taylorian form.

$$V = \frac{273}{T^{0.2}\, f^{0.2}\, d^{0.15}}\ \text{m/min.}$$

f and *d* are in mm. (**Ans.** 165.2 rev./min)

69. A H.S.S. tool runs at 24 m/min. for one hour. Circumstances are such that it becomes desirable to run the tool for 1000 min. Estimate the suitable speed. Take $n = 0.25$ in the Taylor equation.

(15.74 m/min)

70. A bar 76 mm diameter is reduced to 71 mm diameter by means of a cutting tool for which approach angle is 90°, and for which the cutting edge lies in the plane containing the work axis of rotation. The mean length of the cut chip is 73.9 mm and a feed of 0.2 mm/rev. is used. Find the cutting ratio. (0.32)

71. When turning 19 mm diameter bar on an automatic lathe employing sintered carbide tools, the value of n is 0.2 and the value of V_{60} is 104 m/min. At what speed should the spindle run to give a tool life of 6 hours. If a length of 50 mm per component is machined and the feed used is 0.16 mm/rev, what is the cutting time per piece and how many pieces can be produced between tool changes. (1220 rev/min., 16.5 s, 1300)

72. When cutting steel with a H.S.S. cutter, the tool life at a cutting speed of 30 mpm is 60 min. and at a cutting speed of 45 mpm, it is 30 min. What is the tool life equation of the cutter on the material. ($T = 21115/V^{1.724}$)

73. A HSS tool requires regrinding after 3 hours and 20 minutes when machining steel at a cutting speed of 40 m/min. Calculate the tool life if the speed is increased to 70 m/min. (2.7 min.)

74. *C*-steel is machined by a WC cutting tool. The tool has a life of 100 min. between consecutive grinds when operating at 80 m/min. and a life of 33 minutes at a speed of 100 m/min. Determine the values of 'n' and 'C' in Taylor equation. (n = 0.2, C = 201)

75. In Taylor equation, if n = 0.25 and C = 300 m/min., what should be V for a tool life of one hour? (**Ans.** 107.8 m/min)

76. C-40 steel is turned with a H.S.S. tool of back rake angle of 8°. The uncut chip thickness is 0.3 mm, depth of cut (chip width) is 1.5 mm and the cutting speed is 0.6 m/s. One metre long chip weighs 5.7g. The width of the chip remains constant. Calculate the cutting ratio, taking density of material as 7.9 g/cm³. Also determine the chip compression ratio. (**Ans.** 0.625, 1.6)

77. For problem 76, Calculate MRR, power requirement and the cutting force. (270 mm³/s, 891 W, 1485 N)

78. Using Taylor equation and taking n = 0.12, calculate the percentage change in cutting speed required to give an 80% reduction in tool life. (**Ans.** 21.4%)

79. The following data was obtained from the tool-life cutting test of H.S.S. tool material, used to cut die-steel :-

Cutting Speed, m/min :	49.74	49.23	48.67	45.76	42.58
Tool life, min :	2.94	3.90	4.77	9.87	28.27

Determine the constants of the Taylor tool life equation $VT^n = C$. (**Ans.** n = 0.07, C = 54)

80. The power required to turn a medium C-steel is approximately 3.8 W/mm³/s. The power available at the machine spindle is 3.73 kW. Determine :

(a) Maximum MRR. (*b*) Cutting force.

(c) Depth of cut if cutting speed is 36 m/min. and the feed is 0.25 mm / rev.

(**Ans.** 982 mm³/s, 6.216 kN, 6.55 mm)

81. A lathe running idle consumes 325 W. The power input increases to 2.58 kW when cutting an alloy steel at 24.5 m/min. If the spindle speed is 124 rev./min., depth of cut = 3.8 mm and feed is 0.2 mm/rev., determine :

(*a*) Cutting force. (*b*) Torque at the spindle.

(*c*) Specific power consumption. (**Ans.** 5.522 kN, 174 Nm, 7.27 W/mm³/s)

82. In an orthogonal cutting operation on a workpierce of width 2.5 mm, the uncut chip thickness was 0.25 mm and the tool rake angle was zero degree. It was observed that the chip thickness was 1.25 mm. The cutting force was measured to be 900 N and the thrust force was found out to be 810 N.

(*a*) Find the shear strength of the workpiece material. (GATE 1992)

(*b*) If the co-eff. of friction was 0.5, what is the machining constant?

(**Ans.** 227 N/mm², 49.2°, Machining constant = 2ϕ + β)

83. While turning a mild steel bar on a lathe, the following data were obtained.

Machined length	=	150 mm
Speed	=	170 rev./min
Feed rate	=	0.5 mm/rev.
Cutting force	=	2000 N
Dia. of workpiece	=	500 mm

Calculate : Power consumed in machining, machining time and the total amount of heat generated during cutting. *(Nagpur University)*

(Ans: 8.90 kW, 1.765 min., 942.51 kJ)

84. Draw Merchant's circle in order to analyse the mechanies of metal cutting and deduce various force angle relationships. Determine shear angle, friction angle, shear and normal stress on shear plane, shear strain by using the formula data:

Uncut chip thickness	=	0.125 mm
Chip thickness	=	0.250 mm
Width of cut	=	6.5 mm
Cutting speed	=	100 m/min.
Rake angle	=	10°
Cutting force	=	70 N
Thrust force	=	25 N

(Nagpur University)

(**Ans:** 28.33°, 29.65°, 61.24 N/mm², 67.96 N/mm², 2.1866)

85. In an experimental set-up for the tool life of the cutting tool, following data have been noted:

No. of components machined between regrinds	Rev./min.	Feed in mm/min.
700	450	0.25
80	710	0.25

Find out the tool-life equation if the length of the work-piece is 100 mm and the mean diameter at the cutting edge is 25 mm. *(Nagpur University)*

(**Ans**: $VT^{0.14} = 87$)

86. A 50 mm diameter bar of steel was turned at 284 rev./min. and the tool failure occurred in 10 min. The speed was changed to 232 rev/min. and the tool failed in 60 min. of cutting time. Assuming a straight line relationship exists, what cutting speed should be used to obtain a 30 min. tool life? (**Ans:** V_{30} = 39.39 m/min.)

87. For orthogonal cutting with a cutting tool having a rake angle of 10°, the uncut chip thickness is 0.15 mm and chip thickness is 0.4 mm. Determine (*a*) the cutting ratio, (*b*) the shear plane angle and (*c*) the shear strain. (**Ans:** 0.375, 21.55°, 2.734)

88. In an orthogonal cutting operation, the following data have been observed:

Uncut chip thickness,	$t = 0.25$ mm
Chip thickness,	$t_c = 0.75$ mm
Width of cut,	$b = 2.5$ mm
Rake angle,	$\alpha = 0°$
Cutting force,	$F_c = 950$ N
Thrust force,	$F_t = 475$ N

Determine: shear angle, co-efficient of friction, the friction angle, ultimate shear stress of the material. (**Ans:** 18.4°, 0.5, 26.57°, 379.4 MPa)

89. In the machining of mid steel, the following data have been observed :

Uncut chip thickness,	$t = 0.2$ mm
Width of cut,	$b = 2$ mm
Cutting speed,	$V = 200$ m/min.
Rake angle,	$\alpha = 10°$
Co-efficient of friction,	$\mu = 0.5$
Ultimate Shear stress of material,	$t = 400$ MPa.

Determine : Shear angle and the cutting and thrust components of the machining force.

(**Ans:** 36.7°, 420 N, 125 N)

Hint: Use eqn. 14.19 $\phi = 45° + \frac{\alpha}{2} - \frac{\beta}{2}$, to determine, ϕ.

90. Solve the above problem, by assuming the machining constant to be 70° in Merchant's second equation for angles. Also, find out the results by using relation given by Lee and Shaffer relation. (**Ans:** Merchant Theory : F_c = 468.7 N ; F_t = 139.46 N)

Lee and Shaffer theory ; F_c = 463.92 N; F_t = 154.6 N)

91. Determine the cutting force during orthogonal machining of mild steel with an uncut chip thickness of 0.25 mm and width of cut being 2.5 mm. Take a = 0°, μ = 0.5, and τ_x = 400 MPa. (**Ans:** 1000.84 N)

92. Determine the cutting force and the thrust force and the cutting ratio when machining M.S. Take:

Uncut chip thickness = 0.25 mnn

Width of cut = 2 mm

Rake angle = 0°

$\mu = 0.5$

$\tau_x = 400$ MPa

Use Lee and Shaffer relation for shear angle. (**Ans:** 800.65 N ; 400.3 N, 0.333)

93. (*a*) Determine the various components of the machining force when machining a C.I. block on a shaper with depth of cut = 4 mm, feed = 0.25 mm/stroke, Normal rake angle of the tool = 10°, Principal cutting edge angle = 30°. μ = 0.6. τ_x = 340 MPa.

(*b*) Determine the average power consumption if the operation takes with 60 strokes/min. and the length of the job is 200 mm.

Solution: (*a*) Using Lee and Shaffer relation for shear angle.

$$\phi = \frac{\pi}{4} - (\beta - \alpha)$$

Now $\beta = \tan^{-1} \mu = \tan^{-1} 0.6 = 30.96° \cong 31°$

$\therefore$ $\phi = 45° - (31 - 10) = 24°$

Now, cutting force (it acts along velocity vector),

$$F_c = \frac{\tau_x . b . t}{\sec(\beta - \alpha) . \cos(\phi + \beta - \alpha) . \sin\phi}$$

Now, $b = \frac{d}{\cos C_s}$ and $t = f \cos C_s$

$\therefore$ $b . t = d . f = 4 \times 0.25 = 1 \text{ mm}^2$

$\therefore$ $F_c = \frac{340 \times 1}{\sec 21° \times \cos 45° \times \sin 24°} = 1103.8$ N

Thrust force, $F_t = F_c . \tan(\beta - \alpha)$, it acts normal to tool-job interface. (Eqn. 14.18*a*)

$= 1103.8 \times \tan 21° = 423.7$ N

Now feed force component, $F_f = F_t . \sin\lambda$ (normal to the velocity vector)

$= 423.7 \times \sin 60° = 366.92$ N

Normal force component, $F_n = F_t . \cos\lambda = 311.85$ N (normal to the machined surface)

(*b*) Now $\quad$ Work done $= F_c \times$ length of stroke $= 1103.8 \times \dfrac{200}{1000} = 220.76$ J

$$\therefore \quad \text{Average Power} = 220.76 \times \frac{60}{60} = \mathbf{220.76\ watts}$$

94. Calculate the metal removal rate for a 50 mm diameter bar machined at 200 rev./min, with a feed of 0.5 mm/rev. and a depth of cut of 4 mm.(**Ans.** 5.78×10^4 mm³/min)

95. When machining a 250 mm diameter bar at 40 rev./min. with depth of cut of 2 mm and a feed of 0.3 mm/rev., the cutting force at the tool point was 1800 N and the feed force was 400 N. Calculate the power consumption. (**Ans.** 935 W, 0.08 W)

Sol. Power consumption = Force × Velocity

$$\text{and Velocity} = \frac{\pi D_{ave} N}{1000 \times 60} = \frac{\pi \times 248 \times 40}{1000 \times 60} \text{ m/s}$$

$$\therefore \text{Power} = \frac{1800 \times \pi \times 248 \times 40}{1000 \times 60} = 935\ W$$

$$\text{Now velocity of feed force} = \frac{0.3 \times 40}{1000 \times 60}$$

$$\therefore \text{Power consumption} = \frac{400 \times 0.3 \times 40}{1000 \times 60} = 0.08 \text{ W}$$

Which is very negligible as compared to Power consumption for cutting force.

Now MRR = d.f. D_{ave}. π . N

$= 2 \times 0.3 \times \pi \times 248 \times 40 = 1.87 \times 10^4$ mm³/ min.

$$\therefore \text{Specific energy consumption} = \frac{\text{Power}}{\text{MRR}}$$

$$= \frac{935 \times 60}{1.87 \times 10^4} = 3 \text{ W.s/mm}^3$$

96. The resultant force acting at a tool point is 1200 N and its angle of inclination to the horizontal is 54°. The plan approach angle to which it acts perpendicularly is 26°. Determine the three principal component forces.

Solution. $\quad F_c = F_r.\sin 54° = 970.8$ N

$F_f = F_n.\tan 26°$ $\quad$ ($F_n = 635.5$ N; $F_f = 310$ N).

and $\quad F_e^{\,2} = F_c^{\,2} + F_f^{\,2} + F_n^2$

97. A W C cutting tool machining MS gave a life between regrinds of 100 min. when operating at 80 m/min. and 33 min. when operating at 100 m/min. Determine the value of the index and the constant in tool life equation.

Sol.

$$VT^n = C$$

$$\therefore Log\ V + n \log T = \log C$$

$$\therefore Log\ 80 + n \log 100 = \log C \quad ...(1)$$

$$\text{and} \quad \log 100 + n \log 33 = \log C \quad ...(2).$$

from these two equations, $n = 0.2$ and C = 201.

15

DESIGN AND MANUFACTURE OF CUTTING TOOLS

15.1. TYPES OF CUTTING TOOLS

The cutting tools may be classified in different ways. Depending upon the number of cutting points on the tool, the cutting tools are of two types :

1. Single-point cutting tools,
2. Multi-point cutting tools.

A single-point cutting tool has only one cutting point or edge. The tools used for turning, boring, shaping, or planing operations, that is, tools used on lathes, boring machines, shaper, planer, etc. are single point tools. A multi-point tool has two or more than two cutting points, for example, tools used on drilling machines, milling machines, broaching machines etc. A multi-point tool can be considered to be basically a series of single-point tools.

Depending upon the construction of the cutting tool, it is classified as :

1. Solid tools,
2. Tipped cutting tools.

The solid cutting tools are made entirely of the same material, whereas, in a tipped cutting tool, an insert of cutting tool material is brazed or held mechanically to the shank of another material.

Solid type tools are available in square, rectangular, round or some special shapes. Small sized solid tools are held in a tool holder, Fig. 15.1 (*a*), whereas the others are mounted directly in the tool post, Fig. 15.1 (*b*). In tipped cutting tools, the insert is either brazed to a tool shank, Fig. 15.1 (*c*), or is held mechanically in the tool holder, Fig. 15.1 (*d*). The mecianically held insert is called as "Throwaway" or "Disposable insert". The details will be given in next article.

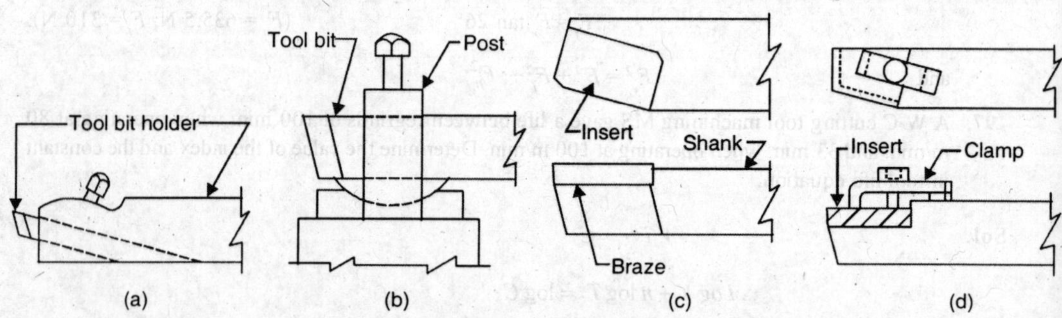

Fig. 15.1. Tool Construction.

The proper selection of tool type depends upon many factors, for example, workpiece material, the machine tool, the type of set up, the power available at the machine, the amount of material to be removed and the kind of operation etc. Cost is the major factor in the use of solid-tools. The tool materials used in solid-tools are less expensive. High carbon steel, Alloy tool steels, H.S.S. are used as solid tools. Due to its high cost, the carbide tool material is used in the form of tips or inserts. Similarly, ceramics are also used as inserts. Cast alloys are used either as tool bits held in tool holder or in the form of tips or inserts. Similarly, diamond is used either as a tip or insert. Polycrystalline boron nitride is used as a throwaway bit in tool holder. UCON is also used as a tip or an insert.

15.2. GENERAL PROBLEMS OF CUTTING TOOL DESIGN

All cutting tools remove a certain layer of material and impart the required shape, size, and surface quality to the workpiece. A great variety of cutting tools have been developed to satisfy the requirements of production. Eventhough the various cutting tools have different shapes and purpose, all of them have :

1. A cutting element, that is, the part of the tool which is actually fed into the workpiece to remove the stock.

2. A mounting or clamping element.

Whatever may be the shape and purpose, the various cutting tools have cutting teeth resembling the point of a single-point cutting tool. Even the individual abrasive grains of a grinding wheel acts as a single-point cutting tool. Due to all this, the various cutting tools have certain common general problems of design.

Design of a cutting tools means the determination of all the dimensions and the shapes of all the elements of a tool, either analytically or graphically. A cutting tool designer performs the following functions :

1. – To determine the forces acting on the cutting surfaces of the tool, on the basis of metal cutting theory.
   – To find the optimum tool geometry.
   – To select the most suitable material for making the cutting element of the tool, and
   – To select a shape for the cutting element that ensures free chip disposal during the cutting operation.
2. – To find the most producible shapes of the cutting and the mounting elements of the tool.
   – To determine the tolerances on the dimensions of the cutting and mounting elements of a tool depending on the working conditions and the required machining accuracy of the job.
3. – To determine the strength and rigidity of the cutting and mounting elements of the tool.
4. – To make a working drawing of the tool and compile the manufacturing specifications.

15.2.1. Cutting Element. To design the cutting element of a tool, it is necessary to know the kinematics of cutting for the given case. Any cutting tool will remove a chip only if its cutting edge moves relative to the work. We know that this relative motion is obtained by adding the absolute motions of the work and the tool. The motions in various metal-cutting machine tools are made up of linear and rotary motions. The actual values of the tool angles depend on the kinematics of the cutting process, as discussed in Chapter 14.

The following design factors should be kept in mind when designing the cutting element :

1. Each kind of tool may have different schemes of load distribution. For example, to cut threads on a shaft on a centre lathe, the tool can either be fed normal to the axis or along one side of the thread profile, (see Fig. 17.4). Each of these methods has advantages and drawbacks. The designer should select the most effective method.

2. Roughing tools should be capable of removing the largest amount of stock at minimum forces and power. Finishing tools should be able to attain the required surface finish.

3. *Sharpening of tools.* When a tool wears out, it is sharpened by grinding off a layer of metal that has become worn during the cutting process. The metal is ground off the point, teeth or blades. The tool is sharpened by grinding either the face or the flank of the tool. The method will depend upon the purpose of the tool and the conditions of the tool operation. The method accepted for sharpening (grinding either the face or flank) determines the principal geometric dimensions of the tooth or blade and its shape. Since cutting tool wear occurs on both the face and flank, a combination sharpening method is used in which both the face and the flank are ground.

4. *Chip disposal.* The free and unrestricted disposal of chips from the cutting edges of the tool and ample chip space are one of the most important design features. Discontinuous chips will not present any problem. In the case of continuous chips, "Chip breakers" are provided on the tool face to break the chips into separate pieces for their convenient removal, see Chapter 14. In the case of drills, the chips are conveyed away from the cutting edges through flutes provided in the body of the drill. Sometimes, the chips are disposed off with the help of a stream of cutting fluid.

5. *Heat disposal.* We know that lot of heat is generated during the cutting process in the deformation of the work metal. This heat can lead to intensive wear of the cutting tool. Therefore, one of the main principles of tool design is to ensure proper heat removal from the cutting edge of the tool being designed. In single point cutting tools, it is achieved by grinding the tool point to certain angles and additionally by an ample supply of cutting fluid to the cutting zone. In the case of complex tools, proper heat removal is achieved by maintaining the body of the cutting tooth to sufficient size and by making channels or passages to deliver the cutting fluid to the cutting edges.

6. *Strength and rigidity.* As discussed in Chapter 14, the variables affecting the cutting process are so complex that it is impossible to do the exact force analysis. However, simplified calculations should be carried out to check the strength of the tool being designed, as discussed in Chapter 14.

It is even more difficult to check the rigidity and vibrational response of a cutting tool. As a rule, in extremely complex and critical cases, several different prototype designs are tested and the best one is selected.

7. *Heat treatment of tools.* We know that heat treatment has a great effect on the strength of a cutting tool. However, high internal stresses developed in hardening a tool may lead to the information of cracks and to failure. Residual stresses get concentrated at sharp corners and at points of abrupt section changes etc. Sharp corners and abrupt section changes cannot be tolerated. They act as stress raisers and greatly reduce the strength of the tool and lead to the formation of cracks and to failure of the tool during heat treatment. Hence, all sharp corners are rounded off in properly designed cutting tools.

8. *Economical utilization of tool materials.* The tool designer should keep the point of economical utilization of tool materials in mind. Tool steels, especially H.S.S. are very expensive than structural steels. Again, cemented carbides are several times more expensive than H.S.S. Due to this, the tool industry follows the economical route of constructing tools in which the cutting element is made of H.S.S. or cemented carbide and the body of the tool is made of structural steel.

9. *Built up tools.* Built up or assembled tools have a great advantage over solid tools, namely, the possibility of size adjustment. For example, a built up reamer with inserted teeth or blades, can be expanded to restore the required diameter after wear. Size adjustment of tools increases their life.

15.2.2. Mounting Element. The mounting element of a tool of a shank-or-arbor-type should be capable of transmitting the power developed by the machine tool spindle to the cutting edge of the tool. If the mounting element is not sufficiently strong or it is incorrectly designed, it will limit the capacity of the cutting tool.

In designing the mounting element, the required accuracy of location of the cutting tool in the machine spindle (twist drills, core drills, reamers etc.) or in the clamping slots of the holder or body (form tools, inserted blades, etc.), should be ensured. For this, the mounting element is manufactured with high degree of accuracy, thereby, ensuring complete interchangeability.

The form of the mounting element should be such that it enables the tool to the clamped with a minimum loss of time in the machine tool. It will still be preferable if this can be done without stopping the machine spindle, for example, rapid tool changing in a drill press.

The most widely applied constructions of mounting elements are given below :

1. A square on a straight cylindrical shank, a taper shank with or without a tang etc.

2. Quick change clamping devices of various construction, etc., for shank-type tools with rotary motion, for example, drill chucks used with drilling machines.

3. A cylindrical mounting hole with a longitudinal key or with drive keys on the end face of a tool, locking devices of various constructions, and taper holes for arbor-type tools with rotary motion.

4. A pull end of shank with a tapered key, rapid change clamping devices of various constructions for tools with a lengthwise motion, for example, pull and push broaches.

15.2.3. Effect of Method of Manufacture. The method of manufacturing a cutting tool should be taken into account while determining the form and dimensions of the cutting tool. If the tool is to be made by casting, the specific features of the various foundary techniques are taken into consideration. If a drill is to be made by forge rolling and twisting, and not by milling the flutes, the distinctive features of the rolling and twisting processes will influence the design of the drill.

All the above mentioned problems are solved inclusively to obtain integrated design of a cutting tool.

15.3. SINGLE POINT CUTTING TOOLS

The solid single point cutting tools have been discussed in detail in last chapter. Here, we shall discuss the tipped single point cutting tools.

Tipped Single Point Cutting Tools. As already discussed the carbides, ceramics, cast alloys, diamond, CBN and UCON are used as tips or inserts which are either brazed into a prepared seat machined on a tough steel tool shank or are clamped to the shank, Fig. 15.1. The second type of tips or inserts are known as indexable inserts or throwaway tips.

1. **Brazed Tipped Tools.** Here suitable shapes of tool material tips or inserts are brazed to a steel shank, Fig. 15.1 (*c*). When the tip or insert gets worn out, it is resharpened with the help of special grinding wheels. For resharpening purposes, the tool will have to be removed from the machine involving a resetting operation. The main drawback of a brazed tip is that because of difference in co-efficients of expansion of tip material and tool shank material, the brazing has to be done very carefully.

2. **Mechanically Clamped Tip Tools.** In these tools, the tips or inserts are clamped mechanically on to the tool shank, Fig. 15.1 (*d*). These tips are known as indexable because these have more than one cutting edges which are used one by one by indexing the tip and these tips are known as throwaway type because once all the edges of the tip have been used, the tip or insert is removed from the tool shank and thrown away or is disposal off (disposable tip). The most common shapes in which these tips are available are : Square, triangular and diamond shapes, Fig. 15.2. The edges of the inserts may be at 90° to the tip face, Fig. 15.2 (*a*) or the edges may be at a small angle to the face, Fig. 15.2 (*b*). In the first case, the tips will provide a negative rake angle because these will have to be clamped on to shank with the seating sloping downwards to provide a clearance angle. Here, the number of cutting points will be twice the number of edges, because when all the edges on the top face have been used, the insert can be turned over to give an additional equal number of cutting edges.

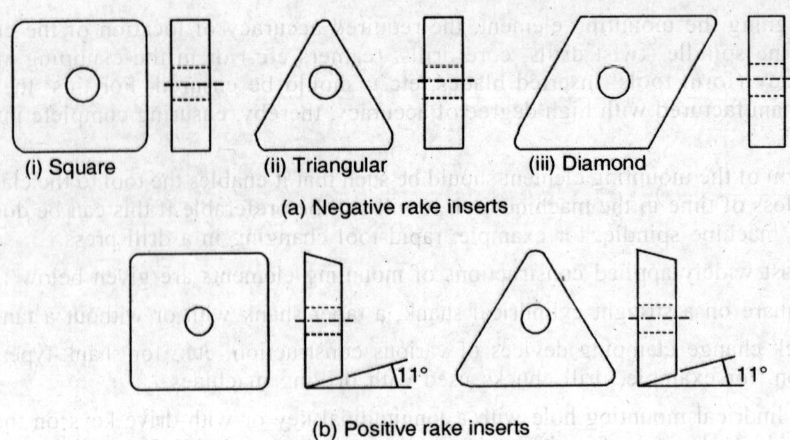

Fig. 15.2. Indexable inserts.

In the second case, Fig. 15.2 (*b*), positive rake is obtained on the tip. Here, the insert cannot be turned over to use the cutting edges on the bottom face, because the small angle provided on the sides of the tip will prevent this.

When a cutting edge on the tip gets worn, the clamp is released and the tip is rotated (indexed) to bring a new cutting edge into the cutting position. When all the edges have been used the tip is thrown away and a new tip is substituted.

The various methods for fastening the tip are : (*i*) screw fastening (*ii*) bridge type clamp (*iii*) pin type clamp.

Chip Breakers. During machining ductile materials, continuous type of chips are produced which are difficult to handle and a sharp, hot chip in motion is a hazard. To handle and dispose off the chips conveniently, the continuous chips are broken up into short segments. This is achieved with the help of chip breakages which give an extra bending to the highly work hardened chip to break it into small pieces. The various types of chip breakers are :

1. A groove may be ground into the top face of the tool after leaving a small land from the tip, Fig. 15.3 (*a*).

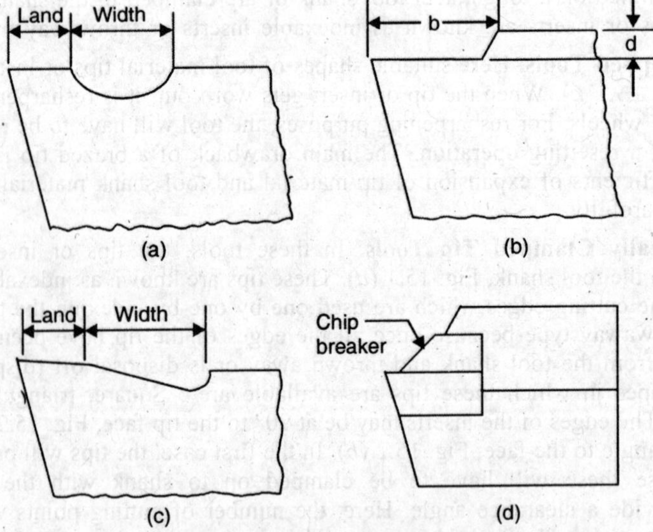

Fig. 15.3. Chip breakers.

2. A step may be ground into the tool, Fig. 15.3 (*b*).

3. By providing a secondary rake angle and chip breaker projection, Fig. 15.3 (*c*).

4. In the case of carbide tipped tools, a chip breaker groove is made all around the boundary or a separate plate or step may be clamped on top of the tool, Fig. 15.3 (*d*).

The typical dimensions of a step ground on a tool are given in table 15.1.

Table 15.1. Dimensions of a chip breaker

Depth of cut, mm	Feed, mm/rev					
	0.2		*0.35*		*0.55*	
	b, mm	*t, mm*	*b*	*t*	*b*	*t*
1	1.5	0.3	2.0	0.4	3.0	0.5
4	2.5	0.5	3.0	0.5	4.0	0.6
9	3.0	0.5	4.0	0.6	4.5	0.6

Dimensions of Tool Shank. The tool shank can be square, rectangular or circular in section. Rectangular section is commonly used since reduction in its strength is less as compared to other sections, when a seat is cut on it for an insert. The dimensions of the tool shank will depend upon the cutting force, overhang of the tool shank from the tool post and the material of the shank, Fig. 15.4. The tool acts as a cantilever. If *Fc* is the cutting force and l is the overhang of the tip of tool, then, bending moment on the tool shank is,

$$M = Fc \times l \qquad ...(15.1)$$

Moment of resistance

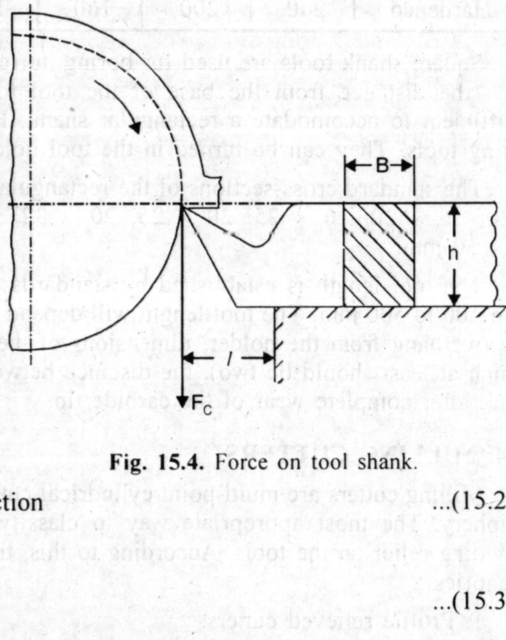

Fig. 15.4. Force on tool shank.

$$= \frac{Bh^2}{6} \times \sigma_b, \text{ for rectangular section} \qquad ...(15.2)$$

$$= \frac{\pi d^3}{32} \times \sigma_b \text{ for circular section.} \qquad ...(15.3)$$

Where *B* is the width and *h* is the height of rectangular tool shank and σ_b is the safe stress in bending for the material of tool shank. The values of *h/B* are taken as follows :

$$h/B = 1.25 \text{ for roughing operations} \qquad ...(15.4)$$

$$= 1.6 \text{ for semifinishing and finishing operations.}$$

The overhang *l* is usually kept as (1 to 1.5) × *h*.

With the help of the above relations, the dimensions of the tool shank can be found out.

The deflection of the tool point should also be checked. For a cantilever, it is given as,

$$\delta = \frac{4F_c l^3}{EBh^3} \qquad ...(15.5)$$

The permissible deflection of the tool may be taken as, 0.10 mm for rough turning

= 0.05 mm for finishing.

The value of Young's modulus of the tool shank material (Carbon structural steel) may be taken as,

$$E = 20 \times 10^4 \text{ to } 22 \times 10^4 \text{ N/mm}^2 \quad ...(15.6)$$

The analysis given above is only an approximate one, since only the plane bending of the tool shank is considered. Also, only the cutting force F_c is taken into consideration. Actually, the tool shank is subjected to combined stresses due to radial and axial forces acting on it in addition to the cutting force. The stresses are also influenced by plan approach angle and the consturction of the tool point. The values for safe bending stress for tool shanks of structural *C*-steel subjected to plane bending (with combined stresses taken into account) are given below :

Value for σ_b N/mm²

	Values of plan approach angle, λ					
Shank	30°	45°	60°	75°	90°	45° (bent shank)
Unhardened	120	100	80	65	55	130
Hardened	240	200	160	130	110	260

Square shank tools are used for boring, turret lathe and screw machine tools and the cases when the distance from the base of the tool to the line of centres of the machine tool is insufficient to accomodate a rectangular shank. Round shanks are used for boring and thread cutting tools. They can be turned in the tool holder to make adjustments.

The standard cross-sections of the rectangular tool shanks are, $B \times h$, 10 × 16, 12 × 16, 12 × 20, 16 × 20, 16 × 25, 20 × 25, 20 × 32, 25 × 32, 25 × 40, 32 × 40, 32 × 50 and 40 × 50 mm.

The tool length is established by standards and depends upon the cross-section. It varies from 100 to 500 mm. The tool length will depend upon the following factors : size of tool point, tool overhang from the holder, dimensions of the tool holder, the number of clamping screws (which at least should be two), the distance between these screws and further utilization of the shank after complete wear of the carbide tip.

15.4. MILLING CUTTERS

Milling cutters are multi-point cylindrical cutting tools with cutting teeth spaced around the periphery. The most appropriate way to classifying the milling cutters is on the method of providing relief on the tools. According to this, the milling cutters are classified into two main categories :

1. Profile relieved cutters.
2. Form relieved cutters.

The profile-relieved cutters are obtained by sharpening a narrow land behind the cutting edge. This narrow land is resharpened by grinding when the cutting edge become dull. Form relieved cutters have a curved relief behind the cutting edge and these cutters have a curved relief behind the cutting edge and these cutters are sharpened by grinding the tool face. There is greater flexibility in adjusting relief angles in profile-relieved cutters since it is fixed in the manufacture of the cutter. However, this type is more suitable for cutters with intricate shapes/ profile since the relief is not changed during resharpening.

The milling cutters can also be classified according to the method of their mounting, for example, arbor type, shank type 'or spindle mounted type'. Most large cutters are arbor type.

Most milling cutters are made as solid of H.S.S., but they are also available with carbide-tipped teeth or with disposable tips of various tool materials.

The milling process is divided into two main types :

1. Peripheral milling, and
2. Face milling.

In peripheral milling, the finished surface is parallel to the axis of the cutter and is machined by cutter teeth located on the periphery of the cutter, Fig. 15.5 (*a*). In face milling, the finished surface is at right angle to the cutter axis and it is obtained by teeth on the periphery and the flat end of the cutter, Fig. 15.5 (*b*).

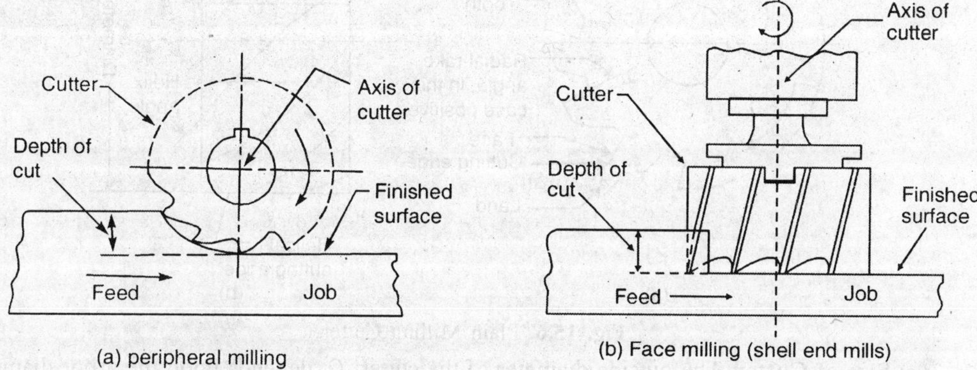

Fig. 15.5. Milling Processes.

Nomenclature : Refer to Fig. 15.6

1. **Body.** It is the part of the solid or tipped cutter exclusive of the teeth or shank.

2. **Arbor.** It is the shaft on which the arbor type cutters are mounted or driven.

3. **Shank.** It is a cylindrical or tapered extension along the axis of the cutter employed for holding and driving shank type cutters.

4. **Cutting edge.** It is the edge formed by the intersection of the face and the circular land or the surface left by the provision of primary clearance.

5. **Land.** The narrow surface of a cutter tooth immediately behind the cutting edge.

6. **Flute or Gash.** The chip and coolant space between the back of one tooth and the face of the following tooth.

7., **Face.** It is the portion of the flute adjacent to the cutting edge on which the cut chips impinge.

8. **Fillet.** It is the bottom surface of the flute.

9. **Helix angle.** The cutting edge angle which a helical cutting edge makes with a plane containing the axis of a cylindrical cutter.

10. **Diameter.** Cutter diameter or the outside diameter of the cutter is the diameter of the circle passing through the peripheral cutting edge.

11. **Relief.** The result of the removal of tool material behind or adjacent to the cutting edge to provide clearance and prevent rubbing.

12. **Primary relief.** It is the relief immediately behind the cutting edge.

13. **Secondary relief or clearance.** The additional space provided behind the relieved land (primary relief) of a cutter to eliminate undesirable contact between the cutter and the workpiece.

14. **Relief angle.** The angle formed between a relieved surface and a given plane tangent to a cutting edge or a point on a cutting edge.

15. **Radial rake angle.** It is the angle in a plane perpendicular to the axis of the cutter, between the face of the tooth and a radial line passing through the cutting ege.

Design Features. The design of a milling cutter will be illustrated by considering a plain milling cutter, Fig. 15.6. The main elements to be considered for design are : Size of cutter, tool angles, number of teeth, flutes, and material etc. These will be discussed below in turn :

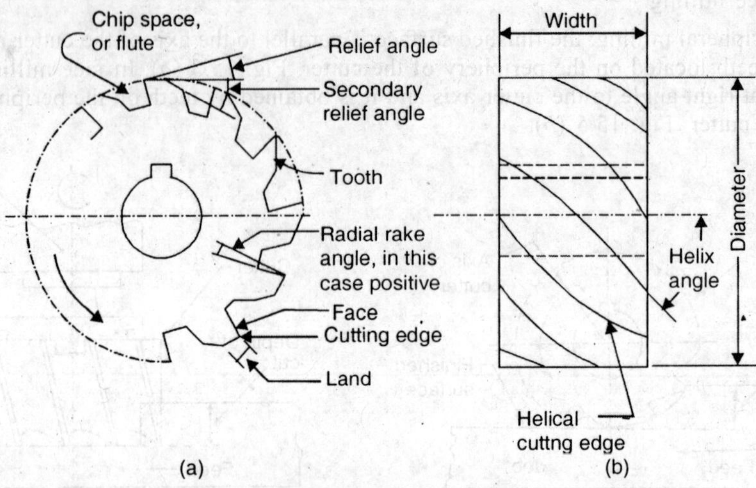

Fig. 15.6. Plain Milling Cutter.

(*a*) **Size of Cutter.** The outside diameter of the cutter, *D*, depends upon the arbor diameter, *d*, thickness of the cutter ring, *t*, and the height of the cutting tooth, *h*, or the depth of the flute. It is given as :

$$D = d + 2t + 2h \quad ...(15.7)$$

A larger cutter diameter will give more flute depth (chip space) resulting in better heat removal, longer tool life and higher cutting rates. However, a larger diameter cutter will require more torque and deflect more. It will also need more tool material, more machining for its manufacture and hence higher tool cost. A smaller cutter will require less torque and deflect less and will also be cheaper. Generally the cutter diameter, *D*, is taken to be about (2.5 to 3) times the arbor diameter. The face width of the cutter should be adequate so as to give sufficient support to the cutting edges.

The arbor is usually selected from the commercially available standard arbor sizes : 16, 22, 27, 32, 40, 50 and 60 mm.

Cutter hub diameter = d + 2 × thickness of body

= (1.5 to 2.5) d

Diameter of Plain milling cutters and the depth of cut are also inter-related as follows :

D = 60 to 90 mm for depth of cut upto 5 mm

= 90 to 110 mm for depth of cut upto 8 mm

= 110 to 150 mm for depth of cut upto 12 mm

Face width of milling cutter = width of workpiece + (2 to 5) mm.

Diameter of face milling cutter is selected depending on the width *B* of the surface to be milled :

D = (1.06 to 1.10) B for H.S.S. cutters

= (1.20 to 1.70) B for Carbide tipped cutters

(*b*) **Tool angles.** A plain milling cutter may have either straight or cylindrical teeth. The helix angle is taken as :

Helix angle = 20° to 30° for plain helical cutters.

= 10° to 15° for side and end mill cutters.

Regarding the tool angles, the same general principles apply for multi-point tools as for single point tools. A compromise has to be made between the efficient cutting action (greater tool angles) and tool strength (smaller tool angles).

Radial rake angle. The radial rake angle varies from 10° to 20°, larger values for cutting softer materials and smaller values for cutting harder materials. Carbide tipped cutters have negative rake angles, – 10° to – 15°.

Relief angles. Relief angle is provided to eliminate heel drag, *i.e.*, to reduce friction between the tooth land and the cutting surface. Relief angles should be small to ensure greater strength of the cutting edge and for better heat dissipation. Larger relief angles reduce the strength of cutting edge leading to its failure and also increase the tendency of chatter. The average relief angles are given in table 15.2.

Table 15.2 Average Relief Angles

Type of cutter	*Tool material*	*Work material*		
		Steel	*Cast iron*	*Non-ferrous and non-metallic*
1. Peripheral	H.S.S.	5–10°	5–10	7–12°
	Carbide	4–6°	4–6	5–10°
	Cast alloy	4–6°	4–6	5–10°
2. Side or end cutting edges	All	1–4°	1–4	2–7°

Larger values of relief angles are used for smaller-diameter cutters and vice-versa. It is clear from Table 15.2, that the relief on side or end cutting edges is much smaller as compared to peripheral teeth, because the problem of heel drag does not exist for these teeth. Usually, relief on side cutting edges $= \frac{1}{4}$ to $\frac{1}{2}$ times the relief on peripheral cutting edges.

Due to repeated sharpenings, the land width will go on increasing, leading to increased tendency of heel drag on the workpiece. To control the land width, a clearance angle (or secondary relief) is ground on the tooth. This secondary relief is approximately twice the relief angle.

(*c*) **Width of land.** To give strength to the cutting point, a narrow land is provided immediately behind the cutting edge. This land is ground to the relief angle. Its values are :

Width of land = 0.127 to 0.254 mm for small end mills

= upto 3.2 mm on large-diameter cutters

= 0.8 mm to 1.6 mm (average)

(*d*) **Number of teeth.** The number of teeth on a milling cutter will depend upon the work material and the surface finish required. For rough cuts, fewer teeth are required, whereas, for finer cuts (to get improved surface finish) greater number of teeth are needed. However, if the number of teeth in contact with the work-material at the same time are large, then the tool or machine will get overloaded. In general, at least, one tooth must be in contact with the work material at all time. Again, if the number of teeth is large, less flute space is available for chips which may lead to jamming of chips and breakage of tool. More chip space is needed (less number of teeth) if the chips are continuous (when milling soft materials). On the other hand, if the chips are well broken, then, less chip space is needed and more number of teeth can be provided resulting in better surface finish, as the feed per tooth can be reduced.

The number of teeth in a milling cutter is given as :

$$n = \frac{f}{f_t \times N} \qquad \text{...(15.8)}$$

where f = feed rate, mm/min

f_t = feed rate per tooth, mm

N = cutter speed, rpm

For H.S.S. plain milling cutters, f_t = 0.05 to 0.6 mm/tooth for milling steel

= 0.1 to 0.8 mm/tooth for milling C.I.

The metal removal rate, mm³/min, is given as,

Metal removal rate, $MRR = w \times h \times f$, mm³/min

where w = width of cut, mm

h = depth of cut, mm

= 3 to 8 mm for roughing

= 0.5 to 1.5 mm for finishing

To prevent overloading the machine motor, the various variables are connected as given below :

$$n = \frac{K \times hp_c}{f_t Nhw} \qquad \text{... (15.9)}$$

where hp_c = horsepower available at cutter

K = machineability factor, mm³/min/hp_c.

The values for '*K*' are given in Table 15.3, for carbide cutters and for feed of 0.250 mm per tooth.

Table 15.3. Values of K

Work material	*cm³/min/hp$_c$*
Aluminium and Magnesium	41 to 65.6
Bronze and Brass, soft	27.85 to 41
Bronze and Brass, medium	16.4 to 23.0
Bronze and Brass, hard	9.85 to 16.4
Cast Iron, soft	24.6
Cast Iron, medium	13 to 16.4
Cast Iron, hard	9.85 to 13
Malleable Iron	14.75
Cold drawn steel	16.4
Forged and Alloy Steel	10 to 14.25
Alloy steel, 300-400 BHN	8.2
Stainless Steel, free machining	18
Stainless Steel, austenitic free machining	13.6
Stainless Steel, austenitic	11.8
Monel metal	9.0
Copper annealed	13.8
Tool-Steel	8.3
Nickel	8.6
Titanium	12.3

According to the cutter diameter, the number of cutter teeth may be taken as given below :

$$n = C\sqrt{D} \quad \text{... (15.10)}$$

For solid cutters :

$$C = 2 \text{ to } 2.8 \text{ for fine tooth cutter}$$
$$= 0.6 \text{ to } 1.05 \text{ for coarse tooth cutter}$$

Solid end cutters :

$$C = 1.2 \text{ for large teeth}$$
$$= 2.0 \text{ for fine teeth}$$
$$= 2.5 \text{ to } 2.8 \text{ for Angle milling cutters}$$
$$= 1.5 \text{ to } 2 \text{ for Form milling cutters}$$
$$= 2 \text{ for Disk milling cutters}$$

For inserted blade cutters,

$$n = 0.04\, D, \text{ for } D \leq 200 \text{ mm}$$
$$= 0.04\, D + 2, \text{ for } D > 200 \text{ mm}$$
$$= 0.10\, D \text{ for cutting C.I.}$$

Another relations is : In eqn. (15.10)

$$C = 0.9 \text{ with helix angle of } 20°$$
$$= 0.8 \text{ with helix angle of } 45°$$

(*e*) **Power Requirements for Milling.** The total horsepower required at the cutter can be found as,

$$hp_c = \frac{\text{Metal removal rate, cm}^3/\text{min}}{\text{K}} \quad \text{...(15.11)}$$

K is taken from Table 15.3.

(*f*) **Flutes.** The flutes of a milling cutter may be straight, helical or angular. Helical flutes are most common, since the entire cutting edge does not come into contact with work material at one time. The helix can either be right handed or left handed.

Example 15.1. *The feed of an 8-tooth face mill is 0.033 cm per tooth at 200 rev/.min. The material cut is 300 BHN steel. Depth of cut is 0.32 cm and the width is 10 cm. Calculate the (a) horsepower at the cutter (b) horsepower at the motor if the efficiency of the machine is 60 per cent.*

Solution. From equation (15.11), the horsepower at the cutter, is

$$hp_c = \frac{\text{Metal removal rate, cm}^3/\text{min}}{\text{K}}$$

Now, Metal removal rate $= w \times h \times f$

where w = width of cut = 10 cm

h = depth of cut = 0.32 cm

f = feed rate, cm/min

$= n f_t N$

where n = number of teeth in the cutter

$= 8$

$$f_t = \text{feed rate per tooth}$$
$$= 0.033 \text{ cm}$$
$$N = \text{Cutter speed, rpm} = 200$$

$\therefore$ $$\text{Metal removal rate} = w\,h\,n\,f_t N \text{ cm}^3/\text{min}$$
$$= 10 \times 0.32 \times 8 \times 0.033 \times 220$$
$$= 170 \text{ cm}^3/\text{min}$$

Now from Table 15.3, $K = 8.2$

(*a*) $\therefore$ $$hp_c = \frac{170}{8.2} = 20.73$$

(*b*) The horsepower at the motor,

$$hp_m = \frac{hp_c}{\eta} = \frac{20.73}{0.6} = 34.55$$

15.5. BROACH DESIGN

A broach is a multi-point cutting tool consisting of a bar having a surface containing a series of cutting teeth or edges which gradually increase in size from the starting or entering end to the rear end. Broaches are used for machining either internal or external surfaces (Sizing of holes and cutting of serrations, straight or helical splines, gun rifling and keyways). The surfaces produced may be flat, circular or of any intricate shape. In broaching, the broach is pushed or pulled over or through a surface of a workpiece. Each tooth of the tool takes a thin slice from the surface. Broaching of inside surface is called 'Internal or hole broaching' and outside surfaces as "Surface broaching".

(*a*) **Details of an internal broach (Hole broach).** A typical internal broach is shown in Fig. 15.7. To machine an internal hole, the broach is gripped by a puller at the shank end. The

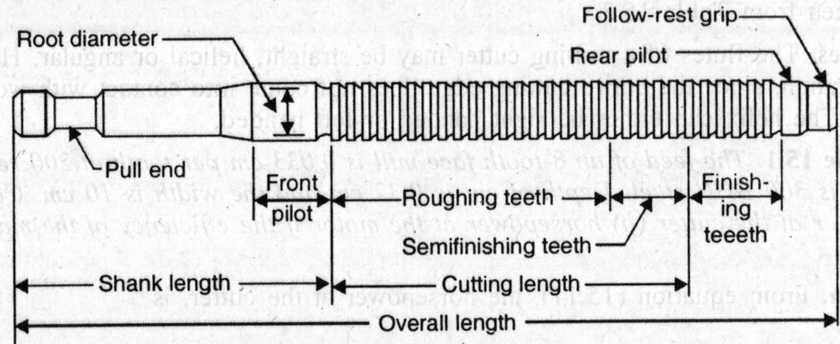

Fig. 15.7. A Typical Internal Broach.

front pilot centres the broach in the hole before the teeth begin to cut. The front taper (5 to 20 mm) facilitates the insertion of the front pilot in the hole. The first set of teeth behind the front pilot, removes most of the material and are called "roughing teeth". These are followed by a few teeth called "semi-finishing teeth" where the depth of cut of individual tooth is quite small. Finally, there are finishing or sizing teeth which are all of the same size and have the shape of the finished hole. Sometimes, a few burnishing teeth may be provided after the finishing teeth. These have no cutting edges but are button shapes and from 0.025 to 0.075 mm larger than the size of the hole. The resulting rubbing action smooths and sizes the hole. They are used primarily on cast iron and non-ferrous metals. The "rear pilot" supports the broach after the last tooth leaves the hole.

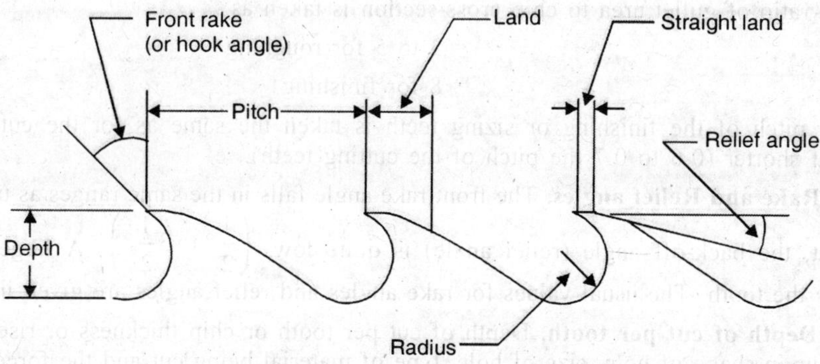

Fig. 15.8. Tooth Shape of a Broach.

(*b*) **Tooth elements.** The tooth shape of a typical broach is shown in Fig. 15.8. The front rake angle (face angle or hook angle) refers to rake angle of a single point cutting tool and the back off angle (relief angle) is provided to prevent rubbing of tool with workpiece.

(*c*) **Material.** High speed steel (H.S.S.) is by far the most widely used material for the broaches. Brazed carbides or disposable inserts are sometimes used for the cutting edges when machining cast iron parts which require close tolerances and production rates. Carbide tools are also used to an advantage on steel casting to offset the damaging effect of local hard spots.

(*d*) **Construction.** A broach may be either solid or assembled or built up from shells, replaceable sections or inserted teeth. Replaceable sections, teeth or shells make a broach easier to repair.

(*e*) **Broaching Allowance.** Or the stock left for broaching is defined as the total thickness of the metal to be removed by broaching. For example, for round broaches, it is the difference between the maximum permissible diameter of the broached hole and the diameter of the hole previous to broaching. The nominal allowance for round holes, machined by drilling or core drilling previous to broaching is,

$$A_b = 0.005\,D + (0.1 \text{ to } 0.2)\sqrt{L}$$

where D = Basic diameter of hole, mm

L = Length of hole to be broached, mm

Broach design is based on the allowance. Formulas are given in handbooks to determine the allowance for other types of broaches, depending upon the shape of the hole. The allowance should also include the manufacturing allowance of the hole.

(*f*) **Pitch of Teeth.** The pitch of a broach is decided in such a manner that any time at least two (preferably three) teeth are cutting in order to keep the cutting operation smooth. The other factors affecting the pitch of the teeth are : depth of cut or chip thickness, length of cut, cutting force required and power of the broaching machine. A good approximation for pitch, P, is :

$$P = (1.25 \text{ to } 1.50)\sqrt{L}, \text{ for plain broaches} \quad ...(15.12)$$

$$= (1.45 \text{ to } 1.90)\sqrt{L} \text{ for progressive broaches.}$$

where L is the length of hole or length being cut. The minimum permissible value for 'P' should be less than 5 mm unless a smaller pitch required for very short cuts to provide at least two teeth in contact simultaneously with the part machined. A smaller value than 5 mm is seldom required in surface broaching,

For surface broaching, the pitch is given as,

$$P = 3\sqrt{\text{length of cut} \times \text{rise per teeth} \times \text{ratio of gullet area to chip cross-section}}$$

The ratio of gullet area to chip cross-section is taken as

$$= 3 \text{ to } 5 \text{ for roughing}$$

$$= 8 \text{ for finishing}$$

The pitch of the finishing or sizing teeth is taken the same as for the cutting teeth or somewhat shorter (0.6 to 0.7 the pitch of the cutting teeth).

(*g*) **Rake and Relief angles.** The front rake angle falls in the same ranges as used for other tools, but, the back-off angle (relief angle) is quite low, $\left(\frac{1}{2}° \text{ to } 3\frac{1}{2}°\right)$. A large relief angle weakness the tooth. The usual values for rake angles and relief angles are given in Table 15.4.

(*h*) **Depth of cut per tooth.** Depth of cut per tooth or chip thickness or rise per tooth(s) depends upon shape of hole, size of hole, type of material being cut and the force available at the machine. It is generally very small, of the order of 0.025 mm for finishing to 0.15 mm for roughing. Its values are given in Table 15.4. In progressive broaching, the cut per tooth can be made more and may be between 0.1 to 0.35 mm or even more.

(*i*) **Width of land.** An average land is 0.12 mm on the roughing teeth and 0.25 to 0.75 mm gradually increasing through the finishing teeth. It is selected as a compromise between tooth strength and chip space. In general, it is taken as,

$$\omega = \left(\frac{1}{4} \text{ to } \frac{1}{3}\right) \times P$$

Land at zero clearance, that is, parallel to the broach axis is taken as 0.13 mm to 0.50 mm.

(*j*) **Depth of cutting teeth.** The size of the depth of cutting teeth (tooth gullet) is directly related to tooth size of pitch. The usual value of the tooth depth is given below :

$$h = (0.35 \text{ to } 0.40) \times P$$

The depth of the tooth gullet can also be found in the following manner :

When the job is getting broached, the metal chips get collected in the gullet space. It is highly desirable that the chips don't get compressed and thereby jam up the entire gullet space. Now, the effective cross-sectional area of the gullet space can be taken as,

$$A_x = \frac{\pi}{4} h^2$$

Table 15.4

Material to be broached	*Depth of cut per tooth, mm*		*Face or Hook angle*	*Relief angle*	
	Roughing	*Finishing*		*Roughing*	*Finishing*
Steel (High Strength)	0.04–0.05	0.01	10–12	1.5–3	0.5–1
Steel (Medium Strength)	0.06–0.10	0.01	14–18	1.5–3	0.5–1
Cast Steel	0.06–0.10	0.01	10	1.5–3	0.5
Malleable Cast Iron	0.06–0.10	0.01	7	1.5–3	0.5
Soft Cast Iron	0.15–0.25	0.01	10–15	1.5–3	0.5

Hard Cast Iron	0.07–0.10	0.01	5	1.5–3	0.5
Zinc die castings	0.10–0.25	0.025	12	5	2
Cast Bronze	0.25–0.60	0.01	8	0	0
Wrought Aluminium	0.10–0.25	0.025	15	3	1
Cast Aluminium alloys	0.10–0.25	0.025	12	3	1
Magnesium diecastings	0.25–0.40	0.025	20	3	1

Now to allow for clearance between the coils of continuous chips and for free formation of the discontinuous chips, the effective gullet section is taken to be from 3 to 6 times the actual chip section. Thus, if the uncut chip thickness is 't', and the job length is L, the chip section area A_c is,

$$A_c = L.t$$

$\therefore$ $A_g = K A_c$, where K is 3 to 6, is the volume factor. Its value depends upon chip thickness and material.

$\therefore$ Depth of the gullet will be gives as,

$$h = \sqrt{\frac{4kLt}{\pi}}$$

Smaller values of K are taken for broaching brittle material where a discontinuous chip is formed.

(*k*) **Tooth fillet radius.** It is provided between the teeth to strengthen the broach tooth. It also helps in the curling of the chips. The various empirical relations to calculate teeth fillet radius r, are :

$$r = (0.20 \text{ to } 0.35) \times P$$
$$= (0.40 \text{ to } 0.60) \times h$$
$$= \frac{1}{3}(\text{land width}) + \frac{1}{2}(\text{tooth depth}) + \frac{1}{4}(\text{Pitch.})$$

(*l*) **Chip Breakers.** Rounded chip breaking grooves are provided at interval along the cutting edges to break up the wide curling chips and preventing them from clogging the chip space, thus reducing the cutting pressure and strain on the broach. These grooves are provided only on roughing teeth. The more ductile the material the wide the chip breaker grooves should be, the smaller the distance between them.

The number of the chip beakers for round broaches can be selected from the following table :

Broach diameter, mm	10 to 13	13 to 16	16 to 20	20 to 25
Number of chip breakers	6	8	10	12

The number of chip breakers for keyway, and slab and contour surface broaches and their dimensions depend upon the concrete conditions in each case. These data can be found in various handbooks on cutting tools.

(*m*) **Total length of Broach.** The total length of a broach is given as,

$$L_b = \text{Length of toothed portion} + \text{length of shank} + \text{length of rear pilot}$$

Now the length of toothed portion will depend upon the number of cutting teeth (roughing teeth plus semi-finishing teeth), the number of finishing teeth and the pitch of the teeth.

Now, the number of cutting teeth can be calculated as,

$$n_c = \frac{T}{s} + (2 \text{ to } 4)$$

where T = metal thickness to be removed

$$= \frac{A_b}{2}, \text{ for round broaches.}$$

The number of finishing teeth varies from 3 to 8, the larger number being for broaches that are to machine more accurate holes. The number of semi-finishing teeth between the roughing teeth and the finishing teeth is usually 3 or 4. The idea of having a large number of finishing teeth is that they are ground many times during the life of a broach and the reserve of finishing teeth provides the last cutting tooth of the same size as the desired dimension of the hole.

Thus, the total number of teeth is,

$$n_t = n_c + (3 \text{ to } 8)$$

Length of toothed portion, $$l = n_t \times P = \left(P \times \frac{T}{s}\right) + (5 \text{ to } 12) \times P$$

(*n*) **Cutting Speeds.** Cutting speeds in broaching process are low and seldom exceed 15 mpm. For steel castings and forgings, the range is 6 to 10 mpm and for C.I., brass and aluminium it can be upto 12 mpm.

(*o*) **Load on Broach.** The load on the broach or the force needed for broaching may be found by the simple relation.

F = Area of metal removed by the teeth in contact with the work × shear strength of material being cut

$= nA\tau_s$

where n = number of teeth cutting at a time

A = cross-sectional area of cut.

The above relation does not consider the fact that the resistance to broaching varies with rise per tooth, being higher for lower values of rise per tooth. Considering this fact, the relation becomes,

$$F = n\,A\,C$$

where C = Specific cutting force, that is, force to remove 1 mm^2 of metal at a given size per tooth.

Considering the blunt broach factor, the expression finally becomes,

$$F = n\,A\,C\,B$$

where B = Blunt broach factor (1.25 to 1.40)

For round broaches, $F = n \times \pi D s \times C \times B$

For surface broaching, $F = n \times s \times b \times C \times B$

For Spline broaching, $F = n \times z \times s \times w \times C \times B$

where D = finish diameter of hole

b = width of contact of each tooth

z = number of splines

w = width of splines

$$\text{Power needed for broaching} = \frac{FV}{1000}, \text{kW}$$

Where F is in newtons and V is the cutting speed is m/s.

The values of C are given in Table 15.5.

$$\text{MRR per pass} = nAV, \text{mm}^3/\text{s}$$

A is in mm^2 and V in mm/s.

Table 15.5. Values of *C*.

Material to be	*Rise per tooth, s, mm*					
	0.03	0.04	0.05	0.06	0.08	0.10
	C, N/mm²					
Mild Steel	42.5	36.5	33.5	31.0	28.0	26.0
Alloy Steel	58.0	49.0	45.0	42.5	38.5	36.0
Grey C.I.	38.0	33.0	30.0	27.0	25.0	23.5
Malleable C.I.	34.0	29.0	27.0	25.0	22.5	21.0

Example 15.2. *The bore of an alloy steel component prior to broaching is* $32.25^{+0.05}_{-0.00}$ *mm. The bore is to be finish broached to* $32.75^{+0.01}_{-0.00}$ *mm. diameter. If the length of bore is 35 mm and the cutting speed is 0.15 m/s, determine the broaching power for broaching and design the broach.*

Solution. (*a*) The relation to determine the broaching force for round broaches is,

$$F = n \times \pi Ds \times C \times B$$

$$n = 2 \text{ or } 3, \text{ say } 3$$

$$D = \text{finish diameter of hole}$$

$$= 32.75 + 0.01 = 32.76 \text{ mm}$$

Let $s = 0.05$ mm

$$B = 1.30$$

From Table 15.5, $C = 45 \text{ N/mm}^2$

$$F = 3 \times \pi \times 32.76 \times 0.05 \times 45 \times 1.30$$

$$= 903.23 \text{ N}$$

$$\text{Broaching Power} = \frac{FV}{1000}, \text{kW}$$

$$= \frac{903.23 \times 0.15}{1000} = 0.1355 \text{ kW}$$

(*b*) **Broach Design :**

Let, the pitch $P = 1.75\sqrt{\text{hole length}}$

$$= 1.75\sqrt{35}$$

$$= 10.35 \text{ mm, say } 10.50 \text{ mm.}$$

This gives a minimum of two teeth cutting at any one time, so, $P = 10.50$ mm is satisfactory.

(*b*) Let Face angle = 10°

Relief angle = 1.5° for roughing, and

= 1.0° for finishing

Depth of cut per tooth, $s = 0.05$ mm

Width of land, $w = 0.3 \times P = 3.15$ mm

Depth of cutting teeth, $h = 0.4 \times P = 4.20$ mm

Tooth fillet radius, $r = 0.3 \times P = 3.15$ mm

(*c*) Length of toothed portion of broach, l, is

$$l = \left(P \times \frac{T}{s}\right) + (5 \text{ to } 12) \times P$$

$$T = \frac{32.76 - 32.30}{2} = 0.23 \text{ mm}$$

$$\therefore \quad \text{No. of cutting teeth} = \frac{T}{s} = \frac{0.23}{0.05} = 5, \text{ say}$$

$$\therefore \quad l = (5+7) \times P = 12 \times 10.5 = \mathbf{126\ mm.}$$

15.6. DRILLS

Drilling is the process of cutting or originating a round hole from the solid material. The tool (drill) and not the workpiece is revolved and is fed into the material along its axis. There are many ways of classifying drills, for example, according to : material, number and type of flutes, drill size, type of shank (straight or taper) and cutting point geometry, etc. However the most common type

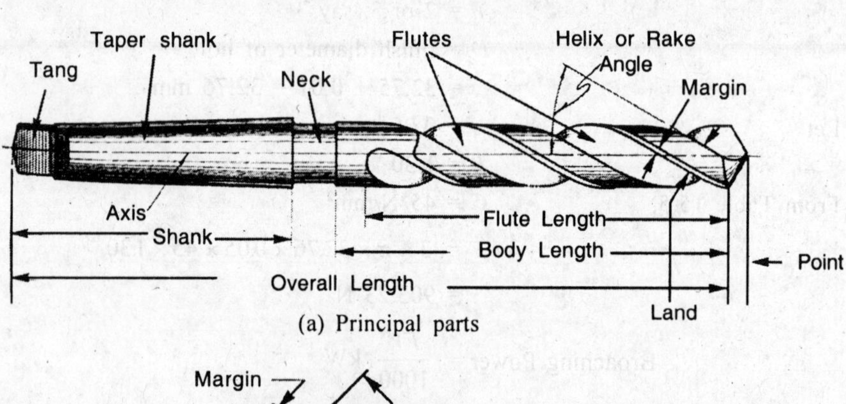

(a) Principal parts

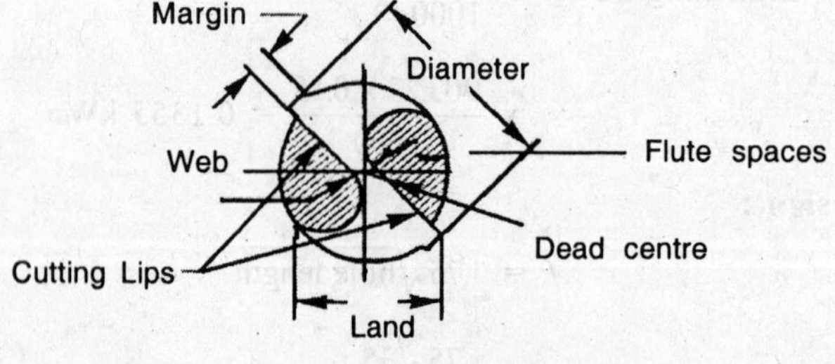

(*b*) The point

Fig. 15.9. A Twist drill

of drill is the fluted twist drill, Fig. 15.9. It is made from a round bar of tool material, and has three principal parts : the point, the body and the shank. The drill is held and rotated by its shank. The point comprises the cutting elements while the body guides the drill in operation. The body of the drill has two helical grooves called "flutes" cut into its surface. The flutes form the cutting surface and also assist in removing chips out of the drilled hole. The two cutting edges are straight and are separated by web thickness of the drill which is provided to strengthen the drill structure. The body of the drill is made slightly less in circumference leaving a narrow "margin" at full nominal diameter along the edge of each flute. This reduces rubbing action between the drill and the hole wall and allows the cutting fluid to reach the point of the drill. The metal cut away to form the margin is known as "body diameter clearance". To help further in reducing the rubbing action, drill bodies are given a slight back taper (About 0.0075 mm per cm of length). The shank can be either straight or tapered (for big drills). Straight shanks are provided for small drills (upto 10 to 12 mm diameter) and are held in chucks.

Nomenclature : See Fig. 15.9

1. **Axis.** It is the imaginary longitudinal centre line of the drill.

2. **Body.** It is the portion of the drill extending from the outer corners of the cutting lips upto the commencement of neck (if present) or shank.

3. **Back taper.** A slight decrease in diameter of the drill from the front end to the back in the body of the drill.

4. **Flutes.** Straight or helical grooves cut or formed in the body of the drill to provide cutting edges, to allow chip removal, and to allow cutting fluid to reach the cutting edges. Regarding the direction of the helical flutes, the twist drills are divided into R.H. cutting (the flute goes up from L to R, the drill cuts when rotated C.C.W), and L.H. cutting (the flute goes up from R to L, the drill cuts when rotated C.W). L.H. cutting drills find limited use.

5. **Land.** The peripheral portion of the drill body between adjacent flutes.

6. **Body clearance.** It is the space provided to eliminate undesirable contact between the drill and the workpiece.

7. **Margin.** It is the cylindrical portion of the land which is not cut away to provide body clearance. It is ground to the diameter of drill. There are 2 margins. Drill guidance and friction losses in drilling depend on the margins.

8. **Drill diameter.** It is the diameter of the drill over the margins measured at the point.

9. **Clearance diameter.** It is the diameter over the cutaway portion of the drill lands.

10. **Web.** It is the central portion of the drill body that connects the lands. The extreme ends of the web forms the chiesel edge on the two-flute drill.

11. **Point.** It is the cutting end of a drill formed by the ends of the lands and the web.

12. **Lips or cutting edges.** These are the cutting edges of a two-flute drill extending from the chiesel edge to the periphery.

13. **Chisel edge.** It is the edge at the end of the web that connects the cutting edges. Its length depends on the drill diameter, being on average 0.13 × drill diameter.

14. **Shank.** The part of the drill by which it is held and rotated.

15. **Tang.** The flattened end of a taper shank which fits a driving slot in a socket.

16. **Lip relief.** It is the axial relief on the drill point.

17. **Lip relief angle.** It is the axial relief angle at the outer corner of the lip. It is the angle formed by the flank and a plane at right angles to drill axis.

18. **Face.** It is the portion of the flute surface adjacent to the lip on which the cut chips impinge.

19. **Flank.** It is the surface on a drill point which extends behind the lip to the following flute.

20. **Heel.** The edges formed by the intersection of flute surface and body clearance. It is the trailing edge of the land.

21. **Point angle.** It is the included angle of the cone formed by the cutting lips.

22. **Helix angle.** It is the angle made by the leading edge of the land with a plane containing the axis of the drill.

23. **Chiesel edge angle.** It is the angle included between the chiesel edge and the cutting lip as viewed from the end of the drill.

24. **Web thickness.** Thickness of web at the point, unless another location is indicated.

25. **Neck.** It is a section of reduced diameter between the body and shank. It is provided to allow the overrun of the grinding wheel in drill manufacture. It carries the marking of the drill.

Design Features

(*a*) **Materials.** Carbon tool steels drills have a low first cost, but these should be used occasionally and at slow speeds. High speed steel drills are the most popular and have good strength. Drills tipped with cemented carbide are economical for high production but are expensive and must be handled carefully to avoid breakage. These are used mainly for drilling of malleable iron castings and non-ferrous metals and alloys such as copper, brass, aluminium, magnesium, zinc and also plastics, hard rubber etc. These are not used for steel components as there is likelihood of their breakage due to high tip pressure.

(*b*) **Speeds and feeds.** These are given in tables 4.2 and 4.3.

(*c*) **Drill sizes.** Standard drills are available in four size series, the size indicating the diameter of the drill body :

(*i*) **Fractional size.** Size range is $\frac{1''}{64}$ to $\frac{1''}{4}$ with increments of $\frac{1''}{64}$.

(*ii*) **Millimeter size.** Size range is 0.5 to 10 mm with increments 0.1 mm.

(*iii*) **Numbered size.** (80 to 10). Size range is 0.0135″ to 0.228″ with very slight increments.

(*iv*) **Lettered size (*A* to *Z*).** Size range is 0.234″ to 0.413″ with very slight increments.

The diameter of the drill should always be slightly smaller than that of the hole it is to drill, since drills always cut oversize.

(*d*) **Lip relief (clearance) angle.** The heel of the drill point is backed off when ground to give relief behind the cutting lips. This will allow the cutting edges to cut without interference. This is equivalent to end relief angle of a single point cutting tool. It is kept 12° to 15°.

(*e*) **Point angle.** The point angle is selected to suit the hardness and brittleness of the material being drilled. It is 116° to 118° for medium hard steel and cast iron, 125° for hardened steel and 130° to 140° for brass and bronze. It is only 60° for wood and fiber. This angle (half) refers to side cutting edge angle of a single point tool.

(*f*) **Helix angle.** This angle is equivalent to back rake angle of a single point cutting tool. It is 24° to 30° for most drills.

(*g*) **Web thickness.** It is an important element of twist drills. If the web is too thin, the drill will not be rigid enough to withstand a high drilling torque. But, with a thinner web, the axial thrust is reduced and the drilling is easier since the chiesel edge becomes shorter. Following are the recommended values for web thickness :

1. **For carbon steel and H.S.S. drills :**

$$\text{Web thickness} = (0.2 \text{ to } 0.25)\,D \text{ for } D = 6 \text{ to } 10 \text{ mm.}$$
$$= (0.13 \text{ to } 0.16)\,D \text{ for } D > 10 \text{ mm.}$$

2. **Carbide-tipped drills :** For these drills, the webs are relatively thicker because the drill body is weakened by the slot for the tip.

$$\text{Web thickness} = (0.27 \text{ to } 0.30)\,D \text{ for } D \text{ upto } 10 \text{ mm.}$$
$$= (0.2 \text{ to } 0.26)\,D \text{ for } D > 10 \text{ mm.}$$

3. **Twist drills with milled flutes.** The web thickness increases by 1.4 to 1.8 mm per 100 mm towards the shank. It increases the strength and rigidity of the drill.

(*h*) **Chisel angle.** This is the angle which the raised line at the dead centre makes with the cutting edge. It is 120° to 135°.

Recommended values of point angle, helix angle and relief angle are given in Table 15.6.

(*i*) **Land width.** Land width affects the strength and rigidity of the drill body. It is usually taken to be equal to the flute width if drill diameter is more than 20 mm. For smaller diameter drills, the land width is made larger than the flute width. As a guide, the following values for land width should be taken :

Land width = 0.62 × drill diameter, for drill diameter 3 to 8 mm
= 0.59 × drill diameter, for drill diameter 8 to 20 mm
= 0.58 × drill diameter, for drill diameter > 20 mm.

In drawings, the land width is indicated perpendicular to a helical flute.

Table 15.6

Material to be drilled	*Point angle*	*Helix angle*	*Lip relief angle*
Steel (Tensile strength up to 700 MPa)	116–118	30	12–15
Steel (Tensile strength 700 to 1000 MPa)	120	25	10–12
Steel (Tensile strength 1000 to 1400 MPa)	125	20	8–10
Stainless steel	120	25	8–12
Hard Manganese steel	130	10	8–10
Cast Iron	118–125	18–30	8–12
Copper	125	35–45	12–15
Bronze	118–125	8–12	12–15
Aluminium alloys	130–140	35–45	12–15
Moulded soft plastics	140	30–40	12–15
Moulded hard plastics	80	12	8–10
Thin moulded hard rubber	80	12	8–10
Marble, Slate	80	12	8–10

(*j*) **Margin.** Margin is usually kept as 0.06 to 0.07 times the drill diameter. Its height is from 0.03 to 0.02 of the diameter.

(*k*) **Back taper.** As mentioned above, the back taper is usually kept as 0.0075 mm per cm of drill body. The usual values are given below :

Back taper (mm per 100 mm length of drill)
= 0.03 to 0.07 for drill diameter 1 to 6 mm
= 0.04 to 0.08 for drill diameter 6 to 18 mm
= 0.05 to 0.10 for drill diameter > 18 mm.

(*k*) **Torque and thrust.** The geometrical analysis of a drill is very complex, because, the inclination angle, the normal rake angle and the effective rake, vary radically as we go from the inner most part of the cutting edge to the outer part. The point of the drill may be considered as consisting of two parts : the cutting edges and the chisel edge. The chisel edge does not cut in the usual sense, but rather displaces the metal sideways as if we were performing the hardness test. The total torque and thrust can thus be divided into two components : (*i*) due to cutting edges (*ii*) due to chisel edge. The contribution of chisel edge to total torque is only small but to total thrust is considerable. The cutting process in drilling is illustrated in Fig. 15.10. The depth of cut '*d*' in drilling from the solid metal is one half the drill diameter. The feed '*f*' is the movement of the drill along its axis in mm per revolution.

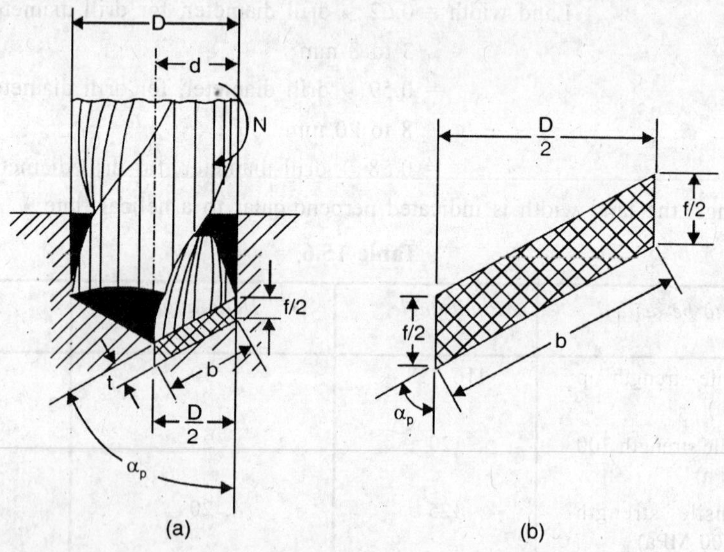

Fig. 15.10. Cutting Process in Drilling.

The chip thickness, $t = \left(\dfrac{f}{2}\right) \cdot \sin \alpha_p$ and, the width of cut, b is

$$b = \frac{D}{2 \sin \alpha_p}, \; D = \text{drill diameter}$$

where $2\alpha_p$ is the point angle of the drill.

Fig 15.11 shows the torque force acting on the cutting edges.

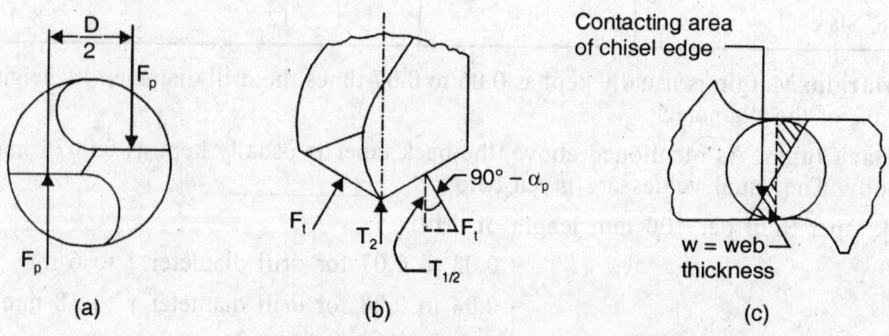

Fig. 15.11. Torque and Thrust in Drilling.

Thus, $$F_p = \sigma_c \times \text{chip ross-section} = \sigma_c \cdot \left(\frac{f}{2}\right)\left(\frac{D}{2}\right)$$

$$= \sigma_c \times b \times t$$

where σ_c is the contact stress, equal to Brinell hardness H_B. The moment due to two forces, F_p separated by half the drill diameter is

$$M = \frac{\sigma_c . D^2 f}{8} = \frac{H_B . D^2 f}{8} \quad ...(15.13)$$

The thrust force T_1 due to cutting edges can be estimated if the mean rake angle is known. However, the effective rake angle goes from a positive value near the outer radius to a negative value near the chisel edge.

Taking $$\frac{F_t}{F_p} = 0.05 \text{ to } 1.0,$$

$$T_1 = (0.5 \text{ to } 1.0)\, 2 \sin \alpha_p \left(\frac{D}{2}\right)\left(\frac{f}{2}\right) \cdot \sigma_c$$

For $$2\alpha_p = 118°$$

$$T_1 = (0.21 \text{ to } 0.42)\ \sigma_c Df \quad ...(15.14)$$

or From (15.13) and (15.14), $$T_1 \approx (1.7 \text{ to } 3.5) \cdot \frac{M}{D} \quad ...(15.15)$$

It is difficult to estimate T_2 (thrust due to chisel edge) as it is very difficult to ascertain the area of contact between the chisel edge and metal. To a very crude approximation, the shaded contact area [Fig. 15.11 (*c*)] is taken as 10 to 20% of the area of a circle inscribed within the web. The web thickness, *w*, is

$$w = 0.2\ D \text{ for } D < 3.2 \text{ mm}$$
$$= 0.1\ D \text{ for } D > 25.4 \text{ mm}$$

As mentioned above, the cutting action of chisel point is very similar to a hardness test, so,

$$T_2 = (0.1 \text{ to } 0.2) \cdot \frac{\pi}{4} w^2 \cdot H_B$$

$$\therefore \quad T = (0.7 \text{ to } 3.5) \cdot \frac{M}{D} + (0.1 \text{ to } 0.2) \cdot \frac{\pi}{4} w^2 . H_B \quad ...(15.16)$$

Power required for drilling by a two-flute drill can be calculated from the following empirical formula :

$$\text{Drilling power, kW} = \frac{1.25\, D^2 . K . N . (0.056 + 1.5 f)}{10^5} \quad ...(15.17)$$

where *K* is the material factor (Table 15.7), and

$$N = \text{r.p.m.}$$

$$\text{Drilling thrust, } kgf = 1.16\ \text{K.D.}\ (100 f)^{0.85} \quad ...(15.18)$$

$$MRR = \frac{\pi D^2}{4} \times f \times N, \text{mm}^3/\text{min.}$$

Table 15.7. Material Factor, *K*

Material to be drilled	*K*
0.2% carbon steel	1.07
0.13% free cutting carbon steel	1.03
0.35% carbon steel	1.15
1.25% Nickel chromium steel	1.56
Chromium molybdenum steel	1.56
2% Nickel molybdenum steel	1.57
3.5% Nickel molybdenum steel	2.04
1% chromium steel	1.69
Chromium vanadium steel	1.84
Cast iron	1.00
Malleable iron	2.03
Aluminium alloy	0.55
Magnesium alloy	0.45
Free cutting Brass	0.55
Alpha Brass	0.73
Manganese Bronze	1.47

Example 15.3. *Estimate the moment, thrust force and power required for 12.7 mm drill having a feed of 0.254 mm/rev, turning at 100 rpm, cutting a steel of Brinell hardness 200.*

Solution. $H_B = 200 \text{ kgf/mm}^2$

$$M = \frac{H_B . D^2 f}{8} = \frac{200 \times (12.7)^2 \times 0.254}{8}$$

$$= 1024.2 \text{ kgf-mm}$$

$$\text{Power} = \frac{2\pi \times 100 \times 1024.2}{4500 \times 1000} = 0.143 \text{ hp.}$$

By formula (15.17),

$$\text{Power} = \frac{1.25 D^2 . K . N . (0.056 + 1.5 f)}{10^5} \text{ kW}$$

Taking $K = 1.10$

$$\text{Power} = \frac{1.25 \times (12.7)^2 \times 1.10 \times 100 \,(0.056 + 1.5 \times 0.254)}{10^5}$$

$$= 0.097 \text{ kW} = \mathbf{0.132 \text{ hp.}}$$

Now

$$T_1 = (1.7 \text{ to } 3.5) \times \frac{M}{D} = (1.7 \text{ to } 3.5) \times \frac{1024.2}{12.7}$$

$$= 137 \text{ to } 282 \textit{ kgf}$$

$$= 209.5 \textit{ kgf} \text{ (Average)}$$

$$T_2 = (0.1 \text{ to } 0.2) \cdot \frac{\pi}{4} \omega^2 \cdot H_B$$

Let $w = 0.14\ D = 1.778$ mm

$$T_2 = 49.66\ kgf \text{ to } 99.32\ kgf$$

$$= 74.50\ kgf \text{ (Average)}$$

$$\text{Average} = T_1 + T_2 = \mathbf{284\ kgf.}$$

By formula (15.18), $\text{Thrust} = 253.4\ kgf.$

15.7. REAMERS

A reamer is a rotary cutting tool generally of cylindrical shape, which is used to enlarge and finish holes to accurate dimensions to a previously formed hole. It is a multiple edge cutting tool, having the cutting edges on its periphery. A reamer consists of three main parts : fluted section, neck and shank, Fig. 15.12. The fluted part consists of chamfer l_1, starting taper l_2, sizing section l_3 and the back taper l_4. Chamfer length or bevel lead length l_1 ensures proper and easy entry of the reamer into the hole. The main cutting action of reamer is done by starting taper l_2. The sizing section serves to guide the reamer and also smooths or sizes the hole. The back taper l_4 (with a difference between the maximum and minimum diameters of from 0.01 to 0.08 mm) reduces friction between the reamer and the hole surface.

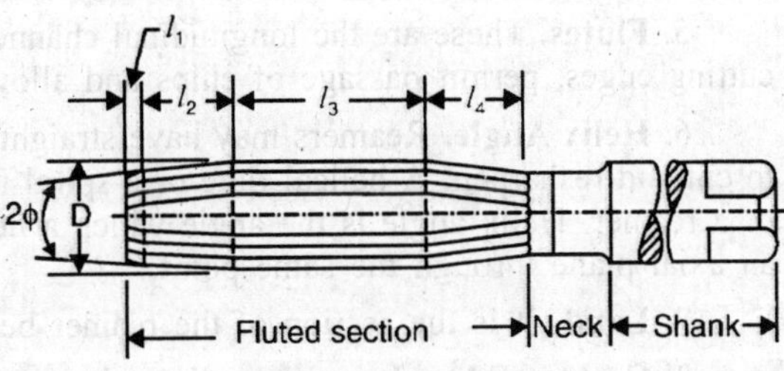

Fig. 15.12. A Reamer.

Types of Reamers

1. **Hand reamers.** These reamers are operated by hand with a tap wrench fitted on the square end of the reamer. The work is held in the vise or vice versa. The flutes may be straight or helical. The shank is straight with a square tang for the wrench.

2. **Machine reamers.** These are similar to hand reamers except that the shank is tapered.

3. **Chucking reamers.** These are machine reamers with shorter flutes. These may be either of the type known as Rose reamers or Fluted reamers. Rose reamers do not cut on the circumference of the flutes but are bevelled off and clearanced to cut on their ends. These are used for heavy roughing cuts, for example, for clearing out cored holes.

4. **Floating reamers.** Here the holders are not rigid but are floating. This permits the reamer to follow the previously made hole naturally and without restraint resulting in a better hole.

5. **Expanding reamers.** These reamers allow slight increase in their size to allow for wear or to remove an extra amount of material. For this, the body of the reamer is bored taper and is slitted. A taper plug runs through the hole and is operated by a screw so that it acts as the expander. The possible variation is generally between 0.15 to 0.50 mm.

6. **Adjustable reamers.** In these reamers, separate blades are inserted into the grooves provided in the body of the reamer. The blades can be moved up or down to increase or decrease the size of the reamer.

7. **Taper reamers.** These reamers are used to finish the taper holes for cutting the taper pins used to secure the collars, pulleys etc. to the shafts.

8. **Shell reamers.** Solid reamers (upto about 20 mm diameter) are usually made of H.S.S. To reduce the cost of larger reamers, the cutting portion is made as separate shells which are mounted on standard shanks made of lower cost sheels. These reamers, are, however, not very rigid and accurate. Inserted teeth or blades in shells will further reduce the cost of the reamer. To increase their production capacity, the reamers can be tipped with cemented carbides.

Nomenclature : See Fig. 15.13

1. **Axis.** It is the longitudinal centre line of the reamer.

2. **Back Taper.** It is a slight decrease in diameter from front to back, in the flute length of the reamer. As pointed out above, it is provided to reduce friction between the reamer and the hole surface. It is also called "longitudinal relief".

3. **Blade.** It is the tooth or cutting element inserted in the reamer body. It may be adjustable and/or replaceable.

4. **Body.** It is the fluted portion of the reamer, inclusive of the chamfer, starting taper and bevel, Fig. 15.12.

5. **Flutes.** These are the longitudinal channels formed in the body of the reamer to provide cutting edges, permit passage of chips and allow cutting fluid to reach the cutting edges.

6. **Helix Angle.** Reamers may have straight flutes or helical flutes. Straight flutes are easy to cut and resharpen. A helical flute or a spiral flute is formed in a helical path around the axis of a reamer. Helix angle is the angle which a helical cutting edge at a given point makes with an axial plane through the same point.

7. **Land.** It is the section of the reamer between adjacent flutes.

8. **Cutting Edge.** It is the leading edge of the land in the direction of rotation for cutting.

9. **Face.** It is the portion of the flute surface adjacent to the cutting edge on which the chips impinge after they are cut from the work.

10. **Heel.** It is the trailing edge of the land in the direction of rotation for cutting.

11. **Chamfer.** It is the angular cutting portion at the entering end of the reamer. It is also called "Bevel" or "Bevel lead".

12. **Neck.** It is the section of reduced diameter connecting the reamer body to the shank.

13. **Shank.** It is the portion of the reamer by which it is held and driven.

14. **Squared Shank.** A cylindrical shank having a driving square on the back end.

15. **Tang.** It is the flattened end of a shank which fits a slot in the socket.

16. **Periphery.** It is the outside circumference of a reamer.

17. **Chamfer Angle.** It is the angle between the axis and the cutting edges of the chamfer measured in an axial plane at the cutting edge. It is also called "Bevel Lead Angle".

18. **Pilot.** A cylindrical portion ahead of the entering end of the reamer body, which is sometimes provided to maintain alignment.

19. **Starting Taper.** It is a slight relieved taper on the front end of the reamer. It facilitates cutting and finishing of the hole. It is also called 'Taper Lead'.

20. **Rake Angle.** It is the angle between the face and a radical line from the cutting edge. It can be zero, Fig. 15.13 (*a*); Negative, Fig. 15.13 (*b*) (describes a cutting face in rotation whose cutting edge lags the surface of the cutting face); Positive, Fig. 15.13 (*c*) (describes a cutting face in rotation whose cutting edge leads the surface of the cutting face).

21. **Hook.** It is the concave condition of a cutting face. The rake angle of a hooked cutting face must be determined at a given point.

22. **Clearance Angle.** Practically, all the cutting action of the reamer is confined to the front tapered portion. Suitable relief angles should be provided to ensure proper cutting action without rubbing. Clearance agles are the angles formed by the primary or secondary clearances and the tangent to the periphery of the reamer at the cutting edge, [Fig. 15.13 (*d*) to (*e*)].

23. **Taper Lead Length.** It is the length of the taper lead measured axially.

24. **Taper Lead Angle.** It is the angle formed by the cutting edges of the taper lead with the reamer axis.

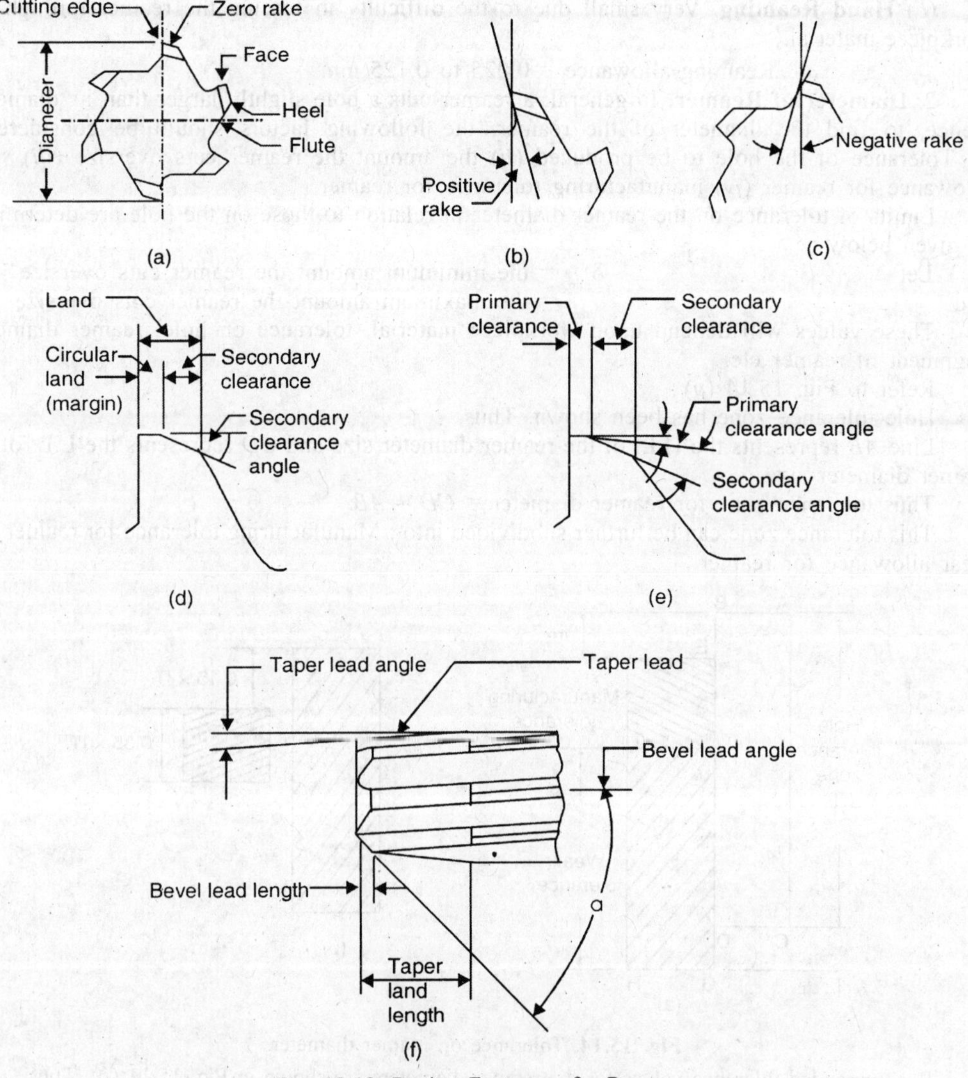

Fig. 15.13. Design Features of a Reamer.

25. **Bevel Lead Length.** It is the length of the bevel lead measured axially.

26. **Bevel Lead Angle.** It is the angle formed by the cutting edges of the bevel lead with the reamer axis.

27. **Margin or circular land.** It is the cylindrically ground surface adjacent to the cutting edge on the leading edge of the land [Fig. 15.13 (*d*)].

Design Features

1. **Reaming Allowance.** Reaming is a finishing operation and hence the reaming allowance or the cutting allowance, that is, the material to be removed is very small. Following are the average values :

(*a*) *Machine Reaming :*

Reaming allowance = 0.125 to 0.25 mm for upto 6.25 mm hole

= 0.25 to 0.375 mm upto 12.5 mm hole

= 0.375 to 0.75 mm upto 37.5 mm hole

(*b*) **Hand Reaming.** Very small due to the difficulty in forcing the reamer through the workpiece material.

Reaming allowance = 0.025 to 0.125 mm.

2. **Diameter of Reamer.** In general, a reamer cuts a hole slightly larger than its diameter. Hence, to find the diameter of the reamer, the following factors should be considered : (*i*) Tolerance of the hole to be produced (*ii*) the amount the reamer cuts oversize (*iii*) wear allowance for reamer (*iv*) manufacturing tolerance for reamer.

Limits of tolerance on the reamer diameter in relation to those on the hole are determined as given below :

Let $\delta_{min.}$ = the minimum amount the reamer cuts oversize

and $\delta_{max.}$ = the maximum amount the reamer cuts oversize

These values will depend upon workpiece material, tolerance on hole, reamer diameter, alignment of reamer etc.

Refer to Fig. 15.14 (*a*) :

Hole tolerance zone has been shown. Thus,

Line *AB* represents the H.L. of the reamer diameter size and *CD* represents the L.L. of the reamer diameter size.

Thus tolerance zone for reamer diameter = *CD* – *AB.*

This tolerance zone can be further subdivided into : Manufacturing tolerance for reamer and wear allowance for reamer.

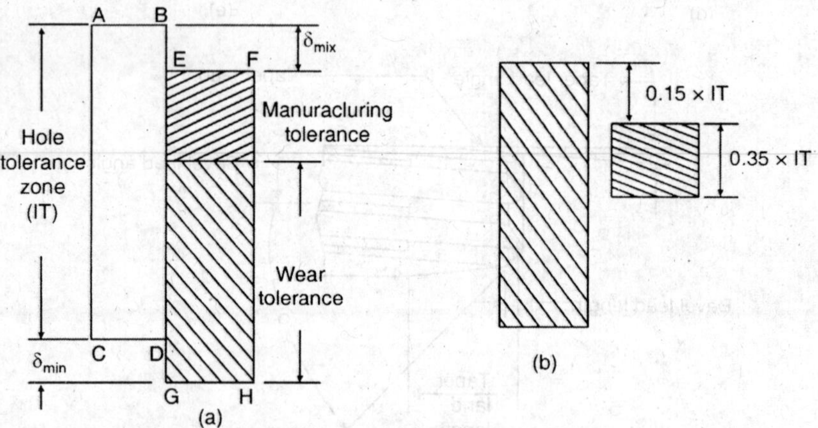

Fig. 15.14. Tolerance on reamer diameter.

Recommended tolerance values for the reamer diameters are shown in Fig. 15.14 (*b*). Thus,

Max. diameter of the reamer = Max. diameter of hole – 0.15 IT

Min. diameter of the reamer = Max. diameter of reamer – 0.35 IT

See Fig. 15.15

Note : The Values of 0.15IT and 0.35 IT are rounded off to the nearest 0.001 mm, towards greater values.

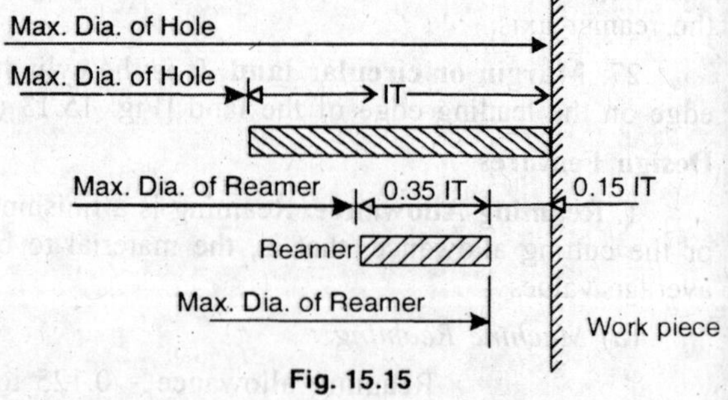

Max. Dia. of Hole
Max. Dia. of Hole
IT
Max. Dia. of Reamer
0.35 IT
0.15 IT
Reamer
Max. Dia. of Reamer
Work piece

Fig. 15.15

3. **Length of body :**

Length of fluted portion ∢ 1.5 × reamer diameter, unless conditions demand so.

Length of guiding or sizing section of reamer = $\left(\frac{1}{4} \text{ to } \frac{1}{3}\right)$ × reamer diameter.

Length of cutting section = (1.3 to 1.4) × reaming allowance × cot ϕ + (1 to 3 mm).

where ϕ = entering angle.

4. **Back taper or relief.** It is a good practice to relieve the reamer with a back taper. It is

provided to reduce friction of the sizing section of the reamer on the surface of the hole being reamed. It is :

(*a*) 0.008 to 0.005 mm for hand reamers.

(*b*) 0.04 to 0.06 mm for rigidly mounted machine reamers

(*c*) 0.06 to 0.10 mm for floating machine reamers.

5. **Front end.** The front end of the reamer should be smaller than the hole diameter by 0.3 to 0.4 times the reaming allowance to ensure the free entry of the reamer into the drilled or bored hole. To further facilitate the entry, a chamfer or bevel is also provided at the front end.

6. **Shape of the teeth.** The reamers may be straight helical-fluted, with straight fluted reamers being more common. Helical-fluted reamers are used for holes with straight slots or with sheet metal. The helix angle varies from 30° slots to 45°. Hand of flute is generally opposite to the hand of cut in order to avoid the screwing in action.

7. **Number of Teeth.** The number of teeth on the reamer can be found out as,

$$Z = 1.5\sqrt{D} + 4, \text{ for cutting brittle materials (C.I. and Bronze)}$$

$$= 1.5\sqrt{D} + 2, \text{ for other metals.}$$

where D = Diameter of the reamer.

It is a general practice to have 'Z' as even, since with odd number of teeth, it is difficult to measure the outer diameter of the reamer. The recommended values of number of teeth are given in Table 15.8.

Table 15.8 : Number of Teeth on a Reamer

Reamer diameter, mm	*Number of teeth*	
	HSS reamer	*Cemented carbide tipped reamer*
6 to 12	6	4
7 to 30	8	6
32 to 45	10	6
45 to 56	12	8
56 to 60	14	8
70 to 75	16	8

Note. Hand reamers have generally more than 6 teeth.

8. **Various Angles :**

(*a*) *Rake angle.* Reamers are generally provided with a zero rake. But proper rake angle can be chosen depending on the workpiece material. Following are the average values of rake angles:

0° to 5°	for Steel
5° to 8°	for Grey C.I.
5°	for Brass
8° to 12°	for Aluminium

(*b*) *Clearance Angles :*

(*i*) *For the cutting section of the reamer*

Primary clearance angle = 5° to 7°

Secondary clearance angle = 10° to 12°

No cylindrical or circular land is provided, Fig. 15.13 (*e*).

(*ii*) *For guiding section.* A circular land is provided for guiding the reamer in the hole.

Circular land or width of margin = 0.08 mm to 0.20 mm for reamers 3 to 10 mm.

= 0.25 to 0.50 mm for reamers 50 to 80 mm.

For very accurate holes, land = 0.02 to 0.03 mm.

There is no primary clearance.

Secondary clearance behind the circular land = 10° to 12°.

(*c*) *Taper lead angle or Plan approach angle.* The cutting edges of the starting taper on a reamer make an angle with the tool axis (point or included angle 2ϕ), Fig. 15.12. The angle is called the taper lead angle or plan approach angle. The agle 2ϕ affects the axial force required in reaming, the larger the angle, the larger the force required. It is taken as :

2ϕ = 1° to 3° for hand reamers

= 30° for machine reamers in reaming through holes in ductile materials (steels)

= 10° for reaming through holes in C.I.

= 60° to 90° for carbide-tipped reamers

= 90° to 120° in reaming blind holes as well as through holes to the 3rd accuracy grade or coarse.

(*d*) *Chamfer or Bevel angle.* The chamfer forms a truncated cone on the starting end of the reamer. It is provided to further facilitate the entry of the reamer into the hole and make the cutting action more convenient. The chamfer angle generally used is 45°.

9. **Materials.** The common materials for reamers are H.S.S. and cemented carbide tipped. Reamers are frequently tipped with cemented carbides to increase their production capacity.

10. **Cutting Speeds and Feeds.**

Cutting speed = 4 to 6 m/min for H.S.S. reamer

= 10 to 15 m/min or cemented carbide tipped reamers

Feed = 0.0375 mm to 0.125 mm per flute per rev.

Example 15.4. *Design a shell-inserted-blade reamer tipped with cemented carbides for reaming a through hole, diameter 55 H7 in a workpiece of structural alloy steel with ultimate strength of 1050 MPa. The diameter of the pre-machined hole is 54.65 mm.*

Solution. 1. *Diameter of reamer :*

Now, 55 H7 is $55^{+0.035}_{-0.000}$ mm

Maximum diameter of hole = 55.035 mm

Minimum diameter of hole = 55.000 mm

Hole tolerance, IT = 0.035 mm

Maximum diameter of reamer = 55.035 – 0.15 IT

= 55.035 – 0.006

= 55.029 mm

Minimum diameter of reamer = Maximum diameter of reamer – 0.35 IT

= 55.029 – 0.35 × 0.035

= 55.029 – 0.013 = 55.016 mm.

2. *Provide the back taper equal to 0.05 mm.*

3. *Let the values for various angles be :*

Rake angle = 5°; Plan approach angle, ϕ = 45°.

Circular land = 0.25 to 0.50 mm.

Secondary clearance angle = 10°.

4. *Length of reamer :*

Length of fluted portion $\nless 1.5 \times 55$

$\nless 82.5$ mm.

Length of cutting portion = (1.3 to 1.4) reaming allowance + (1 to 3 mm)

Let reaming allowance be = 0.18 mm

$\therefore$ Length of cutting section = 0.243 + 2 (say)

$\cong$ 2.25 mm.

Length of guiding section $= \left(\frac{1}{4} \text{ to } \frac{1}{3}\right) \times$ reamer diameter

$= \left(\frac{1}{4} \text{ to } \frac{1}{3}\right) \times 55$

= 13.75 mm to 18.33 mm

$\cong$ 16 mm

5. *Number of teeth :*

$$Z = 1.5\sqrt{D} + 2$$

$$= 1.5\sqrt{55} + 2 = 14$$

Let Z = 10, to secure proper clamping of the blades in the body.

Example 15.5. *Design and draw a single point cutting tool assuming empirical proportions to turn a M.S. bar with a linear cutting speed of 40 m/min, on a lathe equipped with a 10 kW motor. Safe stress for tool material is 250 MPa and the efficiecy of machine tool is 70%.*

Solution. Now cutting power

$$P_c = P_m \times \eta = 10 \times 0.7 = 7 \text{ kW}$$

Now $$P_c = F_c \times V$$

$\therefore$ Cutting force $F_c = \frac{7 \times 1000 \times 60}{40} = 10500$ N

$\therefore$ B.M. on tool shank, $M = F_c \, . \, l$ (Equ. 15.1)

$= \frac{Bh^2}{6} \times \sigma_b$ (Equ. 15.2)

Let, l the overhang be = 1.25 h (Equ. 15.4)

Take h/B = 1.6 (Equ. 15.4)

Now $\sigma_b = 250$ N/mm²

$$\therefore \quad 10500 \times 1.25 \times 1.6B = \frac{B \times 1.6^2 \times B^2}{6} \times 250$$

From here, B, the width of shank = **15.4 mm**

Then h, height of the shank = 1.6 × 15.4

= **24.65 mm**

Shank overhang, l = 1.25 × 24.65

= **30.8 mm**

Note :

1. *Design of Tap* : Refer to Chapter 17, Art. 17.5.2.
2. *Design of Hob* : Refer to Chapter 16, Art. 16.15.

15.8. FORM TOOLS

Form tools are cutting tools which are used to machine complex-shaped surfaces with the cross-section outlined by curves or broken lines. The cutting edge of the form tool is of a shape that produces the desired contour or the workpiece in the turning operation. Form tools are commonly used in large-lot and mass production. Their use ensures :

— a high output

— uniform contour of all workpieces, and

— accurate dimensions.

Classification of Form Tools. Form tools can be classified as follows :

1. According to their type. as circular, flat and end form tools, [Fig. 15.17 and 15.16 (*a*), (*b*)].

2. According to the setting of flat tools in respect to the workpiece, as : Tools with a radial cutting edge, [Fig. 15.16 (*a*)] and tangential tools.

3. According to the position of the tool axis as : tools with the axis parallel to the workpiece axis and tools with an angular location of the axis or mounting surfaces.

4. According to the elements of the contoured surface on the tool, as : circular tools with annular elements, Fig. 15.17, circular tools with helical elements, and flat tools with flat elements, [Fig. 15.16 (*a*)].

Circular and flat form tools with an angular location of the axis or mounting surfaces with respect to the workpiece axis, are seldom used owing to their complex manufacture. They are used only when the shape of the workpiece contour is such that tools with parallel axis cannot be employed.

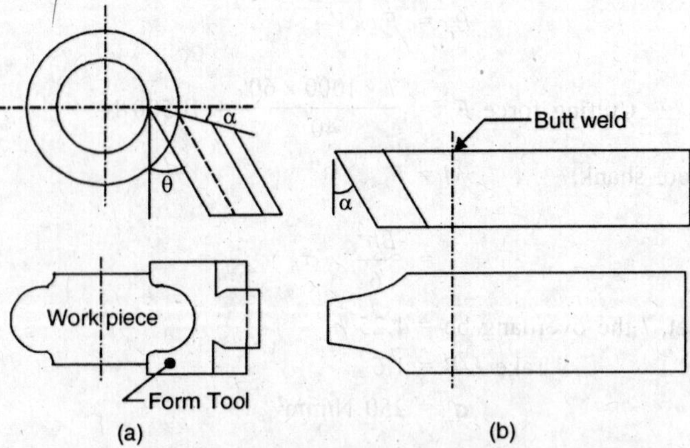

Fig. 15.16. Types of Form Tools.

15.8.1. Design features of form tools. Most form tools are made of H.S.S. However, cemented carbides are increasingly being used for this purpose. The use of contoured cemented-carbide tips for form tools enables the productivity to be raised by 30 to 40 per cent, as compared to H.S.S. form tools.

A form tool should have the proper rake and relief angles, so that, the metal is cut under sufficiently advantageous conditions. Table 15.9 gives the rake angles for the form turning of various materials.

The relief angle depends upon the type of form tool. Relief angle is :

= 10° to 12° on circular form tools

= 12° to 15° on flat form tools

= 25° to 30° on form tools used for relieving form milling cutters.

Table 15.9. Rake Angles for Form Tools

Material	*Rake angle, Degrees*
Aluminium, Copper	20 to 25
Bronze, Leaded brass	0 to 5
Mild steel	25
Medium-hard steel	20 to 25
Hard steel	12 to 20
Very hard steel	8 to 12
Soft cast iron	15
Hard cast iron	12
Very hard cast iron	8

15.8.2. To determine outside diameter of a circular tool. It is determined graphically, (Fig. 15.17). It is explained below for a circular tool of positive rake angle (α > 0). It will depend on the height of the profile to be turned.

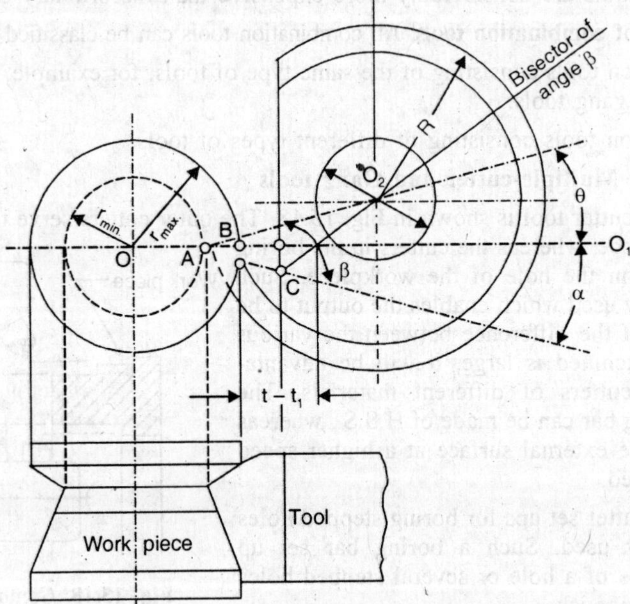

Fig. 15.17. A Circular Form Tool.

1. With O as centre of the workpiece, draw two concentric circles with radii equal to the maximum and minimum radii of the contour to be machined.

2. Through point A, draw a line at the angle, α, representing the trace of the plane ground to produce the tool face. Draw another line through A at an angle equal to the relief angle, θ.

3. Draw a line normal to OO, from a point at a distance of t, from the contact point B. The distance t, is the minimum amount that will permit chip disposal from the tool face. It is taken from 3 mm to 12 m depending upon the chip thickness and the amount of chips to be cut.

4. From the point of intersection C of the vertical line and the line of the tool face, draw a line bisecting angle β.

5. The point of intersection of this bisector and the line drawn at the angle , θ, O_2, is the centre of the circular tool.

6. With O_2 as centre, draw a circle of radius R. Then determine all other dimensions graphically.

7. To determine the diameter of the mounting hole, the wall thickness, t_1, is taken as 6 to 10 mm.

The profile of a form tool, as a rule, does not coincide with the required profile of the workpiece, it must be determined graphically or analytically. However, this is out of the scope of this general book.

15.9. COMBINATION TOOLS

Combination tools are special cutting tools intended for machining several surfaces. A combination tool consists of two or several tools for the same or different kinds. By the use of a combination tool, several operation elements or even operations can be combined into one. The use of combination tools results in the following advantages :

1. There is an increase in productivity because the machining and handling times required for a job get reduced.

2. Simpler machine tools can be employed which results in reduced cost of the operation.

3. Possibility of removing substantial machining allowance by the simultaneous operation of several cutting edges.

Combination tools are considerably more expensive than the ordinary cutting tools.

15.9.1. Types of combination tools. All combination tools can be classified into two groups :

(*a*) Combination tools consisting of the same type of tools, for example, multiple diameter, multiple-cutter and gang tools.

(*b*) Combination tools consisting of different types of tools.

Multiple diameter, Multiple-cutter and Gang tools

(*i*) A multiple cutter tool is shown in Fig. 15.18. The outer cutters serve to turn the external steps on the workpiece, whereas the cutters in the boring bar bore the steps in the hole of the workpiece. Such tools are extensively used which enables the output to be sharply increased. If the difference between the various diameters being machined is large, it will be advantageous to employ cutters of different materials. The cutters on the boring bar can be made of H.S.S., whereas those for turning the external surface at a higher speed can be carbide tipped.

(*ii*) Multiple-cutter set ups for boring stepped holes are also commonly used. Such a boring bar set up enables several steps of a hole or several stepped holes to be bored simultaneously.

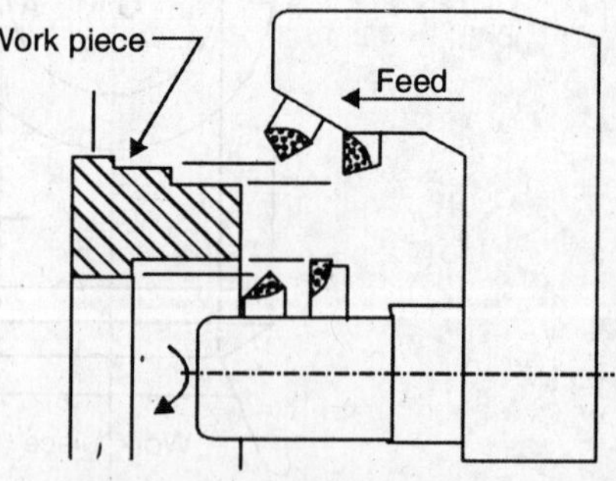

Fig. 15.18. Combination Multiple-Cutter Tool.

(*iii*) A multiple-cutter turner (Box Tool) used on Turret lathes also belongs to this category (Fig. 7.14).

(*iv*) Combination core drills are of built up design and consist of two core drills. The first step is a separate core drill with a tapered shank which fits a corresponding taper socket in the core drill intended for machining the second step. This design facilitates sharpening.

(*v*) Inserted-blade multiple boring heads with size adjustments are available.

(*vi*) A gang milling cutter is a typical example of combination tool.

(*vii*) Combination broaches can consecutively broach a round hold and then cut splines in it.

These are some of the common examples of Combination Tools of the first type. Many other constructions of combination tools are in use.

Combination tools consisting of different kind of tools : Below, we give some common examples of these combination tools :

(*i*) A combination drill and core drill. It can be used to drill and enlarge a hole simultaneously. Here, the drill is inserted in two flute core drill and secured in its body.

(*ii*) A combination drill and reamer can be used for machining through holes. The tool is so designed that the reamer will begin machining only after the hole is drilled through. This ensures proper chip disposal.

(*iii*) A combination drill and tap operates on the same principle as a combination drill and reamer.

(*iv*) A combination stock stop and centre tool is commonly used on Turret lathes, (see Fig. 7.14).

(*v*) Adjustable knee tool is another common combination tool used on Turret lathes, (see Fig. 7.14). This tool is used for turing, combined with drilling, boring or centering.

(*vi*) A combination drill and end mill tool for making elongated holes in track rails is shown in Fig. 15.19. In Fig. 15.19 (*a*), the tool rotates and is fed axially. The drill lips are in operation and they drill the hole. When the hole has been fully drilled, the table feed is switched on. Now the table is fed horizontally perpendicular to the tool axis, Fig. 15.19 (*b*). Now the cutting edges on the cylindrical part of the tool start machining, and the tool acts as an end mill.

(*vii*) A combination coupling boring head is extensively used. This tool combines the facing of the ends, chamfering, boring of the recesses and reaming of the tapers in the couplings in one operation. Such a combination tool is employed with special coupling boring machines in which the coupling rotates in operation. As the head advances, the reamer blades start machining. They machine the internal tapers. As the tool advances into the blank, boring blades

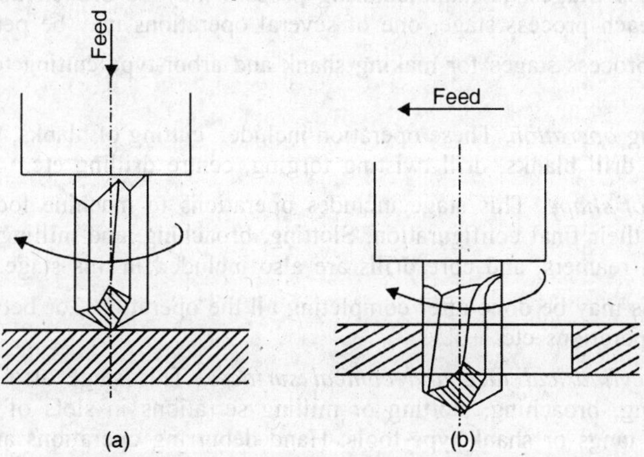

Fig. 15.19. Combination Drill and End Mill.

start to machine the recesses. The facing blades act last and both face the ends and make the chamfers. A great number of holes drilled in the body of the head carry cutting fluid directly to the cutting edges of the tool. The streams of cutting fluids wash away chips.

(*viii*) A combination pipe turning and threading head combines turning and thread cutting.

15.10. MANUFACTURE OF CUTTING TOOLS

In designing any manufacturing process, the aim is to ensure high quality and output of the product at a minimum cost of production.

The quality of any cutting tool is characterised by its reliability. The latter defines the ability of any product to perform its function in accordance with the specification over the whole period of service under the prescribed operating conditions. The main factors of reliability for cutting tools are dimensional and form accuracy, and stable tool life.

Cutting tools should provide for the specified accuracy and surface finish of the parts machined over an optimum tool life. A large number of workpieces machined with one tool between resharpenings as well as upto its total wear point to high quality of the tool. The stability of tool life is the maximum to minimum tool-life ratio in one batch of tools subjected to the test. This ratio should not exceed 2.

High productivity results from adequate machining methods adopted for each stage of manufacture of a given product. The selection of the most productive methods, determined by the type of production, will make for minimum machining cost.

Metal savings are essential in designing initial blanks. The types of blanks to be used (hot rolled, cold drawn or ground bar) will also depend on the type of production, for instance, in high-volume production cold-drawn and ground bars are used on automatic machines, whereas separate blanks are made by forging. As a result, the blanks will be close in shape and dimensions to the finished product.

The user will find a high cost but durable cutting tool profitable as its cost tends to be offset by a larger number of machined workpieces. Such a tool will also be advantageous for the economy as a whole because of reduced consumption of H.S.S. and Cemented Carbides. Thus, the extended tool life is virtually equivalent to production of the tool increased in a direct proportion.

The main demand placed on a manufacturing process is its stability. This property of the process is determined by the capability of all machining equipment to hold the dimensional and form tolerances of the workpieces machined within the prescribed limits during a very long period of time. Such stability is achieved by properly selected and, as far as possible, invariable locating surfaces, and use of appropriate machines, cutting and measuring tools and fixtures.

15.10.1. Process Stages. A manufacturing process may be broken up into a number of process stages. At each process stage, one of several operations may be performed.

For example, process stages for making shank and arbor-type cutting tools can be written as given below :

1. *Blank making operation.* These operation include : cutting of blanks, welding, annealing, hot forge rolling of drill blanks, drill twisting forging, centre drilling etc.

2. *Formation of shape.* This stage includes operations to machine locating surfaces and shape the blanks to their final configuration. Slotting, broaching, and milling keyways in arbor-type milling cutters, reamers, and core drills are also included in this stage.

3. *Making.* This may be done after completing all the operations, or between some process stages or shaping operations etc.

4. *Fluting in cylindrical face and conical surfaces.* This stage also incorporates tooth relieving, slot milling, broaching, slotting or milling serrations in slots of cutter bodies, and milling squares and tangs or shank type tools. Hand deburring operations are also included in this stages.

5. **Formation of additional surfaces.** For example, drilling of spot holes for adjusting screws in round threading dies, drilling and tapping holes for screw fasteners, etc.

6. *Heat treatment.* This stage includes brazing carbide tips to the Cemented-Carbide tools.

7. *Removal of solder and machining main or intermediate locating surfaces for further processing of the tools.* Related to this stage are the operations of the removal of solder from brazed carbide-tipped tools, grinding and lapping mounting holes and faces in arbor-type tools, finishing threaded holes in shanks, grinding and lapping centre holes etc. Intermediate locating surfaces are also shaped at this stage on external surfaces of arbor-type tools to ensure their simple and reliable mounting in work-holding fixtures.

8. Assembly of tool bodies with insertable blades, wedges, and other parts of accordance with the construction of a given tool. This stage is used for inserted-blade cutters.

9. Sharpening tool cutting surfaces and grinding shanks. Grinding out flutes and lands from solid blanks.

The manufacturing process for any cutting tool includes all the above mentioned stages or most of them. The sequence of operations within each stage is different and depends on the design of a given cutting tool and the adopted manufacturing process.

Standardization of Separate Process Stages. A definite operational sequence for tools of similar design can be established within separate process stages. The main is to establish uniform operational sequence, apply optimum overall and operational machining allowances and tolerances, and reduce the number of types and sizes of tooling, jigs and fixtures.

(*a*) **Example 15.6.** *There is a definite processing sequence at the blank-making stage for shank-type welded cutting tools 8 to 80 mm in diameter, for example, drills, reamers, core drills and taps. The sequence of operations is given below :*

1. Cutting-off a blank for the tool body on an abrasive cut-off machine.

2. Cutting-off a blank for the shank on an abrasive cut-off machine.

3. Turning a neck on the body or the shank to equalize the diameters of the portions to be welded together.

4. Cleaning by shot blasting or tumbling.

5. Friction welding on a semi-automatic or electric arc welding.

6. Annealing.

7. Turning off the weld.

8. Cutting the end face on the body. This operation is done to remove the decarburized layer from the body face and also to obtain this specified length of the blank. The end face on reamers, core drills and end mills can be cut by turning it from the periphery towards the centre with a facing tool; and on the body side of twist drills, tapered point is turned.

9. Cutting the end face of the shank on a lathe to the predetermined blank length.

10. Centre drilling on the body end face in a centre drilling machine.

11. Centre drilling on the shank end face.

12. Blank straightening. The operation is done by hand on screw presses, and the radial run out is subsequently checked between centres with a dial indicator. To save operator's physical effort, a hydraulic press should be used.

(*b*) **Example 15.7.** *A standardized process stage can also be explained for arbor-type cutting tools with cylindrical mounting holes. Various types of milling cutters belong to this category, for example, plain, face, side, form, and angle cutters, end mills, disk-type gear milling cutters, and also gear shapping and shaving cutters.*

Blank upto 50 mm in diameter are cut-off from forged bars, and over 63 mm in diameter, by forging. We have the following sequences of operation :

1. Cutting off a separate blank/blank forging.

2. Annealing.

3. Cleaning the blank of scale by shot blasting or tumbling.

4. Turret operation. This consist of the following machining operations on turret lathe : Centre drilling, drilling the central hole, facing, core drilling, hole relief boring, hole edge chamfering and reaming the hole.

5. Machining the second face in a three-jaw chuck on a turret lathe. Here, a layer of stock is removed with the allowance for subsequently grinding the face.

6. Grinding the second face on a surface grinder. The ground face should be parallel to the first face machined on the turret lathe and, hence, square with the hole axis.

7. Making recesses on both sides (if specified by the tool design) with a pilot-guided counter bore on a drill press.

8. Chamfering the hole edge of the second face with a counter sink on a drill press.

9. Broaching a keyway on horizontal broaching machine.

10. Chamfering and removing burrs from the keyway on both sides by hand, with a smooth-cut file.

11. Rough and finish turning of the outside diameter. Several blanks at a time are mounted on a plain mandrel and clamped with a nut. The mandrel is then mounted between the centres of a lathe.

The mounting hole and the faces of any arbor-type milling cutter are used as the locating surfaces both in the process of manufacture of the cutter and during its service. Therefore, the requirements for squareness of the faces to the hole are very strict. For this, the following operations are done :

1. The hole and one of the faces are ground on an internal grinder in one set up.

2. The second face is ground on a surface grinder. The blank is mounted (positioned) on a short mandrel located at the centre of the magnetic chuck.

15.10.2. Manufacture of Drills. The manufacturing process for twist drills differ depending on drill types and sizes. As to their accuracy, both straight and taper-shank drills are divided into precision and general-purpose groups.

(*a*) **Straight shank drills.** Drills with straight shank are usually made upto 12 mm diameter. The starting material (blank) is ground-bar stock. For machining such drills, the outside diameter of the bar stock acts as locating surface. The sequence of operations for making such drills is given below :

1. Cutting off a separate blank on a vertical cut-off automatic with drill point angle (120°) from the body end at right angle from shank end respectively.

2. Hardening and tempering.

3. Rough through-feed grinding on a centreless grinder.

4. Finish through-feed grinding on a centreless grinder.

5. Grinding out the flutes. Grinding out straight and helical flutes in hand-and-machine taps and straight-shank drills upto 12 mm in diameter has recently found use instead of milling. Flutes are ground out on hand-operated, semi-automatic and automatic machine tools.

6. Grinding out the lands.

7. Sharpening. Drills are sharpened on automatic drill grinders.

8. Boiling out and passivation.

9. Rinsing.

10. Marketing the name by rolling on automatic marking machines.

(*b*) **Taper-Shank drills.** For machining straight-and taper-shank drills above 12 mm in diameter, drills are located from a centre hole in the shank end and a tapered point on the body end. The tapered point is made to an included angle of 90° (equal to drill point angle of 118° to 120°) is established for a minimum stock removal in the course of grinding. Blanks with centre-holes are advantageous as these holes, having been ground or even lapped, can provide locating surfaces for subsequent operations. The sequence of operations for producing drills above 12 mm in diameter from blanks made in accordance with a standard process plan (Example 15.6, Art. 15.10.1), is as follows :

1. Turning the body on an automatic lathe.

2. Turning the taper shank. In small lot production, the shank is machined to taper shape on a lathe, against a template. In medium and large-volume production, shanks are machined on specialized semi-automatic lathes fitted with template devices.

3. Facing the end of the shank with a radius-shaped facing tool.

4. Marking the name on the drill neck by rolling on a marking machine.

5. Traverse grinding of the taper shank on a cylindrical grinder to make an adequate locating surface for milling the tang.

6. Traverse or plunge grinding of the drill body on the outside diameter in a cylindrical grinder between centres.

7. Milling flutes and lands on semi-automatic drill-milling machines. First, one flute is milled with a profile sharpened cutter, and, simultaneously, one land. Then, the blank is indexed through 180° and the other flute and land are milled.

8. Milling the tang on horizontal or vertical milling machines, and removing burrs with a smooth-cut file from the lands and the tang. Deburring is done manually, following the respective machining operation.

9. Polishing the flutes on polishers to facilitate a similar operation after heat treatment.

10. Grinding the flanks to relief angle to leave a uniform stock allowance for finish grinding after heat treatment.

11. Heat treatment of the body and the tang, the latter being hardening after tempering the body.

12. Polishing the flutes by hand on polishers.

13. Grinding the centre holes on a centre hole grinder on counter-sinking on a drill press with a carbide tool.

14. Rough traverse or plunge grinding of the body on a cylindrical grinder.

15. Rough grinding of the shank in a similar way.

16. Finish traverse grinding of the shank on a cylindrical grinder.

17. Finish traverse grinding of the drill body on a cylindrical grinder so as to form the back taper.

18. Rough grinding the drill point to an included angle of 118° on tool grinder.

19. Finish grinding of the drill point to 118° on tool grinder.

20. Thinning the web on a special machine or a general purpose tool grinder provided with a special fixture.

In the course of turning, the outside diameter of drill blanks is checked with adjustable snap gauges and vernier calipers. The thickness of the web is checked by means of a micrometer fitted with special spherical tips and by counter gauge. The drill body is checked for radial runout in a special measuring fixture. In grinding operations, before and after heat treatment, the body and shank are measured with a micrometer.

15.10.3. Manufacture of Milling Cutters. The operational sequence for making coarse tooth plain milling cutters is given below :

1. The blank making stage includes forging and the next process stage is shaping. Both these process stage have been explained in Example 15.7, Art 15.10.1.

2. The next process stage is milling helical flutes and clearance surfaces. This is done on general-purpose millers, the number of blanks machined simultaneously being dependent on the rigidity of the taper shank mounting mandrel fitted into the index-centre spindle.

3. Deburring.

4. Heat treatment of blanks.

5. Machining the locating surfaces. The hole and the faces ground as explained after step 11 in example 15.7, Art. 15.10.1.

6. Next is the grinding stage. The profile-sharpened milling cutters are machined in the following sequence :

(*i*) Grinding the tooth faces on a universal tool grinder.

(*ii*) Grinding the outside diameter on a cylindrical grinder.

(*iii*) Grinding the lands to a width not exceeding 0.05 mm.

7. The final operation is marking of the cutter face with a tungsten needle scriber on a pantograph or by the electro-chemical method.

15.10.4. Manufacture of Reamers. The process will be explained for taper-shank H.S.S. machine reamers.

1. The blank-making stage is explained in Example 15.6. Art. 15.10.1.
2. The reamer body is turned in two cuts on a semi-automatic lathe.
3. Taper shank is turned in two cuts on the same lathe.
4. Face the shank end with a radius-shaped facing tool, on a lathe.
5. Grind taper shank to obtain locating surface, on cylindrical grinder.
6. Machine face and 45° chamfer on body end, on centre or semi-automatic lathe.
7. Marking oṇ neck on a marking machine.
8. Mill tang on a horizontal miller to obtain locating surface for next operation.
9. Mill flutes on 8 reamers at a time on semi-automatic miller for reamers.
10. Remove burrs with file manually.
11. Harden and temper.
12. Grind centre holes on centre grinder on drill press.
13. Lap centre holes on a drill press.
14. Grind tooth faces on semi-automatic grinder or on universal tool grinder.
15. Rough grind reamer body to cylinder on cylindrical grinder.
16. Rough grind tapes shank on cylindrical grinder.
17. Finish grind reamer body to cylinder on cylindrical grinder.
18. Finish grind taper shank on cylindrical grinder.
19. Grind body to back taper on cylindrical grinder.
20. Grind chamfer on cylindrical grinder.
21. Grind lands on teeth, leaving margins, on universal tool grinder.
22. Grind chamfer relief sharp on universal tool grinder.
23. Find grind cutting face on universal tool grinder.

24. Find grind outside diameter on cylindrical grinder.

25. Find grind chamfer relief on universal tool grinder.

15.10.5. Manufacture of Broaches. There are a large number of types of broaches. So, it is not possible to give here the manufacturing process for each type. For this reason, the manufacture of a Spline broach is discussed here. Broaches have a large ratio of length of diameter in the range of 13 to 50. We shall consider a broach of 30 to 50 mm diameter. The steps of manufacture are :

1. The blank milling operations are accomplished in accordance with the process plan discussed in Art. 15.10.1.

2. At the shaping stage, a spot for the location in a steady rest is machined in the middle of the blank.

3. Rough turning in two operations, first, on the shank upto the steady rest, and the second, on the body, also upto the rest.

4. Normalizing and annealing of the blank for relieving internal stresses.

5. Both ends are faced, and the centre holes are recovered with a bell-type combined drill and countersink to obtain the locating and protective conical recesses of 60° and 120° respectively. The operation is done on a lathe. One end of the blank is clamped in 3-Jaw Chuck, and the other is supported by a steady rest.

6. The spot for the steady rest is turned for the second time.

7. Now, the front pilot, the neck, and the pull end are finish turned, the latter being chamfered.

8. Finish turning of the back pilot and the sizing portion of the body and the chamfering of the latter are done. Operations 7 and 8 are carried out in a steady rest.

9. The body is turned to a taper against a template, with the use of a steady rest. This completes the shaping stage.

10. The sizing and cutting teeth are cut on a lathe with a steady rest. Here the rake angle is formed.

11. Milling six splines with an allowance for grinding and removing burrs in the course of milling. The side of splines are milled by various methods.

12. Heat treatment of the body and the shank in succession.

13. Workpiece is cleaned chemically or in a shot blaster and straightened on a hand power process.

14. Grinding centre holes on a centre-hole grinder.

15. Rough grinding of faces of the cutting and the sizing teeth on a special broach grinder.

16. Grinding the neck for a steady rest on a cylindrical grinder.

17. Rough grinding of the pull end and the front pilot in a steady rest.

18. Rough grinding of the rear pilot and the sizing tooth, also in a steady rest.

19. Rough grinding for the cut per tooth without a steady rest on a universal cylindrical grinder.

20. Grinding the front taper connecting the neck with the front pilot.

21. Finish grinding of the neck for a steady rest.

22. Finish grinding of the pull end and the front end.

23. Finish grinding of the rear pilot and the sizing teeth.

24. Finish grinding of the cutting teeth for the cut per tooth on a cylindrical grinder.

25. Grinding the clearance surface (with the use of a steady rest) on the cutting and sizing teeth.

26. Finish grinding of the faces of the sizing and the cutting teeth on a special broach grinder.

27. Rough and finish grinding of the spline profile on a special grinder.

28. Engraving the name on a pantograph.

15.10.6. Manufacture of Inserted Blade Cutting Tools. Manufacture of inserted blade cutting tools consists of three stages : Making tool bodies, Making blades and then assembly.

(*i*) **Making tool bodies.** The blank making and shaping stages in the production of bodies for shank- and arbor-type tools are just the same as for solid tools discussed in Art. 15.10. and 15.10.1. Most bodies are marked after the shaping stage.

Then slots are made in place of flutes in the tool bodies, on milling machines. After machining the slots, serrations are broached in the slots, chip grooves are milled on the outside diameter and face, and burr grooves are removed. In inserted-blade cutters, machining of other surfaces is also carried out in accordance with special design features, for example, holes for washers, cams, or screws are drilled in the bodies of inserted-blade reamers, core drills and some types of milling cutters, and in other tools thread is cut.

The next process stage is heat treatment. After that, the cutter bodies are oxidized to obtain a corrosion-proof oxide film. Oxidized bodies are deep-brown or black. The subsequent machining of auxiliary or main locating surfaces is carried out in the same way as with solid tools.

(*ii*) **Making of blades.** The sizes of blades depend on those of the respective tools. The initial blank for blades upto 50 mm wide is sized for three to four blades. Below, we give the manufacturing process for blades used with side milling cutters.The blank is a piece of H.S.S. strip, dimensioned for four blades.

1. Cutting of the initial blank for four blades on an abrassive cut-off machine, a band saw machine, or on an eccentric press.

2. Grinding two planes on a surface grinder.

3. Milling the blank edges in a special fixture on a vertical miller. First the edges are milled to an angle of 25° and then the blanks are placed in another location of the fixture for milling the opposite edges to 90°.

4. Cutting the blanks into separate blades.

5. Milling the back to an angle of 5° on a horizontal miller or on a broaching machine depending upon quantity.

6. Milling Serrations.

7. Heat Treatment.

8. Grinding the face surface on a surface grinder with the wheel periphery.

9. Chemical marking on the blade face.

(*iii*) **Assembly.** Here, the bodies are assembled with blades, wedges and other components incorporated in the design. Special devices and accessories are used to ensure the correct location of blades with respect to the outside diameter and face, and to facilitate the assembly process.

The next process stage is grinding and sharpening. Here the machining methods are identical with those for solid tools. The face surfaces of side and face milling utters, shell and mills, reamers and core drills of all types are ground separately on blades during the process of their manufacture. The face surface location required to form the rake angle is determined by the dimensions of the slot receiving the blade.

Inserted-blade plain milling cutters are rough ground on the outside diameter to obtain an equal projection of the blades out of the body. The face surface of the blades is then ground, and after finish grinding of the outside diameter, the lands on the tooth flanks are ground to form a relief angle.

The rake and relief angles are checked with a special instrument, and the slots, with special contour gauges.

Example 15.8. *A 150 mm long 12.7 mm dia. stainless steel rod is being turned to 12.19 mm dia. on a centre lathe. Spindle speed = 400 rev./min., Axial speed = 203.20 mm/min. Calculate, cutting speed, MRR, Time to cut, Power dissipated and cutting force.*

Solution.

(*i*) Cutting velocity, $V = \dfrac{\pi DN}{1000 \times 60}$

$$= \frac{\pi \times 12.7 \times 400}{1000 \times 60} = 0.26 \text{ m/s.}$$

$$\text{Depth of cut, } d = \frac{12.70 - 12.19}{2} = 0.255 \text{ mm}$$

$$\text{Feed, } f = \frac{203.20}{400} = 0.508 \text{ mm/rev.}$$

(*ii*) Metal removal rate, MRR = Area of cut × V (Art. 14.8)

Now area of cut = $b.t. = d.f.$

$\therefore$ $MRR = d.f.\pi.D_{ave}.N$

$D_{ave.} = 12.445$ mm

$\therefore$ $MRR = 0.255 \times 0.508 \times \pi \times 12.445 \times 400$

$= 2024 \text{ mm}^3\text{/min.}$

(*iii*) Machining time, $T = \dfrac{L}{fN}$, L is tool travel

$$= \frac{150}{0.508 \times 400} = 0.72 \text{ min.}$$

(*iv*) Power $= d \times f \times V \times C$, watts (See P)

d and f are in mm, V is in m/min and C is a constant.

= 56 (From Table on Page 502)

$\therefore$ Power $= 0.255 \times 0.508 \times 0.26 \times 60 \times 56$

= **100.8, watts**

(*v*) Now $P = T.\omega$

$\therefore$ Torque, $T = \dfrac{100.8 \times 60}{2\pi \times 400} = 1.80$ Nm

$\therefore$ Cutting force, $F_C = \dfrac{2T}{D_{ave.}} = \dfrac{2 \times 1.80 \times 1000}{12.445}$

= **290 N**

Note. If we use the values given in Table 14.2. For stainless steel, specific energy is 4 W.s/mm³.

$$\text{Now } MRR = 2024 \text{ mm}^3/\text{min.}$$
$$= 33.73 \text{ mm}^3/\text{s}$$
$$\therefore \quad \text{Power} = 4 \times 33.73$$
$$= \mathbf{134.92 \text{ watts}}$$

Example 15.9. *A 10 mm dia. hole is to be drilled in a block of magnesium alloy. The feed is 0.2 mm/rev. and the spindle speed is 800 rev./min. Calculate : MRR, cutting power and torque on the drill.*

Solution. (*i*)
$$MRR = \frac{\pi D^2}{4} \times f \times N, \text{ mm}^3/\text{min}$$
$$= \frac{\pi \times 100}{4} \times 0.2 \times 800$$
$$= 12568 \text{ mm}^3/\text{min.} = \mathbf{209.47 \text{ mm}^3/\text{s}}$$

(*ii*) Now from Table 14.2, the average cutting power for magnesium alloys is 0.5 W.s/mm³.

$$\therefore \quad \text{Power} = 0.5 \times MRR, \text{ watts}$$
$$= 0.5 \times 209.47$$
$$= 104.735 \text{ watts.}$$

(*iii*)
$$\text{Torque} = \frac{\text{Power}}{\omega}$$
$$\text{now } \omega = \frac{2\pi \times 800}{60} = 83.8 \text{ rad./s}$$
$$\therefore \quad \text{Torque} = \frac{104.735}{83.8} = \mathbf{1.25 \text{ N.m}}$$

Example 15.10. *The following data is given for slab milling of a 300 mm long, 100 mm wide mild steel block :*

Feed = 0.25 mm/tooth; depth of cut = 3.2 mm

Cutter diameter = 50 mm

Number of cutter teeth = 20

Cutter speed = 100 rev./min.

Cutter is wider than the job. Calculate :

(i) MRR

(ii) Power

(iii) Torque

(iv) Cutting time.

Solution.

(*i*)
$$MRR = \text{width of cut} \times \text{depth of cut} \times \text{Table feed}$$
$$\text{Now width of cut} = \text{width of job} = 100 \text{ mm}$$

$$\text{Depth of cut} = 3.2 \text{ mm}$$

$$\text{Table feed} = \text{feed rate, mm/min.}$$

$$= \text{feed per tooth} \times \text{number of teeth in cutter} \times \text{cutter speed}$$

$$= 0.25 \times 20 \times 100$$

$$= 500 \text{ mm/min.}$$

$$\therefore \quad MRR = 100 \times 3.2 \times 500$$

$$= 160000 \text{ mm}^3\text{/min.}$$

$$= \mathbf{2666.7 \text{ mm}^3\text{/s.}}$$

(*ii*) Now from Table 14.2, the average value of power requirement is 6 W.s/mm³.

$$\therefore \quad \text{Power} = 6 \times 2666.7 = 16000, \text{ watts.}$$

(*iii*)

$$\text{Torque} = \frac{16000 \times 60}{2\pi \times 100} = 1527.7 \text{ N.m}$$

(*iv*)

$$\text{Cutting time} = \frac{\text{Total table travel}}{\text{Table feed, mm / min.}}$$

$$\text{Now Total table travel} = \text{length of job} + \text{Added Table travel}$$

$$\text{Added table travel} = \sqrt{Dd - d^2} \quad \text{(See Chapter 4)}$$

$$= \sqrt{50 \times 3.2 - 3.2 \times 3.2}$$

$$= 12.24 \text{ mm}$$

$$\therefore \quad \text{Cutting time} = \frac{300 + 12.24}{500} = \mathbf{37.5s}$$

PROBLEMS

1. Discuss the various types of cutting tools.
2. On what factors the proper selection of tool type depend?
3. Discuss the general problems of cutting tool design.
4. Discuss the design factors to be kept in mind when designing the cutting element of a tool.
5. What is the function of mounting element of a cutting tool?
6. Discuss the most widely used constructions of mounting elements.
7. What is the effect of method of manufacture on the cutting tool design?
8. What is the main drawback of a brazed tipped tool?
9. What is throwaway tipped turning tool and what contribution has it made towards increasing productivity?
10. Discuss the use of chip breakers.
11. Discuss the various types of chip breakers.
12. How does a chip breaker break up a chip?

13. Why large positive rake angles cannot be used on cutting tools?

14. How the tool shank of a single point cutting tool is designed?

15. How the milling cutters are classified?

16. Name the common materials for milling cutters.

17. Discuss the following design features of a milling cutter :

(*i*) Size of cutter

(*ii*) Tool angles

(*iii*) Width of land

(*iv*) Number of teeth

(*v*) Power required for milling

(*vi*) Flutes.

18. Define : Broaching tool and broaching operation.

19. Sketch and discuss a typical internal broach.

20. How do you classify a broaching operation?

21. Sketch the tooth shape of a broach and write briefly about its elements.

22. Name the common materials from which broaches are made?

23. Write briefly on the construction of broaches.

24. Define broaching allowance.

25. How is the pitch of teeth of a broach selected?

26. How the total length of a broach is determined?

27. How will you calculate the power needed for broaching?

28. Discuss the following design features of a broach :

(*i*) Rake and relief angles

(*ii*) Depth of cut per tooth

(*iii*) Width of land

(*iv*) Depth of cutting tooth

(*v*) Tooth fillet radius

(*vi*) Chip breakers

(*vii*) Cutting speeds.

29. What are the various ways of specified the drill sizes?

30. What is point angle of a drill? Give its value for different materials.

31. Sketch a twist drill and write briefly on its following elements :

(*i*) Lip angle, helix angle, chiesel angle and point angle.

(*ii*) Land width, margin and back taper.

32. Write on the materials for drills.

33. How will you determine the drilling torque and hence the drilling power?

34. What is the function of flutes in cutting tools?

35. Discuss the various types of reamers.

36. Discuss the design features of a reamer.

37. What are form tools?

38. How are the form tools classified?

39. Discuss the design features of form tools.

40. What are combination tools?

41. Discuss the various types of combination tools.

42. Discuss the main factors of reliability for cutting tools.

43. Discuss the various process stages for manufacturing a cutting tool.

44. Write in detail about the manufacture of :

(*i*) Drills

(*ii*) Milling cutters

(*iii*) Reamers

(*iv*) Broaches

(*v*) Inserted blade cutting tools.

45. Alloy steel having a hardness of 250 Bhn is to be machined in a milling machine. The depth of cut is to be 6.35 mm, feed is 0.13 mm per tooth and the cutting speed is 1.5 m/s. The milling cutter has 12 teeth and is 25 cm in diameter. The width of the cut is 12.5 cm. Find the horsepower. Take machinability factor as 8 cm^3/min/hpc.

46. Find the drilling power for 50 mm diameter drill having a feed of 0.50 mm/rev. The cutting speed is 0.75 m/s. The material factor for brass is 0.55. Determine also the drilling thrust.

47. A 20 teeth, 10 cm in diameter cutters operating at 80 rev/min with a feed of 16 mm/rev. The material to be cut is medium C.I. The width of the cut is 1.25 cm and the depth of cut is 0.125 cm. Calculate : (*a*) the horsepower at the cutter (*b*) the horsepower at the motor if the efficiency is 40%. Take machinability factor as 15 cm^3/min/hp_c.

48. The following data is given for slabmilling :

Cutter diameter = 150 mm

No. of teeth = 10

Width of cut = 60 mm

Length of job = 500 mm

Depth of cut = 3 mm

Table speed = 0.6 m/min.

Cutter speed = 100 rev./min.

Find : MRR, Power, feed per tooth. The material is high strength Al alloy.

49. A steel bar 100 mm outside diameter is being turned on a centre lathe. The spindle speed is 1000 rev./min. The cutting force is 700 N. Determine : -

(*a*) cutting velocity at outside diameter.

(*b*) cutting velocity when the diameter is 50 mm.

(*c*) power at the two cutting velocities (**Ans.** 5.24 m/s; 2.62 m/s; 3.668 kW; 1.834 kW).

50. A 19 mm bar is being turned on an automatic lathe using Titanium carbide tool for which n = 0.2 and V_{60} = 104 m/min. in Taylor's equation. Find the spindle speed for a tool life of 6 hrs. Find the cutting time for machining a length of 50 mm with feed of 0.16 mm/rev. How many workpieces can be produced between tool changes.

(**Ans.** 72.7 m/s; 1217.95 rev/min; 1315)

51. Calculate the maximum and minimum diameters of a reamer for a 20H7 (20 + 0.021 mm) Hole. (**Ans.** 20.017 mm, 20.009 mm)

52. For an Orthogonal machining operation, Tool Rake angle is 10°, chip thickness is 0.4 mm and the uncut chip thickness is 1.5 mm. Calculate : cutting ratio, shear plane angle and magnitude of shear strain. (**Ans.** 0.375, 21.55°, 2.736)

16

GEAR MANUFACTURING

16.1. INTRODUCTION

Gears are used extensively for transmission of power. They find application in : Automobiles, gear boxes, oil engines, machine tools, industrial machinery, agricultural machinery, geared motors etc. To meet the strenuous service conditions the gears should have : robust construction, reliable performance, high efficiency, economy and long life. Also, the gears should be fatigue free and free from high stresses to avoid their frequent failures. The gear drives should be free from noise, chatter and should ensure high load carrying capacity at constant velocity ratio. To meet all the above conditions, the gear manufacture has become a highly specialised field. Below, we shall discuss the various materials and manufacturing processes to produce gears.

16.2. MATERIALS

The various materials used for gears include a wide variety of cast irons, steels, non-ferrous materials (Bronzes, Aluminium) and non-metallic materials such as : leather, plastic, nylon, texolite etc. The selection of the gear materials depends upon :

(*i*) Type of service

(*ii*) Peripheral speed

(*iii*) Degree of accuracy required

(*iv*) Method of manufacture

(*v*) Required dimensions and weight of the drive

(*vi*) Allowable stress

(*vii*) Shock resistance; and

(*viii*) Wear resistance

1. Cast iron is popular due to its good wearing properties, excellent machinability, and ease of producing complicated shapes by the casting method. However, it is suitable where stresses on gear are not high and where large gears of complicated shapes are needed.

2. Steel is sufficiently strong and highly resistant to wear by abrasion. The teeth can be machined to the required degree of accuracy and also it can transmit heavy loads.

3. Cast steel is used where stress on gear is high and it is difficult to fabricate the gears.

4. Plain carbon steels (medium carbon steels and high carbon steels, C = 0.3 to 0.6%) find application for industrial gears where high toughness combined with high strength, fatigue limit in bending, long life and accuracy of teeth are required.

5. Where high tooth strength and low tooth wear are required, for example, industrial and automobile gears alloy steels (Mn, Ni, Cr, Mo) are used.

6. To combine properly the toughness, gears made of steels are heat treated. The various heat treatment processes employed are : case hardening, through hardening, nitriding etc. High hardness heat treated steels give smaller gears for a given load and also the service life is increased by about ten times without increasing the size or weight of the gears.

(*a*) Case hardened steel with carbon less than 0.2 per cent and alloyed with Ni, Cr and Mo is used for industrial gears where the conditions are of varying loads and shocks.

(*b*) Surface hardening steel (C = 0.4 to 0.6%) and alloyed with Mn, Cr and Mo and flame/induction hardened, is used where tooth surface has to be very hard and core soft, for example, gears used for transmitting shocks loads.

(*c*) Nitrided steel is used where hardness has to be very high (750-800 BHN), for example, aircraft gears, marine propulsion gears and industrial gears etc.

7. Bronzes are very popular in worm gear drives because of their ability to withstand heavy sliding loads and ability to "wear in" to fit hardened steel worms. They are also very useful for corrosive conditions. Like C.I., they can also be easily cast into complex shapes.

8. Aluminium is used where low inertia of rotating mass is desired, for example, instrument gears and appliances.

9. Gears made of non-metallic materials give noiseless operation at high peripheral speeds, but their load carrying capacity is low, for example, gears for instruments and appliances etc.

16.3. METHODS OF MANUFACTURE

The various manufacturing methods used for gears can be listed under the following headings :

1. Casting
2. Metal forming
3. Metal removal

Out of the above methods, metal removal or machining is the best and most widely used method for mass production of high quality gears.

16.3.1. Casting. Gears can be produced by the various casting processes. Sand casting is economical and can take up large size and module, but the gears have rough surfaces and are inaccurate dimensionally. These gears are used in machinery where operating speed is low and where noise and accuracy of motion can be tolerated, for example, farm machinery and some hand operated devices. Sand casting is suitable for one off or small batches. Large quantities of small gears are made by "Die-Casting". These gears are fairly accurate and need little finishing. However, the materials used are low melting ones, such as alloys of zinc, aluminium and copper. So, these gears are suitable for light duty applications only (light loads at moderate speeds), for example, gears used in toys. cameras and counters etc. Gears made by "Investment Casting" may be accurate with good surface finish. These can be made of strong materials to withstand heavy loads. Moderate-size gears are currently being steel cast in metal moulds to produce preforms which are later forged to size. Light gears of thermoplastics are made by "Injection Moulding". This method is suitable for large volume production. However, gear tooth accuracy is no high and initial tool cost is high. These gears find use in instruments, household appliances etc.

For phosphor bronze worm wheel rims, "centrifugal casting" is used far more extensively than any other method. Centrifugal casting is also applied to the manufacture of steel gears. Both vertical and horizontal axis spinners are used. After casting, the gears are annealed or nomalized to remove cooling stresses. They may then be heat treated, if required, to provide the needed properties. Centrifugally cast gears perform as well as rolled (discussed ahead) gears and are usually less expensive. "Shell moulding" is also sometimes used to produce small gears and the product is a good cast gear of sowewhat lower accuracy than one made by investment casting but much superior to sand casting.

16.3.2. Forming Method. Methods included under this heading are : Roll forming, extrusion and cold drawing.

1. **Roll forming.** In roll forming of gears, the gear blank is mounted on a shaft and is pressed against hardened steel rolling dies. The rolls are fed inward gradually during several revolutions which produce the gear teeth from the solid. The forming rolls are very accurately made and the

roll formed gear teeth usually have excellent accuracy. Roll forming is done both hot and cold. In hot/roll forming, the hot-rolled gear is usually cold-rolled which completes the gear with a smooth mirror finish. In cold roll forming, higher pressures are needed as compared to hot rolling. Many of the gears produced by this process need no further finishing. Since the material is strained plastically during the process, it becomes stronger against both tension and fatigue. Since this method can be called as chipless method (no metal removal), there is lot of material saving. Also, there is saving in machining operation, may be 80%. The machinery is very expensive and so the method is suitable only for mass production. Both spur and helical gears are being made by this process.

2. **Extrusion.** Small sized gear can also be made by extrusion process. Other operations like piercing the hole upsetting or heading etc. are also needed to produce the complete gear. At the end, only finish grinding is required to complete the job. There is saving in material and machining time. The method can produce any shape of tooth form and is suitable for high volume production. Gears produced by extrusion find application in watches, clocks, typewriters, etc. Previously, only brass, bronze, aluminium and magnesium alloys were extruded. But now steel bars upto about 60 mm diameter are cold extruded.

16.3.3. Metal Removal. In this method, the metal is removed from the gear blanks to produce the gears. Under this, the main gear cutting methods are :

(*a*) Profiling or Form Cutter Method

(*b*) Generation

Profiling or Form Cutter Method. In this method, the spaces between the teeth are cut by means of a cutter conforming to the gear tooth form. This method necessitates the use of special cutter for each application and so is not very suitable for large volume production. The various methods under this category are :

1. Gear cutting on milling machine with formed disc cutter or end-milling cutter.
2. Gear cutting on planer or shaper with single point formed tool.
3. Cutting external or internal teeth with formed cutter in broaching machine.
4. Gear cutting with form tool blades by shear speed process.

Gear generating process. This process is based on the principle that any two involute gears, or any gear and a rack, of the same module will mesh together properly. If one of the gears (or the rack) is made into a cutter by proper sharpening and it meshes with the gear blank, then the teeth on blank are developed or generated by the relative rolling motion of the cutter and the blank. This process in which the cutter is similar to a mating pinion or a rack is called "Gear Shaping". If the generating tool (cutter) is threaded and gashed, it is called a "hob" and the process is called as "Gear Hobbing".

The major advantages of gear generation is that the same cutter of a particular module can cut gears of different number of teeth without profile deviation. Due to this, this is the most popular method of producing gears in mass.

16.4. GEAR CUTTING BY MILLING

For cutting a gear on a milling machine, the gear blank is mounted on an arbor which is supported between a dead centre and a live centre in the indexing head. The cutter is mounted on the arbor of the milling machine. The geometric centre of the cutter must be aligned exactly vertically with the centre line of the indexing head spindle. The table of the machine is moved upward until the cutter just touches the periphery of gear blank. The vertical feed dial is set to zero. The table is then moved horizontally until the cutter clears the gear blank. The table is then moved upwards by an amount equal to the full depth of the gear tooth. The vertical movement may be less if the gear is to be cut in two or more passes. After this, the longitudinal feed of the table is engaged. The gear blank moves under the rotating cutter and a tooth space is cut. After this, the movement of the table is reversed so that the cutter again clears the gear blank.

The gear blank is then indexed to the next position for cutting the second tooth space. This procedurre is repeated until all the teeth have been milled.

This method is shown in Fig. 16.1 (*a*). The cutter is flat circular disc type cutter. The plane is rotation of the cutter is radial with respect to the blank. The second method, shown at (*b*), makes use of an end-cutter or miller. The cutter rotates about an axis which is set radially with respect to the blank and at the same time the cutter is traversed parallel to the axes of the blank. The cutting edges lie on a surface of revolution, so that any axial cross-section of the cutter corresponds to the shape required for the space between two adjacent teeth on the finished wheel. The milling machine used in this method is vertical milling machine.

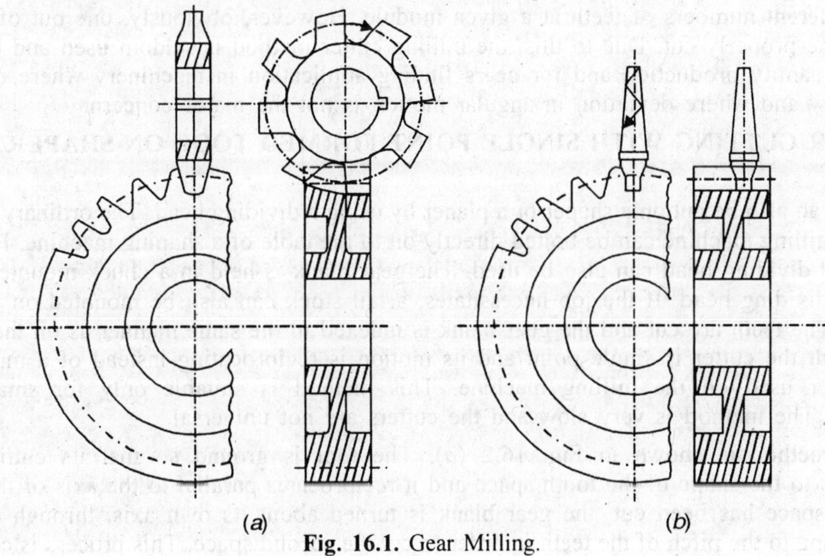

Fig. 16.1. Gear Milling.

The end mill cutter is mounted straight on the milling machine spindle through a chuck. The disk type of cutter is used to cut big spur gear of large pitch, whereas, the end mill type of cutter is employed for the manufacture of pinion of large pitch.

The quality of gear produced by this method is not high since the operation of the indexing device is not precise. Also, this method is very slow since only one tooth is cut at a time. To overcome these two drawbacks, multiple tool shaping cutter head, Fig. 16.2, is used to cut all the tooth spaces of the gear at the same time. The form tools are arranged radially in the cutter head

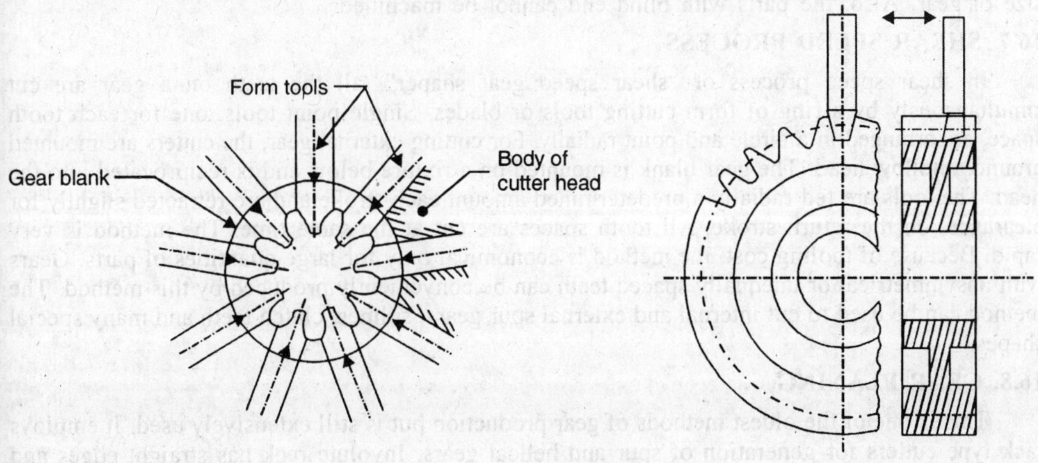

Fig. 16.2. Multiple-tool Shaping Cutter Head. **Fig. 16.2** (*a*)

and their number is equal to the gear-tooth spaces desired. All the tools are fed radially towards the centre of blank by an amount equal to the tooth depth, prior to each cutting stroke. After the cutting stroke (into the plane of paper), all the tools are simultaneously retracted.

Gear milling is a simple, economical and flexible method of gear making. Spur, helical, bevel gears and racks can be produced by this method. The method is versatile in that tooth forms, that are not possible by generation methods, can be produced by milling.

The major disadvantage of this method is that a separate cutter must be used not only for every module but for every number of teeth. If this method is followed, the cost of producing a gear will be very high. To reduce the cost, the usual procedure is to use the same cutter for 8 to 10 different numbers of teeth at a given module. However, obviously, one out of 8 or 10 gears will be properly cut. Due to this, the milling cutter method is seldom used and that only for small quantity production and for gears finding application in machinery where operation speed is low and where deviation in angular motion is not the major concern.

16.5. GEAR CUTTING WITH SINGLE POINT FORMED TOOL ON SHAPER/ PLANER

Gear can also be cut on a shaper or a planer by using a dividing head. The ordinary dividing head on a milling machine can be bolted directly on to the table of a shaping machine. However, a simplified dividing head can also be used. The gear blank is held in a chuck mounted on the spindle of dividing head. If the job necessitates, a tail stock can also be mounted on the table of shaper. The teeth are cut and the gear blank is indexed in the same manner as on the milling machine, but the cutter is single point and its motion is reciprocating instead of a multi-point rotary cutter used on the milling machine. This method is suitable only for small batch production. The method is very slow and the cutters are not universal.

This method is shown in Fig. 16.2 (*a*). The tool is ground so that its cutting edge corresponds to the shape of the tooth space and it reciprocates parallel to the axis of the blank. When one space has been cut, the gear blank is turned about its own axis, through an angle corresponding to the pitch of the teeth in order to cut the second space. This process is continued until all the tooth spaces have been cut.

16.6. BROACHING

Broaching produces highly accurate gears with an excellent surface finish. Internal gears, racks and gear sectors have been produced by broaching. By an operation, called "pot broaching", external gears can be made in one pass by a circular broach having in ward-facing teeth. Broaching is a highly productive method and is suitable for high volume production. The method has the drawbacks : tool is not universal and a separate broach must be used for each size of gear. Also, the parts with blind end cannot be machined.

16.7. SHEAR SPEED PROCESS

In shear speed process or "shear speed gear shaper", all the teeth on a gear are cut simultaneously by a ring of form cutting tools or blades. Single point tools, one for each tooth space, are arranged in a circle and point radially. For cutting external gear, the cutters are mounted around a hollow head. The gear blank is mounted on a fixture below and is reciprocated into the head. The tools are fed radially a predetermined amount each stroke and are retracted slightly for clearance, on the return stroke. All tooth spaces are cut at the same time. The method is very rapid. Because of tooling cost, the method is economical only for large quantities of parts. Gears with unsymmetrical or unequally spaced teeth can be conveniently produced by this method. The method can be used to cut internal and external spur gears, splines, clutch teeth and many special shapes.

16.8. GEAR PLANING

This is one of the oldest methods of gear production but is still extensively used. It employs rack type cutters for generation of spur and helical gears. Involute rack has straight edges and sharp corners and hence can be manufactured easily and accurately. The cutters generate as they cut and as the name implies, the machine cuts the teeth by the reciprocating planing action of the cutter. This is a true generating process since it utilizes the principle that an involute curve can be formed by a straight generator (rack type cutter) when a circle (gear blank) is made to roll without slip relative to the generator, Fig. 16.3.

There are two types of gear planing machines, one based on "The Sunderland Process" and the other on "The Maag Process". Both the methods are identical in principle but differ in machine configuration and detail.

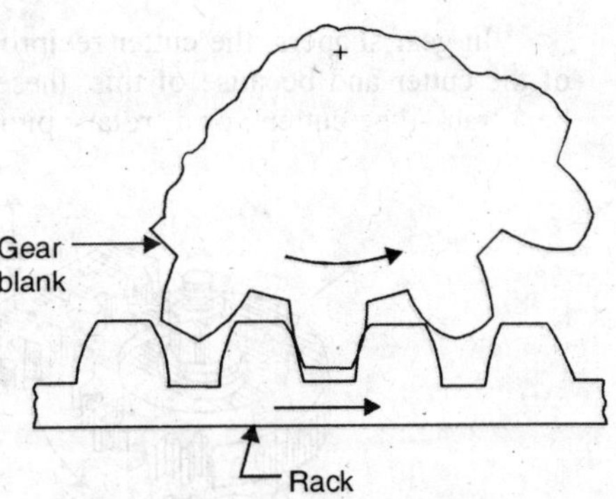

Fig. 16.3. Generation by rack.

The Sunderland process. In this method the work (gear blank) is mounted with its axis horizontal, and the cutter slide is carried on a saddle that moves vertically downwards as cutting proceeds. For cutting spur gears, the cutter reciprocates parallel to the work axis but because it can be swivelled in the vertical plane to any desired angle, the machine is also used for cutting single helical gears. The cutter is gradually fed to the desired depth of teeth after which the depth remains constant. Simultaneously, the gear blank is rotating and the rack is traversed at a tangent, the motion of rack and blank being geared to act in unison on their respective pitch lines. This relative motion brings fresh part of the blank and rack into contact and thus causes the teeth of the cutter to generate wheel teeth of exactly the correct form. Since a long rack to cut a full gear is not practicable, the machine is arranged to index automatically after the rolling motion has occurred for a distance of one or more pitches. The indexing really consists in stopping the rotation of the blank and causing the rack to move back the amount it has advanced and the process is repeated until the blank has completed one revolution.

Fig. 16.4 explains the principle of gear planing. Thus, while the cutter during its cutting stroke is in contact with several teeth at the same time, and with different part of each tooth, it planes comparatively a narrow strip on each tooth at each stroke, and a different part of each tooth is submitted to the action of the cutter at the next stroke.

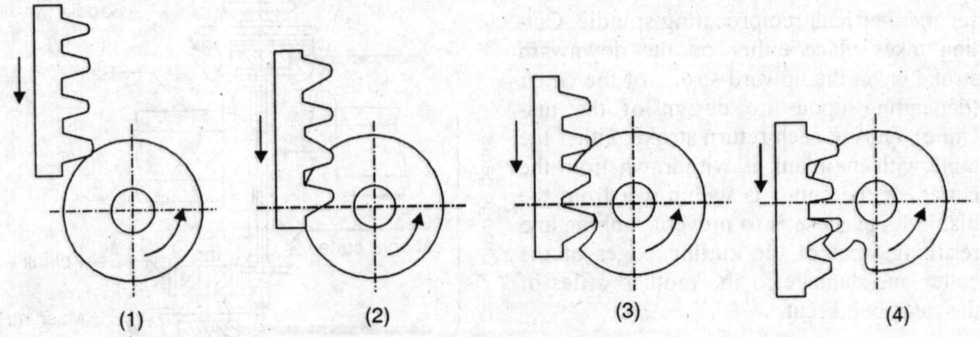

Fig. 16.4. Spur Gear Planing.

To improve the above method regarding the output and quality, two modifications are incorporated :

(*a*) In double cutting method, two blanks are mounted side by side and two cutters are mounted on a duplex cutter holder. The cycle is so arranged that the cutting action of each cutter takes place on alternate strokes.

(*b*) In the double action method, a special box carries two cutters back to back. One cuts during forward stroke and the other during return stroke. One cutter works at the bottom of the tooth space while other finishes the sides. This avoids idle stroke and maintains the cutter condition over a long period.

The Maag Process. In this method, the work is mounted on the machine table with its axis vertical. The rack cutter is carried in a cutter head that is made to move in a vertical plane but the actual direction of motion of the enter in its slide can be set at any desired angle.

16.9. GEAR SHAPERS

In gear shapers, the cutter reciprocates rapidly. The teeth are cut by the reciprocating motion of the cutter and because of this, these machines are called "gear shapers". The cutter can either be a 'rack-type cutter', or a 'rotary pinion type cutter', Fig. 16.5. The main drawback of 'rack-type

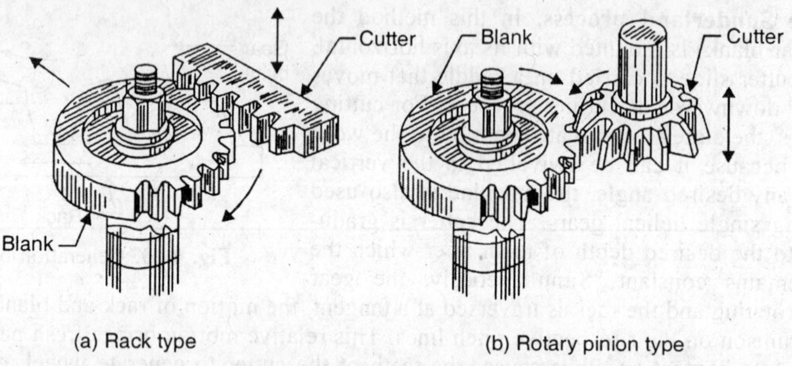

Fig. 16.5. Gear Shaper Cutters.

cutter' is that once the length of rack is covered by the gear blank, the cutting process is interrupted to index the blank back to starting point. In the case of 'rotating pinion type cutter', such an indexing is not required, therefore, this type is more productive and so common. The principle of a gear shaper is shown in Fig. 16.6. The cutter is pinion shaped with a clearance on the tooth face. The gear blank mounted on a vertical spindle and the cutter on the end of a second, vertical, reciprocating spindle. Cutting takes place either on the downward stroke or on the upward stroke of the cutter (depending upon the design of the machine). During each return stroke, either the table with the blank is withdrawn from the cutter, or the cutter is withdrawn from the blank. Its purpose is to prevent rubbing and resulting wear of the cutting edges of the cutter and damage to the tooth profiles of the gear being cut.

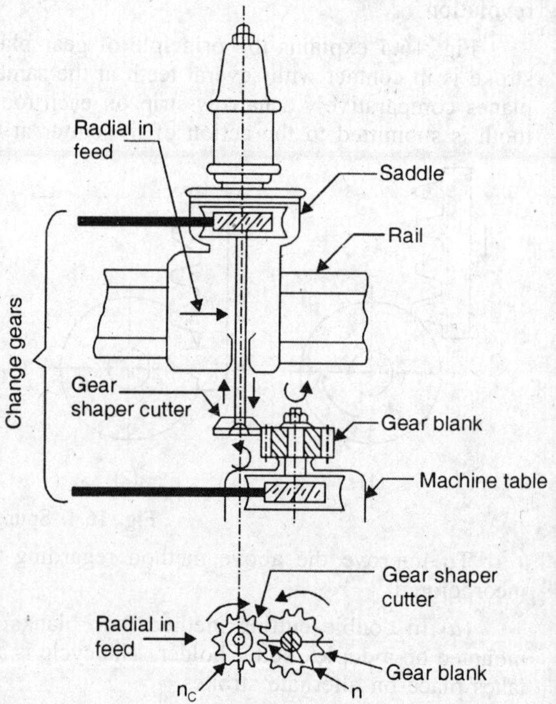

Fig. 16.6. Principle of a Gear Shaper.

To start cutting a gear, the tool is fed into the gear blank before each cutting stroke. When a required depth is reached, the inward feed stops and the cutter and blank slowly start rotating as if they were in mesh. The gear is cut in one complete rotation of the blank. The rotary motions of the cutter and the blank are co-ordinated through change gears so that the surface speeds of the cutter and the blank are the same. The cutter reciprocates at about 100 strokes per minute for an average job and the strokes can be upto 2000 per minute for fine tooth gears. Both straight and helical-tooth gears can be cut on gear shapers. For cutting helical-tooth gears, both the tool and the gear blank are given an oscillating rotational motion, during each stroke of the cutter, turning in one direction during the cutting stroke and in the opposite direction during the return stroke.

Advantages

1. One cutter can be used for cutting all spur gears of the same module, irrespective of number of teeth on the gear.

2. It can cut teeth, up close to a shoulder with only a slight clearance recess.

3. Because the cutting stroke can be adjusted, gear shapers are particularly suitable for cutting cluster gears.

4. The method is versatile and can cut spur, helical, herring-bone, internal and cluster gears, racks, splines, pawls and many others.

5. The cutter has a generated profile which is more accurate in shape than some cutters, so, the gears produced are accurate.

6. The method is suitable for medium and large batch production.

Limitations

1. The cutting action is reciprocating, and cutting takes place only during one half of the stroke, so, only about half the machine time is spent in metal removal.

2. Separate helical guide is required for cutting helical gear of particular helix angle in a particular direction.

Applications. Gears produced by gear shaper find applications in automobiles, machine tools, instruments, machinery, clocks and other equipment.

The above method of gear shaping, in which the work is mounted on a rotating table with its axis vertical and the cutter is carried on a ram that is given a reciprocating vertical motion, is known as 'The Fellows Process'. Another technique of cutting helical gears on such a machine is to use a cutter made with helical teeth and the ram carrying the tool is fitted with helical guides. During the working stroke, the cutter teeth move in a helical path corresponding to the angle of the tooth spirals required on the work.

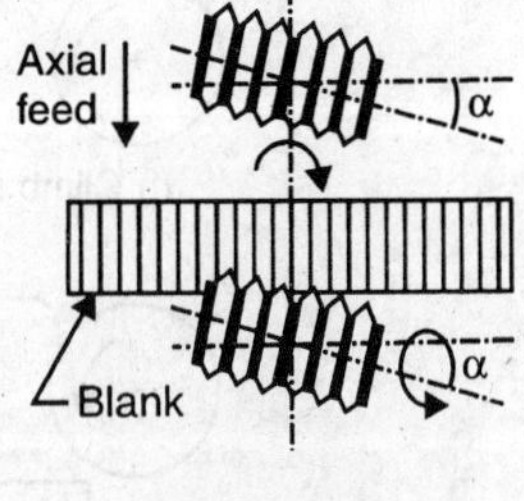

(a) Spur gear

The Sykes Process. In this type of gear shaping machine, the work is mounted with its axis horizontal. The cutter is carried on a ram that reciprocates, horizontally and also rotates uniformly. Spur, single helical and double helical gears can be cut on this machine. For cutting double helical gears, two helical guides are required, one right-and the other left-handed, and one forms a sleeve over the other.

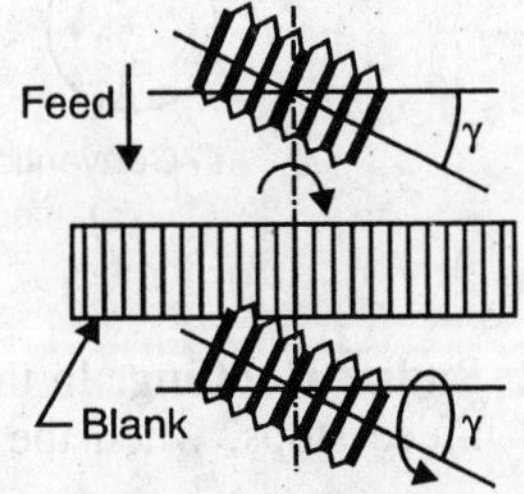

(b) Helical gear

Fig. 16.7. Hobbing Gear Teeth.

16.10. GEAR HOBBING

Hobbing is the process of generating gear teeth by means of a rotating cutter called a "hob". A hob resembles a worm, with gashes made parallel to its axis to provide cutting edges. Relief is provided behind each of the helically arranged cutting faces, for clearance. Gear hobbing is a continuous cutting operation. The hob and the gear blank are connected by means of proper change gears. The ratio of hob and blank speeds is such that during one revolution of the hob, the blank turns through as many teeth as there are starts (threads) on the hob. To start cutting a thread, the hob is made to clear the blank. It is then moved inwards to obtain the required tooth depth. After the tooth depth is reached, the hob is fed in a direction parallel with the axis of rotation of the gear (Axial hobbing). As the gear blank rotates, the teeth are generated and the feed of the hob across the face of the blank extends the teeth to the desired tooth face width. One rotation of the blank completes the cutting unless the blank has a wide face.

To cut spur or helical gears, the hob is set in relation to the axis of the gear blank, so that the threads of the hob facing the gear blank are directed along the axes of the tooth spaces for cutting spur gears and at the helix angle of the teeth while cutting helical gears. This is achieved by setting the hob axis at an angle to the horizontal equal to the helix angle of the hob $\propto$, (for cutting spur gears) and at angle $\gamma = \beta \pm \propto$ for cutting helical gears, where β is the helix angle of the helical gear, Fig. 16.7. Plus sign is to be used if the hands of the helical gears and hob are different (one right hand and the other left hand) and if the hands are the same, minus sign is to be used.

Types of Hobbing. Depending upon the direction of feed of the hob, the gear hobbing may be classified as :

Axial hobbing, radial hobbing and tangential hobbing.

1. **Axial Hobbing.** In this type, firstly the gear blank is brought towards the hob to get the desired tooth depth. The table slide is then clamped. After that, the hob, while rotating, moves along the face of the blank to complete the job, Fig. 16.8 (*a*). Axial hobbing, which is used to cut spur and helical gears can be "climb hobbing" or "Conventional hobbing".

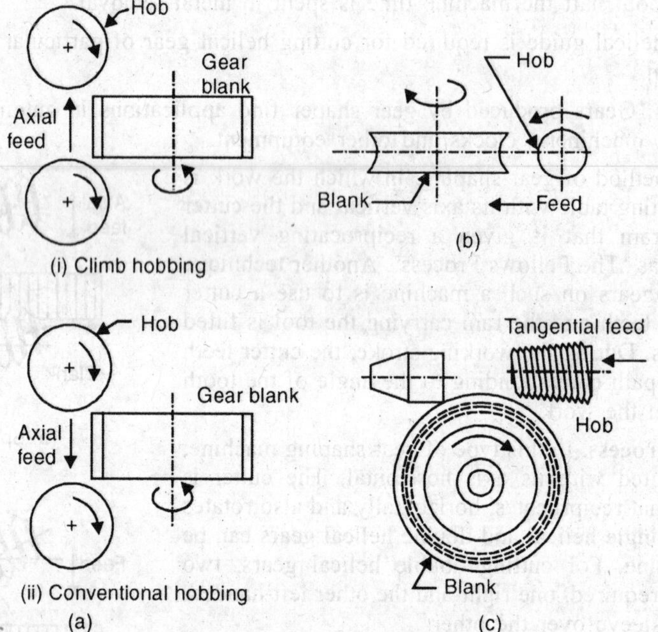

Fig. 16.8. Types of Hobbing.

2. **Radial Hobbing.** In this type, the feed of the hob is radial towards the centre of blank. Radial infeed stops, when the full depth of cut is reached, Fig. 16.8 (*b*).

Radial hobbing has got the following characteristics :

(*i*) Only a small portion of hob length is doing cutting at any time, so, the hob wears non-uniformly. This can affect the tooth profile accuracy of the cut gear.

(*ii*) Again, a comparatively few hob teeth cut each tooth on gear, producing more or less visible flats which also reduces the tooth profile accuracy.

Radial hobbing has a higher production capacity and is used for producing worm wheels whose helix angle does not exceed 6° to 7°.

3. **Tangential Hobbing, Fig. 16.8. (*c*).** In this method, the hob is set at the start of cutting to the full tooth depth and is then fed into the gear blank by an axial feed motion. This method is generally used for generating worm wheels.

Axial hobbing is most commonly used but it has the drawback that there should be enough space for the hob approaching length. When such a condition is not available, the combined "Radial Axial Hobbing" is used. In "Diagonal hobbing", both the axial and the tangential feeds are combined. This method produces a crossed pattern of cutting on the cut teeth, imparting better rolling characteristics to the gear. The wear is uniform all along the hob length which gives longer hob life.

This method is popular for generating tapered gears.

Advantages of Gear Hobbing

(*a*) The method is versatile and can generate spur, helical, worm and worm wheels.

(*b*) Since gear hobbing is a continuous process, it is rapid, economical and highly productive.

(*c*) The method produces accurate gears and is suitable for medium and large batch production.

(*d*) The cutter is universal, because it can cut all gears of same module, irrespective of number of teeth on the gear.

Disadvantages

(*a*) Gear hobbing cannot generate internal gears.

(*b*) Enough space has to be there in component configuration for hob approach.

Applications. The gears produced by gear hobbing are used in automobiles, machine tools, various instruments, clocks and other equipments.

16.11. BEVEL GEAR GENERATING

For generating straight bevel gears, the rolling motions of two pitch cones are employed instead of pitch cylinders (used for cutting spur and helical gears). In cylindrical gears, we have seen that the teeth are generated by tools that simulate or are derived from the basic rack, whereas bevel gear teeth are produced by cutters representing the basic crown wheel (A crown wheel is a bevel gear in which the tooth faces lie in one plane. It is equivalent to the rack in spur gearing).

16.11.1. Straight bevel gear planing. In the production of straight bevel gears by planing, use is made of two straight sided reciprocating tools which shape the profile of the teeth being cut, while the workpiece is given the appropriate rolling motion.

The arrangement to generate straight bevel gears is shown in Fig. 16.9. Tool slides with tools reciprocate along ways provided on the face of the cradle of the machine.

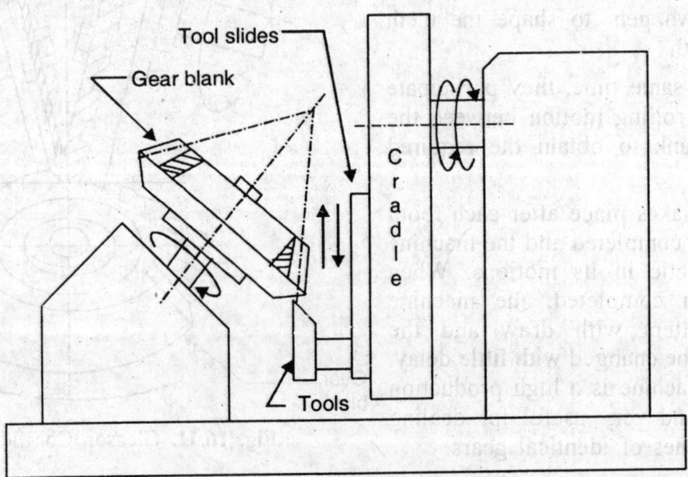

Fig. 16.9. Straight Bevel Gear Planing.

The gear blank is rotated. Also rotated is the cradle with the reciprocating tools. The tools cut with their motion towards the apex of the gear pitch cone. The return stroke is idle, when the tools are withdrawn from the blank to avoid damaging the machined surfaces of the teeth. The two tools represent only one tooth space of the imaginary crown gear.

The complete cycle of generating a tooth is illustrated in Fig. 16.10. At the start, the tools for machining one of the side surface of teeth begins to cut into the blank (positions 1 and 2). Then the second tool, designed to generate the other side surfaces of the teeth, starts to cut the gear (positions 3 and 4). At this point, the first tooth has been completely cut. As the cradle rolls further the tools run out of mesh with the gear blank. So, both gear blank and cradle are reversed. At the end of the reverse rotation, the blank is indexed for the next tooth (positions 5 and 6). This cycle is repeated for each tooth of the gear.

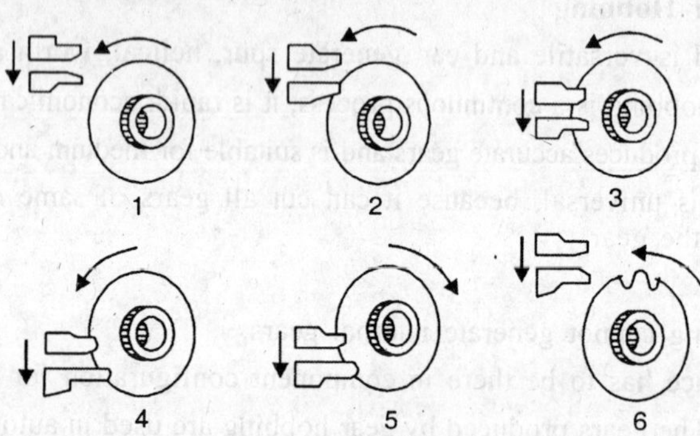

Fig. 16.10. Complete Cycle of Generating a Tooth.

16.11.2. Gleason Method. In this method, two disc milling cutters are employed, Fig. 16.11. The tools form the blanks of a tooth simulating the basic crown wheel. Cutter teeth are inter-meshing and the discs are incined to each other at the required pressure angle (usually 20°). The following motions are involved while cutting a tooth :

1. The rotating cutters revolve about their axes to provide the cutting action.

2. They travel in planes passing through the sides of the teeth on the imaginary crown gear to shape the teeth along their teeth.

3. At the same time, they participate in the relative rolling motion between the cutters and blank to obtain the required tooth profile.

Indexing takes place after each tooth space has been completed and the machine is fully automatic in its motions. When gear has been completed, the machine stops, the cutters with draw and the workpiece can be changed with little delay. This type of machine is a high production rate machine and very useful for dealing with large batches of identical gears.

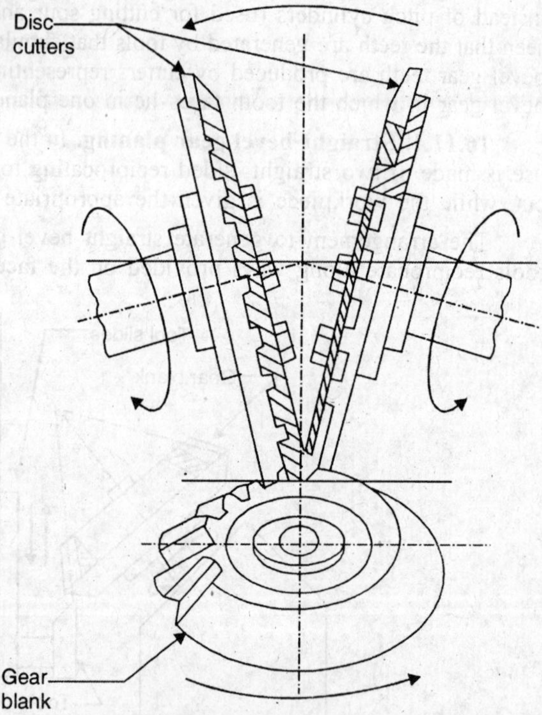

Fig. 16.11. Gleason Coniflex Method.

16.12. MISCELLANEOUS GEAR MANUFACTURING METHODS

Some other methods for making gears can be : Gear stamping, cold drawing, powder metallurgy, electro-chemical matching, electrical discharge machining and HERF.

1. **Cold drawing.** In this method, the material is passed through a number of dies. The shape of the final die corresponds to the desired shape of the tooth. During the passage of the material through the dies, it is squeezed into the shape of the die. Since the material is displaced by pressure, the tooth surface is quite hard and smooth. The materials which can be employed for cold drawing are : low to medium carbon steels, brass, bronze and aluminium. Gears produced by this method find application in : watches, electric clocks, spring wound clocks, typewriters and so on.

2. **Stamping.** Large quantities of gears are made by the method known as "stamping", "blanking" or "fine blanking". The gears are made in a punch press from sheet metal, upto about 12.7 mm (maximum) thick. Such gears find applications in : toys, clocks and timers, watches, household gadgets, water and electric meters and some business equipment. If, after blanking, the gears are shaved, they give best finish and accuracy. The materials which can be stamped are : low, medium and high carbon steels, stainless steel, all brass alloys and aluminium alloys. The method is suitable for large volume production.

3. **Powder metallurgy.** High quality gears (both dimensional and surface quality) can be made by powder metallurgy (P/M) method. The metal powder is pressed in dies conforming to the tooth shape, after which the product is sintered. After sintering, the gear may be coined to increase density and surface finish. This method is usually used for small gears (upto about 25 mm in diameter). Large gears can be made by powder metal forging (P/M forging) in which performs made by powder metallurgy are forged to the final shape. The method is suitable for large volume production.

Sintered gears can be held to close limits of dimensional accuracy with a good surface texture and they require little or no subsequent machining. Sintered gears can be impregnated with a plastic to impart additional wear resistance or with another metal of lower melting point than the sintered material, to give them increased strength.

An example of a suitable powder for producing gears by P/M technique can have the following composition :

Nickel	1.75%
Molybdenum	0.50%
Copper	1.50%
Carbon	0.50%
Iron	Remainder

Accurate gear pump rotors with wear and corrosion resistant surfaces can be made from stainless steels by this method, since stainless steels are very tough to be cut easily by conventional means. Gears made by P/M method find applications in : toys, instruments, appliances, small motor drivers, lawn and garden tractor transmission gears etc.

4. **Hot forging of gears.** In this method, the whole gear is forged from a billet, usually of case hardening steel, in a pair of accurate dies split on the axial plane. It is necessary, therefore, to control the size of the fully machined billet to a fairly close tolerance to reduce the forging flash to a minimum. The dies are produced by a copying technique from a master gear. Dies are expensive to manufacture but the forging process is rapid and is fit for production. This method followed by a cold coining operation is attractive for bevel gear manufacture and is also applicable to spur-gear manufacture. A good grain flow pattern in the teeth, an accuracy in pitch and profile within 0.012 mm and a high surface finish can be obtained by this method.

16.13. GEAR FINISHING OPERATIONS

For the gears to operate efficiently and have satisfactory life, the tooth profile must be very accurate and the teeth should be hard and smooth. Gear made by cold rolling method may not need further finishing, but, gear made by other methods (particularly made by metal removal) often need further finishing operations. The finishing operations can be : shaving and burnishing for untreated gears and grinding, lapping and honing for heat treated gears (hardened).

16.13.1. Gear Shaving. Gear shaving is the most common method for gear finishing. In this method, a very hard gear (gear-shaving cutter) is used to remove fine chips from the gear-tooth profile. The shaving cutter can be : Rotary type or rack type. In rotary shaving, the cutter and the gear are run in mesh. As they rotate, the gear is traversed longitudinally across the shaving cutter or vice versa. The rotary shaving cutter, Fig. 16.12 has a number of peripheral gashes, serrations or grooves, to form a series of cutting edges. The cutter and gear are set up in gear shaving machine with crossed axes in the form of spiral gearing. The usual angles are 10° to 15°. In rack shaving, the cutter is in the form of a rack. During the operation, the gear is rolled in mesh with the cutter. The cutter is reciprocated and at the end of each stroke is fed into the gear.

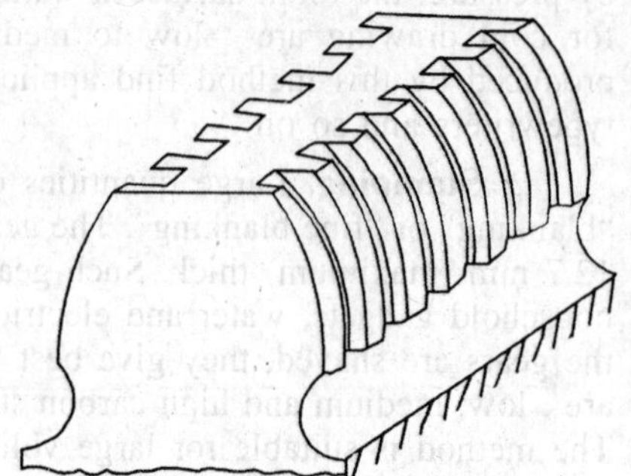

Fig. 16.12. Gear Shaving Cutter.

16.13.2. Gear Burnishing. This is not a common method of gear finishing. In this method, the gear is rolled under pressure with three hardened, accurately formed, burnishing gears. This is a cold working process in which any high points on the tooth surface are plastically deformed to get accurate and finished tooth profile. However, the operation may have a smeared metal surface.

16.3.3. Gear Grinding. Grindling is the most accurate method of gear finishing. The operation can be either by : form grinding or generation grinding. In form grinding, the wheel is dressed to the form of an involute and is similar to the formed cutter used on a milling machine. The cutter is plunged into a tooth space of the gear. The gear is completely finished by indexing after each tooth has been finished by the rotary cutter (disc grinding wheel).

In the generating type, the gear is rolled past the revolving wheel and the finishing is done by the flat face of the wheel. The wheel is wide enough to cover the entire tooth width. so that it does not have to be traversed parallel to the axis of the gear. The drawbacks of the method are : low production capacity and the grinding wheels are expensive and need skilled labour for operation.

16.13.4. Lapping. The lapping process only corrects minute heat treatment distortion errors in hardened gears. The gear to be finished is run in mesh with a gear shaped lapping tool or another mating gear (cast-iron). An abrasive lapping compound is used in between them. Only the tooth contact is improved by this method.

16.13.5. Honing. Honing is employed mainly to remove burrs and nicks and thus for improving the surface finish. In this method, the honing tool (made of a plastic material with abrasive material embedded in it) is run in mesh at crossed axes, with the gear to be finished. The gear is driven at high speeds by the tool and is also reciprocated across the tool. During the cycle, the gear is run in both directions.

16.14. GEAR INSPECTION

For proper and efficient performance of gears, these should be inspected and measured throughout the stage of their manufacture and also afterwards. The various items of the gear to be checked are : tooth profile, tooth thickness, tooth spacing, and so on. The most common method for checking gears in the combination method, in which the gear is run in mesh with a master gear in a gear testing machine.

1. **Gear-tooth vernier calipers.** These calipers can be used to measure the thickness of gear tooth on the pitch circle. A gear-tooth varnier caliper, Fig. 16.13, consists of two beams, with line scales squares with each other. There are two slides vernier scales which move along their

respective beams. The tooth thickness on the pitch circle is measured by the jaws. The procedure is to first calculate for the gear to be tested, the chordal addendum, h, and the tooth thickness at the pitch circle, w, (where h occurs) and then to set the vertical beam of the caliper for 'h' and the horizontal beam for 'w', Fig. 16.14. The dimensions h and w can be derived to be given as :

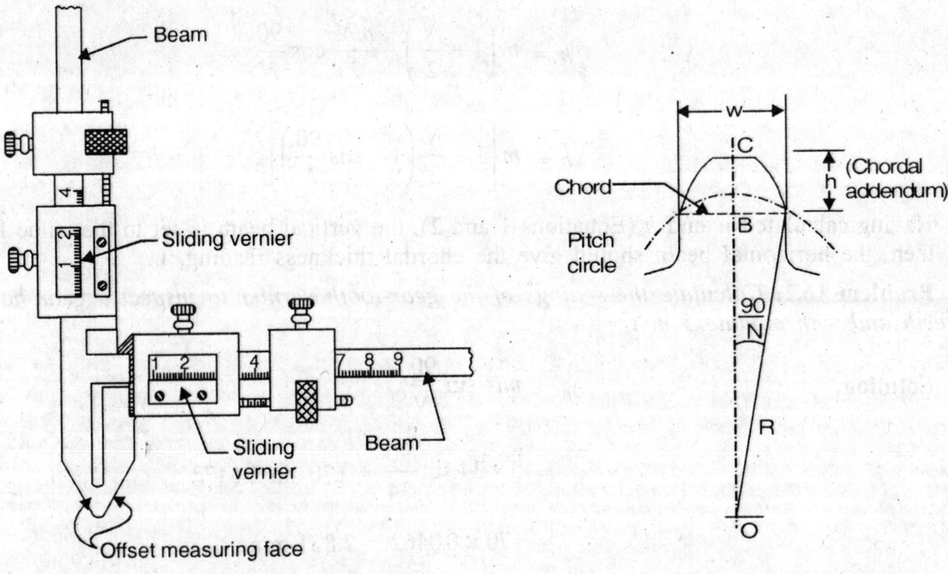

Fig. 16.13. Gear-tooth Vernier Caliper.

Fig. 16.14. Gear Tooth.

It is very clear from the figure that,

$$\frac{w}{2} = R \sin \frac{90}{N}.$$

where N = number of teeth in the gear

and R = pitch circle radius

now $m = \frac{D}{N}$, D is the pitch circle diameter

$$= \frac{2R}{N}$$

$$\therefore \quad \frac{w}{2} = \frac{mN}{2} \sin \frac{90}{N}$$

or $$w = mN \sin \frac{90}{N} \qquad ...(1)$$

Let the point where the vertical axis passing through O meets the face of the tooth be C and where it meets the chord be B. Then,

$$h = OC - OB$$

Now $$OC = R + \text{addendum}$$

$$= R + m = \frac{mN}{2} + m$$

where m = module

$$OB = R\cos\frac{90}{N} = \frac{mN}{2}\cos\frac{90}{N}$$

$$\therefore \qquad h = m\left(1+\frac{N}{2}\right) - \frac{mN}{2}\cos\frac{90}{N}$$

or

$$h = m\left[1+\frac{N}{2}\left(1-\cos\frac{90}{N}\right)\right] \qquad ...(2)$$

Having calculated w and h (Equations 1 and 2), the vertical beam is set to the value for h and, then, the horizontal beam should give the chordal thickness reading, w.

Problem 16.1. *Calculate the settings of the gear-tooth vernier to inspect a gear having 34 teeth and with module 5 mm.*

Solution.

$$w = mN\sin\frac{90}{N}$$

$$= 5\times 34\sin\frac{90}{34} = 170\sin 2.647°$$

$$= 170\times 0.0462 = 7.851 \text{ mm}$$

$$h = m\left[1+\frac{N}{2}\left(1-\cos\frac{90}{N}\right)\right]$$

$$= 5\left[1+\frac{34}{2}(1-\cos 2.647°)\right]$$

$$= 5\left[1+17(1-0.999)\right] = 5\times 1.017$$

$$= 5.085 \text{ mm}$$

2. **Two-wire method.** The method checks the tooth spacing and reference (Pitch circle) diameter. Fig. 16.15 shows a rack tooth symmetrically in mesh with a gear tooth with its straight sides touching the curved sides of the gear tooth at A and B on the lines of action. O is the pitch point. The best wire size is the one that is tangent to the teeth at the pitch line when placed in the gear tooth space (The racktooth is taken as the gear tooth space). In Δ OAC Diameter of wire or plug = 2 × OA = 2 × OC cos α ... (3)

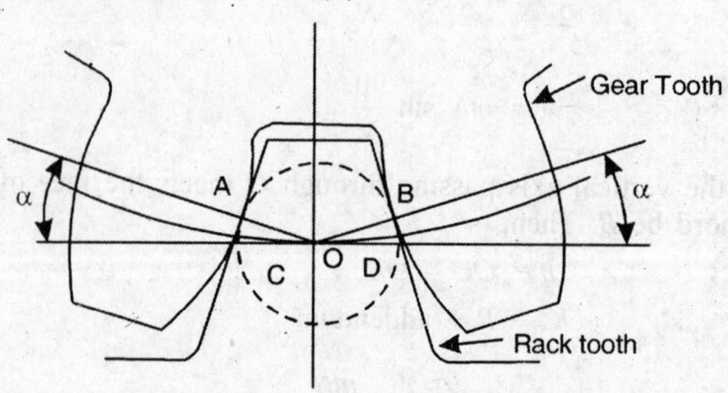

Fig. 16.15. Two-wire Method.

$$= 2 \times \frac{\text{circular pitch}}{4} \cos\alpha \quad \text{... (4)}$$

$$= 2 \times \frac{\pi m}{4} \cos\alpha = \frac{\pi m}{2} \cos\alpha \quad \text{... (5)}$$

This is the diameter of the best wire, roller or plug that will rest in the tooth space and with its centre on the reference (Pitch) circle. With a pair of such plugs placed in diametrically opposite tooth spaces, it is a simple matter to verify the Pitch circle. The accuracy of tooth spacing over any number of teeth may be checked by finding the angle subtended at the centre and comparing with that obtained from a chordal check of the plugs.

3. **Composite Error.** In most of the production shops, instead of measuring individual errors, composite error is measured which consists of pitch error, profile error, tooth alignment error and radial run-out error, For this, the gear to be tested is rolled in mesh with a master gear. The error is indicated on a sensitive dial indicator, recorded on a strip chart. The composite error is representative of the overall operation. The arrangement is shown in Fig. 16.16. The master gear and the gear to be tested are mounted on parallel spindles. The right hand slide is fixed in the required position by a screw and hand wheel to a scale and vernier. The left hand slide is movable and is held against a spring, so that the gears mesh tightly without backlash. When the gears are rotated, there will be variation in centre distance between the two spindles, due to the gear errors. The movable slide will move according to the amount of error. This movement is indicated on a dial indicator or represented on auto-graphic recorder.

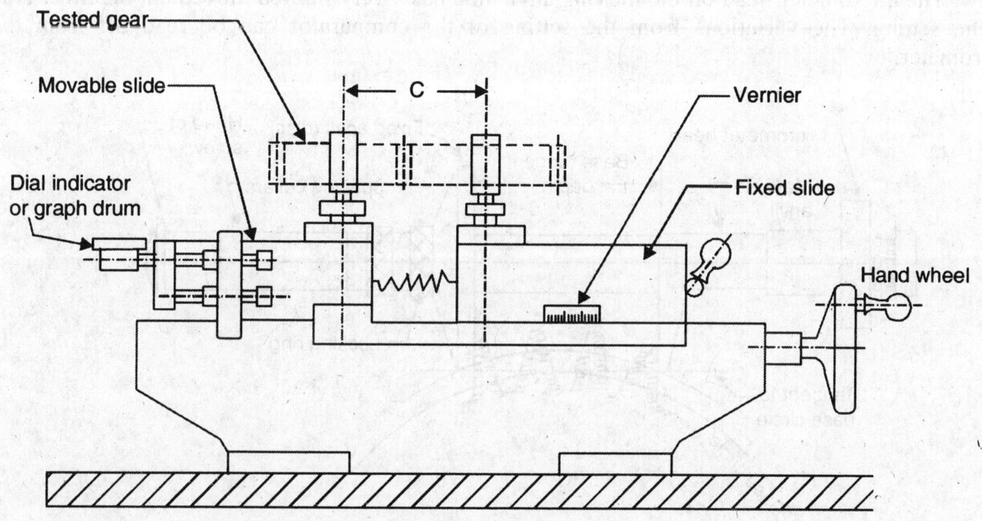

Fig. 16.16. Gear Tester.

4. **Base Tangent Method.** This is a very popular method of checking the tooth thickness. The base tangent length is the distance between two parallel planes which are tangential to opposing tooth flanks, Fig. 16.17. The number of teeth over which the measurement is to be made for a particular gear can be taken from gear handbooks. The base tangent length will consist of one base circular thickness of tooth and several base pitches. Theoretically, the base pitch is given as,

$$\text{Base pitch} = \pi m \cos\alpha \quad \text{...(6)}$$

where α is the pressure angle.

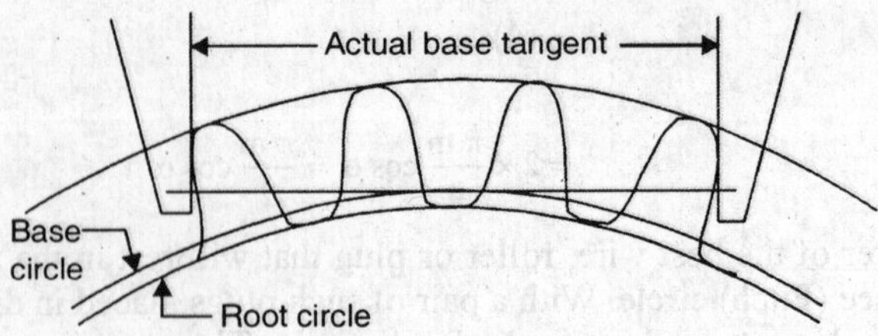

Fig. 16.17. Base Tangent Method.

It is clear that the gauging points (the points at which the two parallel planes are tangent to opposing tooth flanks) lie on the tangent to the base circle. There are two methods for determining the tooth thickness error. In one, the base tangent length is measured with the help of a flanged micrometer. The difference between the actual reading and the theoretical value is the error. The theoretical value can be computed by the following formula :

$$\text{Base Tangent Length} = Nm\cos\alpha\left[\tan\alpha - \alpha + \frac{\pi}{2N} + \frac{\pi N'}{N}\right] \quad ...(7)$$

where N' = Number of tooth spaces contained in the base tangent length being measured.

In the second method, a Base Tangent Comparator is used. It consists of a fixed anvil (disc) and a movable anvil, Fig. 16.18. The base tangent length is calculated with the help of equation (7) and this distance is then set in the tangent comparator by slip gauges after the fixed anvil has been initially set at the desired place by spacing collars put between the disc and the head of screw. The micrometer head on the moving anvil side has a very limited movement on either side of the setting. The variations from the setting of the comparator can be read off from the micrometer.

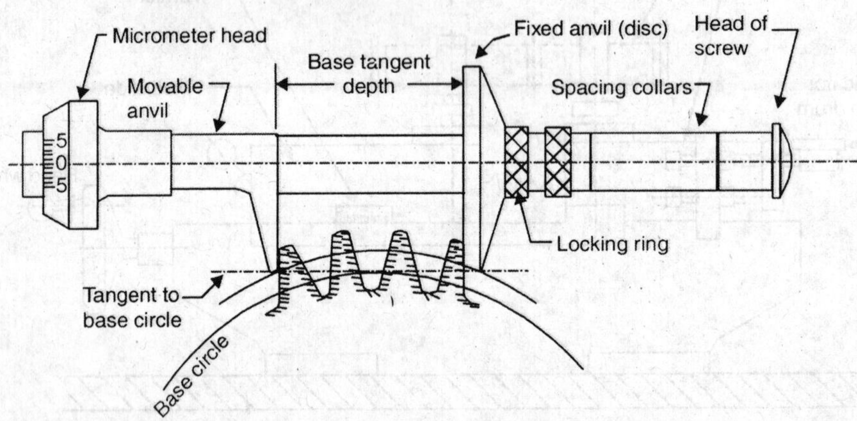

Fig. 16.18. Base Tangent Comparator

This method is more popular as compared to tooth calipers since it uses only one micrometer scale instead of two vernier scales in the case of tooth, calipers. Also, the gear tooth thickness can be measured by this method even during production on the machine, and deviations can be corrected by adjusting the cutting tool to achieve the required tooth depth and hence the tooth thickness.

5. **Pitch measurement.** Pitch is the measurement of two adjacent teeth irrespective of an axis. Pitch error or pitch variation is between any two adjacent teeth, The pitch variation is checked with the help of a pitch measuring machine or instrument. Such an instrument consists of a checking or measuring head and employs two contact points, one fixed and the other movable,

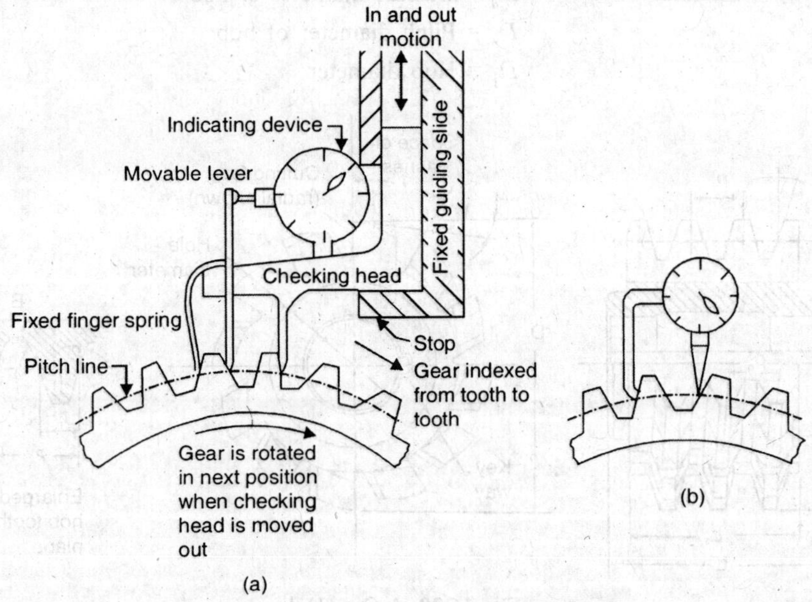

Fig. 16.19. Measurement of Pitch Variation.

Fig. 16.19 (*a*). The fixed contact (fixed finger and stop) is employed for consistent positioning on successive pairs of teeth. The movable tip is in the form of a lever whose one end contacts the gear tooth and the other end actuates the contact point of the dial indicator. The measuring head compares the tooth circular pitch with that of a selected tooth circular pitch setting on two adjacent teeth. The error or variation is displaced on the dial indicator or chart recorder. Readings are influenced by profile variation and run out of the gear as mounted in the checking instrument, Fig. 16.19 (*b*).

16.15. DESIGN OF GEAR HOB

As written earlier, a gear hob is a worm with the difference that it has cutting edges. Gear hobs that are straight sided in a normal section of the thread are extensively used for cutting spur and helical gears.

Fig. 16.20 shows schematically an arbor-type single thread finishing gear hob. The following notation is used in the illustration :

p_a = Axial lead of the thread profile

p_n = Normal lead of the thread profile

ϕ = Pressure angle of the gear to be cut = 20° (standard)

ϕ_n = Normal pressure angle of the thread profile

t_n = Normal tooth thickness

h_1 = Addendum of hob

h_2 = Dedendum of hob

h = Total depth of hob tooth

L = overall length of hob

L_1 = Effective length of hob

α = Rake angle

γ = Relief angle

D_0 = Outside diameter of hob

D_p = Pitch diameter of hob

D_1 = Hob diameter

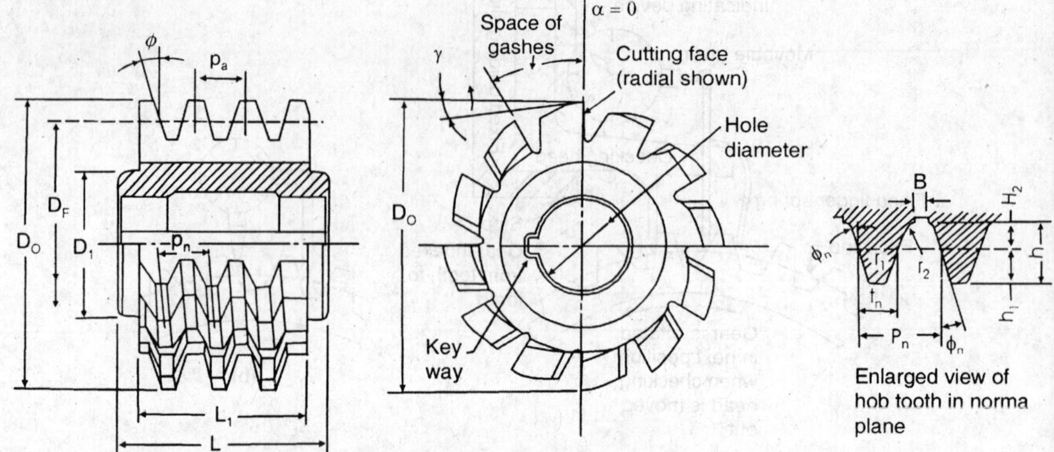

Fig. 16.20. A Gear Hob

The essential dimensions of a hob are : outside diameter, mounting hole diameter, length of hob and number of teeth.

Design features

(*i*) Module *m* is the initial parameter in hob design. For cutting spur gears, module is in the plane of rotation and for cutting helical gears, it is in the normal plane.

(*ii*) The normal pressure angle of the hob profile may be taken equal to the pressure angle of the gear to be cut. But this is not exact. For more accurate gear hob, this angle should be corrected as given below,

$$\tan \phi_n = \tan \phi . \cos \lambda$$

where λ = Lead angle of the hob thread on the pitch cylinder.

(*iii*) The hob addendum, h_1 = Dedendum of gear

= 1.25 *m*

To take into account the clearance between the outside diameter of the gear blank and the root of the hob thread,

Hob dedendum is also taken = 1.25 *m*

Total depth of hob tooth, *h* = 2.5 *m*

(*iv*) For cutting spur gears, the normal lead of the thread profile on hob, $p_n = p_c$, the circular pitch of the gear to be cut.

For cutting helical gears, p_n = Pitch of the gear in the normal plane.

(*v*) Corner radius of hob tooth, r_1 = (0.25 to 0.30) *m*

Root or fillet radius, r_2 = (0.2 to 0.3) *m*

(*vi*) **Outside diameter of hob.** Larger the outside diameter, less will be the effect of profiling error, larger arbor can be used, thereby, reducing vibrations during hobbing. But more H.S.S. will be used and production capacity will be reduced.

The outside diameter, hole diameter and the number of teeth on the hob can be selected from Table 16.1.

Table 16.1. Hob Dimensions

Module	*PrecisionSolid Hobs*			*General-Purpose Solid Hobs*			*General-Purpose Inserted-Blade Hobs*		
mm	*Do* *mm*	*d* *mm*	*z*	*Do* *mm*	*d* *mm*	*z*	*Do* *mm*	*d* *mm*	*z*
1 – 1.25	70	32	16	63	27	12	—	—	—
1.5 – 1.75	80	40	16	63	27	12	—	—	—
2 – 2.25	90	40	14	70	27	12	—	—	—
2.5 – 2.75	100	40	14	80	32	10	—	—	—
3 – 3.75	112	40	14	90	32	10	—	—	—
4 – 4.5	125	50	14	100	32	10	—	—	—
5 – 5.5	140	50	14	112	40	10	—	—	—
6 – 7	160	60	12	125	40	9	—	—	—
8	180	60	12	140	40	9	—	—	—
9	200	60	12	140	40	9	—	—	—
10	225	60	12	160	50	9	180	40	8
11	—	—	—	160	50	9	180	40	8
12 – 14	—	—	—	180	50	9	200	50	8
16 – 18	—	—	—	—	—	—	225	50	8
20	—	—	—	—	—	—	250	60	8

(*vii*) **Hob length and cutting face width**. The cutting face width of a hob L_1, should not be less than the length of the projection of the line of action of the gearing on the base line of the mating rack.

The length of this projection is given as,

$$L_a = h \cot \phi$$

Now hob life can be greatly increased by shifting the hob periodically along the arbor to enable undulled teeth to be put into action. It is good practice to shift the hob by one lead. Moreover, one turn of the thread at each end of the hob may be incomplete in form and should not be used for cutting.

$$\therefore \quad L_{1\,min} > L_a$$

$$> h \cot 20° + \text{some allowance}$$

Putting values of h in terms of module,

$$L_{1\,min} \cong 10\, m$$

The wider the cutting face, the more times the hob can be shifted before sharpening. In fine module hobs, L_1 is taken much greater than $L_{1\,min}$; in coarse module hobs, the face width is near the $L_{1\,min}$. For example, for solid steel gear hobs,

$$L = 44 \text{ mm with } m = 1 \text{ mm}$$
$$= 200 \text{ mm with } m = 11 \text{ mm}$$
$$= 380 \text{ mm with } m = 30 \text{ mm}$$

(*viii*) Rake angle, $\alpha = 0°$

Relief angle at tooth tip, $\gamma = 9°$ to $12°$.

(*ix*) **Cam Relief :** $K = \dfrac{\pi D_o}{z} \tan \gamma$

(*x*) **Pitch diameter :** It is taken as,

$$D_p = D_0 - 2h_1 = 0.3 \text{ K}$$

(*xi*) **Flute depth,** $H_f = h + K + r,$ for hobs with a non-ground profile

$$= h + \frac{K + K_1}{2} + r, \text{ for hobs with a non-ground profile}$$

where K_1 = additional relieving

$= (1.2 \text{ to } 1.5)\, K$

and r = radius of the flute bottom.

(*xii*) Lead angle of the hob thread on the pitch cylinder can be calculated by the formula,

$$\sin \lambda = \frac{p_n}{\pi D_p} = \frac{\pi m}{\pi D_p} = \frac{m}{D_p}$$

where m = module of the gear.

Similarly, $p_a = \dfrac{p_n}{\cos \lambda}$

PROBLEMS

1. List the desirable characteristics of a gear.
2. List the factors on which the selection of material for gear depends.
3. List the various gear materials and give the field of application of each.
4. Give the advantages and limitations of casting for the manufacture of gears.
5. Differentiate between forming and generating.
6. Explain the "Roll forming" method of gear manufacture. Give the advantages of roll forming.
7. Explain the method of cutting gears by milling.
8. Why are not more gears made by broaching?
9. Could a helical gear be cut on a plain milling machine? Why?
10. Explain the process of gear shaping.
11. Why "rack-type cutter" is not used so commonly in gear shaper?
12. List the advantages and limitations of gear shaping.
13. Explain the principle of gear hobbing.

14. Why a gear hobbing machine is more productive than a gear shaper?

15. What is the relation between the axes of a hob and the gear blank for cutting (*a*) spur gears (*b*) helical gears.

16. Give the advantages and limitations of gear hobbing.

17. Write about the following methods of gear manufacture : extrusion, cold drawing, stamping and powder metallurgy.

18. Why gear finishing is required?

19. Discuss the various gear finishing operations.

20. What factors should usually be checked while inspecting gears.

21. With the help of a neat diagram, discuss the design features of a hob.

22. Write the specifications of Gear Making and Finishing Machines (All Types).

Ans.: Module, Gear ϕ.

23. To calculate, for a 40 teeth, 4 mm module and 20° pressure angle

(*a*) Test Plug size (*b*) Distance over plugs in diametrically opposite spaces (GATE : 2009)

Sol. As noted under Art 16.14(2), the best wire or best test plug diameter $= \dfrac{\pi\, m}{2} \cos \alpha$

now m = 4 mm and $\alpha = 20°$

(a) $\therefore$ Test plug size $= \dfrac{\pi \times 4}{2} \times \cos 20° \quad = 2\pi \times 0.9397 = 5.9$ mm

(b) Distance over plugs = P.C.D. + d

= 40 × 4 + 5.9 = 165.9 mm

Here P.C.D. = Pitch circle diameter = m × z = 4 × 40 = 160 mm

Note : Plug size is of a plug (wire or roller) that will rest in the tooth spare and lie with its centre on the Pitch Circle.

24. For the method of cutting teeth or a gear blank on a milling machine, please refer to Appendix IV.

25. Explain as to how a gear cutter actually forms the teeth on a gear blank and state the limitations placed on this type of cutter.

26. Explain the advantages of a helical tooth milling cutter.

27. With the help of a neat diagram, write on "Gear tooth Vernier Caliper"

28. How a vernier caliper is used to check the tooth thickness at the pitch circle ? Write the complete procedure.

29. Write on "Two – wire" method of gear checking.

30. What is "Best Wire Size" ?

31. Write and explain the various terms used in the "Two-Wire" method.

32. What is "composite error"? How is it measured ? Explain

33. What is the "Base Tangent Method" ? For what purpose, this test is employed ? Explain it with the help of a diagrams.

34. With the help of a neat diagram, write a note on "Base Tangent Comparator".

35. With the help of a neat diagram, explain as to how the pitch of a gear is measured and checked ?

36. Write a comprehensive note on "Design of a Gear Hob".

37. Why a gear cutter or a hob is termed as "Arbor type" cutter or hob ?

17

THREAD MANUFACTURING

17.1. INTRODUCTION

Threads are of prime importance to the engineering. These are used as fasteners, to transmit power or motion and for adjustment. The subject of thread manufacture has assumed a great significance because of the ever increasing demand for high precision fastening devices and power transmission devices. At present, the threads are manufactured by the following processes :

1. Casting
2. Thread chasing
3. Thread rolling
4. Die thread and tapping
5. Thread milling
6. Thread grinding

17.2. CASTING

The accuracy and finish of threads made by casting will depend upon the method of casting. Threads made by sand casting are rough and are not used much, except sometimes in vises and rough machinery. Threads made by die-casting and permanent mould casting are very accurate and of high finish, if properly made. However, these can be made only of low melting point non-ferrous metals and, therefore, are not fit for repeated use, being not hard and durable. Lost-wax method can produce highly accurate threads of good finish. But the method is costly and difficult. Sewing machines, vending machines, type writer parts and toys may have their threads cast in place by die-casting and permanent mould casting. Such parts are rarely taken apart, so, the method is very satisfactory. The drawbacks of sand casting can be overcome by using shell-moulding method. Due to the inherent drawbacks of casting method, this has a limited field of application as compared to other methods of thread production.

17.3. THREAD CHASING

The method of cutting threads with a single point tool on a centre lathe and with a multipoint tool on a turret lathe, is called "thread chasing". Thread chasing is a form cutting operation, with the form tool corresponding to the profile of desired thread space.

17.3.1. Thread cutting on a centre lathe. The first step in cutting threads on a lathe is to get an accurately shaped and mounted tool. The form and setting of the tool is checked with the help of a thread template or centre gauge, Fig. 17.1. The job is either mounted between centres or held in a chuck (for external threads) and held in a chuck for internal threads. When mounting the tool in the tool post, it must be ensured that the top of the tool is horizontal and is in line with the axis of rotation of the job, Fig. 17.2. After this, the second step is to establish a specific relationship between the longitudinal movement of the tool parallel to the axis of rotation, and the rotation, of the job. This will determine the pitch or lead of the thread. This is achieved with the help of lead screw and a split nut. The two halves of the split nut are fastened to the carriage. When the nut is closed on to the lead screw, it acts as a complete nut, and the carriage starts moving as the lead screw rotates.

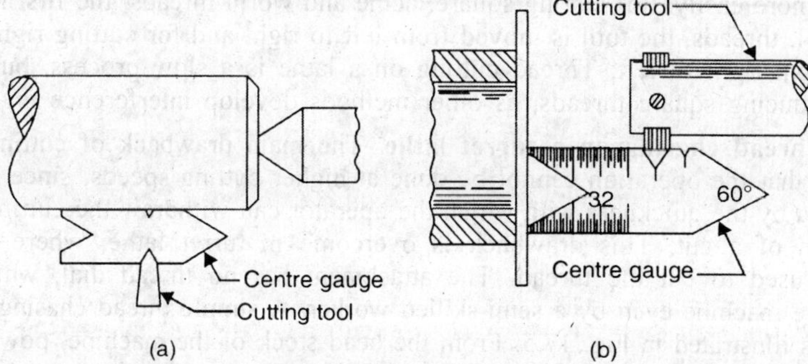

Fig. 17.1. Centre Gauge.

The lead screw is geared to the spindle and the proper speed ratio between the two is set by means of a gear-change box. Therefore, as the lead screw rotates, the carriage will move a predetermined distance (depending upon the pitch or lead of the thread) per revolution of the job. The third requirement is that the split nut must be engaged at an exact predetermined time, for taking successive cuts, so that the tool enters the helical groove of the cut previously produced, otherwise the tool may remove some of the desired thread. This is achieved with the help of a 'thread dial', which is mounted on the carriage and is driven by the lead screw through a worm gear. The face of the thread dial is graduated into an even number of full and half divisions, Fig. 17.3. Whenever the lead screw rotates and the split nut is not engaged, the thread dial rotates. The split nut must be engaged when a particular line on the dial face coincides with the zero line. For cutting even number of threads, the split nut should be engaged when any line on the dial coincides with zero line, and, for cutting odd-number threads, when any numbered line coincides with zero line.

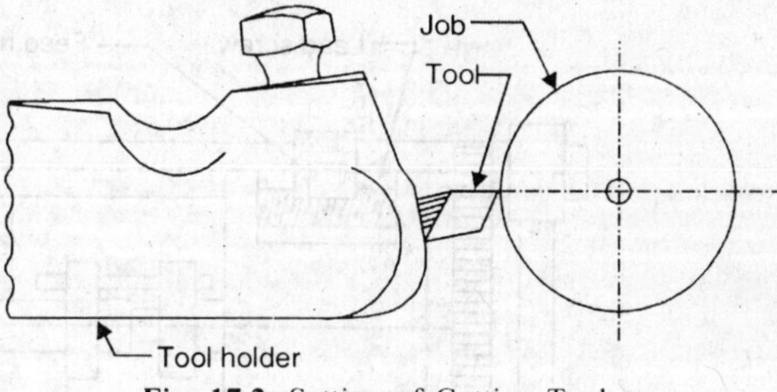

Fig. 17.2. Setting of Cutting Tool.

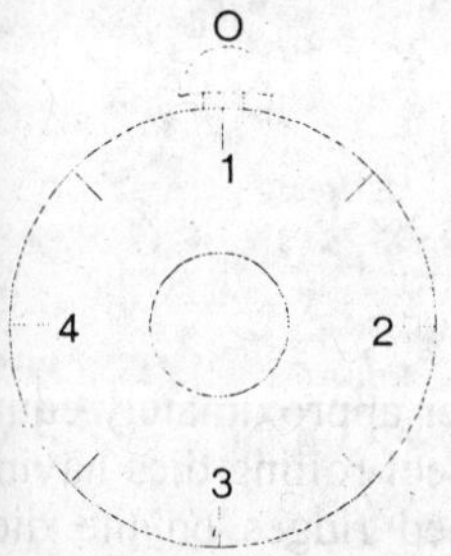

Fig. 17.3. Thread Dial

To start cutting a thread, the tool is fed inward until it first scratches the surface of the job. The graduated dial on the cross-slide is noted or set to zero. The split nut is then engaged and the tool moves over the desired job length. At the end of tool travel, it is quickly withdrawn by means of cross slide. The split nut is disengaged and the carriage is returned to the starting portion, for the next cut. These successive cuts are continued until the thread reaches its desired depth (checked on the dial of cross-slide). The depth of first cut is usually 0.25 to 0.40 mm. This is gradually decreased for the successive cuts until for the final finishing cut, it is usually 0.027 to 0.075 mm. The tool can be fed inward either radially or at an angle of 29 by swivelling the compound rest, Fig. 17.4.

The drawback of the first method is that the absence of side and back rake will not produce proper cutting except on brass and cast iron. In the second method, the cutting mainly takes place on one face of the tool and some side rake can be provided. Also, the

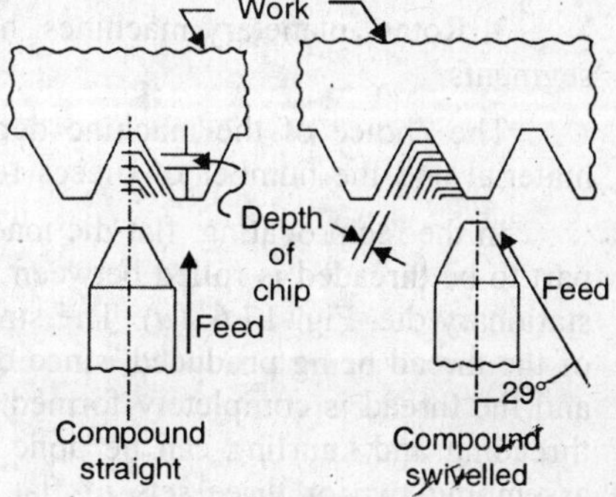

Fig. 17.4. Feeding the Tool into the Job.

chip will curl more easily. For cutting square, acme and worm threads, the first method is used. For cutting L.H. threads, the tool is moved from left to right and for cutting right hand threads, it is moved from right to left. Thread cutting on a lathe is a slow process, but it is the only process of producing square threads, as other methods develop interference on the helix.

17.3.2. Thread chasing on a turret lathe. The main drawback of cutting threads on a centre lathe is that the operation cannot be done at higher cutting speeds, since the permissible speed is limited by the quickness with which the operator can withdraw the cutting tool from the job at the end of a cut. This drawback is overcome in turret lathe, where thread chasing attachment is used to cut the thread. The attachment has no thread dial, which enables the operation of the machine even by a semi-skilled worker. A simple thread chasing attachment for a turret lathe is illustrated in Fig. 17.5. From the head stock of the machine, power is given to a short lead screw, known as the leader, by means of change gears. The feed nut and the tool slide are carried on a shaft, which can be engaged or disengaged to the leader by means of a hand lever. The major advantage of the arrangement is that the feed nut can be engaged to the leader at any portion of the work rotation.

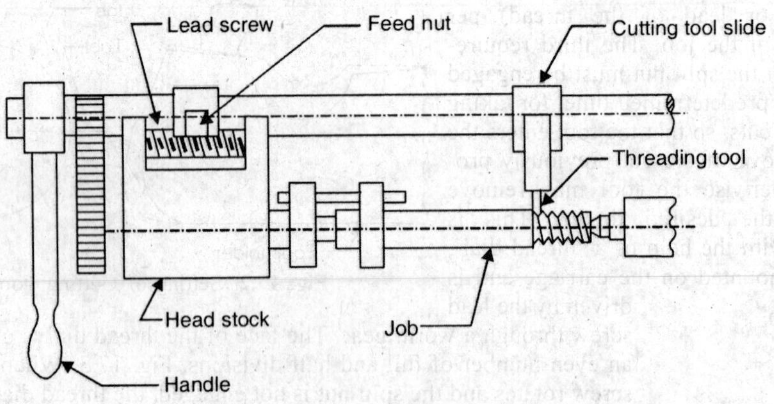

Fig. 17.5. Thread Chasing Attachment.

17.4. THREAD ROLLING

Thread rolling is a cold working process in which a blank of diameter approximately equa to the pitch diameter of the required thread, is rolled between hardened steel rolling dies havin the negative contour of the thread to be produced. As the thread shaped ridges on the die penetrate the blank material, material is displaced from the bottom of the thread and force radially out to form the thread crests. There are three types of thread rolling machines :

1. Reciprocating, flat die machines.
2. Cylindrical-die machines.
3. Rotary planetary machines, having a rotary die and one or more stationary concave-di segments.

The choice of the machine depends upon : size and design of the workpiece, the wor material and the number of pieces to be produced.

In the reciprocating, flat die machine, one die is stationary and the other reciprocating. Th part to be threaded is rolled between the dies, as the moving die reciprocating in reference to th stationary die, Fig. 17.6. (*a*). The stroke of the reciprocating die will depend upon the diamet of the thread being produced, since during one stroke, the blank makes one complete revolutic and the thread is completely formed. This is a highly versatile machine, since at the same tim threading and knurling can be done on a part of right and left hand threads can be rolled, b assembing two or three sets of flat dies. This method is used mainly for the manufacture c commercial bolts and nuts.

In cylindrical die machine, the part to be threaded is rolled between rotating cylindrical dies. The machine can have two round dies located diametrically opposite each other, Fig. 17.6 (*b*), or three round dies usually spaced. This machine is slower than the reciprocating flat die machine and is more suitable for large sized precision threads and for short run production. This machine operates with the following motions :

(*a*) Positive rotation of both the dies (in a two die machine) in the same direction.

(*b*) Radial motion of one of the dies for its rapid approach, infeed and retraction.

This method has the main application of rolling the thread on taps.

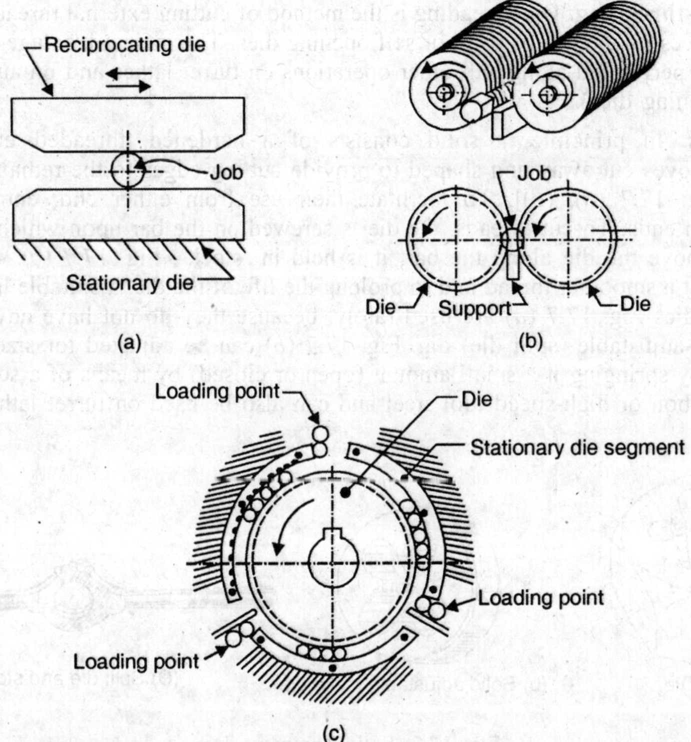

Fig. 17.6. Thread Rolling.

In rotary planetary machine, Fig. 17.6 (*c*), the job is rolled between a central die that rotates continuously about a fixed axis, and one or more concave-shaped die segments located adjacent to the periphery of the rotating die. This being a continuous process, is the fastest method of thread rolling.

Advantages of Thread Rolling

1. It is the fastest method of producing a thread, with production rate more than 2000 pieces per minute.

2. Being a chipless forming process (no material wastage), there is lot of material saving (about 16 to 27 per cent).

3. During thread rolling, the material is strained plastically and is work-hardened, and is, therefore, stronger against both tension and fatigue, especially the latter. Increase in tensile strength is from 10 to 20% and that in fatigue strength is from 10 to 75%.

4. The grain fibres remain continuous and follow the contour of the threaded surface. Due to his, the threads are less easily sheared off than machined threads.

5. The surface of rolled thread is harder than a cut thread, so, wear resistance increases.

6. Surface finish is better as controlled by the rolls.

7. Dimensional accuracy is better, as very little wear occurs on the rolls as it would on a cutting tool.

The major drawback of thread rolling is that it is basically used for producing external threads. It is best suited to diameters upto about 20 mm and fine threads. Thread rolling may also be done from the end slide of a turret lathe.

17.5. DIE-THREADING AND TAPPING

175.1. Die-threading. Die-threading is the method of cutting external threads on cylindrical or tapered surfaces by the use of solid or self opening dies. The main advantage of die threading is that it can be performed alongwith other operations on turret lathes and on automatics (in the case of self-opening dies).

Solid dies. In principle, a solid consists of a hardened, threaded nut with several longitudinal grooves cut away and shaped to provide cutting edges to the remaining portions of the thread, [Fig. 17.7 (*a*), (*b*)]. To facilitate their use from either end, entry chamfers are provided at both ends. To cut threads, the die is screwed on the bar upon which the threads are to be cut. To move the die along the bar, it is held in 'stock'. Fig. 17.7 (*c*), which is rotated manually. To cut a smoother thread and to prolong the life of the die, a suitable lubricant is used. The solid-type dies Fig. 17.7 (*a*) are used rarely, because they do not have any adjustment for wear. The solid-adjustable (split die) dig. Fig. 17.7 (*b*) can be adjusted for size and wear over a small range, by springing it a small amount (open or closed) by means of a screw. These dies are made of carbon or high-speed tool steel and can also be used on turret lathes with suitable holders.

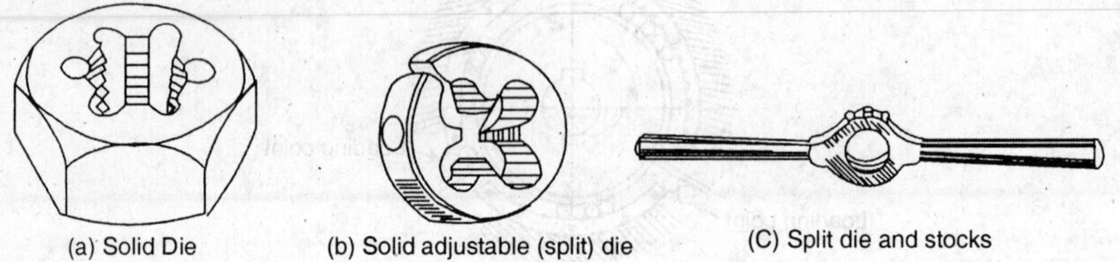

(a) Solid Die (b) Solid adjustable (split) die (C) Split die and stocks

Fig. 17.7. Solid Threading Die.

Self-opening die-heads. The major drawback of solid-type dies is that they must be unscrewed from the workpiece by reversing the machine spindle, to disengage the die from the work. Due to this, these dies are not suitable for use on high speed production machines, for example, turret lathes and automatics. This drawback is overcome by using self-opening die heads. When the required length of thread is cut, the dies open automatically. At the end of the turret slide travel, the front portion of the die-head continues to move forward by a small amount, until the chasers in the die-head are moved radially outwards in the body, under the action of a scroll or a cam. This action clears the chasers from the cut thread and enables the die-head to be withdrawn without reversing the machine spindle. The die-head to be withdrawn without reversing the machine spindle. The die-head, while cutting threads, may advance under its own guidance once it screws itself along the work, until the dies trip opens. However, for better accuracy, there is increasing use of lead screw guides.

Depending upon the type of the chaser, there are three types of die-heads, Fig. 17.8.

1. Radial
2. Tangent
3. Circular

Radial chasers can be more rigidly supported than other types. These are difficult to resharpen and their life is short. Tangential chasers give a long life, because the length of the teeth makes possible a large number of regrinds on the cutting face. Due to this they are very suitable for heavy duty work and large batch production. Circular chasers also have a long working life since these can be resharpened a number of times. All the die-heads can either be stationary or revolving.

When used on automatics, the feed motion of the die-head is controlled by the cam rise, which can be designed accordingly. At the end of the return stroke, the dies are closed automatically when the closing handle strikes a rod. Die-hands are available for cutting threads from 6.35 mm to 114 mm diameter and chasers are available for any thread form.

17.5.2. Thread tapping. Taps are the tools for cutting internal threads. A tap is similar to a threaded bolt, with one to four flutes cut parallel to its axis. The flutes perform three functions :

1. provide cutting edges.
2. conduct the cutting fluid to the cutting region, and
3. act as channels to carry away the chips formed by the cutting action.

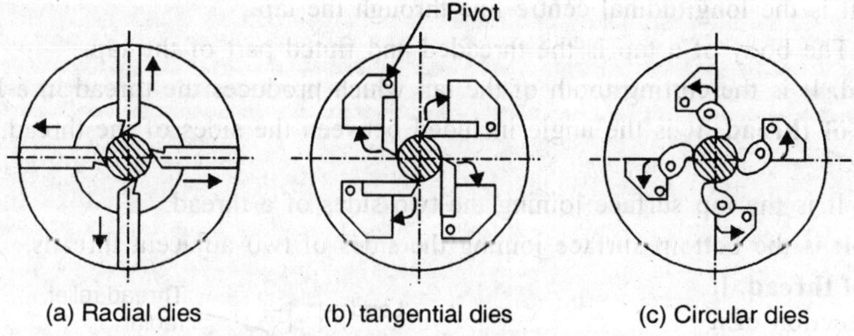

Fig. 17.8. Self Opening Die Heads.

The flutes can be either straight, spiral, helical or spiral pointed. Taps with straight flutes are most commonly used, since it is easier to cut and sharpen these flutes. Tapping can be done manually or on drilling machines, tapping machines, turret lathes and automatics. A hole of diameter slightly larger than the minor diameter of the thread to be cut, must already exist, for thread tapping. The hole can be made by drilling, boring or casting. The two main types of taps are : solid taps and collapsing taps.

(*a*) **Solid taps.** Solid taps are of one piece construction. These taps are usually worked manually but can also be used on machine tools, such as lathes, drill presses and special tapping machines. Taps are made of high carbon or high speed steel. The shank of the taps is kept plain and the end is squared. To operate the tap by hand (Hand taps), it is held at the squared end with the help of a 'tap wrench', which is used to screw the tap into the hole. To cut any particular size, hand taps are available in sets of three : taper, plug and bottoming. The three taps are identical in size and length, but differ in the amount of chamfer at the bottom end. The taper plug has about 8 to 10 threads chamfer at the bottom end, the plug tap has 2 or 3 threads chamfered, whereas, a bottoming tap has no taper threads at its bottom end. The tapered are cut to the full depth gradually, so less effort is required. If a hole is open at both ends, then, after the taper tap, plug is used for finishing the threads as deep into the hole as its shape will permit. Lastly, the bottoming tap is used to finish the entire thread portion. So, the three taps should be used in the order mentioned above. The bottoming tap is the only tap which would nearly reach the bottom of a blind hole. The three taps are shown in Fig. 17.9.

While threading, a combined rotary and axial motion is given to the tap. When using a solid tap on a drill press, a special tapping attachment is used. This makes the tap to rotate slowly as it is fed downward into the job. At the end of tapping, when the spindle is raised, the tap automatically starts rotating in the reverse direction at a higher speed to back the tap out of the hole in a shorter time. On screw machine or turret lathe, a special holder is used for the tap, in which a pin prevents the tap from rotating while it is fed into the job. At the end of travel, the tap pulls the pin so that it is free to rotate with the work. The machine spindle is then reversed in motion and the pin again stops the tap from rotating while it is being backed out of the hole.

(*b*) **Collapsing taps.** For better results, a tap (or a die) should not be backed off the thread it has just produced, because, during backing off, they catch tiny chips which can do damage to the product. So, for good finish and to speed up operations, collapsing taps are used, which collapse inward automatically when the thread is completed. This makes it possible to withdraw the tap from the hole without reversing the machine spindle.

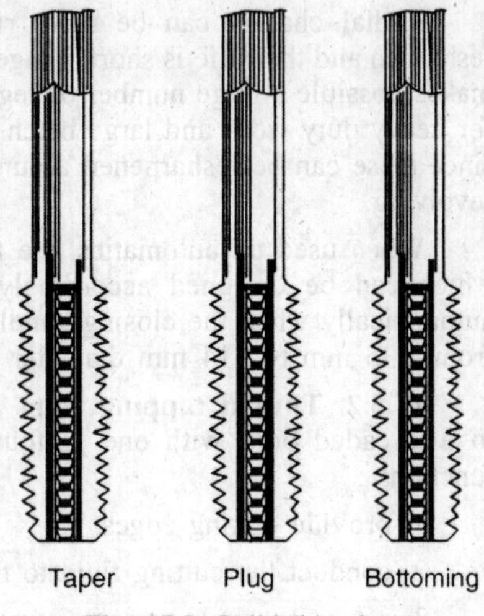

Fig. 17.9. Solid thread taps.

Nomenclature : Refer to Fig. 17.10

1. **Axis.** It is the longitudinal centre line through the tap.
2. **Body.** The body of a tap is the threaded and fluted part of the tap.
3. **Thread.** It is the cutting tooth of the tap which produces the thread in a hole.
4. **Angle of thread.** It is the angle included between the sides of the thread, measured in the axial plane.
5. **Crest.** It is the top surface joining the two sides of a thread.
6. **Root.** It is the bottom surface joining the sides of two adjacent threads.
7. **Base of thread.** It is the bottom section of a thread; the greatest section between the two adjacent roots.
8. **Depth of thread.** The depth of the thread profile is the distance between the top of crest and the base or root of thread measured perpendicular to the axis of the tap.
9. **Side of thread.** It is the surface of the thread which connects the crest with the root.
10. **Land.** It is the threaded web between flutes.
11. **Cutting Face.** It is the front part of the threaded section of the land.

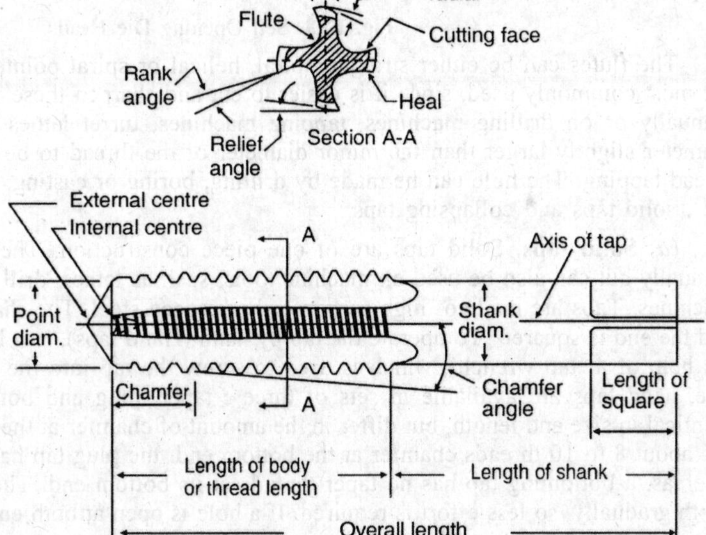

Fig. 17.10. A solid tap.

12. **Hook.** It is the curved undercut of the cutting face of the land.
13. **Heel.** It is the back part of the threaded section of the land.
14. **Chamfer.** The tapered outside diameter at the front end of the threaded section.

15. **Point diameter.** It is the outside diameter at the front end of the chamfered portion.

16. **Flute.** It is the groove providing for the cutting faces of the teeth, chip passage and cutting fluid.

17. **Helix.** It is the curve of an ordinary screw thread.

18. **Helix Angle.** It is the angle made by the helix of the thread at the pitch diameter with a plane perpendicular to the axis.

19. **Shank.** It is the part of the tap behind the threaded and fluted section of the tap. The tap is held or located and driven by the shank.

20. **Square.** It is the squared end of the tap.

21. **Radial Rake Angle.** It is the angle formed in a diametral plane between the face and a radial line from the cutting edge at the crest of the thread form.

22. **Chamfer Angle.** It is the angle formed by the tapered outside diameter at the front end with the top axis.

23. **Web.** The central portion of the tap situated between the roots of the flutes and extending along the fluted section of the tap. Its thickness increases from the front and towards the shank end of the flutes.

24. **Back Taper.** The reduction in diameter of the tap body of the threaded portion from the front end towards the shank end.

25. **External Centre.** It is the cone-shaped end of the tap. It is provided only for manufacturing purposes and only for small taps and usually at the thread end.

26. **Internal Centre.** A small drilled and countersunk hole at the end of the tap, necessary for manufacturing purposes.

27. **Thread Relief.** It is the radial clearance providing a gradual decline in the major, pitch, and minor diameters of the lands, back of the cutting face.

Design features of a Tap. As written above, a tap is essentially a screw that has been fluted to form cutting edges. The cutting end of the tap has a relieved chamfer, which forms the cutting edges and permits it to enter the untapped hole. The design features are illustrated below :

1. **Chamfer diameter.** The chamfer diameter, d_{ch}, of the chamfer at the front end of the tap is made smaller than the minor diameter of the thread as given below :

d_{ch} = Minor diameter of thread — 0.10 to 0.15 mm for diameter upto 18 mm

= " " " — 0.20 to 0.25 " " " from 20 to 39 mm

= " " " — 0.30 to 0.35 " " " from 42 to 52 mm.

2. **Chamfer length and Chamfer angle.** Fig. 17.11 shows the material removal in tapping threads. The cross-hatched area represents the part of the thread groove removed in first revolution of the tap.

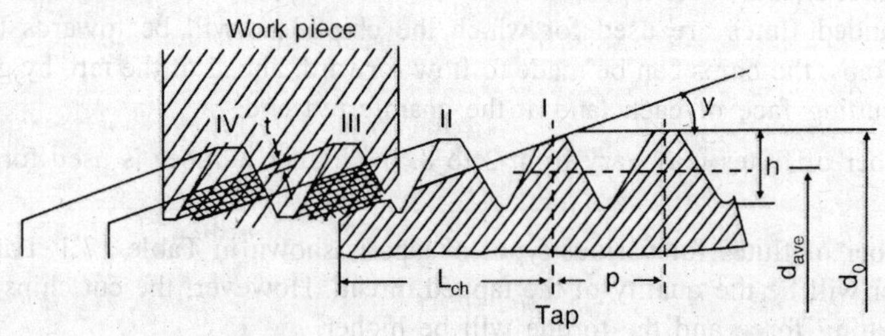

Fig. 17.11. Tap Chamfer Elements.

The uncut chip thickness (measured perpendicular to the tap axis for simplicity) removed by each land is,

$$t = \frac{h}{zf}$$

But $$f = \frac{L_{ch}}{p}; \quad L_{ch} = \text{chamfer length}$$

$$\therefore \quad t = \frac{ph}{zL_{ch}} = \frac{p}{z} \tan\phi$$

It is clear that,

$$L_{ch} = \frac{h}{\tan\phi} \text{ or } = \frac{h}{kz}$$

where h = depth of thread

ϕ = angle of chamfer of the tap

p = pitch of the thread being tapped.

and $k = \frac{t}{p}$ is a characteristic of the construction of a tap.

Its values are :

k = 0.012 to 0.02 for nut taps

= 0.03 to 0.04 for die taps

= 0.06 to 0.10 for hand and machine taps.

The chamfer angle is given as,

$$\tan\phi = \frac{d_0 - d_b}{2L_{ch}} \quad \therefore \quad L_{ch} = \frac{d_0 - d_b}{2} \cdot \cot\phi$$

where d_0 = major diameter of the tap thread

d_b = diameter of blank hole for tapping.

3. **Flutes.** Most taps have straight flutes, but certain special taps have helical flutes. The direction of the chip flow can be changed by changing the hand of the helical flutes on the tap. Left-handed flutes will drive the chips forward, ahead of the tap. Left-handed flutes will drive the chips forward, ahead of the tap, and, so, are used for tapping through holes. For tapping blind holes, right-handed flutes are used for which the chip flow will be towards the shank. With straight fluted taps, the chips can be made to flow forward, ahead of the tap, by grinding a spired point on the cutting face of each land at the chamferred end.

The number of flutes may vary from 2 to 8, the higher number is used for larger diameter taps.

The number of flutes for various types of taps is shown in Table 17.1. Larger the number of flutes, better will be the quality of the tapped thread. However, the cut chips will be thinner, the specific cutting force and the torque will be higher.

Table 17.1. Number of Flutes

Type of Tap	Number of Flutes				
	Major diameter, mm				
	2 to 6	8 to 14	16 to 24	27 to 36	39 to 52
Hand, nut and machinctaps :					
for metric and inch threads	3	3	3 or 4	4	4 to 6
for pipe threads	–	3 or 4	6	6	6
Master taps	3	4	6	6	6 to 8.

4. **Tap geometry :**

(*i*) **Rake angle.** of the sizing and chamfer part is given below, depending upon the type of material to be tapped.

α = 15°, for steel with σ_t < 600 MPa.
= 10°, " " " " 600 to 900 MPa
= 5°, " " " " > 900 MPa
= 5° for Grey C.I.
= 0° for Bronze
= 20° to 30° for Aluminium and its alloys.

(*ii*) **Relief angle.** Relief is provided only on the chamfer length. It is obtained by relieving the thread only on the crests along the length of the chamfer. Its recommended values are :

γ = 8° to 10° for machine taps
= 6° to 8° for hand taps
= 8° to 12° for nut and machine taps
= 3° to 4° for die calibrating taps
= 4° to 8° for taps for light alloys

The relieving over the chamfer length will be given as,

$$K = \frac{\pi d_0}{z} \cdot \tan \gamma$$

There is usually no relief on the sizing section and at the flank of the thread. Relieving reduces the friction between tap and the surface of the hole.

(*iii*) **Back-Taper.** Axial back taper is provided on the tap from the front end towards the shank end to avoid rubbing of the tap with the surface of the hole so as to reduce friction. It is taken as :

= 0.05 mm to 0.10 mm/100 mm for ground taps
= 0.08 mm to 0.12 mm/100 mm for unground taps in which threads are formed by rolling.
= 0.20 mm for tapping especially tough, high strength materials, such as, heat resistant and stainless steels and alloys and tough row-carbon steels etc.

(*iv*) **Chamfer Angle.** The leading edges of a tap is chamfered to help in starting the tap. Smaller the chamfer angle, longer will be the chamfer length. This will result in thinner uncut chips, resulting in increase in cutting force, eventhrough longer chamfer length provides better guiding to the tap and the quality of the thread improves. 17.2 gives the common values taken for chamfer angles.

Table 17.2. Chamfer Angles

Taps in a set	*Type of Tap*	ϕ, *degrees*
1.	Nut	2
2.	Taper	7
	Rougher	7
	Bottoming	20
	Finisher	20
3.	Taper	5
	Rougher	5
	Second	10
	Intermediate	10
	Bottom	20
	Finisher	20

5. **Cutting speeds.** The cutting speeds for machine taps are given in Table 17.3.

Table 17.3. Cutting Speeds

Work material	*Lubricant*	*Tapping speed m/min.*
Aluminium	Kerosene and hard oil	30
Bakelite	Air blast	24
Brass	Soluble or light base oil	42
Cast Iron	Dry or Soluble oil	24
Steel :		
Mild	Soluble or sulphur based oil	18
Medium alloy	Sulphur-base oil	12
Stainless	"	6
Zinc die-cast	Soluble oil	24

6. **Materials.** Taps are usually made of carbon tool steel or H.S.S.

17.6. THREAD MILLING

In thread milling, the threads are cut by a revolving form milling cutter conforming to the shape of the thread to be produced. Both external and internal threads can be cut by this method. Thread milling has got the following characteristics :

1. This is a fast thread cutting method for producing threads usually of too large a diameter for die heads.

2. The threads produced are more accurate than those cut by dies, but less accurate than produced by grinding.

3. Threads running upto a shoulder on the workpiece can be cut without any difficulty.

4. Worms and lead screws which are too large to be cut with a single point tool can be milled.

5. This method is desirable, when the pitch of the thread is too coarse to be cut with a die.

6. The method is more efficient than cutting thread on a lathe, especially when the job is long or when large amounts of metal are to be removed.

For thread milling either single or multiple cutters may be used. A single-form cutter has a single, annular row of teeth, lying in one plane. While thread cutting with a single cutter, it is tilted through an angle equal to the helix angle of the thread to avoid interference while cutting. To start milling the threads, the cutter is fed radially inward equal to the depth of the thread, while the job is stationary, being held between centres of the machine. The job is then rotated slowly and the cutter, while rotating, is also traversed longitudinally parallel to the axis of the job, or vice versa, by means of a lead screw. This operation is stopped when the thread is completed. This method of thread milling is used for cutting coarse (large pitch or multiple-pitch) threads. The threading can be completed in a single cut or roughing and finishing cuts may be used.

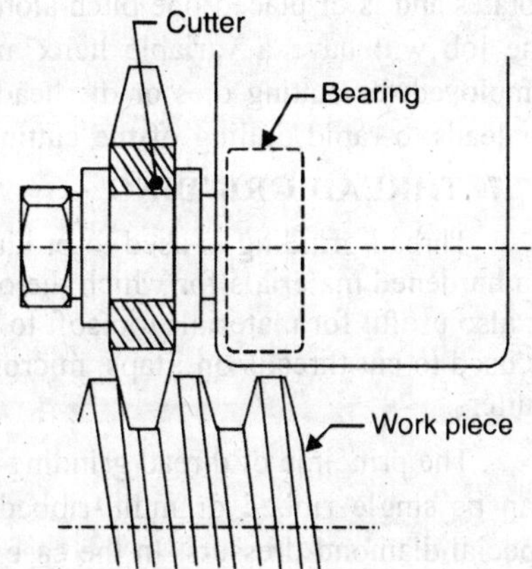

Fig. 17.12. Thread Milling with Single-Thread Cutter.

The method of cutting threads with single-thread or single-rib milling cutters is chiefly employed to cut long threads (chiefly of square and trapezoidal profiles) on various lead screws and worms. Usually, the threads are cut rough by milling and then these are finished by chasing with a single-point tool or a formed grinding wheel. The method is shown in Fig. 17.12.

Multiple cutter is used when the thread to be cut is not too long and it is desired to cut the threads in one revolution of the work. The width of the cutter has to be slightly more than the length of the thread. The cutter is set parallel to the axis of the job and is fed radially inward equal to the depth of the thread while the job is stationary. The job is equal to the depth of the thread while the job is stationary. The job is then rotated slowly, with the cutter moving axially a distance equal to the lead of the thread plus a small overtravel to complete the thread in one pass.

Thread milling with multiple-thread or multiple-rib cutters is illustrated in Fig. 17.13. Fig. 17.13 (*a*) is for cutting external threads.

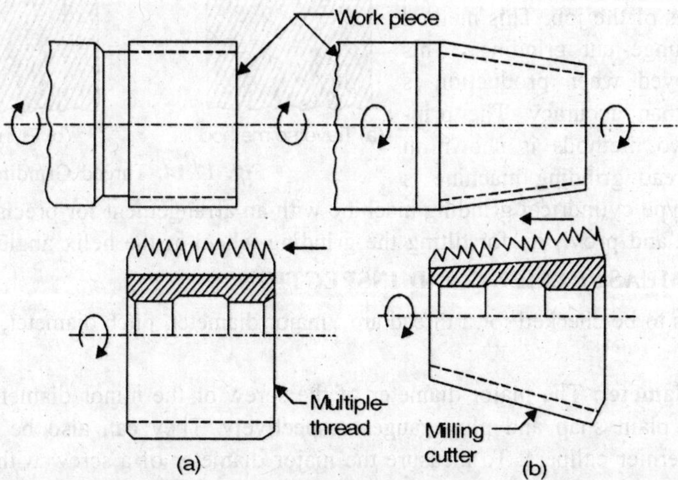

Fig. 17.13. Thread Milling with Multiple-thread Cutters
(*a*) External Threads; (*b*) Taper.

The profile of the cutter teeth should be the same as that of the thread to be cut (Usually vee threads are cut by this method). Taper threads can also be cut by this method, Fig. 17.13 (*b*). Cutters have helical flutes of constant lead, milled on the conical surface of the cutters. The spacing between the flutes is more toward the larger diameter of the cutter. The cutter rotates and is displaced one pitch along an element of the taper as shown. The taper threads cut on the job will have a variable helix angle. Thread milling with multiple-thread cutters is widely employed for cutting dies or die heads, does not provide a sufficient high class of surface finish, or leads to rapid dulling of the cutting edges of the tools.

17.7. THREAD GRINDING

Thread grinding is used to produce very accurate threads. It is also employed to cut threads on hardened materials for which the other methods of thread cutting are not possible. The method is also useful for materials too soft to get a good surface finish by other methods. Thread grinding is used to cut threads on : taps, micrometer screws, lead screws, thread gauges and thread milling cutters.

The principle of thread grinding is similar is principle to thread milling. The grinding wheels can be single ribbed or multi-ribbed, which are shaped (conforming to the thread profile) by special diamond dressers. In the case of single-ribbed wheel, the wheel turns against the rotation of the job. In addition to this rotary motion, a relative axial motion between the wheel and the job is provided with the help of a precision lead screw. The wheel is tilted an angle equal to the helix angle of the thread, to the axis of the job. This method is known as 'Traverse Thread Grinding', and is used to produce long and coarse pitch threads. Also, the pressure on the work and hence the heat generated during grinding is not excessive, resulting in a more accurate thread.

A multi-ribbed wheel, which is slighly longer (one or two threads) than the work, is used to cut the entire threads in one revolution of the work. The wheel is fed into the work to the required depth and moves axially a distance equal to the pitch of the thread while the work revolves through one revolution. The cutter is set parallel to the axis of the job. This method is known as 'Plunge cut grinding'. This method is employed when production is more important than accuracy. The principle of these two methods is shown in Fig. 17.14. A thread grinding machine is similar to centre type cylindrical grinding machine with an arrangement for precise movement of the machine table and provision for tilting the grinding wheel at the helix angle of the thread.

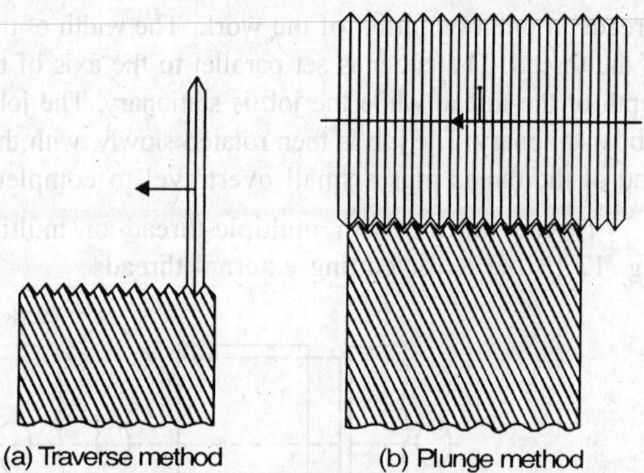

Fig. 17.14. Thread Grinding.

17.8. THREAD MEASUREMENT AND INSPECTION

The elements to be checked for a thread are : major diameter, pitch diameter, pitch and helix angle.

1. **Major Diameter.** The major diameter of the screw or the minor diameter of a nut can be checked by a plain snap and plug gauges respectively. They can also be measured with micrometer and vernier calipers. To measure the major diameter of a screw with a micrometer, the anvils should be of sufficient diameter so as to span two threads. To eliminate the effect of errors between the micrometer screw and the anvil faces, it is always better to first check the instrument on a cylindrical standard of about the same diameter as the screw.

2. **Minor Diameter.** Minor diameter of a screw can be measured with a screw thread micrometer caliper. This instrument is similar to the ordinary micrometer, but instead of usual flat measuring faces, it has specially designed anvil and spindle inserts. The inserts are interchangeable to suit the thread pitch. A simple screw thread micro meter is shown in Fig. 17.15. To check the minor diameter of a screw, two *V*-shaped inserts are used, so that their sharp apexes contact the roots of the screw thread, Fig. 7.16.

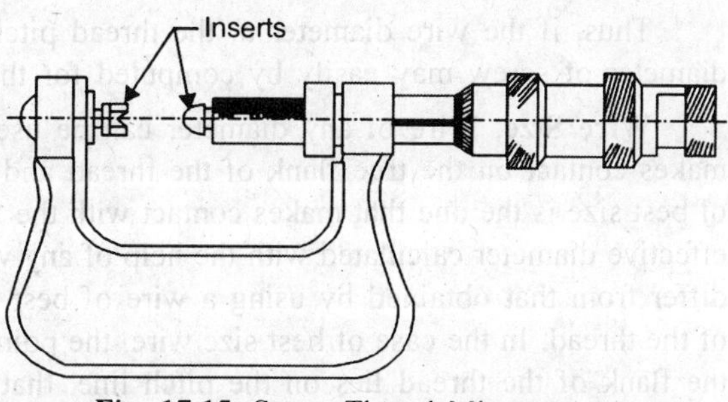

Fig. 17.15. Screw Thread Micrometer.

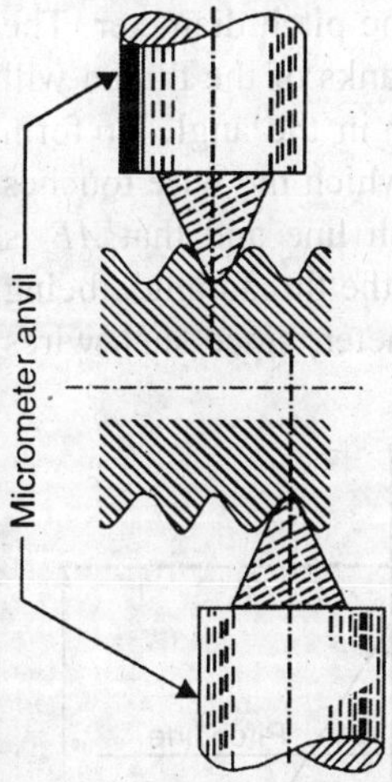

Fig. 17.16. Checking Minor Diameter of a Screw.

To check the pitch diameter, inserts of a type that contact the sides of the screw thread near the pitch diameter are employed. For this, a truncated thread form is used on the inserts.

3. **Pitch Diameter.** One of the most accurate methods for checking the pitch diameter is the 'three-wire method'.

The method consists in placing three small diameter cylinders (three wires of equal and precise diameter) in the thread grooves at opposite sides of a screw and measuring the distance W over the outer surfaces of the wires with an ordinary micrometer caliper having flat measuring faces, [Fig. 17.17 (*a*)]. Three wires are required to prevent misalignment of the measuring faces on the micrometer caliper. The pitch or effective diameter is calculated from the value *W* in the following manner [Refer Fig. 17.17 (*b*)] :

It is clear that,
$$W = P + 2 \times AC + 2 \times \frac{d}{2}$$

where P = pitch or effective diameter

and d = wire size

Now
$$AC = AD - CD = \frac{d}{2}\operatorname{cosec}\frac{\alpha}{2} - \frac{p}{4}\cot\frac{\alpha}{2}$$

where α = thread angle

and p = pitch of threads

After simplification, it can be seen that,

$$W = P + d\left(1 + \operatorname{cosec}\frac{\alpha}{2}\right) - \frac{p}{2}\cot\frac{\alpha}{2}$$

In the case of I.S.O. metric threads, $\alpha = 60°$

$$W = P + 3d - 0.866p$$

$$P = W - 3d + 0.866p$$

Here, the pitch diameter lies 0.3248p inside the crest of the thread that is,

$$P = D - 0.6496p$$

where D = Outside diameter

$$D = W - 3d + 1.5156p$$

Thus, if the wire diameter d, the thread pitch p and the distance W are known, the pitch diameter of screw may easily by computed for the above relations.

Wire Size. Wire of any diameter can be used to measure the pitch diameter, provided it makes contact on the true flank of the thread and provided the thread angle is correct. A wire of best size is the one that makes contact with the flanks of the thread at the pitch diameter. The effective diameter calculated with the help of any wire touching the true flanks of the thread will differ from that obtained by using a wire of best size if there is an error in the angle or form of the thread. In the case of best size wire, the point B [Fig. 17.17 (*b*)] at which the wire touches the flank of the thread lies on the pitch line, that is, BC lies on the pitch line and that AB is perpendicular to the flank position of the thread. If there is a possibility of the thread angle being incorrect, the wire of best size should be used to determine effective diameter, since such wires will be independent of any error in thread angle.

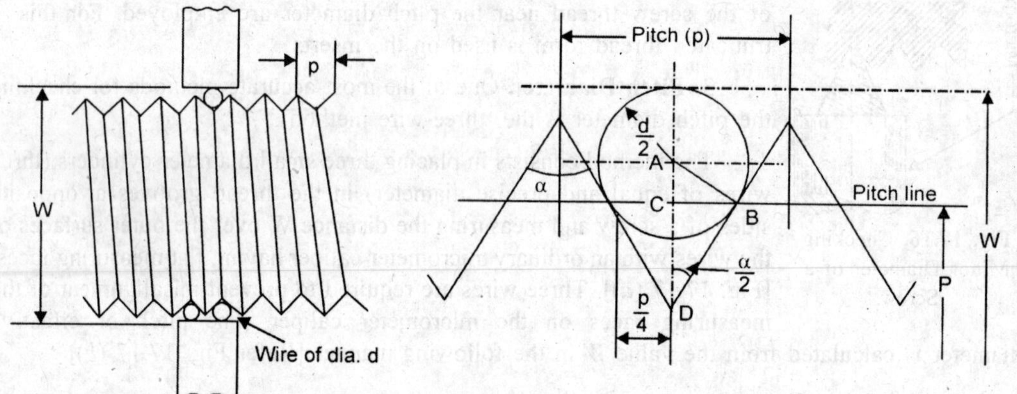

Fig. 17.17 (*a*). Three Wire Method (*b*) Three Wire Measurement.

Now $BC = p/4$

From triangle ABC, $AB = d/2 = BC \sec\frac{\alpha}{2} = \frac{p}{4}\sec\frac{\alpha}{2}$

∴ Best wire size, $d = \frac{p}{2}\sec\frac{\alpha}{2}$

For I.S.O metric threads,

$$d = \frac{p}{2}\sec 30° = 0.5774p$$

Example 17.1. *What is the value of the 'best wire size' for M 80 × 6 thread for measuring effective diameter.*

Solution. $D = 80$ mm, p = 6 mm

$d = 0.5774$ p

$= 0.5774 \times 6 =$ **3.464 mm**.

Example 17.2. *Find the value of the 'best wire size' for M20 × 2 I.S.O. metric thread. Also determine distance over wires.*

Solution. $D = 20$ mm, $p = 2.5$ mm

$$d = 0.5774\ p$$
$$= 0.5774 \times 2.5 = 1.444 \text{ mm}$$
$$W = D + 3d - 1.5156p$$
$$= 20 + 3 \times 1.444 - 1.5156 \times 2.5$$
$$= 20 + 4.332 - 3.789$$
$$= \mathbf{20.543\ mm}$$

4. Pitch. The pitch of a thread is usually measured with 'Screw pitch gauge'. Screw pitch gauges, Fig. 17.18 are sets of flat steel blades (similar to feeler gauges), which are notched on one edge according to various thread pitches represented by the gauge. The blades are pivoted at the end of a holder. To use it, the blade with the required thread pitch is applied to the thread being checked at the radial plane. If the pitch is correct, the gauge will fit tightly to the thread profile and no light will pass between the gauge and the thread profile.

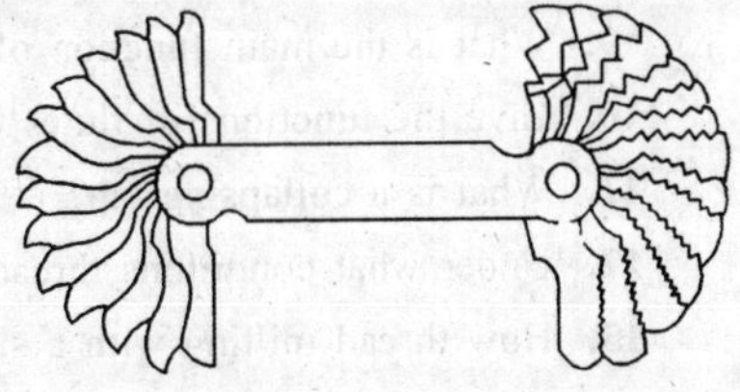

Fig. 17.18. Screw Pitch Gauge

To estimate the values of the screw pitch errors, screw pitch comparators are available. A comparator comprises a frame with two or three rods ending in ball shaped contacts. The rods are linked to a measuring tool, for example, a dial indicator and the ball - shaped contacts are inserted into the thread grooves to be checked. If the comparator has three contacts it will align along the thread axis. A two contact comparator checks the thread pitch in a direction perpendicular to the helix angle. The scale of the dial indicator will indicate the accumulated pitch error over the length of measurement. The average pitch error is estimated by dividing the accumulated pitch error by the number of pitches. This instrument must be set up with gauge blocks to the nominal size of the length of measurement.

Pitch measuring machines are also available to inspect a screw for pitch. The machine consists of a bed with centres at each end (just like a centre lathe) for supporting the screw. Alternate means are also available for holding nuts and sleeves. A head carrying a stylus shaped to fit in the vee of the thread is moved along the bed with the help of an accurate micrometer. The head is provided with an indicator which shows when the stylus is in its lowest position in the groove, that is, bedded home centrally in the groove of the thread. When the head is moved along the bed, the stylus seats successively in each of the threads over the length being examined. The pitch is determined by analysing the micrometer reading.

5. **Angle of the thread.** The screw pitch gauge can also give an indication about the correctness of the thread angle. If the angle is incorrect, light will be seen between the gauge and the thread profile. The thread angle of screws is measured on an optical instrument, the "toolmaker's microscope". Checking of an internal thread is very difficult since moulded copies of the thread profile must be made.

Optical projection methods are very convenient for inspecting the form and angle of a thread. The screw is held between centres provided in the apparatus and tilted to the helix angle so as to get a clear profile of the thread. When a beam of light is thrown on the thread, the magnified image of the thread is projected on to a screen or onto some part of the apparatus and compared with a master template.

The angle of the thread can also be measured/inspected from the projected image with the help of a shadow protractor provided with the apparatus. The blade of the protractor is set to each side of the thread and the angle with the vertical is measured to get the total angle of the thread.

Note. The use of a Tool room microscope for the measurement of screw pitch, pitch diameter of screw and the thread angle, has already been explained in Chapter 12.

PROBLEMS

1. Discuss the casting method for producing threads.

2. In cutting a thread on a lathe, how is the pitch controlled?

3. Explain the function of a thread dial on a lathe.

4. Why thread rolling has become the most commonly used method for thread manufacture?

5. Discuss the various thread rolling machines.

6. List the advantages of thread rolling.

7. Give the advantage of 'self opening die' over 'solid type die'.

8. Discuss the various types of self-opening die heads.

9. What is the main function of a taper tap?

10. Give the functions of flutes on a tap.

11. What is a collapsing tap?

12. Under what conditions thread milling is used to produce the threads.

13. How thread milling with a single cutter and a multiple cutter differ?

14. What are the advantages of producing threads by grinding?

15. Differentiate between 'Traverse thread grinding' and 'Plunge cut grinding'.

16. How will you check major diameter of a screw?

17. How will you check the minor diameter and the pitch diameter of a screw?

18. Explain the 'three wire method' of checking the pitch diameter of a screw?

19. What is a 'Screw pitch gauge'?

20. How the angle of a thread is checked?

21. Discuss the design features of a tap.

22. Write the specifications of threading Machines.

Machine Description		**Main Specifications**
(*i*) Tapping Machines	:	Tap size
(*ii*) Thread Hobbing and Milling Machines	:	Thread ϕ
(*iii*) Bolt and Pipe Threading (Die Type) Machines	:	Thread ϕ
(*iv*) Thread Grinders	:	Swing ϕ
(*v*) Thread Rolling Machines, Thread Whirling Machines	:	Thread ϕ

23. Sketch and write on "Screw thread Micrometer.

24. How a Tool Room microscope is used to check the following features of screw thread : Pitch, Pitch diameter and the Thread angle.

25. Sketch and label a Solid Tap.

26. Why a chamfer angle is provided on a tap ? Write the common values of chamfer angles.

27. What are the usual materials for taps.

28. Write about the common cutting speeds of a tap for the common work materials.

29. Write about the functions of flutes in a tap. What is the common number of flutes in a tap?

30. Write on : (a) Straight flutes (b) Helical flutes (c) Left handed flutes (d) Right handed flutes

31. What are : (a) Rake angle (b) Relief angle (c) Back Taper on Tap geometry ?

32. Describe the relative advantages of self-opening and solid-die heads for threading.

18

DESIGN OF MACHINE TOOL ELEMENTS AND MACHINE TOOL TESTING

18.1. DESIGN OF MACHINE TOOL ELEMENTS

Introduction

The quality of the finished workpiece depends on the relative positions between the tool and the workpiece during the machining operation. This in turn depends upon the holding and guiding devices that determine the positions of those parts on which tools and workpieces are located. These devices must be accurately produced and aligned. Also, their displacements due to play and distortions under the effect of external forces (cutting forces, weight of the workpiece, clamping forces) must also be kept within permissible limits. Furthermore, the machine tool as a whole (including its stationary as well as moving parts) must be rigid enough to prevent their tendency to vibrate and deflect under load, so that the machining accuracy is not impaired.

From above, it is clear that no ensure the required performance, the various parts of a machine tool must possess :

1. Strength to withstand the cutting forces.

2. Stiffness against deformation under load.

3. Rigidity against vibrations.

4. Provisions which ensure that the accuracy of relative location and alignment of all component parts is maintained throughout the working range of the machine tool.

18.1.1. Machine Body. Machine body is the basic structure of a machine tool. The body of the machine tool can be a bed (lathe, grinding machine), a column or an upright (drilling or milling machine) or a combination of both (planer, horizontal borer etc.) On it, the vital parts of the machine are either rigidly fastened in their correct positions or guided along the intended paths of movements in such a manner that they are accurately aligned in relation to each other and kept so during the operation of the machine. So, the main requirement made to the machine body (bed, base or column) of a machine tool is that it maintains the proper relative position of the units and parts mounted on it over a long period of service under all specified working conditions. This is achieved by designing locating datum surfaces on the bed, base or column for the principal parts whose positions remain unchanged during the above mentioned conditions. The locating datum surfaces are called ways or guides. Hence, the machine body must possess shape invariability alongwith strength, producibility, low material requirement and sufficiently low cost. All this will depend upon :

1. Proper selection of the material for the machine body and the manufacturing process.

2. The provision for a static and dynamic rigidity at which the deformation of the bed, base or column, under the action of maximum forces during operation, in within limits conforming with the machining tolerances.

3. A sufficiently high wear-resistance of the ways.

18.1.2. Elements of Design. During machine tool operation, the cutting forces, weight of the stationary and moving parts, weight of the workpiece etc., and in some machine tools, inertia forces act on the bed (base, column, etc.). The stressing conditions of most machine tool bodies are rather complex, but generally a combination of bending and torsion will be found. The bending moment to which a bed is subjected is not as serious as the strain resulting from torsion. The driving force is converted into torsion or a twisting stress, the torque on the work and on the machine body balance one another and must be equal while the actual stresses are usually low, every effort is made to reduce distortion to a minimum. Thus the primary object in the design of a machine body is rigidity (stiffness), because, eventhough the total deflection of a machine is the sum of all its parts, the machine bed constitutes the main unit to which the cutting forces are eventually transmitted. It is impossible to accurately determine the deformation of the bed being designed, especially if it is of complex shape, by calculations. The required rigidity is usually provided for by methods proved in practice.

Section of Machine Body

The configuration of the machine bed (base, column etc.) is determined primarily by :

(*a*) the arrangement of the ways on it for various units of machine tool.

(*b*) the weight, dimensions and length of stroke of the main units and parts.

(*c*) the necessity of housing various mechanisms inside the bed.

(*d*) the necessity of providing various openings, apertures etc. in the bed walls for assembly, disassembly, inspection, adjustment and lubrication of various mechanisms of the machine, and pads, brackets and lugs on the bed walls for mounting various devices.

Two sections are in general use for machine body : Plate or rib section, and the box or tubular section.

1. **Plate or rib section.** The plate is strong in a vertical direction only. It has little strength in a sideways direction and less against torsional strain.

2. **Box section.** The almost completely closed box section gives the best stiffness in torsion, and is also satisfactory in bending. Since producibility cosiderations are also in favour of this shape of cross-section, it is most often the basis for bed design. In respect to strength, for equal weight of metal, a complete box is about 13 times more rigid against torsion and 4 times more rigid against bending than the same amount of metal used in plate form.

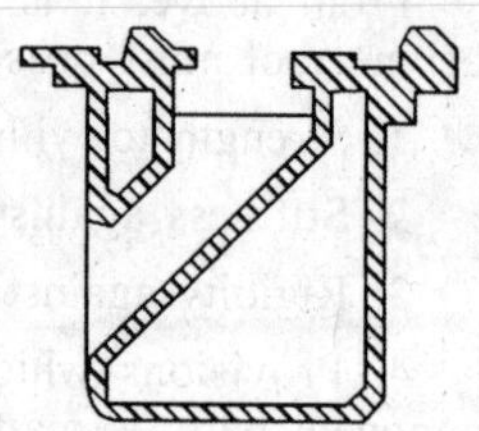

Fig. 18.1. Box-section of Lathe Bed.

3. **Tubular section.** A tubular section has the advantage that there are no internal corners. Fig 18.1 shows the box-section of a lathe bed.

Even though, a completely enclosed (box) section is the best section to resist bending and especially torsion, it is usually impossible to retain such a section over the whole length of the bed because of the necessity for ensuring free chip disposal, and for arranging various mechanisms in the bed etc. It is often necessary, therefore, to provide open sections, or at least openings, in the machine body. This substantially reduces the rigidity of the bed. The weakening must be counteracted by stiffening devices. One way is to retain a longitudinal horizontal stiffening rib (usually trough shaped) unbroken over the full length of the bed. The rib is either inclined or has openings to facilitate chip disposal.

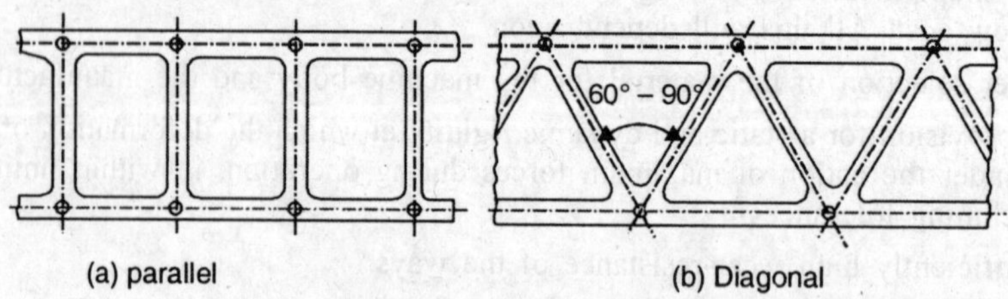

Fig. 18.2. Stiffening Ribs.

A very effective way of attaining the required rigidity is the provision of partitions tying together the longitudinal walls. Transverse partitions are extensively used in machine tool beds, Fig. 18.2. Diagonal partitions are superior as compared to parallel partitions.

The beds of heavy machine tools are often of sectional construction. In designing such a bed, it is necessary to provide means in the construction to obtain a sufficiently high rigidity of the joints between the sections.

The stiffness and rigidity requirements of precision grinding machines are more severe as compared to lathes. The lathe beds are supported on two or three legs which are connected to the foundation with the help of foundation bolts. The foundation, thus, takes its share in stiffening the bed. On the other hands, the beds of grinding machines should not be rigidly fixed to foundation because of possible straining and distortion due to temperature changes. The machine bed should itself provide the necessary rigidity. So, the grinding machine beds are designed as deep box sections, with few openings, to provide the maximum rigidity. The bed is usually freely supported at several points (if possible, only 3) to obtain a statically determinate system. For levelling purposes, adjustable wedge boxes are often used as supports.

Any deviations from straightness in the beds of long grinding an planing machine tools, which have to produce highly accurate work, have to be watched and corrected regularly. Due to this, they are often supported on adjustable wedge boxes at intervals of about 90 cm. Closed box sections are generally used in the uprights of milling or boring machines which are acting as cantilevers clamped at the base and free at the top.

Eventhough a bed must be sufficiently rigid, this alone is inadequate to ensure the rigidity of the machine tool fixture — workpiece complex. The selection of feed and depth of cut that are permissible for the required machining accuracy, the class of surface finished obtained and the specified tool life depend upon the rigidity of the whole above mentioned complex. This is achieved by :

1. Tieing the main parts of the machine tool together, so as to form a closed frame. Fig 18.3 (*a*) shows an open frame construction and Fig 18.3 (*b*) shows a closed frame construction of a knee-type milling machine and a planer. The latter construction increases the stiffness of the structure.

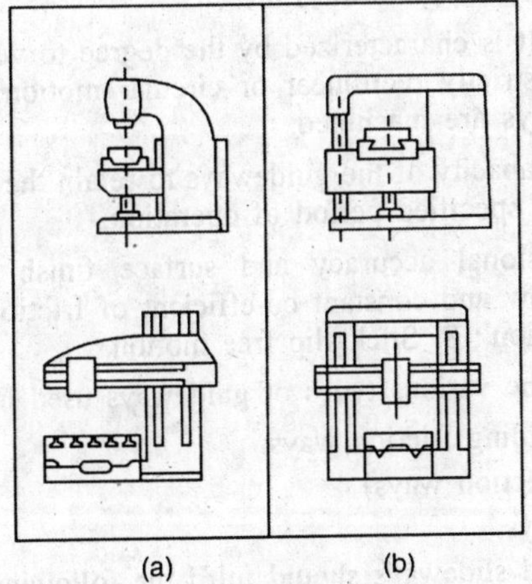

(a) (b)

Fig 18.3

2. Casting the bed integral with the headstock housing.
3. Employing a monoblock construction.

Columns (uprights, stanchions), housing, tables, carriages, slides, cross rails, as well as such components as the knee of milling machines, columns of radial drills, columns of semiautomatic multi-spindle vertical chucking machines, have a great variety of configurations. Their configurations depends upon with what parts of the machine tool these components mate, whether the parts are fixed or movable, the location of the components in the machine tool, the magnitude and direction of the acting forces, and other factors. Just like bed or base of the machine tool, these components have housing-type constructions. The principal requirements concern their rigidity and vibration proof properties. These elements are attached to the machine tool bed or base by welding, bolting or mounting on ways.

These parts are made of the same metals as used for making beds. These housing type parts are made of either cast or welded design. The required rigidity is attained, as in beds, by the box-shape cross-section, and the system of ribbing in cast construction or braces, angle plates and similar stiffening members in welded steel constructions.

The rigidity of such parts as tables and carriages or slides depends to a great extent upon the number of joints or mating surfaces and their arrangement in respect to the acting forces. Fewer the joints or mating parts, the more rigid the construction. In cases where it is not possible to have the number of such joints below a certain limit, it is necessary to enlarge the contacting surfaces in a direction approximately perpendicular to that of the acting force. This reduces the specific pressure and makes provision for firmly and reliably clamping parts which are to be stationary during operation. Similar clamping devices are used to secure the outer column and aim of radial drills, cross-rails of vertical boring mills, planers, planer-type milling machines etc. These clamping devices may be hand operated or powered by an individual motor.

If a housing-type part is traversed along vertical ways by a kinematic train which contains no self-braking transmissions, the part is balanced with a counterweight or spring to facilitate its setting and to prevent it from sliding down when it is unclamped.

The working surfaces of tables have a system of parallel, and sometimes perpendicular, T-slots used to set up and clamp various types of fixtures.

18.1.3. Guideways. The cutting tool or the work, together with the units on which the tool or work is mounted (saddle, cross-slide, table assembly etc.), travels on guideways provided on the machine bed, column or upright. A guideway constraints either the tool or the work to move in a definite path, usually, either a straight line or a circle. To some extent, the workpiece accuracy is dependent on the shape of the guideways used in guiding saddles, cross-slides or tables.

The principal requirements of guideways are :

1. **Accuracy of travel.** It is characterized by the degree to which the actual travel of the units is in compliance with strictly rectilinear or circular motion. It depends mainly on the accuracy with which guideways are machined.

2. **Durability.** It is the capacity of the guideways to retain the initial accuracy of travel of the corresponding unit over a specified period of operation.

3. **Rigidity; 4.** Dimensional accuracy and surface finish; **5.** High wear resistance; **6.** Corrosion resistance; **7.** Low and constant co-efficient of friction, especially at top friction limits and interrupted lubrication.; **8.** Stick-slip free motion.

Types of Guideways. The various types of guideways used in machine tools are :

(*a*) Slideways, that is, sliding friction ways.

(*b*) Antifriction (roller friction ways)

(*c*) Hydro-static guideways.

(*a*) **Slideways.** Slides and slideways should fulfil the following basic requirements :

(*i*) Accurate alignment of parts in question under working load.

(*ii*) Possibility of adjustment to facilitate not only assembly, but also maintenance, especially compensation for wear.

(*iii*) Freedom from unnecessary restraint.

(*iv*) Possibility of efficient lubrication.

(*v*) Prevention of swarf accumulation and ease of swarf removal.

The basic surfaces used for slideways are, (Fig. 18.4) (*i*) Vee-slide (*ii*) Flat slide (*iii*) Dove-tail slide (*iv*) Cylindrical slide.

1. **Vee-slide.** Vee ways are more difficult to manufacture than flat ways. But its advantages are : it provides automatic adjustment for wear in the vertical plane, since gravity acts as a closing force to keep the mating surfaces in contact. So, it is self-compensating for wear in the horizontal plane also. These slides are used for very accurate movement of parts. The upward Vee (external or encompassed) does not allow dirt or swarf to accumulate. Upward Vee is invariably used on lathe beds except in very large machines. This design retains lubricant poorly. The downward Vee (internal or encompassing) found in many large griding and planing machines retains the lubricant, but the danger of a large accumulation of swarf can be overcome only by providing the ways with shields or other protecting devices.

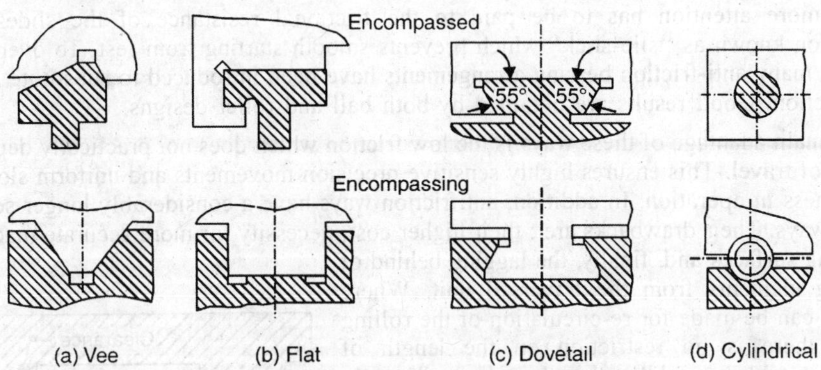

Fig. 18.4. Principal Shapes of Slideways.

Vee ways are made symmetrical, if for example, the load is directed vertically, as from the weight of the travelling unit, so that the force should be shared equally by the two faces of the Vee to maintain alignment. On a lathe, however, most of the surface loading is applied to the left hand end of the inside face of the front shear and to offset this, an asymmetrical form of Vee has been adopted, with the wider side on the inside to meet the combined trust angle more directly.

Drawbacks of Vee guide are : lack of bearing surface, and further the inverted Vees (upward vees) weaken the saddle which has a long unsupported span across the bed.

2. **Flat ways.** With flat guides there is less friction than with Vee guides. They are easier and cheaper to manufacture, particularly on large machines where the fitting and alignment of four Vee faces is not easy. They give plenty of bearing surface and the saddle is well supported while machining operations on the bed are simplified. They are used extensively for heavy duty work and large and heavy machine tables. They separate the guiding actions in the vertical and horizontal direction and thus enable the fitter to align mating surfaces independently in the two planes. However, they require devices for adjusting the clearances, have a tendency to accumulate dirt, and retain the lubricant comparatively poorly when they are of the encompassed type.

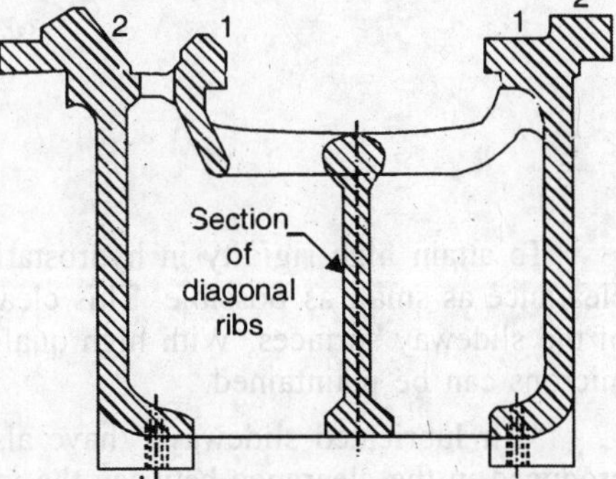

Fig. 18.5. Lathe Bed

Sometimes, a compromise is effected by a design which incorporates both Vee and flat guideways, similar to those used in centre lathes and grinding machines. In centre lathe, two independent sets of slideways are used, one for headstock and tailstock which carries the workpiece, (1) in Fig. 18.5 and the other one for the saddle which carries the tool post, (2) in Fig. 18.5. In grinding machines, the cut is mainly pressing downwards and extremely light when fine boring so that there is practically no tendency to lift the table, and one Vee provides sufficient guiding surface. So far such machines and also for lathe and some planing machines, one Vee and one flat way are used.

3. **Dovetail ways.** These occupy small space and have comparatively simple clearance adjustment by means of a simple taper or flat gib. These are widely used for the guidance of parts producing feed movements such as lathe cross-slides, milling machine tables, saddles and knees and so on. These ways are not suited in cases where the forces try to pull out the guides.

4. **Cylindrical guides.** These are simple to manufacture but require accurate machining because adjustment or fitting by scraping is hardly possible. Their main drawback is the low rigidity due to the fact that they are secured to the bed only at their ends. These are used mainly for axial loading (axially movable spindles of drilling and boring machines).

(*b*) **Anti-friction ways.** As the requirements for accuracy and location of tables and saddles increase, more attention has to be paid to the frictional resistance of the slides and the phenomenon known as "slip-stick" which prevents smooth starting from rest. To overcome this drawback, many anti-friction bearing arrangements have been introduced to substitute sliding by rolling friction. Good results are obtained by both ball and roller designs.

The main adantage of these ways is the low friction which does not practically depend upon the speed of travel. This ensures highly sensitive precision movements and uniform slow motion with lightness in operation. In addition, antifriction ways have a considerably longer service life than slideways. Their drawbacks are : their higher cost, necessity for more accurate machining of the working surfaces and, finally, the lagging behind of the rolling elements from the traversed unit. Where provisions can be made for re-circulation of the rolling elements, there is no restriction on the length of traverse of a table or saddle other than the bed length. Anti-friction guideways are used for precision machine tools.

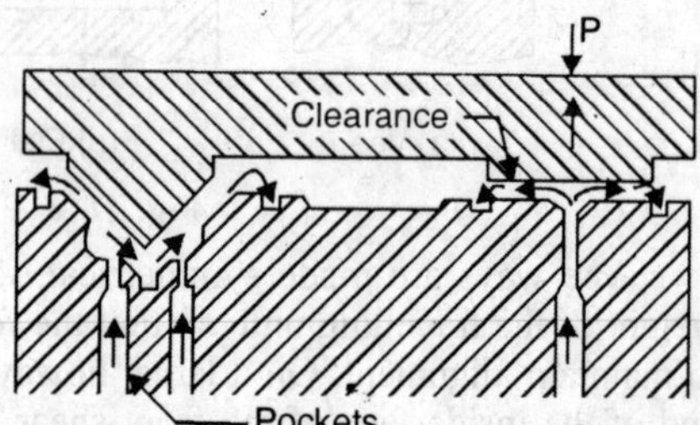

Fig. 18.6. Hydrostatically Lubricated Slideways.

(*c*) **Hydrostatically lubricated slideways.** In these slideways, oil under pressure is delivered between the mating surfaces, so as to produce an oil film over the full contact area, Fig. 18.6. From the pump, oil is delivered under pressure through flow control valves into pockets made in the ways. From the pockets the oil escapes through the clearance between the mating surfaces. The load carrying capacity of these ways can be calculated from the relation :

$$P = p.A.\alpha$$

where p = oil pressure in the pockets

A = area of the slideways

α = factor taking into account the drop in oil pressure in the clearance

$$= \frac{1}{3} \text{ to } \frac{1}{2}$$

To attain high rigidity in hydrostatically lubricated slideways, it is necessary to make the clearance as small as possible. This clearance depends upon the macro-and micro-irregularities of the slideway surfaces. With high quality scrapping, a minimum design clearance of 15 to 25 microns can be maintained.

"Air lubricated slideways" have also been used in machine tools. Here, an air cushion is produced in the clearance between the mating bearing surfaces. Air from compressed air mains passes through a filter and pressure regulating valve and enters the pockets at a pressure of 0.3 to 0.5 N/mm^2 through apertures of small diameter (0.2 to 0.5 mm).

18.1.4. Materials for Machine Body and Guideways. The ideal material for machine tool structures is one which will damp out vibrations and at the same time be wear resistant. The wear of slideways depends to a considerable extent on the materials of the ways of the bed and of the travelling unit, that is, saddle, slide, table etc. A high surface hardness of guideways does not, by itself, guarantee high wear resistance. It has been found that minimum total wear of slideways is attained with different hardness of the mating pair of surfaces due to run-in of the softer material of the pair. In most cases, it is more advantageous to use the harder material for the stationary guideways (bed ways) since their shape is copied by the travelling unit and also, it is more difficult and expensive to repair the bed ways.

When the guideway is made integral with the bed, and correspondingly, with the travelling unit, Grey cast iron is the most commonly used material. The wear resistance of the cast iron slideways can be increased by surface hardening with flame or induction heating. Alloyed cast iron and nitrided cast iron have also been used for machine beds.

Cast iron has many advantages for making machine tool structures :

1. Any complex shape can be cast.

2. Good machinability.

3. Lower production cost in large lot production.

However, it has got the following drawbacks :

1. The casting method takes a long time to make a machine tool due to the making of a pattern, core boxes, aging of the casting before machining and after roughing for stress relieving.

2. Possibility of casting defects and the rejection of castings.

3. Large machining allowance has to be provided.

Welded structures. Welded structures from previously cut pieces of rolled steel are being used as an alternative to castings often with the advantage in regard to cost, lightness and time saving. Welded structures possess the following advantages :

1. The mechanical properties of steel are much higher than those of cast iron. Its modulus of elasticity being about twice that of cast iron, the distortions of steel structures are about half that of cast iron of equal sections and loads.

2. Cast iron is sensitive to shock loads.

A cast bed is often more expedient in large lot production, while a welded steel bed is preferable when it is necessary to make one or several machine tools in short time. Whereas cast iron is more capable of damping vibrations than steel, bed of welded steel construction are not usually inferior to cast iron beds.

If the slideways are not cast integral with the bed or when the bed is a welded structure, then the slideways are made separately and attached to the beds. "Attached ways" are usually of steel, but in some cases they are made of high-quality cast iron. The ways are designed as strips secured to a cast iron bed with screws or welded to a welded steel bed. (Fig. 18.7).

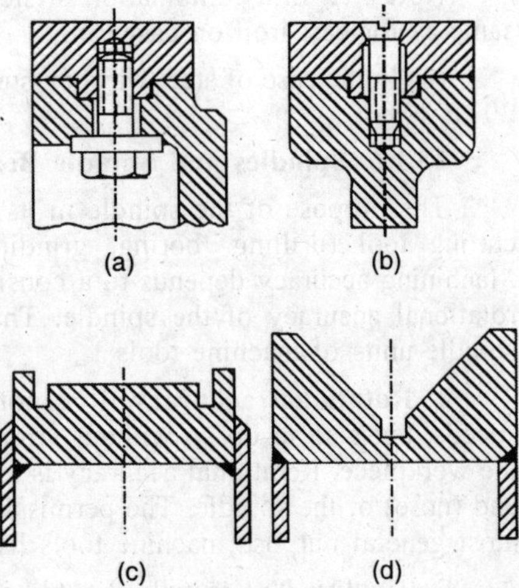

Fig. 18.7. Attached ways.

Plastic guideways. Plastics are promising materials for slideways because of their antiscoring, anticorrosive, reduced friction and excellent wearing qualities. Also, where freedom from lateral play is important they will operate succcessfully with practically no clearance. Laminated fabric strips are used in combination with cast iron for the slideways of heavy machine tools. With plastic material, the metal particles trapped between the table and the bed slides become embedded in the plastic without protruding and do not damage the slideways as will happen in normal slideways. Their drawbacks are : low modulus of elasticity as compared to steel, low co-efficient of thermal conductivity and tendency to swell when they absorb oil. Due to this, it is more advantageous to employ slideways with a thin layer of polymer coating applied by spraying, gluing on a thin film or some other method. Ways of plastics are usually secured by screws but are sometimes glued to the bed.

Sometimes, pads of zinc alloy or of bronze are also used on the slideways. They possess good wear resistance but are expensive.

18.1.5. Slideway Design. One of the most important conditions on which the wear resistance of slideways depends is the pressure distribution over the way surfaces. It should be as uniform as possible, with the average specific pressure not exceeding a certain definite value established on the basis of experience in machine tool operation. The calculations are based on the assumption that the pressure distribution varies according to a linear function lengthwise along the slideway and it is considered to be distributed uniformly across the width of each face of the slideway.

Machine tool industry lists the following permissible values of p_{max}. For cast iron slideways :

(*a*) p_{max} = 2.5 to 30 N/mm^2 at low sliding speeds in the order of the rates of feed (lathes and milling machines).

= 0.8 N/mm^2, at high sliding speeds in the order of the cutting speeds (planers, shapers and slotters).

= For special-purpose machine tools, operating at constant heavy feeds and high speeds, it should be reduced by approximately 25%.

= 1.0 N/mm^2 for heavy machine tools at low sliding speeds, and 0.4 N/mm^2 for high sliding speeds.

= 0.05 to 0.08 N/mm^2 for the slideways of grinding machines.

Note : (*i*) In a combination of steel on the cast iron ways, the values of p_{max} are about the same as for cast iron on cast iron.

(*ii*) In the case of steel ways on steel ways, the permissible values can be increased by about 20 to 30%.

18.1.6. Spindles and Spindle Bearings

The purpose of the spindle in its bearings is to hold securely and transmit motion to the cutting tool (drilling, boring, grinding and milling machines) or the workpiece (lathes). Machining accuracy depends to a considerable degree in many types of machine tools upon the rotational accuracy of the spindle. This imposes the following principal requirements on the spindle units of machine tools :

1. **Rotational accuracy.** It is not sufficient for a spindle to be able to rotate at varying speeds, but also to do so without vibration or irregular running that will cause inaccuracies in the workpiece. Rotational accuracy is usually characterized by the amount of run out of the front end (nose) of the spindle. The permissible spindle nose runout values (both radial and axial) for most general purpose machine tools have been standardized.

2. **Rigidity.** The spindle should be rigid enough to retain its correct position when acted on by various working forces. Excessive deformation of the spindle has a deteremental effect on the machining accuracy and on the service life of the spindle bearings and drive.

3. **Vibration-proof properties.** This is essential for spindles of high speed machine tools, especially those intended for performing finishing operations.

4. **Wear resistance.** Wear resistance of the bearing surfaces is required in cases when the spindle runs in sleeve bearings or when there is relative longitudinal motion of elements of the drive and the spindle (as in drilling, boring and other machines).

18.1.7. Materials for Spindles. One of the main requirements of spindles is sufficient rigidity which depends, in part, on the Young's modulus of spindle material. Since the Young's modulus of various steels is practically the same, the use of alloy steels for spindles is not recommended unless their use is warranted by other requirements. Therefore, the spindles of machine tools are usually made of medium-carbon structural *C*-45 steel which subsequently undergoes a heat treatment (quenching followed by high tempering to a hardness of 22-28 R_c). If the surface of the spindle (or a part of it) must have a high hardness, heat treatment is done to get a hardness of 40-50 R_c. Induction hardening will give better results (hardness of 48-60 R_c) and at the same time, less distortion will take place during heat treatment.

If a very high surface hardness of the spindle journals is needed, then low-carbon case hardening steel is used. The heat treatment consists of : carburization, quenching and tempering to a hardness of 56-62 R_c. Exceptionally high surface hardness with very little deformation can be obtained by nitriding followed by quenching and tempering. Such spindles are used for high-precision machine tools not subject to heavy loads. Spindles of heavy machine tools are made of manganese steel with subsequent normalization (spindles subject to low loads) or hardening followed by high tempering to a hardness of 28-35 R_c.

In specific cases, hollow spindles of large diameter for horizontal boring and other machines can be suitably made of grey cast iron or of high strength nodular cast iron.

18.1.8. Design Features of a Spindle. As noted earlier, a machine spindle performs many functions : provides centering, holds the tool or the workpiece, rotates the tool or the workpiece, and also feeds the tool (as in drilling machine). The design features of a spindle depends upon : the kind of cutting tool or workpiece it is to carry, the fits of the elements of the drive and the type of bearings it is to run in. A typical machine tool spindle is shown in Fig. 18.8.

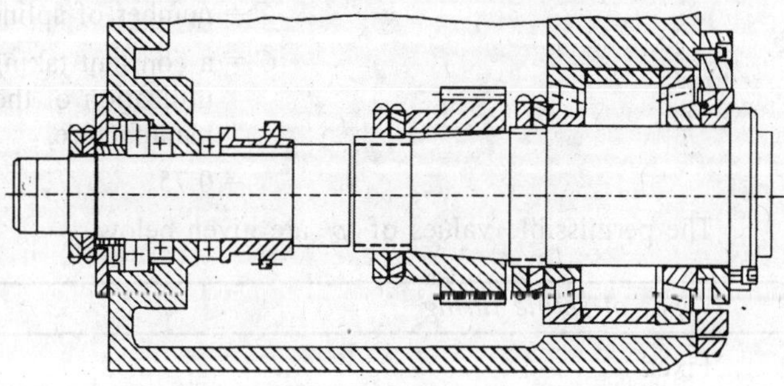

Fig. 18.8. A Machine Tool Spindle.

The front end of the lathe spindle (nose) is shown in Fig. 18.9. The spindle is hollow with a tapered hole at the front end, which is used for internal location (live centre etc.). The taper is Morse taper (approximately 1 : 20) which is self locking to hold the centred part in position, and it is even capable of transmitting a torque (by friction), for example, the cutting torque of the drill. The front outer portion of the spindle, *b*, is threaded for receiving a chuck, collet or face plate. As a screw thread will rarely, if ever provide accurate concentric location, a cylindrical portion '*a*' is provided which acts as spigot for the concentric location of chucks etc. The screw thread holding the chuck axially in position against the shoulder must be free and almost a loose fit so that it will not interfere with the locating action of the cylinder.

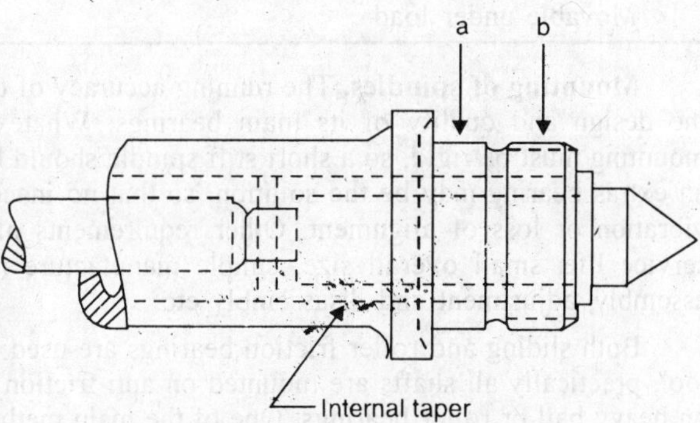

Fig. 18.9. Lathe Spindle nose.

Design calculations. A machine tool spindle is nothing but a shaft supported in bearings. Therefore, the design calculations involve the following features :

(*a*) **Rigidity calculations.** These calculations involve the determination of the deflection in bending and, in some cases, the twist in torsion. The following values of permissible deflection and slope angle are used as tentative norms in machine tool engineering practice :

$$y_{max} \leq 0.0002\, l; \qquad \theta_{max} \leq 0.001 \text{ rad}$$

where l = distance between supports.

(*b*) **Strength calculations.** As noted above, the spindles are subjected to bending moment and also twisting moment. Also, there is a reversal of stress in the fibres due to the rotation of the spindles. Thus, the loaded spindles must be checked for fatigue loading.

(*c*) **Vibration behaviour calculations.** Vibration behaviour calculations including the determination of natural frequency of the spindle to avoid resonance vibrations, should be carried out for high spindles.

(*d*) **Unit pressure calculation.** The specific pressure should be checked on the surfaces of splined sections of spindles. It is given as,

$$p = \frac{8\ Mt}{\left(D^2 - d^2\right).L.Z.C}$$

where Mt = maximum twisting moment.

D, d = outer and inner diameter of the spindle shaft

L = length of engagement

Z = number of splines

C = a constant taking into account the non-uniform utilization of the spline surfaces, due to errors in manufacture,

= 0.75

The permissible values of 'p' are given below :

Types of spline fitting	*p-N/mm²*
Fixed	120 to 200
Movable, but not under load	40 to 70
Movable under load	10 to 20

Mounting of spindles. The running accuracy of the spindle depends to a large extent upon the design and quality of its main bearings. Whatever type of bearing is used, the spindle mounting must be rigid, so a short stiff spindle should be the aim, and if this is not feasible, then an extras bearing may be the solution, so that no inaccuracies are introduced in the product by vibration or loss of alignment. Other requirements of spindle bearings are : sufficiently long service life, small overall size, simple manufacture (sleeve bearings), simple and convenient assembly, adjustment and disassembly etc.

Both sliding and roller friction bearings are used in spindle supports. In a modern machine tool, practically all shafts are mounted on anti-friction bearings, with the main spindle mounted on heavy ball or roller bearings. One of the main methods of increasing the accuracy of running of the spindle is preloading the bearings. With this, any initial slackness that might be effected by the cutting action of the tool is eliminated before machining starts. In addition, this sets up elastic deformation that improves the total rigidity of the spindle unit.

Hydrostatic bearings are being increasingly used. As written under slideways, the clearance between spindle and bearing should be minimum, so that the spindle floats upon a film of oil under pressure. The oil flows out through the clearances between a shoulder of the spindle and the end of the bearing. Ample supply of oil should be supplied through the bearing to restrict temperature rise. These bearings are finding increasing application for grinding machine spindles and high precision machine tools such as fine boring machines and diamond turning lathes. Hydrostatic bearings can operate under fluid conditions even at the lowest speeds of rotation.

Air bearings have less frictional losses but are less rigid as compared to hydrostatic bearings.

We know that the common bearing materials for sliding contact bearings are : cast iron, bronzes and babbits. Some of the plastic materials (nylon, derlin etc.) are proving of value as bearing materials. They have the advantages :

(*i*) After initial running, they can be used without lubrication, an advantageous feature when bearings are mounted in a position difficult to access.

(*ii*) They run quieter and withstand shock loads without producing vibrations.

(*iii*) The low co-efficient of friction enables them to resist wear. Even under abrasive and dirty conditions, they have much longer life as compared to metal bearings.

18.2. MACHINE TOOL TESTING

18.2.1. Introduction. The purpose of the various machine tool tests (acceptance tests) is to heck the accuracy and performance of the machine tool and whether it complies with the manufacturing specifications. Therefore, these tests conclude the process of designing and manufacturing a lot-produced machine tool. As a rule, these tests give a general idea of the quality of the machine tool without taking much time or requiring complex apparatus. Therefore, these tests are to be conducted without fail at the manufacturing plant, and following overhauls and even medium repairs, so that the machine tool meets the requirements of specifications.

According to the present-day manufacturing specifications, the machine tools are subjected to the following acceptance tests :

1. Idle-run tests, checking the quality of manufacture and the operation of all mechanisms in the electric, hydraulic and pneumatic equipment, and in the lubricating and coolant systems, and certificate data checks.

2. Checking whether the machine tool complies with the accuracy standards (that is, checking the geometrical accuracy of the machine tool), and checking the surface roughness, and accuracy of the workpiece being machined.

3. Performance tests under load and determining the rate of production (for special machine tools).

4. Rigidity tests of machine tools.

5. Tests for vibration-proof properties of machine tools in cutting.

The accuracy of the workpiece being machined and its surface roughness may be checked during the load tests.

To conduct these tests, the machine tool is installed on a special foundation (a test stand at manufacturer's works). With the aid of adjusting wedges and/or levelling shoes, positioned in the same manner as for installing the machine for regular operation, it is levelled in the longitudinal and transverse directions to a precise spirit level. After a preliminary trial in which the various mechanisms of the machine tool are switched on to make sure that they are in proper order, the idle-run tests are started.

18.2.2. Idle-run Tests. Idle-run tests are conducted at no load. They determine whether mechanisms are in proper condition and operate correctly when they run without loads. These tests are conducted at different spindle speeds (from minimum to maximum values). The machine should continue too operate at the maximum speed until the spindle bearings reach a steady temperature : 85°C for rolling bearings, 70°C for sliding bearings and 50°C for other mechanisms. At the same time, the operation of the feed mechanisms is checkea at the low, medium and high working feeds, as well as rapid traverse motions.

In addition to this, all the mechanisms are switched on or engaged, the actual spindle speeds are measured, as well as the travel of all units (minimum and maximum) to check whether the characteristics of machine tool conform to the specifications. This tests the proper action of all control devices in engaging, changing over, and transmitting of motions, the action of interlocking devices, reliability of locking facilities and a complete absence of accidental engagements, misalignment and jamming.

Idle-run tests also include :

1. The operation of all automatic devices, stops and indexing mechanisms is checked.

2. Backlash is determined for all the actuating screws of the manual controls.

3. The operation of the mechanisms for clamping the work and cutting tools, of the lubricating and coolant systems, and of the electrical and hydraulic equipment is checked.

4. The dependability of all the protection devices is checked.

5. The effect required to operate the manual control is measured.

6. No-load power at all speeds of the main drive.

Spindle speeds are measured by a 'tachometer' or by a 'revolution counter', in conjunction with a stop watch. The rate of feed of tool slides is measured by counting off 100 spindle revolutions with the revolution counter and measuring the corresponding toolslide travel with a steel rule. Then,

$$\text{feed per revolution} = \frac{\text{Measured travel}}{100}$$

The feed of hydraulically-operated units (mm/min) is determined by measuring the travel in 30 or 60 seconds (using the stop watch). Power measurements are made by a wattmeter or by using a voltmeter and two ammeters.

18.2.3. Accuracy Tests. The geometrical accuracy of a machine tool characterizes the quality of manufacture and assembly of the machine tool. The machining accuracy of a machine tool (machine tool performance), besides depending on the geometrical accuracy of the machine itself, is influenced by such other factors as :

1. The type of cutting tool and its condition.

2. The cutting speed, feed and chip sections.

3. The material to be machined.

4. The shape and size of the workpiece.

5. The machine-fixture-cutting tool-working system, rigidity.

6. The skill of the operator.

Due to all this, the various State Standards stipulate compulsory accuracy tests of machine tools machining work samples including a check of their surface roughness. These tests should be carried out after the preliminary idle running of the machine tool or its load tests. The kind of work sample, its material, and the character of machining for various types of machining tools are given in corresponding standards. For example, in the case of the lathe, the cutting tests cover mainly the turning of cylindrical parts and the cutting of screw threads close to the headstock.

Though the geometrical accuracy of the machine tool cannot qualitatively characterise the accuracy of a workpiece produced by the machine tool, it is one of the important characteristics of machine tool performance in this respect. The geometrical accuracy of each lot-produced machine tool should be checked. Norms of geometrical accuracy were first worked out by Dr. G. Schlesinger in 1927. Typical checks usually include

(*a*) Accuracy tests on the geometrical shape of mounting surfaces (straightness, flatness, out-of-roundness, taper etc.).

(*b*) Relative position of surfaces (parallelism, squareness and alignment), that is, of the principal units and parts.

The aim of accuracy tests is to determine the actual errors in the above mentioned items. Each of these errors should not exceed the permissible values stipulated in the accuracy standards for the given type and size of machine.

Measuring Tools used in Geometrical Accuracy Tests

The principal measuring instruments used in geometrical accuracy tests are :

1. **Straight edges.** (wide-edge type from 500 to 3000 mm long and toolmaker's straight edges upto 300 mm long. See Art. 12.8.1.

2. **Squares.** See Art. 12.9,

3. **Gauge blocks.** See Art. 12.7,

4. **Thickness gauges or Feeler gauges.** See Art. 12.16,

5. **Dial indicators.** See Art. 12.10,

6. **Spirit levels.** See Art. 12.8.4,

The readers should refer to the above mentioned articles for the explanation of their principle of design and operation.

7. **Stub or centre-type test mandrels.** The most widely used inspection tool for acceptance tests of machine tools is the test mandrel. Its quality (especially as far as straightness and roundness are concerned) is of paramount importance for accurate results. Two types of mandrels are used :

(*i*) **Stub mandrel.** It has a cylindrical measuring surface and a taper shank which can be inserted into the taper bore of the main spindle, Fig. 18.10.

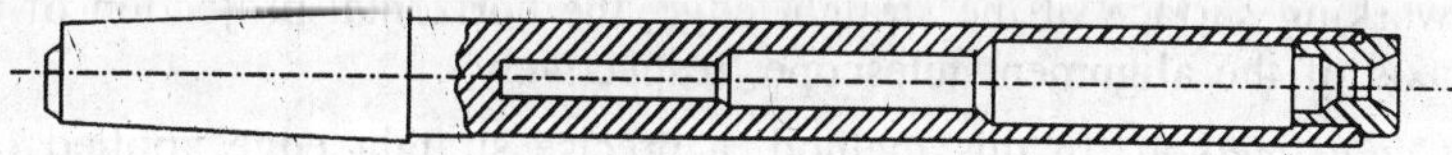

Fig. 18.10. Test Mandrel.

(*ii*) **Cylindrical mandrel.** It can be held between centres.

All mandrels must be hardened, stress-relieved and ground. The measuring length of the cylindrical part of mandrels depends on their purpose. For stub mandrels it varies from 100 to 500 mm. Sagging of the mandrel under its own weight may cause measuring errors. Thus, its diameter must be such that the sag is kept within permissible limits. To reduce their weight, test mandrels are usually made hollow (but not at the cost of their rigidity).

The inserted test mandrel is used to check the true running of the main spindle. It is also to check : the parallelism between spindle axis and saddle sideways, alignment of the tool rest, correct height alignment between head-stock and tailstock of a lathe etc. A test mandrel is always used in conjunction with a dial indicator.

8. **Optical instruments.** Microscope, Telescope, Collimator etc.

18.2.4. Acceptance Tests. Acceptance tests are conducted as per Test Charts for various machine tools (Refer : Testing Machine Tools by G. Schlesinger). The general procedure follows the steps given below :

1. The machine is installed and carefully levelled by the proper use of a spirit level.

2. Testing of straightness, flatness, parallelism and quality of the guiding and bearing surfaces of beds, uprights, base plates.

3. Main spindle is tested for true running-axial slip, and location and position of its axis relative to other axes and surfaces.

4. The movements of other main parts of the machine are checked.

5. The accuracy of the workpieces produced on the machine is checked.

Below, we discuss the various tests for a lathe.

(*a*) **Levelling the lathe bed.** Before the actual acceptance test is started, the machine tool must be carefully levelled, usually by means of a precision spirit level. The levelling is done longitudinally as well as transversely. During levelling, the carriage must be in the middle of the bed. In the case of long beds with more than two legs, it must be between two legs. A spirit level is first put on the rear slideway and has to be checked over the full length of the bed. This slideway is usually plane. The measurements are repeated for the front slideway. This slideway may be intentionally convex, because the cutting torque exerted on the bed will normally tend to press this slideway downwards.

It is advisable to check the levelling in the transverse direction simultaneously with longitudinal levelling. For this, a second spirit level is placed alternatively on either side of the carriage.

(*b*) **Straightness of the slideways.** The slideways must be straight in both horizontal and vertical planes and parallel with the axis of rotation of the spindle. Otherwise, the tool nose, in turning, will not travel along a straight line parallel with the axis of the machine and work of true cylindrical form cannot be obtained.

The straightness of machine tool ways may be measured either by measuring linear values, determining the position of various sections of ways in reference to an initial straight line, or by measuring the location of these sections in reference to one another, consecutively along the whole length of the ways.

In the first case, the common methods are : measurements made by means of a straight edge, gauge blocks and a thickness gauge; measurements along a taut wire; and by an optical sighting method. The initial straight line, in reference to which measurements are made by these methods, is the working surface of the straight edge, the horizontal projection of the taut wire, and the optical axis of the alignment telescope, respectively.

(*i*) *Straight edge method.* In this method, a precise straight edge applied on two gauge blocks of equal height is used. The distance under the straight edge to the ways is measured at various places by a set of gauge blocks and a thickness gauge. If, for example, at various points under the straight edge, thickness gauges of 0.10 mm, 0.05 mm and 0.08 mm will pass and the distance between the blocks supporting the straight-edge equals one metre, then the maximum error per metre will be 0.10 – 0.05 = 0.05 mm. Thus, the maximum error in straightness = Maximum size of thickness gauge – Minimum size of thickness gauge passing between the straight edge and the way.

It is evident that if thickness gauges of the same size pass under the straight edge, all along its length, the way will be as straight as the straight edge.

(*ii*) *Taut wire method.* The microscope is fixed to the saddle and travels along a tight wire. Deviations from straightness of the saddle slideways will be recorded, with the horizontal projection of the taut wire as the reference line, Fig. 18.11.

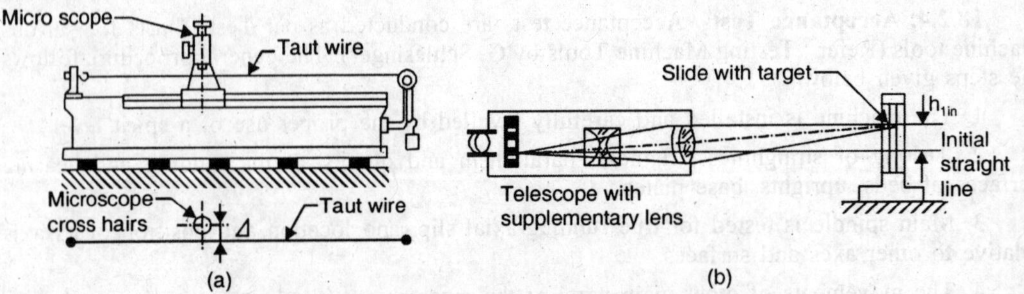

Fig. 18.11. Checking the Straightness of Slideways.

(*iii*) *Optical sighting method.* In this method, a target (glass plate in the centre of which cross hairs are inscribed) which slides on a carriage on the bed is observed through a telescope. The instrument (slide with the target) is applied consecutively at different elementary plane surfaces along the ways. In each position, the deviation of the target from the initial straight line is recorded, Fig. 18.11 (*b*).

In the second case, straightness is checked with a level or by the collimation method. The collimation method has been explained under Art. 12.9.3, section (*a*). These methods are used for machine beds more than 3 m long.

For beds upto 3 m in length, straightness can be checked with the help of precision level. The difference between this test and the levelling of the machine bed is that for the straightness test, the spirit level readings are taken as several positions along the bed (at intervals of about 300 mm).

(*c*) **Flatness of guideways.** Flatness of ways in a horizontal or vertical plane can be checked with an ordinary or frame-type level. The level is moved along the tested surface. Flatness may also be tested with marking compound and a surface plane. A thin layer of marking compound is applied on the surface plate which is then rubbed back and forth with a slight motion on the surface of the way being tested. Flatness is characterized by the number of bearing spots per unit area which remain on the way surface.

(*d*) **Parallelism.** In the case of the centre lathe, parallelism has to be checked :

1. Between the saddle slideway and
   (*i*) the tailstock slideway
   (*ii*) the spindle axis
   (*iii*) the outside diameter of the tailstock sleeve
   (*iv*) the tailstock sleeve taper
   (*v*) the lead-screw axis
2. Between the spindle axis and the tool rest slideway.

The parallelism between the saddle slideway and the tailstock slideway is checked by means of a dial gauge which is clamped to the saddle and travels with it along the bed, Fig. 18.12. This test is best carried out in two positions of the dial gauge, *i.e.*, on each flank of the Vee. Errors in this parallelism would result in tapered workpieces, the amount of taper depending on the tailstock position on the bed.

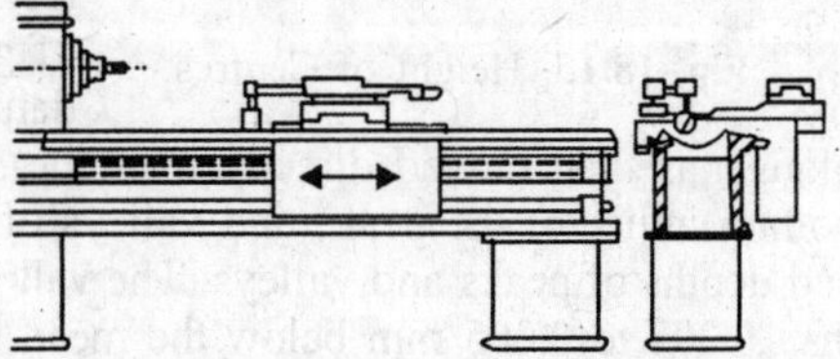

Fig.18.12. Parallelism between Saddle and Tailstock Ways.

Tests 1 (*ii*) to 1 (*v*) are identical in principle and are conducted in the vertical as well as the horizontal plane. For the test 1 (*ii*), the test mandrel is located in the spindle taper and the dial gauge is mounted on the saddle with the plunger touching the test manderel, Fig. 18.13. The spindle is turned into its mean position (mean reading of the dial gauge during one turn of the spindle). The dial gauge is then moved along the test mandrel by traversing the saddle along the bed. Errors in this parallelism will result in tapered workpieces. Measurements are repeated in the vertical plane '*a*' and horizontal plane '*b*'.

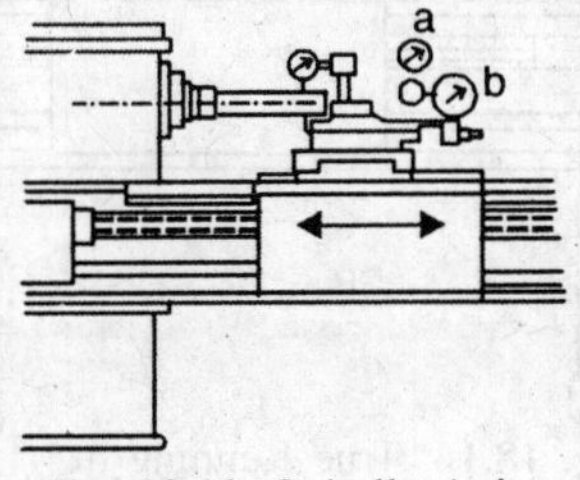

Fig. 18.13. Spindle Axis.

For test 1 (*iii*) and 1 (*iv*), the tailstock sleeve which cannot rotate but can be axially moved, must be clamped during each measurement. For test 1 (*iii*), the dial guage is moved along the tailstock sleeve by traversing the saddle on the bed. For test 1 (*iv*), the sleeve is taken out and a test mandrel is inserted into the sleeve taper. Now the dial gauge will move along the test mandrel.

For measuring the alignment of the leadscrew, test 1 (*v*), the saddle is moved to the middle of the bed and the nut closed. The dial gauge is mounted on a bridge piece which is located by the front Vee of the bed and freely supported on the rear slideway. The dial-gauge plunger touches the outside diameter of the lead screw, (Fig. 18.14). The bridge piece is moved to the right (I to III) and the left (I to II) and the procedure repented both in the horizontal plane *a* and the vertical plane '*b*'.

The test 2, that is, parallelism between tool rest slideway and the spindle axis or the test for the alignment of the tool rest is carried out in the same manner as the test 1 (*ii*) with the difference that now the dial gauge plunger moves along the test mandrel by traversing the tool rest slide on the compound rest of the lathe. This test need be conducted in the vertical plane only, because the tool rest is usually supportd by a swivelling base. This test is carried out to ensure constant height of the tool during setting.

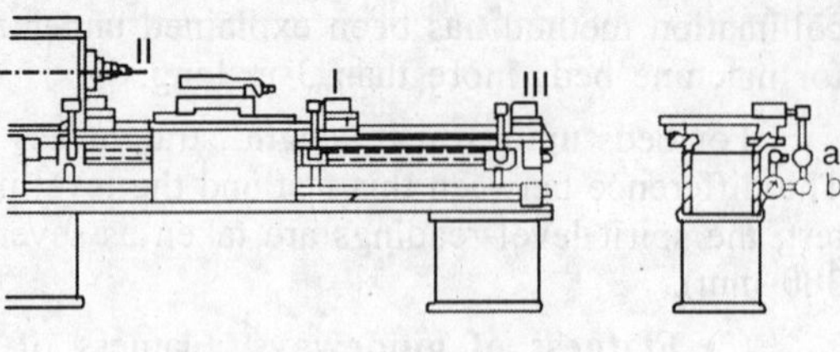

Fig. 18.14. Lead-Screw.

(*e*) **Height alignment of spindle and tailstock sleeve.** For this test, a hollow test mandrel is held between centres. A dial gauge is mounted on the saddle with the plunger touching the top of the mandrel (Fig. 18.15). The saddle is moved along the bed and the indication of the dial gauge noted. The tailstock and the tailstock sleeve must be clamped in position during the tests, that is, they are tested under their normal working conditions. The permissible tolerance allows the tail stock centre, which may in time be lowered due to wear of the tail stock sliding faces, to be above and never below the head stock centre.

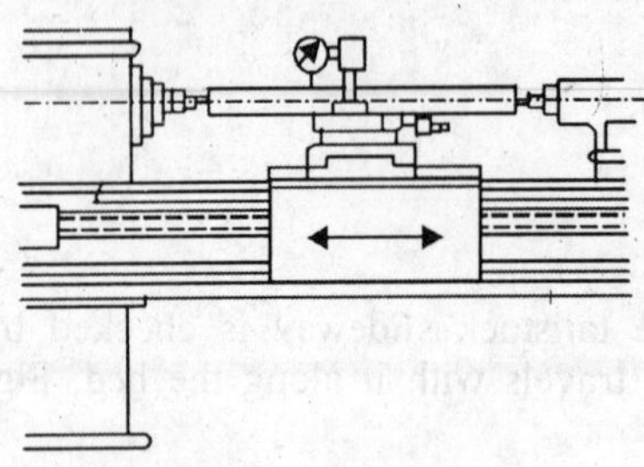

Fig. 18.15. Height of Centres.

(*f*) **Testing the surface quality of slideways and locating surfaces.** Although definite standards concerning flatness, straightness and parallelism of the principal machine ways have been established, no such standards for the surface finish of the slideways have yet been established.

In one method of checking scraped or ground surface, the plunger of a dial gauge is passed over the surface to be tested. The dial gauge support must have a locating surface of generous dimensions and must be guiding along a good datum surface or edge (surface plate or straight-edge). The plunger is then moved directly over the surface to be tested along a series of parallel lines. The pointer indicates the peaks and valleys of the surface. The method determines the average heights and depths of peaks and valleys. The valleys of a well-scraped or ground surface should not be more than 0.002 to 0.005 mm below the mean bearing area.

The waviness meter developed by S.A. Tomlinson also measures and records waviness.

(*g*) **Testing the main spindle.** The main spindle is checked for :

1. True running (roundness, angularity and concentricity) of (*i*) centre (*ii*) the internal taper (*iii*) the external locating cylinder or taper.

2. **Axial float.** The true running of the internal taper is usually measured by means of a dial gauge at the end of a 300 mm long test mandrel which is inserted in the taper, (Fig. 18.16). The dial-gauge plunger rests on the cylindrical surace of the test mandrel and the readings of the dial gauge are taken

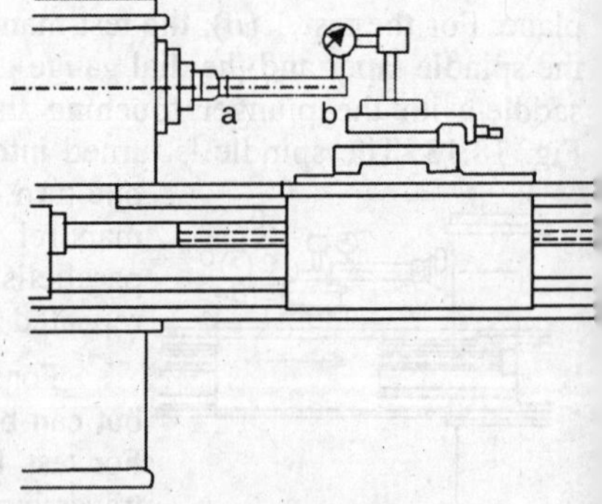

Fig. 18.16. True Running of Internal Taper.

while the spindle is slowly rotated. Readings are taken both near the spindle nose (*a*) and at the end of the test mandrel (*b*).

For checking the centre and external cylinder or taper (used for locating a chuck on a lathe spindle), the dial-gauge plunger should be rested at right angles to the surface to be tested, Fig. 18.17. Readings of the dial gauge are taken while the spindle is rotated slowly. Errors mentioned above cause eccentricity between machined surfaces and centres or cylindrical surfaces clamped in the chuck.

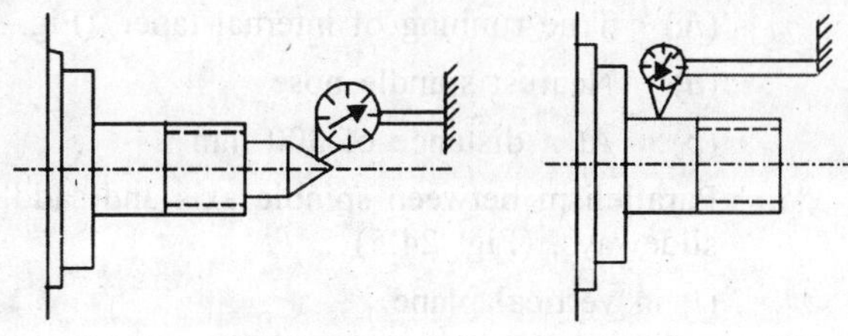

Fig. 18.17. True Running.

Axial float. Axial float or axial slip is different from axial play of the spindle. Axial play is a normal feature of a thrust bearing and had no effect on true axial running as long as the thrust load is applied consistently in one direction. Axial float is due to :

(*i*) errors in the thrust bearing (lack of alignment of the thrust washers).

(*ii*) lack of alignment of the locating shoulder, *i.e.*, the face of the locating shoulder is not perpendicular to the axis of rotation.

(*iii*) the face is irregular.

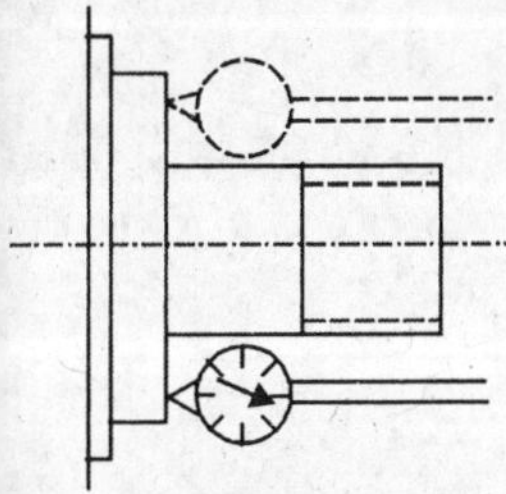

Fig. 18.18. Testing Axial Float.

These errors cause an undesirable axial oscillating movement of the spindle during rotation. Plane surfaces and shoulders cannot be obtained with facing operations if the spindle floats axially.

For the test, after the plunger is placed against the shoulder face of the spindle, (Fig. 18.18), readings are tkaen while the spindle, axially loaded against the thrust bearing, is slowly rotated. Measurements are reported with the dial-gauge plunger resting against the shoulder face at a point diamatrically opposite to that of the first measurement.

In all tests concerning running conditions and alignment of main spindle, the machine must be at its working temperature.

Schlesinger test chart for centre lathe. As per Schlesinger test chart for centre lathe (800 mm swing) acceptance tests are conducted in the following order. The permissible errors are also given.

Test	**Permissible error, mm**
1. Levelling of bed	
(*a*) Longitudinal direction	
(*i*) Front slideway	
(*ii*) Rear slideway	
(*b*) Transverse direction	0 to 0.02 per 1000 mm
	0.02 per 1000 mm
2. Taut wire for straightness of slideways, [Fig. 18.2 (*a*)]	+ 0.02 per 1000 mm, no twist permitted
3. Parallelism between the tail stock slideways and the saddle slideways, (Fig. 18.3).	0.02 per 1000 mm
4. Work-spindle	
(*i*) True running of centre point [Fig. 18.8 (*a*)]	0.02 per 1000 mm
(*ii*) True running of external locating	0.01

cylinder or taper, [Fig. 18.8 (*b*)]

(*iii*) Axial float, (Fig. 18.9)

(*iv*) True running of internal taper, (Fig. 18.7)

(*a*) Nearest spindle nose

(*b*) At a distance of 300 mm

5. Parallelism between spindle axis and saddle slideways, (Fig. 24.5)

(*i*) in vertical plane

(*ii*) in horizontal plane

6. Alignment of the tool rest

7. Tailstock :

(*i*) Tailstock sleeve parallel with bed

(*a*) in vertical plane

(*b*) in horizontal plane

(*ii*) Taper of sleeve parallel with bed

(*a*) in vertical plane

(*b*) in horizontal plane

(*iii*) Axis of centres parallel with bed in vertical plane, (Fig. 28.6)

8. Lead Screw :

(*i*) Accuracy in pitch of load screw is assured within

(*ii*) Check for axial float, (Fig. 18.5)

(*a*) horizontal plane

(*b*) vertical plane

18.2.5. Performance Tests Under Load. These tests are the most important of the tests because they indicate how the machine actually operates under load and its production capacity. In these tests, workpieces are machined at higher speeds and feeds than those employed in regular operation, enabling maximum permissible loads to be attained. As a rule, the machine is loaded 25 per cent in excess of its rated load while employing up-to-date cutting tools of advanced design.

In the load tests, the operation of all the mechanisms and the systems of the machine tool is checked, as well as the operation of all clutches and brakes, and the dependability of the protection and safety devices. The performance tests of machine tools, for which the piece output is indicated in the manufacturing specifications (single purpose, production and automatic machine tools), should be conducted at the maximum output in machining a typical workpiece. Automatics and semi-automatic lathes are tested on a run of several hours, and sometimes several shifts, to determine the production capacity over long periods of time.

The power consumption of the drive motor is also measured. It is determined at medium and at high speeds feeds. Power consumption tests show whether the drive motors have been selected properly. They enable the efficiency of the drive to be determined, as well as the quality of its design and manufacture.

Load tests will how the machine tool operates and what accuracy and surface finish may be obtained on the work it produces.

18.2.6. Rigidity Tests. The rigidity is one of the primary criteria of machine tool performance since it determine the accuracy under load in steady state operation. The less the rigidity of a system, the larger the shape and size errors of the machined workpiece, that is, the lower the machining accuracy.

Rigidity tests are often conducted in the acceptance of a machine tool and consist in determining the capacity of various units for resisting loads that may appear. Machining accuracy depends on the rigidity of the spindle, toolslides, and the machine as a whole. Also, the rigidity values indicate, in some measure, the quality of manufacture and assembly of the machines and may reveal methods of increasing the rigidity.

Practically, rigidity tests consist in applying force on the tested unit and measuring the deflection. The rigidity value is defined as,

$$R = \frac{\text{Force}}{\text{Deflection}} = \frac{F}{Y}, \text{N/mm}$$

The larger this ratio, the higher the rigidity of the unit, or, as they say, the less it will give.

18.2.7. Vibration-resistance Tests. Vibration developed in the operation of machine tools :

(*a*) has a detrimental effect on the surface finish and accuracy of the work produced,

(*b*) it increases wear in the machine parts,

(*c*) it sharply reduces tool life,

(*d*) it, in general, decreases the rate of production,

(*e*) under unfavourable circumstances, vibrations may even be the cause of a break down.

There are many causes of vibrations. They are associated with :

(*i*) unbalanced rapidly rotating parts,

(*ii*) alternating or variable rigidity,

(*iii*) cutting speeds and feeds,

(*iv*) condition and geometry of the cutting tool, and

(*v*) the properties of the material being machined.

Vibration-resistance tests are conducted to develop methods for eliminating vibration. The tests are made during operation, for the most part, and consist in registering the surface finish quality of definite test samples and the absence of chatter marks due to vibration.

Such instruments as the 'vibrograph' or the 'oscillograph' are employed for more accurate vibration resistance research. They measure the frequency and amplitude of vibration of the operative units.

The following practical measures are often applied to increase the vibration resistance :

1. Elimination of excess clearance in the spindle bearings and tool slide ways.
2. Increasing the rigidity of fixed joints.
3. More careful alignment and balancing of rapidly rotating parts including the workpiece.
4. Changing the tool angles, and
5. The application of high-speed cutting methods.

18.2.8. Order of Conducting the Tests. According to many authors, the machine tool tests should be conducted in the following sequence :

(*a*) Idle-run tests; (*b*) Performance tests under load, that is, load tests; (*c*) Accuracy test; (*d*) Rigidity tests of machine tools; (*e*) Tests for vibration-proof properties of machine tools in cutting.

PROBLEMS

1. List the properties which should be possessed by machine tool parts for their proper performance.

2. What is the purpose of machine body?
3. Which is the best section for machine body? Justify your answer.
4. How the machine bodies are stiffened?
5. What is the purpose of guideways?
6. What are the principal requirements of guideways?
7. Name the different types of guideways.
8. Name and discuss the various types of slideways.
9. Write on :

   (*a*) Anti-friction slideways.

   (*b*) Hydro-statically lubricated slideways.
10. Discuss the different materials used for machine body and guideways.
11. Compare the cast and welded structures for machine bodies.
12. What are :

    (*a*) Attached ways.

    (*b*) Plastic guideways.
13. Write briefly on design of slideways.
14. What is the function of machine spindle?
15. Discuss the principal requirements for machine spindles.
16. Write on : Materials for spindles.
17. Write on the design features of spindles.
18. Write on : Mounting of spindles.
19. What is the purpose and importance of machine tools tests?
20. List the various tests to which the machine tools are subjected. Give their sequence.
21. Explain ''Idle-run'' tests.
22. Discuss ''Performance tests under load''.
23. On what factors the machining accuracy of a machine tool depends? Write these.
24. Write about the following tests for slideways :

    (*a*) Straightness (*b*) Flatness (*c*) Parallelism
25. List various instruments used for the geometrical accuracy tests of machine tools.
26. Write on ''Test mandrel''.
27. Write on ''Levelling of the machine tool bed''.
28. Write on the following methods of testing the straightness of slideways.

    (*a*) Straight-edge method (*b*) Taut wire method (*c*) Optical sighting method
29. Write briefly on "Testing the main spindle".
30. Write on the true running of main spindle.
31. What is axial float of main spindle? What are its causes? How is it tested?
32. Write on rigidity tests of machine tools.
33. Write on vibration-resistance tests.
34. Machine tool structures are made .... for high process capability (tough/strong/rigid).

    (GATE 1995) (**Ans :** Rigid)

19

MACHINE TOOL INSTALLATION AND MAINTENANCE

19.1. MACHINE TOOL INSTALLATION

Installation of a machine tool consists in setting the machine tool on its site, levelling and connection to the power supply. The performance of a machine tool and its service life depends to a considerable extent upon its proper installation on the foundation. The foundation enables the load due to the weight of the machine tool and the mounted workpieces to be uniformly transmitted to a larger area of soil and the required position to be quickly imparted to the machine tool in levelling operations. Accuracy in operation may be achieved and retained over a long period of service only if the machine is properly levelled.

Machine tool foundations are subjected to dynamic loads in addition to static loads. While designing a machine foundation, due consideration should be paid to the type of machine, the conditions imposed by it for its safe and efficient working and the type of soil.

All ill designed and/or improperly constructed machine foundation can prove to be a source of trouble and may result in :

(*i*) harm to the machine itself.

(*ii*) poor performance of machine bearings by non-symmetrical wear.

(*iii*) premature fatigue of machine parts.

(*iv*) non-uniform settlement of the foundation, necessitating periodic adjustment and gauging of machines.

Apart from this, a foundation subjected to dynamic loading and vibrations becomes a source of waves propagation through the soil which is harmful to structures, people living within it and the normal operation of the plant and many technological processes (for example moulding and operation of precisions machines).

Due to all the above factors, it is very essential that the design is done rationally and dynamic analysis of the machine foundation is done. The machine foundation should be designed so as to achieve the following objectives :

(*a*) It should provide rigidity and stability for the machine.

(*b*) It should provide sufficient mass to limit the amplitude of oscillation within a prescribed value.

In the satisfactory installation of foundation for new machines, personal experience gained in the design, planning and testing of machine foundations will prove to be an invaluable help. Metal cutting machine tools can be installed on the concrete floor of a shop, on concrete strips or on specially designed foundations arranged on the ground.

1. Machine tools weighing upto 15 tonnes with bed or column of medium and high rigidity (its length to height ratio does not exceed about 2), for example, most lathes, vertical turret lathes, drill presses and other machines of standard accuracy with predominant static loads, make no special requirements to the foundation. As a rule, the foundation of such machine tools is the concrete floor of the shop with a thickness of 150 to 250 mm.

Machine tools operating under dynamic loads such as shapers, slotters etc. and machin tools which are sensitive to vibrations such as Jig borers, should not be installed directly on concrete floor shop. Most machine tools of medium size and normal accuracy (about 90 to 95% used in engineering industries are now installed on common shop concrete floors.

2. If the shop has no concrete floor, separate concrete slabs upto 300 mm thick an 4 m × 4m in plan, can serve as foundations. Quite frequently, particularly in installing specia machine tools united in a production line, concrete strips 1.5 to 3 m wide and upto 6 m long c a continuous concrete strip 300 mm to 400 mm thick can be used as a foundation. Suc foundations are usually checked by determining the deflections at various points for a given loa on the basis of the theory of beams or slabs resting on an elastic base. Then, the pressure on th soil is determined as,

$$p = C.d$$

where C is a co-efficient depending upon the nature of the soil.

The beds or bases of large and heavy machine tools are frequently insufficiently rigid. I such cases, the foundation is secured to the bed to form a common closed circuit, thereb increasing the rigidity of the bed. Various types of foundation bolts are used to secure the bed

3. Installation of a specially designed and previously prepared foundation increases th rigidity and vibration resistance of the machine tools. The machine is placed on a previousl prepared foundation of a depth depending upon the size and weight of the machine and the be is secured to the foundation by various types of foundation bolts. Therefore :

(*a*) Separate foundations are advisable for high speed machines over 2 or 3 tonnes i weight.

(*b*) Special foundations for metal cutting machines of normal accuracy of upto 300 tonne in weight are generally made of concrete, and rarely of rubble concrete or brick.

(*c*) Special foundations are used for machine tools with insufficient rigidity of beds and als for heavy machine tools (over 10 tonnes in weight) which are installed in shops with a concret floor that is not thick enough for machines of such weight. Foundation blocks should extend nc less than 100 mm beyond the base of the machine tool. The thickness of the foundation block is determined as,

$$H_f = K\sqrt{L}, \quad L \text{ is the length of the foundation.}$$

The factor K depends on the type of the machine tool. It is,

K = 0.2, for lathes (including semi-and automatic laths) and for horizontal broaching machines.

= 0.4 for grinding machines.

= 0.6 for gear cutting machines, vertical boring mills, vertical milling machine with rotary tables, horizontal and vertical milling machines, and horizontal boring machines.

= 0.3 for planers and planer-type milling machines,

(*d*) For radial drill presses, shapers, slotters and vertical boring machines, the thickness c concrete foundations lies within a range of 0.6 to 1.4 m. This thickness is increased by 20% fc precision machine tools.

(*e*) Foundations for machine tools weighing over 12 tonnes and also those for machine too subjected to high dynamic loads (such as shapers, slotters etc.) are made of R.C.C. The steel ne is fabricated from round steel bars of 6 to 8 mm in diameter with a cell size c 150 mm × 150 mm. The net is laid under the upper surface of the foundation at a depth of 2 to 30 mm.

(*f*) For light weight machine tools (under 4 tonnes) such as millers, gear cutters, and drilling machines, a 0.25 m thick concrete foundation is used. For these machines, at least 0.5 m thick brick foundation may be laid in dry ground instead of concrete foundations.

(*g*) For precision machine tools and for machine tools in which dynamic loads predominate, the foundation has to perform two functions : to protect the machine against external vibration and to reduce the frequency of natural vibrations of the system by increasing its mass. This reduces the amplitude of forced vibrations of the machine. In such cases, the foundation is in the form of a single monolithic slab from 0.6 m to 0.9 m high.

Machine tools are installed on foundations with the aid of levelling pads, wedges and shoes, with the machine being held down :

(*i*) by means of foundation bolts and its base being grouted with cement mortar.

(*ii*) without foundation bolts but with grouting.

(*iii*) with the use of either levelling wedges or elastic rubber-and metal supports but without foundation bolts and grouting.

The machine tool bed is secured on the foundation by means of foundation anchor bolts, which are placed into special pockets in the foundation and grouted with cement mortar or higher grades of concrete (Fig. 19.1). Pockets in foundations, accommodating anchor bolts, are laid out according to corresponding holes in the machine-tool beds. The distance from a pocket to the outer boundaryy of the foundation should not be less than 120 mm. Size of the foundation blocks is larger than the bed plate of the machine with a minimum all around clearance of 150 mm. Depending upon the material of the foundation, the length of the foundation bolt should be 15 to 20 times the minor diameter of the bolt.

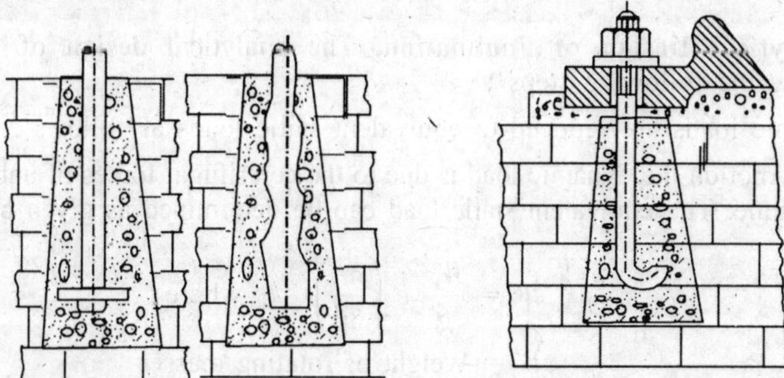

Fig. 19.1. Foundation Bolts.

Vibration Isolation. When a machine tool is operating, then one of the means of increasing the machining accuracy and improving the surface finish, is to achieve two types of vibration isolations :

1. To prevent the transmission of vibrations from the machine tool to the foundation. This is known as "active" vibration isolation.

2. To isolate the machine tool from the vibrations of the foundation base. This is called "passive" isolation.

Rigid installation of a machine tool on its foundations — on wedges with subsequent grouting or on levelling shoes (as discussed above) does not always ensure the necessary vibration isolation.

Elastic vibration isolators and shock mounts are being widely used. Now-a-days in the installation of machine tools that have a dynamic action on the surroundings, in the installation of high precision machine tools susceptible to vibrations of the foundation base, as well as many general purpose models.

Machine tools can be installed much quicker on vibration isolators, a better quality of surface finish is obtained on precision machines, and the noise and dustiness of the air in the shop is reduced. The use of isolators is especially convenient in installing machine tools on upper floors, in rearranging machine tools to suit a changed manufacturing process etc. The vibration isolators are classified according to their material, as :

(*i*) Rubber (sheets and pads) and combined rubber and metal isolators.

(*ii*) All metal isolators and mounts with helical, float or Belleville springs or knitted metal cushions.

(*iii*) Felt pads. These, as a rule, are used for compressive loads and having a high damping effect.

(*iv*) Cord pads. Due to its comparatively high stiffness, is chiefly used for sound proofing.

(*v*) Plastic pads (laminated, impregnated with vinyl plastics and allowing high adjustment.

(*vi*) Pneumatic isolators.

It has been found that combined rubber and metal isolators with the rubber in shear are the most efficient for the installation of machine tools.

Passive vibration isolation is also of prime importance. A special isolated foundation, suspended on springs and having a natural frequency of vibrations of the order of 1.5 to 2 cps, achieves the desired effect but is too expensive, and is therefore applied for unique high precision machine tools. For this reason, the promising and inexpensive method of installing precision machine tools on elastic vibration isolators (as discussed above) has found wide applications. These devices are frequently used for installing machine tools for standard accuracy as well.

19.1.1. Analytical Design of Foundation. The analytical design of a machine tool foundation involves the following steps :

1. The dynamic loads are reduced to equivalent static loads as follows :

(*a*) For rotary motion the dynamic load is due to the centrifugal forces of unbalanced rotating masses of the machine. The equivalent static load can be determined as given below :

$$W_e = \frac{W_r}{g} \cdot e \cdot \left(\frac{\pi n}{30}\right)^2 \cdot k_i, \text{ where}$$

W_r = Weight of rotating masses

e = eccentricity of rotating masses. It can be taken as 0.1 of the workpiece diameter.

n = Speed of the workpiece, rev./min.

k_i = Impact of dynamic co-efficient = 1.5 to 2.

(*b*) For reciprocating motion, the equivalent static load can be taken as equal to (5 to 6) times the value of the cutting force.

2. Knowing the plan dimensions of the machine tool, the plan dimensions of the foundation are determined as discussed under Art. 19.1.

3. The weight of the foundation is determined approximately from the empirical relation :

$$W_f = K_f \cdot W_{mt}$$

where W_{mt} = Weight of the machine tool

K_f = An impirical factor

= 0.6 to 1.5 for machine tools with a static load.

= 2 to 3 for machine tools with a dynamic load.

4. Height of the foundation is determined as :

$$H_f = \frac{W_f}{A_f \cdot w_f}, \text{ where}$$

A_f = Base area of the foundation,

w_f = Specific weight of the material of foundation.

(*a*) The height of the foundation should be sufficient to accommodate the foundation bolts. The length of the foundation bolts is determined from the condition of equivalent strength of the bolt in tension and of a part of the foundation being torn away in the form of an inverted truncated cone. As already mentioned depending upon the material of the foundation.

Length of the foundation bolt = (15 to 20) × minor diameter of bolt.

(*b*) To eliminate the action of foundation on each other due to soil settlement, the angle between the footings of the adjacent foundations, (Fig. 19.2), should be less than the angle of repose of the soil. The angle of repose of the soil is the acute angle which the line of slope makes with the horizontal surface at the base of the embankment.

Angle of repose = 15° to 20° for moist or rich clay

= upto 15° for dry loam.

(*c*) Check the average pressure of the foundation on the soil,

$$p = \frac{W_z + W_f}{A_f}$$

where W_z = Sum of all the vertical forces exerted on the foundation by the machine tool including the dynamic loads (equivalent static loads).

Depending on the soil, its porosity and saturation,

p should be = 1 to 5.5 kgf/cm².

(0.1 to 0.55 N/mm²)

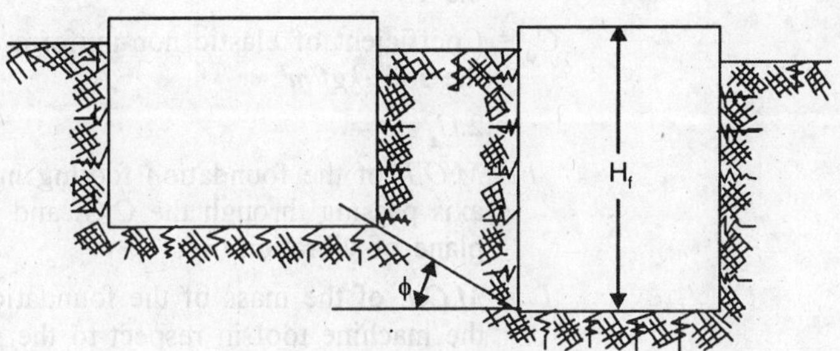

Fig. 19.2

(*d*) The foundation is checked for stability. Calculate the stability factors in respect to overturning about edges *a* and *b*, Fig. 19.3.

$$(k_{sb})_a = \frac{(M_{sb})_a}{(M_{ot})_a} = \frac{(\Sigma W_z + W_f) \cdot b/2}{(M_x + \Sigma W_y \cdot H_f)}$$

$$(k_{sb})_b = \frac{(M_{ab})_b}{(M_{ot})_b} = \frac{(\Sigma W_z + W_f) \cdot a/2}{(M_y + \Sigma W_x \cdot H_f)}$$

where M_{sb} and M_{ot} are the stability and the overturning moments, respectively.

For a stable foundation,

$$K_{sb} \leq 1.8 \text{ to } 2$$

5. Finally, the foundation is checked for resonance. The frequency of the natural vibrations of the foundation alongwith the machine tool is determined from the following formulas, in *cps*,

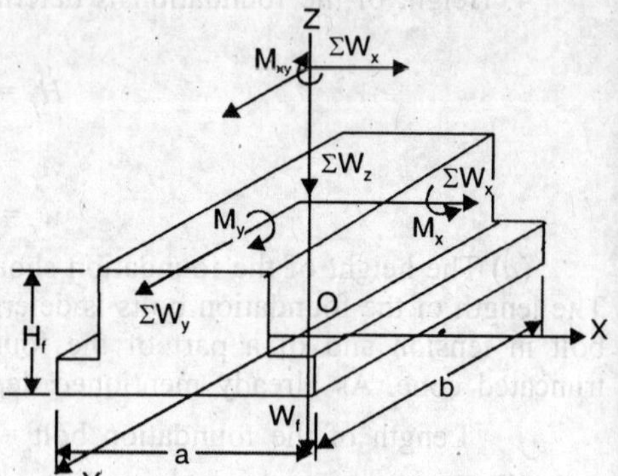

For vertical vibrations,

$$f_{\phi 0} = 0.16 \sqrt{\frac{k_z}{m}}$$

For horizontal vibration,

$$f_{x0} = 0.16 \sqrt{\frac{k_z}{m}}$$

Fig. 19.3

For rocking motion,

$$f_{\phi 0} = 0.16 \sqrt{\frac{k_\phi}{I_m}}$$

where k_z, k_x and k_ϕ are stiffness factors.

$k_z = C_z \cdot A_f$, *kgf* per *m*

$k_x = C_x \cdot A_f$, *kgf* per *m*

$k_\phi = C_\phi . I kgf. m$

where, C_z = Coefficient of elastic uniform compression of the soil, kgf/m^3.

C_x = Coefficient of elastic shear of soil, kgf/m^3.

$= 0.5\ C_z$

C_ϕ = Coefficient of elastic non-uniform compression of the soil, kgf/m^3.

$= 2.C_z$

I = *M.O.I.* of the foundation footing in respect to an axis passing through the C.G. and normal to the plane of vibration, m^4.

I_m = *M.O.I.* of the mass of the foundation together with the machine tool in respect to the same axis, $kgf\text{-}m\text{-}sec^2$.

If f is the frequency of the disturbing periodic forces, than $f / f_0 > \sqrt{2}$, to exclude the danger of resonance.

At $f / f_0 < \sqrt{2}$, the vibration will build up, and at $f = f_0$, resonance will occur.

19.2. MACHINE TOOL MAINTENANCE

1. Aim. The overall aim of maintenance of machine tools is to keep them in working order at optimum efficiency and economy during their expected useful life, without, in any way, affecting the safety and well being of those employed.

2. **Need.** The factors that govern the need for maintenance are :

(*a*) **During operation :**

(*i*) Inadequate or wrong foundations.

(*ii*) Inadequate or wrong lubrications.

(*iii*) Insufficient attention to operating instructions.

(*iv*) Wrong manipulation.

(*v*) Misuse or overloading.

(*vi*) Progressive wear.

(*vii*) Deterioration of some components earlier than others.

(*b*) **Environment effect (direct or indirect).** Under this, we can include the factors such as: vibrations, dust, humidity, heat, corrosion and erosion etc.

3. **Types of maintenance.** The common types of maintenance are given below :

(*a*) *Corrective maintenance.* When the installed machine tools are in operation, a study of repetitive failures may warrant a change in design, materials or working conditions.

(*b*) *Scheduled maintenance.* Well-planned scheduled maintenance entails all operations which are necessary to keep production, machinery and equipment working efficiently. It includes : lubrication, periodic inspection, cleaning and adjustment, periodic repair and repair by replacement and overhaul.

(*c*) *Preventive maintenance.* Preventive maintenance is defined as "maintenance undertaken before the need develops to minimise the possibility of an unanticipated production interruption and breakdown''. Although the operations under this could as well form the part of scheduled maintenance, the philosophy of preventive maintenance differentiates it from other types of maintenance. Under this, the part or accessory of a machine tool is replaced before the end of their useful life is reached. Common examples of preventive maintenance are :

(*i*) Replacement of an Air filter on an Oil filter before the expected time expires.

(*ii*) Changing a bearing of a compressor.

(*iii*) Changing spark plugs of a petrol engine.

(*iv*) Changing carbon brushes of dynamo or starter and so on, before their expected life expires.

(*d*) *Predictive maintenance.* All types of maintenance need fact finding activities to help in their planning. It is necessary to know what is happening to different parts of equipment under actual working conditions. Such a diagnosis of noise or vibrations etc., will help in better assessment of frequency of scheduled maintenance or to know before hand if an unexpected deterioration is taking place in certain parts.

4. **Planned maintenance system.** To prolong the life of a machine tool and to reduce maintenance costs, the periodic maintenance must be carried out according to the periodic maintenance schedule. Periodic maintenance is essential for preventing trouble and accidents and to ensure safety. The maintenance charge is a small price to pay for trouble caused when periodic maintenance is not carried out.

The maintenance of machine tools is accomplished in accordance with a Planned or Preventive Maintenance System (PMS). The system includes all facilities for preserving equipment, such as current (on duty) servicing between repairs, planned periodic inspection and cleaning, periodic accuracy tests, planned minor and medium repairs, and planned general overhauls. The system consists in conducting various preventive measures and maintains the equipment in an excellent operating condition all the time. The PMS can be affected by means of the following methods :

(*a*) *Post-inspection repairs.* This method involves planning of periodic inspections rather than repairs. The time intervals between successive inspections is determined according to the minimum service life of rapidly wearing components. If an inspection confirms that there is no need for repairs and that the machine can operate without these until the next inspection, the repairs are postponed. This method prevents any sudden breakdown of the equipment.

(*b*) *Periodic repairs.* This method involves repairs being done after a given running time.

(*c*) *Standard or compulsory repairs.* This method involves compulsory repairs of equipment at planned intervals, which are standard for each piece of equipment. The PMS includes :

(*i*) *Routing servicing.* This involves normal everyday running of machines, minor repairs and, whenever necessary, adjustment of separate units or members of the machine tool.

(*ii*) *Periodic inspection.* Periodic inspections are conducted according to schedule and involve visual inspection, cleaning and accuracy checks.

(*iii*) Inspections as such consist in exterior checks accompanied by partial disassembly. All the mechanisms are checked in operation and regulated, fasteners are repaired or replaced, the state and wear of the machine tool as a whole and its individual units are assesssed. A report is prepared on the basis of the inspection results, about the mechanical condition of the machine. The date for the next repairs is fixed in accordance with this report. The accuracy check of the machine tool is conducted according to approved standard.

(*iv*) *Scheduled repairs.* These repairs are divided into : minor or routine repairs, medium repairs and general repairs. In routine repairs, separate components or units of the machine are repaired or replaced in accordance with the results of periodic inspection without thorough disassembly of the machine. Medium repairs include all the elements of the routine repairs with the additional restoration of the relative position of the principal units and with partial repairs to basic components. Parts subject to replacement in minor repairs are revealed by minor repairs.

General overhaul involves the complete replacement or repairs of all the basic components, full restoration of the relative position of the principal units and the required accuracy of the machine tool.

The most widely used PMS for machine tools involves the following maintenance cycle : G-I-R-I-R-I-M-I-R-I-R-I-M-I-R-I-R-I-G, where

G : General overhaul, I : Inspection

R : Routine repairs, M : Medium repairs

Thus, the maintenance cycle includes :

9 Inspections

6. Routine repairs, and

2. Medium repairs

19.2.1. Current or On-duty Maintenance or Servicing. As mentioned earlier, the periods of trouble free operation of a machine tool can be substantially prolonged with proper maintenance. On-duty maintenance can be divided into three stages : servicing before starting work, servicing during operation, and servicing after finishing work.

(*a*) **Servicing Before Starting Work.** Before starting to work, each operator or setup man must inquire of his opposite number in the previous shift about all troubles and defects that are discovered. After making oure that the troubles are completely eliminated, he should follow the instructions given below :

1. Inspect the machine.

2. Check whether controls operate without failure.

3. Check whether the clamping devices operate reliably.

4. Cutting tools must always be reliably clamped.

5. Thoroughly lubricate the machine.

6. If the machine has been standing idle for a prolonged period, it should be started and run idle for 10 to 15 minutes.

7. After the machine is switched on for automatic operation, the first few parts should be carefully checked to see whether they conform to the specifications of the detail drawing.

(*b*) **Servicing During Operation.** The operator should carefully observe and see that :

1. all mechanisms operate faultlessly.

2. the temperature of plain bearings does not exeed 60°C and that of the ball/roller bearings does not exceed 75°C.

3. all engagement and disengagements take place without impact.

4. there is no jamming, excess clearance, or misalignment of various units.

In addition, he should :

(*i*) periodically check the devices for feeding lubricant as well as the effectiveness of the oil filters.

(*ii*) see that there are no leaks in the system and cutting fluid is delivered directly to the zones where metal is cut.

(*iii*) see that the guards and shields are not removed and the machine tool operates normally for servicing during operation.

(*c*) **Servicing After Finishing Work.** This exercise consists in :

1. Switching off the machine.

2. Cleaning the machine of any remaining chips.

3. Retracting all operative units to their initial positions.

4. Wiping off the machine with a clean rag.

Systematic and attentive maintenance of machine tools will greatly increase its production capacity and utilisation and reduce the amount of repairs required and the cost of operating the machine. If during servicing, a necessity of adjustments or minor repairs to certain mechanisms arises, such work should be done without any delay, otherwise the defects may lead to serious trouble afterwards.

19.3. MACHINE TOOL LUBRICATION

All moving parts of machinery and machine tools require lubrication. Lubrication means oiling or greasing the parts that move or rotate. Lubrication is of vital importance, because proper lubrication ensures not only trouble free operation and a prolonged service life of a machine tool, but accuracy of performance as well. Lubricant is delivered to the friction surfaces of a machine during operation :

1. To fill the space between the metal parts and act as a cushion.

2. To prevent direct contact of the friction surfaces and thus to reduce friction and wear and to facilitate relative motion of mating parts.

3. To act as a cooling agent by carrying away some of the heat produced by friction.

4. To protect the bearing surfaces from rust and corrosion.

Most parts of a machine tool can be lubricated satisfactorily by grease or by a mineral oil of suitable viscosity. In selecting a grade of lubricant, it is neccessary to take into consideration the speed and pressure between the moving parts to be lubricated. For example, high speed spindle bearings on a grinding machine require a light oil. Heavy oil has been known to ruin the bearing, which means a breakdown and costly repairs. In the circulating system of lubrication, the oil becomes heated in the bearings and gears, and is subject to some churning and spraying. It, thus, repeatedly comes into contact in a finely divided condition with air. The oils tend to get oxidised, resulting in the production of organic acids and sludge. So, an oil which is stable and of good resistance to oxidation should be selected. Moisture can also find its way into lubricating oil systems from atmosphere or by ingress of soluble cutting oils. The oil, thus, should have good "demulsibility" to permit quick separation of water in the sump. For these reasons, solvent-refined oils, with their high stability, are generally recommended. They give long and satisfactory service, provided reasonable care is paid to general cleanliness, filters and shaft seals and to minimizing contamination by metallic dust, dirt, and cutting oil. All reservoirs should be washed and the whole lubrication system blown through each 1.5 or 2 months of operation. In addition, the filters should be cleaned and clean lubricant should be added to the system periodically. As a rule, the service manual of each machine tool includes a lubricating diagram and a chart of instructions which lists the points of lubrication and the amount and grade of lubricant needed for each point.

Proper lubrication is the essence of any good maintenance system. If thoroughly analysed, the cause of most of the breakdowns will be traced to failure of the lubrications. Proper lubrication means :

1. Use of right lubricant.
2. Application of lubricant at right time.
3. Use of right quantity of lubricant, and
4. Proper care of lubricants to prevent their contamination.

19.3.1. Methods of Application. The four main types of the methods of lubricant applications are :

1. **Manual.** Lubrications can be applied directly from an oilcan to oil holes on bearings and slides or by handgun to appropriate oil or grease nipples. This method can give satisfactory results, but it is difficult to ensure that the correct lubricant finds its way in appropriate quantity to the desired spot at the right time, and free of contamination. So, such a system is fast disappearing.

2. **Reservoir System.** In this system, some form of reservoir is used, with simple feed devices, such as drip-feed or siphon-feed lubricators, ring oilers for journal bearings, or oil baths for gear boxes. Gears will be lubricated by a splash or spray method, and worm gearing by arranging the worm in an oil bath. Oil may also be fed by gravity from a reservoir arranged above the points it accommodates. Oil level in the reservoir should be correctly maintained.

3. **Pump System.** In this system, oil is fed under pressure from a lubricator to all the points needing lubrication. Oil reservoirs in conjunction with pump feed to strategic points, are used. A pump connected to the drive system draws oil from a sump and feeds it to a distribution tray, from which it is distributed by pipes and oilways to bearings and gears, and finally drain back to the sump for recirculation.

4. **Automatic System.** Automatic system has been developed from the pump system. Pumps (which may be driven from the gear box or operated at each reciprocation of a machine table) supply filtered oil through a system of metering valves in accurately measured quantities to every point requiring it, that is, to distributors, lubricators or oil baths. Though high in first cost, this system is more satisfactory and is being increasingly adopted. In automatic machine tools and transfer machines, lubricant must be delivered automatically to all points of lubrication before starting operation.

Note. An oiling or lubrication chart comes with each different kind of machine tool. Certain parts require daily lubrication, others should be oiled weekly or monthly as the chart says.

19.4. RECONDITIONING OF MACHINE TOOLS

An discussed under Art. 19.2, one stage in the cycle of machine tool maintenance is general overhaul. A general overhaul consists of complete disassembly, washing of all parts and units, replacement of all defective parts, and a careful restoration or reconditioning of all worn parts. A general overhaul should fully restore the initial accuracy and production capacity of the machine tool. During overhaul, mating parts are measured. This reveals wear and defects and serves to draw up the defect schedule, which, subsequently, is the basis for making repairs.

1. Worn ways of beds, cross-members, slides, and saddles are usually required by scraping and, less frequently, by grinding. The repaired surfaces are checked by surface plates, straight edges, and the mating parts. For example, if bedways are scraped, they are checked finally with the saddle using a marking compound. A thin layer of marking compound is applied on the ways of either the bed or saddle. The saddle is then moved along the bed. If the repair is proper, there will a uniform distribution of bearing marks on the ways. However, if there are "bald spots" without bearing marks whatsoever, or places with a continuous coat, then he scraping has not been done satisfactorily. More metal then must be scraped off where there is much marking compound. This process is repeated until the bearing marks are uniformly distributed in spots over the full length and width of the ways.

2. **Repairs of Spindle Journals and Their Bearings.** If plain bearing are used, the spindle journals are subject to wear and lose their initial form. It will be then necessary to regrind the journals, removing the minimum possible amount of metal. Worn antifriction bearings should be replaced with new bearings. Plain bearings are repaired by scraping and are also checked with marking compound.

3. Excessively worn gears are best to be replaced by new gears.

Repairs Processing Methods. The processing methods employed in repair operations must ensure high quality and the required accuracy. The following are some of the up-to-date repair methods employed to restore and reinforce worn or broken parts.

1. Arc or gas welding, including the welding of cast iron parts.

2. Surfacing of worn places on parts.

3. Flow welding.

4. Electric or gas metal spraying to build up worn places.

5. Building up little-worn vital surfaces by chromium plating.

6. Gluing together certain parts not subject to high heat or loads with carbinol glue (for metals).

7. Lapping-in.

8. Heat treatment.

9. Shot peening : for finishing work surfaces, surface hardening and increasing the fatigue strength.

All these processes have been discussed in detail in the book "A Textbook of Production Technology" by author.

19.5. SAFETY IN MACHINE TOOLS

19.5.1. Introduction. There is an old saying, "A good worker is a safe worker". In a machine shop, a worker will be working with both hand tools and power driven tools or machines (Revolving shafts, spindles, mandrels, drills, reamers, chucks, boring bars, circular saws, milling cutters, grinding wheels, broaching tools and so on). None of these tools and machines is unsafe (They may be dangerous). There are only unsafe workers. The worker must learn to use the equipment in the machine shop in a correct and safe manner. Most accidents are caused by doing the wrong things, or by not following carefully the instructions given.

A good idea is to follow the ABC of safety : "Always Be Careful". The best way for this is to follow carefully the instructions. The operators must be acquainted with the use of all guards and with accepted safety measures. "The correct ways is the only safe way".

19.5.2. General Safety Rules. Some general safety rules are given below :

1. **Dress Properly.** Workers in most machine shops wear either an apron or a shop coat. Unbuttoned, flapping (loose) clothes must not be worn. Sleeves should be tightly buttoned or tied at the wrist. Keep your hair cut reasonably short or wear a head dress.

2. Always protect your eyes. For this, wear safety glasses or an eyeshield, against fine flying chips.

3. Do not attempt to operate a machine until you have received instructions from your instructor or set-up man and have passed a written safety test.

4. Always attend strictly to your work and keep your mind on what you are doing.

5. Always be neat and clean. Everything should be kept in its proper place. Keep the area around the machine clean and free of oil. There should be nothing that the operator may slip on or trip over. Keep tools properly arranged.

6. Make sure that the safety/protective guards are in place, especially after repairs have been done.

7. Never mount or remove the work, replace the tools, clean or lubricate the machine, make any adjustment, take measurements, or remove the swarf, while the machine is running. Always stop the machine before doing any of these operations.

8. Be sure keys, loose levers, hand wheels and wrenches are removed before starting the machine.

9. Before starting the operation, always see that the work and the cutting tools are secured fast.

10. Disengage all the operating levers and place them in neutral position before starting the motor.

11. Keep your hands away from all moving parts, as well as cutting tools.

12. Never try to remove chips with your fingers. Special hooks, brushes or scrubbers should be used to pull them away from the machine.

13. Do not try to stop the chuck with your hands.

14. Do not attempt to lift heavy articles (more than 20 kg) without assistance. Get helper or use a crane.

15. Never stop the spindle with the hands if it still rotates after being disengaged.

16. Never put blanks, finished work, tools, or waste on the machine.

17. Never remove or replace belts, gears or chains without stopping the machine.

18. It is just as important to know how to stop a machine as to start it.

Before Starting the Machine, Check

(*i*) Do the coolant and lubricating systems operate satisfactorily.

(*ii*) Do all chip disposal devices operate properly. Have wire hooks been prepared for pulling out chips.

(*iii*) Are all belts and chains properly tensioned.

After stopping the machine :

(*a*) Switch off the motor, even during the shortest stops.

(*b*) Remove the workpiece.

(*c*) Wipe off the machine and removed all chips caught in the working zone.

19.5.3. Safety Suggestions for Individual Machine Tools.

(*a*) **Drilling Machines**

1. Never leave the chuck key in the chuck.

2. Never hold thin or small work with your hands, use a drill vise.

3. Don't try to stop the spindle after the power is shut off by grasping with your hands.

4. Always ease up on the down-feed pressure as the drill begins to break through the hole. Heavy feed will cause the drill to dig in, the drill may break or pull the work loose.

5. Use the proper drill for a material. For example, never try to cut holes in brass, copper or bronze with a drill ground for steel.

(*b*) **Lathes**

1. When holding work between centres, use the correct size of the centres with good points.

2. Tool bits must be sharp and ground to the correct shape. They should be set at proper height and angle to the work.

3. When work is held on a face plate, turn the face plate by hand for a complete revolution to make certain the work will not strike any part of the lathe.

4. Never leave a chuck wrench in the chuck, even for a moment.

5. Do not change spindle speeds until the lathe comes to a dead stop.

6. Never do file work unless the file has a handle.

7. You should always know in what direction and how fast the carriage or cross feed will move before you turn on the automatic feeds.

8. Be careful to see that the carriage or compound slide does not run into the rotating chuck.

(*c*) **Shaper**

1. The length of stroke and the position of the ram must be correctly set before starting the machine.

2. Always check the speed control levers before starting the machine.

3. Be sure the tool bit is clear of the work before starting the machine, otherwise it will damage the workpiece and will be broken off.

4. Be sure to lock the tool-slide clamp screw after setting the depth of cut, otherwise the tool-slide will start moving downword during the cut.

5. The tool must be fed into the work gradually.

(*d*) **Milling Machines**

1. Do not clean against or rest your hands on the moving table.

2. Never try to reach over a revolving cutter. Stop the machine.

3. It is very dangerous to use rags, cotton waste, or a cleaning brush near a revolving cutter.

4. When removing cutter, always hold a rag or cloth over the cutter to prevent being injured.

5. When mounting the arbor support, keep your fingers away from the bearing hole.

(*e*) **Grinding Machines**

1. When a new wheel is mounted or a grinding wheel is first started, stand aside for a minute. The wheel may be cracked or damaged. If so, it may fly apart when it reaches full speed.

2. Learn how to test a wheel for soundness before mounting it.

3. Never force a work against a wheel roughly.

4. Never attempt to measure a work near a revolving wheel.

5. Make sure the wheel is clear of work before starting the machine.

6. Be sure the correct type of wheel is securely mounted on the spindle.

7. When the work is held on magnetic chuck, see that it is tight. For this, hold the work with your hand to see if it is loose or movable.

8. Magnetic chucks be clean and smooth.

9. Work must never be placed in a grinding machine or on a magnetic chuck when the wheel is revolving.

10. Wheel guards must be placed over the wheels.

11. Never lay or drop tools or workpieces on the accurate table surfaces.

12. When a grinding machine is equipped with feed reverse and trip dogs, always test the movement first by using the hand feed.

19.5.4. Safety and Protecting Devices. As already noted, in modern machine tools a large number of driven elements move at his speeds. Safety from these dangerous elements can be secured by construction of the machine tool, by position of the machine tool and with the help of various types of guards.

(*a*) **Safety by Construction.** When a new machine is being designed and constructed, modifications should be made to eliminate any such dangerous parts by reconstruction, or by placing the source of danger in an alternative and safe position, if possible. Following are some of the examples of carefully through out safety design in the modern machine tools :

1. In the precision grinder or milling machine, motors, belts and pulleys are enclosed in a housing provided with a cast iron door which can be securely fastened.

2. Gear boxes of turret lathes are fully enclosed and guards are provided for the belt and pulley drives.

3. Revolving spindle and shafts can be protected by plain or telescopic sleeves of steel or brass.

4. All gearing should be completely encased with the guard butting tightly against the machine frame so as to prevent fingers being inserted between the two.

5. Give sufficient clearances between control handles and moving tables and between fixed and moving parts.

6. Care should be taken to shroud automatic trip stops for reversing motions as on milling machines.

7. Pedals for clutch control should be placed where they cannot inadvertently be actuated by articles falling on them. Preferably, these should be of the safety self-locking type.

8. Control levers and handles should be placed in convenient positions so as to prevent fatigue which is conductive to accidents. Also, such control should not be too stiff in operation.

9. The finger-light mechanisms of electric, pneumatic and hydraulic devices are far safer then tedious levers and treadles.

10. The risk of carelessness in operation is reduced by the use of different colours for fixed and moving parts.

(*b*) **Safety by Position.** The placing of dangerous parts out of reach, as an equivalent to secure fencing, may sound attractive, but in reality the principle is not easy to maintain. Dangerous parts within reach would be fenced as a matter of course, or guard should be provided for all such parts.

(*c*) **Safety by Guards.** Guards are provided on machines to protect the operator against flying chips and splashing coolant and also to exclude the possibility of the operator's clothes, hair, hands, or fingers, being caught by the rapidly moving parts. Guards should be designed so that :

(*i*) they protect the operator and can be easily removed during servicing.

(*ii*) they should not hinder the operator in the manual operations required for machine control.

(*iii*) they should allow the operator to watch the operation of the machine and the condition of the tools.

There are many types of guards. Fixed guards are legally required in preference to other types. However, if their use is impracticable because of the nature of the operation, other types are considered.

1. **Fixed Guards.** They should prevent all access to the dangerous parts, should be of robust construction to withstand rough usage, and wherever possible should form an integral part and not easily removable from the machine. The materials for the guards can be cast iron sheet steel and expanded metal of small mesh. Cast iron should not be used where it may be subjected to shock. Sheet steel is probably the best material because it is easily adaptable to fabrication methods. Expanded metal with small mesh is useful where a clear view of the moving parts is desired.

2. **The Automatic Guard.** This guard is operated by a moving member of the machine to which it is connected. It is designed to remove the operative away from the dangerous parts before trapping can occur. Automatic guards should not be connected to friction or slipping devices. Positive connections should always be made to a suitable part which moves whenever the machine is started either normally or accidentally. This guard provides protection in the event of a repeat stroke caused, for example, by clutch seizure or other mechanical fault, or due to another accidental operation of the machine. This guard is not suitable on quick-running machines because of the small interval of time available for proper functioning.

3. **The Interlock Guard.** The interlocking arrangement does not allow the machine to be started until the dangerous parts are fully protected by the guard and also these parts cannot be exposed again until the machine has come to rest. There are 2 interlocking arrangements.

(*i*) The moving portion of the guard is connected to a device which prevents the starting-mechanism clutch, fast and loose pulley system, electric motor starter, hydraulic valves, etc., from being manipulated to start the machine until the guard is in the protective position.

(*ii*) Some solid piece connected to the guard is interposed between two moving machine parts so as to prevent the starting-up of the machine until the solid piece is removed by putting the guard member in the protective position.

In semi-aūtomatic lathes, the interlocking arrangement does not allow the machine to be started or spindle rotation engaged if the work is improperly clamped. On automatics, the machine is prevented from being started until all loose handles, levers and handwheels are taken off.

4. **The Trip Guard.** Thsi guard is particularly useful for protecting nips between pairs of revolving parts, or between reciprocating and revolving parts, into which material is fed by hand. The combination of a guard and associated tripping mechanism is designed to bring the machine to rest, or even to reverse it, before full access to the dangerous parts can be attained.

5. **The Positional or Distance Guard.** Such guards are useful when the close-up fencing by fixed or other guards is impracticable and should often be supplmented by a wire or other trip device for quick stopping because access under or over the guard to some machines may be possible during operation. The guard consists of a grill or of bars or rails placed at such a height above the floor or standing place, and distance from the tapping line or area, that arm cannot be extended over the top bar far enough for the fingers to be trapped.

Using Guards

(*i*) Before starting the machine, install all guards and check their fastenings.

(*ii*) In setting up the machine, the guards should not be removed before stopping the machine.

(*iii*) Never touch drive elements or moving components, enclosed by guards, during operation.

(*iv*) After stopping the machine, switch off the power supply before removing the guards.

(*v*) During operation or setting up, eliminate the possibility of chance persons starting the machine.

(*vi*) In semi-automatic lathes, the working zone is usually enclosed by light metal splash guards or shields with transparant windows. These devices must be lifted or shifted to the side each time work is loaded or unloaded. During operation, however, they must completely enclose the working zone.

(*vii*) Current carrying components should always be reliably guarded to prevent contact with operaters. It is absolutely necessary to ground all motor frames, starters and adjacent metal parts, as well as the whole machine.

(*viii*) Electric starting devices should be conveniently located and protected against accidental engagement. In push button stations, the start buttons should be sunk 3 to 5 mm below the face to prevent it from being pushed accidently. Stop buttons should be of red colour and should operate obsolutely without fail from the slightest touch.

PROBLEMS

1. What is meant by machine tool installation?
2. What are the main objectives of machine foundations?
3. Write in detail about the machine foundations.
4. List the various vibration isolators.
5. Write in detail about the analytical design of machine foundations.
6. What is the aim of machine tool maintenance?
7. What is the need for machine tool maintenance?
8. Discuss the various types of maintenance.
9. Write on : Planned maintenance system.

10. What do you understand by servicing? Write on :

(*a*) Servicing before starting work.

(*b*) Servicing during operation.

(*c*) Servicing after finishing operation.

11. Write on : Machine tool lubrication.

12. Write on the various methods of doing machine tool lubrication.

13. Write on : Reconditioning of machine tools.

14. Write the various processing methods employed for repair of machine tools.

15. Write the general safety rules to be followed in a machine shop.

16. Write the safety rules while working on :

(*a*) Drilling machines (*b*) Lathes

(*c*) Shaper (*d*) Grinding machines

17. What is meant by :

(*a*) Safety by construction (*b*) Safety by position (*c*) Safety by guards.

18. Write about the various types of guards used in machine shops.

19. PERT, as discussed in Chapter 23, Art. 23.8.2, can be used in maintenance. As the name implies, it is a method to evaluate a process with respect to the target, quantum of work, inter-dependence of various operation, availability of manpower and machines, optimum time needed for each operation, and the necessity of expediting the operation for achieving the target. The complete cycle of operation is represented by a NET-WORK in the correct sequence and this helps in reviewing the progress of the process. For its correct representation, it is necessary to know the inter-dependence of various operations, i.e., which operation comes first, which comes after, and which operations can be done concurrently.

20. Write a note on Organisation of Maintenance Department.

Every industry, irrespective of the type of production, has a maintenance department to look after and upkeep of the equipment. Its main function is to organise planned preventive maintenance of the plant and machinery and also to carry out routine maintenance works.

For an industry having sizable equipment and many shops, the maintenance work can be grouped into the following categories :-

(a) Control and Coordination

(b) Maintenance Centres for each shop

(c) Central Maintenance workshop

The main functions of the first category, are :-

1. Preparation of annual and monthly preventive maintenance schedules.
2. Conducting enquiry into the accident on the machines
3. Checking equipment for proper care.
4. Spare parts and material requirement and procurement
5. Preparing design/drawing of spare parts on the request of the maintenance engineers of the maintenance centres.

20

DESIGN OF PRODUCT FOR ECONOMICAL PRODUCTION

20.1. INTRODUCTION

The design and production of a product is achieved in three stages :

1. **Conceptual stage.** Here, the functional requirements of the product are established and the designer conceives of an idea for a device that will fulfill the function requirements.

2. **Functional design stage.** Here, the product designer designs the product primarily for function.

3. **Production design or Designing for production.** This means providing a design that can be manufactured economically.

The responsibility for satisfactory functional or product design rests with the product designer or the designing engineer. Therefore at the functional design stage, he is more concerned with functional design and materials, than with the manufacturing processes. But, the major emphasis must be on providing a design that can be manufactured economically. The product must be produced at the lowest possible cost to assure a profit on its sale. The product designer is not always cost concious and often does not have the required knowledge of manufacturing processes to influence his design for low cost production. But, if his knowledge of functional or product design is complete with the experience and knowledge of machines, tools and processes contributed by others in the organization, the ultimate design will be a product that is functionally correct, yet can be economically produced, Low cost production is a responsibility of the "Tool Engineer". He, with his knowledge of the product and its function and his familiarity with methods and processes, is qualified to influence product design for most economical production, thus assisting in product design for most economical production, thus assisting in designing for production. Thus, it is extremely important that the relationship between design and production be given careful consideration throughout the design-production stage. Whatever changes are necessary in the design, must be made before a product is in production, since, changes made afterwards will prove to be very costly. The complete product design, including all details, must be carefully studied, while still in the drawing stage, to determine whether both function and productibility are satisfied. Scale working models are made and tested. This will verify the function and check the production possibilities before the details are released for production.

The recommendations of the tool engineer for changes in design (for economical production) are put into effect only on the authority of the product designer, since, he is responsible for the performance of the product, and he has to ensure that the recommendations do not interfere with the function of the product. If a conflict does arise as to the advisability of a design change a compromise between cost and function can always be arrived at.

Note. Read also chapter 6 (Art. 6.1 and Fig. 6.1)

20.2. SUGGESTIONS FOR DESIGNING FOR PRODUCTION

Machine parts of great variety and complexity, as designed today, involve manufacturing problems which can only be solved by specialists in various branches of manufacturing engineering (Casting, Metal forming, Metal cutting etc.) For this reason, it is advisable, when, designing machine parts, to consult respective production specialists.

There are many suggestions which if followed during the design state, will result in a sound design for a product and in its convenient and economical production. Also, there are many examples where a part will function equally well if designed either of two ways, but will be much cheaper to produce one way than the other. Various suggestions and a few illustrations of the above examples have been discussed below.

20.2.1. Machining. If a part is to be produced by machining, the designer should visualise how the job will be machined and make minor modifications (if necessary) that will permit easy and economical machining. A few examples are illustrated below :

1. When cutting external threads. (Fig. 20.1), no threading tool can work clear up to a shoulder. So, a groove, called 'thread relief', should be provided. Also, the first few cutting threads of the die are ground off to provide a starting lead. Thus the threads close to shoulder

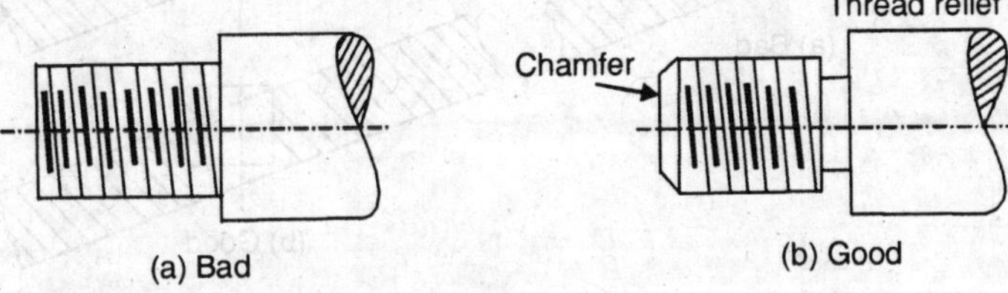

Fig. 20.1. Cutting External Threads.

will be imperfect and not to full depth. The relief or neck at the shoulder will provide full threads on the working length and the mating part will rest fully against the shoulder. The relief does not interfere with the function of the part and can be machined at little cost compared to that for cutting full threads up to the shoulder. The addition of a chamfer at the end of the threaded part, facilitates starting of the die and entering the work in the mating part. The same idea holds for grinding close to a shoulder, (Fig. 20.2). Also, a perfectly sharp inside corner is difficult to produce by any machining method. All cutting tools have a small radius on the cutting tips (for strength reasons). Therefore, in mating parts that must fit flush, a corner relief or an internal chamfer facilitates the production of the parts and their ultimate assembly. Grinding wheels never have sharp corners. Thus when two perpendicular surfaces are to be grounded at once, a relief cut into the corner clears the radius on the wheel and permits the two surfaces to be grounded square with each other. A flush fit with a mating part results.

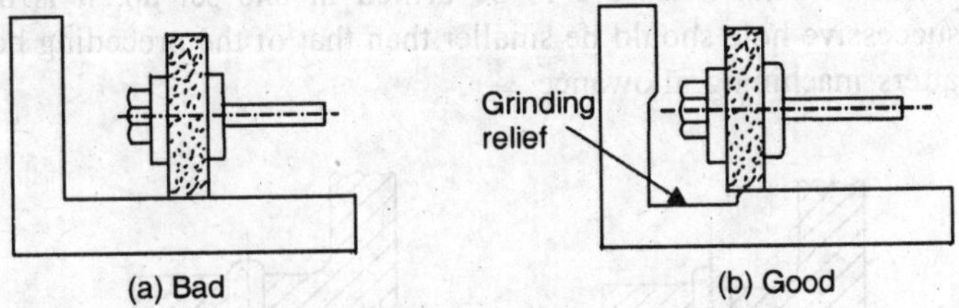

Fig. 20.2. Grinding Close to a Shoulder.

2. The first few threads of a tap are also ground off to provide a lead for the tap to enter a drilled hole. The tapped hole, [Fig. 20.3 (*a*)], is easy to produce, because it goes clear through and all the threads will be of full depth. For a blind hole, a relief should be provided at the bottom, [Fig. 20.3 (*b*)]. It will clear the imperfect threads produced by the lead on the tap, producing full threads on the working length of the hole. This also saves the use of a bottoming tap. If a relief is not provided at the bottom of a blind hole, [Fig. 20.3. (*c*)] then to produce good threads clear to the bottom, a starting tap and a bottoming tap must be used, resulting in one additional operation.

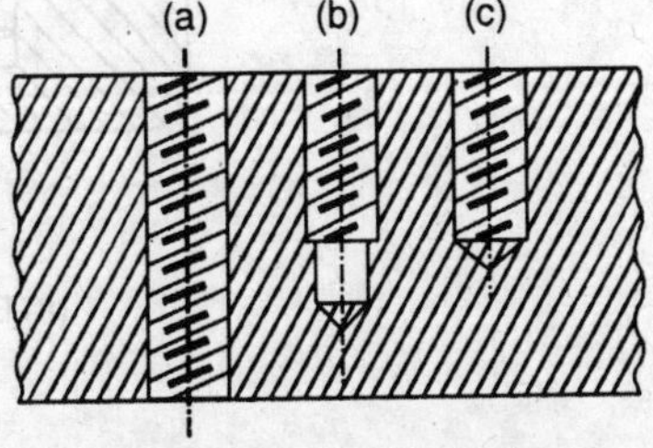

Fig. 20.3. Drilling Holes.

3. Drilled holes should be normal to the surface, top and bottom, (Fig. 20.4). A drill attempting to start at an angle will be deflected, spoiling accuracy and possibly breaking the drill. A drill breaking through a slanting underface is subjected to breakage, because the drill point comes through unevenly.

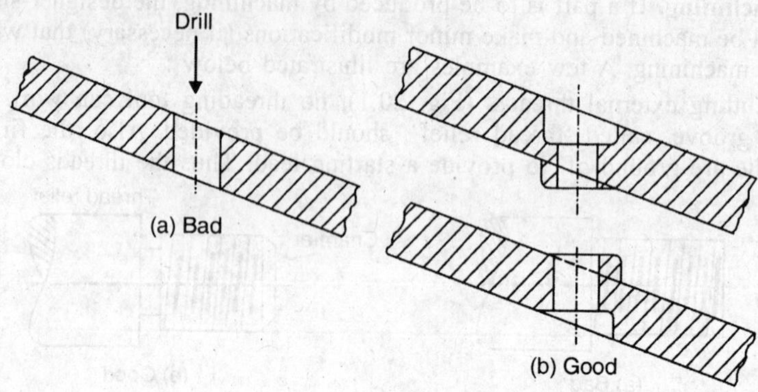

Fig. 20.4. Drilling a Hole (Slanting Surface).

4. Flange holes should not be located very close to a wall otherwise the drill becomes partially embedded in the wall when it breaks through the flange. The hole axis should be spaced a distance $\geq \frac{1}{2} \times$ hole diameter + radius of the rounding between the wall and the flange.

5. Dowels should not be pressed in to blind holes unless necessary. They are harder to insert due to air compressed below them and are much harder to remove. The design at (*b*) (Fig. 20.5) is better for knocking out the dowels.

6. Through holes are preferred to blind ones for easy machining.

7. Accurate blind holes should be provided with a bottom recess to allow tool overrun.

8. Where several co-axial holes are to be drilled in one set up, it is desirable that the diameter of each successive hole should be smaller than that of the preceding hole by an amount greater than the latters machining allowance.

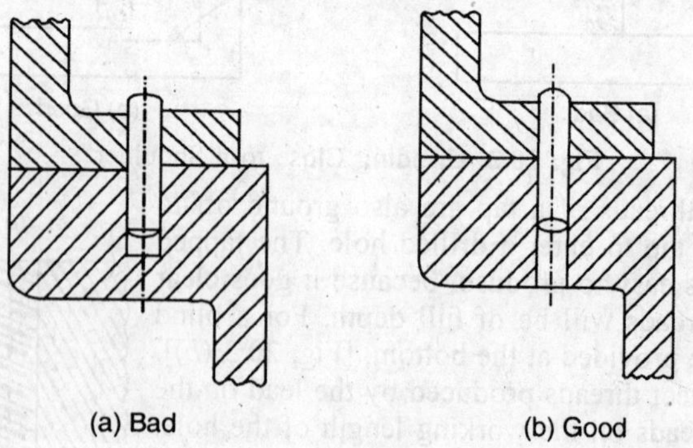

Fig. 20.5. Fixing Dowels.

9. Parts, such as shown in Fig. 20.6 involving excessive material removal, could be more economically made in two or more pieces.

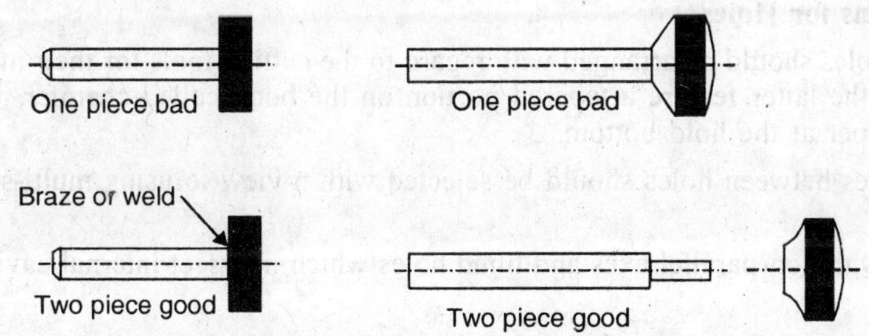

Fig. 20.6. Excessive Material Removal.

10. Grooves, such as oil holes, are easier to machine on outside than inside surfaces. (Fig. 20.7).

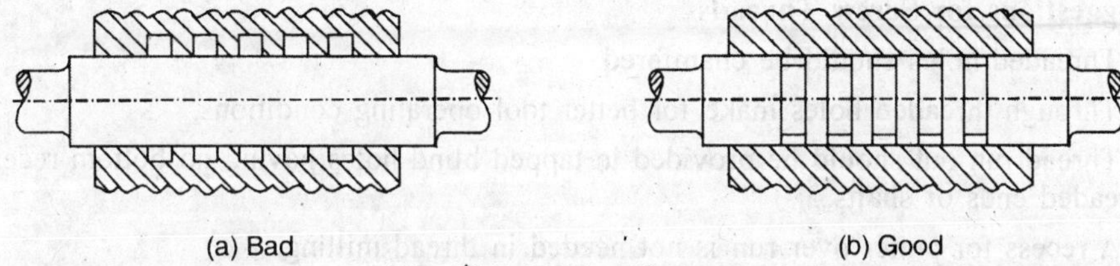

Fig. 20.7. Making Grooves.

11. The example illustrated in Fig. 20.8 are self-explanatory. In Fig. 20.8 (*a*), one more cut is needed to machine the flange for the assembly on the left than to finish the one on the right. [Fig. 20.8 (*b*)], shows a design to withstand cutting action. If a cored hole is to be tapped, particularly in brittle materials, a sharp edge may breakout (left). A countersink to overcome breakage is preferred (Right).

12. The product should be designed so that commercial sizes of standard tools are used for its machining. Special tools cost more than standard tools resulting in additional cost and time delay.

13. Savings in cost also result if a product is so designed that use is made of standard material sizes and shapes.

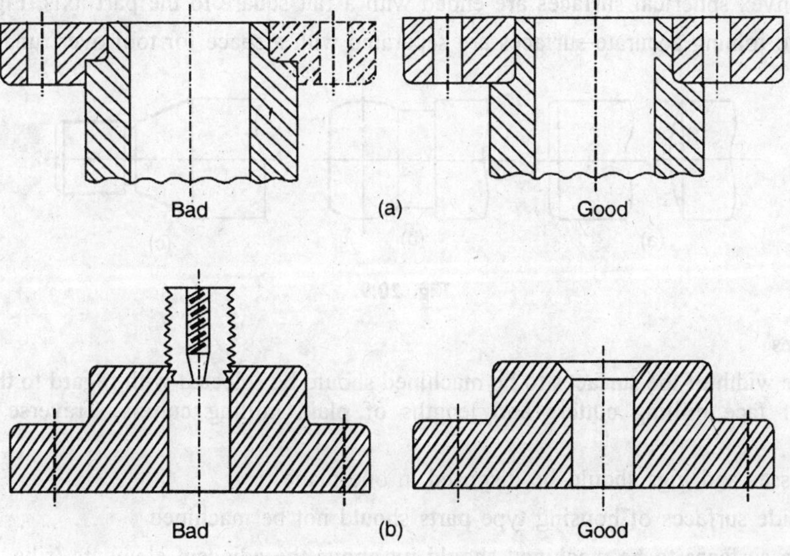

Fig. 20.8. Flange and Assembly.

More Suggestions for Holes

14. Blind holes should be arranged with regard to the cutting tools for their machining (core drills, reamers), the latter feature a tapered portion on the body called chamfer, which forms a corresponding taper at the hold bottom.

15. Distances between holes should be selected with a view to using multi-spindle drilling heads.

16. Holes with non-parallel axes and blind holes which intersect internal cavities should be avoided.

17. Counterboring is better to be replaced with face turning or milling.

18. The boring of recesses in holes on drilling or unit-built machines is not recommended, cast on excusses with a depth of ≥ 1 to 1.5 mm should be introduced instead

More Suggestions for Screw Threads

19. Threaded holes should be chamfered.

20. Through threaded holes make for better tool operating conditions.

21. Thread run out should be provided in tapped blind holes having no bottom recess and at the threaded ends of shafts.

22. A recess for cutter over run is not needed in thread milling.

23. Screw threads should be unified for whole lines of products to be manufactured.

24. Threads under 6 mm in diameter in large parts should be avoided to prevent frequent tap breakdowns.

External Surfaces of Revolution

25. Stepped workpiece portions should have the minimum difference of diameters. Where the difference is large, the steps or flanges are obtained by upsetting or a composite blank is used to lower machining content, and reduce metal losses.

26. Elements rotational parts should be unified to allow the use of the same tooling set ups.

27. Wherever possible, spherical end faces should be replaced with flat ones provided with chamfers, Fig. 20.9a

28. Convex spherical surfaces are ended with a flat square to the part axis, Fig. 20.9b.

29. Two joining accurate surfaces are separated with a space for tool over run, Fig. 20.9c.

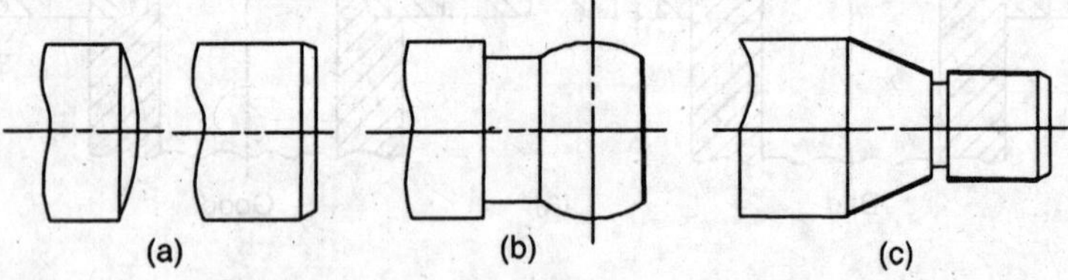

Fig. 20.9.

Flat Surfaces

30. The width of the surfaces to be machined should be selected with regard to the standard diameters of face milling cutters and lengths of plain milling cutters, Traverse milling is preferable.

31. Bosses and lugs should be disposed in one plane.

32. Inside surfaces of housing type parts should not be machined.

33. The surfaces to be machined should jut above the adjacent elements (ribs, lugs, etc.), for simpler machining.

Slots and Seats

34. Slots should allow traverse milling, otherwise the blind end of the slot should be outlined in accordance with the cutter radius.

35. The depth and width of slots are selected with regard to the corresponding dimensions of standard end milling cutters.

36. Slots which allow machining by disc-type milling cutters rather than end milling cutters, are recommended.

37. The rounding radii in recesses or slots should be the same throughout the whole contour and should correspond to the size of standard end mill cutters.

The aforesaid rules are general in character. Additional requirements are placed on parts machined on NC machine tools, unit-built machines and automatic transfer machines, with regard to their special features.

20.2.2. Casting. In the design of castings the basic fact to be kept in mind is that castings contract during cooling and solidification. So, as far as possible, uniform sections should be employed and adequate provisions should be made to avoid excessive shrinkage stresses.

It is necessary to choose the casting method, to determine the position of the casting in the mould, to select the parting plane, to determine the number and disposition of cores, and to specify the thickness of the casting walls. The casting method is selected with regard to the material, shape and accuracy of the casting, the number of castings to be produced and the production schedule. In many instances, castings are the most complex and costly machine parts. In machine tools, i.e. engines, compressors and other machines, the mass of the casting may be as great as 75 to 85% of products' total mass. For this season, the proper selection of the casting method is an important task. If the average cost of grey iron castings is taken as 1, then for other castings, it is: 1.1 for inoculated C.I; 1.3 for malleable C.I.; 1.8 for C steel; 2.5 for low alloy steel; 3 to 6 for non-ferrous alloys and 6 to 8 for high alloy steels.

The shape of the casting should be as simple as possible. That helps to reduce the cast of patterns, cores, shells and moulds. Casting should be made as compact as possible. Large steel castings of complex shapes are well to divide into elements joined by welding.

The position of the casting's surfaces during metal pouring must be taken into account, since gas blow holes may develop on the casting's upper horizontal surfaces. Critical surfaces of castings should be at the bottom part of the mould. Attention must be paid to unhindered filling of the mould with liquid metal. For this, sharp changes in the speed and direction of its flow should be avoided. The casting shape should allow easy cut-off of the gating-system elements and removal of the cores.

The casting design should provide for easy removal of core material and reinforcements and for careful cleaning of the casting recesses.

The minimum height of the bosses should be 5mm, 10 to 15 mm, 20 to 25 mm for the castings overall dimensions of upto 0.5 m, 0.5 to 2 m and over 2 m, respectively.

The following suggestions will result in a sound and economical casting:

1. Design the casting for maximum uniformity of wall thickness. Where it is not possible, light sections should be blended into heavy ones. An abrupt change in section thickness must always be avoided to avoid shrinkage defects and stress concentration at sharp corners. This can be achieved by the use of fillets [Fig. 20.10 (*a*), (*b*)]. It not only facilitates moulding, but also makes the casting much stronger. Cooling and solidification of molten metal start at the surface of the mould cavity and crystals form there first. Then the crystal growth takes place inwards, normal to all surfaces. At the central plane of the casting, the formations from different sides intersect. The metal in this plane is weak because the crystals do not join perfectly or even porosity may result there. In a casting with sharp corners, [Fig. 20.10 (*b*)] a plane of weakness

extends from corner to corner, exactly where stresses are apt to be maximum. In casting with corners rounded off, the plane of weakness runs uniformly through the centre, where the stresses are extremely low.

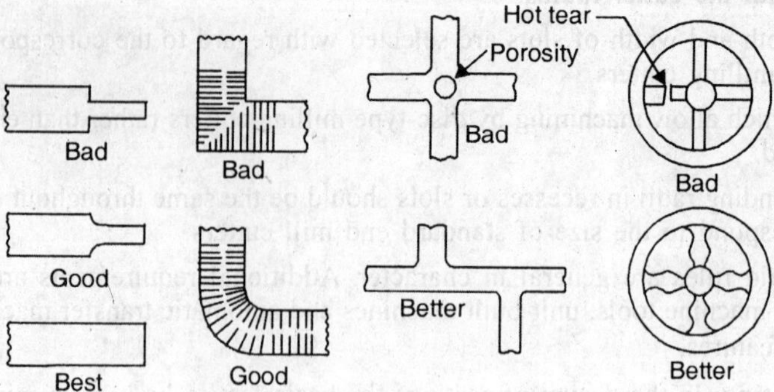

Fig. 20.10. Designing the Casting.

2. The minimum wall thickness for grey cast iron castings can be as low as 3 mm but the average value is 6 mm for parts upto 45 cm in length or diameter and it gradually increases to 18 mm for large and heavy castings. For malleable and ductile iron castings, the minimum thickness may be taken as 3 mm, but the average minimum thickness for heavier castings should be about 6 mm.

Theoretically, the thickness of the casting walls depends on the size and mass of the casting, its material and the casting method. For Grey C.I parts in sand moulds, it may be taken as,

$$t = \frac{L}{200} + 4 \text{ mm}$$

where L = Largest length of the casting

Note: Inner walls are 20% thinner than the external walls.

3. Avoid complex parting lines on the pattern, as far as possible. Parting lines should be in a single plane, if practicable. Irregular parting surfaces increase the cast of moulding.

4. Provide for the required draft $\left(\frac{1}{2}^\circ \text{ to } 3^\circ\right)$ on the wall of the castings to enable the pattern to be withdrawn from the mould. For internal surfaces, the draft values should be higher than for external surfaces.

5. Since cast iron is more strong in compression than in tension (about 4 times), the iron castings should be loaded in compression as far as possible. The tensile and bending stresses can be eliminated or minimised by proper design.

6. Don't use iron castings for impact and fatigue loading.

7. Don't use iron castings at temperatures above 300°C, since its strength starts decreasing.

8. In the case of grey iron castings, the patterns and cores should be simplest and holes should be provided to knock out cores. If possible, the casting should be so designed that no cores are needed.

9. Provide for reinforcing ribs in the casting to increase its resistance to distortion upon redistribution of internal stresses which may be due to non-uniform cooling of parts of the casting, having different sections.

10. If possible, castings should be designed with no undercuts that require special pattern equipment or special moulding techniques, (Fig. 20.11).

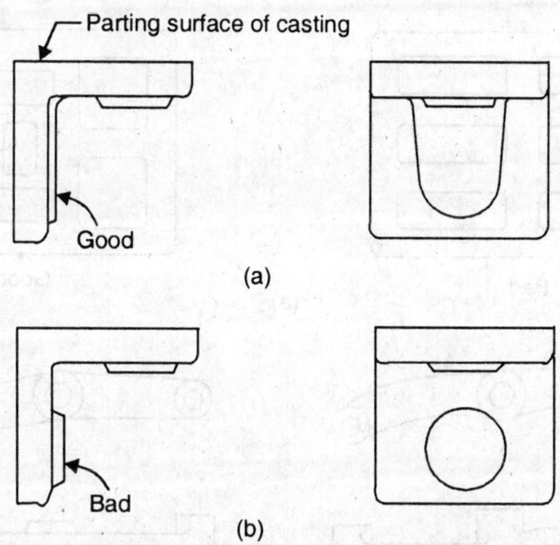

Fig. 20.11. Casting with and without Undercuts.

11. Provide places where holes are to be drilled, to reinforce the walls of the casting, [Fig. 20.12). Design (*a*) is not good, as explained in Fig. 20.4. Design (*b*) is much more satisfactory because it eliminates drill breakage, reduces drilling time and saves lot of material.

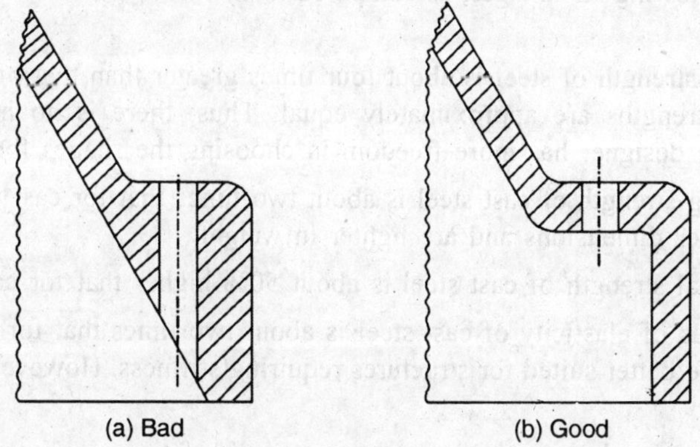

Fig. 20.12

12. [Fig. 20.13 (*a*)] The staggered ribs (right) cause less distortion than the regularly spaced ribs (left). In Fig. 20.13 (*b*) : the design on the right is better, not only because the lines are simpler and the pattern is less costly but, also because a plane parting can be used in the process of moulding.

Steel Casting. The major part of the production steel castings occur in the medium carbon range (0.30 to 0.40% C). About 50% of these castings are used by railroads. Steel is not as fluid as cast iron, therefore, complicated shapes, sharp corners and thin sections can not be cast. The minimum thickness of steel castings is usually taken as 6 mm. However, in small castings, this can be equal to 4.5 mm. All steel castings must be annealed, to relieve them of internal stresses.

Their strength can be increased by subsequent heat treatment. Cast steel has double the contraction, as compared to cast iron. Due to this the size variation is also larger. Therefore, the designer must aim to secure maximum uniformity in contraction which occurs. This can be achieved by using uniform wall-thickness, avoiding concentration of metal etc. However, the cast

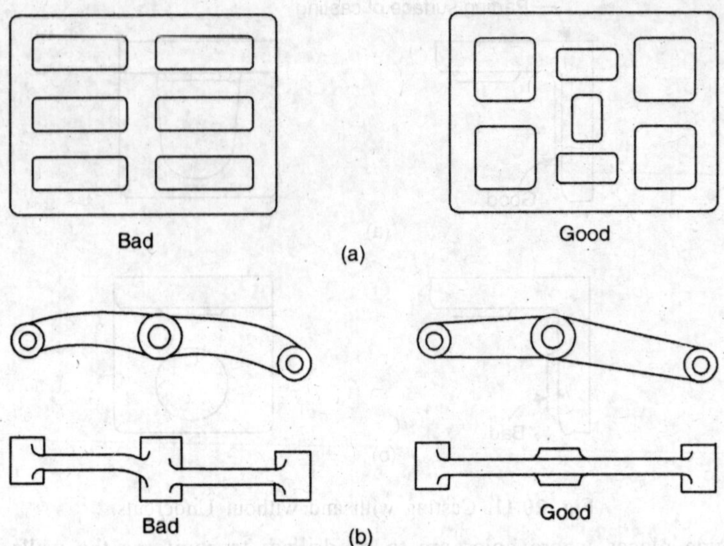

Fig. 20.13

steel has got the following advantages, as compared to cast iron, which will influence the design of steel castings :

1. The tensile strength of steel is about four times greater than that of cast iron. Its tensile and compressive strengths are approximately equal. Thus, there is no need to avoid tensile stresses. Hence, the designer has more freedom in choosing the shapes for steel casting.

2. The bending strength of cast steel is about two times that for cast iron. Therefore, steel castings have smaller dimensions and are lighter in weight.

3. The torsional strength of cast steel is about 50% higher that for cast iron.

4. The modulus of elasticity of cast steel is about two times that for cast iron. Therefore, steel castings will be better suited for structures requiring stiffness. However, cast iron has better damping capacity.

5. By suitable form design steel castings can acquire fatigue resistance.

6. Cast steel has an elongation at fracture of 12 to 20%. Thus steel castings are very suitable for components, which when severely loaded, shall only take on an elastic or permanent formation and shall not suddenly fracture.

7. Cast steel can be used for temperatures above 400°C.

8. Cast steel gives greater toughness and wear resistance, when alloyed with manganese Therefore, it can be used for components which should offer considerable resistance to wear, for example, crushing machines and mechanical shovels.

Die Casting. For the design of castings made by die casting (used mainly for non-ferrous metals and alloys), the following points should be kept in mind :

1. As the light alloys are more sensitive to stress concentration, large fillets should be provided; notches should be avoided and the wall thickness should be kept as uniform as possible The minimum thickness for brass and bronze is 2.3 mm and for Al it is 3 mm.

2. Aim at the simplest possible die design.

3. Allow draft where necessary.

4. Avoid partly closed shapes and closed cavities.

5. Provide for easy flash removal.

6. Provide shrinkage and machining allowances.

7. The light alloys have high co-efficient of linear expansion. They loose their strength at temperatures above 150°C. In no case, the die castings should be used at temperatures above 260°C.

Permanent Mould Casting

1. The shape of the casting should allow its easy removal from the mould.

2. Uniform thickness of the walls is desirable.

3. The thickness of internal walls and ribs is recommended to be 0.6 to 0.7 that of the external walls.

4. The thickness of the casting walls should be as follows :

= .3 mm for silumin

= 8 to 10 mm for steel

= 15 mm for grey cast iron

The above values are for casting walls 30 cm^2 in area.

Investment Casting

Investment casting makes it possible to obtain complex-shape parts with a minimum wall thickness of 1 to 2 mm and a hole diameter of 2 mm. The recommended design factors are :

1. Blind holes should be avoided.

2. Casting without draft can be obtained (a draft angle of 0.5° is needed only for removing the wax from the mould).

3. The minimum radius of edge and corner rounding should be 1 to 3 mm.

4. The uniform thickness of walls in the casting is desirable.

5. Long surfaces on castings should be avoided because ceramic material of the mould is not stiff enough.

Shell Moulding

1. A single parting plane should be provided for the mould.

2. Detachable pattern parts and cores should be avoided.

3. The casting walls of uniform thickness are desirable, the minimum wall thickness being 2 to 2.5 mm.

4. Rounding radii of 2.5 to 3 mm should be used.

5. Draft angles of no less than 1° (preferably 2° to 4°) should be used.

20.2.3 Sheet Metal Working:- Sheet metal working or stamping is done on sheet metal (bands, strips and sheets), and is a group of cold working processes. Band is used for parts from 2 to 2.5 mm thick, strips for parts upto 10 mm thick, and sheets for parts of greater thickness. Stiffening ribs, flanges and like elements formed on stamped parts often make it possible to use stock of smaller thickness. The following points should be kept in mind while designing parts for stamping:-

1. The part out line should allow for layout with minimal losses of stock (See Art 2. 12. 2).

2. For minimum diameter of piercing operation, See Art. 2.11.

3. For bending in dies, the height of the straight portion of the part being bent should be greater than two times the thickness of the stock. A smaller height can be obtained with the aid of a performed groove at the base of the bent wing, or by subsequent machining.

4. In drawing large parts with large flat surfaces, provisions should be made of elongated and criss-crossed stiffening ribs to avoid local distortions.

5. Complex-shape hard-to draw parts should be divided into simple elements which are then joined by welding or soldering.

Note: For more details on 'Stamping' See Chapter 2.

20.2.4. Forging. For the design of forging, two basic factors to be kept in mind are : metal flow and minimizing the number of operations and dies. The various design factors have already been discussed in Chapter 3 (Articles 3.3 and 3.3.1.). In Art. 3.3, it has been discussed that the direction of the fibre flow lines should be kept in mind so that full advantage can be taken of these.

Cutting across the grain or bending at an acute angle is not desirable. It is desirable that the forged blanks should have simple symmetrical shape. Non- simple shapes should be replaced with weldments, comprising simple elements. The lugs and projections on their main surfaces should be avoided.

Other recommendations for a forging to be produced on forging hammers and presses, are discussed below:

1. Pockets and recesses should be minimum to avoid increased die wear.

2. Sections should not be so thin as to restrict the flow of metal. Thin walls in forging reduce die life, since the forging cools rapidly and its resistance to metal flow is increased. This prevents the proper filling of the die-cavity resulting in defective forgings.

3. The parting line on forging, if possible, should divide the forging about in half. Non-uniform shapes and non-uniform drafts result in forces tending to shift the upper and lower dies

4. Maximum flash thickness should not be more than 6.4 mm or less than 0.79 mm, on average.

5. The blank must be so shaped as to allow its easy removal from the die. Thus, the grooves and recesses in blanks can be arranged only along the direction of the die working movement

6. Narrow long projections arranged in the die parting plane or in a plane perpendicular to it are unacceptable, since these form hard to fill recesses in the die and, therefore, cause defects

7. It is expedient to arrange the longitudinal axis of the forging along the grain flow of its material.

8. As discussed in Chapter 3, the blank shape should, as a rule, allow a horizontal parting plane of the dies to avoid side thrusts. This also makes forging and trimmingdies simpler in construction. A broken and especially curved parting surfaces are undesirable. With a broken parting surface, the surface portion should not be inclined in excess of 60° to a horizontal plane for a neatly cut flash.

9. The maximum overall blank dimensions should be located in the die parting plane, which makes it possible to have the dies with the minimum depth of the cavities or impressions, for their proper filling with metal, see Fig. 3.14 (*c*). An exception to this rule is forging shaped as solids of revolution whose length-to-diameter ratio is under 3. These are simpler to forge along the axis, with the bottom die having a deep impression.

10. Great variations in the cross-section of the blank along its length should be ruled out since it makes forging difficult. Here, the blank metal fails to properly fill the die impression. and that results in defective forging.

11. Fins and lugs on the blank should not be near each other, because that interferes with the flow of the metal and filling of the die impression.

12. It is good practice to accommodate the blank in one half of the die, that cuts cost of die manufacture and improves the accuracy of the forging owing to the eliminated effect of the die halves mismatch.

13. A blank symmetical to the die-parting plane and with symmetrical shapes of its projecting walls facilitate die manufacture and the forging process stock allowances are cut and eliminates defects due to mismatch of the two die-halves.

'Forging Machines' are capable of forging blanks of different shape, the most common being solids of revolution with a shoulder or flange at one end, with a through or blind hole :

1. The thickness of the wall in hollow forgings should not be less than 0.15 of the blank outside diameter.

2. The necks along the blank impede the flow of metal during forging and, therefore, should be avoided.

3. Taper shanks are also difficult to forge and should be changed to cylindrical ones, where ever possible.

20.2.5. Welding. Modern machines can be made more producible by the use of welded structures in the design. A complex part can be welded from simple elements obtained by different methods (e.g. stamping, forging, rolling, casting or by cutting off from standard or special rolled stock.). The replacement of steel castings with weldments sometime allows a 20 to 30% reduction in their mass and 30–50% reduction in their machining content. Some examples of complex parts made by welding simple elements (blanks) are blanks for front and rear axles of automobiles, reducing gear housings, propeller shafts etc. Occasionally, cast frames of machine tools (e.g. broaching machines) can be replaced by welded structures, thus saving in material and machining costs. Welding (or brazing) of stampings is used in batch and mass production. In single-piece production, i.e., in building heavy equipment, elements to be welded are made by smith forging, casting or cutting - off from rolled stock. The elements are then joined together by arc welding, ESW. One weldment may comprise elements of different materials.

Welded structures are not machined after welding if their dimensions are not toleranced. Their component elements are machined before welding for positional accuracy in a weldment. After welding, the weldments are only finished. Such a process is advantageous in heavy engineering where it allows a reductions in machining content, because the machine tools used here are often unique, and they impose limitations on the shop productive capacity. In the general engineering, automotive, and other branches of industry, components made by butt welding, arc welding and pressure contact welding are subsequently machined.

When designing welded components, it is necessary to bear in mind the general process plan, to provide for the use of the most efficient welding methods and to take measures which will prevent scrap.

The volume of welding operations should be reduced by substituting a single thick plate for a stack of thin plates (Fig. 20.14a) by replacing welded ribs (stiffening) with stamped ones (Fig. 20.14 b); by employing bending instead of welding at corners (Fig 20. 14c); by using weld joints with minimum included angle between the edges of the welded elements or square butt joints; by designing weld joints without lap straps with the minimum cross-section of welding seams; by using components welded from stamped or cast elements etc.

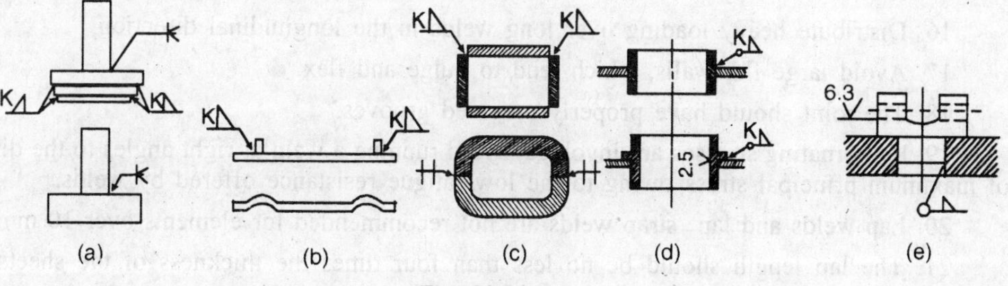

Fig. 20.14

Welded structures containing a great number of elements arranged according to a complex pattern should be devided into simple individual units. The subsequent welding and assembly are thereby facilitated, residual strain is reduced and accruary increased. Priority should be given to weld joints which allow the maximum number of elements to be set in position for welding at a time. Positioning and welding each element in sequence is undesirable.

The machined surfaces of elements to be welded should be protected from damage in welding by arranging the welds at a safe distance from those surfaces. A thin walled bushing being welded to a plate (fig. 20.14d) may get its hole warped, whereas such warpages may be eliminated with a flanged bushing, where fully machined elements are welded to obtain their strictly specified relative position, these should be previously press fitted to each other, and the weld should be located as shown in Fig. 20.14e.

Additional requirements are placed on automatically assembled products. Their components should be simple and symmetical in shape. That facilitates the orientation of the parts as they are fed from the hopper to the working station. The shape of the parts should be such that do not get entangled as they leave the hopper. The maximum use of standard parts is desirable so that assembly equipment of limited variety can be employed.

The following points should also be kept in mind when designing a weldment:

1. Weldments should be designed to require a minimum of weld metal.

2. Thermal contraction of metal, which has been heated by welding, may cause internal residual stresses and distortion. These can be controlled or reduced by :

(*a*) Preheating, (*b*) Minimum number of welds (*c*) Smallest size of weld that fulfills requirements, (*d*) maximum use of intermittent welds, (*e*) slow after cooling.

3. Sharp discontinuities in metal should be kept at a minimum since these cause stress concentration.

4. An important strength weld should not be located where much of it may be removed later by machining.

5. Welds should be located so that adequate strength will be provided at the proper place on a structure or part.

6. As far as possible, a straight line force pattern should be provided.

7. Laps, straps and stiffening angles should be avoided except as required for strength.

8. Where ever possible use butt welds.

9. The ends to be welded should be of equal thickness.

10. Welds at the vulnerable cross-sections should be avoided.

11. The use of welding fixtures should be avoided as far as possible.

12. Welds should not be subjected to bending.

13. A weld should not be located at the point of maximum deformation.

14. Ribs should be designed correctly and these should be used with care.

15. Provide for easy access to welds so that they are accessible for inspection.

16. Distribute heavy loading over long welds in the longitudinal direction.

17. Avoid large flat walls, which tend to bulge and flex.

18. The joint should have properly prepared grooves.

19. If alternating stresses are involved, avoid running a weld at right angles to the directi[on] of maximum principal stress owing to the low fatigue resistance offered by welds.

20. Lap welds and lap strap welds are not recommended for elements over 10 mm thic[k]

21. The lap length should be no less than four times the thickness of the sheets bei[ng] welded.

22. In spot welding, the welded sheets thickness ratio should not exceed 2 : 1, only in some exceptional instances it can be 3:1.

23. In spot welding, two elements should be welded at no fewer than two welds. More than three elements in a joint are not recommended.

24. Critical welds are deposed in places convenient for welding and for usual and special inspection.

25. Some designs can avoid or minimise edge preparation.

26. Horizontal welds should be preferred to vertical, and especially to overhead ones.

27. In order to weld on different surfaces of complex components, these should be allowed to be turned over.

28. Spot welding of parts in mass production should be carried out simultaneously for all the spots.

29. Components should fit properly before welding.

20.2.6 Plastic Parts. Plastic parts are manufactured by : (1) compression moulding of basic materials in the form of powders or tablets with the use of reinforcements or fibre fillers or without these; (2) by injection moulding in special moulding machines; (3) by extrusion to obtain bars, strips, tubes and special shapes, (4) by air-pressure and vacuum forming of lamellar thermoplastics heated to plastic state, (5) by stamping from heated sheets of textolite, micarta, and other thermosetting plastics etc, (6) Machining is employed to obtain parts from bars, tubes, and other shapes, and from individual moulded blanks. The latter are only finished.

While designing the moulded plastic parts, the following design rules should be followed :

1. Allow for shrinkage after moulding.

2. Allow atleast a minimum draft of $\frac{1}{2}$ to a 1 degree for easy withdraw of the parts from the mould.

3. Avoid undercuts whenever possible. They prevent removal unless special mould sections are provided that move at right angles to the opening motion of the main mould halves. Such moulds are costly to construct and to maintain. Alternatively, the undercuts will need cores or split cavity moulds.

4. If possible, the parting line should be located in one plane.

5. Design corners with ample radii or fillets. Provide adequate fillets between adjacent sections also. This will assure smooth flow of the molten material into all sections of the mould and will also eliminate stress concentration at sharp corners. Such a mould will be less expensive to build and also less prone to breakage where thin, delicate mould sections are encountered.

6. The curing time of the product is determined by its thickest section. Thus, thick sections should be preferably kept as nearly uniform in thickness as possible. The minimum wall-thickness depends upon the size of the product and the type of plastic used. It is also limited by the difficulty of removing very thin parts from the mould and also by the high pressures needed to fill at a high width-to-thickness ratio. The minimum recommended wall-thickness is 0.65 mm and it can be 3.2 mm for large parts. Variation in wall thickness of the moulding should not be over 30%.

7. **Stiffening** Ribs should be provided to increase strength and rigidity and to reduce distortion and warping. When extra strength is needed at corners, it is better to provide ribs there, than to have thick corners, which are likely to lead to gas pockets, under curing or creaking. Rib height should not be more than twice its thickness, which should be 0.6 to 0.8 that of the adjoining wall. The ribs should be arranged in the direction of the material flow on the mould.

8. Plastics have low modulii of elasticity. Therefore, large flat surfaces will not be rigid and should be avoided, whenever practicable. However, their strength can be increased by ribbing or doming.

9. Through holes are limited only by the strength of the core pin and are usually held below a length-to-diameter ratio of 8. Blind holes are also made with the help of core pins and these are limited to a depth-to-diameter ratio of 4 for diameter greater or equal to 1.5 mm and to a ratio of 1 for smaller holes. Threaded holes of diameter equal to and greater than 5 mm can be moulded directly. It is better to drill smaller holes. Smaller threads of reasonable strength are best provided by metal inserts. Binding posts, electrical terminals, anchor plates, nuts and many other metallic components are conveniently obtained by moulded-in inserts. Metal inserts (usually made of steel or brass), are held in the plastic only by a mechanical bond, since there is no adhesion between metals and plastics. Therefore, the metal inserts are suitably knurled or grooved so that they are gripped firmly and do not become loose in service. Plastics have much greater thermal expansion as compared to metals. This helps to shrink the plastic onto the insert, but could also cause cracking of a brittle plastic. The wall-thickness around the insert, therefore, must be made large enough to sustain the secondary tensile stresses.

10. Mouldings should be simple in shape for easy removal from moulds.

11. Back tapered features in parts should be avoided since they require complicated moulds and improper moulding condition.

12. It is advisable to reinforce the moulding end faces with shoulders, which prevent the part from cracking. The shoulders are arranged along the end face periphery without interruption.

13. If plastic parts have holes, then the minimum thickness of the wall between the holes should not be less than 0.5 mm for holes 2.5 mm in diameter and no less than 2.5 mm for holes 16 mm in diameter. Minimum distance from the part side edge of a hole.

= 1 mm for holes 2.5 mm in diameter

= 4.5 mm for holes 16 mm in diameter.

14. The *l*/*d* ratio for reinforcement elements in plastic parts (bushing cores, inserts) should not be less than 2 for their reliable fit.

20.2.7. Powder Metallurgy Parts. We know that in P/M process, the parts are made by compressing metal powders in a die. The strength and other properties are added to the parts by subsequent sintering operations :

Design Considerations

1. Strength is closely related to the density of the pressed powder. The many quality attribute of the parts made by P/M is uniform density of their material. Non-uniform density causes stress in the material, and, consequently, warping and cracking of parts. To obtain homogenous strength in a product, it is necessary that uniform pressure or density is obtained throughout the product. For this :

(*a*) An almost uniform cross-section must exist throughout the length of the part.

(*b*) Walls should be of uniform thickness.

(*c*) Part should be relatively short in length in comparison with its diameter, since pressure is not transmitted uniformly through a deep bed of powder. The maximum depth to diameter ratio is practically limited to 2 to 4.

(*d*) Ratio of unpressed length to pressed length should be kept below 2, if possible, and never exceed 3. Select the proper direction of pressing of the parts.

2. The part must be of a shape that can be ejected from the die.

(*a*) Various lugs, bosses, flanges etc. should be arranged in a plane normal to the direction of pressing and as near the upper part of the die as possible.

(*b*) Grooves, cavities and recesses should be arranged so that their axes are along the direction of pressing.

(*c*) Multiple stepped diameters, reentrant holes, grooves and undercuts should be eliminated.

Note. Flexible isostatic compacting moulds permit undercuts or reverse tapers but not ransverse holes.

3. The powder should flow properly into all parts of the die. Therefore, do not design for thin walls, narrow splines or sharp corners, since even under pressure, the powder can not fill very thin sections. The minimum wall thickness should be 1 mm for cylindrical parts and 1.5 mm for parts of other types. The rounding radii for internal and external corners should not be less than 0.3 mm nd 2.5 mm, respectively. This will also result in the construction of strong tooling.

4. Draft should be provided on the part walls perpendicular to the mould parting plane for asy stripping of the part from the mould. The draft angle should range from 5° to 10°. No draft s required for ejection from a lubricated die.

5. Provide wide dimensional tolerances whenever possible. Wide tolerances mean lower iece-part cost and longer tool life.

20.2.8. Heat Treatment Requirements on Part Design: Heat treatment and thermo-hemical treatment of parts carried out at the initial, intermediate and final manufacturing stages nake it possible to drastically improve their functioning and prolong their useful life. That is chieved through improved general properties of the metal and strengthened surface of the parts which results in their greater wear resistance. The following factors should be kept in mind while esigning parts that are to be heat treated:

1. The parts should have a simple and symmetrical shape without sharp edges, thin ribs nd greatly contrasting cross-sections.

2. Grooves, holes and recesses cut before that treatment are undesirable, because during eating and cooling, these may form the pockets of stress concentration and develop cracks.

3. The surface roughness of the parts should be under 10 μm *Ra*. Greater surface roughness ay be conducive to cracking and breakdown of the part.

4. A decarburized surface layer and scales lead to a non-uniform and reduced hardness after eat treatment.

5. In parts to be heated by high-frequency currents, the hardened layer should be deeper an any annular grooves made thereon, for otherwise their fatigue strength falls and they may acture at the grooves.

6. The hardened layer extending to critical (loaded) Zone of the part should be avoided, cause here the hardening stress may add to the working stress, and the part will breakdown.

7. To prevent the melting of the edges on the workpiece end faces and holes, chamfers ould be provided thereon.

8. Threaded parts which undergo thermochemical treatment, should not be hardened because e thread turns out to be excessively hard and brittle.

9. Critical zones of the part (thin walls, ribs etc,) should undergo a local thermochemical atment to prevent cracking in the course of hardening.

10. Parts which tend to warp should be made of alloy steels hardened in oil or in the air.

.3. DESIGN FOR MANUFACTURABILITY (DFM)

The topic "Design of product for economical production" is also known by the name of esign for Manufacturability". DFM, or "Concurrent engineering", C.E.

DFM or CE necessitates that product and process designs be developed simultaneously, rather than sequentially. That means, that all the design constraints, including assembly, material information process and material handling requirements, are included as part of the functional optimization of the design. In this way, the DFM process enables designers or a design team, to consider all aspects of the product design and manufacturing, at early stages of the design cycle, so that design iteration and accompanying engineering changes can be made easily and effectively. This has great advantages, because it leads to few or no manufacturing problems.

A C.E. approach has been illustrated, as shown in Fig. 20.15. It shows that CE strategy requires a parallel iterative team approach — a tiger team'.

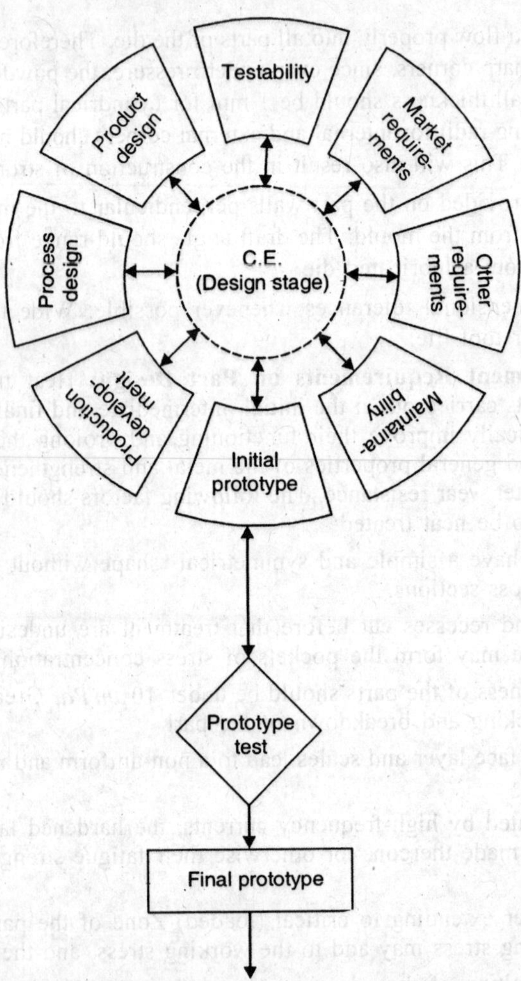

Fig. 20.15. Con-current Engineering Approach.

However, there is one major difference between DFM and CE. Whereas DFM concept is confined within the premises of the company, the concept of CE involves the "Life cycle" of the product, "Life cycle" means that all aspects of a product, such as design, development, manufacturing, distribution, use and its ultimate disposal and recycling are considered simultaneously. On the other hand, the concept of DFM is a comprehensive approach to manufacturing of goods applicable within the walls of the company. It integrates the design process with materials, manufacturing methods, process planning, assembly and quality assurance. Therefore, with DFM concept, the engineering release package will contain not only part drawings, part list, and assembly drawings, but it will also contain the process plan.

PROBLEMS

1. What is meant by Production Design?
2. What are the duties of "Product Designer" and "Tool Engineer"?
3. Discuss the importance of relationship between design and production.
4. State the many reasons as you can for using round corners and fillet radii on manufactured components.
5. Discuss the importance of providing "relief" for various machining operations.
6. Why drilled holes should be normal to the surface, top and bottom? Discuss.
7. Discuss the importance of uniformity of wall thickness for components produced by casting and forging.
8. Discuss the design factors for forgings.
9. Discuss the design factors for weldments.
10. Discuss the design factors for casting.
11. Discuss the design factors for plastics.
12. Discuss the manufacturing requirements for powder metallurgy parts.
13. What is DFM or CE?
14. Give the benefits of DFM.
15. Write about the suggestions for cutting external threads.
16. What is your suggestion for grinding close to a shoulder?
17. Write about the suggestion for locating flange holes.
18. Why a dowel pin should not be pressed in a blind hole?
19. Write about the suggestion about drilling of co-axial holes in a single set-up.
20. Write about suggestions for machining of :-
    (*a*) blind holes
    (*b*) Screw threads
    (*b*) External surfaces of Revolutions
    (*d*) Flat surfaces
    (*e*) Slots and seats
21. What is the significance of minimum wall thickness of a casting?
22. Write the suggestions for designing a casting for:-
    (*a*) Pr. Die-casting
    (*b*) Gravity die-casting
    (*c*) Investment casting
    (*d*) Shell Moulding
23. Write the suggestions for sheet metal working.
24. Write the suggestions for "Heat treatment requirements on Part Design".
25. Why welded structures are preferred in the design of a complex part?
26. How the volume of welding operations is reduced? Explain
27. What is the importance of providing "draft" in the design of a part?
28. What do you understand by the term "Design of a product for economical production"?
29. Sketch the design process by CE approach.
30. What is the major difference between DFM and CE?
31. What is "life cycle"?

21

STATISTICAL QUALITY CONTROL

21.1. INTRODUCTION

Quality means the degree of perfection of excellence when applied to manufactured products. It does not necessarily mean "the best". It should imply "the best for the money". That is, the product should meet the desired requirements at lowest cost. The following facts about quality should be noted :

1. The word "quality" is meaningless unless and until the use of the product is specified. The term "good quality" means that the product serves the purpose well for which it was stipulated.

2. Quality is not absolute but is a relative term. It can only be judged by comparing with some standard. Thus the quality of a unit may be defined as its conformance to standards or specifications.

3. Quality has direct effect on manufacturing costs and on selling price. High quality means high cost and vice versa.

4. Quantity is also affected by quality. The higher the degree of quality. tighter becomes controls and more difficult it becomes to achieve quantity output.

By "quality control" we mean those processes or operations of testing, measuring and comparing the manufactured components with a standard and then determining whether it should be accepted, rejected, adjusted or reworked. It is, of course, more sensible to control the process so that the standards are being maintained and that defective parts will not be produced, instead of accepting or rejecting a part, after it has been manufactured. Thus "quality control" means the systematic control of those variables in a manufacturing process which affect the excellence of the end-product. In addition, cost and labour of inspection must not become excessive. (The difference between quality control and inspection should be clearly understood. Quality control means bringing variables under control. It enlarges the production pile by inspection records and salvage method. Inspection is merely separating good and bad from a lot and it enlarges the scrap pile. Inspection is a part of the quality control programme or it is the principal tool of quality control). Statistical quality control (SQC) procedures, which are based on well established statistical methods, have been developed which provide economical means of maintaining continual analysis and control of processes and products.

SQC applies the theory of probability to sample testing or inspection. The inspector on the shop floor takes controlled random samples of parts. These are then checked by measurement or gauging. The quality of the total batch of components is then judged by the result of the sample. That is, if the number of rejects found in the sample is too large, then the whole batch will be rejected and vice-versa. Thus, the two principal objectives of SQC are :

1. To prevent the production of defective parts by inspecting them and establishing control over manufacturing process during production, instead of waiting to inspect them after they have been completed, when it is too late to take corrective action.

2. To reduce the cost of inspection and at the same time ensure quality of the product by adopting a sampling inspection plan that is based on statistical methods.

In addition to the above two principal objectives, SQC results in the following benefits :

(*i*) It results both in decreased scrap and in less time being spent on ratification of defective components.

(*ii*) Due to uniformly achievable improved product specifications the performance of assembly operations may be improved.

(*iii*) Where inspection has to be by destructive testing, the adoption of SQC is the only reasonable guarantee of quality.

(*iv*) The application of quality control may result in investigations which will lead to improved inspection methods, improved manufacturing methods and improved product specifications.

21.2. STATISTICAL TOOLS FOR QUALITY CONTROL

The main statistical tools used in the quality control jobs are :

1. Frequency distributions.
2. Control charts.
3. Sampling tables.

21.3. FREQUENCY DISTRIBUTION

For any manufacturing method, it is generally recognized that no two parts are exactly alike and also that for every part, there are always variations from established standards. These variations may be large or small. Variations in a product are due to slight differences in successive units of raw material, wear on tools, changed setting on machines, fluctuations in power, changes in temperature, and other numerous causes. All these causes or variables can be grouped into two categories.

1. **Chance Variables.** These variables or causes occur at random and they are difficult to avoid. These causes result in variations which are so small that their elimination can not be justified economically. These causes are :

(*i*) Inherent variations in the raw material, that is, materials not entirely homogeneous.

(*ii*) Imperfections inherent in the design of the machine or in the manufacturing process.

(*iii*) Natural or inherent inaccuracies of inspection instruments.

(*iv*) Feel, eyesight or judgement of the production operator.

2. **Assignable Variables.** These variables or causes result from

(*i*) Trouble in the manufacturing process or improper operation of the machines.

(*ii*) Incorrect sequence of manufacturing operations.

(*iii*) Machines or inspection instruments worn out and in need of repair.

(*iv*) Room temperatures that vary during the day or in different parts of the same room.

(*v*) Variations of plants building affecting the performance of the machines.

Assignable causes result in large variations in the quality of the product and they should be identified and eliminated, if possible.

As long as chance causes are the only factors for part variation, any given characteristic of a product will remain within the limits specified for it and the process is said to be under control. However, if assignables causes come into play, the part variations will exceed the specified limits and the process is said to be out of control. So, the major application of quality control is to determine when an assignable cause of variation enters a manufacturing process in time to take corrective action so that no defective parts are produced.

Recording Part Variations. There are two methods of inspection to determine the variation of a product from established standards :

1. **Inspection by Variables.** Here, the part is checked with a graduated tool or instrument, to determine how much it varies from the ideal value which is specified for it. This method gives a record of the actual measurements of the product.

2. **Inspection by Attributes.** Here, the product is checked with a "go-not go" type of inspection tool to determine if it lies within the limits of variability established for it. This method gives a record of the total number of parts inspected and the number of parts inspected and the number found to be defective.

From the recorded data, a frequency distribution can be established. A frequency distribution is an arrangement of numerical data according to size and frequency. That is, it is tabulation or tally, of the number of times a given characteristic measurement (diameter, length etc.) occurs within the sample of the product being inspected. For example, let the outside diameter of 64 steel pins be recorded, and arranged according to size, as shown in Fig. 21.1. If the pins shown in Fig. 21.1 include all the existing pins of a particular type, this frequency distribution would be referred to as a "population distribution". The dotted curve, which is constructed by drawing a smoothed curve through the maximum value of each group in the frequency distribution, is called the "frequency distribution curve."

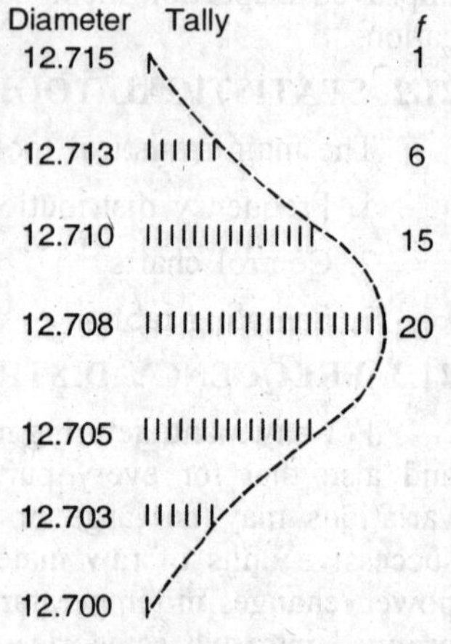

Diameter	Tally	*f*
12.715	I	1
12.713	IIIIII	6
12.710	IIIIIIIIIIIIIII	15
12.708	IIIIIIIIIIIIIIIIIIII	20
12.705	IIIIIIIIIIIIII	
12.703	IIIIII	
12.700	I	

Fig. 21.1. A Frequency Distribution.

Algebraic Measure of Frequency Distribution. The frequency distribution can be described by the following two characteristics :

1. Its central tendency, that is, what is the most representative value.

2. Its spread or dispersion, that is, how much variation is there from the central tendency.

For industrial use, the two most valuable measures of central tendency are — the Average and the Median. The two most useful measures of spread are : the Standard deviation and the Range.

These characteristics are defined below :

(*i*) **Average.** The average is the most useful measure of central tendency. It is obtained by dividing the sum of the values in a series of readings, by the number of readings, that is,

$$\bar{X} = \frac{X_1 + X_2 + X_3 + \ldots + X_n}{n} \qquad \ldots(21.1)$$

where $\bar{X}$ = Average value of the series

$X_1, X_2, \ldots, X_n$ = Value of each reading

n = number of readings

or

$$\bar{X} = \frac{\sum_{1}^{n} Xi}{n} = \frac{\sum X}{n} \qquad \ldots(21.2)$$

Let there be a set of 9 readings : 2.6, 2.7, 2.7, 2.8, 2.8, 2.8, 2.9, 2.9, 3.0. For this,

$$\bar{X} = \frac{2.6 + 2.7 + 2.7 + 2.8 + 2.8 + 2.8 + 2.9 + 2.9 + 3.0}{9}$$

$$= 2.8$$

Where there is a large number of readings in a series, the calculation of the average is greatly simplified by first grouping together readings in suitable cells and then summing up these cells. That is,

$$\overline{X} = \frac{\Sigma fX}{n} \qquad ...(21.3)$$

In the above example

for

$$\begin{aligned} X &= 2.6, f = 1 \\ &= 2.7, \quad = 2 \\ &= 2.8, \quad = 3 \\ &= 2.9, \quad = 2 \\ &= 3.0, \quad = 1 \end{aligned}$$

$$\therefore \quad \overline{X} = \frac{1 \times 2.6 + 2 \times 2.7 + 3 \times 2.8 + 2 \times 2.9 + 1 \times 3.0}{9}$$

$$= 2.8$$

For Fig. 21.1,

$$\overline{X} = \frac{1 \times 12.715 + 6 \times 12.713 + 15 \times 12.710 + 20 \times 12.708 + 15 \times 12.705 + 6 \times 12.703 + 1 \times 12.700}{64}$$

$$= 12.708 \text{ mm}$$

when the values of the average for each scrics of rcadings in a number of series are calculated, it may be desirable to compute the average of these averages. This measure is termed as the "grand average", $\overline{\overline{X}}$ and is given as,

$$\overline{\overline{X}} = \frac{\Sigma \overline{X}}{k} \qquad ...(21.4)$$

where k = number of series, or sub-groups.

(*ii*) **Median.** The median is that value which divides a series of readings arranged in order of the magnitude of their values so that an equal number of values is on either side of the centre or median value. For example, in a set of readings : 11, 12, 13, 15, 16, the value of the median is 13. In another set of readings, 8, 9, 9, 10, 11, 11, 12, 12, 13, 13, 13, 15, the value of the median is 11.5.

(*iii*) **Standard Deviation.** The standard deviation, σ, is the root mean square deviation of the readings in a series, from their average. That is,

$$\sigma = \sqrt{\frac{\left(X_1 - \overline{X}\right)^2 + \left(X_2 - \overline{X}\right)^2 + \ldots + \left(X_n - \overline{X}\right)^2}{n}} \qquad ...(21.5)$$

or

$$\sigma = \sqrt{\frac{\Sigma\left(X - \overline{X}\right)^2}{n}} \qquad ...(21.6)$$

When there is a large number of readings in a series, it is usually convenient to group together readings of the same value into individual cells. Then,

$$\sigma = \sqrt{\frac{\Sigma fX^2}{n} - \overline{X}^2} \quad ...(21.7)$$

(*iv*) **The Range (*R*).** the range is the difference between the lowest and the largest readings in a series. That is,

$$R = X_{max} - X_{min} \quad ...(21.8)$$

For example, with respect to Fig. 21.1,

$$R = 12.715 - 12.700 = 0.015 \text{ mm}.$$

Normal Distribution. There are many types of frequency distributions, for example, Binomial and Poisson, but the distribution which is important to engineers is the normal or the Guassian distribution. It is a continuous distribution having a symmetrical bell shape and is exactly symmetrical about its arithmetic mean, that is, values which are at equal distances from the arithmetic mean have equal frequencies. (Fig. 21.2). The normal distribution curve has been analysed by mathematicians and the percentage of items located at various distances from the arithmetic mean measured in terms of standard deviation, have been determined and tabulated. For example, it is known that 99.73% of the items in a normal distribution, lie within ± 3σ on either side of the arithmetic mean. For these reasons, therefore, the control limits on a quality control chart are usually located at ± 3σ, since then only 0.27% of the parts would be expected not to conform to specifications.

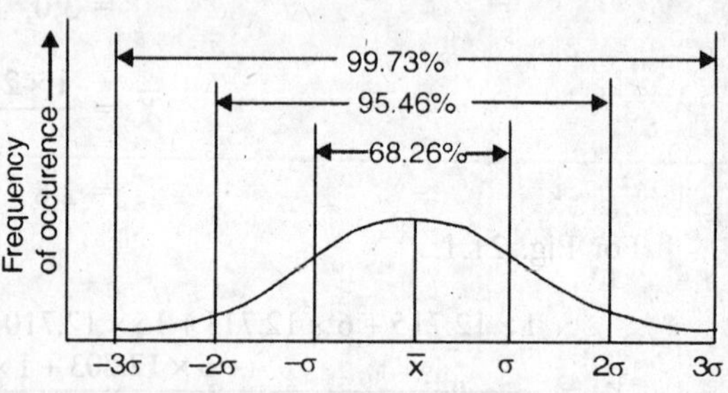

Fig. 21.2. Normal Distribution Curve.

21.4. CONTROL CHARTS

A control chart can be defined as : A chronological (hour-by-hour, day-by-day) graphical comparison of actual product quality characteristics with limits reflecting the ability to produce as shown by past experience on the product characteristics. The concept of the control charts can be understood by the following simple example : Two holes are to be drilled in a plate with tolerance of ± 0.075 mm between drilled hole centres. We know that as long as the causes for variations are only "chance causes or usual causes", it will be possible to maintain these limits. However, if assignable causes come into action, the variations will be unusual, that is, greater than the shopman has learned to expect (in this case ± 0.075 mm). The shop-foreman's immediate reaction will be that something "unusual" has occurred which requires corrective action. Perhaps, the drill is running off-centre, perhaps it is improperly ground, perhaps the drill jig is worn. He will try to apply possible corrective steps. This concept is explained in Fig. 21.3. The circled points are those requiring corrective action. The concept of usual variation limits is carried into the control charts in the form of control limits.

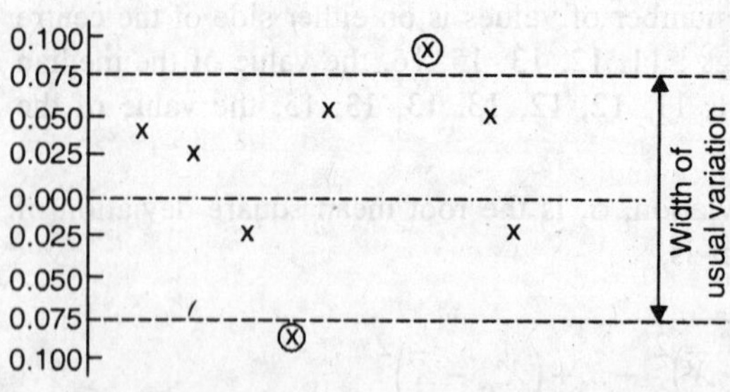

Fig. 21.3. A Control Chart.

21.4.1. Types of control charts:– As noted above, a control chart is a graphic presentation of a product's characteristics. This chart is based on the theory of "Sampling Inspection", See Art 21.5, according to which an adequate sample size drawn at random from a lot, represents the lot.

Control charts have the following functions:

(*a*) Indicate whether the process is under control or not?

(*b*) Detect and determine the extent of variations is a process from the established standard.

(*c*) Ensure product quality.

(*d*) Provide information regarding the proper selection of a process.

Corresponding to the two types of inspection methods, there are two basic types of control charts.

1. When the method of inspection is by variables, the most popular control charts are $\overline{X}$ and *R* charts.

2. When the method of inspection is by attributes, the most popular control charts are fraction or percent defective charts (also called P-charts).

To prepare the above two types of control charts, the approach is similar and it involves the following steps :

1. Select the appropriate quality characteristic to be studied.

2. Record data, taken on a required number of samples, (usually to 10 in number), each composed of an adequate number of parts.

3. Determine the control limits from these sample data.

4. Check if these control limits are economically satisfactory for the job. Are they too wide? Are they too narrow?

5. Plot the control limits on suitable graph papers. Start to record the results of production samples of proper size and at periodic intervals.

6. Take corrective action if the characteristics of the production samples exceed the control limits.

Many times, it is possible that when control limits are first being determined, the process is found to be "out of control" that is, quality characteristics of many samples exceed the control limits. In such cases, causes for the excessive sample variations can be traced down and eliminated. Steps 2 and 3 are repeated until the process becomes controlled.

Control Charts for Variables. These charts are also known as $\overline{X}$ -charts and *R*-charts, since these charts give the control limits for measures of central tendency and spread. The basic principle for computing measurements of control chart limits is similar to that for computing the frequency distribution process, that is, 3-sigma limits. 3-sigma has been chosen in place of 2 or 4-sigma, because, experience has proved 3-sigma value to be most useful and economical for control-chart applications. The reason is that so many frequency distributions, encountered in industry, tend towards normality and as long as only chance variations occur, there is considerable assurance that less than 3 parts in 1000 would fall outside these limits. Sometimes, 3-sigma limits are set for action alongwith 2-sigma limits for warning.

Formulae for computing the control limits for variables, are listed below :

(*i*) **When Range is used as Measure of Spread**

(*a*) **Average.** Lower Control Limit (*LCL*) = $\overline{\overline{X}} - 3\sigma_x$

$$\text{Centre line} = \overline{\overline{X}} \quad \text{...(21.9)}$$

$$\text{Upper Control Limits } (UCL) = \overline{\overline{X}} + 3\sigma_x$$

(*b*) **Range.** $LCL = \overline{R} - 3\sigma_R$

$$\text{Centre line} = \overline{R} \quad \text{...(21.10)}$$

$$UCL = \overline{R} + 3\sigma_R$$

(*ii*) **When Standard Deviation is used as Measure of Spread**

(*a*) **Average.** $LCL = \bar{\bar{X}} - 3\sigma_x$ ...(21.11)

Centre line $= \bar{\bar{X}}$

$UCL = \bar{\bar{X}} + 3\sigma_x$

(*b*) **Standard Deviation.** $LCL = \bar{\sigma} - 3\sigma_\sigma$

Centre line $= \bar{\bar{X}}$

$UCL = \bar{\sigma} + 3\sigma_\sigma$ ...(21.12)

where $\bar{\bar{X}}$ = Grand average

σ_x = Standard deviation of the sample average

$\bar{R}$ = average range

σ_R = standard deviation of the sample range

$\bar{\sigma}$ = average standard deviation

σ_σ = standard deviation of the sample standard deviation.

The use of the above formulae is very inconvenient, since it is a very tedious job to gather a series of samples of small size, determine the values for central tendency and spread for each of these samples and then do the laborious calculations involved in the above formulae. So, the following formulae, in terms of constants, are used in practice :

(*i*) **When Range is used as Measure of Spread**

(*a*) **Average.** $LCL = \bar{\bar{X}} - A_2\bar{R}$

Centre line $= \bar{\bar{X}}$ ...(21.13)

$UCL = \bar{\bar{X}} + A_2\bar{R}$

(*b*) **Range.** $LCL = D_3\bar{R}$

Centre line $= \bar{R}$ ...(21.14)

$UCL = D_4\bar{R}$

(*ii*) **When Standard Deviation is used as a Measure of Spread**

(*a*) **Average.** $LCL = \bar{\bar{X}} - A_1\bar{\sigma}$

Centre line $= \bar{\bar{X}}$ ...(21.15)

$UCL = \bar{\bar{X}} + A_1\bar{\sigma}$

(*b*) **Standard Deviation**

$LCL = B_3\bar{\sigma}$

Centre line $= \bar{\sigma}$ ...(21.16)

$UCL = B_4\bar{\sigma}$

The values of the constants $A_1, A_2, D_3, D_4, B_3,$ and B_4 are given in Table 21.1.

Table 21.1

Sample Size n	A_1	A_2	B_3	B_4	D_3	D_4
2.	3.759	1.880	0	3.658	0	3.268
3.	2.394	1.023	0	2.692	0	2.574
4.	1.880	0.729	0	2.330	0	2.282
5.	1.596	0.577	0	2.128	0	2.114
6.	1.410	0.483	0.003	1.997	0	2.004
7.	1.277	0.419	0.097	1.903	0.076	1.924
8.	1.175	0.373	0.169	1.831	0.136	1.864
9.	1.094	0.337	0.227	1.774	0.184	1.816
10.	1.028	0.308	0.273	1.727	0.223	1.777
11.	0.973	0.285	0.312	1.688	0.256	1.744
12.	0.925	0.266	0.346	1.654	0.284	1.717
13.	0.884	0.249	0.375	1.625	0.308	1.692
14.	0.848	0.235	0.400	1.599	0.329	1.671
15.	0.817	0.223	0.423	1.577	0.348	1.652

Fig. 21.4 shows an example of $\overline{X}$-chart and R-chart. These charts refer to outside diameters of shafts. In the $\overline{X}$ chart, average values of sample are plotted, instead of individual values, because sample averages tend to be more normally distributed than single values. In the $\overline{X}$-chart, each point refers to the average of measured values (diameters) of five shafts. The centre line of $\overline{X}$-chart represents the grand average of the subgroup average $\overline{\overline{X}}$ (eqn. 21.4). Here

$$\text{Control limits} = \overline{\overline{X}} \pm 3\sigma_x$$

where $$\sigma_x = \sqrt{\frac{\sum_1^k \left(\overline{X} - \overline{\overline{X}}\right)^2}{k}}$$

Here, k is the number of subgroups.

The range of values of the R-chart are obtained from the same subgroups of 5 samples each as were in the $\overline{X}$-chart. The centre line, $\overline{R}$, represents the average range of the subgroup ranges, and,

$$\text{Control limits} = \overline{R} \pm 3\sigma_R,$$

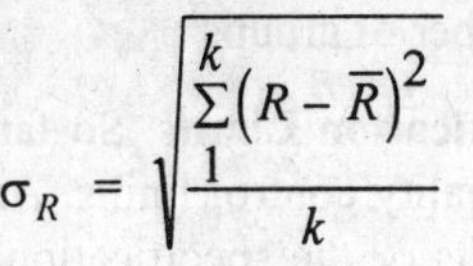

where $$\sigma_R = \sqrt{\frac{\sum_1^k \left(R - \overline{R}\right)^2}{k}}$$

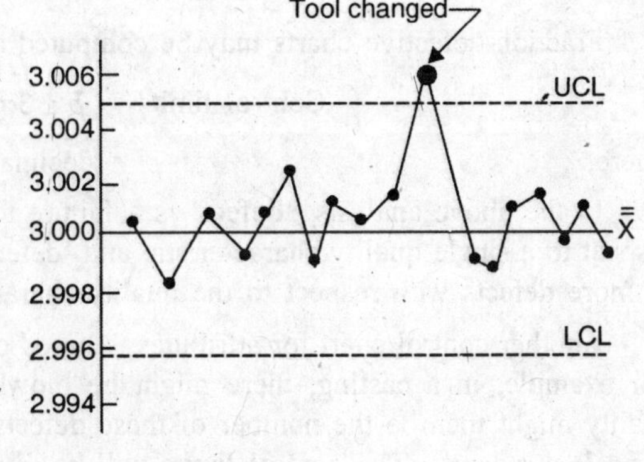

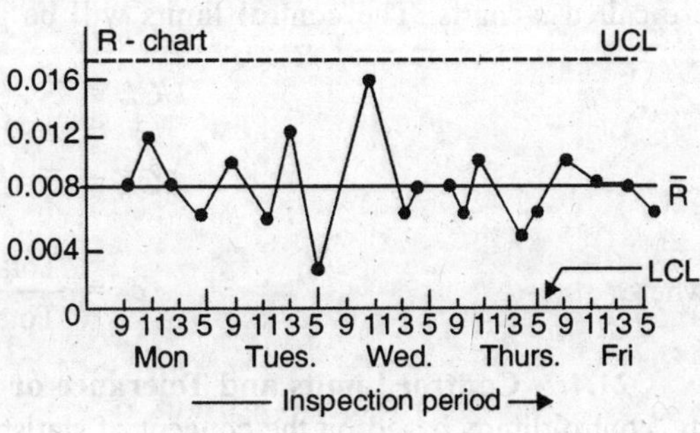

Fig. 21.4. $\overline{X}$ and R-Charts.

Control Charts for Attributes. These charts (*P*-charts) are used when parts are inspected by 'go' and 'not-go' gauges and a unit is classified simply as good or bad. The data is represented in terms of fraction or percent defective.

Fraction defective, $$p = \frac{\text{Number of defectives in a sample}}{n\text{ (sample size)}}$$

Percent defective, $$P = p \times 100$$

Thus, if sample size is 150 and if 6 units are found to be defective, then

$$p = \frac{6}{150} = 0.04$$

and Percentage defective, $$P = 0.04 \times 100 = 4$$

Percent and Fraction defective data can be described fully by means of values for their central tendency and spread. With percent defective data, the average is generally used as the measure of central tendency, expressed as percentage. The standard deviation is the measure of spread in that percentage. That is,

$$\overline{P} = \frac{\text{Total number of defectives found}}{\text{Total number of parts sampled } (\Sigma n)} \times 100 \qquad ...(21.17)$$

and $$\sigma\overline{P} = \sqrt{\frac{\overline{P}(100 - \overline{P})}{n}} \qquad ...(21.18)$$

$$\text{Then control limits} = \overline{P} \pm 3\sigma_{\overline{P}} \qquad ...(21.19)$$

Fraction-defective charts may be computed as below :

$$\text{Control limits} = \overline{p} \pm 3\sigma\overline{p} \qquad ...(21.20)$$

where, $\overline{p}$ = decimal value for average fraction defective.

In the above analysis, 'defect' is a failure to meet a requirement imposed on a unit with respect to a single quality characteristic and 'defective' refers to a defective unit containing one or more defects with respect to the quality characteristic under consideration.

Another control chart for attributes is based on the number of defects per unit of a product. For example, in a casting, there might be blowholes, cracks, sponginees, etc. A measure of quality might then be the number of these defects per unit. Control charts set up on this basis are called *c*-charts. The control limits will be given as,

$$UCL = \overline{c} + 3\sqrt{\overline{c}}$$

$$LCL = \overline{c} - 3\sqrt{\overline{c}}$$

where $$\overline{c} = \frac{\text{Total number of defects}}{\text{Total number of groups}}$$

21.4.2. Control Limits and Tolerance or Specification Limits. So far, we have discussed the control limits based on the concept of statistical quality control limits. Another type of limits we come across in manufacturing is the tolerance limits or the specification limits. These limits are not determined satistically but are fixed when the product is designed, keeping in view the use of the product, as discussed in chapter 9. Tolerance limits are the final test of whether a product will be acceptable or not. If the statistical control limits fall within the tolerance limits, Fig. 21.5 (*a*)), both engineering and customer specifications will be met. Such a case can happen when a product having wide tolerance limits, is being manufactured by a process having little variability. The process is in effect too good for the tolerance limits. The Mean and the Range can be varied considerably without exceeding the tolerance limits. When the statistical control limits fall outside the tolerance limits, (Fig. 21.5 (*b*)), lot of scrap will be produced. This can happen when the product having great variability. The process is too crude for the specifications. One remedy is that the tolerance limits should be relaxed in consultation with engineers or customers. If this can not happen or the process can not be changed, the remedy is to inspect each item. Those parts which fall within the specification limits will be accepted. The others can be retained for rework, salvage or scrap.

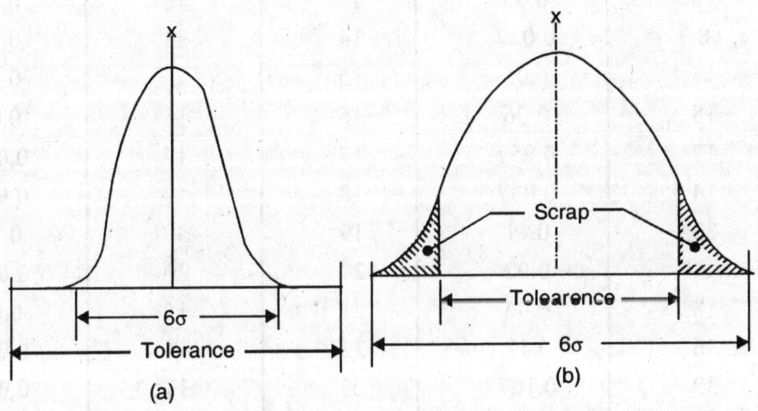

Fig. 21.5. Statistical Limits (*a*) within and (*b*) Outside Tolerance Limits.

The above discussion brings out another important function of control charts. They clearly show whether the process capability is compatible with the specifications or not. This compatibility can be evaluated in terms of 'Relative Precision Index (R.P.I)' which is defined as :

$$R.P.I. = \frac{\text{Tolerance}}{\text{Average Range}} \qquad ...(21.21)$$

The values for R.P.I. are given in Table 21.2

Table 21.2

Sample Size, n	*Low R.P.I.*	*Medium R.P.I.*	*High R.P.I.*
2	< 6	6 to 7	> 7
3	< 4	4 to 5	> 5
4	< 3	3 to 4	> 4
5 and 6	< 2.5	2.5 to 3.5	> 3.5

If R.P.I. is low, the process is unsatisfactory and scrap will result.

If R.P.I. is medium the process is satisfactory.

If R.P.I. is high, the process is very good and the parts will be produced with little variation.

21.4.3. Preliminary and Revised Control Limits. We have noted under Art. 21.4.1., that, when control limits are first being determined, the process may be found to be out of control. In such cases, causes for the excessive sample variations can be traced out and eliminated. Then, the control limits are established. The alternative method is :

Plot the preliminary data, with 3-sigma control limits. If some points fall outside the limits, eliminate these points and determine the revised 3-sigma control limits. This is explained with the help of data given in Table 21.3, which shows the number of defectives found in daily samples of 300, for 24 consecutive production days.

Table 21.3

Day	*No. of Defectives*	*Fraction Defective*	*Day*	*No. of Defectives*	*Fraction Defective*
1	15	0.05	13	12	0.04
2	8	0.027	14	21	0.07
3	15	0.05	15	6	0.02
4	18	0.06	16	15	0.05
5	17	0.057	17	17	0.057
6	14	0.047	18	17	0.057
7	33	0.11	19	39	0.13
8	6	0.02	20	20	0.067
9	18	0.06	21	15	0.05
10	36	0.12	22	14	0.047
11	32	0.107	23	17	0.057
12	23	0.077	24	18	0.06

Total number of parts sampled = 24 × 300 = 7200

Total number of defectives found = 446

$$\therefore \qquad \bar{p} = \frac{446}{7200} = 0.062$$

$$\therefore \qquad \sigma_{\bar{p}} = \sqrt{\frac{\bar{p}(1-\bar{p})}{n}}$$

$$= \sqrt{\frac{0.062 \times 0.938}{300}} = 0.014$$

$$\therefore \qquad UCL = \bar{p} + 3\sigma_{\bar{p}}$$

$$= 0.062 + 0.042 = 0.104$$

$$LCL = 0.062 - 0.042 = 0.020$$

It is clear from Table 21.3, that the data for days 7, 10, 11, and 19 fall outside *UCL*. Therefore, revised control limits will be set as below :

Now, Total number of parts sampled = 20 × 300 = 6000

$$\text{Total number of defectives} = 446 - (33 + 36 + 32 + 39)$$
$$= 306$$

$$\therefore \quad \bar{p} = \frac{306}{6000} = 0.051$$

$$\sigma_{\bar{p}} = \sqrt{\frac{0.051 \times 0.949}{300}} = 0.013$$

$$\therefore \quad \text{Revised } UCL = 0.051 + 0.039 = 0.09$$

$$LCL = 0.051 - 0.039 = 0.012$$

These revised limits reflect the variation due to chance causes.

These are used as standards for checking the future samples.

If any sample falls outside these limits, the cause will be assignable. It will be determined and corrected.

The plot of preliminary and revised control limits is shown in Fig. 21.6.

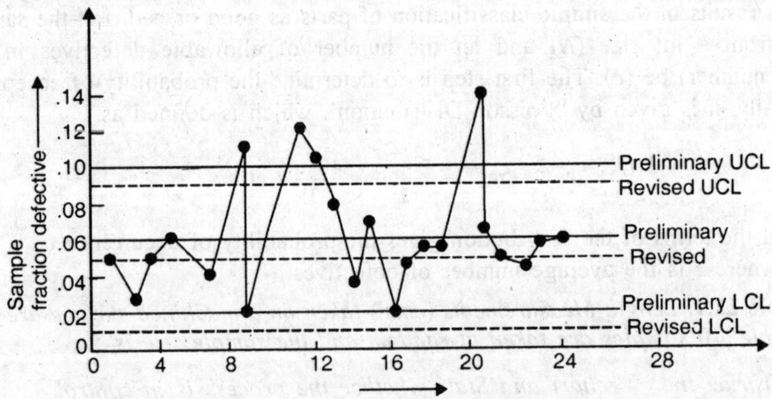

Fig. 21.6. Control Chart for Preliminary and Revised Control Limits.

21.5. SAMPLING INSPECTION

Quality Control Charts may be used in conjunction with either 100% inspection or sampling inspection. In 100% inspection, also known as "Screening", every part in a lot is inspected and good parts are sorted from bad parts. Only by this method, complete assurance can be obtained that all defective parts have been removed from a lot and that there is complete conformity of a product to its specifications. Even then, there are limitations of 100% inspection, as compared with effective sampling inspection carried out on a modern statistical basis. These limitations are :

1. It is expensive since every part must be checked individually, and so is time consuming.

2. Some defective will be overlooked through failure of an inspector or his instruments, due to monotony of repetitive inspection.

3. In cases, where destructive testing is required, 100% inspection is impossible.

In sampling inspection, one or more units are taken from a lot of the product and inspected. The sample is considered to be the representative of the lot. The lot is accepted, inspected further or rejected on the showing of the sample. Sampling inspection is necessary when the testing process destroyed the product and when the cost of 100% inspection can not be absorbed in the selling price of the product.

It is apparent that sampling inspection can not guarantee 100% perfect product to the customer. It will be used when the customer is willing to accept a few defective parts in order to secure lower cost. Also, if the sample size is small, the cost of inspection will be less. But it will increase "consumer's risk" because some bad lots of product will get accepted. Conversely, a large sample will increase the cost but decrease the "Consumer's risk". A "producer's risk" is also involved because some good lots of product may get rejected. So, any sampling inspection plan, involves a compromise between cost and risk.

Whereas "Acceptance Sampling" is applied to a batch of product after it is completed, control charts are applied during production. This means that it is possible with control charts to make adjustments in the manufacturing process if the recorded data indicates that corrections are needed. Acceptance sampling is a procedure in which a sample is drawn from a batch of parts in order to access the quality level of the batch and to determine whether the batch should be accepted or rejected. It is based on the statistical notion that the quality of a random sample drawn from a larger population will be representative of the quality of that population.

21.5.1. Acceptance Sampling by Attributes. As already mentioned, acceptance sampling by attributes results in the simple classification of parts as good or bad. Let the sample size be (n) drawn from a lot size (N) and let the number of allowable defectives in the sample (acceptance number) be (c). The first step is to determine the probability of accepting this lot. Mathematically, it is given by "Poisson Distribution', which is defined as,

$$e^{-z} - ze^{-z} - \frac{z^2 e^{-z}}{2!} - \frac{z^3 e^{-z}}{3!}, \text{ etc.}$$

Each of the terms of the distribution gives the probability of occurrence of 0, 1, 2, 3, etc. defectives, where z is the average number of defectives.

Example 21.1. *Length Measurements (cms.) taken on a machined surface are given in the following table. 10 samples are taken at random and the sample size is 5.*

(a) Construct the $\bar{X}$ chart and State whether the process is in control.

(b) Construct the R chart.

Solution.

Sample No.	Length of each measurement 1	2	3	4	5	ΣX	$\bar{X}$	R
1	11.33	11.18	11.28	11.33	11.25	56.37	11.274	0.15
2	11.30	11.25	11.35	11.15	11.18	56.23	11.246	0.20
3	11.13	11.38	11.23	11.23	**11.05**	56.02	11.204	0.33
4	11.23	11.28	**11.51**	11.40	**11.05**	56.47	11.294	0.46
5	11.23	11.30	11.25	11.28	11.20	56.26	11.252	0.10
6	11.28	11.30	11.28	11.15	11.18	56.19	11.238	0.15
7	11.15	11.20	11.23	11.33	11.35	56.15	11.230	0.20
8	11.30	11.20	11.25	11.20	**11.43**	56.38	11.276	0.23
9	11.28	11.33	**10.90**	11.13	11.40	56.04	11.208	0.50
10	11.23	11.25	11.10	11.35	11.40	56.33	11.266	0.30

(*a*) Mean of each sample $\bar{X}$ is given in column (4).

∴ The mean of the sample means, $\bar{\bar{X}} = \frac{\Sigma \bar{X}}{n} = \frac{112.488}{10}$

$= 11.25$ cm.

The range of each sample is given is column (5)

∴ $\bar{R} = \frac{\Sigma R}{n} = \frac{2.62}{10} = 0.262$

Now from Table 21.1, for sample size five, $A_2 = 0.577$

$$\therefore \quad \text{UCL} = \bar{\bar{X}} + A_2 \bar{R}$$

$$= 11.25 + 0.577 \times 0.262$$

$$= 11.25 + 0.15 = 11.40 \text{ cm.}$$

$$\text{Centre line} = \bar{\bar{X}} = 11.25 \text{ cm.}$$

$$\text{LCL} = \bar{\bar{X}} - A_2 \bar{R} = 11.10 \text{ cm.}$$

With this, the control chart can be drawn for $\bar{\bar{X}}$.

Now, since readings 11.43 and 11.51 lie above UCL and the readings 10.90, 11.05 and 11.05 lie below the LCL, the process is not under control.

(*b*) **R chart:** $\text{LCL} = D_3 \bar{R}$

$$\text{Centre line} = \bar{R}$$

$$\text{UCL} = D_4 \bar{R}$$

From T 21.1, for sample size five, $D_3 = 0$ and $D_4 = 2.114$

$$\therefore \quad \text{Centre line} = \bar{R} = 0.262$$

$$\text{UCL} = 2.114 \times 0.262 = 0.554$$

$$\text{LCL} = 0$$

With this data, the *R* chart can be constructed.

Example 21.2. *In the testing of Piston Pins, 10 samples were taken each of 100 (sample size). The number of defectives found were respectively 3, 2, 3, 5, 3,3, 2, 4, 3 and 2.*

(a) Construct the control charts for fraction defectives.

(b) Construct the percent defect control charts.

Solution.

(*a*) $\Sigma p = 30$

$$\therefore \quad \bar{p} = \frac{30}{10 \times 1000} = 0.03$$

Now, $\sigma_{\bar{p}} = \sqrt{\frac{\bar{p}(1-\bar{p})}{n}}$

$$= \sqrt{\frac{0.03 \times 0.97}{100}} = 0.017$$

∴ Control limits are : $\text{UCL} = \bar{p} + 3\sigma_{\bar{p}}$

$$= 0.03 + 3 \times 0.017 = 0.081$$

$$\text{LCL} = \bar{p} - 3\sigma_{\bar{p}} = -0.0219$$

Since the negative fraction defective is meaningless = 0

(*b*) Now percent defective (Mean)

$$\overline{P} = 100 \times \overline{P} = 3$$

$$\sigma_{\overline{P}} = \sqrt{\frac{\overline{P}(100-\overline{P})}{n}}$$

$$= \sqrt{\frac{3 \times 97}{100}} = 1.7$$

∴ Control limits are :

$$\text{UCL} = \overline{P} + 3\,\sigma_{\overline{P}} = 3 + 5.1 = 8.1$$

$$\text{LCL} = \overline{P} - 3\,\sigma_{\overline{P}} = 3 - 5.1 = -2.1 = 0$$

$$\text{Centre line} = \overline{P} = 3$$

With this, the control charts can be constructed. Since all the points lie within limits, the process is under control.

Example 21.3. *Twenty pistons were selected at random from a process. The number of defects observed in each piston were respectively: 3, 4, 6, 3, 4, 6, 7, 3, 4, 7, 6, 5, 4, 3, 4, 6, 3, 4, 9, 9. Construct the C charts (number of defects per unit).*

Solution.

$$\text{Total } C = \Sigma C = 100$$

∴

$$\overline{C} = \frac{\Sigma C}{n} = \frac{100}{20} = 5$$

∴ Control charts are :

$$\text{UCL} = \overline{C} + 3\sqrt{\overline{C}}$$

$$= 5 + 3\sqrt{5} = 11.708$$

$$\text{LCL} = 5 - 6.708 = -1.708 = 0$$

$$\text{Centre line} = \overline{C}$$

With this, the control charts can be constructed.

Example 21.4. *A lot of 1000 units contain 50 defectives. A random sample of 4 units is taken. Calculate the Poisson probabilities that the sample will contain exactly 0, 1, 2 and 3 defectives.*

Solution. Poisson probabilities are :

$$e^{-z}\left[1 + z + \frac{z^2}{2!} + \frac{z^3}{3!}\right] \text{ for 0, 1, 2, and 3 defectives.}$$

$$z = \frac{50}{1000} \times 4 = 0.2$$

Hence probabilities are $e^{-0.2}\left[1 + 0.2 + \frac{0.04}{2} + \frac{0.008}{6}\right]$

$$= 0.819\left[1 + 0.2 + \frac{0.04}{2} + \frac{0.008}{6}\right]$$

∴ Probabilities for 0, 1, 2 and 3 defectives are : 0.819, 0.1638, 0.01638, 0.00109 respectively.

Operating Characteristic Curve (O.C.C.). The graphical relationship between percentage defective in the lots being submitted for inspection and the probability of acceptance is termed as "Operating Characteristic" of a particular sampling plan. An operating characteristic curve for a particular combination of n and c shows how well the given sampling plan discriminate between good and bad lots. Fig. 21.7 shows a typical O.C.C. for a sampling plan with $n = 50$ and $c = 1$. It gives a clear picture about the probability of acceptance of a lot for various values of percent defectives in the lot. The probability of acceptance of a lot is high for low values of actual percentage defectives and it is low for high values of actual percentage defectives. For example, if the actual percent defectives in a lot are 2, then the probability of its getting accepted is about 73% and that of its getting rejected is about 27%. On the other hand, if the actual percent defectives in a lot are 6, then the probability of its getting accepted is about 17% (quite low). This is the desirable feature of a sampling plan. That is, if the actual quality of a lot is good, there should be high probability of its acceptance, but, if the actual quality is poor, it is desired that the probability of its acceptance must be low.

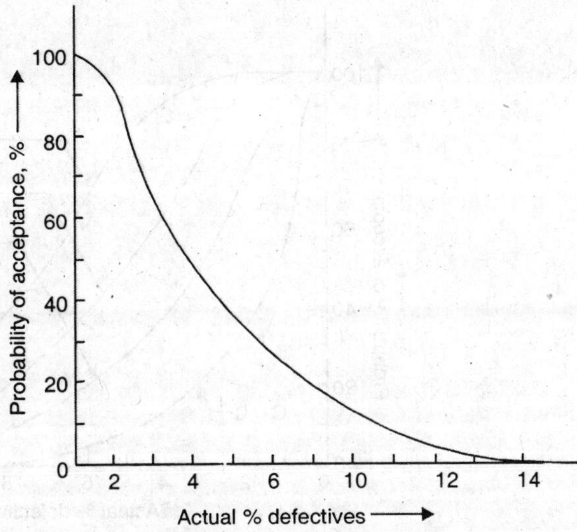

Fig. 21.7. A Typical O.C. Curve ($n = 50$, $c = 1$)

The discriminating power of a sampling plan is greatly dependent upon the sample size. Fig. 21.8 shows typical O.C. curves for sample sizes (n) of 100, 200 and 300 with the acceptance

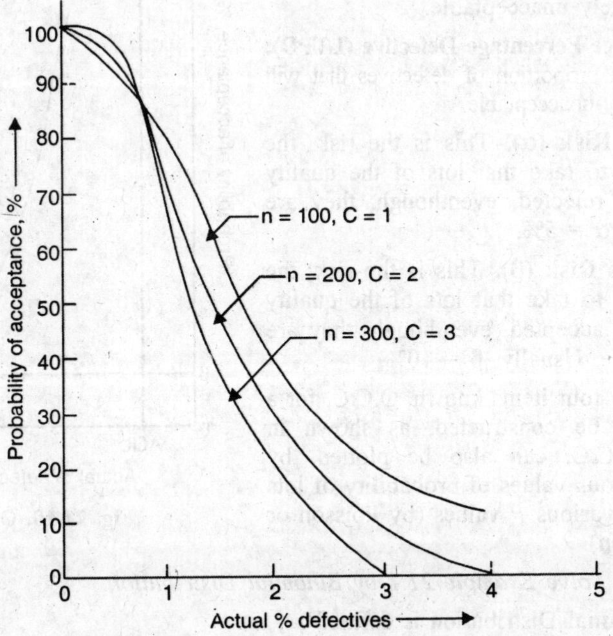

Fig. 21.8. Typical O.C. Curves for Different Sample Sizes. (n).

numbers (*c*) proportional to sample size. It is clear that as the sample size increases, the curve becomes steeper, that is, more discriminating. Fig. 21.9 shows O.C. curves with same sample size (*n*) but different values of acceptance number, (*c*). It is clear that as the acceptance number decreases the sampling plan becomes tighter, that is, more discriminating.

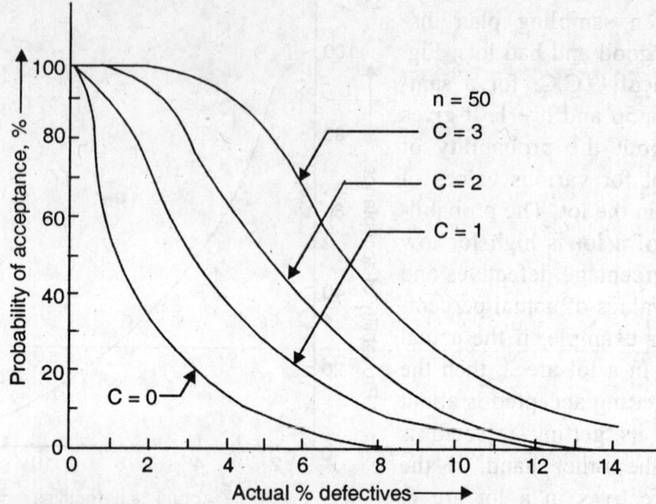

Fig. 21.9. Typical O.C. Curves n = 50, but c increasing 1 to 3.

Construction of O.C. Curve. To develop a sampling plan for acceptance sampling, an appropriate O.C. curve must be selected. To construct an O.C. curve, an agreement has to be reached between the producer and the consumer on the following four points :

1. **Acceptable Quality Level (AQL).** This is the maximum proportion of defectives that will make the lot definitely unacceptable.

2. **Lot Tolerance Percentage Defective (LTPD):** This is the maximum proportion of defectives that will make the lot definitely unacceptable.

3. **Producers Risk (α).** This is the risk, the producer is willing to take that lots of the quality level AQL will be rejected, eventhough, they are acceptable. Usually α = 5%.

4. **Consumer's Risk (β).** This is the risk, the consumer is willing to take that lots of the quality level LTPD will be accepted, eventhough, they are actually unacceptable. Usually, β = 10%.

With the above four items known, O.C.C. for a sampling plan can be constructed, as shown in Fig. 21.10. An O.C.C. can also be plotted, by determining the various values of probability of lots being accepted, for various *P* values (by Poisson or Binomial Distribution).

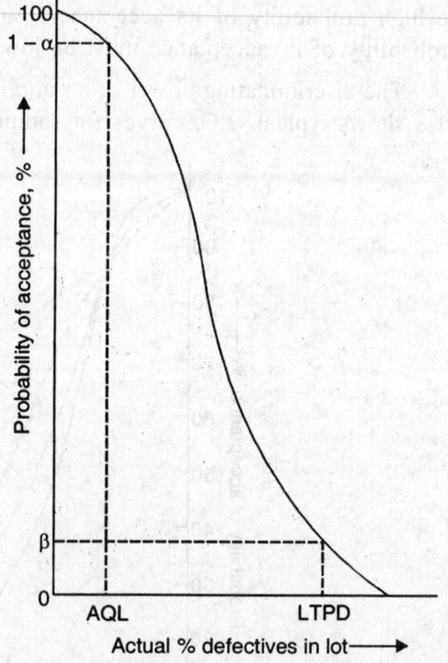

Fig. 21.10. O.C. Curve.

Example 21.5. *Solve Example 21.1 by Binomial Distribution.*

Solution. Binomial Distribution is defined as,

$$(p+q)^n = q^n + \left({}^nC_1\right)q^{n-1}p + \left({}^nC_2\right)q^{n-2}p^2 + \left({}^nC_3\right)q^{n-3}\cdot p^3 + \ldots + p^n$$

Here p = probability of an event happening.

q = probability of an event not happening.

n = sample size.

In the present problem, there are 50 defectives in a lot of 1000 pieces. Therefore, the probability that a single component in a continuous process is defective is,

$$p = \frac{50}{1000} = 0.05$$

$$\therefore \quad q = 1 - p = 0.95$$

Here $n = 4$

∴ Probabilities are :

$$(0.95)^4 + 4 \times (0.95)^3 \times 0.05 + 6 \times (0.95)^2 \times (0.05)^2 + 4 \times (0.95)^1 \times (0.05)^3$$

$$= 0.8145 + 0.1715 + 0.0135 + 0.000475$$

Therefore, the probabilities for 0 defective, 1 defective, 2 defective and 3 defective are, respectively : 0.8145, 0.1715, 0.0135 and 0.000475.

Sampling Plans. Numerous sampling plans are possible. A sampling plan is designed by applying the law of chance to such factors, as lot size, sample size, average outgoing quality, consumer's risk, etc., to determine the "acceptance number" (c) for the sample. Fortunately, sampling schemes have been tabulated and published (Dodge-Roming plans). The basis of any plan is that the inspection cost should be minimum. In D.R. plans, β is taken as 10%. If N (lot size) and other statistical parameters are known, n and c can be read directly from the tables The various sampling plans used in practice :

1. **Single Sampling Plan.** Under this plan, the number of defective in a lot are found. Then the plan is :

If number of defectives is $\leq c$, accept the lot

If number of defectives is $> c$, reject the lot.

In one variation of the single sampling plan, if the number of defectives exceeds the acceptance number, the remainder of the lot is 100% inspected; defective units are replaced with good ones and the lot is accepted. The advantage of single sampling plan is quick decision. However, it needs a large sample and also there is the inherent probability of error in judging the total lot from one sample.

2. **Double Sampling Plan.** To a layman, the idea of "giving it a second chance" is more appealing, as compared to the idea of "sudden death" in single acting plan. In double sampling plan, a first sample is checked. If the acceptance number is not exceeded, the lot is accepted. But if it is exceeded, a second sample is taken at random from the lot, and checked. If the acceptance number for the first and second samples combined is not exceeded, the lot is accepted, otherwise it is rejected.

In one variation, of double sampling plan, there are both acceptance and rejection numbers for a sample. An initial sample is taken and the number of defectives found. This number is then compared to two numbers, c_1 (acceptance number) and c_2 (rejection number), c_2 being greater than c_1.

If number of defectives is $\leq c_1$ accept the lot.

If number of defectives is $> c_2$

If number of defectives lies between c_1 and c_2, take another sample. Now the problem is treated as a single sampling plan. If the total number of defectives in the combined sample

$(n_1 + n_2)$, is less than or equal to c_3 (new acceptance number), accept the lot, otherwise, reject it.

The advantage of double sampling plan is that smaller samples can be taken than single sampling plan. Also, if the whole lot is accepted on the basis of first sample, there will be lot of saving in total inspection.

3. **Sequential Sampling.** In sequential sampling, the idea of double sampling is carried further. For a sequence of samples, there are both acceptance and rejection numbers. After each sample is inspected, the cummulative results are analysed and a decision taken : (*a*) accept the lot (*b*) reject the lot (*c*) delay the decision, that is, take another sample. This process is continued until stage is reached where only two possible decision can be made : acceptance or rejection of the lot. Under this scheme, still smaller samples can be taken. Sequential-Sampling is explained with the help of Table 21.4 and Fig. 21.11.

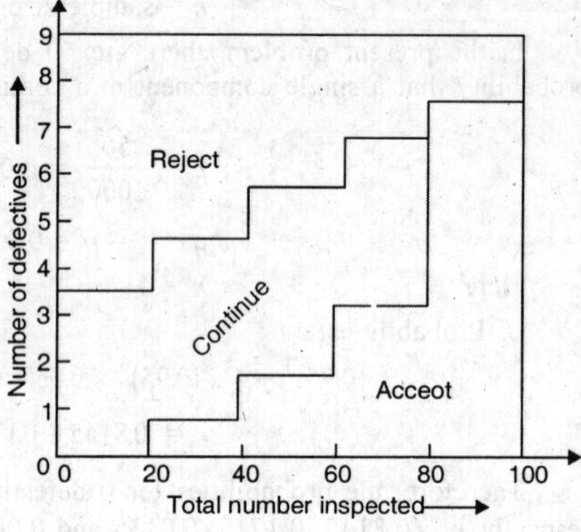

Fig. 21.11. Sequential Sampling.

Table 2.4

Sample Number	*Sample size*	*Commulative Sample Size*	*Commulative Acceptance Number*	*Commulative Rejection Number*
1	20	20	0	4
2	”	40	1	5
3	”	60	2	6
4	”	80	4	7
5	”	100	7	8

(*i*) *First sample is taken.*

If number of defectives is zero, accept the lot.

If number of defectives is ≥ 4, reject the lot.

If number of defectives lies between 0 and 4, take another sample.

(*ii*) *Second sample is taken.*

If number of defectives in the commulative sample is ≤ 1,

accept the lot

” ” ” ” ” ” ” ” ≥ 5, reject the lot

” ” ” ” ” ” ” ” lies between 1 and 5,

take another lot

This will be continued and at the sample number 5, the band of indecision disappears and acceptance or rejection of the lot can be finally decided.

Size of Sample. The determination of sample size is often a function of common sense. In general, it depends upon the economy of gathering the samples, the time available for inspection and the convenience of gathering the samples. The larger the sample, the more representative will be the results obtained from it of the characteristies of the lot. Big samples are better but not proportionately better. A sample twice as big as another sample is not twice as reliable. Moreover, larger samples are more costly to gather and inspect. Therefore, a compromise must be reached to maintain quality control costs at an acceptable level. Industry often uses a sample size of 4 or 5 units in controlling variables with mean and range charts. Sample sizes of about 100 or more are often used in controlling attributes. Eastman Kodak gives the following relation:

$$n = \sqrt{2N}$$

Frequency of Sampling. The question of frequency of sampling is based on judgement and common sense of the inspector. If the process is well under control from analysis of sample data and it remains so for several weeks, the frequency of sampling may be reduced. If the schedule has been one sample every hour, the quality control manager may decide to take one sample each in the morning and in the evening, so long as the process is under control. However, if the process is not under control, more frequent sampling may have to be done to ascertain the rate of going out of control of the process and also to determine the reasons for this. Therefore a compromise on frequency of sampling is : Attainment of acceptable level of quality at a moderate cost.

Example 21.6. *In a single sampling plan, the following data are known.*

$$\text{Lot size} = 500, \quad \text{Sample size} = 71$$

$$\text{Acceptance number} = 1 \quad AQL = 0.005$$

$$LTPD = 0.05$$

Determine produce's risk and consumer's risk and draw O.C.C.

Solution.

(*a*) **Producer's Risk**

The mean number of defects expected per sample,

$$z = np = n \times AQL = 71 \times 0.005 = 0.355$$

To determine the probability of finding 1 or less defective, use Poisson Distribution with $z = 0.355$. This is given as,

$$e^{-z} + ze^{-z} = 0.701 + 0.249 = 0.95$$

$\therefore$ From Fig. 21.8, $\quad 1 - \alpha = 0.95$

$\therefore$ Producer's risk, $\quad \alpha = 0.05 = 5\%$

(*b*) **Consumer's Risk**

Here, $\quad z = np = n \times LTPD$

$$= 71 \times 0.05 = 3.55$$

$\therefore$ Probability for 1 or less defective is,

$$e^{-z} + ze^{-z} = e^{-3.55} + 3.55 \times e^{-3.55} = 0.0287 + 0.102$$

$$= 0.1307$$

$\therefore$ Consumer's risk, $\quad \beta = 13.07$

So, knowing *AQL*, LTPD, α and β O.C.C. can be drawn, as shown in Fig. 21.8.

21.6. AVERAGE OUTGOING QUALITY LIMIT (AOQL)

If rejected lots were always scrapped then the average quality of the products would lie about halfway between *AQL* and LTPD. However, if rejected lots are 100 percent inspected and the defective units in these lots be replaced or repaired, the average outgoing quality can be improved. The worse the quality of lots submitted the better will be average outgoing quality. If majority of the lots are bad, nearly all of them will be rejected. After 100 percent inspection, they will end up nearly perfect. If these are averaged with the good lots that passed plus the few bad lots that got through, we will get a better quality than if most lots were just good enough to pass. With this scheme, a sampling plan gives definite assurance about the average outgoing quality. This is very important when average outgoing quality protection is desired. For any sampling plan, we can plot a curve for the *AOQ* for any level of incoming quality. Let us see how to draw *AOQ* curve for the sampling plan with $n = 50$ and $c = 1$, as shown in Fig. 21.7. If a lot is submitted with percent defective of 2%, these lots would be accepted 73 percent of the time and rejected 27 percent of the time. These lots would be 100 percent inspected and the defective units in these lots would be replaced or repaired.

(*i*) *AOQ* of these lots would be = 27 percent × 0 percent + 73 percent × 2 percent
= 1.46 percent defective.

In this manner, *AOQ* for lots with different values of percent defective, can be determined by referring to Fig. 21.7.

(*ii*) If incoming percent defective is 1%.

AOQ of lots = 12 percent × 0 percent + 88 percent × 1 percent
= 0.88 percent defective.

(*iii*) If incoming percent defective is 3%,

AOQ of lots = 46 percent × 0 percent + 54 percent of 3 percent
= 1.62 percent defective

(*iv*) If incoming percent defective is 5%,

AOQ of lots = 73 percent × 0 percent + 27 percent of 5 percent
= 1.35 percent defective.

(*v*) If incoming percent defective is 8%,

AOQ of lots = 92 percent × 0 percent
+ 8 percent of 8 percent
= 0.64 percent defective

and so on.

With these values, *AOQ* curve is plotted in Fig. 21.12. It is clear that the *AOQ* will never exceed 1.62 percent, whatever may be the incoming quality. This level is called *AOQL*.

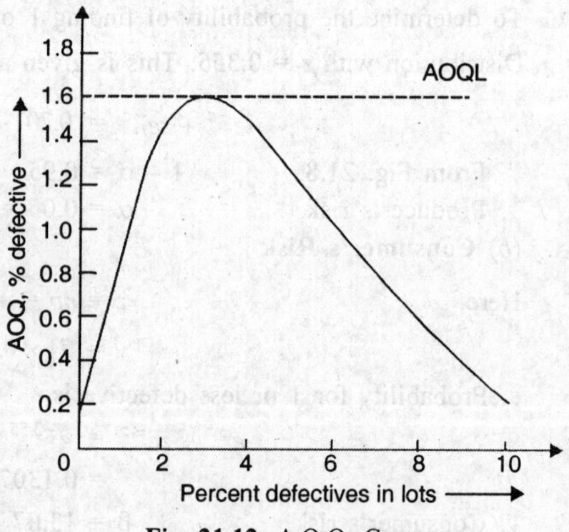

Fig. 21.12. A.O.Q. Curve

Example 21.7. *The data below shows the number of defectives over period of 20 days in a fixed sample size of 200. Determine, whether the data exhibit statistical control ? Evaluate the preliminary and revised control limits for the process.*

Day	No. of defectives	Fraction defective	Day	No. of defectives	Fraction Defective
1	10	0.050	11	21	0.105
2	15	0.075	12	15	0.075
3	10	0.050	13	8	0.040
4	12	0.060	14	14	0.070
5	11	0.055	15	4	0.020
6	9	0.045	16	10	0.050
7	22	0.110	17	11	0.055
8	4	0.020	18	11	0.055
9	12	0.060	19	26	0.130
10	24	0.120	20	13	0.065

Solution. Average fraction defective $= \dfrac{\text{Total number of defectives}}{\Sigma n}$

$$\therefore \qquad \bar{p} = \frac{262}{200 \times 20} = 0.0655$$

$$\sigma_{\bar{p}} = \sqrt{\frac{\bar{p}(1-\bar{p})}{n}}$$

$$= \sqrt{\frac{0.0655 \times 0.9345}{200}} = 0.0175$$

$$\therefore \qquad UCL = \bar{p} + 3\sigma_{\bar{p}}$$

$$= 0.0655 + 0.0525 = 0.118$$

$$LCL = \bar{p} - 3\sigma_{\bar{p}}$$

$$= 0.013$$

Data for days 10 and 19 are out of control. Therefore, for revised control limits,

Total number of defectives = 262 – (24 + 26) = 212

$$\therefore \qquad \bar{p} = \frac{212}{200 \times 18} = 0.0588$$

$$\sigma_{\bar{p}} = \sqrt{\frac{0.0588 \times 0.9412}{200}}$$

$$= 0.0166$$

$$\therefore \qquad UCL = 0.0588 + 0.0498 = 0.109$$

$$LCL = 0.009$$

∴ Preliminary Control Limits are :

$$UCL = 0.118$$
$$LCL = 0.013$$

and Revised Control Limits are :

$$UCL = 0.109$$
$$LCL = 0.009$$

Example 21.8. *A set of assemblies are subjected to inspection. A set consists of 5 assemblies and there are 15 sub-groups. The inspection data for a particular day is given below. Find the control limits for a c-chart.*

Group No.	*No. of defects*	*Group No.*	*No. of defects*
1	*77*	*9*	*45*
2	*64*	*10*	*77*
3	*75*	*11*	*59*
4	*93*	*12*	*54*
5	*45*	*13*	*84*
6	*61*	*14*	*40*
7	*49*	*15*	*92*
8	*65*		

Solution. Total number of defects = 980

$$\therefore \quad \bar{c} = \frac{980}{15} = 65.33$$

$$UCL = \bar{c} + 3\sqrt{\bar{c}}$$
$$= 65.33 + 24.25 = 89.58$$
$$LCL = \bar{c} - 3\sqrt{\bar{c}} = 41.08$$

The data for group numbers, 4, 14, and 15 fall outside these control limits. Therefore, for revised control limits.

$$\bar{c} = \frac{980 - (93 + 40 + 92)}{12} = 62.917$$

$$\therefore \quad \text{Revised } UCL = 62.917 + 3 \times 7.932$$
$$= 86.713$$

and
$$LCL = 62.917 - 23.796$$
$$= 39.121$$

Example 21.9. *The following data are the result of life-tests on 20 samples of 5 each, carried on bulbs manufactured by a Company. The values are in hours.*

1. Is the process in a state of statistical control?

2. Assuming that assignable causes could be discovered and eliminated for all points showing the process to be out of control, what is your best estimate to the capability of the Procéss?

Data. *The formula for control limits based on analysis, past data and various statistics are*

Statistics	*Limits*
Average using σ	$\overline{\overline{X}} \pm A_1\overline{\sigma}$
Average using R	$\overline{\overline{X}} \pm A_2\overline{R}$
Standard deviation	$B_3\,\overline{\sigma}, B_4\,\overline{\sigma}$
Range	$D_3\overline{R}, D_4\overline{R}$

and the values of factors for computing control chart lines for a sample size of 5 are as follows :

Chart of Average : $A = 1.342$ $A_1 = 1.596$ $A_2 = 0.577$

Chart for Ranges : $d_2 = 2.326,$ $d_3 = 0.864$ $D_1 = 0$

$D_2 = 4.918,$ $D_3 = 0,$ $D_4 = 2.115$

Sample No.	$\overline{X}$	*R*	*Sample No.*	$\overline{X}$	*R*
1	*3290*	*560*	*11*	*3220*	*580*
2	*3180*	*410*	*12*	*3590*	*670*
3	*3350*	*200*	*13*	*4270*	*480*
4	*3470*	*300*	*14*	*4040*	*250*
5	*3080*	*90*	*15*	*3580*	*170*
6	*3240*	*650*	*16*	*3500*	*670*
7	*3260*	*890*	*17*	*3570*	*440*
8	*3310*	*410*	*18*	*3560*	*660*
9	*3640*	*1120*	*19*	*2740*	*560*
10	*4110*	*520*	*20*	*3200*	*590*

Solution. $\sum\overline{X} = 69200, \sum R = 10320$

$$\therefore \quad \overline{\overline{X}} = \frac{69200}{20} = 3460$$

$$\overline{R} = \frac{10320}{20} = 516$$

For $\overline{X}$ *Chart :*

$$UCL = \overline{\overline{X}} + A_2\overline{R}$$
$$= 3460 + 0.577 \times 516$$
$$= 3757.732$$
$$LCL = 3460 - 297.732$$
$$= 3162.268$$

For R Chart:

$$UCL = D_4\overline{R}$$
$$= 2.115 \times 516 = 1091.34$$
$$LCL = D_4\overline{R}$$
$$= 0 \times \overline{R} = 0$$

Revised Control Charts

$\overline{X}$ for samples No. 5, 10, 13, 14, and 19 are out of control.

$$\therefore \quad \overline{\overline{X}} = \frac{692000-(3080+4110+4270+4040+2740)}{15}$$

$$= 3397.33$$

$\therefore$ R for sample No. 9 is out of limits for R-chart. Therefore, revised.

$$\overline{R} = \frac{10320-1120}{19} = 484.21$$

$\therefore$ For $\overline{X}$-chart :

$$UCL = 3397.33 + 279.389 = 3676.719$$

$$LCL = 3397.33 - 279.389 = 3117.94$$

For R-Chart :

$$UCL = 2.115 \times 484.21 = 1024.104$$

$$LCL = \mathbf{0}$$

Example 21.10. *A single sampling plan has a sample size of 50 and a rejection number 2. Draw the operating characteristics of this sampling plan and determine the producer's risk at 2% defective and the consumer's risk at 8% defective. Draw also the AOQ curve of this plan. Use 1%, 2%, 4%, 6% and 8% defective as points for construction of the above curves.*

Solution. Using Poisson's distribution, the probabilities for various defective are :

$$e^{-z} + ze^{-z} + \frac{z^2e^{-z}}{2!} \text{ and so on.}$$

Here, sample size $n = 50$

rejection number $= 2$

$$AQL = 0.02$$

$$LTPD = 0.08$$

(*a*) *Producer's Risk :*

The probability of finding 1 or less defective is,

$$e^{-z} + ze^{-z}$$

Here z = mean number of defectives expected per sample

$$= np = n \times AQL$$

$$= 50 \times 0.02 = 1$$

$$\therefore \quad \text{Probability} = e^{-1} + 1 \times e^{-1}$$

$$= 0.7358$$

$$\therefore \quad 1-\alpha = \mathbf{0.7358}$$

$$\therefore \quad \text{Producer's risk} = \mathbf{26.42\%}$$

(*b*) *Consumer's Risk :*

Here $\quad z = np = n \times LTPD$

$$= 50 \times 0.08 = 4$$

$$\therefore \quad \text{Probability} = e^{-4} + 4 \times 10^{-4}$$

$$= 0.09158$$

$\therefore$ Consumer's risk, $\beta = 9.158\%$

AOQ curve

(*i*) Incoming percent defective = 1%

$$\therefore \text{ Average number of defective } z = \frac{50 \times 1}{100} = 0.5$$

$\therefore$ Probability of finding less than 2 defectives is,

$$e^{-z} + ze^{-z} = 0.910$$

$\therefore$ *AOQ* of lot = 9% of 0 percent + 91% of 1%

= **0.91**

(*ii*) Incoming percentage defective=2%

Probability of acceptance = 73.58% (found above)

rejection = 26.42%

$\therefore$ *AO*Q of lot = 26.42 percent × 0 percent + 73.58 percent × 2 percent

= **1.4716**

(*iii*) Incoming percentage defective = 4%

$$\therefore \quad z = \frac{50 \times 4}{100} = 2$$

$$\therefore \quad \text{Probability of acceptance} = e^{-2} + 2 \times e^{-2}$$

= 40.6%

$\therefore$ Probability of rejection = 59.4%

$\therefore$ *AOQ* of lot = 59.4 percent × 0 percent + 40.6 percent × 4 percent

= **1.624**

(*iv*) Incoming percentage defective = 6%

$$\therefore \quad z = \frac{50 \times 6}{100} = 3$$

$$\therefore \quad \text{Probability of acceptance} = e^{-3} + 3 \times e^{-3}$$

= 19.9%

Probability of rejection = 81.1%

$\therefore$ *AOQ* of lot = $0 + 19.9\% \times 6\%$

= **1.194**

(v) Incoming percentage defective = 8%

$$z = \frac{50 \times 8}{100} = 4$$

$\therefore$ Probability of acceptance = 9.158% (found above)

rejection = 90.842%

$\therefore$ AOQ of lot = 0 + 9.158% × 8%

= **0.733**

With the help of above data, AOQ curve can be drawn.

21.7. TOTAL QUALITY MANAGEMENT (TQM)

Introduction. The concepts of 'quality' and of 'quality control' have been discussed. As already stated, the term 'Quality' may be defined as the totality of features and characteristics of a product or service that bear on its ability to satisfy stated or implied needs. The term 'Quality Control' QC' may be defined as the operational techniques and activities that are used to fulfil requirements for quality. In 'QC' the aim is to inspect to prevent defects from occurring and not to find defects after they have occurred. The QC programme would be geared to root out the problems that can cause defective products during production.

Quality evolution phases have been shown in Fig. 21.13. **Operator inspection and Foreman verification phases** focussed on part defect detection through post production inspection. It is concerned with conforming to standards/specifications and sorting out rejects.

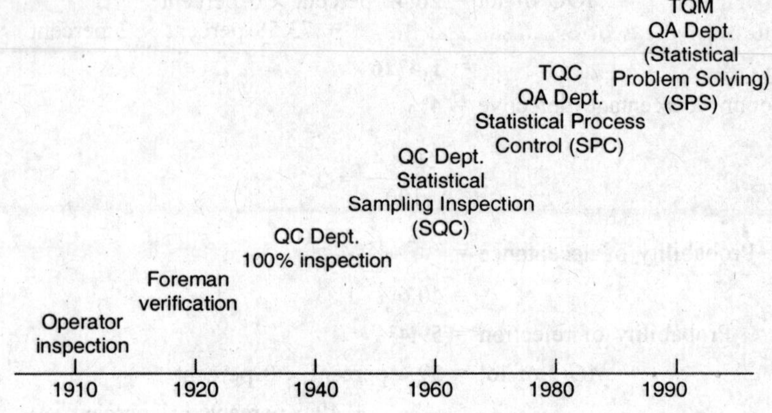

Fig. 21.13. Quality Evolution Phases.

The Statical Quality Control, SQC, concept has been extensively refined during the last 50 years. SQC can be broken into two areas : Acceptance Sampling and Statistical Process Control (SPC). As discussed in Art. 21.5. Acceptance Sampling is the basis of traditional inspection based quality control approaches and is essentially retrospective. It deals with problems of quality after they have been built in. It is still used where there is no guarantee of incoming quality from previous stages or from suppliers. The techniques uses either 100 per cent inspection or sampling inspection where only an 'acceptable number' of defects get through This percentage is set using an O.C.C.

On the other hand, SPC, attempts to monitor and improve things while they are still happening. Statistical methods are employed to detect trends that will ultimately lead to the production of defective parts. The aim is to identify the causes for such trends and then taken the necessary corrective actions before any defective parts are produced. The essence of SPC is that 'if the process is under control, any variations are due only to random causes'. The basic way of checking this out is via Control Charts. Lack of control of a process is indicated by the following conditions :

1. A point falling outside the control lines (UCL and LCL).
2. A run of more than 8 points on the same side of the centre line.
3. A trend with atleast 7 points.
4. 13 out of 15 points falling on the same side of the centre line.

When any of these conditions exists, then corrective actions need to be taken.

Quality Assurance Phase. Quality Assurance, *Q*A, recognizes that inspection is not the answer and that "the entire manufacturing process must be committed to meeting the quality needs of the design". It incorporates all those planned and systematic actions necessary to provide adequate confidence that a product or a service will satisfy the requirements for quality. For effectiveness, *QA* usually requires a continuing evaluation of factors that affect the adequacy of design or specifications for intended application as well as verifications and audits of production, installation and inspection operations. *QA* focusses on procedure compliance and product conformity to specifications through production and operations management, often using SPC as a tracking tool.

The concept of "*QA*" originated in early 1960s. However, by 1969, it had degenerated into an increasingly dogmatic, bureaucratic and specialized function. It mainly became an increasingly bureaucratic set of rules and procedures which suppliers needed to go through to obtain certification. Due to all this, a good idea in principle, the concept of *QA* became a book of rules rather than a live principle.

Total Quality Control, TQC. TQC expands the *QA* philosophy beyond manufacturing operation into other areas of organisational life.

TQC incorporates many of the tools used in *QA*, but the aim is to develop long-term solutions to the problem rather than respnd to short-term variations. Emphasis on direct cost reduction and efficiency is replaced in favour of the pursuit of quality through the elimination of waste and nonvalue added procedures and assuring continuous improvement through the refinement and expansion of quality control systems and procedures. The utilization of computer integrated manufacturing systems (CIMS) or Integrated Manufacturing Production Systems (IMPS) and just in-time (JIT) operations management methods also facilitate this control. Management of the manufacturing system then extends outside the company gate to tie in suppliers, distributors and customers in the "chain-of-quality". According to A.V. Fiegenbaum, TQC is concerned with the integration of all the efforts in the organisation towards quality improvement, quality development and quality maintenance to meet full customer satisfaction at economical levels. Productivity needs to be improved by enhancing quality of products, services and all other activities. *TQ* efforts enhance quality of worklife, employee satisfaction through participation and involvement and consequently the image of the organisation. In TQC, the emphasis is on making it right the first time, and placing the quality in the hand of the workers. To achieve TQC, the following steps have been recommended :

1. **Implementation of Taguchi Methods.** These methods involve the use of a truncated experimental design to determine which process input variables have the greatest effect on the quality or the precision and which have the least effect. Then, those variables which have the greatest effect on precision are set at levels so as to minimize their effect on process variability. The factors having the least effect on precision are adjusted or recentred to achieve the process aim. Thus, the aim of Taguchi methods is to minimize or dampen the effect of the input causes of variability and hence to reduce the total process variability. For example, in machine tools, the input process variables are : machine alignment, the design and rigidity of the workholding device and its set up, the accuracy of the cutting tools, the design of the product, the temperature and the operating parameters (speed, feed and depth of cut etc.). The use of Taguchi methods will determine which of these input variables have the greatest and the least effect on the accuracy and precision of the process.

2. Giving the responsibility of quality to the worker. He is also authorised to stop the work if anything goes wrong. The worker is given the analysis tools he needs to find and expose the problem (control charts and cause-and-effect diagrams). He is encouraged to correct his own errors. For all this, extensive education training of the workers is a must.

3. 100% inspection (often carried out automatically) is the rule and the defective parts are not allowed to be passed on to the next process. The trick is to inspect to prevent defects and the goal is perfection.

4. Get it right first time (control the process while the product is being made).

5. Make quality easy to see (use displays and indicators to highlight progress).

6. First on compliance (quality comes first, output second).

7. Continuous improvement.

An attitude of defect prevention and a habit of constrant improvement are fundamental to TQC and IMPS.

S. Shigo has suggested the following basic ideas to achieve TQC :

1. Make every worker an inspector.

2. Use source inspection techniques that control quality at the stage where defects originate.

3. Use 100% inspection with immediate feedback, rather than sampling.

4. Minimize the time it takes to carry out corrective action.

5. Concentrate on making the process efficient alongwith making the workers and operations more proficient.

Total Quality Management (TQM). During the first half of the twentieth century, customers were expected to pay extra for quality. However, in the competitive business climate of the late 1980's, quality is no longer an option. It is a positive requirement without which an organisation cannot survive. The quality of goods and services has emerged as a central issue in the current business and economic scenario. This is largely due to the need to satisfy the fast changing requirements of the customers which requires enhanced customer focus. In this context, the quality has become everyone's responsibility as against the earlier approach where the QC department was responsible for ensuring quality. Moreover, the management itself has become more leadership oriented rather than just performing a supervisory role. Further, behavioural scientists have proved that the human beings could be made more responsive to the needs of the customer by contributing their hidden potential, initiative/innovation and creativity through recognition and enhanced responsibility. The organisations' outlook on the business front is going through a metamorphosis from the earlier reactive approach to pro-active interaction with the customers. suppliers, including employees and all other stakeholders. Thus, to stay in business and be excellent, situation calls for continuously changing process/methods/procedures that produce goods and services delighting its customers. TQM is a process that creates such an organisation. TQC could not succeed in this, because even though it is supposed to be a company wide movement, but, actually is largely limited to the manufacturing department.

Definition of TQM. TQM is an integrated organisational approach in delighting customers (both external and internal) by meeting their expectations on a continuous basis through everyone involved with the organisation working on continuous improvement in all product/processes alongwith proper problem methodology. In other words, TQM means activities involving everyone in a company—Management and workers in a totally integrated effort towards improving performance at every level. This improved performance is directed towards satisfying cross-functional goals as Quality, Cost, Manpower development, Quality of worklife etc. These activities ultimately lead to increased customer and Employee satisfaction. In short, the definition says "Continuously meeting agreed customer requirements at the lowest cost by realising the potential of all employees". Hence TQM has also been termed as "Continuous Quality Improvement", CQI, by Frazier (1997), Besides "quality", the emphasis is on "Continuity". Quality is a never ending journey (Navaratnam, 1997).

TQM Culture. Total quality is a new approach to improve the effectiveness and flexibility of an organisation as a whole. It basically aims to involve every person in every department of an organisation working together to eliminate errors and prevent waste. It is an organisational culture to ensure things are done right first time.

According to "Ten Compelling reasons for TQM" by Dr. Steve Smith, the results are almost invariable :

1. Committed customers.
2. Improved productivity.
3. Reduced costs.
4. Improved certainly in operations.
5. Improved company image.
6. Dedicated management.
7. Increased employee participation.

The only way to achieve these results is through commencement by management, starting with the top management. TQM must be management lead, company wide in implementation, dedicated to continuous improvement, and the responsibility of every employee.

TQM principles. The guideline principles of TQM are given below :

1. Agree customer requirements.
2. Understand customers/suppliers.
3. Do the right things.
4. Do things right first time.
5. Measure for success.
6. Continuous improvement is the goal.
7. Management must lead.
8. Training is essential.
9. Communicate as never before.

TQM is a process of change in terms of values, beliefs, style and activity. The concepts are essentially simple to understand and yet are difficult to implement.

Commitment. TQM should become a 'way of life'. Once started, it is never ending. Continuous improvement is the goal. As a minimum, the Top Management should demonstrate their commitment to the TQM principles visibly and consistently so that the remainder of the organisation have a new role model to follow. If this does not happen with the honesty and integrity it deserves, the TQM process will never take off as intended. Total in the term TQM means really total. TQM is a process of habitual improvement, where control is embedded within and is driven by the culture of organisation.

Where Applicable. TQM is applicable to all functions as broadly interpreted. Quality means, Quality of work, Quality of service, Quality of information, Quality of process, Quality of division, Quality of people, Quality of system, Quality of company, Quality of objectives etc. Quality has an element of cost consideration in it. Idea is to minimise the cost with quality as the objective.

TQM approach versus Traditional Approach

TQM approach provides a distinctly different way of looking at management style as depicted below :

	TQM Approach	*Traditional Approach*
BELIEFS	No workers, no managers, only facilitators and team members.	Blue collar employees : Workers White collar employees : Managers.
	Employees join voluntarily to look at problems.	Workers participation must be legislated.
	Involvement, not participation. Involvement is participation + Commitment + pride.	Participation, as it is convenient for bargaining.
	Addresses organisational needs, especially those of meeting customer requirements.	Primarily addresses the welfare needs of the employees.
BEHAVIOURS	Integrated, cooperative style	Conflict, win/lose style.
	Workers are empowered and do what is right	Each fights on who is right.
	Family oriented relationship where everyone gives his/her best.	Economic relationship, where everyone gives only as he gets.
VALUES		
	Openness	Secretive
	Trust	Distrust
	Discipline	Lack of discipline
	Patience	Short sightedness
	Respect	Hatred

Fig. 21.14 compares the various quality control phases.

Teamwork for Quality. It is beyond the control of a single person to tackle the complex problems encountered in industry, commerce and services. Such problems can be solved/tackled more conveniently through the use of some form of teamwork. The approach of teamwork has the following advantages over individual effort :

1. The problem is exposed to a greater diversity of knowledge, skill and experience.

2. A variety of problems can be tackled, which are beyond the capability of any one individual or even one department.

3. Interdepartmental or interfunctional problems can be solved more easily.

4. The recommendations of the team are more likely to be accepted and implemented than individual suggestions.

5. The approach is more satisfying to team members and boosts morale.

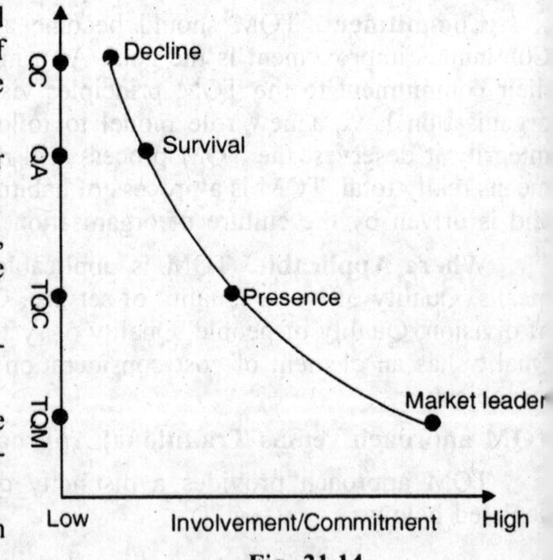

Fig. 21.14

The various companies have evolved many versions of teamwork approach, such as given below :

(*a*) **Quality Improvement Team (QIT).** A quality improvement team is a group of people with the appropriate knowledge, skill and experience who are brought together by management specifically to tackle and solve a particular problem, usually on a project basis. They are cross-functional and often multidisciplinary. The QIT determines policy, establishes direction, provides support, and by example demonstrates commitment to quality improvement. The composition of QIT varies from company to company. It may consist of General Manager, the first line reporting managers and a specialist for facilitating the operations of the quality improvement groups and the corrective action team.

In the past, a somewhat similar team was used to be called as "Task Force". But this used to be at the technology and management levels. But Quality Improvement Teams include the entire production or operating system.

(*b*) **Quality Circles.** Quality circles may be defined as a small group of workers (5 to 10) who do the same or similar works, voluntarily meeting together regularly in their normal working time, usually under the leadership of their own supervisors to identify, analyse and solve work related problems, presenting solutions to management and where possible, implementing the solutions themselves. The Quality Circle is also a means of establishing a better lines of communications between management and workers. The quality circle concept first originated in Japan in the early 1960s, by combining the principles of quality control and small group participation.

The basic cycle of a Quality Circle goes from selection of a problem through analysis, solution generation, presentation to management and finally implementations by management, Fig. 21.15.

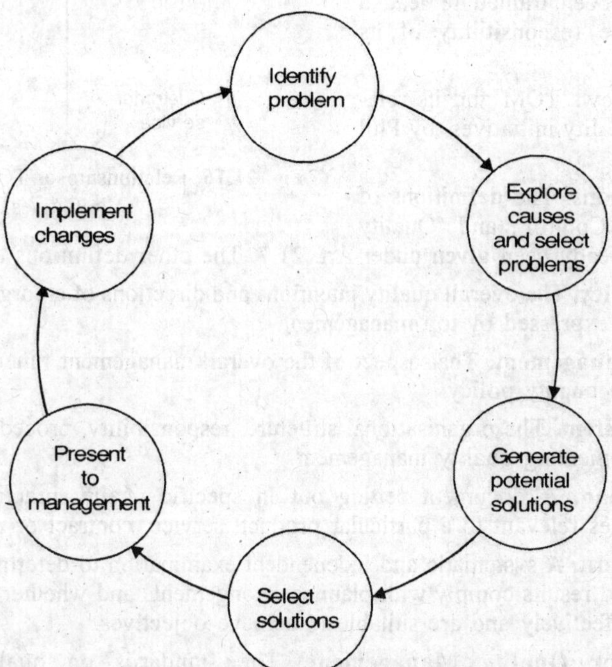

Fig. 21.15. The Quality Circle.

Circle Organization. *The unique feature of quality circles is that people are asked to join and not told to do so. Due to this, the organization of quality circles may vary from company to company. However, any quality circle usually consists of :*

1. Members
2. Leaders
3. Facilitator or Co-ordinator
4. Management

Members are the prime element of the programme. Leaders are usually the immediate supervisors or foremen of the members. The facilitator helps the quality circle started. He is the overall manager of the quality circle programme. He will be responsible for the success of the programme. He is responsible to see that the quality circle continues to operate smoothly and effectively. He coordinates the meetings, the training and energies of the workers and leaders and is the link between the quality circles and the rest of the organisation. Ideally, the facilitator should be an innovative industrial teacher, capable of communicating at all levels and with all departments within the organisation.

The management forms a companywide group called "Steering Committee" that overseas quality circle formation and activity. It consists of department heads or circle participants.

Training. As discussed under TQC, for the successful implementation of any quality control programme, the education and training of workers and other participants is of great importance. The members of the quality circles will be taught the basic problem-solving and quality control techniques. They will possess the ability to identify and solve work-related problems. Leaders will have been trained to lead a circle and bear the responsibility of its success.

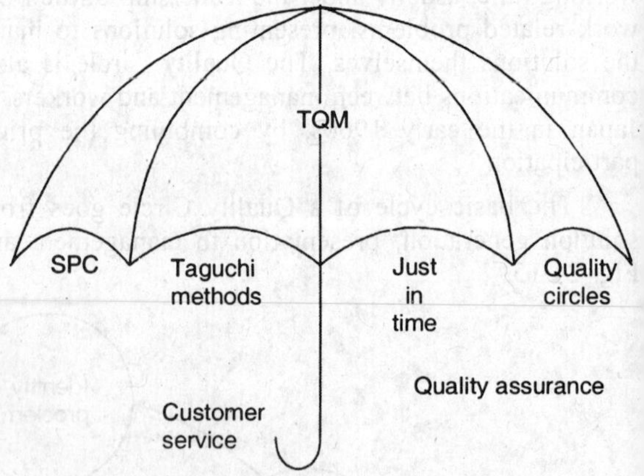

Fig. 21.16. Relationship of TQM to other Quality Initiatives.

Fig. 21.16. shows TQM and its relationship to other quality initiatives, by Phil Atkinson.

Some Definitions. The definitions of 'Quality', 'Quality Control', and 'Quality Assurance' have already been given under Art. 21.7. The other definitions are given below :

1. **Quality Policy.** The overall quality intentions and directions of an organisation as regards quality as formally expressed by top management.

2. **Quality Management.** That aspect of the overall management function that determines and implements the quality policy.

3. **Quality System.** The organisational structure, responsibility, procedures, processes and resources for implementing quality management.

4. **Quality Plan.** A document setting out in specific quality practices, resources and sequence of activities relevant to a particular product, service, contract or project :

5. **Quality Audit.** A systematic and independent examination to determine whether quality activities and related results comply with planned arrangements and whether these arrangements are implemented effectively and are suitable to achieve objectives.

Standards for Quality Management. The standards on quality system are : ISO 9000/IS : 14000.

These standards provide a sound base for TQM. They require the manufacturers "to establish and maintain a documented quality system as a means of ensuring product conformance to specified requirements. The documented quality system embraces all functional areas right from design to after-sales service. Practised in right spirit, it will bring about total standardisation,

Compliance with ISO 9000/IS : 14000 quality system standard provides the foundation for TQM, the four principal elements of which are :

1. The involvement of functions other than manufacturing also in quality activities.
2. The participation of employees at all levels.
3. Continuous quality improvement.
4. Careful attention to customers definition of quality.

Objectives of ISO 9000 :

1. Reducing multiplicity of quality standards.
2. Facilitation of International trade.
3. Facilitate implementation of quality management system.
4. Providing one standard applicable to all countries.

Concepts behind ISO 9000 :

1. PDCA cycle.
2. Self-control.
3. Do it right first time.
4. Customer oriented strategy.
5. Co-operation-Collective wisdom.
6. Transparency and evidence.
7. Continuous improvements.
8. Quality - Everyone's responsibility.
9. Use of Statistical techniques.
10. Customer-Supplier Chains.

Relevant Standards

1. **ISO 9000.** Quality management and quality assurance standards. Guidelines for selection and use.

2. **ISO 9001.** Quality System – Model for quality assurance in design/development, production, installation and servicing.

3. **ISO 9002.** Quality Systems – Model for quality assurance in production and installation.

4. **ISO 9003.** Quality System – Model for quality assurance in final inspection and testing.

TQM Model. As already discussed, TQM provides a distinctly different way of looking at the management style.

– It developes a participative culture where each employee can directly participate in areas relating to his work and decisions concerning his work.

– It is organised through quality circles on voluntary basis and quality improvement teams.

– It builds positive attitude of employees towards quality, organisation and respect for each other leading to a workplace meaningful to be in.

Fig. 21.17 Shows the TQM model.

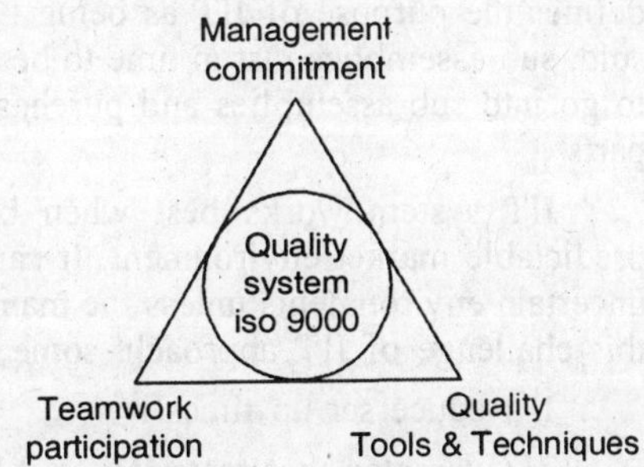

Fig. 21.17. TQM model.

Components of TQM

1. Quality policy and its communications.
2. Teamwork and participation.

3. Problem solving tools and techniques.
4. Standardisation.
5. Design and implementation of Quality System.
6. Quality Costs/Measurement.
7. Process Control.
8. Customer/Supplier integration.
9. Education and Training.
10. Quality and Training.

Just in Time (JIT) approach. In the conventional production planning, the customer orders are converted to a plan and the necessary materials are purchased in order to meet the plan. Stocks of materials are pushed in at one end and products come out of the other. The companies usually resort to over production to meet an order. This is done to keep the machinery utilized or to try to offer better customer service by holding a high level of finishing goods in stock. All this keeps the working capital tied up in inventory and also involves extra storage and handling etc.

Another wasteful practice is to buy high volumes of raw materials at a discount and keeping them in stock and to overcome unforeseen situations. Just in time system of manufacturing attempts at addressing the above problem by aiming at making components available just in time to be used, with the absolute minimum of waste. As Taichi Ohno puts it "There is nothing more wasteful than producing something you do not need immediately and then storing it in a warehouse. Both people and machines are wasted and the warehouse puts your money to sleep".

The primary emphasis of JIT system was originally in the area of inventory management. But later on this approach has been extended to whole manufacturing system. The main aim is to avoid any type of waste. In addition to the two wastes discussed above, the other areas in which there is waste, are :

(*a*) **Waiting time.** Here parts or products are waiting for the next operation, machines and the operators are waiting for the next batch of components to arrive, and so on.

(*b*) **Transport.** Where parts and products are handled and moved around the factory more than necessary.

(*c*) **Processing waste.** The actual process may be wasteful or inefficient. This waste can be reduced by proper maintenance and design for manufacture (DFM) principle.

(*d*) **Quality.** The presence of errors and defects lead to physical waste in the form of scrap, time and inventory.

JIT systems attempt to minimize inventory by only making components available just in time for them to be used and apply this principle right along the supply line. This is an ideal situation, but it is a target towards which continuous improvements can be made. Schonberger defines the purpose of JIT as being : To produce and deliver finished goods just in time to be sold, sub-assemblies just in time to be assembled into finished goods, fabricated parts just in time to go into sub-assemblies and purchased material just in time to be transformed into fabricated parts.

JIT system works best when batch sizes are fairly large and where there is a fairly predictable market environment. It can run into problems when it has to try to react quickly in uncertain environments unless the manufacturing system is geared up for rapid response. To meet this challenge of JIT approach, some way is needed to :

1. Reduce set up times.
2. Guarantee that materials will be there without holding excess inventory.
3. Guarantee machine and reliability.

4. Ensure easy and rapid availability of tools, jigs and fixtures etc.
5. Guarantee incoming quality with zero defects.
6. Ensure smooth flow through the plant.
7. Reduce inventory without the risk of running out of stock.
8. Make the whole plant responsive and agile.

It is clear from Fig. 21.16, that JIT is an important feature of TQM, and for this the essential requirement is the availability of defect-free parts at all stages in manufacture (point 5 above). This requires an awareness of the quality problem, which is shared and owned by everyone and for which everyone carries responsibility (TQM). This gives rise to the principle of "Quality at Source" as distinct from "Quality Control" as a separate function. Everyone is responsible for checking quality, and for not passing on anything less than 100 per cent perfect to the next 'customer'.

Bibliography :

1. Managing *Advanced Manufacturing Technology* by J. Bessant.
2. TQM : NPC, New Delhi.
3. *Materials and Manufacturing Processes*, 7th edition, E.P. De Garmo.

PROBLEMS

1. Define the term 'quality' as used in manufacturing science.

2. Explain, what is meant by "Quality Control".

3. Differentiate between Quality Control and Inspection.

4. List the primary objectives of SQC.

5. How SQC is used to control the 'quality'.

6. Name the main tools used in SQC.

7. Differentiate between chance and assignable variables. Explain their significance from the point of view of quality control.

8. Distinguish between 'inspection by variables' and 'inspection by attributes' Show why the distinction is important?

9. Define 'frequency distribution'.

10. Distinguish between "population distribution' and "sample distribution'.

11. Define the following terms. How these are computed?

(*i*) Average

(*ii*) Median

(*iii*) Standard deviation

(*iv*) Range.

12. Explain "Normal frequency distribution'.

13. What is a Control Chart?

14. Name the various types of Control Charts.

15. List the steps for preparing Control Charts.

16. What is the significance of ± 3 Sigma' in establishing the control limits on a quality control chart.

17. Write the formulas for control limits on charts for variables, when :

(*a*) Range is used as measure of spread.

(*b*) Standard deviation is used as measure of spread.

18. Define : Grand average and Average range.

19. Name the control charts for variables. How these are set up?

20. Name the control charts for attributes. How these are set up?

21. Distinguish between statistical control limits and tolerance limits.

22. Define "Relative Precision Index".

23. Distinguish between "Primary control limits and "Revised control limits.

24. What is "sampling inspection'. Compare it with 100 percent inspection.

25. What is "acceptance sampling". Define "acceptance number".

26. What is 'operating characteristic' of a sampling plan?

27. How the 'operating characteristic' is affected by sample size and acceptance number?

28. Define :

(*a*) AQL

(*b*) LTPD

(*c*) Producer's risk.

(*d*) Consumer's risk.

29. How an O.C. curve is constructed for a sampling plan?

30. Name the various sampling plans? How they work?

31. Describe the differences among single sampling plans, double sampling plans and sequential sampling plans.

32. How the size of a sample is decided?

33. List the factors affecting the frequency of sampling.

34. Explain terms :

(*a*) AOQ

(*b*) AOQL.

35. How AOQ curve is plotted?

36. A process, which is under statistical control, has a population mean of 10,000 cm and a population standard deviation 0.002 cm. Determine : (*a*) the natural tolerance limits of the process, (*b*) If the specification limits are 10.002 ± 0.004 cm, what percentage of the product is defective, assuming that the process out put deviations are normally distributed?

37. For a given operation the special dimension is 12.02 cm. The sum of averages for a group of 10 samples of 5 each is 120.210. The value of ax for these same sample is 0.0024, $\overline{R}$ = 0.003 cm. Individual measurements of five sample groups of 5 pieces each are :

Sample	1	2	3	4	5
	12.020	12.020	12.017	12.022	12.020
	12.022	12.018	12.022	12.020	12.020
	12.022	12.020	12.020	12.022	12.021
	12.024	12.020	12.022	12.021	12.022
	12.018	12.020	12.020	12.020	12.022

Draw the $\overline{X}$ -chart and the *R*-chart and plot the average and range values for each sample. Is the process under control.

38. Determine the probability of accepting a lot whose incoming quality is 4% defective. Sample size is 30 and acceptance number is 1. (0.663)

39. Table below shows the data for successive lots of spindles that are coming out of a machine shop. The spindles are subjected to inspection for defects. They are inspected in samples of 100 each. Presence of a single or more defect is not tolarated. Compute and construct a *p*-chart, showing ± 3σ limits for this data and comment on your results :

Sample No.	*Sample Size*	*Defectives*	*Sample No.*	*Sample Size*	*Defective*
1	100	1	11	100	3
2	100	1	12	100	2
3	100	2	13	100	1
4	100	1	14	100	2
5	100	0	15	100	0
6	100	1	16	100	2
7	100	0	17	100	7
8	100	1	18	100	1
9	100	2	19	100	2
10	100	3	20	100	0

40. Define : SPC. How it deffers from SQC?

41. Define : QA.

42. Discuss : TQC.

43. What is TQM?

44. Define TQM. Write a detailed note on TQM.

45. How TQM differs from TQC?

46. Enumerate TQM principles.

47. What are Quality circles?

48. What is JIT approach?

49. Checking the diameter of a hole using Go, NO-GO Gauges is an example of inspection by .... (Variables/attributes). (GATE 1995) (**Ans.** Attributes)

50. What do you mean by Statistical Quality Control ?

51. List the steps needed to meet the challenge of JIT approach.

52. Define the following terms : Quality policy, Quality management, Quality system, quality plan and Quality audit.

53. What is QIT?

54. What is "Process Capability Index (PCI)"?

PCI, like RPI (equ. 21.21), is a way of measuring the suitability of a process. It is defined as,

$$PCI = \frac{\text{Tolerance}}{6\sigma}$$

The two indices are compared as:-

RPI	PCI	Remarks
> 8 σ	> 1.33	High Relative Precision
6 σ to 8 σ	1 to 1.33	Medium Relative Precision
< 6 σ	< 1	Low Relative Precision

When PCI = 1, a process matches the required tolerance. Machine capability (MC) is defined as,

$$MC = \frac{6\sigma}{\text{Tolerance}} \times 100(\%)$$

If MC = 133%, capability is excellent. At MC = 100%, the process is just acceptable and at MC < 100%, it is not acceptable.

PC is often expressed by the reciprocal of MC. An acceptable process has the capability of 1 (100%) or less.

22

KINEMATICS OF MACHINE TOOLS

22.1. INTRODUCTION

Machine tools are precise and complex machines which are used to produce various types of components by metal cutting, that is, by removing metal in the form of chips. The workpiece is held in a machine tool with the help of various types of work-holding devices such as chucks, collets, face plates, mandrel etc, or held between centres (centre lathes, grinders etc) or clamped on a machine table (planers, shapers, slotters, milling machines, surface grinding machines etc.). The metal cutting tool (cutter) is held in various types of tool holders, for example, a tool post of a centre lathe, an arbor of a milling machine, spindle of a drilling machine and so on.

Motions in Machine Tools. For removing the metal from a workpiece, a relative motion is necessary between the tool and the job. The various motions characteristics of machine tools are : working motions and Auxiliary motions.

'Working motions' affect the process of chip removal. These are transmitted either to the cutting tool, or to the work or to both simultaneously. Working motions include :

1. Primary cutting motions
2. Feed motions

Primary cutting motion provides for cutting the chip from the job at the cutting speed (defined in Chapter 4) which determines the rate at which the chips are formed. Primary cutting motions most commonly used are : rotation and straight line reciprocation. The primary cutting motions of certain machine tools may be quite complex but it can be described as a combination of rotary and reciprocating motions. Rotary motions may either be transmitted to the job (lathes), or to the tool (milling cutters, drills etc) or to both simultaneously (cylindrical grinding). Straight line reciprocating primary motion is employed in planers, shapers, slotters, broaching machines, power hacksaw machines etc. This motion can either be transmitted to the tool (shapers, slotters) or to work (planners). The feed motions (defined in chapter 4) are the movements either of the tool or of the work in reference to each other. This motion enables the cutting operation to be extended to the whole surface of the workpiece to be machined. This motion may be rectilinear or curvilinear. It may be in a longitudinal direction or in a cross direction. In the example of turning a cylindrical job on a lathe, the cutting motion is obtained by the rotation of the job between centres and the feed motion is obtained by the movement of the tool parallel to the axis of the job and normal to the cutting motion.

'Auxiliary motions' prepare the machine, workpiece and tool for carrying out the cutting process and check whether the movements have been properly made. These motions include : loading and clamping the job, removing the finished work, clamping and swivelling units on which the work or tools are mounted, rapid approach and withdrawl of units carrying the cutting tools, measuring the workpiece and other operations. Since all these motions are non-cutting motions, these should be performed as fast as possible to reduce the total time to produce one component and thereby increase the rate of production of the machines. It is also desirable that these motions are combined with each other or with cutting motions to further increase the rate of production (Discussed in detail in chapters 6, 7 and 8).

22.2. DRIVES IN MACHINE TOOLS

The primary cutting motions of machine tools are power driven. Similarly, feed motions are also power driven except on small machines, where these may be performed manually. Auxiliary motions on manually operated machine tools are performed manually by the operator. However, these are power operated on automatic machine tools.

The operating cycle of machine tools, including both working and auxiliary motions, is obtained by means of a drive and definite units and mechanisms. The drive of a machine tool consists of :

1. A source of energy, and

2. Devices for transmitting power from the source of energy to the operating elements (spindles, slides, tables etc.) for producing the cutting motions and feed motions.

Machine tools are driven almost universally by electric motors. Each machine may be driven individually by its own motor or driven by belt from a line shaft furnishing power to other machine tools as well, this being called "Group Drive".

Group Vs Individual Machine Tool Drive. The choice between the two depends on :

1. Comparative first cost.

2. Total annual operating expenses.

3. Such minor advantages and disadvantages from the production point as can be foreseen from experience in similar installations.

(*a*) *Group Drive.* Group drive motors are often mounted overhead and the machine tools are driven in groups through line shafts and belts. This limits the size of the motor to about 75 kW or preferably not over 37.5 kW as they are unwidely to replace in case of failure. This drive is most suitable where power consumption of individual machines is extremely variable, with occasional brief high peaks. Group drive is usually more economical in fixed charges, power consumption and maintenance.

(*b*) *Individual Drive.* Such a drive should be used in the following situations :

(*i*) in areas requiring over head crane service.

(*ii*) On machines which would require countershafts if grouped and are likely to be moved frequently as activity in department varies.

(*iii*) On machines that require considerable power, say about 17.5 kW or more, operating at a fairly full constant load.

(*iv*) On a few machines scattered over a large area.

(*v*) In instances where the requirements of production or materials handling are best met by locating machines at odd angles.

(*vi*) Machines, requiring wide speed variations also are best driven by individual drive.

(*vii*) In complex machines, various movements are better synchronized electrically than mechanically.

Individual drives are slightly more expensive but are more flexible, permit better plant layout and changes to facilitate the flow of work through the plant, help maintenance and allow cleanliness and better lighting. Working hazard is thus reduced.

Each drive should be selected on its own merit, but the individual motor drive has largely superseded the group drive. The motion and power is transmitted from the drive motor to the various units by devices called 'transmission elements'. In 'electrical drive' the direct motor drive drives the machine drive shaft through direct coupling. In "Mechanical Drive", the transmission elements include : belts, chains, toothed gearing or some multi-or variable speed transmissions. A 'hydraulic drive' comprises a pump, a distributing device and an operating cylinder with piston rod.

The transmission elements between the input and output shafts can perform the following functions :

1. Convert rotary motion into translatory motion and vice-versa.

2. Convert rotary motion into rotary motion.

3. Convert translatory motion to translatory motion.

22.2.1 Mechanical Drive. In a mechanical drive, the transmission elements will depend upon the type of conversion needed between the drive shaft and the driven shaft.

1. **Conversion of rotary motion into rotary motion.** The following transmission elements are used to convert rotary motion of the drive shaft (input shaft) to the rotary motion of the driven shaft (out put shaft).

(*a*) *Belt drives.* In such drives, a pulley is mounted on the drive shaft and another on the driven shaft and the motion is transmitted from the drive pulley to the driven pulley with the help of belts. The section of the belts can be round, rectangular (flat), or V. The drive can be : open or crossed (V-belts are not crossed). In open drive, the driver and the driven shafts rotate in the same direction, whereas in crossed drive, their direction of rotation is reversed.

$$\text{Transmission ratio of the drive, } i = \frac{\text{RPM of driven Shaft}}{\text{RPM of driver Shaft}} = \frac{n_2}{n_1}$$

Now since $\pi D_1 n_1 = \pi D_2 n_2$ (no slip)

$$i = \frac{n_2}{n_1} = \frac{D_1}{D_2} \qquad \text{...(22.1)}$$

(*b*) *Chain Drivers.* Roller chain and silent chain are employed in machine tool drives. The direction of the rotation of the driving sprocket and driven sprocket is the same.

$$i = \frac{\text{Number of teeth on the driving sprocket}}{\text{Number of teeth on the driven sprocket}} = \frac{Z_1}{Z_2}$$

$$\therefore \quad n_2 = in_1 = \frac{Z_1}{Z_2} n_1 \qquad \text{...(22.2)}$$

(*c*) *Toothed gearing.* Toothed gearing serves for transmitting motion between parallel, intersecting and crossed shafts. Straight tooth and helical tooth spur gears are used to connect parallel shafts. For intersecting shafts, straight or spiral bevel gears are used and for crossed shafts, worm gearing is used.

$$i = \frac{Z_1}{Z_2} = \frac{n_2}{n_1} \quad \text{for spur and bevel gears} \qquad \text{...(22.3)}$$

and $$i = \frac{k}{Z} = \frac{\text{number of threads or starts on the worm}}{\text{number of teeth on the worm wheel}} = \frac{n_2}{n_1} \qquad \text{...(22.4)}$$

for worm gearing.

If a number of drives are arranged in series, the total transmission ratio,

$$i = i_1 \times i_2 \times i_3 \times \ldots \qquad \text{...(22.5)}$$

2. **Conversion of Rotary motion into rectilinear reciprocating motion.** The following kinematic links are used to convert rotary motion into rectilinear motion and often vice-versa, Fig. 22.1.

(*i*) *Rack and Pinion.* If the pinion [Fig. 22.1 (*a*)] is rotated on the driving shaft the rack will travel in a straight line. Conversely, if the rack is reciprocated as the driving element, the pinion will rotate as the driven element. For one revolution of the pinion, the travel of the rack will equal the number of teeth on the pinion, that is, travel of the rack, S will be given as,

$$S = pZn \text{ mm per min} \quad ...(22.6)$$

where p = pitch of rack teeth, mm

Z = Number of teeth on the pinion

n = Speed of pinion, rev/min

Now $p = m\pi$, where m is module of a gear

$\therefore$ $S = \pi m\ Zn$ mm per min.

This drive is commonly used in the movements of slides (Turrets, Automatics), table of a Planer.

(*ii*) *Worm and Rack* [Fig. 221.1 (*b*)]. Upon rotation of the worm, the travel of the rack will be given as,

$$S = pnk \quad ...(22.7)$$

where k = number of threads or starts on the worm.

(*iii*) *Screw and Nut* [Fig. 22.1 (*c*). Rotation of a screw, fixed in an axial direction, will cause a nut, held against rotation, to travel in a straight line along the screw. If, on the opposite, the nut is fixed axially and rotated, the screw will travel axially when it is held against rotation.

The travel of the nut or the screw (depending upon which element is rotated and which travels axially) can be determined by the equations :

$$S = pn \text{ mm per min} \quad ...(22.8)$$

where n = speed of the screw or nut in rev/min

The common example of this drive is the lead screw of a lathe and other machine tools.

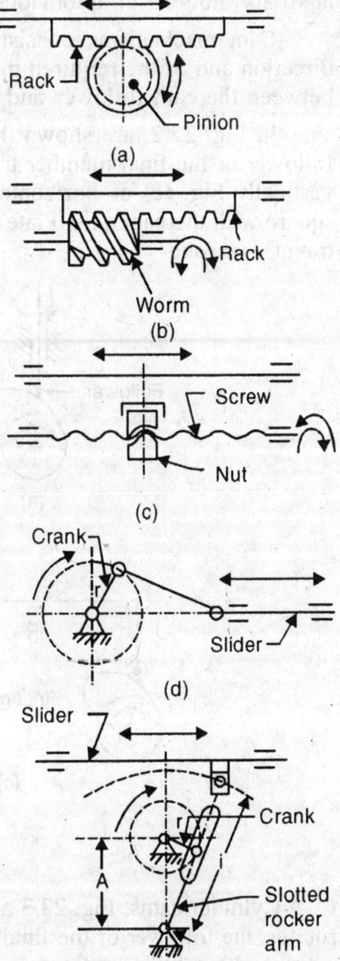

Fig. 22.1. Converting Rotary Motion into Rectilinear Motion

(*iv*) *Slider and Crank* [Fig. 22.1 (*d*)]. The number of strokes per min of the slider-crank mechanism equals the speed in rev/min of the driving crank, while the slider travel is calculated from the following relation :

$$S = 2r, \text{ mm} \quad ...(22.9)$$

where r = radius of crank, mm ...(22.10)

(*v*) *Crank and Slotted rocker arm* [Fig. 22.1 (*e*)]. The travel of the ram or slider is determined from the formula,

$$S = 2.L/A.r, \text{ mm} \quad ...(22.11)$$

where r = crank radius, mm

L = rocker arm length, mm

A = distance from the rocker arm pivot to the crank centre, mm.

An example of this drive is the movement of ram in a shaper.

(*vi*) *Cams.* Plate (disc) and cylinder (barrel) cams are widely used to obtain working and auxiliary motions in automatics and semiautomatics.

Cam mechanisms consist of a cam and a follower. To obtain, motion in the required direction and of the required magnitude, intermediate linkages such as levers, are often arranged between the cam follower and the final member.

In Fig. 22.2, are shown the plate cam actuated movements. Upon rotation of the cam, the follower or the final member travels up and down (Fig. *a*), swings about a pivot (Fig. *b*), travels vertically Fig. (*c*) or horizontally Fig. (*d*). Plate cams may be suitably employed for motions square with the cam axis. Plate cams are of one piece design and are used of comparatively short travel or stroke.

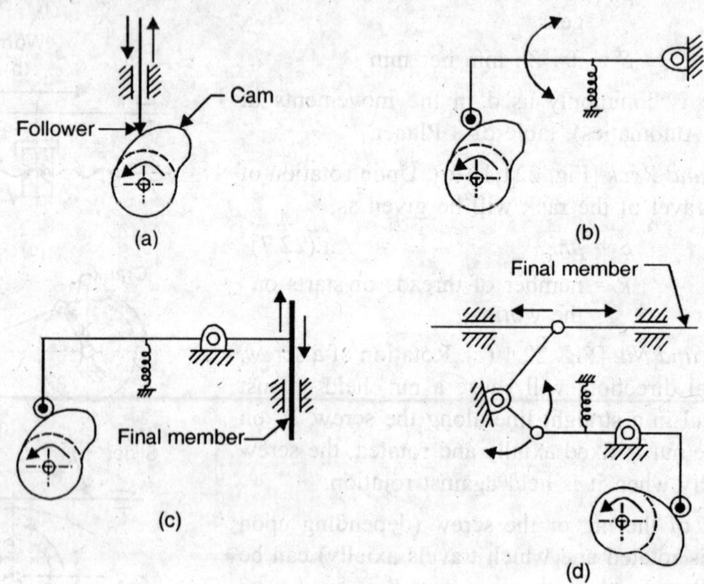

Fig. 22.2. Plate Cams.

Cylinder cams, Fig. 22.3 are drums having helical slots or cam members. When the cylinder rotates, the follower or the final member travels horizontally or vertically. These cams are usually employed to obtain motions parallel with the cam drum axis. Such cams are of short travel. For longer travel, cam plates or members are attached by screws to the drum. The magnitude, direction and speed of travel of the operative element or follower, may be varied by imparting a particular form to the cams. Cams discussed above are single edged cams. These actuate the follower in one direction only and depend on some external force produced by a spring or weight to return

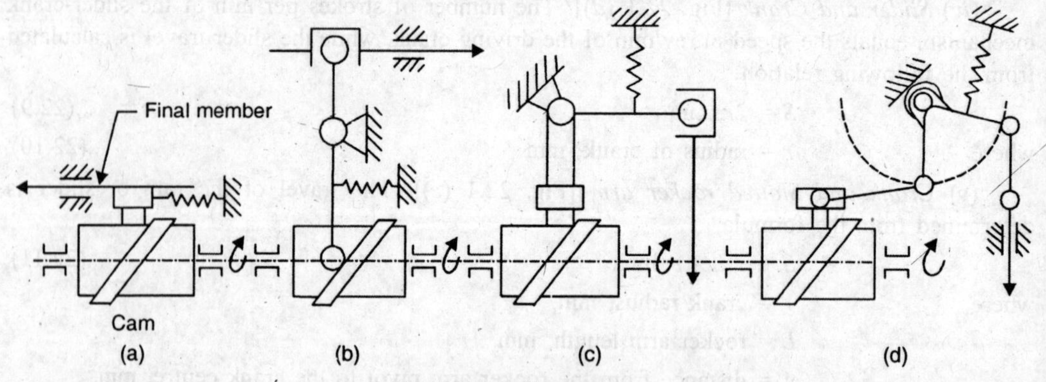

Fig. 22.3. Cylinder Cams.

the follower to its starting point. Two edged cams are positive motion cams and provide for positive follower travel in both directions. Such cams have a curvilinear groove which accommodates the follower roller.

3. *Mechanisms for Periodic (intermittent) Rotation.* Periodic rotation mechanisms include a stop with pins, Fig. 22.4 (*a*); Ratchet devices combined with crank or cam mechanisms [Fig. 22.4 (*b*), (*c*)] and Geneva mechanisms (discussed in Chapter 8).

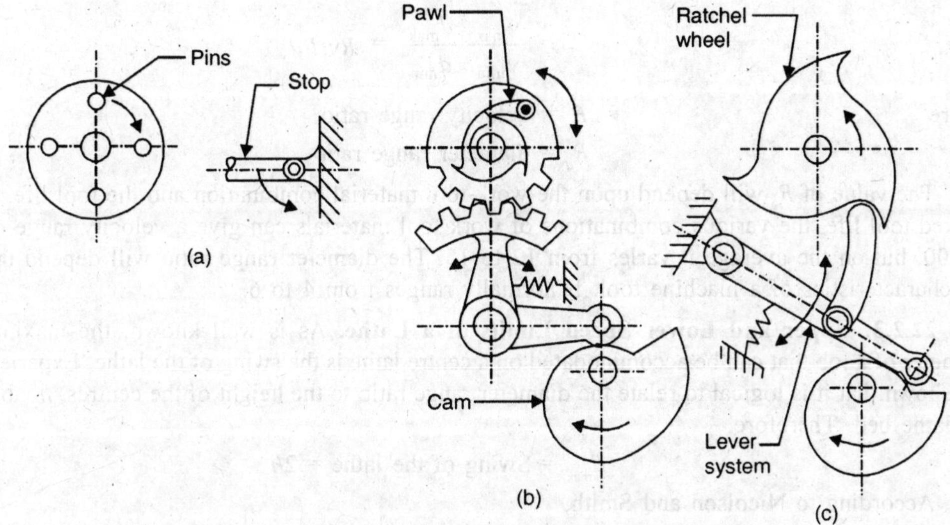

Fig. 22.4. Mechanisms for Periodic Motion.

22.2.2. Selecting the Maximum and Minimum Cutting Speeds and feeds. The cutting speed depends chiefly on the material to be machined, the material of the cutting tools, the depth of cut and the rate of feed (See Chapter 4). Maximum and minimum cutting speeds and feeds, for machining blanks of the maximum and minimum sizes that are to be accommodated, are selected by analyzing the manufacturing process. The usual range of cutting speeds and feeds for various combinations of work material and tool material can be found in standard tables. The higher the cutting speed, the larger the rate of production. The application of cutting speeds, higher than the recommended values, will decrease the life of the cutting tools which will be subject to premature wear or failure. The rate of feed is selected to suit the machining allowance, provided for the given operation, and the required accuracy and surface finish. Naturally, the maximum feed rate will be for the roughing cut for the hardest tool-softest work material combination and the minimum feed rate corresponds to the finishing cut. Other conditions being equal, the rate of production will increase in direct proportion to an increase in the rate of feed. Therefore, to reduce the total machining time so as to increase the rate of production, it is desirable to increase time so as to increase the rate of production, it is desirable to increase both cutting speed and rate of feed within feasible limits.

Spindle Speeds for Rotary Motion. After determining the extreme values of the cutting speed V_{max} and V_{min} and the extreme values of diameters of jobs d_{max} and d_{min} which can be accommodated on a machine tool, the extreme values of spindle speeds n_{max} and n_{min} can be determined by the formula :

$$n = \frac{1000\nu}{\pi d}, \text{ rev / min} \qquad ...(22.12)$$

where n = cutting speed, m/min

d = job or cutter diameter, mm

$$\therefore \qquad n_{max} = \frac{1000\nu_{max}}{\pi d_{min}} \qquad ...(22.13)$$

$$n_{min} = \frac{1000\nu_{min}}{\pi d_{max}}$$

The ratio of n_{max} and n_{min} is called the range ratio of the spindle speed variation, that is, the range ratio is :

$$R_n = \frac{n_{max}}{n_{min}}$$

$$= \frac{\nu_{max}}{\nu_{min}} \cdot \frac{d_{max}}{d_{min}} = R\nu . Rd$$

where R_v = velocity range ratio

and R_d = diameter range ratio

The value of R_v will depend upon the work-tool material combination and the tool life. For a fixed tool life, the various combinations of work-tool materials can give a velocity range of 4 to 100, but on the average it varies from 12 to 15. The diameter range ratio will depend upon the characteristics of a machine tool. It normally ranges from 4 to 6.

22.2.3. Upper and Lower Speed Limits of a Lathe. As is well known, the maximum diameter of a job that can be accommodated on a centre lathe is the swing of the lathe. Experience has shown that it is logical to relate the diameter range ratio to the height of the centres, *h*, above the lathe bed. Therefore,

$$d_{max} = \text{Swing of the lathe} = 2h$$

According to Nicolson and Smith,

$$d_{min} = h/8$$

$$\therefore \quad R_d = \frac{d_{max}}{d_{min}} = 16$$

now

$$n_{max} = \frac{1000\nu_{max}}{\pi d_{min}} = \frac{1000 \times \nu_{max} \times 8}{\pi \times h}$$

$$\therefore \quad n_{max} = 2546.15\left(\frac{\nu_{max}}{h}\right) \quad \text{...(22.14)}$$

Nicolson and Smith have also suggested a formula for n_{min}. According to them, ν_{min} should be selected such that a finishing cut at a cutting speed of 15 f.p.m. (4.5 m/min) can be taken on a work piece diameter of slightly less than the swing of the lathe, with a H.S.S. tool. They suggested $d_{max} = 2\left(h - \sqrt{h}\right)$, *h* in inches

$$\therefore \quad n_{min} = \frac{12 \times 15}{\pi 2\left(h - \sqrt{h}\right)} = \frac{28.65}{h - \sqrt{h}} \text{ rev/min.}$$

$$= \frac{733.45}{h - 5\sqrt{h}}, \text{ rev / min } (h \text{ in mm})$$

For carbide tools they suggested a finishing cut at about 100 f.p.m. (30 m.p.m.). Thus

$$n_{min} = \frac{4890}{h - 5\sqrt{h}} \text{ for carbide tools.}$$

The common range ratio of spindle speeds for various machine tools are given in Table 22.1.

Table 22.1. Common Values of Rn

Machine Tools	*Rn*
Planer, Shaper, Slotter	6 to 10
Centre lathe	40 to 60
Turret lathe (ram type)	80 to 100
Automatic lathes	8 to 10
Semi-automatic lathes	15 to 25
Drilling machines	15 to 30
Boring machines	40 to 60
Milling machines	30 to 50
Grinding machines	1 to 12

22.2.4. Stepped and Stepless Drives. With a constant speed motor, there is a need for some method of varying the speed over this range (n_{max} and n_{min}). The infinitely variable (stepless) speed variation is possible with suitable mechanical, hydraulic and electrical drives. However, the torque/speed characteristics of available stepless drives do not meet the requirements of spindle drives which demand an increased driving torque to the spindle at lower output speeds in order to maintain a constant rate of metal removal. The stepless drives which do possess the required torque/speed characteristics, are limited by the speed range over which these characteristics can be maintained. In order to provide for a wide range of operating speeds together with adequate torque at lower spindle speeds, it is necessary that the spindle speed range be convered in a number of discreet steps (Stepped drive). This is achieved by a constant speed motor used in conjunction with conepulleys (almost out of choice now) or a gear box (most commonly used) which provides for a series of spindle speeds in a mechanical stepped drive. The number of speed steps provided in a machine tool is likely to represent a compromise between what is desirable for efficient operation and the cost of drive (it will increase with the number of steps) which can be justified in given circumstances.

22.2.5. Characteristics of Mechanical Stepped Drive. If the work piece diameter d and the optimum cutting speed v are given or known, then the corresponding spindle speed can be determined with the help of equ. (22.12). Cutting speed charts of the rectilinear type are often used to quickly determine the spindle speed for a give cutting speed, Fig. 22.5. Each slanting line or "ray" represents a definite step, in rpm, of the spindle speed of the given machine tool. A logarithmic chart of the type, Fig. 22.6, is often used instead of the rectilinear chart to determine spindle speeds.

Construction of Ray diagram. It is clear from equ. (22.12),

$$n = \frac{\pi dn}{1000} \text{ m/min}$$

then $v \propto d$, with n being constant.

This relation is represented graphically by a straight line passing through the origin, because $v = 0$ when $d = 0$. To draw the straight line, we need another point. The number of straight lines or rays emerging from the origin for a given cutting speed range will be equal to the number of steps in this range. This is achieved as follows :

$$v = \frac{\pi n}{1000} \cdot d \approx \frac{n}{318} \cdot d$$

$\therefore$ when $$d = 318 \text{ mm}, \; v = n$$

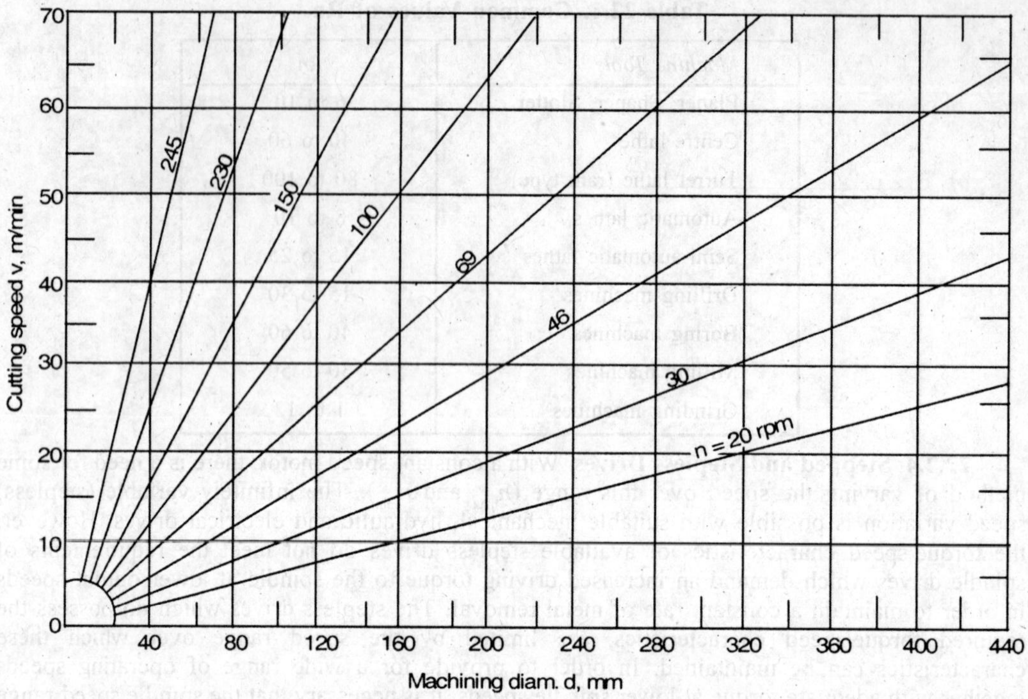

Fig. 22.5. Cutting Speed Chart.

i.e. the cutting speed is equal in value to the spindle speed. As is clear from Fig. 22.5, diameters are plotted along X-axis and cutting speed along Y-axis. Refer to Fig. 22.5A, A vertical line at d = 318 mm is drawn, and horizontal lines for different values of v (equal in value to n) are found. For example, let us take the speed steps are :

$$n_1 = 45, \quad n_2 = 63, \quad n_3 = 90, \quad n_4 = 125, \; n_5 = 180$$

$$n_6 = 250, \quad n_7 = 355, \; n_8 = 500 \text{ and } n_9 = 710 \text{ rev/min.}$$

∴ The corresponding cutting speeds will be : 45, 63, 90, 125, 180, 250, 355, 500 and 710 m/min respectively. From points corresponding to these cutting speeds on Y-axis, horizontal lines are drawn. Their points of meeting the d = 318 mm lines are plotted. By joining all these points to the origin, the ray diagram will be completed.

It is clear that due to limited scale of the diagram, it is not possible to draw the ray lines for 500 and 710 rev/min. In such cases, it is convenient to take d = 31.8 mm and vertical line corresponding to this diameter drawn. It is clear that the cutting speeds corresponding to 500 and 710 rev/min will be 50 and 71 m/min respectively. So, the rays can be drawn as explained above. If needed, d = 159 mm can also be used. In this case, it will be equal in value to n/2. The most common use of ray

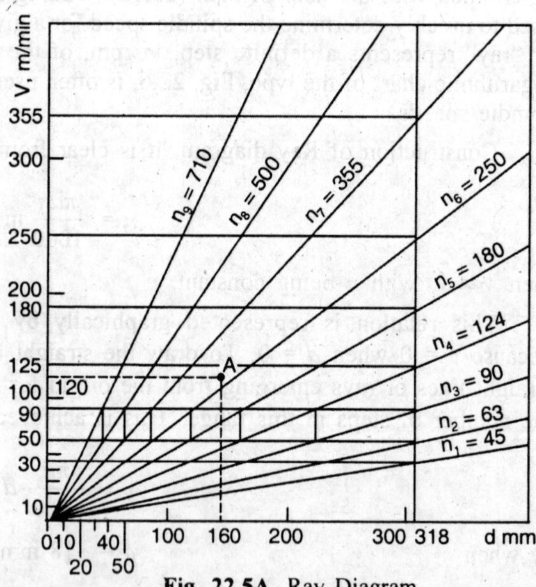

Fig. 22.5A. Ray Diagram.

diagram is to find the optimum speed of rotation of given values of v and d. For example, if v = 120 m/min and d = 160 mm, the corresponding point on the diagram, Fig. 22.5 A, will be A. This point lies in between rays for $n_6 = 250$ and $n_7 = 355$ rev/min. It is better to take $n = 250$ rev/min to increase the life of the tool.

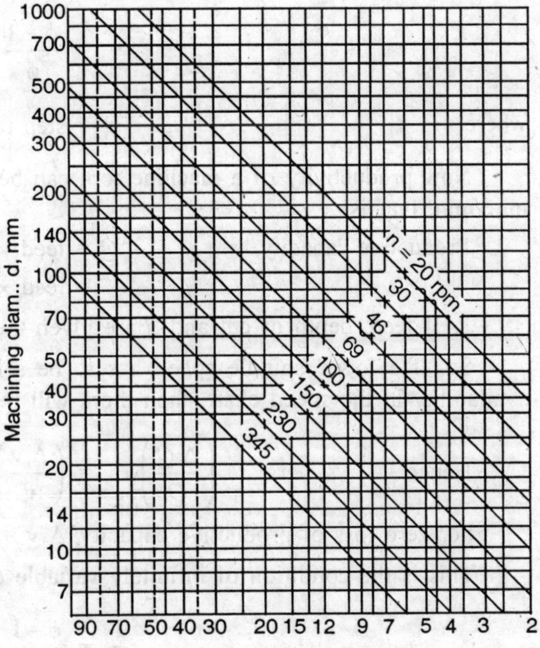

1. *Series of Spindle Speeds for Machine Tools.* Before we discuss the various series, let us study about the loss of productive capacity. Let the maximum value of the optimum cutting speed be v. From the speed chart, Fig. 22.7 let d_{j-1} be the diameter that can be machined at spindle speed n_{j-1} and dj the diameter at the spindle speed n_j. Let there be a workpiece of diameter d lying between d_j and d_{j-1}. It is clear from the Fig. 22.7 that for this machining diameter, there is no spindle speed ray, the speed rays n_{j-1} and n_j being the two definite steps for a given machine tool. So, 'd' being greater than d_j, the spindle speed to be selected will be n_{j-1}, which is lower than the calculated one (from equ. 22.12) for diameter 'd'. So, there is a speed loss of Δv, which is given as,

Fig. 22.6. Logarithmic Speed Chart.

$$\Delta v = v_a - v_b = \pi dn - \pi dn_{j-1}$$

It is clear that maximum speed loss will be when $d = d_j$.

Then,
$$(\Delta v)_{max} = \pi d_j n_j - \pi . d_j . n_{j-1}$$

$$\therefore \quad \text{Maximum \% loss of speed, } A = \left(\frac{\Delta v}{v}\right)_{max}$$

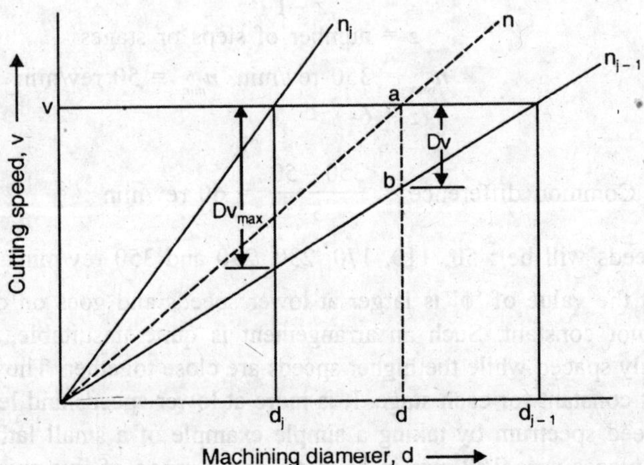

Fig. 22.7. Speed with Constant Cutting Speed

$$= \frac{\pi d_j \left(n_j - n_{j-1}\right)}{\pi d_j n_j} \times 100$$

$$= \frac{\phi - 1}{\phi} \times 100$$

where ϕ = step ratio = $\frac{n_j}{n_{j-1}}$

Now productivity of a machine tool can be written in terms of the metal removal rate in mm³/min, that is,

Productive capacity, Q = feed × depth of cut × v × 1000 mm³/min

= feed × depth of cut × πdn, mm³/min

∴ If feed, depth of cut and d are given then $Q \propto n$.

So, if a lower spindle speed has to be selected than the calculated one due to its non availability in the speed chart, then there will be loss in the productivity capacity.

Hence $$\left(\frac{\Delta Q}{Q}\right)_{max} = \left(\frac{\Delta v}{v}\right)_{max} = \frac{\phi - 1}{\phi} \times 100 \quad ...(22.15)$$

For zero loss of productive capacity, $\Delta v = 0$, therefore $\phi = 1$.

This is the condition of infinitely variable drive. For 50% efficiency,

$$\frac{1}{2} = \frac{\phi - 1}{\phi}. \therefore \phi = 2$$

So, the value of ϕ should preferably lie between 1 and 2.

Series of Spindle Speeds. As discussed above the spindle speeds vary over a discreet number of stages in a stepped drive. These steps usually form a series. The series can be A.P., G.P. or Logarithmic Progression (L.P.).

(*i*) *Speed Spectrum in Arithmetic Progression.* As is clear, in this series, there will be a constant common difference between any two consecutive spindle speeds. The common difference will be given as :

$$\text{Common difference} = \frac{n_{max} - n_{min}}{z - 1}$$

where z = number of steps or stages

Let n_{max} = 350 rev/min, $n_{min.}$ = 50 rev/min

and $z = 6$

$$\therefore \quad \text{Common difference} = \frac{350 - 50}{5} = 60 \text{ rev/min}$$

∴ Spindle speeds will be : 50, 110, 170, 230, 290 and 350 rev/min.

It is clear that the value of 'ϕ' is larger at lower speeds and goes on decreasing at higher speeds, and so is not constant. Such an arrangement is quite unsuitable, because, the lower speeds are too widely spaced while the higher speeds are close together. The productive capacity loss is thus also not constant for each stage. It is more at lower speeds and less at higher speeds. Let us draw the speed spectrum by taking a simple example of a small lathe. It is intended to accommodate workpieces upto 100 mm in diameter and a range of five spindle speeds is to be provided. Cutting speed is 24 m/min.

Let us consider workpieces of $d_{min} = 20$ mm with $d_{max} = 100$ mm

$$\therefore \quad n_{max} = \frac{1000 \times 24}{\pi \times 20} = 382 \text{ rev/min}$$

$$n_{min} = \frac{1000 \times 24}{\pi \times 100} = 76.4 \text{ rev/min}$$

$$\therefore \quad \text{Common difference} = \frac{382 - 76.4}{4} = 76.4 \text{ rev/min}$$

$\therefore$ n_2, n_3, n_4 will be 152.8, 229.2, 305.6, rev/min

$$\therefore \quad d_2 = \frac{1000 \times 24}{\pi \times 152.8} = 50 \text{ mm}$$

$$d_3 = 33.33 \text{ mm}$$

$$d_4 = 25 \text{ mm}$$

Now with the help of the above data, a speed spectrum can easily be drawn, Fig. 22.8. Firstly, a horizontal line corresponding to cutting velocity of 24 m/min is drawn. Then vertical lines from diameters d_{min}, d_2, d_3, d_4 and d_{max} are drawn to meet the speed line. The meeting points are joined to the origin to get the speed rays. Then by drawing the test line, that shows the speed loss as a 'saw diagram' for A.P. series can be drawn.

(*ii*) *Speed Steps in G.P.* We know the properties of a G.P. series. The common ratio between two consecutive steps is equal and constant, that is, the series can be written as,

$$n_1, \phi n_1, \phi^2 n_1, \phi^3 n_1, \ldots \phi^{z-1} n_1$$

where $n_1 = n_{min}$, $\phi^{z-1} \cdot n_1 = n_{max}$

and z = number of spindle speed steps

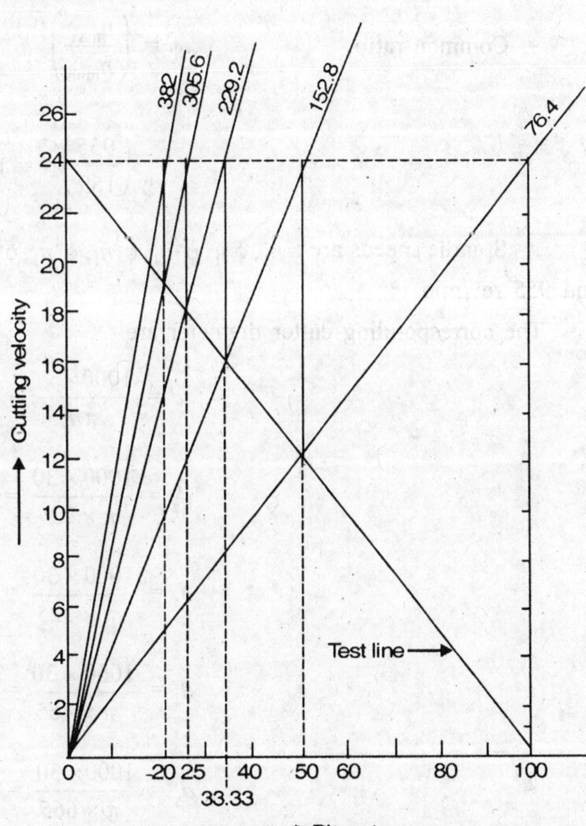

Fig. 22.8. Speed Spectrum and Saw Diagram for A.P.

$$\because \quad n_{max} = \phi^{z-1} \cdot n_{min}$$

$$\therefore \quad \phi^{z-1} = \frac{n_{max}}{n_{min}} = Rn$$

$$\therefore \quad \phi = (R_n)^{\frac{1}{z-1}} \quad \ldots(22.16)$$

$$\text{and} \quad z = 1 + \log Rn / \log \phi = \frac{\log (Rn.\phi)}{\log \phi}$$

If Rn and ϕ are given, z can be calculated from the above formula. It is then rounded off to a whole number and then the corresponding value of Rn is found out.

It is clear from equ. (22.15) that since ϕ is constant, the percent loss of cutting speed and productive capacity (at any optimum cutting speed) is constant for all speed steps. In addition to its economical advantages, geometrical series of spindle speeds have other advantages that are of great importance in designing the machine tool drive. For all these reasons, G.P. series has found wide applications in machine tools. The speed spectrum and saw diagram can be drawn with the help of the following example :

Example 1. *A machine spindle is to operate on ferrous metals at 30 m/min and is required to have six speeds. The spindle can accommodate H.S.S. cutters ranging from 10 mm to 60 mm. Determine the spindle speeds. Plot a graph between cutting velocity and cutter diameter for each spindle speed and find the range of cutting velocity for 12 mm cutter and 36 mm cutter.*

Solution. Now,

$$n_{min} = \frac{1000 \times 30}{\pi \times 60} = 159 \text{ rev/min}$$

$$n_{max} = \frac{1000 \times 30}{\pi \times 10} = 955 \text{ rev/min}$$

$\therefore$ Common ratio,

$$\phi = \left(\frac{n_{max}}{n_{min}}\right)^{\frac{1}{z-1}}$$

$$= \left(\frac{955}{159}\right)^{\frac{1}{5}} = 1.43$$

$\therefore$ Spindle speeds are : $n_1, \phi n_1, \phi^2 n_1, \phi^3 n_1, \phi^4 n_1, \phi^5 n_1$, that is, 159, 227, 235, 325, 465, 665 and 955 rev/min.

The corresponding cutter diameter are :

$$d = \frac{1000 \times v}{\pi n}$$

$$\therefore \quad d_2 = \frac{1000 \times 30}{\pi \times 227} = 42 \text{ mm}$$

$$d_3 = \frac{1000 \times 30}{\pi \times 325} = 29.4 \text{ mm}$$

$$d_4 = \frac{1000 \times 30}{\pi \times 465} = 20.5 \text{ mm}$$

$$d_5 = \frac{1000 \times 30}{\pi \times 665} = 14.4 \text{ mm}$$

With the help of above data, the speed spectrum can be drawn, Fig. 22.9, as explained earlier.

To find the range of cutting speed for cutter diameters of 12 mm and 36 mm, draw vertical lines from these lines to meet the speed rays corresponding to the diameter steps within which these two diameters lie. Read the corresponding values of cutting velocity. It is clear from the diagram, that for 12 mm cutter, v_{max} = 36m/min and $v_{min.}$ = 24.5 m/min.

$\therefore$ Range of cutting speed = 36 – 24.5 = 11.5 m/min.

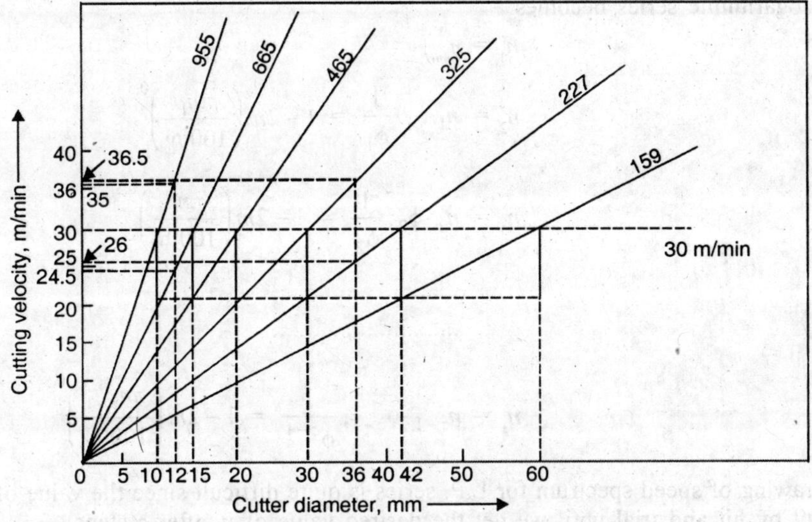

Fig. 22.9. G.P. Series Speed Spectrum.

Similarly for 36 mm cutter, v_{max} = 36.5 min, v_{min} = 26 m/min

∴ Range of cutting speed = 36.5 – 26 = 10.5 m/min.

Let us take the case when we write machining diameter in place of cutter diameter. We write the machining diameter and the corresponding spindle speed :

d = 60 42 29.4 20.5 14.4 10 mm

n = 159 227 20.5 465 665 995 rev/min

Now, in moving from a speed n_1 = 159 to n_2 = 227 rev/min, a material of 18 mm must be removed from the workpiece. This can be achieved in 3 passes, if we take maximum depth of cut equal to 3 mm. Now let us consider the last two speed steps. In moving from 665 to 955 rev/min, a material of 4.4 mm diameter will have to be removed, which can be very easily done only in one pass (with a cut equal to only 2.2 mm). Therefore, to make the performace of the machine tool equal in the whole speed spectrum, the low speeds should be brought closer, whereas the higher speed values can be widened a little.

(*iii*) *Logarithmic Series.* This series is based on the plea that the diameter step or range is a function of the diameter. Refer to Fig. 22.5, the diameter range is,

$$\Delta d_{j-1} = d_{j-1} - d_j$$

Now Δd_{j-1} is a function of d_{j-1} and is given as

$$\Delta d_{j-1} = 2m\left(d_{j-1}\right)^p \qquad ...(22.17)$$

According to Knonenberg, $p = 0.5$

$$\therefore \quad \Delta d_{j-1} = 2m\sqrt{d_{j-1}}$$

$$\therefore \quad d_{j-1} - d_j = 2m\sqrt{d_{j-1}} \qquad ...(25.18)$$

$$\therefore \quad \frac{1000v}{\pi n_{j-1}} - \frac{1000v}{\pi n_j} = 2m\sqrt{d_{j-1}}$$

From here,

$$\frac{n_{j-1}}{n_j} = \frac{1}{\Phi_{j-1}} = 1 - 2m\left(\frac{\pi n_{j-1}}{1000v}\right)^{0.5} \qquad ...(22.19)$$

∴ Logarithmic series becomes :

$$n_1 = n_{min}$$

$$n_2 = n_1.\phi_1, \frac{1}{\phi_1} = 1-2m\left(\frac{\pi n_1}{1000v}\right)^{0.5}$$

$$n_3 = n_2 \cdot \phi_2, \frac{1}{\phi_2} = 1-2m\left(\frac{\pi n_2}{1000v}\right)^{0.5}$$

....

....

$$n_z = n_{z-1} \cdot \phi_{z-1}, \frac{1}{\phi^{z-1}} = 1-2m\left(\frac{\pi n_{z-1}}{1000v}\right)^{0.5} \qquad ...(22.20)$$

The drawing of speed spectrum for L.P. series is quite difficult since the value of 'm' is to be found out by hit and trial until we get the desired value of n_z after z steps.

If a speed spectrum is drawn for L.P. series, then it has been found that the values lie in between those for A.P. and G.P. series. So, L.P. series is most suitable from the operational point of view. But due to the many advantages of G.P. series, it is commonly used in machine tool drives.

2. **Standard values of the Ratio ϕ.** Standard values of the ratio ϕ for standard series of spindle speeds have been based on the following : The values of ϕ are normalized which makes it possible to normalize speed and feed ranges, thereby, simplifying the kinematic design of machine tool :

(*a*) Generally, two speed 3ϕ electric motors are employed in the spindle speed drives of machine tools. The ratio of their synchronous speeds equals 2, for example, 3000/1500 or 1500/750 rpm. Hence if in the G.P. series, there is a spindle speed n_x, then there must also be a speed n_y, so that

$$n_y = 2\, n_x$$

Also, n_x and n_y must belong to G.P. series, therefore,

$n_y = n_x \cdot \phi^{E_1}$, where E_1 is a whole number or a fraction

$$\therefore \qquad \phi = \left(\frac{ny}{nx}\right)^{\frac{1}{E_1}} = (2)^{\frac{1}{E_1}} \qquad ...(22.21)$$

(*b*) For parameters in machine tools, the series of preferred numbers are in the form of G.P. whose constant ratio must satisfy the condition :

$$\phi = (10)^{\frac{1}{E_2}} \qquad ...(22.22)$$

where E_2 is a whole number or a fraction.

Thus, standard values of ϕ must satisfy the conditions :

$$\phi = (2)^{\frac{1}{E_1}} = (10)^{\frac{1}{E_2}}$$

$$\therefore \qquad E_2 \log_2 = E_1 \log_{10}$$

or $$3E_2 = 10E_1$$

$\therefore$ $$E_1 = 3E' \text{ and } E_2 = 10\,E',$$

where E' is any whole number.

and $$E_2 = 40,\ 20,\ 10,\ 5 \text{ and so on.}$$

$\therefore$ The standard values of ϕ are given as :

$$(10)^{\frac{1}{40}},\ (10)^{\frac{1}{20}},\ (10)^{\frac{1}{10}},\ (10)^{\frac{1}{5}}, \text{ that is,}$$

$$1.06,\quad 1.12,\quad 1.26,\quad 1.58$$

With the addition of the values $\phi = \sqrt{2}, \phi = \sqrt[3]{2}$ and $\phi = (10)^{\frac{1}{4}}$, that is, 1.41, 2 and 1.78 respectively, the series of the standard values of ϕ are given in T 22.2.

Table 22.2. Standard Values of ϕ

ϕ	1.06	1.12	1.26	1.41	1.58	1.78	2
Maximum Speed loss $\frac{\phi - 1}{\phi} \times 100$	5	10	20	30	40	45	50
Average Speed loss, $\frac{\phi - 1}{\phi + 1} \times 100$	3	5	12	17	22	28	33

For calculating average speed loss, it is clear from Fig. 22.7 that the average speed loss will be $\frac{1}{2}(\Delta\, v)_{\max}$ and

$$\text{Average speed will be} = \frac{1}{2}\pi d_j \times \left(n_{j-1} + n_j\right)$$

now $$\frac{1}{2}(\Delta\, v)_{\max} = \frac{1}{2}\pi dj\left(n_j - n_{j-1}\right)$$

$\therefore$ $$\text{Average \% loss of speed} = \frac{n_j - n_{j-1}}{n_j + n_{j-1}}$$

$$= \frac{\phi - 1}{\phi + 1} = \text{constant} \qquad \ldots(22.23)$$

It is clear from Table 22.2 that both maximum speed loss and average speed loss increase with an increase in the value of ϕ. Actual spindle speeds may differ from the tabular values by not more than $\pm$ 10 $(\phi - 1)$ percent.

The arrangement of feeds varies considerably between different types of machine tools. These may be in G.P., but more usually they increase in A.P. This holds true of reciprocating primary cutting motions and for feed series.

3. *Selecting the optimum value of ϕ and number of Steps*. For a given Rn, it is clear from equ. (22.16) that the number of steps, z, increases rapidly as the value of ϕ is reduced. Thus, in

selecting the values of ϕ and z, it is necessary to find an economically optimum compromise between the effort to reduce losses in cutting speeds by having more steps and thereby complicating the drive, and the effort to reduce the cost of the drive/machine tool by keeping it as simple as possible. The following points should be considered while selecting ϕ and z.

(*i*) In majority of the general purpose machine tools, satisfactory operation is achieved by the use of a ratio $\phi = 1.26$ or 1.41.

(*ii*) For automatic and semi-automatic machine tools intended for lot or mass production and when speed changes are to be made in the drive gear train by means of change gears, then $\phi = 1.12$ or 1.26 proves satisfactory.

(*iii*) A G.P. series has an excessively large number of steps in the higher speed range (used mainly in machining small diameters). Indeed, the maximum diameter interval, x_j (Fig. 22.10) accommodated at a constant cutting speed, v, by two adjacent steps of spindle speed series is :

$$x_j = d_{j-1} - d_j \qquad ...(22.24)$$

$$= \frac{v}{\pi}\left(\frac{1}{n_{j-1}} - \frac{1}{n_j}\right)$$

$$= \frac{v}{\pi\, n_{j-1}}\left(1 - \frac{n_{j-1}}{n_j}\right)$$

$$= d_{j-1}\left(1 - \frac{1}{\phi}\right) = d_{j-1}A \qquad ...(22.25)$$

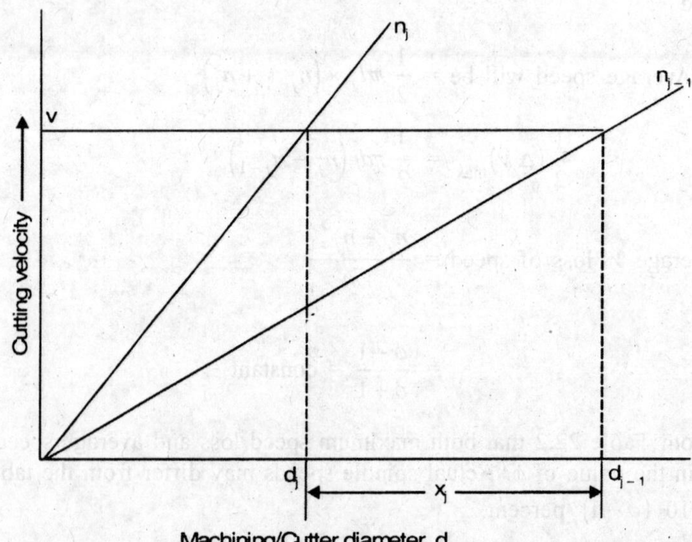

Fig. 22.10

If, to avoid a large number of steps in the series, the diameter interval is selected so that it is not less than the interval $\Delta\, d$ of standard bar stock diameter or of tool sizes (for example, drills), then

$$xj = d_{j-1}\, A \geq \Delta\, d$$

$$\therefore \quad d_{j-1} \geq \frac{\Delta d}{A} = \frac{\phi}{\phi - 1} \cdot \Delta d \qquad ...(22.26)$$

This equation enables to find the minimum diameter of bar stock to be machined or of tools to be used, with various values of ϕ, and conversly, the minimum values of ϕ that can be assigned to machine tools with various minimum diameters of work to be machined. Thus large values of ϕ = 1.58 and 1.78 are used in small machine tools which accommodate small work diameters, while smaller values of ϕ = 1.26, 1.12 and sometimes 1.06 are used in heavy machine tools.

(*iv*) It is good practice to select a number of speed steps, *z*, having the factors 2 and 3, so that

$$z = 2^{E_1} \cdot 3^{E_2} \qquad ...(22.27)$$

This requirement is met by the values of

$$z = 2, 3, 4, 6, 8, 9, 12, 16, 18, 24, 27, 32 \text{ and } 36$$

The most frequently used values are :

$$z = 3, 4, 6, 8, 12, 18 \text{ and } 24.$$

The values of *Rn*, *z* and feed variation R_s and steps of feeds z_s, may vary within quite large numbers. For each type and size of machine tool, their values depend upon purpose of machine tool, nature of the process, type of cutting tool to be used and more importantly upon the required degree of versatility. The more versatile a machine tool is to be, the more different types of tools that are to be used, the larger the range ratios *Rn* and R_s must be for efficient operation.

In most cases, $z \leq 36$ in machine tools with a rotary primary cutting motion. The same is true for z_s. However, centre lathes designed to cut threads of various pitches have feed mechanisms with z_s = 48 to 60 or more.

For machine tools with a reciprocating primary cutting motion, the range ratios of the number of strokes per min, *Rn*, or of the working stroke speeds *Rv*, are narrower than the range ratios *Rn* in machine tools with a rotary primary motion.

In specialized and in particular, special machine tools, the values of *Rn*, R_s, *z*, z_s should be much smaller as compared to general purpose machine tools.

22.2.6. Designing Layout for Mechanical Stepped Drives. As already discussed, a constant speed A.C. is used as the power source and the common mechanisms employed in the design of variable speed transmission are — Step-cone pulley drive, and all geared drive.

1. *Step-cone pulley drive.* The step-cone pulley and belt drive, Fig. 22.11 is one of the oldest and most widely used forms of stepped drive system. It gives definite steps of speed ranging from driven speeds considerably less than that of the driver to speeds much higher than that of the driver. The system has been popular for line-shaft drives and where the power is not very great, for individual gears. The driving shaft and hence the cone pulley mounted on it is driven from the electric motor with the help of a belt drive.

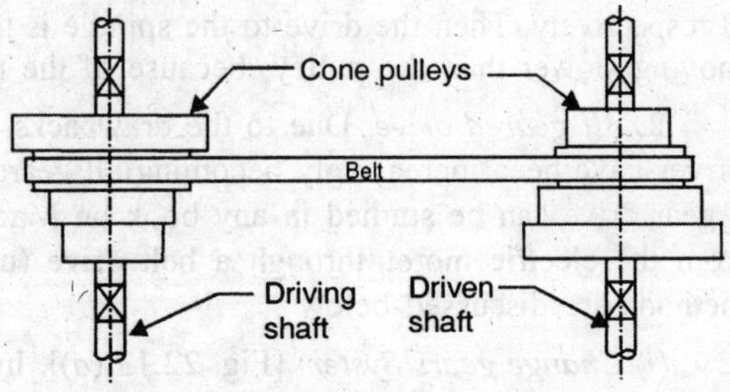

Fig. 22.11. Step-cone Pulleys Drive.

Flat-belt drives (open and crossed) as well as *V*-belts are used in this system. The shifting of *V*-belts from one set of pulleys to the other, is, however, more difficult than in the case of flat belts. But, frequently it is possible to mount one set of pulleys on a very short countershaft or on the motor shaft which can be easily moved while the belts are being changed.

The drive is very simple, and cheap in design but it occupies lot of space and the number of speed steps are limited. Also, belt drives (flat) are not positive drives. Shifting of belts also takes time affecting the rate of production.

Fig. 22.11 is a 4-step cone pulley drive. The number of steps can be increased by using a back gear arrangement.

Back-gear Arrangement. Fig. 22.12 shows a 3-step cone pulley drive with a back gear. The cone pulley is driven from a similar pulley on the countershaft which is driven at constant speed by the main shaft. Three speeds will be obtained by the cone pulley. The other 3 speeds are obtained by the back gear as follows — Gear *A* is keyed to and revolves with the cone pulley

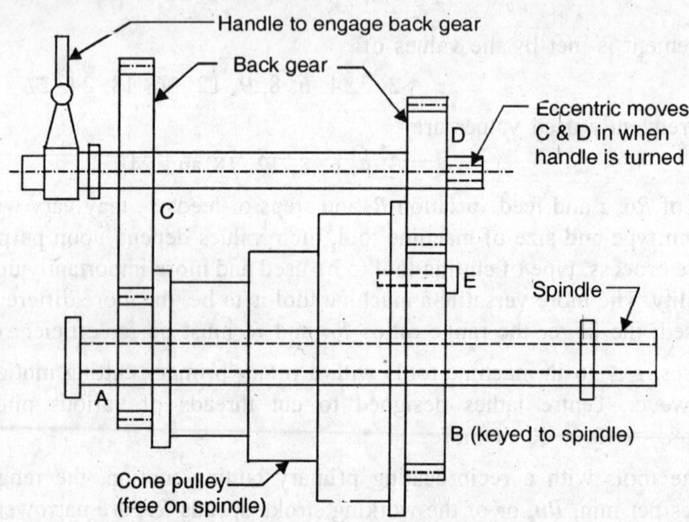

Fig. 22.12. Step Cone Pulley with Back Gears.

and gear *B* is keyed to the spindle. Gears *C* and *D* form one unit, move freely on the back shaft and may be moved in or out of mesh with gears *A* and *B*. If the locking pin *E* is withdrawn, the cone pulley is disengaged from gear *B* and is free on the spindle, otherwise the pulley is attached to B and hence to the spindle. The first three speeds (without the back gears) are obtained by engaging *E* and the cone pulley drives the spindle via the gear *B*. To get the 3 back gear speeds, *E* is withdrawn and the backshaft is moved so that the gears *C* and *D* mesh with gears *A* and *B* respectively. Then the drive to the spindle is transmitted from *A* to *C* and *D* to *B*, the spindle moving slower than the pulley, because of the reduction effect of the gears.

2. *All geared drive.* Due to the drawbacks of the step-cone pulley drive, the machine tool drives have been increasingly becoming all geared drives. The advantages and disadvantages of a gear drive can be studied in any book on Machine Design. The driving shaft is again driven from the electric motor through a belt drive (usually *V*-belt drive). Different speed changing methods are discussed below :

(*i*) *Change gears System* [Fig. 22.13 (*a*)]. In the selection of change gears, mounted on two shafts, the sum of the number of teeth on the various pairs of gears must remain constant. The gearing ratio of change gears is usually not over 2 : 1 and not less than 1 : 4.

Change gears are usually installed on the outer overhanging ends of the shafts. With this, they can be removed and a new pair installed with a minimum time loss. During operation, the change gears should not slip off. For this, they are installed securily on the shafts. Change gears are replaced only after switching off the drive of the machine tool. This system is used in spindle drives where speeds are to be changed infrequently it takes time to install change gears.

(*ii*) *Clutch type drive* [Fig. 22.13 (*b*)]. In this method, a jaw clutch is manually switched over which engages freely mounted gears to the driving or driven shafts. One or more than one clutch may be used. In the Fig. (*b*), when the clutch is shifted towards the left, gears 1 and 3 will engage and gears 2 and 4 will run idle and when the clutch is shifted to the right, the opposite will happen. So, with one driving speed, two driven speeds can be obtained. The toothed clutch can only be engaged at rest or when the relative speed of rotation does not exceed 0.7 m/s.

(*iii*) *Sliding gear type drive* [Fig. 22.13 (*c*)]. If a double cluster gear (mounted on the shaft with the help of splines) is shifted so that one of the gear rims of the cluster gear meshes with one of the gears on the second shaft, one of the two available speeds of the driven shaft may be obtained. Sliding gears may be shifted only with the rotation of the driving shaft switched off or disengaged and the setting is such that one set of gears must get completely disengaged before the other set starts getting engaged. However, speed changes without stopping can be permitted if surface speed of gears is less than 0.5 m/s. A cluster of 2, 3 or occasionally 4 gears can be used. This drive is very commonly used in machine tools.

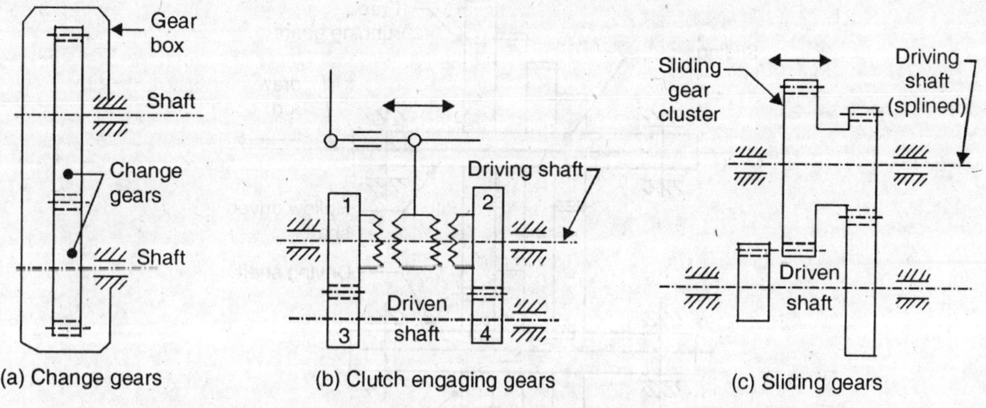

Fig. 22.13. Gear Drives

(*iv*) *Tumbler-gear or Norton gear-box drive.* Sec Fig. 22.14. The tumbler gear can slide on shaft S_1. It can mesh with any gear on shaft S_2 through an intermediate gear which is located on a swinging and sliding lever so that it can engage gears 1 to 8 of different diameters, on shaft S_2. The lever can be fixed in any desired ratio position with the help of a stop pin. The

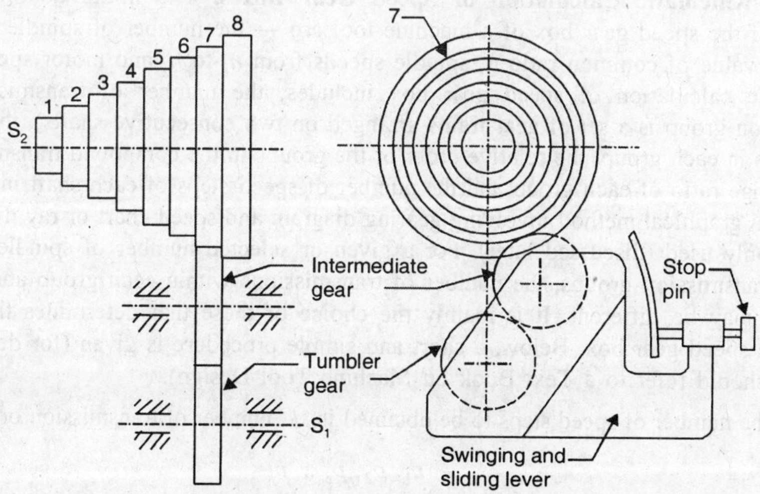

Fig. 22.14. Norton Type Gear Box.

main advantage of this drive (which is used as a feed drive) is the less number of gears and short length of the feed drive) is the less number of gears and short length of the shaft. However, the drawbacks are : low rigidity, requires a long odd shaped opening in the gear box wall for the operation of the swinging lever, difficult to protect the mechanism against dust and difficult to ensure proper lubrication. Closed type of Norton gear box eliminates some of these advantages.

(v) *Sliding key gear box* [See Fig. 22.15]. In this drive, a number of gears run free on a shaft (driven). Anyone of these can be engaged with the shaft by a sliding key which can be axially moved to engage one of the gears. An equal number of gears are permanently connected to the driving shaft. When a gear on the driven shaft is made to connect the driving shaft by a sliding key, a preselected ratio will be obtained. Main advantages are : compact construction, short length of assembly, control of engagement of a gear pair with a single lever. The major drawbacks are : long and deep keyways weaken the shaft, even those gears which do not transmit and power are in mesh and due to certain clearance between the key and keyway, local bearing area of the key is limited. Hence this type is used for transmitting relatively small torques.

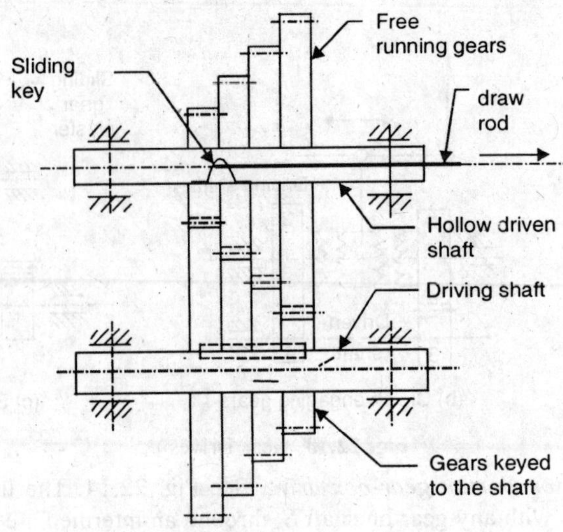

Fig. 22.15. Sliding Key Gear Box.

22.2.7. Kinematic Calculation of Speed Gear Boxes. The initial data in kinematic calculation of the speed gear box of a machine tool are — the number of spindle speed steps, z; a definite value of common ratio φ, spindle speeds from n_1 to n_z and motor speed n_m. Then the kinematic calculation of speed gear box includes; the number of transmission groups (a transmission group is a set of gear trains arranged on two consecutive shafts), the number of transmissions in each group, the relative order of the groups in the compound transmission train the speed range ratio of each group and the number of speed steps of each shaft in each group For this, semi graphical method involving gearing diagram and speed chart or ray diagram is the most commonly used (discussed later). For a given or selected number of spindle speeds, the number of transmission groups, the number of transmissions within each group and the group arrangement may be different. It is mainly the choice of these that determines the kinematic layout of the speed gear box. Below, a short and simple procedure is given (for detailed study the readers should refer to a Text Book on Machine Tool Design).

If z is the number of speed steps to be obtained in 'k' number of transmission or stages, then

$$z = p_1, p_2, p_3 \cdots p_k$$

where p_1 = number of speed steps in the first transmission group and so on. Refer to Fig. 22.16. In Fig. (*a*), the three gear sliding cluster of shaft S_1 and three gears to mesh with it on shaft S_2 form the first transmission group. The second transmission group consists of two gear sliding cluster on shaft S_2 and two gears on shaft S_3.

$$p_1 = 3,\ p_2 = 2$$

$$\therefore \quad z = p_1 \times p_2 = 3 \times 2 = 6$$

For Fig. (*b*), $\quad p_1 = 2$ and $p_2 = 3$ and $z = 2 \times 3 = 6$

For obtaining a particular number of speed steps using a minimum number of gears, it is necessary that

$$p_1 = p_2 = p_3 \dots p_k$$

Therefore, the number of speed steps in each transmission group can be found from the relationship :

$$p = (z)^{\frac{1}{k}}$$

For example, if $z = 8$ and $k = 3$, then

$$p = (8)^{\frac{1}{3}} = 2$$

If $(z)^{\frac{1}{k}}$ does not come out to be a whole number, then, as discussed already, equ. (22.27), speed steps should be divided such that,

$$z = 2^{E_1} . 3^{E_1}$$

For the most common values of z, the following variants may be used :

$z = 4 \quad = 2.2$

$z = 6 \quad = 2.3 = 3.2$

$z = 8 \quad = 2.2.2 = 2.4 = 4.2$

$z = 12 \quad = 2.2.3 = 2.3.2 = 3.2.2 = 3.4 = 4.3$

$z = 16 \quad = 2.2.2.2 = 4.2.2 = 2.4.2 = 2.2.4 = 4.4$

$z = 18 \quad = 2.3.3 = 3.2.3 = 3.3.2$

$z = 24 \quad = 3.2.2.2 = 2.3.2.2 = 2.2.3.2 = 2.2.2.3 = 2.3.4 = 2.4.3 = 3.2.4$
$= 3.4.2. = 4.2.3 = 4.3.2$

The optimum variant is selected by following the rules discussed ahead.

Note. As is clear from above, the maximum number of gears on the shafts for one group should be maximum equal to 4 and that too in exceptional cases. In general, this number should be upto 3.

In machine tools with G.P. series for spindle speed range, group transmission ratios constitute a G.P. series with the common ratio ϕ^x, where

x = group characteristic
= total speed steps number of transmission groups kinematically preceding the given group.

The general set up equation for group transmission may be expressed as :

$$i_1 : i_2 : i_3 : \dots i_p = \phi^x : \phi^{2x} : \dots \phi^{(p-1)x}$$

For the drive arrangement of Fig. 22.16 (*a*), the formula for the structure of the drive will be :

z = 3[1].2[3], where the digits in the square brackets denote the group characteristics.

The first transmission group is called the main transmission group and the subsequent ones as extension groups. It is clear that for the main group, $x = 1$ and for the extension group $x_1 = p_1 = 3$, hence the formula.

For Fig. 22.16 (*b*), z = 2[1].3[2]

As discussed in the last article, the basic unit of speed change device is the 2-axis drive. Due to space limitations or when the number of teeth on gear or the surface velocity of the gears have to be limited, it becomes necessary to limit the transmission ratio. The following can be taken as a guide :

$$\frac{1}{4} \leq i \leq \frac{2}{1}, \text{ for spur gears} \qquad ...(22.28)$$

For helical gears; i_{max} can be taken as 2.5.

For feed gear boxes which have slow speed gearing and small diameter gears, the ratio can be :

$$\frac{1}{5} \leq i \leq 2.8, \qquad ...(22.29)$$

Thus for a 2-axis gear box, the limiting maximum range ratio, of the group

$$i_g = (R_n)_{max} = \left(\frac{i_{max}}{i_{min}}\right)_{lim} = \frac{2}{1/4} = 8 \qquad ...(22.30)$$

Gear boxes with larger speed ranges and with greater number of steps can be designed by arranging several 2-axes drives in series.

Some other points to be kept in mind while designing a gear box are :

1. The minimum number of teeth in a set of gears should be greater than 17 (preferably equal to 20) to avoid interference.

2. Since the centre distance between two shafts is constant, therefore, in sliding gears, the sum of the number of teeth of mating gears must be constant.

3. The minimum difference between the number of teeth of adjacent gears must be 4.

The layout of a gear box with more than 2 axes can be shown graphically on a ray diagram or speed chart. Ray diagrams for a 6 spindle speed gear box are shown in Fig. 22.16. Speeds are plotted vertically on a logarithmic scale and the shafts are shown as vertical parallel lines at equal distance from each other. Distance between consecutive speed lines is constant because of G.P. series, as given below :

$$\frac{n_2}{n_1} = \frac{n_j}{n_{j-1}} = \phi$$

$$\therefore \qquad \log n_j - \log n_{j-1} = \log \phi = \text{constant} \qquad ...(22.31)$$

Thus the various speed values on log scale appear at equal intervals (equal to log ϕ See Fig 22.6) and transmission ratios between 2 axes are indicated by the vertical distances between the corresponding speed values. As the distance between the axes are shown equal, the slope of the lines joining the speed values on different axes represent the transmission ratios. The ray diagrams show at a glance, the kinematic arrangement, shaft speeds at different stages and torques at various speeds. An optimum ray diagram is the one which will result in a compact gear box with less number of gears, shafts, bearings and shifting levers resulting in reduction of overall cost. For this, the following points should be kept in mind when drawing a ray diagram :

1. Input point should be preferably located towards the higher speed to get lower transmission ratios between the motor shaft and the input shaft.

2. The speed on the intermediate shaft should be as high as possible in order to reduce the size of gear box.

3. Tool high torques and too high (drastic) gears reductions should be avoided by suitably selecting the speed of the driving spindle.

4. Number of gears on the last shaft (spindle shaft) should be minimum possible.

The following constraints should be taken into account when deciding the optimum diagram :

1. *Ray constraint.* To reduce the size of gear box, the maximum number of teeth on a gear should not be more than 120. Since, the minimum number should be 20, therefore,

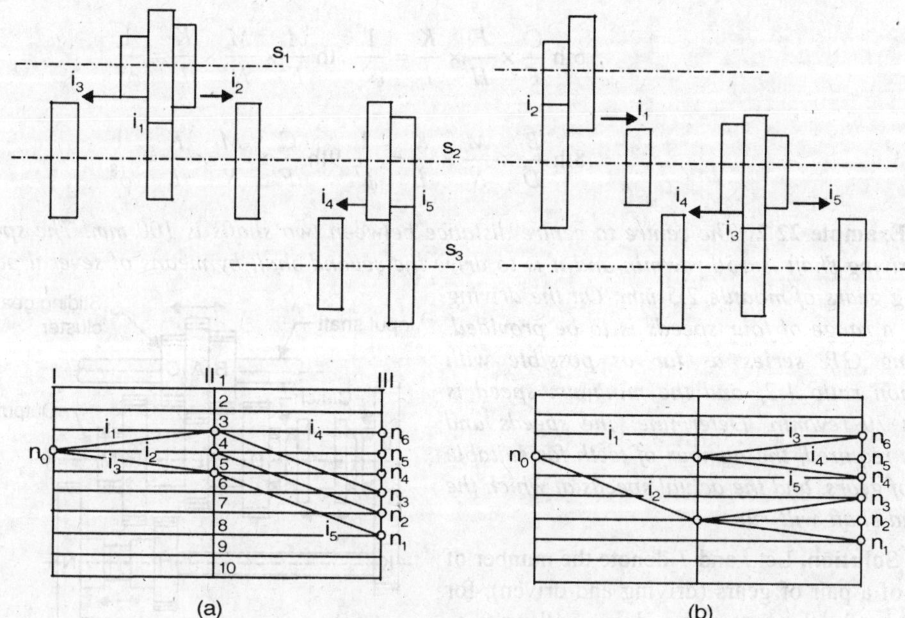

Fig. 22.16. Ray Diagram

$$(\phi)^m_{\max} = \frac{120}{20} = 6$$

where m = number of intervening spaces, that is, the number of intervals spanned by a ray,

2. *Stage constraint.* We have already seen, that

$$i_g \leq 8, \therefore \left(\frac{i_{\max}}{i_{\min}}\right)_{\lim} = (\phi)^{(p-1)} x_{\max} \leq 8 \qquad \text{...(22.32)}$$

where $x_{\max}$ = characteristic of last extension group.

and p = number of transmissions in this group.

3. **Node method of optimization.** Nodes in a ray diagram are the points from where a ray starts or terminates. These are numbered from the maximum speed towards the minimum speed. The number of nodes in ray diagrams are added and compared. The ray diagram which gives the least sum of the nodes will be the best layout. For example in Fig. 22.16, Layout (*a*) is better, because,

For (*a*) $\sum \text{nodes} = 4+5+8 = 17$

For (*b*) $\sum \text{nodes} = 4+7+8 = 19$

Fig. 22.16 (*a*) gives cross-layout, while Fig. 22.16 (*b*), gives an open layout.

Fig. 22.17 shows a 9-spindle speed gear box, utilising two sliding gear clusters. The speed changes are obtained as under :

$$1\text{st } \frac{A}{E} = \frac{1}{1}, 2nd\ \frac{B}{D} = \frac{1}{\phi}\ 3\text{rd} = \frac{C}{F} = \frac{1}{\phi^2}$$

$$4\text{th } \frac{A}{E} \times \frac{F}{H} \times \frac{K}{L} = \frac{1}{\phi^3}\ 5\text{th } \frac{B}{D} \times \frac{F}{H} \times \frac{K}{L} = \frac{1}{\phi^4}$$

$$6\text{th } \frac{C}{F} \times \frac{F}{H} \times \frac{K}{L} = \frac{1}{\phi^5}\ 7\text{th } \frac{A}{E} \times \frac{M}{N} \times \frac{K}{L} = \frac{1}{\phi^6}$$

$$8\text{th } \frac{B}{D} \times \frac{M}{N} \times \frac{K}{L} = \frac{1}{\phi^7}\ 9\text{th } \frac{C}{F} \times \frac{M}{N} \times \frac{K}{L} = \frac{1}{\phi^8}$$

Example 22.2. *The centre to centre distance between two shafts is 100 mm. The speed of the driving shaft is 150 rev/min and it is to drive the second shaft by means of several pairs of sliding gears of module 2.5 mm. On the driving shaft, a range of four speeds is to be provided, forming G.P. series as far as possible with common ratio 1.2, and the minimum speed is about 70 rev/min. Determine: the speeds and ratios required, the number of teeth for suitable pair of gears, and the actual speeds at which the second shaft will run.*

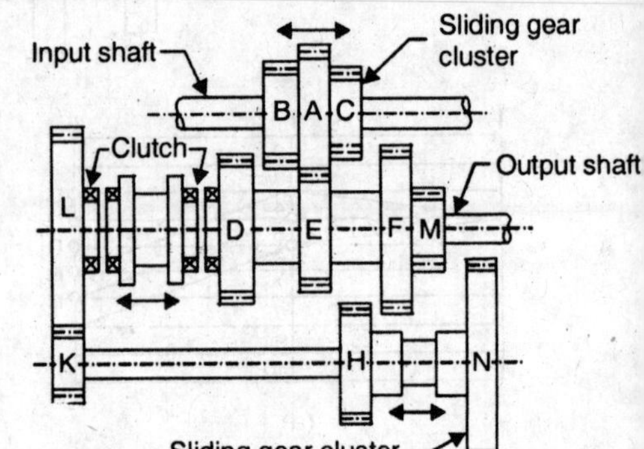

Fig. 22.17. Nine-Spindle Speed Gear Box.

Solution. Let t and T denote the number of teeth of a pair of gears (driving and driven), for example, for the lowest speed, we will write t_1, T_1 and speed n_1.

now $\frac{1}{2}$(sum of diameters of pair of gears)

= 100 mm

∴ Diameter of driver gear + diameter of driven gear = 200

∴ $m(t+T) = 200$, m = module

∴ $t + T$ = sum of teeth of a pair of gears $= \frac{200}{2.5} = 80$

now $n_1 = 70$ rev/min

∴ $n_2 = \phi n_1 = 1.2 \times 70 = 84$ rev/min

$n_3 = \phi^2 n_1 = 100.8$ rev/min

and $n_4 = \phi^3 n_1 = 120.96$ rev/min

now $$i = \frac{\text{Speed of driven shaft}}{\text{Speed of driving shaft}}$$

$$\therefore \quad \frac{t_1}{T_1} = \frac{70}{150} = \frac{7}{15}, (t_1 + T_1 = 80)$$

$$\therefore \quad \text{From here, } t_1 = 25\frac{5}{11} \text{ and } T_1 = 54\frac{6}{11}$$

$\therefore$ For nearest practicable ratio : number of teeth on driver = 25

and number of teeth on driven gear = 55

Similarly $$\frac{t_2}{T_2} = \frac{84}{150}$$

From here, $$t_2 = 28\frac{28}{39} \text{ and } T_2 = 51\frac{11}{39}$$

$\therefore$ Nearest approximation is : Driver = 29 teeth, Driver = 51 teeth.

And like this, we can obtain :

$$t_3 = 32, T_3 = 48$$

$$t_4 = 36, T_4 = 44$$

$\therefore$ The actual running speeds of the driven shaft will be :

$$n_1 = 150 \times \frac{25}{55} = 68.18 \text{ rev / min}$$

$$n_2 = 150 \times \frac{29}{51} = 86 \text{ rev/min}$$

$$n_3 = 150 \times \frac{32}{48} = 122.73 \text{ rev/min}$$

and $$n_4 = 150 \times \frac{36}{44} = 122.73 \text{ rev/min.}$$

Example 22.3. *A gear box of a drilling machine is to be designed to give speed variation between 100 and 180 rev/min in 6 steps. The speed of the input shaft is 225 rev/min. The intermediate shaft is to have 3 speeds.*

Solution. $$\phi = (R_n)^{\frac{1}{z-1}}$$

now $$z = 6, Rn = \frac{n_{max}}{n_{min}} = \frac{180}{100} = 1.8$$

$$\therefore \quad \phi = (1.8)^{0.2} = 1.124, \text{ say } 1.12 \text{ (standard in G.P.)}$$

$\therefore$ the six spindle speeds are : 100, 112, 126, 142, 160, 180 rev/min.

There can be many designs of gear box. But a suitable one is shown in Fig. 22.18. The three speeds of intermediate shaft are :

142, 160 and 180 rev/min.

Now from the rule of sliding gears :

$$T_a + T_b = T_c + T_d = T_e + T_f$$

and
$$T_h + T_j = T_g + T_i$$

Also
$$\frac{n_a}{n_b} = \frac{T_b}{T_a} = \frac{225}{142} = 1.58$$

Again
$$\frac{T_d}{T_c} = \frac{n_c}{n_d} = \frac{225}{160} = 1.40$$

and
$$\frac{T_f}{T_e} = \frac{n_e}{n_f} = \frac{225}{180} = 1.25$$

Now
$$\frac{T_b}{T_a} = 1.58 = \frac{32}{20} = \text{(taking minimum number of teeth as 20).}$$

$$\therefore \quad T_b = 32,\ T_a = 20$$

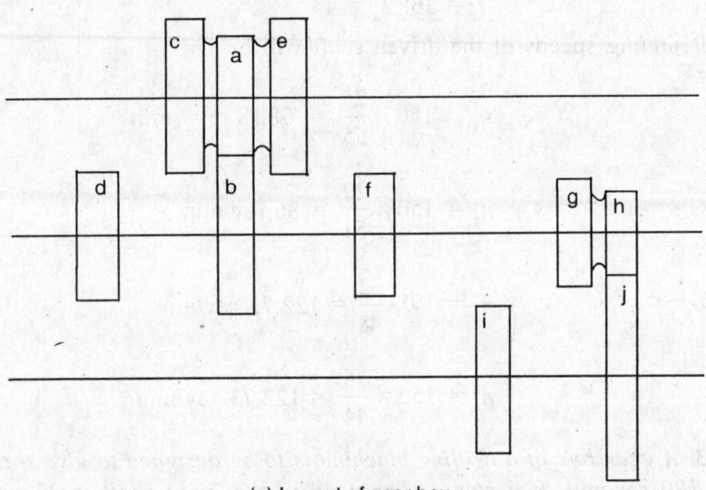

(a) Layout of gear box

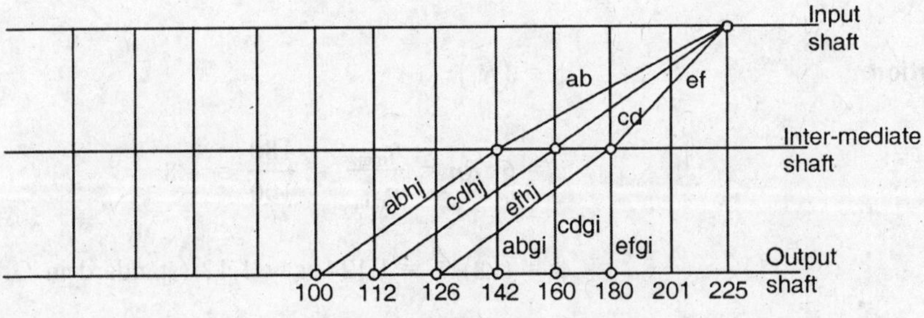

(b) Ray Diagram

Fig. 22.18. A 6-Speed Gear Box.

now $$\frac{T_d}{T_c} = 1.40 \text{ and } T_c + T_d = 52$$

$\therefore$ $$T_c = 22 \text{ and } T_d = 30$$

$$\frac{T_f}{T_e} = 1.25 \text{ and } T_e + T_f = 52$$

$$T_e = 23 \text{ and } T_f = 29$$

Again $$\frac{T_h}{T_j} = \frac{n_j}{n_h} = \frac{100}{142} = 0.7$$

and $$\frac{T_i}{T_g} = \frac{n_g}{n_i} = \frac{142}{142} = 1$$

Also $$T_h + T_j = T_i + T_g$$

$$\frac{T_h}{T_j} = 0.7 = \frac{19}{27}$$

$\therefore$ $$T_h = 19 \text{ and } T_j = 27$$

$\therefore$ $$T_i = T_g = 23$$

Example 22.4. *A gear box for a drilling machine is to give six spindle speeds. If the size of the drills to be used is from 5 to 20 mm diameter and the cutting speed is 21 m/min. Calculate the common ratio for G.P. series of speeds and hence the various spindle speeds.*

Solution. Let us first calculate the maximum and minimum spindle speeds

$$n_{\text{max}} = \frac{1000v}{\pi\, d_{\text{min}}} = \frac{1000 \times 21}{\pi \times 5} = 1336.90 \text{ rev/min}$$

$$n_{\text{min}} = \frac{1000\, v}{\pi d_{\text{max.}}} = \frac{1000 \times 21}{\pi \times 20} = 334.23 \text{ rev/min}$$

$\therefore$ $$\phi = \left(\frac{n_{\text{max}}}{n_{\text{min}}}\right)^{\frac{1}{z-1}} = \left(\frac{1336.90}{334.23}\right)^{0.2} = 1.32$$

Various spindle speeds are :

Now $$n_1 = 334.23 \text{ rev/min}$$

$\therefore$ Spindle speeds are : $n_1, \phi n_1, \phi^2 n_1, \phi^3 n_1, \phi^4 n_1$ and $\phi^5 n_1$ that is, 334.23, 441.2, 582.36, 768.72, 1014.78 and 1336.9 rev/min.

Example 22.5. *The maximum speed at the input shaft of a gear box is to be 1400 rev/min. The gear box is to provide six speeds based upon the 'R10' series. There are three shafts and three changes are required between the input and intermediate shafts, and two changes between the intermediate and output shaft. Determine :*

(a) the gear ratios between the shafts using a speed diagram

(*b*) *the number of teeth in each gear*

(*c*) *the actual value of the six output speeds*

Solution. The '*R* 10' series in G.P. means that the common ratio,

$$\phi = (10)^{\frac{1}{10}} = 1.26$$

(*a*) Refer to Speed diagram, Fig. 22.18 A

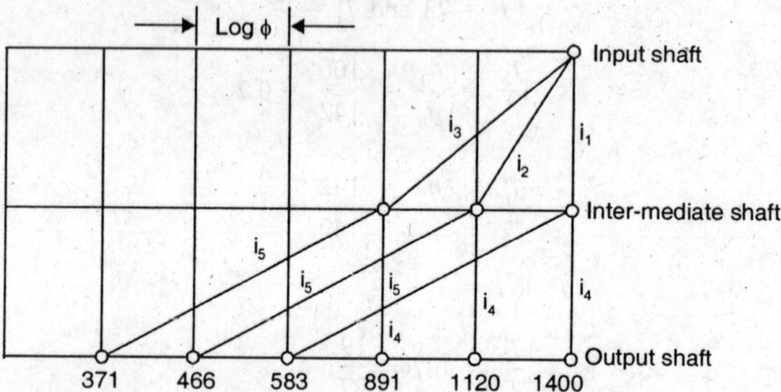

Fig. 22.18A. A Speed Diagram.

The three gear ratios between the input and intermediate shafts are :

$$i_1 = \frac{1}{1},\ i_2 = \frac{1}{1.26},\ i_3 = \frac{1}{1.26^2},$$

The two ratios between the intermediate and output shafts are :

$$i_4 = \frac{1}{1},\ i_5 = \frac{1}{1.26^3}$$

(*b*) To obtain the number of teeth on each gear :

(*i*) For input to intermediate shaft, theoretical

$$L.C.M. = (1 + 1)(1 + 1.26)(1 + 1.26^2)$$

Let $$\frac{1}{1.26} = \frac{4}{5} \text{ (approx) and } \frac{1}{1.26^2} = 0.63 = \frac{7}{11} \text{ (approx.)}$$

Thus $$LCM = (1+1)(4+5)(11+7) = 2 \times 9 \times 18$$

Let us select, 18×3 as the sum of teeth on any pair of gears, that is equal to 54.

∴ Number of teeth on the three pairs of gears will be :

$$\frac{27}{27}, \frac{24}{30}, \frac{21}{33}$$

(*ii*) For the intermediate shaft to output shaft :

$$\text{theoretical LCM} = (1 + 1)(1 + 1.26^3)$$

Now $$\frac{1}{1.26^3} = \frac{5}{12} \text{ (approx.)}$$

$\therefore \quad$ theoretical LCM $= (1+1)(12+5) = 2 \times 17 = 34$

Now it is better to take minimum of teeth on any gear as 20, so,

$$\frac{5}{12} = \frac{20}{48},$$

$\therefore$ Total number of teeth on two pairs of gears = 68

$\therefore$ Number of teeth on two pairs of gears will be :

$$\frac{34}{34} \text{ and } \frac{20}{48}$$

(c) *Output speeds* :

$$n_6 = \frac{27}{27} \times \frac{34}{34} \times 1400 = 1400, \quad n_5 = \frac{24}{30} \times \frac{34}{34} \times 1400 = 1120$$

$$n_4 = \frac{21}{33} \times \frac{34}{34} \times 1400 = 891, \quad n_3 = \frac{27}{27} \times \frac{20}{48} \times 1400 = 583$$

$$n_2 = \frac{24}{30} \times \frac{20}{48} \times 1400 = 466 \text{ and } n_1 = \frac{21}{33} \times \frac{20}{48} \times 1400 = 371 \text{ rev/min.}$$

22.3. STEPLESS MECHANICAL DRIVES

Stepless or infinitely variable main and feed drives have found considerable application in modern machine tools. Their main advantages are : the possibility of setting up the optimum cutting conditions (speeds and feeds) with higher accuracy than with a stepped drive and the possibility of changing speeds of the main drives or feeds without stopping the machine. The stepless, drives can be : mechanical, electrical, hydraulic or combined drive. They have their own advantages and disadvantages. The selection will depend upon : purpose of the machine (general purpose, or special); for roughing, finishing or microfinishing: range ratio, the power required and the cost.

Mechanical stepless drives are of : friction type and positive type. The operation of friction type drives involves friction losses. Other drawbacks of mechanical drives are : non-rigid kinematic characteristics and the variations in the maximum transmitted power when the speed is changed.

22.3.1. Friction Type Stepless Mechanical Drives. These drives are based on the principle : the driven link contacts the driving link either directly or through some intermediate element (roller, disc, ring or belt). The driven and driving elements are held tightly together and the friction force developed will cause one element to rotate when the other is rotated. If the diameter of contact on both the elements (driving and driven) or on at least one of them is varied, then the transmission ratio of the drive will vary accordingly. There are many designs of friction type devices. Some of these will be discussed below :

1. *Roller and disc drive.* This drive, Fig. 22.19 is a very elementary friction drive. A single control lever permits smooth variation in speed ratios over a wide range from zero when the roller is over the recessed portion of the disc to a maximum when the roller contacts the disc at its outer end. The transmission ratio being given as :

$$i = r/R \text{ (R being variable)} \quad \text{...(22.32)}$$

Direction of rotation can be completely reversed by bringing the friction roller into contact

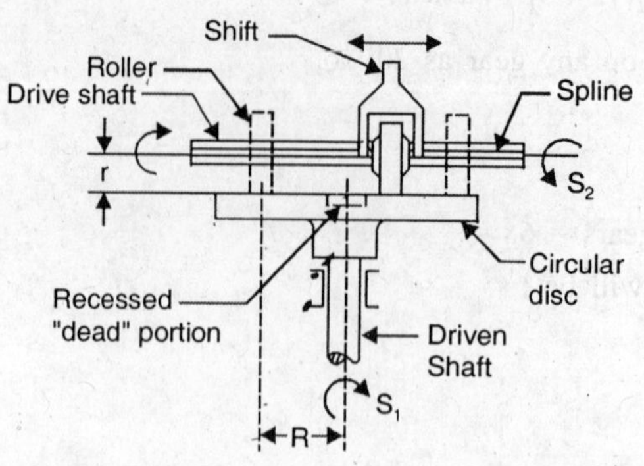

Fig. 22.19. Disc and Roller Drive.

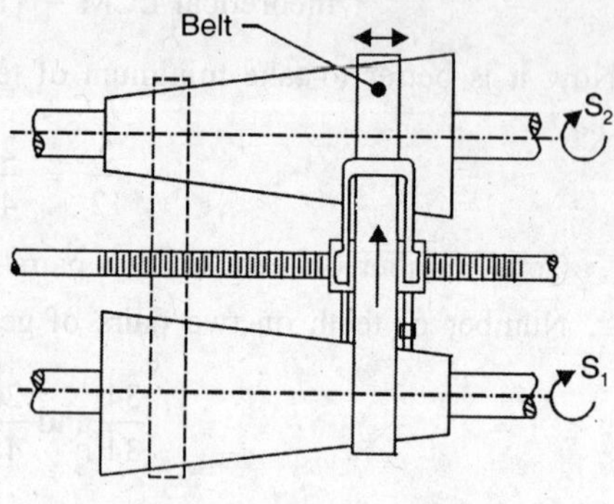

Fig. 22.20. Friction Cone Pulley Drive.

with the disc on either side of the recessed centre portion. The roller must be mounted on a splined shaft to move it across the face of the disc, while being positively driven by the shaft. Drawbacks of this drive are : uneven wear of disc and the rapid wear of the roller. Maintaining proper pressure between the contact surfaces is another cause for trouble.

2. *Friction cone with flat belt drive, Fig. 22.20.* The principle of this drive is the same as that of the stepped cone pulley and belt systems, except that there are no fixed steps but/ rather it provides the possibility of very slight changes in speed ratios over a wide range of adjustment. Adjustment of speed is obtained by means of a belt shifting device shown. To prevent the belt from slipping, the cone angle is usually 10°. The drive is not suitable for large powers due to the flexibility of the belt.

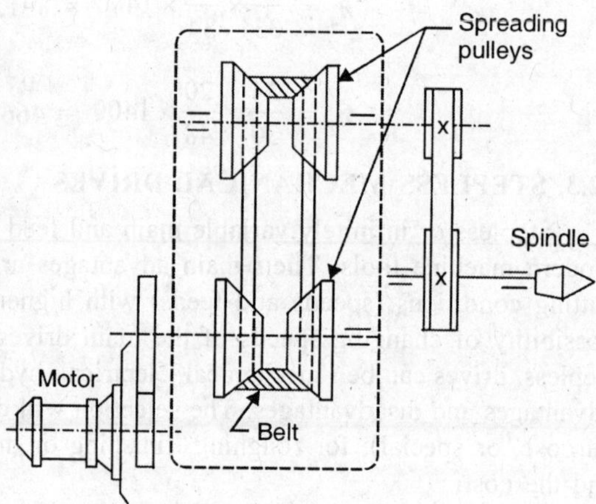

Fig. 22.21. Spreading Pulleys Drive.

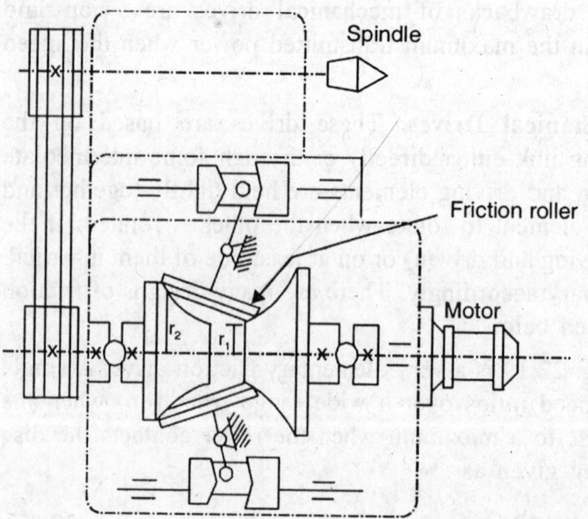

Fig. 22.22. Conical Discs and Friction Roller Drive.

3. *Spreading Conical pulleys drive, Fig. 22.21.* The distance between the pairs of driving and driven cone pulleys can be changed by axially moving one member in each pair. With this the diameter of contact of the belt with driving or driven pairs of cone pulleys can be hardened metal ring as the frictional member can be used.

4. *Conical discs and friction roller drive, Fig. 22.22.* Friction rollers, arranged between spherically-shaped cones or discs on the driving and driven shafts may be inclined in different positions. This will change the radii of contact between the roller and the driving and

driven discs and thereby the transmission ratio. In the position shown in the figure, (*n*) driven (*n*) driver (r_2 being r_1).

When the axes of the top and bottom rollers coincide vertically, the speeds of driving and driven pulleys will be same. The speed of the driven shaft is transmitted to the spindle through a *V*-belt drive. By changing the position of the rollers along the spherical surfaces of cones (driving and driven), spindle speeds can be infinitely varied from the minimum to the maximum values. The range can be 4 to 8. By combining this friction drive (Svetozarov drive) with a 3-stage geared headstock, a still wider range of infinitely variable speeds can be obtained.

Positive Drive. The positive infinitely variable (P.I.V.) drive is a variant of "Spreading conical pulleys friction drive" in which a chain is used in place of a belt and the conical faces of the alloy-steel wheels are grooved. The self tooth forming chain engages with these grooves. The wheels mounted on splined shafts are free to move laterally. A control handwheel moves the control levers about central pivolts to change the ratio of the effective wheel diameters. The chain links are slidable transferse slots that form the power transmitting teeth. A shoe mechanism, with spring tension, applies pressure to the slack side of the chain to keep it in adjustment. All moving parts run in oil with automatic splash lubrication.

The range ratio of mechanical stepless drives depends upon the principle and construction of the device. It may range from about 2 to 4 (with wide *V*-belts and adjustable sheaves) to about 10 to 25 (chain and ball type drives). In most cases, range ratio = 4 to 6.

22.4. HYDRAULIC DRIVES

Hydraulic drives are widely used to obtain infinitely variable rates of rectilinear motion in machine tools. Mostly, it refers to feed motions but in some machine tools (Planers, Shapers, Slotters and Broaching machines) main drive speeds are varied in this way. The advantages of hydraulic drives are :

1. Rapid and infinitely variable adjustment obtainable during operation, for the length, speed and direction of travel of a machine tool unit.

2. Faster, reverse and acceleration rates are possible because of less inertia and cushioning effect of the fluid.

3. The drive is smooth and reverses without shock.

4. The drive stops when fluid pressure reaches a preset maximum and tool breakage is less likely than with a mechanical drive.

5. Automatic protection against overloads.

6. Ability to stall against an obstruction without damage to the tool or to the machine.

7. Convenience of remote control and its automation.

8. Self lubrication of the system.

Drawbacks :

1. Insufficiently flat characteristic curve resulting from leakage.

2. Effect of the temperature on the oil viscosity.

3. At slow speeds (12 to 15 mm/min), the operation of the drive becomes unstable.

A hydraulic drive for straight line motion consists of :

(*a*) A pump, which delivers oil under pressure

(*b*) An operating cylinder with a piston and piston rod

(c) Control devices.

Either the piston rod or the cylinder is attached to the operative unit of the machine tool and the other is stationary. The speed of travel of the operative unit will depend on the volume of oil delivered by the pump in a unit of time. This speed can be adjusted in two ways:

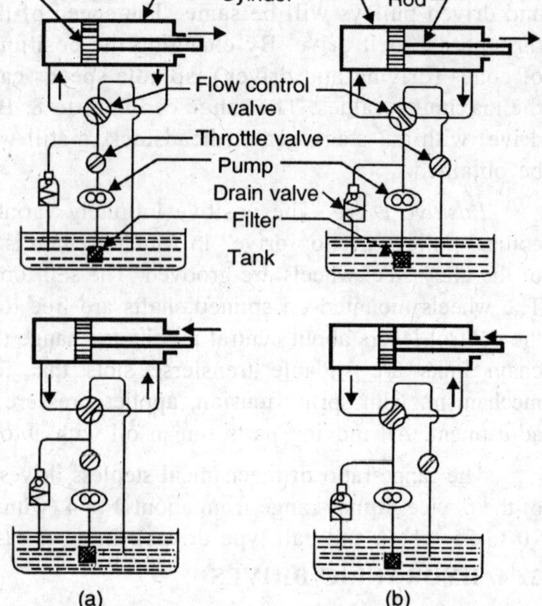

Fig. 22.23. Throttle Type Speed Control.

(i) by using a variable delivery pump which can deliver different amounts of oil per unit of time.

(ii) by using a constant delivery pump and adjusting the speed with the help of a throttle valve.

The variable delivery pumps are more expensive and complex than constant delivery pumps. No excess oil is pumped, except for a small amount of slippage and leakage. This system is more efficient than a throttle type system. However, in many cases, the difference is not large and the saving in operating costs alone does not justify the more expensive equipment.

There are two ways of controlling the speed in the throttle type system :

(a) *Metering in Control, Fig. 22.23* (a). The throttle valve is mounted in the pipeline through which oil is delivered to the cylinder.

(b) *Metering-out Control, Fig. 22.23* (b). Here, the throttle value governs the discharge oil from the opposite end of the cylinder. The amount of oil passing through the adjustable throttle valve per unit time, will determine the piston speed. If the amount of oil delivered to the working end of the cylinder is too large, surplus oil is drained to the tank through a relief valve.

Pumps. As mentioned above, the pumps used in the hydraulic drive can be :

1. Constant delivery pumps (non-adjustable)
2. Variable delivery pumps (adjustable)

According to their principle of operation and design, pumps are classified as,

(i) Gear type pump (ii) Vane type pump (iii) Piston type pump

(a) **Gear Pump.** It is the simplest and the most commonly used type of pumps. It is a constant displacement pump, and is designed for operating pressure upto 25 atm. It consists of two spur gears, Fig. 22.24 (a), of equal diameter. One of the gear is driven by an electric motor. They operate by the suction of oil through suction pipeline by the rotating gears. The oil is carried in the tooth spaces to the discharge pipeline. The discharge pressure can be increased by increasing the speed of the pump and the number of teeth of the gears.

(b) **Vane Pumps.** In this pump a rotor (whose axis is eccentric to that of the assembly) revolves in a housing. The rotor carries vanes located in slots of the rotor. These pumps, Fig. 22.24 (b), also have a constant displacement and are designed for working pressures upto 65 atm.

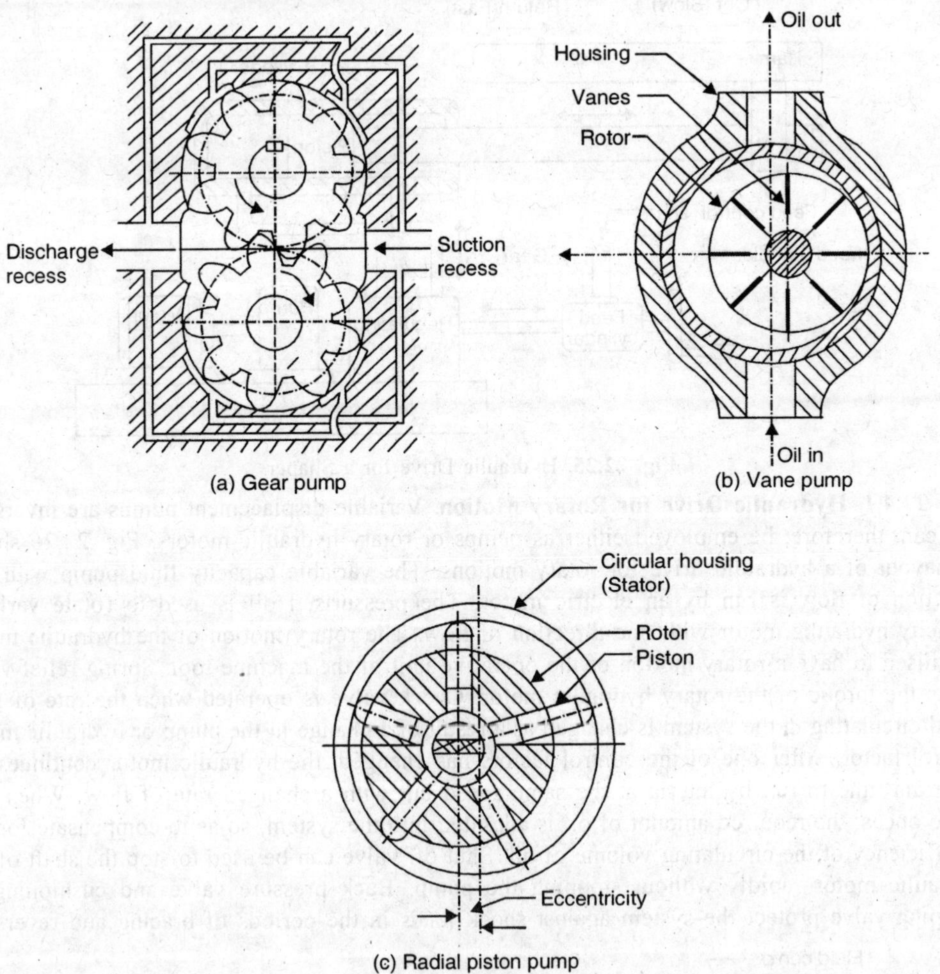

Fig. 22.24. Pumps.

(*c*) **Rotary piston pumps.** These pumps may be designed with either radially or axially arranged pistons. They may operate at pressures upto 100 atm. In the radial arrangement Fig. 22.24 (*c*), the cylinders rotate centrically in the core of a circular housing. The oil inlet and outlet ports are located near the centre of the rotating cylinder block. When the rotating pistons move outward, oil will be sucked in the rotating cylinder and it will be discharged out when the piston moves inwards. Radial piston pumps may contain from 5 to 126 pistons arranged in one, two, three or four rows, depending on the pump delivery. These pumps have delivery ranges of 0.2 to 4 *l*/min for small size models and 18 to 600 *l*/min for large size models.

Hydraulic drives with a power rating of ≥ 0.45 kW are economically sound. A hydraulic drive is advisable for developing high torques and pulling forces. Its cost is less as compared to electric drive of the same rating. The application of hydraulic drive in copy system (copy turning, copy milling) and flow forming lathes is also of great importance.

Fig. 22.25 shows a hydraulic drive for a shaper.

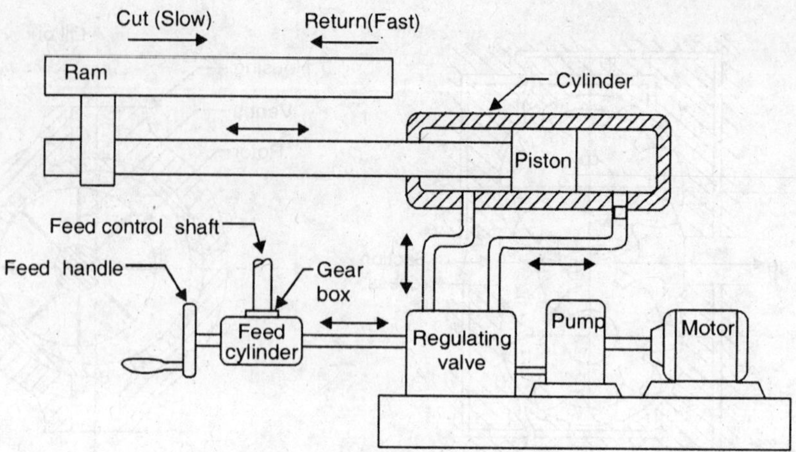

Fig. 22.25. Hydraulic Drive for a Shaper.

22.4.1. Hydraulic Drive for Rotary Motion. Variable displacement pumps are inversible and can, therefore, be employed either as pumps or rotary hydraulic motors. Fig. 22.26 shows the layout of a hydraulic drive for rotary motions. The variable capacity fluid pump with one direction of flow is run by an electric motor. The pressurised oil is used to rotate variable capacity hydraulic motor with one direction of flow. The rotary motion of the hydraulic motor is utilised to have a rotary motion of the operative unit of the machine tool. Spring relief valve limits the torque of the rotary hydraulic motor. Check valve is operated when the rate of flow of oil circulating in the system is changed as a result of a change in the pump or hydraulic motor control factor. After one of the control factors has changed, the hydraulic motor continues for a certain time to run by interia at the same speed but with a changed rate of flow. When this valve opens, the required amount of oil is admitted into the system, so as to compensate for the insufficiency of the circulating volume of oil. Shut off valve can be used to stop the shaft of the hydraulic motor rapidly without stopping the pump. Back pressure valve and cushioning or damping valve protect the system against shock loads in the periods of bracing and reversal.

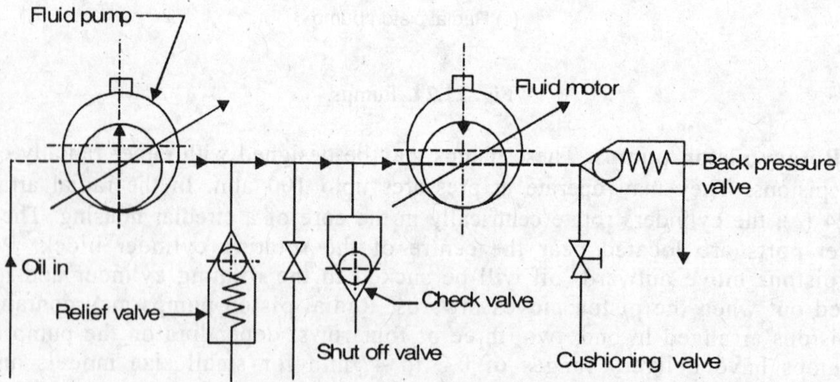

Fig. 22.26. Hydraulic Motor.

The motors for rotary hydraulic drives are particularly complex, expensive and of low efficiency after wear and are not common.

22.5. PNEUMATIC DRIVES

Compressed air is already available in most plants and can be put to work with rather inexpensive equipment. Air flow is fast, but the use of compressed air has several disadvantages that limit to light service. Pressures available are usually not high. A compressed air system by itself is hard to control because of the compressibility of air. Feeds and speeds are inclined to vary too much as the load changes and the equipment may not stop and reverse within desired limits. An air driven but hydraulically controlled circuit, Fig. 22.27, mitigates some of the shortcomings of air, Compressed air is admitted into the oil tank at a pressure of 0.4 to 0.5 N/mm^2. Consequently, oil will pass through the pipe line and check valve to the left of the cylinder. This moves the piston to the left and withdraws the toolslide from the work. Air from the right end of the cylinder will be released to the atmosphere. When the tool slide is to be advanced to the work, a three way valve is turned so that compressed air is admitted to the right hand end of the cylinder and not to the oil tank. The speed of the piston, however, will be restricted by the oil draining from the cylinder as it can pass only through the throttle valve. Thus, the speed of the piston (toolslide) depends on the size of the slit in the throttle valve through which oil flows.

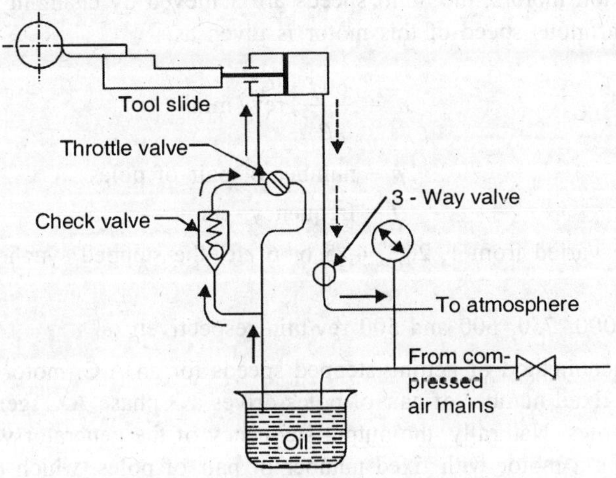

Fig. 22.27. Hydro-pneumatic Drive.

22.6. ELECTRICAL DRIVES

The trend in the development of machine tools drives has been towards more complete motorisation. The electric derives of machine tools comprise one or several drive motors and devices for their control. Up to date electric motors can be reversed or braked. The motorisation of machine tools and an ever wider use of electric controls will continue to be one of the principal factors in their improvement, in increasing the rate of production and reliability of machine tools and in reducing operator fatigue.

Three phase squirrel cage motors for a power supply of 220/380 or 500 V, 50 c/s are most frequently employed in machine tools. Such motors have a flat characteristics, that is, their speed falls only slightly with an increase of load. Three phase electric motors have speeds in rpm near to the following series : 3000, 1500, 1000, 750, 600 and 500.

Shunt wound direct current motors also have a flat characteristics. The speeds of these motors are usually infinitely variable in a speed range ratio of 3 or 4. The main advantage of these motors is that their speeds are infinitely variable (practically with a very small difference 4 to 5 percent between adjacent steps).

Electric motors permit momentary overloads upto 1.5 to 2 times the rated values. This feature is used in starting a motor since a considerable overload is experienced at this time.

(*a*) *Stepped Electrical Drives.* The electric motors used in mechanical and hydraulic drives are ordinary three - phase electric motors having only one shaft speed. For stepped electrical drives, multi-speed squirrel cage induction motors and shunt wound D.C. motors are used. These are used invariably with mechanical gear boxes (Electro-Mechanical regulation). The use of multi-speed electric motors enable the mechanical part of the drive to be simplified. However, the multi-speed motors are more expensive. The possibility of changing speeds while the machine tool is running is a great advantage of multi-speed motor. Hence, they are often used in the drives of small machine tools in conjunction with transmissions that can be readily changed over without stopping the machine tool, for example, gearing in which speeds are changed by the engagement of mechanical, electro-magnetic or hydraulic friction clutches, variable speed transmission etc. Such an arrangement reduces the handling time when the machining time is very short, and also to automatically change spindle speeds and feeds during the working cycle in automatic machine tools of various sizes.

In A.C. induction motors, the multi speeds are achieved by changing the number of pairs of poles. The synchronous speed of this motor is given as,

$$n_s = \frac{60f}{p}, \text{ rev / min} \qquad ...(22.33)$$

where p = number of pair of poles

f = frequency, c/s

Therefore p is varied from 1, 2, 3, 4, 5 or 6 etc. the stepped synchronous speeds of the motor will be :

3000, 1500, 1000, 750, 600 and 500 rev/min respectively.

There is another method of getting stepped speeds for an A.C. motor drive. A standard 3 phase motor with a fixed number of pair of poles drives a 3 phase A.C. generator with variable number of pair of poles. Naturally, the output frequency of the generator will be variable. This generator then drives a motor with fixed number of pair of poles, which ultimately drives the machine operative unit. The synchronous speed of this motor, Fig. 22.28 (*a*) will be given as :

$$n_s = \frac{60f}{p}\left(\frac{p_g}{p_m}+1\right) \qquad ...(22.34)$$

where f is the frequency of the main supply, and p, p_g and p_m are the number of pairs of poles in the driving motor (final), generator and the standard A.C. motor respectively.

The various stepped speeds in a D.C. motor can be obtained by :

(*i*) by changing the voltage drop in the armature circuit.

(*ii*) by varying the field flux.

(*iii*) by varying the power circuit voltage.

(*b*) **Stepless Electrical Drives.** Electrical variation is achieved by varying the speed of the electric motor which drives the corresponding train of the machine tool. D.C. electric motors with shunt adjustment are used chiefly in heavy machine tools. Most convenient, in this case, is a generator-motor drive (Ward-Leonard drive) with a range ratio of 10 to 15, Fig. 22.28 (*b*). An A.C motor drives a D.C. generator which supplies D.C. power to a D.C. motor. By changing the resistences in the field windings of the generator and the D.C. motor stepless speed variations can be obtained within the maximum and minimum limits. The speed range can be substantially increased by the use of rotary amplifiers in this drive. Such drives are suitable for machine tools requiring very large range ratios, in the order of 500, 1000 of even more.

Stepless electrical drives can be easily automated. Their main drawback is large overall size and costs and also low efficiency (about 0.65).

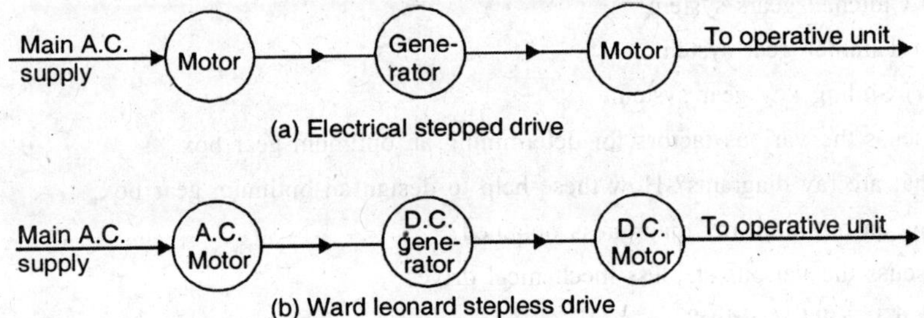

Fig. 22.28. Electrical Drives.

PROBLEMS

1. Name the various motions of a machine tool.
2. Define primary cutting motions, feed motions and auxiliary motions.
3. Compare group drive and individual drive.
4. Discuss the various elements for converting a rotary motion into a rotary motion in a machanical drive.
5. Discuss the various mechanisms for converting rotary motions into rectilinear motions in a mechanical drive.
6. Discuss the various cam mechanisms.
7. Discuss the various mechanisms for obtaining periodic rotary motion.
8. How the maximum and minimum cutting speeds and feeds are selected?
9. Define : spindle speed range ratio, velocity range ratio and diameter range ratio.
10. List the usual values of speed range ratio for various machine tools.
11. How the upper and lower speed limits of a centre lathe are determined?
12. What are 'Stepped' and 'Stepless drives?
13. What is step ratio? What are its extreme values?
14. Derive an expression for percentage loss of productive capacity.
15. Name the commonly used series of spindle speeds.
16. Draw Speed Spectrum in A.P. Series.
17. Derive expressions for Step ratio and number of steps in G.P. Series.
18. Draw speed spectrum in G.P. Series.
19. Why G.P. Series is the most preferred series for speed steps of machine tools.
20. How the standard values of step ratio are determined? List the standard values.
21. How the optimum values of step ratio and number of steps are determined?
22. Derive an expression for finding the minimum machining diameter or diameter of cutter to be used for various values of step ratio.
23. List the optimum number of step sizes.
24. With the help of neat sketch, discussed the following stepped mechanical drives :
    (*i*) Step-cone pulley drive.
    (*ii*) Step-cone pulley drive with back gears
    (*iii*) Change gears system.
    (*iv*) Sliding gears system

(*v*) Clutched gears system

(*vi*) Tumbler gear system

(*vii*) Sliding key gear system

25. Discuss the various factors for determining an optimum gear box.

26. What are ray diagrams? How these help to design an optimum gear box.

27. How an optimum ray diagram is obtained?

28. Discuss the various stepless mechanical drives.

29. What is a P.I.V. drive?

30. List the advantages and disadvantages of hydraulic drives.

31. How the speed in a hydraulic drive can be varied?

32. What are 'Metering-in' and 'Metering out' controls?

33. Discuss the various types of pumps used in hydraulic drive.

34. What is a 'Hydraulic Motor'? Why it is not commonly used?

35. Why pneumatic drives are not commonly used for machine tool drives?

36. What is an 'Hydro-pneumatic drive'? Sketch and explain it.

37. Name the types of motors used in electrical drives.

38. How the stepped electrical drives are achieved?

39. Explain a stepless electrical drive.

40. Determine the 6 spindle speeds of a turret lathe with maximum and minimum spindle speeds of 900 and 100 rev/min.

41. Design a gear box for a drilling machine to give speed variation between 100 to 250 rev/min in nine steps. The driving shaft is to run at a constant speed of 300 rev/min. Assume G.P. series.

42. Show that for speed steps arranged in G.P. series, the speed loss is constant.

43. A machine spindle is required to have 8 speeds in geometrical progression and to operate on ferrous metals at 28m/min. If the spindle accommodates HSS cutters ranging from 10 mm to 50 mm diameter, determine the spindle speeds and rationalise them in accordance with the preferred number series. Plot a graph of cutting velocity against cutter diameter for each spindle speed. Find the spindle speed for operating with cutters of (*a*) 12.5 mm diameter (*b*) 40 mm diameter. Find, for each cutter, the fall in cutting speed caused by dropping to the next lower spindle speed.

44. Determine the various speed steps for a lathe, which has to have a minimum speed of 18 rev./min. and a maximum speed of 1050 rev./min. The number of speed steps desired is 18.

45. Calculate the intermediate speeds of a head stock spindle of a lathe, given the following data :

Minimum speed = 45 rev./min.

Maximum speed = 2000 rev./min

Number of speeds = 18

46. Design a gear box to give speed variations between 100 and 300 rev./min in four steps. The driving shift is to run at a constant speed of 176 rev/min. Assume that the spindle speed are in G.P.

47. A drilling machine is to be designed to have 6 spindle speeds ranging from about 110 rpm to about 650 rpm. Assuming a proper series for the layout of the speeds, determine the values of all these 6 spindle speeds. Modify the computed valves so as to render them acceptable as standard. (GATE 1995) (**Ans** : 110, 155.1, 218.7, 308.35, 434.8, 650 rpm)

23

PRODUCTION PLANNING AND CONTROL

23.1. INTRODUCTION

Production can be defined as the transformation of raw materials by manufacturing methods into useful and valuable things needed by society. Production planning is the very basis of manufacturing. Production planning and control is the process of planning production in an industry in advance of actual production. It consists of the following activities :

1. Determining the practicability of a product design.
2. Analysing the product for best methods of production.
3. Determining the economic lot size.
4. Selecting the best equipment for any manufacturing process.
5. Determining the best sequence of operations to manufacture each individual item, part, or assembly (routing).
6. Designing or supervising the design of tools, jigs, fixtures, or other devices to best assist the manufacturing equipment to function as planned.
7. Estimating expenditure for equipment and tools and help making cost estimates on new jobs.
8. Setting starting and finishing dates for each important item, assembly and the finished product (Scheduling).
9. Controlling the inventory of raw materials and in-process parts.
10. Despatching materials, tooling and equipment to the plant locations specified by the schedule (Despatching).
11. Keeping track of progress of production. (Follow-up).
12. Providing programme evaluation review.

Thus, production planning and control is the function of looking ahead, anticipating difficulties and taking steps to remove the causes before they materialize.

No work should be done without planning. It is an old business axiom that time is money. Therefore, planning future events and scheduling (putting the plan into the time frame of calender) them so that they are accomplished with a minimum of time delay is an important part of production process. For large production projects, detailed planning and scheduling is a must. Computer based methods for handling the large volume of information have become common place.

23.1.1. Principles of Production Planning and Control. Production planning and control includes the investigation, coordination and evaluation of manufacturing capabilities and requirements that ensure timely production of projects through efficient and optimum use of facilities, that is, men, machines, money and materials (the four M's of production). These principles of Production planning and Control are discussed below :

1. Investigation. This activity includes the following functions .

(*i*) Organization and interpretation of information received from the selling (sales forecasts), purchasing (inventory details) and engineering departments (product design and drafting).

(*ii*) What production steps have to be performed as well as where and when to perform them.

(*iii*) Calculation of material requirements.

(*iv*) Development of detailed production schedules.

The product designer determines the shape, size, strength and material of the part to be made. The process engineer prepares the manufacturing plan, establishing th sequence and type of operation involved, selecting the tools and equipment required, determines the amount of material to be used for each part and in what form the material is to be purchased, for example, bar stock, plate or strip stock etc.

Determination of production steps and material requirements involves receiving and analyzing the above complete information and integrating it with manufacturing plans for other products or parts to assure maximum utilization of manpower and equipment.

2. Co-ordination. This activity consists of initiating action which brings the needed materials, tools etc. to the proper machines and at the scheduled times.

3. Evaluation. This activity consists of constantly reviewing production materials, methods, tooling, operating times etc., so that the planned manufacturing results are realized in terms of quantity, quality, time and location.

23.1.2. Objectives of Production Planning and Control. From the above discussion, it is clear that there are basically two major and important objectives of production planning and control :

1. Planning the activities of production :

(*i*) Collect and organize the pertinent production information generated by the selling, purchasing and engineering departments.

(*ii*) Translate the information collected above into production schedules such that the overall plant facilities are economically used and the production requirements of each department satisfied.

2. Controlling the production activities :

(*i*) Release the production orders in conjunction with the production schedules (Despatching).

(*ii*) Review production progress and initiate remedial action. If it becomes necessary to review production scheduling (Follow-up).

23.1.3. Production Planning. Production planning is the very basis of manufacturing. It takes a given product and organizes in advance the men, materials, machines and money required for a predetermined output in a given period of time.

Production planning starts with a product that can be manufactured and consists of the following functions :

1. Selection of process by which the product can be produced.

2. A breakdown of parts and materials.

3. A decision to either produce or purchase.

4. Planning the machines, equipments and tools etc.

5. The development of machining operations, that is, preparing detailed process operation sheets.

6. The development of time standards or estimates.

7. Determine the path the raw materials will take through the organization to produce the finished part (Routing).

8. Make up production schedules.

The quantity of parts to be produced and the frequency of reorders play an important part in decisions rendered by the planner. The type of order dictates the method of manufacture to be utilized and in turn the type of machine shop (batch production, mass production etc.).

Duties of a production planner. In short, the duties of a production planner are :

1. Originate, plan and evaluate manufacturing processes and tooling equipments to develop and/or improve production methods and manufacturing techniques.

2. Collaborate or design-fabrication problems with the engineering project group.

Qualifications of a production planner. Broad knowledge of all shop operations, including machine shop practices, assembly and fabrication methods, processing and finishing operations, knowledge or scheduling, product design, tool design and drafting techniques.

Objectives of production planning. The consumer acceptance of a product is based on appearance, convenience, durability, purchase price and maintenance cost. Production planning aims at meeting the above conditions of product cost and also ensuring a profit to the enterprise. Thus, the main objectives of production planning are :

1. Operating the plant at a predetermined level of efficiency.
2. Obtaining a prescribed level of profit.
3. Utilizing available plant facilities.
4. Reducing manufacturing costs through R & D.

23.1.4. Production Control. Production control function directs and controls numerous instructions to all parts of the factory to produce the item desired and at the time specified. Production control has been likened to a nervous system, sending directions to all parts of a factory to achieve production schedule.

The functions of production control are performed with the help of the following :

1. Despatching. This function involves moving materials, tools and equipment. Moving work to machines on operation-to-operation basis. Liason with scheduling.

2. Follow-up. This function includes expediting materials, in-process parts, and assembly.

Objectives of Production Control

1. The prime purpose and objective of production control is to get out the desired products economically and on time.
2. Minimizing idleness of man and machine.
3. Meeting promises to customers.
4. Maximizing inventory turn over.
5. Improving quality of the product.

Production control and production planning must work in reciprocal fashion. each assists the other in the performance of its duties. Routing and scheduling rely on planners for assistance in preparing schedules. Planners rely on despatching and follow-up for tooling and material assistance in tool tryout programmes.

23.1.5. Advantages of Production Planning and Control. The development of a more formal and systematic procedure or planning and controlling production, seeks to accomplish the following aims :

1. Maximum quality production at minimum cost through even distribution of work to available equipment and personnel.
2. Added flexibility in equipment and personnel to meet unavoidable emergencies.
3. Harmony and co-operation between departments.
4. Achieve a high standard or reliability of delivery.

The achievement of the above aims results in the following advantages :

(*a*) To the Investors or Financiers :

1. Security of money.
2. Adequate return on investment.

(*b*) To the Producers :

1. Company can earn more money and pay better wages to the workers.
2. Stable employment.
3. Job security to the workers.
4. Improved working conditions.
5. Increased satisfaction to the workers.
6. Increased productivity.

(*c*) To the Customers :

1. Better values.
2. Delivery on time.

(*d*) To the Nation :

1. Economic and social stability.
2. Security.
3. Prosperity.

23.2. TYPES OF PRODUCTION

From the point of production planning and control, there are two basic types of productions : production for stock and production for customer order.

Production planning and control principles and objectives are the same for both types, but the systems for achieving the objectives differ in two types of productions. Both types of productions can occur simultaneously in a company.

1. Production for stock. As the name implies, the finished parts are stocked and the customers are served from an inventory. For this, a forecast of future customer demand is required and the parts are produced by keeping this forecast in view.

The parts produced under ''production for stock'' are usually the standardized items that have a continued demand which need not be a constant demand. The forecast is usually prepared by the sales department. The forecast submitted by the representatives of sales, manufacturing and finance are discussed and the final forecast is approved before issuing the forecast to production. The forecast is then compared with the finished goods inventory and the difference is the net production requirement.

Once the net production requirements are determined for a period, the requirements of total manpower, material and machines are arrived at by using the conventional tools of standard times, operation sheets, bill of materials and operating budgets. These requirements are matched against the available facilities and the necessary steps are taken to meet the additional requirements of men, material and machines.

2. Production for customer order. Under this system, the process of production planning and control starts only after a confirmed customer order is received. However, such an approach will be impractical. If no work were done prior to the receipt of an order, it will be very difficult to meet the due date of the customer. To overcome this problem, it is customary to stock a certain amount of raw material, and in most cases, common parts are maintained in a finished or semi-finished state.

In this system, frequently all customer's orders received during a given period (a week or a month) are grouped for batch production of common parts in order to achieve greater production economies.

The system for controlling production orders for finished goods has to provide the eventual control of individual customer orders. This usually means a more costly control system if production output does not approximately coincide with customer delivery requirements.

On the basis of quantity of parts to be manufactured, three types of productions are distinguished in engineering :

1. Jobbing production or Piece production.
2. Batch production.
3. Mass production.

(*a*) **Jobbing production.** This type of production deals with a great variety of products made in fairly small quantities, often one-of-a-kind products. The same type and size of product either is made never again or repeated over indefinite periods of time. The distinctive feature of this type of production is a variety of operations carried out at each working place. The objects of jobbing production are unique items made to special orders (that is, custom built), for example, giant hydroturbines, machine tools, rolling mills and other heavy equipment, shipbuilding, manufacture of aeroplanes and oil field equipments etc.

Because of its nature, the jobbing production should be universal and flexible to be able to cope with diverse tasks. Hence, mainly general-purpose machines, standard cutting tools and universal measuring facilities are employed. Such machines require the smallest capital outlay, but the workers have to be highly skilled and the unit costs are high.

The principle of interchangeability is not complied with in machining the workpieces. Hence, fitting is resorted to in assembly operations.

(*b*) **Batch production.** Batch production or medium production or lot production is characterized by the manufacture of parts and products in lots which are regularly repeated in definite periods of time. The total production of one type of component may range from 25000 to 100,000 per year. The total quantity is not sufficient to keep a production unit continuously engaged. So, the total continuing demand is met by producing parts in lots or batches. The batch size or lot size is the number of parts produced in an uninterrupted run. Tentatively, the following can be taken as the lot size for the different kinds of lot production :

(*i*) Small-lot production — upto 50 parts/lot.

(*ii*) Medium-lot production — 50 to 300 parts/lot.

(*iii*) Large-lot production — More than 300 parts/lot.

Batch production is the commonest type of production and deals with a variety of products made in significant quantities. After a batch of parts of one design has been manufactured, the plant facilities are used to produce a batch of parts of another design. Examples of batch production are : machine tools, stationary internal-combustion engines, pumps, compressors, equipments for food processing, printing of books etc.

In batch production, a considerable part of the equipment consists of general purpose machine tools equipped with universal adjustable and sectional built up jigs, fixtures and tools. NC machine tools are also employed for bach production. In large batch production, programmable automatics are also very economical. Principles of interchangeability are to be strictly complied with. All this enables the labour input and the cost of production to be substantially reduced as compared to jobbing production.

(*c*) **Mass production.** It is characterized primarily by an established and stable object of production. The parts are produced in large quantities either intermittently or continuously, but are not dependent upon individual orders. The quantity is usually 100,000 parts per year. However, the most characteristic feature of mass production is not the quantity of products being manufactured; it is an arrangement whereby a single specific operation is continuously performed at most of the working places. A mass production enterprise deals with standardized products of limited variety, such as consumer durables, and products for industrial use (automobiles, tractors, bicycles, electric motors, sewing machines, nuts, bolts, screws, washers, pencils, matches and engine blocks etc.).

Other features typical of mass production are : an extensive use of specialized (usually permanently set-up) and single purpose machine tools and mechanization and automation of production processes with strict compliance of the principle of interchangeability. The latter greatly reduces the time required for assembly operations. In mass production type of production, semi-skilled or even unskilled workers are needed to operate the machines. This type of production is capital intensive, but the unit cost of production is low.

The most advanced type of mass production is "Continuous flow production", whose main feature is that the time required for each operation of the production line is equal to or a multiple of the set standard time all along the line. This enables work to be done without producing stock piles and in strictly definite intervals of time. Examples includes : oil refineries and continuous chemical plants.

23.3. SALES FORECASTING

"Forecasting" is a calculated economic analysis. It is, in fact, perceiving the shape of things to come. Virtually, all the management and planning decisions depend upon forecasts. Managers study the "Sales forecasts" for the following purposes :

1. To know the likely demand of the finished goods in the market.

2. To know the requirements of the raw materials necessary in the production and of the inventory levels.

3. To make decisions on working capital needs.

4. To make decisions about the size of work force.

5. To take decisions related to production planning and control.

6. To take decisions about the location of facilities, the amount of advertising and sales promotion and the need to change prices etc.

7. To determine the most economical production design of products, processes, equipment, tooling, capacities and layouts.

Although, "forecasts" are critically important for making economically sound decisions, they are never as accurate as managers would like. This is because of the "Law of Multiplicity of Causes". This law could never make deeper dent than at present. This is because of knowledge explosion. As such, forecasting will never be perfect. Yet it is neither a guess work nor a pessimistic/optimistic thinking.

23.3.1. Methods of Sales Forecasting. "Sales forecasting" are based on a number of factors that when properly considered lead to a reasonably accurate determination of the volume of future sales. Most sales forecasting is concerned only with short run projections for established products. If the product is well established in its field or only a change of model is concerned, the sales forecast may be based on experience with information drawn from past behaviour of sales, with allowance made for normal business increase or decrease after all business conditions have been considered from existing markets. If the product is new on the market, the forecast, which has been made on market analysis, business trends, seasonal fluctuations, and various other factors, must be more carefully considered. This calls for greater ingenuity and expense. There is no previous sales history and no empirical evidence. The forecast is built from scratch.

23.3.1.1. Sales Forecasting for Establishing Products. This type of forecasting takes many forms : personal judgement and experience and algebraic formulas. Both methods can be used as cross checks. In short, sales forecasting for this category means the use of past sales figures to estimate the future demand.

Patterns of demand for specific products might vary widely, but in general can be reduced to five components : average demand, trends in the average, seasonal effects, cyclic effects and random variations. Cyclic variations are beyond our scope. So, we need to be able to state a forecast for the upcoming period which takes account of the four components : average demand, trend, seasonal effects and random variations.

In all the methods of forecasting for established products, we take the help of time-series data. Time series is a sequence of observations of values of a variable obtained at regular time intervals. A trend component indicates a long term tendency of the series and may be the result of population changes, inflation, government regulatory policies or other factors that produce gradual changes over time. The general trend of a time series is often the principal basis of a forecast.

The main forecasting methods are discussed below :

1. Regression analysis. One way to access the general trend of the future demand is the method of Regression analysis also known as the "Method of Curve Fitting". In this method, the data on past sales (along vertical axis) is plotted against time (along horizontal axis), and the best curve called the "Trend Line" or "Regression Line" or "Trend Curve" is fitted to these data points, as shown in Fig. 23.1. The forecast is obtained by extrapolating this trend line or curve. The trend line or curve is determined by using the Statistical method of least squares curve fitting.

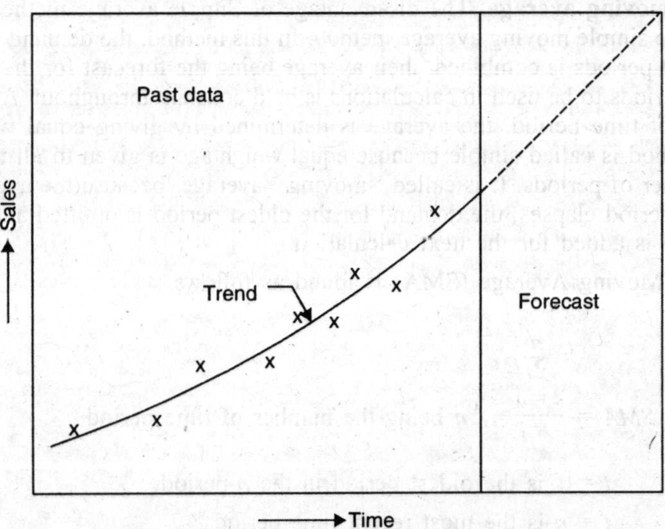

Fig. 23.1. Regression Analysis.

The main drawback of the regression analysis is that we would have to assume some linear or quadratic (sometimes cubic or exponential and so on) model that might reflect trend. That is, we will have to make an assumption that might not be true. Also, the computations are more complex than other methods, the method requires much data, which is costly to store. However, when data are linear and stable, linear regression can sometimes be used.

2. Averaging models. The methods use past data (time-series data) to calculate an average of past demand. This average is then used as a forecast. The various methods of calculating the average are discussed below :

(*i*) **Simple average.** In this method, the average of the complete data in the time-series is calculated. That is,

$$\text{Simple average,} \qquad SA = \frac{\sum_{t=1}^{n} D_t}{n} \qquad \text{...(23.1)}$$

where D_1 = the demand in the most recent period.

D_n = the demand that occurred in n periods ago.

It is clear, that in this method, all the previous demands enter into calculations and are equally weighted. Its advantage is that since all the previous demands enter into calculations, the effects of randomness are minimized. However, there is one major disadvantages. If the demand pattern changes overtime, the forecast may not be representative of the future. This is so because although the time-series data for many past periods may not be indicative of recent trends, they are given the same weight as the most recent data.

Example 23.1. *The total demand for Onida T.V. sets at Electroage shop has been 90, 100 and 80 each of the last quarters. Calculate the forecast on the basis of simple average.*

Solution.

$$SA = \frac{D_1 + D_2 + D_3}{3}$$

$$= \frac{90 + 100 + 80}{3}$$

$$= \mathbf{90}$$

So, the forecast will be 90 T.V. sets per quarter.

(*ii*) **Simple moving average.** The disadvantage of Simple average method is overcome to some extent by the simple moving average method. In this method, the demand data from several of the most recent periods is combined, their average being the forecast for the next period. The number of past periods to be used in calculations is held constant throughout. After selecting this suitable number of time period, the average is determined by giving equal weightage to each demand. The method is called simple because equal weightage is given to all the demands over the selected number of periods. It is called "moving" average, because the average moves over time. After each period elapses, the demand for the oldest period is omitted and the demand for the newest period is added for the next calculation.

The Simple Moving Average (SMA) is found as follows :

$$SMA = \frac{\sum_{t=1}^{n} D_t}{n}, \text{ } n \text{ being the number of time period} \qquad ...(23.2)$$

where $t = 1$ is the oldest period in the n-period.

$t = n$ is the most recent time period.

The value of 'n' is selected so as to get greater smoothing effect on the data.

Example 23.2. *Usha Lexis has experienced the following demand for storage water heaters for the last six months :*

Time	*Demand*
July	*300*
August	*350*
September	*400*
October	*500*
November	*600*
December	*700*

Prepare a forecast by the SMA method, to forecast January sales.

Solution. Under this method, the expected demand (forecast) for the next planned period is equal to the simple average of the actual demand of the previous n-periods. The formula (23.2) can also be written as,

$$F_t = \frac{\sum_{i=1}^{n} D_{t-i}}{n}$$

If $n = 3$, that is, by using 3-month moving average,

$$F_t = \frac{D_{t-1} + D_{t-2} + D_{t-3}}{3}$$

$$F_t \text{ for January} = \frac{700 + 600 + 500}{3}$$

$$= 600$$

(*iii*) **Weighted moving average.** A general accepted principle in forecasting is that the recent data contain more information than old data. Therefore, under this method, different weightings are given to the actual demands of every one of the previous *n*-periods. Usually, more weighting is given to actual demands for the most recent periods. According to this method,

$$F_t = \sum_{i=1}^{n} W_i D_{t-i} \qquad \text{...(23.3)}$$

where W_i is the relative weight of the actual demand for the previous *i*th period, and

$$0 \le W_i \le 1.0$$

and

$$\sum_{i=1}^{n} W_i = 1.0$$

Example 23.3. *Solve Example 23.2 by WMA method, to forecast the demand for January using 3-period model with the most recent period's demand weighted twice as heavily as each of the previous two period's demand.*

Solution. $W_1 = 0.50,\ W_2 = W_3 = 0.25$

$$\therefore \quad F_t = \sum_{i=1}^{n} W_i D_{t-i}$$

$$= W_1 D_{t-1} + W_2 D_{t-2} + W_3 D_{t-3}$$

$$= 0.50 \times 700 + 0.25 \times 600 + 0.25 \times 500$$

$$= 625$$

(*iv*) **Exponential-Smoothing.** This method differs from the WMA method by the special way it weights each of the past demands in calculating an average. The pattern of ''Weighting'' is non-linear (exponential in form). The ''weighting'' is assigned in such a manner that an observation becomes more remote (farther back in time from the current time), it receives less and less weight.

According to this method,

Forecast of next periods' demand = $\alpha \times$ Most recent demand + $(1 - \alpha) \times$ Most recent forecast

that is,

$$F_t = \alpha D_{t-1} + (1 - \alpha) F_{t-1} \qquad \text{...(23.4)}$$

where α = smoothing parameter, lies between 0 and 1.

Example 23.4. *In a General Hospital, the demand for disposable plastic tubing in General Surgery department for the last two months has been : November – 250 units, December – 300 units. Using 200 units as the November forecast and a smoothing co-efficient of 0.7, calculate the forecast for January.*

Solution. First, calculate as to what the forecast for December would have been,

$$F_t \ (t = \text{December}) = \alpha \ D_{t-1} + (1 - \alpha) \ F_{t-1}$$

$$\therefore \quad F_t = 0.7 \times 250 + 0.3 \times 200$$

$$= 175 + 60$$

$$= 235$$

$\therefore$ Forecast for January would be (t = January),

$$F_t = \alpha . D_{t-1} + (1 - \alpha) . F_{t-1}$$

$$= 0.7 \times 300 + 0.3 \times 235$$

$$= 210 + 70.5$$

$$= 280.5, \text{ say } \mathbf{281 \text{ units.}}$$

23.3.1.2. Forecasting Demand for New Product. As discussed earlier, if the product is new on the market, the forecast, which is based on market analysis, business trends, seasonal fluctuations and various other factors, must be carefully considered. Even under the best conditions, it is difficult to predict how the public will accept a new product. One method to know the customer's response is ''market sampling''. The new product is introduced in say one representative area in every sales region of the company. The customer reponse to the product is carefully assessed to obtain estimates of sales demand. Another metod can be : Opinion poll approach.

23.3.2. Techniques of Prediction. A few important techniques of prediction are discussed below :

1. Visionary prophecy. In this method, a top seasoned executive of the company who possesses deep knowledge, insight and judgement in business and technological affairs makes a prediction of future events. This prediction is based on a combination of factual knowledge, instinct and imagination. This prediction cannot be relied upon entirely. At best, it can be used as a check on statistical and other forecasts which may have missed to take into account some emerging factors in business and technology.

2. Estimates by salesmen. In this method, a district sales manager estimates item-wise future demand (for a month or a year) based on the data provided by the salesman. The regional sales manager collects the data from all the districts and modifies it, if necessary. Finally, the estimates of regional sales managers are collected by the General Sales Manager at the headquarters, and modified if necessary.

3. Delphi Method. The Delphi Method as already been discussed for ''Cost estimating''. In this method, a questionnaire is carefully prepared by a co-ordinator, regarding future trends of specific events, for example, future state of technology, trends in design, consumer demand preference, materials, competition and so on. This questionnaire is sent to a number of experts. Their answers are received and a summary of the answers is made. This summary alongwith a new questionnaire comprising aspects on which difference of opinion is fund amongst the experts, are sent back to all the experts. This process is repeated until a consensus is reached.

23.3.3. Types of Forecasts. The forecasts to predict the future sales activity of the firm's product are of three types :

1. Long range forecasts. These forecasts are for a period greater than five years. These forecasts involve plant construction, equipment procurement etc.

2. Medium range forecasts. (1 to 2 years). Such forecasts are regarding long lead time material planning.

3. Short range forecasts. (3 to 6 months). Such forecasts involve hiring personnel and production scheduling etc.

23.4. ECONOMICAL BATCH QUANTITY

Many plants produce parts for customer order or for internal use. As already noted, "Batch production" is the most common type of production. After a batch of components of one design is produced, the machine is used to produce a batch of another design. Now if the entire year's requirement of a component is produced in a single batch, great deal of money will be tied up in the finished parts (inventory) and storage costs. This is so, because the customer's orders will be received at intervals or the parts will be used internally at a fairly constant rate throughout the year. So, it will be an uneconomical production to manufacture the total requirement in a batch, Due to all this, the total annual requirement is produced periodically throughout the year in a number of batches or lots to replenish the stock. Typically, they are produced at a rate that is higher than the use rate. Now the question arises as to what should be the optimum batch quantity or lot size. Minimum cost concept is used to determine the optimum (economical) lot size. That is, the economical lot size is the number of parts produced per set-up which results in a minimum total cost per part.

As discussed in Chapter 5, calculating economical lot size is not an easy task. It is based on a number of recognizable factors. However, the most important of these are :

1. The time required to set-up the machine.
2. The amount of capital invested in the finished parts.
3. Cost of available storage space and handling.

Making a machine ready for starting the production of parts is called as "Setting up" of the machine. A lot of time is used to set-up a machine. This cost of setting up a machine is called the set-up cost and it occurs each time a lot is produced. This cost is essentially constant, regardless of quantity produced in a lot. Consequently, the set-up cost per unit varies inversely with the size of the batch. So, large lot sizes are justified to reduce the set-up cost per unit. But, as the lot size increases, a number of costs increase, for example, the amount of storage space

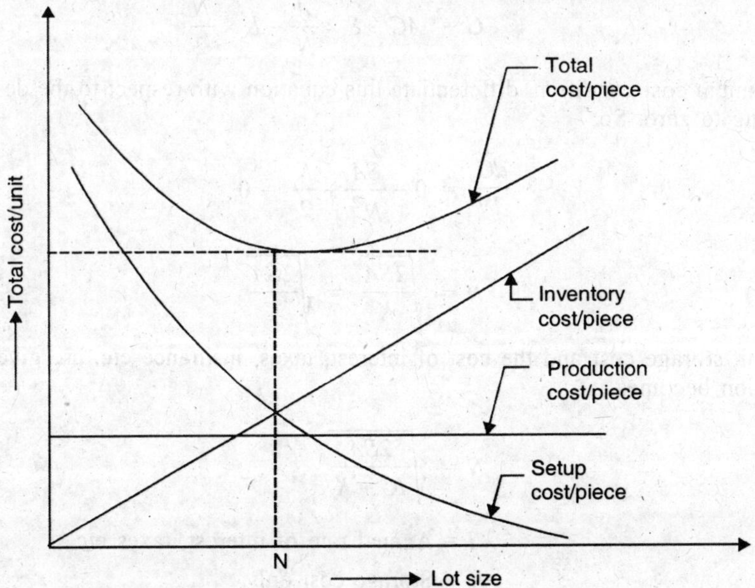

Fig. 23.2. Variation of Components of Production Cost

and its cost, the interest on the average inventory in the stock, insurance and taxes will increase directly with the lot size. Also, large lot sizes suffer more from shelf wear and increase the danger of loss from obsolescence. So, the above two cost factors are conflicting in nature and in determining the most economical lot size, we have to strike a balance between the two. The cost of production per unit is independent of lot size.

We will use the minimum-cost concept to determine the economical lot size. See Fig. 23.2, which shows the variation of the three cost factors discussed below :

(*i*) If the stock size is zero, when the new batch is started :

Let A = Annual requirement of parts (at a constant rate).

C = Unit cost of item.

S = Set-up cost per lot.

K = Carrying cost per unit per year.

Here, the carrying cost factor accounts for interest, taxes, insurance, storage costs, obsolescence etc. It is usually expressed as a percentage of unit cost, that is,

$$K = iC \quad \text{where } i = 5 \text{ to } 30\%$$

If N = most economic lot size, then

$$\text{Average stock inventory} = \frac{N}{2}$$

$$\text{Number of batches or lots per year} = \frac{A}{N}$$

$$\therefore \quad \text{Annual set-up cost} = S \times \frac{A}{N}$$

$$\text{Annual carrying cost} = K \times \frac{N}{2}$$

Now, Total Annual Cost = Production Cost + Set-up Cost + Carrying Cost

$$\therefore \quad G = AC + S \times \frac{A}{N} + K \times \frac{N}{2}$$

For minimum cost condition, differentiate this equation with respect to the design variable. N and equating to zero. So,

$$\frac{dG}{dN} = 0 - \frac{SA}{N^2} + \frac{K}{2} = 0$$

$$\therefore \quad N = \sqrt{\frac{2SA}{K}} = \sqrt{\frac{2SA}{iC}} \qquad \text{...(23.5)}$$

If the unit storage cost and the cost of interest, taxes, insurance etc. are given separately, then the relation becomes :

$$N = \sqrt{\frac{2SA}{IC + K}} \qquad \text{...(23.6)}$$

where I = Annual rate of interest, taxes etc.

and K = Storage cost only.

(*ii*) In the above analysis, it is assumed as if the entire lot is produced instantaneously, when the stock becomes zero. But that can never be possible. Also, some parts are being used before the production of the entire lot is completed. So, a correction factor must be applied to account for this condition, see Fig. 23.3.

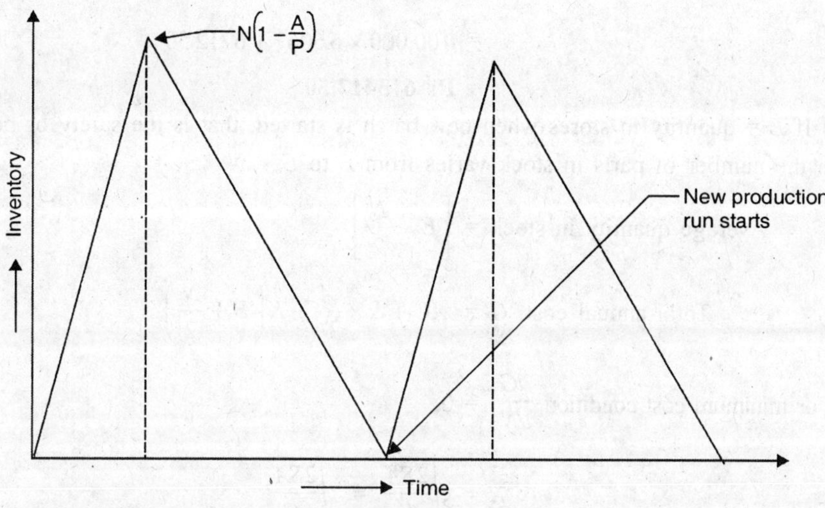

Fig. 23.3

If P = the production rate per year, with production at a constant rate, then the average quantity will be $= \dfrac{N}{2}\left(1 - \dfrac{A}{P}\right)$.

Using the concept expressed while deriving eqn. (23.5), the most economical lot size to produce is :

$$N = \sqrt{\frac{2S}{[(1/A)-(1/P)]K}} \qquad \text{...(23.7)}$$

Example 23.5. *Determine the most economical lot size for the following conditions :*

Setting up cost per lot = *Rs 600*
Cost of each item = *Rs 6.00*
Annual demand for the item = 100,000 pieces
Annual carrying charges = 25% of average inventory
What is the total cost of producing and storing 1 year's worth of this commodity ?

Solution. $N = \sqrt{\dfrac{2SA}{K}}$, S = Rs 600, C = Rs 6.00

A = 100,000

K = 0.25 × 6 = Rs 1.50, per unit/year.

$$\therefore \quad N = \sqrt{\frac{2 \times 600 \times 100{,}000}{1.5}}$$

= **8950 pieces**

$$\text{T.C.} = AC + S \times \frac{A}{N} + K \times \frac{N}{2}$$

$$= 100{,}000 \times 6 + 600 \times \frac{100{,}000}{8{,}950} + 1.50 \times \frac{8{,}950}{2}$$

$$= 100{,}000 \times 6.6067 + 6712.50$$

$$= \textbf{Rs 613412.50}$$

(*iii*) If E = quantity in stores when new batch is started, that is the safety or buffer stock then the number of parts in stock varies from E to $E + N$

$$\therefore \quad \text{Average quantity in stock} = \left[E + \frac{N}{2}\right]$$

$$\text{Total annual cost, } G = AC + S \times \frac{A}{N} + K\left(E + \frac{N}{2}\right)$$

$\therefore$ For minimum cost condition, $\frac{dG}{dN} = 0$,

and $$N = \sqrt{\frac{2SA}{K}} = \sqrt{\frac{2SA}{iC}} \quad \text{...(23.8)}$$

23.5. PRODUCTION PLANNING AND CONTROL FUNCTIONS

As discussed above, the various production planning and control functions are : routing, scheduling, despatching and follow-up. Frequency, routing and scheduling are known as planning functions and despatching and follow-up functions are known as control functions. These functions are discussed below :

23.5.1. Routing. Routing includes the planning of : what work shall be done on the material to produce the product or part, where and by whom the work shall be done. It also includes the determination of path that the work shall follow and the necessary sequence of operations which must be done on the material to make the product.

Routing procedure consists of the following steps :

1. The finished product is analysed thoroughly from the manufacturing stand point, including the determination of components if it is an assembly product. Such an analysis must include :

(*i*) Material or parts needed.

(*ii*) Whether the parts are to be manufactured, are to be found in stores (either as raw materials or worked materials), or whether they are to be purchased.

(*iii*) Quantity of materials needed for each part and for the entire order.

One method of determining raw material requirements for production is by explosion technique. The technique consists of breaking down of scheduled quantities of end products into the required quantities of component parts or raw materials. These are two common methods of doing this :

(*a*) **Synthesis.** In this method, a list is set-up for each part, showing all the future production schedules on which it is expected to be used. This includes schedules for optional subassemblies as well as final end products. The approximate requirements for each part are then determined by summarizing the expected usage for future production schedules. This method is often used when producing to stock and any overruns on a component part will be readily absorbed in future production.

(*b*) **Analysis.** In this method, the production parts requirements for each end product is determined by breaking it down into subassemblies and/or parts. Eventually, all subassemblies are broken down to their component parts and the exact usages of each component part can be computed. This method is more often used when producing on order, and the expense of scrapping any unused components is high.

It is important to recognize that much of this work is done by the engineering department or process engineering group. The function of production planning and control department is to study the above data and check this form the production stand point.

2. The fixing of the sequence of completion in manufacture that one part or piece of material bears to another, in order that all may be brought together as needed in the process of manufacture.

3. The determination of operations which must be performed at each stage of manufacture (that is, their sequence) and the place where these shall be performed. This is usually illustrated with the help of a "Route Sheet" and "Operation Sheet" which indicate the sequence of operations that a part is to follow through the production process. These also include the name of machines on which the operations are to be performed and the departments where these are to be done. Tools, jigs, fixtures needed and the time analysis are also included in these sheets.

4. To determine the economic lot size to meet the total requirements of the product quantities.

23.5.1.1. Routing documents. The two important routing documents are : Route Sheet or Route Card and Operation Sheet. These documents specify the process sequence through the plant.

1. Route sheet. A route sheet or route card lists manufacturing operations in proper sequence and associated machine tools for each parts. It also indicates the departments in which a particular operation is to be done on a part/parts, and to which department the parts must go to for the next operation. A route sheet travels with the parts which move in batches between the processes from one point in the plant to another. That is why, it is also known as "traveller". A typical route sheet is shown in Table 23.1.

Table 23.1. Route Sheet

Name of Part________	Part No.__________	Drawing No.______		
Product__________	Product No.________	Quantity__________		
Material__________	Economic lot size____			
Order No.__________	Due Date__________			
Operation No.	**Description of Operation**	**Machines Or Equipment**	**Department**	**Tooling**

2. Operation sheet. The operation sheet lists the various operations in sequence, required for producing a part. Operation sheets vary greatly as to details. A simpler operation sheet specifies only the operation and the machines to be used. Cutting speeds, and depths of cut are left to the discretion of the operator, particularly if he is skilled and small quantities of parts are involved. However, in more common types of operations sheets, complete details are given regarding cutting speeds, feeds, depth of cut and tools. Often, the time analysis for each operation is also included in the sheet. A typical operation sheet is shown in Table 23.2.

Table 23.2. A Typical Operational Sheet

Part Name ________				Part No. ________			
Material ________				Quantity ________			
Ope. No	Operation	Machine Tool	Cutting Tool	Cutting Speed	Feed mm/rev	Depth of cut, mm	Time Analysis

Bill of materials. Bill of materials is a listing of all the component parts and/or raw materials required to make a product with associated descriptions and required quantities of each component. All components are identified by name and number. The bill of materials is generally prepared by the engineering department (product design department) or the process planning engineer. This information is essential for Inventory Control. A typical bill of materials is shown in Table 23.3.

Table 23.3. Bill of Materials

Name of Product ________			Product No. ________	
Compiled by ________			Checked by ________	
Approved ________			Date ________	
S. No.	Part No.	Part Name	Material Specification	Weight/Quantity

23.5.2. Scheduling. As already noted, the scheduling function consists of putting the production plan into the time frame of the calendar. It is the activity of setting the starting and completion dates for processing products for manufacturing so that congested manufacturing conditions at one time and idle machines at another time may be prevented. So, the main objectives of scheduling function are :

(*a*) Maximization of resource utilization, chiefly men and machines.

(*b*) Delivery of orders on or before due date.

(*c*) Minimization of costs and waiting times.

There are two types of schedules used : Master Schedules and Shop or Production Schedule.

1. Master schedule. The first step in scheduling is to prepare the *Master Schedule.* A master schedule specifies the products to be manufactured, the quantity to be produced and the delivery date to the customer. It also indicates the relative importance or manufacturing orders. The scheduling periods used in the master schedule are usually months. Whenever a new order is received, it is scheduled on the master schedule taking into account the production capacity of the plant. Based on the master schedule, individual components and sub-assemblies that make up each product are planned and :

(*i*) Orders are placed for purchasing raw materials to manufacture the various components.

(*ii*) Orders are placed for purchasing components from outside vendors.

(*iii*) Shop or production schedules are prepared for parts to be manufactured within the plant.

To schedule a manufacturing order, the first step is to determine the delivery dates or shipping dates from which the starting dates may be calculated. These dates are calculated by scheduling the order on master schedule based on the following elements :

1. Classification or type of equipment.
2. Customer's demand or due date.
3. Existing scheduled load.
4. Normal delivery time.

When the delivery date is determined, the order requests are sent to the manufacturing nformation section. Here, the product is subdivided into parts and sub-assemblies. By cheduling backward from the established shipping date, the starting dates for the manufacture of parts and sub-assemblies are quickly determined.

There are two methods for scheduling jobs :

(*a*) The manufacturing of all the parts is started on the same date. Since the standard time f each part will be different, the part requiring the longest standard time will determine the date n which all the parts may be assembled into a finished product. The parts requiring less standard imes will each await assembly according to their finish dates.

(*b*) The various parts are successively started for production as per their individual nanufacturing times. In this way, all the parts will be ready for assembly on the same date. This s the preferred method of scheduling.

Thus, the master schedule is the composite plan for completing all programmes. As noted bove, it is generally prepared monthly and extends into the future as far as orders are booked. he master schedule continues to roll on a monthly basis; as one month is completed, a new one s added at the end and the schedule adjusted, if necessary, for blind-schedule conditions or sales orecasts.

Objectives of master schedule. The objectives of master schedule are :

1. It helps in keeping a running total of the production requirements.

2. With its help, the production manager can plan in advance for any necessity of shifting om one product to another or for a possible overall increase or decrease in production equirements.

3. It provides the necessary data for calculating the back log of work or load ahead of each ajor machine.

4. After an order is placed in the master schedule, the customer can be supplied with obable or definite date of delivery.

2. Shop or production schedule. After preparing the master schedule, the next step is to epare shop or production schedule. This includes the department machine and labour-load schedules, and the start dates and finish dates for the various components to be manufactured within the plant.

A scheduling clerk does this job so that all processing and shipping requirements are relatively met. For this, the following are the major considerations to be taken case of :

(*i*) Due date of the order.

(*ii*) Whether and where the machine and labour capacity are available.

(*iii*) Relative urgency of the order with respect to the other orders.

Orders are scheduled according to either machine capacity or labour capacity. The first method is more common especially when the machines are independent of each other. Machines are listed according to types and operations, with their standard-time values that can be performed on each. The second method is perferred when the machine tools can be designated individually, such as when a mattery of machies are dependent upon each other for the completion of a parts. In this case, the types of labour are established and their total capacity designated in available labour units.

The machine type of loading is more accurate, but the labour type of loading is more flexible and simpler to handle. It allows quick changes in schedule and with this, its accuracy can become almost equal to that of the machine-load type.

In very large plants, the preparation of production schedules is usually complex due to the following factors :

1. The individual parts being manufactured concurrently may run into hundreds or thousands.

2. Each part may have its own separate process route through the plant.

3. The parts are routed through a large number of machines in different departments.

4. The machines may be limited in number, they perform different operations, and have different features, capacities and capabilities.

5. Customer's orders may be of different priorities.

6. Parts become defective during manufacturing, cycle times vary, machines breakdown etc

Machine Loading. Allocating the job to work centres is referred to as "Machine Loading" while allocation of jobs to the entire shop is called "Shop Loading". The production planners can safeguard the production schedule and unit cost through the use of a machine loading system. Machine loading is an important tool of management of securing the most efficient use of manpower and equipment. The load capacity of a machine may be expressed in terms of pieces for a given length of time or in time for a given number of pieces. In either case, the capacity may be determined very readily from the standard time values of the operation performed by the machine.

Machine loads change daily, and the records must reflect the change. Each new job assigned increases the future load on a machine and each finished job decreases the load already assigned. The work backlog is reflected accurately from these records. Loading procedure should attempt to assign and schedule a machine or group of machines so that the total time to complete a job is as short as possible.

Machine loads should be received regularly for overload or unload conditions. This will indicate any need for change in machine or labour capacity, overtime, policing, or number of shifts employed per day.

A machine load chart is a chart for showing the work ahead for various machines and processes. A typical machine load chart is shown in Fig. 23.4. Here, the load is expressed in terms of the number of hours for a given number of pieces. Such a chart is known as 'Bar Chart' or 'Gantt Chart'. A bar represents a task. It is shown along the horizontal axis which indicates time scale.

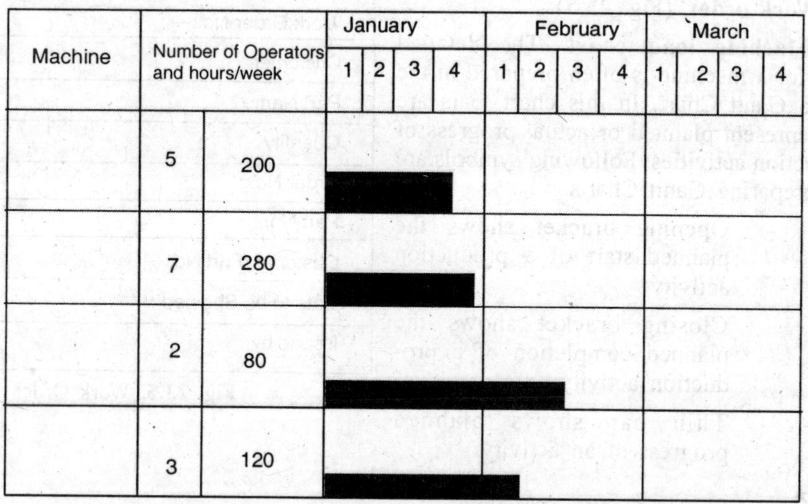

Fig. 23.4. A Typical Machine Load Chart

Objectives of Production Schedule :

1. It meets the output goals of the master schedule and fulfils delivery promises.
2. It keeps a constant supply of work ahead of each machine.
3. It puts manufacturing orders in the shortest possible time consistent with economy.

23.5.3. Despatching. After the schedule has been completed, the production planning and control department makes a master manufacturing order with complete informations including routing, the desired completion dates within each department or on each machine and the engineering drawings. From this master manufacturing order, departmental manufacturing orders can be made up giving only the information necessary for each individual foreman. These include inspection tickets and authorization to move the work from one department to the next when each department's work is completed. When a foreman of a particular department receives the manufacturing order, he is authorized to begin production in his department. The despatching of these orders and instructions at the proper time to the proper people is usually done by a person known as "Despatcher".

So, "despatcher" function consists of issuing the orders and instructions which sets production in motion in accordance with production schedules and routings. This function is purely a clerical function and requires voluminous paper work.

Duties of a despatcher. The detailed duties of a despatcher are :

1. Initiate the work by issuing the current work order instructions and drawings to the different production departments, work stations, machine operators or foremen. The various documents despatched include : detailed machine schedcules, route sheets, operations sheets, materials requisition forms, machine loading cards, move or material ticket etc.

2. Release materials from stores.

3. Release production tooling, that is, all tools, jigs, fixtures and guages for each operation before operation is started.

4. Keep a record of the starting and completion date of each operation.

5. Getting reports back from the men when they finish the jobs.

Works order documents. The usual formats of various works order documents used by the despatcher are given below. The route sheet (card), operation sheet and machine loading chart have already been discussed.

1. Work order. (Fig. 23.5).

Work Order No.:—	
Customer	Date
Part Name :	
Quantity :	
Order No. :	
Part No. :	
Customer Part No. :	
Date to be Shipped :	
Issued by :	

Fig. 23.5. Work Order

2. Machine load chart. The detailed schedule for a machine is often prepared in the form of a Gantt Chart. In this chart, bars are used to represent planned or actual progress of the production activities. Following symbols are used in preparing Gantt Charts :

[– Opening bracket shows the planned start of a production activity.

] – Closing bracket shows the planned completion of a production activity.

—— – Thin bar shows planned progress of an activity.

☒ – Shows change over time.

Gantt chart for Machine loading is shown in Fig. 23.6.

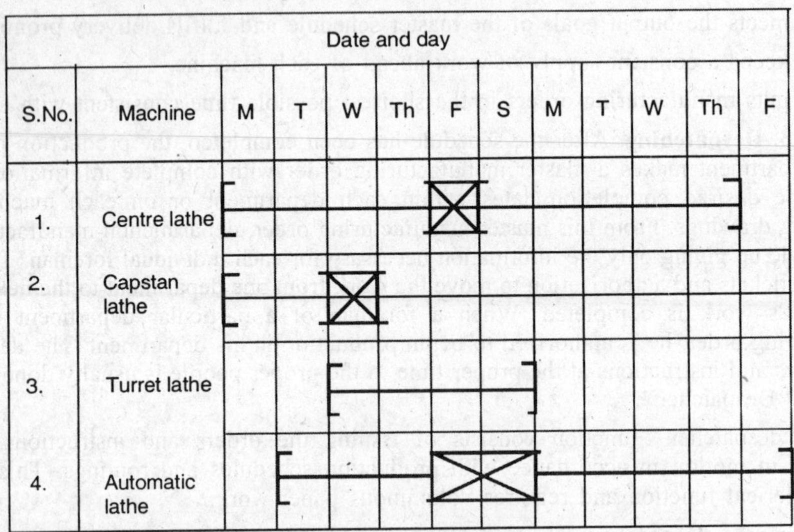

Fig. 23.6. Machine Loading Chart

3. Material requisition form. This form specifies the material required for the schedule job and the necessary quantity to be issued, Fig. 23.7.

MATERIAL REQUISITION

Part :	Part number :	Drawing number :
Product :	Product number :	Quantity :
Material :	Material Code :	
Quantity authorized :		
Quantity issued :		

Fig. 23.7. Materials Requisition Form

4. Move ticket. This is a form attached to a part or a group of parts which authorizes transfer or movement of materials or parts from the store room to the department where the first operation is to be performed and then between operations or stations or departments, Fig. 23.8.

MOVE TICKET

Date :	Order number :
From Dept. :	To dept.
Quantity delivered :	Quantity accepted :
Foreman :	Foreman :
Moved by :	
Remarks :	

Fig. 23.8. Move Ticket.

5. **Inspection ticket.** This is used to report the quantity of parts passed and rejected at each inspection operation, Fig. 23.9.

INSPECTION TICKET

Deptt. :			Date :
Order number :			Charge to :
Quantity	Material	Number Rejected	number accepted
Cause of rejecton :—			
Operator's fault :		Last Operation :	
Disposition :			
			Inspector

Fig. 23.9. Inspection Ticket

The term "despatching" is not much heard in decentralized control where the foreman passes out the jobs. It is used mainly with centralized control where the production control's branch office (despatch office) in each department tells men what jobs to work on. In this system, one shop order copy, known as "traveller"; circulates through the shop with the parts from operation to operation. This copy has the progress report denoted by the foreman. Another copy of shop order is kept in the despatch office. This copy is maintained by a clerk who duplicates the reporting from the foreman. Material is returned to the despatch office after operations are completed in any one department. Material is held and dispersed to the next operation according to the sequence of operations. This system of centralized despatching permits easy tracking and expediting the material. After all work is completed, the order is closed out.

23.5.4. Progress Reporting and Follow-up. Shortage of parts and failure of production foremen to meet production schedules are common. There are numerous reasons for these occurrences, for example, errors in routing, scheduling and despatching; lack of material; labour difficulties; breakdown of equipments; lack of proper tools, jigs and fixtures; and excessive rejections etc. They cannot be pinpointed on any one individual but rather on a build up of system folleis and lack of communication or follow through. Progress reporting is essential to the prevention of these delays.

When the manufacturing within a department is completed, an inspection is made to check the quality and conformity with the engineering specification. The inspection ticket prepared by the inspector is sent back to the production control department. From the information on this ticket, the progress of the order can be recorded on the master control chart. In this way, the production control department can follow every job without too much difficulty.

Gantt charts were the first to be used to control production. As discussed earlier, a Gantt chart is a bar chart. The activities are listed in the vertical direction, and elapsed time is recorded horizontally, Fig. 23.10. The horizontal axis is a time scale and the desired or required start and finish times for each task may be conveniently displayed on the chart. If somebody takes the time and trouble to update the chart at regular intervals, indicating (perhaps by shading the bars) the stage of completion of all tasks at a particular point in time, the Gantt chart affords the manager an excellent visual aid for assessing production progress. The main drawback of Gantt chart for planning and control purposes is its lack of adequate depiction of interrelationships between the separate tasks, that is, how the ability to start one task depends upon the successful completion of other tasks.

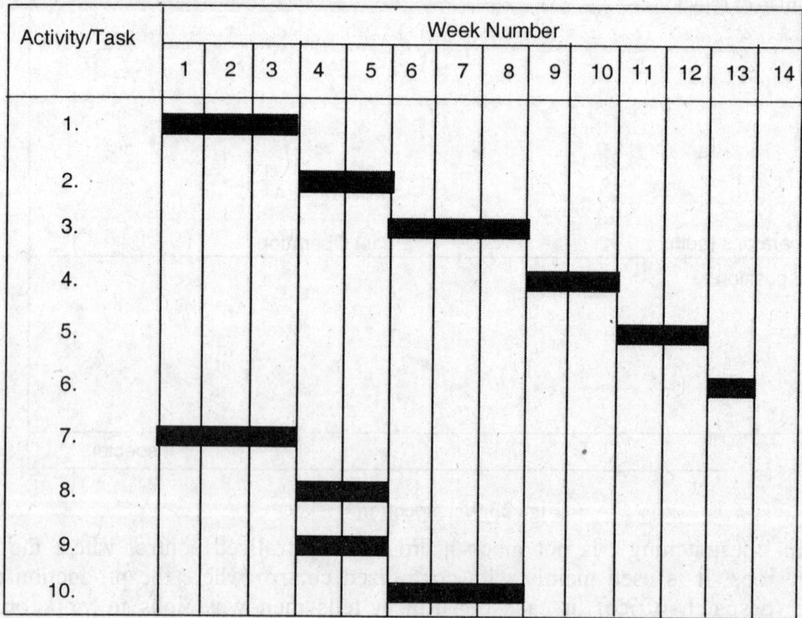

Fig. 23.10. Gantt Chart for Scheduling.

Today, industry has modernized the Gantt charts into modern graphic portrayals of production status and control, for example, Production Control Board is a pegboard arrangement, Sched-U-graph boards etc. These are used to help management to visually observe work status. These visual progress charts must be kept current to be effective.

Another chart to aid scheduling, fabrication, and assembly is the manufacturing and assembly flow chart, Fig. 23.11. It depicts the flow of detail parts into one subassembly. The accumulation of several subassemblies results in one finished assembled product. The illustration also reflects the standard hours of the individual assembly and the accumulated assembly time over two or more subassemblies. The cost of the product can be determined at any in progress station. The chart can also be utilized for other purposes, such as the determination of lead time. Lead time is the time interval required to produce a part to a fixed reference point. The common reference generally used is the shipping point in a plant.

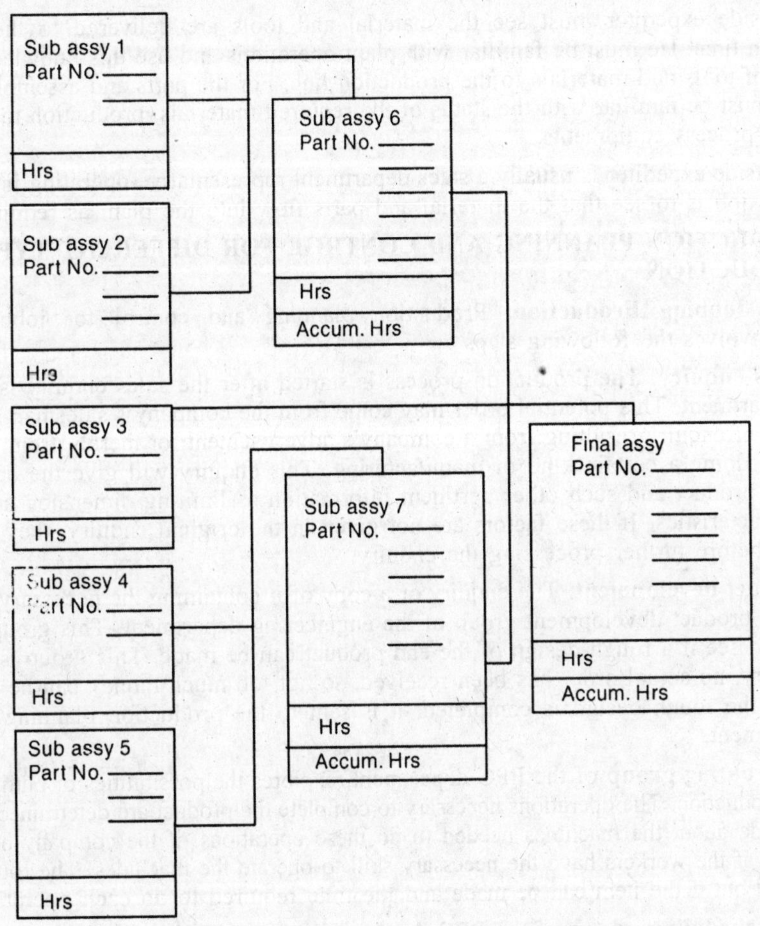

Fig. 23.11. Typical Manufacturing Flow Chart.

Follow-up. Progress reports prompt use of follow-up men also known as stock chasers or expediters to locate lost jobs and push late jobs through to completion. They attempt to foresee and eliminate further delays. So, the follow-up function or expediting consists in determining the current status of a part which is on other or a production order which is in process and initiating efforts to speed up operations when failure to meet schedules appears likely. It regulates the progress of material and parts through the production process. The cause of delays or shortages may also be investigated and an attempt made to prevent their recurrence. This includes the physical tracing of the work in the plant and the contacting of vendors who provide outside parts. For all this a follow-up section is set up in the production planning and control department.

Follow-up function can be carried out in two ways :

1. When organized on the basis of the customer's order, each follow-up man follows an order or a group orders all the way through the plant.

2. When organized on the basis of the related departments, each follow-up man follows all orders in his respective unit or department.

Follow-up or expediting is commonly divided into two categories :

(*a*) Production operations inside the plant.

(*b*) The supply of the parts and materials from outside the plant.

The inside expediter must see the material and tools are delivered to their seheduled destination in time. He must be familiar with plant operations and use this knowledge to ensure a free flow of tools and materials to the production line. For the parts and assemblies under his control, he must be familiar with the status of the required materials, production tooling, cutting tools and in-process equipments.

The outside expediter is usually a sales department representative operating in the plants of vendors. His job is to see that the material and parts flow into the plant as required.

23.6. PRODUCTION PLANNING AND CONTROL FOR DIFFERENT TYPES OF PRODUCTION

23.6.1. Jobbing Production. Production planning and control for jobbing type of production involves the following steps :

1. Sales enquiry. The production process is started after the sales enquiry is received by the sales department. This potential order may come from the company's salesman, from agents, from a letter of enquiry resulting from a company's advertisement, or merely from a company's reputation of doing a certain kind of manufacturing. This enquiry will give the desired use of the potential product and such other pertinent information as limiting dimension and necessary strength characteristics. If these factors are not given in the original enquiry, they will have to be obtained before further processing the enquiry.

2. Product development. The enquiry or a copy of it containing the necessary information is sent to the product development group of the engineering department. This group checks the information to see if a rough design of the end product can be made. This group is working on a sales enquiry, no actual order has been received, so not too much money can be spent at this stage. When the rough design is completed, it is sent to the production planning and control (PPC) department.

3. The routing group of the PPC department, explores the possibilities of putting the rough design into production. The operations necessary to complete the product are determined. This group will also decide about the machines needed to do these operations, if the company has necessary machines, and if the workers have the necessary skill to operate the machines. The routing group's duty is to find out if the item can be made and the time required to do each operation.

4. The scheduling group of the PPC department must then decide when the item can be made. This group determines this by referring to the load chart and to the approximate time needed for each operation. This load chart shows the amount of work scheduled by the company and the exact time for which it has been scheduled. Knowing the plant capacity, the scheduling group can readily determine the capacity available and determine the earliest possible completion date for the proposed product.

5. The cost control group examines the rough design of the product and from the information provided by the routing group and scheduling group, is able to determine the cost of manufacturing the product (see Chapter 4).

6. The sales department adds to the cost estimate, the necessary profit margin dictated by the company's policy to establish the selling price. It is then the duty of the sales department either by correspondence or by personal contact, to give all this information to the potential buyer and endeavour to obtain his order.

7. Engineering drawings. When the actual order is received from the customer, the company begins its work in earnest. The product development group is notified that its rough design has been approved by the customer. The engineering department must then transpose the rough design into manufacturing drawings. This includes making individual drawings of every part. An assembly drawing showing the individual parts assembled has to be made for the workers who assemble the product. The engineering department prepares a parts stock list. This list contains the standard parts (nuts, bolts, etc.) and similar items that manufacturing companies purchase from outside sources and carry in their stockroom. A committee determine manufacturing specifications, which include the product quality requirements and specify any tests that the sales, PPC, or engineering departments feel necessary to ensure an acceptable finished product.

8. All the informations developed by the engineering department is sent to the PPC department, which determine first all the items necessary for the finished product. These include stock items and basic raw materials necessary. The procurement of the needed items is initiated. The list of items is sent to the purchasing department which is to have them available on a certain date. This date is determined by the PPC department after referring to the desired delivery date.

9. Now the actual functions of the PPC department come into play, that is, routing, scheduling, despatching and follow-up.

In jobbing type of production, since low cost general purpose machinery is used, costs of equipment breakdown are not serious. However, the total manufacturing process require to be closely coordinated and progress carefully monitored in order to meet delivery dates and control costs.

23.6.2. Mass Production. The production planning and control procedure in a mass production company is radically different from that in jobbing company. Because the company manufactures a standardized product, the production control procedure is simplified. When the standard product is selected, the plant layout group plans the production process. It decides what machinery is necessary and how it should be arranged in the plant in the best possible manner, for example, a straight line arrangement to eliminate all back-tracking of the work. The material handing group investigates the best method of transporting the material from operation to operation. A production forecasting committee studies overall business conditions and similar factors affecting the company and endeavours to forecast the quantity of the standardized product that should be produced over a future time period.

The product development group of the company works on improving the product. All successful improvements are adopted.

In mass production companies, the sales orders usually take the place of sales enquiry. Sales come in directly because the standard product is well-known. When the company receives the order, it is sent to the shipping room where the standard product is taken from the finished goods inventory and shipped to the customer. In actual filling of the order in this type of company, the PPC department is not directly involved. However, the PPC department is responsible for having the item available in the stock room. Through its production forecasting, it determines what quantity is to be manufactured throughout a definite time period. The amount of goods in finished goods inventory is a result of time decision. Since the company is manufacturing a standard item for stock, the forecasting group is able to determine the daily rate of production, and the purchasing department ordinarily knows what raw materials are needed for each day. A routing sheet is not necessary because the routing is established and the plant layout is fixed. The scheduling group schedules the daily production determined by the forecasting group. The despatching group notifies each foreman of his daily production output.

A record may be kept showing the future daily production output. A record may be kept showing the future daily production and the rae of present daily production. This daily production progress report can be made from the reports of the inspectors who check the product to see if it meets specifications.

So it is clear that in the mass production company, all drawings, raw materials and specifications are standard, because the product is standard. The routing is standard, and the scheduling, despatching and follow-up are routine. All this depends upon proper production routing, layout and capacity when the product was first put into production. However, the main problem is that of plant maintenance, because breakdown of even one machine or workcentre is liable to result in stoppage of production on the whole line.

23.6.3. Batch Production. Production planning and control for batch type of production is most problematic. A variety of products is being manufactured at the same time in the company. The demand of products usually fluctuates. So, the raw material requirements are not constant. Moreover, these may be known only a short time before actual production is to commence. The main problems in batch production are — demand forecasting, short-term capacity planning, scheduling of a large number of different job orders at short notice, purchasing and inventory control. High machine and labour utilization and production quality has to be ensured.

23.7. INVENTORY CONTROL

The term "inventory" can be defined as the stock of raw materials, parts and finished products held by a business company. It is a comprehensive list of materials, resources and goods that are idle at a given point of time.

Inventory is the life blood of a production system. It represents a major portion of a manufacturing facility's assets. Inventories in a typical manufacturing unit comprise of the following five general categories of materials :

1. Production inventories. These include the items which go into final product, that is, raw materials, purchased components etc.

2. MRO inventories. These include the items which do not form a part of the final product but are consumed in the production process, for example, maintenance, repair and operating supplies like spare parts and consumable items (oil, grease, cotton waste, cutting fluids etc.).

3. In-process inventories. These include semi-finished products at various stages of production.

4. Finished-goods inventories. These include the finished products ready for despatch.

5. Miscellaneous inventories. These arise out of the above four types of inventories and include — scrap, surplus and obsolete items, which are to be disposed off.

Excessive inventories represent locked-up capital and waste storage space. Even though, the cost of carrying large inventories is substantial, they are needed for the following reasons :

(*i*) *Fluctuation in demand.* Finished-goods inventories are needed to meet the fluctuating demand. During the periods of low demand, excess production is stored in the warehouse, and during periods of increased demands goods are supplied from the stock.

(*ii*) *Fluctuation in supply.* Materials and parts are not always available for buying whenever, wherever and in whatever quantity they may be needed.

(*iii*) *Quantity purchasing.* It is usually cheaper to buy in large quantities because of reduced cost of ordering, reduced unit price (called quantity discounts) and reduced cost of transportation.

(*iv*) In case of production delay due to process breakdown, lost batches or defective lots, finished goods inventory reduces chance of delays in meeting sales delivery dates.

(*v*) In-process inventories are needed to maintain production on other machines if one machine in the line breaks down.

(*vi*) Finished parts and products-inventories are required for reducing overall production cost in batch production.

Inventory control endeavours to achieve a balance between too little inventory (with possible stock-out of raw materials) and too much inventory (with investment and storage tied up).

23.7.1. Objectives of Inventory Control. From above, it is clear that the objectives of inventory control are :

1. To try not to run out of materials and thus to minimize production delays owing to non-availability of raw materials and parts.

2. To keep the capital blocked up in inventory to the minimum point in consistence with the operating, sales and financial requirements of the company.

3. To minimize the overall production costs by keeping a balance between production and operating costs on one hand and the customer service on the other.

4. To ensure sufficient supply of required materials, parts, supplies and other items to maintain most efficient level of operations to meet the customers demand, that is, stabilizing production levels.

5. To minimize losses in inventory deterioration, obsolescence, damage or wastage.

23.7.2. Functions of Inventory Control. The functions of an effective inventory control system to achieve the above objectives are as follows :

1. To develope policies, plans and standards essential to achieve the objectives.

2. To build up a logical and workable plan of organisation for doing the job satisfactorily.

3. To develop procedures and methods that will produce the desired results economically.

4. To provide the necessary physical facilities.

5. To maintain overall control by checking results and taking corrective actions.

23.7.3. Selective Inventory Control. Inventory in a medium size company comprises of thousands of items in stock. The control of all these items creates a serious problem to the management in keeping track of each and every item and having the same extent of control on each of the item. In general, it can be said that overstocking and stockouts do not effect all the items in store. It is quite common to come across some items which move so fast that it is quite difficult to keep them replenished, while in the next bin, there are other items that have been collecting dust or rust for weeks or even months and tying up a good amount of capital. Therefore, in order to execute proper control, it is necessary to take a selective approach and find out the attention required for each item according to its importance. This is essential for achieving maximum benefits with minimum efforts.

***A B C* Analysis.** The common and important of the selective inventory control of *ABC* analysis. *ABC* analysis is done for items on stock and the basis of analysis is the annual consumption in terms of money value. If the annual consumption of every item is worked out, it may be noticed that in general only about 10% of the items account for 70% of total annual consumption (*A* item), about 20% of the items account for 20% of the total annual consumption (*B* item) and the rest 70% of the items account for only 10% of total annual consumption (*C* item). These percentages are only guidelines for segregation. They may vary from company to company.

The pattern of *ABC* analysis can be clearly presented on a curve with percentage of items of X-axis and percentage of total annual usage on Y-axis, Fig. 23.12.

Control of *A* item. Maximum savings can be expected to result if stock levels and buffer stocks of the *A* class items are carefully controlled. For this :

1. These items must be ordered frequently and scheduled to arrive just before use, to reduce capital tied up. Annual contract for supplies with staggered deliveries is economical.

2. Minimum safety stock or even fluctuating safety stock should be kept by maintaining good vendor relationships, speculation of market condition etc.

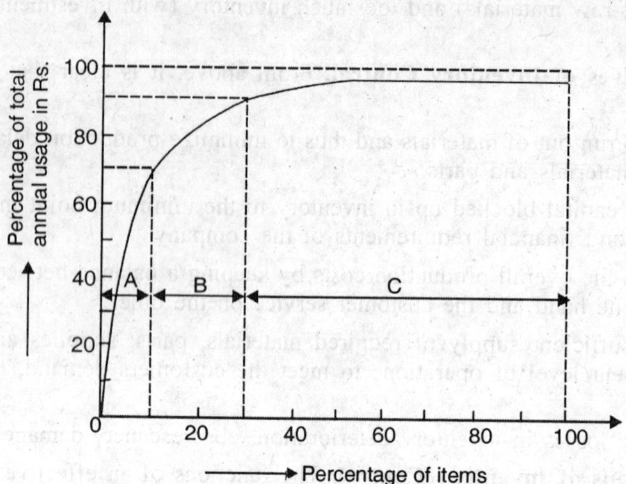

Fig. 23.12. A, B, C – Curve.

3. Stock positions, consumption patterns, ordering point, ordering quantity and safety stock should be serviced frequently.

4. Precise quality specifications and material standards should be evolved.

5. Purchasing must take maximum efforts to buy these items at competitive prices.

6. Cost reduction measures to be adopted to find out cheaper substitute, better source of supply, to reduce scrap, rejects, rework and substandards and to reduce overall costs ultimately.

7. If possible, supplies may be encouraged to store and supply materials at the factory site at their own cost.

8. Stores records must be kept meticulously.

Control of *C*-items :

1. It is usually sufficient to buy in the largest economic quantities and therefore maintain a comparatively large buffer stock.

2. Purchasing costs can be minimized through single tender system, blanket contract, clubbing of similar items in one purchase order, travel orders etc.

3. Two bin system can be followed very conveniently.

Control of *B*-items. The policy of control of *B* items lies midway between that of *A* and *C* items. Order quantities, reorder points and safety stocks on the basis of formulae work more effectively for *B* items and reviews are not necessary very often. Six monthly or quarterly contracts are suitable in general but purchaser has to put quite some efforts for high value items in *B* group.

23.7.4. Inventory Management Systems. Having excessive inventories in stock is a painless way of avoiding production bottlenecks and a luxury. Business cannot afford to have capital tied up in the stock room. Controlling the stock on hand and ordering replacements whenever they are needed are must for a successful business. The rate of a parts use and the lead time necessary to replace it must be considered. It is necessary to know when stocks will run out and how long it will take to secure a replacement in order to determine how far ahead to place an order for more of each item. Regular and periodic reordering are essential to maintain proper inventory balance. A popular method of control is minimum-maximum, Fig. 23.13.

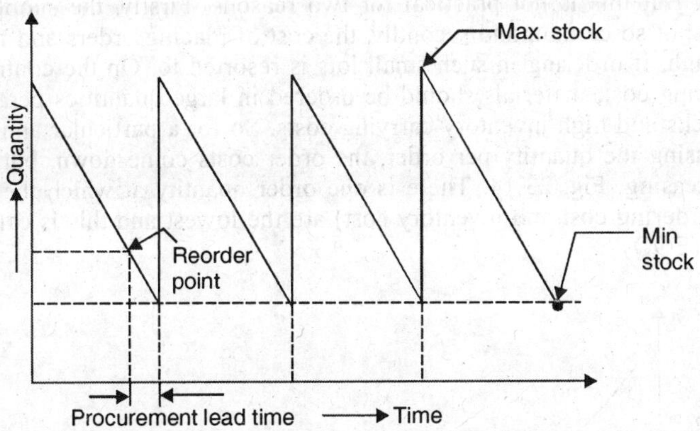

Fig. 23.13. A Simple Inventory Management System

(*a*) *Minimum Inventory.* Minimum inventory or buffer stock is needed to take care of any temporary unpredictable increase in the part usage or in procurement lead time.

(*b*) *Reorder point.* Reordering point is sufficiently above the minimum inventory to allow for issuing the purchase order and for delivery by a vendor. Reorder point stock level is equal to the minimum stock plus the expected consumption during the procurement lead time.

(*c*) *Reorder quantity.* This is the fixed quantity of item for which order is placed everytime the stock drops to the reorder point. This quantity is fixed either on the basis of experience or calculated as discussed in the next Article.

(*d*) *Procurement lead time.* This comprises the time required for preparing the purchase order, the time gap between placing an order and receiving supplies and time required for inspection etc.

(*e*) *Maximum inventory.* It is approximately the sum of the order quantity and the minimum inventory. It will exactly equal the sum of these two quantities if the ordered material is received just when the minimum stock is reached.

23.7.5. Economic Order Quantity (EOQ). One of the basic problems of inventory management is to find out the quantities to be ordered for each of the supplies so that it is most economical to the company from overall operational aspects. For this the minimum cost concept is used as discussed in Article 23.4. The various inventory costs are :

1. **Ordering cost.** When the supply is exhausted or diminished to a minimum level, a new order is placed to replenish the stock. This cost of placing the order is called the ordering cost and it occurs each time an order is placed. The ordering cost includes — the cost of originating, placing and paying for an order. This will be essentially constant, regardless of the quantity ordered. Thus, the order cost per unit varies inversely with the size of order, Fig. 23.14.

2. **Carrying cost.** When an item is procured and kept in stock, it introduces a large number of costs for the company, namely — amount of storage space and its rent or cost, interest on the average inventory in stock, insurance, depreciation on storage and handling facilities, obsolescence, deterioration and wastage etc., salaries and wages of stores personnel and so on. It is clear that all these costs increase directly with the order quantity. Fig. 23.14.

3. **Item cost.** The cost per piece of the item or material purchased is, in many cases, independent of order size; Fig. 23.14.

It is very clear from above that the ordering cost and carrying cost or holding cost are conflicting in nature and in determining the most economical quantity to order, we have to strike a balance between the two. From the point of view of the carrying cost, in an ideal condition there should be no stock at all. Every item should arrive just before use and in quantity just enough for a day or so. But this is not practical for two reasons. Firstly, the supplies and production programmes are not so certain, and secondly, the cost of placing orders and follow up work will shoot up very high, if ordering in such small lots is resorted to. On the contrary, from the point of view of ordering cost, materials should be ordered in large quantities per order. But this will result in big stocks and high inventory carrying costs. So for a particular annual consumption as we go on increasing the quantity per order, the order costs come down, but inventory carrying costs go on increasing, Fig. 23.14. There is one order quantity at which the total costs of item (material cost, ordering cost and inventory cost) are the lowest and this is called the "Economic Order Quantity (EOQ)".

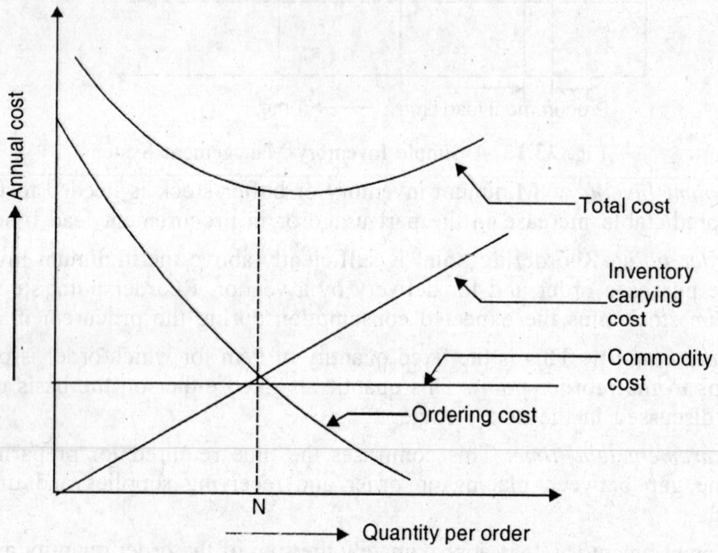

Fig. 23.14. Economic Order Quantity

The relation for "EOQ" can be derived mathematically as below, ignoring the quantity discounts :

Let A = Annual requirement of parts

C = Unit cost of part

R = Ordering cost per lot

K = Carrying cost per unit per year

N = Most economical ordering quantity per lot

$$\text{Average inventory} = \frac{N}{2}$$

$$\text{Number of orders per year} = \frac{A}{N}$$

$$\therefore \quad \text{Annual carrying cost} = K \times \frac{N}{2}$$

$$\text{Annual procurement (ordering) cost} = R \times \frac{A}{N}$$

$$\text{The annual commodity cost} = C \times A$$

The cost of procuring and storing one year's worth of the commodity.

$$G = CA + K \times \frac{N}{2} + R \times \frac{A}{N}$$

For minimum cost condition, $\frac{dG}{dN} = 0$

$$\therefore \quad 0 + \frac{K}{2} - \frac{RA}{N^2} = 0$$

$$\therefore \quad EOQ = N = \sqrt{\frac{2RA}{K}} \quad ...(23.9)$$

If N_s = Safety or buffer stock

then average inventory = $N_s + N/2$

However, the relation of "EOQ" will be still equ. (23.9).

It is clear that equation (23.9) resembles equation (23.5) for economical batch quantity with the difference that in place of set up cost per lot, we have ordering cost per lot.

Example 23.6. *Determine the most economical order quantity when annual usage is 8000 parts. Unit commodity cost is Rs 60 and the cost of placing an order is Rs 150 and the annual inventory carrying cost is 30% of the average inventory.*

Solution. $N = \sqrt{\frac{2RA}{K}}$

A = 8000 parts

C = Rs 60

R = Rs 150

i = 30%

$K = iC = 0.30 \times 60$ = Rs 18

$$\therefore \quad N = \sqrt{\frac{2 \times 8000 \times 150}{18}} = \textbf{365 units.}$$

23.8. NETWORK TECHNIQUES

The normal Gantt or bar chart as used for production scheduling, has two serious limitations:

1. It does not clearly indicate details regarding the progress of activities.
2. It does not give a clear indication of interrelationship between the separate activities.

These deficiencies may be eliminated to a large extent by showing the interdependence of various activities by means of connecting arrows. This procedure leads directly to network techniques. There has been considerable growth in application of network techniques in the field of planning and scheduling, which have proved to be quite effective and versatile tools. The two most commonly used techniques are—CPM (Critical Path Method) and PERT (Programme Evaluation and Review Technique).

These techniques were developed to aid in the management of large engineering projects. Now, production management differs from Project Management in the sense that whereas Project Management is generally a non-repetitive work, Production Management deals with work of permanent nature of repetitive type. However, CPM and PERT can be applied to Production Management also, particularly for jobbing production and servicing work orders, planning and launching of new products and services, changeover to new products or new models etc.

Difference between CPM and PERT. The basic difference between CPM and PERT lies in the amount of uncertainties existing in a project. CPM is used in production management mainly for the jobs which are repetitive in nature, where the activity time estimates can be predicted with considerable certainty due to the existence of past experience. PERT, on the other hand, is utilised for non-repetitive type of jobs (research and development work), where the time and cost estimates tend to be quite uncertain. Hence, this technique uses probabilistic time estimates.

Basic tool of CPM and PERT. The basic tool of CPM and PERT is an arrow network diagram. The chief elements of this diagram are :

1. **An activity.** It is a time consuming effort that is required to perform part of a work. An activity is shown on an arrow diagram by a line with an arrow head pointing in the direction of progress in completion of the work.

2. **An event.** An event is a point in time where one or more activities start and/or finish. It is a part of accomplishment and/or decision. A circle is used to designate an event.

The events are usually numbered as 10, 20, 30, 40 and so on, when the network is first drawn. The network progresses from left to right. Early events are given low numbers and later events are assigned higher numbers. If it becomes necessary to alter the network by adding more events, these can be given the numbers 11, 12, ..., 19, 21, 22, ... 29 and so on. Activities are usually referred to by showing the events that they connect, for example, activity 10–20, activity 20–30, activity 30–40 and so on. Sometimes, the events may also be numbered as 1, 2, 3, ... and activities as *A, B, C,* ... and so on.

An arrow network is an orderly representation of activities and events. This network is prepared according to logical sequence, that is, what would preceeds and what should succeed. The various logic restrictions to constructing the network diagram are :

(*i*) An activity cannot be started until its tail end is reached. Thus in,

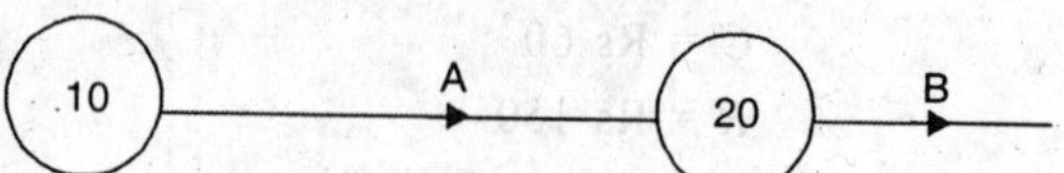

activity *B* cannot be started until activity *A* (10–20) has been completed.

Similarly in

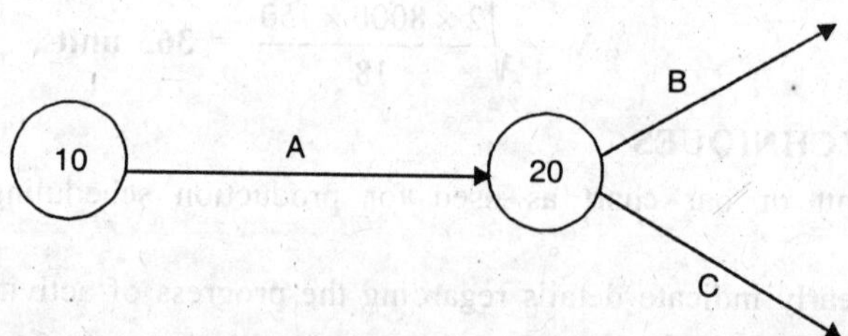

activities *B* & *C* cannot be started until activity *A* has been completed.

(*ii*) An event cannot be reached until all activities leading to it are complete. Thus in,

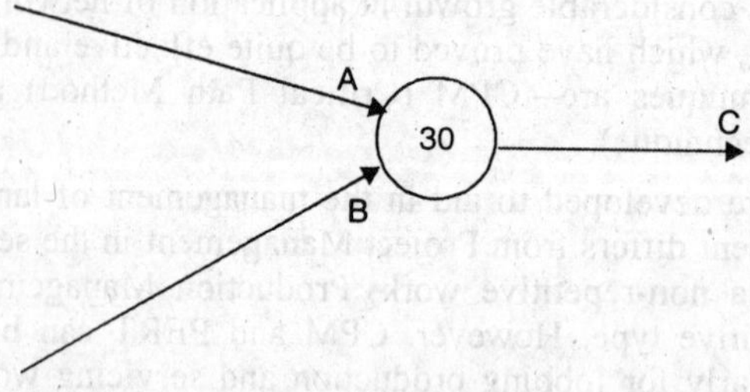

activities *A* and *B* must preceed *C*.

(*iii*) Sometimes an event is dependent on another event, preceding it even though the two events are not linked by an activity. To establish the relationship between such two events without the consumption of time and resources in a network, we call it as a dummy activity which is represented by the dotted line and the arrow as its end (head). For example, in fig. 23.15, activities *A* and *B* must both the completed before activity *D*, but activity *C* depends only on activity *A* and is independent of activity *B*.

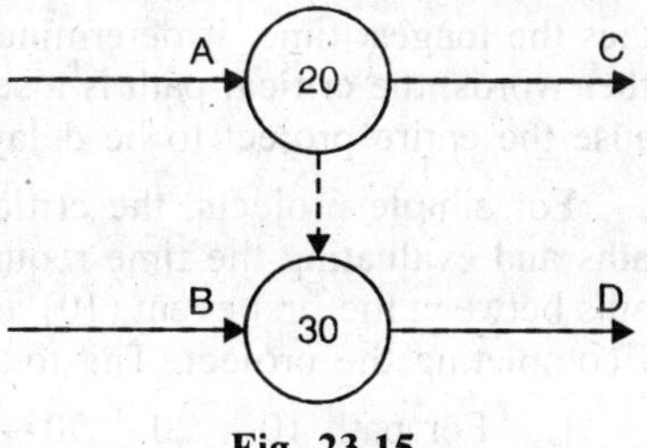

Fig. 23.15

(*iv*) All activities must both start and finish with an event.

(*v*) Any event, except the first and the last, must have at least one activity leading into it and at least one activity leading out of it.

(*vi*) Events should be numbered in the time sequence in which they occur.

(*vii*) Network, should be drawn so that event numbers increase progressively from left to right.

Network drawing. The basic concept of network drawing is nothing but a diagram of events and activities arranged in logical sequence. The first requirement is to define the end project date, then break up the project into various activities which could be done by piecemeal or breaking up the whole project into the small activities right in the beginning. The major events that must be accomplished towards the achievement of the objective must then be determined and laid out with their associated activities in network form. Then arrange the activities in a logical way showing their interdependence. Once the configuration is prepared, we can go ahead with the job of finding out the time estimates.

Fig. 23.16 shows activities and events that are required for carrying out a Plant Installation Project. The various activities are described below :

1. *A* (10 – 20) : Partial building construction
2. *B* (20 – 50) : Complete building construction
3. *C* (10 – 30) : Purchase of equipment
4. *D* (30 – 60) : Installation of equipment
5. *E* (10 – 40) : Recruitment of workers
6. *F* (40 – 60) : Training of workers
7. *G* (50 – 60) : Safety inspection
8. *H* (60 – 70) : Pilot production

Numbers on the activity lines denote activity duration in weeks.

Activity *I* (20 – 30) is the dummy activity which shows that activity *A* (10 – 20) must be completed before activity *D* (30 – 60) can start.

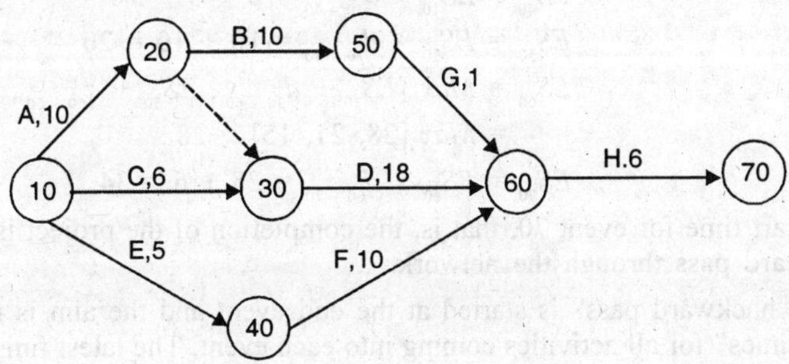

Fig. 23.16. Network Diagram

23.8.1. Critical Path Method (CPM). The Critical Path through the network is the path that takes the longest time. It determines the minimum time in which a project can be completed. In other words, the critical path is a series of connected activities and events which, if delayed, will cause the entire project to be delayed.

For simple projects, the critical path can be determined quite quickly by enumerating all paths and evaluating the time required to complete each. Referring to Fig. 23.15 there are four paths between the first event (10) and the last event (70). Each of these paths must be undertaken in completing the project. The total time along each path is determined below :

1. For path 10 – 20 – 50 – 60 – 70 — 27 weeks
2. For path 10 – 40 – 60 – 70 — 21 weeks
3. For path 10 – 30 – 60 – 70 — 30 weeks
4. For path 10 – 20 – 30 – 60 – 70 — 34 weeks

Thus the paths 10 – 20 – 30 – 60 – 70 is the critical path.

For more complex problems, a methodology must be established. The critical path is determined by a two-phased procedure as explained below :

The "forward pass" through the network starts at the initial node and proceeds towards the end node. Conversely, the "backward pass" moves from the end node to the starting node.

(*a*) In the forward pass, a number is determined at each node that represents the earliest occurrence of the respective event. As already written, activities are usually described in terms of an $(i - j)$ index, where $i < j$.

Let, ES_i = the earliest start times for all activities starting from event i

ES_j = the completion time

when $i = 1$, that is, the initial event,

$E_1 = 0$

The earliest start time for the jth event in the network is calculated by the relationship —

ES_j = Maximum values of $(ES_i + d_{ij})$ for all activities commencing from event i and leading to event j.

where d_{ij} = duration of activity $(i - j)$

For the network shown in Fig. 23.15,

$$ES_{10} = 0$$

$$ES_{20} = ES_{10} + d_{10-20} = 0 + 10 = 10$$

$$ES_{30} = Max\,[ES_{20} + d_{20-30},\ ES_{10} + d_{10-30}]$$
$$= Max\,[10, 6] = 10$$

$$ES_{40} = ES_{10} + d_{10-40} = 0 + 5 = 5$$

$$ES_{50} = ES_{20} + d_{20-50} = 10 + 10 = 20$$

$$ES_{60} = Max\,[ES_{30} + d_{30-60},\ ES_{50} + d_{50-60},\ ES_{40} + d_{40-60}]$$
$$= Max\,[28, 21, 15] = 28$$

$$ES_{70} = ES_{60} + d_{60-70} = 28 + 6 = 34$$

The earliest start time for event 70, that is, the completion of the project is 34 weeks. This completes the forward pass through the network.

(*b*) Now, the "backward pass" is started at the end event and the aim is to determine the "latest completion times" for all activities coming into each event. The latest finish time for event i is denoted by LF_i and for the end event,

$$LF_i = ES_i$$

The latest finish times for any given event i is determined as :

LF_i = Min. value of $(LF_j - d_{ij})$ for all activities commencing from event i to event j.

Hence,

$$LF_{70} = ES_{70} = 34$$
$$LF_{60} = LF_{70} - d_{60-70} = 34 - 6 = 28$$
$$LF_{50} = LF_{60} - d_{50-60} = 28 - 1 = 27$$
$$LF_{40} = LF_{60} - d_{40-60} = 28 - 10 = 18$$
$$LF_{30} = LF_{60} - d_{30-60} = 28 - 18 = 10$$
$$LF_{20} = Min.\ of\ [LF_{50} - d_{20-50},\ LF_{30} - d_{20-30}]$$
$$= Min.\ of\ (27 - 10,\ 10 - 0) = 10.$$
$$LF_{10} = Min.\ of\ [LF_{20} - d_{10-20},\ LF_{30} - d_{10-30},\ LF_{40} - d_{10-40}]$$
$$= Min.\ of\ (10 - 10,\ 10 - 6,\ 18 - 5) = 0.$$

This completes the "backward pass".

Critical path. The critical path in the network is that sequence of activities and events where there is no "Slack". Slack is defined as the difference between latest event time and earliest event time, that is

$$Slack = LF - ES$$

So, for Zero Slack,

$$ES_i = LF_i$$
$$ES_j = LF_j$$

The events lying along this path are "Critical" and the connecting activities are called "critical activities" in the sense that their occurrence can't be delayed if the scheduled completion time is to be met. When the slack for an event is zero, it means that any delay in the occurrence of that event will cause delay in starting subsequent activities. A positive slack indicates that the activity could be delayed by that amount without causing any delay in the total project.

Table 23.4 summarizes the ES_i and LF_i times and the amount of slack associated with each event.

Table 23.4 Slack Times for each Event

Events i	*ES*	*LF*	*Slack (Weeks)*
10	0	0	0
20	10	10	0
30	10	10	0
40	5	18	13
50	20	27	7
60	28	28	0
70	34	34	0

So, the critical path is : 10 – 20 – 30 – 60 – 70.

Critical events are shown by double circles and critical activities are shown as thick lines as illustrated in Fig. 23.16.

23.8.2. PERT. The PERT technique uses the same ideas as the CPM; but instead of using just the most likely estimates, it uses a probabilistic estimate of time for completion of any activity. Whereas CPM uses only a single time estimate called the "Point estimate" for all activity durations, PERT uses three probable time estimates, namely,

t_o = an optimistic time estimate if everything goes smoothly.

t_p = a pessimistic time estimate if everything goes badly.

t_m = the most likely time estimate and is bracketed between t_o and t_p.

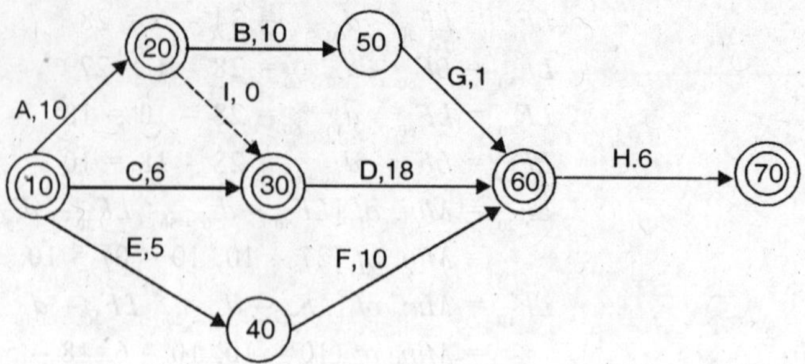

Fig. 23.17. Critical Path.

Then the expected time t_e, is found by assuming that the time estimates follow a beta frequency distribution, that is,

$$t_e = \frac{(t_o + 4t_m + t_p)}{6} \quad \text{...(23.10)}$$

The expected time is computed for each activity and the expected times are used to determine the "Critical Path" and the boundary times.

The expected time for each activity also has a standard deviation which describes its scatter and is given by

$$\sigma = \frac{t_p - t_o}{6} \quad \text{...(23.11)}$$

The variance of the expected time is obtained with the help of the equation

$$\text{variance} = \sigma^2 = \left(\frac{t_p - t_o}{6}\right)^2 \quad \text{...(23.12)}$$

The variance of the expected time of the project, σ_{cp}^2, is obtained by adding the variance of the expected times of all activities along the critical path, that is,

$$\sigma_{cp}^2 = \Sigma(\sigma)_i^2 \quad \text{...(23.13)}$$

The expected time of the project is the sum of the expected times of all activities lying on the critical path, that is,

$$t_{cp} = \Sigma t_e \quad \text{...(23.14)}$$

Now assuming that the completion time for the project has a normal distribution about the expected completion time, t_{cp}, with a standard deviation of σ_{cp}, the probability that the project will be completed in a given time, t_d, is given by

$$p = \varphi\left(\frac{t_d - t_{cp}}{\sigma_{cp}}\right) \quad \text{...(23.15)}$$

$$= \varphi(z)$$

where $\varphi(z)$, denotes the cumulative distribution function for the variable z corresponding to a standardised normal distribution, Fig. 23.17.

If $z = 0$, there is a 50% probability of completing the project on the scheduled time. If $z = -0.05$, there is 30% probability and so on.

23.8.3. Other uses of CPM and PERT. In addition to their use in production planning and control, CPM and PERT can also be used for economic studies. Once a critical path is determined, the time required to complete a project can be shortened by shifting resources or by changing procedures. However, such activities usually involve some increases in costs. Such costs are called as "Crash Costs". And there usually is a range of crash costs that can be incurred. So the economic study is : balancing costs verus revenues. CPM and PERT can also be used for another class of solution — least-cost solution using costs associated with the normal time to complete the activities.

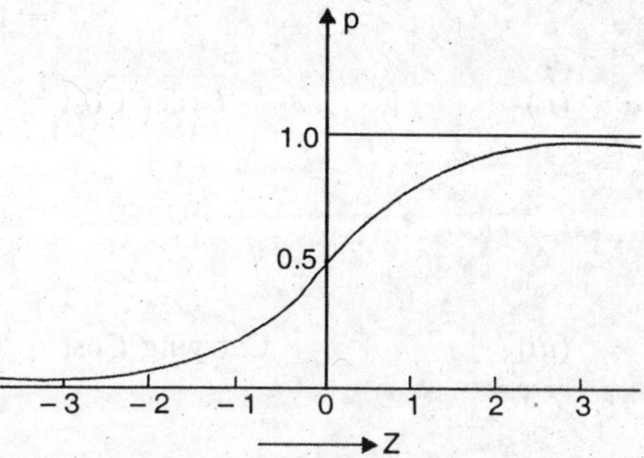

Fig. 23.18. Cumulative Distribution of $\varphi(z)$

Example 23.7. *Determine the most economic lot size for the data given below :*

Annual requirement = 12000 parts

Unit cost of part = Rs 5

Set up cost per lot = Rs 60

Production rate per year = 18750 parts

Inventory carrying cost = 20% of unit cost per year.

Solution. Given :

$$A = 12000;\ C = \text{Rs } 5$$

$$S = \text{Rs } 60;\ P = 18750$$

Now
$$N = \sqrt{\frac{2S}{\left[(1/A) - (1/P)\right] K}} \qquad \text{...(23.7)}$$

Now
$$K = iC = 0.2 \times 5 = 1$$

$$N = \sqrt{\frac{2 \times 60}{\left[\frac{1}{12000} - \frac{1}{18750}\right] \times 1}}$$

$$= \textbf{2000 parts}$$

Example 23.8. *The annual requirement of raw material of a company is 15625 units. The cost per unit is Rs 12. The ordering cost is Rs 60 per order and the inventory carrying cost is Rs. 1.20 per unit. Determine :*

(*i*) *EOQ* (*ii*) *Ordering Cost* (*iii*) *Carrying Cost* (*iv*) *Total Inventory Cost.*

Solution. (*i*)
$$EOQ = N = \sqrt{\frac{2RA}{K}}$$

Now
$$A = 15625 \text{ units};\ R = \text{Rs } 60$$

$$C = \text{Rs } 12;\ K = iC = \text{Rs } 1.20$$

$$\therefore \quad EOQ = \sqrt{\frac{2 \times 60 \times 15625}{1.20}}$$

$$= \textbf{1250 units}$$

(*ii*) $$\text{Order Cost} = R \times \frac{A}{N}$$

$$= 60 \times \frac{15625}{1250} = \text{Rs. } 750$$

(*iii*) $$\text{Carrying Cost} = K \times \frac{N}{2}$$

$$= 1.20 \times \frac{1250}{2} = \text{Rs } 750$$

(*iv*) Total Inventory Cost = Rs 750 + Rs 750 = Rs. 1500.

Example 23.9. *The annual requirement of raw material of a company is 50 tonnes. The cost of raw material is Rs. 500 per tonne. Ordering cost is Rs 100 per order. The company is offered a discount of 2 per cent of the purchase cost if the order per lot is 25 tonnes. Should the company accept the offer ? The inventory carrying cost is 20% of unit cost per annum.*

Solution. Economic Ordering Quantity is,

(*i*) $$EOQ = \sqrt{\frac{2RA}{K}}$$

R = Rs 100, A = 50 tonnes; $K = iC = 0.2 \times 500$ = Rs 100

$$EOQ = \sqrt{\frac{2 \times 100 \times 50}{100}} = 10 \text{ tonnes}$$

$$\text{Ordering Cost} = R \times \frac{A}{N} = 100 \times \frac{50}{10} = \text{Rs } 500$$

$$\text{Inventory carrying cost} = K \times \frac{N}{2} = 100 \times \frac{10}{2} = \text{Rs } 500$$

$\therefore$ Total Inventory Cost = Rs 500 + Rs 500 = Rs 1000

(*ii*) If the order per lot = 25 tonnes,

then $$\text{Ordering Cost} = 100 \times \frac{50}{25} = \text{Rs } 200$$

$$\text{Inventory carrying cost} = K \times \frac{N}{2}$$

$$= 100 \times \frac{25}{2} = \text{Rs } 1250$$

Total Inventory Cost = Rs 200 + Rs 1250 = Rs 1450

$\therefore$ Increase in Inventory Cost = Rs 1450 – Rs 1000 = Rs 450

$$\text{Discount offered} = \frac{2}{100} \times 500 \times 50$$

$$= \text{Rs } 500$$

∴ Offer is worth accepting.

Example 23.10. *The requirement of a manufacturer for the next year which will have 250 working days, is 1000,000 components. The storage cost is Rs 4 unit per year, and the cost per order is Rs 32. Safety stock has to be two days requirements and the lead time for the supplier is 4 days. Calculate EOQ if the order is placed at the end of the day and the delivery is also at the end of the day. Also, determine the reorder point.*

Solution. (*i*)
$$EOQ = \sqrt{\frac{2RA}{K}}$$

$$R = \text{Rs } 32;\ A = 1000{,}000;\ K = \text{Rs } 4$$

∴
$$EOQ = \sqrt{\frac{2 \times 32 \times 1000{,}000}{4}}$$

$$= \textbf{4000 components}$$

(*ii*)
$$\text{Re-order point} = \text{Safety stock} + \text{Average daily usage} \times \text{Lead time}$$

$$\text{Now Safety stock} = \frac{1000{,}000}{250} \times 2 = 8000 \text{ components}$$

∴
$$\text{Re-order point} = 8000 + 4000 \times 4$$

$$= \textbf{24000 components}$$

Example 23.11. *Solve Example 23.6 for the variable price-schedule given below :*

Lot Size	*Unit Price*
1 – 200	*Rs 63*
201 – 500	*Rs 60*
501 and above	*Rs 55.50*

Solution. Given data is

$$\text{Annual usage, } A = 8000 \text{ parts}$$

$$\text{Cost per order} = \text{Rs } 150$$

Annual inventory carrying cost = 30% of the average inventory. Such problems are usually solved by multiple-step procedure, because there is never any assurance that the theoretical *EOQ* in any price range will fall within the lot size range for the particular price, or that this lot size will give the minimum total cost.

The problem can be solved on the same lines as Example 5.42, or we can solve it as given below :

(*a*) When lot sizes of 1 to 200 are considered, purchase orders/year will be 40 or more.

$$\text{The minimum total cost will be} = CA + K \times \frac{N}{2} + R \times \frac{A}{N}$$

With purchase orders equal to 40, *N* will be = 200

$$\text{Min, total cost} = 63 \times 8000 + 0.30 \times 63 \times \frac{200}{2} \times \frac{8000}{200} \times 150$$

$= 504000 + 1890 + 6000$

$= \text{Rs } 511{,}890.00$

(*b*) For lot sizes of 201 to 500,

Orders placed per year will be = 16 to 39.

With orders = 16, *N* will be 500

$$\text{Total cost} = 60 \times 8000 + 1.8 \times 250 + 150 \times 16$$
$$= 480{,}000 + 450 + 2400 = \text{Rs } 482{,}850.00$$

with orders = 39

$$\text{Total cost} = 480{,}000 + 1.8 \times \frac{8000}{39 \times 2} + 39 \times 150$$
$$= 480{,}000 + 184.60 + 5850 = \textbf{Rs 486034.60}$$

Total cost for the second price schedule will vary between Rs 482,850 and Rs 485,034.60. or, we can proceed like as given below :

$$EOQ = N = \sqrt{\frac{2RA}{K}} = 365 \text{ units (See Ex. 23.6)}$$

∴ This is a feasible quantity in the price of Rs 60 per unit as it lies between 201 and 500.

$$\text{Min. T.C.} = 480{,}000 + 1.8 \times \frac{365}{2} + 150 \times \frac{8000}{365}$$
$$= 480{,}000 + 328.50 + 3287.70$$
$$= \textbf{Rs. 483616.20}$$

(*c*) Order of 501 and above,

∴ Number of orders = 16 or less

Now with number of order = 16 and lot size = 501

$$\text{T.C.} = 55.50 \times 8000 + 1.665 \times \frac{501}{2} + 16 \times 150$$
$$= \text{Rs } 444000 + 417 + 2400$$
$$= \text{Rs } 446817.00$$

With number of lots = 14

T.C. will be = Rs 45087.00

With number of lots = 10

T.C. will be = Rs 45216.00, that is, T.C. is increasing.

Min. T.C. is Rs 446817.00

$N = 501$

No. of lots = 16

PROBLEMS

1. Define the term : Production planning and control.
2. List the activities involved in production planning and control.
3. Discuss the principles of production planning and control.
4. Write the objectives of production planning and control.

5. Write the functions included under the activity of production planning.
6. Write the duties of a production planner.
7. Write the qualifications of a production planner.
8. Write the objectives of production planning.
9. What is production control function?
10. Write the objectives of production control.
11. Write the advantages of production planning and control.
12. Discuss : Production for stock and Production to order.
13. Discuss the three types of production — Jobbing, Batch and Mass Production.
14. Write the purpose of Sales forecasting.
15. Discuss the following methods of sales forecasting :
    (*a*) Regression analysis.
    (*b*) Simple averaging method.
    (*c*) Simple moving average method.
    (*d*) Method of Weighted Moving Average.
    (*e*) Exponential Smoothing.
16. How will the demand for a new product be forecasted?
17. Discuss "Economical Batch Quantity".
18. Derive an expression for economical batch quantity.
19. What is "Routing" function?
20. Discuss the steps of routing procedure.
21. Discuss the following routing documents :
    (*a*) Route sheet
    (*b*) Operation sheet
    (*c*) Bill of materials.
22. Define scheduling function.
23. Write the main objectives of scheduling function.
24. What is Master Schedule?
25. Write the objectives of a Master Schedule.
26. How a Master Schedule is prepared?
27. What is a Shop Schedule?
28. How a Shop Schedule is prepared?
29. Discuss : Machine loading.
30. Draw a typical Machine loading chart.
31. Write the objectives of Shop schedule.
32. What is "Despatching function"?
33. Write the duties of a despatcher.
34. What are works order documents? Write on :
    (*a*) Work order form
    (*b*) Material requisition form
    (*c*) Move ticket
    (*d*) Inspection ticket.

35. What is progress reporting?
36. Write on Gantt Chart and Manufacturing flow chart.
37. What is "Follow-up" function?
38. Discuss the two types of follow-up functions.
39. Discuss the procedure for production planning and control for :
    (a) Jobbing type of production
    (b) Mass production
    (c) Batch production
40. Write the various types of inventories used during production process.
41. Why large inventories are essential in a plant?
42. Define : Inventory Control.
43. Write the objectives of inventory control.
44. Write the functions of inventory control.
45. Write on Selective inventory control.
46. Discuss a simple inventory management system.
47. What is Economic Order Quantity?
48. What are Ordering costs and carrying costs?
49. Show the variation of Ordering costs and carrying costs with order quantity.
50. Derive an expression for *EOQ*.
51. Name the two common network techniques used for production scheduling and control.
52. What is the difference between CPM and PERT techniques.
53. Define : Activity and Event.
54. Discuss the various logic of restrictions for constructing a network diagram.
55. Define Earliest start time, Latest finish time and slack.
56. Discuss the procedure of CPM.
57. Discuss the procedure of PERT.
58. The annual requirement of a company is 8000 parts. The cost of each part is Rs 20. The ordering cost per order is Rs 15. The carrying cost is 15% of the average inventory per year. Determine the economic order quantity.
59. A plant can produce 800 hydraulic valves per month. The annual sale is 3000 valves. The selling price of each valve is Rs 250. The ordering and set up cost is Rs 300 per order. The inventory carrying cost is 20%. Determine the quantity the warehouse should order from the plant.
60. An automobile industry purchases a lot of 1000 axles which is a 3 month supply. The cost per axle is Rs 500 and the ordering cost per lot is Rs 600. The inventory carrying cost is 20% of unit cost. Calculate the total annual inventory cost. Determine the EOQ for the company and the money saved by the company per year by employing EOQ.

    (Rs. 52400; 215; Rs 30487.20)
61. A plastic moulding company produces 24000 teflon bearing inserts per year. Setting up cost is Rs 85 and the weekly production rate is 1000 pieces. Cost per unit is Rs 5.50 and the carrying cost is Rs 0.50 per unit. Determine the production during each production run.

    (3895 units)
62. The date for a component to be produced by machining process is given below :

    Annual requirement = 4800 pieces

    Machine rate = 20 pieces per shift

Number of shifts in a year = 320

Set up cost = Rs 400

Production cost per piece = Rs 100

Inventory carrying cost per year = 12% of unit cost.

Determine EBQ. (1131 units)

63. A bearing company is committed to supply 24000 bearings per year to a fan company on a steady daily basis. The set up cost for each bearing manufacture run is Rs. 650 and the inventory carrying cost is 20 paise per unit per month. Determine :

(*a*) the optimum run size for bearing manufacture.

(*b*) time interval between two consecutive bearing manufacture runs.

(*c*) minimum inventory carrying cost. (3600 units, 54 days, Rs 4320)

64. Determine the most economical size of a purchase order for the following conditions :

Order preparing cost = Rs 350

Annual demand = 200,000 parts

Annual carrying charges = 25% of average inventory

Cost of each item = Rs 8.00

65. Consider the following data for a product :

Demand = 10,000 units/year

Order Cost = Rs. 40/ order

Holding Cost = 10% of the unit cost/unit yr.

Unit Cost = Rs. 550

(*a*) What is the EOQ?

(*b*) Under the EOQ, what is the number of annual orders? (GATE 1995)

(**Ans.** 126, 80)

66. Determine the number of production runs and also the total incremental cost in a factory for the data given below :-

Annual requirement = 15,000 units

Preparation cost per order = Rs. 25

Inventory holding cost = Rs. 5/unit / year

Production rate = 100 units / day

Number of working days = 250 (GATE 1997)

Sol. From equ. 23.7

$$N = \sqrt{\frac{2R}{[(1/A)-(1/P)]K}}$$

$$\text{optimum time interval} = \frac{N}{A} = \sqrt{\frac{2RP}{KA(P-A)}}$$

R = Ordering cost = Rs. 25 P = Production rate per year = 100 × 250 = 25000

A = Annual requirement = 15000 K = Inventory holding cost / unit/year = Rs. 5

$$\therefore \text{ Optimum time interval} = \sqrt{\frac{2\times 25\times 25000}{5\times 15000\times 10000}} = 0.040825 \text{ years}$$

$$\therefore \text{ Number of production runs/ year} = \frac{1}{0.040825} = 25$$

24

MANUFACTURING SYSTEMS AND AUTOMATION

24.1. INTRODUCTION

A system can be defined as the entire combination of hardware, information and people, necessary to accomplish some specified mission. A large system is usually divided into subsystems, which are, in turn, made up of components. Over the years, the manufacturing of consumer and durable goods has become very complex due to : proliferation of the number and variety of products, increased variety of materials, requirements for more precision and better quality at reduced cost, product reliability and on-line delivery of parts etc. The manufacturing thus consists of a series of interrelated activities and operations involving design, material selection, quality assurance etc. Due to all this, manufacturing has grown to become a ''System'' with many components that interact in a dynamic manner. A manufacturing system can be defined as a system organized to manufacture parts and products. The system takes inputs and produces products for the customer.

The inputs to a manufacturing system are : men, materials, information, and energy. These inputs are given to a complex set of elements known as ''Machines" or machine tools, processes, tools etc. These elements are known as "Converting Equipment". Here, the materials are processed and they gain value. The output of a manufacturing system consists of : consumer goods or producer goods (inputs to some other processes). Thus in a manufacturing system, the four M's of the system, that is, machines, materials, money and manpower alongwith the information, are co-ordinated so as to manufacture the parts of desired quality at the minimum possible cost, resulting in profit to the organization. A manufacturing system consists of : a group of machines, material handling, storage and control devices.

24.2. MATERIAL MOVEMENT

The movement of materials, parts, tools and products, is an essential element of any manufacturing system. Therefore, before studying the various types of manufacturing systems, one must know about the movement of these items in a manufacturing system. Material movement in any manufacturing system consists of the following activities :

1. Movement from the receiving area to the incoming inspection area.
2. Movement from the incoming inspection area to stores.
3. Movement from stores to shop floor.
4. Movement between production units on the shop floor.
5. Movement from the shop floor to the warehouse.

In this chapter, while discussing the various manufacturing systems, we will be interested only in the material movement on the shop floor between production units. It has been observed that about 95% of the total manufacturing time (in batch type production) is utilized in moving the parts from one place to another or in waiting for something to happen. Even out of the remaining 5% of the time, the actual machining time consumes only about 30%. The remaining time gets utilized in loading, unloading, positioning, clamping and gauging etc. Hence, to increase the productivity of a manufacturing system, the methods of material movement, loading, positioning, clamping, gauging and unloading of parts and tools must be improved. The material handling equipment used in any manufacturing system serves two functions : to move parts between machines and to orient and locate these parts for processing at the machines.

The material movement can be : attended material movement or mechanized material movement.

24.2.1. Attended Material Movement. In this method, the operators can move the material etc., but this method is least efficient and most costly. Manually operated trucks (1, 2, 3 or 4 wheeled) can be used. Use of pallets, platforms or trays facilitate quick and easy loading and unloading. Fork lift trucks are a flexible and highly maneourable vehicle. But it needs unobstructed passageways. Overhead cranes do not need floor space and ensure flexibility, but may interfere with one another.

24.2.2. Mechanized Material Movement. Mechanized material movement or automated material handling is an essential component of a modern integral automation industry. Such an industry requires automatic loading, unloading and movement of material from one machine to the next. For this, automatic conveying, storing, feeding and work-handling devices must be provided. The workpieces mounted on pallets or fixtures move through the system by means of towlines combined with shuttle system, roller conveyors, drag chains or automatically guided vehicles, AGVS. Combined with shuttle system, industrial robots, located suitably with respect to a group of machine tools, transfer workpieces from one machine to the next.

1. On-line Automatic Conveyors (OLAC's). On-line automatic conveyors are driven by an endless chain driven by geared hydraulic servo-unit which provides the required velocity and acceleration. OLAC gets the instructions from the Central Computer. The conveyor can also carry a cutter magazine for interchanging tools on the various NC machine tools.

2. Automatic Guided Vehicle System (AGVS). This is a very flexible material movement system on the shop floor. The vehicle can move along any one of several paths, under inductive guidance provided by wire or cable guides embedded in the shop floor. The path of a vehicle can be readily programmed and reprogrammed to meet the changing production needs and for optimizing the path of each vehicle. The movement of the vehicles can be stopped when they encounter an obstruction, with the help of sensors. The AGV system can be easily expanded to includes miles of factory shop floor.

3. Robots. To realize complete computer controlled manufacturing, general purpose programmable work handling devices have been developed. Robot is such a device. So, a robot can be defined as an operator-less work handling device which is programmable. According to the Robot Institute of America, a robot is a ''reprogrammable, multifunctional manipulator designed to move material, parts, tools, or special devices through variable programmed motions for the performance of a variety of tasks''. This definition puts the condition that the mechanical arm of the robot must be computer controlled. However, Japanese usually include ''fixed-sequence robots'' in their robot statistics. The mechanical arm of these machines is controlled not by a computer but by a series of electro-mechanical switches.

Industrial robots consist of two major component systems. Firstly, there are the moving parts, chiefly comprising the arms, wrist and hand elements, actuated by drive mechanisms. Secondly, it is the ''brain'', that is, a computer memory that would store and operate on the list of instructions that a robot needs to tell it how to go about doing a particular job.

The key to deciding if a machine is robotic is that it must be ''reprogrammable''. In order for a robot to carry out a particular task, it has to be given a ''program'', that is, a list of instructions is changed, then the robot will do a different task. Making the change is called ''reprogrammable''. This enables the robot to switch quickly from one simple task to another. Robots and robotic system relieve man from mindless drudgery of performing tedious, repetitive and dangerous jobs.

The automobile industry is the leading employer of robots, variously described also as ''dumb servants'', or ''mindless mechanical workers''. In this industry, assembly line production is replete with simple, repetitive jobs that the robots can do so well. Here, robots are being put to do the following work : loading and unloading conveyors, welding car bodies together and spray-painting the finished product.

Smaller manufacturers are also using robots to perform tasks like loading and unloading moulds and presses. Robots are also performing the cornerstone of ''flexible manufacturing system''. These are combinations of robots and computer-controlled machine tools which bring the benefits of automation to the production of small batches of parts. Here, the ability of the robot to switch from one task to another, counts. By varying the order in which parts are transferred between different machine tools, the robot gives such systems the flexibility to handle a variety of jobs. It means that one need not produce any single component in large quantities to justify the costs of automation.

Robots can boost productivity with their untiring speed, and boost quality with their mindless ability to do the same job in exactly the same way every time.

The two most important applications of robots are : (*i*) Part handlers (*ii*) Tool handlers.

Part handlers are helpers. Their job is to boost the productivity of other machines by feeding them with parts quickly and smoothly. They have to work in systems of machines. Because they are only cogs in a system, part handlers are often more difficult to put to work successfully. Their brainlessness means they cannot cope with errors made by other parts of the system. The machines with which they work must be organized to present parts to them in the same manner everytime. They must suit their pace to work to that of the rest of the system. If the machines with which robot works operate at different pace, or a human is required to run a machine, no amount of robot efficiency will guarantee higher productivity. Tool handlers are more independent. They do skilled jobs. Although, they, too are often integrated into production systems, their work requires more individual ''machine brain power'' than that done by part handlers. Their skills allow them to fit into assembly line jobs. They can also work in more or

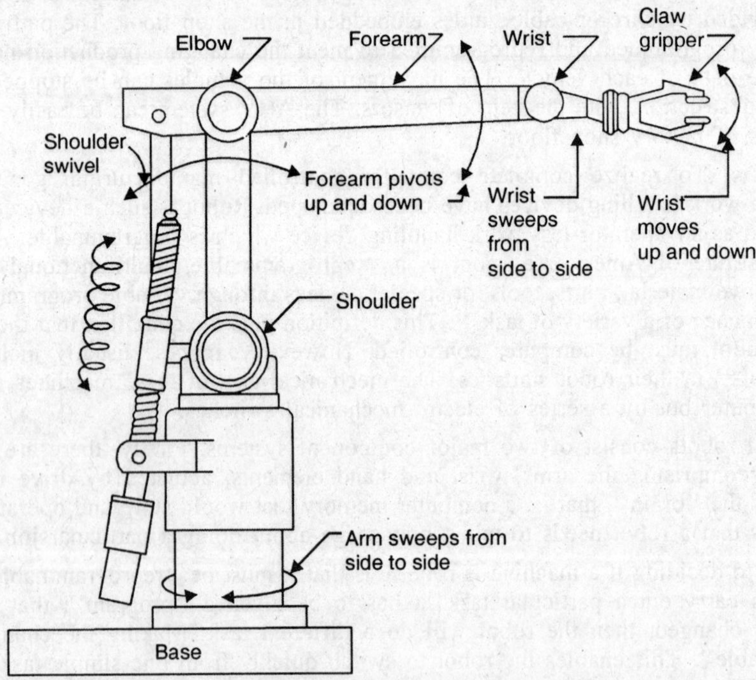

Fig. 24.1. A Typical Robot (with claw type gripper).

less self-contained areas like paint-spraying booths. Their problem is to acquire the skills they need to do. A robot can have several axes of motion and can handle components weighing, from 1 kg to as much as 900 kg. The range of a robot can vary from 75 mm to 600 cm. Fig. 24.1 shows a typical robot. The degrees of freedom of a robot correspond to those of human arms, hands and fingers. A typical robot can translate as a whole in two directions. Its arms can translate in the third dimension with respect to the body. Also, its hands are capable of swivelling around its wrists and bending about two mutually perpendicular directions above its wrist. Further, the robot's hands are provided with jaws or electromagnetic or suction devices for grasping and holding.

Generations of robots. The stages of development of robots are referred to as ''generations''. The term ''generation'' is used here often conditionally. Robot generations do not change each other historically. New robot generations come to life, while the first generation continues to live and find even increasing application. Robot design is perfected and performance improved. Different types of robots of each generation find application in industry.

(*i*) **First generation robots.** They are quite basic machines. They have a simple brain and are good at doing exactly what they are told to do. Such robots can perform not only the relative simple tasks of loading and unloading, die casting and plastic moulding machines, forging pressess, welders and other machine tools, but they can also guide a welding gun over a spatial track to weld, say pipes into a T-joint. To illustrate, the robot picks up a workpiece, loads the press, unloads the press, moves the workpiece to another place, returns to the initial position. For this purpose, use is often made of a two-arm robot. Then picking up a workpiece by one hand coincides with unloading the press by second, next one hand loads the press and the other moves the workpiece to another place, while the press operates. Both hands return to the starting position and the operation repeats. First generation robots have been widely used since late 70s. Their main limitation is that they have no feedback, that is, no sensing ability to tell them that something is wrong and what they should do about it. The robot used in car factories to weld body panels and spray the car shells for example, would continue to operate, even though a fault on the assembly line had delayed the components. A welding machine will continue to make spot-welds, regardless of whether the components to be joined are in correct position. Thus, all parts and tools must be located in exact positions.

(*ii*) **Second generation robots.** The second generation or supervisory controlled robots have feed-back facility. They are provided with rudimentary sensors at the wrists or gripper units of robots : tactile, force, locational, visual and heat. These sense will allow them to complete a particular range of tasks within their specific environment. Since it is possible to send and receive signals to and from the associated equipment, the robot has the ability to control operation, depending upon the received information and thus avoid mismatch between the programmed performance and the actual performance of robots. So, these robots will be able to tell whether there is a car shell in front of them or not. The sensors will help to locate the exact position of holes etc. into which a robot has to insert some part, for example, workpiece into a chuck or a tool bar into a turret etc. For this, a robot has to possess certain adaptivity which is essential in motion, operations, involving parts with undetermined orientation, assembly, welding etc.

The tactile and force feedback comes from the touch sensors provided in the grippers of a robot. The sensors may consist of microphones bedded in foam rubber pads on the tips of the gripper. The microphones pick up the very faint noises which are produced when the foam is crushed. These noises can tell the robot that it has touched the job. And, as sound will be created by the object slipping from the gripper, the robot will know whether or not there is enough pressure in its grip. Strain gauges which can measure the amount of pressure that is being used, will be able to tell it just how much force it is applying on the object.

Infra-red sensors will allow a robot to see hot or living objects in the dark. Microwave devices can detect movement and speed, again in total darkness. When waves bounce off a moving object, they come back with a change in pitch. The 'note' is higher or lower depending on which direction the object is moving in. By measuring this shift in pitch, it is possible to work out how fast the object is moving. In ultrasonic sensors, sound waves are sent out which bounce off objects in front of them and then come back. By knowing how fast sound waves travel through air, we can work out how far an object is. A robot fitted with ultrasonic or microwave sensors (or a combination of the two) would, therefore, be able to sort out two things. First, it would tell how far away the objects were, and secondly what speed and direction (if any) they are moving in.

(*iii*) **Third generation robots.** The third generation robots are endowed with "artificial intelligence". They possess richer means of sensory perception, pattern recognition, decision making and motion implementation. Visual perception, digital simulation of the environment and complex use of sensors assume herein a special significance. Robot systems using TV cameras have been developed which make use of camera vision to acquire un-oriented workpieces for example, turbine blade forgings from bins, orient them and transport them to the desired locations.

A robot has three main parts :

(*a*) Drive to power the robot.

(*b*) Mechanical arm.

(*c*) Hands to be attached to the end of the arm.

(*a*) **Power drives for robots.** There are three types of drives which can be selected for a robot :

(*i*) **Pneumatic drives.** Here, compressed air is used to move the robot's mechanical arm. They are light-weight, fast and relatively inexpensive, but, they cannot provide much strength.

(*ii*) **Hydraulic drives.** These use compressed fluid to drive the mechanical arm. They are much stronger, but, are more expensive and are prone to leaking.

(*iii*) **Electric drives.** These are the strongest and the least energy consuming robots. However, they are the most expensive — both to buy and to maintain.

To stop the robot, one of the following two methods can be used :

(*i*) Once the robot's arm has reached a desired position, the drive will be cut off and the brakes will be slammed. Such robots are called non-servo-controlled robots. They don't employ feed-back control for their drive units.

(*ii*) A servo-controlled system can be used. This will begin to slow down the arm as it nears the desired end point, then use the drive's own power to completely stop it. It can even reserve the drive to the arm if it overshoots the mark. This system is more sensitive, but slower and more expensive.

(*b*) **Mechanical arm.** One out of the following four designs, can be selected, for the arm of a robot, Fig. 24.2.

(*i*) **Anthropomorphic or revolute arms.** Like human arms, these can bend and swivel at the shoulder and bend at the elbow. These are the most flexible or robotic arms and can reach any nook and corner. However, they are difficult to control properly and are less able to carry heavy loads.

(*ii*) **Cylindrical arms.** These consist of two basic parts : an extendable arm and the central pole on which it is mounted. The arm can move horizontally out from the pole, swivel round it and move vertically up and down along it.

(*iii*) **Polar arms.** These consist of the similar two parts as for cylindrical arms. Like a cylindrical arm, a polar arm can swivel round the central pivot. However, instead of moving up and down it, the arm tilts to reach out above or below the level at which it is mounted.

(*iv*) **Cartesian arms.** In this design, a gripper moves along three different, perpendicular tracks. One track controls its height, the second its width and the third its depth of operation. Here, we get great positional accuracy but the arm becomes slow.

For loading and unloading a machine tool, a cylindrical arm driven by a hydraulic motor is most suitable. The hydraulic drive gives the arm the necessary strength and speed. For paint spraying, the requirements are different. The robot must be fast, agile and accurate but not necessarily strong. Here, pneumatically driven, anthropomorphic arms are popular. For spot-welding auto bodies, a polar robot might be the best choice. The main task here is to stretch out to the car body and grip it with the welder. Because the body areas to be welded are often not horizontal, the polar robot's ability to tilt and reach is handy.

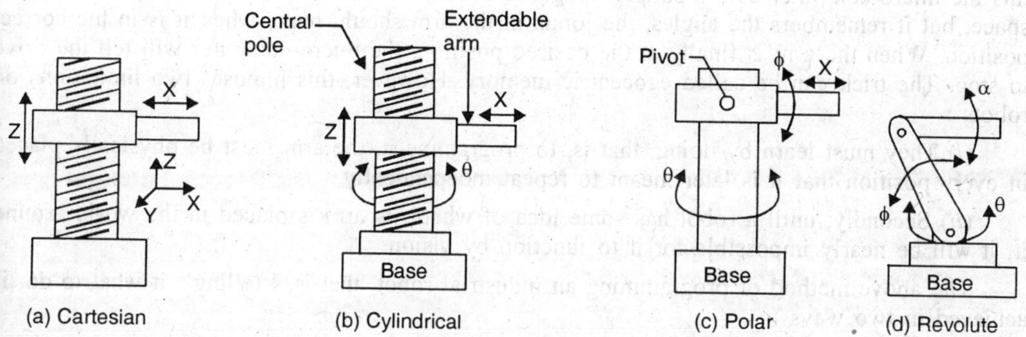

Fig. 24.2. Arm Geometry.

(*c*) **Hands or grippers.** There is a wide variety of hands to be attached to the end of the rm. They may range from claw gripper, magnetic gripper, suction gripper or cup, ladles, or inchers, to tools like grinders, paint spray guns or welding torch. The "claw type" is perhaps ie most commonly used. This gripper comes in many forms. Claws can have tactile feed back. uction type gripper works in much the same way as a vacuum cleaner does. They are ideal for icking light weight plastic items quickly. They are not really strong enough for heavy objects. Magnetic grippers are used in factories where sheet metal parts are made. On many industrial obots, the gripper is replaced by what is known as "end units". These include welding torches, aint spray guns, drilling devices and the like. They are either bolted or temporarily fixed to the rist of the robot's arm. The various grippers are shown in Fig. 24.3. The claw type gripper has een shown in Fig. 24.1.

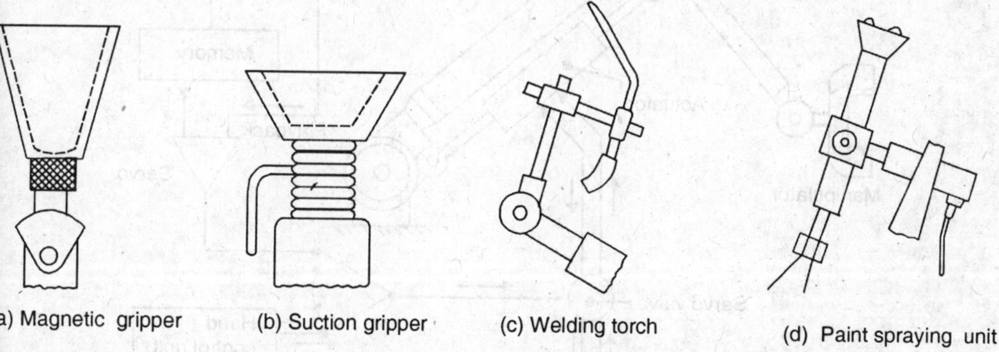

Fig. 24.3. Robot Grippers.

Software. As mentioned above, the present day's robots are endowed with computer emory. The computer memory or computer "brains" come is two categories :

(*a*) The pick-and-place robots or the loaders and unloaders robots. These are the dumbest odels and their microcomputer controller merely tells them to move their arms from one place another. It does not specify the precise path the arm should take in doing so. Its path is termined by the architecture of the arm.

Such robots are called Point-to-Point (PTP) robots.

(*b*) For many jobs, the arm must follow regular and well-defined paths, for example, the arm of a robot meant to do paint spraying. If it does not, the paint job will be patchy. For such jobs, more control is needed and a so-called continuous path (CP) robot is required. Here, the whole of the arm's path is carefully controlled from start to finish.

In either case, the brain of a robot must send signals to the drive of the robot's arm telling the drive to switch the arm until its actual position matches the one in the brain's memory. For this the micro-computer uses a simple dodge. It does not 'visualize' the position of the arm in space, but it remembers the angles, the joints in the arm should make when it is in the correct position. When the arm is finally at the desired position, the micro-computer will tell the drive to stop. The trick can be called egocentric memory. However, this imposes two limitations on robots :

(*i*) They must learn by doing, that is, to programme it, the arm must be physically placed in every position that it is later meant to repeat independently.

(*ii*) Secondly, until a robot has some idea of where its arm is placed in the world around it, it will be nearly impossible for it to function by vision.

The above method of programming an industrial robot, that is, "telling" it what to do, is achieved in two ways :

(*i*) **Walk through method.** In this method, the robot is guided manually through a sequence of operations. The successive positions of all the robot's joints are stored in the memory, of the micro-computer. By switching from the "tell" or "teach" to the "playback" mode, stored positions are repeated during production.

(*ii*) **Lead through method.** In this method, the robot is guided through a sequence of operations, by using a hand controller or a teach pendent, (Fig. 24.4), instead of doing it manually. The other procedure is the same as above. This method is more popular as compared to walk-through method because of its ease and convenience.

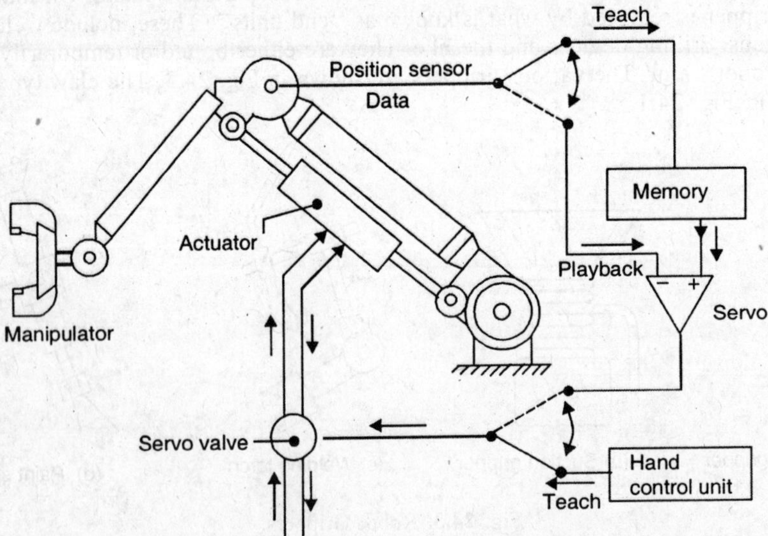

Fig. 24.4. Robot Programming.

Another method of robot programming is :

(*iii*) **Off-line programming.** This method is just similar to NC part programming. The program for a particular job is prepared and stored in the robot memory for use during the production cycle. There is no need to "teach" the robot, resulting in saving in production time.

Also, off-line programming can be completed for a new job while the robot is still working for the old job.

The various robot programming languages, which have been developed are : -

HELP	:	High Level Procedural Language
AML	:	A Manufacturing Language (by IBM)
AL	:	A Language
RCCL	:	Robot control 'C' Library
RPL	:	Robot Programming Language
VAL	:	Versatile Assembly Language (created by Unimation)
AR-Basic	:	By American Robot Corporation
ARMBASIC	:	By Microbot, Inc.

Advantages of robots. In addition to the advantages already discussed, the robots have the following advantages :

1. They can perform a variety of tasks with only a change in program and gripper hand or end effector, *i.e.*, greater flexibility.

2. They can work around the clock. They don't take tea or coffee breaks and don't participate in trade union activities.

3. They don't take any kind of leave.

4. They don't get hurt and don't complain about the working conditions, that is, heat, cold, lust or fumes.

5. They quickly learn reasonably complex task.

Robots have proved exceptionally popular in the following areas : spot welding, arc velding, spray painting, die casting, injection moulding, machine tool loading/unloading and also naintenance of nuclear power plants and foundaries.

Some other benefits from robot use are listed below :

1. Improved product quality and consistency.
2. Lower labour costs alongwith increased productivity.
3. Improved working conditions and safety resulting in better labour relations.
4. Greater reliability and hence reduced downtime.
5. Better management control.
6. Increased technical expertise.
7. Lower material costs and less wastage.
8. Less capital tied up in WIP (work-in-process).

4.3. CLASSIFICATION OF MANUFACTURING SYSTEMS

The two most important factors that will determine the choice of a manufacturing system e :

1. The total number (volume) and variety of parts to be manufactured.

2. The rate of production, that is, the number of parts per unit period (hour, day, month or ar).

These two factors together will influence the economies of machinery/equipment and oling.

In addition to volume, variety and the rate of production, the choice of a manufacturing stem also depends on part geometry and accuracy.

Conventional manufacturing systems have been marked by one of the two distinct features :

(*i*) **Job shop type of manufacturing system.** Such a system is capable of producing a riety of products, but the unit cost is high. Product variety is high.

(*ii*) **Mass production systems.** Mass production systems, or transfer equipment can produce rge quantities of a product at a reasonable cost. However, these systems being inflexible, are nited to the production of one, two, or very few different parts (limited variety).

In between these two extremes, comes the *batch production* (medium production) systems.

One way of classifying manufacturing systems is :

1. Fixed (Hard wired) systems, and
2. Flexible (Soft wired) systems.

This classification is normally applicable to both the processing equipment (machine tools etc.) and the material handling system.

Fixed automated manufacturing systems are typically used for the production of high volume parts and are usually limited to the manufacturing of a single product, or, at the most, a few different products. That is, the variety of products is very much limited. The example can be of a dedicated transfer line which performs a variety of manufacturing operations automatically to manufacture a product in large numbers.

Each process or machine usually consists of dozens of mechanisms that create the required relative motions to complete an activity. These mechanisms include : cams, linkages, ways and slides, pistons (pneumatic and hydraulic), vibratory devices, push rods and screw mechanisms. For a new product, the change in all these systems become very tedious, time consuming and costly. So, such a system can be called as a "Fixed programmed manufacturing system".

On the other hand, the flexible manufacturing systems are easily programmable and are used in the production of a variety of products automatically in low-to-medium volume batch quantities. Advent of NC and Robotics has provided us with reprogrammable capabilities at the machine level with minimum set up time. NC machines and robots provide the basic physical building blocks for reprogrammable manufacturing systems. Manufacturing systems are being designed that not only process the parts automatically but also move the parts from machine to machine and sequence the ordering of operations in the systems.

24.3.1. Job Shop Type of Manufacturing Systems. This type of manufacturing system has already been discussed in Chapter 23 under Art. 23.2 (Jobbing production). The manufacturing equipment is grouped according to the general type of manufacturing process, for example, lathes in one department, drilling machines in another and so forth. Such a layout enables a wide variety of products to be manufactured. Each product is routed through the respective departments in the proper order as per its operation sheet. Forklifts and handcarts are used to move materials from one machine to the next. The production rates in conventional type of job shop manufacturing system are quite low.

For job shop type of manufacturing system, the stand alone NC machines are also best changeovers. They have proved to be economical for this type of manufacturing. They are highly flexible, that is, they can be conveniently reprogrammed to accommodate changes in product design or even product changeovers. However, the production rates of these machines are low.

24.3.2. Mass Production Manufacturing System. This manufacturing system has already been discussed in Chapter 23 under Art. 23.2. For mass production, automated dedicated transfer lines are very efficient manufacturing systems when a product is to be manufactured in large volumes at high production rates. But their main limitation is their inflexibility. That is, if there is change in part/product design, the line has to be shut down and retooled (which cannot be allowed in the case of mass production) and if design changes are extensive, the line may be rendered obsolete. So, automated transfer lines are most suitable for mass production where the part variety has to be minimum possible.

Mass production manufacturing system is organized on a flow-line principle. The output pace is the main design characteristic which is used to arrange all individual workplaces into a single manufacturing system. The uninterrupted operation of flow lines is ensured by a rationally planned manufacturing process, uniform feeding of blanks, and rhythmical functioning of all units of the manufacturing system. Transfer lines must be carefully balanced to equalize the output of various stages, otherwise the most time-consuming station would slow down the entire line. To prevent interruptions in flow line operation due to emergency shut downs of individual in-line machines, some stockpiles of workpieces are provided at separate workplaces.

Transfer Machines. A transfer machine can be defined as a large automatic installation consisting of several individual machining heads or units which are fastened together by conveying units. Workpieces are loaded at one end and are automatically transferred alongwith their fixtures, from station to station. At each station, machining operations are performed on the workpieces and the completed workpieces leave at the other end of the machine. Thus, a transfer machine is a combined material handling and material processing machine. These are basically special purpose machines and are often the most suitable method for a continuous manufacture of identical or very similar parts in mass production. Thus a transfer machine is economically justified only if the continuous production of a workpiece is met by an equal demand for it.

The machining stations on a transfer machine consist of powerhead production units. A powerhead unit consists essentially of :

1. Powered spindle/spindles mounted in suitable bearings, which are usually driven from a self-contained motor though a gear box.

2. A means for power feed to traverse the head along the slideways while the cutting tools work upon the components. The power feed may be mechanical or hydraulic.

3. A frame or bed to support the limit.

Type of transfer machines. The two most common types of transfer machines are :

1. In-line transfer machines.

2. Rotary transfer machines.

1. In-line transfer machines. In this type, the machining heads are arranged in a line and the component is automatically transferred from one machining station to the next by one of the following three methods :

(*a*) By pulling along supporting rails by means of an endless chain conveyor.

(*b*) By pushing along continuous rails by air or hydraulic pistons.

(*c*) By moving by an overhead chain conveyor, which may lift and deposit the work at the machining stations. This method is used only for lighter workpieces.

The use of transfer pallets or fixtures for holding the components will depend upon the size, rigidity and design of the components. If these can be avoided by suitably designing the component, it will result in saving of their cost and also greater freedom of machine layout is provided. The varius geometric arrangements of machining heads, in in-line transfer machines are :

Straight line, *L, U,* Square or Rectangular.

If transfer pallets are used, arrangements have to be made to return the empty pallets to the starting point of the machine. The method commonly used is a rapidly moving return conveyor parallel to the main transfer line. If enough floor space is not available, then, *L, U,* Square or Rectangular pattern can be adopted. In the case of square or rectangular pattern, there is no need of return conveyor, since the pallets are automatically returned to the loading point.

The size of a straight in-line machine, Fig. 24.5, can be almost unlimited (if floor space is available) and such machines have been employed, for example, to perform all the machining operations on an automobile-engine cylinder block. Such machines also include inspection and assembly operations in addition to the machining operations.

When transfer pallets are used, they contain locating holes or points that mate with retracting pins or fingers at each machining station. The pallets are then located and then clamped in proper positions. When pallets are not used, locating bosses and points are designed into the workpiece itself.

2. Rotary transfer machines. The rotary transfer machine, Fig. 24.6, is used when, only 6 to 10 or fewer machining stations need to be employed. The machining heads are arranged radially along the periphery of a rotary indexing table. The table rotates about a vertical axis and its movement may be continuous or intermittent. For indexing the table, a 'Geneva type indexing mechanism' discussed in Chapter 7, can be employed. This type of machine is very compact and there is saving in floor space.

The operations performed on a transfer machine include : drilling, boring, counterboring, reaming tapping, countersinking, chamfering, face milling, spotting, hollow milling, trepanning, gauging (air or electrical), blow-out or dumping chips and rolling over or revolving work to reposition for the following operations.

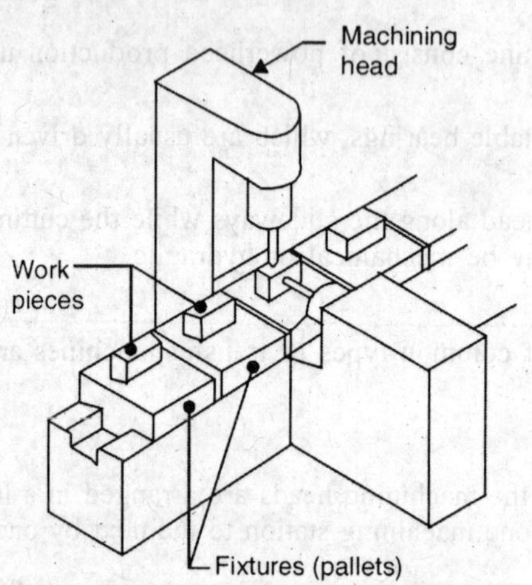

Fig. 24.5. In-line Transfer Machine.

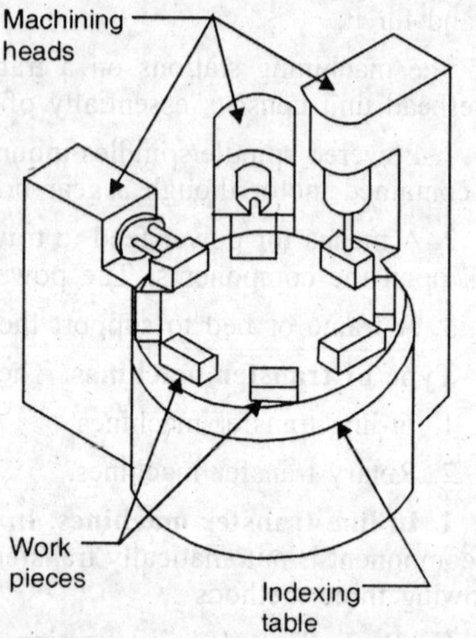

Fig. 24.6. Rotary Transfer Machine.

The most frequent machine tools associated with a transfer machine are : drilling machines, reaming machines, single point boring machines, milling machines, tapping machines, inspection equipment. Less frequently used machine tools include : broaching machines, boring machines, polishing machines and turning machines.

24.3.2. Advantages of transfer machines. 1. The machining operations are speeded up, thereby reducing the production cycle.

2. Fewer operators are needed.

3. Secure higher output.

4. Manual handling of the component is avoided, except loading and unloading.

5. A chip conveyor can be used for the removal of chips.

6. The alignment of the job at each machining station is simplified and automized.

7. Greater accuracy of the job is achieved.

8. Lower cost of product.

9. Better use of floor space.

Disadvantages

1. This system is justified only for high production of components.

2. Initial cost is very high.

3. The whole set up is to be changed if the design of the component changes.

4. The breakdown of one machine will stop all the machines.

5. Electrically, the system is very complex.

Incorporating flexibility in transfer lines

Increasing global competition, frequent changes in product design to meet the rapidly changing customer demands, makes it imperative that even the mass production manufacturing systems have to become more and more flexible. Some of the modifications in this direction are given below :

1. By replacing fixed machine tools by "Powerhead production units" incorporating many interchangeable attachments. This will facilitate the carrying out of various machining operations (turning, drilling, tapping, milling etc.) as per need. This aspect has already been given while discussing the various types of transfer lines (machines). Sometimes even CNC machining centres are incorporated.

2. Quick-change tool holders complete with preset tools can be used.

3. The flow line is grouped into "sections" (5 to 12 stations) with a smaller buffer storage in between. This will ensure that breakdown, tool change, or setting in one section does not stop the whole line.

4. **Linked Lines.** In a linked line system, the general purpose and/or special purpose machine tools are linked together usually by a powered belt conveyor. The components are transferred from the conveyor and loaded into the machine, when the work on this machine is done, the component is transferred back to the conveyor to be taken to the next machine. The transferring and loading can be done automatically or manually. The production in this system is based on the flow line principle as in a transfer machine, but this system is more flexible and cheaper. The linked line can take any route : straight or curved. Unlike a transfer machine, the production in this system is not interrupted if one machine breaks down, because every machine is an independent unit.

While setting up and laying out a linked line, either of the two methods can be adopted :

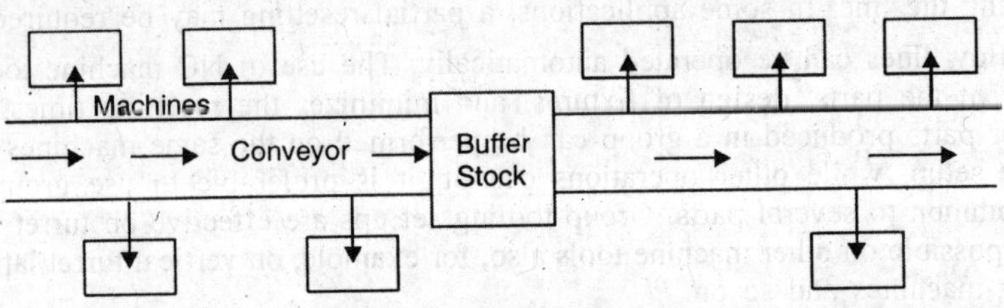

Fig. 24.7. Linked Line

(*i*) Line balancing, so that the processing time of each machine is approximately the same so as to reduce the cycle time.

(*ii*) The machines can work at their rates, with a buffer, stock, between each machining station. The size of the buffer stock should be decided carefully after considering the cost of storage and the effect of breakdown. Fig. 24.7 shows a schematic diagram of a linked line.

24.3.3. Batch Production System. This manufacturing system has already been briefly discussed in Chapter 23 under Art. 23.2. In the manufacturing system traditionally used for batch production, the machine tools are grouped by types, for example, turning, milling, grinding, or gear cutting machines. Lathes, drilling machines, milling machines, grinders and so on, are clustered so that they remain inside a single department. This system is known as the "Process-oriented or Functional layout system", and is one of the most common machine layouts used in industry. Batches of workpieces pass through some or all of these groups according to a preplanned route. Each machine is attended by an operator and the work has to be fed continuously to each machine to achieve its full utilization. Parts are moved by some flexible means, *i.e.*, manually, by overhead cranes (conveyors), prolift trucks and so on, from machine to machine. Unfortunately, functional layout also requires that the product flows throughout the entire system in somewhat random manner. The other main drawback of this system is the high cost of workhandling and temporary storage. Also, after completing a batch of products of a given design, the production machines are reset for starting the production of the next batch of another design. This means lost production time.

The drawbacks of "Functional layout" can significantly be reduced if parts requiring similar operations are grouped together into a "Part family", and the machines required to manufacture the family are organized into a "Cell". The flow of the product through the system is much more direct and material handling can be reduced significantly. Such a layout is called as "Cellular layout" or "Group Technology (GT) layout" . Group technology has been defined as :

Group technology is the realization that many problems are similar, and that by grouping similar problems, a single solution can be found to a set of problems thus saving time and effort.

For manufacturing purposes, the basis of "GT layout" is part-family formations. Components requiring similar operations/processes are grouped together into the same family. Components that are not similar in shape may still require similar manufacturing processes. For example, a dowel pin and a small shaft may be very similar in appearance but different in function. Spur gears of different sizes need the same manufacturing processes and vary only in size.

By using "Cellular layout", the structural principles of flow lines (used for mass production) can be introduced into batch production (medium production). This "GT layout" is also known as "Group flow-line" or "Unlinked flow-line system". The specially planned manufacturing processes under this layout are called as "Group manufacturing process" or as "GT processes" implemented on group flow lines.

In a group flow line, the equipment is laid out in accordance with the flow routes of various parts, similar in configuration and design, alloted to the line. All the parts dealt with by the line are processed alternately in separate batches and the general operation mode resembles that of a continuous-flow production line. Changeover from one to another size of part is possible without restting the line. In some applications, a partial resetting may be required.

Group flow lines can be operated automatically. The use of NC machine tools simplifies the grouping of the parts, design of fixtures, and minimizes the resetting times. When some operations for parts produced in a group can be performed on the same machines and with the same tooling setup, while other operations cannot, it is preferable to use group set ups on operations common to several parts. Group tooling set ups are effective on turret lathes. These can be made possible on other machine tools also, for example, on vertical turret lathes or boring mills, milling machines and so on.

Benefits of GT principle

1. Set up times are minimized.
2. The total quantity of production of all parts belonging to a given family becomes large enough to warrant mass production. Jigs and fixtures, therefore, become economical for facilitating production.
3. Tooling costs can be reduced by standardization.
4. Reduced production cycle time, shorter lead times and faster response to customer needs.
5. Better production planning and control.
6. Reduction in variety and quantity of starting material and also in the in-process inventory.
7. The operations become less labour intensive, semi-skilled workers can be used and the unit cost is lowered.

24.3.4. Flexible Manufacturing System (FMS). Some of the recent trends in manufacturing have been :

1. Parts have been subjected to a short product life.
2. Frequent changes in the product design.
3. Small in-process inventory restrictions, that is, the just-in time (JIT) approach.

The end result has been that over 90% of manufactured parts are produced in lot sizes of less than 50. Due to this, the dedicated production lines (transfer lines) which were so effective in mass production of inexpensive parts/products are in the process of being eliminated. This has resulted in the emergence of flexible or programmable manufacturing systems to produce low-to-medium volume batch quantities. To complete in the international market, it has become mandatory to manufacture goods quickly and to keep the inventory to a minimum. So, the need is of a small batch dynamic manufacturing system. For this, the two pre-requisites are : (*i*) Automation, and (*ii*) Flexibility.

Automation provides good quality and low cost and flexibility is necessary to adopt to changes of the product and demand and for producing a variety of products in low-to-medium volume batches. The solution is adoption of computer aided manufacturing at the shop floor. The use of NC machines and robotics have provided potential solutions to many flexibility problems. The two principles whereby flexibility is achieved are : integration and reprogrammability. Numerical Control (NC) and robotics provide reprogrammable capabilities at the machine level with minimum set up time. NC machines and robots provide the basic physical building blocks for reprogrammable or Flexible Manufacturing Systems (FMS).

Thus, a flexible manufacturing system is a reprogrammable manufacturing system capable of producing a variety of products in low-to-medium volume batches, automatically. FMS is also known as Variable Mission Manufacturing (VMM) and computerized manufacturing system. The economic objective of FMS is to approach the efficiency of mass production for low to moderate production. The concept of GT is the heart of FMS. Hence, when a product is to be manufactured in the midrange volume between that for transfer lines and stand alone NC machines (for jobbing production), and a flexibility is also needed to accommodate product variety, neither transfer lines nor the stand alone NC machines are suitable. This gap is filled by FMS.

Thus a FMS can be defined as a manufacturing system which consists of a group of NC machines connected together by an automated material handling system and operating under computer control. Each FMS will be different in design as per the production requirements of the user. FMS is the most automated form of manufacturing system for the manufacture of discreet products.

Basic components of FMS. The basic components of a FMS are given below :

1. **Machine tools and the related equipment.** It is clear from the definition of FMS, that the machines must be automated and reprogrammable to accommodate a large variety of products. For this reason, the majority of equipment consists of standard CNC machines (CNC turning centres and CNC machining centres), special purpose machine tools, tooling for these machines, inspection stations or special inspection probes used with these machine tools. These machines are capable of accommodating a variety of tooling via an automatic tool changer and tool storage system.

2. **Material handling equipment.** The material handling equipment used in a FMS serves two functions : to move parts between machines and to orient and locate these parts for processing at the machines, automatically. The workpieces mounted on pallets or fixtures move through the system by means of powerful handling systems such as towlines combined with shuttle system, roller conveyors, drag chains or automatically guided vehicles, AGVS. Combined with shuttle system, industrial robots, located suitably with respect to a group of machine tools, transfer workpieces from one machine to the next. A robot is normally only capable of addressing one or two machines and a load-and-unload station. Thus, a FMS may consist of a number of robotic workstations.

3. **Computer control system.** An FMS is a complex network of equipment and process that must be controlled via a computer and network of computers. It consists of control of machines, control of material handling system, to monitor the performance of the system and to schedule production.

4. **Human labour.** Even though FMS is a highly automated manufacturing system, involvement of human labour is needed to run the system. The various human labour may include : system manager, tool setter, load/unload man, fixture set up and lead man, electrical technician, mechanical/hydraulic technician and robot operator.

Types of Flexible Manufacturing System. There is a large range of definitions which the people use in the flexible, manufacturing technology. Thus the name 'FMS' has been equally applied to a single, relatively simple computer controlled machining centre and a 30-machine tool factory. In this connection, the following definitions would be appropriate :

(*i*) **Flexible Manufacturing Unit (FMU).** An FMU consists of a single, multifunction CNC machine tool and is the most simple flexible manufacturing system that can be constructed. It consists of a processing machine (CNC machine tool), a load/unload area and a material handler (a robot). The parts that move down a conveyor are loaded into the machine by a robot. After that, the robot is retracted and the processing begins. After the machining has been completed, the robot takes the part off the machine and moves it to the output bin.

(*ii*) **Flexible Manufacturing Cell (FMC).** This flexible system consists of two or more CNC machine tools alongwith one or more robot work stations, but not under DNC-linked control.

(*iii*) **Flexible Manufacturing System (FMS).** This system consists of a number of CNC machine tools under supervisory computer control via some form of DNC linkage.

(*iv*) **Flexible Manufacturing Transfer Line (FML or FTL).** It consists of a multimachine layout including several CNC machine tools and other specialist pieces of equipment all under supervisory computer control. This system is an alternative to a dedicated transfer line and is used for high volume production.

FMS, covering the moderate or medium production volume and medium part variety belongs to the family of CIM systems (discussed in Chapter 25). CIM can be further classified into finer systems depending upon the different levels of compromises between flexibility and product quantity as : Manufacturing Cell, FMS and Special Manufacturing System. In the above three categories, a manufacturing cell is the most flexible, special manufacturing system is the least flexible and FMS comes in between these two extreme positions. The product quantity per product per year and product variety for these systems may be as given below :

(*a*) **Annual product quantity per part :**

Special system = 1500 to 15,000

FMS = 40 to 2000

Manufacturing cell = 15 to 500

(*b*) **Product variety :**

Special system = 2 to 8

FMS = 4 to 100

Manufacturing cell = 40 to 800

This has been shown in Fig. 24.8.

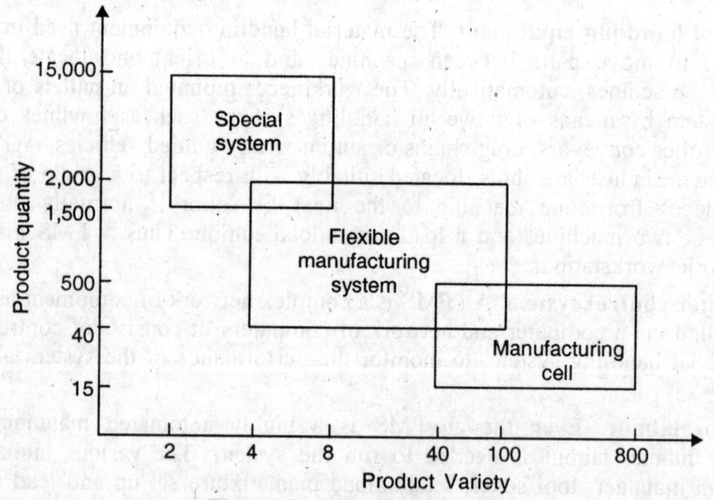

Fig. 24.8. Categories of FMS.

It is very clear from above that because of least product variety in special manufacturing system, specialized machine tools would not be uncommon in this system. Also, this system will incorporate a highly integrated interconnecting work part handling system. On the other hand, a manufacturing cell might consist of a number of stand alone NC machines without any interconnecting workpart handling system. In FMS, parts are loaded and unloaded at a central location and pallets are used to transfer parts between machines.

FMS results in the following benefits : Increased machine utilization, reduced direct and indirect labour, reduced manufacturing lead time, lower in process inventory and scheduling flexibility. Also, quality improves because human error is eliminated.

24.4. CHARACTERISTICS OF MANUFACTURING SYSTEMS

Table 24.1 gives the summary of the characteristics of the various manufacturing systems and the processing machines.

Table 24.1. Characteristics of Manufacturing Systems

Type of System/ Processing machines	*Volume of Production*	*Part Variety*	*Unit Cost*	*Flexibility*
Transfer line	Highest	1 or 2	Minimum	Lowest
Special systems	High	1 to 12	Low	Low
Stand alone automatic machines	Low	Many	Moderate	High
FMS	Moderate	Large	Low	V. High
Stand alone manual machines	Lowest	Most	Highest	Highest

Special systems such as an automatic screw machine system are specially designed systems that can be altered (via alternate cams, lever systems, adjustable screw settings etc.). Stand alone automatic machines include NC-machines. Several factors other than volume and part variety such as part geometry and accuracy affect the choice of the manufacturing system.

24.5. PRODUCTION SYSTEM

A production system helps a manufacturing system to produce the goods. These are many types of manufacturing systems (job shop, flow line etc.), but all the manufacturing systems are derived by a production system. A production system serves the manufacturing system/systems and the individual processes to manufacture goods, without itself manufacturing products. Most production systems are functionally organized. A classical production system may include the following typical functional elements :

Product design engineering
Purchasing
Sales and Marketing
Finance
Accounting
R&D
Personnel procurement including recruitment training, labour relations and safety
Production planning and control
Process engineering (process planning, cost estimating etc.)
Inventory control
Quality control
Plant engineering

24.6. AUTO-MATION

The word "automatic" is derived from the Greek and it means "self-moving" or "self thinking". The word "automation" has been used to refer to a type of manufacturing system in which the various manufacturing functions such as material processing, material/part handling and inspection are carried out by self operating machines without human intervention.

Auto-motion has been developed out of the need for higher productivity, lower cost and more precise manufacturing. Avoidance of human intervention, omission of conventional tooling and fixturing and quick change capability of these elements are the primary factors, which were considered for automation. Hence, there is an optimization of cutting tool life and quality of jobs, possibility of making parts which are impossible in conventional machining, and quick and more accurate inspection and detection of error in design and fabrication.

There are mass production machine tools, which are highly specialized but inflexible, because of their longer set up times while changing over to new jobs as these machines need cams, templates, stops, electrical trip dogs etc. Also, there are general purpose machine tools which are highly flexible, but are not suitable for mass production. Thus, a great need of automation was felt which could bridge the gap between the highly specialised machine tools and the general purpose machine tools.

Hence, Automation is done in case of following cases : -

1. Loading and Unloading of parts
2. Automatic production lines
3. Automatic tool changing
4. FMS
5. Factory automation
6. The emphasis on reducing cycle time.
7. Reduction of idle time.
8. Complex parts
9. Frequently changing design.

As already noted, the involvement of computers (CNC and DNC machines) are used to increase flexibility, reliability, programmability and repeatability.

The aim of any manufacturing system is to manufacture a product in the most economical manner possible. Whatever may be the form of automation, it should be economically justifiable. In general, automation results in :

- Reduction in direct labour cost due to less operator skill required.
- High quality of product
- Increased productivity
- Increased shop efficiency
- Prevention of shut down
- Reduced in-process inventory
- Reduced material handling cost and time
- Increased manufacturing control
- Increased safety and
- Versatility
- Greater accuracy and repeatability and hence less rejections.
- Reduction in lead time and set up time lowers the production cost.
- High production rates as the machining conditions are optimized.
- Better machine utilization due to reduced idle time.
- Changes in part design can be incorporated very easily and at a low cost.
- Lower tooling costs as expensive jigs and fixtures are not required.

- Reduced cycle time and increased tool life.
- Automatic table indexing, automatic pallet changing and automatic tool changing.
- Ability for higher levels of integration such as :
  ÇAD, CAM, CAE, CIM, DNC, FMC, Adaptive Control (AC)

Disadvantages :

1. Very high initial Cost
2. High maintenance Cost
3. Skilled software knowledge specialists needed

Automation is also the answer to the reduction of problems such as : operator fatigue, carelessness, and other human frailities.

There is difference in the terms "Mechanization" and "Automation". Mechanization means the replacement of manual labour with machines. For example, the movement of a tool slide with the help of a cam. However, there is no provision of feedback. So, this is an "open loop" system. On the other hand, automation means a "closed loop" control system in which there is a provision of feedback.

Basically, automation is of two forms :

1. Hard or fixed automation.
2. Soft or flexible automation.

Hard or fixed automation uses mechanical or pneumatic/hydraulic controls. Mechanical controls consists of dozens of mechanisms that create the required relative motion to complete an activity. These mechanisms include : cams, levers, ways and slides, push rods, screw mechanisms etc. Automatic screw machines and many such machines still use such a control system. Mechanical control is difficult to fabricate and reprogram (in the event of the change of product) and is subject to wear. A group of such semi-automatic or automatic machines are linked together by an "intrasystem" material handling system.

Pneumatic or Hydraulic control is also used for fixed automation, because reprogramming is not feasible. The system consists of compressed air/pressurised oil, pistons and cylinders, valves, switches and various mechanisms and pipe lines. Such a system is slow and is subject to wear.

In electro-mechanical control, use is made of relay devices (switches, relays, timers, counters etc.). Since current is used instead of air or oil, the system is faster and more flexible.

Electronic control is similar to electro-mechanical control except that the moving mechanical components in electro-mechanical control device are replaced by electronic switches. These are more faster and are more reliable.

Soft and flexible automation system uses tape, a hand-held control box, microprocessors, programmable logic controllers (PLC) or computers. The system can be reprogrammed very easily, simply by changing the software.

Fixed automation is most appropriate for high-volume long-life products. Flexible automation system has brought automation to some relatively low-volume products.

The concepts of fixed automation and flexible automation have already been discussed while discussing the various forms of manufacturing systems, for example, transfer lines (fixed automation), FMS (flexible automation) and so on.

1. Lean Manufacturing. Lean manufacturing can be defined as: A systematic approach to identify and eliminate waste (non-value added activities) through continuous improvement by flowing the product at the pull of the customer in pursuit of perfection.

Thus the sole aim of the concept of Lean Manufacturing is of cutting wastes of all kinds (labour, equipment, machinery, management personnel), leading to world class manufacturing with lower costs of production without compromising with quality, so as to compete globally.

The concept aims at continuously improving the efficiency and profitability of the enterprise by eliminating all kinds of wastes, without cutting back resources. The word 'Lean' in "Lean Manufacturing" is very apt because is this system, the emphasis is on cutting the 'fat' or waste in the manufacturing process. Waste is defined as anything that does not add value to the customer, or anything the customer is unwilling to pay for.

The fundamental principle of Lean Manufacturing is 'Flow-Line Manufacturing' which is a time based process that pulls the material through a production system without any interruption. In this system (Pull system), the production is driven by the real consumer demand and not on the sales forecasts (Push system), see Art. 26.15. This has been made feasible by the recent trends in manufacturing where computer has been involved in every activity of product design and manufacturing, e.g., CAD, CAM, CIM, CAPP, GT, FMS, JIT etc. The success of the system will depend entirely on complete understanding and co-operation and team work between the management and work force.

Lean manufacturing has resulter is shorter lead times and the production of high-quality products economically, in lower volumes and making these available to the customer faster than the mass production systems.

2. Agile Manufacturing. Lately, the manufacturing paradigm has shifted from labour intensive, high volume and low variety production (Mass production) to high variety and low volume production. Mass markets are fragmenting into niche markets. Also, markets are becoming more dynamic and global and the Global business competition is intensifying. Customers expect low volume, high quality and custom products. Due to all this, very short development time and production and lead times are mandatory. Traditional manufacturing systems (where the main emphasis is on the productive use of resources) are no longer optional for this type of environment. This has led to the concept of "Agile Manufacturing", which aims at developing the capability of surviving and prospering in a competitive environment of continuous and unpredictable change by reacting quickly and effectively to changing markets, driven by customer-designed products and services. Thus, the concept of "Agile Manufacturing" involves "agility" and hence flexibility on the part of an enterprise so that it can respond to changes in product demand and customer needs.

The concept of "Lean Manufacturing" and "Agile Manufacturing" are different. Lean manufacturing is a response to compititive pressures with limited resources, Agile manufacturing is a response complexity brought about by constant change. Lean manufacturing is a collection of operational techniques focussed on productive use of resources. On the other hand, Agile manufacturing is an overall strategy focussed on thriving in an unpredictable environment. However, all the basic fundamentals of Lean manufacturing are incorporated in the concept of Agile manufacturing. The concept requires highly intgrated and flexible technologies of production, not necessarily high tech, but highly capable. The company's work force must be highly educated and trained, and significantly empowered within the constraints of a clear vision and delineated company principles and goals. The company itself must have the ability to affect changes rapidly, have highly flexible management structures, and comprehensive methods of introducing change and prospering from it.

It should be borne in mind that 'Agility' is not a "magic wand" to solve all the problems of the present day manufacturing. It is built upon the firm foundations of World class or Lean manufacturing methods, coupled with an organization that is physically, technologically and managerially established for rapid and unpredictable change.

PROBLEMS

1. Define a manufacturing system.
2. Write the inputs to and outputs of a manufacturing system.
3. Which are the four M's of a manufacturing system?
4. What constitutes a manufacturing system?

5. Write the activities performed by 'Material movement' in any manufacturing system.
6. Define : Attended material movement and Mechanical material movement.
7. Write brief notes on : OLAC's and AGVS.
8. Define a "Robot".
9. Discuss the various generations of Robots.
10. Discuss the various power drives for Robots.
11. Discuss the various designs of Robot arms.
12. Sketch and discuss the various Robot grippers.
13. Discuss the various methods of programming a Robot.
14. Write the advantages of Robots.
15. How the manufacturing systems are classified?
16. Define : Job shop type of manufacturing system, Mass production system and Batch production system.
17. Define : Fixed manufacturing systems and Flexible manufacturing systems.
18. Write detailed notes on :
    (*i*) Job shop type of manufacturing system.
    (*ii*) Mass production manufacturing system.
    (*iii*) Batch production systems.
19. Define a transfer machine.
20. Discuss the types of transfer machines.
21. How flexibility can be incorporated in transfer machines.
22. Write the advantages and disadvantages of transfer machines.
23. Write a note on "Linked lines".
24. Define "Functional layout system". Write its drawbacks.
25. Define : "Part family" and "Cell".
26. Define : "Cellular layout".
27. Write a brief note on "Group flow lines".
28. Write a advantages of GT principle.
29. What is Flexible Manufacturing System (FMS)?
30. Discuss the basic components of FMS.
31. Write about the various types of flexible manufacturing systems.
32. List the characteristics of manufacturing systems.
33. Define : Production system.
34. Define : Automation.
35. What are :
    (*i*) Hard or fixed automation.
    (*ii*) Soft of flexible automation.
36. Write the advantages of automation.
37. What is the difference between 'Mechanization' and 'Automation'?
38. Cellular manufacturing is suitable for : -
    (*a*) A single product in large volume
    (*b*) One-off production of several varieties.
    (*c*) products with similar features made in batches.
    (*d*) large variety of products in large volumes. (GATE 2000) **[Ans. : (*c*)]**
39. Specifications of Transfer machines (Rotary, Inline):
    No. of stations, Product

25

COMPUTER INTEGRATED MANUFACTURING

25.1. INTRODUCTION

Modern manufacturing facilities use a computer for a variety of manufacturing, monitoring and control functions. The modern factory environment has become an example of computer-controlled manufacturing. Facilities exist where,

1. Parts are created on a computer (CAD system).

2. Production plans are created from the CAD data base using an automated process-planning system (CAPP).

3. N-C part programmes are created using the tool-path requirements on a CAD system.

4. Parts are manufactured under the control of a computer.

Computer integrated manufacturing (CIM) can be defined as : The integration of computer based monitoring and control of all aspects of the manufacturing process, drawing on a common data base and communicating via some form of computer network.

The main activities under CIM are :

(*a*) Computer-Aided Design (CAD).

(*b*) Computer-Aided Manufacturing (CAM).

(*c*) Interfacing of CAD and CAM (CAD/CAM).

(*d*) Computer-Aided Production Management (CAPM).

25.2. COMPUTER-AIDED DESIGN (CAD)

Design is a key feature of any production system and its importance is growing. It is becoming a major determinant of competitiveness due to emphasis on design and product innovation stressing noval features, frequent changes, greater response to customer requirements and more rapid new product development cycles. All this necessitates the use of not only new design tools such as computer aided design (CAD), but also approaches to integrate design with manufacturing.

Definition of CAD. Computer-aided design can be simply defined as : using a computer in the design process.

The more comprehensive definitions can be :

"By CAD, we mean the development and use of special computer programmes (software) to help the designer in carrying out routine computations for the design of products and processes."

"CAD can be defined as the use of computer systems to assist in the development, analysis, modification, and optimization of an engineering design. It also includes storing and communication of design information."

"By CAD is meant the use of Interactive Computer Graphics (ICG) programmes to develop assemblies, parts lists, computer models and mathematical results. The output includes the working engineering drawings."

CAD helps in getting the analytical results very quickly. This will enable the designer to evaluate more than one design alternatives which would otherwise not be possible. Also, optimum design solutions can be obtained by using sophisticated programmes. This will result in significant savings in unit costs. Other advantages include improved information access and manufacturing data creation.

In CAD, the total design work is divided between the designer and the computer for which each is best suited. For example, the engineering knowledge, creative skill and intuitive analysis skill (imagination and judgement) of the designer are combined with the following traits of the computer, to obtain better and better designs in short time:

1. The extraordinary large number of functions that can be performed by the computer.
2. The great speed at which each function can be performed.
3. The accuracy and capacity for repetitions of operations.
4. The memory or storage system.

The designer and the computer are in constant interaction throughout the period of the project. For this, the designer should have a thorough knowledge of software and hardware tools.

The following data and information can be stored in the computer memory, which will help and guide the designer at every stage of the design process:

(*i*) Expert knowledge of experienced designers can be acquired and stored in the computer memory.

(*ii*) Knowledge based expert systems (computer programmes based on Artificial Intelligence, AI) can be written. These programmes are based on systematic reasoning and will help the designer to get guidance from the stored knowledge.

(*iii*) Standard data from Handbooks and other sources can be stored in data bases. These are available to the designer at any time. To help the designer to interact with this database, software tool known as "Database Management System" can be designed.

CAD is now firmly and profitably established in Aerospace, Ship design, Chemical engineering, Nuclear engineering, Mechanical engineering and Structural engineering etc.

All the above has been made possible by two significant developments :

1. **Interactive Computer Graphics.** With this, the designer is able to interact with the computer using a probe or a light pen on a graphics terminal. The analytical design is also displayed on the same terminal, by the computer. If the designer wants a change in the computed solution, he can do so by sketching the desired modification on the graphics terminal with his light pen. The modified design is analysed by the computer and the results are again displayed on the graphics terminal.

2. **NC drafting machines.** These machines enable part and assembly drawings to be produced, based on the computational results of CAD.

When the designer is finally satisfied with the design, it is assigned a classification number. All the data concerning the product, namely, the component and part numbers, detail and assembly drawing numbers, material for each part, dimensions, tolerances and surface finish requirements are stored in a master file.

25.2.1 The Components of a CAD System. A typical CAD system consists of three main parts :

1. **Hardware.** It consists of the computer and the Input-Output (I/O) devices. Input devices are generally used to transfer information from a human or storage medium to the computer. A keyboard is the standard input device used to transmit alpha numeric data to the system. The standard output for CAD is a CRT (Cathode Ray Tube), that is, the monitor. The other input devices can be : function keyboard, graphics tablet, light pen, joy stick and mouse etc. The other output devices can be : hard copy unit, printer, plotter, video tape, COM (Computer output micro-film), CAM (Computer-aided manufacturing). In many cases, the design may never be produced

on paper. The data generated by a CAD system can be directly utilized by a CAM system, for example, it may be transmitted to CNC programming routines used to produce components.

2. **Operating system software.** It is the interface between the CAD application software and the hardware.

3. **Application software.** It is the heart of a CAD system. It consits of programmes that do: 2–D, 2.5–D, or 3–D geometric modelling; engineering analysis; and drafting.

(*a*) *Geometric modelling.* Geometric modelling is an important part in CAD. It allows the creation of a geometric model to represent the size and shape of the component. It corresponds to the synthesis phase of the design process.

For geometric modelling, the designer gives the following commands to the computer :

(*i*) Commands for generating basic geometric elements such as points, lines and circles etc.

(*ii*) Commands for scaling, rotation or other transformations of the geometric elements.

(*iii*) Commands which join the various elements into the desired shape of the part being created.

The computer converts these commands into a mathematical model, stores it in the computer data file and displays it as an image on the CRT screen.

The geometric models may be :

(*i*) 2-D for flat objects.

(*ii*) 2.5-D for parts with constant section and having no side-wall details.

(*iii*) 3-D for generalized part shape.

The wire frame geometric modelling is the simplest and most common. It is easy to create and requires less computer time and memory. However, it is difficult to interpret the image. The perception of wire frame modelling can be enhanced by : colour graphics, dashed lines to portray invisible edges, removal of hidden lines and surface models. Surface models define part geometry precisely and can be used to produce NC machinery instructions automatically. However, these models don't give information about the part being solid or hollow and about the internal features of the part. The most advanced technique of geometric modelling is 3-D solid modelling.

(*b*) **Engineering analysis.** All the engineering design problems require some type of analysis such as :

(*i*) Stress-strain calculations.

(*ii*) Heat transfer computations.

(*iii*) Fluid flow problems.

(*iv*) Lubrication problems.

(*v*) Static and dynamic analysis of complex structures such as aircraft, cars, dams, bridges, etc. and so on.

The Finite Element Method (F.E.M.) is probably the most powerful and widely used tool in CAD.

Computer-Aided Process Analysis. Until about mid sixties, the analysis of metal working processes was based on analytical methods, such as 'The Slab Method', 'The Slip-line field Method', and 'Upper Bound Method', without the help of computers. The main aim used to be of predicting the loads to perform a metal working operation. Then, with the wide spread use of digital computers, these analytical methods were increasingly computerized. This led to an increase in the accuracy and the efficiency of predicting the results by these methods. Also, it became possible to obtain progressive solutions once the initial equations and boundary conditions have been set up. Inspite of all this, these methods do not represent the boundaries of the real materials, based as they are on any simplifying assumptions.

More recently, fully numerical techniques such as 'Finite Difference Method', 'Finite Element Method' and 'Matrix Method' have been developed for the analysis of metal working processes, with an emphasis on the behaviour of real materials. Due to this, the research activities in this field have shifted from simply predicting the forming loads to improvement, adaptation and extension of the analytical and numerical method; the residual stresses; interrelation between the flow of material, friction and design of tools; limits and defects of deformation and the optimization of the process.

(*c*) **Design review and evaluation.** It has already been discussed in Art. 25.2 under "Interactive computer graphics". Computer programmes based on "optimization Techniques" are used to get the best design.

(*d*) **Automated drafting.** See Art 25.2. Automated drafting increases productivity four times over manual drafting. The quality of the drawings is improved and consistent. In addition, automated drafting allows automatic dimensioning, generation of cross-hatched areas, sectional views, scaling and enlarged views of particular part details.

Other benefits of the system are :

(*i*) Drawings can be stored and recalled easily. Storage space needed is less. Revision on a later date can be done much faster as compared to traditional techniques.

(*ii*) By means of a coding system, similar part designs can be grouped into classes and the similarities related. By this, the designer can recall a particular design and modify it to suit a different application. This will be much more convenient and cheaper than always designing new parts. This feature is very useful in linking the CAD system with the CAM system.

Another powerful feature of CAD systems is "Simulation". Basically, the principle behind simulation is that it is cheaper, faster and easier to explore the behaviour of products and systems before they are made or used, through simulation in a CAD system. A variety of software techniques and mathematical models are used for this purpose. By FE analysis and simulation techniques, the final product design can be progressively improved.

For many companies, a key design problem is the storage and retrieval of information to wide and old product ranges which are still active and for which orders may be expected. CAD systems allow the use of computer storage to hold electronic information and the use of database management techniques to radically improve the speed and efficiency of access. Another different to a part already held in the library, the library drawing can be modified quickly to create the new design. This cuts the overall response time.

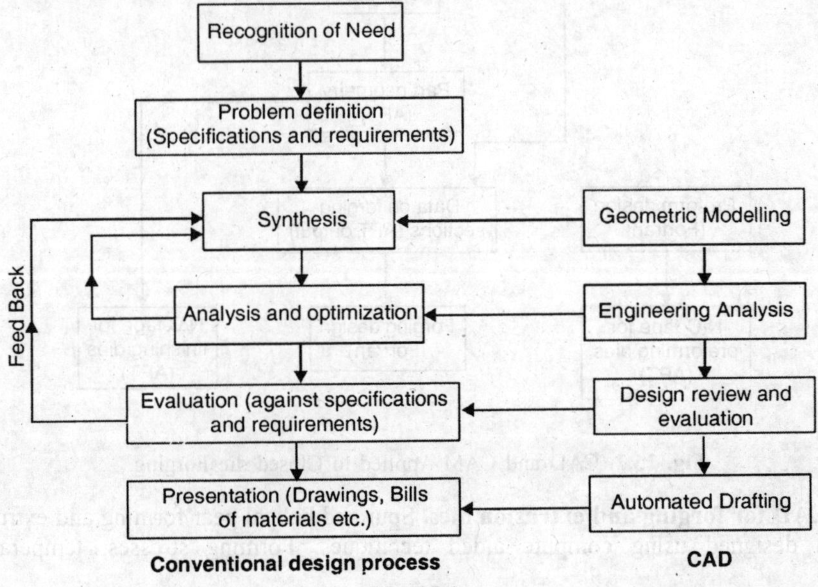

Fig. 25.1

Thus, a modern CAD system consists of the following four functional areas :

1. Geometric modelling.
2. Engineering analysis.
3. Design review and evaluation.
4. Automated drafting.

Fig. 25.1 illustrates the relationship between Conventional design and CAD. The left hand side portion of the Fig. 25.1 depicts the conventional design process. The complete fig. 25.1 illustrates the assistance of the computer in the total design process.

25.2.2. Examples. (*i*) **Design of a double column machine tool structure.** CAD has been used for the design of a double column machine tool various design parameters, the structure is examined visually on the graphics terminal. Both static and dynamic operational conditions are investigated and the corresponding response of the structure is then analyzed.

(*ii*) **CAD for forged part design and forging dies.** To design a forged part, the data base comprises : material coding, shape coding, relative material costs, operating costs and die costs etc. The input to the programme is a part design and the output is the total forging cost. Alternative part designs are investigated for obtaining the optimal design. CAD is also used to establish the proper design for performing and finishing dies in closed die forging. CAD has been applied to rib-web type airframe forgings and to airfoil shapes, but it can be applied to any class shape. Starting with a drawing of the final part, the CAD system defines this geometry in terms of points, planes, cylinders and other regular shapes using the APT computer language. Next, the co-ordinates of the various cross-sections of the forging are determined and these are used to perform design calculations to establish such factors as the location of the neutral surface, the shape difficulty factor, the cross-sectional area, volume, the flash geometry, the stresses, the loads and the centre of loading. An important aspect of this system is that is takes the part geometry and flash dimensions and generates the N/C tape for machining the electrodes in the sinking of the finishing dies by electric discharge machining (EDM). Thus this system also involves CAM. CAM is also used to machine the perform dies. This is explained in Fig. 25.2.

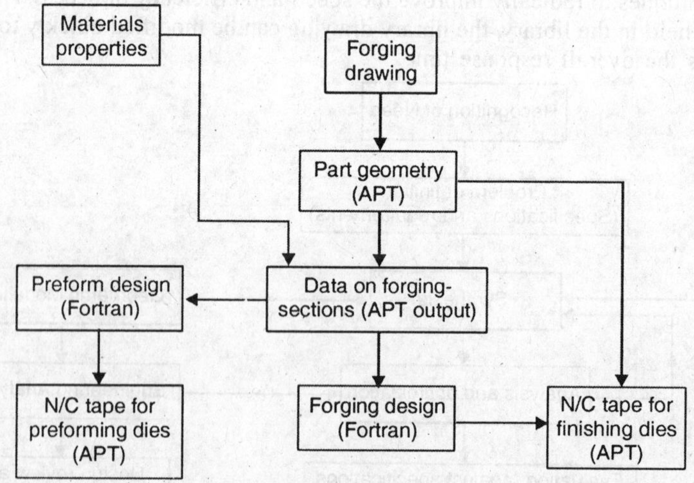

Fig. 25.2. CAD and CAM Applied to Closed die Forging.

(*iii*) **CAD for forging and extrusion dies.** Spur and helical gear forming and extrusion dies have been designed using computer-aided techniques. Forming stresses, temperature, die deflections and bulk shrinkage etc. are taken into account to calculate the perform volume and die-geometry.

(*iv*) **CAD for roll profiles for cold roll forming.** For given dimensions of the required product such as a tube, the CAD software enables computation of optimal pass-schedule and roll profiles.

(*v*) **CAD for preform design in plain strain rolling.** The front and rear ends of plates formed during ingot rolling are usually irregular and therefore defective. The Finite Element Method (FEM) can be used to predict the shapes of the ignot ends so as to minimize these defects.

Thus the CAD system performs the following two activities :

1. Product design.
2. Production preparation, that is, part drawings, assembly drawings, bill of materials etc.

25.3. COMPUTER-AIDED MANUFACTURING (CAM)

CAM can be simply defined as "the application of computers in manufacturing". Most elaborately, CAM refers to the use of comptuers in the control of production machines and ancilliary operations, for process optimization and control, process planning, process management, materials management, material movement (including transfer lines, robots, etc.), production scheduling and monitoring etc. These activities under CAM are written in detail as given below. Numerical control can be considered as the beginning of CAM.

1. **Process Planning.** Process selection, process design, process parameters, group technology, NC parts programming, tool and fixture design (including dies, jigs, gauges, etc.), quality control.
2. **Process *R* & *D*.** Process choice, optimization, modelling.
3. **Processing.** Manufacturing of parts; sensing and corrective action; storing, moving and handling of materials, parts, tooling, jigs and fixtures; and assembly.
4. **Production Planning and Control.** Routing, scheduling, follow up (tracking), machine load monitoring, inventory (parts, materials, in-process), purchasing, receiving, maintenance, quality assurance (standards, inspection, etc.). Thus, CAM centres around four main areas : NC, process planning, robotics, and factory management.

These activities will be discussed in brief as below :

Numerical control has been discussed in the book "A Text Book of Production Technology" by the author.

25.3.1. Process Planning. In computer-aided process planning (CAPP), the logic, judgement and experience required for process planning are incorporated into computer programmes. Based on the characteristics of a given product or component. The programme automatically generates the manufacturing operation sequence which is obtained in the form of computer listing. The main advantages of CAPP are : it reduces lot of routine paper work of manufacturing engineers and it helps in getting a rational, consistent and even optimal manufacturing sequences. Two approaches have mainly been used :

(*i*) **Group Technology Approach.** In batch production which no doubt forms a major portion of all manufacture, the usual method is of working out separate sequences of production operations for every single part of every product. It even often happens that completely different sequences are specified for closely similar or even identical parts. Therefore time, talent and money are unnecessarily spent and paperwork increased. Because production sequences are worked out for relatively small quantities of production, little use is made of special tooling set-ups or of high productivity attachments.The situation can be remedied by using Group Technology (GT).

Group technology is based on the general principle that many problems are similar and by grouping similar problems, a single solution can be found to a set of problems, thus saving time and effort. This principle can be applied to any branch of engineering.

In manufacturing, group technology is a means of achieving variety reduction, allowing manufacture on "part families". By classifying components in accordance with their shapes or their technological features, and not according to their functions, it is possible to arrange them in groups which can be machined in "families". For this, groups of machines are formed to manufacture these families of components on a flow-line principle. The "families of components" consist of various components each of which is required only in small number but which can be grouped together to form economical batches.

The groups of machines are formed so that all the components in one family can be manufactured by one machine group. These machine groups can be arranged in two ways :

1. The 'Group-layout system'.
2. The 'Group flow-line system'.

In the first system, the machines are arranged into groups in such a manner that each group can carry out all the machining operations needed for the family of components. For example, a particular family of components requires machining operations on a lathe, a drilling machine, a milling machine and a lapping machine. These four machines can be grouped into a cell and located in one small area of the floor space.

In the second system, the machines are arranged in the sequence of the production operations and are usually linked by a conveyor arrangement.

Group technology is also known as "part-family manufacture".

Advantages of GT

1. Because of large production quantities involved in each family of parts, it becomes economical to use a wide range of high productivity equipment.

2. Process planning paper work gets greatly reduced, because, instead of, say, the several hundred production sequences for an equal number of parts needed in conventional manufacture, of parts needed in conventional manufacture, only a few (10 to 20) production sequences would be required in GT for an equal number of families.

The data on the various families and the corresponding sequence of operations, machines, tools, jigs and fixture is stored on a magnetic tape. Whenever a part is to be produced, its classification number is fed as data to the computer and the computer prints out the required sequence of operations, machine group number, cutting speeds, depths of cut and feed rates for individual operations in the machining sequence.

(*ii*) **Geometrical Elements Approach.** In this approach, the part to be produced is described in terms of its geometrical elements (*e.g.,* cylindrical surface, plane surface, external and internal taper, threaded portions, holes, splines, keyways etc.), their dimensions, required dimensional accuracy and surface finish. The computer is provided with a technological data file in its back-up storage containing the required list of operations, machine tools, jigs and fixtures etc. capable of producing each geometrical element. When the geometrical elements of a part are fed as data to the computer, it displays the alternative sequences of operations on the terminal. If the process planner is satisfied with the computer-sequences, other data such as production times and costs are also displayed or printed out.

CAPP is also known as CAM-I automated process planning system. CAM-I stands for Computer-Aided Manufacturing-International, a non-profit industrial research organization.

Advantages of CAPP. In addition to the advantages of CAPP mentioned in the beginning of this article, CAPP has the following advantages :

1. It can reduce the skill required of a process planner.
2. It can reduce the process planning time.
3. It can reduce both process-planning and manufacturing costs.
4. It can increase productivity.

25.3.2. Computer-Aided Production Management (CAPM). The traditional production planning and control as discussed in Chapter 23 suffers from the following bottlenecks :

Its inability to cope up with plant capacity problems and suboptional production scheduling, inadequate inventory control and low work centre utilization.

Again, new and important considerations have become a part of the modern production systems, for example, increase in complexity of the products manufactured resulting in stringent specifications and tolerances, more consumer awareness and expectations, increased competition and reduction in overall time.

Due to all the above factors, the volume of information to be handled has become very large. For this, CAPM is finding increasing favour with companies.

CAPM can be subdivided into three main areas :

1. Input information and influences of manufacturing.

2. Production planning.

3. Output information for the realization and control of production including feedback from production itself.

The input information for CAPM is the output information of the design department and other planning departments of the company. The information from the design and engineering department includes, for instance, shape, size, shape elements, accuracy and surface quality. Input informations from other planning departments is especially economic data, that is, lot sizes, limitations of costs and planning data, for instance, assortment plan, time schedule etc.

The tasks under production planning cover the following essential areas :

1. Preparation of manufacturing processes resulting in the preparation of operational sheets and NC punch tapes etc.

2. Technological design for reconstruction of equipment and effective application of technical capacity.

3. Technological design and manufacturing of special machine tools, fixtures, special tools and all kinds of technical aids.

4. Technological normalization of work, for the further development of company time standards, and carrying out of time studies and work-scientific analyses.

For the solution of the above mentioned tasks, information stores such as data banks and suitable programmes are necessary. Such data bases are necessary for machine tools, tools, materials, recommended values for machining data and technological times, fixtures and other aids.

The output information in the field of production planning is coincident with the primary data for the entire factory organization. It is used for the technological and time control of production sequences.

CAPP discussed earlier forms a part of CAPM. Other functions of CAPM are discussed below :

1. **Capacity planning.** Capacity planning forms the second principal step in the production system, the Product and Service design step being the first. The term "Capacity" of a plant is used to denote the maximum rate of production that the plant can achieve under given set of assumed operating conditions, for instance, number of shifts and number of plant operating days etc.

Capacity planning is concerned with determining labour and equipment capacity requirements to meet the current master production schedule and long term future needs of the plant.

Short term capacity planning involves decisions on the following factors :

(*a*) employment levels

(*b*) number of work shifts

(*c*) labour overtime hours

(*d*) inventory stock piling

(*e*) order back logs

(*f*) subcontracting jobs to other plants/shops in busy periods.

Long term capacity planning involves decisions on the following factors :

(*i*) investment in new machines/equipments

(*ii*) new plant construction

(*iii*) purchase of existing plants

(*iv*) closing down/selling obsolete facilities.

2. **Cost planning and control.** Cost planning and control system consists of the data base to determine expected costs to manufacture each of the products of the firm. It also consists of the cost collection and analysis software to determine what the actual costs of manufacturing are and how these actual costs compare with the expected costs.

Cost planning answers the first question, that is, manufacturing and selling costs.

Cost control is concerned with actual manufacturing and selling costs and the differences between actual and expected costs. The reasons for the differences could be due to machine breakdown, increase in price of raw material, deviation of actual process from route sheet etc. A document is prepared which projects actual product costs and variances from standard costs.

3. **Inventory management.** The functions of good inventory management are : to keep the investment in inventory low without affecting good customer service.

The inventory management module of CAPM consists of two functions :

(*a*) **Inventory accounting.** This function involves the following activities : inventory transactions and inventory records which include : receipts, disbursements or issue returns etc. The "Item Master File" describing the computerized inventory record file contains three segments.

(I) **Item master data segment.** This segment gives part's identification, lead time, cost and order quantity.

(II) **Inventory status segment.** This segment indicates future changes occurring against inventory status alongwith the current level of inventory.

(III) **Subsidiary data segment.** This segment contains, miscellaneous information pertaining to purchase orders, scraps etc.

(*b*) **Inventory planning and control.** This function involves the following activities :

(*i*) determination of economic lot sizes

(*ii*) determination of safety stock levels

(*iii*) determination of reorder parts

(*iv*) automatic generation of requisitions for purchasing

(*v*) analysis of usage rates for lot size calculations etc.

Using updated historical records on the requisitioning of each different material, part, tool or machine etc.; optimum procurement lead times, order points, reorder quantities are periodically worked out by the computer using software based upon simulation or some other operation research technique.

4. **Production scheduling.** With the receipt of a job order, the problem of production scheduling arises. The computer is supplied with data on product classification number and quantities of production. It uses the data of Design Master File to explode the product into parts

and components. Next it computes the quantities of parts and components to be procured from outside suppliers as well as the number of parts and components to be produced in the plant. It then refers the Inventory Master File to find which materials and parts are available in stock and which are not. If they are, necessary authorizations are printed out. On the other hand, if requisite materials and parts are not available in stock, necessary supply orders are printed for postage. Also with the help of Process Master File, the computer identifies the possible sequence of operations required for various parts and components. Finally with reference to the required delivery schedule and present machine loading available in its back-up storage, the computer identifies which machines are available, selects the most suitable machine for the operation, computes the optimum process parameters and finalizes or updates the detailed schedule. Rescheduling of several machines may sometimes be necessitated by such reasons as acceptance of rush jobs, breakdown of some machines or non-availability of certain materials.

5. **Material requirement planning (MRP).** MRP function is a computational technique with the help of which the master schedule for end products is converted into a detailed schedule for raw materials and components used in the end product. The concept of MRP is quite simple but the data to be banded is so huge that the application of the technique becomes quite complicated.

Inputs to MRP

(*i*) Master production schedule

(*ii*) The bill of materials

(*iii*) Inventory records relating to raw materials, purchased components, parts to be delivered, in process materials and parts, finished products and tools and maintenance supplies.

With this informations as the input, MRP manages the inventories with due regard to the timing of the material requirements.

Output of MRP

(*i*) Order release notice

(*ii*) Future order release notice

(*iii*) Scheduling and rescheduling notices

(*iv*) Cancellation notices

(*v*) Inventory status reports

(*vi*) Inventory forecasts, performance reports, deviation from schedules etc.

A good MRP system results in :

(*a*) Reduced lead times

(*b*) Minimum inventory

(*c*) Faster response to customer requests

(*d*) Increased productivity

6. **Shop floor control (SFC).** SFC function consists of three computer software modules :

(*i*) *Order release.* Its purpose is to provide necessary documentation that accompanies as it is processed through the shop : route sheet, material requisitions, job cards and parts list.

(*ii*) *Order schedule.* Its purpose is to provide assignments of orders to various machines in the factory. It is assignment of specific jobs to specific centres. It consists of machine loading and job sequencing. Job sequencing or priority sequencing involves the determination of the order in which a given line of jobs should be processed through a given machine. It is governed by priority rules. Some of the more important priority rules are :

(*a*) First come, First served rule of First-in, First-out. It is a fair rule but is not in general the best as regards the minimization of waiting time, set up time or inventory time.

(*b*) Highest priority given to "Earliest Delivery Date" jobs.

(*c*) Highest priority given to "Shortest Processing Time" jobs.

(*d*) Highest priority given to "Least Slack" job.

Slack = Time remaining till delivery data : Process time remaining.

(*e*) Highest priority given to "Lowest Critical Ratio" job.

(*iii*) *Order progress.* Its function is to monitor shop order status, maintain information on work in process, and production output data for capacity control.

The data to be collected would be :

— piece counts
— scraped/rework parts
— machine breakdowns
— labour time turned in against a job.

7. **Inspection and testing.** In a computer controlled inspection machine, the component is suitably mounted on the machine. The measuring probe or sensor is automatically moved (under numerical control) to the selected positions on the component, one after the other, where required dimensions are to be checked. The motion of the probe (or probes) are controlled by an "inspection programme" that contains the necessary information about the inspection points and the tolerance limits on each measured dimension. If a measurement indicated by the probe is within the tolerance limits, the probe moves to the next inspection location. Otherwise, the computer prints out the job number, the inspection location and the corresponding error. In addition to checking the dimensions, the computer carries out a statistical analysis of measured dimensions on each batch of parts. The output of this analysis is either displayed or automatically used to reset tools.

A computer vision system has been designed to inspect components. The component is held vertically, back lit and rotated in front of a solid state camera. A three dimensional image is re-constructed by the computer, which is then compared with the specifications of the component.

Computer aided inspection is also carried out on completed assemblies. A number of probes work at different locations on the assembly, to check or measure different variables. For example, for the testing of aircraft engines, hundreds of measuring probes, work at the same time to measure temperature, pressure, strain rate etc., at critical locations. The measurements are relayed to the computer which analyses the data and prints or displays the results.

8. **Assembly.** In a typical computer aided assembly, the numerically controlled mechanical hand or an industrial robot picks up a part from a storage point, carries it to the assembly point, locates it at the proper position, inserts it or attaches it to another part and fastens them together by pressure, sliding or turning etc.

9. A further improvement in CAPM is "Manufacturing Resource Planning" also called MRP-II or closed loop MRP, which integrates a complete manufacturing control system. It performs the following functions :

— Production planning
— Master scheduling
— Capacity requirement planning
— Function necessary for executing the production plan (including vendor schedules and dispatch lists)

Provisions are made for continuous updating. The management part can be integrated through a computer network independent of a CAM system. It then is called a "Computer-Integrated Production Management System", CIPMS.

25.4. COMPUTER-INTEGRATED MANUFACTURING (CIM)

Many definitions have been given for CIM in literature by various authors. But for our purpose, we will refer to our earlier definition of CIM given in Art. 25.1 as :

"The integration of computer based monitoring and control of all aspects of the manufacturing process, drawing on a common data base and communicating via some form of computer network."

CIM includes not only manufacturing functions (CAM), but also engineering functions (CAD) and business functions. The activities under CIM include :

1. CAD
2. CAM
3. Market research, market forecast, product concept and sales (order processing).
4. Customer service.
5. Shipping (inventory, invoicing and accounting etc.).

25.4.1. The Trend Towards Integration. The two factors which have influenced the emergence of Advanced Manufacturing Technology (AMT) are : Information Technology (IT) and Integration. By integration, the separate functions of the manufacturing process are brought together. It accelerates the process. Through a synthesis of different elements, the whole becomes greater than the sum of its parts. Manufacturing is regarded as a system with interdependent large variety of activities (sub-systems). These sub-systems have become specialities on their own. The integration of these sub-systems offer significant improvements in quality, flexibility, productivity and so on. Integration within functional areas towards integration between areas of activities and finally integration of all areas of activities of a manufacturing system leads to CIM. This is briefly discussed below :

1. **Integration within areas of activities** : (*a*) *Design.* In the field of design, the various functions associated with this complex process have gradually integrated into computer-aided design systems. The traditional engineering department drawing office has been replaced by a computer-aided design facility in which the designers work with computer terminals. The result is integrated CAD workstations.

(*b*) *Manufacturing.* In the field of manufacturing, we have moved from originally NC, then CNC/DNC, the FMS and now AI (Artificial Intelligence). Via this process of integration, we have moved from a single machine tool to a complete manufacturing system over a period of about 150 years. We have : integrated cells, CNC machinery and so on.

(*c*) *Co-ordination.* Such as integrated stock control, MRP, MRP2. Integration of the various functions of CAPM results in CIPMS (Computer Integrated Production Management System).

2. **Integration between areas of activities.** Such as CAD/CAM, FMS. Since CAD systems make use of information coded in electronic form, it follows that other systems—such as those for manufacturing and co-ordination—which also use such information—can be linked in via some form of network. Such an integration leads to sharing a common pool of information. This is the basis of CAD/CAM on the manufacturing side, and increasingly of CAPM systems on the co-ordination side. Benefits arising from such an integration are enormous : significantly reduced lead times, improved quality, better machine utilization and much improved customer service.

CAD/CAM. The real benefits of CAD and CAM can be fully realized only by integrating them and creating which is usually referred to as CAD/CAM, computer-aided design and manufacturing (also called CAE, computer-aided engineering), Fig. 25.3.

CAD/CAM systems essentially take the output of the design process and instructions are generated and sent to the machine tools and other devices which actually manufacture the parts. Information flows in both directions (CAD $\leftrightarrow$ CAM). This ensures that parts and assemblies will

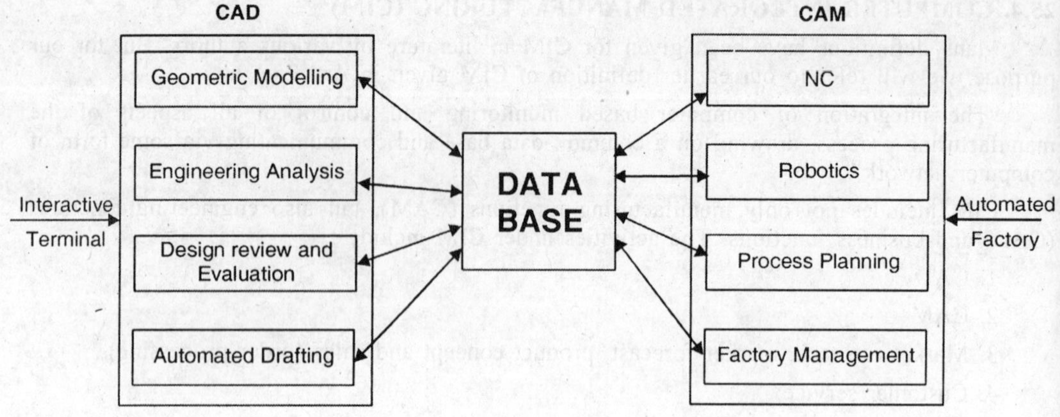

Fig. 25.3. Integrated CAE System.

be designed with the capabilities and limitations of materials and manufacturing process in mind. In the conventional design and manufacturing functions, the interface between them is called "Concurrent Engineering (CE)", See Chapter 20, Art. 20.3.

Benefits of CAD/CAM Systems

(*i*) Superior products can be created with improved quality.

(*ii*) New products are developed more rapidly and at a lower cost, that is, reduction in lead times and savings in materials.

(*iii*) Better customer service.

(*iv*) Reduced drawingroom costs.

(*v*) Simple production planning

(*vi*) The system forces a review and improvement of existing design and manufacturing practices and production planning.

And since CAD systems hold all the information about the products to be made, and fixtures, tools, materials etc. needed, this information can be used to generate the basic information needed to order materials and schedule work for the factory.

3. **Integration of all areas of activities.** Since CAD systems, CAM systems, CAPM systems and business functions are all IT (Information Technology) based systems, they can be integrated to communicate with each other via some form of network. This is the basis of Computer-Integrated Manufacture (CIM). It means the integration of all areas of activities of a modern production system. In CIM, a product is designed on a CAD system which then provides all the information for both the physical production and the management of that process, including business functions. In other words, the factory behaves as if it were a giant, complex but integrated machine.

25.5. COMPUTERIZED INFORMATION SYSTEM

As discussed above, the Information Technology (IT) has played a significant role in the emergence of Advanced Manufacturing Technology (NC, Robotics, CAD, CAM, CAPM, FMS and so on). The different components of a computerized information system (sometimes referred to as Information Storage & Retrieval System or Management Information System) include a Data Bank consisting of several master data files, software comprising of a package of programmes, means of Data Acquisition and Updating, Information Display and On line control.

(*i*) **Master Data Files.** Several master data files have to be created and maintained. The chief among them are the following :

(*a*) Engineering data master file contains design data on all products, *e.g.*, their constituent components and parts, all relevant dimensions of the different parts, material specifications, dimensional accuracy, surface finish and heat treatment, classification of parts into families.

(*b*) Process data master file stores data on the different feasible sequences of operations required for producing different geometrical elements or parts belonging to different families or groups alongwith corresponding dimensional accuracy and surface finish, particulars of machine tools in the plant on which the operations can be performed, cutters or tools and the recommended jigs, fixtures and machining conditions.

(*c*) Inventory master file contains data on all materials, brought-out parts, components, tools, cutters, machines and consumable currently held in stock, on order or being processed for ordering, stock already allocated to specific jobs, free stock, list of suppliers, unit purchase price, quantity discounts etc.

(*d*) Sales master file stores data on all orders currently on hand, order quantities, promised delivery dates, historical sales records of individual items, forecast quantities etc.

(*e*) Schedule master file contains the detailed schedule for every machine or workcentre, shop load, scheduled quantities and completion dates regarding parts, components and assemblies for each machine and shop, quantities actually produced etc.

(*f*) Current status master file maintains updated data on all employees and production facilities, absentee workers, machines under repairs and those awaiting repairs etc.

(*ii*) **Software.** Special computer programmes based upon statistical, mathematical and operational research techniques are developed for computing and problem solving. For example, the design office would use a library of subroutines for analyzing alternatives for each part, component or product. The process design function would require programmes for process planning specially tailored to the machines installed in the plant and a library of part programs for use on the NC machines of the plant. Sales planning function requires forecasting and other statistical programs for market research. The production planning and scheduling, and inventory functions make use of linear programming simulation and other OR programmes. Accounts function requires data processing programmes for recording all transactions and the preparation of various statements of account.

(*iii*) **Data acquisition and updating.** It is imperative that all data files are speedily updated. For example, data on new product designs and processes must be entered on the respective files by means of remote terminals on punched cards. Similarly each stores transaction such as issue, allocation or receipt of a given material is used to update the available stock levels. Also acceptance of a new job order or delivery of a pending order should be reflected in the sale and schedule master files. Data on operator attendance and machine status should likewise be fed to the current status file.

(*iv*) **Information display.** In a plant, a certain decision has to be taken by the particular manager charged with that responsibility. To arrive at a correct decision in an efficient manner, it is essential that pertnent facts be made available to the manager. Because the computer has ready access to a vast amount of data in its memory store, the manager is able to obtain the needed data in the required format on the remote terminal in his office by relaying a request to the central computer. This facility can also be used to over-ride certain analytical decisions obtained through the computer software. For example the production planner may decide to alter the computed schedule in response to a situation unforeseen in the original scheduling programme.

(*v*) **On-line control.** As explained earlier, the production schedule and the current plant status is available in appropriate master files. On the basis of this information, the computer can transmit commands to the appropriate peripherial computers regarding production of specific parts or assembly of specific products. Each peripherial computer is supplied with a library of part programmes in its back-up storage. Obeying the central computer, it transmits appropriate commands to individual production machines, materials handling devices, industrial robots, assembly, inspection and testing machines. The peripheral computer also receives information back from machines etc. through appropriate sensors regarding completion of parts and products, their quality and machine status. It transmits some of this information further on to the central computer. The peripherial computer also analyzes information on quality and if necessary modifies future manufacturing instructions to the production machines in the light of this analysis.

25.6. THE AUTOMATIC FACTORY

From the previous paragraphs, it is clear that computerization is today well advanced in individual sub-systems (*e.g.*, product design, process planning, machine tool operation, materials handling, inventory control, sales, etc.) of a manufacturing system. The various sub-systems can be integrated with one another by means of an on-line computer and several peripheral computers. Such an integrated system would be able to work with minimal human intervention in the form of managerial decision-making, supervision and physical tasks. One may even visualize an integrated fully automatic manufacturing system in which the prime input is the information about human needs for some products and whose outputs are finished products. The system comprises of a combination of software and hardware, concerning product design, production operations. The system has flexible automation and is self-optimizing. Several manufacturing system have actually been installed which are integrated to varying degrees. Some system are in existence which operate in 3 shifts, out of which operations in 2 shifts are carried out in almost total absence of human supervision or any other kind of human intervention. It may take a few more decades for the installation of a completely automatic factory.

It is not difficult to visualize an integrated computer automated factory. The raw materials entering the factory will be automatically transferred and stored. The identification and location of each item in storage will be retained in computer files. When required, the automated materials handling system will move the materials to the manufacturing location in the shop. There, the raw materials will be processed under the control of computer through :

1. A closely coordinated schedule according to the proper sequence of machines.
2. The optimal routing for the different workpieces.
3. In-process inspection to ensure the quality of the parts.
4. Adaptive control feed back to ensure that all the tools are in good working order.
5. The use of Robots to help in many of the above operations.

After the processing of the parts is completed, the next steps will be :

(*i*) Automated transportation of the parts to automated assembly lines.

(*ii*) Automatic testing of the assemblies.

(*iii*) Then transportation to an automated warehouse for storage until shipment to the customer.

The role played by operators in an automated factory consists of functions requiring higher skill and intellect and training. These functions include :

(*a*) Computer-aided planning.

(*b*) Programming and operating computers.

(*c*) CAD, CAM and computer-aided repair.

(*d*) CAPM and supervision.

(*e*) Computer aided diagnosis of machine breakdowns.

(*f*) Computer prepared preventive maintenance schedules.

Benefits of computer controlled and integrated automatic factory :

1. Increased productivity.
2. Increased human productivity because many human decision-making functions are replaced by or assisted by computers.

3. More efficient planning.
4. Enable small batches of products to be manufactured at competitive costs.
5. More efficient use of machines and production facilities.
6. More efficient use of raw materials.
7. Less scrap.
8. Less in-process inventory.
9. Shorter lead times.
10. Faster response to market demands.
11. Less direct and indirect labour costs.

This computer controlled automated factory will differ from the automotive transfer line in that it will be a flexible manufacturing system capable of producing a wide variety of parts under computer control.

25.7. SIMULTANEOUS ENGINEERING

The concept of integrating design and manufacturing at the start of the design process has already been discussed in Chapter 20 under Art. 20.3. This concept is known as "Concurrent Engineering", CE. The concept of Simultaneous Engineering (SE) is much broader. It also involves the people responsible for market analysis and those concerned with sales and distribution. Many authors consider CE and SE as the same.

Fig. 25.4 shows a SE model. The overall aim of SE is to consider all aspects of the product design and manufacturing at the very beginning of a project so as to minimize the costs of changes in design late in the product development cycle. SE is an integrated approach in which the design is refined and developed on the basis of "real-time" interaction so that it is constantly evolving and improving.

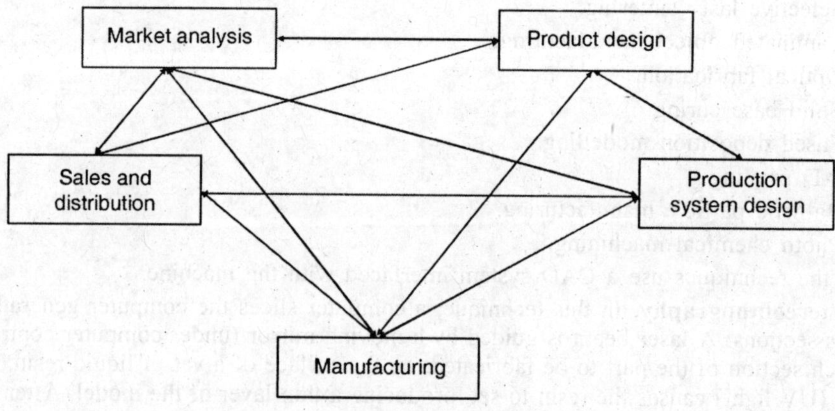

Fig. 25.4. A SE Model.

Benefits :

1. Improved product design.
2. Increased product quality.
3. Reduced manufacturing cost.
4. Rapid product development cycle.
5. Increased market share through reduced time to introduce the product in the market.

25.8. RAPID PROTOTYPING TECHNIQUES

In the 'design-manufacturing' cycle, after the "detailed design stage", the next step is prototype fabrication and testing. A prototype is an original working model of the product, and in a sense, is the first stage of manufacturing. The traditional methods of fabricating a prototype (shaping, forming, machining etc.) take a long time (anywhere from weeks to months) depending upon the product complexity. This greatly enhances the cost of the product.

As discussed under Art. 25.2, Computer-Aided Design (CAD) has recently replaced the conventional design process. The two important constituents of a CAD system are : computer graphics and computer modelling. The tremendous advances in these two fields have led to a desire to quickly convert the computer model of product design into a 3-D prototype. This has resulted in the development of various "Rapid Prototyping Techniques (RPT)". These techniques rely on CAD/CAM and various consolidation process (using polymers or metal powders) such as resin curing, sintering, deposition and solidification etc. The techniques have been developed and generally used for producing prototypes in the form of solid physical model of a component rapidly and at a lower cost, directly from a 3-D CAD drawing. These techniques are being further developed so as to be used for low-volume production of components. The process is also called as "Desk-top manufacturing", "Free form manufacturing" or "Layered manufacturing". It is called "Layered manufacturing" because the product is built up from a collection of 2-D profiles joined together by some means to fabricate a 3-D part. RPT can significantly reduce the costs of prototype fabrication and product development cycle times.

25.8.1. Various Rapid Prototyping Techniques. The input for all the rapid prototyping techniques is a 3-D computer model of the part to be fabricated. An important stage in all these techniques is the slicing operation performed on the CAD representation to generate the geometrical description of each layer to be formed. The coordinate information which defines the layers provides the instructions which drive the computer controlled prototype system. The part is built up in a series of thin layers.

The various RPT are :

1. Stereolithography.
2. Selective laser sintering.
3. Laminated object manufacturng.
4. Optical fabrication.
5. Solid base curing.
6. Fused deposition modelling.
7. 3-D printing.
8. Ballistic particle manufacturing.
9. Photo chemical machining.

All the techniques use a CAD system interfaced with the machine.

1. **Stereolithography.** In this technique, a computer slices the computer generated design into cross-sections. A laser beam is guided by a moving mirror (under computer control) which traces each section of the part to be fabricated, on the surface of a vat of liquid resin. The laser radiation (UV light) causes the resin to set, producing a thin layer of the model. After one layer is completed, it drops slightly below the surface of the resin bath, so that the next layer can be built on top of it. Like this, the complete model is fabricated. To ensure complete curing of the part, it is subjected to post curing with UV light or by controlled heat curing. This technique can convert a computer generated 3-D model of the design into a 3-D plastic model within hours.

Automotive, aerospace, electronics and medical industries are using this technique as a rapid and inexpensive method of producing prototypes.

One major application is in the field of making patterns and moulds and dies for casting and injection moulding. For example, in precision investment casting (lost wax process), wax patterns can be replaced with accurate resin patterns made directly by this RPT, using CAD data and without the use of dies, at a fraction of the time and cost of dies for making wax patterns. The maximum part size which can be made by stereolithography is 0.5 m × 0.5 m × 0.6 m.

RPT have made a very valuable contribution to Concurrent Engineering (CE). A rapid prototyping model with drawing avoids misinterpretation of drawing.

Advantages :

(i) Well proven technique.

(*ii*) High and better dimensional accuracy as compared with other techniques and best surface finish.

(*iii*) Unattended system.

Disadvantages :

(*i*) The technique requires support structure.

(*ii*) The components can get warped.

(*iii*) Resin is expensive.

(*iv*) Resins and solvents can be environmentally hazardous.

2. **Selective Laser Sintering.** Here, discreet metal powder particles are joined together into a solid body by selective laser sintering. A computer controlled high powered laser beam (based on a 3-D CAD programme of the part to be fabricated) is focussed on a layer of the powder, tracing and sintering the defined shape into a solid section. The loose powder supports the sintered layer. The table is then lowered by an exact amount and another layer of powder is then deposited on top and the cycle is repeated. A new layer of solid mass is produced which gets fused to the lower layer. In this way, the entire 3-D model is slowly built up. The completed part is recovered by shaking off the loose particles. Unless, it is ceramic, the part does not need post curing. However, it needs firing to remove the binder.

Advantages :

(*i*) Good range of materials can be used in this process, including polymers, wax, metals and ceramics with suitable binders.

(*ii*) No post curing unless the material is ceramic.

(*iii*) Relatively low cost of materials.

Disadvantages :

(*a*) Long time to heat up and cool down (to avoid part break up dice to residual heat).

(*b*) Porous component. However, it can be made 100% dense by filling it with copper powder. Copper is sintered by refiring.

(*c*) When binder is removed by firing, thin walled sections may become fragile.

(*d*) Higher initial cost.

(*e*) Powders require careful swing to avoid "ball-ups".

(*f*) Some portions of the part may require support.

3. **Laminated Object Manufacturing.** Laminated object manufacturing (LOM) process consists of fabricating a part by bonding together layers of sheets (such as metal foil, paper or plastic) in a stack. The machine consists of a platform over which an adhesive backed paper (say) is stretched with the help of two rolls (one feed up roll and the other take up roll) one on either side of the platform. The laser then traces the required shape of the part on the paper. The unused portions are cross-hatched by the laser so that they can be easily removed after the model is completed. After this, the table descends, the paper rolls on, placing the next layer on top of the just completed. The laser performs its function on the new layer. A steam roller is then used to adhere this layer to the previous layer. The cycle is completed until the model is completed.

Advantages :

(*i*) The method is one of the simplest and cheapest because of the low cost of raw materials.

(*ii*) The material type can be easily changed.

(*iii*) Materials used are not hazardous.

(*iv*) Supports are not needed.

Disadvantages :

(*i*) The material (paper) suffers from hydration. So, it needs to be lacquered to seal the surface, to avoid distortion or expansion of the part.

(*ii*) The quality of the surface (finish) is not of high order.

(*iii*) Slender components need careful handling.

4. **Optical fabrication.** The technique is similar to stereolithography. It uses a visible light argon ion laser. The part is made on a stationary platform. Point-by-point solidification of the liquid takes place at the intersection of two laser beams at right angles to each other, that is, by holographic interference.

5. **Solid base curing.** This method uses photosensitive polymers. The resin in each thin layer is covered with a photomask which is then exposed to UV light and is cured in a few seconds. The unexposed liquid is then removed and the voids are filled with molten wax to support the next layer. The cycle is repeated until the entire part is built up.

6. **Fused deposition modelling.** In this technique, the model is fabricated by depositing very thin layers of molten thermoplastic or wax on top of each other. The filament of the material is fed through a heated extruding heat. The molten filament is deposited on a platform to form the part. The process is controlled by a computer which guides a robotic arm to form the 3-D part layer by layer. The fabricated part needs no further curing.

7. **3-D Printing.** This technique is similar to the ink-jet printing. A piston supports the power bed. The printer head moves across the bed emitting a continuous stream of droplets of a binder, selectively binding the powder. The piston is lowered incrementally and with each step, a layer is deposited and joined by the binder. After the part is completed, it is removed from the powder bed and the loose powder is removed. It is then heated to remove the binder and is finally sintered. The powders used are : Al_2O_3, SiC, SiO_2 and Zirconia. The ink-jet head is guided by a three-axis robot.

8. **Ballistic particle manufacturing.** In this technique, streams of materials, such as plastics, ceramics, metals and wax are ejected (using a ink-jet mechanism) through a small orifice, at a surface. The part is completed with layers of material deposited on top of each other. Again, the ink-jet head is guided by a 3-axis robot.

9. **Photo-chemical machining.** This process is similar to stereolithography but uses two lasers beams intersecting each other to form the part. One beam moves in the X-Y plane and the other in the Y-Z plane. The process is more versatile and the layer by layer technique is not needed to print the part.

25.9. ENTERPRISE RESOURCE PLANNING (ERP)

As has been discussed earlier, the two factors that have contributed to Advanced Manufacturing Technology (AMT) are : Information Technology (IT) and integration.

Until recently, MRP-II package (See Art. 25.3.2. Point 9) has been the ultimate solution which controls all aspects of manufacturing through feed back, that is, a closed-loop solution. However with tremendous growth of global trade a flexible business information system that quickly responds to rapid changes and customer needs (due to technology development which has greatly shortened the product-life cycle), if it has to compete in the world market. A successful enterprise will be the one which is able to most effectively gather the vital information and quickly act upon it. Hence with globalisation and technology development, the entire enterprise must be managed within a more global, tightly integrated closed-loop solution. To be successful in the present and future world market, a high level of interaction and co-ordination along the entire supply line is essential. This is the basis of ERP. The various ERP software packages, which have been developed, cater to this need.

Both MRP-II and ERP are closed-loop, totally integrated and on-line solutions. But the basic difference between the two is that, whereas MRP-II is essentially manufacturing set up oriented (It is capable of final production scheduling, monitoring actual results in terms of performance and output, and comparing those results against the master production schedule), ERP caters to the information requirements of the entire enterprise. Hence, MRP-II is basically on automation solution, which in many cases may not end up as a business solution. On the other hand, ERP tries to provide a "total solution", that is, it is a business solution. MRP-II has become a subset of ERP.

ERP packages can be applied to all types of enterprises (manufacturing, service, banking, insurance etc.) Just like TQM, ERP needs the total commitment of the management, which must also take appropriate follow up actions to maximize gains.

Benefits of ERP :

1. Reduced cycle times (upto 80%) and lead times (upto 60%).
2. On-time delivery (upto about 99%).
3. Work-in-process (WIP) reduced to about 70%.
4. Manifold increase in business.
5. Improved and better customer satisfaction and vendor performaı.ce.
6. Greater flexibility.
7. More efficient and economical utilization of resources, resulting in increased productivity.
8. More reliable feed back.
9. Reduced quality costs.
10. Improved decision taking capabilities.

Following are companies in the field of ERP : Baan (Based in Putten, Netherlands and Menlo Park California, U.S.A.), SAP (a German company), IFS (Industrial and Finance Systems) Qad and Ramco systems.

PROBLEMS

1. What do you mean by computer controlled manufacturing?

2. Define CIM. List the functions performed under CIM.

3. Define CAD.

4. How does CAD influence engineering design?

5. Write on N.C. drafting machines and computer graphics used in a CAD system.

6. List and discuss the components of a CAD system.

7. What is geometric modelling? Discuss the various geometric models used in a CAD system.

8. Write the benefits of Automated Drafting used in a CAD system.

9. Give the benefits of a CAD system.

10. Define CAM.

11. List the activities performed by a CAD system.

12. List the functions performed by a CAM system.

13. What is CAPP? Write the benefits of CAPP?

14. What is Group Technology? What are its main advantages?

15. What is CAPM? How is it better than the traditional production planning and control?

16. Write short notes on :

(*a*) Capacity planning.

(*b*) Cost planning and control.

(*c*) Inventory accounting.

(*d*) Inventory planning and control.

(*e*) Production scheduling.

(*f*) Material Requirement Planning (MRP).

(*g*) Shop floor control.

(*h*) Computer-aided inspection and testing.

(*i*) Computer-aided assembly.

17. What is MRP-II?

18. Explain how computer is used for .

(*a*) Production scheduling

(*b*) Inventory management.

19. Discuss the two factors which have helped in the emergence of Advanced Manufacturing Technology (AMT).

20. What is integration in manufacturing technology?

21. What is a CAD/CAM system? Write the benefits of such a system.

22. Write a short note on computerized information system used in CIM.

23. Write a brief note on "An Automated Factory". List its benefits.

24. What is Simultaneous Engineering? Write its benefits.

25. What do you understand by Rapid Prototyping?

26. What are Rapid Prototyping techniques?

27. What are the advantages of Rapid Prototyping techniques?

28. Write short notes on the following Rapid Prototyping techniques :

(*a*) Stereolithography.

(*b*) Selective Laser Sintering.

(*c*) Laminated object manufacturing.

(*d*) Optical fabrication.

(*e*) Solid base curing.

(*f*) Fused deposition modelling.

(*g*) 3-D printing.

(*h*) Ballistic particle manufacturing.

(*i*) Photo-chemical machining.

29. Define ERP.

30. How MRP-II and ERP differ from each other?

31. Write the benefits of ERP.

32. For planning the procurement or production of dependent demand items, the technique most suitable is .... (MRP/EOQ)

(GATE 1995) (**Ans :** MRP)

26

PLANT LAYOUT

26.1. INTRODUCTION

Plant layout refers to the arrangement of various plant facilities, such as, Equipment, Machines, Materials, Manpower, etc., and services of the plant within the area of the plant site.

26.2. DEFINITION

Many definitions of Plant Layout have been put forward by various authors. Some of these are listed below :

(*i*) Plant layout is the arrangement of machines within a factory, so that each operation is performed at the point of greatest convenience.

(*ii*) Plant layout is a technique of locating machinery and plant services within the plant site so that the greatest possible output of high quality at the lowest possible total cost can be achieved. A relative arrangement of the various departments of a plant within the selected plant site is determined as also the relative arrangement of machines and equipment within each department.

(*iii*) **Mallick and Gandreau :** Plant layout planning and practice. It is a floor plan for determining and arranging the desired machinery and equipment of a plant, whether established or contemplated, in the best place, to permit the quickest flow of material at the lowest cost and within the least amount of handling in processing the product, from the receipt of raw material to the shipment of finished product.

(*iv*) **Apple : Plant and material handling.** It is planning the path each component part of the product is to follow through the plant, co-ordinating the plant of the various parts so that the manufacturing process may be carried out in the most practical and economical manner, then preparing a scale drawing, or other representation of the arrangement, and finally seeing that the plant is put into effect.

(*v*) **Sansonneti and Mallick : Factory Management.** It is placing the right equipment, coupled with right method, in the right place, to permit the processing of a product unit in the most effective manner, through the shortest possible distance, and in the shortest possible time.

So, in short, a plant layout should be such that it results in greatest co-ordination and efficiency of men, machines and materials in a plant.

Thus, Plant layout is a plan or the act of planning optimum arrangement of industrial facilities, including personnel, operating equipment, and all other supporting services, alongwith the design of the best structure to contain these facilities.

26.3. NECESSITY OF PLANT LAYOUT

The need or necessity for planning a plant layout may arise due to one of the following reasons :

1. Setting up of a new plant.

2. Change in the design of a product. This may result in change in the manufacturing operations or sequence of operations.

3. Expansion of existing department(s).

4. Relocation of existing department(s).

5. Addition of new department(s).

6. Addition of a new product to the existing facilities.

7. Replacement of existing machines/equipment by modern and more efficient machines/equipment.

8. Improving the existing plant layout.

26.4. IMPORTANCE OF PLANT LAYOUT PLANNING

Plant layout is an important aspect of the flow of components and materials throughout the manufacturing cycle. The arrangement of production machinery and material handling equipment should be orderly and efficient. The time and distances required for moving parts and materials should be minimised, and storage areas and service centres should be organised accordingly. Hence a well-planned plant layout is very essential because : Relative location of the various departments within the plant site will affect the inter-departmental material handling costs. Also, the size and complexity of buildings and therefore the first cost of the plant depends upon the plant layout.

A good plant layout :

1. Avoids congestion and underutilisation of plant site.
2. Minimises material/part handling and movement.
3. Minimises work in process (WIP) inventory.
4. Minimises workers movements and hence worker fatigue.
5. Ensures convenience and safety of workers.
6. Improves quality of production.
7. Minimises cost of production.
8. Ensures optimum utilisation of plant facilities (men, machines, materials, etc.)
9. Ensures efficient control over the various production processes.

It is very true that the costs and efforts of careful plant layout planning pay for itself many times over in the form of reduced operating costs and other benefits. Whereas, the costs of improper plant layout planning are incurred daily.

26.5. OBJECTIVES OF PLANT LAYOUT

A good plant layout is one which provides satisfaction to everybody involved in the company or the enterprise, that is, management, shareholders and employees. Thus, the objectives of a good plant layout can be summarised as given below :

It should

1. Provide overall satisfaction to all concerned.
2. Reduce the part and material handling costs.
3. Provide for worker convenience, and better working conditions.
4. Promote job satisfaction.
5. Promote safety.
6. Provide high work-in-process (WIP) turnover.
7. Utilize the plant site effectively, that is, no congestion and underutilisation.
8. Help in effective utilisation of employees, machines and services.
9. Avoid unnecessary investment of capital.
10. Increase operator output and reduce fatigue.
11. Simplify control of production.
12. Assist supervision.
13. Reduce inventory.
14. Reduce manufacturing time.

15. Reduce production delays
16. Increase productivity.
17. Be flexible so that it is easier to expand and diversify.
18. Result in reduced cost of production.

26.6. ADVANTAGES OF GOOD PLANT LAYOUT

Some of the advantages of good plant layout have been written under Art. 26.4. The detailed advantages as summarised by **"Mallick and Gandreau : Plant Layout Planning and Practice"** are given below :

A. To the Worker :

1. Reduces the effort of the worker.
2. Reduces the number of handling.
3. Extends the process of specialisation.
4. Permits working at maximum efficiency.
5. Produces better working conditions by eliminating congestion.
6. Reduces the number of accidents.
7. Provides better employee service facilities.
8. Provides basis for higher earning.

B. In Labour Cost :

1. Increases the output per man hour.
2. Reduces set up time involved.
3. Reduces the number of workers (operators).
4. Reduces the number of handlers.
5. Reduces the length of hauls.
6. Reduces lost motions between operations.
7. Converts operator into a producer instead of a handler by eliminating the various unnecessary motions.

C. In other Manufacturing Costs :

1. Reduces the cost of expense supplies.
2. Decreases maintenance costs.
3. Decreases total replacement costs.
4. Effects a saving in power loads.
5. Decreases spoilage and scrap.
6. Eliminates some of the wastage in raw material consumption.
7. Improves the quality of the product by decreasing handling.
8. Provides better cost control.

D. In the Manufacturing Cycle :

1. Shortens the moves between work-centres.
2. Reduces the manufacturing cycle in each department.
3. Reduces the length of the travel by the product.
4. Reduces the overall time of manufacturing the product.

E. In Production Control :

1. Facilitates receipts, shipments and delivery.
2. Provides adequate and convenient storage facilities.
3. Permits the maximum possible output.
4. Paces production.
5. Determines production flow.

6. Makes production time predictable.
7. Makes scheduling and dispatching automatic.
8. Sets up production centre.
9. Permits straight-line layout by products for mass production.
10. Permits layout by process for job lot manufacturing.
11. Moves work in process by most direct lines.
12. Reduces the number of lost or mishandled parts.
13. Reduces the paper work for production control.
14. Reduces the number of stock chasers.
15. Reduces production control expense.

F. In Supervision :

1. Tends to ease the burden of supervision.
2. Determines the supervisory control.
3. Reduces the cost of supervision.
4. Reduces cost of piece counts.
5. Decreases the amount of inspection.

G. In Capital Investment :

1. Holds permanent investment at a minimum.
2. Adopts plant to present manufacturing methods.
3. Provides for changed methods and future expansion.
4. Keeps the plant from becoming obsolete before it is worn out.
5. Reduces the investment in machinery and equipment by
   (*a*) Increasing the production per machine.
   (*b*) Utilising idle machine time.
   (*c*) Reducing the number of operations per machine.
6. Maintains a proper balance of departments.
7. Eliminates wasted aisle space.
8. Reduces the floor space and shop areas required for manufacturing.
9. Reduces the amount of material handling equipment required.
10. Reduces the eliminates elevator service.
11. Reduces the inventory of work in process and of finished products.

26.7. FACTORS INFLUENCING PLANT LAYOUT

The following are the factors which mainly influence the plant layout :

1. **Type of Industry.** The various industries are generally grouped into the following four categories depending upon the process of manufacturing the product :

(*i*) *Analytical Process.* In this process, the end products are obtained by breaking the input material. The best example is petroleum refineries.

(*ii*) *Synthetic Process.* It is just the opposite of the analytical process. Here, two or more input materials are mixed together to get the end product, for example, production of self-lubricated bearings by Power Metallurgy process, where powders of iron, tin and graphite are mixed together alongwith some lubricant, compacted and sintered and impregnated with oil to get self-lubricated bearings.

(*iii*) *Extraction Process.* The various metallurgical processes, for example, extraction of metals from their ores.

(*iv*) *Conditioning Process.* Here, the form of the raw material is changed to get the end product, for example, Textile mills, Hosiery plants etc.

2. **Type of Product.** The plant layout is also influenced by the type of product : Light or heavy, big or small, solid or liquid and so on.

3. **Volume of Production and its Variety.** Based on these two factors, production can be classified as follows :

(*i*) Jobbing production

(*ii*) Batch production

(*iii*) Mass production

These types of production have been thoroughly discussed in Chapter 23 (Art. 23.2) and Chapter 24 (Art. 24.3).

(*iv*) *One-off production.* One-off production or project construction refers to the production of a single, large and compex product. Examples are : Erection of a Bridge, Building of a Dam, Construction of a Power Plant or Hospital etc.

4. **Process.** The complete process used to convert raw material(s) into end product also greatly influences the planning of the plant layout.

26.8. TYPES OF PLANT LAYOUT

On the basis of type of industry, volume of production and variety of production, we have the following four types of plant layouts :

(*i*) **Product Layout.** This layout is also called flow-line layout, line layout or production line layout. In this layout, the machines, equipment and work centres are arranged in a straight or curved line, in the order in which they have to be used, that is, according to the sequence of operations needed to manufacture a product. To justify the product layout, the product must be standardised and manufactured in large quantities. Hence, this system is best suited for mass production. Examples are : automobile assembly lines, bottling plant and so on. The raw material enters at one end of the line and moves from one machine to another in the line without back-tracking or cross-movements and finally the end product leaves from the other end of the line, Fig. 26.1.

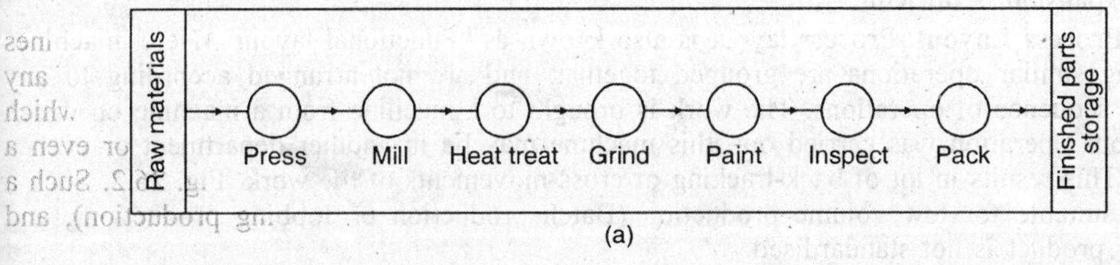

(a)

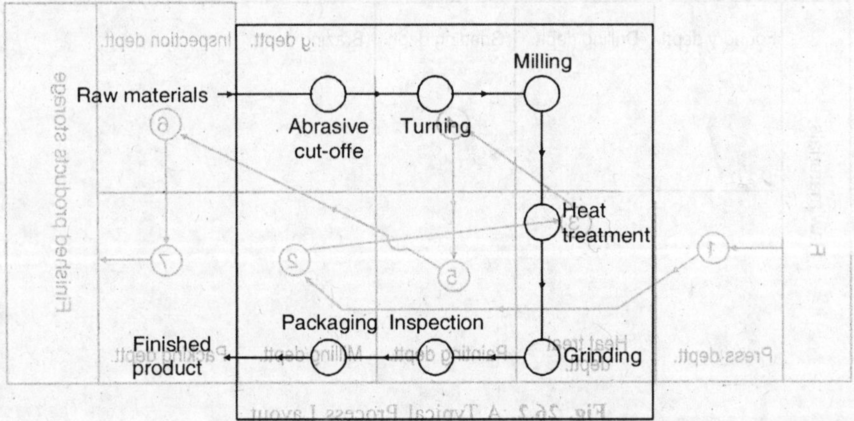

(b)

Fig. 26.1. A Typical Product Layout

Advantages of Product Layout :

1. Reduced total production time.
2. Minimum of handling and transportation resulting in lower total materials handling cost.
3. Less floor area needed per unit of production.
4. There in less work-in-process (WIP), that is, lower stocks.
5. Reduced delays due to the flow of work in the forward direction.
6. Higher productivity.
7. Easy supervision, easy inspection and easier co-ordination.
8. Better utilisation of machines and workers.
9. Greater simplicity of production control leading to lower accounting cost and need for fewer controls and records.
10. Offers greater incentive for groups of workers to raise productivity.
11. **Work simplified.** Due to increased mechanisation and the possibility of breaking work down into simple tasks, less skilled workers can be employed resulting in lower labour costs.

Disadvantages :

1. The layout is fixed, that is, if the product changes, the whole line will have to be rearranged, that is, lower flexibility.
2. The break down of a single machine in the line leads to shut-down of the whole production line.
3. The manufacturing cost rises if the volume of production falls.
4. If one or more lines are running light, there is considerable machine idleness.
5. Line balancing is important, so that the material being processed, does not have to wait for the next operation.
6. High capital investment.
7. Expansion is difficult.

(*ii*) **Process Layout.** Process layout is also known as "Functional layout". Here, machines performing similar operations are grouped together, and are not arranged according to any particular sequence of operations. The work is brought to a machine from a machine on which the previous operation was carried out, this machine may be in another department or even a building. This results in lot of back-tracking or cross-movements of the work, Fig. 26.2. Such a layout is suitable for low volume production (Batch production or Jobbing production), and where the product is not standardised.

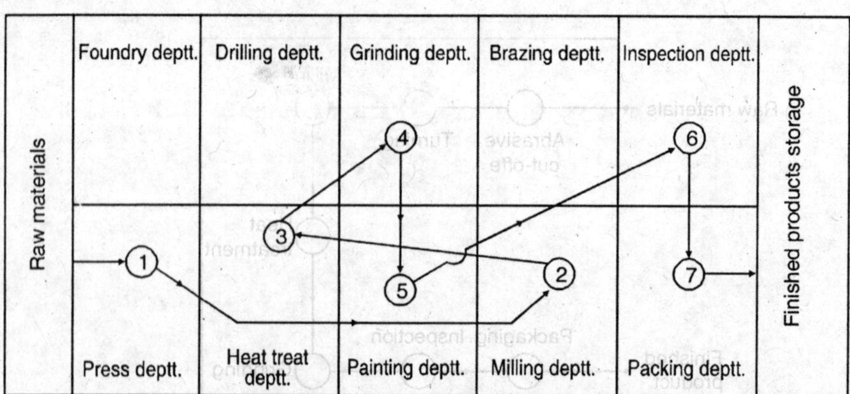

Fig. 26.2. A Typical Process Layout

Advantages of Process Layout :

1. Greater flexibility of production. Change in product design can be easily accommodated.
2. Lower initial investment in machinery because of less duplication of equipment.

3. Break down of one machine will not shut-down the production as the work of that machine can be transferred to another machine or worker.

4. If offers greater incentive to individual workers to raise level of performance.

5. If offers better and more efficient supervision through specialisation as similar jobs are produced on similar machines.

6. Better control of complicated or precision processes, especially where much inspection is needed.

7. Better utilisation of high production equipment.

8. Ease in maintenance

Disadvantages :

1. Generally, more floor space is required.

2. More handling costs because of back-tracking and cross-movements of work, resulting in chaotic material movement.

3. More labour cost, since the workers have to be skilled to work or general purpose machines.

4. Specialisation creates monotony.

5. Inspection is costlier.

6. Longer production cycles.

7. High in-process inventory.

8. Damage to material and equipment.

9. Loss of materials.

10. Higher accident rates.

Note. Also read Art. 24.3.3.

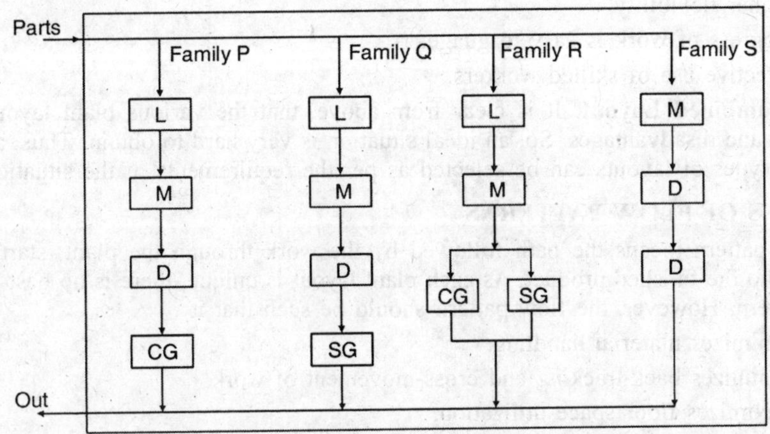

Fig. 26.3. G.T. Layout

(*iii*) **Group Technology Layout.** The principle of "Group Technology (G.T.)" which has been mainly applied to machined components, has been explained in Art. 24.3.3. Here, similar parts or those requiring similar sequence of operations are grouped together in a family. All the equipment needed to manufacture a family of parts is grouped into a "Cell". That is why, this layout is also called "Cellular layout". It is used for batch production and overcomes the drawbacks of conventional batch production by Process Layout.

The parts are transferred from one unit to another with minimum movement and wastage of time. A number of line layouts are established, one for each family of G.T. parts, Fig. 26.3.

Note. The models for product and process layouts are distinctly different : Product models generally focus on minimising idle labour time through line-balancing techniques and process models generally minimise load-distance moved relationships.

(*iv*) **Fixed-position Layout.** This type of layout is used for products which are very massive (in weight or size or both). It is very inconvenient or is not desirable to move them. Here, the work remains in a fixed position. The machines, materials, equipment and workers needed for production are brought to the site of work, Fig. 26.4. The examples are : Shipbuilding industry, Air-craft production etc., and also in assembly operation industries.

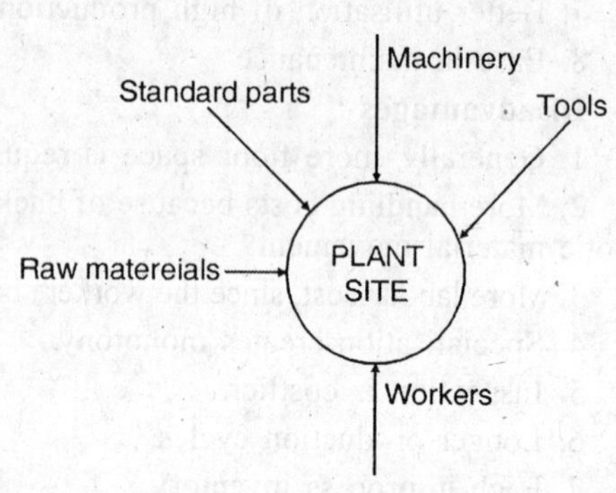

Fig. 26.4. Fixed-position Layout

Advantages :

1. Capital investment in plant layout is minimum.

2. Greater flexibility, allowing the change in

(*a*) Product design.

(*b*) Product mix.

(*c*) Production volume of demand.

3. Because, a worker does more than one type of job, his skills get enhanced which is known as "Jobs enlargement".

4. The worker begins to take some pride as he identifies himself with the product.

5. Higher flexibility.

6. Progress of work is easy to guage.

7. Effective use of skilled workers.

(*v*) **Combined Layout**. It is clear from above, that the various plant layouts have their advantages and disadvantages. So, an ideal situation is very hard to obtain. Thus, a combination of various types of layouts can be selected as per the requirements of the situation.

26.9 TYPES OF FLOW PATTERNS

Flow pattern means the path followed by the work through the plant, starting from raw material upto the finished product. As each plant layout is unique, there is no best way to layout a flow pattern. However, the flow pattern should be such that it :

1. Minimizes material handling.
2. Minimizes back-tracking and cross-movement of work.
3. Maximizes floor space utilization.
4. Results in most flexible layout.

We have the following six basic flow patterns :

(*i*) **Straight Line.** Most suited for product layout, Fig. 26.5 (*a*).

(*ii*) ***L*-Flow.** Similar to straight line (I-flow), but economizes on floor space, Fig. 26.5 (*b*).

(*iii*) ***U*-Shaped : Fig 26.5 (*c*).** The raw materials enter and the finished products leave the plant at the same end. This pattern results in better space utilization and easier inspection.

(*iv*) ***S*-Flow.** Still better space utilization and simpler inspection. However, the material enters the plant at one end and the finished products leave the plant from the other end, Fig. 26.5 (*d*). Used when production line is very long.

(*v*) **Circular or O-Flow.** Fig. 26.5 (*e*). Easier supervision and minimum back-tracking of work. This flow is suited where the operations are carried out on rotary tables or where the starting point and the finishing point are the same, for example, mechanised foundry.

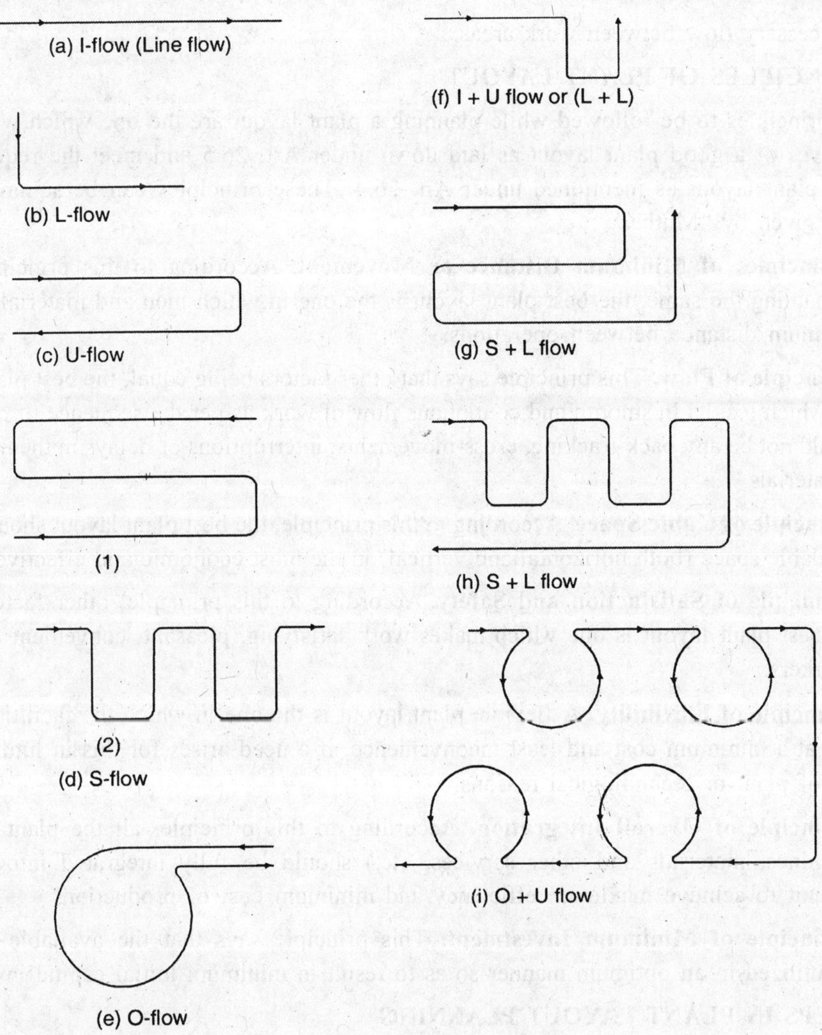

Fig. 26.5. Basic Horizontal Flow Patterns

(*vi*) **Convoluted.** Fig. 26.5 (*d*, 2). This pattern has the same characteristics as the S-flow pattern.

Factors affecting Flow Pattern. Above, the main types of flow patterns have been discussed. We have also mentioned the conditions to be satisfied by an ideal flow pattern. However, the final choice will be affected by the following factors :

1. External transportation facilities.
2. Number of parts to be handled.
3. Number of operations on each part.
4. Number of sub-assemblies made up ahead of assembly line.
5. Number of units to be processed.

6. Amount and shape of space available.

7. Necessary flow between work areas.

26.10 PRINCIPLES OF PLANT LAYOUT

The principles to be followed while planning a plant layout are the one which will satisfy the objectives of a good plant layout as laid down under Art. 26.5 and meet the requirements of a good plant layout as mentioned under Art. 26.4. These principles can be summarized as follows as given by "Muther" :

1. **Principles of Minimum Distance or Movement.** According to this principle, other factors remaining the same, the best plant layout is the one in which men and materials have to move minimum distance between operations.

2. **Principle of Flow.** This principle says that other factors being equal, the best plant layout is the one which results in smooth and continuous flow of work as per the sequence of operation. There should not be any back-tracking, cross-movements, interruptions or delays in the movement of work/materials.

3. **Principle of Cubic Space.** According to this principle, the best plant layout should utilize all the available space (both horizontal and vertical) in the most economic and effective manner.

4. **Principle of Satisfaction and Safety.** According to this principle, other factors being equal, the best plant layout is one which makes work satisfying, pleasant, convenient and safer for the workers.

5. **Principle of Flexibility.** A flexible plant layout is the one in which the facilities can be rearranged at a minimum cost and least inconvenience, if a need arises for this in future due to expansion of plant or technological reasons.

6. **Principle of Overall Integration.** According to this principle, all the plant facilities (men, machines, materials and other services, etc.) should be fully integrated into a single operating unit to achieve maximum efficiency and minimum cost of production.

7. **Principle of Minimum Investment.** This principle says that the available facilities should be utilized in an optimum manner so as to result in minimum initial capital investment.

26.11. STEPS IN PLANT LAYOUT PLANNING

Guiding Fundamentals :

1. Plan the whole and then the details.
2. Plan the ideal and then the practical.
3. Follow the cycle of layout development and overlap the phases.
4. Plan the process and machinery around the material requirement.
5. Plan the layout around the process and the machinery/equipment.
6. Plan the building around the layout.
7. Plan with the aid of clear visualization.
8. Plan with the help of others.
9. Check the layout.
10. Sell the layout.

Plant layout planning is a specialized task and it should be carried out systematically. Before the planning of the plant layout is undertaken, the following steps of the Design Phase in the life cycle of a manufacturing system must be completed so as to provide the basic data needed for planning the plant layout. Life cycle of a manufacturing system consists of the activities as shown in Fig. 26.6.

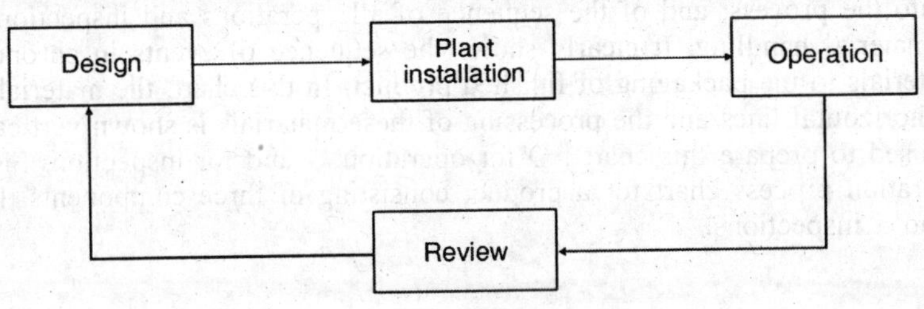

Fig. 26.6. Life Cycle of a Manufacturing System

(*a*) Detailed design of products and services proposed to be manufactured in the plant. This will provide the data for products and services to be produced.

(*b*) Maximum rate of production of goods and services for which facilities have to be provided in the plant, are calculated, that is, production volume is determined.

(*c*) **Overall Process Planning.** Under this phase :

(*i*) the optimum sequence of production operations, including parts manufacture, assembly and service operations needed for each different product or service, are determined :

(*ii*) the different production machines and equipment are selected.

(*iii*) material handling equipment is selected.

(*iv*) best sequence of production operations is selected.

(*v*) line balancing is decided (for mass production layout).

(*vi*) Make or buy decision is made.

(*vii*) Bill of materials is prepared.

(*d*) **Detailed Process Planning.** During this step, the optimum method for carrying out each operation is determined alongwith the corresponding tools, workplace layout and process variables.

(*e*) **Organizational Planning.** During this step, the plant facilities and activities are organized into departments and jobs. Functions of each department are determined. The content, wages and incentives of each job are also decided.

(*f*) The site for the proposed manufacturing plant is selected. This will also determine the topology of the site.

With the above information available, the planning of the plant layout is undertaken. The following are the major steps in the planning of a plant layout :

1. **Analyse Manufacturing Requirements.** For this, the following basic data is available from the information obtained above :

(*i*) Complete list of parts to be manufactured and assembled.

(*ii*) Their drawings.

(*iii*) Operation sequence and method of manufacture of each part.

(*iv*) Estimated process time for each operation.

All this data is available in the form of "operation planning sheet" prepared by a Process Planning Engineer (Chapter 6) under the stage of Process Planning of Design Cycle.

To help visualize manufacturing and assembly operations, charts should be prepared. An operation process chart is most suitable for this purpose, since at this preliminary stage, a more detailed analysis of work flow is not desirable.

Operation Process Chart. It is a graphic representation of the points at which materials are introduced into the process, and of the sequence of all operations and inspection except those involved in material handling. It clearly shows the sequence of events in chronological order from raw materials to the packaging of finished product. In this chart, the materials entering are indicated by horizontal lines and the processing of these materials is shown vertically. Only two symbols are used to prepare this chart : O for operation □ and for inspection. Fig. 26.7 shows a typical operation process chart for a product consisting of three components. It involves 17 operations and 5 inspections.

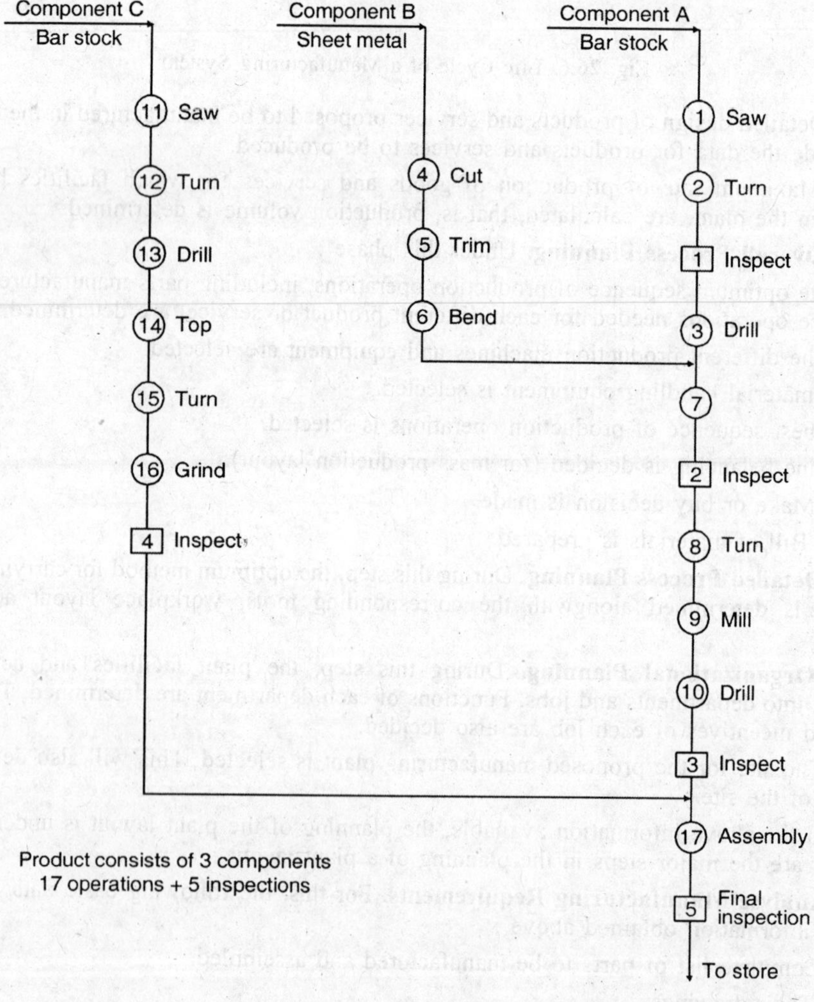

Fig. 26.7. A Typical Operation Process Chart

2. **Determination of Space Requirements.** The total space required for various departments and supporting services is determined as given below :

(*i*) *Production Departments.* To determine the space required for each production department, the following data is already available :

(*a*) Activities to be carried out in each department (from organizational planning).

(*b*) What operations are to be performed on which machines and in which department (Process planning).

(*c*) Standard times for individual operations are then calculated as explained in Chapter 4, and listed in operation planning sheets (Chapter 6).

Then, the required number of machines of each type are calculated as under :

Let annual output of parts = N

Expected scrap = S%

No. of working hours/year = A

$$\text{Output required for } N \text{ good parts} = N \times \frac{100 + S}{100}$$

$$\therefore \quad \text{Parts required per hour} = N \times \frac{100 + S}{100} \times \frac{1}{A}$$

Let estimated time per part = t, seconds

$$\text{Output from one machine per hour} = \frac{3600}{t} \times \frac{\eta}{100}, \text{ parts}$$

where η = production efficiency of the machine.

$$\text{Machines required} = \frac{\text{Parts required per hour}}{\text{Output per machine per hour}}$$

Example 26.1. *Let N = 1,00,000 parts; S = 2%; t = 105 seconds; η = 80%; A = 2300 hours.*

Solution. $\text{Parts required per hour} = 100{,}000 \times \frac{102}{100} \times \frac{1}{2300}$

$= 44.4$

$$\text{Now output from one machine per hour} = \frac{3600}{105} \times \frac{80}{100} = 27.4 \text{ parts}$$

$$\text{No. of machines required} = \frac{44.4}{27.4} = 1.62$$

Hence, if the machine is to be used exclusively for the part considered, two machines would be required.

In this manner, the total number of machines required in a given department can be calculated, since we know the different operations which are to be carried out in that department.

Space needed for each machine can then be determined with the help of "Machine Data Cards" and the requirements of other equipment needed. Machine data cards give information about capacity, foundations, space needed, handling devices etc.

The other equipment needed include : material handling equipment, work racks, tool racks, work benches (fitting tables, surface plates, etc.). The space needed should also include the space for : worker, clearance space for work, stairways, gangways etc. All this will help in designing the work station. "Work Station" or "Work Place" is defined as the floor space occupied by the machine or machines and the worker. In this manner, the space needed for each department is determined. Extra space for future expansion is also set apart for each department.

(*ii*) **Space for utilities.** The common utilities for a production plant include : Power house, electrical sub-station, pump house, fire-fighting station, compressed air plant, heating and cooling plant, telephone exchange, vehicle storage, maintenance shops etc.

(*iii*) **Space for production related activities.** These activities include : Engineering department, Quality control, Production planning and control and stores.

Stores include : Stores for raw-materials, Parts and consumables, Tools and gauges, Component and sub-assembly, Spares warehouse, Finished products warehouse.

(*iv*) **Space for Administration Offices.** These include : offices of chairman, Board of directors, and M.D. (or C.E.O. or President), a conference room, departments of accounts, sales, purchase, *R* and *D*, personnel, reception, a data processing centre etc.

(*v*) **Space for Personal Services.** These include : sick bay, workers canteen, officers canteen, drinking water, public telephone booths, toilets, parking areas, time clocks, locker rooms etc.

3. **Choice of Plant Layout.** The various types of plant layouts have been discussed under Art. 26.8 In most of the cases, a choice has to be between a product layout or a process layout or to use a combination of the two. As already discussed, in product layout (which is used for mass production), the manufacturing equipment is arranged in the same sequence as the operations performed on the work. Each operation should be capable of processing work at the rate required for assembly and so far as possible the output for each operation should be balanced so that all the parts match in step to final assembly (line balancing).

In process layout (associated with jobbing and batch production), the equipment will be grouped together so that all machines of a similar type are in the same section of a company.

4. **Determination of Flow of Materials.** The smooth flow of work into, through and out of the factory depends, to a large extent, on the type of production and on the success of the layout chosen. In flow production (product layout), the work movement will be predetermined, with the equipment arranged in operation sequence and the components and subassemblies flowing into the main assembly line at the correct position. The type of flow pattern adopted will depend upon the space available (Art. 26.9).

For process layout, it is possible to establish only a general direction of work movement, due to the difference in operation sequence of different parts of the product. A work movement diagram can be drawn for each part by superimposing the path of movement of each part on a floor plan of the factory. But with so many parts being manufactured, the total picture will become very confusing. A better way is to take the help of Travel Charts.

Travel Charts. Travel charts which are also known as "Cross-Chart" or "From-To-Chart", can be used to represent the frequency of movement between departments in concise form. By analysis of travel

From \ To	A	B	C	D	E	Total
A						
B	3		1	3	4	11
C	—	5		4	1	10
D	4	2	4		1	11
E	—	—	2	1		3
Total	7	7	7	8	6	35

Fig. 26.8. A Typical Travel Chart

charts, it is possible to improve upon the first tentative positioning of departments and achieve a better arrangement. Fig. 26.8 shows a Typical Travel Chart. It is in the form of a matrix. Back tracking is shown below the main diagonal and the forward movement above the main diagonal. Back tracking is considered undesirable, so it should be eliminated if possible, minimized if not. The departments are arranged in different ways with respect to another, until the above aim is achieved. The number of forward movements and back tracking are indicated in squares.

The desired flow of materials between the various operations and departments in the plant can be depicted with the help of "Work-Flow Diagram" and "Flow Process Chart".

Work-Flow Diagram. It is the plan view of the work-place under study, to a certain scale. A line diagram indicates the path followed by the object under study. The object may be materials, men or equipment. In the case of work/material movement, it represents sequence of operations. The flow diagram provides a compact overall view of an existing or proposed process and is useful in making improvements. A typical flow diagram is shown in Fig. 26.9. Symbols used are :

○ : Operation; □ : inspection; ⇨ : transport;

D : delay; ▽ : storage;

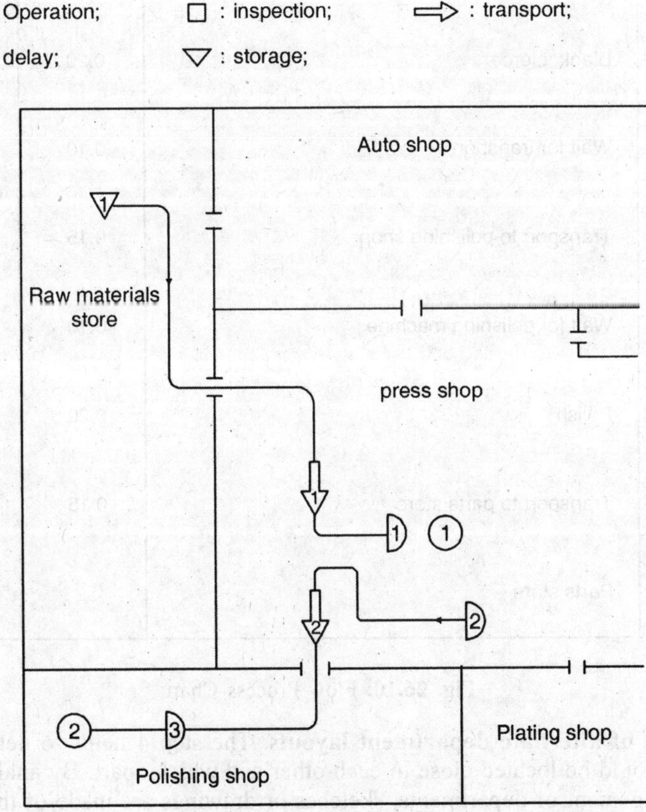

Fig. 26.9. Work-Flow Diagram

Flow-Process Chart. A flow process chart is more useful than a work-flow diagram as it incorporates more details than a flow diagram, such as operation times, distances travelled, transportation and so on. It has been defined as follows : Flow process chart is a graphic representation of all operations, transportation, delays and storage occurring during a process of procedure and includes information considered desirable such as distance moved and time required. A systematic study and analysis of these charts can result in improvements in : methods of manufacturing, sequence of operations, plant layout etc. Fig. 26.10 shows a typical flow process chart.

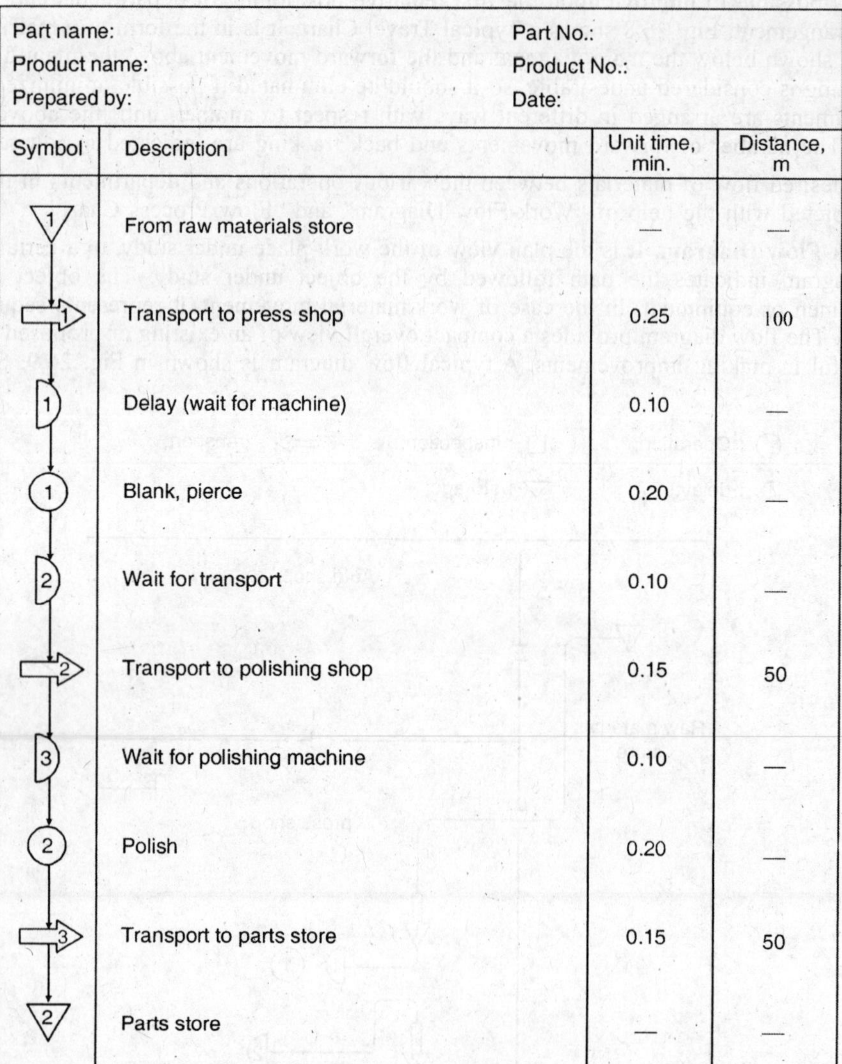

Part name: Part No.:
Product name: Product No.:
Prepared by: Date:

Symbol	Description	Unit time, min.	Distance, m
▽ 1	From raw materials store	—	—
⇨ 1	Transport to press shop	0.25	100
D 1	Delay (wait for machine)	0.10	—
○ 1	Blank, pierce	0.20	—
D 2	Wait for transport	0.10	—
⇨ 2	Transport to polishing shop	0.15	50
D 3	Wait for polishing machine	0.10	—
○ 2	Polish	0.20	—
⇨ 3	Transport to parts store	0.15	50
▽ 2	Parts store	—	—

Fig. 26.10. Flow Process Chart.

5. **Synthesis of alternate department layouts.** The step 4 helps to determine which pair of departments should be located close to each other and which apart. By analysing travel charts for different arrangement of departments, sketches or drawings are made of the possible relative arrangement of the various departments. In this manner, we get a number of rough layouts of departments considering the shape and size of each department. These rough layouts are modified after taking into account the topology of the plant site. For this exercise, it is very convenient to prepare templates. "Templates" are the scaled wooden or plastic strips showing the floor space occupied by each department. These templates are placed over a floor plan of the plant site according to the rough layouts. These are then arranged and rearranged manually. until a satisfactory layout is obtained. In this manner, we get a number of satisfactory layouts corresponding to the number of rough layouts.

6. **Planning stores and service areas.** The size and position of the various store areas are determined.

(*i*) Normally, raw materials stores are located at one end of the company and the finished products warehouse at the other, with a component store between the manufacturing and assembly areas.

(*ii*) Their sizes will depend upon the type of production, with mass production, the stores need keep a few days' stock only, but with batch production, the inventory may be equal to an average stock of several weeks or even months.

(*iii*) Planning is done about the tool and process materials stores. In the case of large factories, they are usually sub-divided so as to be near to the production areas they serve.

(*iv*) The sizes of the various offices directly connected with the factory should be worked out with the departmental managers concerned, and they should be sited as near as possible to the factory activity they are intended to control.

(*v*) Commercial offices and the personnel department should located be near to the factory entrance so that the general public is kept away from the production departments.

7. **Choice of the best overall layout.** The satisfactory layouts obtained under step 5 will have their own plus points and minus points. To select the best out of these layouts, the following ways can be followed :

(*a*) *Annual cost method.* As discussed in Chapter 5 under Art. 5.2.2, the alternative which has the least equivalent annual cost is chosen (based on initial investment and operating expenses).

(*b*) *Comparison of materials handling costs.* The total materials handling cost for moving the material from one department to another, can be found out by the following relation :

$$C_T = \sum_{i=1}^{n} \sum_{j=i+1}^{n} C_{ij} \, D_{ij}$$

where, $C_{ij} = W_{ij} \times C_{ij}^u$

W_{ij} = number of loads to be moved between departments i and j per unit time

C_{ij}^u = cost of moving one load per unit distance between departments i and j

D_{ij} = distance of one move between departments i and j

and n = number of departments.

The alternative which results in the minimum total materials handling cost is selected.

8. **Detailed factory layout planning.** The next and the last step is to plan the detailed layout of each department. This means that individual items of equipment have to be located and the work stations planned for each department. Space has to be allowed for : operators, gangways, machines, feeding work to the machine, that is, for conveyor stations, work racks, pallet stands etc.

This is done by considering each department separately and determining the best arrangement for machines, equipment and facilities within the department. Steps 4, 5 and 7 of the overall plant layout can be used for the detailed layout planning of each department. Work-flow diagram (Fig. 26.9) and flow process chart (Fig. 26.10) can be very useful for this purpose. The building plan of the department may be viewed as the plant 'site' and each machine, equipment or facility as a 'department'.

Use of Templates and Scale Models

The use of templates in the layout of the various departments on the plant site has been discussed above under step 5. In planning the layout of individual departments, the templates will show the floor space of the department occupied by individual pieces of equipment. These templates are placed over a floor plan of a department and then arranged and rearranged until the most satisfactory layout is obtained. The floor plan of the department is mounted on a board and the templates are secured to the plan by mapping pins.

Use of scale models is also useful particularly for large projects. As they are 3-D models, they create a more realistic impression. Also, they are useful for checking clearances, particularly where over-head cranes are used.

When the detailed plans of all the departments have been decided, the overall plant layout is finalised by adding to it roads, entry gates, Lawns, boundary walls, etc. Finally a layout drawing is made and copies of it issued for action.

Summary of the procedure of Plant Layout :

1. Collect data.
2. Analyse the data.
3. Determine the machinery/equipment required.
4. Select the material handling equipment.
5. Draw the plan of site.
6. Determine the general flow of materials.
7. Plan the overall layout.
8. Design the individual work station.
9. Fit individual layouts into overall layout.
10. Determine the storage space needed.
11. Make floor diagrams.
12. Plan and locate service areas.
13. Make Master layout.
14. Visualise with the help of visualisation aids.
15. Check the final layout.
16. Present and get approval.
17. Install the approved layout.
18. Follow up.

26.12. VISUALISATION AIDS OF THE LAYOUT ENGINEER

It is clear from Art. 26.11, that the various tools and techniques used by a layout engineer for planning the plant layout are :

1. Operation process chart.
2. Travel chart.
3. Work-flow diagram.
4. Flow process chart.
5. Templates.
6. Scale models.
7. Machine data cards.

All these tools have been discussed in Art. 26.11.

26.13. SINGLE STOREY VERSUS MULTI-STOREY BUILDING

Both single storey construction and multi-storey construction have their own advantages as given below :

Advantages of single-storey buildings :

1. Detailed layout planning is simplified.

2. Provides greater flexibility in affecting layout changes.

3. Materials handling is easy and its cost is very low.

4. Vibration effect are minimum. So, there is a greater floor bearing capacity and heavy equipment can be placed anywhere on the floor area.

5. No special treatment for foundations is needed, so construction costs are lower.

6. Provides better use of natural light and ventilation.

7. Overall operating costs are lower.

8. Better and effective supervision is provided.

9. Less risk of damage due to fire and accident.

10. Future plant expansion can be obtained by constructing upper floors.

Advantages of multi-storey building :

1. More efficient use of land and space, particularly, in situations where land available is limited and costly.

2. More compact layout.

3. Cheaper to heat because heat loss/gain through the building is smaller due to less floor and roof area.

4. Upper storeys are free from noise and dust etc.

5. More fire-resistant because, generally, R.C.C. construction is used.

6. Lower cost of construction per unit of floor area.

Hence, depending upon the situation and taking into account the respective advantages of single-storey and multi-storey building, a final decision can be taken. However, in general and in practice, the production departments are single-storeyed and office buildings are multi-storeyed.

26.14. COMPUTERIZED TECHNIQUES FOR PLANNING A PLANT LAYOUT

Many computer programmes have been developed to help the plant layout engineer in planning the best possible plant layout. Some of the widely used softwares (as per Apple : "Plant Layout and Materials Handling") are given below :

1. **CORELAP.** It means "Computerized Relationship Layout Planning". According to this technique, the most related departments are located close to one another. Then other facilities are added progressively on the basis of relative closeness desired.

2. **CRAFT.** (Computerized Relative Allocation of Facilities Technique). This technique determines the desirable relative arrangement of production departments so as to minimise the total materials handling cost. The input data is : inter-departmental flow per unit time, unit load, material handling cost per unit distance, initial block programmes.

Then, when the total materials handling cost has been minimized and no further improvement is possible, the computer provides print outs of block-type layout of the location of each department and facility.

3. **ALDEP.** (Automated Layout Design Program). In this technique, the first activity/facility is randomly selected and located. Then, the subsequent activities/facilities, in desired size, are selected and located as per relative closeness desired, or at random if no particular inter-relationships are desired. Alternate layouts are generated and compared.

4. **PLANET.** (Plant Layout and Evaluation Technique). Here, the "penalty" cost associated with separating departments are computed on the basis of inter-departmental flow data. Three

heuristic algorithms are used for generating alternative configurations to be manually evaluated and adjusted.

5. **COFAD.** (Computerized Facilities Design). This technique combines CRAFT-based layouts with materials handling selection based on various criteria.

6. **CAN-Q.** (Computer Analysis of Network of Queues). This technique is used to analyse the work flow in a production system. It is quite easy to use, very efficient and versatile.

26.15. EFFECT OF "PUSH" OR "PULL" TYPE OF REPETITIVE MANUFACTURING SYSTEM ON PLANT LAYOUT

We know that repetitive manufacturing is used for mass production of units of one product or of several models of one basic product, for example, automobiles, appliances and toys etc. We have already discussed that it is a "Flow-line" system of manufacturing. Plant layout is greatly affected by the manufacturing approach (Push or Pull) adopted.

The traditional approach is the "Push system". Here, a predetermined production schedule is established from anticipated demand for the product. Once the schedule is set in motion, the raw materials, fabricated parts and sub-assemblies are continuously pushed along the flow line until the final assembly occurs. Since the schedule is based on anticipated demand, we need storage for purchased materials, storage for finished products with high inventories between stages along the flow line (cushion inventories). In this system, also known as "Assembly-to-schedule", a relatively fixed assembly line using dedicated single purpose machines with high output capabilities will be used. Materials handling equipment shuttles materials and parts to work centres from supplying departments or storage depots when scheduled by the materials control staff. The workers perform their specialized tasks repetitively. Long production runs avoid expensive set up and changeover costs.

In the "Pull system" (very popular in Japan), also known as "Assembly-to-order", it is recognised that the actual demand will vary from what was anticipated. The upstream activities (sub-assemblies, fabrication of parts, purchasing of materials) are geared to match the final assembly needs. The production is adjusted to meet the variations in the final demand, that is, the sub-assemblies, fabricated parts etc. are pulled along the flow line to meet the actual end item demand, instead of continuously pushing these along the flow if components are not needed now, do not produce them now or ahead of the time. When they are needed, be prepared to create them rapidly in required quantity. This is JIT approach or Stockless production. A very close co-ordination is must between the work centres and the system has to be very efficient. The machines needed are less expensive, smaller and adaptable. Intricate tools and attachments permit rapid equipment changeovers as needed. The work centres are much closer to one another. This gives station-to-station visibility work progress and it eliminates in-process inventories, inventory storage areas, inventory conveying systems and materials staff.

It is clear from the above discussion, that, in the "Push system" a straight-line layout will be suitable, whereas for the "Pull system" where the emphasis is on simplicity and flexibility, a U-shaped flow line is preferred.

PROBLEMS

1. Define plant layout.
2. What is the necessity of plant layout ?
3. Discuss the importance of plant layout planning.
4. Enumerate the objectives of plant layout.
5. Discuss in details the advantages of a good plant layout.
6. Discuss the various factors influencing plant layout.
7. Distinguish between project, jobbing, batch and mass production.

8. Discuss the four types of plant layout, namely,

   (*i*) Product layout

   (*ii*) Process layout

   (*iii*) Group technology layout

   (*iv*) Fixed position layout

9. Discuss the advantages and disadvantages of the four types of plant layouts.
10. Give examples of plants where the four types of plant layouts are used.
11. What is a flow pattern ? Discuss the various flow patterns used in plant layouts.
12. Discuss the principles of plant layout.
13. Describe the various steps in plant layout planning.
14. List the major activities for which the space has to be provided in a production plant.
15. Why are stores needed in a production plant ? Where they should be located within the plant ?
16. Compare the advantages and disadvantages of single-storey and multi-storey building.
17. How the required number of machines of each type are determined ?
18. What is a travel chart ? Explain it.
19. List the tools of a plant layout engineer.
20. Discuss the use of the various plant layout tools for planning the plant layout.
21. Write briefly on :

   (*i*) Work flow diagram.

   (*ii*) Flow-process chart.

   (*iii*) Templates.

   (*iv*) Scale models.

   (*v*) Machine data cards.

   (*vi*) Operation process chart.

22. Discuss the various techniques for computerized planning of plant layout.
23. Fill in the blanks :-

   (a) ———— type of plant layout is also known as flow-line layout :-

   (*i*) Product layout (*ii*) Process layout

   (*iii*) Fixed position layout (*iv*) Combined layout

   (b) In ----------- type of layout, the equipment is arranged in order in which they have to be used.

   (*i*) Product (*ii*) Process

   (*iii*) Fixed Position (*iv*) Combined.

   (c) In --------------- type of layout, the machines are arranged in order of their functions :-

   (*i*) Product (*ii*) Process

   (*iii*) Fixed (*iv*) combined

24. Which type of layout has more material handling :-

   (*i*) Product (*ii*) Process

   (*iii*) Fixed (*iv*) Combined

25. True or False

   (a) Product layout is best suited for Mass production

   (b) Process layout is used for shipbuilding industry

   (c) Process layout has more material handling as compared to Product layout.

27

PRODUCTION AND PRODUCTIVITY

27.1. DEFINITION

27.1.1 Production. The word "Production" is often used interchangeably with the word "Manufacturing". Whereas the term "Manufacturing Engineering" is widely used in U.S.A., the equipment term "Production Engineering" is prevalent in Europe and Japan. The word "Manufacturing" is derived from the Latin words 'manus' and 'factus' meaning hand and made respectively, that is, the literal meaning is 'made by hand'. However, in the modern sense, the word 'production' or 'manufacturing' means making of goods or articles from raw materials by hand and/or machinery by following a well-defined plan for each activity required. Thus "Manufacturing Engineering" or "Production Engineering" may be defined as the art of making components and machines of specified quality on the planned production scale with the minimum consumption of materials and maximum productivity of labour. The word 'Product' means something that is produced. Production does not mean the making of only goods but in general it includes services also. Quantitatively, the word 'production' means quantity produced of goods or services, that is, volume of production.

27.1.2 Productivity. For any manufacturing enterprise, the input resources are : men, money, materials, machinery, time, energy and space etc. The output is : the manufactured goods and/or services. The main aim of the management is to earn profits. So, the company endeavours to produce the goods at the minimum possible competition. Hence, the management is always concerned with obtaining the most productive utilization of input resources. The degree of effectiveness of utilization is often measured by the familiar equation :

$$\eta = \frac{\text{Out put}}{\text{Input}}$$

In manufacturing and production systems, we use the term "Productivity" in place of efficiency. Thus,

Productivity can be defined as the ratio between the output of goods or services and the input of resources. That is,

$$\text{Productivity Index} = \frac{\text{Value or quantity of goods and / or Services produced}}{\text{Value or quantity of given input resources}}$$

Thus productivity refers to a comparison between the quantity of goods or services produced and the quantity of resources employed in turning out these goods or services. When the same resources that were employed in the past now produce more than they did before, it means that productivity has increased. Hence, to establish whether and how much productivity is changing, take into account the relationship between what comes "out" of production and what goes "in" to production, that is, the ratio between output and input. So,

$$\text{Productivity Index} = \frac{\text{Output}}{\text{Input}}$$

Thus higher productivity means producing more with the given input resources or producing a given quantity with lesser input resources. Thus productivity basically measures operating efficiency.

It follows from the definition that higher the productivity, lesser will be the cost of production for a given amount of input resources. Thus, there is reciprocal relationship between productivity and cost of manufacturing. For getting higher productivity, the wastage of input resources must be eliminated if possible or minimized, if not.

So, productivity may be defined as the optimum utilization of all the resources of organization : men, money, materials, machinery, energy, space and technology, etc. Output per employee per hour, in all phases must be maximized. The manufacturing enterprise must constantly strive for higher productivity. Economy and efficiency in every aspect of the production process is the key to higher productivity. Two of the major resources that should be optimally planned are : human resources and equipment.

27.2. DIFFERENCE BETWEEN PRODUCTION AND PRODUCTIVITY

The two terms are entirely different from each other and this difference should be clearly understood. Whereas production is an absolute concept, productivity is a relative concept. Production reflects the volume of production of goods/services by an enterprise, whereas productivity reflects the operating efficiency of the enterprise. While defining production, no reference is made to the quantity or quality of input resources. But, once the concept of efficiency with which the input resources are utilized, is incorporated, we enter the field of productivity.

Increased productivity should not be confused with increased production. Production can be increased by increasing the input resources. But productivity may remain the same, or decline due to inefficient use of inputs. Similarly, productivity can be improved by the economic and efficient use of input resources, but the production may not increase.

27.3. IMPORTANCE OF PRODUCTIVITY

The prosperity of any society or country is directly dependent upon the productivity of every activity of economic development. Higher the productivity, greater will be the prosperity and vice-versa. Every country has limited input resources. So, in a world of competition, the only golden rule to survive is higher production and higher productivity with the given input resources. Hence, the given input resources must be utilized efficiently and economically.

Higher productivity leads to many benefits :

1. Cost of production decreases which makes the goods available to the public, cheaper.
2. Lower cost of production and increased production increase profits of a company.
3. Alleviation of poverty.
4. Brings prosperity to a society/nation due to economic growth and social progress.
5. Increased productivity in a growing economy will ultimately help in increasing employment by stimulating the development of industry.
6. Employees can get better wages and improved working conditions.
7. Increased productivity at the national level would mean that our meagre foreign exchange reserves could be better conserved and the targets set in the plans could be achieved at a lower cost.
8. Our exporting companies would be in a better position to compete in foreign markets and earn more foreign exchange.

27.4. MEASUREMENT OF PRODUCTIVITY

Productivity has been defined as,

$$\text{Productivity} = \frac{\text{Output}}{\text{Input}}$$

Now, output can be measured in physical terms, such as, volume of production or quantity of goods or services produced. But, there is a problem of measuring input in physical terms. The input resources are varied and are not inter-convertible. So, to compare output and input, the better way is to convert both into money terms. Therefore, the more appropriate definition of productivity, as far as its measurement is concerned, is :

$$\text{Productivity} = \frac{\text{Value of goods or Services produced}}{\text{Value of input resources}}$$

There are two types of "Productivity index" when we deal with the measurement of productivity :

1. Partial Productivity
2. Total Productivity

In partial productivity, we consider the effect of only one input resource at a time. So, we have : labour productivity, material productivity, machine productivity, energy productivity and so on. Thus,

$$\text{Labour Productivity} = \frac{\text{Value of Total output}}{\text{Value of Total man - hours worked}}$$

$$\text{Machine Productivity} = \frac{\text{Value of Total output}}{\text{Value of Total machine - hours worked}}$$

$$\text{Material Productivity} = \frac{\text{Value of Total output}}{\text{Value of Total material input}}$$

$$\text{Energy Productivity} = \frac{\text{Value of Total output}}{\text{Value of Total energy input}}$$

and so on.

When defining "Total Productivity" or "Overall Productivity" we consider the contribution of all the input resources. Thus,

$$\text{Total Productivity Index} = \frac{\text{Value of Total output (goods and / or Service)}}{\text{Value of Total inputs (Labour + Materials + Machinery + Money + Energy etc.)}}$$

Labour Productivity. Out of all the partial productivity indices, it is the labour productivity, which is most commonly used. It is so because it is labour productivity that ultimately determines the standard of living of a nation. Further, labour is least likely to change and is most sensitive to any change in the other factors.

One concept of labour productivity is to compare the total output with the simple sum of all the hours of labour-the "Man-hours"-spent in production. This concept takes into account not only the number of persons engaged but also the hours put in by each of them. This concept is known as "Output per man-hour" or "Production per man-hour".

The second concept that is used for the measurement of labour productivity is generally referred to as OHP, that is, "Operating Hours for Unit Production". This concept is more common than the "Production per man-hour", because of its easy analysis. When the raw material comes out as a finished product after passing through the various production departments, the OHP for each department can be separately assessed and the sum of the separate OHP's will give the operating hours for the product. The concept is thus additive and it enables us to locate the areas of low productivity of the different departments of a company from time to time. It also enables us to compare different units producing the same product and to study trends in productivity in industry. While, it is a useful measure of productivity for an individual factory, it is not a suitable measure for a country as a whole.

Total productivity. Here, the total output is compared with total input, that is, with the aggregate of all the resources employed in production. Now, the relative contribution of various input factors to total productivity is different in different industrial situations. It depends upon a number of considerations such as, the level of technological development, the level of skill of the labour, the relative cost of labour and equipment, the cheapness or otherwise of the various materials available, and so on. Hence, while measuring total productivity, each input resource is appropriately weighted.

Total productivity is usually measured as follows : The total input resources include : manpower measured simply by number of man hours but also the intangible capital invested in education to improve the quality of labour and the tangible capital invested in plant, equipment, materials, land, inventory, etc. Thus,

Total input = Weighted man-hours + Weighted machine-hours.

Here machine-hours is short hand for the hours worked by all kinds of tangible capital. Weighted man-hours differ from the unweighted man-hours used in calculating output per man-hour, because an hour of high quality labour-a highly paid man hour of work-counts for proportionately more in the weighted aggregate than a low quality hour of labour-a lower wage man hour. In this way, we take into account the differences in skill, education, length of experience and other factors determining the quality of labour. Similarly, the weighted machine hours are also calculated.

In the above procedure, we have considered labour and all other input resources in the form of tangible and intangible capital. So, "Total Productivity" is also known as "Output per unit of labour and capital". Thus, total productivity measures efficiency in the use of labour and capital. However, the method is not exactly accurate, because the services of certain types of intangible capital are omitted from the total input because we can not count them.

Labour productivity requires less information and is easier to calculate. Total productivity is a more complex measurement because it requires more information.

27.5. FACTORS AFFECTING PRODUCTIVITY

As discussed above, for any manufacturing enterprise, the following are the input resources for the production of goods and/or services : men, machines, materials, money, time, space, energy, technology, etc. The productivity of a concern will depend upon how well these resources are managed and utilized. All the factors affecting productivity are of two types :

(*a*) Internal factors.

(*b*) External factors.

Internal factors. These factors are of an individual factory, and the main factors are given below :

1. Environmental factors within the company, whether they are conducive to increasing productivity or not. These may include, incentive schemes, working conditions, etc.

2. Level of mechanisation.

3. Technical and managerial skills available. These will depend upon the selection, training and number of manpower.

4. Use of materials and processes.

5. Application of productivity techniques such as work, study, method study, product design, plant layout, quality control, material handling and management techniques, etc.

6. Type of industrial relations existing in the factory.

External factors. 1. Economic factors, such as, available of capital, raw materials, power and market.

2. The level of competition.

3. Government rules and regulations.

4. Policies followed by a government.

5. Sociological factors. Resistance to change is a well-known phenomenon. Workers are afraid that a change (to achieve higher productivity) may not be for their betterment.

Productivity in its wider concept is not only an economic concept but also a sociological concept. It is not only a method of working in economic terms, but also a way of thinking and acting. It is an attitude of mind. The environment in which we live and work, our mutual relationships, as groups and as individuals have a profound influence on the effectiveness of our work.

Factors 3 and 4 above alongwith the effect of attitude of people towards asceticism on productivity, are aptly illustrated by the following examples :

(*i*) In United States and Western countries, the economy is free and is not under the entire control of the state. Moreover in these countries, worldly-success and ever increasing material prosperity have been accepted as desirable social objectives individually as well as collectively. They have been following these objectives relentlessly and with no reservations. All this has contributed to the growth of national wealth and a better standard of life for everyone.

(*ii*) In communist countries (the earstwhile Soviet Union and Eastern European countries), eventhough the goal is material prosperity, but because of the political system, it is collective well-being and not individual prosperity. Also the economies have been under the control of the state. Hence, there has been no individual motivation to achieve higher productivity. Due to this, the economies of all these countries have been sluggish.

(*iii*) Japan is a small country with inadequate natural resources. They have to import most of their raw materials and also food. The only way to do so is to maintain a high level of exports. For this, one has to compete in the world market. This they have achieved by obtaining a high level of productivity, thereby raising the standard of living of their people.

27.6. TECHNIQUES FOR IMPROVING PRODUCTIVITY

A high level of productivity is not an accident. It is the result of : (*i*) planned, effective and economical utilization of input resources, (*ii*) the deliberate adoption of methods and processes most suitable in a given situation, and (*iii*) the application of scientific knowledge to the problems of industry.

Any failure to make full use of any of these factors reduces productivity proportionately.

Increase in productivity depends largely upon improving the means of production. It depends upon the growth and development of technology and the ability to invent and apply new ideas and innovations.

Only the constant introduction of technological changes and scientific knowledge as and when they are available combined with the use of productivity techniques will result in increasing productivity.

Also, for creating and maintaining high productivity, the active co-operation of workers and trade unions is vital.

All the industrial activities concerning both design and production should be planned, controlled, coordinated and integrated for achieving a high level of productivity.

The various techniques adopted to raise productivity are briefly discussed below :

1. **Work-study.** This technique was originally developed by F.W. Taylor and has been

known by the time "Scientific Management". We have two aims in this technique : one is to discover the best way of doing a job and the time taken to do it efficiently. This is achieved by breaking down the job into its various elements, eliminating all unnecessary movements and assessing the time taken to do this job with the help of a stop-watch. The second aim is to ensure that all workers engaged in the job are trained to do it the best way. This technique coupled with a piece rate system would ensure high output, high wages for the workers and low costs of production.

2. **Product design.** A good design of product which helps in the economical and convenient manufacturing (designing for production) will minimize scrap and reduce the cost of production.

(*i*) *Product simplification* means the reduction to the minimum of the number of separate components for a product. This will result in decrease in the cost of production and increase in reliability. More complex a product, higher will be the cost of production. Greater the number of separate parts, less will be the reliability of the product. This is due to the fact that longer the chain of parts greater are the chances of failure of one or more parts in the chain. The shapes of the separate parts of the product should be simple. This will also reduce the cost of manufacturing.

(*ii*) *Product standardisation.* The purpose of standardisation is to establish mandatory or obligatory norms or standards to which the different types, grades, parameters, quality characteristics, test methods and so on, should conform. Its aim is to control variety of materials, tools, machine tools, parts, components and all types of machinery and equipment. All this will result in interchangeable parts which can be manufactured 'by mass production', thereby, reducing the costs.

3. **Cost control.** Cost of production can be reduced by efficient and economical utilization of all the input resources, elimination of all types of wastages, and by product design, product simplification and product standardisation as discussed above. Cost will be also reduced by better utilization of machinery through proper maintenance and balancing of production.

4. **Plant layout.** The benefits of a good plant layout have been listed in Chapter 26. All these benefits will result in lower cost of production and higher productivity. Plant layout also includes 'material handling' which leads to overall reduction in cost of production.

5. **Ergonomics.** Ergonomics is the study of considerations which should be taken into account while devising working conditions for workers engaged in production. A typical production situation comprises a man-machine system. This system must be designed and operated in a manner that ensures high productivity as well as safety of workers.

6. **Market research.** Market research is carried out to :

(*a*) determine the actual requirements of the product.

(*b*) assess popularity of the company's products.

(*c*) test reaction of consumers and competitors to any proposed increase in the price of the products.

(*d*) identify and tap new markets.

(*e*) estimate the effect of any new emerging factors, such as, technologies, materials, competition etc. on the projected demand.

Market research is a scientific tool which enables a company to earn maximum while giving the customer, the maximum satisfaction.

7. **Research and development.** Every company who believes in development and higher productivity, has a strong Research and Development (R&D) department. R and D department carries out the following activities to increase productivity and reduce cost of production :

(*i*) Using latest scientific and technological knowledge available for designing new products.

(*ii*) Improving the existing processes or devising new ones for improving quality of production and reducing cost of production.

(*iii*) Constantly improving the technical performance of the existing products and reducing their costs, thereby pushing up the sales.

(*iv*) Evaluating new available materials for use in existing designs.

(*v*) Standardising parts and materials used in design and production.

8. **Process planning.** As discussed in Chapter 6 and Chapter 26, the function of the process planning is to plan the process of manufacturing a product, select proper equipment and tools and establish a proper sequence of operations to manufacture the product. The total planning is based on the Principle of Economics so as to reduce the cost of production.

9. **Production planning and control.** Production planning and control includes the investigation, coordination and evaluation of manufacturing capabilities and requirements that ensure timely production of parts through efficient and optimum use of input resources (Chapter 23). The two network techniques, CPM and PERT have proved to be quite effective and versatile tools in planning and scheduling.

10. **Quality control.** Quality control function includes inspection of incoming and outgoing materials, parts and products, inspection of work-in-progress and prevention of poor quality by timely warning. All this results in improving the quality and increasing the sale. The rejections and scrap are minimized, thus reducing wastage and cost of production.

11. **Inventory control.** Inventory is the life blood of a production system. Inventory control endeavours to achieve a balance between too little inventory (with possible stock-outs of raw materials) and too much inventory (with investment and storage tied up). Thus inventory control fulfills the following objectives which leads to higher productivity :

(*i*) To minimize the overall production costs by keeping a balance between production and operating costs on one hand and the customer service on the other.

(*ii*) To ensure sufficient supply of required materials, parts, supplies and other items of maintain most efficient level of production to meet the customer demand.

(*iii*) To minimize losses in inventory deterioration, obsolescence, damage or waste.

12. **Operation Research** is a technique for taking decisions which searches the optimum in purity with the overall objectives and constraints of the organization.

13. **Automation.** Mechanization means the replacement of human labour by machines. But machines still are directed and controlled by humans. On the other hand, automation means self control of the machine so that it can work for a relatively long time without human intervention. Both these techniques have increased the speed and accuracy of production. These techniques are beneficial if the benefits of increased productivity are equitably shared with workers.

14. **Human factor.** The importance of human factor in production has been increasingly realised in recent times. It has been fully realised that the man behind the machine is more important than the machine itself. However efficient the machine might be, unless the man operating it was efficient, properly trained and willing, productivity could not be increased. In fact, as technology progresses, the human factor becomes more important, not less. Because, as the productive capacity of the machine increases, the responsibility of the man behind the machine also increases. Any carelessness, ignorance or unwillingness on his part would have a considerable adverse effect on productivity. This realisation has resulted in the establishment of better human relations within the factory, to the improvement in the working conditions within the factory as well as improved conditions of living and to the provisions of psychological incentives in order to maximize productivity.

The behaviour of a worker inside the organization is conditioned by the environment within the organization, his home and outside world and society. The effectiveness of his work, and consequently productivity, is largely influenced by these factors.

The social system in which we live also has a tremendous impact on productivity. The distinguishing features of a socio-economic system, such as, the division of labour, rewards and punishment, authority and prestige are some of the relevant factors. For having maximum productivity, one has to be satisfied with the prevalent system.

The following techniques are adopted to increase productivity by taking into account the factors (social as well as economic) as discussed above :

(*a*) **Incentive schemes.** The various incentive schemes help increase productivity. The incentives can be non-financial or financial.

The non-financial incentives include : job title, importance in the organization, vacations, good working conditions, good canteens, dispensaries, schools, clubs, recreation facilities, etc.

The financial incentives include : various bonus schemes.

(*b*) **Workers' participation in management.** This helps in developing mutual respect and co-operation between workers and the management. They can act in unison to increase productivity.

(*c*) **Management by objectives (MBO).** This is a new style of management. In this method, emphasis is on the achievements of results expressed in terms of objectives. These are decided jointly and in a participative manner by seniors and subordinates in the organization. The contribution of each individual is assessed jointly towards achieving the objectives. This technique greatly motivates the people, leads to better understanding between supervisors and subordinates and provides better co-ordination between different departments. All this helps to increase productivity.

(*d*) **Job enrichment and job enlargement.** These two techniques are very effective in increasing the motivation and morale of the workers.

Repetitive jobs are usually monotonous. Many workers would like a variety in their job to prevent monotony and boredom. This can be done by the following methods :

(*i*) The workers are allowed to periodically switch over from one job to another within the organization. This is known as job rotation.

(*ii*) The workers are allowed to participate in the planning, organizing and controlling their job. This is known as "Job enrichment".

(*iii*) The workers are allowed to work on large jobs, that is, those involving a greater number of operations and longer cycle times. This is known as "Job enlargement".

(*e*) **Flexitime.** This is a new technique which is different from the conventional one where the rigid time schedules are to be followed. The workers are given the freedom to schedule their working hours subject to a minimum number of hours per week. During a period as 'core hours' all the workers are required to be present in the organization. Flexitime reduces tardiness, short time absentism and overtime, thereby increasing productivity.

(*f*) **Total quality management (TQM).** TQM means activities involving everybody in an enterprise-Management and Workers-in a totally integrated effort towards improving performance at every level. This is a truly participative management approach in which everybody in an organisation, from worker to top management, is involved in achieving an overall excellence. TQM is discussed in detail in Chapter 21 (Art. 21.7).

Quality circles and Just in Time (JIT) approach are also parts of TQM.

27.6.1 Involvement of Computers in Design and Manufacture. Computers are being increasingly used in the design/production cycle of a product. The use of computers in manufacturing has result in :

1. Improved productivity.
2. Improved quality.
3. Improved equipment utilization.
4. Reduced inventory.
5. Faster delivery.

Computers are being used in every aspect of manufacturing, for example :

1. Computer-Aided Design (CAD).

2. Computer-Aided Manufacturing (CAM).

This includes : (*i*) Use of NC (Numerical Control) of production machines and processes, CNC (Computer Numerical Control), and D-N-C (Direct Numerical Control).

(*ii*) Computer-Aided Process Planning (CAPP), including Group Technology (GT) approach.

(*iii*) Industrial Robots.

(*iv*) Inspection and Testing

(*v*) Assembly.

3. Computer-Aided Production Management (CAPM). The activities included under this are:

(*i*) Capacity planning.

(*ii*) Cost planning and control.

(*iii*) Inventory management.

(*iv*) Production scheduling.

(*v*) Material requirement planning (MRP).

(*vi*) Shop-floor control (SFC).

(*vii*) Inspection and testing.

(*viii*) Assembly.

(*ix*) Manufacturing resources planning (MRP-II).

4. Computer-Aided Quality Control (CAQC).

5. Enterprise Resource Planning (ERP).

Computer-Integrated Manufacturing (CIM). CIM includes not only the CAD and CAM functions, but also the following activities :

(*i*) Market research, market forecast, product concept and sales (order processing).

(*ii*) Customer service.

(*iii*) Shipping (inventory, invoicing and accounting, etc.)

Note : For details, please refer to Chapter 24 and Chapter 25.

PROBLEMS

1. Define the terms "Production" and "Productivity".
2. How production and productivity differ from each other?
3. Discuss the importance of productivity.
4. How productivity is measured?
5. What is 'Partial productivity' and what is 'Total productivity'?
6. Define : Labour productivity, machine productivity, material productivity and energy productivity.
7. Discuss the factors affecting productivity.
8. Discuss the various techniques used for increasing productivity.
9. Discuss the role of human factor in raising productivity.

APPENDIX – I

A. Economy of Cold-Working. The total cost of producting a component can be split into five parts, as given below:

$$\text{Total Cost of a product} = \text{Material Cost} + \text{Tooling Cost} + \text{Set-up Cost} + \text{Production Cost} + \text{Overhead Cost}$$

(*a*) **Material Cost.** Compared to machining and other fabrication process, including hot working, cold working produces lot of saving in material. There is direct material saving due to less wastage. Indirect material savings are achieved due to the following factors :

1. Lighter components can be used due to their increased strength because of work-hardening.

2. Due to very good surface finish and dimensional tolerances obtained, the subsequent machining operation is either eliminated or minimised, thus saving material.

The actual material saving varies from component to component and may be as high as 50 to 60% in some cases. Thus, the material cost is invariably lower for cold working.

(*b*) **Tooling Cost.** The tooling cost for cold forming is in general more than for competitive process. However, when large volumes of components are to be produced, cold forming will become competitive compared to other processes. The batch size at which the cold forming (say cold heading) becomes economical depends upon the component and in general, it is about 20,000 to 30,000 work-pieces.

(*c*) **Set-up Cost.** The set-up cost is also higher in cold forming as compared to other methods. Since, the set-up cost is only a small percentage of the total production cost, this may not play a very important role in the economy.

(*d*) **Production Cost.** Due to high rate of production, the production cost is very small compared to other methods. This results in a large amount of saving.

As mentioned earlier, the initial capital expenditure for cold forming is higher than for other processes. But, due to high rate of returns, cold forming results in greater savings more quickly.

B. Characteristics of Hammers :

1. **Gravity Drop hammer :**

Total blow energy = K.E. of ram

$$\therefore \quad E_T = \frac{1}{2} m V^2 = \frac{1}{2} \cdot \frac{W}{g} \cdot V^2 = W.H$$

where,

m = mass of dropping ram

V = Velocity of ram at the start of deformation

W = Weight of ram

H = Height of ram fall.

2. **Power drop hammer :**

$$E_T = \frac{1}{2} m V^2 = (W + pA)H$$

where p = air, steam or oil pressure acting on the ram cylinder in the downstroke.

A = Surface area of the ram cylinder.

3. **Counter-blow hammer :**

$$E_T \text{ per blow} = 2 \times \frac{1}{2} mV^2 = \frac{W}{g} V^2$$

If V_t = actual velocity of blow of the two rams

$= 2V$

Then,
$$E_T = \frac{W.V_t^2}{4g}$$

During the working stroke, the total nominal energy, E_T, of a hammer is not entirely transformed into full energy available for deformation, E_A. Some small amount of energy is lost in overcoming friction of the guides and a significant portion is lost in the form of noise and vibrations to the environment.

Thus, blow efficiency of hammer,

$\eta = E_A/E_T$ is < 1

= 0.8 to 0.9 for soft blows (small load and large displacement)

= 0.2 to 0.5 for hard blows (high load and small displacement)

C. Inter-relation of Grit Size of Absrasive and Surface Roughness Value:

(a) Lapping Process:

Grit Size	Abrasive	Ra, µm
320	SiC	0.6 – 0.8
400	"	0.5 – 0.6
500	"	0.1 – 0.5
800	SiC	0.1 – 0.5
	Al_2O_3	0.05 –0.08
900	Al_2O_3	0.02 –0.05

(b) Honing Process:

Grit Size	Ra, µ m
150	0.05
280	0.3
320	0.25
400	0.125
600	0.05

Appendix I(D)

TOOL AND DIE STEELS

Tool and die steels for cold work

IS : 3749–1966

Designation of steel	C %	Si %	Mn %	Cr %	Mo %	V %	W %	Brinell hardness number HB, max	Typical applications
T 50	0.45-0.55	0.10-0.35	0.60-0.90	—	—	—	—	240	Covers the requirements for plain carbon and alloy tool and die steels in the form of bars, blanks, rings and other shapes for cold work, capable of being hardened and tempered. These are used for the making tools and dies for blanking, trimming, shaping and shearing.
T60	0.50-0.60	0.10-1.35	0.60-0.90	—	—	—	—	240	
T70Mn 65	0.65-0.75	0.10-0.35	0.50-0.80	—	—	—	—	240	
T80Mn65	0.75-0.85	0.10-0.35	0.50-0.80	—	—	—	—	240	
T90	0.85-0.95	0.10-0.30	0.20-0.35	—	—	—	—	200	
T103	0.95-1.10	0.10-0.30	0.20-0.35	—	—	—	—	200	
T133	1.25-1.40	0.10-0.30	0.20-0.35	—	—	—	—	210	
T90V 23	0.85-0.95	0.10-0.30	0.20-0.35	—	—	0.15-0.30	—	200	
T118Cr45	1.10-1.25	0.10-0.30	0.20-0.35	0.30-0.60	—	0.30 max	—	200	
T105Cr1Mn 60	0.90-1.20	0.10-0.35	0.40-0.80	1.00-1.60	—	—	—	230	
T140W4Cr 50	1.30-1.50	0.10-0.35	0.25-0.50	0.30-0.70	—	—	3.50-0.35	250	
T55Ni2Cr 65Mo30	0.50-0.60	0.10-0.35	0.50-0.80	0.50-0.80	0.25-0.35	—	—	255	
T105W2Cr60V25	0.90-1.20	0.10-0.35	0.25-0.50	0.40-0.80	0.25 max	0.20-0.30	1.25-1.75	230	
T110W2Cr1	1.00-1.20	0.10-0.35	0.25-0.50	0.90-1.30	—	—	1.25-1.75	230	
T90Mn2W50 Cr45	0.85-0.95	0.10-0.35	1.25-1.75	0.30-0.60	—	0.25 max	0.40-0.60	230	
T215Cr12	2.00-2.30	0.10-0.35	0.25-0.50	11.0-13.0	0.80 max	0.80 max	—	260	
T45Cr1Si95	0.40-0.50	0.80-1.10	0.55-0.75	1.20-1.60	—	—	—	230	
T55Cr70-V15	0.50-0.60	0.10-0.35	0.60-0.80	0.60-0.80	—	0.10-0.20	—	230	
T55Si2Mn90 Mo33	0.50-0.60	1.50-2.00	0.80-1.00	—	0.25-0.40	0.12-0.20	—	230	
T40W2Cr1V18	0.35-0.45	0.50-1.00	0.20-0.40	1.00-1.50	—	0.110-0.25	1.75-2.25	230	
T50W2Cr1V18	0.45-0.55	0.50-1.00	0.20-0.40	1.00-1.50	—	0.10-0.25	1.75-2.25	230	

Note : Prefix 'T' is for Tool Steels

Appendix I(E)

TOOL AND DIE STEELS

Tool and die steels for hot work IS : 3748–1966

Designation of steel	C %	Si %	Mn %	Cr %	Mo %	V %	W %	Brinell hardness (annealed) HB, max	Typical applications
T33W9Cr3V 38	0.25-0.40	0.10-0.35	0.20-0.40	2.80-3.30	—	0.25-0.50	8.00-10.0	241	Used for extrusion dies, hot swaging dies, forging die inserts, brass forging dies, hot shear blades, trimmer dies , die-casting dies for copper etc.
T35Cr5Mo1V30	0.30-0.40	0.80-1.20	0.25-0.50	4.75-5.25	1.20-1.60	0.20-0.40	—	229	
T35Cr5MoVI	0.30-0.40	0.80-1.20	0.25-0.50	4,.75-5.25	1.20-1.60	1.00-12.0	—	229	
T35Cr5MoW1V30	0.30-0.40	0.80-1.20	0.25-0.50	4.75-5.25	1.20-1.60	0.20-0.40	1.20-1.60	229	
T55W14Cr3V45	0.50-0.60	0.10-0.35	0.20-0.40	2.80-3.30	—	0.30-0.40	13.0-15.0	248	

Steels for die blocks for drop forgings

Designation of steel	C %	Si %	Mn %	Ni %	Cr %	Mo %	Brinell hardness HB		Typical applications
							Annealed max	Hardened and tempered	
T60	0.55-0.65	0.15-0.35	0.50-0.80	—	—	—	209	212-269	Steel for die blocks in square, rectangular and sections for drop forging
T60Nil	0.55-0.65	0.15-0.35	0.50-0.80	1.0-1.4	—	—	209	212-269	
T55NiCr65	0.50-0.60	0.15-0.35	0.50-0.80	1.25-1.65	0.50-0.80	—	230	235-302	
T50NiCr35	0.48-0.53	0.15-0.35	0.45-0.65	0.80-1.00	0.80-1.00	0.30-.40	255	269-477	

Note : Prefix 'T' is for Tool Steel

Appendix I(F) – FUNDAMENTAL TOLERANCES OF GRADES I T01 TO IT 16

Diameter Steps in mm		Values of Tolerances in Microns																	(1 micron = 0.001 mm)
		Tolerance Grades																	
		0.1	0	1	2	3	4	5	6	7	8	9	10	11	12	13	14*	15*	16*
To and inc	3	0.3	0.5	0.8	1.2	2	3	4	6	10	14	25	40	60	100	140	250	400	600
Over To and inc	3 6	0.4	0.6	1	1.5	2.5	4	5	8	12	18	30	48	75	120	180	300	480	750
Over To and inc	6 10	0.4	0.6	1	1.5	2.5	4	6	9	15	22	36	58	90	150	220	360	580	900
Over To and inc	10 18	0.5	0.8	1.2	2	3	5	8	11	18	27	43	70	110	180	270	430	700	1100
Over To and inc	18 30	0.6	1	1.5	2.5	4	6	9	13	21	33	52	84	130	210	330	520	840	1300
Over To and inc	30 50	0.4	1	1.5	2.5	4	7	11	16	25	39	62	100	160	250	390	620	1000	1600
Over To and inc	50 80	0.8	1.2	2	3	5	8	13	19	30	46	74	120	190	300	460	740	1200	1900
Over To and inc	80 120	1	1.5	2.5	4	6	10	15	22	35	54	87	140	220	350	540	870	1400	2200
Over To and inc	120 180	1.2	2	3.5	5	8	12	18	25	40	63	100	160	250	400	830	1000	1600	2500
Over To and inc	180 250	2	3	4.5	7	10	14	20	29	46	72	115	185	290	460	720	1150	1850	2900
Over To and inc	250 315	2.5	4	6	8	12	16	23	32	52	81	130	210	320	520	810	1300	2100	3200
Over To and inc	315 400	3	5	7	9	13	18	25	36	57	89	140	230	360	570	890	1400	2300	3600
Over To and inc	400 500	4	6	8	10	15	20	27	40	63	97	155	250	400	630	970	1550	2500	4000

* Up to 1 mm, Grades to 16 are not provided

APPENDIX – II

PROCESS PLANNING SHEETS FOR SOME MECHANICAL COMPONENTS

II.1. AUTOMATED PLANT FOR PISTON MANUFACTURE

The following steps are involved in an automated plant for the manufacture of pistons for internal combustion engines :

1. Melting of aluminium alloy pigs in an electric furnace and pouring into permanent moulds mounted on a multiple-station pouring machine.
2. Cutting of gates and risers with the help of a special machine.
3. Loading the castings of the pistons into containers in which they are annealed for about 5½ hours by a stream of hot air from an air heater.
4. Transferring the pistons to an automatic hopper. This is a storage bank which provides a reserve, used when the pouring machine is shut down.
5. For machining the piston, the locating surfaces are machined on the piston. One locating surface is on the centre of piston head in the form of a small hollow cylinder which is counter sunk to provide a locating surface for the centre of the machine. This small piece is machined off after all the machining operations have been completed on the piston. The second locating surface is a bored portion of the open end of the piston. To provide proper seating surface to the locating device, the open end/skirt end of the piston is faced and chamfered.
6. Then, each piston is automatically mounted on a plate-fixture on which it is conveyed through the machining department. The machining department consists of an automatic transfer machine consisting of seven machine tools. The lathes in the department are 4-spindle, vertical, unit built machines. In each of the two lathes, one for roughing and the other for finishing, turning is performed by two carriages. One carriage has a cross-slide for facing the end, and for turning the grooves and the second carriage feeds longitudinally and turns the outside diameter. The carriages are automatic hydraulic units.
7. At the end of the machining operations, the fixture runs on a lift table. Here, they are removed from the pistons, lowered to the chain conveyor and returned to the station where locating surfaces are machined on the piston.
8. After passing through the washing machine, the pistons are transferred to the automatic hopper or storage tank.
9. From the storage tanks, the pistons are delivered for boring, reaming and finishing wristpin hole.
10. Washing
11. Inspection and sorting.
12. Slushing
13. Wrapping in waxed paper.
14. Accumulating the piston size in sets.
15. Packing in Cardboard cartons.
16. Closing the cartons with sealing tape.

II.2 PROCESS PLANNING SHEET FOR CONNECTING ROD OF AN I.C. ENGINE

INTRODUCTION

Connecting rod is a link between the piston and the crank shaft. It converts the reciprocating motion of piston into the rotary motion of crank shaft. The connecting rod has small end through which the piston pin is mounted and a big end that is split and bolted around the rod journal bearing of the crank shaft. The two ends are joined by a rigid link, most typically of an I-beam section.

As connecting rod has anoscillatory motion, it can lead to the fatigue of the connecting rod. To avoid this it is always in a forged form.

Material : AISI (American Iron and Steel Institute) 1078 steel.

Form : Forging.

Supplier : Bharat forge Ltd.

PROCESS PLANNING SHEET

Sr. No.	*Operation*	*Machine/Equipment*
1.	Semi finsih both sides	Duplex Milling Machine
2.	Semi finish bore and chamfer Small end.	Twin spindle Special Purpose Machine (SPM)
3.	Location pad turning	SPM
4.	Milling of nut seat and bolt seat, big end	Machine centre
5.	Gundrill bolt holes, Big end	Pneumatic gun.
6.	Counter boring of bolt holes, big end	4 Spindle SPM
7.	Split theBig end of rod.	Multi spindle, SPM.
8.	Finish grindmatching faces, Big end.	Horz. milling machine
9.	Finish bore and chamfer bolt holes	Surface grinding machine
10.	Rough assembly	Twin spindle boring machine.
11.	Rough bore of big end	Hydraulic press.
12.	Finish crank and pin bore	Single Spindle SPM
13.	Finish turn crank width and chamfer	Twin spindle SPM
14.	Dismantle and deburr	Facing and chamfering machine
15.	Mill lock slot.	Pneumatic gun and manual de-burring
16.	Gun drill oil hole, deburr, chamfer	Horizontal milling machine
17.	Oil holes and press bush	Twin spindle SPM and hydraulic press
18.	Lap mating faces, deburr and assemble.	Lapping machine.
19.	Mill Taper both sides at pin end.	Taper milling machine.
20.	Mill balance weights for weight correction.	CNC milling machine
21.	Deburring	Manual
22.	Finish hone crank bore	Honing machine
23.	Bush boring	Boring SPM.
24.	Washing	Washing booth.
25.	Crack testing.	Magnaflux crack testing m/c

Function of Location Pad

To rest or clamp the connecting rod for future operations.

GUNDRILLING

In case of drilling long holes, lubrication of cutting area becomes difficult. To overcome this, gundrilling is used, in which drill having continuous oil hole is used. It passes oil from spindle to cutting area and reduces friction.

LAPPING

The purpose of lapping is :

a. To produce geometrically true surface.
b. To imrpove dimensional accuracy.

c. To correct minor surface imperfections.

For Connecting rod, lapping of mating faces is done.

Powder - Silicon carbide.

Solution - 0.5 kg. Powder + 2 lit SAE 40 oil + 5 lit of kerosene.

This solution is poured drop wise on lapping face.

HONING

It is an abrading process used to remove grinding or tool marks left on surface to obtain smooth surface finish. Abrasive sticks of aluminium oxide or silicon carbide mounted in a mandrel or fixture are used. The honing tool is given a slow reciprocating motion along with rotational motion. Machine used is vertical spindle honing machine.

II.3 PROCESS PLANNING SHEET FOR CAM-SHAFT OF I.C. ENGINE

INTRODUCTION:

It is the brain of the engine and it gives reciprocating motion to valves and injector plunger. Supplying accurate quantity of fuel on right time and at regular time interval is very critical factor in case of any diesel engine. As camshaft is the part which actuates intake valves, exhaust valves and also injector, it is very critical part of diesel engine.

Material - AISI 1078 steel.

Camshaft is basically a shaft with cam lobes and journals. It receives rotary motion from crankshaft through gear train. It is rotating with half the speed of crankshaft. The roller followers pivoted on a shaft ride on the lobes of camshaft. The vertical motion of followers is transmitted to rockers levers through push tubes. The rocker levers pressed on crosshead facilitate opening and closing of valves or injector.

Lubricating oil is supplied under pressure to the bearing via drilled passage in the rear of cylinder block. The manufacturing process involves following steps:

1. CAM MILLING

CNC M/C

- The turned bar stock received from supplier is held between two hydraulically operated chucks and cutter moves in both X and Y directions simultaneously to produce profile cam lobe. The cutteruses 40 carbide inserts with speed of 110 rpm.
- The machine has 4 axes:

  a. X : movement of table towards and away from operator in the horizontal plane and parallel to the column of the machine. Thus, the longest motion of the machine is in the direction of X-axis.

  b. Y : movement of workpiece in the horizontal plane towards and away from machine column.

  c. Z : movement of cutter towards and away from operator, in the vertical plane. The Z-axis, perpendicular to both the X and Y axes, is always parallel to the axis of cutting tool.

  d. C : axis of rotation of the workpiece.

  The three axes X, Y, Z are mutually perpendicular to one another.

- The machine can machine length of cam shaft with high production rate and without bending.
- On this machine vary cam is used, this technique is nothing but carrying the feed as per the cam i.e. feed is slow when there is uneven material removal and is fast when there is even material removal. With the use of this machine cycle time is reduced by 30% and tool life is increased by 15%.

2. DEBURRING :

Pneumatic Grinder :

- To remove the burr and improve surface finish.

3. **HEAT TREATMENT**

INDUCTION HARDENING MACHINE:

- The temperature of heating is about 850-900°.
- Case depth of 130-140 thou.
- the hardness achieved is 60 RC.

4. **TEMPERING**

- Tempering is done at 210° for 3 hours.
- This removes residual stresses, reduce brittleness and for refining the grain.

5. **STRAIGHTENING**

Hydraulic Press :

- During heat treatment shafts get bent (Allowed bend - "0.005" T.I.R.)
- Cam lobes governing opening and closing of valves.

6. **JOURNAL GRINDING**

Journal Grinding Machine

- Ordinary cylindrical grinder is used.
- For all bearings to be of the same size, LVDT (Linear variable Differential Transducer) is used. When exact size is obtained, Marposs sends signal causing withdraw of wheel.

7. **KEY-WAY MILLING**

Key Way Milling Machine

- On the cam-shaft cam gear is mounted with the help of woodruff key with reference to keyway, timing marks are obtained on gear.
- Thus gear is fixed relationship between camlobes and keyway.

8. **CAM GRINDING**

CNC Machine

- The grinding wheel moves towards or away from shaft each degree rotation according to programmed valves.
- First rough grinding is done and then they are finish grinded.

9. **BURN TEST**

SPM based on Electrolytic Etching

- It is a non-destructive test for testing grinding burns and treatment defects.
- It is used to identify Untempered Martensite from tempered Martensite.

Procedure

- First camshaft is cleaned by Sodium Hydroxide (NaOH) solution.
- The job is cleaned with water.
- It is then etched by HCL till uniform black column is obtained, with current of 10A.
- The job is cleaned with water and Citric Acid is applied to remove black colour.
- If back stain remains, the portion is said to be burnt.

10. **CRACK TEST:**

Magnaflux Testing Machine

11. **LAPPING**

Centre Lathe :

- It is done by abrassive cloth forming lapping belt. It is driven by separate motor.
- Due to rubbing of belt on the lobe smooth surface finish of 15-20 microns is achieved.

12. **TUMBLING**

Vibrating Container Machine

- It is done to remove sharp cutting edges on camshaft
- It also provides cross hatch surface pattern for better oil retention.

II.4 MANUFACTURE OF PISTON RINGS

1. Manufacturing Tolerances :

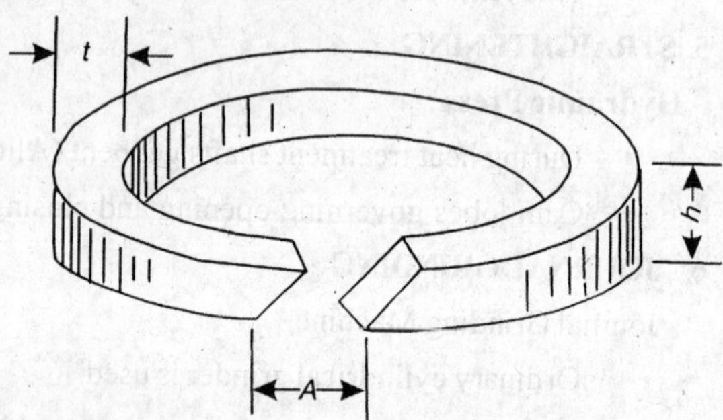

Fig. II.1: PISTON RING

(i) Thickness of ring, $h = \left(\frac{1}{30} \text{ to } \frac{1}{35}\right) D$; D = cylinder diameter. It should be less than the groove height by about 0.01 mm, so that, it is neither tight nor loose in the groove.

(ii) Width of ring, t = (0.6 to 1.0) h
It should be less than the groove depth by 1.5 – 2 *mm*

(iii) A = (0.0015 – 0.003) D, for Steam cylinders
= 0.004D, for compressors
= (0.005 – 0.008) D, for automobiles
= 0.05 mm for hydraulic and pneumatic cylinder upto 120 mm ϕ
= 0. 10 mm for hydraulic and pneumatic cylinder from 120 – 150 mm ϕ

(iv) **Surface Finish :**

End faces : Ground
I.D. : Machined
O.D. : Fine Machined or ground

2. Material : The most commonly used material for piston rings is C.I. Typical Composition are given below, in Table :

S.No.	*Application*	*Composition*						
		C	*Si*	*Mn*	*P*	*S*	*Cr*	*Ni*
1	Aviation	2.8	1.6	1.2	0.4	0.1	1.35	0.40
		3.3	2.0	1.6	0.6	–	–	–
2	Automobiles	3.5	2.3	0.6	0.3	0.1	0.3	–
		3.7	2.6	1.0	0.7	–	–	–
3	Tractor Building	3.75	2.4	0.5	0.45	0.05	0.25	–
		3.90	2.6	0.75	0.50	–	0.35	–
4	Diesel	2.9	1.4	0.7	0.4	0.12	0.3	0.4
		3.2	1.6	0.9	0.8	–	–	–
5	Locomotive	2.9	1.5	1.0	0.4	0.12	0.2	0.2
		3.2	1.9	1.5	0.8	–	–	0.5
6	Compressor, Hydraulic & Pneumatic Cylinders	2.9	1.5	0.7	0.4	0.12	0.4	–
		3.2	1.9	1.0	0.8	–	–	–

3. Manufacturing Method : The method involves the following steps :

(i) Suitable size hollow *C.I.* cylinder (obtained by centrifugal casting) is taken and its O.D. and *I.D.* are turned to D' and d' respectively. In order to achieve the required springing action, firstly, the *O.D.* of the ring should be turned and then its *I.D.*,

$$D' = D + 1.5\,t$$

$$d' = d + 0.5\,t$$

Where d is the I.D. of finished ring.

(ii) Rings of thickness t' are now parted off from the cylinder, where,

$$t' = t + (0.5 \text{ to } 1.0) \text{ mm}$$

(iii) Rings are diagonally slit at 45° as shown in Fig. the ends of the ring are brought together and joined by soldering /Brazing.

(vi) The ring O.D. is finish - turned on special mandrel.

(v) The ring I.D. is finish turned in special mandrel.

Note: In the absence of special mandrels, operations (iv) and (v) can be carried on the lathe chuck itself.

(iv) The end faces of the ring are now ground to size on a surface grinding machine.

(viii) Soldered/Brazed ends of the ring are made free by melting away the filler material.

II.5 FABRICATION OF BUSH BEARINGS

Bi-Metalling : it is economical to use bi-metal bushes with steel/C.I. body and babbit/bronze lining. The use of such bush bearings is recommended for all sizes.

(1) Bronze Lining by Centrifugal Casting : Fig. II. 2(a) shows a Bush (C.I./steel body) with bronze lining.

For repairing the worn-out internal surface or lining (bi-metalling) the new one, the steel/C.I. body of the bush bearing is rough bored (preferably threaded with torn/rough threads) by 3–6 mm more than the desired nominal diameter so that bronze layer of 1.5–3 mm (depending on the bush dia.) is left on the finally machined bush.

Depending upon the heating method, bi-metal lining can be obtained by any one of the following methods :

(a) Furnace heating: The C.I./steel body is pre-heated (to about 300°C), if necessary, and a cover welded at one end of it. The bronze powder/chips in the required quantity are now placed in the bush and another cover welded on the other end of it, Fig. II.2(b). For the out let of gases, a small hole 4–6 mm ϕ is drilled in one of the covers.

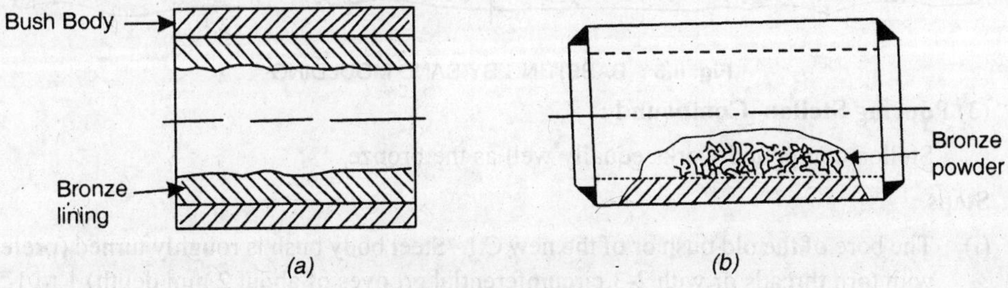

Fig. II.2: A BUSH

This bush with the bronze powder/chips in now heated in a furnace to a temp. of 1150°–1180°C, at which bronze become viscous and starts flowing. This is now held in special chuck of the lathe and

and rotated at about 100 rev/min for 10–15 minutes under the centrifugal force, the bronze is uniformly spread on the internal surface of the bush and adheres there. Subsequently, this is bored on a lathe to the required diameter.

(b) Electric arc heating : In this method, the bronze powder/chips are heated by an electric arc struck between two electrodes. The remaining procedure is the same as for furnace heating. The electrodes form a part of the total assembly which is rotated on a centre lathe.

(2) Babbiting by Sand Moulding :

(i) The steel/ C.I. body is roughly bored as explained under the previous method.

(ii) The sand mould is now prepared with the internal core, as shown in Fig. II.3

(iii) The molten babbit (m.pt. about 450° C) is now poured in the mould.

(iv) It is allowed to cool and solidify.

(v) The sand mould is broken and the lined bush taken out.

(vi) It is cleaned

(vii) It is then bored on a lathe to the required diameter.

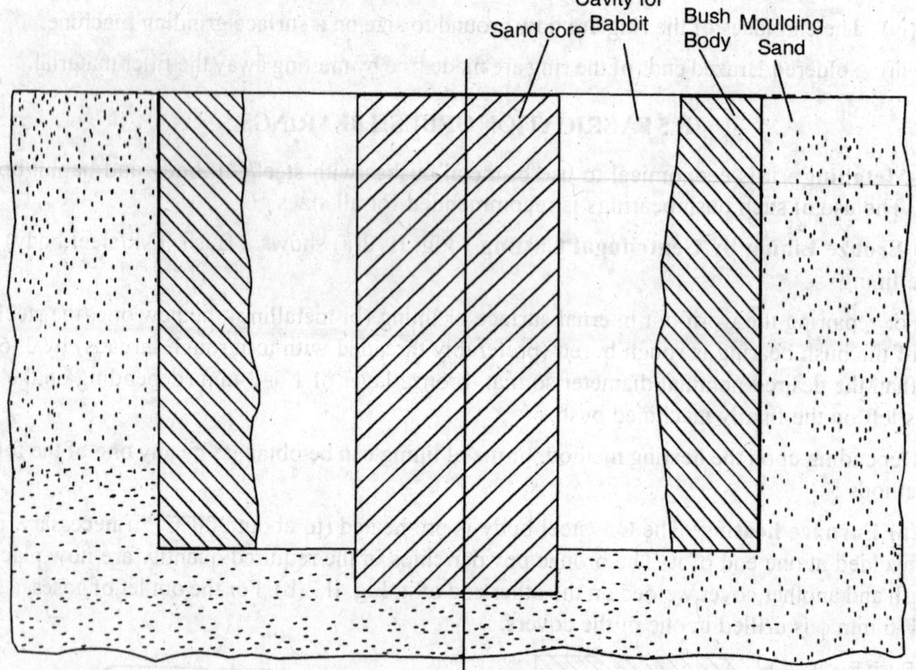

Fig. II.3 : BABBITING BY SAND MOULDING

(3) Pouring Stellon Compound :

Stellon compound works equally well as the bronze.

Steps :

(i) The bore of the old bush or of the new C.I. /Steel body bush is roughly turned (preferably with torn threads or with 2-3 circumferential grooves of about 2 mm depth) 1 to 1.5 mm leaving end collars, Fig. II.4.

(ii) One end of the bush is closed by pasting a paper at that end.

(iii) The bush is mounted on a lathe and rotated at about 20 rev./min.

(iv) Stellon compound is now poured through a paper cone from the open end while the bush is rotating. Due to centrifugal force, the compound will evenly spread and adhere to the bush walls. The operation is continued from about an hour and a 2 to 3 mm thick compound wall is allowed to be formed. The bush is now removed from the lathe and allowed to set for 12 hours.

(v) The bush is next bored to the required size, with the final compound to be 1 to 1.5 mm thick.

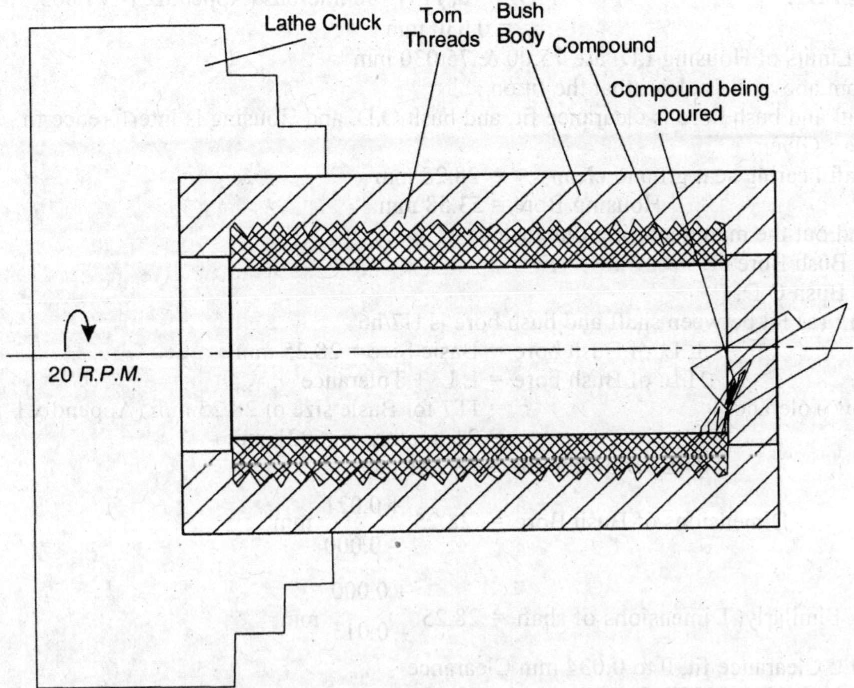

Fig. II.4 : BIMETALLING A BUSH WITH STELLON COMPOUND

II.6.1 FIT ON BUSH BEARING

Note: (a) Fit between the shaft and the bush bore is a sliding fit, H7/h6.

(b) Fit between the bush outside diameter and the housing inside Diameter is push fit, H7/js6.

Example: Given:

Nominal bush bore diameter = 48 mm

Nominal bush outside diameter = 75 mm

Determine : Tolerance on shaft, bush and the housing.

Solution: Fit between, shaft and bush is H7/h6

Shaft: For shaft 'h' F. D. = 0

∴ H.L. of shaft = Basic size = 48 mm

L.L. of shaft = H.L. of shaft – Tolerance

Tolerance, IT6 = 16 microns (See Table Appendix I – F, P 865)

= 0.016 mm

∴ L.L. of shaft = 48 – 0.016 = 47.984 mm

Bush Bore for Hole 'H', F.D. = 0

∴ L.L. of Bush bore = Basic (Nominal size) = 48 mm.

H.L. of Bush bore = L.L. + Tolerance

Tolerance, I T 7, for nominal size of 48 mm

= 25 microns (Appendix I–F, P865).

= 0.025 mm

∴ H.L. of Bush bore = 48.025 mm

Bush O.D. and Housing : Nominal size = 75 mm

Fit. H7/js6

Bush O.D. : For js6, Tolerance = ± 0.0095 mm

∴ **Limits are :** 74.9905 and 75.0095 mm

Housing I.D. : F.D. = 0, IT7 = 30 microns (Appendix I-F, P865)

= 0.030 mm

∴ Limits of Housing I.D. are 75.00 & 75.030 mm

From above, it is clear, that the fit on :

shaft and bush bore is clearance fit, and bush O.D. and Housing is interference fit.

Example : Given,

Shaft bearing seat ground clean = 28.25 mm

Housing Bore = 53.58 mm

Find out the manufacturing dimensions of

(a) Bush Bore

(b) Bush O.D.

Solution : (a) Fit between shaft and bush bore is H7/h6

∴ L.L. of Bush bore = Basic Size = 28.25 mm

H.L. of Bush bore = L.L. + Tolerance

Now Tolerance = IT7 for Basic size of 28.25 mm (Appendix I–F, P865)

= 21 microns = 0.021 mm

∴ H.L. of bush bore = 28.271 mm

$$\text{Dimensions of Bush Bore} = 28.25^{+0.021}_{-0.000} \text{ mm}$$

$$\text{Similarly, Dimensions of shaft} = 28.25^{+0.000}_{-0.013} \text{ mm}$$

Fit. : Clearance fit, 0 to 0.034 mm Clearance

(b) Fit is H7/Js6, Basic Size = 53.58 mm

∴ **Dimension of Housing Bore :**

L. L. = 53.58 mm

H.L. = L.L. + Tolerance

Tolerance, IT7 for Basic Size of 53.58 mm

= 30 microns (Appendix I–F, P865)

∴ H.L. of Housing Bore = 53.61 mm; i.e. $53.58^{+0.030}_{-0.000}$ mm

Similarly, for Bush O.D. = $53.58^{+0.0095}_{-0.0095}$ mm, i.e. 53.5705 to 53.5895 mm

Fit is interference : – 0.0095 mm interference to + 0.0395 mm clearance

∴ Manufacturing dimensions are :

(a) Bush Bore : $28.25^{+0.034}_{-0.000}$ mm

(b) Bush O.D. : $53.58^{+0.0095}_{-0.0395}$ mm

APPENDIX – III

CUTTING FORCE MEASUREMENT – DYNAMOMETERY

III.1 Introduction: As has been discussed in Art. 14.8, the force analysis of the cutting process is quite complex and has been simplified and idealised with the help of many assumptions. Due to this, the actual investigation of any real cutting process is mainly dependent upon the results of experimental work. The main aim of any experimental work is to determine the cutting forces. Tools used for this purpose are known as "Force Dynamometers".

Any cutting force dynamometer consists of two parts:

1. An elastic element/member, which bears the force to be measured, and gets deflected/deformed/strained.
2. A transducer or sensing element which picks up the deflection/strain of the elastic element and converts it into a measurable signal – Mechanical, Electrical, Electronic, Hydraulic, Pneumatic etc. The deformation is magnified by an amplifier to enhance sensitivity.

III. 2 Requirements of Dynamometers:- The requirements of force dynamometers are the same as of any measuring instrument, that is,

1. It should not alter the rigidity of the Machine – Tool – Work Piece (MTWP) system beyond the permissible limits.
2. Its incorporation in the system should not affect the dynamic characteristics of the system.
3. It should have high resolution, i.e., large response to small variations in cutting force.
4. The sensitivity of the transducer should not be affected by time and environmental conditions like temperature and humidity.
5. The frequency response should be as wide as possible. For accurate results, the natural frequency of dynamometers should be at least 4 times higher than that of the cyclic variations in the forces to be measured. This implies less weight and high rigidity of the elastic member. Often, a compromise is made between rigidity of the elastic element and the desired resolution.
6. As far as possible the readings should bear a linear relationship with the forces to be measured.
7. The adjustment of zero points and meauring ranges should be non-critical.
8. The cross-effects should be minimal.

III. 3. Types of Force Dynamometers: Depending upon the type of transducer used, the force dynamometers are of the following types:-

(i) Mechanical Dial - Gauge type.

(ii) Hydraulic pressure measuring type.

(iii) Pneumatic type

(iv) Capacitive type

(v) Inductive type

(vi) Piezo-electric type

(vii) Resistance Strain - Gauges dynamometer.

1. **Mechanical Dial - Gauge type dynamometer:-** The deflection or strain in the elastic elements supporting the cutting tool is measured by a sensitive dial gauge. The main drawback of this type is that, due to vibrations of the elastic member, the reading of the dial gauge is difficult to measure accurately.

2. **Hydraulic pressure measuring type:-** Here, a membrane or piston - cylinder system is employed to transmit the cutting forces from the tool, in the form of hydraulic pressure, to a pressure measuring meter (Manometer), Fig. III. 1. The limitation of this type is that the frictional forces between the piston and the cylinder can seriously affect the accuracy of readings.

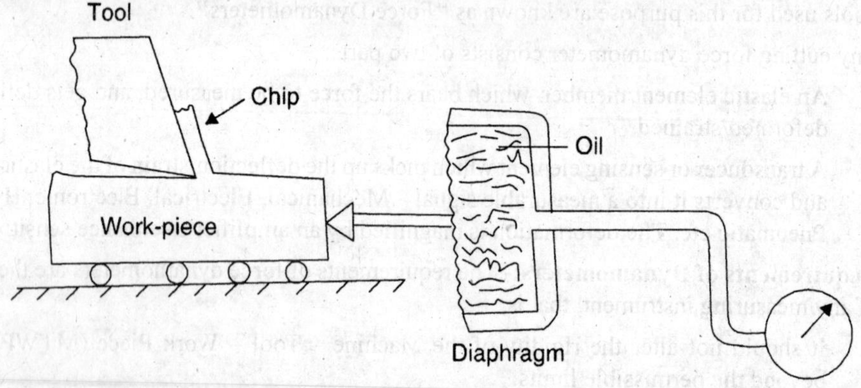

Fig. III.1 Hydraulic pressure measuring type Dynamometer

3. **Pneumatic type Pick-up: -** Pneumatic heads may be used to measure the deflection of elastic elements (diaphragms, bellows etc.), Fig. III. 2.

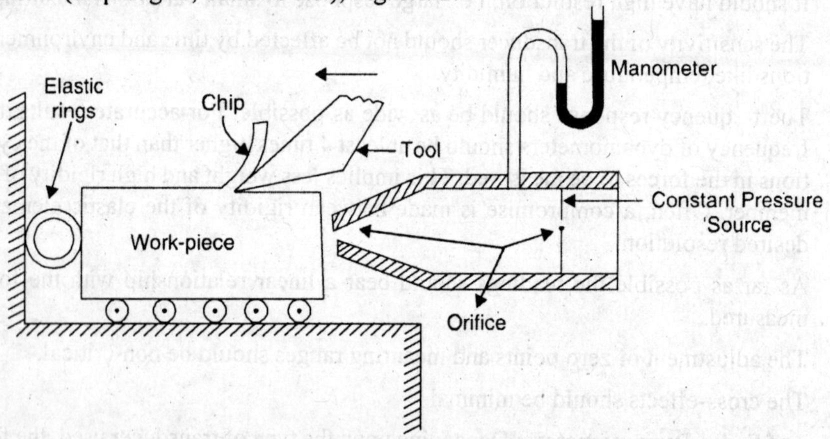

Fig. III. 2 Pneumatic head type pick-up

4. **Capacitive type Pick-up :-** A capacitor consists of two parallel plates with an air - gap in between. the capacitance of the set up is given as, Fig. (III. 3):-

$$C = k.\frac{A}{d}$$

where A = Effective area of plates constituting the capacitor d = Distance between the two plates.

k = a constant

In capacitive type pick up, the deformation/deflection of the plates under the action of cutting forces, changes the air-gap, d, thereby the value of C.

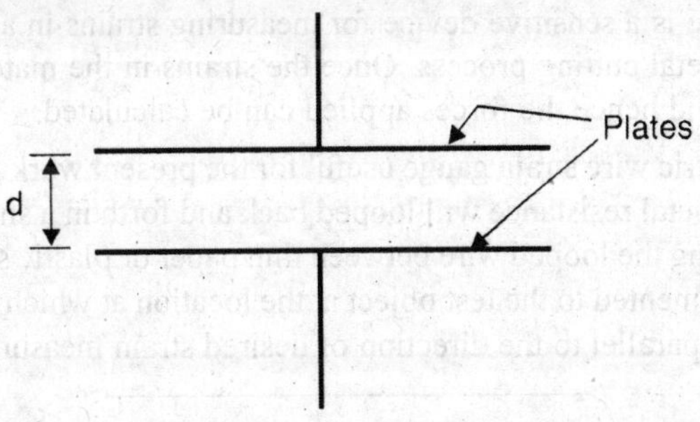

Fig. III. 3 Capacitive pick-up

The major disadvantage of this type of pick up is its relatively large output impedence which needs careful shielding and short connections to subsequent stages.

5. **Inductive type Pick – up** : - The inductance of a coil is given as:

$$L = n^2 \mu F$$

where n = number of turns of the coil

μ = permeability of the material in and around the coil

F = Form factor

The inductive type pick-up is based on the change in 'L' with change in any one of the three factors n, μ and F. The easy and common method is the change of the inductance of the magnetic path.

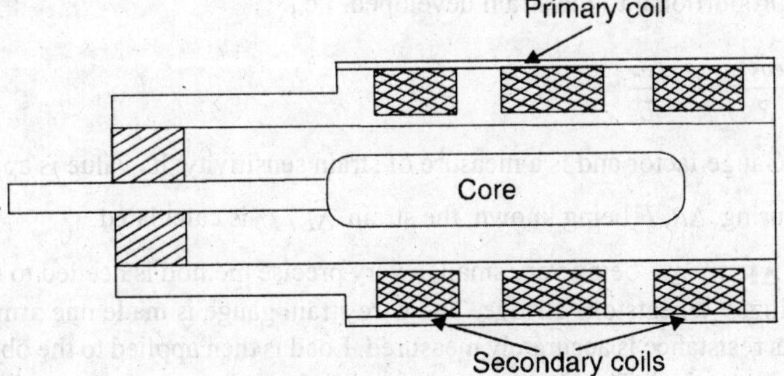

Fig. III. 4 An Inductive Pick up

One type of inductive pick up dynamometer is shown in Fig. III. 4. It consists of a primary coil and two secondary coils. A high frequency supply is fed to the primary coil. This induces e.mf. in the secondary coils. A movable iron cone (core) placed within the coils pick up the displacements/strains of the elastic member. Any change in the position of the iron cone changes the induced e.m.f. in the secondary coils. The secondary coils are connected in opposition, so that there is no output when the core is symmetrically placed with respect to the secondary coils. Such a pick up can be designed for measuring very large displacements and hence the cutting forces.

6. **Piezo-electric type pick-up:-** This pick-up is based on the piezo-electric effect: If the flat ends of a crystal (quartz crystal) deform under the action of cutting force (contract or elongate), an electric potential is induced in the normal direction. The charges thus appearing on the surfaces are collected and transformed into an analogue voltage by a charge amplifier.

7. **Resistive Strain Gauge:-** The electric resistance wire strain gauge, also known as the bonded resistance wire gauge is a sensitive device for measuring strains in a wide variety of circumstances, including metal cutting process. Once the strains in the material are determined, the stresses developed and hence the forces applied can be calculated.

   Working:- The electric wire strain gauge useful for the present work is a bonded strain gauge. It consists of a fine metal resistance will looped back and forth in a single plane, fig. III.5. It is insulated by cementing the looped wire between thin paper or plastic strips. This 'sandwiched' wire gauge is then cemented to the test object at the location at which strain is to be measured, with the wire length parallel to the direction of desired strain measurement.

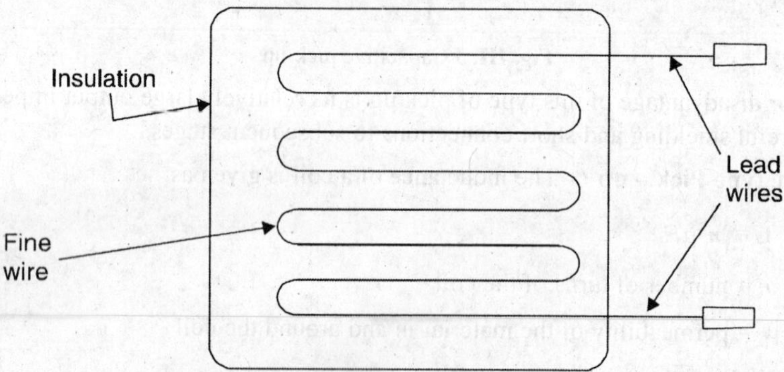

Fig. III.5 Resistance wire Strain Gauge

When a load is applied to the test object, it deforms and the wire gauge bonded to it deforms equally with it. This slightly alters the electrical resistance of the wire. the change in resistance is precisely proportional to the strain developed, i.e.,

$$\frac{\Delta R}{R} = F \cdot \frac{\Delta L}{L}$$

where F = Gauge factor and is a measure of strain sensitivity. Its value is approximately 2.

So, by measuring ΔR, F being known, the strain $\Delta L / L$ is calculated.

Measuring ΔR : - ΔR being very small, a very precise method is needed to measure it. Such a method utilizes Wheatstone Bridge. The wire strain gauge is made one arm of the four-arm bridge and its resistance is accurately measured. Load is then applied to the object to which the strain gauge is attached. The Gauge gets deformed. Its resistance changes which is again measured by balancing the bridge. ΔR is then calculated, substituted in the gauge factor equation and strain determined.

The relationship between the resistances of the bridge,

Fig. III. 6, is

$$\frac{R_1}{R_4} = \frac{R_2}{R_3} \quad \text{or} \quad \frac{R_1}{R_2} = \frac{R_4}{R_3}$$

Any imbalance in the bridge circuit caused by a change of resistance in R_1, can be removed by adjusting the ratio R_2 / R_3 or R_4 / R_3. The amount of adjustment is read in terms of R_1.

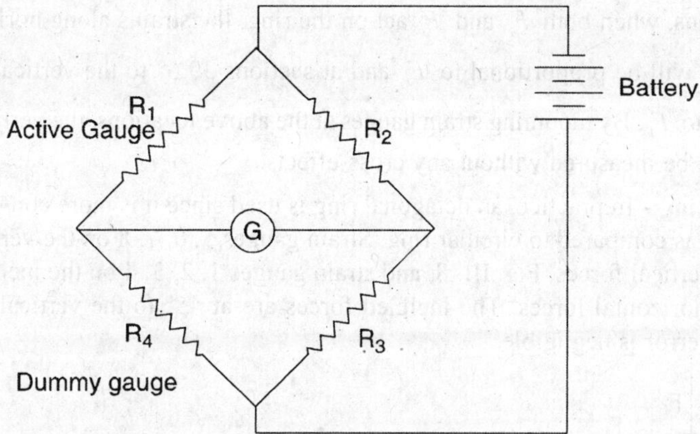

Fig. III. 6 Wheatstone Bridge

In practice, use is made of a strain indicator instrument which directly indicates the strains caused by various loadings. The indicator read-out scales are calibrated. The strain indicator is nothing but a wheat stone bridge with provision for making external resistances two arms of the bridge, and with a signal amplification arrangement.

The resistance of wire strain gauge also changes due to temperature. This will introduce error unless compensated. For this, a dummy gauge is located close to the active strain measuring gauge, Fig. III. 6. It is free of any strain. The effects of temperature on both active and dummy gauges will be equal and cancel each other. This is due to the fact that changes in resistance of like sign are in effect substractive in adjacent arms of the bridge.

Strain Gauge Dynamometers:- Strain gauge dynamometers use either a circular or an octogonal ring. When a circular ring is subjected to load, Fig. III. 7 (a), there are locations on the ring where the strain is maximum, while at some other locations, the strain is minimum or zero. These locations are known as "Strain nodes". If strain gauges are fixed at these locations, they would give the maximum or minimum response according to the nature of the nodes. The cross effects are also minimum on these nodes.

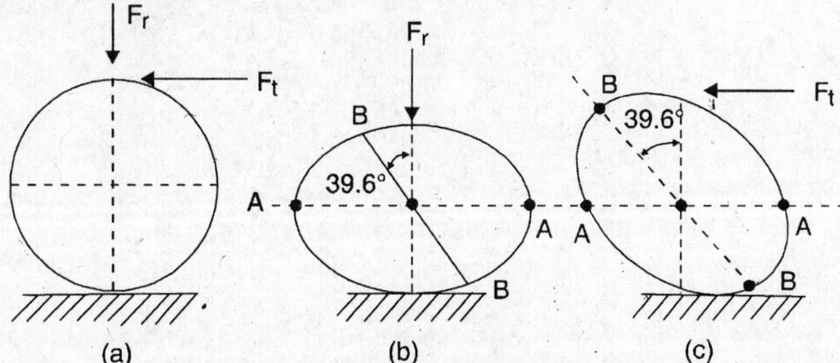

Fig. III. 7. Circular ring subjected to load

It has been analyzed that if a circular ring is subjected to a radial load, Fr, passing through its centre, Fig. III. 7 (b), maximum strains occur on sections lying on horizontal diametral plane,

A – A. Whereas there is no strain on sections located at 39.6° from the vertical, that is B – B. Similarly, if a horizontal force (Tangential force), F_t, lying in the plane of the ring acts on the ring at the heighest point Fig. III 7(c), the strains on A – A will be zero and it will be maximum on B – B. Thus, when both F_r and F_t act on the ring, the strains along horizontal diametral plane A – A, will be proportional to F_r and at sections 39.6° to the vertical, B – B, will be proportional to F_t. By mounting strain gauges at the above locations, the vertical and horizontal forces can be measured without any cross-effects.

Octagonal Ring:- In practice, an octagonal ring is used since it is more convenient to fix and is more rigid as compared to circular ring. Strain gauges 5, 6, 7, 8 on the vertical faces of the ring are for vertical forces, Fig. III. 8, and strain gauges 1, 2, 3, 4 on the inclined faces of the ring are for horizontal forces. The inclined forces are at 45° to the vertical as compared to 39.6°, so the error is negligible.

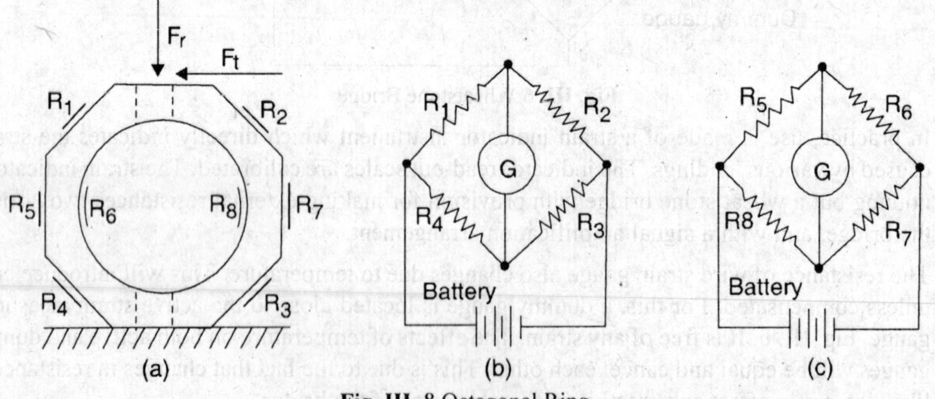

Fig. III. 8 Octagonal Ring

Materials for Rings:- The materials for rings should have stable mechanical properties and should be free from creep, hysteresis or age hardening. The common materials for rings are:- Stainless steel for large cutting forces and Aluminium for small cutting forces.

APPENDIX – IV

GEAR MANUFACTURING

Gear manufacturing has been discussed in Chapter 16. Here, we will discuss the various steps involved in the cutting of gears by different machining methods.

1. Spur Gears: Gear cutting by milling has been dealt with under Art. 16.4. Fig. 16.1 (a) shows the milling of spur gears on a horizontal milling machine with the help of a flat circular disc type cutter and Fig. 16.1(b) shows the gear milling on a vertical milling machine by making use of an end- cutter or miller.

(a) **Spur gear cutting on Horizontal Milling Machine**: The following steps are involved in cutting spur gears on Horizontal Milling Machine with involute disc gear cutter :

(i) For every module, a set of involute disc gear cutters are available as given below :

Cutter Number	No. of gear teeth (z) for which cutter is used
8	12–13
7	14–16
6	17–20
5	21–25
4	26–34
3	35–54
2	55–134
1	135 and above

From module, the number of teeth of the gear is determined as below :

Outer diameter = P. C. D. + 2x addendum

∴ Do = $m.z. + 2m = m\ (z + 2)$

∴ Z = Do/m – 2

∴ For the given outside diameter of the gear blank, Do, the number of gear teeth, z, is determined for the desired module.

(i) For z number of teeth, the gear cutter of suitable number is selected from the above table.

The cutter is mounted on the arbor of the horizontal milling machine.

(iii) The gear blank is mounted on the mandrel or between centres of a dividing head fixed on the machine table, and made true.

(iv) By trial and error, the cutter is brought just over the gear blank which is checked by equal cutting of the blank on both the sides.

(v) The table is now moved back to clear the blank. It is then moved vertically up (by elevating screw) by 2.25 m (addendum + dedendum). The longitudinal feed of the table is engaged for automatic cutting.

(vi) When the cutting of one tooth space is completed, the table is moved back.

(vii) The gear blank is indexed $\frac{40}{z}$ rotation by means of the dividing head, using simple Indexing procedure.

(viii) Following the above steps, the complete gear in milled.

(b) Spur gear cutting on a Gear Shaper: See Art 16.9 P 590.

As explained, the principal motions involved in the gear shaping are as follows :

(i) Reciprocating and Rotary motion of the cutter.

(ii) Rotary motion of the work spindle on which the gear blank is mounted.

(iii) Infeed cutter motion into the blank by cam rotation, to get depth of cut.

(iv) Gear blank relief movement during return stroke of the cutter (to avoid their rubbing with each other).

(v) Indexing motion : Slow speed continuous rotation of the cutter spindle and work spindle to provide circular feed, the two speeds being regulated through the change gears such that against each rotation of the cutter, the blank rotates through Zc/Z rotation, where,

Z = No. of teeth to be cut or the gear blank

Zc = No. of teeth on the cutter.

The cutter head moves towards right (Fig. 16.6) to achieve radial infeed, with the help of infeed cam. At least 2 or 3 passes are needed to obtain the required depth of cut. The indexing and reciprocating motions (of spindle) continue until the required number of teeth are cut all along the periphery of the gear blank.

On completion of cycle, the cutter is retracted back completely, the cut gear in unloaded and a new gear blank mounted on the work spindle.

(c) Spur gear cutting on a Hobbing machine : See Art 16.10 P 591.

It involves the following steps :

(i) The gear hob of the desired module is mounted on the machine spindle.

(ii) The angle of hob, as punched or written on the hob is noted, say α°h (right). The cutter head with the hob is rotated by α°h to the left.

(iii) The gear blank is mounted on the table and centred.

(iv) The table and hob speed ratio is adjusted so that for every single rotation of the hob, the table turns by *n/z* rotation, where n is the number of starts of the hob cutter.

Note : The hobs are commonly 75 to 150 mm ϕ. Two and three start hobs are also made. They cut faster but are less accurate.

(v) Suitable feed and speed are selected.

(vi) The hob is advanced to the blank, so that it just starts cutting.

(vii) The tool head is now moved up, above the blank so as to clear it.

It is them moved towards the gear blank by (2,25 m). For larger module, several cuts may be required.

(viii) The machine is now started on automatic feed and gear cut. See Fig. 16.8 a (ii) p 592.

(ix) Several gears can be cut at a time is one setting.

2. Helical Gears

(a) Gear cutting on a Hobbing Machine : The method differs from that for spur gear cutting, in the following two aspects :

(i) The tool head alongwith the hob is turned by $\beta \pm \alpha h$, where β is the helix angle of the helical gear. Plus sign in to be used, if the hands of the helical gear and hob are opposite (one R.H. and the L.H.) and if the hands are the same, then, minus sign is to be used.

(ii) Table speed, Rev/min $= \frac{n}{z}$(Hob speed, Rev./min.) $\pm \frac{\text{Tan}\,\beta}{\pi.m.z} \times$ vertical feed of the hob in mm/min.

If the table speed comes out to be equal to (whole numbers + a fraction), the extra rotation is achieved with the help of differential gearing.

(b) **Gear Cutting on Gear Shaper :** See Art. 16.9 P590.

(c) **Helical Gear Cutting on Universal Horizontal Milling Machine :** To facilitate the cutter to cut helical grooves, the workpiece must be given two simultaneous motions normal to each other : the linear motion along its axis and a rotary motion. The workpiece is held in the dividing hear (mounted on the machine table) which is connected by means of gears to the table feed screw of the machine. When the table feed screw rotates, the spindle of the dividing head and hence the work will rotate through change gears, worm and worm wheel of the dividing head. Also when the table feed screw rotates in the nut, the machine table and hence the dividing head has linear motion. The milling cutter is mounted on the arbor of the machine. To cut the helix, it is necessary to swing the machine table through an angle equal to the helix angle, on the vertical axis. Due to this, the spiral grooves are cut on a universal milling machine, Fig. IV.1.

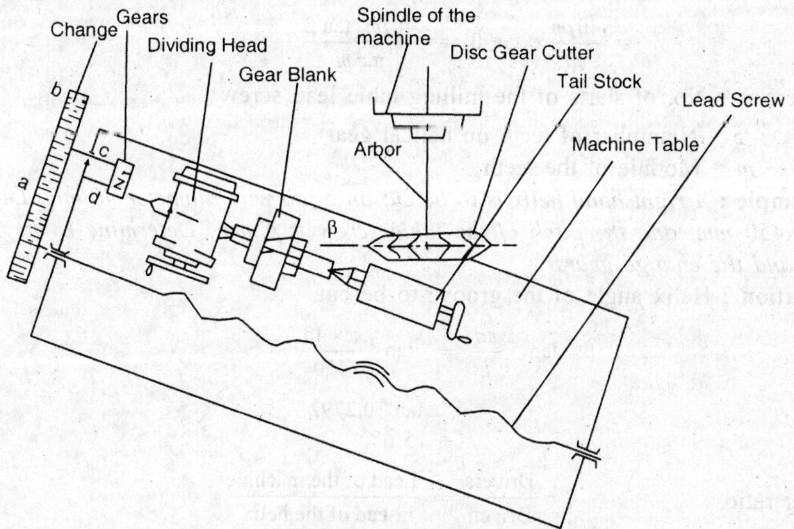

Fig. IV.1 : MILLING OF SPIRAL GROOVES.

The helix angle of the groove on the job in given as,

$$\tan\beta = \frac{\pi d}{L}$$

where d is the diameter of the job and L is the lead of the helix. The helix angle is the angle between the tangent to the thread helix on the pitch cylinder and the axis of the job.

Supposing the gear ratio between the table feed screw and the worm shaft (of the dividing head) is 1 : 1. Then, one revolution of table feed screw will rotate the worm shaft through one revolution. The dividing head spindle and hence the work will rotate through $\frac{1}{40}$ of a revolution.

That is, for the work to complete one revolution, the worm shaft and hence the table feed screw will make 40 revolutions. If "p" is the pitch of the table feed screw, then, the distance travelled by the table feed screw for one revolution of the work will be 40 p. This is called the "Lead of the machine". Hence, the lead of the machine is defined as the distance travelled by the machine table corresponding to one complete revolution of the dividing head spindle (and hence of the work) when the gear ratio between the table feed screw and worm shaft is 1 : 1. Thus,

$$\text{Lead helix will be} = \text{Lead of machine} = 40\text{ p.}$$

The diameter of the job and lead of the helix to be cut will determine the setting of the machine and the gear ratio. The hand of the helix will determine whether an idler is needed or not in the gear train. The work must rotate left handed for left hand helix and right handed for right hand helix. Thus, for right handed helix, the table feed screw and the worm shaft should rotate in the same direction and for left handed helix, they should rotate in the opposite directions.

$$\text{Gear ratio } \frac{a}{b}\times\frac{c}{d} = \frac{\text{Drivers}}{\text{Driven}}$$

$$= \frac{\text{Lead of the machine}}{\text{Lead of the helix to be cut}}$$

$$= \frac{40\,p}{L}\times n$$

$$= \frac{40\,pn}{\pi d}\times \tan\beta = \frac{40\,pn\ \tan\beta}{\pi.z.m}$$

Where n = No. of starts of the milling table lead screw

z = Nummber of teeth on helical gear

m = Module of the teeth

Example : *A right hand helix is to be cut on a job with diameter 40 mm. Lead of helix to be cut is 450 mm, and the pitch of the table screw is 6 mm. Determine is the setting of the machine and the change gears.*

Solution ; Helix angle of the groove to be cut,

$$\beta = \tan^{-1}\frac{\pi d}{L} = \tan^{-1}\frac{\pi\times 40}{450}$$

$$= \tan^{-1} 0.2792$$

$$= 15.6°$$

Gear ratio, $$i = \frac{\text{Drivers}}{\text{Driven}} = \frac{\text{Lead of the machine}}{\text{Lead of the helix}}$$

$$= \frac{40p}{450} = \frac{40\times 6}{450}$$

$$= \frac{8}{15} = \frac{2\times 4}{5\times 3}$$

$$= \frac{40\times 32}{100\times 24}$$

It is a compound gear drive.

Use a gear of 40 teeth on table feed screw, gear of 24 teeth on worm, 100 teeth gear on First gear on stud and Second gear of 32 teeth on stud, and no idler gear.

After mounting the gears, the job and the cutter, the machine feed table should be swung at an angle of 15.6°

The following steps are involved in cutting helical gears on Universal Horizontal Milling Machine :

(i) The disc gear cutter number corresponding to virtual number of teeth ($z_v = z/\cos^3 \alpha$: α pressure angle) and of given module is selected from the table given under Art IV. 1. It is then mounted on the arbor of the milling machine.

(ii) The gear blank is mounted on the mandrel or between the centres of the dividing head fixed on the machine table. The motion of the dividing head and of the machine lead screw are connected through suitable changeable gears (*a, b, c, d*) as shown is Fig. IV. I, so that for one rotation of the gear blank, the table advances by an amount equal to the pitch of the helix.

(iii) The machine table or the spindle head of the universal machine is rotated through an angle β of the gear helix, as shown in Fig. IV.1.

(iv) The remaining steps are exactly similar to the ones for milling spur gears.

2a. Herringbone Gears / Double Helical Gears : are two helical gears, having the same helix angle but of opposite hand.

Milling each gear is done either on a special machine or more commonly on a gear hobber / universal horizontal milling machine. The two gears are separately turned and milled, and then assembled together.

For the purpose of proper assembly, one section of the gear is turned with a projection and the other section with the corresponding locating surface. Both the sections are then locked with locking screws, Fig. IV. 2.

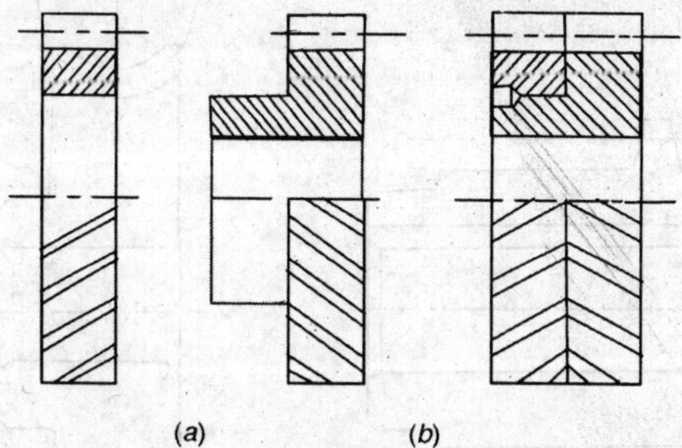

(*a*) (*b*)

Fig. IV. 2: HERRINGBONE GEARS

This was the original method of manufacturing herringbone or double helical gears. But now a days, herring bone gears are produced as a single unit on special machines which cut the teeth in two directions at one time.

3. Manufacturing of Bevel Gears : See Art. 16.11 p 593

4. Worm and Worm - Wheel Cutting :

4. 1 Worm Cutting : The worm can be cut on a thread milling machine, or more frequently cut on a lathe. The method involves the following steps :

(i) After the blank has been turned, the machine is set for the axial thread pitch of the worm ($p = \pi.\ m.\ n$; m = module, n = number of starts of the worm).

(ii) The tool is ground to the required profile and the worm cut with the usual threading operation for the given number of starts. After the first thread has been completed, the second thread is cut by advancing the compound slide by $\pi.\ m$. For more number of starts, the compound slide is advanced by $\pi.\ m.$ after the completion of each thread.

(iii) During the process, the thread thickness at a certain depth is checked to see whether the process in under control or not.

4.2 Worm - Wheel Cutting: On gear Hobbing Machine, Fig. IV.3.

The steps are :

(i) The hob cutter should have the same number of starts as that of the worm and diameter equal to or slightly greater (upto 5 mm) than that of the worm. This cutter is mounted on the machine.

(ii) If the angle of the hob is αh (R.H.) and helix angle is β (L.H.), then the tool head must be rotated to the left by an angle = $\beta + \alpha h$, and so on.

(iii) The hob cutter and table speeds are adjusted such that,

$$\frac{\text{Hob speed}}{\text{Table speed}} \text{ (Rev./min.)} = \frac{n}{z}$$

(iv) The hob cutter is adjusted centrally to the gear blank (by trial and error) so that it cuts equally both the edges.

(v) The hob is now advanced till it just starts cutting in the centre. From this position, the hob is advanced by (2.2 m) in steps. At every one complete rotation of the table, the cutter is fed till the required depth of cut is obtained.

(iv) During the process, the axial distance 'A' should be checked, which serves as a control for the proper cutting.

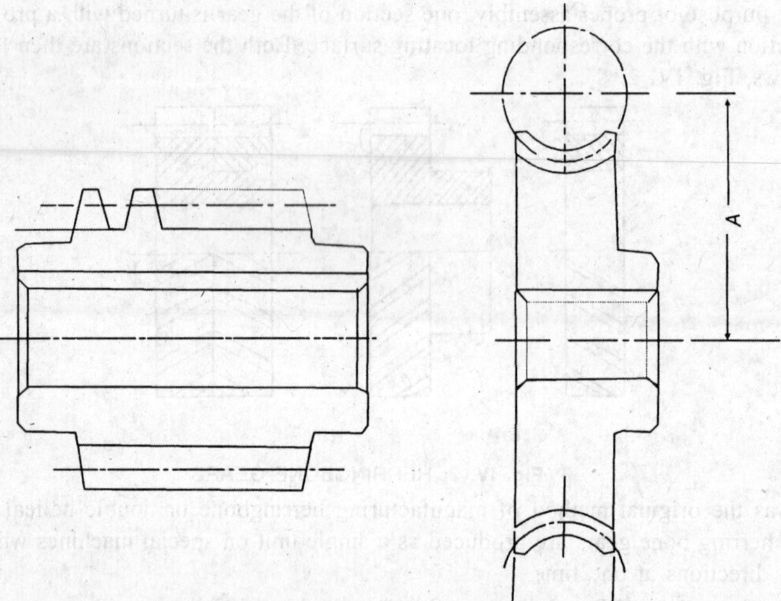

Fig. IV. 3: WORM AND WORM WHEEL

APPENDIX – V

PROBLEMS FROM COMPETITIVE EXAMINATIONS

1. Cylindrical bars of 100 mm diameter and 576 mm length are turned in a single-pass operation. The spindle speed used is 144 rpm and the total feed is 0.2 mm/rev. Taylor's tool life relationship is $VT^{0.75} = 75$, where V is the cutting speed (m/min.) and T is the tool life (min.). Calculate:–
   (*i*) Time for turning one piece.
   (*ii*) The average total change time per piece given that it takes 3 minutes to change the tool each time, and
   (*iii*) The time required to produce one piece given that the handling time is 4 min.

[*GATE 93*]

Sol. (*i*) $$T_m = \frac{L}{fN} \text{ (See equ. 4.10)}$$

$$= \frac{576}{0.2 \times 144} = 20 \text{ min.}$$

(*ii*) Taylor equation is,

$$VT^{0.75} = 75$$

Now $$V = \frac{\pi DN}{1000} = \frac{\pi \times 100 \times 144}{1000} = 45.24 \text{ m/min}$$

∴ From here, $T = 1.962$ min

∴ Number of tool change during the cutting of one piece = $\frac{20}{1.962} \approx 10$ times

∴ Total tool change time = 10 × 3 = 30 min.

∴ Total time to machine one job= 20 + 30 = 50min

Now handling time = 4 min.

∴ Total time needed to produce one piece

= 50 + 4 = 54 min.

2. If under a given condition of plain turning, the life of the cutting tool decreases by 50% due to increase in the cutting velocity by 20%, then by what percentage will the life of that tool increase due to reduction in the cutting velocity by 20% from its original value?

[*GATE 1995*] (**Ans.** 44%).

3. While turning a C-15 steel rod of 160 mm diameter at 315 rpm, 2.5 mm depth of cut and feed of 0.16 mm/rev. by a tool of geometry 0°, 10°, 8°, 9°, 15°, 75°, 0 (mm), the following observations were made:–

   Tangential component of the cutting force = 500 N

   Axial component of the cutting force = 200 N

   Chip thickness = 0.48 mm.

   Draw schematically the Merchant's circle diagram for the cutting forces in the present case.

[*GATE 1995*]

4. A 50 mm diameter disk is to be punched out from a carbon steel sheet 1.0 mm thick. The diameter of the punch should be :–
(a) 49.925 mm (*b*) 50.00 mm (c) 50.075 mm (*d*) None of these
(GATE 1996) (**Ans:**– a).

5. Five holes of diameter 10 mm each are to be punched in a sheet 3 mm thick at a pitch of 25 mm. What should be the minimum capacity of the press (in tonnes) if the yield point of the material is 50 MPa and
(*a*) one hole is punched per stroke
(*b*) five holes are punched in a single stroke
Show with a neat sketch the arrangement of punches so as to avoid eccentric loading (i.e. provide balanced loading). It may be noted that no shear is provided on the punches.
[*GATE 1996*] (**Ans:** 0.48 Tonnes, 0.40 Tonnes)

6. A project requires an initial investment of Rs. 500,000 and returns of Rs. 200,000/– at the end of each year for five years with no terminal salvage. What is the undiscounted pay back period for the project? If the interest rate is 20%, compute the discounted pay back period.
[*GATE 1996*] (**Ans.** 2.5 years, 5 years).

7. In a typical metal cutting operation, using a cutting tool of positive rake angle of 10°, it was observed that the shear angle was 20°. The friction angle is
(*a*) 45° (*b*) 30° (*c*) 60° (*d*) 40°
[*GATE 1997*] (**Ans.** c)

8. A cutting tool has a nose radius of 1.8 mm. The feed rate for a theoretical surface roughness of $R_a = 5\ \mu m$ is
(*a*) 0.36 mm/rev. (*b*) 0.187 mm/rev. (*c*) 0.036 mm/rev. (*d*) 0.0187 mm/rev.
(**Ans.** b).

9. Find the number of production runs and also the total incremental cost in a factory for the data given below:–
Annual requirement = 15,000 units
Preparation cost per order = Rs. 25
Inventory holding cost = Rs. 5/unit/year
Production rate = 100 units/day
Number of working days = 250 [*GATE 1997*].

10. In a turning trial using orthogonal tool geometry, a chip length of 84 mm was obtained for an uncut chip length of 200 mm. The cutting conditions are: V ≈ 30 m/min, uncut chip thickness = 0.5 mm, Rake angle = 20°. Cutting tool is of H.S.S. Estimate the shear plane angle ϕ, chip thickness and the shear plane angle for minimum chip strain.
[*GATE 1997*] (**Ans.** 24.7°, 1.19 mm).

11. Estimate the reduction in piercing load for producing circular hole of 50 mm diameter in a 3 mm thick steel strip, when the punch was provided with a shear of 1mm. Assume 30% penetration and shear strength of steel as 400 N/mm^2. [*GATE 1997*] (**Ans.** 99.21 kN).

12. In an orthogonal machining operation, the chip thickness and the uncut thickness are equal to 0.45 mm. If the tool rake angle is 0°, the shear plane angle is
(*a*) 45° (*b*) 30° (*c*) 18° (*d*) 60°
[*GATE 1998*] (**Ans.** a)

13. Choose the correct statement:
(*a*) A fixture is used to guide the tool as well as to locate and clamp the work piece.
(*b*) A jig is used to guide the tool as well as to locate and clamp the work piece.
(*c*) Jigs are used on CNC machines to locate and clamp the work piece and also to guide the tool.
(*d*) No arrangement to guide the tool is provided in a jig. [*GATE 1999*] (**Ans.** b)

14. What is approximate percentage change in life of a tool with zero rake angle used in Orthogonal cutting when its clearance angle is changed from 10° to 7°?

[*GATE 1999*] (**Ans.** *b*).

(*a*) 30% increase (*b*) 30% decrease (*c*) 70% increase (*d*) 70% decrease

Hints. Wear rate is proportional to tan γ.

15. In an orthogonal cutting experiment with a tool of rake angle of 7°, the chip thickness was found to be 2.5 mm when the uncut chip thickness was set to 1 mm.
(*a*) Find the shear angle.
(*b*) Find the friction angle assuming that Merchant's formula holds good.

[*GATE 1999*] (**Ans.** 22°–39'; 51°– 42')

16. The lives of two tools governed by equation $VT^{0.125} = 2.5$ and $VT^{0.25} = 7$ respectively in certain machining operation where V is the cutting speed in m/s and T is the tool life in seconds.
(*a*) Find out the speed at which both the tools have the same life. Also calculate the corresponding tool life.
(*b*) If you have to machine at a cutting speed of 1 m/s, which one of these tools will you choose in order to have less frequent tool changes? [*GATE 1999*].

Sol. Let V_0 = speed at which both the tools have the same life.

Now for the first tool,

$$V_1 T_1^{0.125} = 2.5$$

$$\therefore \quad T_1 = \left(\frac{2.5}{V_1}\right)^8$$

Similarly for the second tool,

$$T_2 = \left(\frac{7}{V_2}\right)^4$$

Now $\quad T_1 = T_2$ at $V_1 = V_2 = V_0$

$$\therefore \quad \left(\frac{2.5}{V_0}\right)^8 = \left(\frac{7}{V_0}\right)^4$$

From here, $\quad V_0 = 0.893$ m/s.

(*b*) At $\quad V_1 = V_2 = 1$ m/s

$$T_1 = (2.5)^8 = 1525.88 \text{ seconds}$$

and $\quad T_2 = (7)^4 = 2401$ seconds

Since $\quad T_2 > T_1$

$\therefore$ Second tool will be selected

17. A 5 mm thick M.S. plate is cut in a shearing machine and the length of cut is 500 mm. The shearing strength of material is 300 MPa. Find (a) the force required (b) the force required if the cutting blade is inclined at 1°, and the penetration is 40%. [*GATE 2000*]

[**Hints:** (*a*)

$$F_{max} = L.\, t.\, \tau_s$$
$$= 500 \times 5 \times 300 = 750 \text{ kN}$$

(*b*)

$$F = \frac{Kt}{I} \times F_{max} \quad \text{(Eqn. 2.7)}$$

I is in degree, so this equation can not be used.

$$\text{Length to be cut} = \frac{t}{\sin\alpha} = \frac{5}{\sin 1°} = 286.533 \text{ mm}$$

$$\therefore \quad F = 286.533 \times 5 \times 300 = 429.8 \text{ kN.}$$

18. Side cutting edge angle (according to American System) of a single point tool is 30° and this cutting edge lies in a plane parallel to the base of the tool. The effective rake angle in a plane perpendicular to both the base and the cutting edge is 20°. Find the back rake and side rake angles. *[GATE 2000]*

Hint:

$$\lambda = 90 - 30 = 60°$$

$$i = 0,\ \alpha = 20°$$

$$\tan\alpha_b = \cos\lambda\tan\alpha + \sin\lambda\tan i$$

$$\tan\alpha_s = \sin\lambda\tan\alpha - \cos\lambda\tan i$$

19. A 15 mm diameter HSS drill is used at a cutting speed of 20 m/min. and a feed rate of 0.2 mm/rev. Under these conditions, the drill life is 100 min. The drilling length of each hole is 45 mm and the time taken for idle motions is 20s. The tool change time is 300s. Calculate
(*a*) Number of holes produced using one drill
(*b*) Average production time per hole. *[GATE 2000]*

Sol.

$$V = \frac{\pi DN}{1000}$$

$$\therefore \quad N = \frac{1000 \times 20}{\pi \times 15}, \text{ rev/min}$$

$$\therefore \quad T_m = \frac{L}{fN} = \frac{45 \times 15\pi}{0.2 \times 1000 \times 20} = 0.53 \text{ min.}$$

$$= 31.8 \text{ s.}$$

(*a*) $\therefore$ Number of holes produced using one drill $= \dfrac{300}{31.8}$

$$= 9.43 \text{ say } 9$$

(*b*) Average production time per hole

= machining time + idle motion time + tool change time

$$= 31.8 + 20 + \frac{300}{9} = 85.1\text{s}$$

20. A company places orders for supply of two items A and B. The order cost for each of the items is Rs. 300/order. The inventory carrying cost is 18% of the unit price per year per unit. The unit prices of the items are Rs. 40 and Rs. 50 respectively. The annual demands are, 10,000 and 20,000 respectively. (a) Find the economic order quantities and the minimum total cost. (b) A supplier is willing to give a 1% discount on price, if both the items are ordered from him and if the order quantities for each item are 1000 units or more. Is it profitable to avail the discount? *[GATE 2000]*

Sol.

$$N_1 = \sqrt{\frac{2RA}{K}} = \sqrt{\frac{2 \times 10000 \times 300}{0.18 \times 40}}$$

$$= 912.87; \text{ say } 913$$

$$N_2 = \sqrt{\frac{2 \times 20000 \times 300}{0.18 \times 50}} = 1154.7 \text{ say } 1155$$

$$\therefore \quad (\text{Total Cost})_1 = 10000 \times 40 + 300 \times \frac{10000}{913} + 7.2 \times \frac{913}{2}$$

$$= 400{,}000 + 3285.87 + 3286.80$$

$$= \text{Rs. } 406572.67$$

$$(\text{Total Cost})_2 = 20000 \times 50 + 300 \times \frac{20000}{1155} + 9 \times \frac{1155}{2}$$

$$= 100{,}000 + 5194.80 + 5197.50$$

$$= \text{Rs. } 1010392.30$$

$$\text{Total Cost} = \text{Rs. } 1416964.97$$

(b) $N_1 = N_2 = 1000$

$C_1 = \text{Rs. } 39.60,\ C_2 = \text{Rs. } 49.50$

$$\therefore \quad (\text{T.C.})_1 = 10000 \times 39.60 + 300 \times \frac{10000}{1000} + 7.2 \times \frac{1000}{2}$$

$$= 396000 + 300 + 3600$$

$$= \text{Rs. } 402600$$

$$(\text{T.C.})_2 = 20000 \times 49.50 + 300 \times \frac{20000}{1000} + 9 \times \frac{1000}{2}$$

$$= 990000 + 6000 + 4500$$

$$= \text{Rs. } 1000500$$

$$\text{T.C.} = \text{Rs. } 1403100$$

∴ It is profitable to avail the offer.

21. The actual observed time for an operation was 1 minute per piece. If the performance rating of the operator was 120 and a 5 per cent personal time is to be provided, the standard time in minutes per piece is:

(*a*) 1.00 (*b*) 1.20 (*c*) 1.250 (*d*) 1.260

[*GATE 1993*]

(**Ans.** *d*)

Sol. Performance rating = 120

∴ Performance factor, $R = 1.20$

Now $R = \dfrac{T_a}{T_e}$ ... (eqn. 4.7)

Now observed or estimated time = 1 minute

∴ Actual needed time, $T_a = 1 \times 1.20 = 1.20$ minute

Now standard time = Actual needed time or Normal time × Allowance factor

= 1.20 × 1.05 = 1.26 minutes

22. Match four correct pairs between list I and list II:

List I (Measuring Instruments)	*List II* (Applications)
(*a*) Talysurf	1. T–Slots
(*b*) Telescopic gaugue	2. Flatness
(*c*) Transfer calipers	3. Internal diameter
(*d*) Auto-Collimator	4. Roughness.

[*GATE 1995*]

Ans. (*a*) – 4; (*b*) – 3; (*c*) – 1; (*d*) – 2.

23. (*a*) Schematically draw a single point turning tool which is designated as: 10°, 12°, 7°, 5°, 20°, 50°, 0 (mm) and show the angles in that sketch.

(*b*) Also determine the values of normal rake and normal clearance of the above mentioned tool. [*GATE 1995*].

Hint: (*a*) The ASA designation of single point cutting tool is:–

$$\alpha_b, \alpha_s, \theta_s, C_e, C_s, R$$

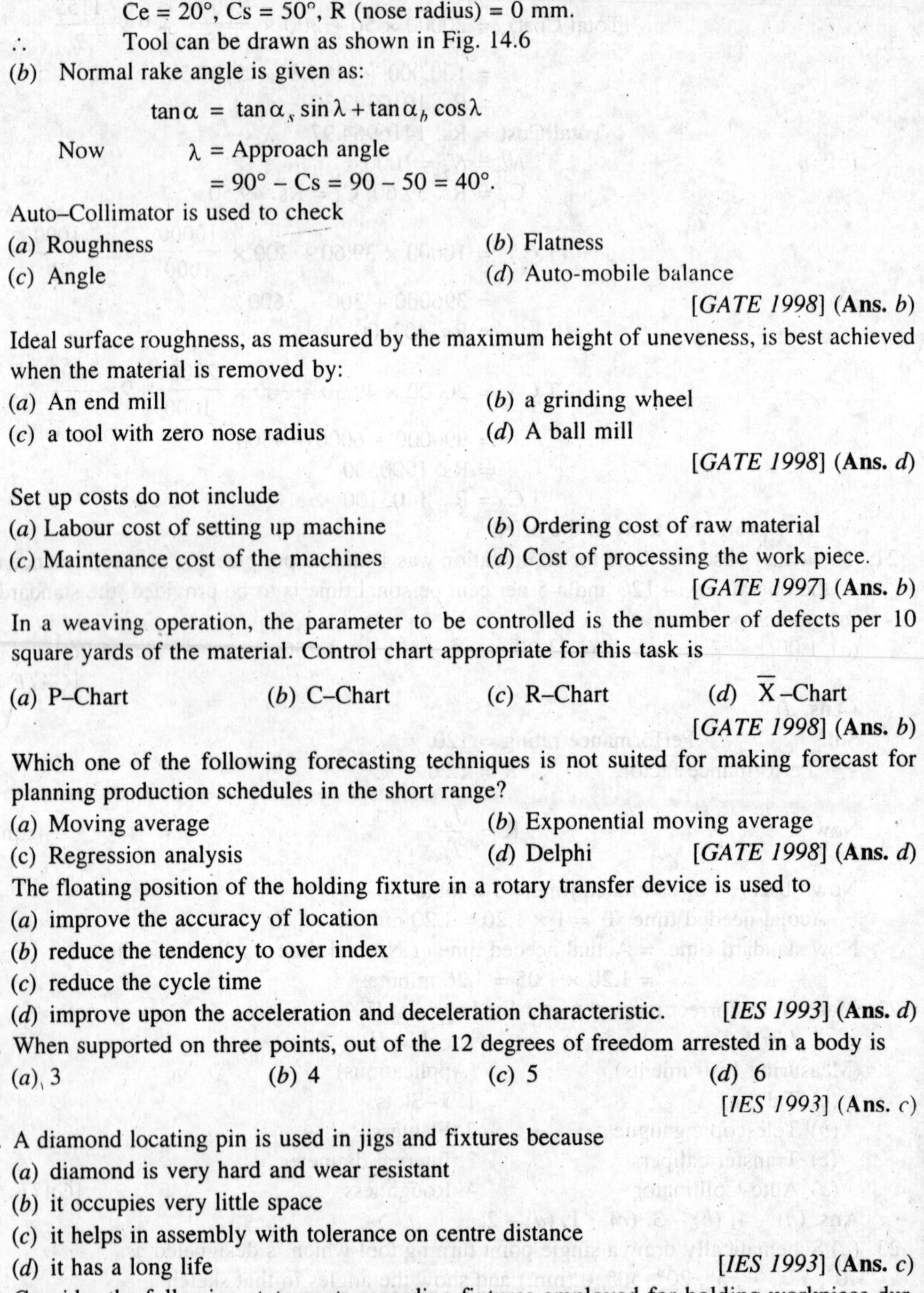

$\therefore$ $\alpha_b = 10°, \alpha_s = -12°, \theta_e = 7°, \theta_s = 5°$

Ce = 20°, Cs = 50°, R (nose radius) = 0 mm.

$\therefore$ Tool can be drawn as shown in Fig. 14.6

(*b*) Normal rake angle is given as:

$$\tan\alpha = \tan\alpha_s \sin\lambda + \tan\alpha_b \cos\lambda$$

Now λ = Approach angle

= 90° – Cs = 90 – 50 = 40°.

24. Auto–Collimator is used to check
    (*a*) Roughness (*b*) Flatness
    (*c*) Angle (*d*) Auto-mobile balance
    [*GATE 1998*] (**Ans.** *b*)

25. Ideal surface roughness, as measured by the maximum height of uneveness, is best achieved when the material is removed by:
    (*a*) An end mill (*b*) a grinding wheel
    (*c*) a tool with zero nose radius (*d*) A ball mill
    [*GATE 1998*] (**Ans.** *d*)

26. Set up costs do not include
    (*a*) Labour cost of setting up machine (*b*) Ordering cost of raw material
    (*c*) Maintenance cost of the machines (*d*) Cost of processing the work piece.
    [*GATE 1997*] (**Ans.** *b*)

27. In a weaving operation, the parameter to be controlled is the number of defects per 10 square yards of the material. Control chart appropriate for this task is
    (*a*) P–Chart (*b*) C–Chart (*c*) R–Chart (*d*) $\overline{X}$–Chart
    [*GATE 1998*] (**Ans.** *b*)

28. Which one of the following forecasting techniques is not suited for making forecast for planning production schedules in the short range?
    (*a*) Moving average (*b*) Exponential moving average
    (c) Regression analysis (*d*) Delphi [*GATE 1998*] (**Ans.** *d*)

29. The floating position of the holding fixture in a rotary transfer device is used to
    (*a*) improve the accuracy of location
    (*b*) reduce the tendency to over index
    (*c*) reduce the cycle time
    (*d*) improve upon the acceleration and deceleration characteristic. [*IES 1993*] (**Ans.** *d*)

30. When supported on three points, out of the 12 degrees of freedom arrested in a body is
    (*a*) 3 (*b*) 4 (*c*) 5 (*d*) 6
    [*IES 1993*] (**Ans.** *c*)

31. A diamond locating pin is used in jigs and fixtures because
    (*a*) diamond is very hard and wear resistant
    (*b*) it occupies very little space
    (*c*) it helps in assembly with tolerance on centre distance
    (*d*) it has a long life [*IES 1993*] (**Ans.** *c*)

32. Consider the following statements regarding fixtures employed for holding workpiece during machining.
    1. The location is based on the 3-2-1 principle
    2. The numbers refers to the pins employed in three mutually perpendicular planes to arrest all the degrees of freedom

3. Fixture also provides good guidance. Of these statements
(*a*) 1, 2 and 3 are correct (*b*) 2 and 3 are correct (*c*) 1 and 3 are correct (*d*) 1 and 2 are correct

[*IES 1994*] (**Ans.** d)

33. Match List I with List II and select the correct answer using the codes given below the lists:

List I (Task)	List II (Recommendation)
A Three components in a straight line should be worked in one loading	1. Clamp with a floating pad
B Unloading of clamp element from jig is essential	2. Quick action nut
C Clamping of rough surfaces	3. Cam clamp
D Need for heavy clamping force	4. Equalising clamp
	5. Strap Clamp

Codes:	A	B	C	D		A	B	C	D
(*a*)	5	2	3	4	(*b*)	4	2	1	5
(*c*)	1	4	2	3	(*d*)	4	1	5	3

[*IES 95*] (**Ans.** *d*)

34. If the diameter of the hole is subjected to considerable variation, then for locating in jigs and fixtures, the pressure type of locator used is
(*a*) Conical locator (*b*) Cylinderical locator
(*c*) Diamond pin locator (*d*) Vee locator [*IES 1995*] (**Ans.** *a*)

35. Consider the following statements:
The cutter setting block in a milling fixture
1. Sets the cutting tool with respect to two of its surfaces.
2. Limits the total travel required by the cutter during machining.
3. Takes location from the location scheme of the component.
(a) 1, 2, 3 are correct (*b*) 1 and 2 are correct
(*c*) 2 and 3 are correct (*d*) 1 and 3 are correct [*IES 1996*] (**Ans.** b)

36. One of the pins in a dual pin locator of a jig or fixture is shaped as a "diamond pin locator" because
(*a*) diamond does not wear fast
(*b*) it is easy to clamp
(*c*) any variation between the centres of the hole is taken care of
(*d*) it will be easy to machine afterwards when the locator is worn out. [*IES 1998*] (**Ans.** c)

37. Diamond pin location is used in a fixture because
(*a*) it does not wear out
(*b*) it takes care of any variation is centre distance between two holes
(*c*) it is easy to clamp the part on diamond pins
(*d*) it is easy to manufacture. [*IES 1999*] (**Ans.** *b*)

38. In sheet metal work, the cutting force on the tool can be reduced by
(*a*) grinding the cutting edges sharp (*b*) increasing the hardness of tool
(*c*) providing shear angle on tool (*d*) increasing the hardness of die

[*IES 1993*] (**Ans.** *c*)

39. In sheet metal blanking, shear is provided on punches and dies so that
(*a*) press load is reduced (*b*) good cut edge is obtained
(*c*) warping of sheet is minimised (*d*) cut blanks are straight

[*IES 1994*] (**Ans.** *a*)

40. For obtaining a cup of diameter 25 mm and height 15 mm by drawing, the size of the round blank should be approximately:
(*a*) 42 mm (*b*) 44 mm (*c*) 46mm (*d*) 48 mm

[*IES 1994*] (**Ans.** c)

41. For 50% penetration of work material, a punch with single shear equal to thickness will
    (*a*) reduce the punch load to half the value
    (*b*) increase the punch load to half the value
    (*c*) maintain the same punch load
    (*d*) reduces the punch load to quarter load. [*IES 1997*] (**Ans.** *a*, See. equn. 2.7)
42. A cup of 10 cm height and 5 cm diameter is to be made from a sheet metal of 2 mm thickness. The number of deductions needed will be:
    (*a*) one (*b*) two (*c*) three (*d*) four
    [*IES 1997*] (**Ans.** *c*; See table 2.9)
43. Consider the following statements:
    Earing in a drawn cup can be due to non-uniform
    1. speed of the press 2. clearance between tools
    3. material properties 4. blank holding
    Which of these statements are correct?
    (*a*) 1, 2 and 3 (*b*) 2, 3 and 4 (*c*) 1, 3 and 4 (*d*) 1, 2 and 4
    [*IES 1999*] (**Ans.** *d*)
44. A hole is to be punched in a 15mm thick plate having an ultimate shear strength of 3 kN/mm^2. If the allowable crushing stress in the punch is 6 kN/mm^2, the diameter of the smallest hole which can be punched is equal to
    (*a*) 15 mm (*b*) 30 mm (*c*) 60 mm (*d*) 120 mm
    [*IES 1999*] (**Ans.** *b*)
45. In a blanking operation to produce steel washer, the maximum punch load used is 2×10^5 N. The plate thickness is 4 mm and percentage penetration is 25%. The work done during this shearing operation is
    (*a*) 200 J (*b*) 400 J (*c*) 600 J (*d*) 800 J
    [*IAS 1994*] (**Ans.** *a*)
46. Consider the following factors
    1. Clearance between the punch and the die is too small.
    2. The finish at the corners of the punch is poor.
    3. The finish at the corners of the die is poor.
    4. The punch and die alignment is not proper.
    The factors responsible for the vertical lines parallel to the axis noticed on the outside of a drawn cylindrical cup would include
    (*a*) 2, 3 and 4 (*b*) 1 and 2 (*c*) 2 and 4 (*d*) 1, 3 and 4
    [*IAS 1994*] (**Ans.** *d*)
47. In blanking operation the clearance provided is
    (*a*) 50% on punch and 50% on die
    (*b*) On die
    (*c*) On punch
    (*d*) On die or punch depending upon designers choice [*IAS 1995*] (**Ans.** *c*)
48. Which one of the following manufacturing processes requires the provision of "gutters"?
    (*a*) Closed die forging (*b*) Centrifugal casting
    (*c*) Investment casting (*d*) Impact extrusion [*IES 1993*] (**Ans.** *a*)
49. Which of the following pairs of process and draft are correctly matched?
    1. Rolling ..... 2 2. Extrusion .... 50 3. Forging 4
    Select the correct answer using the codes given below: Codes:
    (*a*) 1, 2 and 3 (*b*) 1 and 2 (*c*) 1 and 3 (*d*) 2 and 3
    [**IES 1994**] (**Ans.** a)

50. Which one of the following processes is most commonly used for the forging of bolt heads of hexagonal shape?
    (*a*) Closed die drop forging (*b*) Open die upset forging
    (*c*) Closed die press forging (*d*) Open die progressive forging
    [*IES 1998*] (**Ans.** *d*)

51. Consider the following operations involved in forging a hexagonal bolt from a round bar stock, whose diameter is equal to the bolt diameter:
    1. Flattening 2. Upsetting 3. Swaging 4. Cambering
    The correct sequence of these operations is
    (*a*) 1, 2, 3 and 4 (*b*) 2, 3, 4, 1 (*c*) 2, 1, 3, 4 (*d*) 3, 2, 1, 4
    [*IES 1999*] (**Ans.** *a*)

52. The time (in minutes) for drilling a hole is given by

$$t = \frac{\text{Depth of hole} + \text{h}}{\text{Feed} \times \text{RPM}}$$ where '*h*' is the

    (*a*) Length of the drill (*b*) drill diameter
    (*c*) flute length of the drill (*d*) cone height of the drill
    [*IAS 1994*] (**Ans.** *d*)

53. Stroke of a shaping machine is 250 mm. It takes 30 double strokes per minute. Overall average speed of operation is
    (*a*) 3.75 m/min (*b*) 5-0 m/min (*c*) 7.5 m/min (*d*) 15-0 m/min
    [*IAS 1994*] (**Ans.** *c*)

54. A drill bit of 20 mm diameter rotating at 500 rpm with a feed rate of 0.2 mm/rev. is used to drill a through hole is a M.S. plate of 20 mm thickness. The depth of cut in this drilling operation is
    (*a*) 100 mm (*b*) 20 mm (*c*) 10 mm (*d*) 0.2 mm
    [*IAS 1995*] (**Ans.** *b*)

55. Production cost refers to prime cost plus
    (*a*) factory over heads
    (*b*) factory and administration over heads
    (*c*) factory, administration and sales overheads
    (*d*) factory, administration, sales overheads and profits. [*IES 1995*] (**Ans.** *b*)

56. A grinding wheel of 150 mm diameter is rotating at 3000 rpm. The grinding speed is
    (*a*) 7.5π m/s (*b*) 15π m/s (*c*) 45π m/s (*d*) 450π m/s
    [*IES 1996*] (**Ans.** *a*)

57. The standard time of an operation has been calculated as 10 minutes. The worker was rated at 80%. If the relaxation and other allowances were 25%, then the observed time would be
    (*a*) 12.5 min. (*b*) 10 min. (*c*) 8 min. (*d*) 6.5 min
    [*IES 1999*] (**Ans.** *b*)

58. Money required for the purchase of stores, payment of wages etc. is known as:
    (*a*) Block Capital (*b*) Reserved Capital
    (*c*) Authorised Capital (*d*) Working Capital [*IAS 1994*] (**Ans.** *d*)

59. Fixed investment for manufacturing a product in a particular year is Rs. 80,000/–. The estimated sales for this period is Rs. 2,00,000/–. The variable cost per unit for this product is Rs. 4/–. If each unit is sold at Rs. 20/–, then the break even point would be
    (*a*) 4,000 (*b*) 5,000 (*c*) 10,000 (*d*) 20,000
    [*IAS 1994*] (**Ans.** *b*)

60. For a small scale industry, the fixed cost per month is Rs. 5000/–. The variable cost per product is Rs. 20/– and the sales price is Rs. 30/– per piece. The break even production per month will be:
(*a*) 300 (*b*) 400 (*c*) 500 (*d*) 10000
[*IES 1995*] (**Ans.** c)

61. Two alternative methods can produce a product. The first method has a fixed cost of Rs. 2000/– and variable cost of Rs. 20/– per piece. The second method has a fixed cost of Rs. 1500/– and a variable cost of Rs. 30/–. The break even quantity between the two alternatives is
(*a*) 25 (*b*) 50 (*c*) 75 (*d*) 100
[*IES 1996*] (**Ans.** *b*)

62. Process I requires 20 units of fixed cost and 3 units of variable cost per piece, while process II required 50 units of fixed cost and 1 unit of variable cost per piece. For a company producing 10 pieces per day
(*a*) Process I should be chosen
(*b*) Process II should be chosen
(*c*) either of the two processes could be chosen
(*d*) a combination of process I and process II should be chosen [*IES 1997*] (**Ans.** a)

63. Interchangeability can be achieved by
(*a*) standardisation (*b*) better process planning
(*c*) simplification (*d*) better product planning
[*IES 1993*] (**Ans.** a)

64. A multispindle automatic performs four operations with times 50, 60, 65 and 75 seconds at each of its work centres. The cycle time (time required to manufacture one work piece) in seconds will be
(*a*) 50 + 60 + 65 + 75 (*b*) (50 + 60 + 65 + 75)/4
(c) 75/4 (*d*) 75 [*IAS 1994*] (**Ans.** *a*)

65. In a single spindle automatic lathe, two tools are mounted on the turret, one form tool on the front slide and the other, a parting tool on the rear slide. The parting tool operation is much longer than form tool operation and they operate simultaneously (overlap). The number of cams required for this job is
(*a*) one (*b*) two (*c*) three (*d*) four
[*IES 1994*] (**Ans.** a)

66. Which one of the following steps would lead to interchangeability?
(*a*) Quality control (*b*) Process planning (*c*) Operator training (*d*) Product design
[*IES 1994*] (**Ans.** *a*)

67. If the chip-tool contact length is reduced slightly by grinding the tool face, then
(*a*) both cutting force and interface temperature would decrease.
(*b*) both cutting force and interface temperature would increase.
(*c*) the cutting force would decrease but the interface temperature would increase
(*d*) the cutting force would increase but the interface temperature would decrease.
[*IES 1993*] (**Ans.** *c*)

68. Tool life in the case of a grinding wheel is the time
(*a*) between two successive regrinds of the wheel
(*b*) taken for the wheel to be balanced
(*c*) taken between two successive wheel dressings
(*d*) taken for a wear of 1 mm on its diameter. [*IES 1993*] (**Ans.** a)

69. For achieving a specific surface finish in single point turning, the most important factor to be controlled is
(*a*) depth of cut (*b*) cutting speed (*c*) feed (*d*) tool rake angle
[*IES 1993*] (**Ans.** *c*)

70. In ASA system, if the tool nomenclature is 8–6–5–5–10–15–2mm, then the side rake angle will be
(*a*) 5° (*b*) 6° (*c*) 8° (*d*) 10°
[*IES 1993*] (**Ans**. *b*)

71. Consider the following characteristics :
1. The cutting edge is normal to the cutting velocity
2. The cutting forces occur in two directions only
3. The cutting edge is wide than the depth of cut.
The characteristics applicable to orthogonal cutting would include
(*a*) 1 and 2 (*b*) 1 and 3 (*c*) 2 and 3 (*d*) 1, 2 and 3
[*IAS 1994*] (**Ans.** a)

72. Thrust force will increase with increase in
(*a*) Side cutting edge angle (*b*) tool nose radius
(*c*) rake angle (*d*) end cutting edge angle
[*IAS 1995*] (**Ans.** *a*)

73. In an orthogonal cutting, the depth of cut is halved and the feed rate is doubled. If the chip thickness ratio is unaffected with the changed cutting conditions, the actual chip thickness will be
(*a*) doubled (*b*) halved (*c*) quadrupled (*d*) unchanged
[*IAS 1995*] (**Ans.** *a*)

74. In a single point turning operation with a cemented carbide and steel combination having a Taylor exponent of 0.25, if the cutting speed is halved, then the tool life will become
(*a*) half (*b*) two times (*c*) eight times (*d*) sixteen times
[*IAS 1995*] (**Ans.** *d*)

75. The angle between the face and the flank of the single point cutting tool is known as
(*a*) rake angle (*b*) clearance angle (*c*) lip angle (*d*) point angle
[*IES 1995*] (**Ans.** *c*)

76. Single point thread cutting tool should ideally have
(*a*) Zero rake (*b*) positive rake (*c*) negative rake (*d*) normal rake
[*IES 1995*] (**Ans.** *a*)

77. Consider the following statements about nose radius
1. It improves tool life 2. It reduces the cutting force
3. It improves the surface finish. Of these statements
(*a*) 1 and 2 are correct (*b*) 2 and 3 are correct
(c) 1 and 3 are correct (*d*) 1,2 and 3 are correct
[*IES 1995*] (**Ans.** *d*)

78. Match list I with List II and select the correct answer using the codes given below the lists

List I (*Wear type*)	List II (*Associated mechanism*)
A Abrasive wear	1. Galvanic action
B Adhesive wear	2. Ploughing action
C Electrolytic wear	3. Molecular transfer
D Diffusion wear	4. Plastic deformation
	5. Metallic bond

Codes: A B C D A B C D A B C D A B C D
(a) 2 5 1 3 (b) 5 2 1 3 (c) 2 1 3 4 (d) 5 2 3 4
[*IES 1995*] (**Ans.** *a*)

79. Crater wear is predominant in
(*a*) Carbon tool steel (*b*) tungsten carbide tools
(*c*) high speed steel tools (*d*) Ceramic tools [*IES 1995*] (**Ans.** *b*)

80. Consider the following work materials:

1. Titanium   2. Mild steel   3. Stainless steel   4. Grey cast iron

The correct sequence of these materials in terms of increasing order of difficulty in machining is

(*a*) 4, 2, 3, 1   (*b*) 4, 2, 1, 3   (*c*) 2, 4, 3, 1   (*d*) 2, 4, 1, 3

[*IES 1995*] (**Ans.** *a*)

81. The primary tool force used in calculating the tool power consumption in machining is the

(*a*) radial force   (*b*) tangential force   (*c*) axial force   (*d*) frictional force

[*IES 1995*] (**Ans.** *b*)

82. Chip equivalent is increased by

(*a*) an increase in side cutting edge angle of tool

(*b*) an increase in nose radius and side cutting edge angle of tool

(*c*) increasing the plan area of cut

(*d*) increasing the depth of cut [*IES 1996*] (**Ans.** *a*)

83. Notch wear at the outside edge of the depth of cut is due to

(*a*) abrasive action of the work hardened chip material

(*b*) oxidation

(*c*) slip-stick action of the chip

(*d*) chipping [*IES 1996*] (**Ans.** *c*)

84. Which of the following indicate better machinabilities?

1. Small shear angle   2. Higher cutting forces

3. Longer tool life   4. Better surface finish

(*a*) 1 and 3   (b) 2 and 4   (c) 1 and 2   (d) 3 and 4

[*IES 1996*] (**Ans.** *d*)

85. Which of the following forces are measured directly by strain gauges or force dynamometers during metal cutting?

1. Force exerted by the tool on the chip acting normally to the tool fake.
2. Horizontal cutting force exerted by the tool on the work piece.
3. Frictional resistance of the tool against the chip flow action along the tool face.
4. Vertical force which helps in holding the tool in position.

(*a*) 1 and 3   (*b*) 2 and 4   (*c*) 1 and 4   (*d*) 2 and 3

[*IES 1996*] (**Ans.** *b*)

86. Specific energy requirements in a grinding process are more than those in turning for the same metal removal rate because of the

(*a*) specific pressures between wheel and work being high

(*b*) size effect of the larger contact areas between wheel and work

(*c*) high cutting velocities

(*d*) high heat produced during grinding [*IES 1996*] (**Ans.** *d*)

87. In orthogonal cutting, the depth of cut is 0.5 mm at a cutting speed of 2 m/s. If the chip thickness is 0.75 mm, the chip velocity is

(*a*) 1.33 m/s   (*b*) 2 m/s   (*c*) 2.5 m/s   (*d*) 3 m/s

[*IES 1997*] (**Ans.** *a*)

88. Consider the following elements:

1. Nose radius   2. Cutting speed   3. Depth of cut   4. Feed

The correct sequence of these elements in decreasing order of their influence on tool life is

(*a*) 2, 4, 3, 1   (*b*) 4, 2, 3, 1   (*c*) 2, 4, 1, 3   (*d*) 4, 2, 1, 3

[*IES 1997*] (**Ans.** *a*)

89. The rake angle in a twist drill
   (*a*) varies from minimum near the dead centre to a maximum value at the periphery
   (*b*) is maximum at the dead centre and zero at the periphery
   (*c*) is constant at every point of the cutting edge
   (*d*) is a function of the size of the chisel edge. [*IES 1997*] (**Ans.** c)

90. Consider the following forces acting on a finish turning tool:
   1. Feed force 2. Thrust force 3. Cutting force
   The correct sequence of the decreasing order of the magnitudes of these forces is
   (*a*) 1, 2, 3 (*b*) 2, 3, 1 (*c*) 3, 1, 2 (*d*) 3, 2, 1
   [*IES 1997*] (**Ans.** *c*)

91. In turning operation, the feed could be doubled to increase the metal removal rate. To keep the same level of surface finish, the nose radius of the tool should be
   (*a*) halved (*b*) kept unchanged (*c*) doubled (*d*) made four times
   [*IES 1999*] (**Ans.** *b*)

92. The radial force in single-point tool during turning operation varies between
   (*a*) 0.2 to 0.4 times the main cutting force (*b*) 0.4 to 0.6 times the main cutting force
   (*c*) 0.6 to 0.8 times the main cutting force (*d*) 0.5 to 0.6 times the main cutting force
   [*IES 1999*] (**Ans.** *a*)

93. Consider the following criteria in evaluating machinability:
   1. Surface finish 2. Type of chips
   3. Tool life 4. Power consumption
   In modern high speed CNC machining with coated carbide tools, the correct sequence of these criteria in decreasing order of their importance is
   (*a*) 1, 2, 4, 3 (*b*) 2, 1, 4, 3 (*c*) 1, 2, 3, 4 (*d*) 2, 1, 3, 4
   [*IES 1998*] (**Ans.** *c*)

94. Internal gears are made by
   (*a*) hobbing (*b*) shaping with pinion cutter
   (*c*) shaping with rack cutter (*d*) milling [*IES 1993*] (**Ans.** *b*)

95. The blank diameter used in thread rolling will be
   (*a*) equal to minor diameter of the thread
   (*b*) equal to pitch diameter of the thread
   (c) a little larger than the minor diameter of the thread
   (*d*) a little larger than the pitch diameter of the thread. [*IES 1993*] (**Ans.** *b*)

96. While cutting helical gears on a non-differential gear hobber, the feed change gear ratio is
   (*a*) independent of index change gear ratio
   (*b*) dependent on speed change gear ratio
   (*c*) interrelated to index change gear ratio
   (*d*) independent of speed and index change gear ratio [*IES 1995*] (**Ans.** *c*)

97. Consider the following processes of gear manufacture
   1. Milling with form cutter 2. Rack type gear shaper (gear planer)
   3. Rotary gear shaper (gear shaper) 4. Gear hobbing
   The correct sequence of these processes in increasing order of accuracy of involute profile of the gear is
   (*a*) 3, 2, 4, 1 (*b*) 2, 3, 4, 1 (*c*) 3, 2, 1, 4 (*d*) 2, 3, 1, 4
   [*IES 1996*] (**Ans.** a)

98. Gear cutting on a milling machine using an involute profile cutter is a
   (*a*) gear forming process (*b*) gear generating process
   (*c*) gear shaping process (*d*) highly accurate gear producing process
   [*IES 1996*] (**Ans.** *a*)

99. For the manufacture of full depth spur gear by hobbing process, the number of teeth to be cut = 30, module = 3 mm and pressure angle = 20°. The radial depth of cut to be employed should be equal to

(*a*) 3.75 mm (*b*) 4.50 mm (*c*) 6.00 mm (*d*) 6.75 mm

[*IES 1996*] (**Ans.** *d*)

100. Which of the following motions are not needed for spur gear cutting with a hob?

1. Rotary motion of hob
2. Linear axial reciprocatory motion of hob
3. Rotary motion of gear blank
4. Radial advancement of hob

Select the correct answer using the codes given below:

(*a*) 1, 2 and 3 (*b*) 1, 3 and 4 (*c*) 1, 2 and 4 (*d*) 2, 3 and 4

[*IES 1997*] (**Ans.** *d*)

101. Match List I (Gear component) with List-II (Preferred method of manufacturing) and select the correct answer using the codes given below the lists:

List I	*List II*
A. Gear for clocks	1. Hobbing
B. Bakelite gears	2. Stamping
C. Aluminium gears	3. Powder Compacting
D. Automobile transmission gears	4. Sand Casting
	5. Extrusion

Codes: A B C D A B C D A B C D A B C D

(a) 2 3 5 1 (b) 5 3 4 2 (c) 5 1 2 3 (d) 2 4 5 3

[*IES 1997*] (**Ans.** *a*)

102. A straight teeth slab milling cutter of 100 mm diameter and 10 teeth rotating at 200 r.p.m. is used to remove a layer of 3 mm thickness from a steel bar. If the table feed is 400 mm/min, the feed per tooth in this operation will be:

(*a*) 0.2 mm (*b*) 0.4 mm (*c*) 0.5 mm (*d*) 0.6 mm

[*IES 1999*] (**Ans.** *a*)

103. Consider the following processes for manufacture of gears:

1. Casting
2. Powder metallurgy
3. Machining from bar stock
4. Closed die forging

The correct sequence is increasing order of bending strength of gear teeth is

(*a*) 1, 2, 3, 4 (*b*) 1, 2, 4, 3 (*c*) 2, 1, 4, 3 (*d*) 2, 1, 3, 4

[*IES 1999*] (**Ans.** d)

104. The ratio between two consecutive spindle speeds for a six-speed drilling machine using drills of diameter 6.25 to 25 mm size and a cutting velocity of 18 m/min is

(*a*) 1.02 (*b*) 1.32 (*c*) 1.62 (*d*) 1.82

[*IES 1994*] (**Ans.** *c*)

105. A PERT network has three activities on critical path with near time 3,8 and 6 and standard derivation 1, 2 and 2 respectively. The probability that the project will be completed in 20 days is

(*a*) 0.50 (*b*) 0.66 (*c*) 0.84 (*d*) 0.95

[*IES 1993*]

Sol. The standard deviation of all activities on critical paths

$$\sigma_{cp} = \sqrt{\sigma_1^2 + \sigma_2^2 + \sigma_3^2} = 3$$

The probability that the profit will be completed in a given tıme is given by

$$p = \phi\ (Z) \qquad \text{... (23.15)}$$

$$\therefore \quad Z = \frac{t_d - t_{cp}}{\sigma_{cp}}$$

$$t_d = 20 \text{ days}$$

$$t_{cp} = \sum te = 3 + 8 + 6 = 17 \text{ days}$$

$$\therefore \quad Z = \frac{20-17}{3} = 1$$

From Fig. 23.17, for Z = 1, p = 0.84

∴ (**Ans.** *c*)

106. The following activities are to be performed in a particular sequence for routing a product:
1. Analysis of the product and breaking it down into components.
2. Determination of the lot size.
3. Determination of operation and processing time requirement.
4. Taking make or buy devision.
The correct sequence of these activities is
(*a*) 1, 2, 3, 4 (*b*) 3, 1, 2, 4 (*c*) 3, 1, 4, 2 (*d*) 1, 4, 3, 2
[*IAS 1994*] (**Ans.** *a*)

107. If orders are placed once a month to meet an annual demand of 6000 units, then the average inventory would be
(*a*) 200 (*b*) 250 (*c*) 300 (*d*) 500
[*IAS 1994*] (**Ans.** *d*)

108. Which of the following characteristics are more important in the equipment selected for mass production shops?
1. Fast output 2. Low tooling cost 3. Lower labour cost 4. Versatility
[*IAS 1995*] (**Ans.** 1 and 3)

109. Which one of the following charts gives simultaneously, information about the progress of work and machine loading?
(*a*) Process chart (*b*) Machine load chart
(c) Mass-machine chart (*d*) Gantt chart
[*IAS 1995*] (**Ans. c**)

110. There are two products A and B with the following characteristics:

Product	Demand (in units)	Order cost (Rs/order)	Holding cost (Rs/unit/year)
A	100	100	4
B	400	100	1

The economic order quantities (EOQ) of products A and B will be in the ratio
(*a*) 1 : 1 (*b*) 1 : 2 (*c*) 1 : 4 (*d*) 1 : 8
[*IES 1994*] (**Ans.** *a*)

111. Which one of the following methods can be used for forecasting the sales potential of a new product?
(*a*) Time series analysis (*b*) Jury of Executive opinion method
(*c*) Sales force composite method (*d*) Direct survey method
[*IES 1995*] (**Ans.** *d*)

112. Classifying items in A, B and C categories for selective control in inventory management is done by arranging items in the decreasing order of
(*a*) total inventory costs (*b*) item value
(*c*) annual usage value (*d*) item demand [*IES 1995*] (**Ans.** *a*)

113. In manufacturing management, the term "Despatching" is used to describe
(*a*) despatch of sales order
(*b*) despatch of factory mail
(*c*) despatch of the finished product to the user
(*d*) despatches of work orders through shop floor. [*IES 1995*] (**Ans.** *d*)

114. Which of the following factors are to be considered for production scheduling?
1. Sales forecast 2. Component design
3. Route sheet 4. Time standards
Codes:
(*a*) 1 2, 3 (*b*) 1, 2 and 4 (*c*) 1, 3 and 4 (*d*) 2, 3 and 4
[*IES 1995*] (**Ans.** *d*)

115. Consider the following statements:
1. ABC analysis is based on Pareto's principle
2. FIFO and LIFO policies can be used for material valuation in materials management.
3. Simulation can be used for inventory control
4. EOQ (Economic Order Quantity) formula ignores variation in demand pattern.
Of these statements.
(*a*) 1 alone is correct (*b*) 1 and 3 are correct
(*c*) 2, 3 and 4 are correct (*d*) 1, 2, 3 and 4 are correct
[*IES 1995*] (**Ans.** *d*)

116. In inventory control theory, the economic order quantity is
(*a*) average level of inventory
(*b*) optimum lot size
(*c*) lot size corresponding to break-even analysis
(*d*) capacity of warehouse. [*IES 1995*] (**Ans.** *b*)

117. The routing functions in a production system design is concerned with
(*a*) man power utilisation
(*b*) machine utilisation
(c) quality assurance of the product
(*d*) optimising material flow through the plant. [*IES 1996*] (**Ans.** *b*)

118. A production line is said to be balanced when
(*a*) there are equal number of machines at each work station
(*b*) there are equal number of operators at each work station
(*c*) the waiting time for service at each station is the same
(*d*) the operation time at each station is the same. [*IES 1997*] (**Ans.** *d*)

119. Annual demand for a product costing Rs. 100 per piece is Rs. 900. Ordering cost per order is Rs. 10 and inventory holding cost is Rs. 2 per unit per year. The economic lot size is
(*a*) 200 (*b*) 300 (*c*) 400 (*d*) 500
[*IES 1997*] (**Ans.** *d*)

120. Match List I (Methods) with List II (Problems) and select the correct answer using the codes given below the lists:

List I	*List II*
A Moving average	1. Assembly
B Line balancing	2. Purchase
C Economic lot size	3. Forecasting
D Johnson algorithm	4. Sequencing

Codes: A B C D
(*a*) 1 3 2 4
(*b*) 1 3 4 2
(*c*) 3 1 4 2
(*d*) 3 1 2 4
[*IES 1998*] (**Ans.** *d*)

121. Consider the following statements:
Dispatching
1. is the function of operation, planning and control
2. releases work to the operating directions
3. conveys instructions to the shop floor of these statements
(*a*) 1, 2 and 3 are correct (*b*) 1 and 2 are correct
(*c*) 2 and 3 are correct (*d*) 1 and 3 are correct [*IES 1998*] (**Ans.** *b*)

122. Which of the following are the benefits of assembly line balancing? 1. In minimises the m-process inventory 2. It reduces the work content 3. It smoothens the production flow 4. It maintains the required rate of output.
Select the correct answer using the codes given below:
Codes:
(*a*) 1, 2 and 3 (*b*) 2, 3 and 4 (*c*) 1, 3 and 4 (*d*) 1, 2 and 4
[*IES 1998*] (**Ans.** *d*)

123. Estimated time Te and variance of activities 'V' on the critical path in a PERT network are given in the following tables.

Activity	Te (days)	V $(days)^2$
a	17	4
b	15	4
g	8	1

The probability of completing the work in 43 days is
(a) 15.6% (b) 50.0% (c) 84% (d) 90.0%
[*IES 1998*] (**Ans.** *c*)

124. The earliest occurrence time for event '1' is 8 weeks and the latest occurrence time for event '1' is 26 weeks. The earliest occurence time for event '2' is 32 weeks and the latest occurrence time for event '2' is 37 weeks. If the activity time is 11 weeks, then the total float will be
(*a*) 11 (*b*) 13 (*c*) 18 (*d*) 24
[*IES 1998*]

Sol. Total float = 37 – 8 – 11 = 18 days: (**Ans.** *c*)

125. Which of the following cost elements are considered while determining the Economic lot size for purchase?
1. Inventory carrying cost 2. Procurement cost 3. Set up cost
Select the correct answer using the codes given below:
Codes:–
(*a*) 1, 2 and 3 (*b*) 1 and 2 (*c*) 2 and 3 (*d*) 1 and 3
[*IES 1998*] (**Ans.** *a*)

126. Consider the following costs:
1. Cost of inspection and return of goods 2. Cost of obsolescence
3. Cost of scrap 4. Cost of insurance
5. Cost of negotiations with suppliers
Which of these costs are related to inventory carrying cost?
(*a*) 1, 2 and 3 (*b*) 1, 3 and 4 (*c*) 2, 3 and 4 (*d*) 2, 4 and 5
[*IES 1999*] (**Ans.** *d*)

127. A company intends to use exponential smoothing technique for making a forecast for one of its products. The previous years forcast has been 78 units and the actual demand for the corresponding period turned out to be 73 units. If the value of the smoothening constant α is 0.2, the forecast for the next period will be
(*a*) 73 units (*b*) 75 units (*c*) 77 units (*d*) 78 units
[*IES 1999*] (**Ans.** c, see equn. 23.4)

128. Which one of the following statements is correct in relation to production, planning and control?
 (*a*) Expediting initiates the execution of production plans, whereas dispatching maintains them and sees them through to their successful completion.
 (*b*) Dispatching initiates the execution of production plans, whereas expediting maintains them and sees them through to their successful completion.
 (*c*) Both dispatching and expediting initiate the execution of production plans.
 (*d*) Both dispatching and expediting maintain the production plan and see them through to their successful completion. [*IES 1999*] (**Ans.** *b*)

129. Time estimates of an activity in a PERT network are: Optimistic time t_0 = 9 days; pessimistic time tp = 21 days most likely time t_m = 15 days
 The approximate probability of completion of this activity in 13 days is
 (*a*) 16% (*b*) 34% (*c*) 50% (*d*) 84%
 [*IES 1999*] (**Ans.** *a*)

130. Consider the following situations
 1. Loads are uniform
 2. Materials move relatively continuously
 3. Movement rate is variable
 4. Routes do not vary
 For material transportation, conveyors are used when the prevailing conditions include
 (*a*) 1, 3 and 4 (*b*) 1, 2 and 4 (*c*) 1, 2 and 3 (*d*) 2, 3 and 4
 [*IAS 1994*] (**Ans.** *b*)

131. Transfer machines can be defined as
 (*a*) material processing machines
 (*b*) material handling machines
 (*c*) material handling and material processing machines
 (*d*) component feeders for automatic assembly [*IES 1999*] (**Ans.** *c*)

132. Which of the following data are needed for MRP?
 1. Master production schedule 2. Inventory position
 3. Machine capacity 4. Bill of materials
 Select the correct answer using the codes given below:–
 Codes:
 (*a*) 1, 2 and 3 (*b*) 2, 3 and 4 (*c*) 1, 2 and 4 (*d*) 1, 3 and 4
 [*IES 1998*] (**Ans.** *b*)

133. Process capability of a machine is defined as the capability of machine to
 (*a*) produce a definite volume of work per minute
 (*b*) perform definite number of operations
 (c) produce job at a definite spectrum of speed
 (*d*) hold a definite spectrum of tolerances and surface finish. [*IES 1993*] (**Ans.** *d*)

134. Production scheduling is simpler, and high volume of output and high labour efficiency are achieved in the case of
 (*a*) fixed position layout
 (*b*) process layout
 (*c*) product layout
 (*d*) a combination of line and process layout [*IES 1993*] (**Ans.** *c*)

135. Consider the following advantages
 1. Lower in process inventory
 2. Higher feasibility in rescheduling in case of machine breakdown.
 3. Lower cost in material handling equipment.

When compared to process layout, the advantages of product layout would include

(*a*) 1 and 2 (*b*) 1 and 3 (*c*) 2 and 3 (*d*) 1, 2 and 3

[*IAS 1994*] (**Ans.** *b*)

136. To avoid excessive multiplication of facilities, the layout preferred is

(*a*) product layout (*b*) group layout (*c*) static layout (*d*) process layout

[*IAS 1995*] (**Ans.** *d*)

137. Consider the following situations that would warrant a study of the layout:

1. Change in the work force 2. Change in production volume
3. Change in product design4. competitions in the market.

The situation (s) that would lead to a change in the layout would include:

(*a*) 1, 2, 3 and 4 (*b*) 1, 3 and 4 (*c*) 3 alone (*d*) 2 alone

[*IES 1994*] (**Ans.** *c*)

138. Which of the following charts are used for plant layout design:

1. Operation process chart 2. Mass machine chart
3. Correlation chart 4. Travel chart

Select the correct answer using the codes given below:

Codes:–

(*a*) 1, 2, 3 and 4 (*b*) 1, 2 and 4 (*c*) 1, 3 and 4 (*d*) 2 and 3

[*IES 1995*] (**Ans.** *b*)

139. The type of layout suitable for use of the concept, principles and approaches of 'group technology' is:

(*a*) product layout (*b*) job-shop layout
(*c*) fixed position layout (*d*) cellular layout [*IES 1999*] (**Ans.** *a*)

140. A systematic job improvement sequence will consist of

1. Motion study 2. Time study
3. job enrichment 4. jobs enlargement.

An optimal sequence would consist of

(*a*) 1, 2, 3 and 4 (*b*) 2, 1, 3 and 4 (c) 3, 1, 2 and 4 (*d*) 3, 4, 1 and 2

[*IAS 1994*] (**Ans.** *a*)

141. Production cost per unit can be reduced by

(*a*) producing more with increased inputs (*b*) producing more with the same inputs
(*c*) eliminating idle time (*d*) minimising resource waste

[*IAS 1995*] (**Ans.** *b*)

142. Procedure of modifying work content to give more meaning and enjoyment to the job by involving employees in planning, organisation and control of their work, is termed as

(*a*) job enlargement (*b*) job enrichment (*c*) job rotation (*d*) job evaluation

[*IES 1996*] (**Ans.** *b*)

Directions: The following questions consist of two statements, one labelled the 'Assertion A' and the other labelled the 'Reason R', you are to examine the two statements carefully and decide if the Assertion A and Reason R are individually true, and if so, whether the Reason is a correct explanation of the Assertion. Select your answers to these questions using the codes given below and mark your answer sheet accordingly.

(*a*) Both A and R are true and R is the correct explanation of A
(*b*) Both A and R are true and R is not a correct explanation of A
(*c*) A is true but R is false
(*d*) A is false but R is true.

143. *Assertion (A):* A workpiece with rough unmachined surface can be located in a jig or fixture on three supporting pins.

Reason (R): Indexing is made accurate by supporting on three pins. [*IES 1996*] (**Ans.** c)

144. *Assertion (A):* The first draw in deep drawing operation can have upto 60% reduction, the second draw upto 40% reduction and, the third draw upto 30% reduction only.
Reason (R) : Due to strain hardening, the subsequent draws in a deep drawing operation have reduced percentages. [*IES 1998*] (**Ans.** *a*)

145. *Assertion (A) :* In drop forging besides the provision for flash, provision is also to be made in the forging die for additional space called gutter.
Reason (R): The gutter helps to restrict the out ward flow of metal thereby helping to fill thin ribs and bases in the upper die. [*IES 1997*] (**Ans.** *a*)

146. *Assertion (A):* Plastic deformations in metals and alloys is a permanent deformation under load. This property is useful in obtaining products by cold rolling.
Reason (R): Plastic or permanent deformation in metal or alloy is caused by movement of dislocations. [*IES 1998*] (**Ans.** *c*)

147. *Assertion (A):* It is possible to have more than one break-even point in break even charts.
Reason (R): All variable costs are directly variable with production. [*IES 1999*] (**Ans.** *d*)

148. *Assertion (A):* In a multispindle automat, the turret is indexed to engage each of the cutting tool mounted on it.
Reason (R): Turret is a multiple tool holder so that the machining can be continued with each tool without the need to change the tool. [*IAS 1995*] (**Ans.** *d*)

149. *Assertion (A):* In a Swiss-type automatic lathe, the turret is given longitudinal feed for each tool in a specific order with suitable undexing.
Reason (R): A turret is a multiple tool holder to facilitate machining with each tool by indexing without the need to change the tools. [*IES 1995*] (**Ans.** *b*)

150. *Assertion (A):* A large margin of safety in break-even analysis is helpful for management devisions.
Reason (R): If the margin of safety is large, it would indicate that there will be profit even when there is a serious drop in production. [*IES 1997*] (**Ans.** *c*)

151. *Assertion (A):* In case of control charts for variables, if some points fall outside the control limits, it is concluded that the process is not under control.
Reason (R): It was experimentally provided by Shewhart that averages of four or more consecutive readings from a universe (population) or from a process, when plotted, will form a normal distribution curve. [*IES 1999*] (**Ans.** *a*)

152. *Assertion (A):* For a negative rake tool, the specific cutting pressure is smaller than for a positive rake tool under otherwise identical conditions.
Reason (R): The shear strain undergone by the chip in the case of negative rake tool is larger. [*IES 1993*] (**Ans.** *a*)

153. *Assertion (A):* An increase in depth of cut shortens the tool life.
Reason (R): Increase in depth of cut gives rise to relatively small increase in tool temperature. [*IAS 1995*] (**Ans.** *a*)

154. *Assertion A:* Tool wear is expressed in terms of flank wear rather than crater wear.
Reason R: Measurement of flank wear is simple and more accurate. [*IES 1994*] (**Ans.** *a*)

155. *Assertion A:* Machine tool beds are generally made of C.I.
Reason R: C.I. possess good self-lubricating properties. [*IES 1994*] (**Ans.** *a*)

156. *Assertion A:* Planning and scheduling of job order manufacturing differ from planning and scheduling of mass production manufacturing.
Reason R: In mass production manufacturing, a large variety of products are manufactured in large quantity. [*IES 1994* (**Ans.** *c*)

157. *Assertion A:* Special purpose machine tools and automatic machine tools are quite useful for job shops.
Reason R: Special purpose machine tools can do special types of machining work automatically. [*IES 1996*] (**Ans.** *d*)

158. *Assertion A:* Generally PERT is preferred over CPM for the purpose of project evaluation.
Reason R: PERT is based on the approach of multiple time estimates for each activity.
[*IES 1996*] (**Ans.** *a*)

159. *Assertion (a):* Product layout is more amenable to automation than process layout.
Reason (R): The work to be performed on the product is the determining factor in the positing of the manufacturing equipment in product layout. [*IES 1995*] (**Ans.** *a*)

160. *Assertion (a):* Job shop production leads to large work in-process inventory.
Reason (R): Jobbing production is used to manufacture medium demand variety production.
[*IAS 1994*] (**Ans.** *a*)

161. A dealer sells a radio set at Rs. 900 and makes 80% profit on his investment. If he can sell it at Rs. 200 more, his profit as percent age of investment will be
(*a*) 160 (*b*) 180 (*c*) 100 (*d*) 120
[*IES 1999*] (**Ans.** *d*)

162. Given that average cutting speed = 9 m/min, the return time to cutting time ratio is = 1 : 2, the feed rate = 0.3 mm/stroke the clearance at each end of c ut = 2 mm and that the plate is fixed with 700 mm side along the direction of tool travel, the time required for finishing one flat surface of size 700 × 30 mm in a shaper will be
(*a*) 10 min (*b*) 12.5 min (*c*) 15 min (*d*) 20 min
[*IES 1994*] (**Ans.** *b*)

163. The following parameters determine the model of continuous chip formation:
1. True speed 2. cutting velocity
3. chip thickness 4. Rake angle of the cutting tool.
The parameters which govern the value of shear angle would include:
(*a*) 1, 2 and 3 (*b*) 1, 3 and 4 (*c*) 1, 2 and 4 (*d*) 2, 3 and 4
[*IES 1994*] (**Ans.** *b*)

164. What is the correct sequence of the following parameters in order of their maximum to minimum influence on tool life?
1. Feed rate 2. Depth of cut 3. Cutting speed
Select the correct answer using the codes given below:
Codes:
(*a*) 1, 2, 3 (*b*) 3, 2, 1 (*c*) 2, 3, 1 (*d*) 3, 1, 2
[*IES 1994*] (**Ans.** *d*)

165. Tool geometry of a single point cutting tool is specified by the following elements:
1. Back rake angle 2. Side rake angle 3. End cutting edge angle
4. Side cutting edge angle 5. Side relief angle
6. End relief angle.
7. Nose radius. The correct sequence of these tool elements used for correctly specifying the tool geometry is
(*a*) 1, 2, 3, 6, 5, 4, 7 (*b*) 1, 2, 6, 5, 3, 4, 7 (*c*) 1, 2, 5, 6, 3, 4, 7 (*d*) 1, 2, 6, 3, 5, 4, 7
[*IES 1994*] (**Ans.** *b*)

166. The ratio of cutting force to thrust force is nearly 2.5 in
(*a*) turning (*b*) broaching (*c*) grinding (*d*) plain milling
[*IES 1994*] (**Ans.** *c*)

167. Which one of the following operations is carried out at the minimum cutting velocity if the machines are equally rigid and the tool work materials are the same?
(*a*) turning (*b*) grinding (*c*) boring (*d*) milling
[*IES 1994*] (**Ans.** *d*)

168. Match the following quality control objective functions with the appropriate statistical tools:

Objective functions	*Statistical Tools*
(*A*) A casting process is to be controlled with respect to hot tearing tendency	(P) X-chart
(*B*) A casting process is to be controlled with respect to the number of blow holes, if any, produced per unit casting	(Q) c-chart
(*C*) A machining process is to be controlled with respect to the diameter of shaft machined	(R) Random sampling
(*D*) The process variability in a milling operation is to be controlled with respect to the surface finish of components.	
	(S) p-chart
	(T) Hypothesis testing
	(U) $\overline{X}$, σ-charts

[*GATE 1992*] (**Ans.** A–S, B–Q, C–P, D–U).

169. In PERT, the distribution of activity times is assumed to be
(*a*) Normal (*b*) Gamma (*c*) Beta (*d*) Exponential
[*GATE 1995*] (**Ans.** *c*)

170. Statistical quality control was developed by
(*a*) Frederick Taylor (*b*) Walter Shewhart (*c*) George Danzig (*d*) W.E. Deming
[*GATE 1995*] (**Ans.** *b*)

171. Match four correct pairs between List I and List II for

List I *(Problem Areas)*	*List II* *(Techniques)*
(*A*) JIT	1. CRAFT
(*B*) Computer assisted layout	2. PERT
(*C*) Scheduling	3. Johnson's rule
(*D*) Simulation	4. Kanbans
	5. EQQ rule
	6. Monte Carlo

[*GATE 1995*] (**Ans.** A– 4, B – 1, C – 2, D – 6)

172. The individual human variability in time studies to determine the production standards is taken case of by
(*a*) Personal allowances (*b*) Work allowances
(*c*) Rating factor (*d*) none of the above
GATE 1996) (**Ans. :** *c*)

173. A dummy activity is used in PERT network to describe
(*a*) precedence relationship (*b*) necessary time delay
(*c*) resource restriction (*d*) resource idleness
[*GATE 1997*] (**Ans.** *a*)

174. Match four correct pairs between List I and List II

List I	*List II*
(*a*) Grinding	1. Surface for oil retention
(*b*) Honing	2. Surface for Max. load capacity

(c) Super finishing — 3. Surface of limiting friction
(d) Burnishing — 4. Surface of matte finish
5. Surface for pressure sealing
6. Surface for interference fit [GATE 1997]

175. A project is given below:–

Activity	Time duration in weeks	Predecessors
A	2	None
B	2	None
C	7	A
D	12	A
E	10	B
F	3	D, E
G	4	C, F

(a) Construct a PERT network
(b) Find the critical path and estimate the project duration [GATE 1997] (**Ans.** 21 days)

176. In inventory planning, extra inventory is unnecessarily carried to the end of the planning period when using one of the following lot size decision policies:–
(a) lot–for–lot production
(b) Economic Order Quantity (EOQ) lot size
(c) Period Order Quantity (POQ) lot size
(d) Part period total cost balancing [GATE 1998] (**Ans.** c)

177. In a time study exercise, the time observed for an action was 54 seconds. The operator had a performance rating of 120. A personal time allowance of 10% is given. The standard time for the activity, in seconds, is
(a) 54 (b) 60.8 (c) 72 (d) 58.32
[GATE 2000] (**Ans.** c)

178. Consider the following machine tools:–
1. Hobbing machine 2. Gear shaping machine
3. Broaching machine
The teeth of internal spur gears can be cut in
(a) 1, 2 and 3 (b) 1 and 2 (c) 1 and 3 (d) 2 and 3
[IES 1994] (**Ans.** d)

179. In a transfer line:
(a) all the machine tools must be automatic
(b) the work stations must form a closed loop
(c) the cycle time is total time taken by all the machining operations
(d) all the machine tools must be of conventional and general purpose type.
[IES 1994] (**Ans.** a)

180. Consider the following approaches normally applied for the economic analysis of machining:
1. Maximum production rate 2. Maximum profit criterion
3. Minimum cost criterion
The correct sequence in ascending order of optimum cutting speed obtained by these approaches is:
(a) 1, 2, 3 (b) 1, 3, 2 (c) 3, 2, 1 (d) 3, 1, 2
[IES 1999] (**Ans.** c)

181. A 100 mm sine bar is used to find the included angle of a plug gauge. The height of the slip gauges below the lower cylinder is 25 mm.
(A) If the angle is 30°, what is the height of the slip gauges below the upper cylinder?
(B) What would be the height of slip gauges below the upper cylinder if the length of the sine bar was 100.005 mm and the upper cylinder 0.005 mm larger. **(GATE 1992)**

Solution:- Refer Fig. 12.31 (b)

$$h_2 = 25\,\text{mm},\quad h_1 = ?,\ 2\theta = 30°$$

$$\text{Now Sin } 2\theta = \frac{h_1 - h_2}{L}$$

$$\therefore h_1 - h_2 = 100 \times \sin 30° = 50$$

$$\therefore h_1 = 50 + h_2 = 50 + 25 = \mathbf{75mm}$$

(b) If d_1 = diameter of upper cylinder, and

d_2 = diameter of lower cylinder, then

$d_1 - d_2 = 0.005$ mm

$$\therefore \text{Sin } 2\theta = \frac{h_1 + \frac{d_1}{2} - h_2 - \frac{d_2}{2}}{L}$$

Now $L = 100.005$ mm

$$\therefore \frac{1}{2} \times 100.005 = h_1 - 25 + \frac{1}{2}(d_1 - d_2)$$

$$\therefore h_1 = 50.0025 + 25 - 0.0025 = \mathbf{75mm}$$

182. A sine bar and a set of gauge blocks (slip gauges) are used to measure the angles of tapered components. Neglecting errors due to the use of slip gauges, prove that a given sine bar achieves greater accuracy while measuring smaller angles than while measuring larger angles. (GATE 1993)

Solution:- Refer Fig. 12.31 (b)

Let the included angle be $= \theta$

$$\therefore \text{Sin } \theta = \frac{h_1 - h_2}{L} = \frac{h}{L}$$

$$\therefore d\theta = \frac{1}{\sqrt{1 - \left(\frac{h}{L}\right)^2}} \cdot \frac{dh}{L}$$

Now slip gauges are standard thickness gauge blocks. From the above equation, it is clear that the introduction of the smallest thickness gauge block with *dh* = 1.001 mm results in larger $d\theta$ when 'h' is more (*i.e.* θ is more) and smaller $d\theta$ when *h* tends to zero (*i.e.* θ tends to zero) It is clear thus that angles can be measured more precisely with the smallest thickness slip gauge, when h is less, *i.e.*, lower angle θ.

183. A single point cutting tool made of HSS has the values of constants $C = 80$ and $n = 0.2$ in the basic tool life equation. If the tool cost per regrind is Rs. 2 and the machine hour rate is Rs 30, determine the most economical cutting speed. The tool cost includes the cost of time spent on changing. (GATE 1994)

Solution:-

Refer Eqn. 14.31 (Page 512)

$$V = C\left[\frac{C_m}{C_t}\cdot\frac{n}{1-n}\right]^n$$

$$C = 80,\ n = 0.2,\ C_m = \text{Rs. } 30,\ C_t = \text{Rs. } 2$$

$$\therefore V = 80\left(\frac{0.2}{0.8}\right)^{0.2}\cdot\left(\frac{30}{2}\right)^{0.2}$$

$$= \mathbf{104.2\ m/min.}$$

184. Balls of diameter 30 mm and 15 mm were used to measure the taper of a taper ring gauge. During inspection, the ball of 30 mm diameter was protruding by 2.5 mm above the top surface of the ring. This surface was located at a height of 50 mm from the top of the 15 mm diameter ball. Calculate the taper angle. (GATE 1997) (**Ans.**: 21.6°)

185. The standard time of an operation while conducting a time study is

(A) Mean observed time + allowances

(B) Normal time + allowances

(C) Mean observed time × rating factor + allowances

(D) Normal time × rating factor + allowances.

(GATE 2002) (**Ans.**: C)

186. An item can be purchased for Rs. 100.00. The ordering cost in Rs. 200 and the inventory carrying cost is 10% of the item cost per annum. If the annual demand is 4000 units, the E.O.R. (in units) is

(A) 50 (B) 100 (C) 200 (D) 400

(GATE 2002) (**Ans.** D)

187. In a blanking operation, the clearance is provided on

(A) The die (B) Both punch and die equally

(C) The punch (D) Neither the punch nor the die. (GATE 2002) (**Ans.** C)

188. A tube of 35 mm 0.D. was turned on a lathe and the following data was obtained:-

Rake angle = 35°, Cutting speed = 15 m/min.

Feed rate = 0.1 mm/rev.

Length of continuous chip in one revolution = 60 mm:

Cutting force = 200 N, Feed force (Thrust force) = 800 N

Calculate the chip thickness, shear plane angle, velocity of chip along tool face and co-efficient of friction. (GATE 2002)

Solution: $l_c = 60$ mm, $l = \pi D = 35\pi$

$$\therefore r = l_c / l \quad \text{[equ. 14.5]}$$

$$= 0.545$$

$$\phi = \tan^{-1}\frac{r\cos\alpha}{1 - r\sin\alpha} \quad \text{(Equ. 14.4)}$$

$$= 33°$$

$V_c = r \times V = 8.175 \, m/\text{min}.$

μ (equ. 14.15) = **1.528**

189. **While measuring the effective diameter of an external metric screw thread gauge of 3.5 mm pitch, a 30.5 mm diameter cylindrical standard and 2 mm diameter wires were used. The micrometer reading over the standard and wires was 13.3768 mm. The corresponding reading over the thread gauge and wire was 12.2428 mm. Calculate the thread gauge effective diameter.** (GATE 2002)

190. **The cutting force in punching and blanking operations mainly depends upon**
(A) The modulus of elasticity of metal. (B) the shear strength of metal
(C) the bulk modulus of metal (D) The yield strength of metal.
(GATE 2001) (**Ans.** B)

191. **Allowance in limits and fits refers to**
(A) Maximum clearance between shaft and hole.
(B) Minimum clearance between shaft and hole.
(C) Difference between maximum and minimum size of hole.
(D) Difference between maximum and minimum size of shaft. (GATE 2001 (**Ans.** B)

192. **Production flow analysis (PFA) is a method of identifying part families that uses data from**
(A) Engineering drawings (B) Production schedule
(C) Bill of materials (D) Route Sheets (GATE 2001) (**Ans.** A)

193. **When using a simple moving average to forecast demand, one would**
(A) Give equal weight to all demand data.
(B) Assign more weight to the recent demand data.
(C) Include new demand data in the average without discarding the earlier data.
(D) Include new demand data in the average after discarding some of the earlier demand data.
(GATE 2001) (**Ans.** D)

194. **3-2-1 method of location in a jig or fixture would collectively restrict the workpiece in *n* degrees of freedom, where the value of *n* is**
(A) 6 (B) 8 (C) 9 (D) 12 (GATE 2001) (**Ans.** C)

195. **For rigid perfectly plastic material, negligible interface friction and no redundant work, the theoretically maximum possible reduction in the wire drawing operation is**
(A) 0.36 (B) 0.63 (C) 1.00 (D) 0.72
(GATE 2001) (**Ans.** A)

196. **During orthogonal cutting of mild steel with a 10° rake angle tool, the chip thickness ratio obtained was 0.4. The shear angle (in degrees) evaluated from this data is**
(A) 6.53 (B) 20.22 (C) 22.94 (D) 50.00
(GATE 2001) (**Ans.** C)

197. **Fifty observations of a production operation revealed a mean cycle time of 10 min. the worker was evaluated to be performing at 90% efficiency. Assuming the allowances to be 10% of the normal time, the standard time (in seconds) for the job is**
(A) 0.198 (B) 7.3 (C) 9.0 (D) 9.9
(GATE 2001) (**Ans.** D)

198. **A number of cold rolling passes are required in a two-high rolling mill to reduce the thickness of a plate from 50 mm to 25 mm. the roll diameter is 700 mm and the co-efficient of friction at the roll-work interface is 0.1. It is required that the draft in each pass must be the same. Assuming no front and back tension, determine**
(A) The minimum number of passes required (B) The draft in each pass.
(GATE 2001) (**Ans.** 8, 3.125 mm)

IES - 2003

199. Which one of the following is assumed for timing the activities in PERT networks?

(A) α - distribution (B) β - distribution
(C) Binomial distribution (D) Erlangian distribution **(Ans.** B)

200. The three time estimates of a PERT activity are : optimistic time = 8 min, most likely time = 10 min. and pessimistic time = 14 min. The expected time of the activity would be

(A) 10.00 min (B) 10.33 min (C) 10.66 min (D) 11.00 min. **(Ans.** B)

201. A shop owner with an annual constant demand of "A" units has ordering costs of Rs. 'P' per order and carrying costs Rs. 'I' per unit per year. The conomic order quantity for a purchasing model having no shortage may be determined from

(A) $\sqrt{\dfrac{24P}{AI}}$ (B) $\sqrt{\dfrac{24AP}{I}}$ (C) $\sqrt{\dfrac{2AP}{I}}$ (D) $\sqrt{\dfrac{2AI}{P}}$ **(Ans.** C)

202. Economic Order Quantity is the quantity at which the cost of carrying is

(A) Minimum (B) Equal to the cost of ordering
(C) Less than the cost of ordering (D) Cost of over-stocking **(Ans.** B)

203. Why do we need inventory? Explain why we need to optimise the order quantity. The demand for a component is 10000 pieces per year. The cost per item is Rs. 50 and the interest cost is 1% per month. The cost associated with placing an order is Rs. 240. What is the EOQ?

(Ans. 894)

204. The proper sequence of activities for material requirement planning are

(A) Master production schedule, capacity planning, MRP, and order release.
(B) Order release, master production schedule, MRP and capacity planning.
(C) Master production schedule, order release, capacity planning and MRP.
(D) Capacity planning, master production schedule, MRP and order release. **(Ans.** D)

205. In the tolerance specification 25D6, the letter D represents

(A) Grade of tolerance (B) Upper deviation
(C) Lower deviation (D) Type of fit **(Ans.** D)

206. In a machine tool gear box, the smallest and the largest spindles are 100 rpm and 1120 rpm respectively. If there are 8 speeds in all, the fourth speed will be

(A) 400 rpm (B) 280 rpm (C) 800 rpm (D) 535 rpm **(Ans.** B)

207. The value of surface roughness 'h' obtained during the turning operation at a feed 'f' with a round nose tool having radius 'r' is given as

(A) $\dfrac{f}{8r}$ (B) $\dfrac{f^2}{8r}$ (C) $\dfrac{f^3}{8r}$ (D) $\dfrac{f^3}{8r^2}$ **(Ans.** B)

208. In economics of machining, which one of the following costs remains constant?

(A) Machining cost per piece.
(B) Tool chaging cost per piece
(C) Tool handling cost per piece
(D) Tool cost per piece. **(Ans.** C)

209. Match list I (Files in MRP) with list II (Inputs required) and select the correct answer using the codes given below the lists:

List I (Files in MRP)	List II (Inputs required)
A. Master Production Schedule	1. Scheduled receipts
B. Bill of materials	2. Unit costs and discounts
C. Inventory records	3. Production capacity
	4. Product structure

Codes:

	A	B	C
(a)	4	1	3
(b)	3	4	2
(c)	3	4	1
(d)	4	3	1

(Ans. b)

210. Match list I (limits in normal distribution) with list II (Population covered) and select the correct answer using the codes given below the lists:

List I (Limits in normal distribution)	List II (Population covered)
A $\pm$ 3σ	1. 0.3413
B $\pm$ 2σ	2. 0.6826
C $\pm$ 1σ	3. 0.9973
	4. 0.9545

Codes:

	A	B	C
(a)	3	4	2
(b)	3	2	4
(c)	4	2	3
(d)	4	3	2

(Ans. a)

211. In a drilling operation under a given condition, the tool life was found to decrease from 20 min to 5 min due to increase in drill speed from 200 r.p.m. to 400 r.p.m. What will be the tool life of that drill under the same condition if the drill speed is 300 r.p.m.?

Solution:- $VT^n = C$

$\therefore\ 200\,(20)^n = 400\,(5)^n$

$\therefore\ (4)^n = 2 \quad \therefore\ n = 0.5$

$\therefore\ C = 200 \times (20)^{0.5} = 894.4272$

Now, $300.(T)^{0.5} = 894.4272$

$\therefore\ T =$ **8.9 min.**

212. How much force will be required to pierce a circular hole of diameter 20 mm in a 2 mm thick plate of mild steel with the help of flat ended die and punch in a press tool? The shear strength of the work material is 350 MPa.

Solution: F = Perimeter $\times\ t \times fsu$

$= \pi D t \ fsu$

$= \pi \times 20 \times 2 \times 350$

$=$ **43.98 k.N.**

213. In the case of a sine-bar, show that the accuracy of the angle of set is a function of the accuracy & the centre distance between the rollers and of the setting height.

Solution: We know that

$$\text{Sin } \theta = \frac{h}{L}$$

By partial differentiation,

$$Cos\theta.\Delta\theta = \frac{\Delta h}{L} - \frac{h \Delta L}{L^2} = \frac{\Delta h}{L} - \sin\theta.\frac{\Delta L}{L}$$

$\therefore$ Total error in θ, $\Delta\theta = \frac{\sec\theta}{L}.\Delta h - \tan\theta.\frac{\Delta L}{L}$

It is clear that:

1. The Higher the value of L, the greater the accuracy of angular setting, other things remaining constant.
2. The higher the value of θ, the lower the setting accuracy, since as $\theta \to 90°$ $\sec\theta \to \infty$ and $\tan\theta \to \infty$.

GATE - 2003

214. The dimensional limits on a shaft of 25h7 are
(A) 25.00, 25.021 mm (B) 25.000, 24.979 mm
(C) 25.000, 25.007 mm (D) 25.000, 24.993 mm (**Ans.** B)

215. When a cylinder is located in a Vee-block, the number of degrees of freedom which are arrested is
(A) 2 (B) 4 (C) 7 (D) 8 (**Ans.** C)

216. The symbol used for Transport in work study is
(A) $\Rightarrow$ (B) T (C) ▭ (D) ▽ (**Ans.** A)

217. Quality screw threads are produced by
(A) thread milling
(B) thread chasing
(C) thread cutting with single point tool
(D) thread casting (**Ans.** A)

218. A metal disc of 20 mm diameter is to be punched from a sheet of 2mm thickness. The punch and die clearance is 3%. The required punch diameter is
(A) 19.88 mm (B) 19.94 mm (C) 20.06 mm (D) 20.12 mm (**Ans.** B)

219. A shell of 100 mm diameter and 100 mm height with the corner radius of 0.4 mm is to be produced by cup drawing. The required blank diameter is
(A) 118 mm (B) 161 mm (C) 224 mm (D) 312 mm (**Ans.** C)

220. A brass billet is to be extruded from its initial diameter of 100 mm to a final diameter of 50 mm. The working temperature is 700°C and the extrusion constant is 250 MPa. The force required for extrusion is
(A) 5.44 MN (B) 2.72 MN (C) 1.36 MN (D) 0.36MN

Hint. $P = k.A_0.ln\frac{A_0}{A_f}, k = 250\ MPa$ **(Ans.** B)

221. A threaded nut of M16, ISO metric type, having 2 mm pitch with a pitch diameter of 14.701 mm is to be checked for its pitch diameter using two or more number of balls or rollers of the following sizes.

(A) Rollers of 2mm ϕ (B) Rollers of 1.155 mm ϕ

(C) Balls of 2mm ϕ (D) Balls of 1.155 mm ϕ **(Ans.** D)

222. Two machines of the same production rate are available for use. On machine 1, the fixed cost is Rs. 100 and the variable cost is Rs. 2 per piece produced. The corresponding numbers for the machine 2 are Rs. 200 and Re 1 respectively. For certain strategic reasons, both the machines are to be used concurrently. The sale price of the first 800 units is Rs. 3.50 per unit and subsequently it is only Rs. 3.00. The break-even production rate for each machine is

(A) 75 (B) 100 (C) 150 (D) 600 **(Ans.** B)

223. Market demand for springs is 8,00,000 per annum. A company purchases these spring in lots and sells them, the cost of making a purchase order is Rs. 1200. The cost of storage of springs is Rs. 120 per stored piece per annum. The economic order quantity is

(A) 400 (B) 2828 (C) 4000 (D) 8000 **(Ans.** C)

224. A cylinder is turned on a lathe with orthogonal machining principle. Spindle rotates at 200 rpm. The axial feed rate is 0.25 mm per revolution. Depth of cut is 0.4 mm. The rake angle is 10°. In the analysis, it is found that the shear angle is 27.75°.

1. The thickness of the produced chip is

(A) 0.511 mm (B) 0.528 mm (C) 0.818 mm (D) 0.846 mm **(Ans.** A)

2. In the above problem, the co-efficient of friction at the chip tool interface obtained using Earnt and Merchant Theory is

(A) 0.18 (B) 0.36 (C) 0.71 (D) 0.98 **(Ans.** D)

225. A project consists of activities A to M shown in the net in the following figure with the duration of the activities marked in days.

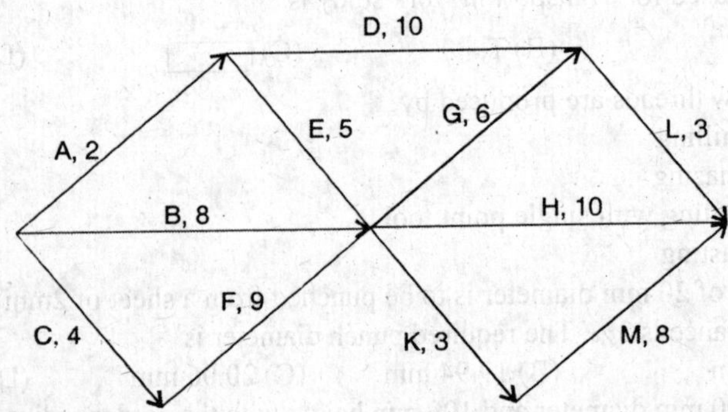

The project can be completed

(A) between 18, 19 days (B) between 20, 22 days

(C) between 24, 26 days (D) between 60, 70 days **(Ans.** C)

226. A part shown in the figure is machined to the sizes given below:-

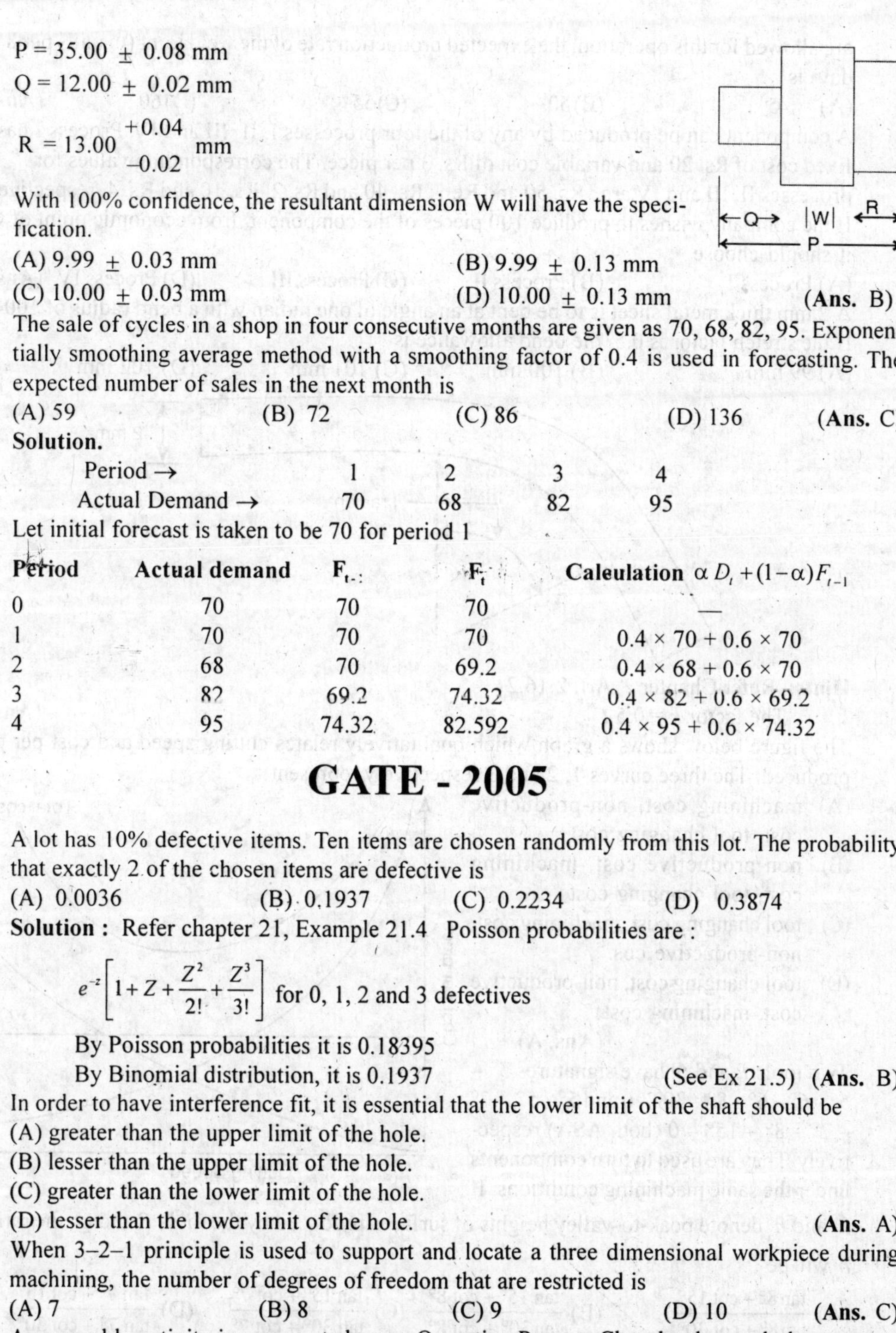

P = 35.00 ± 0.08 mm

Q = 12.00 ± 0.02 mm

R = 13.00 $^{+0.04}_{-0.02}$ mm

With 100% confidence, the resultant dimension W will have the specification.

(A) 9.99 ± 0.03 mm (B) 9.99 ± 0.13 mm

(C) 10.00 ± 0.03 mm (D) 10.00 ± 0.13 mm **(Ans.** B)

227. The sale of cycles in a shop in four consecutive months are given as 70, 68, 82, 95. Exponentially smoothing average method with a smoothing factor of 0.4 is used in forecasting. The expected number of sales in the next month is

(A) 59 (B) 72 (C) 86 (D) 136 **(Ans.** C)

Solution.

Period →	1	2	3	4
Actual Demand →	70	68	82	95

Let initial forecast is taken to be 70 for period 1

Period	Actual demand	F_{t-1}	F_t	Calculation $\alpha D_t + (1-\alpha)F_{t-1}$
0	70	70	70	—
1	70	70	70	0.4 × 70 + 0.6 × 70
2	68	70	69.2	0.4 × 68 + 0.6 × 70
3	82	69.2	74.32	0.4 × 82 + 0.6 × 69.2
4	95	74.32	82.592	0.4 × 95 + 0.6 × 74.32

GATE - 2005

1. A lot has 10% defective items. Ten items are chosen randomly from this lot. The probability that exactly 2 of the chosen items are defective is

(A) 0.0036 (B) 0.1937 (C) 0.2234 (D) 0.3874

Solution : Refer chapter 21, Example 21.4 Poisson probabilities are :

$$e^{-z}\left[1+Z+\frac{Z^2}{2!}+\frac{Z^3}{3!}\right] \text{ for 0, 1, 2 and 3 defectives}$$

By Poisson probabilities it is 0.18395

By Binomial distribution, it is 0.1937 (See Ex 21.5) **(Ans.** B)

2. In order to have interference fit, it is essential that the lower limit of the shaft should be

(A) greater than the upper limit of the hole.

(B) lesser than the upper limit of the hole.

(C) greater than the lower limit of the hole.

(D) lesser than the lower limit of the hole. **(Ans.** A)

3. When 3–2–1 principle is used to support and locate a three dimensional workpiece during machining, the number of degrees of freedom that are restricted is

(A) 7 (B) 8 (C) 9 (D) 10 **(Ans.** C)

4. An assembly activity is represented on an Operation Process Chart by the symbol

(A) □ (B) A (C) D (D) O **(Ans.** D)

5. A welding operation of time-studied during which an operator was pace-rated as 120%. The operator took, on an average 8 minutes for producing the weld-joint. If a total of 10% allow-

are allowed for this operation, the expected production rate of the weld joint (in units per 8 hour day) is

(A) 45 (B) 50 (C) 55 (D) 60 (**Ans.** A)

6. A component can be produced by any of the four processes I, II, III and IV. Process I has a fixed cost of Rs. 20 and variable cost of Rs. 3 per piece. The corresponding values for processes II, III and IV are : Rs. 50 and Re 1; Rs. 40 and Rs. 2; Rs. 10 and Rs. 4, respectively. If the company wishes to produce 100 pieces of the component, from economic point of view it should choose

(A) Process I (B) Process II (C) Process III (D) Process IV (**Ans.** C)

7. A 2 mm thick metal sheet is to be bent at an angle of one radian with a bend radius of 100 mm. If the stretch factor is 0.5, the bend allowance is

(A) 99 mm (B) 100 mm (C) 101 mm (D) 102 mm

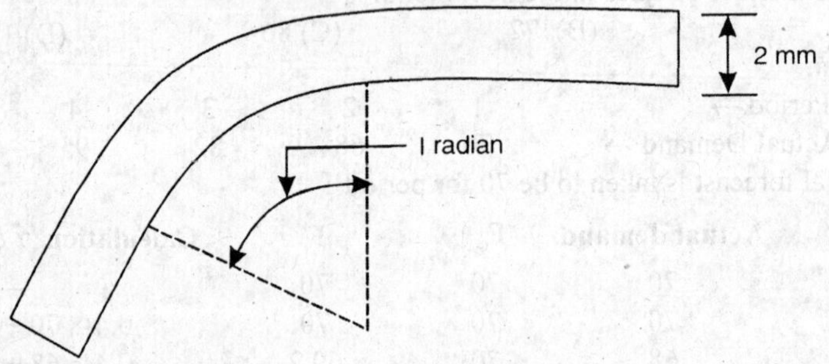

Hints: Refer Chapter 2, Art. 2.16.2

The factor $k = 0{,}5\ t$ (**Ans.** C)

8. The figure below shows a graph which qualitatively relates cutting speed and cost per piece produced. The three curves 1, 2 and 3 respectively represent

(A) machining cost, non-productive cost, tool changing cost

(B) non-productive cost, machining cost, tool changing cost

(C) tool changing cost, machining cost, non-productive cos.

(D) tool changing cost, non-productive cost, machining cost

(**Ans.** A)

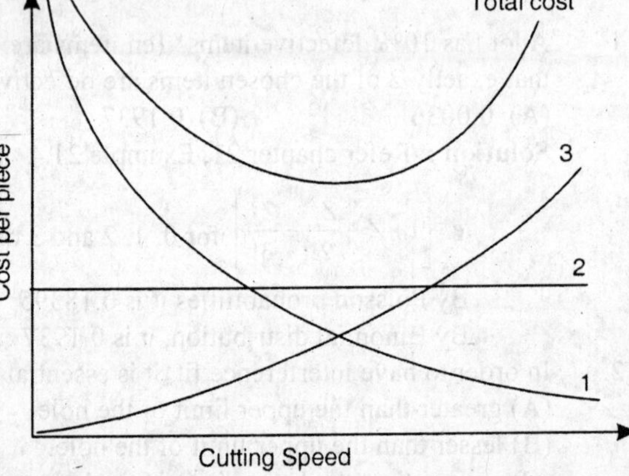

9. Two tools P and Q have signatures 5° – 5° – 6° – 6° – 8° – 30° – 0 and 5° – 5° – 7° – 7° – 8° – 15° – 0 (both ASA) respectively. They are used to turn components under the same machining conditions. If h_p and h_q denote peak-to-valley heights of surfaces produced by the tools P and Q, the ratio h_p/h_q will be

(A) $\dfrac{\tan 8° + \cot 15°}{\tan 8° + \cot 30°}$ (B) $\dfrac{\tan 15° + \cot 8°}{\tan 30° + \cot 8°}$ (C) $\dfrac{\tan 15° + \cot 7°}{\tan 30° + \cot 7°}$ (D) $\dfrac{\tan 7° + \cot 15°}{\tan 7° + \cot 30°}$

(**Ans.** B)

Hint : See chapter, 14, Art. 14.14.

10. The sales of a product during the last four years were 860, 880, 870 and 890 units. The forecast for the fourth year was 876 units. If the forecast for the fifth year, using simple exponential smoothing, is equal to the forecast using a three period moving average, the value of the exponential smoothing constant α is

(A) 1/7 (B) 1/5 (C) 2/7 (d) 2/5
(Ans. C)
Hint : See chapter 23, Art 23.3

IES - 2004

1. Process *X* has fixed cost of Rs. 40, 000 and variable cost of Rs. 9 per unit whereas process *Y* has fixed cost of Rs. 16, 000 and variable cost or Rs. 24 per unit. At what production quantity, the total cost of *X* and *Y* are equal?
(A) 1200 units (B) 1600 units (C) 2000 units (D) 24000 units(**Ans.** B)
2. Which one of the following information combinations has lowest break-even point?

	Fixed cost (in Rs.)	Variable cost / unit (in Rs.)	Revenue / unit (in Rs.)
(A)	30,000	10	40
(B)	40,000	15	40
(C)	50,000	20	40
(D)	60,000	30	40

(**Ans.** A)
3. Consider the following fits :
1. I.C. engine cylinder and piston
2. Ball bearing outer race and housing
3. Ball bearing inner race and shaft
Which of the above fits are based on the shaft basis system?
(A) 1 and 2 (B) 2 and 3 (C) 1 and 3 (D) 1, 2, and 3 (**Ans.** B)
4. Consider the following alignment tests on machine tools :
1. Straightness
2. Flatness
3. Run out
4. Parallelism
Which of the above alignment tests on machine tools are common to both lathe and shaper?
(A) 1 and 2 (B) 2 and 3 (C) 3 and 4 (D) 1 and 4 (**Ans.** A)
5. In a machining operation, chip thickness ratio is 0.3 and the back rake angle of the tool is 10°. What is the value of the shear strain?
(A) 0.31 (B) 0,13 (C) 3.00 (D) 3.34 (**Ans.** D)
6. The rake angle of a cutting tool is 15°, shear angle 45° and cutting velocity 35m /min. What is the velocity of chip along the tool face?
(A) 28. 5 m/min. (B) 27.3 m/min. (C) 25.3 m/ min. (D) 23.5 m/min(**Ans.** A)
7. Consider the following statements :
1. As the cutting speed increases, the cost of production initially decreases, then after an optimum cutting speed, it increases.
2. As the cutting speed increases, the cost of production also increases and after a critical value it reduces.
3. Higher feed rate for the same cutting speed reduces cost of production.
4. Higher feed rate for the same cutting speed increases the cost of production.
Which of the statements given below is/are correct?
(A) 1 and 3 (B) 2 and 3 (C) 1 and 4 (D) 3 only (**Ans.** D)
8. Consider the following statements :
During the third stage of tool-wear, rapid deterioration of tool edge takes place because
1. Flank wear is only marginal
2. Flank wear is large

3. Temperate of the tool increased gradually
4. Temperature of the tool increased drastically.

Which of the statements given above are correct ?

(A) 1 and 3 (B) 2 and 4 (C) 1 and 4 (D) 2 and 3 (**Ans.** B)

9. It is given that the actual demand is 59 units, a previous forecast 64 units and smoothing factor 0.3. What will be the forecast for the next period, using exponential smoothing.

(A) 36.9 units (B) 57.5 units (C) 60.5 units (D) 62.5 units (**Ans.** D)

10. The demand for a product in the month of March turned out to be 20 units against an earlier made forecast of 20 units. The actual demand for April and May turned to be 25 and 26 respectively. What will be the forecast for the month of June, using exponential smoothing method and taking smoothing constant α as 0.2 ?

(A) 20 units (B) 22 units (C) 26 units (D) 28 units (**Ans.** B)

11. Consider the following statements :

1. Preparation of Master production Schedule is an iterative process.
2. Schedule Charts are made with respect to jobs while load charts are made with respect to machines.
3. MRP is done before master production scheduling.

Which of the statements given above are correct ?

(A) 1, 2 and 3 (B) 1 and 2 (C) 2 and 3 (D) 1 and 3 (**Ans.** B)

12. Consider the following statements with respect to PERT :

1. It consists of activities with uncertain time phases.
2. This is evolved from Gantt Chart.
3. Total slack along the critical path is not zero.
4. There can be more than one critical path in PERT network.
5. It is similar to electrical network.

Which of the statement given above are correct?

(A) 1, 2 and 5 (B) 1, 3, and 5 (C) 2, 4, and 5 (D) 1, 2 and 4 (**Ans.** A)

13. Match list-I (PPC functions) with list-II (Activity) and select the correct answer using the codes given below the lists :

List–I (PPC functions)	**List–II** (Activity)
A. Capacity Planning	1. Listing products to be assembled and when to be delivered.
B. Shop floor control	2. Rescheduling orders based on production priorities.
C. Master Production Schedule	3. Closure tolerances
D. Material Requirement Planning	4. Monitor progress of orders and report their status.
	5. Planning of Labour and equipment

Codes :

	A	B	C	D
(a)	1	4	3	2
(b)	5	2	1	4
(c)	1	2	3	4
(d)	5	4	1	2

(**Ans.** d)

14. Consider the following statements :

1. Single Machine tool

2. Manual materials handling system
3. Computer control
4. Random sequencing of parts to machines

Which of the above characteristics are associated with flexible manufacturing system ?

(A) 1, 2 and 3 (B) 1 and 2 (C) 3 and 4 (D) 2, 3 and 4 **(Ans.** C)

15. Distinguish between a jig and fixture, with the help of diagrams for at least five each.
16. Distinguish between unilateral and bilateral tolerances for the dimensions. Indicate the tolerance zones for each by taking an example.
17. What is the hole basis and shaft basis of fits ?
Which one if preferred in practice and why ?
18. Clearances have to be provided on the press tools. What is the order of clearances for shearing operation and deep drawing operation ? On what tool will you provide the clearance in
(i) Punching operation (ii) Blanking operation
(Ans. : 10% to 20% of t, 2.5t, Die, punch)
19. What are the velocities which come into existence when a metal is cut orthogonally. Show these velocities graphically on a velocity diagram and determine the mathematical relationship in terms of shear and rake angles.
20. What is the basis for Selective Inventory Control ? Mention four applications of the same in the Inventory Management.
21. What is the distribution followed by the activity durations of a PERT network? Mention its mean and standard deviation.
22. What do you understand by tracking signal in forecasting? How is it computed?
23. Distinguish between P-system and Q-system in Inventory control.
24. What are the characteristics of Cellular manufacturing system.
25. The demand for a product during the last 10 years is given below :
Estimate the demand for the next two years by the method of regression

Year	1	2	3	4	5	6	7	8	9	10
Units	124	135	145	150	167	157	161	170	187	168

26. For *xyz* company, the annual requirement of an item is 2400 units. Each item costs the company Rs. 6. the supplier offers a discount of 5% if 500 or more quantities are purchased. The ordering cost in Rs. 32 per order and the average inventory cost is 16%. It is advisable to accept the discount ? Comment on the result. **(Ans.** : Yes)
27. Match list–I (study) and list–II (Related factors) and select the correct answer using the codes given below the lists :

List–I (Study)	**List–II** (factors)
A. Job enrichment	1. Gilbreth's principle
B. Job evaluation	2. Movement of limbs by work factor system
C. Method study	3. Herzberg motivators
D. Time study	4. Jacques time span of discretion

Codes :

	A	B	C	D
(a)	2	1	4	3
(b)	3	4	1	2
(c)	2	4	1	3
(d)	3	1	4	2

(Ans. b)

INDEX